Artificial Life IX

Artificial Life IX

Proceedings of the Ninth International Conference on the Simulation and Synthesis of Artificial Life

edited by Jordan Pollack, Mark Bedau, Phil Husbands, Takashi Ikegami, and Richard A. Watson

A Bradford Book

The MIT Press
Cambridge, Massachusetts
London, England

MIT Press books may be purchased at special quantity discounts for business or sales promotional use. For information, please email special_sales@mitpress.mit.edu or write to Special Sales Department, The MIT Press, 5 Cambridge Center, Cambridge, MA 02142.

This book printed and bound in the United States of America.

Library of Congress Control Number: 2004108085
ISBN: 0-262-66183-7

10 9 8 7 6 5 4 3 2 1

Contents

Preface..xii

Self-Organization

1. **The Chemoton: A Model for the Formation of Long RNA Templates**........... 1
   Chrisantha Fernando & Ezequiel Di Paolo

2. **Connecting Transistors and Proteins**... 9
   Claudio Mattiussi & Dario Floreano

3. **Designed and Evolved Blueprints For Physical Self-Replicating Machines**... 15
   Efstathios Mytilinaios, Mark Desnoyer, David Marcus & Hod Lipson

4. **Emergent Robustness and Self-Repair through Developmental Cellular Systems** 21
   Can Öztürkeri & Mathieu S. Capcarrere

5. **On Self-referential Shape Replication in Robust Aerospace Vehicles**.......... 27
   Mikhail Prokopenko & Peter Wang

6. **An Evolving and Developing Cellular Electronic Circuit**............................. 33
   Daniel Roggen, Yann Thoma & Eduardo Sanchez

7. **An Environment for Simulating Kinematic Self-Replicating Machines** 39
   William M. Stevens

8. **Towards an evolutionary-developmental approach for real-world substrates** 45
   Shivakumar Viswanathan & Jordan Pollack

9. **Stepwise evolution of molecular biological coding**......................... 51
   Peter R. Wills

Robotic Studies

10. **Once More Unto the Breach: Co-evolving a robot and its simulator** 57
    Josh C. Bongard & Hod Lipson

11. **Closing the loop: Evolving a model-free visually guided robot arm**............. 63
    Thomas Buehrmann & Ezequiel Di Paolo

12. **Performance Evaluation of Neural Architectures for Sequential Tasks** 69
    Genci Capi & Kenji Doya

13. **EvolGL: Life in a Pond**.. 75
    Santi Garcia Carbajal, Martin Bosque Moran & Fermin Gonzalez Martinez

14. **Evolving Plastic Neural Controllers stabilized by
Homeostatic Mechanisms for Adaptation to a Perturbation** 81
Thierry Hoinville & Patrick Hénaff

15. **A Stigmergic Cooperative Multi-Robot Control Architecture** 88
Thomas G. Howsman, Daniel O'Neil & Michael A. Craft

16. **Crawling Out of the Simulation: Evolving Real Robot Morphologies
Using Cheap, Reusable Modules** 94
Ian Macinnes & Ezequiel Di Paolo

17. **Functional Freeform Fabrication for Physical Artificial Life** 100
Evan Malone & Hod Lipson

18. **Quadrupedal Locomotion: GasNets, CTRNNs and
Hybrid CTRNN/PNNs Compared** 106
Gary McHale & Phil Husbands

19. **Evolving Simulated Mutually Perceptive Creatures for Combat** 113
Michael J. T. O'Kelly & Kaijen Hsiao

20. **Information Trade-Offs and the Evolution of Sensory Layouts** 119
Lars Olsson, Chrystopher L. Nehaniv & Daniel Polani

21. **Evolving Flying Creatures with Path Following Behaviors** 125
Yoon Sik Shim, Sun Jeong Kim & Chang Hun Kim

22. **Behavioural Categorisation: Behaviour makes up for bad vision** 133
Emmet Spier

23. **The Evolution of Control and Adaptation in a 3D Powered
Passive Dynamic Walker** 139
Eric D. Vaughan, Ezequiel Di Paolo & Inman R. Harvey

Language, Brain, Culture

24. **Evolving Imitating Agents
and the Emergence of a Neural Mirror System** 146
Elhanan Borenstein & Eytan Ruppin

25. **A Comparison of Population Learning and Cultural Learning
in Artificial Life Societies** 152
Dara Curran & Colm O'Riordan

26. **A Computation Framework to Simulate
the Coevolution of Language and Social Structure** 158
Tao Gong, Jinyun Ke, James W. Minett & William S-Y Wang

27. **Boom and Bust: Environmental Variability Favors the
Emergence of Communication** 164
Patrick Grim & Trina Kokalis

28. **Expressing and Understanding Desires in Language Games** 170
Michael Klein, Hans Kamp, Guenther Palm & Kenji Doya

29. The Evolution of Affect-Related Displays, Recognition and Related Strategies 176
Robert Lowe, Lola Cañamero, ChrystopherL. Nehaniv & Daniel Polani

30. Language, Altruism and Docility: How Cultural Learning Can Favour Language Evolution 182
Marco Mirolli & Domenico Parisi

31. Evolution of Plastic Sensory-motor Coupling and Dynamical Categorization 188
Gentaro Morimoto & Takashi Ikegami

32. Song Grammars as Complex Sexual Displays 194
Kazutoshi Sasahara & Takashi Ikegami

33. Analogies between Genome and Language Evolution 200
Luc Steels

34. The Effects of Learning on the Evolution of Saussurean Communication 208
Edgar E. Vallejo & Charles E. Taylor

35. Minimum cost and the emergence of the Zipf-Mandelbrot law 214
Paul Vogt

Formal Games and Automata

36. Modcling Multicellular and Tumorous Existence with Genetic Cellular Automata 220
Armand Bankhead III, Nancy Magnuson, & Robert B. Heckendorn

37. Whatever Emerges should be Intrinsically Useful 226
Hugues Bersini

38. Shaping collective behavior: an exploratory design approach 232
Pablo Funes, Belinda Orme & Eric Bonabeau

39. Updating Schemes in Random Boolean Networks: Do They Really Matter? 238
Carlos Gershenson

40. Reducing Prejudice: A Spatialized Game-Theoretic Model for the Contact Hypothesis 244
Patrick Grim, Evan Selinger, William Braynen, Robert Rosenberger, Randy Au, Nancy Louie, & John Connolly

41. Quasi-Stable States in the Iterated-Prisoner's Dilemma 250
Philip T. Mueller

42. Evolving Memory: Logical Tasks for Cellular Automata 256
Luis Mateus Rocha

43. Complex Genetic Evolution of Self-Replicating Loops 262
Chris Salzberg, Antony Antony & Hiroki Sayama

44. **Redrawing the Boundary between Organism and Environment** 268
Tim Taylor

45. **Imitation and Inequity in Avoiding the Tragedy of the Commons** 274
Terry Van Belle & David H. Ackley

46. **The Quantum Coreworld: Competition and Cooperation
in an Artificial Ecology** 280
Alexander Wait

47. **Self-reproduction by glider collisions: the beehive rule** 286
Andrew Wuensche

Evolutionary and Adaptive Dynamics

48. **The value of death in evolution: A lesson from Daisyworld** 292
Matthew Bardeen

49. **The Flexible Balance of Evolutionary Novelty and Memory
in the Face of Environmental Catastrophes** 297
Andrew Buchanan, Mark Triant & Mark A. Bedau

50. **Kin-Selection: The Rise and Fall of Kin-Cheaters** 303
Sherri Goings, Jeff Clune, Charles Ofria & Robert T. Pennock

51. **Homeostasis and Rein Control: From Daisyworld to Active Perception** ... 309
Inman Harvey

52. **Measuring Biological Complexity in Digital Organisms** 315
Wei Huang, Charles Ofria & Eric Torng

53. **The Role of Nearly Neutral Mutations in the Evolution
of Dynamical Neural Networks** 322
Eduardo Izquierdo-Torres

54. **Sustained Evolution from Changing Interaction** 328
George Kampis & Laszlo Gulyas

55. **See How She Runs: Towards Visualising
Artificial Red Queen Evolution** 334
James A. R, Marshall & Simon Tokumine

56. **Sexual reproduction and Muller's ratchet in digital organisms** 340
Dusan Misevic, Richard E. Lenski & Charles Ofria

57. **Chaotic Population Dynamics and the Evolution of Aging** 346
Joshua Mitteldorf

58. **The Role of Non-Genetic Change in the Heritability, Variation
and Response to Selection of Artificially Selected Ecosystems** 352
Alexandra Penn & Inman Harvey

59. **Ecolab, Webworld and self-organisation** 358
Russell K. Standish

60. **Tierra's missing neutrality: case solved**.................................... 364
Russell K. Standish

61. **Drastic Changes in Roles of Learning in the Course of Evolution**............. 369
Reiji Suzuki & Takaya Arita

62. **Niche Construction and the Evolution of Complexity**.......................... 375
Tim Taylor

Cellular Development

63. **A Model for Exploring Genetic Control of Artificial Amoebae** 381
Barry Drennan & Randall D. Beer

64. **Asymmetric cell division and its integration with other developmental processes for artificial evolutionary systems**..... 387
Peter Eggenberger Hotz

65. **A Functional Model of Cell Genome**.. 393
Alessandro Fontana & Walter Steven Fraccaro

66. **Asynchronous Dynamics of an Artificial Genetic Regulatory Network**..... 399
Jennifer Hallinan & Janet Wiles

67. **Small World and Scale-Free Network Topologies in an Artificial Regulatory Network Model**.................................. 404
P. Dwight Kuo & Wolfgang Banzhaf

68. **Inertia of Chemotactic Motion as an Emergent Property in a Model of an Eukaryotic Cell**...................................... 410
Shin I. Nishimura & Masaki Sasai

69. **Phenotypic Variability in Canalized Developmental Systems** 415
Sean T. Psujek & Randall D. Beer

70. **Self-repairing and Mobility of a Simple Cell Model**.......................... 421
Keisuke Suzuki & Takashi Ikegami

71. **Evaluating an Evolutionary Approach for Reconstructing Gene Regulatory Networks**........................... 427
Dion J. Whitehead, Andre Skusa & Paul J. Kennedy

Artificial Chemistries

72. **Bonding as an Emergent Phenomenon in an Abstract Artificial Chemistry**.................................... 433
Dominique Chu & Rune Vabø

73. **Evolution of Robust Developmental Neural Networks**.......................... 438
Alan Hampton & Christoph Adami

74. **A Functional Self-Reproducing Cell in a Two-Dimensional Artificial Chemistry** 444
Tim J. Hutton

75. **Metabolic closure in (*M,R*) Systems** 450
Juan-Carlos Letelier, Jorge Soto-Andrade, Flavio Guíñez-Abarzúa,
Athel Cornish-Bowden & María-Luz Cárdenas

76. **Flows of information in spatially extended chemical dynamics** 456
Kristian Lindgren, Anders Eriksson & Karl-Erik Eriksson

77. **Cellular Dynamics in a 3D Molecular Dynamics System**
with Chemistry 461
Duraid Madina & Takashi Ikegami

78. **Lipidia: An Artificial Chemistry of Self-Replicating Assemblies**
of Lipid-like Molecules 466
Barak Naveh, Moshe Sipper, Doron Lancet & Barak Shenhav

79. **Towards the Simulation of Reaction Networks in Astrochemistry** 472
Pierre Philippe, David Weiss Solís, Tom Lenaerts & Hugues Bersini

80. **Homochirality as Fixed Point of Prebiotic Chemistry** 478
Raphaël Plasson, Hugues Bersini & Auguste Commeyras

81. **An Evolvable Artificial Chemistry**
Featuring Continuous Physics and Discrete Reactions 484
Thomas E. Portegys

82. **The Role of RNA Editing in Dynamic Environments** 489
Luis Mateus Rocha & Chien-feng Huang

83. **A Tangled Hierarchy of Graph-Constructing Graphs** 495
Chris Salzberg, Hiroki Sayama & Takashi Ikegami

84. **Graded Artificial Chemistry in Restricted Boundaries** 501
Barak Shenhav, Ran Kafri & Doron Lancet

85. **Spacial Representation for Artificial Chemistry**
Based on Small-World Networks 507
Hideaki Suzuki

Art and Philosophy

86. **Behavioral Adaptive Autonomy. A milestone in the Alife route to AI?** 514
Xabier Barandiaran

87. **A Physiological Approach to the Generation of Artificial Life Forms** 522
Marc Cavazza, Simon Hartley, Louis Bec, François Mourre,
Gonzague Defos du Rau, Remy Lalanne, Mikael Le Bras, & Jean-Luc Lugrin

88. **Mechanistic and ecological explanations**
in agent-based models of cognition 528
Jason Noble & Manuel de Pinedo

89. **Empiricism in Artificial Life** 534
Eric Silverman & Seth Bullock

90. Using the Universal Similarity Metric to Model Artificial Creativity and Predict Human Listeners Response to Evolutionary Music.................. 540

Nils Svangård, Jon Klein, & Peter Nordin

91. Playing Music by Conducting BOID Agents................................. 546

Tatsuo Unemi & Daniel Bisig

Agents Ants and Swarms

92. An Evolutionary Approach to Complex System Regulation Using Grammatical Evolution.................................... 551

Saoirse Amarteifio & Michael O'Neill

93. Analyzing Evolved Fault-Tolerant Neurocontrollers 557

Alon Keinan

94. Tracking Information Flow through the Environment: Simple Cases of Stigmergy 563

Alexander S. Klyubin, Daniel Polani & Chrystopher L. Nehaniv

95. Ant Foraging Revisited... 569

Liviu A. Panait & Sean Luke

96. Learning Ant Foraging Behaviors.. 575

Liviu A. Panait & Sean Luke

97 Systems Biology Thought Experiments in Human Genetics Using Artificial Life and Grammatical Evolution 581

Bill C. White & Jason H. Moore

Author Index 587

Preface

Artificial life is the study of the organizational principles of life, rather than the study of carbon-based life as it exists on Earth. Highly interdisciplinary across physics, biology, computer science, and complex systems, some of the fundamental questions of artificial life are:

- What are the principles of evolution, learning and growth that can be understood well enough to simulate as an information process?

- Can robots be built faster and cheaper by mimicking biology than by the product design process used for automobiles and airplanes?

- What kinds of constraints should be placed on sciences, such as "Wet ALife," which works with self-replicating elements?

- What components of physics and chemistry support emergence and automatic discovery of physical and cognitive mechanisms of life forms?

- How can we unify theories from dynamical systems, game theory, evolution, computing, geophysics, and cognition?

Ten years ago, the ALIFE 4 conference in Boston heralded breakthroughs such as Karl Sim's co-evolving computer graphic creatures. We have nearly 100 papers and posters within, and results of workshops in a companion volume, and while we can't predict which are the most significant breakthroughs (we can only tell in retrospect), we hope you join us in celebrating how the field has blossomed.

When the President of the Society for Artificial Life approached me to chair the next Artificial Life Conference (and hold it in Boston), it was immediately after I'd given a keynote at the European Conference on Artificial Life, and was enjoying all the youthful energy, idealism and late night socializing of the many young researchers attending this great conference in Dortmund. Although I resolved never to run a conference again after 1996 SAB dissipated my summer energy without reducing entropy, there was a co-arms race between my brain and my mouth, and my mouth won and said "YES."

It was October 2003, we didn't have an organizing committee or a program committee, a date, a venue, or sponsors, or speakers, or conference schwag. What other patsies could I locate to be on an organizing committee with only 11 months? I owe great thanks to Richard Watson in Boston, Mark Bedau on the west coast, Phil Husbands in the UK, and Takashi Ikegami in the far east.

The first thing we needed was, of course, a logo and a poster and a Website. We owe a great debt to Cliff Pickover for donating to the conference reproduction rights for his mathematical artwork "Telopodite Fractal 1" to use as our logo, which signifies the chaos in organizing a conference. Cliff's art really improved the website I designed myself.

The second thing we needed was a date and a place to hold it. After calling a few of the usual downtown venues only to find them booked or too expensive, I located the Boston Convention Bureau. Julie Bennett Taylor put our conference profile out as a fax to the catering sales offices of every hotel in Boston and its suburbs. After being overwhelmed with offers, we settled on a great downtown location with wonderful space, great food, and free in-room internet access, and I would like to thank the staff of the Wyndham Tremont Hotel, in advance, for their hospitality.

The third thing we needed were distinguished keynote speakers, and we are indebted to George Whitesides, Eors Szathmary, and Murata Satoshi for their willingness to come to the meeting even before it was organized.

The fourth thing we needed were papers. To get the papers, we only had to send out a self-reproducing email virus. But, to select the papers…

The fifth thing we needed was a distinguished program committee to contribute their time and energy in review. Thanks to the following people for doing the artificial selection of the memes that will drive the field forward:

Hussein Abbass
Chris Adami
Takaya Arita
Wolfgang Banzhaf
Eric Bonabeau
Mark Bedau
Randy Beer
Richard Belew
Hugues Bersini
Seth Bullock
Rafael Calabretta
Angelo Cangelosi
Alastair Channon
Sung-Bae Cho
Thomas Christaller
Dave Cliff
John Collier
Kerstin Dautenhahn
Ezequiel Di Paolo
Marco Dorigo
Alan Dorin
Sevan Ficici

Dario Floreano
Paulien Hogeweg
Gregory Hornby
Phil Husbands
Takashi Ikegami
Brian Keeley
Graham Kendall
Jan T. Kim
Simon Kirby
John Koza
Kristian Lindgren
Hod Lipson
Paul Marrow
Alcherio Martinoli
John McCaskill
Jon McCormack
Barry McMullin
J.J. Merelo
Alvaro Moreno
Satoshi Murata
Chrystopher Nehaniv
Jason Noble
Stefano Nolfi

Naoaki Ono
Domenico Parisi
Daniel Polani
Jordan Pollack
Tom Ray
Eytan Ruppin
Hiroki Sayama
Moshe Sipper
Andre Skusa
Russell Standish
Luc Steels
Hideaki Suzuki
Charles Taylor
Tim Taylor
Guy Theraulaz
Peter Todd
Richard A. Watson
Barbara Webb
Michael Wheeler
Claus Wilke
Peter Wills
Andy Wuensche
Jens Ziegler

Conferences get disorganized over reviewing. But, not to worry, with the latest in electronic reviewing provided by Titus Brown, this conference would be self-organizing! While Titus Brown is not one of the editors of this volume, his software for paper submission and review relieved the paperwork which would have required three humanoid robots. Nevertheless, extra kudos go to Richard Watson for juggling the lion's share of the proceedings grunt work between his job interviews at seemingly every university in the UK.

Seventh, we needed sponsors. We talked to large and small businesses. Still they seemed artificially devoid of life in the post bubble age. We ended up with a Best Student Paper award thanks to Dave Cliff at HP Labs, Bristol, as well as a corporate sponsorship from Icosystem, thanks to Eric Bonabeau and Paul Edwards.

We thank the International Society for Artificial Life for the seed funding which made organization possible. And we thank Brandeis University as the major university sponsor, providing my time, my students' time, the CS department offices, Sponsored Programs Accounting, the Theatre Box Office which has the only credit card machine on campus, and the eight hours of administrative time involved with acceptance of the liability of contracts. Brandeis has also contributed the energy of Ms. Myrna Fox, without whom ALife9 couldn't happen at all.

Jordan B. Pollack
Chairman, ALife9.org
April 14th 2004

The Chemoton: A Model for the Origin of Long RNA Templates.

Chrisantha Fernando and Ezequiel Di Paolo
Centre for Computational Neuroscience and Robotics.
Department of Informatics
University of Sussex
Brighton, BN1 9QH
{ctf20,ezequiel}@cogs.susx.ac.uk

Abstract

How could genomes have arisen? Two models based on Ganti's Chemoton are presented which demonstrate that under increasingly realistic assumptions, template replication is facilitated without the need of enzymes. It can do this because the template state is stoichiometrically coupled to the cell cycle. The first model demonstrates that under certain kinetic and environmental conditions there is an optimal template length, i.e. one which facilitates fastest replication of the Chemoton. This is in contradiction to previous findings by Csendes who claimed that longer templates allowed more rapid replication. In the second model, hydrogen bonding, phosphodiester bonding and template structure is modeled, so allowing dimer and oligomer formation, hydrolysis and elongation of templates. Here, monomer concentration oscillates throughout the cell cycle so that double strands form at low monomer concentrations and separate at high monomer concentrations. Therefore, this simulation provides evidence that a protocell with Chemoton organization is a plausible mechanism for the formation of long templates, a notorious problem for studies of the origin of life.

Introduction

The Chemoton, invented by Ganti (Ganti, 1971), and largely ignored since, is a model for a minimal protocell that provides a compelling hypothesis for the origin of long RNA template replicators, without the need for enzymes. Briefly, the problem of long template replication is that, beyond a certain length of double stranded RNA, double strands remain stuck together and cannot replicate. In modern cells, protein enzymes catalyze the separation of these strands. However, the length of the strand required to produce this enzyme is greater than the length of the strand that can spontaneously separate (Szathmáry, 2000). Therefore, some mechanism is required to facilitate strand separation before such protein enzymes came to be. This lead Eigen to propose the hypercycle, whereby RNA templates might act as enzymes to facilitate the replication of other RNA templates (Eigen, 1971). However, long templates were prevented from forming due to Eigen's paradox which resulted from the assumption that specific sequences of RNA template had different replication rates. This assumption does not hold in the Chemoton, because Ganti wanted to conceive of a minimal system that was capable of replication without any enzymes. In the simplest Chemoton it is only the length of template strands that effect replication rate. The plausibility of hypercycles as a solution to Eigen's paradox has been discussed elsewhere (Szathmáry and Maynard-Smith, 1997). Ganti proposed that the Chemoton could delicately control nucleotide concentrations, producing oscillatory conditions where at low levels, double strands formed, and at high concentrations, strand separation took place, so allowing template replication without enzymes. This paper demonstrates the mechanism of Ganti's proposal in simulation.

A review of the Chemoton is beyond the scope of this paper, and readers are referred to "The Principles of Life" (Ganti, 2003), "Chemoton Theory Vol I and II" (Ganti, 2004), and to Tibor Csendes' simulation of the Chemoton (Csendes, 1984). Briefly, the Chemoton is the simplest biologically plausible model of a primitive cell, which couples metabolism, membrane, and RNA templates, such that growth and division can occur under reasonable assumptions. In the simplest model, the sequence of templates is irrelevant, only the length carries information, i.e. regulates, the other two subsystems. Each subsystem is autocatalytic, i.e. regenerates itself, in strict coordination with the other two subsystems because the chemical reactions of each subsystem are coupled stoichiometrically. In the Chemoton, the high error rate of non-enzymatic template-directed template replication is of no concern, since it is only the length of templates, and later on in evolution, the ratio of nucleotides, that effect replication rate. This is because Ganti proposed that the original role of templates was merely as a "sink" to soak up waste products of metabolism, so regulating the rate of metabolism, and osmotic pressure. Thus, the Chemoton model provides an intermediate step in an evolutionary tale of the origin of genomes we know and love today. Just as feathers in birds might have initially been for warmth, and only later co-opted for flight, in the Chemoton, what seems to be the simplest possible coupling

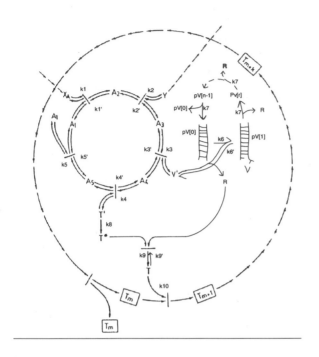

Figure 1: Adapted from Fig 4.6 of Chemoton Theory Vol I. The Chemoton consists of three stoichiometrically coupled auto-catalytic chemical subsystems, a reversible metabolic subsystem based on the formose reaction, an irreversible membrane subsystem capable of growth and division, and an irreversible template polycondensation subsystem which serves as an informational control system potentially capable of unlimited heredity by a set of plausible evolutionary steps.

between genotype (template properties) and phenotype (metabolism and membrane) is proposed as the origin of long genomes.

Figure. 1 shows the chemical reactions of the Chemoton. Let us consider the metabolic subsystem first. A high energy precursor molecule, X, permeable to the membrane, T_m, reacts with metabolite, A_1, producing A_2. A_2 forms waste molecule Y (which diffuses out of the Chemoton), and another molecule A_3. A_3 produces V', an impermeable template precursor monomer, and also produces A_4. A_4 produces T', an impermeable membrane precursor molecule, and also produces A_5. A_5 produces two copies of A_1, thus completing the autocatalytic cycle. All members A_1 to A_5 are impermeable to the membrane. Metabolism is reversible, but the rate of reverse reaction is less than the forward reaction as free energy of the precursor X is greater than the free energy of the products Y, V', and T' of the cycle.

Next we consider the template subsystem. This system

replicates by template directed synthesis in which a pre-formed polynucleotide directs and catalyses the synthesis of its complementary form, from mono- or oligonucleotide building blocks. Monomer molecule V' comes together in a polycondensation reaction to form pVn (double stranded polymerised V' molecules of length n/2). This occurs only when V' is present in a concentration greater than the polycondensation threshold, $[V']*$ [1]. So, $[V']$ increases until $[V']*$ whereupon it begins to be bound to the polymer, also releasing molecule R, a hydrophilic component necessary for incorporation with $T*$ to form phospholipids, T, for the membrane. After N molecules of V' have bound to the double stranded polymer, the polymer splits up to produce two semi-conserved double stranded polymers, again of length n/2.

Finally, we consider the membrane subsystem. The T' molecule reacts irreversibly to produce $T*$ which binds to R to produce the phospholipid, T, which is spontaneously incorporated into the membrane. In each metabolic cycle, precisely equal quantities of T' and V' are produced. As the membrane grows, it has been shown experimentally that, if the volume does not increase rapidly enough to maintain the cell as a sphere, the cell divides into two daughter cells.

The three sub-systems regulate each other by feedback due to the existence of reversible reactions. When $[V']$ is less than $[V']*$ (as long as no residue $T*$ and R are present) the volume does not increase because T cannot be formed. Thus, as the metabolic cycle functions, increasing $[V']$ hinders the operation of the metabolic cycle, i.e. the Chemoton slows down. The increasing concentration of intermediates creates an increase in the osmotic pressure within the Chemoton. Above $[V']*$ the volume starts to increase as R is produced, but the number of osmotically active internal particles can never maintain the Chemoton as a sphere because V' is produced only in proportion to R. T is produced in proportion to surface area, at the same rate as V', and so V' cannot increase in proportion to the volume. Osmotic equilibrium can be reached only when the volume of the Chemoton has been reduced from that of a sphere. When the surface area of the sphere is doubled, the spherule divides into two spherules of approximately equal size by a well studied self-organizing process (Bachmann et al., 1992).

Model 1: Methods

The basic Chemoton described above is modeled using a set of standard kinetic equations with reaction rates the same as in Csendes' model and initial conditions which approximate the conditions at the start of cell division.

---
[1] [X] means the concentration of X.

Metabolic sub-system

The metabolic subsystem is described by the same differential equations as used by Csendes. See Figure. 1 to see the meaning of the rate constants, k.

$$\frac{dA_1}{dt} = 2(k_5A_5 - k_5'A_1A_1) - k_1A_1X + k_1'A_2 \quad (1)$$

$$\frac{dA_2}{dt} = k_1A_1X - k_1'A_2 - k_2A_2 + k_2'A_3Y \quad (2)$$

$$\frac{dA_3}{dt} = k_2A_2 - k_2'A_3Y - k_3A_3 + k_3'A_4V' \quad (3)$$

$$\frac{dA_4}{dt} = k_3A_3 - k_3'A_4V' - k_4A_4 + k_4'A_5T' \quad (4)$$

$$\frac{dA_5}{dt} = k_4A_4 - k_4'A_5T' - k_5A_5 + k_5'A_1A_1 \quad (5)$$

Consider equation 1 for example. A_1 is produced by the reaction of an A_5 molecule to produce $2A_1$ molecules, hence the term $2k_5A_5$. Two A_1s are used up in the reverse of this reaction hense the term $2k_5'A_1A_1$. A_1s are also used up in reaction with X at rate k_1 and are produced in the reverse of this reaction at rate k_1'. By comparing the separate terms in each equation with the metabolic cycle in diagram 1, one can see they describe the stoichiometric equations correctly.

Template sub-system

We were unable to provide a meaningful physical interpretation for Csendes' model of template replication, in which he uses an "auxiliary variable" to represent template state. This implementation of template polycondensation is designed afresh, see Eqns. 6 to 10 and Figure. 1. The initial conditions consist of double stranded templates of length N/2 at concentration 0.01. Initiation is a reversible reaction taking place only at concentrations above [V']*. Propagation was an irreversible reaction, taking place also only when $[V'] > [V']^*$, on templates onto which at least one V monomer had already been bound. New full length double stranded templates,$(pVn[0])$, were formed when a final V' monomer was bound to a strand of length $pV(n-1)$. Thus, we modeled the concentration of polymers at each discrete stage of replication, with a separate differential equation.

$$\frac{d(pVn[0])}{dt} = 2k_7pVn[N-1]V' + k_6'pVn[1]R - k_6pVn[0]V' \quad (6)$$

$$\frac{d(pVn[1])}{dt} = k_6pVn[0]V' - k_6'pVn[1]R - k_7pVn[1]V' \quad (7)$$

$$\frac{d(pVn[r])}{dt} = k_7pVn[r-1]V' - k_7pVn[r]V'$$
$$: n-1 \geq r \geq 2 \quad (8)$$

$$\frac{dV'}{dt} = k_3A_3 - k_3'A_4V' + k_6'pVn[1]R -$$

$$k_6pVn[0]V' - \sum_{1}^{n-1}k_7pVn[r]V' \quad (9)$$

$$\frac{dR}{dt} = k_6pVn[0]V' - k_6'pVn[1]R + k_9'T -$$
$$k_9T^* - k_9T^*R + \sum_{1}^{n-1}k_7pVn[r]V' \quad (10)$$

Membrane sub-system

The membrane precursor, T', is irreversibly converted into T^* which can then react with R (the by-product of polycondensation) in a reversible reaction to produce T. T then reacts in an irreversible reaction with the membrane, T_m, producing T_{m+1} in proportion to surface area S.

$$\frac{d(T')}{dt} = k_4A_4 - k_4'A_5T' - k_8T' \quad (11)$$

$$\frac{d(T^*)}{dt} = k_8T' - k_9T^*R + k_9'T \quad (12)$$

$$\frac{d(T)}{dt} = k_9T^*R - k_9'T - k_{10}TS \quad (13)$$

$$\frac{d(S)}{dt} = k_{10}TS \quad (14)$$

Calculation of Volume.

The calculation of volume, Q, is the least realistic calculation of the current model. It is simply calculated as $Q = S^{3/2}$, with an initial volume of 1.0. Thus, we assume that the Chemoton remains a sphere. Although this assumption is valid before $[V']^*$, after threshold it will not be accurate as the volume will be less than that expected of a sphere of that surface area, because the Chemoton has deformed due to decreasing internal osmotic pressure, thus causing water to leave the cytoplasm. At each time step the ratio, Q(t)/Q(t+1), is multiplied by each of the values found for Eq 1-13 above, thereby adjusting the concentrations of the substrates, so that if volume is decreasing, the concentration of the substrates will be increased.

A time-step of 0.0001 is used, and the fastest reaction rate constant is 100. This model is deterministic and continuous.

Model I: Results

The dynamics of model I are shown in Figure. 2. The concentration of metabolites, A_1 to A_5, increases as X is metabolized. $[V']^*$ is exceeded in the second half the cell cycle. After replication, residues of T, T^*, R, and T remain, and this results in slow growth immediately after cell division. Templates at all stages of replication are present at the point of cell division, and are passed onto the daughter cells. However, template replication does not continue immediately after cell division as $[V]^*$ is not yet reached. The Chemoton

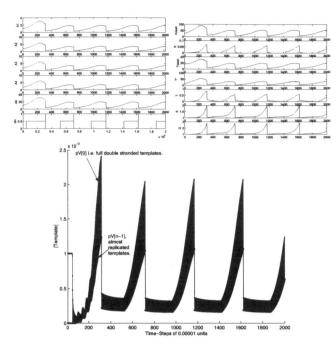

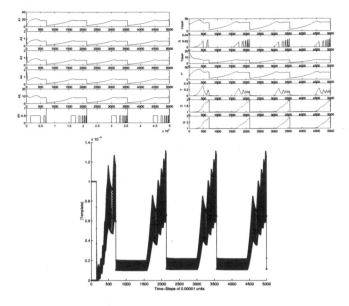

Figure 2: Top Left (1-5): Concentrations of A1-A5 over 4 cell cycles. Top Left (6): div = 0.5 when $[V'] > [V']^*$, and div = 1.0 at division. Top Right(1-6): V', R, T', T, Surface Area(S), and Volume (Q). Bottom: 25 lines, each showing the concentration of polymers (of length $N = 25$) at different stages of replication, i.e. $pV[0]$, is the full double stranded polymer, $pV[1]$ has one extra V' bound, and $pV[n-1]$ requires only one more V' to be bound for separation into 2 semi-conserved $pV[0]$ polymers once again.

settles into a stabalized generation time, with a stable composition of molecules being passed to the progeny.

Many of Csendes' findings are confirmed. If the rate constants for the backward reactions of the metabolic cycle are made equal to the forward reactions, as would be the case if the free energy of X were not greater than the free energy of the products of the metabolic cycle, $[V']^*$ is not reached. Even if the Chemoton is initialized without V', T' and T^*, after a brief transient state a stable cell cycle is re-established, demonstrating the stability of the Chemoton to changes in initial state. Even if only A_1 is present, after a long transient a normal cycle is re-established. Csendes' findings that generation times are slightly longer for higher $[V']$ are confirmed.

If $[X]$ is reduced to 10, replication time increases from 0.455 to 0.65, and when $[X] = 1.0$ replication time is 2.4, that is slower by a factor of only 5.3 compared to the replication time at $[X] = 100$. The explanation given by Bekes (Bekes, 1975) for this non-linear adaptability of the Chemoton was that the concentration of $[A1]$ showed a compensating increase such that the product k_1XA_1 was maintained at a high value. A_1 buildup would be caused by the slowness

Figure 3: Chemoton functioning at $[X] = 1.0$, compared to previously where $[X]$ was 100.0. Since $[V']$ cannot be maintained above threshold permanently, (see top Left(6)),R is produced in spikes.

of the reaction incorporating X into the metabolic cycle, at low $[X]$. We indeed observed that when $[X] = 1.0$, $[A1]$ increased to a maximum of 20.0 from a previous maximum at $[X] = 100$ of 2.5. All other membrane chemicals had decreased in concentration. Also, at low $[X]$, the Chemoton functions by fits and starts because $[V']$ cannot be maintained above $[V']^*$ continuously (see Figure. 3).

Template Length and its Effect on Replication Rate.

Csendes claimed that Chemotons containing longer templates could more effectively compensate for decreases in nutrient concentration. If this is true, it would indicate a selection pressure for longer templates. However, using model I, for both high and low $[X]$ values, we found replication time increased with increasing template length, see Figure. 4. What accounts for these contradictory findings? Ganti writes, "A longer template molecule (i.e. larger N) requires more molecules V' for its reproduction, thus decreasing the $[V']$ and allowing the cycle to function more intensively. Conversely, a shorter template molecule allows the cycle to function more slowly and less intensively thus the template molecule pVn carries information concerning the system." So, what is happening to $[V']$ in our simulation? Examining the details of replication for a Chemoton with $N = 1000$ and $N = 3$ respectively, we see that $[V']$ responds in a manner opposite to that predicted by Ganti. At $N = 3$, $[V']$-max is apx. 35, for $N = 25$, $[V']$ max is apx. 85, and for $N = 1000$, $[V']$ max is apx. 1000. So, longer templates are not 'mopping up' the V' molecules more rapidly, rather the

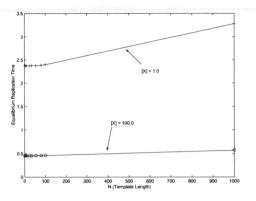

Figure 4: Replication time increases with longer templates, for both high and low nutrient concentration. Although difficult to see, the increase in replication time with increasing template length is monotonic.

opposite is the case.

In Csendes' model, and in ours so far, the rate of propagation of template polycondensation, ($k_7 = 10.0$) , is equal to the rate of initiation of polycondensation ($k_6 = 10.0$), therefore propogation proceeds at the same rate as initiation. According to these kinetics, it is merely the concentration of templates, irrespective of template length which will effect the rate of incorporation of [V']. What is more, with some consideration, it is predicted that template concentration will be less with increased template length, because a lower number of long templates are required to store all the V produced before cell division. This is indeed the case, for at $N = 3$, $pV[0]$ max = 0.4, at $N = 25$, $pV[0]$ max = 0.0025, and at $N = 1000$, $pV[0]$ max = 0.000007.

Are there any kinetic conditions for the basic Chemoton in which longer templates allow more rapid replication? What if $k_7 >> k_6$, such that the rate limiting step to polycondensation is initiation? We would expect that longer templates would have to undergo fewer rate limiting initiation steps before undergoing rapid proliferation, although, the same problem of lower template concentrations with longer templates would still persist. The same experiment as above was conducted but with $k_7 = 100.0$, see Figure. 5. The effect of template length in both conditions was very small. However, there was a qualitatively different profile compared with the previous experiment. At low [X], templates of length 50 are optimal. This could be explained by a trade off between maintaining high template concentration and reducing the number of rate limiting initiation events.

Of-course these findings are crucially dependent upon our particular choice of template dynamics. Were it the case that longer templates provided a greater number of simultaneously accessible binding sites for proliferation, then the ef-

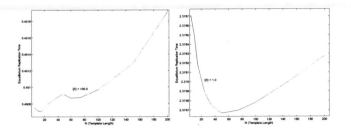

Figure 5: Left: At [X] = 100.0 the Chemoton exhibits slower replication with increased template length. The bump before this is unexplained. Right: At [X] = 1.0 the Chemoton shows a non-linear response to template length, with most rapid replication occuring at apx $N = 50$.

fect of reduced concentration could be counteracted. Alternatively, if the rate of proliferation of longer templates were greater, then the finding would also be reversed. The experiment above is conducted only on the most basic Chemoton without accounting for variable template length, enzymatic effects of specific template sequences upon metabolism, nor for the ability of templates composed of differing nucleotide ratios to simultaneously regulate alternative metabolic cycles. Thus, the relevance of such small effects of template length on replication rate may be of-course be irrelevant if other properties of long tamplates were to confer fitness.

Model II: Methods

In order to see whether the Chemoton could still function with a slightly more realistic model of template replication, a simulation was written that was loosely based on models by Breivik (Breivik, 2001), Kanavarioti and Bernasconi (Kanavarioti and Bernasconi, 1990) and Wattis and Coveney (Wattis and Coveney, 1999). This model consists of monomeric units, V', interacting by both hydrogen (Watson-Crick type base pair bonds) and phosphodiester bonds. The probability of bond formation and breakage respond differently to environmental fluctuations in monomer concentration. Hydrogen bonds link strands to form double strands, and phosphodiester bonds link monomers lengthwise along a single strand. We assume that a phosphodiester bond is formed with very high probability, (0.5), in the configuration shown in Figure. 6. In all other configurations, the formation of hydrogen bonds is assumed to be more probable than phosphodiester bond formation. Also, we assumed that hydrogen bonds tend to break at fairly low monomer concentrations, whereas phosphodiester bonds are much more stable to high monomer concentrations.

The model is a discrete probabilistic model coupled to the continuous deterministic model described previously. The polymers are modeled as a variable size array of data structures shown in Figure. 6, and the probabilities of reactions

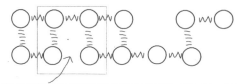

P(p bond formation) = 0.5 here because of configuration shown in box.

Figure 6: The data structure for a single polymer is a 2*M array, containing monomers, h-bonds and p-bonds. Potential bond sites can be deduced. The box shows the configuration necessary for high probability formation of a p bond, that is, the two monomers between which the p-bond can form must be hydrogen bonded to two monomers opposite. The two monomers opposite must already have a p-bond linking them to each other.

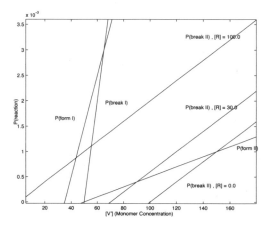

Figure 7: The dependency of the reaction probability of hydrogen and phosphodiester reactions upon $[V']$ and $[R]$ is shown. Bond I = hydrogen bond, and Bond II = phosphodiester bond. Where the lines for P(bond formation) and P(bond breakage) cross, there is an equilibrium.

are shown in Figure. 7.

For each site, the probability of reaction is $P(r) = 1 - e^{dE_r} : dE > 0$ and dE is calculated as shown in Eq. 15-19.

$$-dE(BondI\,form) = R_{hf}([V] - k_{hf}) \qquad (15)$$

$$-dE(BondI\,break) = R_{hb}([V] - k_{hb}) \qquad (16)$$

$$-dE(BondII\,form) = R_{pf}([V] - k_{pf}) \qquad (17)$$

$$-dE(BondII\,break) = R_{pb}([R] + [V] - k_{pb}) \qquad (18)$$

Where $R_{hf} = 0.0001, R_{hb} = 0.0002, k_{hf} = 35.0, k_{hb} = 50.0, R_{pf} = 0.00001, R_{pb} = 0.00002, k_{pf} = 50.0, k_{pb} = 100.0$. Thus, elongation and hydrolysis of chains can occur. However, several unrealistic features remain. We have not modeled the interactions between chains fully, i.e. reactions between polymers in separate data structures do not occur, so that spontaneous association of single-stranded oligomers are not modeled between data structures, only within data-structures. Detachment of oligomers is modeled, see below. The probabilities of binding and breakage,

except for the configuration in Figure. 6, are not dependent upon any structural feature of the polymer, e.g. a h-bond in the center of a double stranded polymer is equally likely to break as one on the edge of the polymer (i.e. no stacking reaction). Only one nucleotide, V' has been modeled. This reduces the problem of a combinatorial explosion in polymer configurations. Hydrolysis and dimerization of activated V' monomers is ignored, although dimers of V' can be formed by cleavage of longer strands. These decisions will clearly effect the findings, and future work must be to make these polymer dynamics more biologically plausible.

The simulation is initiated with 500 double stranded polymers of length 3. At each time-step, as well as calculating the membrane and metabolism equations, we calculate a probability of formation and breakage of type I & II bonds for each actual and potential binding site in each individual polymer. Based on these probabilities, the state of each polymer is changed. If two or more separate strands exist in the same data-structure, the detached strands (except one) are moved to new empty data-structures. Shorter oligomers dissociate faster than longer ones since more h-bonds need to be broken. The net change in $[V']$ caused by binding or release of V' from polymers is then calculated and incorporated into the differential equations for $[V']$ & $[R]$, after multiplication by a scaling factor (10). A sufficiently large number of polymers must be simulated to obtain a smooth net change in $[V']$. We assume R is produced in proportion to the incorporation of V' onto strands[2].

Model II: Results.

Figure. 8 shows the Chemoton initialized with 500 double stranded polymers of length 3, and with $[X]$ changed from 100 to 1.0 half way through the trial. The average polymer size drops from 6.0 to 5.5. The number of polymers (pV) is doubled rapidly when strands split, and this corresponds to a decrease in the average number of h-bonds. New p-bonds are formed immediately following the incorporation of monomers onto the single strand by the high probability p-bond formation when strands are in the configuration shown in Figure. 6. At low $[X]$, splitting of double strands by h-bond breakage is staggered since $[V']$ cannot be maintained above 50.0 for very long. Elongation takes place only rarely, since $[V']$ remains below 50.0 for much of the time.

---

[2]Future models will only have R produced when V' joins to the strand by a p-bond.

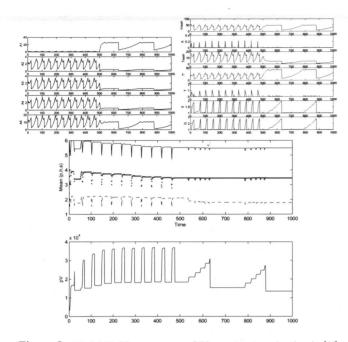

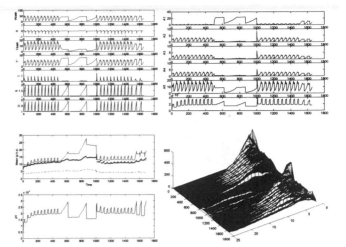

Figure 8: Model II: The response of Chemoton to reduction in $[X]$ is shown. Bottom(1): Mean (p,h,s) refers to mean number of p-bonds (dark line), mean number of h-bonds (dashed line) and the mean polymer size (no. of monomers in a polymer). Bottom(2): pV refers to the total number of monomers in the genome.

Figure 9: At First, $[X] = 100.0$, it is then decreased to 1.0, increased to 50.0 and reduced again to 10.0 $k_{pf} = 40.0$. Bottom Left: An equalibrium template length of 7 is reached with $[X] = 100.0$. When $[X]$ is reduced to 1.0, templates elongate further. Bottom Right: Absolute frequency of templates of different lengths, over the whole run.

Equilibrium template length depends on $[X]$ and p/h-bond probabilities.

Could changes in the relative probabilities of p-bond and h-bond formation effect the tendency for template elongation? The probability of p bond formation at low $[V']$ was increased by making $k_{pf} = 40.0$. Figure. 9 shows that, at high $[X]$ (100.0), elongation does occur until an equilibrium mean monomer number of 14 is reached, corresponding to double strands of length 7. At decreasing $[X]$ (1.0), there was a tendency for strands to elongate to longer lengths because $[V']$ did not exceed 50.0 as often, so that h-bonds were broken at a lower rate. This corresponds to a decrease in the total number of polymers being passed to each daughter cell.

Externally imposed $[V']$ oscillation verses Chemoton controlled $[V']$.

Figure. 10 (Top) shows the distribution of polymer sizes obtained in a trial with $k_{pf} = 35.0$ and $[X] = 50.0$. Compared to a trial in which oscillation of $[V']$ is imposed externally as a sine function, we see that a similar distribution of template sizes is obtained, see Figure. 10 (Bottom). However, to achieve this tight control of $[V']$ experimentally is extremely difficult, and such an experiment has not been conducted. The Chemoton provides a parsimonious explanation of how such oscillatory concentrations could be achieved.

Conclusions

In model I, it was demonstrated that under certain kinetic conditions, an optimal template length exists. Below this optimum length, template initiation is rate limiting, but above this length, although fewer initiation steps are required, the template concentration is decreased, so that propagation becomes rate limiting. In model II it was demonstrated that the equilibrium template length at any given $[X]$ was dependent upon the relative probabilities of h-bond and p-bond formation. When p-bonds could form at low $[V']$, at low $[X]$ there was a tendency for templates to grow as double strands and not separate as often, so decreasing the total template number passed onto each daughter cell. Thus, the length and concentration of templates conveyed information about the environmental conditions experienced by the parents. Finally, model II demonstrated that the tight coupling of the template state to the cell cycle in the Chemoton resulted in a different distribution of template sizes compared to that obtained when [V'] was oscillated by an external experimenter.

Further work will considerably improve the biological plausibility of the template dynamics, e.g. adding more realistic interactions between polymers in separate data-structures, adding the stacking reaction, and altering binding probabilities on the basis of experimental data. The hope is of providing a simple and viable alternative for experimentalists, (Rasmussen et al., 2003), of how an evolvable genotype-phenotype coupling could occur in a real protocell.

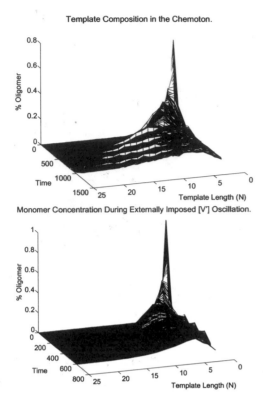

Figure 10: Top: Distribution of templates of average length N, calculated by summing the number of monomers in a polymer and dividing by 2, for the Chemoton, undergoing 4 divisions. Bottom: Distribution of templates for an in vitro study where $[V'] = 30 + 23 * \sin(500)(0.032)(0.00001).t$, is imposed externally to roughly match the $[V']$ observed in the Chemoton, but where $[R]$ is maintained at 0. $[V']$ is thus, un-reactive to template state. In practice, achieving this sine wave would be extraordinarily difficult. Slightly longer templates are obtained by the Chemoton.

Acknowledgements

Special thanks to Eors Száthmary for discussions, advice, references, and the idea for this paper. Also thanks to Simon McGregor, Inman Harvey, Phil Husbands, Naoaki Ono, and Tim Hutton. Thanks also to my parents and kind family in Sri-Lanka where this paper was written.

Appendix

Initial conditions in model 1 : [A1] = 1.0, [A2] = 1.8, [A3] = 1.9, [A4] = 1.7, [A5] = 10.0, [V'] = 26.0, [T'] = 17, [T*] = 14, [T] = 0.0, [R] = 0.0, [X] = 100.0, [Y] = 0.1, [pVn] = 0.01, Surface Area (S) = 1.0, Volume (Q) = 1.0, Polycondensation Threshold ([V']* = 35.0), Template Length (N) = 25.

Reaction Rate Constants in model 1: k1 = 2.0, k2 = 100.0, k3 = 100.0, k4 = 100.0, k5 = 10.0, k6 (if $[V'] > [V']*$) = 10.0 else k6 = 0, k7 = 10.0, k8 = 10.0, k9 = 10.0, k10 = 10.0 k1' = k2' = k3' = k4' = k5' = 0.1, k6' = 1.0, k9' = 0.1

References

Bachmann, P., Luisi, P., and Lang, J. (1992). Autocatalytic self-replicating micelles as models for prebiotic structures. *Nature*, 357:57–59.

Bekes, F. (1975). Simulation of kinetics of proliferating chemical systems. *Biosystems*, 7:189–195.

Breivik, J. (2001). Self-organization of template-replicating polymers and the spontaneous rise of genetic information. *Entropy*, 3:273–279.

Csendes, T. (1984). A simulation study of the chemoton. *Kybernetes*, 13:79–85.

Eigen, M. (1971). Self-organisation of matter and the evolution of biological macromolecules. *Naturwissenshaften*, 58:465–523.

Ganti, T. (1971). *The Principle of Life(In Hungarian)*. Budapest: Gondolat.

Ganti, T. (2003). *The Principles of Life*. Oxford University Press.

Ganti, T. (2004). *Chemoton Theory Vol I and II*. Kluwer.

Kanavarioti, A. and Bernasconi, B. (1990). Computer simulation in template-directed oligonucleotide synthesis. *Journal of Molecular Evolution*, 31:470–477.

Rasmussen, S., Chen, L., Nilsson, N., and Abe, S. (2003). Bridging living and non-living matter. *Artificial Life*, 9:269–316.

Szathmáry, E. (2000). The evolution of replicators. *Philosophical Transactions of the Royal Society of London. B.*, 355:1669–1676.

Szathmáry, E. and Maynard-Smith, J. (1997). From replicators to reproducers: the first major transitions leading to life. *Journal of Theoretical Biology*, 187:555–571.

Wattis, J. and Coveney, P. (1999). The origin of the RNA world: A kinetic model. *Journal of Physical Chemistry B.*, 103:4231–4250.

Connecting Transistors and Proteins

Claudio Mattiussi[1], Dario Floreano[1]

[1] Autonomous Systems Laboratory, ASL-I2S-STI-EPFL, 1015 Lausanne, Switzerland
claudio.mattiussi@epfl.ch

Abstract

We connect transistors and proteins in two ways. The first is by showing that they have much in common as fundamental devices of electronics and life. The second is by describing how an evolvable wiring of electronic devices can parallel the wiring of proteins into genetic regulatory networks. We then transform this connection into a methodology for the study of the evolutionary properties of circuits. The approach is based on the use of analog electronic circuit simulators. We present an example of implementation with the first results obtained.

Introduction

Many functions within living cells are performed by proteins in their role as catalysts (Alberts et al. 2002, Creighton 1993). In the simplest scenario, a chemical substance generically called the substrate must be converted into another substance called the product. The free energy of the substrate is higher that that of the product. Hence, the former would convert spontaneously in the latter. However, the conversion requires the passage through a less favorable transition state. In the absence of the catalyst, the barrier constituted by the transition state keeps the reaction rate low. The effect of the catalyst is to lower the barrier and thus accelerate the reaction rate.

The operation of active semiconductor devices such as transistors is conceptually similar. For example, a bipolar junction transistor (BJT) is composed by three adjacent regions of semiconductor having different physical characteristics (Cooke 1990). These regions are called the emitter, the base, and the collector. In the typical circuit configuration, the voltages applied to emitter and collector, make energetically favorable the flowing of current carriers from emitter to collector. This current, however, must pass through the base. When the base is left unconnected, it acts as a barrier to the current flow, which is therefore small. A suitable voltage applied to the base lowers this barrier with the effect of increasing the current flowing from emitter to collector.

These descriptions reveal a striking analogy in the operation of proteins and transistors (Figure 1). Evolution designed the basic devices of life just as engineers designed the basic devices of electronics. Both kinds of devices permit the variation of the rate of some physical process. In other words, they are the key to the implementation of constraints to the spontaneous dynamics of those physical processes. As argued by Pattee (Pattee 1995), natural selection leads indeed to the formation of structures whose presence influences the dynamics of the surrounding space-time in ways that favor the persistence and, eventually, the self-reproduction of these structures. If Pattee's intuition is correct, we should therefore expect to observe the emergence of devices performing this kind of function within our synthetic experiments on the evolution of life. Thus, we can take a major evolutionary shortcut if we adopt directly these structures as our basic building blocks. Note that this does not spoil our inquiry since, as we will argument below, there remains to study the crucial aspect of the establishment of the connectivity between the structures.

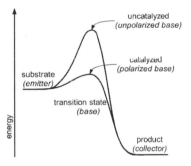

Figure 1: The analogy between the stages of a chemical reaction, and the regions of a bipolar transistor. The vertical axis represents the free energy of the substances (substrate, transition compound, product) in the course of the chemical reaction or the energy of the current carriers in different regions (emitter, base, collector) of the transistors body. The presence of the catalyst decreases the height of the barrier that hinders the transformation of the substrate into the products, just as a suitable polarization of the transistors base decreases the barrier that hinders the flow of current carriers from the emitter to the collector.

If we follow this approach, we can take advantage of the existence of analog electronic circuits simulators (Vladimirescu 1994). In these simulators, the physics of the devices and of their interaction is modeled at high level, through a set of algebraic and differential equations, which embed the relevant conservation laws of physics. The resulting implementation is efficient and physically sound. Besides, by using the models of energy storing components such as capacitors, this approach allows the modeling of delays to the propagation of signals, a phenomenon that affects also chemical signals that must diffuse across spatially extended structures in cells.

Signals and connections

A collection of unconnected devices performs no function. The task of the engineer and of evolution consists in finding how to connect the available components to obtain the desired behavior.

In the biological case the connectivity corresponds to the network of interactions between the elements. For genetic regulatory networks and omitting many details (for the full picture see for example Alberts et al. 2002) the interactions can be schematized as follows (Figure 2). A protein interacts with a stretch of a DNA, and activates a transcription machine called RNA polymerase. Each cell contains many copies of a few types of these transcription machines. The output of the machine is a molecule of RNA, which, after a number of further steps, leads to the synthesis of a protein. This protein can in turn interact with the DNA, to activate or repress the transcription of another sequence of DNA, and so on. The identity of the connected elements and the strength of the interaction depend on the chemical nature of the participants in a way that we will describe below.

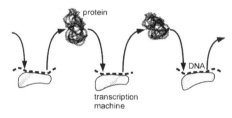

Figure 2: A very schematic representation of the interactions that compose a genetic regulatory network. The mediation is due to proteins that can activate or repress the functioning of specialized transcription machines.

In the case of analog electronic circuits, the connection between devices is determined by conducting wires that guide the signals. The strength of the connection can be varied by changing the value of resistance of the connection, from a minimal value of zero, which corresponds to the maximum strength of the interaction, to the absence of direct interaction, which corresponds to a virtual infinite-valued resistance (Figure 3).

Inspired by this similarity, we could thus imagine to evolve an electronic circuit by determining the connections between the electronic devices in a way that reminds that used in biological systems. Since in the biological case it is

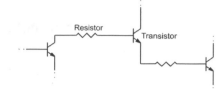

Figure 3: The interactions between electronic devices are mediated by conducting wires connecting the terminals. The value of resistance of the connection determines the strength of the interaction.

the characteristics of two DNA sequences that determine the existence and the strength of the interaction (Figure 2), we could imagine to associate a sequence of characters to each terminal of the circuit components. Then we could define a mapping of pairs of sequences in order to determine the existence of a connection and its strength. By coding those sequences in an artificial genome, we could then parallel the process of evolution of biological circuits (Figure 4).

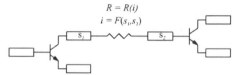

Figure 4: The strength of the connection between devices can be assigned by associating sequences of characters to the terminals, defining a mapping $i = F(s_1, s_2)$ from pairs of sequences to values of connection strength, and a further mapping $R(i)$ that gives the values of resistance.

The protein folding objection

Before we proceed to detail the nature of the mapping outlined above, there is a major objection to the whole program that must be addressed. After the transcription of the DNA sequence into an RNA molecule, the latter undergoes a series of transformations that convert it into a chain of amino acids. To become a functional protein, this chain must fold into a precise three-dimensional shape (Alberts et al. 2002). In living cells the folding usually proceeds effortlessly. However, the simulation of this process appears computationally daunting. Thus, if the decoding of sequences into interaction strengths requires the computation of an equivalent of the folding process, our suggested approach becomes computationally impractical.

Some authors (for example, Conrad 1999) have argued that the characteristics of the folding process are unique in determining the evolvability of living systems. We can think of this processing as a mapping from the space of sequences to the much higher dimensional space of protein shapes. This mapping provides redundancy to the evolutionary process, thanks to its being potentially many-to-one; it brings a degree of smoothness to the discrete universe of DNA sequences, but still allows abrupt discontinuities in shape with only a few nucleotides substitutions. Moreover, it determines the shaping of proteins that gives them their specificity, and leads to the phenomenon of allostery (a phenomenon that consists in the change of shape of the protein in presence of physical or chemical signals).

Fortunately, molecular biologists have discovered that at the level of genetic regulatory network, things seem to be simpler than was previously imagined (Ptashne and Gann 2002). In the case that interests us, the process of transcription proceeds as illustrated schematically in Figure 5. Here a single protein (called activator) is assumed to be in charge of the activation of the transcription of a certain DNA sequence. This protein recognizes a sequence of nucleotides along the DNA and binds to it in a well-defined position. Then it recruits the transcription machine to the DNA through another binding interaction, which is sufficient for the transcription to start and proceed autonomously. It turns out that in most cases, the activator does not need to alter – as was instead previously imagined – the transcription machinery, for example with some complicated allosteric interaction. The only specific interaction is the readout of the sequence of DNA that binds the activator. The surprise of molecular biologist at this finding is witnessed by the following extract (Ptashne and Gann 2002, p. 176)

That so much of the specificity of regulation and hence so much of development and evolutionary change depends on simple binding interaction is (or we think should be) hard to swallow. It certainly is for us. We, and we suspect many others, had expected that the meanings of biological signals would have been, somehow, more solidly based.

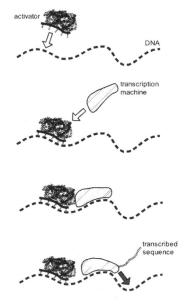

Figure 5: In many cases, the regulation of the transcription of DNA sequences depends on simple binding interactions which correspond to the recognition of a sequence of nucleotides, and to a generic adhesive recruitment of the transcription machinery.

The consequence of this finding is that we can hope to obtain an equivalent of the protein-mediated interaction in terms of a mapping from pairs of sequences which does not imply the complexity of protein folding. Note that we are not saying that protein folding, three-dimensional shape and allostery play a minor role in the existence of living beings. These phenomena are obviously essential in a world where significant physical and chemical signals, the energy sources, the strength of materials, the dynamics of motion, and many other essential aspects, are imposed from the outside and must be complied with in order to survive. What we observe is merely the contingent fact that where living systems "talk to themselves" and are free to define their own language for example in exchanging internal signals across genetic regulatory networks they appear to employ forms of interaction where allostery and three-dimensional shape play at most a generic role, and where the communication can be interpreted as a sequence-to-sequence correspondence (Figure 6).

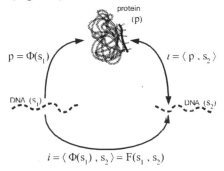

Figure 6: The interaction between two DNA sequences within genetic regulatory networks can be schematized by composing the mapping $F(s_1)$ that converts a sequence s_1 into a folded protein p, and the mapping $\langle p,s \rangle$ representing the interaction of the protein p with another DNA sequence s_2. Since the resulting process is based on simple binding interactions, we can hope to model it with a computationally tractable mapping $F(s_1,s_2)$ that gives the strength of interaction i.

Defining and decoding the genome

We can now proceed to the definition of an evolutionary system based on the ideas presented in the previous section. The first thing that we need to specify is the structure of the genome. The genome must contain at least the description of the devices and the sequences of characters associated with the terminals of the devices, which determine the strength of the connections (Figure 4). It is useful to have also the possibility to evolve the value of some parameters associated with the devices, for example the capacitance value of a capacitor, or some parameter of a transistor.

To fulfill these requirements, we use a genome constituted by a sequence of characters. Each kind of device is identified by a token of a few characters, for example "NBJT" for an NPN BJT, and "CAPA" for a capacitor. We define two other tokens: one relative to the terminals, for example "TERM", and one to the parameters of the device, for example "PARM". Note that we chose human-readable tokens just to facilitate the visual inspection of the genomes.

The decoding proceeds as follows. We search in the genome the first token identifying a device, which signals the start of the fragment of genome coding for that device. Each kind of device has a characteristic number of terminals and evolvable parameters, and we search that number of terminal and parameter tokens in that fragment. If all the required tokens are found before the end of the chromosome (or before the next device token, if no overlap of device descriptors is allowed), a device – for the moment, unconnected – is created in the circuit. The sequences of characters delimited by the tokens are associated with the terminals and parameters of the device (Figure 7). When the terminals of the device are not interchangeable, an order is specified for them and the association of extracted sequences follows that order. Once a device has been decoded we proceed to search the next device token in the genome, and so on until all the genome has been examined.

Genome fragment

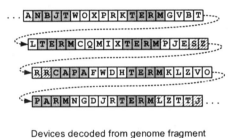

Devices decoded from genome fragment

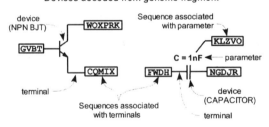

Figure 7: A fragment of genome (top) and the corresponding devices decoded from it (bottom). A series of tokens (shaded) identify the start of coding regions and delimit the sequences of characters associated with the terminals and evolvable parameters of the devices. The hatched characters correspond to noncoding genome.

Connecting the evolved components

The result of the process just described is a collection of devices with sequences of characters associated with their terminals and parameters. To connect the devices we proceed as follows. We define a mapping $F(s_1, s_2)$ that transforms pairs of sequences in a scalar value i that represents an abstract interaction strength. For each pair of terminals

of the collection of devices extracted from the genome, we calculate the interaction strength determined by their associated sequences. Then, we transform i into a resistance value with a predefined mapping $R(i)$, and we insert in the circuit a resistor connecting the two terminals and having $R(i)$ as resistance value (Figure 4). We will give below an actual example of the mappings $F(s_1, s_2)$ and $R(i)$.

An analogous process assigns the values to the parameters of the devices. With each evolvable parameter is associated a sequence s of characters extracted from the genome (Figure 7). We define a fixed sequence v that will be used for all the devices parameters, we evaluate $i = F(s, v)$ and transform the result into the parameter value with another mapping, for example $C(i)$ for the capacitance of a capacitor.

External connections

A living system is connected to the external world, to absorb energy and matter, expel waste, exchange signals. The evolution of these interactions is actually a topic of major interest for ALife (Bedau et al. 2001). In the case of an electronic circuit, this corresponds to the presence of external devices or circuits, such as power supplies, signal generators and output loads. We must thus specify how our decoded circuit connects to these external parts and how these connections can evolve. The simplest solution consists in associating predefined fixed sequences to the terminals of the external devices that must connect to the evolved circuit (Figure 8), so that the connection strategy described for the devices decoded from the genome can be extended to the external devices, and evolution of these connections is possible. For more complex approaches to the establishment of external connections, see (Mattiussi and Floreano 2004).

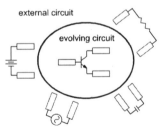

Figure 8: By associating sequences of characters to the terminals of the external devices, connections from the circuit specified by the genome (shaded region) to the external devices can be subjected to evolution.

Compartments, modules and hierarchies

With the decoding strategy described above, the strings associated with the terminals implicitly determine the connections between all the terminals in the external and decoded circuit. This frees the genome from the necessity of specifying explicitly all those connections. However, this comes at the cost of calculating the value of the mapping $F(s_1, s_2)$ for all pairs of terminals. The number of evaluations grows quadratically with the number of devices in the circuit. At the same time, the function $F(s_1, s_2)$ cannot be too simple without compromising the evolvability of the system. Therefore, the computational cost of the decoding could become intolerable as the complexity of circuits grows.

A solution to this problem is the inclusion in the tokens for terminals, of an evolvable marker for the compartment to which the terminal belongs. In this way, only the connections for pairs of terminals belonging to the same compartment would have to be considered. At the same time, the system would have the possibility of evolving a compartmentalized or modular architecture, with all the advantages that this entails (Kazadi et al. 2000). Minor elaborations on this strategy, may allow the evolution of hierarchical and multicellular structures.

Genetic operators

The genome as defined above can be composed of several distinct sequences of characters that we can call chromosomes. The structure of the genome and of the decoding process permits the execution of many genetic reorganization operations that are known to apply to biological genomes (Graur and Li 2000) but are seldom used in artificial evolution experiments because they usually make the genome undecodable. In our case, besides the usual substitution of single characters, we can perform operations such as insertion and deletion of them; operations on chromosome fragments, such as duplication, deletion, transposition, recombination of pairs of chromosomes, and insertion of component descriptors; operations on whole chromosomes, such as duplication and deletion; and the duplication of the whole genome. The possibility of performing such operations is important, since they are assumed to play a crucial role in the evolution and complexification of living systems (Graur and Li 2000). Note that from the point of view of the genetic operators each chromosome is just a sequence of characters where the tokens for devices, terminals and parameters (Figure 7) have no special meaning. Therefore, the tokens are *not* protected from the action of the genetic operators, whose action can invalidate any device descriptor present in the genome, making that particular descriptor undecodable.

An example of implementation of the mapping

So far, we have described only in abstract terms the mapping $F(s_1, s_2)$ that transforms pairs of sequences into interaction strengths i, and the function $R(i)$ that gives the value of connecting resistors. We will describe now briefly the characteristics of that mapping for an actual implementation of the system, along with some results obtained performing evolutionary runs with the implementation.

The genome is composed by sequences of uppercase alphabetic ASCII chars. To derive a connection strength from pairs of sequences extracted from the genome, we use local

sequence alignment (Sankoff and Kruskal 1983). The basic idea is that subsequences of one sequence can be put in correspondence with subsequences of the other through operations of insertion, deletion and substitution of characters (Figure 9). To each operation is assigned a score that rewards close matches and the absence of insertions and deletions. The value of the local alignment score $i = F(s_1, s_2)$ of two sequences s_1 and s_2 is defined as the maximum value of the sum of the scores that can be obtained putting in correspondence a subsequence of s_1 with one of s_2. Some favorable properties of this mapping are its high redundancy, the possibility to operate with sequences of variable length, and the possibility due to the locality of the alignment of matching several distinct sequences with a single one.

The values of i obtained are non-negative integers. These are transformed in resistance values through a table of correspondences. This means that there is a finite set of possible values, but this is not a serious limitation; for example, the number of commercial values available to engineers for their designs is also finite. A whole range of values of i corresponds to the zero-valued resistor (direct connection), and another one to the infinite-valued resistor (no direct connection).

Figure 9: The local alignment of sequences is based on the establishment of correspondences between the subsequences of two sequences of characters, using operations of insertion, deletion, and substitution of characters.

Experiments

We ran a first series of experiments of circuit evolution using SPICE as simulator (Vladimirescu 1994). The experiments were targeted at the synthesis of a circuit giving a constant voltage as output, in presence of a variable input voltage and environment temperature. This problem is interesting in an ALife perspective, since the solution implies the evolution of capabilities of measurement and control (Pattee 1995). We obtained good results, while observing biologically evoking phenomena such as phenomena of gene overlapping (Graur and Li 2000)) and the appearance of vast zones of noncoding genome.

A logarithmically distributed set of resistance values was used to connect the components using the string alignment technique. This resistance set covers 6 decades with 8 values per decade, from 1Ω to $1M\Omega$. The 1Ω resistance value corresponds to an alignment scores of 20, whereas the $1M\Omega$ value is associate with a score of 68. The whole range of scores below 20 is associated with an infinite-valued resistor (no connection) and that above 68 is associated with a zero-valued resistor (direct connection).

The external circuit is the one represented in Figure 10, where the voltage of the power supply (left) can vary from 4V to 6V, the source resistance is 1kΩ, and the load resistance (right) is 1kΩ. The goal is the generation of a constant 2V voltage across the load when the temperature varies from 0°C to 100°C. To this end, the decoded circuits where simulated with a power supply voltage varying from 4V to 6V in steps of 0.1V, with simulation temperatures of 0°C, 25°C, 50°C, 75°C, and 100°C. For each power supply voltage and circuit temperature, the square of the difference between the actual voltage on the load and the required output voltage was computed. The fitness was defined as the opposite of the sum of all these squares, so that the goal was the maximization of the fitness, with optimal value zero.

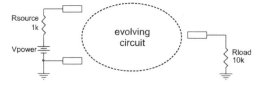

Figure 10: The components of the external circuit for the voltage reference evolutionary experiment.

The genome of all the individuals of the initial population was constituted by one chromosome containing as devices 10 NPN BJT descriptors. The terminal sequences for all the devices had an initial length of twenty characters, randomly filled with elements of the genetic alphabet. We used a genetic algorithm with tournament selection, tournament size of 5, and elitism. The size of the population was 100. The probabilities of nucleotide insertion, deletion, and substitution, those of chromosome duplication and deletion and the probability of genome duplication were set to 0.001. The probabilities of chromosome fragment duplication, deletion, transposition, and that of chromosome single point crossover were set to 0.01. Chromosome fragment reorganization was performed by selecting two random points in the source chromosome to define the fragment, and one random point in the target chromosome, when required. Figure 11 illustrates the course of the best of four runs of 10000 generations evolution. The evolved circuit gives an output voltage that stays within ±1.5% of the prescribed value in the whole temperature and input voltage range. For further details see (Mattiussi and Floreano 2004).

Conclusions

We have presented a methodology to genetically represent and evolve collections of interconnected elements. The technique allows the variation in the course of evolution of the number of elements and of the number and strength of the connections between them. The approach is biologically motivated by the interaction of genes and proteins within genetic networks but does not imply the implementation or the mimicking of the details of protein folding and chemical

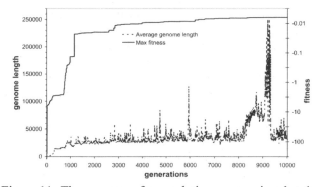

Figure 11: The progress of an evolutionary run aimed at the synthesis of a voltage reference circuit.

reactions. The resulting genome tolerates drastic reorganizations such as duplications and transpositions, which appear instrumental to the open-endedness of the evolutionary process. The first results obtained with this representation witness the evolutionary potential of the proposed approach.

Acknowledgments

This work was supported by the Swiss National Science Foundation, grant no. 620-58049.

References

Alberts, B, et al. 2002. Molecular Biology of the Cell (4th edition). New York : Garland Science.

Bedau, M.A. et al. 2001. Open Problems in Artificial Life. Artificial Life 6:363-376.

Conrad, M. 1999. Molecular and evolutionary computation: the tug of war between context freedom and context sensitivity. Biosystems 52: 99-110.

Cooke, M.J. 1990. Semiconductor Devices. New York: Prentice Hall.

Creighton, T.E. 1993. Proteins: Structures and Molecular Properties (2nd edition). New York: W.H. Freeman.

Graur, D., and Li, W.-H. 2000. Fundamentals of Molecular Evolution. Sunderland, Mass: Sinauer.

Kazadi, S. et al. 2000. Levels of Compartmentalization in Alife. In Proceedings of the 7th International Conference on Artificial Life, 81-89. Cambridge, Mass: MIT Press.

Mattiussi, C., and Floreano, D. 2004. Evolution of Analog Networks using Local String Alignment on Highly Reorganizable Genomes. in Proceedings of the 2004 NASA/DoD Conference on Evolvable Hardware. Forthcoming.

Pattee, H.H. 1995. Artificial Life Needs a Real Epistemology. In F. Moran et al. eds. Advances in Artificial Life, 23-38. Berlin: Springer.

Ptashne, M. and Gann, A. 2002. Genes & Signals. Cold Spring Harbor, New York: Cold Spring Harbor Lab. Press.

Sankoff D., and Kruskal J.B. eds. 1983. Time Warps, String Edits, and Macromolecules: The Theory and Practice of Sequence Comparison. Reading, Mass: Addison-Wesley.

Vladimirescu, A. 1994. The SPICE Book, New York: Wiley.

Designed and Evolved Blueprints For Physical Self-Replicating Machines

Efstathios Mytilinaios, Mark Desnoyer, David Marcus, and Hod Lipson

Computational Synthesis Lab, School of Mechanical & Aerospace Engineering,
Cornell University, Ithaca NY 14853, USA
hod.lipson@cornell.edu

Abstract

Self-replication is a process critical to natural and artificial life, but has been investigated to date mostly in simulation and in abstract systems. The near absence of physical demonstrations of self-replication is due primarily to the lack of a physical substrate in which self-replication can be implemented. This paper proposes a substrate composed of simple modular units, in which both simple and complex machines can construct and be constructed by other machines in the same substrate. A number of designs, both hand crafted and evolved, are proposed.

Introduction

Self-replication is a process critical to both natural and artificial life, but has been investigated to date mostly in simulation and in abstract systems (Sipper and Reggia, 2001). Even in formal systems, self-replication has been difficult to define consistently (Nehaniv and Dautenhahn, 1998; Sanchez *et al* 1997; McMullin, 2000). Self-replication has been examined empirically numerous times, mostly in the context of replicating programs, cellular automata, and artificial chemistries (e.g. Arbib, 1965; Burks, 1970; Moore, 1970; Langton, 1984; Lohn and Reggia, 1997). In contrast, physical machines capable of self-reproduction have been scarcely discussed.

The practical potential of physical self-reproduction was recognized in the early 80's as a possible method for remote colonization (Freitas and Gilsbreath, 1981) but was abandoned due to many unresolved technical difficulties. More recently, interest is being revived in this area as a fabrication paradigm for both macroscopic space applications and for micro- and nano-scale systems (Freitas, 2003). At small scales, such as at the molecular level, self-replication can occur through *stochastic* self-assembly processes catalyzed by a self-replicating entity. At large scales, such as multicellular organisms and robots, self-replication through stochastic processes is energetically implausible, and explicit reproduction must take place.

Artificial physical self-replication at macro-scale was first demonstrated using stochastic tumbling blocks with special geometries (Penrose, 1959). Deterministic self-reproduction of robotic systems has only recently been demonstrated by Chirikjian (2002), where a Lego™ robot composed of three modules was able to assemble three other modules into a new identical robot. To do so, the base module of the robot followed a path drawn on the ground, pushing the other modules and joining them into an assembly using magnetic connections or with the assistance of an external passive joining rig.

Chirikjian's work demonstrated a physical, deterministic self-reproducing machine for the first time, but left more to be desired. If a robot composed of two components is able to reproduce by actively assembling these two components together, then self-replication has indeed occurred, but in a way nowhere as impressive as, say, a machine able to reproduce itself from raw materials or as biological life that is able to reproduce from amino acids.

The apparent existence of different *levels* of self-replication has led us to abandon the view of self-replication as a binary quality, simply existing or not. Instead, we see self-replication (or self-reproduction) as a continuum, quantifiable based on the amount of information being replicated. For example, the amount of information needed to reproduce a two-component robot *given* these two components is less than the amount of information needed to reproduce a robot given many lower-level components. Similarly, the amount of information needed to reproduce an inaccurate copy of a machine is less than the amount of information necessary to produce a more exact duplicate. The amount of information needed to assemble a machine that is independently likely to spontaneously appear in a domain is less than the information needed to assemble a unique machine unlikely to appear by chance. The amount of information involved in the replication is dependent on the definition of the replicating system and the contribution of the environment to the replication process. Based on these assumptions, we have formally defined a domain-independent metric of self-replication (Adams and Lipson, 2003). The metric compares the probability of spontaneous emergence of a system in a given environment, to the probability of emergence of a system given that one instance of the system is already present in the environment. This metric not only avoids some of the difficulties of earlier definitions (how to deal with trivial replicators, and how to apply it to non-formal systems), but also provides a graded value that is more amenable to evolution, if a self-replicating machine is to be evolved directly.

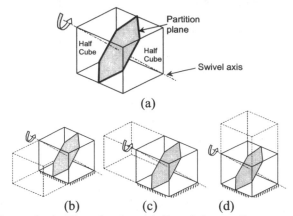

Figure 1: A 3D molecube: (a) Two halves split across the (1,1,1) plane. (b,c,d) Swiveling the top half causes an adjacent block to cycle into new configurations.

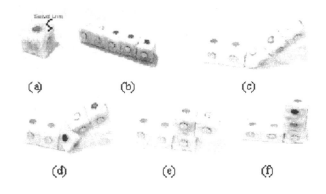

Figure 2: Physical model of a 3D molecube: (a) A single block, with magnets and swivel line. (b-f) a sequence of swivels that transform an object from a line to a Z-shape to an L-shape.

Following this formal definition, we have sought a physical substrate in which a variety of self-reproducing machines can systematically be constructed, ranging from simple to complex replicators. We also seek a substrate that is both physically plausible and faithfully simulatable, so that questions about the origin of physical self-replication can be studied.

This paper reports on some of our progress towards this goal. We first describe a physically-plausible substrate, both in its two-dimensional and three-dimensional versions. We then show a number of hand-designed machines and evolved machines, and the way in which they were evolved. Finally, we show a preliminary physical implementation of this substrate.

A space for physical self-replication

The space of machines we put forward is based on a single cubical building block – a *molecube*. Each cube, shown in Figure 1 below, can connect to its adjacent neighbors. Patterns on the surface of the cubes assure that the assemblies are perfectly aligned when joined. Specific bonding and patterning mechanisms depend on the scale of the implementation, and can be magnetic, electrostatic, or hydrophilic/hydrophobic, for example.

A cube is split into two parts along a plane that is perpendicular to its long diagonal (e.g. 1,1,1); the plane is shaded in Figure 1a. One half of the cube can swivel about the long axis in increments of 120 degrees, each time cycling the faces of the cube. If magnets on all three faces on one half have a polarity of north, and magnets on the three faces of the other half have a polarity of south, then polarities are retained after swiveling and then the structure will always have consistent global polarities.

Swiveling a cube while other cubes are attached to it causes a reconfiguration of the structure. For example, the cube shown in Figure 1b has another cube on its left. Swiveling this cube once causes the adjacent cube to move to the back (Figure 1c). Swiveling once more causes the

adjacent cube to move to the top (Figure 1d). Swiveling again will restore the original configuration (Figure 1b). Each cube thus has 3 possible swivel states, and 4 possible orientations for the swivel axis.

Figure 2 shows a physical model of a set of five building blocks. Two of these blocks are swiveled 120 degrees in turn, causing the entire structure to reconfigure into a Z-shape and then into an L-shape. Other swivels may yield three-dimensional configurations.

Structure topologies may have loops and branches, and multiple blocks can swivel simultaneously. However, while swiveling, a structure needs to go through intermediate states. Due to collisions some of these configuration transformations may not be possible. Similarly, bonding will occur if two cubes become adjacent during reconfiguration and their magnets are in attractive polarities, and the structure will lock. Other physically-realistic constraints may be placed on the structure depending on its environment, such as collision with the ground, gravitational stability, actuation torque limits and motion dynamics. However, because the end positions of the cubes lie at regular grid locations, actuation sequences can be calculated rapidly and simulated without accumulation of error.

A second form of actuation is the ability to change polarities of the faces, so that adjacent faces attract, repel, or are passive. If magnetic bonding is used, then electromagnets can switch between 'north', 'south' and 'off' states. Each cube thus has 3^6=729 possible states. Transitioning states allows a block to pick up, hold and drop other blocks or groups of blocks, as well as grip and climb other structures.

Control of the machine is specified using a sequence of swiveling and bond-state switching commands, executed in open loop (without feedback). It is possible to envision more elaborate controllers that incorporate sensing, branching, memory, and stochastic elements, as well as distributed control where cubes and groups of cubes execute programs locally.

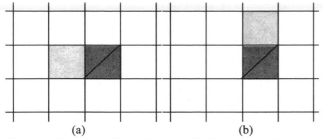

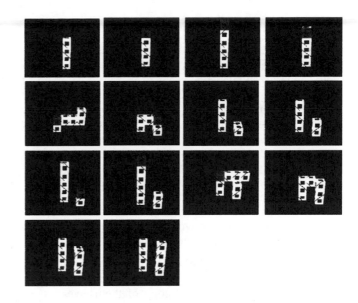

Figure 3 (above): A 2D molecube: (a) Two halves split across the diagonal. (b) Swiveling the top half causes any adjacent blocks to cycle into new configurations.

Figure 4 (right): A self-assisted reproduction sequence (hand designed): Top-left: The original structure contains four cubes. New cubes are dispensed from the top. The machine positions these cubes and the newly formed structure reconfigures during the reproduction to assist in its own construction.

To maintain physical plausibility, power for actuation and bonding is provided through the ground, and transmitted via local connections. Battery operation is possible though inconvenient for physical implementation due to excessive weight and short operation runs.

A two dimensional space

A simpler version of this space exists in two dimensions. In this case each cube is a square; the square is split along its diagonal. A swivel of the square causes two faces to switch as shown in Figure 3. Each square thus has 2 possible swivel states, 2 possible swivel axes and $3^4=81$ possible bonding states. Since the swiveling motion causes the squares to go out of plane during transition, no intermediate collisions exist. Because of this reduced space size and simpler physics, a two-dimensional molecube space is amenable to fast simulation.

The molecube space can also be generalized to other dimensionalities, though it is more difficult to visualize. For example, each four dimensional unit would have 4 possible swivel states, 6 possible swivel axes, and 3^8 possible bonding states.

Possible modes of self-replication

There are a number of ways self-replication may occur in a molecube space. Since any replication process requires material, we assume some grid positions may act as *dispensers*, where new cubes reappear when removed from that location. A machine is considered replicated only when the new copy is *identical* and *detached* from its parent.

- **Direct reproduction**: A machine reconfigures to pick cubes from a dispenser and place them in a new location, gradually building a copy from the ground up.

- **Tandem reproduction**: Multiple machines are required to produce a single copy. One machine may place cubes while the other reorients the constructed machine.
- **Self-assisted reproduction.** The machine being constructed reconfigures during the construction process to facilitate its own construction.
- **Multi-stage reproduction.** Intermediate constructions are required before the target machine can be made. The intermediate machine is then discarded as a waste product, or can be used to catalyze the production of additional machines.

Various combinations and extensions of these modes of operation are possible, and one may imagine an ecology of competing and cooperating machines.

Manually-designed replicators

We initially explored this space manually, looking for possible self-replicating designs. Exploration of the three-dimensional space was carried out using a simulator, able to simulate arbitrary 3D molecube structures and execute sequences of swivel and bonding commands. The simulator accounts for collisions during transformation, loops and locked structures, as well as incompatible bonding polarities and maximum torque loads due to gravity and moment arms.

A number of self-reproducing machines (structure + control) were found. Figure 4 shows one of the simplest machines containing 4 cubes. New cubes are dispensed from the top. The machine positions these cubes, and the newly formed structure reconfigures during the reproduction to assist in its own construction. This is a form of self-assisted reproduction. Larger and more complex machines were found, including a 9-piece machine.

Evolved physical replicators

A more difficult challenge is to *evolve* self-replicators, rather than design them manually. Ideally, such replicators would emerge spontaneously out of a primordial soup of cubes, where – as in nature – self-replication is an *implicit* fitness criterion in itself. However at this initial stage we experimented with direct evolution of replicators where replication is an *explicit* fitness criterion.

Treating self-replication as a binary criterion would not provide any gradient for the evolutionary process to follow, and so would be unlikely to yield any viable solution in this vast space of machines, including both structure and control. Instead we broke down the evolutionary process into two stages, and used the graded definition of self-replication to produce a gradient:

- **Stage 1: Evolve morphologies** of machines that are mechanically capable of reaching an area large enough to contain a detached copy of themselves. The percentage of coverage provides a gradient.

- **Stage 2: Evolve controllers** that would make a given morphology pick cubes from dispensers and place them at the correct position. The number of dispensers needed provides a gradient.

We carried out the initial experiments on the two-dimensional version of molecube as it is faster to simulate and provides a smaller search space. In this particular experiment we also required that each cube be either a swiveling block with permanent magnets or a non-swiveling block with switchable electromagnetic (an 'end effector'), but not both. This restriction was placed both for practical consideration for physical implementation, and also to rule out the trivial solution of a single cube sitting at the dispenser location (this was of course one of the first 'unintended' solutions to be found).

At the initial morphology-search stage, the fitness function first exhaustively mapped out the area that the end-effectors covered, while pruning illegal configurations due to collisions and self-locking. This step can be done in polynomial time using convolution of reconfiguration steps. Once the coverage of the machine was obtained, then a copy of the machine was tried exhaustively to be fitted within that space in any of the four orientations. This step can be done in linear time. The maximum amount of the original structure that would fit in the mapped area provided the fitness for the first stage.

Morphologies were represented as a series of code-pairs. Starting with a cursor at the origin, the first code moves the cursor in one of the four cardinal directions, while the second code defines the type of block to try to place. If there is already a block at that position, the new block is ignored and the cursor continues to move as defined by the next element in the array. Morphology strings may be of variable length, but were limited to under 20 units. Variation was achieved through crossover (p=0.9) and three types of mutation: change (p=0.001), addition(p=0.06), and removal (p=0.0005). The population

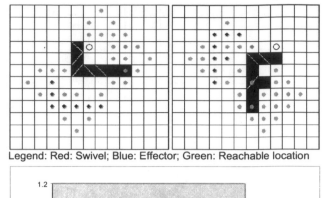

Legend: Red: Swivel; Blue: Effector; Green: Reachable location

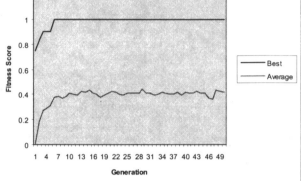

Figure 5: Evolutionary progress for morphology stage (top) Two evolved shapes (morphologies) that can cover an area that contains themselves. (bottom) fitness vs. generations for the run that produced these two results.

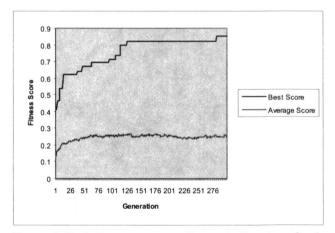

Figure 6: Evolutionary progress for controller stage for the F-Shape.

contained 400 individuals and underwent generational fitness-proportionate selection.

At the second controller-search stage, the fitness function evaluated a control sequence for the given morphology by executing that sequence and measuring the percentage of the potential duplicate that was covered.

Controllers were represented as a series of code-triplets, describing a set of commands to be executed in sequence. Each triplet first described a command ('Swivel', 'Attach', or 'Detach'), and a block number. For the 'Attach' command, the third parameter also specified which of the four sides a new block should attach to. The attach operation also implicitly defined where dispensers are expected to exist – a factor that influences fitness.

Control strings may be of variable size, but were limited to under 300 commands. Variation was achieved through crossover (p=0.9) and three types of mutation: change (p=0.002), addition (p=0.12), removal (p=0.001). Population contained 1000 individuals and underwent generational fitness-proportionate selection for 300 generations. Progress of this stage is shown in Figure 6 for the F-morphology.

The final fitness function used was

$$g(cmds) = 0.8p + 0.2\frac{c-s+1}{c} - 0.03e$$

where

p = weighted fraction of the goal location covered
c = number of blocks inside the model
s = number of dispensers assumed
e = number of excess blocks beyond those needed

Of the three morphologies found by the evolutionary process in the first stage, only two were successful at the second stage, yielding a functional, physically-plausible self-replicating machine. The morphologies and sequences of reconfigurations associated with these two machines are shown in Figure 7 and Figure 8.

Physical Realization

Physical realization of the molecube environment involves several challenges, including reliable transfer of power, information and moments among the units. In particular, since each cube involves six sides with four possible orientations for each side, the number of interface contacts can become impractical and unreliable very quickly.

Our design involves two concentric rings on each face of the cube. The inner ring is an electromagnet, and the outer ring is a conductor. Together, they transfer both power and force, and their circularity avoids orientation issues. Control information is passed on top of the power line as a power modulation signal, readable using conventional power-line modem (PLM) components. The power feeds from the base unit into the rest of the machine.

To date we have constructed two physical prototypes, shown in Figure 9. These are 4"x4" ABS cubes, and contain a PIC microcontroller, power modulation and decoding circuitry, electromagnets, power management circuitry, and a servomotor. Though preliminary, they demonstrate that the key challenges in transmission of power, force and data are feasible, and a realization of a simple machine such as the one shown in Figure 4 is within reach.

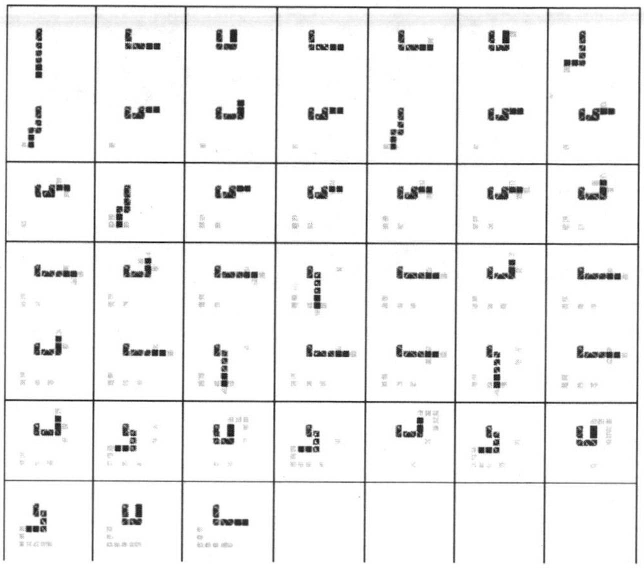

Figure 7: Evolved replication sequence for the L-morphology. Red: Swivel blocks; Blue: Effector blocks; Gray: New material. Note that the sequence contains several redundant cycles.

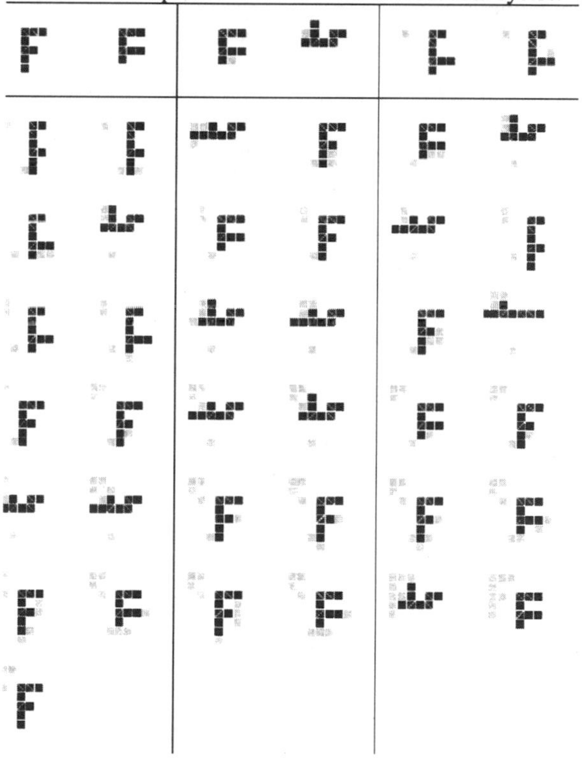

Figure 8: Evolved replication sequence for the F-morphology. Red: Swivel blocks; Blue: Effector blocks; Gray: New material. Note that the sequence contains several redundant cycles.

Figure 9: Prototype realization of substrate. (a) The 4"x4" swivel cube, (b) the cube is able to connect, hold, and transmit power and data to adjacent block.

Conclusions

This paper proposes a physical substrate composed of simple modular units, in which both simple and complex machines can construct and be constructed by other machines in the same substrate. A number of designs, both hand crafted and evolved, are proposed, and the key challenges in physical realization of this substrate have been overcome.

Our purpose in this research is to identify a rich substrate in which physically-feasible self-replication can be investigated both in simulation and in reality. The substrate we have identified and propose here is both realizable and easy to simulate quickly and faithfully due to its discrete nature. We have demonstrated preliminary results showing how various self-replicating machines can be found *systematically*. Although our evolutionary processes required user guidance embedded in the fitness function, we expect that future steps along this path will allow us to examine more ways for self-replication to emerge from first principles, thus examining another critical dimension of artificial-life.

Acknowledgments

This paper was funed by US department of energy (DOE), grant #DE-FG02-01ER45902.

References

Adams B., Lipson H., "A metric for self-replication phenomena", In proceedings of the European Conference on Artificial Life, ECAL 2003.

Arbib, M. A., Comments on Self-Reproducing Automata In: Hart, J. F. and Takasu, S, Systems and Computer Science (1965) 42-59

Chirikjian, G.S., Zhou, Y., Suthakorn, J., Self-replicating Robots for Lunar Development, In: IEEE/ASME Trans. on Mechatronics 7. 4 (2002) 462-472

Freitas, R.A., Gilsbreath, W.P., A self-replicating, growing lunar factory In: Proceedings of the Fifth Princeton/AIAA Conference May 18-21 (1981)

Freitas, R.A., Manufacturing Systems for Molecular Nanotechnology: Kine-matic Self-replicating Machines In preparation (2003)

Langton, C. G., Self-Reproduction in Cellular Automata in Physica 10D (1984) 134-144

Lohn J.D. Reggia J.A. (1997). Automatic discovery of self-replicating structures in cellular automata. IEEE Trans. Evolutionary Computation, 1(3):165-178

McMullin, B, John von Neumann and the Evolutionary Growth of Complexity: Looking Backward, Looking Forward In: Artificial Life 6 (2000) 347-361Sanchez,

Moore, E. F., Machine Models of Self-Reproduction in Burks, A. W., Essays on Cellular Automata (1970) 187-203

Nehaniv C., Dautenhahn K., Self-Replication and Reproduction: Considera-tions and Obstacles for Rigorous Definitions. Abstracting and Synthesizing the Principles of Life, Verlag Harri Deutsch, pp. 283-290, 1998.

Penrose, L.S., Self-reproducing machines. In: Scientific American, 200 (6) (1959) 105-114

Sanchez, D., et al., Phylogeny, Ontogeny, and Epigenesis: Three Sources of Biological Inspiration for Softening Hardware. In: T. Higuchi, et al, editors, Proceedings of ICES96, Lecture Notes in Comp. Sci., Vol. 1259, (1997) 35-54.

Sipper, M., Reggia, J.A., Go forth and replicate. In: Scientific American 285/265(2), August 2001, 35-43

Smith, A., Turney, P., and Ewaschuk, R., JohnnyVon: Self-Replicating Automata in Continuous Two-Dimensional Space. (2002) Technical Report ERB-1099, Institute for Information Technology, National Research Council Can-ada.

Von Neumann, J, completed and edited by Burks, A. W., Von Neumann's Self-Reproducing Automata In: Burks, A. W., Essays on Cellular Automata (1970) 4-65

Hutton T.J., "Evolvable self-replicating molecules in an artificial chemistry", Artificial Life Vol. 8. No. 4, 2002

Bratley P. Millo J. ``Computer recreations: Self-reproducing programs." Software Practice and Experience, Vol. 2, pages 397-400, 1972

Ray, T. S. (1992), An Approach to the Synthesis of Life, in C. G. Langton, C. Taylor, J. D. Farmer & S. Rasmussen, eds, `Artifical Life II', pp. 371-408

Wilke, C. O, J. Wang, C. Ofria, R. E. Lenski, and C. Adami. (2001). Evolution of digital organisms at high mutation rate leads to survival of the flattest. Nature 412:331-333

Emergent Robustness and Self-Repair through Developmental Cellular Systems

Can Öztürkeri and Mathieu S. Capcarrere

Kent University
Natural Computation Group, Computing Laboratory
Canterbury, Kent CT2 7NF, United-Kingdom
{co24, M.Capcarrere}@kent.ac.uk
http://www.cs.kent.ac.uk/research/groups/aii/ncg/

Abstract

Fault-tolerance and, even more, self-repair remain elusive properties in computing systems. In contrast, natural systems are often cited as examples of flexible, self-repairable systems. Such capabilities rely on many different aspects, but our hypothesis is that (adaptive) growth and cellularity are at the heart of these properties lacking so dearly in human artifacts. In this paper, we propose a simple cellular developmental system to back-up through experimental results this hypothesis. First, we show that it is possible to evolve such systems to do specific tasks. Second, and more importantly, that these systems exhibit *emergent* robustness and self-repair capabilities through their own nature rather than specific design or directed evolution.

Introduction & Previous Research

Fault-tolerance and, even more, self-repair remain an elusive property in computing systems. If redundancy is a necessity in a physical system, in abstract or software systems, first detection of errors and then reconstruction are the two main issues in designing robust and self-repairable systems. This reconstruction process is all the easier if the system is able to construct itself in the first place. It can be hypothesised thus, that if the system stabilises its self-construction process into a useful, working state, then any perturbations, any errors, would create an unstable state that should redevelop, restabilise, reconstruct into the desired stable working state. This approach, though still rather undeveloped, has been hinted at to a greater or lesser extent in a certain number of previous studies, (De Garis, 1999; Miller, 2003; Streichert et al., 2003; Capcarrere, 2004; Mange et al., 2000; Macias and Durbeck, 2002).

Natural organisms are often cited as examples of flexible, self-repairable systems and are often at the base of the works previously cited. Though this self-repair capacity seems to diminish with increasing complexity, examples of reconstruction are not rare in nature. Such capabilities rely on many different aspects, but (adaptive) growth, and especially re-growth, is certainly at the heart of these properties lacking so dearly in human artifacts. Growth is itself dependent on a fundamental structure of living being: cellularity. While this cellular aspect has been adopted in some of the previous studies, its necessity for self-repair has often been neglected. In earlier works (Righetti et al., 2003) however, we used only that cellular, decentralised aspect, without growth, as a means of getting fault-tolerant computation.

In this paper, we thus propose a simple cellular developmental system to gain robustness and self-repair capabilities. Rather than study growth in itself, as has been the case in most studies previously quoted, our aim here is to back the above-mentioned hypothesis about self-repair ability through experimental results. Nevertheless, we also demonstrate that by using a reasonably simple framework, it is possible to evolve practical solutions to two problems often confronted in works concerned with these issues. First we obtain stable growth. This may sound a very basic necessity, but is in fact extremely hard to obtain. Many of the previous works that studied growth highlighted that "cancer", uncontrolled growth is one of the main problems when developing these systems (Miller, 2003; Streichert et al., 2003; Capcarrere, 2004). In this paper, we show that by using a technical, practical, trick, it is possible to evolve perfectly "stable" growth, in the sense that perturbed system restabilises more often than not in their working state. Moreover we also evolve highly decentralised computation, a task that has been shown to be non-obvious even for simple problems (Capcarrere, 2002).

In the first section we present the system framework and the evolutionary approach taken to obtain the results presented in the following section. These results, while not evolved specifically for self-repair, exhibit very good emergent robustness and self-repair capabilities. We then conclude on the many open paths for future research.

A Developmental Cellular System

While nature is the starting point for our inspiration, our purpose here is not bio-realism as is the case in (Streichert et al., 2003). The system developed for this work aims at extracting some quintessential abstract principles of the growth process while remaining computationally tractable. The eventual goal here is to gain fault-tolerance and self-repair capabilities in computing artifacts.

Our system can be divided into two main parts and we will describe these separately in the next two sections. The first part is what could be described as the *engine* or the physics of the system. It is directly inspired by the work of Miller (Miller, 2003) but is simplified. This is where the novelty of this specific system lies. The second part describes briefly the evolutionary algorithm used, and most importantly, the parameters used to allow for replication of the experiments.

The engine

The system developed and studied in this paper is based on the model of cellular automata. More exactly the cellular system lies on and develops along a two-dimensional discrete topology and each cell seperately executes the same program. According to its surrounding states and its own, the cell changes its state and decides whether or not to "replicate" in a free space of its neighbourhood.

Each cell lies at the vertex of a 2-dimensional lattice. It possesses a x-bit state that is readable *and modifiable* by any one of its 8 neighbouring cells, where x is dependent on the given problem. This neighbourhood matches Moore neighbourhood (North, N.-East, East, S.-East, South, S.-West, West, N.-West). Though not stored as a lookup table, the memoryless cell program will deterministically act according to the $9x$ bits input from all the neighbourhood cell states and its own. Like a classical CA, the result of this internal program execution will determine the new state of the cell. However, unlike the CA model, this cell may also alter the state of its 8 neighbouring cells.

Hence this internal program could be described as a $9x$-bit input, $9x$-bit output function. It is important to note that while our model is close to the one developed previously by Miller, it does not include any chemicals nor any chemicals diffusion, and, is thus, in our opinion, simpler.

The grid used is non-toroidal. To the program, the "virtual" cells beyond the borders appear to have a 0^x state that cannot be altered. Experiments showed that this border is very useful to obtain a *stable* growth. Obviously it contains growth, but it also provides, unexpectedly, stability. The theoretical questions posed by the problem of "cancer" evoked earlier is not solved as such, but this border allows us to practically use the advantages of growth in terms of self-repair without its disadvantages.

The neighbour's state rewriting property obviously entails a series of questions on the order of update of the cells. It forbids a fully parallel update. A single model of asynchronous updating was studied. It is a fully sequential deterministic model where each cell is updated along a line sweep order (Giacobini et al., 2003). This is what was used in previous works (Miller, 2003). A more interesting model from a research viewpoint, but less practical from a hardware viewpoint, is a random asynchronous update mode. This is to be studied in future works. One time step of the simulation is the evaluation of n cells, where n is the number of cells in the environment.

The evolutionary framework

We used an adaptation of Cartesian GP (CGP). CGP was first designed to evolve digital circuits (Miller and Thompson, 2000) and is well suited for problems involving binary inputs/outputs. They present the great advantage of not suffering from the bloat problem and of keeping the representation compact. At the same time they allow for quick and efficient data manipulation and work with small populations. However, as a drawback, they do require a large number of generations, but one has to keep in mind that the number of sampled solutions is still extremely small compared to the size of the search space. We will not go into the details of the description of CGP here, but simply highlight the idiosyncrasies of the encoding, the selection and the genetic operator we used.

The encoding of our program is as described in (Miller, 2003). The inputs to the program, as highlighted in the above section, are the bits from the states from all the cells in the neighbourhood. Each bit is considered individually as a possible input. We also use the same operations as were used in previous works by Miller. These consist of only four kinds of basic operations: if (input0) then output (input1) else output (input2); if ($\overline{input0}$) then output (input1) else output (input2); if (input0) then output ($\overline{input1}$) else output (input2); if (input0) then output (input1) else output ($\overline{input2}$). This allows for any boolean function of three variables. Each of these operators is to be called a node henceforth. The size of the evolved program is evaluated in terms of number of nodes.

The population size used in all experiments is only 5. Only the best individual is selected, and 4 new mutants are created from it to maintain the population size. Hence it is an extremely elitist algorithm. This is in great part imposed onto us by the very small population size. This was chosen as our experiments showed that reasonably bigger populations were not sensibly decreasing the necessary number of generations per run, thereby increasing greatly the time needed. It is highly probable that a significantly bigger population (5000 to 50000) would certainly greatly decrease the number of generations, but this has not been tested to date. Mutation is the only genetic operator used. It mutates an equal percentage of the following: Inputs to the CGP nodes, the function of the nodes, and the output table which is separate. The only thing worth mentionning here is that the mutator makes sure a cell gets its inputs from previous cells or system inputs, entailing a loop free, easily executable organism.

Emergent Robustness Through Growth

In this section, we present the results obtained. Very interestingly, we observe, as expected, that they exhibit *emergent* fault-tolerance through self-repair, thereby entailing lasting safe behaviour. Two experiments are described here: – first, an experiment that displays visually this emergent self-repair property; – second, a functional experiment, surely more promising in terms of potential future research and applications.

A "visual" example

The first set of experiments aims to demonstrate the qualities of robustness and self-repair obtainable in growing systems. To do so "dramatically", we chose a morphogenetic example where the reconstruction process is visually interesting. The aim of the task is to get a system that grows *and stabilises*. in the shape of a French flag, i.e., a rectangular shape, made up of three different kinds of cell, red, blue and white, themselves arranged in rectangles. As the growth process is not our main focus for this research, we chose to use a shape that has been shown to be evolvable in the past (Miller, 2003).

We use an 8x8 non-toroidal environment in which we aim at evolving systems that develop into a 6x3 horizontal shape, made of 3 blocks of 3x2 cells each. Each cell has got a 5-bit state, the first two bits encoding the colour displayed. The aim is to get a stable configuration in which the first two bits of the cells' state are: 00 (encoding black) outside the rectangular shape, 01 (blue) for the cells in the first block, 10 (white) in the second block and 11 (red) in the third block (see fig 1). The workings of the system are no different than the model described in the above section but there is a specific constraint on the internal workings of the program. If the first two bits of the state of the cell are 00 then that cell's program does nothing. The interpretation of that constraint could be that a cell with a 00-starting state is dead, or environmental, and the other cells are alive. Each cell being able to change its own and its surrounding states, it can thus "decide" to die, reproduce or kill a neighbouring cell.

The CGP evolutionary algorithm used was as described earlier with a population of 5 individuals. The starting configuration is an empty grid, all cells with 00000 state, except for the central cell's state set to 10000. Fitness at time step t, f_t, is defined as the number of cells whose first two bits are in the expected final state. The other three bits which may be, and are, used for the computation are not evaluated directly in the fitness. The fitness used for the evolutionary purpose, F, is defined as the simple aggregate of the fitnesses after 7 and 15 time steps, $F = \frac{f_7 + f_{15}}{2}$. The two fitnesses are employed to discard any unstable growth strategy. The hurdle of unstable and unlimited growth highlighted earlier has been solved here using *both* this fitness constraint and the non-toroidal environment described earlier.

Successful evolution of "organisms" growing and stabilising in the desired shape took on average 300000 generations. Perfect (64/64 cells) or excellent individuals (63/64 cells) in terms of fitness were obtained on 54 out of 100 runs. Though growth is not the focus of this paper, it may be noted that these results compare favourably with past studies, both in terms of success rate and on the stability of the growth of the evolved organism. The size of the successful organisms' program uses less than 100 nodes, which is reasonably small. Further investigations are necessary however to determine exactly how many of these were necessary for the task, as "junk DNA" constitutes, a priori, a non negligible part.

Emergent robustness: The purpose of this paper is to back the hypothesis that cellularity and growth are sufficient conditions for robustness and self-repair. Therefore we tested out the solutions evolved in a *safe* environment. It is important to note that the self-repair capability is thus *emergent* and as such can be deemed to be a consequence of the developmental and cellular nature of the system. The error model used here is a one-time strike model. From the seed configuration, the organism is run for 20 steps. After 20 time steps, an error occurs, once, and the system is run again for 100 more time steps. The experiments described below are on one specific individual but our tests, still preliminary, seem to show that these properties can be roughly found in the other organisms successfully evolved.

Experiment 1: After 20 steps, one of the cells of the grid is reset to 00000, i.e., both the visible and invisible bits are reset. One should note also that this "kills" the cell subject to the error and thus it cannot repair itself. In 95% (61/64) of all the possible cases, the organism perfectly recovered, reconstituting to its perfect working state.

Experiment 2: In this experiment, 10 cells at random are reset in one go. Obviously it is impossible to test the behaviour exhaustively on all C_{64}^{10} possible cases, but the experiment was repeated randomly 1000 times, giving a good overview of the behaviour of the system in the case of dense, but non block, errors. The system resisted rather well and reconstituted itself perfectly in 65.4% of the tests. An example of such a perfect recovery may be seen in figure 1. The system reconstituted itself but for one cell in a further 18.8% of the case.

Experiments 3,4,5: In these experiments, we test block rather than sparse errors. We reset, respectively, all the cells in the white block, red block and blue block. In each of these experiments the system recovers completely in less than 10 time steps. In a *sixth experiment*, we removed both the red and white blocks at the same time, and the system still recovered completely.

However, the system is not indestructible and

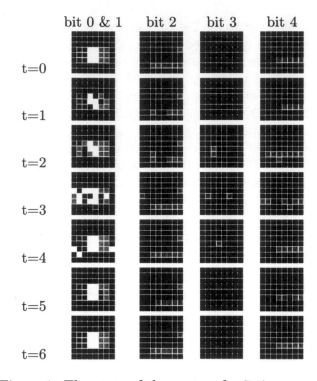

Figure 1: The state of the system for 7 time steps. The growth phase is not shown here, and at time 0 we see the system in its mature, stable state. At time 1 the errors occur. Time 2 to 6 shows the recovery. Time flows downward. Horizontally one can see the state of the two first bits of each cell (the "visible" state), and then of each of the following bits of its state.

when the error block is too large or of horizontal shape the recovery is mitigated or even impossible.

In *experiments 7 and 8*, respectively, the red and white block, and the blue and white block is removed. The system does reach a stable state and a complete French flag is reconstituted, but spurious visible states remain and the original state is never recovered. Hence the self-repair process cannot be judged satisfactory.

Finally, in *experiments 9,10,11*, respectively, the horizontal top, middle and bottom line is removed. The middle line removal causes no trouble, and the system fully recovers. The top and bottom line, on the other hand, creates total chaos, and the system only recovers pieces of the French flag shape, and its stable state includes lots of spurious cells. The recovery in these cases is beyond its ability.

As can be seen, while not perfect, the emergent self-repair capabilities in this visual example are striking. This should be nuanced by the fact that our error model, though harsh in its extent in space, is rather kind time-wise. This aspect will be investigated in future studies. Nevertheless the self-repair process is very efficient and takes usually less than 10 time steps. Hence, the assumptions behind a strike-once model are not unreasonable.

If self-repair capabilities are to be of any use, it should be applied to a functional rather than a visual example. This is investigated in the next section.

A "functional" example

As the French-flag example illustrated, systems based on cellularity and development may display interesting self-repair capabilities. To go beyond this "dramatic" example, we explore the possibility of making a digital circuit on a cellular structure configured through a developmental process and then test its self-repair capabilities, if any. The idea to have a cellular, developmental digital circuit to increase robustness is not new in itself (Mange et al., 2000). However, our system is simpler in its deployment as it can be evolved. More importantly, it allows at the same time for "automatic" error detection (no special signal is needed), and recovery from more bulky errors than the embryonics system, developed by Mange *et al.*. Of course, there is a price to pay, which is the lack of guarantee, before testing, on the exact recovery abilities. Another interesting approach is the cell matrix approach (Macias and Durbeck, 2002) which is rather close to our approach of self-assembling and reconfiguration. Besides, their real hardware approach is certainly interesting for future applications. However, unlike our system, it requires active detection of errors and must be hand designed.

We adopted a very abstract approach. Our circuit is made-up of a grid of $n \times n$ multiplexers, topologically identical to our $n \times n$ cell grids. Each cell's state encodes which type of multiplexers is to be at this position and and what are its connections. The states of all the cells could be interpreted as the configuration string of a basic homogeneous FPGA. All the external inputs of the system are connected by default to false, except for two (or more) of them which are connected to the inputs of the problem. Identically, one output (or more) is designated as the output of the system. The resulting circuit can then be evaluated to see if on given inputs it gives the expected output(s).

The first two bits of the state of the cell determine which of the four types of multiplexer is used. All the multiplexers have three inputs and one output. The first input determines which of the second or

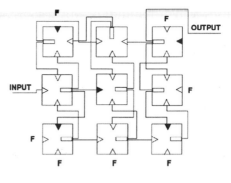

Figure 2: Example of a 3x3, one-input, one-output circuit with its connections. The triangles on the multiplexers are its three inputs, and the bar is its output. Black inputs are inverted. The evaluation of the circuit is done right to left, top to bottom 4 times before the global output of the circuit is read.

third input is connected to the output. In the first type no input is inverted, in the second type input 1 is inverted, in the third input 2 is inverted and finally in the fourth input 3 is inverted. The multiplexer output is always potentially connected the 4 adjacent multiplexers, however it can have four directions, each one being the 90 degrees rotation of the previous, which will determine which multiplexer is connected to which input (see Figure). The next two bits of the cell's state encode this. One may note that this may create meaningless circuits. Being in an abstract model, this is not a problem, and the circuits not working are naturally eliminated from the population evolved. There are a further three bits available in the cell's state that are not used to directly code the multiplexers configuration. The evolutionary process used is exactly as in the previous experiments. The fitness here is quite simply how many inputs to the circuit give a correct output. Thus, unlike in the previous experiment, there is no direct constraint on the cells' state from the fitness. This is a rather crude fitness, but that proved sufficient to evolve successful system. It is important to note that unlike the last experiment in this set of experiments all cells are "alive", i.e., the internal program is executed in every cell, whatever its state.

Results: Using this scheme, it was possible using 5×5 grids to evolve AND and XOR gates. In both cases the inputs were connected to cells (0,1) and (0,3), and output to cell (4,1)[1]. Obviously this

---

[1]Each multiplexer in the $n \times n$ grid is numbered by a pair (i, j), where i is the column number and j, the row number

scheme uses a high redundancy of resources. However it should be noted that this increase in the size of the search space does not increase the complexity of the evolution. Actually, basing ourselves on early experiments, we can even say that it simplifies the evolutionary process greatly. We obtained perfect evolution in 100% of the runs. Here the 100 node CGP cell's program was evolved in less than 5000 generations on average, two orders of magnitude less than for the French flag experiment. Though our work is still in an early stage, we were able to evolve also more complex systems, such as a full binary adder using a 9×9 grid.

Emergent robustness: The XOR gate was tested for robustness using the same error model as for the French flag example. While it is obvious that is extremely redundant to use a 5×5 grid to make an XOR circuit, it is important to note that this is *not* what makes the circuit fault-tolerant. As for the French flag, after an error occurred, the circuit reconstitutes itself, i.e., reconfigure itself to exactly the original configuration. Hence all of the 25 multiplexers are of the "correct" type in the "correct" position, even though some of them are not necessary to the workings of the circuit. In a first experiment, a unique cell is reset after 20 time steps. The circuit recovered fully and completely in all the 25 cases possible. The self-repair ability seeming good, the circuit was further tested with a 5 random cell error. It recovered fully in 70.8% of 1000 experiments. The error was only on one bit of only one cell's state in a further 23.2% of the experiments. And the biggest error was no more than 7 bits out of the 175 (25×7) bits of all the states of the cell. This very good recovery property of this circuit is further increased if one takes only into account the functionality of the circuit, and thereby uses its inherent redundancy. Then the XOR circuit gives the correct output in 100.0% of 1000 5-cell error and 1000 10-cell error experiments.

Concluding Remarks

In this paper we have shown that it was possible to evolve systems exhibiting self-repair capabilities thanks only to their workings. The only direct constraints on the system are that it should have a cellular structure and should develop into its desired final state. The "error-detection" and self-repair properties are the emergent consequence of these constraints. These encouraging results, most notably the fact that perfect repair occurs in a vast majority of cases, call for more research in these directions.

Obviously the functionalities explored were still basic, and the error model used, while not unrealistic, is rather kind. Future works should explore further error models and aims at more complex tasks. Attention should also be devoted to understanding the workings of the developmental process so as to establish clearly the limits of the repair abilities.

References

[1] Capcarrere, M. S. (2002). *Cellular Automata and Other Cellular Systems: Design & Evolution*. Phd Thesis No 2541, Swiss Federal Institute of Technology, Lausanne (EPFL).

[2] Capcarrere, M. S. (2004). An evolving ontogenetic cellular system for better adaptiveness. *BioSystems*, (To Appear).

[3] De Garis, H. (1999). Artificial embryology and cellular differentiation. In Bentley, P., ed., *Evolutionary Design by Computers*, pages 281–295, Morgan Kaufmann.

[4] Giacobini, M., Alba, E., and Tomassini, M.. (2003). Selection intensity in asynchronous cellular evolutionary algorithms. In et al., E. C.-P., ed., *GECCO 2003 proceedings*, vol. 2723 of *LNCS*, pages 955–966. Springer-Verlag.

[5] Macias, N. J. and Durbeck, L. K. (2002). Self-assembling circuits with autonomous fault handling. In Stoica, A., Lohn, J., Katz, R., Keymeulen, D., and Salem Zebulum, R., editors, *Proceedings of the 2002 NASA/DoD Conference on Evolvable Hardware*, pages 46–55.

[6] Mange, D., Sipper, M., Stauffer, A., and Tempesti, G. (2000). Towards robust integrated circuits: The embryonics approach. *Proceedings of the IEEE*, 88(4):516–541.

[7] Miller, J. F. (2003). Evolving developmental programs for adaptation, morphogenesis, and self-repair. In W. Banzhaf et al, ed, *ECAL'03 proceedings*, vol. 2801 of *LNAI*, pages 256–265, Springer-Verlag.

[8] Miller, J. F. and Thompson, P. (2000). Cartesian genetic programming. In *EuroGP'00 proceedings*, vol. 1802 of *LNCS*, pages 121–132, Springer-Verlag.

[9] Righetti, L., Shokur, S., and Capcarrere, M. S. (2003). Evolution of fault-tolerant self-replicating structures. In W. Banzhaf et al,, ed., *ECAL'03 proceedings*, vol. 2801 of *LNAI*, pages 278–288, Springer-Verlag.

[10] Streichert, F., Spieth, C., Ulmer, H., and Zell, A. (2003). Evolving the ability of limited growth and self-repair for artificial embryos. In W. Banzhaf et al, ed., *ECAL'03 proceedings*, vol. 2801 of *LNAI*, pages 289–298, Springer-Verlag.

On Self-referential Shape Replication in Robust Aerospace Vehicles

Mikhail Prokopenko and Peter Wang

Centre for Intelligent Systems Design, CSIRO Information and Communication Technologies Centre
Locked bag 17, North Ryde 1670, Australia, {mikhail.prokopenko, peter.wang}@csiro.au

Abstract

We describe a multi-cellular shape replication mechanism implemented in a sensing and communication network, motivated by robust self-monitoring and self-repairing aerospace vehicles. In particular, we propose a self-referential representation (a "genome"), enabling self-inspection and self-repair; an algorithm solving the problem for connected and disconnected shapes; and a robust algorithm recovering from possible errors in the "genome". The presented mechanism can replicate combinations of predefined shapes and arbitrary shapes that self-organise in response to occurring damage.

Introduction

NASA's goal of robust aerospace vehicles requires structures that are capable of self-assessment and self-repair. Previous work in the joint CSIRO-NASA Ageless Aerospace Vehicle (AAV) project developed and examined concepts for integrated sensing and communication networks which are expected to detect and react to impact location and damage over a wide range of impact energies, ranging from micro-particles to meteoroids (Price et al., 2003; Prokopenko et al., 2004; Lovatt et al., 2003). One of the most important design principles distinguishing an intelligent vehicle health management system from other sensing systems, is the requirement for continued functionality in the presence of damage, and, ultimately, the ability to carry out repairs.

In this paper we investigate a possible first step towards the self-repairing ability, focussing, in particular, on the need for a robust self-replication of multi-cellular shapes. This *shape replication* ability should cover both "standard" and "non-standard" shapes. In other words, we expect that not only a standard shape predefined by an available structural "blueprint" can be produced when required, but also that any non-standard and unpredictable shape covering a damaged region can be dynamically replicated on demand. Importantly, we investigate the self-replication mechanism that would allow us to combine "standard" and "non-standard" shapes if necessary. Repair actions, such as shape replication, might be progressing in the environment where further impacts are likely to occur, and therefore, there is a need for *robust* shape replication algorithms.

The next section will briefly describe the notion of emergent impact boundaries, used to uniquely encode a shape that might be replicated. We follow by setting the relevant background on self-replication architectures. An algorithm for shape replication is then presented and illustrated. The algorithm incorporates a self-referential representation, and solves the problem even if the shape is "disconnected" in certain sense. Finally, we consider the case when shape replication progresses in adverse circumstances, and new impacts damage some parts of the shape being replicated.

Stable Impact Boundaries

The developed hybrid Concept Demonstrator models a two-dimensional array of cells: some cells existing in dedicated hardware (two micro-processors per cell) and some residing within inter-connected personal computers (a number of cells per PC) (Price et al., 2003). We also used a stand-alone AAV Simulator capable of simulating some simple environmental effects such as particle impacts of various energies. In the AAV Simulator, cells are represented as objects (squares) on a two-dimensional plane (e.g., Figure 1), where they asynchronously interact *only* with their immediate neighbours in von Neumann neighbourhood, through connected (geometrically overlapping) communication ports. This approach uses the idea of localised algorithms, in which simple local behaviours lead to *self-organisation* of spatiotemporal multi-cellular patterns, achieving a desired global objective.

Typically, the damage on the AAV skin caused by an impact is most severe at the point of impact (an epicentre). However, not only the cells at the epicentre are destroyed, but the communication capability of the neighbouring cells is reduced. Multiple impacts result in overlapping damaged *impact-surrounding regions* with quite complex shapes.

Let us briefly describe multi-cellular *impact boundaries*, self-organising in presence of cell failures and connectivity disruptions. On the one hand, it is desirable that an impact boundary, enclosing damaged areas, forms a continuously connected closed circuit. This circuit may serve as a reliable communication pathway around the impact-surrounding region within which communications are compromised. Every cell on a continuously connected closed circuit must always have two and only two neighbour cells, designated as the circuit members (circuit-neighbours of this cell). On the other

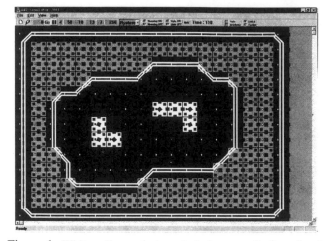

Figure 1: White cells are destroyed, dark-grey cells form "scaffolding", black cells form the "frame". Boundary links are shown as white double-lines.

hand, a continuously connected closed impact boundary provides a template for repair of the impact-surrounding region, uniquely describing its shape (Figure 1). Both these functionalities of impact boundaries can be contrasted with non-continuous "guard walls" investigated by Durbeck and Macias (Durbeck and Macias, 2002) that simply isolate faulty regions of the Cell Matrix, without connecting elements of a "guard wall" in a circuit.

In order to serve either as a communication pathway or a repair template, an impact boundary should be robust to communication failures caused by proximity to the epicentre. The algorithm producing such circuits and the metrics quantitatively measuring their spatiotemporal stability are described elsewhere (Foreman et al., 2003). In this paper, we assume that our impact boundary is a stable continuously connected closed circuit. It is sufficient to mention here the following spatial self-organising layers:

- *scaffolding* region, containing the cells that suffered significant communication damage;

- *frame boundary* — an inner layer of normal cells that are able to communicate reliably among themselves;

- *closed impact boundary*, connecting the cells on the frame boundary into a continuous closed circuit by identifying their circuit-neighbours.

The "frame" separates the scaffolding region from the cells that are able to communicate to their normal functional capacity. These internal layers (scaffolding, frame and closed boundary) completely define an impact-surrounding region as a layered spatial hierarchy. In general, the impact-surrounding region can be seen as an example of annular spatial sorting: "forming a cluster of one class of objects and surrounding it with annular bands of the other classes, each band containing objects of only one type" (Holland and Melhuish, 1999). It could be argued that, as an emergent structure, the impact-surrounding region has unique higher-order properties, such as having *an inside* and *an outside*.

Self-replication: background and motivation

In this section we attempt to position our shape replication scheme with respect to some well-known approaches. Theoretical foundations for artificial self-replicating systems were laid down by Von Neumann, who proposed two central elements: a Universal Computer and a Universal Constructor. A program Π encoded in the Universal Computer directs the behavior of the Universal Constructor. The latter is used to manufacture both another Universal Computer and a Universal Constructor. The program Π is then copied into the newly manufactured Universal Computer. It is possible to develop self-replicating automata which do not require universality. For example, a well-known self-replicating structure is a Langton's loop constructed in two-dimensional, cellular space. The loop is a closed data path, capable of transmitting data in the form of signals. These signals not only encode the loop's "genome", but serve also as the instructions for replication. In executing the instructions the loop extends itself and folds into a daughter loop, also containing the genome and capable of self-replication (Langton, 1984).

As pointed out by (Sipper, 1998), Langton's loop and its various extensions as well as other self-replicating automata based on Von Neumann architecture, can be thought of as unicellular organisms: there is a single genome describing and contained within the entire automaton. Another class of self-replicating automata includes artificial *multi-cellular organisms*, where each of the several cells comprising the organism contains a copy of the complete genome. One well-advanced approach exploiting such artificial multi-cellular organisms is the embryonic electronics (*embryonics*), aimed at very large scale integrated circuits with self-repair and self-replication capabilities (Sipper et al., 1997; Mange et al., 2000). Essentially, embryonics employs three biologically inspired principles: multi-cellular organisation, cellular differentiation, and cellular division. Cellular differentiation takes place by having each artificial cell compute its coordinates (i.e., position) within a one- or two-dimensional space, after which it can extract the specific gene within the artificial genome responsible for the cell's functionality. Cellular division occurs when a "mother cell" arbitrarily placed within the grid, multiplies to form a new multi-cellular organism. In addition to self-replication of the original circuit in case of a major fault, this artificial organism also exhibits self-repair capabilities, allowing partial reconstruction in case of a minor fault. In summary, the embryonics approach models multicellular organisms that *ontogenetically* develop in order to perform useful tasks.

Another relevant concept is *self-inspection*. In some cases, the genome is predetermined and simply needs to be replicated. This would be the case when a "blueprint" of a standard shape is available. Sometimes, however, there is a need to dynamically construct the genome describing a non-standard shape for its subsequent replication. Moreover, sometimes self-inspection should proceed concurrently with the interpretation of the genome (Laing, 1977).

The shape replication algorithms developed in the context of AAV and presented in the following sections are based on the principles of multi-cellular organisation, cellular differentiation, and cellular division as well — similarly to the embryonics approach. A desired shape is encoded when an emergent impact boundary self-inspects itself and stores the "genome" in a "mother" cell. The genome contains both data describing the boundary and a program of how to interpret these data. The mother cell is then seeded in a new place outside the affected AAV array. Executing its program initiates *cell-replication* in the directions encoded in the genome. Each cell-replication step involves copying of the genome (both data and the program) followed by differentiation of the data: an appropriate shift of certain coordinates. Newly produced cells are capable of cellular division, continuing the process until the encoded shape is constructed.

In order to provide a unifying view on the inter-related concepts briefly described above, we informally characterise the shape replication process in self-referential terms, employing two logical levels: an object level and a meta-level. It is well- known that self-replication of a system can be characterised by emergent behaviour and *tangled hierarchies* exhibiting Strange Loops: "an interaction between levels in which the top level reaches back down towards the bottom level and influences it, while at the same time being itself determined by the bottom level" (Hofstadter, 1989). In terms of shape replication, one may argue that the genome encodes the shape in each cell together with the meta-level instructions of how to replicate it. In other words, each cell contains a model of the whole multi-cellular shape, unfolding it at every cell replication and differentiation step. In addition, we shall illustrate that self-inspection of an emergent impact boundary can be mirrored by self-inspection of the genome inside each cell, at every cell-replication step. Similarly, we shall demonstrate that self-repair of the overall damaged impact-surrounding region can be reflected in self-repair of the code embedded in each cell.

Self-referential Shape Replication

Shape Structure

In this section we provide formal definitions of an impact boundary and internal scaffolding, and draw a clear distinction between "connected" and "disconnected" cases. A two-dimensional AAV array can be represented by a *planar grid graph* $G(V, E)$: the product of path graphs on m and n vertices, where the vertices $V(G)$ are any set of points on the planar integer lattice. The edges $E(G)$ connect vertices at unit distances. Given an impact, all cells that are located within the impact-surrounding region can be represented by an *impact* subgraph A of G (Figure 2).

First of all, we identify all the vertices $S \subseteq A$ which have precisely 4 edges each. Then we define the *scaffolding subgraph* Υ_A of A as a subgraph induced on the impact graph A by the set S: i.e., as the set of the vertices S together with any edges $E(A)$ of the impact graph A whose endpoints are both in the subset S.

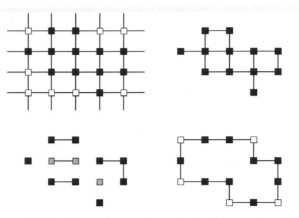

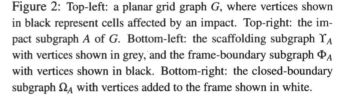

Figure 2: Top-left: a planar grid graph G, where vertices shown in black represent cells affected by an impact. Top-right: the impact subgraph A of G. Bottom-left: the scaffolding subgraph Υ_A with vertices shown in grey, and the frame-boundary subgraph Φ_A with vertices shown in black. Bottom-right: the closed-boundary subgraph Ω_A with vertices added to the frame shown in white.

Secondly, we identify the *frame-boundary subgraph* Φ_A of A as a subgraph induced on the impact graph A by the set-complement $A \setminus \Upsilon_A$: i.e., as the set of the vertices $A \setminus \Upsilon_A$ together with any edges $E(A)$ of the impact graph A whose endpoints are both in the subset $A \setminus \Upsilon_A$. Figure 2 (bottom-left) illustrates the case when both the scaffolding subgraph Υ_A and frame-boundary subgraph Φ_A are disconnected.

Finally, the *closed-boundary subgraph* Ω_A of G is defined as follows. We intend to add to the frame-boundary subgraph precisely those elements from G which provide a shortest path (outside the scaffolding subgraph) between components of the possibly disconnected frame-boundary subgraph. Formally, we employ the graph-theoretic definition of convex sets, according to which a set of vertices X in a connected graph is called convex if for every two vertices $u, v \in X$, the vertex set of *every shortest path* between u and v lies completely in X. We now identify the *graph-theoretic convex hull* H_A of the set of frame-boundary vertices $V(\Phi_A)$ in G but not in scaffolding subgraph Υ_A, as the smallest graph-theoretic convex set in $V(G) \setminus V(\Upsilon_A)$ containing $V(\Phi_A)$. A subgraph induced on the graph G by the set H_A is the desired closed-boundary subgraph Ω_A (Figure 2: bottom-right). It can be shown that the closed-boundary subgraph Ω_A is always *cyclic*.

These definitions are not constructively used by the decentralised boundary formation algorithms, localised within each cell (Foreman et al., 2003). The graph-theoretic notions require global information, used, for example, in specifying all the vertices of the impact graph in advance, or finding shortest paths and convex hulls. In reality, autonomous cells asynchronously deal with unreliable communication messages, while trying to determine whether they belong to scaffolding, frame or closed boundary. We introduced here the formal definitions in order to give a graph-theoretic semantics to these emergent structures, and in particular, to distinguish between *connected* and *disconnected* scaffolding sub-

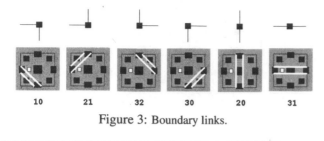

Figure 3: Boundary links.

x	0	0	0	1	2	3	3	4	4	4	3	2	2	1
y	0	1	2	2	2	2	1	1	0	-1	-1	-1	0	0
λ	32	20	30	31	31	10	32	10	20	21	31	32	10	31

Table 1: An example boundary genome.

graphs. It is precisely the second case that presents some difficulties for shape replication.

Given the planar grid topology, each cell on the closed impact boundary may have 6 boundary links, connecting ports "left-right", "left-top", etc. Enumerating four communication ports from 0 to 3 ("bottom" to "right" clockwise) allows us to uniquely label each boundary link with a two-digit number λ, e.g., "32" would encode a link between the "right" and "top" ports (Figure 3). Then, the whole impact boundary can be encoded in an ordered list of these labels. For instance, the boundary depicted in the Figure 2 can be simply represented by the list $\{32, 20, 30, 31, 31, 10, 32, 10, 20, 21, 31, 32, 10, 31\}$. However, in order to replicate the bounded shape, filling it cell by cell, we need to introduce a system of coordinates relative to a cell containing the shape list. More precisely, the boundary genome is a list of triples (α, β, λ), where (α, β) are relative coordinates of a cell with the boundary link λ. The boundary genome for our example is shown in the Table 1. The instructions of how to interpret these data can be easily represented in an assembler-like language, with each "program" triple encoding two operands and an instruction type.

An Algorithm for Disconnected Scaffolding

The first phase is *self-inspection of the impact boundary*, producing the genome, e.g., the genome in the Table 1. The process starts with a selection of a mother cell (any cell s_0 on the boundary), and involves the following steps:

- the mother cell inserts the triple $(0, 0, \lambda_0)$ into the empty genome, where λ_0 is the boundary link maintained by the cell s_0, and sends the incomplete genome to the neighbour in a specific direction (e.g., counter-clockwise);

- each boundary cell receiving the genome determines the relative (a, b) coordinates of the message sender given the port of the incoming message: e.g., if the message comes on the bottom port (labelled as 0), then the sender's relative coordinates are $(0, -1)$ (the possibilities are encoded in a look-up Table 2);

- the cell increments all (x, y) coordinates in the genome as follows: $x = x + a, y = y + b$;

- the cell appends the triple $(0, 0, \lambda_i)$ to the genome, where λ_i is the the boundary link maintained by this cell;

- if this cell is not the mother cell, it sends the genome to the counter-clockwise neighbour, otherwise, the process terminates.

port	0	1	2	3
a	0	-1	0	1
b	-1	0	1	0

Table 2: The look-up table of directions and coordinates.

The next phase is *shape replication per se*. It starts when the mother cell is seeded in some available space. The process involves cell-replications carried out by not only new boundary cells, but also by new scaffolding cells. The cell-replication program encoded in each cell has the following steps (starting with the seed at the beginning):

- the cell iterates through the genome and determines whether there is a triple $(0, 0, \lambda)$, for some λ;

- if such a triple is found, a Boolean flag ω is set to true ("boundary cell"); otherwise $\omega = $ false;

- the cell iterates through all possible directions π, where $0 \leq \pi \leq 3$, doing for each π the following:

   1) retrieve from the look-up Table 2 the coordinates (a, b) for construction in the direction π;

   2) check if there is a cell at the relative location (a, b): if there is, then the direction π should not be used, and the cell moves to the step (6), otherwise it continues with the following steps;

   3) check if both $\omega = $ true and the two-digit number λ does not include the digit π;

   4) if the condition (3) is satisfied (meaning a boundary cell is considering to produce a scaffolding cell), then
   a) fix the vertical "strip" $x = a$ and, by varying the y coordinate across the genome, compute the number of times n_+ and n_- the boundary fully crosses this "strip", above and below $y = b$ respectively (this computation is described below);
   b) check if either n_+ or n_- is odd and neither is 0;

   5) if either the condition (3) is not satisfied, or the conditions (3) and (4.b) are both satisfied (a boundary cell produces a scaffolding cell), then
   a) construct a cell in the direction π;
   b) copy the genome to the constructed cell;
   c) decrement all coordinates in the constructed cell's genome as follows: $x = x - a, y = y - b$;

   6) increment the direction π;

- the process stops when all directions have been checked.

In order to compute the numbers n_+ and n_- relative to the location (a, b), the cell iterates through the genome for $x = a$, varying $y > b$ and $y < b$ respectively. The number n_+ is initiated to 0, and is incremented each time either a) an entry $\lambda = 31$ is encountered, or b) $\lambda = 32$ is encountered, followed (not necessarily immediately) by $\lambda = 10$, or c) $\lambda = 21$ is encountered, followed (not necessarily immediately) by $\lambda = 30$. The number n_- is computed similarly with the λ pairs in (b) and (c) reversed. The numbers n_+ and n_- determine whether the location (a, b) is *inside* or *outside* the shape.

Let us exemplify the cell-replication phase, and in particular cell-differentiation occurring at step (5). We continue with the example genome (Table 1), and assume that the seed $(0, 0, 32)$ starts the process from the bottom-left corner of the boundary. Let us start with the direction $\pi = 0$. The look-up table suggests the coordinates $a = 0, b = -1$, the shift down. The location $(0, -1)$ is free. The condition (3) is satisfied as the number $\lambda = 32$ does not have $\pi = 0$ in it. However, the vertical "strip" on which a possible new cell

x	0	0	0	1	2	3	3	4	4	4	3	2	2	1
y	-1	0	1	1	1	1	0	0	-1	-2	-2	-2	-1	-1
λ	32	20	30	31	31	10	32	10	20	21	31	32	10	31

Table 3: The genome updated after cell-differentiation.

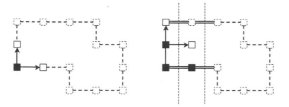

Figure 4: Shape replication. Boundary cells encoded in the genome but not yet produced are shown with dashed lines. Left: a black cell (seed) produces two white cells, indicated by arrows. Right: Two more cells are being produced: one of them is a scaffolding cell, pointed to by the horizontal arrow. The inside direction is recognised by the vertical strip being "crossed" above and below the considered location.

would be located is not "crossed" by the boundary at all, so $n_+ = n_- = 0$, and there is no need to produce a scaffolding cell (which would be outside of the desired shape). The next direction $\pi = 1$ is similar. The direction $\pi = 2$, however, "belongs" to the number $\lambda = 32$, triggering the production of another boundary cell with the coordinates $a = 0, b = 1$. A cell is constructed in the direction 2 (top) relative to the seed; the genome is copied to the newly constructed cell which is differentiated by updating all coordinates as follows: $x = x - 0$, $y = y - 1$. In other words, all y coordinates are decremented, resulting in genome shown in Table 3.

The (seed) cell that produced the copy is encoded in the copy's genome by triple $(0, -1, 32)$, i.e., the seed and the copy have different representations of the same shape. The last direction $\pi = 3$ is similar, and another boundary cell is produced to the right. The shape replication process is now driven by these two newly produced cells. For example, the cell produced to the top of the seed considers 4 directions π. It excludes $\pi = 0$, because there is a cell already in the place indicated by $\pi = 0$. The direction $\pi = 1$ is excluded because it's outside of the shape, as recognised by $n_+ = 0$ and $n_- = 0$. The direction $\pi = 2$ is selected because 2 is within $\lambda = 20$, triggering the cell-replication and cell-differentiation process similar to the one described previously. The direction $\pi = 3$ is interesting now because the corresponding location $a = 1, b = 0$ is inside the shape as recognised by $n_+ = 1$ and $n_- = 1$. In other words, there are two places where the boundary "crosses" the strip $x = a = 1$: one above the level $y = b = 0$, and one below (Figure 4:right).

The process terminates precisely because boundary cells distinguish between inside and outside, and do not replicate outside the desired shape. The scaffolding cells can only reach the boundary from inside: if a scaffolding cell produces a new cell whose genome has a triple $(0, 0, \lambda)$ after an update, then this replica recognises itself as a boundary cell, and does not replicate outside. If the scaffolding subgraph was always connected, a simpler algorithm would be possible. It would involve a) boundary cells building *only* other

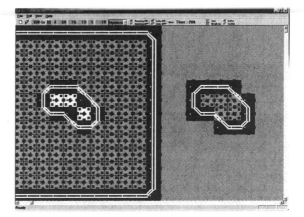

Figure 5: Completed shape replication.

boundary cells; and b) seeding a single scaffolding cell. It would, furthermore, avoid any need to verify whether replicated cells are inside or outside the shape (i.e., the numbers n_+ and n_- would not be needed). In either case (connected or disconnected scaffolding) the shape replication is driven by multiple cells, progressing in parallel (Figure 5).

Robust Shape Replication

The shape replication process described above assumes that at any cell-replication step there are no errors. However, there may be cases when some fragments of the genome are damaged due to copying process or processors/memory failures. If the genome is not repaired then the shape replication process would not terminate at the boundary, and some cells would be replicated beyond the desired shape by going through missing boundary cells.

In this section we consider an advanced robust algorithm designed to recover from such errors. More precisely, we consider the case when some triples (x, y, λ) in the genome are corrupted. The main challenge is not only to recognise corrupted data, but to avoid a replication that may produce an incorrect shape. The proposed solution involves 1) self-inspection of the genome within a cell, determining the endpoints of disconnected boundary fragments, 2) self-repair of the genome within a cell, adding the triples between the disconnected fragments. Thus, self-inspection of the genome on the cell level *mirrors* the boundary self-inspection, while self-repair of the genome on the cell level *mirrors* the repair of the overall shape. We believe that these are examples of Strange Loops because on the one hand, events on the multi-cellular shape's level trigger cellular transformations (e.g., a global repair activating the ontogenetical development), while on the other hand, the actions carried out by each cell (e.g., self-repair of the encoded genome) affect the higher level where the shape is replicated.

Each phase of *self-inspection of the genome within the cell* determines a pair of triples (x_1, y_1, λ_1) and (x_2, y_2, λ_2) such that the triple (x_1, y_1, λ_1) is the last before the break in the counter-clockwise direction, and (x_2, y_2, λ_2) is the first after the break counter-clockwise. If either of these triples is not found, then the genome damage is not repairable. Otherwise, a *self-repair* phase follows with the following steps:

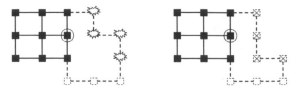

Figure 6: The cell shown inside a circle attempts self-repair. Left: the corrupted triples are shown with the "star"-like signs. Right: the repaired triples are marked with crosses.

- for the start triple (x_1, y_1, λ_1): a) select a counter-clockwise direction π_1 in λ_1; b) retrieve from the look-up Table 2 the coordinates (a_1, b_1) for repair in the direction π_1; c) set $x_1 = x_1 + a_1$ and $y_1 = y_1 + b_1$;

- for the end triple (x_2, y_2, λ_2): a) select a clockwise direction π_2 in λ_2; b) retrieve from the look-up Table 2 the coordinates (a_2, b_2) for repair in the direction π_2; c) set $x_2 = x_2 + a_2$ and $y_2 = y_2 + b_2$;

- approximate the line between the points (x_1, y_1) and (x_2, y_2) and its slope μ;

- set $x = x_1$, $y = y_1$, and continue an iterative process until the points (x, y) and (x_2, y_2) are the same:

  1) determine coordinates (a, b) such that the line between $(x + a, y + b)$ and (x_2, y_2) has a slope closest to μ; if the genome contains a triple $(x + a, y + b, \lambda^*)$ for some boundary link λ^*, then the algorithm terminates and the repair fails because a projected fragment intersects an existing boundary;

  2) retrieve from the look-up Table 2 the direction π corresponding to the coordinates (a, b);

  3) set the direction π_1^* opposite to π_1 by using modulo 4, i.e. $\pi_1^* = (\pi_1 + 2) \bmod 4$;

  4) form λ by concatenating π and π_1^* in decreasing order;

  5) insert (x, y, λ) and set $x = x + a$, $y = y + b$, and $\pi_1 = \pi$;

- when the points (x, y) and (x_2, y_2) are the same, "seal" the break:

  1) set the direction π_2^* opposite to π_2 as $\pi_2^* = (\pi_2 + 2) \bmod 4$, and the direction π_1^* opposite to π_1 as $\pi_1^* = (\pi_1 + 2) \bmod 4$;

  2) form λ by concatenating π_1^* and π_2^* in decreasing order;

  3) insert (x, y, λ);

The genome is partially repaired (Figure 6) within each cell which detected a discontinuity. Although the repaired genome does not cover all the missing cells, it does not introduce any cells which were not in the original shape, exhibiting soundness but not completeness property. In other words, the repaired boundary is always contained within the original shape. Importantly, there is a redundancy in the shape replication process: other cells which did not suffer any damage would successfully replicate the parts not encoded in the partially repaired genomes.

The described algorithms handle both standard ("blueprint") and non-standard shapes, self-organising in response to damage. Moreover, it is possible to combine these types. For example, structural data can be encoded in the form of triples, and a given genome can be extended in run-time with the data produced by self-inspecting emergent boundaries. Similarly, the self-repair phase within a cell which detected an anomaly in the genome may draw some data from the structural "blueprints" rather than approximate segments between disconnected fragments.

Conclusions and Future Work

We investigated a multi-cellular shape replication mechanism, implemented in a sensing and communication AAV network. The main algorithm solves the problem for connected and disconnected scaffolding. The underlying self-referential representation enables self-inspection and self-repair — contributing to a robust algorithm that can recover from possible errors in the "genome". The presented mechanism can replicate predefined standard shapes, arbitrary emergent shapes and their combinations. One future direction would involve treating the program in the same way as the data, so that *reprogrammable* cells may redirect and improve an ongoing shape replication process.

References

Durbeck, L. and Macias, N. (2002). Defect-tolerant, fine-grained parallel testing of a cell matrix. In Schewel, J., James-Roxby, P., Schmit, H., and McHenry, J., editors, *Proceedings of SPIE ITCom 2002 Series, Vol. 4867*.

Foreman, M., Prokopenko, M., and Wang, P. (2003). Phase transitions in self-organising sensor networks. In Banzhaf, W., Christaller, T., Dittrich, P., Kim, J., and Ziegler, J., editors, *Advances in Artificial Life - Proceedings of the 7th European Conference on Artificial Life (ECAL)*, volume 2801 of *LNAI*, pages 781–791. Springer Verlag.

Hofstadter, D. R. (1989). *Gödel, Escher, Bach: An Eternal Golden Braid*. New York: Vintage Books.

Holland, O. and Melhuish, C. (1999). Stigmergy, self-organization, and sorting in collective robotics. *Artificial Life*, 5:173–202.

Laing, R. (1977). Automaton models of reproduction by self-inspection. *Journal of Theoretical Biology*, 66:437–456.

Langton, C. (1984). Self-reproduction in cellular automata. *Physica D*, 10:135–144.

Lovatt, H., Poulton, G., Price, D., Prokopenko, M., Valencia, P., and Wang, P. (2003). Self-organising impact boundaries in ageless aerospace vehicles. In Rosenschein, J., Sandholm, T., Wooldridge, M., and Yokoo, M., editors, *Proceedings of the 2nd International Joint Conference on Autonomous Agents and Multi-Agent Systems*, pages 249–256. ACM Press.

Mange, D., Sipper, M., Stauffer, A., and Tempesti, G. (2000). Towards robust integrated circuits: The embryonics approach. *Proceedings of the IEEE*, 88:516–541.

Price, D., Scott, A., Edwards, G., Batten, A., Farmer, A., Hedley, M., Johnson, M., Lewis, C., Poulton, G., Prokopenko, M., Valencia, P., and Wang, P. (2003). An integrated health monitoring system for an ageless aerospace vehicle. In *Proceedings of the Fourth International Workshop on Structural Health Monitoring*. Stanford University.

Prokopenko, M., Wang, P., Foreman, M., Valencia, P., Price, D., and Poulton, G. (2004). On connectivity of reconfigurable impact networks in ageless aerospace vehicles. *Journal of Robotics and Autonomous Systems*, Special Issue:in press.

Sipper, M. (1998). Fifty years of research on self-replication: An overview. *Artificial Life*, 4:237–257.

Sipper, M., Mange, D., and Stauffer, A. (1997). Ontogenetic hardware. *BioSystems*, 44:193–207.

An Evolving and Developing Cellular Electronic Circuit

Daniel Roggen, Yann Thoma* and Eduardo Sanchez*
Autonomous Systems Laboratory, Institute of Systems Engineering
*Logic Systems Laboratory, Institute of Computing and Multimedia Systems
EPFL, Lausanne, Switzerland
http://asl.epfl.ch, *http://lslwww.epfl.ch
E-mail: name.surname@epfl.ch

Abstract

A novel multi-cellular electronic circuit capable of evolution and development is described here. The circuit is composed of identical cells whose shape and location in the system is arbitrary. Cells all contain the complete genetic description of the final system, as in living organisms. Through a mechanism of development, cells connect to each other using a fully distributed hardware routing mechanism and differentiate by expressing a corresponding part of the genetic code thereby taking a specific functionality and connectivity in the system. The configuration of the system is found by using artificial evolution and intrinsic evolution at the schematic level is possible. Applications include the approximation of boolean functions and the evolution of a controller capable of navigating a Khepera robot while avoiding obstacles. The circuit is suited for a custom chip called POEtic, which is a generic platform to implement bio-inspired applications.

Introduction

New approaches to the creation of electronic circuits have been explored in the last 10 years. Evolvable Hardware (EHW) (Higuchi et al., 1993) consists of using artificial evolution (e.g. genetic algorithms) to create electronic circuits. This approach showed that more efficient circuit implementations than those obtained with traditional techniques may be found (Coello et al., 2000; Vassilev et al., 2000). Evolved circuits may also operate using different principles than those which are obtained by design. They may exploit physical characteristics of devices to implement rich dynamical behaviour from simple building blocks, so better use of the hardware resources may be achieved (Thompson, 1997). EHW is believed to have a lot of potential (Yao and Higuchi, 1999), for instance in adaptive hardware (Kajitani et al., 1998), or in fault-tolerant hardware (Keymeulen et al., 2000), and it becomes more of an industrial reality (Higuchi et al., 2000).

Another approach is to mimick the way living organisms develop from a single cell to form a complete multi-cellular organism. In Embryonics (Marchal et al., 1994; Mange et al., 1996), cellular decomposition and development are key features which are used to provide a self-repairing substrate for the implementation of electronic circuits, as illustrated by the self-repairing BioWatch (Stauffer et al., 2001). As embryonic cells contain the complete description (genetic code) of the circuit, mechanisms such as self-reproduction and self-repair become possible. Self-repair is an alteration of the development process in which faulty cells are avoided. Faults may be detected by duplication of computational unit (Mange and Tomassini, 1998, 251–258). However the cellular paradigm allows new ways to perform fault-detection. Immunotronics takes inspiration from the way the immune system is capable of discriminating between the normal and anormal behavior of a living cell and transposes such concept to electronic circuits (Bradley et al., 2000). Cellular system also open new possibilities in genotype to phenotype mappings which can improve evolvability of circuits.

Motivated by the previous points, we are interested in the combination of evolution and development in hardware. Such a circuit may combine efficient implementations thanks to evolution with the advantages of the cellular and developmental paradigm to explore bio-inspired hardware. Novel in this paper is the development mechanism combined with evolution, and their implementation on a prototype of the custom chip *POEtic* (Tyrrell et al., 2003). POEtic is a generic platform to implement bio-inspired applications comprising mechansisms such as evolution (Phylogenesis), development (Ontogenesis) and learning (Epigenesis). Retaining the POEtic terminology we call the circuit described here a *PO circuit*. The PO circuit consists of cells which all contain the complete genetic description of the final circuit. Cells can be of any size/shape and can be located anywhere in the circuit. The development process uses a fully distributed dynamic hardware routing mechanism which is available in POEtic. Cells connect to each other at run-time and get to know which part of the genetic code they must express and take a specific connectivity and functionality in the system. The functionality of the system is found by evolving the circuit genotype using genetic algorithms. Applications include the approximation of boolean functions (adder and multiplexer) and the evolution of a controller capable of navigating a Khepera robot while avoiding obstacles. Self-repair and self-replication issues are discussed. The next section describes the POEtic hardware on which the PO circuit is implemented. Afterwards the cell structure and the development mechanism are discussed, followed by a section showing how the circuit can be evolved. Finally the results are discussed before concluding.

POEtic hardware

The multi-cellular PO circuit is implemented on a novel substrate: the POEtic chip (Tyrrell et al., 2003). The POEtic chip is a platform to test bio-inspired mechanisms, such as mechanisms of evolution (Phylogenesis), development (Ontogenesis) and learning (Epigenesis). Its architecture is summarized below and described extensively in (Thoma et al., 2003).

The POEtic chip is composed of a CPU and an *organic subsystem*. The CPU is a 32-bit RISC with bit manipulation and random number generation capabilities which make it suited to run evolutionary algorithms.

The organic subsystem is where multi-cellular PO circuits are implemented. It is composed of two layers: a layer of *molecules* and a *routing layer* (see fig. 1, right).

Molecules contain a 16 bits lookup table (LUT), a flip-flop and a switchbox for local routing. The molecule output can be registered or combinational. Molecules can operate in different modes. In the *3/4-LUT* mode, the molecule computes a logic function of 3- or 4-inputs. Modes *input* and *output* allow molecules to interact with the routing layer. Mode *reconfigure* is used to reconfigure another molecule. This is extensively used for the development mechanism of the PO circuit.

The routing layer establishes long distance connections between molecules (e.g. for inter-cell communication) and interfaces molecules to physical pins of the circuit. It is capable of *dynamic routing* to create connections between molecules automatically and at *run-time*. The routing layer is composed of routing units (RU) which are identified by a 16-bit ID and tagged as either *sources* or *targets*. When dynamic routing is triggered, logic within the RU connects sources and targets which have the same ID using a breadth-first search algorithm (Thoma et al., 2003; Moreno Aróstegui et al., 2001). Molecules can change the identifiers in the RU and retrigger the routing mechanism. Therefore new connections can be built at runtime. This may be used to respond to environmental changes, for example changes in the location of sensors or actuators in a robot. Note that this feature does not exist in FPGAs where physical connections are mapped at *design-time* to the FPGA.

Because the final POEtic chip will be available by the summer, prototyping is done by implementing the key parts of POEtic on an FPGA. The main difference is that here the organic subsystem is serially configured whereas in the final POEtic chip it is mapped in the address space of the CPU and accessed in parallel. The architecture of the system is

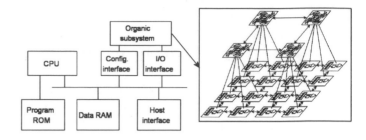

Figure 1: Left: the architecture of the system implemented on the FPGA. Right: the organic subsystem is composed of an array of molecules and an array of routing units.

shown in fig. 1, left. A 4K-word program ROM and an 8K-word RAM is interfaced to the CPU. The configuration interface takes care of the serial configuration and the I/O interface is used to read and write to the I/O of the organic subsystem, to set control bits and to read its status. The host interface allows the CPU to send data to the host (the PC hosting the FPGA board) for example to monitor the status of evolution.

Synthesis has been done for a Xilinx Virtex XC2V3000-4 FF1152. Ressource usage are around 5% of the LUTs for the CPU and 66% for an organic subsystem composed of 80 molecules, while some other logic serves to connect both together. Placement and routing of the system ended up with a 97% ressource usage, and 51% of RAM blocks usage. Placement could succeed with 100 molecules but routing could not. After place and route no theoretical maximum speed could be calculated, as there are combinational paths because of the molecules switch boxes. However, successful execution could be achieved at 10MHz.

One could argue that 80 molecules is a very small number compared to a Xilinx that owns more than 14000 slices, but we have to keep in mind that an FPGA is not designed to emulate another FPGA. For instance, as 76 configuration bits define a molecule, at least 76 flip-flops are needed, only for this purpose. Therefore, a hardware realization of POEtic is the only way to obtain a efficient system.

Circuit development

Cell architecture

The basic unit of the PO circuit is the cell. Key characteristics are that all cells are identical and contain the complete genetic description of the whole organism. A development mechanism is used to build the multi-cellular circuit from initially unconnected and undifferentiated cells. Following the view introduced in (Tyrrell et al., 2003) the cell can be viewed as the three layered structure illustrated in fig. 3. The genotype is a memory storing the genetic code of the entire circuit. The mapping layer is where development mechanisms take place. The phenotype layer is the functional part of the cell. In this implementation it has three inputs and one output and can compute any logic function of the inputs with a lookup table.

A cell composed of 40 molecules has been designed (fig. 2) which allows to fit 2 cells in the system. The molecules are composed of three different blocks, corresponding to the three layers depicted in fig. 3:

• The genotype layer is implemented with 16 molecules which store the genome. Each cell of the organism needs 4 molecules. Therefore the same cell can be used for organisms of up to four cells.

• The mapping from genotype to phenotype is done by 18 molecules: 8 molecules are responsible for the growth process, and 10 molecules used to differentiate cells.

• The phenotype layer is implemented by 5 molecules. Three of them serve as inputs, one as the functional part of the cell (a 3-LUT), and one as the output.

Development mechanism

Development maps the functionality of the cell from its genotype. It operates in two phases: growth and differentiation. The mechanism is based on a unique identifier for each cell, encoded in one-hot (a single bit is 1 at any time and indicates the ID) to reduce the cell size (binary encoding is possible but takes more space). For a system of n cells (n being a number in [1,16]), the first cell has an ID of n-1. For instance, if n=4, its ID would be 0...01000 (in this prototype n=2). IDs are 16-bits.

Growth The growth phase lets the organism grow from one cell to the whole organism. It is initiated by the CPU which first configures the entire organic subsystem with the cell description, including the genome. All cells are identical, totipotent, but are unconnected and undifferentiated. Cells have an *input* molecule waiting for a connection, with address 1...111. An external agent (e.g. the CPU) starts a dynamic routing, by configuring an I/O routing unit to be a source, with address 1...111. Dynamic routing creates a path to the closest cell. Once the routing process is over, the external agent sends serially the ID of the cell (number n-1). The cell stores it and computes the ID of the next cell by shifting its own ID by 1 (if cell is 0...01000, then next one is 0...0100). Afterwards, an output of the cell, with address 1...111 launches a new routing process, to connect to the nearest available cell. It then transmits the newly calculated ID, and the process continues until a cell receives the ID 0...01. It recognizes it is the last one, and does not start a new routing process.

The end of the growth phase is detected by the way of a global enable. This special feature of POEtic allows molecules which are sensitive to this signal to act on a global enable. In our system, the molecules involved in the growth process are not sensitive to this enable, but the others are. Once a cell has received its ID, it sets its global enable line, and so, the differentiation phase can not start before the or-

ganism is totally built.

Differentiation Cells know their ID and can differentiate to express the corresponding part of the genome. The genome codes the functionality (3-LUT content) and the connectivity (three inputs addresses) of each cell. Hence the topology on the phenotype layer needs not be the same as that of growth which is topologically linear on the mapping layer. The topology of the phenotype is given by the genotype and can be anything. For example, at the phenotype level, cells could be connected to form an array.

Four shift-memory molecules are required to store the genome of a cell. The partial reconfiguration capabilities of POEtic are fully exploited in the differentiation phase, where the 3 input molecules and the 3-LUT (functional) molecule are reconfigured with the corresponding part of the genome.

During the differentiation phase, the genome is shifted, the output being redirected to the input. Based on the ID of the cell, counters are used to enable the partial reconfiguration of the phenotype at the right time, that is when the output of the genome corresponds to the current cell. The reconfiguration lasts 4x16=64 clock cycles for each cell, but as the genome has to shift entirely it makes the differentiation longer.

At the end of the differentiation phase, every cell has its functional part ready, but cells are not yet inter-connected at the phenotype level. Therefore, every cellular input launches a routing process, until all cells are connected on the phenotype layer. Finally, the organism is ready to operate, the microprocessor can apply the inputs, retrieve the outputs, and calculate the fitness of this newly created individual.

Circuit evolution

In the PO concept, circuits are *evolved* rather than *designed*. The behaviour of the PO circuit is determined by its *configuration*: how cells are interconnected and what functionality cells take. This information is stored in the genotype layer of cells and can be evolved using genetic algorithms (GA) to obtain the desired functionality.

To demonstrate that the PO circuit can be evolved, two applications are considered. The first consists in implementing logic functions (adders and multiplexers) in the PO circuit, and the second consists in evolving a Khepera robot controller to perform navigation with obstacle avoidance using proximity sensors. In the latter case the relation between inputs and outputs which give the desired robot behaviour may potentially be difficult to determine which makes evolution well suited for such problems (Harvey et al., 1993).

Evolved "organisms" (PO circuits) are implemented in the organic subsystem of the POEtic hardware. A PO circuit is composed of two three-input cells (see 4, left). It has six inputs and two outputs which are connected by design to the output of the cells. Cells can compute any logic function of three inputs by the mean of a lookup table. The inputs of

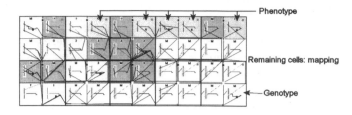

Figure 2: The cell is composed of 10 by 4 molecules. The complete circuit contains two such cells. Within a molecule is drawn its configuration. Five molecules form the phenotype, 16 molecules store the genotype and the remaining cells are used for the growth and differentiation.

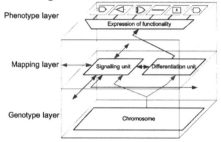

Figure 3: The 3-layer decomposition of a cell. The genotype layer holds the genetic code. The mapping layer handles development. The phenotype layer is the functional part of the cell, selected among a given repertoire of functionalities.

the cells can be connected to one of eight possible locations: one of the six inputs, or the output of one of the two cells. Input and outputs of the PO circuit can be accessed by the CPU which is in charge of running the GA and computing the circuit fitness. The physical location of the inputs and outputs in the organic subsystem are shown on the right of fig. 4. The outputs are placed close to the output molecule of the cells to minimize the use of RU. The only consideration regarding the placement of the inputs is to avoid the leftmost column of routing units, which are used by the development mechanism.

Evolution is applied to the input connectivity of the cells and to the content of their lookup table. The genetic code is a compact version of what is stored in the genotype layer of the cells to reduce the size of the search space. Eight bits are used to encode the content of the LUT and three bits are used for each cell input to encode the connectivity. Therefore the complete genetic code takes 34 bits (17 bits per cell).

Evolving logic functions

The PO circuit is evolved to implement logic functions. Two functions are considered: a multiplexer and a full adder. Three inputs are used (inputs 0 to 2) while inputs 3, 4 and 5 are set at all time to constant values 0, 1 and 0 respectively. The multiplexer uses one output whereas the adder uses two which represent the sum and the carry out. Circuits

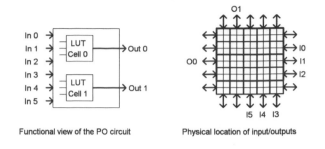

Figure 4: Left: functional view of the PO circuit with two cells which can compute logic functions of three inputs. The circuit has two outputs and five inputs. The outputs are connected by design to the outputs of the cells. The connections of the inputs of the cells are evolved together with the logic function. Right: physical location of the input and outputs of the circuit.

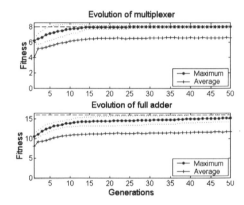

Figure 5: Maximum and average fitness over the generations when evolving logic functions. The horizontal dashed line represents the maximum fitness. The dotted lines either side the maximum represent the standard deviation between 20 runs.

are evolved by a GA with the following parameters: population size of 200, rank selection of the 20 best individuals, 5% of mutation rate, one-point crossover rate of 30% and elitism.

The fitness is evaluated by comparing the output of the circuit with the desired output for all possible inputs. It is equal to the number of times the outputs take the correct value. The maximum fitness is 8 and 16 for the multiplexer respectively the full adder. Fig. 5 shows the fitness over the generations averaged on 32 runs for the multiplexer and full adder. Evolution managed to implement the multiplexer in 31 of the 32 runs (in one run the maximum fitness obtained is 7). The full adder is evolved in 20 of the 32 runs (remaining runs achieve a maximum fitness of 14). As expected the multiplexer is easier to evolved than the full adder because it is a simpler circuit which can be implemented in only one cell whereas the full adder needs two.

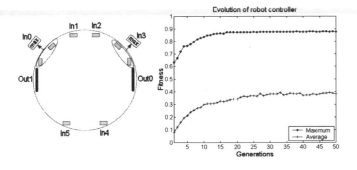

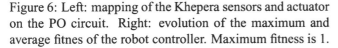

Figure 6: Left: mapping of the Khepera sensors and actuator on the PO circuit. Right: evolution of the maximum and average fitnes of the robot controller. Maximum fitness is 1.

Evolving a robot controller

The PO circuit is evolved to control the navigation of a Khepera robot to avoid obstacles, that is to map sensory inputs to motor commands. The Khepera is a two-wheeled robot with 8 proximity sensors (K-TEAM, 1999). The proximity sensors are connected to the inputs of the PO circuit, and the motors of the Khepera are controlled by the output of the cells. Fig. 6 left shows the mapping of the sensors and motors to the PO circuit. Some sensors are grouped by taking the value of the most active sensor. The robot has a sensory motor period of 100ms during which the speed of the wheels remain constant. At the end of the period, the outputs of the PO circuit update the speed of the wheels. An output of 1 corresponds to a wheel speed of +80mm/s, while an output of 0 corresponds to a speed of -80mm/s. Obviously wheels can never stop. This is a limitation imposed by the small amount of cells which fit on the FPGA. Afterwards, obstacles are sensed and the inputs of of the circuit are set for the next sensory motor period.

The fitness of the controller is determined from the behaviour of the robot: straight motion should be maximized while minimizing contacts with the walls. The fitness function is the sum of the speeds of the wheels at each sensory-motor step when both wheels spin forward (Floreano and Mattiussi, 2001). It favors obstacle avoidance because the wheels of a robot which is stuck against an obstacle do not spin due to the friction with the ground.

The circuit has been evolved using the same GA parameters as previously. Fig. 6 right shows the evolution of the fitness over the generations. A good obstacle avoidance behaviour is already obtained after about 10 generations. Note that evolution was performed in simulation to speed up experiments and the best individual tested successfully on a real robot.

Discussion

The development mechanism is novel in many respects. An important difference compared to embryonics development is that cells can be of any shape and can be physically placed anywhere in the organic subsystem even at irregular intervals. This may be interesting for example if parts of the organic subsystem are damaged: cells may be placed on the functional molecules (an off-line test may reveal the damaged locations), and dynamic routing takes care connecting cells in a transparent way.

Compared to classic unconstrained evolution which consists of manipulating directly the configuration bits of an FPGA (Thompson, 1997), evolution could be performed at a higher level thanks to dynamic routing. Indeed the approach could be classified as *intrinsic schematic* evolution. Connections are evolved by encoding identifiers of source cells (like in a net-list), rather than by encoding the configurations of many switchboxes. Consequently the genetic coding is more compact and evolution may be faster. Note that the genetic coding resembles Cartesian Genetic Programming (Miller, 1999) which also encodes the functionality and connectivity of every cell. In particular circuits which were evolved in simulation in the latter paper can be intrinsically evolved in the PO circuit.

Several features unique to the POEtic chip have been used to realize the PO circuit. The close interaction between the CPU and the organic subsystem allow fast reconfiguration of the organic subsystem by the CPU when running evolutionary algorithms. The POEtic CPU containes features such as a hardware random number generator which may speedup the execution of evolutionary algorithms. Dynamic routing and the possibility of one molecule to reconfigure another molecule to achieve self-reconfiguration are also at the core of the development mechanism.

The POEtic chip implemented on the FPGA has been changed in a number of ways compared to the final POEtic chip to reduce the space taken. Notably the final POEtic chip will contain an AMBA bus to interface with internal and external peripherals. Additional peripherals such as timers and hardware multipliers will be available. Also several chips may be cascaded to form a larger organic subsystem.

Conclusions

An evolving and developing circuit has been implemented on a novel POEtic chip. The circuit is composed of identical cells, all of them containing the complete genetic description of the final system, as in living organisms. A hardware development mechanism using specific features of the POEtic chip is used to build the circuit starting from unconnected and undifferentiated cells through a growth and differentiation process. Evolution has been used successfully to find suitable configurations of the circuit in tasks such as the evolution of logic functions and the evolution of a robot controller.

The circuit described here can be improved in a number of ways. Self-repair although mentionned has not yet been explored. Self-repair requires a means to detect faults in the circuit. This can be done by functional redundancy within the cells or by following the immunotronics approach. Upon detection of a defective cell, it would go offline and a new development process could be triggered which would make use of spare cells placed in the circuit.

Cells contain the genetic code of the complete circuit, however they were not designed to transfer the genetic code to other cells during or after development. As such, circuit self-replication is not yet possible. Modifications to allow self-replication may be explored in the future.

A direct genotype to phenotype mapping has been used and this is known to lead to scalability issues when evolving complex circuits. Indirect genotype to phenotype mappings can be implemented in the mapping layer of the cells. In particular a *morphogenetic coding* has been developed taking inspiration from the way inter-cellular chemical signalling regulate the functionality of cells. It has been designed to remain simple and suited for hardware implementations (Roggen et al., 2003). Further work may explore the combination of evolution and development using such indirect genetic codings.

Acknowledgments

This project is funded by the Future and Emerging Technologies programme (IST-FET) of the European Community, under grant IST-2000-28027 (POETIC). The information provided is the sole responsibility of the authors and does not reflect the Community's opinion. The Community is not responsible for any use that might be made of data appearing in this publication. The Swiss participants to this project are funded by the Swiss government grant 00.0529-1.

References

Bradley, D., Ortega-Sanchez, C., and Tyrrell, A. (2000). Embryonics + immunotronics: A bio-inspired approach to fault tolerance. In Lohn, J. et al., editors, *2nd NASA/DoD Workshop on Evolvable Hardware*, pages 215–223, Los Alamitos, California. IEEE Computer Society Press.

Coello, C. A., Aguirre, A. H., and Buckles, B. P. (2000). Evolutionary multiobjective design of combinational logic circuits. In Lohn, J. et al., editors, *2nd NASA/DoD Workshop on Evolvable Hardware*, pages 161–170, Los Alamitos, California. IEEE Computer Society Press.

Floreano, D. and Mattiussi, C. (2001). Evolution of spiking neural controllers for autonomous vision-based robots. In Gomi, T., editor, *Evolutionary Robotics IV*, pages 38–61. Springer-Verlag, Berlin.

Harvey, I., Husbands, P., and Cliff, D. (1993). Issues in evolutionary robotics. In Meyer, J.-A. et al., editors, *Proc. of the 2nd Int. Conf. on Simulation of Adaptive Behaviour*, pages 364–373. MIT Press/Bradford Books.

Higuchi, T. et al. (1993). Evolving Hardware with Genetic Learning: A First Step Towards Building a Darwin Machine. In Meyer, J.-A. et al., editors, *Proc. of the 2nd Int. Conf. on Simulation of Adaptive Behaviour*, pages 417–424, Cambridge, MA. MIT Press-Bradford Books.

Higuchi, T., Iwata, M., Keymeulen, D., et al. (2000). Real-world applications of analog and digital evolvable hardware. *IEEE Transactions on Evolutionary Computation*, 3(3):220–235.

K-TEAM (1999). *Khepera User Manual.* K-TEAM S.A., Préverenges, Switzerland (http://www.k-team.com).

Kajitani, I. et al. (1998). A Gate-Level EHW Chip: Implementing GA Operations and Reconfigurable Hardware on a Single LSI. In Sipper, M. et al., editors, *Proc. of the 2nd Int. Conf. on Evolvable Systems (ICES 98)*, pages 1–12, Berlin. Springer.

Keymeulen, D., Zebulum, R., Jin, Y., and Stoica, A. (2000). Fault-tolerant evolvable hardware using field-programmable transistor arrays. *IEEE Transactions on Reliability*, 49(3):305–316.

Mange, D., Goeke, M., Madon, D., Stauffer, A., Tempesti, G., and Durand, S. (1996). Embryonics: A new family of coarse-grained field-programmable gate array with self-repair and self-reproducing properties. In Sanchez, E. and Tomassini, M., editors, *Towards Evolvable Hardware*, pages 197–220, Berlin. Springer.

Mange, D. and Tomassini, M. E. (1998). *Bio-Inspired Computing Machines.* PPUR, Lausanne.

Marchal, P., Piguet, C., Mange, D., Stauffer, A., and Durand, S. (1994). Embryological development on silicon. In Brooks, R. and Maes, P., editors, *Proc. of the 4th Int. Conf. on the Synthesis and Simulation of Living Systems*, pages 365–370. MIT Press.

Miller, J. F. (1999). On the filtering properties of evolved gate arrays. In Stoica, A. et al., editors, *1st NASA/DoD Workshop on Evolvable Hardware*, pages 2–11, Los Alamitos, California. IEEE Computer Society Press.

Moreno Aróstegui, J.-M., Sanchez, E., and Cabestany, J. (2001). An in-system routing strategy for evolvable hardware programmable platforms. In Keymeulen, D. et al., editors, *3rd NASA/DoD Workshop on Evolvable Hardware*, pages 157–166, Los Alamitos, California. IEEE Computer Society Press.

Roggen, D., Floreano, D., and Mattiussi, C. (2003). A Morphogenetic Evolutionary System: Phylogenesis of the POEtic Tissue. In Tyrrell, A. M. et al., editors, *Proc. of the 5th Int. Conf. on Evolvable Systems (ICES 2003)*, pages 153–164, Berlin. Springer.

Stauffer, A., Mange, D., Tempesti, G., and Teuscher, C. (2001). A self-repairing and self-healing electronic watch: The biowatch. In Liu, Y. et al., editors, *Proc. of the 4th Int. Conf. on Evolvable Systems (ICES 2001)*, pages 112–127, Berlin.

Thoma, Y., Sanchez, E., Moreno Arostegui, J.-M., and Tempesti, G. (2003). A dynamic routing alogrithm for a bio-inspired reconfigurable circuit. In *Proc. of the 13th Int. Conf. on Field Programmable Logic and Applications (FPL'03)*, pages 681–690, Berlin. Springer Verlag.

Thompson, A. (1997). An evolved circuit, intrinsic in silicon, entwined with physics. In Higuchi, T. et al., editors, *Proc. of the 1st Int. Conf. on Evolvable Systems (ICES 96)*, pages 390–405, Berlin. Springer.

Tyrrell, A. M., Sanchez, E., Floreano, D., Tempesti, G., Mange, D., Moreno, J.-M., Rosenberg, J., and Villa, A. (2003). POEtic Tissue: An Integrated Architecture for Bio-Inspired Hardware. In Tyrrell, A. M. et al., editors, *Proc. of the 5th Int. Conf. on Evolvable Systems (ICES 2003)*, pages 129–140, Berlin. Springer.

Vassilev, V. K., Job, D., and Miller, J. F. (2000). Towards the automatic design of more efficient digital circuits. In Lohn, J. et al., editors, *2nd NASA/DoD Workshop on Evolvable Hardware*, pages 151–160, Los Alamitos, California. IEEE Computer Society Press.

Yao, X. and Higuchi, T. (1999). Promises and challenges of evolvable hardware. *IEEE Transactions on Systems, Man, and Cybernetics*, 29(1):87–97.

An Environment for Simulating Kinematic Self-Replicating Machines

William M. Stevens

Open University, England
william.stevens@open.ac.uk

Abstract

A simulation framework is described in which a collection of particles moving in continuous two-dimensional space can be put together to build machines. A self replicating machine has been designed in this environment. It is proposed that an environment such as this may facilitate the fabrication of self-replicating manufacturing systems.

Introduction

Artificial self-replicating manufacturing systems may offer technological advantages in areas ranging from computer systems engineering to space exploration (Freitas 1981, Drexler 1985), but no system offering such advantages has yet been fabricated.

Research into artificial self-replicating systems began in the late 1940s with von Neumann's proposal for a universal constructor and computer embedded in a cellular automaton (CA) array, which could be programmed to build arbitrary objects in the CA array, with self-reproduction as a special case (von Neumann 1966).

Since then, several other self-replicating systems have been devised for a variety of reasons. Sipper gives a comprehensive list of these systems (Sipper 1998,2003). Artificial self-replicating systems typically possess one or more of the following attributes:

a. Simple implementation (e.g. Langton 1984)
b. Constructional capability (e.g. von Neumann 1966)
c. Computational capability (e.g. Tempesti 1995)
d. Fast operation (e.g. Byl 1989)
e. Physical realism (e.g. Penrose 1959)
f. Evolutionary potential (e.g. Sayama 1999)
g. Reliable operation
h. Similarity to living systems

Any artificial self-replicating *manufacturing* system must at the very least possess attributes b and e. That is, it must be capable of being instructed to manufacture something other than itself, and it must actually exist in the physical world.

The simulation framework described in this paper was originally explored because it offers a greater degree of physical realism than the cellular automata frameworks that are often used to investigate self-replicating systems. In addition, the self-replicating system that has been devised in this framework may have the potential for limited constructional and computational capabilities, though these capabilities have not yet been demonstrated.

The simulation framework and the self-replicating machine described in this paper are offered as catalysts for exploring attributes b and e in the above list, in the belief that simulations of self-replicating systems having these attributes can act as a bridge towards the fabrication of self-replicating manufacturing systems.

The NODES System

NODES is a system in which circular particles called nodes interact with each other in a two-dimensional universe. There are several different types of node. Each type performs a specific function. There are types that perform logical functions, types that join nodes together, and types that exert forces on nodes. Nodes have four terminals which are used to send and receive signals to and from other nodes.

The two-dimensional space in which nodes move is continuous and has no boundaries, each node has two spatial coordinates and an orientation. The motion of nodes is governed by Newtonian-like laws. In addition to these laws, there are also rules that determine how signals can pass between nodes, how nodes can be connected together and how they should behave when they are connected.

Machines can be constructed from collections of nodes connected up in an appropriate way.

An Example

To help clarify all of this, here is an example. This example uses three different types of node.

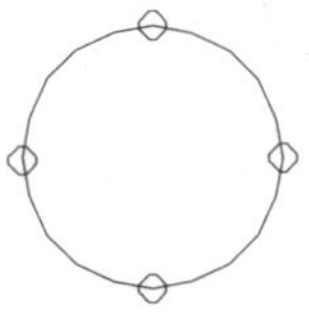

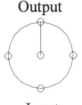

Figure 1: The 'Insulator' type has no function. It simply obeys the laws of motion and can be connected up to other nodes, but has no inputs or outputs.

Output

Input

Figure 2: The 'Not' type functions as an inverter. When it receives no input signal, it outputs a signal. When it receives an input signal, it outputs nothing.

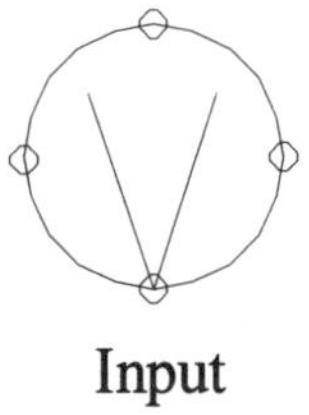

Input

Figure 3: The 'Thrust' type exerts a force on itself when it receives an input signal. The force acts in the direction from the centre of the node towards its input terminal.

On the left hand side of figure 4, a 'Not' node is connected to a 'Thrust' node so that the output signal from the 'Not' node feeds into the 'Thrust' node. Two 'Insulator' nodes are connected to the other side of the 'Not' node. On the right hand side of the figure is a vertical line of 'Insulator' nodes, connected together.

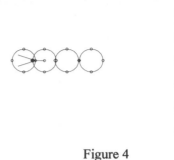

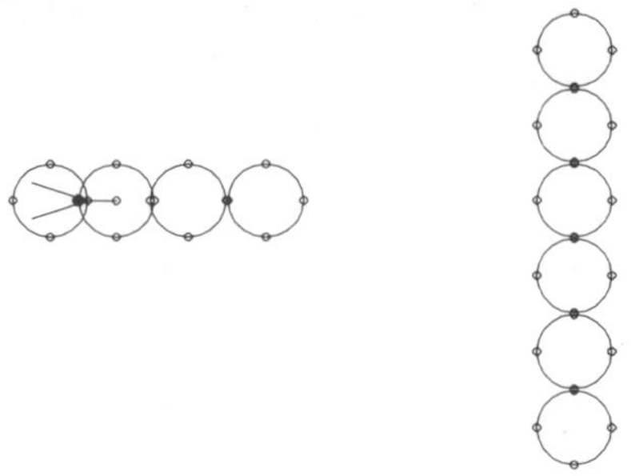

Figure 4

When the arrangement shown in figure 4 is simulated, the output from the 'Not' node activates the 'Thrust' node, which exerts a force on itself. This force ultimately gets transmitted along all of the nodes in the structure to which the 'Thrust' node belongs, and so the whole structure moves towards the right and collides with the vertical line of 'Insulator' nodes. Figure 5 shows this happening.

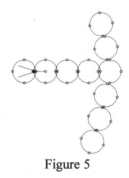

Figure 5

The vertical line of 'Insulator' nodes is pushed aside and comes to rest. Figure 6 shows the result.

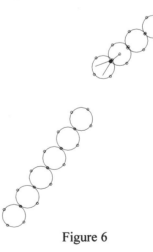

Figure 6

A Concise Description of NODES

In NODES, time is discrete, and moves forward in steps of 1 unit. At a given instant of time, a single isolated 'Insulator' node can be described by its position, orientation, velocity and angular velocity. Every node has a mass of 1 unit and a moment of inertia of 1 unit.

All nodes in the universe experience a frictional force proportional to their velocity, and an angular frictional force proportional to their angular velocity.

Nodes have a radius of 1 unit and there is a repulsive force between any pair of overlapping nodes.

Nodes can be connected to other nodes. When two nodes are connected, they exert forces on each other in such a way as to bring themselves into proximity and alignment. Every node has four terminals evenly spaced around its edge which can be used to make connections with the terminals of other nodes. A line of connected nodes is called a filament.

Node Types and Signals

Three node types have already been introduced – these were the 'Insulator', 'Not' and 'Thrust' types. Signals were mentioned in the informal descriptions of these types.

The terminals which are used to connect nodes together are also used as inputs and outputs for passing signals between nodes. Nodes do not need to be connected in order for signals to pass between them. If an output terminal is near an input terminal, then any signal it is outputting will be received by the input terminal. Each terminal is either an input or an output. If a terminal has no explicit definition, it is effectively an output producing no signal.

Signals are 32-bit integer values. The absence of a signal corresponds to a value of zero.

A node's type determines how it responds to input signals, and whether it produces any output signals.

Table 2 gives an informal description of 22 commonly used node types. In table 2 the letters N,S,E and W (for North, South, East and West) are used to refer to terminals and also to indicate directions. The context should indicate which usage is meant.

The notation used for expressions in table 2 is that used by the C programming language, summarized in table 1.

Operator	Name and meaning
+	Plus Sum of operands
*	Times Product of operands
==	Equals 1 if operands are equal, zero otherwise
!=	Not Equals zero if operands are equal, 1 otherwise
!	Logical Not 1 if operand is zero, zero otherwise
&&	Logical And 1 if both operands are non-zero, zero otherwise
||	Logical Or 1 if any operand is non-zero, zero otherwise

Table 1: Operators used in table 2

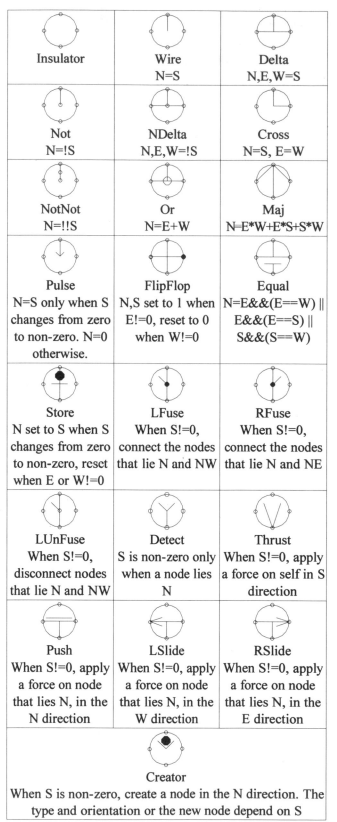

Insulator	Wire N=S	Delta N,E,W=S
Not N=!S	NDelta N,E,W=!S	Cross N=S, E=W
NotNot N=!!S	Or N=E+W	Maj N=E*W+E*S+S*W
Pulse N=S only when S changes from zero to non-zero. N=0 otherwise.	FlipFlop N,S set to 1 when E!=0, reset to 0 when W!=0	Equal N=E&&(E==W) || E&&(E==S) || S&&(S==W)
Store N set to S when S changes from zero to non-zero, reset when E or W!=0	LFuse When S!=0, connect the nodes that lie N and NW	RFuse When S!=0, connect the nodes that lie N and NE
LUnFuse When S!=0, disconnect nodes that lie N and NW	Detect S is non-zero only when a node lies N	Thrust When S!=0, apply a force on self in S direction
Push When S!=0, apply a force on node that lies N, in the N direction	LSlide When S!=0, apply a force on node that lies N, in the W direction	RSlide When S!=0, apply a force on node that lies N, in the E direction

Creator

When S is non-zero, create a node in the N direction. The type and orientation or the new node depend on S

Table 2: Common node types

A Self Replicating Machine in NODES

The node types described in the previous section have been used to make a self replicating machine (SRM). This SRM is made from several smaller machines, which will be referred to by name. The names of these machines are: 'Instruction Tape', 'Tape Reader', 'Tape Copier', 'Dragger', 'Releaser' and 'End Finder and Reader' (EFR).

The SRM would be far more complex were it not for the 'Creator' node type. This type allows new nodes to appear from nowhere and is the least physically realistic part of the SRM.

Figure 7 shows the machine shortly after it starts. The Tape Reader is advancing along the Instruction Tape, creating a filament of nodes. Several filaments will be created, and will assemble themselves to form a 'Tape Copier', two 'End Finders and Readers' a 'Releaser' and a 'Dragger'. The Tape Copier makes a copy of the instruction tape. The Dragger takes the end of this copy, and pulls it to a new position in the universe. The Releaser disconnects the Dragger from the copy. Then, one of the 'End Finder and Reader' machines finds the end of the copy and starts the cycle again at the tape reading step. The second 'End Finder and Reader' machine does the same with the original tape. Figure 8 shows the system part way through the first replication cycle.

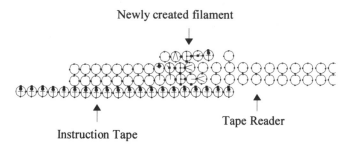

Figure 7: The SRM shortly after it is started. Note that only the rightmost part of the instruction tape is shown.

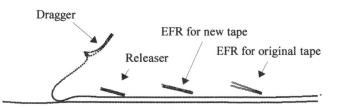

Figure 8:The SRM part way through the first replication cycle. The Dragger is dragging the new tape away from the original tape. The Tape Reader and the Tape Copier that worked on the original tape have finished their jobs and are outside the frame of this picture.

The following sections briefly describe the components of the self replicating machine. There is not room here for a complete description of these components. Further details can be discovered by obtaining the software required to simulate the system, details of which are given near the end of this paper.

The Instruction Tape

The instruction tape is a filament of Store nodes, the outputs of which lie along one side of the filament, as shown in figure 7.

The Tape Reader

The Tape Reader is a very simple component used only in the parent self-replicating machine. For all descendants of this parent, the function of the Tape Reader is included in the EFR components.

The job of the Tape Reader is to move along the Instruction Tape, passing the signals it encounters from the Store nodes in the Instruction Tape to a Creator node to create a sequence of nodes, which it joins together to form a filament. The Tape Reader also outputs a signal to act as a trigger to any filament that needs it.

The Tape Copier

The Tape Copier is the first component to become active after the Tape Reader has finished its task. The Tape Copier moves from its initial position towards the rightmost end of the Instruction Tape. When it touches the Instruction Tape it changes direction and begins to move along the tape, creating new Store nodes and connecting them together to make a second Instruction Tape.

Figure 9: The Tape Copier in action, copying the leftmost end of the original Instruction Tape. The V-shaped component just above Tape Copier is the Dragger, not yet fully assembled. To the right is the Releaser, assembled and waiting to be activated.

The Dragger

The Dragger is designed so that the time it takes to assemble is longer than the time taken for the Tape Copier

to finish its task. When it is active, the Dragger starts moving towards the leftmost end of the copy of the Instruction Tape. It connects itself near the end of this tape and then starts pulling it away from the original tape, as shown in figure 8. As the tape is pulled, it touches and activates the Releaser and the two EFRs.

The Releaser

After the Releaser is activated by contact with the second Instruction Tape, it proceeds to move along this tape, heading towards the Dragger. It moves very slowly, but eventually catches up with the Dragger. When the Releaser touches the Dragger, the Dragger disconnects itself from the tape.

The End Finder and Reader for the Copy

As the second Instruction Tape is dragged away from the original Instruction Tape, it brushes against this EFR. Before it is fully assembled, this component detects a signal from a Store node near the end of the Instruction Tape, and in response it connects itself to the end of this tape. It is then dragged along with the tape. When the Releaser causes the Dragger to let go of the tape, this EFR can finish assembling and then begin its task of reading the second Instruction Tape, starting another replication cycle.

The End Finder and Reader for the Original

Once all of the components needed for the child SRM have moved away from the original Instruction Tape, reading of the original tape can begin again. The second EFR component initiates this. After it is assembled, it moves slowly towards the end of the original Instruction Tape, where it starts reading the tape.

The SRM in Action

Figure 10 illustrates the SRM in action. The original machine has produced three children, and is part way through making a fourth. The first child has produced a child of its own.

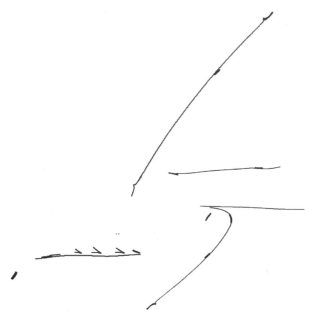

Figure 10: One SRM has produced several others. The scparation between the SRMs has been reduced here to fit them all on the page.

Conclusion

A system has been demonstrated in which a self-replicating machine can be constructed. The laws of motion used in this system are based on Newton's laws. The particles in this system are simple, except for the 'Creator' type which can produce other particles out of nowhere. With further work, it may be possible to improve on the self replicating machine described here with respect to attributes b and e defined in the introduction.

It seems reasonable to suppose that the SRM described here can be modified so as to construct things other than copies of itself. To devise a SRM in NODES which is not dependant on the 'Creator' type seems much more challenging.

Obtaining NODES for your System

Source code, examples (including the self replicating machine described here) and a user guide for NODES are available by following the URL:

http://willsthings.mysite.freeserve.com/SRM/presentation.htm

NODES is written in C++ and has been compiled under Linux and under DOS (Using DJGPP)

References

Byl J. 1989. Self-Reproduction in small cellular automata. Physica D, Vol. 34, 295-299

Drexler K. E. 1985. Engines of Creation: The Coming Era of Nanotechnology. London: Fourth Estate. URL: http://www.foresight.org/EOC/index.html

Freitas R. A. Jr. 1981. Report on the NASA/ASEE summer study on advanced automation for space missions. Journal of the British Interplanetary Society, Vol. 34, September 1981: 407-408.

Langton C.G. 1984. Self-reproduction in cellular automata. Physica D, Vol. 10: 135-144.

Penrose L.S. 1959. Self-reproducing machines. Scientific American, Vol. 200, No. 6: 105-114

Sayama H. 1998. Constructing Evolutionary Systems on a Simple Deterministic Cellular Automata Space. Ph. D Dissertation, Department of Information Science, Graduate School of Science, University of Tokyo.

Sipper, M. 1998. Fifty years of research on self-replication: An overview. Artificial Life, Vol. 4, No. 3: 237-257.

Sipper M. 2003. The Artificial Self-replication page. URL: http://www.cs.bgu.ac.il/~sipper/selfrep/

Tempesti G. 1995. A new self-reproducing cellular automaton capable of construction and computatioon. In F. Morán, A. Moreno, J.J. Merelo, and P. Chacón, editors, ECAL'95: Third European Conference on Artificial Life, volume 929 of Lecture Notes in Computer Science, pages 555-563, Berlin, Springer-Verlag.

von Neumann, J. 1966. Theory of Self-Reproducing Automata. Edited and completed by A.W. Burks. Urbana, Illinois: University of Illinois Press.

Towards an evolutionary-developmental approach for real-world substrates

Shivakumar Viswanathan and Jordan Pollack
DEMO Lab, Computer Science Dept., Brandeis University, MA 02454
{shiva,pollack}@cs.brandeis.edu

Abstract

Extending "body-brain" evolution to the real-world presents a number of difficulties due to conflicting idealizations between evolutionary and constructional models. Toward addressing this gap, we develop a simple model system to analyze the effects of undoing these idealizations. Preliminary experiments with this system show that high variability developmental substrates can influence evolutionary dynamics by causing ambiguities in selection. Furthermore the substrate can enable the evolution of adaptive responses to non-deterministic developmental effects.

Introduction

Background

An important approach towards Artificial Life is via the investigation of the system principles of interaction and self-organization by building robots(Brooks, 1992). A key principle being to design the robots such that their adaptive behavior is achieved with a parsimony of global knowledge models (Brooks, 1991; Harvey et al., 1997; Pfeifer, 1996). The rationale being that the structure of the systems required to achieve the desired behavior under this constraint could provide insights into the nature of interaction-based adaptive behavior both in biological as well as engineering systems. This rationale in turn suggests that the same principle of global knowledge parsimony could be applied to the design of the very processes by which the entire robots (i.e. "body and brain") themselves come to be (Sims, 1994; Pollack et al., 1999).

We can distinguish two interpretations of the general question of how complete adapted robots could be obtained by interaction based processes. From a *functional* perspective, the explanandum is of how adapted robots could be evolved from unadapted ones with respect to a particular behavioral context. From an *embodiment* perspective, the explanandum is how particular physically embodied robots could be constructed (or transformed) from simpler precursors by such processes. When the solution concept of interest is in terms of interaction-based processes that can gener-

ate actual embodied robots performing the desired behavior[1] both these aspects require to be addressed.

Motivation

In industrial engineering practice, knowledge intensive protocols of abstract specification and validation enable the functional and embodiment aspects to be addressed with a significant degree of independence, as problems of design and manufacturing respectively. Applying the principle of knowledge parsimony, the question then is how could these aspects be addressed in an integrated way without making the knowledge intensive specification/implementation distinction. Even though such a distinction is not involved in biological evolution, we believe that achieving this integration in an artificial evolutionary system presents some basic conceptual difficulties.

In evolutionary algorithms with an explicit developmental phase[2], the embodiment aspects are treated as a deterministically unfolding process that satisfies the Genotype-Phenotype map abstraction. Though this abstraction is a key axiom in Evolutionary Computation, development as a generative representation of the phenotype is far removed from the issues involved in the real-world construction of complex structures.

A more relevant, even if simplistic, characterization of problems related to embodiment have been discussed by (Simon, 1962) and (Crane, 1950). In the parable of the two watchmakers, Simon (Simon, 1962) discusses how problems caused by the instability of intermediate stages in constructing complicated structures depends on the structure of the construction process even though the specifications of the final structure are known to both watchmakers. Similarly, Crane(Crane, 1950) discusses the problem of error accumulation in the assembly of structures involving a large number of parts and how the specificity of interactions dur-

---

[1] As differ from a solution concept in terms of build-able specifications for such robots as in(Funes and Pollack, 1998; Lipson and Pollack, 2000; Hornby and Pollack, 2002)

[2] See(Hornby and Pollack, 2002; Stanley and Miikkulainen, 2003) for extensive reviews

ing construction has an impact on this accumulation. However, in both these examples the functional aspects are neglected. There is an implicit idealization of a goal structure where processes unable to produce this specific outcome are categorically maladapted. Furthermore there is no notion of how these processes may be systematically varied to retain these properties while producing different final structures.

This suggests the existence of a gap in the way the way that the functional and embodiment aspects of biological systems has been conceived, where each is typically studied by idealizing the other. The conundrum is that with respect to the problem of producing complete robots neither of these idealizations is entirely valid. In this paper we discuss preliminary investigations toward identifying concepts required to bridge this functional-embodiment divide within an evolutionary developmental framework.

Problem definition

Basic model

In order to model both functional as well embodiment aspects, we adopt a simple evolutionary model and introduce modifications to the developmental phase to accommodate issues relevant to embodiment processes. This is described in an implementation independent manner below.

Development is considered to be a series of transformations starting with the zygote and terminating at the adult organism which enters the reproducing population on which selection acts. The process occurs in a particular environmental context ζ and Φ_ζ is the set of all phenotypes. The zygote is defined as $\phi_{start} \in \Phi_\zeta$, and the net effect of the development process is to produce a series of intermediate states of the embryo $\phi_{start} \to ...\phi_t \to \phi_{t+1} \to ...$ till it halts at the "adult" phenotype ϕ_{final}. This temporally ordered sequence of intermediate phenotypic states from ϕ_{start} to ϕ_{final} constitutes the ontogeny of ϕ_{final}. The final configuration ϕ_{final} obtained on a given execution instance is taken to be the individual which enters the reproducing population. The behavioral measure of interest (i.e. the fitness function) is defined to be $\mathbf{e} : \Phi_\zeta \to \mathbb{R}$.

The zygote has associated with it a genotype $g \in \mathcal{G}$ and the cellular interpreter machinery $\mathbf{i} \in I$ for the genotype ($\mathcal{G}$ is the set of all genotypes and I is the set of all interpreters). The interaction of g and $\mathbf{i}$ results in "actions" that effect the present state of the developing embryo. The various interactions involved in development are assumed to be effected by ambient variability that has a stable pattern particular to the operative physics and ζ. So development with the same zygote ϕ_{start} would not necessarily result in the same outcome.

Consequently, rather than a genotype-phenotype map we have a relation $\mathcal{D} \subset \mathcal{G} \times \Phi_\zeta$, where a pair $(g,\phi) \in \mathcal{D}$ indicates the possible genotype and phenotype combination for a given individual. So each genotype g can be considered to be associated with a finite set (sample space) Ω_g of possible adult phenotypes where $\Omega_g \subset \Phi_\zeta$. A phenotype $\phi \in \Phi_\zeta$

is in Ω_g iff ϕ has a non-zero probability of being produced as ϕ_{final} with g. Formally the development process can be described as a function $\delta_{\phi_{start}} : \mathcal{G} \to \tilde{\Omega}$ where $\tilde{\Omega}$ is the set of pairs (Ω_g, P_g) for all $g \in \mathcal{G}$, with P_g being the probability distribution on Ω_g.

Selection for specificity

Consider the sample space Ω_g associated with a genotype g. *Specificity* is used here with respect to the maximum possible difference in fitness between two phenotypes in Ω_g. Suppose e_{min} and e_{max} are the minimum and maximum values of $\mathbf{e}$ over the phenotypes in Ω_g. A closed interval $I_{\Omega_h} = [e_{min}, e_{max}]$ is defined on the real-number line using these two values, where e_{min} and e_{max} are the greatest lower bound (g.l.b) and least upper bound (l.u.b) of I_{Ω_g} respectively($e_{max}, e_{min} \neq \pm\infty$). So the specificity associated with g is the length of the interval $s = |I_{\Omega_g}| = e_{min} - e_{max}$

So we can say that the developmental process has high specificity if s is a "small" value and low specificity if s is a "large" value. In general terms, a process that could produce individuals having high fitness values with high specificity even in the presence of ambient variability is very desirable as a solution.

With respect to an evolutionary process though, it is a higher-order property as compared to the fitness function $\mathbf{e}$ based on the behavior of the individuals in the population. So a question of interest is whether there would be an implicit selection for genotypes that have greater *specificity* of fitness outcomes especially when associated with high values of both e_{min} and e_{max}. This is also question of interest with respect to biology where developmental processes are known to have high reproducibility in their outcomes. It raises a chicken-egg question of whether developmental specificity is due to properties unique to the living or whether the specificity was an implicitly evolved property. Here we use evolution on a simple toy-system to demonstrate how developmental substrates capable of "measurement" could influence the evolution of specificity(Viswanathan and Pollack, 2004).

Development with a tiling machine

Implementing genetic "actions" The toy system described in this section instantiates the model described in the previous section. The "actions" resulting from the interpretation of the genotype, take the form of assembly actions. These are performed by a tiling machine modeled as a gantry robot that is restricted to movement in two dimensions (see Figure 1).

This machine has a head that can be moved under programmable control to locations in the workspace identified by their (x,y) coordinates, by a series of horizontal and vertical translations. The workspace, which is the equivalent of the environmental context ζ, is a square partition of a two dimensional plane such that it can perfectly accommodate

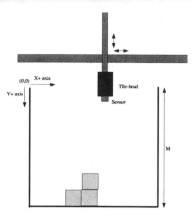

Figure 1: Tiling machine in workspace ζ

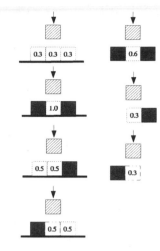

Figure 2: Tile interference physics

$M \times M$ identical square tiles without any gaps. The machine has an explicit spatial existence so it cannot move through tiles.

The atomic operations performed at the head are a **release** operation to release tiles individually, and a **sense** operation by means of a sensor that returns a binary outcome based on whether the position immediately in front of it in the $Y+$ direction is occupied or not with a **1** or **0** respectively. A finite number of tiles are available to the machine which can be released individually at the head under programmable control.

This programmed control is achieved by means of a program g (the genotype) which is executed with a fixed interpreter $\mathbf{i}$ embedded in the tiling machine, where a program is a finite sequenced list of (x,y) locations in the workspace. From this we can see that configuration of tiles in the workspace at time t, is the equivalent of the state of the embryo ϕ_t. And with the execution of the program g, the tile configurations also change resulting in a developmental process[3]

Tile physics The physics governing the behavior of a tile on being released at the tile-head is similar to the Tetris game. There is a constant velocity "diffusion" acting downward along the $Y+$ direction such that a tile released at a particular location moves at the rate of one tile length per time step. The stochasticity arises from the interference when a released tile comes in contact with other tiles or with the edges of the workspace and continues till the tile comes to rest.

---

[3]The motion from the origin to a point (x_a, y_a) takes place by horizontal motion along the $y = 0$ edge till the x_a coordinate is reached, followed by downward motion till the location is reached if unobstructed. This path is retraced to return to the origin. A movement between consecutive locations occurs indirectly by first returning to the origin.

The interference is modeled as occurring as defined in Figure 2. The hatched square is the tile that is descending. It's possible position in the next time step are shown as squares with dotted outlines. The numbers indicate the probability with which the tile will come to rest at that position. the particular values chosen are arbitrary. The dark line indicates a tile or a wall at that periphery and the tile cannot proceed through it.

Interpreters The focus in this setup is on the effect of features in the genotype-interpreter relation. Here we consider two interpreters $\mathbf{i}_{coupled}$ and $\mathbf{i}_{state}$. The interpreter $\mathbf{i}_{coupled}$ is such that for each location in the program, the tiling-head releases a tile at that location and on completion does not return to the origin but continues to release tiles till the sensor indicates a tile occupying the location immediately in front of it. At this point it returns to the origin and the next command in executed. Due to this, the execution of the program is now coupled with the developing configuration.

The interpreter $\mathbf{i}_{state}$ is identical to $\mathbf{i}_{coupled}$ except that it maintains a state variable that is incremented each time it detects a tile in front of it and shifts to execute the next command.

Evolutionary setup

The fitness function

A behavior evaluation function $\mathbf{e} : \Phi_\zeta \to \mathbb{R}$ is defined as:

$$\mathbf{e}(\phi) = \sum_{i=1}^{n} (d(T_i, A) + d(T_i, B))$$

where $\phi \in \Phi_\zeta$, T_i is the location of the ith tile in ϕ, and $d(p,q)$ is the non-linearized version of the Manhattan distance between two points p and q in ζ where $d(p,q) = |x_p - x_q| +$

$|y_p - y_q|$ if $|x_p - x_q| \leq M/2$; and $d(p,q) = M/3 + |y_p - y_q|$ otherwise.

In words, the value $\mathbf{e}(\phi)$ is the sum of the distances (as defined by d) of every tile in a tile-configuration ϕ from two pre-chosen points A and B in the workspace ζ. The points chosen here are $A = (1, M)$ and $B = (M, M)$ where M is the length of each side of the workspace. These points correspond to the lower left and lower right corners of ζ in Figure 1. The fitness of a configuration is related to the presence of specific features of the configuration. Configurations that have tiles concentrated close to $X = 5$ axis, and away from the $Y = M$ edge would tend to have higher values as compared to one where the tiles are randomly distributed around ζ. Construction of these features require tight correlations between the locations of the tiles that are released, and the fitness drops rapidly with the lack of correlation.

Algorithm

Each genotype is represented as a tape consisting of a string of integer values specifying a series of (x, y) coordinate pairs in the workspace ζ. These are treated as being of fixed length defining n locations. So the genotype space G consists of $(M^2)^n$ programs where M is the length of a side of the workspace ζ. Here $M = 10$ and $n = 20$. The maximum number of tiles available for release is also equal to 20. The maximum value of $\mathbf{e}$ for these parameter values is equal to 414.64.

The algorithm used here is similar to a canonical evolution algorithm having the structure

$$... G_i \rightarrow P_i \rightarrow P_i' \rightarrow G_i' \rightarrow G_{i+1} \rightarrow P_{i+1} \rightarrow P_{i+1}' ...$$

where P_i represents the ith generation of individuals with phenotypes on which selection acts, and G_i represents the gene pool of the generation.

The gene pool G_0 is initialized with a random sampling of G. The size of the gene pool is maintained constant at 30 throughout.

Contingent development $(G_i \rightarrow P_i)$: Each of the genotypes in this gene pool generates a maximum of N individuals by a developmental process involving the interpreter $\mathbf{i}$. Only a fraction k of these N is assumed to come to maturity and enter the reproductively viable population P_i. Here $N = 10$ and $k = 0.5$. So, the population always consists of $30 \times 5 = 150$ individuals. The developmental context i.e. the interpreter and the environmental noise model is assumed to be the same for the entire population and throughout the evolutionary process.

When the interpreter is $\mathbf{i}_{coupled}$, the first kN developed individuals are added to the population without any exclusion. However with the interpreter $\mathbf{i}_{state}$, whether an individual enters the population depends on the state accumulated during the developmental process of the individual. Here we pick a hand-designed value for this threshold to be 25% of the total number of tiles, so an individual enters the population

only when the accumulated state crosses this threshold. If no instance crosses this threshold after $N - 1$ trials, the last individual always enters the population.

Fitness assignment $(P_i \rightarrow P_i')$: Based on the behavior of the individuals as determined by $\mathbf{e}$, the population P_i is assumed to take on a reproductive viability in relative proportion to their behavior. This redistribution of P_i in terms of their relative fitness is represented as P_i'.

Selection $(P_i' \rightarrow G_i)$: In proportion to their relative fitness in P_i', a subset of the individuals in the population and hence their genotypes G_i' are selected to be reproduced. Due to the fixed size of the gene-pool, 30 individuals in the population are selected for asexual reproduction in each generation. The selection is elitist with the default selection of the top 2 individuals with highest fitness.

Variation $(G_i' \rightarrow G_{i+1}')$: With the exception of the genotypes corresponding to the elite individuals, the selected genotypes differ from the genotypes that constructed the parents. This difference is implemented explicitly with a variation operation to produce the gene pool of the next generation G_{i+1} from G_i'. The variation is restricted to single-locus mutations of the parental genotype, where the probability of a mutation occurring is uniform over the entire parental genome. A mutation results in a random change in a location specified within a local neighborhood $\mathcal{N}$ to the extent that it lies within ζ. The neighborhood of a location (x_0, y_0) is given as $\mathcal{N}_{(x_0, y_0)} = \{(x, y) : (|x - x_0| \leq r, |y - y_0| = 0) \vee (|x - x_0| = 0, |y - y_0| \leq r)\}$ where $r = 2$.

Results

The results from a representative run with the interpreter $\mathbf{i}_{coupled}$ are shown in Figures 3 and 4. Firstly, we see that the fitness of the best individual does not change much throughout even in the first generation. However this has a pronounced effect on the fitness values in the population as the mean fitness can be seen to increase rapidly with the first 15 generations, but then remains flat after that. However the stable value of the mean fitness is significantly below the best fitness due to the large variation in fitness values for each genotype as can be see in Figure 4.

In comparison, the fitness of the best individual and the mean fitness with $\mathbf{i}_{state}$ (Figure 5) is of the same order as $\mathbf{i}_{coupled}$. However, there is a significant difference in the interval lengths of the individuals entering the population with $\mathbf{i}_{state}$ (Figure 6). As the best and mean fitness are of the same order as $\mathbf{i}_{coupled}$, this is due to an increase in the value of e_{min} of the genotypes in the population though without an increase in e_{max}. This reduction in the interval length can be seen as the downward shift in the scatter plot as compared to that of $\mathbf{i}_{coupled}$.

The higher specificity with $\mathbf{i}_{state}$ can be attributed to the manner in which the maintained state is used to selectively exclude individuals from the population based on the contingencies of their particular ontogenies.

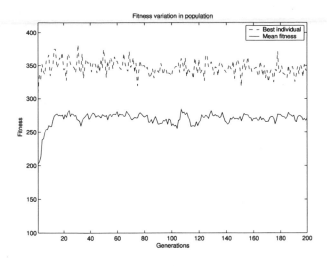

Figure 3: Fitness variation with $\mathbf{i}_{coupled}$

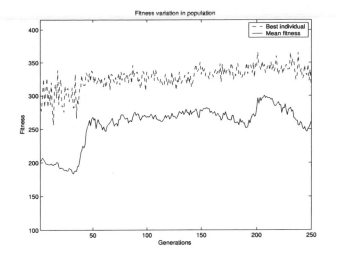

Figure 5: Fitness variation with $\mathbf{i}_{state}$

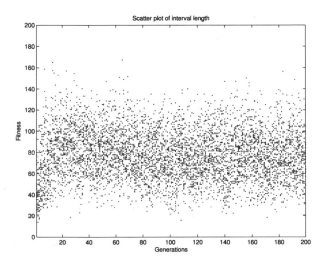

Figure 4: Scatter plot of interval length with $\mathbf{i}_{coupled}$

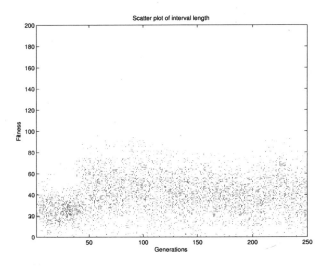

Figure 6: Scatter plot of interval length with $\mathbf{i}_{state}$

The rationale is that at different stages in development there may exist low-dimensional measurable properties and events that provide information about the global state. So developmental elements whose behavior is contingent on the "measurements" of this low dimensional information could influence subsequent states in the ontogenic trajectory. However, developmental interactions are hidden from the selective pressure acting on the population. So on evolutionary time scales these contingent responses can be "calibrated" by evolution such that they are related to the fitness of the outcomes.

In the case of the tiling-machine, a sensor response of **1** at the head occurring immediately after release indicates that the tile was at the very location it was released at. On the other hand a sensor response of **0** after releasing the tile indicates uncertainty as it provides no information about where the tile would eventually come to rest. So a larger number of **1**s would indicate greater certainty about the outcomes of the tile release operations though it says nothing about the fitness of the outcome. As can be seen in Figure 7 that plots the accept rate i.e. number of individuals accepted relative to those that were obtained, the accept rate is about 50% as compared to the 100% in the case of $\mathbf{i}_{coupled}$.

The existence of such low-dimensional measurable properties containing global-state information is tested empirically by determining whether evolution can find a way to use the additional state maintained by $\mathbf{i}_{state}$ to influence the outcome distribution. This "calibration" to exclude individuals that correspond to low-fitness individuals is clearly observable in this case.

Discussion

The tiling machine system described here is primarily used here as a way to provide a transparent and qualitative

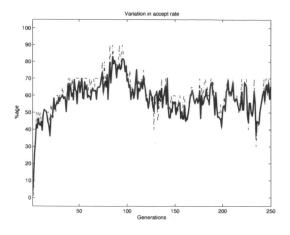

Figure 7: Accept rate

demonstration of the issues to be addressed toward a unified functional-embodiment approach to the evolution of complex real-world systems.

The fundamental difference that arises with the presence of non-genetic variation in the population is that selection of genotypes is now an ambiguous process as the observed fitness differences in the individuals in the population may not correspond to heritable variation. More importantly, this variation is not due to lack of precision in evaluating the fitness but involving systemic variations in the structure and behavior between the different developmental outcomes. So there is no "true" phenotype associated with a genotype with the other phenotypic variants being the effect of noise. As a result the key problems are related to how order and structure can be extracted from the high variability involved. Here we have used a simple experiment to demonstrate the critical role that developmental substrates can play in this context by enabling the evolution of specificity. A more comprehensive analysis of this issue is presented in (Viswanathan and Pollack, 2004).

By incorporating interactional specificity into construction but without reference to a "goal" structure and showing the relation of embodiment processes to the behavior rather than structure of the outcomes, this paper presents an incremental step toward bridging the gap between the functional and embodiment perspectives.

Conclusion

This paper identifies a class of problems arising due to the idealizations in functional and embodiment perspectives of evolutionary development. Toward addressing this issues, a preliminary abstraction for the analysis of evolvable developmental substrates is described and demonstrated using a simple toy-system. Experiments with this system suggest that the developmental substrate could exert adaptive influences on the ontogeny at the interactional level.

References

Brooks, R. (1991). Intelligence without reason. In Myopoulos, J. and Reiter, R., editors, *Proceedings of the 12th International Joint Conference on Artificial Intelligence (IJCAI-91)*, pages 569–595, Sydney, Australia. Morgan Kaufmann publishers Inc.: San Mateo, CA, USA.

Brooks, R. (1992). Artificial life and real robots. In Varela, F. and Bourgine, P., editors, *Toward a practice of autonomous systems: Proceedings of the First European Conference on Artificial Life*, pages 3–10. MIT Press.

Crane, H. R. (1950). Principles and problems of biological growth. *The Scientific Monthly*, 70(6):376 – 389.

Funes, P. and Pollack, J. B. (1998). Evolutionary body building: Adaptive physical designs for robots. *Artificial Life*, 4(4):337–357.

Harvey, I., Husbands, P., Cliff, D., Thompson, A., and Jakobi, N. (1997). Evolutionary robotics: the sussex approach. *In Robotics and Autonomous Systems*.

Hornby, G. and Pollack, J. (2002). Creating high-level components with a generative representation for body-brain evolution. *Artificial Life*.

Lipson, H. and Pollack, J. B. (2000). Automatic design and manufacture of robotic lifeforms. *Nature*, 406(6799):974–978.

Pfeifer, R. (1996). Building fungus eaters: Design principles of autonomous agents. In *From Animals to Animats 4: Fourth International Conference on Simulation of Adaptive Behavior*, pages 3–12. MIT Press/Bradford.

Pollack, J. B., Lipson, H., Funes, P., Ficici, S. G., and Hornby, G. (1999). Coevolutionary robotics. In Koza, J. R., Stoica, A., Keymeulen, D., and Lohn, J., editors, *The First NASA/DoD Workshop on Evolvable Hardware*. IEEE Press.

Simon, H. A. (1962). The architechture of complexity. *Proceedings of the American Philosophical Society*, 106:467–482.

Sims, K. (1994). Evolving 3D morphology and behavior by competition. In Brooks, R. and Maes, P., editors, *Artificial Life IV*, pages 28–39. MIT Press.

Stanley, K. O. and Miikkulainen, R. (2003). A taxonomy for artificial embryogeny. *Artificial Life*, 9:93–130.

Viswanathan, S. and Pollack, J. (2004). On the evolvability of replication fidelity in stochastic construction. Technical Report CS-04-248, Brandeis University.

Stepwise evolution of molecular biological coding

Peter R Wills[1,2]

[1] Department of Physics, University of Auckland, Private Bag 92019,
Auckland 1020, New Zealand
[2] Fraunhofer Gesellschaft, Biomolecular Information Processing (BioMIP),
Schloss Birlinghoven, D-53754 Sankt Augustin, Germany
p.wills@auckland.ac.nz

Abstract

Two principles that embody necessary characteristics of self-sustaining physical systems are espoused. These principles are then used to analyze the dynamics of coding self-organization in the process of nucleic acid sequence-dependent protein synthesis. An artificial system is constructed in which catalysis of codon to amino-acid assignments is embedded hierarchically in protein sequence space. An example is provided of a system that evolves through execution of a coarse-grained binary code to execution of a more refined quarternary code. General implications for the construction of ALife systems are considered.

Introduction

We take as our starting point the fact that all molecular biological systems involve the translation of genetic information. Strings of nucleotides (RNA copies of DNA genes) serve as templates for the construction of strings of amino-acids that fold to give functional proteins. During translation, molecular recognition of a nucleotide triplet ("codon") causes the addition of a specified amino-acid onto a chain which, when completed, is a protein molecule. A set of proteins which are themselves products of the synthetic process, the amino-acyl synthetases, catalyze the physical process that assigns a particular amino-acid to a particular codon. The universal genetic code is often represented as a table listing a set of unique assignments from each of the 64 codons $X \equiv \{A, B, ...\}$ onto the set of 20 standard amino-acids $y \equiv \{a, b, ...\}$. When any of the 1216 assignments $X \rightarrow y$ not found in the coding table is made during translation, as occurs rarely in reality, it is considered by human observers to be an error.

Three basic problems confront our understanding of how such a complex self-contained information-processing system can maintain physical stability and accuracy, let alone have emerged from molecular disorder in the first place. These problems arise in one form or another in any consideration of systems that are autocatalytic – systems whose components are involved obligatorily in their own production.

The first problem is the "error catastrophe" predicted to arise due to amplification of dysfunctionality. Macromolecules that turn out to be partially dysfunctional due to random errors in their synthesis will participate in the synthetic processes required to produce further macromolecules that will be even more likely to be incorrectly synthesized and dysfunctional. Although first outlined by Orgel (1963) in relation to the process of translation, the essential solution to the error catastrophe problem was provided by Eigen (1971) who showed how Darwinian selection can lead to the accumulation of genetic information, rather than its loss, in populations of replicating macromolecules. Self-organization occurs when the accuracy of symbol replication is above a certain system-determined threshold.

Eigen (1971) did not solve the second problem, the instability of error-prone translation, but Hoffmann (1974) was able to show that stable translation can be achieved if the specificity of the adaptors assigning codons to amino-acids is likewise above some threshold. Bedian (1982) provided the first demonstration of thermodynamically driven coding self-organization. Wills (1993) and Nieselt-Struwe and Wills (1997) subsequently established that any translational mapping from nucleic acid to protein sequences can be established in practice only when codon-to-amino-acid assignment capability (adaptor functionality) is distributed among proteins in certain defined ways. That is to say, the way in which catalytic capability is embedded in protein sequence space, the so-called "protein structure-function relationship" must meet certain formal requirements before translation can ever emerge.

The third problem is the establishment and maintenance of functional coupling between the individually error-prone processes of information replication and translation. It is one thing to demonstrate, as Eigen (1971) has, that Darwinian selection takes place when the means of genetic replication is already provided, or as Bedian (1982) and Wills (1993) have, that coding self-organization occurs spontaneously when appropriate genetic information is provided, but it is quite another thing to demonstrate that the two processes can be mutually self-sustaining when they are coupled together and the effect of errors can seep from one process to the other. Only recently have Füchslin and McCaskill (2001) shown that simple reaction-diffusion coupling is sufficient to allow disordered processes of both kinds to condense into a fully-fledged operational system of self-encoded translation in a cell-free system involving genetic replication.

None of the theories that have been proposed for these three problems is complete enough to give a plausible description of the pathway or timescale of prebiotic events. This is due in large part to the complexity and extreme specificity displayed by the final products of that early stage of evolution: 64 codons mapping onto a 20-member set of amino-acids. On the other hand, the universal genetic code displays regularities that offer potential clues to not only the history of its evolution but also some basic features of how the function of a protein varies with its amino-acid sequence, its primary structure. It has long been noted that the distribution of redundancy in the coding table is far from random and that amino-acids with similar properties are represented by similar codons. These regularities are at least partly explained in terms of chemical and structural similarities among codons or amino-acids. On this basis, a group theoretical decomposition of the coding table was provided by Hornos and Hornos (1993), but Nieselt-Struwe and Wills (1997) reframed the question to ask whether progressive decomposition from relatively coarse-grained to increasingly fine-grained translation might be a feature of the real-world historic events that gave rise to the universal code. In this paper I take the argument one step further by asking under what conditions dynamic processes could have led to the progressive refinement of the genetic code, all the way to the final form that has since survived to our own very late evolutionary epoch.

The concept of structure-function decomposition has been presented in several separate contexts (Nieselt-Struwe and Wills 1997; Wills and Henderson 2000; Wills, 2001). The basic idea is that the macromolecules that perform any function are selected in a progressive fashion such that inexact information is first used to specify a general class of catalysts from which those whose functions are more optimally refined emerge through a process of mutual selection. There is a close analogy to the progressive increase in Shannon information that Adami *et al.* (2000) report in the Avida replication system, except that we are concerned here with the way in which components that carry out elementary algorithmic operations cooperate to optimize their function, rather than the overall competitive advantage being seized by a whole quasi-species of algorithms.

I begin by enunciating two basic principles of complex autocatalytic systems and then show how those principles guide the construction of an artificial system in which the self-organization of genetic coding takes place in a stepwise fashion.

Theory

Scientific principles, like Newton's three laws of motion, express empirically valid generalizations that seem logically self-evident when the relevant phenomena are perceived from within some rigorously prescribed framework. The virtue of principles lies in the extent to which they limit the terms of phenomenological description but capture some universal aspect of the empirical domain under consideration. Here we are interested in how genetic information sustains the existence of complex molecular biological systems or their algorithmic counterparts.

The *principle of reflexivity* expresses a truism: all of the complex components of a self-sustaining system must be constructible as a result of executing the operations that their structures functionally enable; or, all of the elementary chemical reactions required within living systems must be catalyzed by the molecular components that are produced by execution of those reactions. The principle applies to all autocatalytic systems, but it has special application to organisms because their existence relies on genetically encoded information: the genetic sequences in a living cell must, *via* translation, be reflexive *vis-à-vis* the catalytic functions of the products of their translation, especially coding assignments. Let us consider the process of coding self-organization (Wills 1993).

Suppose there existed a very primitive mechanism, a completely indiscriminate process analogous to ribosomal protein synthesis, whereby proteins could be produced from RNA templates by joining amino-acids together in strings as a result of some sequential tri-nucleotide docking process. Such indiscriminate "translation" of sequences of triplet codons would produce completely random proteins. However, suppose that proteins with particular structures exerted an influence on the process, some proteins capable of biasing the assignment of particular amino-acids to particular codons during the synthetic process. For example, some proteins could cause a bias in the assignment of GYN, UUN and AUN codons to hydrophobic amino-acids. [The 4 standard nucleotide units, A, C, G and U, are categorized such that N stands for any of them and Y stands for the pyrimidines C and U. The particular codons mentioned here are actually mapped onto hydrophobic amino-acids in the universal coding table.] Suppose that the ability to cause such a bias in the assignment of amino-acids to codons depended on a preponderance of hydrophobic residues at particular positions, perhaps relative positions, in the proteins responsible for the assignment bias.

The presence of RNA templates with a preponderance of the specified codons in appropriate relative positions would allow specific "biasing" proteins to generate themselves autocatalytically in the manner of coding self-organization (Wills 1993). A primitive code in which GYN, UUN and AUN codons were assigned predominantly to hydrophobic amino-acids could evolve as a result of dynamic symmetry-breaking, induced by the matching of RNA template sequences to the structure-function properties of proteins and not in any way dependent on the chemical mechanism of peptide bond formation or direct chemical interactions between amino-acids and the codons to which they became assigned. The maintenance of coding would be based on the presence of specifically selected molecules with special functions, namely, proteins that guaranteed the required assignment bias by virtue of the reflexive relationship of their amino-acid sequences, *via* translation, to the available genetic information.

How might the primitive code in such a system become refined so that the genetic information was interpreted in a more differentiated fashion and the assignment of amino-acids to codons was more selective? All we need do to answer this question is apply the principle of reflexivity again. Suppose in our example that the selective presence of alanine rather than other hydrophobic amino-acids at some positions in proteins conferred the function of biasing the assignment to alanine of GCN codons relative to other GYN, UUN and AUN codons. [This particular choice is made by way of example because GCN codons happen to stand for alanine in the universal coding table.] In the presence of genetic information with GCN codons in appropriate relative sequence positions, the concentration of the alanine-containing proteins could be autocatalytically amplified relative to the overall concentration of the original set of generally hydrophobic proteins. GCN codons would then be assigned to alanine as a result of the dynamic state of the system.

The recapitulation of reflexivity at increasing levels of functional specificity constitutes the *principle of decomposition*: a class of structures can become effectively decomposed into narrower categories through the operation of more refined recognition processes only if the products of those processes have the specified capability of selective recognition. Only when the effect of some new bias in functional specificity results in the synthesis of components that reinforce that bias can a system evolve as a result of autocatalytic selection. When the appropriate condition is met, more refined translation of the genetic information supplied to a system produces assignment catalysts that carry out translation assignments more selectively. In the example considered above, the emergence of the GCN-to-alanine from the GYN-to-hydrophobic assignment depended on the supposition that proteins containing alanine at sequence positions corresponding to GCN codons in the available genetic information were instrumental in biasing the placement of alanine at those positions in the sequences of proteins being produced.

Model System

I now describe a self-organizing coding system in which the embedding of assignment functions in the protein sequence space decomposes in a rigorous hierarchical fashion. The system first evolves to execute the assignment functions defined by a binary code that subsequently decomposes into a quarternary code. The design of the system is based on the principles of reflexivity and decomposition. First, the information supplied to the system is constructed so that components that carry out the assignments of a code are produced when they themselves are used exclusively for the translation process. Second,

the requirements of decomposition are met by constructing a hierarchical embedding of assignment functions in the protein sequence space.

The first step in the construction of the system is to specify classes $\{K, L\}$ of codons and $\{k, l\}$ of amino-acids. In keeping with previous practice (Wills, 1993; Füchslin and McCaskill, 2001), single points in protein sequence space are chosen at random to represent "catalytic centers", optimal sequences for the performance of each of the 4 binary assignment functions $\{K, L\} \rightarrow \{k, l\}$. These arbitrary sequences are shown in the second column of Table 1.

K→k	llkkkklkkllk	A→a	ddbbabcaadca*
		A→b	cdbbbbcabcda
		B→a	dcaaaadaaccb
		B→b	cdababdbadca*
K→l	kkklllllkkll	A→c	baacddddbadc
		A→d	abacdddcbbdc
		B→c	bbbcdcccbacc
		B→d	baadcddcbbdc
L→k	lkklkklllkkl	C→a	dbadabcddbbc
		C→b	dabdabdcdabd
		D→a	daacbbcccaad
		D→b	cabdbbdcdbac
L→l	klkkkkkklllk	C→c	bcbaabbbccdb*
		C→d	bcababaaddcb
		D→c	bdbbabbaccdb
		D→d	acabbabaddca*

Table 1. Protein sequences for optimal assignment activity. The nominal binary assignments, $\{K, L\} \rightarrow \{k, l\}$, were first embedded in the nominal binary sequence space of length 12. The sequences of the 4 optimal catalysts, chosen at random, are shown in column 2. The 16 possible quarternary codon to amino-acid assignments, $\{A, B, C, D\} \rightarrow \{a, b, c, d\}$, were then embedded in the defined subspaces of the quarternary sequence space. Assignments and optimal catalytic sequences for the code $A \rightarrow a$, $B \rightarrow b$, $C \rightarrow c$ and $D \rightarrow d$ are indicated*.

The process of protein synthesis requires that amino-acids be assigned sequentially to codons in genes that are supplied to the system. Which amino-acid is assigned to a particular codon in the process of stringing together amino-acids to form a protein depends on the catalytic capabilities of proteins already in the system. When supplied with the two gene sequences *LLKKKKLKKLLK KLKKKKKKLLLK* a population comprised exclusively of proteins with the sequences *llkkkklkkllk klkkkkkklllk* corresponding to the assignment activities $K \rightarrow k$ and $L \rightarrow l$ for a binary code can clearly maintain its absolute purity because no $K \rightarrow l$ or $L \rightarrow k$ assignments can be made in a system lacking proteins with those catalytic capabilities.

Error-proneness is introduced into the model system by allowing that sequences near to optimal centers for each assignment function are assigned relative catalytic capability R that diminishes rapidly with Hamming distance h from that center according to a Gaussian profile

$$R = \exp(-h^2 / h_0^2) \qquad (1)$$

where h_0 is a constant. This also means that a protein with a randomly chosen sequence is most likely to catalyze all assignment functions in the set $\{K, L\} \rightarrow \{k, l\}$ at an equally minimal rate. Coding self-organization from an initially random population of minimally active proteins in a system of this sort has been extensively documented (Wills 1993).

The second step in the construction of the system is to assume that the classes of codons $\{K, L\}$ and amino-acids $\{k, l\}$ are decomposable into more refined categories, specified as $\{A, B\} \equiv K$ and $\{C, D\} \equiv L$ for codons, and $\{a, b\} \equiv k$ and $\{c, d\} \equiv l$ for amino-acids, generating 16 refined assignment functions, $\{A, B, C, D\} \rightarrow \{a, b, c, d\}$. In keeping with the supposition that a and b are alternative forms of the amino-acid of sort k, and in like manner for all of the decomposed codons and amino-acids, we should expect that the optimal sequence for catalysis of any one of the functions $\{A, B\} \rightarrow \{a, b\}$ will be a specialized example of *llkkkklkkllk* that is optimal for the assignment $K \rightarrow k$. So, the optimal protein sequences for catalysis of assignments $\{A, B\} \rightarrow \{a, b\}$ were chosen by randomly substituting a or b for k and c or d for l in the optimal $\{k, l\}$-level sequence for $K \rightarrow k$ assignment catalysis. The arbitrary $\{a, b, c, d\}$-level sequences generated in this way are shown in the fourth column of Table 1. Suboptimal catalytic activities of sequences were calculated from the Gaussian profile described above with $h_0 = 0.7$ and with h defined in terms of digital representations $a \equiv 00$, $b \equiv 01$, $c \equiv 10$ and $d \equiv 11$.

Results

Simulations of protein synthesis involving the quarternary assignments $\{A, B, C, D\} \rightarrow \{a, b, c, d\}$ were carried out using the methods described in Wills (1993). At each timestep each gene in the system was used as a template for translation to produce a new protein which, when added to the population, contributed to assignment catalysis according to its capabilities as defined by its Hamming distance from each of the optimal sequences (Eq. 1). The genes supplied to the system had sequences *DDBBABCAADCA CDABABDBADCA BCBAABBBCCDB ACABBABADDCA*. When translated according to the assignments $A \rightarrow a$, $B \rightarrow b$, $C \rightarrow c$ and $D \rightarrow d$ constituting the quarternary code, these genes generate proteins with sequences *ddbbabcaadca cdababdbadca bcbaabbbccdb acabbabaddca* that are optimal for those same coding assignments. Note that these genetic sequences are refinements of the $\{K, L\}$-level genetic sequences needed for elementary coding self-organization, as discussed above.

In an illustrative example, the system started with 3000 proteins whose sequences had been chosen at random. It underwent well-separated transitions after about 3000 and 10000 generations. The binary code first evolved from the initially random assignments of codons to amino-acids (Figure 1). In the initial situation all possible assignments $\{A, B, C, D\} \rightarrow \{a, b, c, d\}$ were carried out at equal rates and random proteins were synthesized. Specified coding assignments, $\{A, B\} \rightarrow \{a, b\}$ and $\{C, D\} \rightarrow \{c, d\}$ representing $K \rightarrow k$ and $L \rightarrow l$, became selectively amplified as a result of the reflexive information supplied to the system for translation. Correspondingly, components that catalyze non-coding assignments, $K \rightarrow l$ and $L \rightarrow k$, initially equally represented, effectively disappeared from the system.

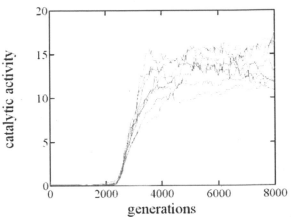

Figure 1. Transition to a binary code. The 8 quarternary assignment functions $\{A, B\} \rightarrow \{a, b\}$ and $\{C, D\} \rightarrow \{c, d\}$, equivalent to the two binary coding assignments $K \rightarrow k$ and $L \rightarrow l$, are selected from the complete set of 16 possible assignments $\{A, B, C, D\} \rightarrow \{a, b, c, d\}$. Each of the traces represents the net rate at which one codon-to-amino-acid assignment $X \rightarrow y$ is catalyzed in the system.

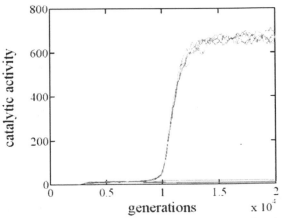

Figure 2. Further transition to a quarternary code. After the first transition to a binary code, the 4 assignment functions $A \rightarrow a$, $B \rightarrow b$, $C \rightarrow c$ and $D \rightarrow d$ are selected from the binary coding sets $\{A, B\} \rightarrow \{a, b\}$ and $\{C, D\} \rightarrow \{c, d\}$. Each of the traces represents the net rate at which one translation assignment $X \rightarrow y$ is catalyzed in the system.

A second stage of evolution resulted in selection of the quarternary code, based on the refined assignments of codons from decomposed sets, $\{A, B\} \equiv K$ and $\{C, D\} \equiv L$ to amino-acids from subclasses $\{a, b\} \equiv k$ and $\{c, d\} \equiv l$ (Figure 2). Specified coding assignments, $A \rightarrow a$, $B \rightarrow b$, $C \rightarrow c$ and $D \rightarrow d$, became selectively amplified and the concentration of components that catalyze the other four quarternary assignments, $A \rightarrow b$, $B \rightarrow a$, $C \rightarrow d$ and $D \rightarrow c$, still represented in the coarse-grain binary code, were reduced to a relative minimum.

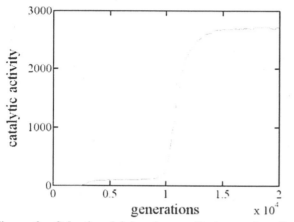

Figure 3. Selective driving force. During each coding selection transition there is a radical increase in total catalytic activity, a measure of the general rate at which proteins are synthesized.

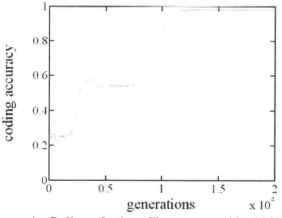

Figure 4. Coding selection. The accuracy with which the final code $A \rightarrow a$, $B \rightarrow b$, $C \rightarrow c$ and $D \rightarrow d$ is executed undergoes two transitions. The initial probability of approximately 0.25 corresponds to the random assignment of codons to amino-acids, the intermediate probability of about 0.5 corresponds to the binary code $\{A, B\} \rightarrow \{a, b\}$ and $\{C, D\} \rightarrow \{c, d\}$ and the final probability close to 1.0 corresponds to the fully-fledged quarternary code.

The two transitions in the dynamics of the system are evident in further results. Figure 3 displays what is in effect the thermodynamic driving force behind the transitions, the improvement in the overall rate of assignment catalysis derived from the selection of catalytically active components that mutually stimulate protein synthesis.

Figure 4 demonstrates evolution from initially random protein synthesis, through the partially resolved binary stage to the quarternary code. In the final stationary state most of the proteins in the system are highly active for the four assignments that constitute the full code.

Discussion

One of the criticisms voiced in the community of researchers interested in the origin of life is that the systems studied are artificially constructed so as to display exactly the behavior being sought as an indication of evolutionary self-organization. Thus, it is argued, rather than enlightening us as to how life actually evolved out of disordered chemical events and processes, we are left with no more than a description of how a system behaves when it is designed to behave that way, the biochemical equivalent of a tautology. Such criticism must be taken seriously when assessing the significance of hypothetical prebiotic scenarios. It is also true that ALife systems are completely hypothetical constructs conforming to preconceived design principles. However, the description and study of ALife systems can demonstrate quite generally that certain preconditions must necessarily be fulfilled before life can evolve. The principles of reflexivity and decomposition describe rather obvious preconditions of that sort, but they have not been generally applied to our understanding of the accumulation of biological information and its exploitation in the precise specification of biochemical function. This is because it is very difficult, after the event of evolution, to reconstruct either the context within which information has promoted the circularity of processes needed for the maintenance of autocatalysis or the classes of primitive processes that were subsequently refined as biochemical complexity increased. Further discouragement from seeking explanations for the origin of biological complexity comes from the confident neo-Darwinian retort "Selection takes care of all of that!" The retort may be true, but it is not complete.

The principle of reflexivity is significant because the complexity and specificity of the most fundamental molecular biological structures and functions is such that their sudden appearance out of a morass of undifferentiated chemical processes and events defies simplistic explanations. As has been shown elsewhere (Nieselt-Struwe and Wills 1997; Wills and Henderson 2000) the structure-function relationship of catalysts must be such as to contain the possibility of reflexive closure between components and operations in the first place. Of course it is important that there exist components that carry out selected operations, but it is also important that the functional specificity of the generated population of inexactly synthesized components keeps the system above the error-catastrophe threshold. Whether this is possible depends in the end on how operational function varies with structure.

From a design point of view, the structure-function relationship of catalytic polymers is created by embedding catalytic capabilities in the relevant polymer sequence space, as in Table 1. It has been possible to design systems that support the stable execution of molecular biological translation (Wills, 1993) by ensuring that the embedding of assignment catalysts in the protein sequence space is consistent with the possibility of constructing genetic sequences that are reflexive *vis-à-vis* the production of a coding set of assignment catalysts *via* translation. In the present paper it is shown how this basic principle can be extended to produce nested realizations of reflexivity that support a gradual pathway for the development of complex structures and functions. The principle of decomposition is more subtle than its parent principle (Wills 2001) because it relies on the description of components of the system at different levels: "coarse-grained" like $\{K, L\}$ and $\{k, l\}$, "fine-grained" like $\{A, B, C, D\}$ and $\{a, b, c, d\}$, *etc*. In ALife we have the luxury of defining hierarchical levels *a prior*, but it is unclear how they can be chosen adequately to describe natural biological systems. The formalized principle of decomposition may have more application to the design of artificial systems than the description of living systems found in nature. Its general significance is apparent in the work of Watson and Pollack (). It remains to be seen whether systems constructed according to the principle of decomposition enlighten intermediate stages of evolution when coding self-organization is coupled to the error-prone replication of the required genetic information. The study of Füchslin and McCaskill (2001) could be extended in this direction.

A philosophical note is appropriate in closing. The Central Dogma of molecular biology (Crick 1957) specifies that molecular sequence information can flow between nucleic acids and into protein but "once it is in protein it cannot get out again" (Crick 1970). The principle of reflexivity complements the Central Dogma by requiring that the information specifying an organism be adapted to the structure-function relationship of proteins such that coordinated execution of the informationally specified functions of proteins produces the very proteins that carry out those functions. In many ways the non-algorithmic embedding of biological functions in the folded structure of proteins is a more characteristic feature of biological systems than the one-way flow of sequence information from DNA to RNA to protein. Algorithmic information flow, translation essentially, guarantees the stability of the processes of inheritance and the operation of natural selection. On the other hand, the non-algorithmic embedding of function within structure space can be exploited to produce the basic phenomena of life, replication, variation and selection, without informational encoding, as the existence of prions demonstrates (Wills 1988). Further articulation and discussion of these ideas is likely to contribute significantly to our understanding of bioinformatics as we attempt to build models of the relationship between complex assemblages of genes and the characteristics of the organisms to which they belong.

Acknowledgments. Support from the Alexander von Humboldt Foundation and helpful advice and encouragement from John McCaskill are appreciated. It is the author's wish that no person or agency should ever derive military or purely commercial benefit that depends in any way on the publication of this paper.

References

Adami, C., Ofria, C. & Collier, T. C. 2000. Evolution of Biological Complexity. Proc. Natl. Acad. Sci. USA, 97:4463-4468

Bedian, V. 1982. The possible role of assignment catalysts in the origin of the genetic code. Orig. Life 12:181-204

Crick, F. H. C. 1958. On protein synthesis. Symp. Soc. exp. Biol. 12:138-163

Crick, F. H. C. 1970. Central Dogma of Molecular Biology. Nature 227:561-563

Eigen, M. 1971. Selforganization of Matter and the Evolution of Biological Macromolecules. Naturwiss. 58:465-523

Füchslin, R. M. & McCaskill, J. S. 2001. Evolutionary self-organization of cell-free genetic coding. Proc. Natl. Acad. Sci. USA, 98:9185-9190

Hoffmann, G. W. 1974. On the Origin of the Genetic Code and the Stability of the Translation Apparatus. J. Mol. Biol. 86:349-362

Hornos, J. E. M. & Hornos, Y. M. M. 1993. Algebraic model for the evolution of the genetic code. Phys. Rev. Lett. 71:4401-4404

Nieselt-Struwe, K., & Wills, P. R. 1997. The Emergence of Genetic Coding in Physical Systems. J. Theor. Biol. 187:1-14

Orgel, L. 1963. The Maintenance of the Accuracy of Protein Synthesis and its Relevance to Ageing. Proc. Natl. Acad. Sci. USA, 49:517-521

Watson, R. A. & Pollack, J. B. 2001. Symbiotic Composition and Evolvability. In Proceedings of Advances in Artificial Life, 6th European Conference, (ECAL 2001), Lecture Notes in Computer Science 2159, 480-490 (Springer)

Wills, P. R. 1988. Genetic Information and the Determination of Functional Organization in Biological Systems. Systems Research 6:219-226

Wills, P. R. 1993. Self-organisation of genetic coding. J. Theor. Biol. 162:267-287

Wills, P. R. 2001. Autocatalysis, Information and Coding. BioSystems 60:49-57

Wills, P. R. & Henderson, L. 2000. Self-organisation and information-carrying capacity of collectively autocatalytic sets of polymers: ligation systems. In Unifying Themes in Complex Systems: Proceedings of the First International Conference on Complex Systems, 613-623 (Perseus Books)

Once More Unto the Breach[1]:
Co-evolving a robot and its simulator

Josh C. Bongard and Hod Lipson
Sibley School of Mechanical and Aerospace Engineering
Cornell University, Ithaca, New York 14850
[JB382|HL274]@cornell.edu

Abstract

One of the major challenges facing evolutionary robotics is crossing the reality gap: How to transfer evolved controllers from simulated robots to real robots while maintaining the behavior observed in simulation. Most attempts to cross the reality gap have either applied massive amounts of noise to the simulation, or conducted most or all of the evolution onboard the physical robot, an approach that can be prohibitively costly or slow. In this paper we present a new co-evolutionary approach, which we call the *estimation-exploration algorithm*. The algorithm automatically adapts the robot simulator using behavior of the target robot, and adapts the behavior of the robot using the robot simulator. This approach has four benefits: the process of simulator and controller evolution is automatic; it requires a minimum of hardware trials on the target robot; it could be used in conjunction with other approaches to automated behavior transferal from simulation to reality; and the algorithm itself is generalizable to other problem domains. Using this approach we demonstrate a reduction of three orders of magnitude in the number of evaluations on a target robot (thousands compared to only five).

Introduction

An evolutionary robotics experiment requires an evolutionary algorithm to optimize some aspect of a simulated or physical robot in order to generate some desired behavior. Because it is difficult and very time-consuming to perform the thousands of fitness evaluations required by an evolutionary algorithm on a physical robot, most or all of evolutionary robotics experiments are performed in simulation.

Evolution in simulation raises a major challenge: The transferal of evolved controllers from simulated to physical robots, or 'crossing the reality gap' (Jakobi, 1997). There are several approaches to this challenge, including: adding noise to the simulated robot's sensors (Jakobi, 1997); adding generic safety margins to the simulated objects comprising the physical system (Funes and Pollack, 1999); evolving directly on the physical system ((Thompson, 1997), (Floreano and Mondada, 1998) and (Mahdavi and Bentley, 2003)); evolving first in simulation followed by further adaptation on the physical robot ((Pollack et al., 2000) , (Mahdavi

and Bentley, 2003)); or implementing some neural plasticity that allows the physical robot to adapt during its lifetime to novel environments ((Floreano and Urzelai, 2001), (DiPaolo, 2000), (Tokura et al., 2001)).

Another approach that can be used in lieu of, or in addition to the above-mentioned approaches is outlined in this paper. Our approach uses a co-evolutionary algorithm to automatically evolve a robot simulator so that the simulated robot acts more like the target robot. Instead of only evolving a controller in simulation and transferring it to the target robot, sensory data recorded by the target robot is returned to the algorithm and used to evolve the *morphological* aspects of the simulated robot. This approach has two benefits: it automates the process of adapting a simulation to a particular physical system; and it accomplishes this with a minimum of hardware trials.

Most evolutionary robotics experiments evolve controller parameters for a robot with a fixed controller topology and fixed morphology (examples include (Floreano and Mondada, 1998) and (Reil and Husbands, 2002)), whether the evolution is performed in simulation or the real world. Other approaches have widened evolution's control over the design process by subjugating the controller topology and/or the robot's morphology to modification as well (eg. (Sims, 1994), (Adamatzky et al., 2000), (Lipson and Pollack, 2000), (Bongard and Pfeifer, 2001) and (Hornby and Pollack, 2002)) with the aid of simulation.

The ability to evolve a simulated robot's morphology, in addition to its controller, has become much easier recently due to the advent and availability of physics-based simulators, which allow for faster than real-time evaluation of different physical systems. In the next section we present our algorithm, which relies on the physical simulator to update the morphology of the simulated robot to better approximate the morphology of a 'physical' robot. In the work presented here, the 'physical' robot is also simulated, but is different than the simulated robot in ways unknown to the algorithm. In future work we plan to replace the simulated target robot with a physical target robot. The section after that presents some results; the penultimate section provides some discussion regarding the generality and applicability of our approach for future physical evolutionary robotics ex-

---

[1] Shakespeare, W. *King Henry V*, Act III, Scene 1. **Breach:** [n] an opening (especially a gap in a dike or fortification). Also [n] a failure to perform some promised act or obligation.

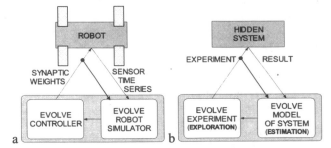

Figure 1: **Schematic view of the proposed algorithm. a:** The cyclical flow of the estimation-exploration algorithm, as applied to evolutionary robotics. The exploration phase evolves controllers in simulation and transfers them to a target robot; based on sensor time series and the evolved controller, the estimation phase improves the simulation. **b:** The general layout of the algorithm for inferring the structure of some hidden target system.

periments; and the final section offers some concluding remarks.

Methods

The algorithm presented in this paper takes as input a target robot, and an initial approximate simulator of the target robot. In other words, the initial simulated robot is morphologically different, in some unknown way, from the target robot. After the algorithm is completed, two structures are returned. The first is a robot simulator that better describes the target robot. The second is an evolved controller that should produce behavior in the target robot that is very similar to the behavior that was observed in the robot simulator, using the same controller. Figure 1 shows the algorithm flow schematically, as well as the flow of the algorithm for any partially hidden, nonlinear system.

The Algorithm. The estimation-exploration algorithm is comprised of two phases: the exploration phase, which in this case evolves neural network controllers; and the estimation phase, which in this case evolves improvements to the initial approximate robot simulator, given pairs of evolved controllers and resulting sensor data obtained from the target robot. The algorithm is cyclical. First, using the default simulator, the exploration phase evolves a controller that allows the robot to move as far forward in its environment as possible, during a fixed time period. Then, the best evolved controller is applied to the target robot, and the sensor values over time are recorded.

Using this evolved controller/sensor data pair (plus any previously evolved controller/sensor data pairs obtained during previous cycles), the estimation phase evolves simulator modifications that allow the simulated robot to better reproduce the observed sensor data, given the evolved controller. The best evolved simulator (that which best reproduces the i behaviors observed during the previous i passes through the algorithm) is then passed to the exploration phase, and the cycle continues for a number of times. A more detailed explanation of the algorithm is given after the robot simulator and the robot is itself are described.

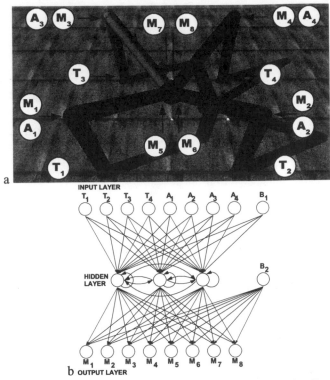

Figure 2: **a:** The morphology of the robot, including the distribution of its four touch sensors (T1-T4), four angle sensors (A1-A4), and eight motorized joints (M1-M8). **b:** The neural network controller of the robot, which connects the eight sensors to the eight motors via a single hidden layer, and an additional two bias neurons (B1-B2).

The Robot. In this work a quadrupedal robot was simulated and used to test the algorithm. The robot simulator is based on Open Dynamics Engine, an open-source 3D dynamics simulation package[1]. The simulated robot is composed of nine three-dimensional objects, connected with eight one-degree of freedom rotational joints. The joints are motorized, and can rotate through $[-\pi/4, \pi/4]$ radians away from their starting angles. The robot is shown in Figure 2a. The robot also has four binary touch sensors, and four angle sensors that return values in $[-1, 1]$ commensurate with the angle of the joint to which they are attached. In this work the target robot is not a physical robot out in the world, but a separate, simulated robot which is identical to the default simulated robot except for some unknown morphological differences. The task of the algorithm is to indirectly infer these differences, and modify the default simulator to reflect them accurately.

Both robots (simulated and target) are controlled by a partially recurrent neural network, which receives sensor data at the beginning of each time step of the simulation into its input layer, propagates those signals to a hidden layer containing three hidden neurons, and finally propagates the signals to an output layer. The neural network architecture is shown in Figure 2b. Two types of sensors are used: touch

---
[1]opende.sourceforge.net

sensors and angle sensors. The touch sensors are binary, and indicate whether the object containing them is in contact with the ground plane or not. The angle sensors return a value commensurate with the flex or extension of the joint to which they are attached. At each time step the values arriving at the eight motor neurons are treated as desired joint angles. The difference between the joint's actual angle and the desired angle output by the corresponding motor neuron is computed, and a torque commensurate with this difference is applied to the joint. All neuron values and the 68 synaptic weights are scaled to lie in the range $[-1.00, 1.00]$. A threshold activation function is applied to the hidden and motor neurons. Once the torques have been applied to the joints, the other external forces acting on the robot's body parts (gravity, friction, and inertia) are summed, and the positions, velocities and accelerations are updated at the next time step. Evaluation continues for 1000 time steps.

Algorithm Application. Our algorithm is quite generalizable: to date, we have applied to the algorithm to other system identification problems such as gene network inference (Bongard and Lipson, 2004b) and function recovery for damaged robots (Bongard and Lipson, 2004a), and other problem domain applications are planned. In order to apply the algorithm to a particular problem (which in this paper is the automated evolution of a robot simulator in order to cross the reality gap) four steps must be followed: **initialization**, setting up the **exploration phase**, setting up the **estimation phase**, and finally setting the **termination criteria**.

1) Initialization: The algorithm begins with the default robot simulator described above, and starts to evolve a controller in the exploration phase.

2) Exploration Phase: The exploration phase uses a given robot simulator to evolve a set of synaptic weights that, when applied to the neural network shown in Figure 2b, causes the robot to move forward as far as possible in a fixed time period. During the first pass through the exploration phase, the default simulator is used; in subsequent passes through this phase, the best evolved simulator from the estimation phase is used.

Genomes in the exploration phase are strings of 68 floating-point values in $[-1, 1]$, representing the 68 synaptic weights of the network. The fitness of a given genome is simply the z component (forward distance) of the robot's center of mass (in meters) during the first time step of the simulation subtracted from the z component of the center of mass at $t = 1000$.

Once all of the genomes in the population have been evaluated, they are sorted in order of decreasing fitness, and the 50 least fit genomes are deleted from the population. Fifty new genomes are selected to replace them from the remaining 50, using tournament selection, with a tournament size of 3. Each floating-point value of a copied genome has a 1% chance of undergoing a point mutation. Each value selected for mutation is either: nudged up or down by 0.0001 (50% probability); or replaced with a new random value in $[-1, 1]$. Of the 100 copied and mutated genomes, 24 pairs are randomly selected and undergo one-point crossover. The

Body Part or Sensor	Simulated Body Part	Target Body Part	Simulated Sensor	Target Sensor
1	5.0 (kg)	6.5 (kg)	0 (t)	15 (t)
2	1.0	2.4	0	7
3	1.0	2.0	0	19
4	1.0	3.0	0	4
5	1.0	2.7	0	10
6	1.0	1.9	0	5
7	1.0	2.8	0	18
8	1.0	2.7	0	16
9	1.0	2.4		

Table 1: The set of differences between the default simulated robot and the target robot. Note that the physical characteristics of the target robot (third and fifth columns) are hidden from the algorithm.

population is evolved for 30 generations.

When this phase terminates, the controller with the best fitness is downloaded to the target robot, and the resulting sensor values are recorded. Both the evolved controller and resulting sensor time series are passed into the estimation phase.

3) Estimation Phase: The genomes in the estimation phase encode modifications to be made to the default robot simulator such that the resulting simulated robot mimics as well as possible the observed behavior of the target robot.

The experimenter must choose sets of morphological characteristics that may differ between the simulated and target robot, or those characteristics which greatly affect behavior. For the work here, two such characteristics were chosen: mass distribution and sensor time lags. Each genome then in the estimation phase is a string of 17 values. The first nine values indicate mass changes, in kilograms, to be made to the default masses of the robot's nine body parts. The last eight values correspond to the robot's eight sensors, and indicate, in time steps, how long it takes the recording of a sensor signal to reach the corresponding input neuron. Each genome encodes 17 values in $[-1, 1]$. During parsing the first nine values are scaled to lie in $[0, 3]$, indicating a mass increase in 0 to 3kg for that body part (we assume we know that the target robot is heavier in some way from the default simulated robot). The last eight values are scaled to $[0, 20]$, to indicate that a sensor may have no time lag up to a 20 time step time lag. Table 1 indicates the morphological settings for the robot in the default simulator, and the hidden morphological settings for the target robot.

For these initial experiments, we assume that the target robot only differs from the default simulation in those characteristics that we have chosen to place under evolutionary control. From similar experiments (Bongard and Lipson, 2004a) we have found that in some cases even an approximate simulator that does not refine all the differing physical characteristics between the simulated and target robot allows for adequate transfer of behavior.

The quality of a candidate simulator is given by the ability of the simulated robot to mimic the observed behavior of the

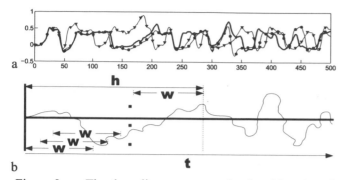

a

b

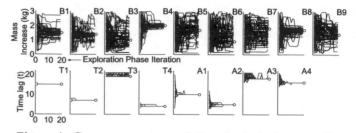

Figure 4: **Convergence toward the physical characteristics of the target robot.** Each pass through the estimation phase produces a set of mass changes for each of the nine body parts of the robot (top row) and a set of time lags for each of the eight sensors (bottom row). Each trajectory represents the best guess output by each pass through the estimation phase of a single run. A total of 50 runs were executed. The open circles indicate the actual differences between the target robot and the starting default simulated robot (for example the first body part of the target robot is 1.5kg heavier than the corresponding body part of the default simulated robot).

Figure 3: **a**: The three lines represent the time histories of one of the simulated robot's angle sensors: each time history corresponds to three different simulated robots using the same controller. Only the first 500 time steps of the 1000 time step evaluation period are shown. The thick line indicates the angle sensor's values when one of the robot's motors is weakened by 10%; the line with circle markers when the same motor is weakened by 20%; and the line with triangle markers when one of the touch sensors is weakened by 50%. **b**: Outline of the rolling mean fitness metric.

target robot. More specifically, given i previously evolved controllers tested on the target robot, the simulated robot should mimic as closely as possible the i sets of sensor time series produced by the target robot. However, quantitatively comparing sensor data from two highly coupled, highly non-linear machines like the robot used here is very difficult: slight differences between the two machines rapidly leads to uncorrelated signals. To address this, we have formulated a comparison metric called the *rolling mean metric*. This metric has been shown to achieve a high value for morphologically dissimilar robots, a lower value for similar robots, and zero for identical robots (Bongard and Lipson, 2004a).

More specifically, the fitness of a genome g_p in the estimation phase is given by:

$$f(g_p) = 1 - \frac{\sum_{i=1}^{c}\sum_{j=1}^{n}\sum_{k=w/2}^{h-w/2} d(i,j,k,p)}{cn(h-w)}, (1)$$

$$d(i,j,k,p) = \frac{\sum_{t=k-w/2}^{t=k+w/2}|s_{\text{act}}^{(i,j)}(t) - s_{\text{obs}}^{(i,j,p)}(t)|}{w}. (2)$$

In this formulation i is the number of evolved controllers tested on the target robot so far; j is the number of sensors contained in the robot (here $j = 8$); h is some header length, which indicates how much of initial sensor time series data to use (here $h = 20$); and w indicates the width of a time window within this time header (here $w = 5$). $d(i,j,k,p)$ indicates the differences between the sensor values of the target robot and the simulated robot when it is morphologically modified by the values encoded in genome g_p. $d(i,j,k,p)$ computes a piecewise comparison between short time periods of sensor activation, namely time periods described by $[t-w/2, t+w/2]$. $s_{\text{act}}^{(i,j)}(t)$ indicates the activation of the jth sensor at time t obtained from the target robot when it is using controller i, and $s_{\text{obj}}^{(i,j,p)}(t)$ indicates the activation of

the jth sensor at time t from the simulated robot using controller i and modified by genome g_p. We refer to this metric as the rolling mean metric because it compares sets of average sensor activations over a short initial time period. The rolling mean helps smooth the sawtooth activation patterns produced by the touch sensors, easing comparison.

Figure 3 depicts this metric graphically. Figure 3a shows that three robots with slightly different morphologies, using the same controller, produce sensor time series that rapidly diverge. However the sensor data for the two similar robots (indicated by the thick line and the line with circle markers) are similar for a longer initial time period (divergence around $t = 250$), compared to the divergence of the different robot from the two similar robots (around $t = 50$). The rolling mean metric then gives small values for morphologically similar robots, and higher values for morphologically different robots.

The genetic algorithm operating in the estimation phase is the same as that operating in the exploration phase, except for a few differences. This phase also evolves a population of 100 genomes, but the genomes are strings of 17 values (the simulator modifications) compared to strings of 68 values as in the exploration phase (the synaptic weights). Second, during subsequent passes through this phase, the initial population of random genomes is seeded with a copy of the genome with the best fitness (equation 1) obtained during the previous pass. Third, genome evaluation is different: the default simulation is modified according to the genome; the simulated robot is evaluated using the i previously evolved controllers; i sets of sensor time series are thus produced; and the genome's rolling mean is calculated. Selection, mutation and crossover are the same as in the exploration phase. The population of the estimation phase is also evolved for 30 generations during each pass, and outputs the best evolved simulator to the exploration phase, which begins again with

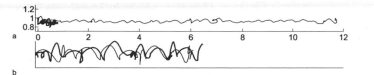

Figure 5: **Behavior recovery after controller transferal.** After the first pass through the exploration phase, the best evolved controller was used by the default simulated robot. The trajectory of its center of mass is given by the thin line in **a**. The same controller was then supplied to the target robot, and the resulting trajectory of its motion is given by the thick line in **a**. The movement of the simulated robot in the updated simulator after the 20th pass through the exploration phase (using the new best evolved controller) is given by the thin line in **b**. The motion of the target robot using the same controller is given by the thick line in **b**. The horizontal axis indicates forward distance, and the vertical axis indicates height (both are in meters).

a random population.

4) Termination Criteria: The algorithm iterates through the cycle shown in Figure 1a 20 times, starting at the exploration phase with the default simulator. This produces 20 evolved controllers (the fittest genome from each pass through the exploration phase), and 20 simulator modifications (the fittest genome from each pass through the estimation phase). The choice of 20 is arbitrary in this case; generally the algorithm continues until satisfactory behavior is attained on the target system, or until no simulator adaptations can explain the observed behavior. This latter failure can occur if either the evolutionary search cannot find a sufficiently accurate simulator, or if no combination of modifications can accurately describe the target system, implying that there is no point continuing in simulation.

Results

Fifty independent runs of the algorithm were conducted against the target robot outlined in Table 1. Figure 4 shows the 50 series of 20 best simulator modifications output after each pass through the estimation phase.

One of the runs was selected at random, and the gait of the simulated robot was compared against the gait of the target robot, when both used the same evolved controller. Figure 5a indicates the change in behaviors when the first evolved controller was transferred, and Figure 5b shows the behavior change when the 20th evolved controller was transferred, during the last iteration through the algorithm's cycle. Finally, for each pass through the exploration phase, the distance traveled by the simulated robot was averaged over all the 50 runs. Similarly, the distance traveled by the target robot using the same controller was averaged over the 50 runs. The results are shown in Figure 6.

Discussion

Figure 4 makes clear that for all 50 runs, the algorithm was better able to infer the time lags of the eight sensors than the

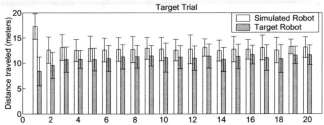

Figure 6: **Average transferal success after each target trial.** The light gray bars indicate the average distance traveled by the simulated robot using the best evolved controller output by that pass through the exploration phase, over all 50 runs. The dark gray bars indicate the average distance traveled by the target robot using the same controller, during each target trial. Error bars indicate two units of standard deviation.

mass increases of the nine body parts. This is not surprising in that the sensors themselves provide feedback about the robot. In other words, the algorithm automatically, and after only a few target trials, deduces the correct time lags of the target robot's sensors, but is less successful at indirectly inferring the masses of the body parts using the sensor data.

Convergence toward the correct mass distribution, especially for body parts 1, 2, 4 and 8 can be observed. This is most likely due to the relatively short time lags of the sensors near to those body parts, providing more information to the rolling mean metric (equation 1).

Even with an approximate description of the robot's mass distribution, the simulator is improved enough to allow smooth transfer of controllers from simulation to the target robot. Using the default, approximate simulation, there is a complete failure of transferal, as indicated by Figure 5a: the target robot simply moves randomly, and achieves no appreciable forward locomotion.

After 20 iterations through the algorithm, an improved simulator is available to the exploration phase, which evolves a controller that allows the simulated robot to move forward, although not as far as the original simulated robot (indicated by the shorter trajectory in Figure 5b compared to Figure 5a). Also, the new gait causes the robot to hop (indicated by the large vertical curves of the robot's center of mass in 5b) instead of walk (indicated by the steady trajectory of Figure 5a). In contrast to the first pass, the target robot exhibits very similar behavior to the simulated robot when it uses the same controller: both travel a similar distance (about 6.5m), and both move in the same way (both exhibit a hopping gait that produces trajectories with similar frequencies and amplitudes).

Finally, Figure 6 shows that this improvement in behavior transferal success is a general phenomenon. On average, over the 50 independent runs, there is a drop by 50% in the distance traveled by the target robot, compared to the default simulated robot. After about five iterations through the algorithm's cycle there is only a statistically insignificant decrease in distance traveled between the two robots. Although not shown in Figure 6, this similar distance is matched in all of the cases viewed by a qualitative similarity in gait patterns, as shown for a single run in Figure 5b.

Conclusions

We have described a new algorithm—the estimation-exploration algorithm—for automatically improving a robot simulator using an evolutionary algorithm. The improvement is shown to be sufficient to allow for successful transferal of an evolved controller from simulation to a target robot, which exhibits some unknown differences compared to the default initial simulated robot.

This algorithm has several benefits: it is automatic; it achieves successful transferal after only a minimum of trials on the target robot; the sensors onboard the robot are used to generate behavior and provide indirect evidence about the morphological characteristics of the robot; and the algorithm is generalizable, as it has already been applied in other problem domains such as systems biology (Bongard and Lipson, 2004b) and robot damage diagnosis and recovery (Bongard and Lipson, 2004a).

Future experiments are planned in which the target robot, which is currently also simulated, will be replaced by an actual physical robot. Second, initial results (Bongard and Lipson, 2004a) have indicated that based on sensor feedback the virtual environment of the simulated robot can be automatically updated to reflect the target robot's physical environment. An interesting next step would be to equip the robot with rudimentary visual sensors, such as sonar, that could be used to automatically construct virtual analogues of obstacles or other agents in the robot's environment. We are currently formalizing methods for applying the algorithm to an arbitrary hidden nonlinear system. Source code, executables and documentation for verifying the experiments and extending the results reported here are available at http://www.people.cornell.edu/pages/jb382/.

Acknowledgements

This work was supported by the U.S. Department of Energy, grant DE-FG02-01ER45902. This research was conducted using cluster computing resources of the Cornell Theory Center.

References

Adamatzky, A., Komosinski, M., and Ulatowski, S. (2000). Software review: Framsticks. *Kybernetes: The International Journal of Systems & Cybernetics*, 29:1344–1351.

Bongard, J. and Pfeifer, R. (2001). Repeated structure and dissociation of genotypic and phenotypic complexity in Artificial Ontogeny. *Proceedings of The Genetic and Evolutionary Computation Conference (GECCO 2001)*, pages 829–836.

Bongard, J. C. and Lipson, H. (2004a). Automated robot function recovery after unanticipated failure or environmental change using a minimum of hardware trials. In *Proceedings of The 2004 NASA/DoD Conference on Evolvable Hardware*.

Bongard, J. C. and Lipson, H. (2004b). Automating genetic network inference with minimal physical experimentation using coevolution. In *Proceedings of The 2004 Genetic and Evolutionary Computation Conference*.

DiPaolo, E. A. (2000). Homeostatic adaptation to inversion of the visual field and other sensorimotor disruptions. In Meyer, J. A., Berthoz, A., Floreano, D., Roitblat, H. L., and Wilson, S. W., editors, *From Animals to Animats 6*, pages 440–449. MIT Press.

Floreano, D. and Mondada, F. (1998). Evolutionary neuro-controllers for autonomous mobile robots. *Neural Networks*, 11:1461–1478.

Floreano, D. and Urzelai, J. (2001). Neural morphogenesis, synaptic plasticity, and evolution. *Theory in Bioscience*, 120:225–240.

Funes, P. and Pollack, J. (1999). Computer evolution of buildable objects. In Bentley, P., editor, *Evolutionary Design by Computer*, pages 387–403. Morgan Kauffman, San Francisco.

Hornby, G. S. and Pollack, J. B. (2002). Creating high-level components with a generative representation for body-brain evolution. *Artificial Life*, 8(3):223–246.

Jakobi, N. (1997). Evolutionary robotics and the radical envelope of noise hypothesis. *Adaptive Behavior*, 6(1):131–174.

Lipson, H. and Pollack, J. B. (2000). Automatic design and manufacture of artificial lifeforms. *Nature*, 406:974–978.

Mahdavi, S. H. and Bentley, P. J. (2003). An evolutionary approach to damage recovery of robot motion with muscles. In *Seventh European Conference on Artificial Life (ECAL03)*, pages 248–255. Springer.

Pollack, J. B., Lipson, H., Ficici, S., Funes, P., Hornby, G., and Watson, R. (2000). Evolutionary techniques in physical robotics. In Miller, J., editor, *Evolvable Systems: from biology to hardware*, pages 175–186. Springer-Verlag.

Reil, T. and Husbands, P. (2002). Evolution of central pattern generators for bipedal walking in a real-time physics environment. *IEEE Transactions on Evolutionary Computation*, 6(2):159–168.

Sims, K. (1994). Evolving 3D morphology and behaviour by competition. *Artificial Life IV*, pages 28–39.

Thompson, A. (1997). Artificial evolution in the physical world. In Gomi, T., editor, *Evolutionary Robotics: From intelligent robots to artificial life (ER'97)*, pages 101–125. AAI Books.

Tokura, S., Ishiguro, A., Kawai, H., and Eggenberger, P. (2001). The effect of neuromodulations on the adaptability of evolved neurocontrollers. In Kelemen, J. and Sosik, P., editors, *Sixth European Conference on Artificial Life*, pages 292–295.

Closing the loop: Evolving a model-free visually guided robot arm

Thomas Buehrmann[1], Ezequiel Di Paolo[1]
[1]Centre for Computational Neuroscience and Robotics
Department of Informatics, University of Sussex
{t.buehrmann, ezequiel}@sussex.ac.uk

Abstract

Dynamic neuro-controllers are incrementally evolved for reaching and tracking movements by a physically simulated robot arm. An active vision system capable of controlling gaze direction and focus replaces the need for internal models of the robot. It is shown that closing the feedback loop allows for robot control being robust to changes in environment, sensors and robot-morphology.

Introduction

The task of controlling a robotic arm can be defined as the movement of the end-effector to visually identified positions or along particular trajectories in workspace. For visually guided movements, computational approaches traditionally consist of the following stages: i) a visual observation of the target is translated to desired position and orientation of the robot's end-effector by using visual pre-processing, feature extraction, inverse perspective projections or other computer vision based methods; ii) inverse kinematics is used to translate the desired end-position into a set of desired joint angles; iii) trajectory planning: a path, i.e. a series of intermediate points is calculated in joint space along which the robot moves from its current state to the desired state; iv) actuator commands (e.g. torques) are calculated which when applied to the arm move the end-effector along the desired intermediate points (dynamics problem). All of these transformations depend on using a correct model of the robot. For accurate control, these models must be constructed and calibrated so that they correspond with the parameters of the real robot. This introduces several problems. First, supervised learning schemes (Massone, 1998; Jordan and Wolpert, 1999) are difficult to apply if obstacle avoidance or other non-trivial behaviors are to be taken into account. Representative input-output samples describing the correct behavior are hard to produce in this case. Secondly, these controllers rely on static configurations of their environment as well as their sensorimotor interfaces. Because they use models calibrated to a particular environment and robot morphology, they can't adapt to changing environments or bodily reconfigurations. This rigidity leads to high

costs of maintenance and calibration. Lack of flexibility can be solved either by re-calibrating the internal model during the robot's lifetime, or alternatively by foregoing the use of a model at all. Instead, the robot-environment feedback loop can be closed and a controller constructed which translates directly from sensory input to joint dynamics (e.g. (van der Smagt, 1995)).

This report deals with the second alternative. It shows how a minimalistic approach inspired by evolutionary robotics and the dynamical systems perspective allows for flexible and robust robot control without using prior knowledge about the robot and its environment in form of internal models. In summary, neural controllers are evolved to use low-level active vision for the guidance of a physically simulated robot arm. By using a genetic algorithm, the robot controller is evolved to directly translate from the visual domain to joint dynamics in order to reach for and track objects in its environment. Since the controller does not use any models of the robot or its sensors, there is no need for calibration. Instead of computationally expensive processes involving filtering, edge finding, feature extraction, flow field analyses or transformation of data into world-based frames-of-references, this system relies on active and egocentric low-level vision only.

Methods

Robot platform

The robot arm in this project is an articulated model that consists of three segments linked by 1-dimensional rotational joints which are controlled by angular motors. The motors apply torque to a joint in order to pivot it at a desired speed. In the experiments described, torque limits are set and neural networks control the arm by specifying the desired velocities for each joint. Also, the range of motion of two connected bodies is limited by setting stops on the joint (maximum and minimum angles). Calculation of the physics (including gravity and inertia) is provided by ODE (http://opende.sourceforge.net). Figure 1 shows the configuration of the robot arm. To avoid both complex image processing (as e.g. in triangulation from stereoscopic

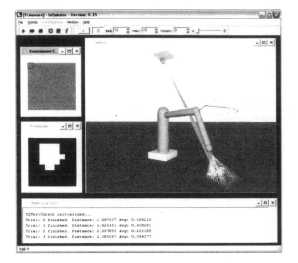

Figure 1: Robot arm, laser range sensors, and camera images.

images) as well as incorporation of prior knowledge (as in heuristics on the relation between image- and actual size of objects) a rather atypical sensor system was used to provide 3-dimensional visual information: a minimal camera system consisting of a two-dimensional array of "laser range sensors" (figure 1). Each individual sensor is a light-ray whose output is proportional to the distance of its collision with an object in the world. A number of such rays is arranged in a rectangular grid. The rays all originate at the same focal point in space and the angle between them determines the camera's field-of-view. By changing this angle, a camera effectively has an adjustable focus and thus control over image resolution[1]. In fact, two such cameras were used in this project. The first one is a world-based camera mounted above the robot arm. Its position is fixed, but it is able to move in two degrees-of-freedom: pan around the vertical axis and tilt around its local x-axis. The second camera is mounted to the end-effector (eye-in-hand, or egocentric setup). The advantage of this camera is that the information it provides is relative to the end-effector and can be used in the final approach to target objects even when the arm occludes the first camera.

In addition to visual input, controllers also have access to proprioceptive information. Angles of the three robot joints as well as the orientation of the camera can be used by the modular neurocontrollers.

Neuro-Controller

Continuous-time recurrent neural networks are used for the neurocontrollers. The state of each node is described by

$$\tau_i \dot{y}_i = -y_i + \sum_{\forall j} w_{ji} \phi_j(y_j + \vartheta_j) + g\mathcal{I}_i(t) \qquad (1)$$

---

[1]A similar active vision system was evolved in (Kato and Floreano, 2001) to perform shape discrimination

where y_i is the cell potential of that neuron, τ_i its time constant, w_{ji} the weights of incoming synapses, ϕ the sigmoidal function $\phi(x) = 1/(1 + e^{-x})$ calculating the firing rate, ϑ the threshold of the neuron and $g\mathcal{I}$ gain-scaled input respectively. The parameters for each neuron are obtained from appropriate scaling of elements in the genotype (distributed over the range [0,1]). Weights and biases were scaled to the interval $[-4, 4]$, time constants to $[0.1, 10]$ and input gains to $[0, 10]$. The Euler method with a time step of 0.2 was used for integrating the differential equations.

Genetic Algorithm

A genetic algorithm was used to evolve fixed network architectures. A linear ranking selection scheme and stochastic universal sampling were used for reproduction. Also, elitism was applied by always keeping the best individual of each generation. Recombination is realized through an ordinary two-point crossover operator. The particular form of mutation used is a variation of the *creep* operator determined by two parameters: one specifies the maximal amount of mutation for all components while the other one determines the probability of mutation for individual components. Mutated values are clipped to the interval $[0, 1]$.

An incremental approach to evolution was used in three different ways. First, the desired behavior was decomposed into several independent behavioral competencies. The overall system was then partitioned into sub-modules which are individually evolved to produce one of the more basic behaviors. Secondly, some of the modules were evolved to produce solutions to a series of increasingly complex evaluation tasks. This was done to avoid local minima of unsatisfying solutions when the initial search space was too big. Finally, evolutionary parameters (such as mutation probability) sometimes were interactively decreased throughout evolution to allow for the population to converge on and optimize the best solution it had found so far.

Fitness Evaluation

The performance measure to be maximized by the controllers consists of mainly two terms. First, those controllers receive higher fitness that reduce the Euclidean distance between the robot end-effector and the target from the beginning of a trial to its end. From the distance at time t (d_t) and the distance at the start of the trial (d_0) this fitness value is given by $f_d = \frac{1}{T} \sum_{t=0}^{T}(1 - \frac{d_t}{d_0})$ where T is the total number of time steps per trial (in the case of the camera trying to center the target, angular distance between direction of view and direction of target was the measure used). The second fitness term tries to minimize movement at the end of the trial so that the robot arm finally comes to a stop: $f_v = 1 - \frac{1}{\dot{\theta}_{max}} \sum_{i=1}^{3}(\dot{\theta}_i)$, where $\dot{\theta}_i$ are the absolute velocities of all joints and $\dot{\theta}_{max}$ the maximum velocity. Although breaks will be used in some experiments at the end of movement, smooth deceleration towards targets is more desirable. Usually, this term is multiplied with the distance measure at the end of a trial, such that high fitness values can be achieved only if the robot arm simultaneously gets close to the target and has minimum velocity.

Each individual in the population is tested in several trials of fixed length. At the beginning of each trial the robot arm and the external camera are initialized to their resting positions, while the target is placed randomly in a cubic volume within the arm's workspace. The overall fitness of a controller is calculated after its last trial by averaging its individual trial fitnesses.

Experiments

Preliminary experiments without the visual system had shown that a single feedforward CTRNN can be evolved to move the robot arm in ways appropriate for reaching and tracking (using Cartesian coordinates of end-effector and target as well as joint angles as inputs). When cameras were included however, the system was broken down into four neuro-controllers which were evolved successively. The first one enables the external camera to find and centre on objects within its visual field. The second module controls the horizontal orientation of the robot arm and makes it align with the camera's direction of view. The third controller is responsible for finding and closing in on the target in the vertical plane, while keeping the orientation of the end-effector such that the target could actually be grasped. An optional fourth controller is connected to the eye-in-hand camera, and allows the arm to track an object even when it is occluded by the arm on the image of the external camera.

External Camera

The external camera has a global view on the scene, and if it's able to centre on the target, its angular position and the distance information of its sensors can be used to inform the arm about the location of the target. The neural network controlling this camera (figure 2) was designed taking into account the symmetries (vertical and horizontal) of the task as well as the sensor arrangement. Through this 'quadlaterally symmetric' architecture, incorporating knowledge about the task (not the robot or its environment), the control problem is simplified and evolutionary search made easier. Parameters of the 104 neurons in the controller were encoded by 42 real numbers (in addition to symmetrical connections, biases and time constants were shared extensively). The initial angle of the camera's focus and the random positions of the target were initialized such that the target was always located within the camera's field-of-view. However, since the space between sensory rays increases with distance from the focal point, objects sometimes lay in between the rays, thus not producing any input. An individual's fitness was equal to the averaged fraction of

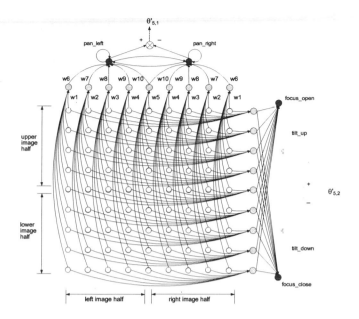

Figure 2: The network has one input neuron per distance sensor of the camera. All input neurons in a given row or column project to a hidden neuron using the same weight. Connections from the left half of the image are mirrored by connections from the right. Also, rows and columns share the same weights. The hidden neurons symmetrically project to three pairs of output neurons which control the camera's pan and tilt angle as well as its focus. To allow for richer dynamics the pairs of output neurons are interconnected fully recurrent.

angular distance covered from the beginning to the end of a trial (f_d). After only a few generations valid solutions were found for reaching for static targets. Near optimal fitness was achieved by generation 100. The best individuals from the last 100 generations on average received 98.2% of the maximum fitness, the population average 89.1%. For static targets without interference of the arm, the camera within a few steps reduces the angular distance to less than 0.1°. The focus is used by the controller to adjust the camera's resolution to the size of the target. While the camera has a tendency to close the focus, the outer sensory rays are used to interrupt this behavior. Consequently, the rays optimally cover the surface of the target object. This behavior is independent of the size or the shape of the target. The tendency to reduce the angle between individual rays has another advantage. If a target initially is positioned between some of the rays, the focusing behavior makes it likely that one of the rays will eventually intersect the object. The evolved focus mechanism turned out to be essential to achieve a high precision centering response, as well as a robust way of avoiding distraction through the robot.

Controlling Arm Orientation

Orienting the robot arm is easy if the active external camera is centred on the target. In this case, all the neurocontroller has to do is to use the angles of the robot and the camera to reduce the distance between both. The basic network evolved for this task is shown in figure 3. A power-switch

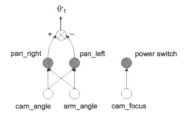

Figure 3: Module controlling rotation around vertical axis.

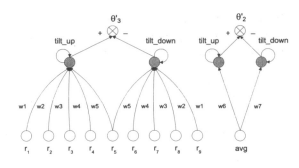

Figure 4: Neuronal modules controlling second and third joint.

is used to allow movement (horizontal rotation) of the robot arm only if the output of the corresponding neuron exceeds a threshold of 0.5. The controller can thus decide when to start and stop moving depending on its input (e.g. the camera's focus). With this information available, controllers can evolve which wait until the camera has found the target before making the arm interfere with its sensors. Given a population size of 60, a valid solution was existent in the first randomly initialized population. Using the distance between arm and target as well as the arm's velocity at the end of a trial for the fitness function ($F = f_{d_a} * f_v$), the averaged performance of the last 10 generations was 99.94% of the maximum for the best individual and 96.93% for the population average. In 100 trials, the best controller of the last generation on average reduces the angular error to $0.56°$ (with a standard deviation of $0.14°$). The velocity of the arm at the end of a trial is negligible. It can also be observed, that the arm only starts to move after the camera has focused on the target.

Reaching
In the next step, the eye-in-hand camera was included for approaching the target in the vertical plane. The task to be solved consists of two parts. First, because in the initial state targets will most likely not be located within the field-of-view of the eye-in-hand camera, the arm has to be moved to a position from which the object can be perceived in the first place. Such a behavior necessitates spontaneous internal dynamics of the controller, since no sensory information will be present at this stage (possible in CTRNNs through (self-)recurrent connections and non-zero biases). Once the object intersects the internal camera's sensors, the arm can be guided towards the target position. After experimenting with different neural architectures, it became apparent that the two joints can be assigned different roles. While the third joint (controlling the limb to which the camera is attached) mainly has to centre the target on the camera's field-of-view, the second joint can and must be used to approach the target. This task-decomposition inspired the neural architecture shown in figure 4. The

idea is analogous to the controller for the external camera. Instead of having an array of input neurons however, a hidden neuron directly receives the average of all sensory activations from one row of the camera array. Also, because of the task decomposition employed, all information needed for controlling joint j_2 is the relative distance of the target. Directional information is only needed for centering the target(j_3). Hence, the module controlling θ_2 only receives the overall average of sensor activation.

The controller was evolved using a fitness function consisting of several different terms. First, throughout the trial an individual was rewarded for maximizing the average sensor activation so as to favour individuals which came as close to the target as possible. Secondly, controllers were rewarded for reducing the angular distance to the target. This term enforced the arm to approach the target full frontal rather than from an angle which would not allow to actually grasp the object. This was necessary because there were no additional degrees of freedom in the wrist which could be used to independently orient the hand relative to the target. Third, controllers were punished whenever their joint angles reached their limits. This way, individuals were filtered out which took over the population by producing stereotypic movements. Finally, at the end of a trial the product of the terms for reducing angular distance and minimizing velocity was added to the fitness. Individuals which did no move at all or got stuck on the robot base or the camera fixture were assigned a fitness of zero.

A first evolutionary step produced controllers which made the arm centre the target on the image of the camera but failed in actually approaching it. Also, it produced small oscillations around the direction of the target. To make the arm approach the target, in a second incremental step the best solution so far was further evolved using a fitness function having larger weights on the absolute distance covered and the final velocity. Additionally, another power-switch was included which depending on the averaged activation level of the camera could stop the second and third limb from moving.

From the trajectory in figure 5 and figure 6 it can be seen that the distance is now minimized in each of the three

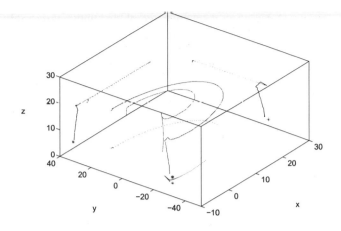

Figure 5: Positioning task: trajectories of end-effector and elbow joint and projection on three planes. Dots mark the position of the target. End-effector and target are initially positioned on opposite sides of the robot base along the y-Axis.

dimensions. In 100 trials, the average Euclidean distance at the end of a trial was 0.24 (which is roughly 1% of the length of the arm). The behavioral strategy is as follows: after the target is found ($t \approx 90$), the arm rotates and lifts its second limb while slowly lowering the third limb. As soon as the target object enters the eye-in-hand camera however ($t \approx 220$), the second limb is now lowered in order to approach the target, while the third limb centers the object (oscillating movements). The change in camera activation during approach leads to deceleration of the robot arm towards the target.

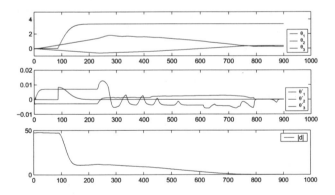

Figure 6: Joint angles(top), joint velocities(middle) and distance from the target for the same positioning task.

At time step 800 the controller uses the power-switch to stop the arm from moving.

Tracking
Since the arm is almost perfectly aligned with the external camera and thus occludes the target object whenever it is close to it, object tracking can not be guided by the external camera. Instead, the pan angle has to be determined by

the eye-in-hand camera. Consequently, to evolve tracking behavior, another controller was introduced. Its connectivity is similar to the controller depicted in figure 4 on the left. It gets as inputs the averaged activations from individual columns in the sensory array. This controller however has feedforward connections only, and as a consequence of its symmetry does not produce any output if no input is available. This property in turn allows for a simple (hand-coded) override mechanism coordinating the two modules which influence the pan angle (j_1). As long as there is no output from the new controller (in absence of the target), the corresponding angle is controlled by the external camera. As soon as the eye-in-hand camera has found the target however, the resulting input drives the new controller and its output is used instead. The final evolved behavior of the robot arm trying to track moving objects is shown in figure 7. Clearly, the arm is now able to follow moving targets in all three dimensions.

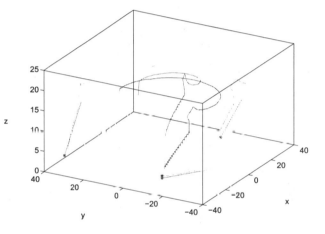

Figure 7: Trajectories and projections for a tracking task.

Robustness
The experiments described so far showed that modular neural networks can be evolved to implement robot controllers which are able to position the end-effector appropriately for grasping visually identified objects. The main advantage of such a controller is that it is not based on calibrated models of the robot, the sensors or the environment. Hence, it should be robust to all kinds of alterations in these factors without the need to re-evolve. In order to check for this property, the final controller was tested on a series of different conditions. Table 1 summarizes the average results from 13 tests of 100 trials each. In the first three tests the diameter of the spherical target object was varied. Obviously, the smaller the target, the higher the precision. Two reasons can be given for this result. First, a smaller target leads to a smaller angle of the external camera's focus, and hence to a better resolution and precision in its centering response. Secondly, if all sensors are equally activated (very close to

Condition	μ_{end}	σ_{end}	μ_{avg}	σ_{avg}
$\varnothing_t = 2.0$	0.35	0.69	13.76	2.62
$\varnothing_t = 1.5$	0.24	0.3	11.26	1.73
$\varnothing_t = 1.0$	0.22	0.16	11.50	2.34
no switch	0.43	1.12	11.81	2.48
$1/2 * \dot{\theta}_{max}$	0.86	1.61	20.12	3.02
$1/2 * \dot{\theta}_{max}$, g/9.0	0.2	0.15	22.52	4.34
$2 * \dot{\theta}_{max}$	6.17	5.51	12.32	5.51
IO-input	0.3	0.59	11.13	1.75
$|l_3| + 3.0$	0.27	0.24	10.9	1.19
$|l_3| + 6.0$	0.47	0.32	10.75	1.11
$|l_2| + 10.0$	0.44	0.29	18.05	1.79
$|l_2|, |l_3| + 10.0$	0.57	0.40	16.66	0.88
$|l_2| + 10.0, |l_3| + 5.0$	0.50	0.36	16.96	1.36
g/9.0	0.31	0.27	12.8	3.12

Table 1: Performance measured over 100 trials. μ_{end} and σ_{end} denote the mean and standard deviation of the distance measured at the end of the trials. μ_{avg} and σ_{avg} are the corresponding values averaged over all time steps per trial.

the target), the arm exhibits no centering response anymore and is being lowered until some rays do not intersect the target anymore. Now, the bigger the object, the bigger the range of movement the arm can produce until the centering response is elicited again. Consequently, there is a bigger variance in positioning and hence a lower precision (if the target gets too small however, the external camera will have problems finding it). In another test, the functionality of the power-switch was deactivated. The result is an increase in the variance of the final position, because the arm starts producing small oscillations around the target again. Next, the output-gains determining the maximal velocity of the joint motors ($\dot{\theta}_{max}$) were changed[2]. A robot arm producing only half of the velocity usually produced during evolution, performs somewhat worse than under normal circumstances. The reason is identified by looking at the next test. Having the same gain but less gravity ($g/9.0$), restores performance to the expected level. Hence, it is likely that having a decreased maximum velocity, and thus less force because ODE applies an amount of force that is needed to achieve the desired velocity, it is harder for the arm to compensate for gravitational force. Doubling the velocity gain, in contrast, leads to movements so erratic that the arm can not reliably center on the target anymore and thus often misses the target completely. To test for independence from details of the sensors, in one test the inputs to the module controlling the centering response were changed to binary mode rather than continuous values encoding the distance of intersection (IO-input). Thus a sensor's response is 1 if intersection occurs and 0 otherwise. As can be seen from the table, there is no considerable decrease in performance. This setup could be useful when implementing a real robot

___
[2]The length of the trial was also changed to allow the robot to move the same distance.

system. A sampling mechanism combined with a threshold function applied to a traditional camera image could provide the same kind of visual feedback as used in the simulation. Another set of tests varied the length of the limbs ($|l_i|$). Since the usual lengths were 16 and 9 units for the second and third limb respectively, adding 10 units to each limb means a lengthening by 110% for limb 3 and 62.5% for limb 2. However, even in the most extreme cases the performance decreases only gradually. The accuracy would still be good enough to actually grasp the object. Finally, reducing gravity to a ninth of the usual value does not affect performance either.

Conclusions

Robotic sensorimotor-coordination was re-formulated as the problem of designing embodied, situated and adaptive controllers which are dynamically coupled to an ever changing environment. This was seen in contrast to classical approaches in which a series of internal representations is generally constructed. It was shown that simple modular neural networks can be evolved as model-free controllers for visually-guided robot motion. Spatial coordination (alignment of gaze direction and arm orientation) and temporal coordination (delaying movements until target is identified) were achieved by coupling individual modules through proprioceptive feedback. In this approach, neither were explicit coordinate-transformations necessary, nor the learning of robot models or sensor calibration. The resulting system is able to reliably position and track objects in its environment and is robust to changes in sensory-, environmental- and morphological parameters. Its behavior is general enough to allow for the desired outcome even without the need for adaptation. For validation of these results, future experiments will aim at evolving a similar control scheme on a real robot in order to compare it with the simulation. Also, the particular form of end-effector trajectories can be improved by adding additional costs to the fitness function which are analogous to well-known trajectory optimization principles like minimum variance, minimum torque-change or minimum jerk (Jordan and Wolpert, 1999).

References

Jordan, M. and Wolpert, D. (1999). Computational motor control. In Gazzaniga, editor, *The Cognitive Neurosciences*. Cambridge, MA: MIT Press.

Kato, T. and Floreano, D. (2001). An evolutionary active-vision system. In *Proceedings of the Congress on Evolutionary Computation*. Piscataway, IEEE-Press.

Massone, L. (1998). Sensorimotor learning. In Arbib, M., editor, *Handbook of Brain Theory and Neural Networks*. MIT press.

van der Smagt, P. (1995). Visual robot arm guidance using neural networks. Ph.D. thesis, University of Amsterdam.

Performance Evaluation of Neural Architectures for Sequential Tasks

Genci Capi[1], and Kenji Doya[2]

[1] Faculty of Information Engineering
Fukuoka Institute of Technology
3-30-1 Wajiro-Higashi, Higashi-ku, Fukuoka, 811-0295, Japan
[2] CREST, JST, ATR, Computational Neuroscience Laboratories
"Keihanna Science City", Kyoto, 619-0288, Japan
capi@fit.ac.jp

Abstract

This paper considers a sequential task where the agent has to alternatively visit two rewarding sites to obtain food and water after first visiting the nest. To achieve a better fitness, the agent must have a working memory to reach the target position, must ignore irrelevant sensory inputs, and at a higher level, it has to deal with the non-Markovian order of sequential task in which the preceding state alone does not determine the next action. We compare the performance of neural control architectures in different environment settings, and analyze the neural mechanisms and environment features exploited by the agents to achieve their goal. Simulation and experimental results using the Cyber Rodent robot show that a specific architecture outperformed the general recurrent controller.

Introduction

The performance of sequential actions requires the storage and update of internal states. In connectionist studies, supervised learning algorithms were applied to recurrent neural networks (Doya, 2002). Despite encouraging early results in training simple neural oscillators and state machines (Doya et al. 1989), it has turned out to be very difficult to train a network to process complex sequences with long-range interactions (Bengio, 1994) by fine tuning the architecture and parameters, though not impossible (Hochreiter, 1997). Reinforcement learning algorithms have also been applied to recurrent neural networks for sequential behaviors (Nakahara et al. 2000).

Evolution of recurrent networks has also been explored for sequential behaviors (Urzelai et al. 2001). For example, Urzelai et al. 2001 evolved neural controllers for a Khepera robot to travel back and forth between lighted and dark target areas, by utilizing the proximity, light and visual sensors.

Yamauchi et al. (1996) showed that continues time recurrent neural networks capable of sequential behavior and learning can be evolved to display reinforcement learning-like abilities. The task studied was generation and learning of short bit sequences based on reinforcement from the environment. Blynel further extended this work to an artificial agent task where a simulated Khepera robot has to navigate first in a simple and then a double T-Maze

(Blynel 2003). Tuci et al. (2002) has demonstrated that it is possible to evolve an integrated dynamic neural network that successfully controls an agent engaged in a simple learning task where the robot learn the relationship between the position of a light source and the location of its goal. The networks have fixed synaptic weights and leaky integrator neurons.

In this work, we consider a more complex non-Markovian sequential task in which the memory of a preceding event alone does not determine the next action to be taken. Therefore, in addition to ordinary working memory of the most recent event, the agent must store and update a higher-order working memory. We compare the performance of a feedforward neural network (FFNN) controller and two recurrent neural networks, globally-recurrent neural network (GRNN), also known as Elman's network (Elman, 1990), and locally-recurrent neural network (LRNN), which has memory units with self-excitation and lateral inhibition. We used a real-valued GA to set the connection weights as well as the number of hidden and memory units.

The task is for a Cyber Rodent (CR) robot (Capi et al. 2002) to navigate between three places, marked by different colors, in a given order: food-nest-water-nest-food-nest-water-nest-... The results show that FFNNs performed fairly well when the locations of food, water, and nest were fixed throughout the course of evolution. Therefore, it did not need to rely on explicit internal memory. However, in the second environment, when the positions of the food and water were changed during agent life, agents with FFNN performed badly. GRNN also failed to capture the higher-order structure of the sequence; LRNN gave the best performance by successfully switching between two-levels of memory and by utilizing local recurrent connections. The output of memory units showed a strong relation with the agent intention. When the reward units were substituted by one touch sensor, the task was not completely solved. However, the LRNN neural controller performed better trying to focus its intention by memory units.

The neural controllers evolved in simulation were then tested in the real hardware of the CR robot. Despite their differences, especially with regard to timing, the robot performed the sequential navigation task well.

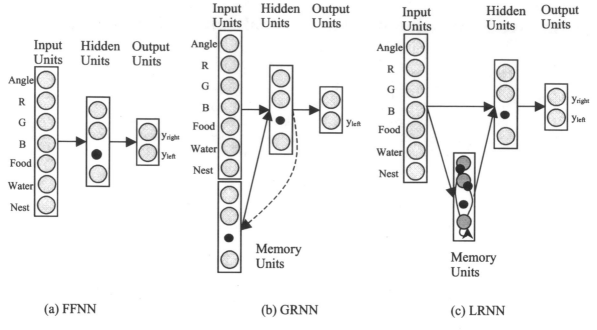

(a) FFNN   (b) GRNN   (c) LRNN

Figure 1. Neural architectures.

Evolving Dynamic Networks

Neural Architectures

The FFNN, GRNN and LRNN control architectures are shown in Figures 1(a), (b) and (c), respectively. The FFNN has one hidden layer. The GRNN architecture is similar to Elman's recurrent network, where the hidden units activation values are copied back and used as extra inputs in the next time step. The LRNN contains a memory layer where neurons with self-excitation and lateral inhibition receive inputs from the input layer. The neural architectures do not have bilateral symmetry.

The inputs of neural controllers are the angle to the nearest landmark (food, water or nest), three units (R, G, B) for color-coding of the nearest landmark, and three reward units (food, water or nest), each of which is activated only when the agent reaches the landmark according to the sequential order. The angle to the nearest landmark varies from 0 to 1 where 0 corresponds to 45° to the right and 1 to 45° to the left.

The hidden, memory and output units use sigmoid activation functions. The output of memory units of LRNN is 1 if the activation is larger then threshold and 0 otherwise. The output units directly control the right and left wheel angular velocities

$$\omega_{right} = \omega_{max} * y_{right}, \qquad (1)$$
$$\omega_{left} = \omega_{max} * y_{left}, \qquad (2)$$

where ω_{max} is the maximum angular velocity and y_{right} and y_{left} are the neuron outputs. The maximum forward velocity of CR robot is considered to be 0.5 m/s.

Genetic Encoding

The genome of every individual of the population is designed to encode the structure of each neural controller. In addition to the weight connections, we also optimized the number of hidden and memory units varying between 1 and 5. The maximum genome length of the FFNN controller is 46. The first 35 genes encode synaptic weights of input-to-hidden connections, while the next 10 genes encode those of hidden-to-output connections, and the last gene encodes the number of hidden units. The respective genome lengths of GRNN and LRNN are 76 and 147. The GRNN genome includes weight connections (35 for input-to-hidden, 10 hidden-to-output and 25 memory units-to-hidden), initial values of memory units (5) and the number of hidden units (1). The LRNN genome encodes synaptic weights (35 for input-to-hidden, 10 hidden-to-output and 35 input-to-memory, 25 interconnected memory units, 25 memory-to-output), thresholds (5) and initial values of memory units (5). In addition, one gene for each memory unit determines whether the neuron has self and lateral connections. The last two genes encode the number of hidden and memory units. The range of input, self and lateral connection weights are from -10 to 10, 0 to 10 and -10 to 0, respectively.

Rather than using variable-length genotypes to allow varying numbers of hidden and memory neurons, we use fixed-length genotypes with the maximum number of hidden and memory nodes. When the number of hidden or memory units is smaller than the maximum, some parts of the genome are ignored in constructing a network.

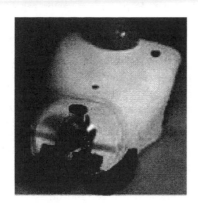

Figure 2. Cyber Rodent robot.

Non-Markovian Foraging Task

Cyber Rodent Robot

In our experiments, we used the CR robot, which is a two-wheel-driven mobile robot as shown in Figure 2. The CR is 250 mm long and weights 1.7 kg. The CR is equipped with:

- Omni-directional C-MOS camera.
- IR range sensor.
- Seven IR proximity sensors.
- 3-axis acceleration sensor.
- 2-axis gyro sensor.
- Red, green and blue LED for visual signaling.
- Audio speaker and two microphones for acoustic communication.
- Infrared port to communicate with a nearby agent.
- Wireless LAN card and USB port to communicate with the host computer.

Five proximity sensors are positioned on the front of robot, one behind and one under the robot pointing downwards. The proximity sensor under the robot is used when the robot moves wheelie. The CR contains a Hitachi SH-4 CPU with 32 MB memory. The FPGA graphic processor is used for video capture and image processing at 30 Hz.

Food-Nest-Water-Nest task

The task for the CR robot is to alternately visit food and water locations after first visiting the nest. The agent must learn low-level reactive responses. In addition, the agent must have working memory to reach the specific landmark according to the sequential order. When the agent visits the nest, a higher memory level is needed to remember if the previous visited landmark was food or water.

The environment is a square of 5 m x 5 m. The food, water and nest are placed randomly in the environment, where the distance between them varies from 1.8 m to 2.3 m. We considered the performance of neural controllers in different environmental settings. In the first environment,

food, water and nest are fixed during the agent's life and throughout generations. In the second setting, the food and water positions change after one third of the evaluation time and then change back after two thirds. In simulated environments, food, water and nest positions are considered squares of 0.05 x 0.05 m and their respective colors are red, blue and green, enabling the agent to distinguish them.

Evaluation and evolution

The CR robot is initially placed in a random position and orientation. Each individual of the population controls the agent during a lifetime of 100 seconds (50 ms x 2000 time steps). The agent receives a positive reward of 1 if the visited position matches the sequential order, and is penalized by -1 otherwise. The individuals are selected for crossover and mutation operations based on their performance.

A real-value GA was employed in conjunction with the selection, mutation and crossover operators. Many experiments comparing real-value and binary GA show that real-value GA generates superior results in terms of the solution quality and CPU time (Michalewich, 1994). The GA selection, mutation and crossover functions and their respective parameters are given in Appendix. Initially, 500 individuals were randomly created. The evolution terminated after 100 generations.

Table 1 Real-number GA functions and parameters.

Function Name	Parameters
Arithmetic Crossover	2
Heuristic Crossover	[2 3]
Simple Crossover	2
Uniform Mutation	4
Non-Uniform Mutation	[4 GNmax 3]
Multi-Non-Uniform Mutation	[6 GNmax 3]
Boundary Mutation	4
Normalized Geometric Selection	0.08

Results

Fixed environment

Figure 3 shows the performance of the agent controlled by FFNN. Under this environment setting, where the food nest and water are aligned almost linearly, the FFNN with three hidden neurons achieves fairy good performance. Due to the lack of memory, the agent reaches any landmark that enters the field of its visual sensor. After moving to the food and then to the nest positions, the agent reaches the water place that happens to be within its field of vision. When the visual sensor detects no landmarks, the agent rotates clockwise, due to strong connection weights to the left wheel. After reaching the water position, the agent rotates clockwise and finds the food first. The agent then continues to rotate clockwise, and as soon as the nest becomes visible, it heads for this landmark. If the angle between the food and nest seen from the water location is larger than the agent's view angle, the FFNN-controlled agent heads to the food and cannot complete the task according to the sequential order.

The GRNN and LRNN also solve the sequential task in fixed environments. Figure 4 shows the behavior of the agent controlled by GRNN. The agent does not find irrelevant landmarks after reaching the food position nor after reaching the nest coming from the food location. In addition, the agent reaches the nest coming from the food location in such a way that the food location would be visible only for a very short time, and the agent starts moving toward the water location. The agent ignores the food after reaching the water location.

In a fixed environment, the easiest evolved strategy is to avoid seeing landmarks that do not match the sequential order. When avoidance is not possible, the agent ignores them.

Changing environment

In changing environment, the evolved FFNN and GRNN neural controllers contained four units in the hidden layer, while LRNN had three units in the hidden layer and four units in the memory layer. Three of the four units in the memory layer had lateral and self-connections, and one unit was not interconnected.

The population average of fitness value of the three neural controllers is shown in Figure 5. The fitness during the first generations is low because there are agents, which have no attraction to any landmark or only rotates passing through the same landmark. These agents are removed from the population during the course of evolution. The fitness of FFNN controller improved quickly during the first few generations and converged after 20 generations. The fast convergence of FFNN was due to the short length of the genome and it's evolved behavior. The average fitness of GRNN is higher than that of the FFNN controller. The LRNN fitness improves slower compared to other architectures, but achieved the best final performance. The GA already converges after the 80 generations.

The agent controlled by FFNN controller, due to its lack of memory, has the worst performance. The agent rotates in the environment and approaches whichever landmark that appears in its visual sensor. The trajectories of evolved GRNN and LRNN controllers are shown in Figure 6 (a) and (b), respectively. Three different environments with different food and water positions are shown. The GRNN controller performs better than FFNN. In the first and third environments, the agent successfully ignores the water and nest locations when it should reach the food. However, in the second environment, the agent reached the water before

moving to the nest. Furthermore, after reaching the nest from the food moves again to the food place.

Figure 6b shows the trajectory of the agent controlled by LRNN. The agent follows the sequential order in three different environments. When the food and water positions change, the agent that was heading toward the food ignores the water location.

Figure 7 illustrates the unit activation values of LRNN. This figure shows how the agent utilized the angle and color of the nearest landmark and output of memory units in order to reach or avoid them. The activation of memory neurons shows that the first memory neuron is active when the agent has to visit the nest place. The second neuron is higher in the hierarchical level and deals with the higher-order structure of the sequence. The third and forth units specify whether to go to food and water, respectively.

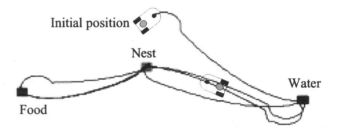

Figure 3. Performance of FFNN in fixed environment.

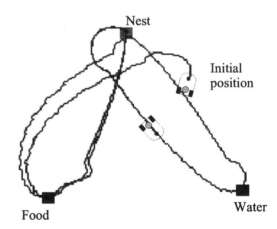

Figure 4. Performance of GRNN in fixed environment.

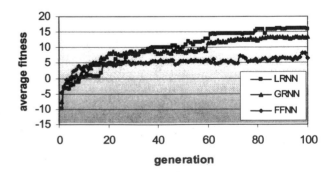

Figure 5. Average of fitness value.

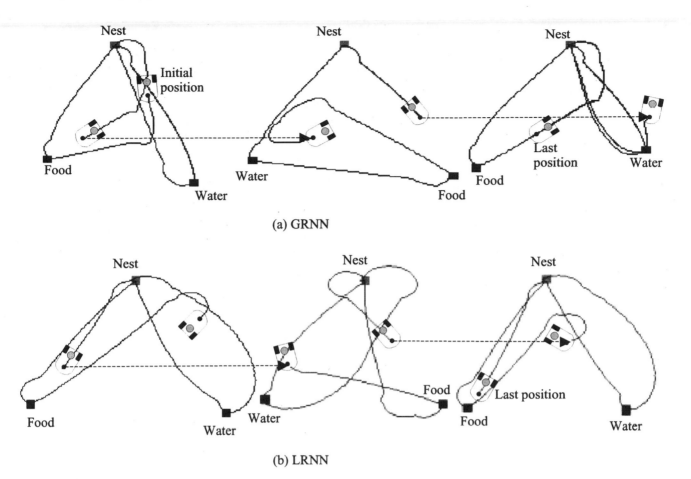

Figure 6. Performance of neural controllers.

Figure 7. Neurons activations of LRNN.

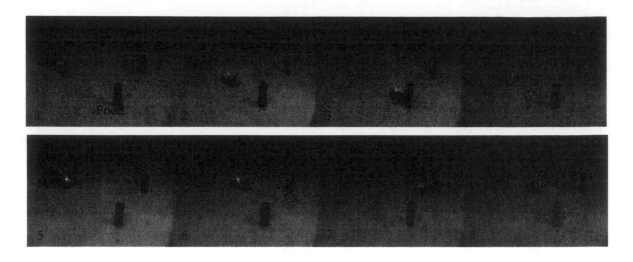

Figure 8. Video capture of hardware implementation.

We implemented the evolved LRNN controller with reward units, on the real hardware of the CR robot (Figure 8), using the visual and proximity sensors. The visual and proximity sensors in front of the CR robot are utilized to activate the input units when reaching the food, water or nest location according to the sequential order. Despite some differences due to the task and CR hardware specifications, the neural controller performed well. In addition, because there were three different colors in the environment, the FPGA needed time to analyze the captured image. Therefore, the sampling time needed to obtain the sensor data and calculate the wheels velocity was 90ms. Despite these differences, the CR robot completed the sequential task well.

Conclusions

In this paper, we presented the performance of three evolved neural controllers for a sequential task, in which the agent had to obtain food and water after first visiting the nest. We evaluated the controllers' performance with simulation and hardware implementation using the CR robot. Based on these results, we conclude:

- The agent performance is strongly related to neural network size, structure and architecture.
- LRNN is more suitable for complex sequential tasks in dynamic environments, because the self-excitation and lateral inhibition of memory units has a great effect on the long term working memory and appropriate time switching among different agent intentions.
- The training environment influences the agent's evolved behavior.
- The evolved neural controller performed well when implemented onto the CR robot.

In the future, we plan to combine learning end evolution in order to deal with environmental changes.

References

Bengio, Y., Simard, P., & Frasconi, P. (1994). Learning long-term dependencies with gradient descent is difficult, *IEEE Transactions on Neural Networks*, 5(2):157-166.

Blynel, J., (2003). Evolving reinforcement learning-like abilities for robots. In A. Tyrrell, P.C. Haddow, and J. Torresen: *Proc. of 5th Int. Conference on Evolvable Systems: From Biology to Hardware*, (pp. 320-331).

Capi, G., Uchibe, E., & Doya, K., (2002). Selection of neural Architecture and the environment Complexity. *Proceedings of International Symposium on Human and Artificial Intelligent Systems* (pp. 231-237), Fukui, Japan.

Doya, K., (2002). Recurrent networks learning algorithms, in: The Handbook of Brain Theory and Neural Networks.

Doya, K. & Yoshizawa, S., (1989). Adaptive neural oscillator using continuous-time backpropagation learning. *Neural Networks*, 2:375–386.

Elman, J., (1990). Finding structure in time, *Cognitive Science*, 14:179-211.

Hochreiter, S., (1997). Recurrent neural net learning and vanishing gradient. *Proceedings of Fuzzy-Neuro Workshop*, Soest, Germany.

Nakahara, H., Doya, K., & Hikosaka, O., (2001). Parallel cortico-basal ganglia mechanisms for acquisition and execution of visuomotor sequences-A computational approach. *Journal of Cognitive Neuroscience*, 13:626-647.

Michalewich, Z., (1994). Genetic algorithms + data structures = evaluation programs. Springer-Verlag.

Tuci, E., Quinn, M., & Harvey, I., (2002). An evolutionary ecological approach to the study learning behavior using a robot-based model. *Adaptive Behavior*, 10:201-221.

Urzelai, J., & Floreano, D., (2001). Evolution of adaptive synapses: robots with fast adaptive behavior in new environments. *Evolutionary Computation*, 9:495-524.

Yamauchi, B., & Beer, R., (1996). Sequential behavior and learning in evolved dynamical neural networks. *Adaptive Behavior*, 2(3):219-246.

EvolGL: Life in a Pond.

Santi Garcia Carbajal[1], Martin Bosque Moran[2] and Fermin Gonzalez Martinez[2]

[1]Computer Science Department, Unuiversity of Oviedo, Spain
[2]INDRA Graphics Department, Alcala de Henares, Madrid, Spain

Abstract

In this work we present the first version of Evolgl, an artificial environment for the development and study of 3D artificial lifeforms. In this first phase on the development of the project we have focused in setting up a virtual world governed by its own laws, whose state had direct influence upon the artificial beings that inhabit it. Starting from the definition of this virtual world, we have designed a basic type of creature (Evolworm), and the genetic coding of its main characteristics. Evolutionary techniques are then used to evolve the morphological features and behavioral aspects of Evolworms. They must learn to be unfolded inside the world, escape from their enemies, find couple, and obtain food. All of this in absence of an explicitly defined fitness function. In the future we are using this environment to study some classical techniques in the evolutionary computation field, like niche programming, and promotion of junk code (introns). GA-P techniques are used to code the external appearance of the individuals (the texture), to let evolution end up with individuals adapted to be invisible in some zones of the world. The artificial system of vision, and the implementation of the worms' behavioral mechanisms so that their actions are provoked exclusively by the sensory information are still under development. At this moment, we have obtained distinct forms of evolworms, as well as different bosses of behavior that we describe in this article.

Introduction. Related work. GA-P algorithms.

In this section we review some existing works from different authors that have worked in similar artificial life projects. Most of them focus in obtaining highly realistic graphical results, or in the development of an accurate physic model of the environment. We have discarded the development of a physics-based virtual marine world due to the high computational cost of such a model, to focus our work in the behavioral aspects of the evolution, and deserve some calculation possibilities to enhance the artificial vision system of the creatures. We include also a brief explanation of GA-P algorithms, an hybrid between Genetic Programming and Genetic Algorithms that we have used to code some features of the artificial creatures that live in Evolgl.

Related work.

M. Komosinski and Szymon Ulatowsky (1998) have developed during the last years the *Framsticks Project*, where some artificial living organisms called *Framsticks* learn to maximize any given criteria (total displacement over the environment, speed, etc). Framsticks are constructed from sticks, three kinds of receptors, muscles, and a neural controller for the resulting body. These authors use a coevolutive approach to obtain concurrently the best adapted body and the optimal controller in order to reach a pre-stated goal. Complete information about this project is maintained at `http://www.frams.poznan.pl/`. Karl Sims' works are broadly known in the evolutionary community. We have mainly looked at *Blockies*(1992), because of their ability to evolve complex behaviors from a numerical fitness function. Despite their visual realism, Blockies are not thought to be built, becoming real objects. Since the evolutionary computation process generates both the physical architecture and the control programs, these works fall into the body-brain coevolution category, like Framsticks. Xiaoyuan Tu and D. Terzopoulos (1994, 1995) have developed a realistic model of a virtual marine world, where three kinds of fishes swim and interact. The result of their work is very realistic. The mental state and habits of the artificial fishes are modeled through three mental state variables (Hunger, Libido, Fear), and habit parameters. An intention generator choses a behavior routine from the sensory information. The implementation described runs on a Silicon Graphics R4400 Indigo2, with 10 fishes, 15 food particles, and 5 static obstacles, obtaining 4 frames/sec. Our hardware is very modest, and we need to work with populations of great size to expect that the evolution take place. Therefore, we carried out a simulation of a very simplified marine world. Technosphere is an internet-based artificial life simulator allowing people to create their own creatures, and receive communications from them as they grow, evolve, and die in a 3d virtual environment. Users create artificial lifeforms and follow their interactions in the virtual environment, via e-mail and a suite of world wide web based interface tools. Full information about this project is available at http://www.technosphere.org.uk/

GA-P algorithms.

GA-P technique (Howard and D'Angelo, 1995) is an hybrid between genetic algorithms and genetic programming, that was first used in symbolic regression problems. Individuals in GA-P have two parts: a tree based representation and a set of numerical parameters. Different from canonical GP, the terminal nodes of the tree never store numbers but linguistic identifiers that are pointers to the chain of numbers (see figure .) The behavior of the GA-P algorithm is mainly due to its crossover operator. Either both parts of the individual can be selected and crossed. We have employed GA-P algorithms in the identification and control of complex dynamical processes, and in classification problems (Garcia and Gonzalez, 1999). In this work, we use the GA-P approach to code and evolve some of the features of the living beings that inhabit evolgl. The most important feature we have modeled with GA-P is the Texture Composition System, to obtain the final texture of each individual from a composition of the basic textures. We discuss this point in section

Why Evolgl?

As we previously mentioned, there is a number of works where some kind of artificial creature is modeled, with extremely realistic results, like Artificial Fishes, or Framsticks. Other works obtain artistically interesting results, like some works of Karl Sims. Unfortunately, accurate mechanics, a complex hydrodynamic model, and large population of artificial creatures are not possible when genetic operators (selection, crossover, mutation) are not

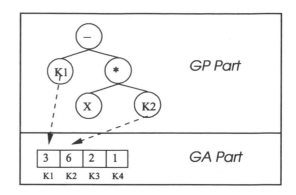

Figure 1: Representation of a generic individual in GA-P techniques. GA-P individuals consist of two parts: a tree and set of numerical constants.

implemented in the standard way in Evolutionary Computation Techniques. In an environment like Evolgl, where generations do not exist, and the fitness function is not explicitly defined, the main attention and computational effort is dedicated to the behavioral aspects of evolution. Our aim is to obtain, in the next phases of this project, an artificial life simulator where we can test some classical techniques in the evolutionary computation field, like niche programming, or observe the effects on evolution of the absence or presence of junk code in the internal representation of the individuals. There is no spectacular result from this work. Evolgl is ready now, and we plan to include the stereoscopic vision system in the next months, to test the importance of the Texture Combination System we have developed for the Genotype-Phenotype mapping process.

Evolgl.

The world.

Evolgl is a virtual world consisting of two zones: the exterior is a forest in which life takes place in a pond. The function of the forest is strictly aesthetic. All the events, calculations, the evolution itself, occurs below the surface of the pond. It is limited by four rocky walls, and warmed up by the action of four light sources (red, yellow, white, and green) with different pulses. See figure 2. The state of each light source causes the apparition of vegetation in greater or smaller measurement in its neighborhood. The plants generate directly the balls of energy Evolworms are fed with. This way, at every point of the pond, there is a temperature and a vegetation density. Plants are born and die as the nearest light source oscillates in its intensity. We introduce the different pulses and intensity for each light source to simulate periodicity in the virtual world and force Evolworms to develop the ability to store enough food to survive the "cold" periods. Moreover,we made the consumption of energy of Evolworms to depend on the temperature of the environment. Figure 3(left) shows the external aspect of the pond, and a view taken from into the pond (right). Figure 2 (right) shows a plot of the wire frame model of Evolgl, and the situation of the four light sources.

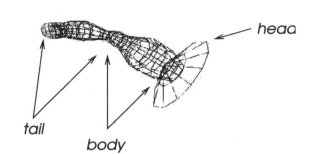

Figure 4: An Evolworm. There is three components: head, body, and tail.

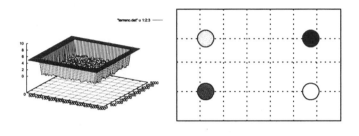

Figure 2: Evolgl. a plot of the wireframe model of the pond (left). Situation of the four light sources(right).

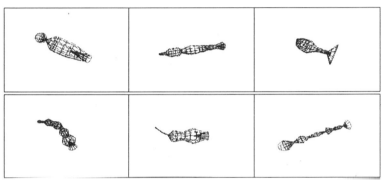

Figure 5: Evolworms. Wire frame model.

Figure 3: External view of the pond (left). A view from inside the pond (right).

There is three different components:

1. Head. The shape and size of the head determines the vision angle of the worm, and the penetration factor into the water. A creature with a big head will have a good perception of the environment, but will move slower than the same individual with a thinner head.

2. Body. The volume of the body determines the storage ability of an individual. This feature lets some Evolworms to be less dependent of the stational variations of the environment than other. Of course, an excessive growth of this part will lead to slow individuals, easier to hunt.

3. Tail. The number of segments of the tail, and their lengths determine the self-protection, and attack abilities of Evolworms, since the way to kill another individual is by hitting him repeatedly with the tail. Besides, the consumption of energy per time unit is partially determined by the total volume of the individual.

Evolworms.

Evolworms are the artificial living organisms that inhabit Evolgl. We choose this name because of their appearance, that reminds the one of some kinds of real aquatic worms. In this section we describe the morphological, behavioral and aesthetic aspects of this creatures, and explain the genotypic representation we created to simulate the evolutive process in the pond.

Physical structure and appearance. Figure 4 shows the generic structure of an Evolworm. Figure 5 shows some Evolworms in wireframe model.

The "brain". The transition from one state of the automata to another is partially determined by the weight of the arc that connects both states. This way, over a generic representation of the automata, many different kinds of individuals (aggressive, coward, food seekers) can exist. Some transitions are possible only under certain conditions. For example, an evolworm will never attack another if its color says "I am an enemy". The list of enemies is used to escape from possible predators, and to identify food (energy balls, or other individuals). It will evolve to store the best set of colors in order to survive into Evolgl. Figure 6 shows the behavior automata of an Evolworm.

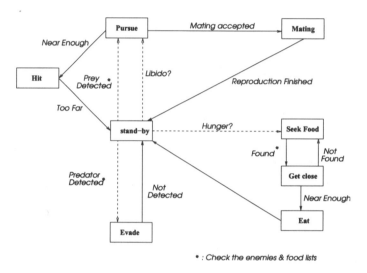

Figure 6: The brain (simplified). Discontinuous traces mean transitions that are executed probabilistic.

Genotype. Figure 7 shows the basic genotypic representation of Evolworms. It consists of the following elements:

1. The behavior automata (the "brain" of an individual).

2. A list where the individual stores the colors of his enemies.

3. A list containing the colors of any kind of food (energy balls, other individuals)

4. The structural features of the individual: number of segments, radius of each segment, etc.

5. The genetic code for the vision system.

6. Additional features like color detection tolerances, re-activity, etc.

7. A GA-P expression that codes the external appearance of the "skin".

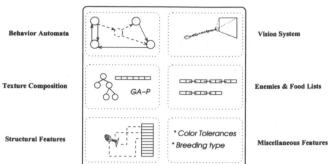

Figure 7: Evolworms: genotype codes the structural and behavioral features of each individual.

Vision system. In this first version of Evolgl, the vision system is quite simplified. Due to the computational costs of keeping large populations of evolworms wandering around, fighting and mating, we had to make a simple and static vision system. Evolworms have a viewing range and angle of view. These characteristics are part of genotype. From the 2D projection of what an Evolworm can see at each moment, it decides if there is another individual detected, food, or anything. To do this, the individual musts check the enemies and food lists, and use the tolerance parameters coded in the genotype to decide if an object is an enemy, food, or nothing relevant. See figure 8. To date, a more complex vision system is under development. Briefly, the idea is to introduce in the genotype a GP evolved function that receives as input information from an standard stereoscopic vision system, and returns a vector of three components that determines the advance direction of the individual, together with an indication of the kind of object detected.

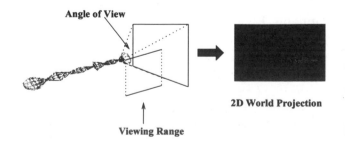

Figure 8: Artificial vision system.

Genotype-Phenotype mapping : The Texture Combination System. Most of the structural features stored in the genome (number of segments, radius, vision angle, etc) can be translated to their visual representation in a straightforward manner. We explain in detail the method we use to code and evolve the external texture applied on the wire frame model of evolworms, because we think it will be very important in the evolutive process when the vision system we are using now will be replaced with a more complex one. The coding of the texture in a way that let the evolution process end up with creatures adapted to be "invisible" in some zones of the pond is based in a GA-P representation (Howard and D'Angelo, 1995) of the external texture of an individual. We adapt the basic GA-P scheme by keeping a structural part (GP Part) that represents a set of operations made on and between different basic textures. The GA part stores the set of textures for an individual. We adapted the genetic operators to work over such a scheme. GP crossover and mutation are the standard ones defined by Koza (1994) . GA crossover and mutation are adapted to perform recombination between 64 per 64 pixel images instead of between numerical parameters. See figure 9.

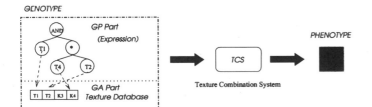

Figure 9: GA-P representation of the texture of an individual. Texture Combination System maps an expression into a resulting texture.

Experiments.

Hardware settings.

We have used a 24-node cluster based on MOSIX (Barak et al., 1999) to carry on the evolution process. All the machines included in the cluster run Red Hat 7.2. when the evolution process is quite advanced, Evolgl is visualized on a intel-based Pc running W98 with a ELSA Gladiac MX graphics card. We chose this configuration because of the poor performance of this card under Linux.

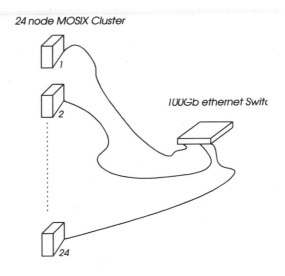

Figure 10: MOSIX cluster.

Niche Programming Model.

Results.

We distinguished three kinds of individuals with respect of the kind of food they are able to eat (energy balls, another individuals, both things), and named each specie Herbivorous, Carnivorous, and Omnivorous. We restricted the mutation operator to not let an individual change his specie, and crossover is permitted only between individuals of the same specie. The aim of these experiments was test and tune the ranges for all the parameters involved in the evolution process, check the maximum number of individuals Evolgl was able to deal with, and debug all the components of the system. Table 1 resumes the parameters of each run. We observed that the environment can manage over 600 individuals, so we restricted the experiments to work with accumulative populations always under this number.

Table 1: Evolgl. Preliminary results.

Experiment	Species	Individuals
A	Herb. + Carn.	200 + 200
B	Herb. + Omn.	200 + 200
C	Carn. + Omn.	200 + 200
D	H. + O. + C.	150 +150 + 150

Figure 11 shows some Evolworms obtained after leaving the environment evolve during a period of 24 hours.

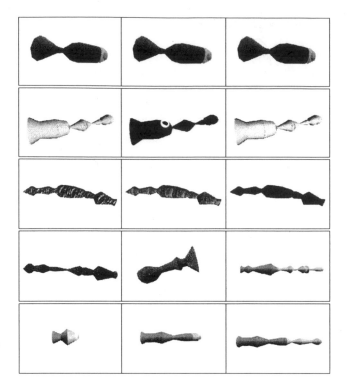

Figure 11: Some pictures from Evolworms.

Conclusions and future work.

We have given the first step toward the development of an artificial world were living organism can evolve and learn complex behaviors. There is still a lot of work to do, because we hope that the final model of evolworms will comprise:

1. An artificial vision system based on the subjective view of the world that an individual has, that provokes directly changes on the behavior automata, and generates the actions of the individuals.

2. A set of functions genetically induced that generate the movement of each part of a creature only from a set of *sensations* like temperature, oscillations produced by other creatures, etc.

References

[1] Barak, A., La'adan, O., and Shiloh, A. (1999). Scalable cluster computing with mosix for linux.

[2] Garcia, S. and Gonzalez, F. (1999). Evolving fuzzy rule based classifiers with GAP: A grammatical approach. In Poli, R., Nordin, P., Langdon, W. B., and Fogarty, T. C., editors, *Genetic Programming, Proceedings of EuroGP'99*, volume 1598 of *LNCS*, pages 203–210, Goteborg, Sweden. Springer-Verlag.

[3] Howard, L. M. and D'Angelo, D. J. (1995). The GA–P: A genetic algorithm and genetic programming hybrid. *IEEE Expert*, 10(3):11–15.

[4] Komosinski, M. and Ulatowski, S. (1998). Framsticks - artificial life.

[5] Komosinski, M. and Ulatowski, S. (1999). Framsticks: Towards a simulation of a nature-like world, creatures and evolution. In *European Conference on Artificial Life*, pages 261–265.

[6] Koza, J. R. (1992). *Genetic Programming: On the Programming of Computers by Means of Natural Selection*. MIT Press, Cambridge, MA, USA.

[7] Sims, K. (1991a). Artificial evolution for computer graphics. Technical Report TR-185, Thinking Machines Corporation.

[8] Sims, K. (1991b). Artificial evolution for computer graphics. *ACM Computer Graphics*, 25(4):319–328. SIGGRAPH '91 Proceedings.

[9] Sims, K. (1992a). Interactive evolution of dynamical systems. In Varela, F. J. and Bourgine, P., editors, *Toward a Practice of Autonomous Systems: Proceedings of the First European Conference on Artificial Life*, pages 171–178, Paris, France. MIT Press.

[10] Sims, K. (1992b). Interactive evolution of equations for procedural models. In *Proceedings of IMAGINA conference, Monte Carlo, January 29-31, 1992*.

[11] Sims, K. (1993a). Evolving images. Lecture. Lecture presented at Centre George Pompidou, Paris on March 4, 1993. Notebook. Number 5.

[12] Sims, K. (1993b). Interactive evolution of equations for procedural models. *The Visual Computer*, 9:466–476.

[13] Terzopoulos, D. and Rabie, T. (1995). Animat vision: Active vision in artificial animals.

[14] Terzopoulos, D. and Tu, X. (1994). Artificial fishes: Autonomous locomotion, perception, behavior, and learning in a simulated physical world. *Artificial Life*, 1(4):327–351.

Evolving Plastic Neural Controllers stabilized by Homeostatic Mechanisms for Adaptation to a Perturbation

Thierry Hoinville and Patrick Hénaff

LIRIS-UVSQ-CNRS, 10-12, avenue de l'Europe, 78140 VÉLIZY, FRANCE

thierry.hoinville@liris.uvsq.fr

Abstract

This paper introduces our ongoing work consisting of evolving bio-inspired plastic neural controllers for autonomous robots submitted to various internal and external perturbations: transmission breaking, slippage, leg loss, etc. We propose a classical neuronal model using adaptive synapses and extended with two bio-inspired homeostatic mechanisms. We perform a comparative study of the impact of the two homeostatic mechanisms on the evolvability of a neural network controlling a single-legged robot that slides on a rail and that is confronted to an external perturbation. The robot has to achieve a required speed goal given by an operator. Evolved neural controllers are tested on long-term simulations to statistically analyse their stability and adaptivity to the perturbation. Finally, we perform behavioral tests to verify our results on the robot controlled with a sinusoidal input while a perturbation occurs. Results show that homeostatic mechanisms increase evolvability, stability and adaptivity of those controllers.

Introduction

Evolving neural controllers to control autonomous robots has been successfully applied to various problems (Meyer et al., 2002; Ijspeert et al., 1998; Nolfi, 1997; Gallagher et al., 1996). However, the majority of the proposed methods produces solutions that are efficient in constant environmental conditions. Recently, evolutionary robotics raise the issue of attempting to build fault-tolerant controllers *adaptive* to internal and environmental *perturbations* (Floreano and Urzelai, 2000; Dittrich et al., 1998). The IRON[1] project, is intended to increase the autonomy and the robustness of robots by confronting evolutionary adjusted neural controllers with different kinds of perturbations: transmission breaking, adherence losses, leg loss, material wear, *etc*. A promising way to do this is to incorporate into neural controllers some plasticity mechanisms inspired by biology (Floreano and Urzelai, 2000). Nevertheless, this plasticity

---

[1]Implémentation RObotique de Neuro-contrôleurs adaptatifs (http://www.liris.uvsq.fr/iron/Iron.html). This project, initiated by J.-A. Meyer from the AnimatLab, is supported by the ROBEA program of the CNRS (http://www.laas.fr/robea).

might not be sufficient because it tends to *destabilize* those controllers.

In this paper, we will perform a *comparative* study of effects of two bio-inspired homeostatic mechanisms on the evolvability of plastic neural controllers embodied in a single-legged robot. Subsequently, we will analyse control stability and its adaptivity to an *external* perturbation.

The paper is organized as follows: the first section briefly presents two homeostatic mechanisms from a biological point of view, the second is devoted to the formalization of these mechanisms in our neural model and its application to make a controller for a simulated robot in an evolutionary way. The third section shows statistical and behavioral results, after which results and mechanism modelling are discussed. In the final section, we conclude by giving further developments within our project.

Homeostatic mechanisms in biological neurons

Historically, research in neurophysiology initially centered on synaptic plasticity (Hebb, 1949). The mechanisms controlling this plasticity are commonly considered as the main vector of information storage in neural networks and synaptic connection refinement during cerebral development. Thus, by establishing correlation between simultaneous active neurons, Hebbian rules allow neural circuits to adapt to received information. Nevertheless, these flexibility mechanisms are sources of instability (Miller and MacKay, 1994). Recent studies (Turrigiano, 1999), show that they are associated with homeostatic rules which regulate intrinsic properties of each neuron. In this paper, we are interested in two kinds of these: one regulating neuronal excitability and the other stabilizing total synaptic input strength of each neuron.

Regulation of excitability

Excitability of a neuron, i.e. its propensity to transmit action potentials according to information it gets, depends on concentrations of various molecules which are present in its cell body and in its close vicinity. Ion channels inserted in its membrane *actively* regulate these concentrations. According to (Desai et al., 1999), this regulation seems to be driven by the average activity of the cell. Thus, when activity of a neuron is high, its excitability decreases to return to a functional firing rate. Conversely, if the cell tends to be silent, its excitability increases until its firing rate gets back to a functional range (top of figure 1).

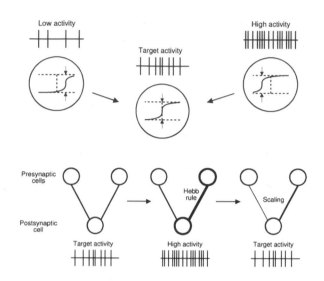

Figure 1: Two homeostatic mechanisms regulating intrinsic properties of neurons. Circles represents nerve cells. (These figures are directly inspired from (Turrigiano, 1999)) Top: Regulation of excitability. The cell's activation threshold is regulated according to its own activity. Bottom: Multiplicative scaling of all synaptic inputs strengths. After an hebbian potentiation of a synapse, the scaling mechanism induces synaptic competition on postsynaptic cell's inputs. Bold circles symbolize activated neurons. Line widths indicates the strength of the corresponding connection.

Stabilization of total synaptic strength

During development and learning, the number of synapses and their properties are submitted to marked changes. These modifications can severely alter activity patterns of neurons. According to (Turrigiano et al., 1998), to preserve their functionalities, neurons regulate their total synaptic input strength. Thus, the excitation amplitude remains in a relevant domain (bottom of figure 1). Experimental observations tend to indicate that this regulation is applied *multiplicatively* and globally to all synaptic inputs of a related cell. By its multiplicative nature, this mechanism ensures relative weights of the different connections. Furthermore, far from neutralizing individual synaptic plasticity, this process induces *competition*: if some connections are potentiated, the strengths of the others must decrease.

Methods

Neuron and synapse model

In the following, we propose extensions to a classical neural model that implement the two homeostatic mechanisms presented above. The extensions can be turned on or off, independently of each other, so it allows comparative study of these mechanisms. It should be noted that these mechanisms, naturally dynamic and activity-dependent, are formalized here in a *static* way. Since the time constant of these mechanisms is much higher than those of the activation and learning processes, we assume this is an acceptable simplification.

Classical model (CM) Nerve cells composing our neural controller are based on a *leaky integrator* model (Beer, 1995):

$$\tau_i^\star \frac{dy_i}{dt} = -y_i + \frac{1}{N_i} \sum_{j=1}^{N_i} w_{ij} o_j + I_i \qquad (1)$$

$$o_i = \frac{1}{1 + e^{-\alpha_i(y_i - \theta_i)}} \qquad (2)$$

where y_i represents the mean membrane potential of the neuron i and o_i its activity. w_{ij} is the synaptic strength of the connection from neuron j to neuron i, N_i is the number of synaptic inputs of the neuron i and $\tau_i^\star$ denotes the time constant of the membrane potential. I_i corresponds, in the case where the cell i is a sensory neuron, to an external excitation coming from a sensor. Finally, α_i is a gain determining the slope of the sigmoidal activation function and θ_i is the threshold of the neuron's activity.

As we want to study neural controllers with intrinsic plasticity, we use the *adaptive synapses* model taken from (Floreano and Urzelai, 1999) in which one local adaptation rule among four is assigned to each connection, as suggested by biological observations. The synaptic weights are updated at each sensory-motor cycle according to the following expression:

$$\widetilde{w}_{ij}^{t+\Delta t} = \widetilde{w}_{ij}^t + \frac{\Delta t}{\tau_{ij}^\odot} \Delta \widetilde{w}_{ij} \qquad (3)$$

where $\widetilde{w}_{ij}$ means $|w_{ij}|$, $\tau_{ij}^\odot$ is the time constant of the adaptation rule (comparable to the learning rate, η, in the Floreano's model) and $\Delta \widetilde{w}_{ij}$ is one of the four adaptation rules:

Plain Hebb rule

$$\Delta \widetilde{w}_{ij} = (1 - \widetilde{w}_{ij}) o_j o_i \qquad (4)$$

Pre-synaptic rule

$$\Delta \widetilde{w}_{ij} = (1 - \widetilde{w}_{ij}) o_j o_i + \widetilde{w}_{ij} o_j (o_i - 1) \qquad (5)$$

Post-synaptic rule

$$\Delta \widetilde{w}_{ij} = (1 - \widetilde{w}_{ij})o_j o_i + \widetilde{w}_{ij}(o_j - 1)o_i \qquad (6)$$

Covariance rule

$$\Delta \widetilde{w}_{ij} = \begin{cases} (1 - \widetilde{w}_{ij})\delta(o_j, o_i) & \text{if } \delta(o_j, o_i) > 0 \\ \widetilde{w}_{ij}\delta(o_j, o_i) & \text{otherwise} \end{cases} \qquad (7)$$

where $\delta(o_j, o_i) = \tanh(4(1 - |o_j - o_i|) - 2)$ is a measure of the difference between o_j and o_i. Note that, in this model, values of synaptic weights are constrained in the interval [-1,1].

Center-crossing mechanism (CC) In order to model the mechanism that regulates neuronal excitability, we use the model of *center-crossing neural networks* proposed in (Mathayomchan and Beer, 2002). This paradigm consists of determining the ideal activation threshold of a neuron according to its synaptic input weights.

$$\theta_i = \frac{1}{2}\sum_{j=1}^{N_i} w_{ij} \qquad (8)$$

In this way, the operating range of each neuron is centered about the most sensitive region of its activation function. Indeed, due to the sigmoid asymmetry (about the x-axis), the excitation range of a neuron can be shifted according to the weight values of its synaptic inputs. Formally, this concept can be reduced to the use of a symmetric activation function like the hyperbolic tangent. In our model, we don't use eq. (8) to replace θ_i in eq. (2) but we adapt it according to the following expression:

$$\tau_i^\star \frac{dy_i}{dt} = -y_i + \frac{1}{N_i}\sum_{j=1}^{N_i} w_{ij}(2o_j - 1) + I_i \qquad (9)$$

Thus, we preserve the same parameterization between the CM and the CC models. Also, the difference of the CC model is that the activation threshold of a neuron becomes totally independent of its synaptic input strengths.

Normalization of synapses mechanism (NS) On the other hand, the mechanism that regulates total synaptic input strengths is modeled by a multiplicative normalization of $\| \vec{w}_i \| = \sqrt{\sum_{j=1}^{N_i} w_{ij}^2}$ (Gerstner and Kistler, 2002). To implement this mechanism, eq. (1) and eq. (3) are altered as

follows:

$$\tau_i^\star \frac{dy_i}{dt} = -y_i + \frac{1}{\sqrt{N_i}}\sum_{j=1}^{N_i} w_{ij}o_j + I_i \qquad (10)$$

$$\widetilde{w}_{ij}^{t+\Delta t} = \frac{\widetilde{w}_{ij}^t + \frac{\Delta t}{\tau_{ij}^\odot}\Delta \widetilde{w}_{ij}}{\sqrt{\sum_{j=1}^{N_i}\left(\widetilde{w}_{ij}^t + \frac{\Delta t}{\tau_{ij}^\odot}\Delta \widetilde{w}_{ij}\right)^2}} \qquad (11)$$

Finally, the above extensions allows the instantiation of four models:

- *center-crossing* model with normalized synapses (CCNS),

- *center-crossing* model (CC),

- normalized synapses model (NS),

- classical model (CM).

Application problem

The aim of the IRON project is to provide a *multi-legged* robot with a neuro-controller, synthesized by evolution, that can adapt its behavior to perturbations. These perturbations can be external (environmental changes) or internal (mechanical or electrical faults). However, a main issue of evolutionary robotics, called the *scalability* problem, is the application of its methods to systems that show a high degree of complexity. Also, to compare the four neural models described before, we apply our approach to a single-legged robot simulation[2].

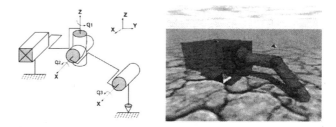

Figure 2: Morphology of the single-legged robot. Left: Kinematic model. Right: Simulation view.

The robot (figure 2) is composed of a body and a leg endowed with three degrees of freedom: two for the hip and one for the knee (see table 1 for mass and geometrical parameters). A binary contact sensor is fixed on the leg tip and on each joint, a servo-motor ordering its angular position is simulated.

---

[2]Based on a simulator called Open Dynamic Engine (http://q12.org/ode/).

Part	Shape	Mass [kg]	Dimension [cm]
body	box	3	$20 \times 20 \times 10$
hip	sphere	0.5	$\varnothing 8$
thigh	capped	0.5	$15 \times \varnothing 4$
shank	cylinder		

Table 1: Mechanical parameters of the robot

A prismatic link connecting the robot body to the ground, constrains its movements by guiding it on $\overrightarrow{X}$ axis. A viscous friction force is applied to the robot. This force F_{fr} is equal to $-k_{fr}V_{eff}$ where V_{eff} is the effective speed of the robot body's centre of mass and k_{fr} is the viscous coefficient.

The task of the robot is to respect a desired walking speed V_{des} and to simultaneously offset a potential perturbation consisting of varying the coefficent k_{fr}. This perturbation, external from the leg's point of view, could simulate, from a multi-legged robot's point of view, an internal perturbation as a mass growth or a disruption of another leg.

The performance p, of a controller is evaluated at the end of a simulation of $T = 10$ sec. according to the following expression:

$$p = \frac{1}{T} \int_0^T |V_{des} - \widetilde{V_{eff}}| dt$$

where $\widetilde{V_{eff}}$ is the global walking speed of the robot[3].

During evolution, each controller is evaluated through three successive simulations with different scenarios. As we can see on table 2, a scenario is defined by temporal variations of V_{des} and k_{fr} parameters. The scenario A corresponds to a simple control behavior, the scenario B rewards the capacity of inhibiting robot gait and the scenario C favours adaptation to the perturbation.

	Scenario A	Scenario B	Scenario C
V_{des}	0.4 0 — T/2 — 0	0.4 0 — T/4 — 3T/4	0.3
k_{fr}	10	10	20 10 — T/4

Table 2: Three evaluation scenarios

The global fitness of an individual is the quadratic combination of the three elementary scores obtained from these evaluations (the lower the fitness is, the better the controller behaves).

$$fitness = \sqrt{p_A{}^2 + p_B{}^2 + p_C{}^2}$$

This kind of combination restricts the compensation effect produced by a classical average and supports behaviors that include the three qualities described by the scenarios.

---

[3] Temporal average calculated by application of a second order low-pass filter on V_{eff}. This value is preferred to V_{eff}, subject to high amplitude variations at the time of each stride.

Neural controller structure and genetic encoding scheme

Figure 3 represents an example of an evolved controller. We arbitrarily fix the network size to eight neurons. Two neurons receive sensorial information from the environment, the first one is excited with the consign error $V_{des} - V_{eff}$, the second one is connected to the contact sensor. Three motoneurons drive angular positions of servomotors. The three remaining neurons form the hidden layer of the network.

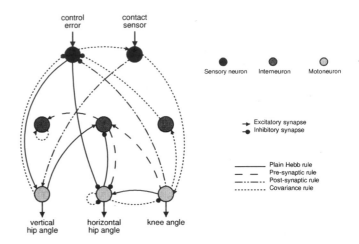

Figure 3: Example of the phenotype structure of an evolved controller.

Controllers are genetically encoded with numerical and symbolic alleles strings divided in eight neuronal blocks. Figure 4 shows structure of the neuron genotype. A neuronal block is composed of a list of intrinsic neuronal parameters ($\tau^{\star}$, α et θ) and a list of eight synaptic blocks defining properties of cell's output connections with others neurons (including itself). The first synaptic block gene condition the network structure by activating or not the related connection. The next gene indicates its excitatory or inhibitory mode. The two remaining genes dictate the connection dynamic by associating to it one of the four learning rules showed above and the related time constant, $\tau^{\odot}$. The value of each numerical gene is taken from a five length allele set. Table 3 shows these allele sets.

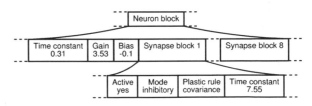

Figure 4: Structure of the neuron genotype.

From a complexity point of view, each genotype is provided with $8 \times (3 + 8 \times 4) = 280$ genes and the size of the

Gene	Alleles set				
$\tau^\star$	0.02,	0.165,	0.31,	0.455,	0.6
α	2.46,	3.53,	5.34,	9.43,	31.26
θ	-0.2,	-0.1,	0,	0.1,	0.2
$\tau^\odot$	0.2,	2.65,	5.1,	7.55,	10

Table 3: Alleles sets of numerical genes

genotype search space is $\left(5 \times 5 \times 5 \times (2 \times 2 \times 4 \times 5)^8\right)^8 = 3.74 \times 10^{138}$ possibilities.

Controllers are evolved by a generational and elitist genetic algorithm. Genetic operators are the allelic mutation ($P_{mut} = 0.001$) and the uniform crossover ($P_{cross} = 0.6$). Individuals are selected by the stochastic universal sampling algorithm (Baker, 1987) according linearly to their rank in the population (Goldberg, 1989) (with the best individual producing an average 1.1 offspring).

Results

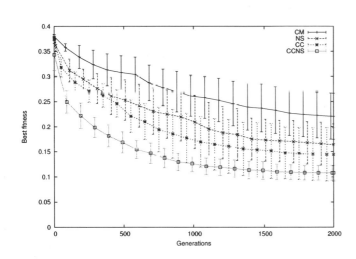

Figure 5: Best fitness for each neural model averaged over 10 runs with different random initializations of the population.

For each neuronal models, we performed 10 evolution runs with different random initializations of populations of 200 individuals. Populations are evolved during 2000 generations. Figure 5 shows statistical results of this experiment in term of best fitness. From these data, four main results can be observed:

- Both homeostatic mechanisms (CC and NS) clearly improve evolvability of the controllers, either in final solution or in evolution speed.

- With the CC controller model, results of evolutionary process strongly depends on the initial random population.

- The NS model is, on average, less effective than the CC model but the standard deviation of the results in the NS case is smaller than in the CC case.

- The CCNS model has the best performance and always results in efficient controllers.

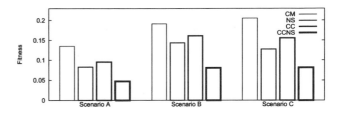

Figure 6: Average fitness of the best controllers (one from each evolution run), according to a scenario and a neural model. Fitness are evaluated during a simulation of 100 sec.

To verify relevance of these results, we evaluated, for each neuronal model and scenario, average fitness of the 10 best controllers taken from the above 10 evolution runs. To test long-term stability of these controllers, each is evaluated during a $T = 100$ sec. simulation. Resulting data are presented on Figure 6. These results corroborate our previous observations except for relative performances of the CC and NS models. This point will be discussed in the next section.

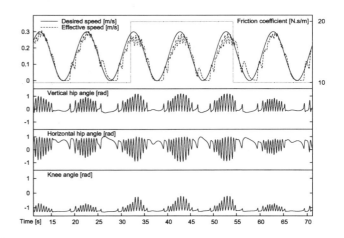

Figure 7: Time-plots of V_{des}, $\widetilde{V_{eff}}$, k_{fr} and joint commands for a CCSN controller. The robot is perturbated from $t = 32$ sec during about 22 sec. Robot strides cause light oscillations of $\widetilde{V_{eff}}$ (attenuated by the low-pass filter).

As a final step, we did a behavioral analysis of the best controllers. During this experiment, a sinusoidal desired speed is indicated to the robot while a perturbation occurs. Figure 7 shows an example of time-plots obtained for a simulation of a CCNS controller. Friction coefficient and joint commands given by the neural network ouputs are also plotted. From these plots, we can see that the control task is satisfied even when the perturbation occurs. Indeed, in the case of a non-adaptive controller, an increase of 100% of the viscous friction coefficient should cause decrease of 50% of its speed. Yet, figure 7 shows that the perturbation does not significantly alter the robot gait, the control task error remaining relatively small. Moreover, perturbation adaptation is confirmed by time-plots of joint commands. Increasing of the friction force causes amplification of signals send to servo-motors (particularly obvious on the knee joint). Note that walking frequency is not altered by the perturbation.

Discussion

As the above results show, when simulation time is increased, NS controllers are more efficient than CC controllers. By analysing long-term behavior of both models, it can be seen that normalized synapses contribute to stabilizing neural network function. Indeed, CC and CM controllers produce less robust oscillatory patterns. Preliminary obervations indicate that these neural networks cannot reconfigure their synaptic strengths after an inhibition or a perturbation of the robot gait. We suppose that the synaptic normalization constraint favours dynamical stability of neuronal systems. Also, this could explain the low standard deviation of NS and CCNS evolution results (figure 5).

Either during evolution (figure 5) or during post-evolutionary evaluations (figure 6), statistical results show that the CCNS model is clearly more efficient than the CC and NS models. Since their associated use is more salient than their individual application, we may conclude that there is a *synergy* between the two mechanisms.

In our model, the synaptic normalization is expressed in a static way. This assumption is made due to the fact that, from a biological point of view, this mechanism is relatively slow. However, it should be interesting to study its dynamic modelling as in Oja's learning rule (Oja, 1982). Moreover, by its local nature, this rule is more biologically plausible. In the same way, a dynamical regulation of neuron excitability could allows a wider range of dynamics.

Conclusion and further work

In this paper, we show that evolvabilty, in terms of both speed and final result, of plastic neural controllers is improved by couping them with bio-inspired homeostatic mechanisms that regulate neuronal excitability. Evolved controllers show increased stability and robustness. These results support the hypothesis that constraining neuronal architectures with homeostatic mechanisms at the micro-level induces robust behaviours at the macro-level (Di Paolo, 2002).

On the other hand, since our neuronal model is not specific to the task we test it on, our results suggest that these mechanisms could be beneficial to other applications or methodologies (back-propagation, reinforcement learning, *etc*). Indeed, in the case of back-propagation, the gradient descent on an error could include a multiplicative normalization mechanism of synaptic weights as in eq. (3). Moreover, the center-crossing mechanism should not need any adaptations to fit classical connexionism methods.

Within the scope of the IRON project, the aim of our current work is to extend our approach to robots of several morphologies undergoing various kind of perturbations. To this end, the addition of a dynamic system, driving intrinsic neuronal properties and synaptic plasticity, could improve neural controllers' adaptivity by allowing them to reconfigure themselves. A potential issue is to integrate to our neural model some bio-inspired paradigms based on chemical messengers (Eggenberger et al., 1999; Husbands et al., 1998).

References

Baker, J. E. (1987). Reducing bias and ineffiency in the selection algorithm. In Associates, L. E., editor, *Proc. of the second Int. Conf. on Genetic Algorithms and their Applications*, Hillsdale.

Beer, R. (1995). On the dynamics of small continuous-time recurrent neural networks. *Adaptive Behavior*, 3(4):469–509.

Desai, N., Rutherford, L., and Turrigiano, G. (1999). Plasticity in the intrinsic excitability of neocortical pyramidal neurons. *Nature Neuroscience*, 2(6):515–520.

Di Paolo, E. A. (2002). Fast homeostatic oscillators induce radical robustness in robot performance. In *SAB'2002*. MIT Press.

Dittrich, P., Buergel, A., and Banzhaf, W. (1998). Learning to move a robot with random morphology. In Husbands, P. and Meyer, J.-A., editors, *Evolutionary Robotics, First European Workshop, EvoRob98*, pages 165–178. Springer, Berlin.

Eggenberger, P., Ishiguro, A., Tokura, S., Kondo, T., and Uchikawa, Y. (1999). Toward seamless transfer from simulated to real worlds: A dynamically-rearranging neural network approach. In Wyatt, J. and Demiris, J., editors, *Proc. of the Eighth Eur. Workshop on Learning Robots*, pages 4–13, EPFL, Lausanne, Switzerland.

Floreano, D. and Urzelai, J. (1999). Evolution of neural controllers with adaptive synapses and compact genetic encoding. In *5th European Conf. on Artificial Life*.

Floreano, D. and Urzelai, J. (2000). Evolutionary robotics: The next generation. In Gomi, T., editor, *Evolutionary Robotics III*, Ontario (Canada). AAI Books.

Gallagher, J., Beer, R., Espenschied, K., and Quinn, R. (1996). Application of evolved locomotion controllers to a hexapod robot. *Robotics and Autonomous Systems*, 19:95–103.

Gerstner, W. and Kistler, W. (2002). *Spiking Neuron Models*, chapter 11. Cambridge Univ. Press. (`http://diwww.epfl.ch/~gerstner/BUCH.html`).

Goldberg, D. E. (1989). *Genetic algorithms in search, optimization, and machine learning*. Addison-Wesley, Reading, MA.

Hebb, D. (1949). *The Organization of Behavior: A Neurophysiological Theory*. John Wiley and Sons.

Husbands, P., Smith, T., Jakobi, N., and O'Shea, M. (1998). Better living through chemistry: Evolving gasnets for robot control. *Connection Science*, 10(3-4):185–210.

Ijspeert, A. J., Hallam, J. C. T., and Willshaw, D. J. (1998). From lampreys to salamanders: Evolving neural controllers for swimming and walking. In *From Animals to Animats: Proc. of the Fifth Int. Conf. of the The SAB*. MIT Press.

Mathayomchan, B. and Beer, R. (2002). Center-crossing recurrent neural networks for the evolution of rhythmic behavior. *Neural Computation*, 14:2043–2051.

Meyer, J.-A., Doncieux, S., Filliat, D., and Guillot, A. (2002). Biologically inspired robot behavior engineering. In Duro, R., Santos, J., and Graña, M., editors, *Evolutionary Approaches to Neural Control of Rolling, Walking, Swimming and Flying Animats or Robots*. Springer-Verlag.

Miller, K. and MacKay, D. (1994). The role of constraints in hebbian learning. *Neural Computation*, 6:100–126.

Nolfi, S. (1997). Evolving non-trivial behaviors on real robots: a garbage collecting robot. *Robotics and Autonomous System*, 22:187–198.

Oja, E. (1982). A simplified neuron model as a principal component analyzer. *Math. Biol.*, 15:267–273.

Turrigiano, G. (1999). Homeostatic plasticity in neuronal networks: the more things change, the more they stay the same. *Trends in Neuroscience*, 22(5):221–228.

Turrigiano, G., Leslie, K., Desai, N., Rutherford, L., and Nelson, S. (1998). Activity-dependent scaling of quantal amplitude in neocortical pyamidal neurons. *Nature*, 391:892–895.

A Stigmergic Cooperative Multi-Robot Control Architecture

Thomas G. Howsman[1], Daniel O'Neil[2], Michael A. Craft[1]

[1]Dynamic Concepts, Inc.
P.O. Box 97, Madison, AL 35758
(thowsman@dynamic-concepts.com)

[2]NASA/MSFC, Marshall Space Flight Center, AL 35812
(dan.oneil@msfc.nasa.gov)

Abstract

In nature, there are numerous examples of complex architectures constructed by relatively simple insects, such as termites and wasps, which cooperatively assemble their nests. A prototype cooperative multi-robot control architecture which may be suitable for the eventual construction of large space structures has been developed which emulates this biological model. Actions of each of the autonomous robotic construction agents are only indirectly coordinated, thus mimicking the distributed construction processes of various social insects. The robotic construction agents perform their primary duties *stigmergically*, i.e., without direct inter-agent communication and without a preprogrammed global blueprint of the final design. Communication and coordination between individual agents occurs indirectly through the sensed modifications that each agent makes to the structure. The global stigmergic building algorithm prototyped during the initial research assumes that the robotic builders only perceive the current state of the structure under construction. Simulation studies have established that an idealized form of the proposed architecture was indeed capable of producing representative large space structures with autonomous robots. This paper will explore the construction simulations in order to illustrate the multi-robot control architecture.

Introduction

NASA's long term goals include construction of large space structures (Figure 1) which will require the assembly of structural elements numbering in the thousands, and will potentially be performed primarily by robotic means. Advantages of using robotic agents rather than astronauts for the assembly process include astronaut safety, construction efficiency, and cost savings. Comparisons between robotic assembly and human construction (via EVA) of structures have been made (Lake 2001), with conclusions stating that for very large structures (greater that 100m in dimension) some form of robotic assembly is required.

Although the need for robotic assembly in space has been established, many aspects of the required systems have yet to be developed. The research discussed herein focuses upon the overall assembly strategy from a distributed control point of view (Mataric 1995). In other words, how can the efforts of multiple robotic construction agents be coordinated so that the desired structure emerges from the assembly effort? Taking a cue from social insect behavior (Theraulaz *et.al.* 1995), a form of decentralized coordination based upon stigmergic principles appears promising. In this scenario, each individual agent's behavior is controlled by stimuli provided by the common environment of the emerging structure–a form of indirect communication. Rather than using direct communication between agents, each individual communicates with its fellow agents via the small changes each one makes to the structure under construction. In fact, the emerging structure itself serves as a form of external memory (Beckers *et.al.* 1994).

An advantage of stigmergic cooperative behavior is that massive redundancy is automatically built in to the system. Failure of a particular individual building agent (robot) will only slow the building process, not stop it. The required computer processing power of the individual agents is reduced by the fact that the overall design is not maintained, or even understood, by the agents (Valckenairs *et. al.* 2001). Rather, the final design is an emergent property of the stigmergic assembly algorithm. An individual agent in the system transitions from one state (e.g., platform assembly) to another (e.g., antenna building) based upon sensor inputs which monitor the local environment, or structure.

The initial phase of the research focused on the formulation of stigmergic building algorithms suitable for the assembly of large space structures. A software simulation of the cooperative building process was developed, allowing for visualization of the assembly process. An internet enabled version of the design and simulation tools is also available. Hardware demonstrations of the developed algorithms are currently underway using relatively standard robotic hardware.

Building Algorithm

During the assembly process, each individual assembly robot moves about the work volume by making discrete, but continuous, transitions from one cell to another. From the robotic agent's point of view, each

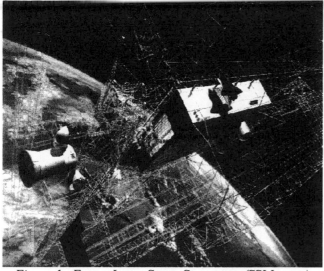

Figure 1. Future Large Space Structure. (SSI Image.)

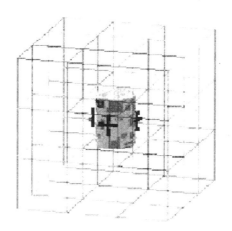

Figure 2. Cubic Sensor Lattice.

agent is centered in a 3×3×3 lattice of cubic cells, as illustrated in Figure 2. The agent occupies the center cell of the 27 cells in the lattice, and can move to any of the 26 neighboring cells that are unoccupied. An alternate hexagonal description of the work volume is also available for used by the algorithm.

The construction agent is assumed to possess sensors capable of detecting the presence of building objects or other robotic agents within the local 3×3×3 lattice. For the purposes of this research, we allowed for twenty distinct types of building blocks. It is assumed the agent can distinguish between the different colored blocks (blue, green, etc.) as well as other building agents which might be present within his sensor lattice.

The data visualization and rule building process is much easier if we utilize a two dimensional display map of the three dimensional sensor lattice (Theraulaz et. al. 1995). An example of this mapping is shown in Figure 3. As can be seen in the figure, the front face of the sensor cube maps to cells (13, 14, 15), (4, 5, 6), and (22, 23, 24).

As the agent moves about the work volume, the sensor patterns associated with the local neighborhood are monitored. If the pattern matches that of a given "building rule," then the rule is activated and the corresponding building action is taken.

Figure 4 illustrates a simple example in which two rules are utilized to build a column structure. The first rule can be interpreted as, "if a blue (dark) block is the only object within the sensor lattice, and it is directly above the building agent, then deposit a green (light) block." Similarly, the second rule also calls for the deposit of a green block, but only if a green block is detected directly above the agent. Finally, as can be seen in the figure, a total of eight building agents were active in the assembly of the column. Each builder acts independently, and is essentially unaware of the activities of other builders, except when a new block is

deposited (detected as change in sensor lattice) or if two builders attempt to move into the same physical location – which triggers the collision avoidance mechanism. A simplified block diagram of the assembly procedure is shown in Figure 5. The basic algorithm repeats the pattern .. move building agent .. check for rule match .. deposit block (if rule satisfied) .. move building agent, and so on. The building agents, which can vary in number, must be initially located within the work volume. It is assumed that the initial position of the building agents is not in the immediate vicinity of the building seed block, which by default is always located at the origin. As indicated in the flowchart, agent locations are updated and the sensor patterns are compared to the set of assembly rules. Note that the agents are only allowed to move into a neighboring cell (one of the 26 surrounding cells in the agent's local 3 × 3 × 3 cell lattice) during a single iteration.

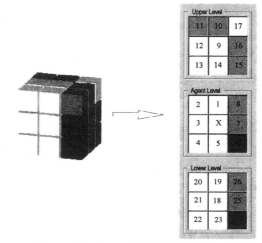

Figure 3. Sensor Lattice Mapping.

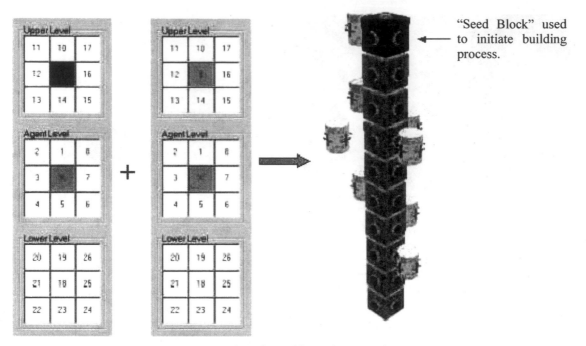

Figure 4. Simple Builder Rule Example.

Initially, the agents are located some distance from the seed block. One possible method of searching the work space for cell locations in which the agent's sensor patterns match a rule is to simply allow the agent to perform a random walk through the work volume. Unfortunately, such a scheme is very inefficient. Consider a work volume that extends 100 cells along each axis. This corresponds to a volume containing 1,000,000 cells. Initially, only the 26 cells bordering the seed block are candidates for block deposition, and typically only one of the 26 will satisfy a building rule. The statistics of this type of 3-dimensional random walk are quite complex; however, it is safe to say that on average, many thousands of iterations would be required to locate the correct initial position for depositing a building block. This fact was borne out with a series of simulation trials. In 19 of 25 simulation cases, the agent performing a random walk in the work volume failed to locate the seed block within 10,000 iterations. The earliest the seed block was found using the random walk approach was in approximately 4,000 iterations. Clearly a more effective search procedure is required.

In nature, insects utilize a variety of methods to provide them with a sense of direction and location (e.g., sunlight angle of incidence, chemical markers, etc.). If we allow the seed block to somehow publish its approximate location to the building agents, then the size of work volume that must be searched can be drastically reduced. An example of how this could be accomplished would be to place a light source upon the seed block, and instrument the robotic agents with light detectors. Essentially, the agents would simply move toward the light until they are in close proximity of the seed block. Once the agents are within a predetermined range of the seed block (perhaps determined

by the measured intensity of the light source), they could begin searching for cell locations corresponding to sensor patterns that match a rule in the rule set.

While interesting structures can be assembled with a totally homogeneous population of agents, greater efficiencies can be achieved by allowing for specializations within the population. This can be accomplished by allowing different members of the population to follow different rule sets, or in effect, creating population subclasses. Delaying the introduction of certain classes into the work volume has also proved beneficial in controlling the coherence of the building process. A coherent set of assembly rules is defined as a rule set that disallows ambiguity in the building process, and thereby generates identical structures given different initial agent conditions (Mason 2002). Significant improvements in the efficiency of the building process were obtained by using a contact, or crawling, algorithm which restricts the searching agents from leaving the surface of the structure once located.

Simulation Results

In the preceding sections, the architecture of an assembly strategy for the autonomous robotic construction of large space structures was developed. Results obtained by applying the algorithms and software to the assembly of a test structure will now be examined. A planar platform penetrated by a central beam was the architectural design goal for the test case. This structure notionally represents a Solar Power Satellite, but the underlying algorithms are applicable to a wide variety of assembly tasks.

Figure 7 illustrates the building process of the targeted structure with selected simulation screenshots and

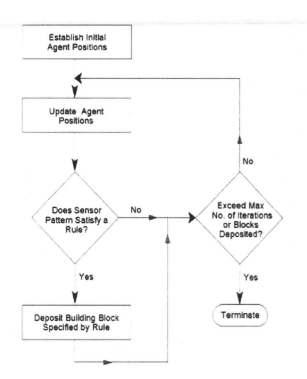

Figure 5. Simplified Building Algorithm.

the corresponding rule most recently triggered by the local configuration. This structure was assembled using a building rule set containing 43 individual rules. For this example, the agent population was completely homogeneous in that each agent operated with the exact same rule set. As can be seen in the figure, there were 8 building agents present, and after 17 iterations the first block was deposited (the triggered rule is adjacent to the sub-figure). The central planar section was complete after 40 iterations, and by iteration 175 the central beam was beginning to emerge. The complete planar platform was complete by iteration 402, after which only the central beam continued to be assembled.

Figure 6 illustrates a more complicated structure under construction. A non-homogeneous agent population was utilized for this example, with three different agent classes defined by three different rule sets. A phased introduction of the agents into the work space was also used in this example to simplify the coordination of the assembly process, (e.g., rim building agents were not introduced until the spoke builders had completed their task).

It is important to note that the number of building agents taking part in the assembly process does not effect the final geometric shape (morphology) of the structure. In general, the build time, or number of iterations, is greater when fewer agents are working. Conversely, the more agents present, the quicker the build time.

Hardware Demonstrations

Several hardware demonstrations of the assembly procedures are underway at the present time. In order to minimize hardware requirements and complications, a global vision system is being employed. This system utilizes an object detection algorithm to examine the entire work volume. The global object location information is transformed into the local sensor lattice of each agent, followed by pattern matching logic to assist in the all important task of deciding whether or not a rule has been satisfied. This global to local transformation of the configuration state of the structure is illustrated in Figure 8.

Summary

Construction of future large space structures will potentially be performed primarily by robotic means; however, many aspects of required robotic systems have yet to be developed. The research presented herein focuses upon the development of an autonomous control architecture in which individual robotic assemblers cooperatively construct the target structure. This form of decentralized coordination based upon biologically inspired principles appears promising, as was demonstrated by the simulations.

The control strategy employed by the individual robots utilizes short range sensing in conjunction with a relatively simple set of construction rules to govern the building process. Direct communication between the robotic building agents is minimal, or even non-existent. Coordination of the building process is an inherent property of the system since the construction rules are activated by local sensing of structural configuration. An advantage of this type of cooperative behavior is that massive redundancy is automatically built in to the system. Failure of a particular individual building agent (robot) will only slow the building process, not stop it. The required computer processing power of the individual agents is reduced by the fact that the overall design, or blueprint, is not maintained, or even understood, by the agent. The stigmergic building algorithm produces the final design as an emergent system property.

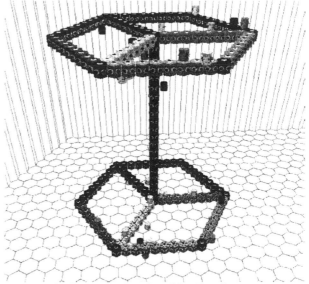

Figure 6. Structure Produced by Non-homogeneous Population of Agents.

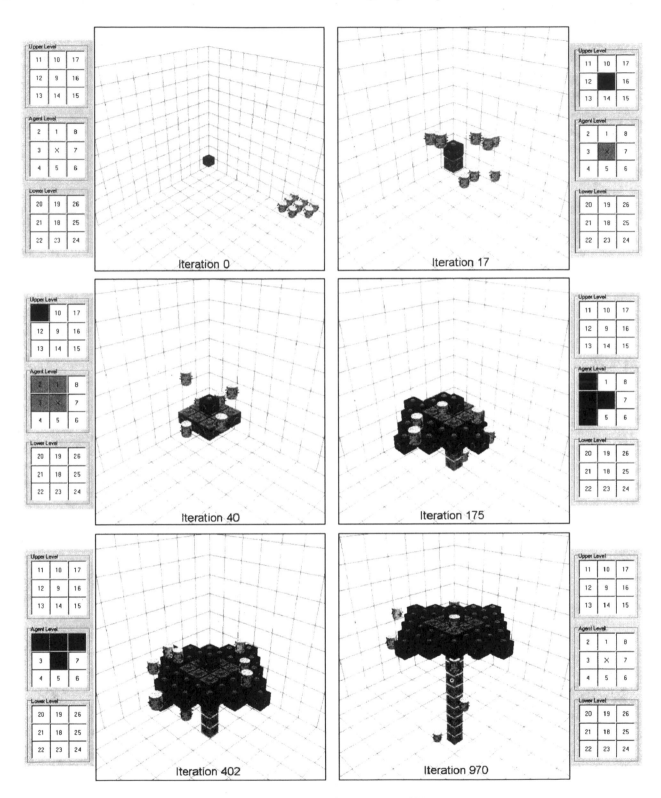

Figure 7. Structural Assembly Process of SSP Prototype.

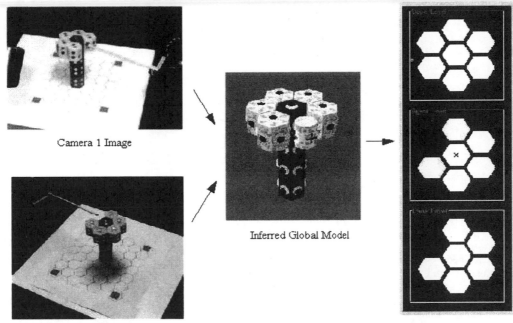

Camera 1 Image

Camera 2 Image

Inferred Global Model

Agent Local Senor Model

Figure 8. Global to Local Object Transformation Used in Hardware Studies.

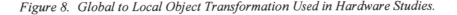

Acknowledgement

The authors would like to thank NASA for sponsoring this research.

References

Beckers R., Holland O. E., and Deneubourg J.-L., "From Local Actions to Global Tasks: Stigmergy and Collective Robotics". In Brooks R. and Maes P., editors, Proc. of the Fourth Workshop on Artificial Life, Boston, MA, July, 1994, The MIT Press, pp. 181-189.

Lake, Mark S., "Launching a 25-Meter Space Telescope: Are Astronauts a Key to the Next Technically Logical Step After NGST?", IEEE Aerospace Conference, Big Sky, MT., March 10, 2001, Paper No. 460.

Mason, Z., "Programming with Stigmergy: Using Swarms for Construction," Artifcial Life VIII, Standish, Abbass, Bedau (eds), MIT Press, 2002, pp 371-374.

Mataric, M.J., "Issues and Approaches in the Design of Collective Autonomous Agents," Robotics and Autonomous Systems, v. 16, 1995, p. 321-331.

Theraulaz, G. And Bonabeau, E., "Modeling the Collective Building of Complex Architectures in Social Insects with Lattice Swarms," Journal of Theoretical Biology, v. 177, 1995, p. 381.

Valckenaers, P., Kollingbaum, M., Van Brussel, H., Bochmann, O., Zamfirescu, C., "The Design of Multi-Agent Coordination and Control Systems using Stigmergy." Proceedings of the IWES'01 Conference, Bled, Slovenia, March 12th - 13th, 2001.

Crawling Out of the Simulation: Evolving Real Robot Morphologies Using Cheap, Reusable Modules

Ian Macinnes[1] and Ezequiel Di Paolo[2]

Centre for Computational Neuroscience and Robotics, University of Sussex, Brighton, U.K.

[1]I.A.Macinnes@sussex.ac.uk [2]ezequiel@sussex.ac.uk

Abstract

A current issue in evolutionary robotics involves the co-evolution of robot controllers and body morphologies built from modular parts. As part of ongoing research, a model for the evolution of the morphologies and neural network controllers of robots is described. Several robots are evolved for locomotion in simulation built from modules representing cheap, preexisting parts and one is physically built that has comparable behaviour with its original simulated version. The behaviour in simulation of such example robots is described. A brief comparison is made between the behaviour of a simulated robot whose design and behaviour has been evolved and its physically instantiated counterpart.

Introduction

A current line of research in evolutionary robotics involves the co-evolution of robot controllers and body morphologies. Much of the existing research, going back to the work of Sims (Sims, 1994b; Sims, 1994a), is either concerned with exploring arbitrary evolved structures in simulation (Komosinski and Ulatowski, 1999) or evolving controllers for a given complex morphology (Ijspeert, 2001), or building self-repairing modular robots that alter their morphology during the lifetime of the agent within constraints (Murata et al., 2001), or exploring appropriate generative mappings to develop body structures out of genotypic data, some of which are then tested in the real world (Pollack et al., 2001; Hornby, 2003; Funes, 2001). The different (but related) research issues of these lines involve questions of modular design, evolvability, and the relation between embodied structures and behavioural performance.

The present paper is part of a research direction within this area with two central objectives: the development and understanding of more complex behavioural capabilities beyond basic movement and locomotion using co-evolved robot bodies and controllers, and the achievement of this first aim in the real world by means of cheap and re-usable components. The long term aim is to explore the feasibility of machines capable of engaging in simple tasks of self-repair and partial self-assembly for which the two previous objectives are prerequisites.

We have started exploring the first objective using physical simulations and involving tasks such as orientation, positive and negative taxis, and pushing (Macinnes, 2003). Our purpose in this paper concerns now the second objective: the development of a feasible evolutionary scheme capable of handling evolved structures made out of cheap, re-usable real-world components using a standardised encoding itself capable of re-use in a variety of contexts. Such components cannot be subject to arbitrary changes (e.g., elongations, continuous displacements) and will constrain the evolutionary search by their fixed properties and discrete spatial arrangements.

The idea is to find an appropriate combination of the power of evolutionary search and well-established engineering principles such as standardisation of modules, re-usability and low part and assembly costs. We return to simpler behaviours like locomotion because our emphasis is on solving the issues involved in encoding components using a genotype that contains information about their physical properties and potentiality of assembly. This genotype must also be evolvable under the constraints of using modules with fixed properties and must generate a convenient plan for the resulting structures so that they can be assembled and tested in the real world.

This work represents a change in emphasis from previous work in evolvable morphology for various reasons. Firstly, the use of a flexible library of pre-existing components from which the robot is to be constructed. This describes both the physical properties and other constraints such as the maximum number of modules. An advantage of this is that we can easily restrict the robot to evolve from and utilise the parts we have currently available. It also allows specification of a minimal subset of essential parts that the robot must contain to be viable, such as a controller block and portable power source. Another advantage is that parts can be added or removed from the library according to changing availability, or the set of components can be replaced to produce a completely different kind of robot.

Secondly, the physical properties of the actuators used are modelled as part of the dynamics of the robot, along with the other essential parts such as the controller block that are necessary to build a self-contained autonomous robot. We hope that this will allow the evolutionary process to take advantage of the physical dynamics of all its constituent parts and hence build more adapted robots.

Thirdly, sensory feedback from the actuators are fed into the controller, creating a closed sensorimotor loop and allowing the robot to modulate its behaviour. This is preliminary to adding more sophisticated sensors such as cameras that has already been done in simulation to direct and modulate behaviour (Macinnes, 2003).

In summary, the morphologies and controlling neural networks of a population of thirty robots are co-evolved over time via the use of an evolutionary algorithm. This is performed in simulation by a cluster of PC's. An example robot is then physically instantiated using the design provided by its genotype. The following sections describe details regarding the real robots, the simulated robots, the neural network controller, the genotype and evolutionary algorithm, and an example of an instantiated robot.

Real Robots

We define a robot to consist of a controller block (Figure 1), a variable number of Lego bricks, a variable number of servomotors (Figure 2), and a servo-driver block for each servomotor. The on-board controller is a Motorola 68332 microprocessor running at 25MHz. The servomotors have been adjusted to provide the position of the disk back into the controller. The controller runs a continuous time recurrent neural network (CTRNN) that has a genetically determined structure. The feedback from the servomotors is fed as input into the network. The output generated by the network is used to specify the power supplied to the servomotors. The control loop for the robot is:

1. Feed the disk positions of the servomotors into the neural network.

2. Perform an iteration of the neural network.

3. Set the positions of the servomotors as specified by the neural network.

Simulated Robots

The physical parts from which the real robots are constructed need to be simulated by a physics engine[1] so we can have an estimation of how the robot would behave if it were physically constructed. The approach taken is to describe the shape and other physical properties of preexisting parts such as servomotors together with the various ways they can be connected together. This information is stored

---

[1]The Open Dynamics Engine (ODE) is used, available from http://opende.sourceforge.net

Figure 1: The controller has Lego parts glued to it allowing it to be attached to other parts.

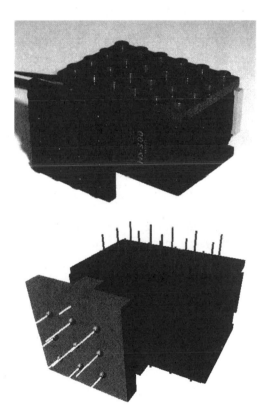

Figure 2: The servomotors have Lego parts glued onto them which allow them to be connected to other parts. They are simulated by approximating their shape with blocks. The anchor point positions are indicated by spheres and the anchor point directions are indicated by lines. It takes four Lego studs to make a connection with another block as can be seen by comparing the above servomotor with its simulated representation below it.

in the *parts library*. An XML schema was developed to facilitate this[2]. The shape and other physical attributes of each part was approximated by reducing it into a series of blocks. Each block has two principal properties, the block dimensions and density. The possible variations for these values are defined in the parts library.

Each block can have associated an unlimited number of possible *anchor points*. An anchor point describes a possible location for a connection with another block. When two blocks are to be connected, an anchor point is chosen on each and the blocks are moved and rotated so that both anchor points occupy the same position and opposing directions. The anchor point properties used for this are:

1. Coordinates specifying a position, usually lying on the edge of the block although it need not be within the body of the block at all. It may be relative either to the centre of the block or to the block edge.

2. A vector specifying the direction of the anchor point allowing the relative orientation of each block to be specified in two dimensions.

3. A set of angles specifying the rotations of two connected blocks, allowing the relative orientation in a third dimension.

Neural network controller

The controller used is a continuous time recurrent neural network (CTRNN) (Beer, 1996). This is co-evolved in conjunction with the morphology - that is, over time, the neural network and physical morphology co-evolve together as a single integrated entity. This means that the controller and the robot with be closely coupled together as a single dynamic system, with feedback going both ways between the controller and the servomotors. The potential in each neuron is given by:

$$\tau_i \frac{dy_i}{dt} = -y_i + \sum_j w_{ji} z_j \qquad (1)$$

where τ_i is the time constant of neuron i, y_i is its potential, and w_{ji} is the weight from i to j. The activation of the neuron z_j is given by:

$$z_j = tanh(b_i - y_i) \qquad (2)$$

where b_i is the genetically determined bias value for neuron **i**.

At each time-step, the input neurons of the controller are set to the positions of the servomotors, which are normalised to the range -1..1. An iteration of the neural network is performed. Then output neurons are normalised to a range 0..255, and are used to set the desired position of the servomotors.

---

[2]Both XML schema and library in XML format are available from *http://www.cogs.susx.ac.uk/users/ianma/alife9*

The Genotype

Each robot has a genotype describing how it is constructed. The genotypes that construct the most successful robots are more likely to be selected to have offspring in the next generation. Rank selection is used to pick genotypes.

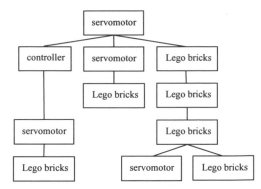

Figure 3: The genotype is constructed as a tree structure where each node represents a modular part and each branch represents a joint.

A direct genotype-to-phenotype mapping is used. For each attribute in the genotype there is a corresponding attribute in the robot. The parts library describes how each block can vary. A block's dimensions and density may be limited to a set of discrete values or be constrained within a range. Offspring are copies of their parents with numerical values altered and/or small structural changes (Tables 2 and 3). Recombination is not used so offspring have only a single parent.

Block	Anchor point	Neural network
dimensions	position	neuron bias
density	axis direction	neuron time constant
		neuron weight

Table 1: There are various numerical attributes within the genotype that can mutate during the production of offspring. There are also a number of structural mutations such as the random addition or deletion of neurons or modular parts such as servomotors or Lego bricks.

The Trial

The fitness of a robot is provided by the minimum distance moved by any of its constituent blocks during a period of one minute. The measuring period starts after thirty seconds to avoid the strategy of the robot growing top heavy and falling over, hence unfairly gaining a respectable fitness (Sims, 1994b; Macinnes, 2003).

Minimal simulations is used to facilitate the transfer to reality (Jakobi, 1998). It requires that the block dimensions

Neural network mutation	probability of occurring
discrete numerical mutation	0.5
adding a random weight	0.1
randomly removing a weight	0.1
moving a weight's endpoints	0.1
adding a random neuron	0.1
randomly removing a neuron	0.1

Table 2: Different mutations can occur to the neural network controller with various probabilities during reproduction of the genotype.

Morphology mutation	probability of occurring
inserting a random part	0.05
adding a random part	0.05
randomly removing a part	0.05
dimension or density	0.3
joint moving	0.3
change in feedback sensor	0.3

Table 3: The morphology can mutate in various ways. Offspring are tested to make sure they can be properly constructed, for example that no block occupies the same space. If their offspring is not viable, its genotype is discarded and another offspring is produced. This is repeated until a viable offspring is generated.

and densities specified by the genotype are subject to noise when the robot is constructed in simulation. The parts library specifies the magnitude of the noise for each block and it often in the order of plus or minus ten percent. Therefore the same genotype will construct differing robots for each trial. This is performed to help cope with inaccuracy in the simulation and to help produce more robust robots. It should be noted then that we do not attempt to simulate the robot as it will exist in reality, but rather a noisy artifact that we will use to guide the evolution of the population of genotypes.

Two robots are created using the genotype per trial. The fitness assigned to the genotype is the average of the fitness of both robots.

Results

Several robots were evolved in simulation (Figures 4 and 5) and one was transfered to reality. A robot (Figure 6) was built from a genotype in generation 2149. It evolved a locomotion strategy of actively using two of its four servomotors, the other two being locked into fixed positions[3]. It uses servomotor c to extend limb a, resulting in a redistribution of its weight causing it to lean forward. Servomotor b has an edge on the ground. It turns, pushing the robot forward. The

[3] Quicktime movies of this robot moving may be downloaded from http://www.cogs.susx.ac.uk/ianma/alife9.html

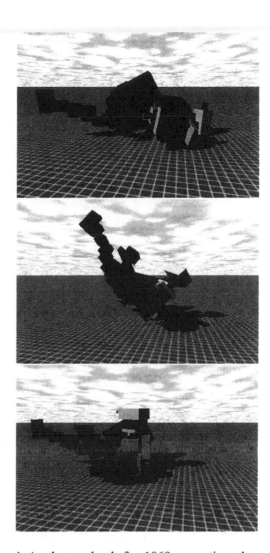

Figure 4: A robot evolved after 1869 generations that moved by throwing its two limbs up in the air and across. One of its limbs developed a larger moment of inertia by increasing its length and mass. It used its servomotors to fling both limbs so the body tilted forward. The smaller limb changed its angle so the robot fell in a different position from when it started. The robot then fell forward and by the end of the movement cycle had changed its position.

robot then draws in limb b, shifting the weight causing the robot to lean back. The servomotors other edge is now on the ground and it turns in the other direction, again pushing the robot forward.

This strategy takes advantage of the movement of servomotor b when it turns in either direction. It depends on coordinating the movement of servomotors b and c, varying its centre of mass and continuously redistributing its weight.

There are obvious differences in the behaviour of the simulated and real robot. The motions of the real robot are far more jerky that than the simulated robot. Although the

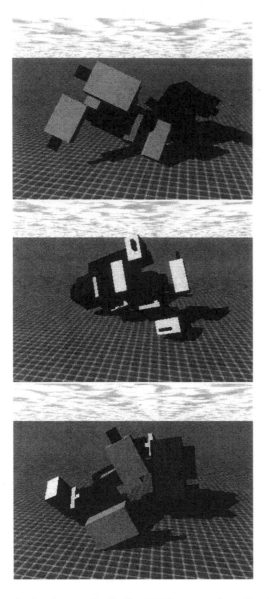

Figure 5: A robot evolved after 2166 generations that tumbles and rolls along. The robot starts perched on its servomotors. It moves its controller up above, which unbalances it, causing it to fall over onto its back. Then it twists the servo connecting the controller and tumbles over, moving its position along.

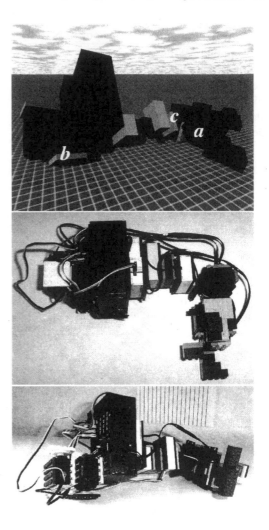

Figure 6: The locomotion strategy of this robot depends closely on its close coupling between its morphology and controller. Above is the simulated robot and below the view of its real counterpart from above and the front. The real robot's behaviour is comparable with its original simulation.

strategy used is clearly the same, the simulated robot travelled approximately 55cm per minute whereas the real robot moved approximately 14cm per minute. One reason was found to be because updating the neural network in the simulation was always assumed to be done within a single time-step of 0.05 seconds. The microprocessor in the real robot's controller was set to use this as a minimum period but took longer than this to perform a single update of the neural network. As a result, it took one minute forty seconds for the real robot to perform the same number of updates. For the measuring period of one minute, there are (60 / 0.05 seconds) 1200 discrete neural network updates. The real robot moved a distance of 23cm using a measuring period defined by the number of updates rather than time. It is expected that the time taken for the real robot to perform a neural network update will vary depending upon the number of neurons and speed of the controller. This may result in it extending beyond any fixed update time used in the simulated robot. We conclude that the neural network update time should be subject to *minimal simulation* noise to result in a better reality transfer.

Conclusions

It is possible to evolve the morphologies and controllers of simulated robots to perform locomotion and successfully transfer them to reality using cheap, reusable, and modular parts. The model shown does not use continuous values within the physical morphology section of the genotype but discrete values from a predefined set of preexisting components. The resulting robots successfully employ a variety of strategies that depend on the close coupling of their physical body and their controller. This is facilitated via careful use of *minimal simulations* (Jakobi, 1998), which requires considerable inter-trial noise.

The next step is reducing the difference between the simulated and real robots, and to produce the more complicated sequential behaviour that has already been produced in robots with evolved morphologies made from genotypes with continuous values (Macinnes, 2003). It is probable that an indirect genotype-to-phenotype mapping would improve the evolvability of the methodology. More complicated mappings have previously been explored (Hornby and Pollack, 2001a; Hornby and Pollack, 2001b) and would seem to be a fruitful area of study combined with the technology described here.

Acknowledgements

We'd like to thank Bill Bigge of the Autonomous Systems Laboratory at the University of Sussex for help at various times on the hardware of the robots.

References

Beer, R. (1996). Towards the evolution of dynamical neural networks for minimally cognitive behaviour. In *From animals to animats 4: Proceedings of the Fourth International Conference on Simulation of Adaptive Behaviour*, pages 421–429. MIT Press.

Funes, P. (2001). *Evolution of Complexity in Real-World Domains*. PhD thesis, Brandeis University, Waltham, Massachusetts, USA.

Hornby, G. (2003). *Generative Representations for Evolutionary Design Automation*. PhD thesis, Brandeis University, Dept. of Computer Science, Boston, MA, USA.

Hornby, G. and Pollack, J. (2001a). The advantages of generative grammatical encodings for physical design. In *Proceedings of the 2001 Congress on Evolutionary Computation CEC2001*, pages 600–607, COEX, World Trade Center, 159 Samseong-dong, Gangnam-gu, Seoul, Korea. IEEE Press.

Hornby, G. and Pollack, J. (2001b). Body-brain co-evolution using L-systems as a generative encoding. In Spector, L., Goodman, E. D., Wu, A., Langdon, W. B., Voigt, H.-M., Gen, M., Sen, S., Dorigo, M., Pezeshk, S., Garzon, M. H., and Burke, E., editors, *Proceedings of the Genetic and Evolutionary Computation Conference (GECCO-2001)*, pages 868–875, San Francisco, California, USA. Morgan Kaufmann.

Ijspeert, A. (2001). A connectionist central pattern generator for the aquatic and terrestrial gaits of a simulated salamander. *Biological Cybernetics*, 85(5):331–348.

Jakobi, N. (1998). *Minimal Simulations for Evolutionary Robotics*. PhD thesis, University of Sussex.

Komosinski, M. and Ulatowski, S. (1999). Framsticks: Towards a simulation of a nature-like world, creatures and evolutions. In Floreano, D., Nicoud, J., and Mondada, F., editors, *Advances in Artificial Life - Proceedings of the 5th European Conference on Artificial Life (ECAL)*, pages 261–265. Springer-Verlag.

Macinnes, I. (2003). Visually guided physically simulated agents with evolved morphologies. In W.Banzhaf, T.Christaller, P.Dittrich, J.T.Kim, and J.Ziegler, editors, *Advances in Artificial Life - Proceedings of the 7th European Conference on Artificial Life (ECAL)*, volume 2801 of *Lecture Notes in Artificial Intelligence*, pages 821–828. Springer Verlag Berlin, Heidelberg.

Murata, S., Yoshida, E., Kurokawa, H., Tomita, K., and Kokaji, S. (2001). Self-repairing mechanical systems. *Autonomous Robots*, 10(1):7–21.

Pollack, J., Lipson, H., and Funes, P. (2001). Three generations of automatically designed robots. *Artificial Life*, 7(3):215–233.

Sims, K. (1994a). Evolving 3D morphology and behaviour by competition. In Brooks, R. and Maes, P., editors, *Artificial Life IV Proceedings*, pages 28–39, MIT, Cambridge, MA, USA. MIT Press.

Sims, K. (1994b). Evolving virtual creatures. In *Computer Graphics, Annual Conference Series, (SIGGRAPH 1994 Proceedings)*, pages 15–22.

Functional Freeform Fabrication for Physical Artificial Life

Evan Malone, Hod Lipson

Mechanical and Aerospace Engineering, Cornell University
evan.malone@cornell.edu

Abstract

Solid freeform fabrication (SFF) allows 3D-printing of arbitrarily shaped structures, directly from computer-aided design (CAD) data. SFF has traditionally focused on printing passive mechanical parts. Advances in this technology and developments in materials science make it feasible to begin the development of a single, compact, robotic SFF system – including a small set of materials - which can produce complete, active, functional electromechanical devices - mobile robots, for instance. We are advancing steadily toward this goal, and successes thus far have included the freeform fabrication of zinc-air batteries, conductive wiring, flexure joints, and combinations of these with thermoplastic structures. Several essential functionalities – actuation, sensing, and control electronics - still remain to be realized before complete electromechanical systems can be produced via SFF. Conducting polymers (CP) are a class of materials which can be used to produce all of these. Several SFF-compatible CP processing methods have been identified, and actuators produced via one of these have been demonstrated. When coupled in a closed-loop with an evolutionary design system, the ability to produce robots entirely via SFF becomes a bridge between the physical and the simulated, giving artificial evolution a complete physical substrate of enormous richness to explore with little or no human involvement.

Introduction

Solid freeform fabrication (SFF) is the name given to a class of manufacturing methods which allow the fabrication of three-dimensional structures directly from computer-aided design (CAD) data. SFF processes are generally additive, in that material is selectively deposited to construct the part, rather than removed from a block or billet. Most SFF processes are also layered, meaning that a geometrical description of the part to be produced is cut by a set of parallel surfaces (planar or curved) and the intersections of the part and each surface – referred to as slices or layers – are fabricated sequentially. Together, these two properties mean that SFF processes are subject to very different constraints than traditional material removal-based manufacturing. Nearly arbitrary part geometries are achievable, no tooling is required, mating parts and fully assembled mechanisms can be fabricated in a single step, and multiple materials can be combined, allowing functionally graded material properties. New features, parts, and even assembled components can be "grown" directly on already completed objects, suggesting the possibility of using SFF for the *repair and physical adaptation of hardware!* On the other hand, a deposition process must be developed and tuned for each material, geometry is limited by the ability of the deposited material to support itself and by the (often poor) resolution and accuracy of the process, and multiple material and process interactions must be understood. SFF has traditionally focused on printing passive mechanical parts in a single material, and the emphasis of research has been on improving the quality, resolution, and surface finish of parts, and on broadening the range of useable materials. Especially the latter effort has met with remarkable success: the range of materials amenable to freeform fabrication has been greatly expanded from original soft plastics to engineering polymers, metals (Mazumder et al. 1997, Rabinovitch 2003), ceramics (IBID, Wang and Krstic 1998), electronics (Church et al. 2000), and even biological tissues (Sun and Lal 2002).

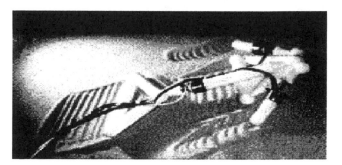

Figure 1. Evolved GOLEM robot with freeform fabricated structure

It has not escaped the notice of some that the diversity of materials and SFF processes is nearing the point at which almost any conventionally manufactured part could be made entirely via SFF processes, and that a synthesis of compatible processes may permit the fabrication of entire functional assemblies (Weiss and Prinz 1998; Safari et al. 2000) and even complete functional systems. We have pioneered the application of SFF to research in artificial life with the GOLEM robots (Figure 1) (Lipson et al. 2000), which are evolved in simulation, then constructed through a combination of freeform fabrication (of structure and articulation) and manual assembly (of actuators, control and power sources). We are now actively pursuing the development of a "basis set" or library of functionalities which are essentially geometry-independent, accurately modeled, and mutually compatible. Such a set can in principle be used to build arbitrary electromechanical systems. One possible set would be structure, articulation, actuation, sensing, power source, logic/control, and signal/power interconnection. In pursuit of this goal, we have designed and built a multi-material-SFF research platform, and successfully demonstrated the freeform fabrication of zinc-air batteries, conductive wiring, elastomer flexure joints, and combinations of these with thermoplastic structures (Malone et al. 2003) – four of the seven functionalities on the list. The next phase of research - adapting a new class of materials, called conducting polymers (CP), to solid freeform fabrication - has begun with the identification of several SFF-compatible processing methods, and the demonstration of actuators produced via one of these.

Fabrication Platform

In order to provide maximum freedom for experimentation with a wide variety of materials and processes, a custom robotic platform, control software, and two material deposition tools have been designed and constructed.

Positioning System

The initial requirement employed in the design of these tools can be stated briefly as follows: the motion control platform should provide maximal parametric freedom to the deposition processes, and permit the sequential use of many deposition processes in the course of fabricating a given object.

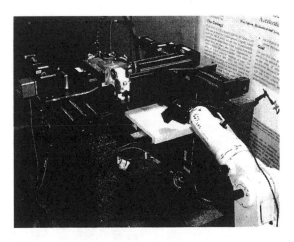

Figure 2. SFF Positioning System

Positioning is limited to three Cartesian axes, and an emphasis is placed on velocity regulation, path-following and positioning accuracy, and high acceleration to achieve fine features while printing at a constant material feed rate. A Cartesian gantry robot configuration (Figure 2) with brushless-motor driven ballscrew stages has been selected for this application for its relatively simple

control, large payload capacity and rigidity, and positioning/path-following performance.

Software

A software application has been created to manage path planning and control. Multi-material objects are defined in computer-aided design (CAD) software as assemblies, then exported as multiple STL (stereolithography) files, each file describing a single part of the assembly. Within our SFF software, each component is graphically assigned a material. Each material is associated with a deposition tool and material properties governing layer thickness, deposition width and deposition rates. Geometry slicing and path generation algorithms construct unique perimeter contour- and fill raster-paths based on the tools' and materials' parameters, and combine the layers into a fabrication sequence with increasing height. Special care is needed to prioritize layers of similar heights according to interaction among materials.

Deposition Tools

For the experiments conducted thus far, two separate material deposition tools have been designed and built—one for plastics and metals, and another for liquid chemicals and pastes.

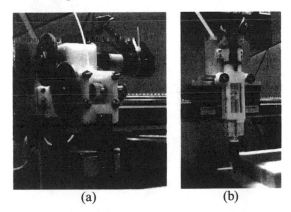

(a)   (b)

Figure 3. (a) Wire-fed tool, (b) syringe extruder

The former (Figure 3a) is an extruder that feeds material in wire-form (0.050" – 0.070" diameter) through an actively air-cooled metal guide-tube and into a heated liquefier block containing a nozzle (Swanson et al. 1999). The tool has been successfully employed with solid core Pb-Sn solder wire, as well as ABS (acrylonitrile-butadiene-styrene) thermoplastic wire. A separate tool is used to deposit liquids and chemical pastes (Figure 3b). This tool accepts standard commercial 10cc Luer-lock syringe barrels and plungers. The plunger position is actuated by a linear stepper-motor capable of exerting 50 lbf. Materials are changed by substituting a different syringe barrel and plunger in the plunger-driver component of the tool.

Freeform Fabrication Experiments

Zn-Air Battery

The fabrication of a functional battery was selected as a comprehensive test of the multiple material capabilities of the previously described platform.

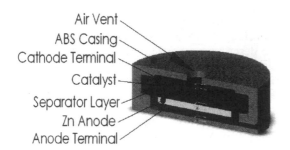

Figure 4. Cut-away model of freeform Zn-air cell

There is little literature on the production of freeform, three-dimensional energy storage devices, though a planar thin-film cell has recently been demonstrated by Power Paper Ltd. (2003) for low-power applications.

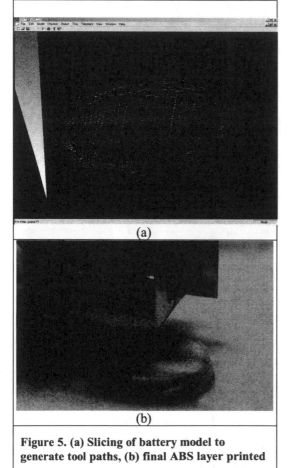

Figure 5. (a) Slicing of battery model to generate tool paths, (b) final ABS layer printed

For the purpose of these experiments, a cell is defined as a device that converts the chemical potential energy between its anode and cathode materials into electrical energy by

means of redox reactions: reduction (electron gain) at the cathode, oxidation (electron loss) at the anode.

A battery comprises one or more connected cells. The essential components of a Zn-air cell are shown in Figure 4. Zinc-air cell chemistry was the first considered because of its simplicity and high energy density. Several series of experiments were performed to establish the effects on cell performance of electrolyte concentration, current collector composition and geometry, cell structure and construction, and cathode catalyst composition. After the initial tests were conducted, and the fundamental chemical and geometrical characteristics understood, the research shifted to adapting materials and a battery design to solid-freeform fabrication processes (Malone et al. 2003).

The first successfully freeform fabricated cells were unenclosed stacks of the active materials. Later experiments focused on achieving encased cells. Figure 4 shows a cross-sectional view of a successfully fabricated, encased cell design. The internal structure is very similar to the unenclosed cells. The main functional difference in the designs is that the case limits air diffusion into the cell, reducing the peak power output (~2.5mW for several hours), but prolonging the life. Figure 5 highlights some key steps in the successful freeform fabrication of this encased cell design. Producing these cells involves deposition of 5 separate materials. Polymeric materials, e.g. poly(vinyl alcohol), are being investigated as a means of producing a thinner separator with lower ionic resistance, and also as a means of stabilizing the electrolyte to control evaporation and flow. These improved materials will allow exploration of less conventional cell geometries, and should improve performance as well.

Elastomer Flexure Joints

As a demonstration of a multi-material functional mechanical assembly, a flexure joint was fabricated using ABS as the rigid end members and a 1-part, room-temperature vulcanizing (RTV) silicone as the flexible connection (Figure 6).

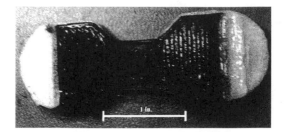

Figure 6. ABS and silicone flexure joint

The silicone is filled with carbon black to make it freestanding upon extrusion. The combination of electrical and mechanical functionality within the same freeform fabricated part is highly desirable and a number of approaches have been investigated (Ting et al. 2001, Safari et al. 2000). Methylcellulose (MC) gels loaded with powdered metals are used as current collectors in our batteries. Qualitative experiments with applications of these materials outside of batteries have revealed that dehydration of the MC gel can lead to shrinkage and cracking, and that adhesion to substrates also suffers. A rudimentary electromechanical assembly was freeform fabricated as lines of silver / MC paste embedded in an ABS and silicone flexure joint. This device successfully carried sufficient current to light an LED (~10mA), but was too delicate to survive much mechanical use due to cracking and detachment of the conductive paste.

Conductive Wiring

Solder alloys are being investigated as a means of depositing wiring into components, and fabricating metal parts (Priest et al. 1997), and at least one commercial process exists which is capable of depositing solder alloy wiring onto a wide variety of substrates (Hayes et al. 1998). We have experimented with using a solid-core, Pb-Sn solder wire was used as the feedstock for the wire-fed extrusion tool.

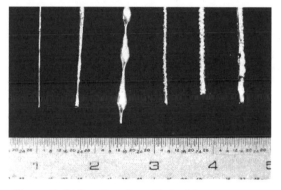

Figure 7. PbSn alloy deposited wiring

Figure 7 depicts the results of these tests. It was found that there are small, separate regions of the deposition control parameter space in which it is possible to form lines of overlapping frozen droplets of approximately 2-4mm in diameter (Figure 7, right), or a very thin but continuous wire of approximately 250µm-1mm in diameter (Figure 7, left). Ink-jet deposition of molten solder will be investigated as well.

Conducting Polymers

Conducting polymers are long chain organic molecules in which successive carbon atoms along the axis of the chain are bound alternately by one shared electron pair (single, or σ-bond) and by two shared electron pairs (double, or π-bond). Like most organic polymers, CP's are normally poor electrical conductors. By adding or removing electrons from the CP chains, and embedding positive or negative ions, respectively, into the chains to maintain electrical neutrality, it is possible to make the locations of the π-bonds less well defined (MacDiarmid 2002). Classically speaking, this allows some of the electrons

more freedom to move from atom to atom, increasing electrical conductivity. Through this process known as redox doping, the conductivity of CP's can be varied continuously from insulating to semi-conducting to metallic. This allows the fabrication in CP of many of the electronic devices currently made with doped silicon, including photovoltaic cells (Arango et al. 2000), chemical sensing (Bhat et al. 2003), optical sensing (Pede et al. 1998), force sensing (Spinks et al. 2003), transistors / redox conductivity switching (Gelink et al. 2000), organic LEDs and displays (Smela et al. 1998), and passive electronics including capacitors, resistors and inductors (Murphy et al. 1999).

Beyond these essentially electronic applications, CP's have been discovered to function as mechanical actuators, with actuation performance comparable to that of human skeletal muscle (Madden et al. 2001). If the CP is placed in contact with an electrolyte and an electrical potential is applied between the polymer and another electrode immersed in the electrolyte, free ions (and attached solvent molecules) from the electrolyte diffuse into or out from between the tangled polymer chains to compensate for the changed charge of the polymer, thereby causing the overall volume of the polymer to change (Bar Cohen 2001). Actuator applications of conducting polymers to date have been restricted primarily to planar (Madden et al. 1999, Madden et al. 2001, Sansinena et al. 1997, Spinks et al. 2003, etc.) and tubular (Ding et al. 2003, Hutohioon et al. 2000) thin-film devices made from polypyrrole (PPy) and polyaniline (PA), which seem to demonstrate the best combinations of mechanical strength, actuation stress, and actuation strain. The majority of CP research has involved liquid electrolytes, because liquids allow high ion mobility, hence faster actuation, but a few air-operable CP actuator designs have been investigated (Hutchison et al. 2000, Madden et al. 1999, Sansinena et al. 1997). These provide a good starting point for the selection or development of solid electrolytes or encapsulation methods.

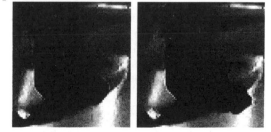

Figure 8. Actuation of solution-cast P3OT; elapsed time 5s

Unfortunately, one method of producing CP films dominates actuator-related CP research: electrochemical synthesis, which is a laboratory chemistry process and not readily adaptable to solid freeform fabrication. Our investigations thus far have identified some promising approaches to freeform deposition of CP. Deposition of solubilized forms of doped CP, via ink-jet or syringe for instance, followed by evaporation of solvent, is one promising approach. This solution casting method has been demonstrated to produce films with reasonably good mechanical properties in at least two conducting polymers (poly(3-octylthiophene) (P3OT), and polyaniline (PANI)) (Dahman 1999), and actuation of solution-cast P3OT in liquid electrolyte has been reported by Chen (Chen and Inganas 1995). We have cast films of P3OT using the methods of Dahman, and verified their actuation (Figure 8) in liquid electrolye. Work on production of air-operable, solution-cast P3OT and PANI actuators is underway. Other approaches under consideration include photo-initiated polymerization of a CP monomer or precursor (Murphy et al. 1999), and embedding of CP in a matrix of more easily processed material (Vidal et al. 2003). With a successful method of producing freeform actuators in hand, we will extend and validate existing electrochemical and electromechanical models (Madden et al. 2001) to more complex and non-planar actuator designs.

Beyond actuators, the adaptation of conducting polymers to solid freeform fabrication opens up the possibility of tapping the rapidly expanding field of polymer electronics for high-performance circuit, sensor and optoelectronic device designs. This should permit rapid completion of the library of freeform functionalities that we are seeking. It will be possible to integrate these with actuation, structure and power sources to create entirely new classes of devices of great commercial, as well as academic interest.

Summary

A traditional approach to evolutionary robotics requires robots to be manually assembled from commercial components or custom-made components which are available in a limited range of geometries and performance levels. The ability to produce complete electromechanical systems via freeform fabrication renders the space of achievable physical robots essentially continuous. As a physical substrate in which to explore artificial evolution, freeform electromechanical systems may offer a far less rugged fitness landscape at far lower cost and effort than manually assembled or even reconfigurable robots. Building upon the successful work with the GOLEM robots, we intend to couple this technology to an evolutionary design system in a closed loop. Simulated evolution will incorporate physically accurate models of freeform functionalities. Fitness evaluations will periodically take place with physically realized individuals to refine simulation accuracy. This will permit a radical acceleration in the evolution of physical artificial systems, and will simultaneously allow a drastic reduction in the human biases contained in these physical realizations.

Acknowledgments

This work was supported in parts by the U.S. Department of Energy, grant DE-FG02-01ER45902.

References

Arango, A.C., Johnson, L.R., Bliznyuk, V.N., Schlesinger, Z., Hörhold, H., and Carter, S. A., 2000, "Efficient titanium oxide/conjugated polymer photovoltaics for solar energy conversion." *Advanced Materials* **12**(22): p.1689-1692.

Bar Cohen Y., 2001, "Electroactive polymer actuators as artificial muscles" Washington: SPIE Press.

Bhat, N. V., Gadre, A. P., and Bambole, V. A., 2003, "Investigation of electropolymerized polypyrrole composite film: Characterization and application to gas sensors." *Journal of Applied Polymer Science* **88**(1): p.22-29

Chen, X. and O. Inganas, 1995, "Doping-induced volume changes in poly(3-octylthiophene) solids and gels." *Synthetic Metals* **74**(2): p. 159-164.

Church, K.H. Fore, C., and Feeley, T., 2000, "Commercial applications and review for direct write technologies" Materials Research Society Symposium - Proceedings, **624**: p. 3-8

Dahman, S.J., 1999, "The Effect of Co-Dopants on the Processability of Intrinsically Conducting Polymers." *Polymer Engineering and Science* **39**(11): p. 2181-2188.

Ding, J., Liu, L., Spinks, G. M., Zhou, D. and Wallace, G. G., 2003, "High performance conducting polymer actuators utilising a tubular geometry and helical wire interconnects." *Synthetic Metals* **138**(3): p. 391-398.

Gelinck, G. H., Geuns, T. C. T., and de Leeuw, D. M., 2000, "High-performance all-polymer integrated circuits." *Applied Physics Letters* **77**(10): p. 1487-1489.

Hayes, D.J., Cox, W.R., and Grove, M.E., 1998, "Micro-jet printing of polymers and solder for electronics manufacturing." *Journal of Electronics Manufacturing* **8**(3-4): p. 209-16.

Hutchison, A.S., Lewis, T.W., Moulton, S.E., Spinks, G.M., and Wallace, G.G., 2000, "Development of polypyrrole-based electomechanical actuators." *Synthetic Metals* **113**(1-2): p. 121-127.

Lipson, H., and Pollack J. B., 2000, "Automatic Design and Manufacture of Artificial Lifeforms." *Nature* 406: p. 974-978.

MacDiarmid, A.G., 2002, "Synthetic metals: a novel role for organic polymers." *Synthetic Metals* **125**(1): p. 11-22.

Madden, J.D., Cush, R.A., Kanigan, T.S., Brenan, C.J., and Hunter, I.W., 1999, "Encapsulated polypyrrole actuators." *Synthetic Metals* **105**(1): p. 61-64.

Madden, J.D.W., Madden, P.G.A., and Hunter, I.W., 2001, "Characterization of polypyrrole actuators: modeling and performance." in Smart Structures and Materials 2001: Electroactive Polymer Actuators and Devices, Yoseph Bar-Cohen, Editor, Proceedings of SPIE **4329**: p.72-83.

Malone, E., Rasa, K., Cohen, D.L., Isaacson, T, Lashley, H., and Lipson, H., 2003, "Freeform fabrication of 3D zinc-air batteries and functional electro-mechanical assemblies." *Rapid Prototyping Journal* **10**(1): p. 58-69.

Mazumder, J., Choi, J., Nagarathnam, K., Koch, J., and Hetzner, D., 1997, "Direct metal deposition of H13 tool steel for 3-D components." *JOM* **49**(5). p. 55-60.

Murphy, O., Hitchens, G., Hodko, D., Clarke, E., Miller, D., and Parker, D., 1999, "Method of Manufacturing Passive Elements Using Conductive Polypyrrole Fomulations." United States Patent 5855755.

Pede, D., Smela, E., Johansson, T., Johanssen, M., and Inganäs, O., 1998, "A General Purpose Conjugated Polymer Array for Imaging." *Advanced Materials* **10**(3): p. 233-237.

Priest, J., Smith, C., and DuBois, P., 1997, "Liquid Metal Jetting for Printing Metal Parts", in Proceedings of the Solid Freeform Fabrication Symposium, Bourell, D. L., Beaman, J. J., Marcus, H. L., Crawford, R. H., and Barlow, J. W., Editors, Austin, TX.

Safari, A., Danforth, S.C., Jafari, M., Allahverdi, M., Jadidian, B., and Mohammadi, F., 2000, "Processing and properties of piezoelectric actuators developed by fused deposition technique", in ISAF 2000: Proceedings of the 2000 12th IEEE International Symposium on Applications of Ferroelectrics 1, p. 79-82.

Rabinovich J, 2003, "Net shape manufacturing with metal alloys." *Advanced Materials and Processes* **161**(1): p. 47-86.

Sansiñena J.M., Olazábal V., Otero T.F., da Fonseca C.N., and DePaoli, M.A., 1997, "A solid state artificial muscle based on polypyrrole and a solid polymeric electrolyte working in air." *RSC Chemical Communications* (22): p. 2217-2218.

Smela, E., Kaminorz, Y., Inganäs, O., and Brehmer, L., 1998, "Planar Microfabricated Polymer Light Emitting Diodes." *Semiconductor Science and Technology* **13**(4): p. 433-439.

Spinks, G., Wallace, G., Liu, L., and Zhou, D., 2003, "Conducting Polymers Electromechanical Actuators and Strain Sensors." *Macromolecular Symposia* **192**(1): p. 161-170.

Sun, W., and Lal, P., 2002, "Recent development on computer aided tissue engineering - A review." *Computer Methods and Programs in Biomedicine* **67**(2): p. 85-103.

Swanson, W. J., and Hopkins, P. E., 1999, "Thin-wall tube liquifier", United States Patent 6004124.

Ting, F.P.Y., Gibson, I., and Cheung, W.L., 2001, "Study on selective laser sintering components with electrically conductive channels." in Proceedings of the Solid Freeform Fabrication Symposium, Bourell, D. L., Beaman, J. J., Marcus, H. L., Crawford, R. H., and Barlow, J. W., Editors, Austin, TX.

Vidal, F., Popp, J.F., Plesse, C., Chevrot, C., and Teyssié, D., 2003, "Feasibility of conducting semi-interpenetrating networks based on a poly(ethylene oxide) network and poly(3,4-ethylenedioxythiophene) in actuator design." *Journal of Applied Polymer Science* **90**(13): p. 3569-3577.

Wang, G., and Krstic, V.D., 1998, "Rapid prototyping of ceramic components – review." *Canadian Ceramics* **67**(3): p. 52-58.

Weiss, L. and Prinz, F., 1998, "Novel Applications and Implementations of Shape Deposition Manufacturing," *Naval Research Reviews 1*, Office of Naval Research.

Yamaura, M., Higiwara, T., and Iwata, K., 1988, "Enhancement of electrical conductivity of polypyrrole film by stretching: counter ion effect", *Synthetic Metals*, **26** (3): p. 209-224.

Quadrupedal Locomotion: GasNets, CTRNNs and Hybrid CTRNN/PNNs Compared

Gary McHale[1] and Phil Husbands[2]
[1]Department of Informatics, University of Sussex, Brighton BN1 9QH, UK
[2]Centre for Computational Neuroscience and Robotics, University of Sussex, Brighton BN1 9QH, UK
cglife@hotmail.com

Abstract

Evolutionary Robotics seeks to use evolutionary techniques to create both physical and physically simulated robots capable of exhibiting characteristics commonly associated with living organisms. Typically, biologically inspired artificial neural networks are evolved to act as sensorimotor control systems. These networks include; GasNets, Continuous Time Recurrent Neural Networks (CTRNNs) and Plastic Neural Networks (PNNs). This paper seeks to compare the performance of such networks in solving the problem of locomotion in a physically simulated quadruped. The results in this paper, taken together with those of other studies (summarized in this paper) help us to assess the relative strengths and weaknesses of the these three different approaches.

Introduction

Evolutionary Robotics has two broad goals; to gain an insight into biological systems through suitable abstractions of these systems, and secondly to seek to discover techniques that are of economic value in the development of robots and physically simulated agents. The underlying methodology involves the use of Genetic Algorithms (GAs) to evolve Artificial Neural Networks (ANNs) that act as sensorimotor control systems in real and simulated robots.

There is an increasing body of work that has successfully applied these techniques to the evolution of ANNs suitable for controlling bipedal and quadrupedal locomotion. This includes earlier work (Reil and Husbands, 2002) which demonstrates that it is possible to evolve a bipedal motor control in a physically simulated agent using a conventional Dynamic Recurrent Neural Network (DRNN) without sensor input. Bongard and Paul have evolved bipedal locomotion in a physically simulated agent through genetic encoding that comprises morphological as well as ANN parameters (Bongard and Paul, 2001). Researchers have evolved bipedal locomotion in a physically simulated robot that incorporates a model of neuromodulation (Ishiguro et al., 2003). Billard and Ijspeert have been successful in evolving Quadrupedal locomotion in a real robot (Billard and Ijspeert, 2000).

Unfortunately, there are remarkably few comparative studies that enable us to judge which of the diverse approaches taken by different researchers is the most expedient when applied to a specific problem. It is hoped that this paper provides a useful insight into the relative performance of some of the more common neural network varieties. This paper is intended as a complementary study to work carried out by the authors (McHale and Husbands, 2004) which sought to identify the relative performance of 14 different network varieties applied to bipedal locomotion.

Network Descriptions

Put in the simplest terms, Continuous Time Recurrent Neural Networks (CTRNNs) (Yamauchi and Beer, 1994) represent the "plain vanilla" form of DRNN's, GasNets represent an approach to incorporate neuromodulation into a form of DRNN (Husbands et al., 1998), and Plastic Neural Networks (PNNs) seek to incorporate Hebbian dynamics (Floreano and Mondada, 1996). The particular variants used in this experiment are described in detail in this following section.

One thing that should be noted is that for each network, network morphology has been constrained to correspond more closely to that associated with coupled-oscillator circuitry. All networks comprise a total of 16 nodes or cells. An initial population is seeded with networks that have a single symmetry axis, such that we have two subnetworks of 8 nodes, each with identical parameters. In the case of GasNets (where nodes have a physical location in a 2 dimensional plane) the position of each node from one subnetwork is mirrored in an axis that divides the plane. This is shown more clearly in Figure 1. The mirrored nodes are interconnected via mutually inhibitory connections. Whilst initial populations comprise symmetrical networks, mutation and crossover results in the introduction of asymmetries over a period of time (symmetry is only enforced in the initial population).

Symmetry Axis

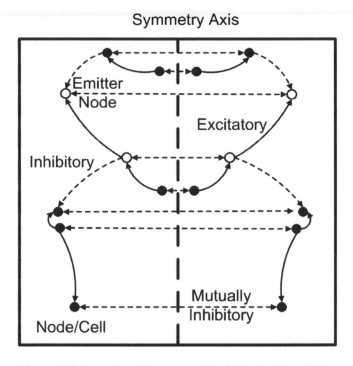

Figure 1: Schematic of the symmetrical distribution of Gas-Net nodes.

Center-Crossing CTRNNs

The characteristic equation of the conventional CTRNN (Beer, 1995) is shown below;

$$y_i^{t+1} = y_i^t + \frac{T}{\tau_i}\left(-y_i^t + \sum_{j=1}^N \omega_{ji}\sigma(y_j^t + \theta_j) + I_i\right) \qquad (1)$$

$$i = 1,2,...,N$$

Where:

$y_i t + 1$ is the activation of the i'th node at time $t + 1$.

y_i^t is the activation of the i'th node at time t.

τ_i is the time constant for the i'th node calculated according to equation 18.

I_i a sensor input to the i'th node where I is either 1 (in contact with the floor) or 0 (not in contact with the floor).

θ_j a bias term for the j'th node where $\theta \in [-2,2]$.

T is the time slice (in this case T is set to 1).

ω_{ji} is the weight of the output from the j'th node to the i'th node where $\omega \in [-4.0, 4.0]$.

σ is the logistic activation function.

$$\sigma(z) = \frac{1}{(1 + e^{-z})} \qquad (2)$$

The network is fully interconnected. Node connection weights and bias are under evolutionary control. This study uses a variant of the conventional CTRNN, referred to as the Center-Crossing CTRNN (Mathayomchan and Beer, 2002), where initial biases are calculated such that:

$$\theta_i = \frac{-\sum_{j=1}^N \omega_{ji}}{2} \qquad (3)$$

Mathayomchan and Beer suggest that populations seeded with center-crossing networks may be more likely to yield a wider range of dynamics than a population of random networks.

The incorporation of a single symmetry axis (as used in this study) results in a final form, as described by;

$$y_i^{t+1} = y_i^t + \frac{T}{\tau_i}\left(-y_i^t + \sum_{j=1}^N \omega_{ji}\sigma(y_j^t + \theta_j - z_i^t) + I_i\right) \qquad (4)$$

$$i = 1,2,...,N$$

Where:

z_i^t is the activation of the corresponding i'th node in the symmetrical subnetwork at time t.

N is the number of nodes in each subnetwork (in this case 8).

GasNets

GasNets are an example of a class of ANNs that seek to model aspects of neuromodulation. A key attribute of the GasNet model is that the transfer characteristics of network nodes are modified via the influence of diffused gases (modeled in a 2-dimensional plane). This network model is inspired by the action of Nitric Oxide in biological systems (Husbands et al., 2001). Earlier work has shown that Gas-Nets are more *evolvable* than comparable networks that do not incorporate gas modulation, in simulation and when used in real robots (Smith et al., 2003).

In GasNets, node transfer functions can be modulated by local gas concentrations in the vicinity of the node. Nodes can also act as chemical emitters, under either gas or electrical stimulation. GasNet nodes exist in a geometric plane where internode distances determine gas concentrations and (in conjunction with additional genetic parameters) network connectivity. Under typical evolutionary parameters the GasNet connectivity rules result in a sparsely connected network.

$$y_i^{t+1} = \tanh\left[k_i^t\left(\sum_{j \in C_i} \omega_{ji}\sigma(y_j^t + I_i)\right) + b_i\right] \qquad (5)$$

Where:

k_i^t is a time-varying transfer function modulator. The value of k varies with gas concentrations at the i'th node, see equation 9.

C_i is the set of all nodes that have an input to the i'th node.

I_i a sensor input to the i'th node.

b_i a bias term for the i'th node where $b_i \in [-2, 2]$.

The original GasNet diffusion model (upon which this implementation is based) is controlled by two genetically specified parameters, namely the radius of influence r and the rate of build up and decay s. Spatially, the gas concentration varies as an inverse exponential of the distance from the emitting node with a spread governed by r, with the concentration set to zero for all distances greater than r (Equation 6). The maximum concentration at the emitting node is 1.0 and the concentration builds up and decays from this value linearly as defined by Equations 7 and 8 at a rate determined by s.

$$C(d,t) = \begin{cases} e^{-2d/r} \times T(t) & d < r \\ 0 & \text{else} \end{cases} \quad (6)$$

$$T(t) = \begin{cases} H\left(\frac{t-t_e}{s}\right) & \text{emitting} \\ H\left[H\left(\frac{t_s-t_e}{s}\right) - H\left(\frac{t-t_s}{s}\right)\right] & \text{not emitting} \end{cases} \quad (7)$$

$$H(x) = \begin{cases} 0 & x \leq 0 \\ x & 0 < x < 1 \\ 1 & \text{else} \end{cases} \quad (8)$$

where C(d,t) is the concentration at a distance d from the emitting node at time t. t_e is the time at which emission was last turned on, t_s is the time at which emission was last turned off, and s (controlling the slope of the function T) is genetically determined for each node. The total concentration at a node is then determined by summing the contributions from all other emitting nodes (nodes are not affected by their own concentration, to avoid runaway positive feedback).

For mathematical convenience, in the basic GasNet there are two 'gases', one whose modulatory effect is to increase the transfer function gain parameter (k_i^t from equation 5) and one whose effect is to decrease it. It is genetically determined whether or not any given node will emit one of these two gases (gas 1 and gas 2), and under what circumstances emission will occur (either when the 'electrical' activation of the node exceeds a threshold, or the concentration of a genetically determined gas in the vicinity of the node exceeds a threshold. Note these emission processes provide a coupling between the 'electrical' and 'chemical' mechanisms). The concentration-dependent modulation is described by Equation 9, with transfer parameters updated on every time step as the network runs.

$$k_i^t = k_i^0 + \alpha C_1^t - \beta C_2^t \quad (9)$$

where k_i^0 is the genetically set default value for k_i, C_1^t and C_2^t are the concentrations of gas 1 and gas 2 respectively at node i at time t, and α and β are constants. Both gas concentrations lie in the range $[0, 1]$. Thus the gas does not alter the electrical activity in the network directly but rather acts by continuously changing the mapping between input and output for individual nodes, either directly or by stimulating the production of further virtual gas. The concentration dependent modulation can, for instance, change a node's output from being positive to being zero or negative even though the input remains constant. Any node that is exposed to a non zero gas concentration will be modulated. This set of interacting processes provides the potential for highly plastic systems with rich dynamics.

CTRNN/PNN Hybrid

One of the underlying concepts associated with Plastic Neural Networks is that there is value in evolving artificial neural networks that are capable of exhibiting learning through ontogenetic change (Floreano and Mondada, 1996). Let us first start with a description of a basic PNN (Urzelai and Floreano, 2000). A key characteristic of PNN's is that connection weights vary over time based on Hebbian learning rules given by:

$$\omega_{ji}^t = \omega_{ji}^{t-1} + \eta \Delta\omega_{ji} \quad (10)$$

Where η is a learning rate ($0.0 < \eta < 1.0$) and ω_{ji} is the connection weight of the input to node i from node j. The adaptation rule $\Delta\omega_{ji}$ is genetically determined for each node. All inputs to a given node are subject to the same adaptation rule (referred to as node encoding by the original authors).

Where x is the activation of node j, which is an input to node i (which has an output activation of y), the adaptation rule is one of:

Plain Hebb Rule

$$\Delta\omega_{ji} = (1 - \omega_{ji})x_j y_i \quad (11)$$

Post-Synamptic Rule

$$\Delta\omega_{ji} = \omega_{ji}(-1 + x_j)y_i + (1 - \omega_{ij})x_j y_i \quad (12)$$

Pre-Synaptic rule Rule

$$\Delta\omega_{ji} = \omega_{ji}x_j(-1 + y_i) + (1 - \omega_{ji})x_j y_i \quad (13)$$

Covariance Rule

$$\Delta\omega_{ji} = \begin{cases} (1 - \omega_{ji}) & \text{if } F(x_j, y_i) > 0 \\ (\omega_{ji})F(x_j, y_i) & \text{otherwise} \end{cases} \quad (14)$$

Where:

$$F(x_j, y_i) = \tanh(4(1 - |x_i - y_j| - 2)) \quad (15)$$

All nodes in the PNN are fully interconnected. The rate of learning η can only assume one of four values (0.0, 0.3,

0.6, 0.9). The characteristic equation for the PNN is shown below:

$$y_i^{t+1} = \sigma\left(\sum_{j=1}^{N} \omega_{ji}^t(y_j^t)\right) + I_i \quad i = 1, 2, \ldots, N \quad (16)$$

Where:

ω_{ji}^t is the adaptive weight for the j'th input to the i'th node.

σ is the standard logistic activation function.

I_i a sensor input to the i'th node where I is either 1 (in contact with the floor) or 0 (not in contact with the floor).

The CTRNN/PNN Hybrid used in this study is a variation on the conventional PNN, where activation signals are further modified by a node based time constant under evolutionary control (in a similar fashion to conventional CTRNNs). This model was first introduced by the authors of this paper in an attempt to create a PNN that exhibited richer frequency dynamics than those of the conventional PNN (McHale and Husbands, 2004). The range of y_i is $[0,2]$ for input neurons and $[0,1]$ for hidden and output neurons (Blynel and Floreano, 2002). It is modified again here such that the network comprises two symmetrical sub-networks, with mutual inhibition;

$$y_i^{t+1} = y_i^t + \frac{T}{\tau_i}\left(-y_i^t + \sum_{j=1}^{N} \omega_{ji}^t \sigma(y_j^t + \theta_j - z_i^t) + I_i\right) \quad (17)$$

$$i = 1, 2, \ldots, N$$

Where:

ω_{ji}^t is the adaptive weight for the j'th input to the i'th node.

z_i^t is the activation of the corresponding i'th node in the symmetrical subnetwork at time t.

N is the number of nodes in each subnetwork.

Genetic Algorithm

The Genetic Algorithm (GA) used in this experiment is the same as that used in the aforementioned comparative study. The population comprises a 2-dimensional grid of 100 individuals. A distributed steady-state GA is used, with a small tournament size corresponding to three individuals. A *principal* is selected, followed by two *neighbors*. These neighbors are selected based on a random walk (of length in the range [1,4] grid cells) originating at the principal. If the principal is fitter than both neighbors individual, then the weakest individual is replaced by a mutated version of the principal. If not, then the weakest member of the tournament group is replaced by the fitter two individuals genes, using single-point crossover, followed by mutation. The replacement of 100 individuals corresponds to a single pseudo-generation. A total of 14 evolutionary runs were carried out for each network type. Each trial was allowed to continue for 50 pseudo generations.

The genetic encoding strategy follows a similar approach for all networks. Network parameters are stored on a node or cell basis. Each gene comprises a list of real valued and integer parameters (comprising 16 parameters per node for a conventional GasNet for example). Connection weights (where relevant) are also stored on a per node basis.

Mutation takes place either after recombination, or after cloning of the principal tournament member (as described earlier). Mutation takes place at 20 percent of the nodes (rounded to 3 in a 16 cell network) selected at random. A single mutation event will result in the mutation of a single real or integer parameter in each of the randomly selected nodes. The magnitude of this mutation corresponds to 4 percent of the real valued parameters range with a probability of 0.2, and 1 percent of the parameters range with a probability of 0.8. In the case of integer parameters we follow a similar strategy of small mutations with a probability of 0.8 and large mutations with a probability of 0.2. These mutation parameters were chosen in preliminary experiments to avoid premature convergence and maintain a reasonable degree of phenotypic diversity across the different network varieties during evolution.

CTRNNs undergo further mutation. Each randomly selected cell has all of its weights mutated (again by a factor of 4 percent with a 20 percent probability and 1 percent with an 80 percent probability).

Time constant initialization was devised to yield a wide range of of values. An exponent f was randomly selected from the set:

$$f \in [-10, -8, -6, -4, -2, 0, 2, 4, 8, 10]$$

A second random variable $r \in [0.0, 1.0]$ was then used to scale the value such that the time τ constant is calculated from:

$$\tau = 1.0 + r(10^f) \quad (18)$$

The time constant mutation operator increments or decrements the exponent by 1 with a probability of 0.2, and generates a new value of $r \in [0.0, 1.0]$ with a probability of 0.8.

Experimental Setup

A screen shot of the quadruped used in this experiment is shown in Figure 2. Whereas the previous (bipedal locomotion) study used a physics package called AutoSim, this study uses an open source package called Open Dynamics Engine (ODE).

The quadruped torso is simulated with 6 physical degrees of freedom (unlike the previous biped study, where the biped was physically incapable of falling sideways). The quadruped comprises 9 rigid bodies, two rigid bodies for

Figure 2: Physically Simulated Quadruped

each leg, and a single rigid body for the quadruped torso. Lower limbs are connected to the the upper limbs via a limited hinge joint with a single rotational degree of freedom. Upper limbs are connected to the torso again with a limited hinge joint with one degree of freedom. The angular limits are shown in a scale diagram of the quadruped in Figure 3.

The assessed fitness of each individual is simply taken as the absolute distance traveled by the quadruped in a fixed time interval. The neural network is updated at half the frequency of the physics simulation, for a total of 5000 updates (approximately 20 seconds of real-time simulation).

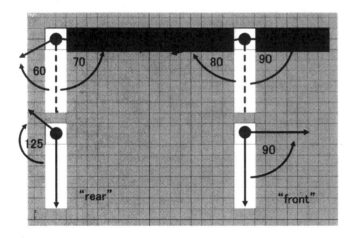

Figure 3: Scale drawn diagram showing joint angular limits (in degrees)

Table 1: Sensor Nodes

Limb	Sensor Nodes
Right Rear	0,1
Left Front	5,6
Left Rear	8,9
Right Front	12,13

Sensor input to the neural network comprises simple contact sensors associated with each lower limb. When a lower limb is in contact with the ground, the sensor value is 1, at all other times it is zero. Each contact sensor is connected to two network nodes as shown in Table 1.

Motor output nodes are shown in Table 2. The output signal of each motor node is mapped linearly into the hinge angular range. This becomes a target angular displacement. A velocity value for the joint is then calculated based on the difference between the current angular displacement and this target displacement. The physics engine than applies torque necessary to arrive at this joint velocity, constrained by a maximum torque value.

Table 2: Motor Nodes

Joint	Motor Nodes
Hip Right Rear	0
Knee Right Rear	2
Hip Left Front	4
Knee Left Front	6
Hip Left Rear	8
Knee Left Rear	10
Hip Right Front	12
Knee Right Front	14

Results

The results of each evolutionary run are shown in Table 3. The distance traveled is normalized by the body length of the quadruped so as to present the data in a more intuitive fashion. A distance traveled of 1.3 body lengths simply corresponds to the quadruped falling forwards. Between 2 and 3 body lengths, typically one or two steps have been taken. Distances greater than 4 body lengths usually correspond to a slow or unstable gait. Distances greater than this correspond typically to cyclical gaits. The global fitness peak is likely to be around 14 body lengths.

Although the results for GasNets and the hybrid CTRNN/PNN are very similar, there are differences in the stability of evolved gaits, with those of the GasNets exhibiting greater stability. Over all, the fittest individual was evolved using GasNets, however the CTRNN/PNN achieved a marginally higher average fitness measure. The results for the Center-Crossing CTRNN were generally poor, however one of the runs did discover the same gait as the fittest Gas-Net and CTRNN/PNN. As a consequence the fittest Center Crossing CTRNN individual attained a fitness very close to that of the other two networks considered.

Discussion

Although not obvious from Table 3, the quality of motion exhibited by the quadruped varied substantially with different networks. The fittest CTRNN driven quadruped exhibits motion that is similar to what we might expect from a Central Pattern Generator. The gait is symmetrical and the frequency of oscillation appears to be relatively stable.

Table 3: Distance Traveled by the Fittest Individual (normalized to quadruped body length). Letter superscripts correspond to distinct gaits described in the text.

Run Index	CTRNN	GasNet	CTRNN/PNN
1	1.4	7.2^b	4.2
2	1.5	4.6	5.7^e
3	1.3	2.0	1.3
4	1.3	3.0	1.6
5	11.7^a	2.0	1.3
6	1.3	3.4^d	7.2^f
7	1.4	2.5	6.2
8	1.4	5.2^c	1.5
9	1.4	2.7	1.3
10	1.3	2.0	4.9
11	1.3	1.7	6.2^e
12	1.3	2.6	13.6^a
13	1.4	13.7^a	1.1
14	1.4	3.2	1.3
Average.	2.1	4.0	4.1
Median.	1.4	2.8	2.9
Maximum.	11.7	13.7	13.6

The CTRNN produced a stable gait that continues for a prolonged time period after the end of a trial (if allowed to continue).

In contrast, the gaits generated by the CTRNN/PNN hybrid appear to be highly reactive, with little evidence of forced oscillations (excluding that of the fittest CTRNN/PNN individual). Although the order of stepping may assume a regular pattern, there is considerable variability in the speed of subsequent steps. In this respect motion closely resembles irregular passive dynamic walking. The GasNet demonstrates gaits which exhibit aspects of reactive behavior together with forced oscillations (producing the fasted quadruped within the evaluation time period).

The GasNet and CTRNN/PNN Hybrid exhibited the widest range of gaits. The CTRNN/PNN exhibited some gaits that were not discovered by GasNet, although the CTRNN/PNN gaits were relatively unstable. If we consider the form of the CTRNN/PNN network, it is clear that connection weights will gradually decline if there is a lack of coincident activity. In such a dynamic environment, rich external sensory input may play a more significant role, than it would in networks that exhibit strong intrinsic dynamic activity (such as self-oscillation). This may well go some way to explaining why the gaits exhibited by the CTRNN/PNN appear to be more reactive, but seem to lack strong oscillatory activity.

The evolved quadrupeds exhibit a variety of of the gaits, and body configurations. The quality of the motion varies from driven-oscillatory to ballistic-reactive. Some of the most distinctive patterns are described below. The letters

correspond to that which appears next to the fitness value in Table 3).

a The front legs hit the ground together, then the back legs, corresponding to the bound gait.

b The quadruped jumps from its rear limbs, stopping itself with its fore-limbs, before returning to a squatting position. This cycle then repeats. This does not correspond to any of the commonly observed animal gaits.

c In this case the left fore-limb remains in a forward position, whilst the right fore-limb remains in a rearward position. Rear limbs push off from the ground in a coordinated fashion. Motion resembles that of the three-legged bound gait.

d In some runs, the quadruped assumes a crawling configuration. Early in the evolutionary run a suspended walking gait is evident. Two diametrically opposed limbs are always in contact with the ground.

e This motion pattern most closely resembles ballistic walking. It is an highly irregular gait, with little evidence of regular oscillatory movement.

f This is another bounding gait, however it makes use of "elbows" rather than "hands/feet" in its forelimbs.

Comparison with Prior Study

In the previous study, which considered bipedal locomotion, GasNets appeared to offer the best solution (the only network to achieve cyclical bipedal locomotion) followed by Center-Crossing CTRNNs. For detailed analysis of GasNet dynamics and performance, the reader is referred to (Smith et al., 2003) and (Philippedes et al., 2002). In this respect the results for the GasNet are broadly in line with those of the previous study.

In this study, only one of the CTRNN runs resulted in locomotion. This is in keeping with prior work (Reil and Husbands, 2002) where CTRNNs were evolved for bipedal locomotion control. Only 10% of runs generated oscillatory activity that resulted in bipedal locomotion. No stable cyclical gaits were generated for the biped in the previous comparative study using CTRNNs.

Compared to the previous study, the biggest difference is in the relative performance of the Hybrid CTRNN/PNN. Whilst at best mediocre in the prior study, the results when applied to quadrupedal locomotion are comparable with those of the GasNet. There are two possible explanations that spring to mind. Firstly, modifying the original Hybrid CTRNN/PNN so that it more closely resembles a coupled-oscillator, may result in dynamic activity that is more suitable to oscillation and locomotive control. Note that in the previous study networks were single heterogeneous networks with no axis of symmetry. Secondly, quadrupedal locomotion may be more amenable to reactive solutions than the intrinsically less stable problem of bipedal locomotion.

CPGs are currently the dominant motor control paradigm. Work has shown that it is possible to model all the common quadrupedal gaits using a network of eight cells (Buono and Golubitsky, 2001). The results described here may lead us to question whether or not reactive responses are just as important in generating locomotive activity.

Conclusion

In conjunction with previous studies, GasNets appear to provide us with a reliable approach to evolving locomotion in physically simulated agents. However ANNs that exhibit substantially different dynamics (such as CTRNN/PNNs), may yield alternative solutions that are comparable in certain cases. Future work will investigate issues of gait stability and directed motion.

References

Beer, R. (1995). On the Dynamics of Small Continuous-Time Recurrent Neural Networks. *Adaptive Behavior*, 3(4):469–509.

Billard, A. and Ijspeert, A. K. (2000). Biologically Inspired Neural Controllers for Motor Control in a Quadruped Robot. In Amari, S., Giles, C., Gori, M., and Piuri, V., editors, *IEEE-INNS-ENNS International JointConference on Neural Networks (IJCNN 2000)*, volume 6, pages 6637–6641.

Blynel, J. and Floreano, D. (2002). Levels of Dynamic and Adaptive Behaviorin Evolutionary Neural Controllers. In Hallam, B., Floreano, D., Hallam, J., Hayes, G., and Meyer, J.-A., editors, *From Animals to Animats 7: Proceedings of the Seventh International Conference on Simulation of Adaptive Behavior (SAB 02)*, pages 272–281. MIT Press.

Bongard, J. and Paul, C. (2001). Making Evolution an Offer it Can't Refuse: Morphology and the Extradimensional Bypass. In Keleman, J. and Sosik, P., editors, *Proceedings of the Sixth European Conference on Artificial Life (ECAL 2000)*, pages 401–412.

Buono, P.-L. and Golubitsky, M. (2001). Models of central pattern generators for quadruped locomotion. *Journal of Mathematical Biology*, 42:291–326.

Floreano, D. and Mondada, F. (1996). Evolution of Plastic Neurocontrollers for Situated Agents. In Maes, P., Mataric, M., Meyer, J.-A., Pollack, J., and Wilson, S., editors, *From Animals to Animats 4: Proceedings of the Fourth International Conference on Simulation of Adaptive Behavior (SAB 96)*. MIT Press.

Husbands, P., Phillipedes, A., Smith, T., and O'Shea, M. (2001). Volume Signalling in Real and Robot Nervous Systems. *Theory in Biosciences*, 120:253–269.

Husbands, P., Smith, T., O'Shea, M., Jakobi, M., Anderson, J., and Phillipedes, A. (1998). Brains Gases and Robots. In et al, N. N., editor, *Proceedings of the 8th International Conference on Artificial Neural Networks (ICANN98)*, pages 51–64. Springer-Verlag.

Ishiguro, A., Fujii, A., and Eggenberger Hotz, P. (2003). Neuromodulated Control of Bipedal Locomotion Using a Polymorphic CPG Circuit. *Adaptive Behavior*, 11:7–17.

Mathayomchan, B. and Beer, R. (2002). Center-Crossing Recurrent Neural Networks for the Evolutionof Rhythmic Behavior. *Neural Computation*, 14:2043–2051.

McHale, G. and Husbands, P. (2004). GasNets and other Evolvable Neural Networks applied to Bipedal Locomotion. From Animals to Animats 8: Proceedings of the Eigth International Conference on Simulation of Adaptive Behavior (SAB 2004) - Accepted.

Philippedes, A., Husbands, P., Smith, T., and O'Shea, M. (2002). Fast and Loose : Biologically Inpsired Couplings. In Standish, R., Bedau, M., and Abbass, H. A., editors, *Artificial Life VIII: Proceedings of the Eight International Conference on Artificial Life*, pages 292–301. MIT Press.

Reil, T. and Husbands, P. (2002). Evolution of central pattern generators for bipedal walking in a real-time physics environment. *IEEE Transactions on Evolutionary Computation*, 6:159–168.

Smith, T., Husbands, P., Philippedes, A., and O'Shea, M. (2003). TemporallyAdaptiveNetworks:Analysis of GasNet Robot Controllers. In Standish, R., Bedau, M., and Abbass, H. A., editors, *Artificial Life VIII: Proceedings of the Eight International Conference on Artificial Life*, pages 274–282. MIT Press.

Urzelai, J. and Floreano, D. (2000). Evolutionary Robots with Fast Adaptive Behavior in New Environments. In Thompson, A., editor, *Third International Conference on EvolvableSystems: From Biology to Hardware (ICES2000)*. Springer Verlag.

Yamauchi, B. and Beer, R. (1994). Integrating Reactive, Sequential and Learning Behavior Using Dynamical Neural Networks. In Cliff, D., Husbands, P., Meyer, J., and Wilson, S., editors, *From Animals to Animats 3: Proceedings of the Third International Conference on Simulation of Adaptive Behavior (SAB 94)*, pages 205–224. MIT Press.

Evolving Simulated Mutually Perceptive Creatures for Combat

Michael J.T. O'Kelly[1], Kaijen Hsiao[2]

[1] Department of Physics, Massachusetts Institute of Technology, Cambridge, MA 02139
[2] Computer Science & Artificial Intelligence Laboratory, Massachusetts Institute of Technology
Contact: mokelly@mit.edu

Abstract

A fundamental obstacle in evolutionary simulations is the necessity of designing more complex simulations to elicit more complex behaviors. We use a combat-based fitness measure to attempt to circumvent this problem. We have designed a simulation that simultaneously evolves the brains and bodies of creatures for one-on-one combat in a three-dimensional environment with realistic physics. By giving the creatures a rich sensorium, we allow them to react sensibly to each others' actions. We discuss the effective, elegant and diverse simulated fighters that emerge, and examine whether qualitatively greater evolutionary complexity arises.

Introduction

The purpose of a Sims-style (Sims 1994a) simultaneous evolution of creature bodies and brains is to allow the greatest possible flexibility in evolution's solution to a given fitness problem. Sims found that a goal as simple as rapid locomotion elicited a variety of wildly different morphologies, with neural circuitry finely tuned for bounding across the virtual Serengeti.

In (Sims 1994b) the same environment and creature representation was used to evolve participants in a kind of contact sport. Creatures had to grab a box in front of them and then figure our how to keep it away from their opponents. This simple task elicited an evolutionary arms race as each generation's tactics were foiled by the next.

Subsequent work of this kind has gone in two directions. More complex and subtle genetic representations, such as L-systems (Hornby & Pollack 2001) and genetic regulatory networks (Brongard 2002), have been used to evolve articulated robots with realistic physics--but only to accomplish simple tasks such as locomotion. Co-evolution and arms races in warring animats has been considered in, for instance, (Cliff & Miller 1996) and (Floreano & Nolfi 1997), which simulate the chase between predator and prey with highly simplified models of the world.

(Nolfi & Floreano 1998) discuss the circumstances in which arms races may occur, concluding that both environmental and sensory richness encourage this phenomenon. We therefore embed our creatures in a rigorously simulated physical environment and provide them with a full suite of kinesthetic and external senses. Our creatures are then tested on their ability to engage in physical combat. Pairs of creatures are placed in an arena and allowed to fight. A specific part of each body is designated a target; the winner is the first creature to touch its opponent's target.

Sims' competing creatures could see only the box they sought, but were blind to their opponents. Our combat-based fitness measure benefits from the moving fitness landscape of all co-evolving systems and the phenotypic complexity of a Sims-style simulation, while also requiring our creatures to be non-deterministic and responsive in dealing with opponents. We hope to observe an improvement in the resultant complexity of our organisms.

Creature Morphology

As in Karl Sims' work, our creatures are three-dimensional, articulated creatures whose morphologies are determined by directed graphs. Unlike his creatures, ours are composed entirely of spheres, connected by motors that spin along the axis of attachment, much like a wheel and axle. All body parts originate from a root node that also functions as the target for opponents' attacks. Each sphere has a set of sensors associated with it, along with an embedded neural structure that connects sensors to the joint motor via a network of computational neurons.

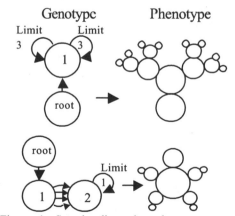

Figure 1: Sample directed-graph genotypes and the resulting phenotypes. Limits show how many more links will be followed.

Examples of two directed graph genotypes and the morphologies they represent are shown in Figure 1. A genotype is translated into a phenotype by starting from the graph's root node and following all of the outgoing links; each time a link is traversed, a new sphere is attached in the phenotype. Links specify the radius of the first sphere in a limb, a scaling factor for subsequent spheres, and other parameters such as the position of attachment. Nodes can be linked recursively back to themselves or in larger cycles, and a single node can link multiple times to its child.

The total number of spheres allowed in any creature is limited globally, forbidding them from becoming too large to simulate. To prevent the representation of infinite-length limbs, nodes have recursive limit parameters that set how many more downstream links may be traversed by the translation algorithm. A separate set of links flagged as extremities can be followed when a recursive limit is reached, so that structures like hands may be attached to the end of recursively-built arms.

These features make our directed graph representation very expressive. Small mutations in the genotype can result in drastic changes in morphology, facilitating evolutionary exploration of the fitness landscape.

Sensors

Each part of a creature's body has sensors that feed it input about its environment so that it can react to changing conditions. One sensor indicates whether the body part is touching either the ground or any part of its opponent. Three sensors indicate the spherical coordinates of the enemy's root node, and three more indicate the spherical coordinates of the nearest enemy sphere. The last two sensors report the position and velocity of the sphere's joint motor. All values returned by sensors are given in the reference frame of the sphere to which the sensor is attached.

In giving our creatures sensors, it would be ideal to provide complete information about the posture and movement of the opponent--information a human would glean from watching a video of a battle. However, there is a tradeoff between having more information available to enable more sophisticated tactics, and keeping the neural structure simple enough for evolution to find useful control structures (Martin 2001). The sensors we used proved more than adequate.

Neural Structure

We want to provide our creatures with the most general possible way of responding to the information provided by their sensors. Ideally, the creature should be able to mull over past events, combine information from throughout its body, and maintain an internal state.

To this end, we use directed graphs to specify our neural circuitry. Our method is comparable to that of (Sims 1994a), with some elaborations. Each node of the morphologic genotype has an embedded graph directing how information will be processed in the phenotype. Nodes in the embedded graphs, referred to as "neurons," represent a specific operation that takes some number of inputs and computes a single output. Links of the graph

indicate which outputs are to be fed to which inputs. Self-loops and cycles in the graph represent computational feedback loops, and make our representation Turing complete.

Sensors are represented as neurons that take no inputs. A lone motor neuron in each body sends its input to the attached axle motor. Our motors are velocity controlled, interpreting the motor neuron's signal as an intended velocity; a positive signal indicates counterclockwise motion, while a negative number indicates clockwise motion.

The rest of the neuron types perform computations:
- standard arithmetic operations +, -, *, /
- unary operations sin, cos, atan, log, exp, & sigmoid
- logical operations <, >, & if-greater-than
- integration, differentiation, smoothing & interpolation
- min, max, sum-threshold
- sinewave & sawwave
- constants (fixed output)

The sinewave and sawwave neurons are time-dependent, varying with simulation time. Other neurons, such as the constant neurons, are dependent on evolved parameters in addition to any variable inputs.

At each time step of the simulation, the neurons calculate their outputs in parallel based on their current inputs. It therefore takes more time for the organism to make a complex calculation than one that is simple.

Figure 2 illustrates the neural structure for a simple behavior, a proportional controller. The spherical coordinate denoting the angle in the xy plane of the opponent's target (φ) is multiplied by -0.1 and fed to the motor neuron. In simulation, this body part would rotate to keep the opponent's target at a constant bearing.

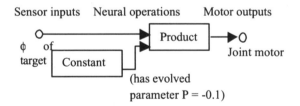

Figure 2: A basic neural structure encoding a proportional controller.

Advanced Neural Structure

The basic neural circuitry provides connections only between sensors and motor neurons on the same body part. In order to allow a creature to coordinate behavior across its body—for instance, to swing a limb in attack or flex away in defense—we want a general framework for the transmission of information from one body part to another.

The root node is the ideal location in our creature for evolving a seat of cognition. It has no motors to control, but it is the most vulnerable part of the body. Its sensors are situated to recognize danger and communicate states to the rest of the body. We therefore allow every neuron in the body to take input from any neuron in the root node.

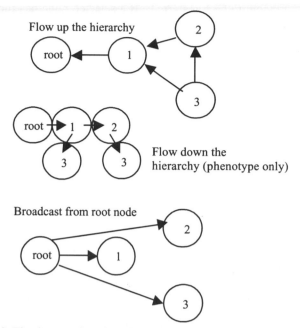

Figure 3: The three modes of communication across the body. Any neuron may draw input from neurons in a child node or in the root node. To take data from a parent node in a way that can be specified unambiguously, special "spinal neurons", which remain unconnected until instantiated in the phenotype, must be used.

Two other modes of linking are provided for the genotype to unambiguously specify how data flows between adjacent body parts. They are illustrated in Figure 3.

Combat

We test our creatures' relative fitness by pitting them against each other in pairs. To create the world in which our creatures are embodied, we use the open-source physics simulation engine Open Dynamics Engine (ODE). ODE models the creatures' bodies, joints, and motors in a physically realistic way, as well as the floor and forces such as gravity and friction. Collision detection and handling is also done by ODE, using a hash space collision detection algorithm for efficiency.

Combat works as follows: two creatures to be compared are embodied in the simulated arena, made to face each other at a fixed distance, and awakened. The first creature to touch its enemy's root node is deemed the winner. Before the contest begins, the creatures are made to relax for a short time without the use of their joint motors, so that they are in a settled configuration when the competition starts. We discovered that without this tweak our creatures evolve only toward extreme height, fighting by falling upon their enemies. Forcing them to relax before combat obligates them to discover a means of locomotion.

We tuned our numerical fitness measure to encourage discovery of basic behaviors such as movement and tracking in early generations. The first creature to hit the other's root node gets the maximum possible fitness value, MAX_FITNESS. The loser is then assigned a fitness value inversely proportional to its distance from the enemy's root node, as a consolation prize for at least having gotten close to its opponent. Without this consolation fitness our creatures found it too risky to approach each other, and never evolved to meaningful engagement. If neither creature hits the other's root node within a fixed amount of time, the game is over and both creatures earn a fitness value inversely proportional to the distance between the two. The actual fitness value is ½ MAX_FITNESS * (initial distance − final distance). Fitness can never be less than zero, however, so if a creature wanders off into the distance, its fitness will be zero. This is to prevent an otherwise fit creature from being eliminated because its opponent went romping off to the middle of nowhere.

Evolution

Selecting Parents

In order to apply evolutionary pressure to our creatures, we require that those that are better at combat reproduce with greater likelihood than those that are worse. However, in our combat-based fitness measure, and in fact any fitness measure derived from direct competition, it is possible for Player A to beat B, B to beat C, and C to beat A. How do we determine who is most deserving?

Since our fitness measure assigns a numerical fitness value to both participants in a fight, an obvious method would be to pair off every creature in a round-robin tournament, and let those with the greatest accumulated fitness reproduce. However, simulated combat runs very slowly, requiring ~1e10 floating point operations per test. A round-robin tournament requires N^2 tests, which becomes untenable with reasonable population sizes.

Instead, we perform selection via miniature round-robin tournaments, a variation on traditional tournament selection. A handful of creatures (we used a tournament size of 4) are picked at random. This subset of the full population plays a round-robin tournament with all possible pairings, and the winner is the player with the greatest accumulated fitness. That player is handed over to the algorithm creating the next generation of players.

This selection algorithm is a good compromise between a desire to rank players fairly and the constraint of reasonable runtime. In essence, any player that can defeat at least three of its peers is considered good enough for reproduction.

(Sims 1994b) discusses a variety of other approaches such as competition between species or against the previous generation's champion. Brief experiments showed little qualitative difference in our results.

Mutation

Directed graphs lend themselves to elegant mutation operators. A genotype selected for mutation may have nodes and links randomly added and removed, or the targets of links randomly changed. Parameters such as scaling factor and recursive limit may be randomly changed. The effect of mutation on the phenotype can be as small as a slightly larger radius and as large as a completely altered morphology. The embedded neural graphs are subject to a similar mutation operation.

Crossover

Our general directed graphs do not have as natural a means of crossover as would, say, a tree graph. The presence of cycles makes it difficult to choose a particular subset of a graph for transplantation. Instead, an artifact of our implementation is used to align parts of two parent graphs by similarity. Each node and link in a genotype is given a unique ID upon its creation. The descendants of a particular genotype also inherit the ID numbers of their components. This results in the genotypes of related creatures having the same ID's for components of similar function.

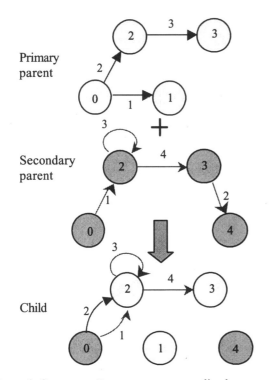

Figure 4: Crossover. Parent genotypes are lined up according to link and node ID's, and some material from the secondary parent is randomly chosen to replace that in the primary.

The crossover operation is illustrated in Figure 4. To perform this operation, a primary parent and secondary parent are selected from the population. Up to half of the nodes and links of the secondary parent are selected at random to be transferred. If a selected node has the same ID number as a node already present in the primary parent, that node is overwritten, along with its underlying neural graph. If the ID is not already present, the node is simply added to the genotype. Links with already present ID's are likewise overwritten, possibly resulting in changes to the graph's topology. Links with distinct ID's are added in the same manner as with nodes.

If the sexually reproducing parents share a common ancestor, as is likely in our small populations, this form of crossover is less brittle than a completely random exchange of genetic material would be. This allows innovations discovered by two different creatures to be shared; in particular, improvements to the neural circuitry are easily shared.

Genotype Validation and Garbage Collection

Before a genotype resulting from crossover or mutation is introduced to the general population, it is checked for physical validity. A genotype could specify a phenotype that self-overlaps, which is forbidden by our physics model. Invalid offspring are destroyed and the reproduction operators are applied again until a valid offspring is discovered.

After mutation or crossover is performed, genotypes are often left with nodes that can no longer be reached from the root node, and therefore are not expressed, as well as links that no longer point to valid nodes. Changes to the directed graph topology will also make many neural connections invalid, such as when one node's neuron tries to draw input from another node that is no longer its child.

The garbage collector works in two stages. First it identifies all unreachable nodes and hanging links, and removes them from the genotype. Then it checks all the remaining neurons to verify that their inputs remain valid. Invalid inputs are reassigned, drawing randomly from the set of all currently valid inputs.

Putting It All Together

We can now describe a complete evolutionary experiment. An initial population is created by randomly generating small directed graphs, checked for physical validity. We use a typical population size of 40. To form subsequent generations, we select 40 parents using the tournament selection rule described above. Duplicate selections are allowed; indeed, the best fighters are usually selected many times. Half of the parents reproduce by mutation, and the other half are paired off to reproduce by crossover, trading off the roles of primary and secondary parent.

The process of selection and reproduction is iterated as many times as desired; our longest runs have about 400 generations.

Results

Our evolutionary runs resulted in a wide variety of successful combat strategies and methods of locomotion.

Most of the creatures we discuss emerged between 20 and 100 generations. While many runs resulted in fairly homogeneous populations by the time they were stopped, each run produced completely different final creatures. A selection of the most successful creatures is shown in Figure 5.

Our creatures typically discovered a successful method of locomotion early on, since spherical structures and the absence of motor joint limits made it easy to roll. The first motion discovered in most evolutionary runs was spinning rapidly in place—an easy way to protect the root node, but not useful for attack.

Usually by the 10th generation a consistent method of locomotion was discovered that remained a theme throughout the rest of the experiment. Panel 1 shows a pair of worms that arch their backs and roll toward their opponents. Panel 2 shows a pair of "breakdancers" that hopped back and forth rapidly from one end to the other; in the frame shown, one is completely in the air while the other has one end on the ground. The breakdancers were rare in that they learned to track their opponents despite being hopping creatures. Behaviors could usually be grouped into wild, defensive behaviors that reacted little to opponents, and more deliberate behaviors tuned to move towards the opponent wherever it went.

Panel 4 shows a creature making a successful adaptation to trump its peers. In a population of creatures that use large defensive arms to shield and attack, the "Spanker" evolved a long extremity that sits inactive until its opponent touches one of its body parts. The creature then swings its long arm around to smack its opponent's protected target. This improvement led the "spanker" to immediate domination in the next generation.

Panels 3 and 5 show more advanced (~60 generations) creatures combining solid attack and defense. The "Proboscis" creatures use a long thin probe both for a cautious rolling attack and as a pivot to sweep at its opponent with its whole body. The "Buggy" rolls toward its opponent, has a front bumper-shield to protect its root node from being casually hit, and uses a mace-like arm to hit an opponent's protected root node.

Our most successful creature is shown in Panel 6. We named it "The Hedgehog" because of its prickly structure and whole-body rolling attack. The Hedgehog has four interchangeable limbs. Two of them touch the ground at any given time, rolling in the direction of the opponent, while the remaining two twirl in the air. The Hedgehog is good at tracking an opponent and its limbs provide some passive defense. What's most impressive, though, is that the limbs constantly move to keep the vulnerable root node as far from the enemy as possible. In response to a sudden attack the Hedgehog's entire body flexes to flip its root node away from harm and land its upper limbs on the attacker's body.

The Hedgehog is also remarkable for the simplicity of its representation. Its genotype is specified with just two nodes, linked recursively; and its neural circuitry consists of just one link between an enemy positional sensor and a motor neuron. (No other neurons are used to any effect.) The Hedgehog's sophisticated behavior is entirely emergent from the geometry of its body and the local interaction of sensors and motors.

Some of our more ambitious intentions did not pan out. With experiments requiring days to simulate a population of just 40 individuals to 400 generations, it remains unclear whether our combat-based fitness measure allows ever-escalating evolutionary complexity or if the plateau has just not been reached. Population diversity tended to bottom out after an initial period of morphological variation lasting around 100 generations.

(Nolfi & Marocco 2002) suggest one possible explanation. They argue that neural structure beyond simple sensory-motor coordination is difficult for evolution to discover because two necessary components--useful internal variables, and connections from them to sensory-motor flow--must arise simultaneously and spontaneously, since no advantage to the creature comes from either alone. Our neural mutation operators were not designed with facilitating this sort of discovery in mind. Evolution of complex neural structure might have consequently required many more generations than a comparably complex morphology. This agrees with our observation that the use of advanced neural structure was surprisingly rare, given the ease of its discovery by our mutation operators.

Also, the fact that fitness depended explicitly on the other members of the population had the unanticipated effect of often encouraging our creatures to be opportunistic. Rather than evolving to outsmart the best of their peers, they instead developed strategies to take advantage of weaklings. A significant fraction of each generation was crippled by mutation or crossover gone awry, and the most successful creatures were those that dealt rapidly and savagely with the defenseless.

These problems are not insurmountable. The first might be handled by designing a new neural representation or reproduction operator, or by evolving one as in (Teller 1996). (Nolfi & Floreano 1998) discuss methods of tournament selection that may better promote arms races. What is most needed, however, is longer experiments on larger populations. (Cliff & Miller 1995) distinguish open-ended from cyclic evolution by testing whether the latest generation of individuals can reliably defeat prior generations. Though cyclic behavior was not observed in our short-lived experiments, neither can open-ended evolution yet be confirmed.

Videos and source code for our experiments can be found online at www.simons-rock.edu/~towk/alife/bubblegene.html .

Conclusion

We successfully evolved creatures for combat exhibiting diverse morphologies and behaviors. Our creatures engaged in an evolutionary arms race, discovering feints, attacks, and dodges once the simpler skills of spinning in place or moving straight forward no longer sufficed. We conclude that our reasons for choosing a combat-based fitness measure—it allows the creatures themselves to determine their fitness landscape, and it forces them to react sensibly to opponents' actions—were validated.

Another facet of our evolutionary framework is the complex behaviors and strategies that can emerge from very simple structures. The best example of such emergent behavior was seen in the Hedgehog, a creature that was able to track its opponent, actively defend its target node by rotating it out of reach of its opponent, and continually attack its opponent with flailing arms. An observer would expect that the Hedgehog must possess a complex internal structure to demonstrate such a sophisticated strategy, whereas in reality all of the above behaviors emerged from a staggeringly simple design.

Acknowledgements

This research was supported by the Fannie & John Hertz Foundation and by the National Science Foundation.

References

Bongard, J. C. 2002. Evolving Modular Genetic Regulatory Networks. In *Proceedings of the IEEE 2002 Congress on Evolutionary Computation (CEC2002)*, vol. 2, pp. 1872-1877.

Cliff, D., & Miller, G. F. 1995. Tracking the Red Queen: Measurements of Adaptive Progress in Co-evolutionary Simulations. In *Advances in Artificial Life: Proceedings of the Third European Conference on Artificial Life*, Berlin.

Cliff, D., & Miller, G. F. 1996. Co-evolution of Pursuit and Evasion II: Simulation Methods and Results. In *Animals to Animats IV: Proceedings of the Fourth International Conference on Simulation of Adaptive Behavior.*

Floreano, D. & Nolfi, S. 1997. God Save the Red Queen! Competition in Co-Evolutionary Robotics. In *2nd Conference on Genetic Programming*, San Mateo, CA.

Hornby, G. S. & Pollack, J. B. 2001. Evolving L-Systems to Generate Virtual Creatures. *Computers and Graphics.* 25:6, p. 1041-1048.

Martin, M. C. 2001. The Simulated Evolution of Robot Perception. Ph.D. thesis, School of Computer Science, Carnegie Mellon University.

Nolfi S., & Floreano D. 1998. Co-evolving Predator and Prey Robots: Do 'Arm Races' Arise in Artificial Evolution? In *Proceedings of Artificial Life IV*, 311-335.

Nolfi, S., & Marocco, D. 2002. Evolving Robots Able To Integrate Sensory-Motor Information Over Time, In *Biologically Inspired Robot Behavior Engineering*, Berlino, Springer-Verlag.

Sims, K. 1994a. Evolving Virtual Creatures. *Computer Graphics. SIGGRAPH Proceedings*, 24-29.

Sims, K. 1994b. Evolving 3D Morphology and Behavior by Competition. In *Proceedings of Artificial Life IV*, p. 28-39.

Teller, Astro. 1996. Evolving Programmers: The Co-evolution of Intelligent Recombination Operators. In *Advances in Genetic Programming 2*. Cambridge, MA: The MIT Press.

Figure 5: A selection of evolved creatures

Panel 1: Worms

Panel 2: Breakdancers

Panel 3: Proboscis

Panel 4: Spanker (left)

Panel 5: Buggy with Mace (right)

Panel 6. Hedgehog (left)

Information Trade-Offs and the Evolution of Sensory Layouts

Lars Olsson, Chrystopher L. Nehaniv, Daniel Polani
Adaptive Systems Research Group
School of Computer Science
University of Hertfordshire
College Lane, Hatfield Herts AL10 9AB
United Kingdom
{L.A.Olsson, C.L.Nehaniv, D.Polani}@herts.ac.uk

Abstract

In nature, sensors evolve to capture relevant information needed for organisms of a particular species to survive and reproduce. In this paper we study how sensor layouts may evolve in different environments and under pressure of different informational constraints. To do this we evolve sensor layouts for different environments and constraints using a fitness measure with weighted terms for *redundancy* and *novelty*, using, respectively, mutual information and Crutchfield's information metric. The results show how different sensor layouts evolve depending on the structure and complexity of the environment but also how selective pressure for redundancy or novelty might affect the design.

Introduction

Nature has produced a wide variety of sensory organs that are well adapted to the specific animals and their respective environments (Dusenbery, 1992): For example, consider the amazing echolocation capabilities exhibited by bats, dolphins, and some species of whales. Sound waves are emitted by the animal and it then listens for the echo, which is used to perceive predators, prey, and objects in the environment. Another example is some birds and bees that probably can use the horizontal component of the magnetic force around the earth to determine their direction. The senses and their interpretation in the brain can also develop during the lifetime of an individual. One example is humans that for some reason go blind and then learn to use other senses to navigate in place of vision.

But how are the different sensoric channels used by a particular species selected for to begin with and how do they evolve over time? These are some of the questions pondered in the field of *sensor evolution* (Dautenhahn et al., 2001). In contrast with natural systems, sensors of artificial systems are often, due to practical and historical reasons, seen as something that is "given" and fixed. Only recently has there been a strong research focus on building artificial systems where the sensors can evolve and adapt, be combined, or the sensory channels created or selected, as this may lead to more autonomous, adaptive, and powerful systems as well as

better understanding of how sensors evolve in nature (Cariani, 1998; Dautenhahn et al., 2001; Nehaniv et al., 2002).

In this paper we discuss how the positioning and *informational coverage* of similar sensors, for example vision sensors, can evolve and adapt depending on the environment. This has been studied in for example (Stryker et al., 1978) where kittens were restricted to seeing either only vertical or horizontal lines. It was found that the cortical cells selective for orientation in the kittens preferred to fire mostly in response to the orientation that they had been exposed to. Evolution of sensory layout in an artificial agent was considered in (Jung et al., 2001) in relation to learning, where it was shown that longer learning periods lead to better suited sensors. In this paper we also consider the selective pressure posed by the trade-off between redundancy and novelty in sensoric input. If two sensors of the visual system of an agent are completely uncorrelated it is hard to find structure that for example can be used to compute optic flow (Gibson, 1986). However, if the sensors are completely correlated the information they transmit is redundant and one sensor is enough, unless they are used in an noisy environment where the redundancy may be used to filter noisy input, and thus provide robustness. Thus, there is a trade-off between similarity of information and the novelty of information between pairs and groups of sensors. To study the environmental impact on sensor layout but also the trade-off between redundancy and novelty, we apply evolution to sensor layouts where the redundancy and novelty can be weighted to be more or less important. In the experiment we consider the evolution of layout of visual sensors in different environments, ranging from an environment consisting of only vertical lines to a complex environment and a random environment without structure. The results show that the layout of the sensors depends on the environment in which the agent is situated in but also the selective pressure for redundancy or novelty.

The remainder of this paper is organized as follows. The next section discusses methods for computing differences between sensors and especially information-theoretic measures. One example with real world data from a robot using the described methods is also discussed. Then we describe our experimental model, the results, and their interpretation. Finally we summarize the paper and discuss possible applications and future directions of the presented work.

Information Distance between Sensors

In order to discuss the effectiveness and layout of sensors it is fruitful to be able to quantify the functional and informational distances between sensors. To do this a number of different methods can used, e.g., the Hamming distance and frequency distribution distance (Pierce and Kuipers, 1997). In (Olsson et al., 2004) these distance metrics are compared with the *information metric*, which was defined and proved to be a metric in (Crutchfield, 1990). The distance between two information sources is there defined in the sense of classical information theory (Shannon, 1948) in terms of conditional entropies. To understand what the information metric means we need some definitions from information theory.

Let X be the alphabet of values of a discrete random variable (information source, in this case a sensor) X with a probability mass function $p(x)$, where $x \in X$. Then the entropy, or uncertainty associated with X is

$$H(X) = -\sum_{x \in X} p(x) \log_2 p(x) \qquad (1)$$

and the conditional entropy

$$H(Y|X) = -\sum_{x \in X} \sum_{y \in Y} p(x,y) \log_2 p(y|x) \qquad (2)$$

is the uncertainty associated with the discrete random variable Y if we know the value of X. In other words, how much more information do we need to fully predict Y once we know X.

The *mutual information* is the information shared between the two random variables X and Y and is defined as

$$I(X;Y) = H(X) - H(X|Y) = H(Y) - H(Y|X). \qquad (3)$$

To measure the dissimilarity in the information in two sources Crutchfield's *information metric* (Crutchfield, 1990) can be used. The information metric is the sum of two conditional entropies, or formally

$$d(X,Y) = H(X|Y) + H(Y|X). \qquad (4)$$

Note that X and Y in our system are information sources whose $H(Y|X)$ and $H(X|Y)$ are estimated from the time series of two sensors as described in the next section using (2).

It is worth noting that two sensors do not need to be identical to have a distance of 0.0 using the information metric. What an information distance of 0.0 means is that the sensors are completely correlated. As an example, consider two sine-curves where one is the additive inverse of the other. Even though they have different values in almost every point

is the distance 0.0 since the value of one is completely predictable from the other. In this case, the mutual information, on the other hand, will be equal to the entropy of either one of the sensors.

Redundancy and Novelty

We will in this paper use mutual information as a measure of redundancy and the information metric to measure novelty between pairs of sensors. Redundancy can be seen as robustness to noise while novelty is important to capture as much different information as possible about the environment. Thus, there is a trade-off between capturing redundant and novel information.

An agent may optimize the relation between redundant sensors and sensors that detect novel things, differently according to selective pressure for redundancy or novelty and the environment. The *informational coverage* achieved by a set of sensors S can be used to calculate how sensors should be selected and is defined as

$$ic(S) = \sum_{X \in S} \sum_{Y \in S} (w_{mi} I(X;Y) + w_{im}(H(X|Y) + H(Y|X))) \qquad (5)$$

where $X \in S$ and $Y \in S$ are the sensors of individual S, w_{mi} a weight associated with the mutual information and w_{im} a weight associated with the information metric. Obviously, if $w_{im} = w_{mi} = 1$, then $ic(S) = \sum_{X \in S} \sum_{Y \in S} H(X,Y)$, since $H(X) - H(X|Y) + H(X|Y) + H(Y|X) = H(X) + H(Y|X) = H(X,Y)$, which is the joint entropy. The weights w_{mi} and w_{im} are used to specify how much emphasis that is put on redundancy and novelty.

As an example, consider Figure 1, which is a scatter plot with mutual information on the y-axis and the information distance on the x-axis for all pairs of sensors including the visual sensors of a SONY AIBO[1] robot dog.

The upper left corner of Figure 1 contains sensors with large mutual information, and hence redundancy, but a small informational distance. The lower right corner, on the other hand, is where we find pairs of sensors with a large informational distance but little redundancy. Finally, the upper right corner is the interesting part where pairs of sensors both share a large amount of mutual information but also have a large informational distance. This is implies that these sensors must have a high individual entropy. In Figure 1 the vision sensors are the sensors in the diagonal cloud where the upper left corner consists of vision sensors that are physically close together on the AIBO, since neighbouring pixels have much redundancy. One application of a scatter plot like this is that related sensors can be grouped together from raw uninterpreted sensor data, in a similar way to the sensory reconstruction method developed in (Pierce and Kuipers, 1997) and extended in (Olsson et al., 2004).

---

[1] AIBO is a registered trademark of SONY Corporation.

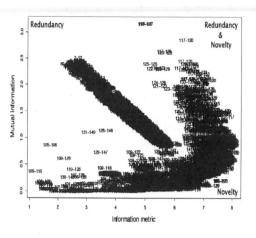

Figure 1: Scatter plot of all sensors and 100 vision sensors (pixels) from a SONY AIBO. The pixels are the diagonal linear cloud.

Experimental Model

An individual will have to select and layout sensory channels so as to maximize robustness and novelty of the information they provide. On a 200 x 200 pixel environment a 10 x 10 individual can move a maximum of one pixel per time step in the x-direction and a maximum of 1 pixel in the y-direction and hence dx and $dy \in \{-1, 0, 1\}$, but not both 0 at the same time. This means that there are 8 possible directions. Each time step there is a 15% probability that either dx or dy, or both, change value by -1 or 1. The 10 x 10 body has 100 possible positions to place sensors, numbered starting with 1 in the upper left corner and 100 in the right lower corner.

Each sensor saves for each time step the current value of the underlying pixel, where the value is in the range $[0, ..., 255]$ and then normalized in the range $[0.0, .., 1.0]$. Thus each sensor has an associated time series, with one element of data for each time step.

In the experiments four different environments were used, see Figure 2. The first environment with only vertical stripes in Figure 2(a) is similar to the experience of the kittens in (Stryker et al., 1978). The one in Figure 2(b) has stripes in two perpendicular orientations, while in the third one in Figure 2(c), each pixel has an equal probability of having any value in $[0, ..., 255]$. Finally, Figure 2(d) is a realistic image of some rocks on a beach. This image will give more piecewise smooth effects of movement on sensory features than the other images, something that is important for biological systems (Gibson, 1986) as well useful when building models of sensory features in artificial systems (Pierce and Kuipers, 1997; Olsson et al., 2004).

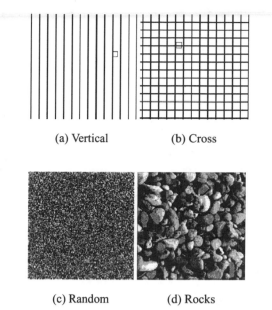

(a) Vertical   (b) Cross

(c) Random   (d) Rocks

Figure 2: The different environments used to evolve the sensor layouts.

Effects of Environments on Information Distances between Sensors

To illustrate the effects of environments on the information distances between sensors we first consider an individual with 100 sensors placed in a 10 x 10 grid that moves around in a simple environment like Figure 2(a). After 1000 time steps, or frames, a metric projection (Olsson et al., 2004) is created using the information metric to compute the distance between all pairs of sensors. A metric projection is a map that represents, in a small number of most relevant dimensions (in this case two), the distances between sensors so that the distance between two sensors is an approximation of their informational distance. The metric projection is shown in Figure 3(a). As we can see there are 10 groups of sensors with 10 sensors in each group. This is due to the fact that in a simple vertical environment it is not possible to detect vertical movement and since there are ten vertical lines in the vision layout all sensors in one line will be grouped together since they always have the same input. Now assume that after 1000 time steps the agent moves to the more rich environment of Figure 2(d). In Figure 3(b) to Figure 3(f) metric projections are shown after 1400 to 3000 time steps. As the individual experience more of the complex environment the informational distance of the sensors increases. After 3000 time steps the real layout of the sensors has been found[2].

---

[2]Since the raw input data contains no directional information it is impossible to find the absolute layout, and only the relative positions can be computed, see (Olsson et al., 2004).

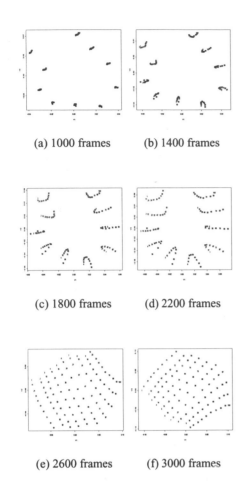

(a) 1000 frames   (b) 1400 frames

(c) 1800 frames   (d) 2200 frames

(e) 2600 frames   (f) 3000 frames

Figure 3: Metric projections of informational distances between visual sensors. After 1000 time steps the agent moves to a richer environment from the vertical environment and the distances between the sensors increase until their layout has been found.

Evolution of Sensor Layouts

Suppose that an individual in an evolving population has enough resources to select only ten sensors from amongst the 100 possible to lay them out as to maximize informational coverage, where the layout is genetically determined by the genome. To evolve the layout of the 10 sensors a Microbial genetic algorithm (GA) (Harvey, 2001) was used, with a population size of 20, a mutation rate of 5%, and crossover probability of 50%. The Microbial genetic algorithm uses tournament selection and was chosen for its extreme simplicity, speed, and the fact that it seems to perform just as well as many other GAs without much tweaking. Each individual's body is a 10 x 10 square with 10 sensors placed somewhere on it. The genome thus encodes a list of 10 positions within that square. The number of possible

genomes with 10 genes in the range [1,..,100] is

$$\frac{n!}{k!(n-k)!} = \frac{100!}{10!(100-10)!} \approx 6.2 * 10^{19}.$$

Each position can only be used once in each agent and thus care is taken during mutation and crossover not to let the same position occur twice. If this occur during crossover that particular position is not copied, while in mutation a new random position is selected.

Each individual i moves around the image for t time steps, in our experiments $t = 4000$, and is then evaluated according to its informational coverage, Equation (5). Note that the fitness of one sensor position is dependant on the other positions in the genome, since the fitness is calculated over all pairs of sensors in the genome. In the first experiment the weights were $w_{mi} = 2$ and $w_{im} = 1$, which means that the theoretical maximum value for the mutual information term and the information metric is the same. In the second experiment $w_{mi} = 1$ and the information metric weight $w_{im} = 4$, to reward novelty more than redundancy.

Figure 4 shows the evolved sensors after 10000 generations of evolution of a typical run using the Microbial GA. First consider the case where the mutual information and information metric is given equal importance, i.e., $w_{mi} = 2$ and $w_{im} = 1$, displayed in Figure 4(a), 4(c), 4(e), and 4(g). In Figure 4(a) all sensors are aligned in a single vertical line. This is an example of maximizing the mutual information since all sensors in every time step will extract the same value from the vertical environment. In general for simple environments with variation in only one dimension, like this one, a sensor layout that maximizes the mutual information between sensors will be a spatial representation of the environment.

In Figure 4(c) and Figure 4(g), we find that the sensors are grouped together in both dimensions. This is due to the fact that these environments are two-dimensional, where the environment with only vertical lines informationally can be seen as a one-dimensional environment. In Figure 4(e) and Figure 4(f), the random environments, we find that there is no structure in the sensor layout. An analysis of the fitness landscapes for this environment reveals that there is no structure, and almost all configurations have exactly the same fitness. This is an obvious result since each pixel has the same probability of being in any state, and thus prediction or finding structure is, at least in practice, impossible.

Now consider the case where there is a strong selective pressure for maximizing the distance between sensors (novelty), the case where $w_{mi} = 1$ and $w_{im} = 4$. In Figure 4(b) the layout is completely different from Figure 4(a). Now the sensors are all placed in different horizontal positions, with some redundancy in the vertical positions. By placing each sensor in a different column the distance between the sensors has been maximized, and the fact is that the position in different rows does not make a difference at all in this en-

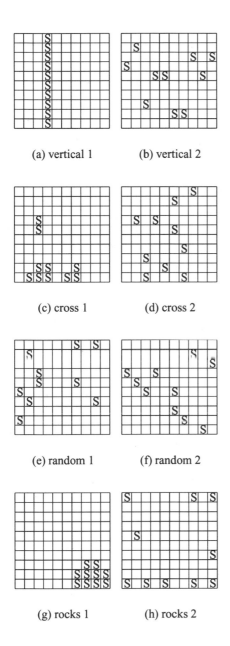

(a) vertical 1 (b) vertical 2

(c) cross 1 (d) cross 2

(e) random 1 (f) random 2

(g) rocks 1 (h) rocks 2

Figure 4: Evolved sensor layouts for the different environments. In Figure 4(a), 4(c), 4(e), and 4(g) the weights in the fitness function are $w_{mi} = 2$ and $w_{im} = 1$. In Figure 4(b), 4(d), 4(f), and 4(h) $w_{mi} = 1$ and $w_{im} = 4$.

vironment, so that coverage of the dimension giving rise to novelty is complete. In Figure 4(d) we find that the number of collinear sensor pairs and the number of horizontal and vertical lines is maximized by maximizing information coverage (with a high weight for novelty). In Figure 4(h) we find that the sensors are spread out over the grid, which maximizes the informational distance between the sensors.

What would happen if several sensors were allowed in the same position? In the second case where the informational distance is important nothing would change. But, if we consider the case where the mutual information is more important, we would find that all sensors would be placed on the same position. This would be the position with the highest individual entropy, since the mutual information between this position and any other position must be lower, due to the fact that $I(X;Y) \leq H(X)$.

Conclusions

This paper has presented some initial results regarding some of the trade-offs associated with evolving sensor layouts and sensory channel selection that an agent can face. We use formal, information-theoretic, measures of redundancy (mutual information) and novelty (information distance) between pairs of sensors. One trade-off is between redundancy, something that might be important in noisy and/or static environments, and novelty, which is more important in dynamic and complex environments. Some redundancy is necessary in vision since vision is a spatial sense, and without any relation between different vision sensors (eye cells) no spatial information can be found. To study this problem four visual environments were used. These range in complexity from a very simple one with only vertical stripes to a realistic image of rocks and a completely random image. In these environments agents evolve their sensor layout where the fitness of the agent depends on the weights associated with mutual information (redundancy) and the informational distance (novelty) between the sensors. The fitness associated with one sensor's position depends on the positions of the other nine sensors. The results show that the evolved layouts depend both on the environment and the kind of coverage rewarded (in our studies, a weighted combination of redundancy and novelty).

One interesting topic that we will consider in our future work is the relevance of information acquired by a certain sensor to the agent. This notion of relevant information was introduced in (Nehaniv, 1999) and formalized in relation to utility in (Polani et al., 2001) by associating the relevance of information with the utility for a particular agent acting in its environment. By measuring the relevant information acquired by different sensors it should be possible to compute the distance between sensors regarding their relevance to an agent performing a certain task. Equipped with the notion of relevant information and a way to measure the distance between sensors using a metric we expect it to be easier to design sensoric systems that can adapt to different conditions and tasks to be performed, and also to understand how the selection of sensory channels happens in nature. This is especially important since using each sensory and effector channel has an associated cost and this has to be traded against the utility of using that channel (Nehaniv, 1999; Polani, 2003). For example, in some cases it might be more effective from the agent's point of view to accept a slight decrease in overall utility depending on the expenditure for attaining the information. Thus, it is interesting and important to develop a predictive theory of how an agent can most effectively select, combine, or integrate channels to solve a certain task.

Finally, is it important to note that what you do in the world determines what you can distinguish (active perception). Conversely, the world also to some extent determines what you can know about your sensors.

References

Cariani, P. (1998). Epistemic autonomy through adaptive sensing. In *Proceedings of the 1998 IEEE ISIC/CIRA/ISAS Joint Conference*, pages 718–723.

Crutchfield, J. P. (1990). Information and its Metric. In Lam, L. and Morris, H. C., editors, *Nonlinear Structures in Physical Systems – Pattern Formation, Chaos and Waves*, pages 119–130. Springer Verlag.

Dautenhahn, K., Polani, D., and Uthmann, T. (2001). *Artificial Life (Special Issue on Sensor Evolution)*, 7(2).

Dusenbery, D. B. (1992). *Sensory Ecology*. WH Friedman & Co, first edition.

Gibson, J. J. (1986). *The Ecological Approach to Visual Perception*. Lawrence Erlbaum Associates, first edition.

Harvey, I. (2001). Artificial evolution: A continuing saga. In Gomi, T., editor, *Evolutionary Robotics: From Intelligent Robots to Artificial Life: Proc. of 8th Intl. Symposium on Evolutionary Robotics (ER2001)*, pages 94–109. Springer Verlag.

Jung, T., Dauscher, P., and Uthmann, T. (2001). Some effects of individual learning on the evolution of sensors. In Kelemen, J. and Sosík, P., editors, *Advances in Artificial Life, 6th European Conference, ECAL 2001, Prague, Czech Republic, September 10-14, 2001, Proceedings*, Lecture Notes in Computer Science, pages 432–435. Springer Verlag.

Nehaniv, C., Polani, D., Dautenhahn, K., te Boekhorst R., and Cañamero, L. (2002). Meaningful information, sensor evolution, and the temporal horizon of embodied organisms. In Standish, R. K., Bedau, M. A., and Abbass, H. A., editors, *Artificial Life VIII: Proceedings*, pages 345–349. MIT Press.

Nehaniv, C. L. (1999). Meaning for observers and agents. In *IEEE International Symposium on Intelligent Control / Intelligent Systems and Semiotics, ISIC/ISIS'99*, pages 435–440.

Olsson, L., Nehaniv, C. L., and Polani, D. (in press, 2004). Sensory channel grouping and structure from uninterpreted sensor data. In *2004 NASA/DoD Conference on Evolvable Hardware June 24-26, 2004 Seattle, Washington, USA*. IEEE Computer Society Press.

Pierce, D. and Kuipers, B. (1997). Map learning with uninterpreted sensors and effectors. *Artificial Intelligence*, 92:169–229.

Polani, D. (2003). Trade-offs in sensoric information acquisition and processing. In Miller, J., Polani, D., and Nehaniv, C., editors, *Evolvability and Sensor Evolution Symposium Abstracts*. EPSRC Network on Evolvability in Biological & Software Systems/University of Hertfordshire Technical Report 384.

Polani, D., Martinetz, T., and Kim, J. (2001). An information-theoretic approach for the quantification of relevance. In Kelemen, J. and Sosík, P., editors, *Advances in Artificial Life, 6th European Conference, ECAL 2001, Prague, Czech Republic, September 10-14, 2001, Proceedings*, Lecture Notes in Computer Science, pages 704–713. Springer Verlag.

Shannon, C. E. (1948). A mathematical theory of communication. *Bell System Tech. J.*, 27:379–423, 623–656.

Stryker, M. P., Sherk, H., Leventhal, A. G., and Hirsch, H. V. (1978). Physiological consequences for the cat's visual cortex of effectively restricting early visual experience with oriented contours. *Journal of Neurophysiology*, 41(4):896–909.

Evolving Flying Creatures with Path Following Behaviors

Yoon Sik Shim[1], Sun Jeong Kim[1], and Chang Hun Kim[1]
[1]Korea University, Seoul, Republic of Korea
necromax@kebi.com

Abstract

We present a system which evolves physically simulated 3D flying creatures and their maneuvers. The creature is modelled as a number of articulated cylinders connected by triangular patagia in between. A creature's wing structure and its low-level controllers for straight flight are generated by an evolutionary algorithm. Then a feed-forward neural network is attached to the low-level controllers, and the connection weights of the network for a given trajectory are found by a genetic algorithm. We show that a control system sufficiently effective to allow aerial creatures to follow a complicated path can be achieved by two-step evolution process.

Introduction

To date, artificial evolution researchers have found only a few applications for virtual embodied agents. Sims (Sims, 1994) pioneered an innovative approach in generating both the morphology and the behavior of physics-based articulated figures by artificial evolution. His "blockies" creatures moved in a way that looked both organic and elegant. Other recent work such as Sexual Swimers (Ventrella, 1998), Framsticks (Komosinski and Ulatowski, 1999), and Virtual Pets (Ray, 2000), have produced various applications in their own way. The Golem project (Lipson and Pollack, 2000) even took this technique a step further by evolving simulations of physically realistic mobile robots and then building real-world copies.

Relatively little work has addressed the simulation of flying behavior. The interaction between a flapping wing and the air is very complicated. The forces generated by this interaction are chaotic and their simulation is often unstable because of high sensitivity. Computational fluid dynamics (CFD) methods have sufficient accuracy, but they are too time-consuming to build into an evolution framework. Reynolds (Reynolds, 1987) developed a stochastic model of the flocking behavior of birds, although he did not animate individual bird animations. Ramakrishnananda and Wong (Ramakrishnananda and Wong, 1999) presented a physically based bird model using two segmented airfoils but the entire motion of both wings is controlled by predesigned functions and only overall lift and thrust were calculated from

aerodynamics. Wu and Popović (Wu and Popović, 2003) developed a realistic model of a bird with flexible feathers and successfully generated flying motions corresponding to a given trajectory by optimizing the control parameters for each wingbeat using simulated annealing.

In terms of virtual creatures, our prior work (Shim and Kim, 2003) has proposed a new developmental model for evolving wing morphology and behavior with a traditional nested graph structure, but it was difficult to generate higher-level behaviors, such as following targets, because of the instability of the air surroundings and unnecessarily complex controller networks.

We now present a two-step evolution process for evolving path-following flying creatures. We describe an improved evolutionary design for creature generation and simplified aerodynamics for use in the time-consuming evolution process and to make convenient of genetic encoding. Since the activity of flying is very sensitive to the characteristics of the environment(low density of fluid, hard to maintain balance against the gravity), we use a more robust periodic controller for motor control. Since the new evolved wing structure has many more degrees of freedom of joint than the predesigned wings used in previous work, searching through all the control parameters at each wingbeat is too computationally expensive and time-consuming process. We therefore exploit a continuous map of the neural network structure so that the optimization is only required once for an entire simulation.

Overview

The entire process is divided into two stages. In the first stage, our creature evolution system generates the creature's body and its low-level controllers that support straight forward flight. After this basic forward-flying creature has been generated, a neural network is connected to its controllers, and the connection weights of the network are optimized by a genetic algorithm to make the creature to follow given trajectory. During the optimization, a set of arrays which contains the weight vectors for the neural network is used as the population. At each time step, the network receives the position and velocity of the creature's central body, the feedback

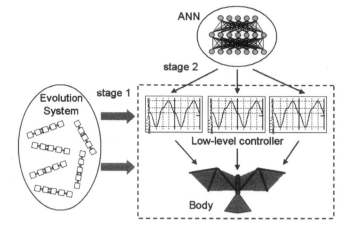

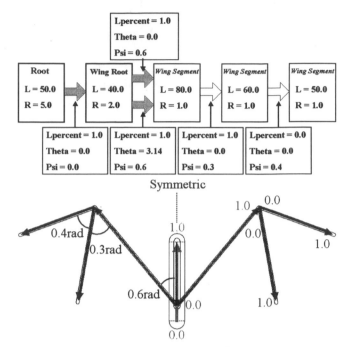

Figure 3: An example of a genotype and the resultant skeletal structure. In the lower picture, the parameters of the left wing and right wing describe **Psi** and **Lpercent** respectively.

Figure 1: Overview of the entire system. The evolution system creates straight-flying creature, and a neural network controls the creature so that it will follow a given trajectory.

of the neural network output, and the target position which is calculated in realtime based on the creature's velocity and the given trajectory. The signals coming out of the network are fed to the creature's controllers, and vary their behavior so that the motion of the wings is changed.

The concept can be explained at an intuitive level by comparing it to three elements of a real life form: the body, the motor nerves of the cerebellum (evolved low-level controllers) which is in charge of motor control, and the cerebrum (a neural network) which gives high-level control to the cerebellum. In the next sections, we describe each process in more detail.

Genotype Representation and Creature Development

The creature is constructed from a basic model which has truss-shaped wings (Shim and Kim, 2003). The genotype encoding is accomplished using a multi-connection list. This allows complete symmetry of shape and wing motion. In this form, variations of the genotype, such as mutation or

Figure 2: A genotype structure. The wing-root has double connections for two symmetric wings.

mating, because of its linear structure. The Root represents the central body (fuselage) of a creature and the wing-root represents the root of wings, which lies inside the central body. These first two nodes and their connections exist in all individuals by default; but they are followed by an arbitrary number of nodes for wing segments. Only the wing-segment nodes are deleted or added during genetic variation.

Each wing cylinder has a joint which connects it to its parent. The wing joint can have two degrees of freedom, allowing both dihedral and sweep motions. The corresponding geometric primitive is a capped cylinder which is aligned along the local z-axis. Each cylinder's two end points (the centers of each hemi-spherical end) are known as the start and the end point. The start point is defined as position 0 on the cylinder and the end point corresponds to limit parametric distance along the cylinder's local z-axis. A node connection specifies how and where to attach each cylinder to the parent. **Lpercent** indicates the position on a parent cylinder where the child's 0 point will be placed, and the value is set to 0 or 1 to obtain a truss-like wing structure. **Theta** (θ) is an initial sweep angle rotated about the parent's local z-axis. The wing grows in the same direction with making an acute angle of **Psi** (Ψ) which is the angle between the local z-axes of parent and child. **Psi** and **Theta** are set as neutral values of each joint at the time of initialization and varies over time during the simulation. After two cylinders have been attached to each other, a zero mass rigid film is combined between the two cylinders. Since this film is attached to a child cylinder, its linear and angular velocities can be calculated from that of the child cylinder. The whole of a creature's wing is thus composed of a sequence of child-parent pairs with attached films, and it can be considered as a webbed structure like a kite or a hang-glider, and each film is subject to the appropriate aerodynamic forces. The model has a fan-shaped tail which is attached to the rear of the cen-

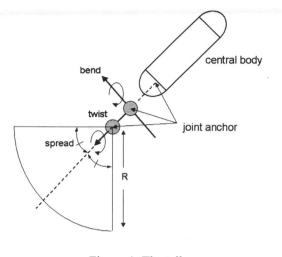

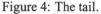

Figure 4: The tail.

Figure 5: An example of a completed creature.

tral body cylinder. The tail has three degrees of freedom which are: bend, twist, and spread. The tail plays an important role in balancing and in turning movements. The tail radius is also included as a parameter of the genotype.

Controller parameters

The controller is defined by a combination of piecewise sinusoidal functions. The output value of the controller is a desired value of the proportional-derivative controller which is used for calculating the actual motor torque. The controller parameter includes the duration of a cycle and the number of control points. Each control point can have arbitrary values between -1 and 1, but all controllers operate with a cycle of the same duration. Hence, all joint motions can be controlled synchronously. We can write the controller function U as:

$$U(t) = \frac{p_n + p_{n+1}}{2} + \frac{p_n - p_{n+1}}{2} cos(\frac{Nt}{D} - n)\pi \quad (1)$$

$$n = 0, 1, 2, ..., N-1 \qquad p_N = p_0 \qquad 0 \le t \le D,$$

where N is the number of control points, and d is duration of a cycle. The control function is constructed by smooth interpolation between each pair of neighboring control points. The last control point is connected back to the first one, which makes the function to be periodic. These parameters are also coded into each node of the genotype. Using a

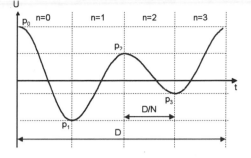

Figure 6: Low-level controller.

periodic sinusoidal function has two advantages. First, most locomotion (including flapping flight) of real animals can be expressed using a sinusoidal function. Second, the sinusoidal function has an infinite derivative continuity so that unexpected fluctuations of a joint motor can be minimized.

Dynamics Simulation

The movement of a physical creature is calculated by an articulated rigid body dynamics. Resistance forces on patagia are calculated by simplified aerodynamics, with gravity. We used the Open Dynamics Engine by Russell Smith (Smith, 1998) which is an open-source articulated body simulator. The integrator is a first order semi-implicit integrator (Stewart and Trinkle, 1996), where the constraint forces are implicit, and the external forces are explicit. A medium speed ($O(n^3)$, where n is the number of links) Lagrange multiplier method (Anitescu and Potra, 1997) is used to calculate the articulated body dynamics.

Force on the Triangular Film

The forces acting on a surface depend on its area and the angle of attack with respect to the velocity of the airstream. A film is divided into several patches, and the calculated forces are considered to be exerted on the center of mass of each patch. The stream velocity for patch i is simply a negation of the velocity of the patch's center of mass. The

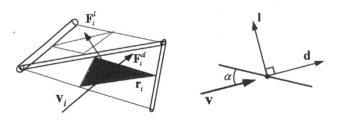

Figure 7: Force on a triangular patch.

lift and drag forces are due to the lift coefficient C_L and the drag coefficient C_D which are functions of the angle of attack α. Drag acts in the direction of the airstream velocity. The direction of lift is perpendicular to that of the drag, so it produces not only lift but also thrust tangential to a patch during the flapping motion of the wings. Each force on the patch i can be simply written as:

$$\mathbf{F}_L^i = \frac{1}{2}\rho C_L(\alpha)A_i\|\mathbf{v}_i\|^2\mathbf{l}_i \tag{2}$$

$$\mathbf{F}_D^i = \frac{1}{2}\rho C_D(\alpha)A_i\|\mathbf{v}_i\|^2\mathbf{d}_i, \tag{3}$$

where A is area of the patch, and ρ is the density of the air (Fox and McDonald, 1976). We synthesized functions to produce lift and drag coefficients at any particular angle based on the typical plots for a real bird (Withers, 1981; Mueller, 2001). After the forces on each patch have been calculated, the resulting force and the torque on the skeleton are obtained by:

$$\mathbf{F}_{total} = \sum_{i\in N}\mathbf{F}_L^i + \mathbf{F}_D^i \tag{4}$$

$$\mathbf{T}_{total} = \sum_{i\in N}(\mathbf{F}_L^i + \mathbf{F}_D^i) \times \mathbf{r}_i \tag{5}$$

where N is the set of triangular segments covered by film and the $\mathbf{r}$ is the relative vector from the cylinder's center of mass to the center of the each segment. The forces on the tail can

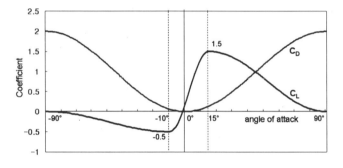

Figure 8: Lift and drag coefficients.

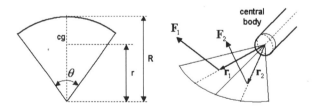

Figure 9: Forces on the tail

be calculated in the same way as those on a patch. Since the tail is fan-shaped, it is simply divided into two sub-fans to accommodate tail twist. The center of mass of the tail can be calculated using the length from a fan's vertex to the center of mass.

$$\|\mathbf{r}\| = \frac{4R\sin\frac{\theta}{2}}{3\theta} \tag{6}$$

R is the tail radius, and θ is the angle of a fan.

Force on the Cylinder

The forces on a central body cylinder are also calculated using existing method (Wejchert and Haumann, 1991). Instead of calculating the forces on every surface of the cylinder, the total resistance is divided into two terms that act at its center of mass: these are linear resistant force and resistant torque. In case of the linear resistance, the area that receives force

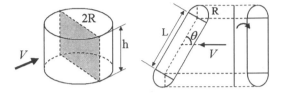

Figure 10: Idealized linear resistance of a cylinder.

is idealized as the projection of the capped cylinder on to a surface normal to the velocity vector. Therefore, the linear resistance of the cylinder can be written as:

$$\mathbf{F} = \alpha_n(2RL\sin\theta + \pi R^2)v\mathbf{v}. \tag{7}$$

The resistant torque is calculated by dividing a cylinder into slices and integrating each over all cylinder slices. For convenience, we approximated a capped-cylinder by a standard cylinder with the length of L+2R.

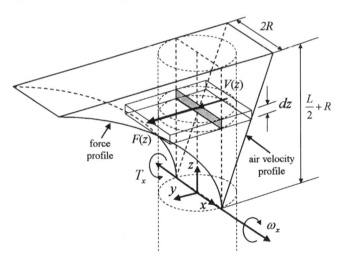

Figure 11: The angular resistance of a cylinder.

$$T_x = -4\alpha_n\omega_x^2 R \int_0^{\frac{L}{2}+R} z^3 dz \tag{8}$$

$$T_y = -4\alpha_n\omega_y^2 R \int_0^{\frac{L}{2}+R} z^3 dz \tag{9}$$

$$T_z = -2\alpha_n\omega_z^2(L+2R) \int_0^R r^3 dr \tag{10}$$

The angular velocities and torques described above are measured about the local axis of each cylinder.

Evolution Process

Each creature is initially placed at a given altitude, and then the simulation is started. The fitness f is defined by the sum of the hovering time (t_h), the flight speed (in the direction of $\mathbf{d}_f$), with the addition of a penalty for unnecessary shaking of the central body.

$$f = k_a t_h + \int_\zeta \{k_b(\mathbf{v}(t) \cdot \mathbf{d}_f) - k_c \|\omega(t)\|\} dt \qquad (11)$$

The terms $\mathbf{v}(t)$ and $\omega(t)$ are the linear and angular velocities of the central body respectively. The symbol ζ denotes the simulation period and k is the weight constants for each term. In the early stage of evolution, hovering time is crucial for survival, because most creatures fall to the ground before the end of the simulation period. The simulation speed is improved by prematurely aborting the simulation of any individual that is not performing well. A creature is also considered to blow-up if the speed of its body parts exceeds some value (typically $1000m/s$), and that creature is aborted. Additionally, interim fitness measuring is performed after a quarter of the total evaluation period has elapsed by removing individual whose fitness is less than one fifth of that of the least fit creature of the previous generation. After fitness evaluation, a typical genetic operation (mutation, crossover, grafting) (Sims, 1994) is performed to produce the next generation.

Combining Controllers with Neural Network

The creature control is done by giving variation to the evolved controller of a creature using a neural network. Since we do not have the training samples necessary for path following control, nor the gradient information of numerical physics simulator needed for the gradient-based training methods such as backpropagation, the network weight is found by a genetic algorithm.

Neural Network Parameters

Two control values are assigned to a single controller which are: scaling s and phase transformation p. Since the tail has no controller, each degree of freedom of the tail is directly controlled by the network (θ^B, θ^T, θ^S). The duration of a cycle D is the same for all the controllers. The scaling signal is simply multiplied to the controller output, and a phase transformation performs time warping (we use t' instead of t, see figure 14) in a way similar to (Wu and Popović, 2003). At the start of each cycle, new values of scaling, phase, and duration are obtained by sampling the continuous output of the network, and these values are retained until the cycle ends. The output of the previous state becomes the input of the network, together with other external states of the creature such as the target position(T), the velocity of the central body(v), and the values from a simplified gyrosensor (G) (Shim and Kim, 2003). All vectors (T,v) are transformed to the local

coordinates of the central body. The bipolar sigmoid which has a range of -1 to 1 is used as the activation function of a neuron.

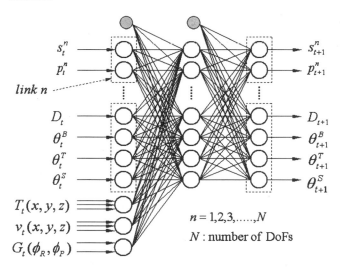

Figure 12: Input and output signals of a neural network.

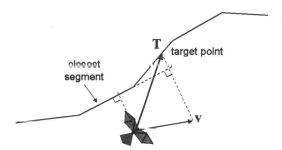

Figure 13: Obtaining the target position on the trajectory.

The path trajectory is given as a piecewise linear curve. At each simulation time, the desired target point is obtained as a function of the current velocity of the central body similar to a heading-based approach (Ribals et al., 1993).

Learning for Given Trajectory

The length of a chromosome for the genetic algorithm is H×(4N+17), including connection weights from the bias term, where H is number of hidden nodes and N is the total number of degrees of freedom in all joints. We use 1 to 2 times as many hidden nodes as the number of input nodes. The fitness is given by

$$f = \int_\zeta \{k_1 d_{min}(t) + k_2 v_t(t) + k_3 a_T(t)\} dt, \qquad (12)$$

where d_{min} is the minimum distance from the trajectory, v_t is the speed of the creature in the current path direction, and a_T is the acceleration towards the target point. Lower values represent better fitness. Each time the fitness evaluation

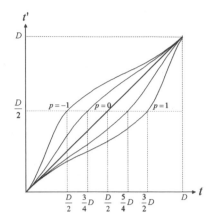

Figure 14: Time warping for phase transformation.

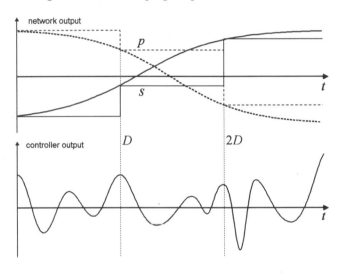

Figure 15: Neural network output(upper) and resultant controller signal(lower).

interval ($0 \leq t \leq t_c$), which we call the "checktime". That is because, if we check these criteria in all simulation periods, few creatures would be survived in the first group(especially most of the creatures will commit the heading failure). The checktime is initially set to some fraction (α) of the full duration, and increases gradually as a function of the number of survivors in the first group

$$t_c^{g+1} = \zeta\{(1-\alpha)\frac{S^g}{P} + \alpha\}, \qquad (14)$$

where g is the generation, S is the number of survivors, and P is the population size.

Results

Each creature was placed in the air at a height of $1000m$ from the ground, and the simulation was started at an initial push of $50m/s$. Typical settings for our runs were as follows. In the first stage (evolving simple forward flight creatures), the population size was generally 200, and runs lasted for 100 to 200 generations. The top 20% of genotypes were used as elites, and the remaining 80% of the new generation were created by selecting a couple of parents by roulette wheel selection, and copying one of them (choosen by tournament selection with a 90% probability of selecting the fitter) with probability 30%. The rest of the offsprings are made up of 30% crossovers, 30% grafting, and 10% random individuals. Mutations were then applied stochastically to the newly generated genotypes. The integration step was generally 0.05 seconds, and the evaluation period was around 500 simulation steps.

In the network evolution stage, the population size was 200-300, and the number of generations was 300-500. The simulation period was set to 1000-2000 time steps, depending on the speed of a creature in straight flight. The setting for the genetic algorithm were the same as those for creature evolution, but the number of crossover points was in the range of 1 to 4. The running time was generally 4-6 hours for creature evolution and 6-10 hours for path-following, running on a single Pentium-IV 2.53Ghz PC.

As expected, most of the creatures fell to the ground during the simulation in the early generations. A creature capable of staying airborne for longer than 500 time steps did not appear until the first quarter of the total running period had almost elapsed. This suggests that air is a harsh environment for evolved creatures, compared with water or land, especially in terms of maintaining balance. Since the creatures have to retain balance against the disturbance of gravity through their life time, this seems to limit the range of motion patterns that evolve. In terms of path-following behaviors, there were some local optima such as hovering near the trajectory with low velocity, or continuing to move straight ahead when the variation of path direction is small.

is complete, the population is divided into three groups according to fitness, giving a strong penalty to last two groups which show unexpected behaviors. The first such behavior is a blow-up of the simulation, and the other occurs when the heading direction of central body deviates seriously from the current path direction ($h \cdot d_t \leq 0$, where h is the heading direction and d_t is desired direction). A creature which does not commit any of these misdemeanor (called a survivor), is placed in the first group of the population. Heading failures go into the second group, and blow-ups go into the worst group. In those two worse cases, the simulation is aborted and the fitness is assigned by following metric:

$$f = P, Q + \frac{1}{t_a}\int_0^{t_a}\{f(t)\}dt. \qquad (13)$$

P and Q are sufficiently huge numbers to distinguish each group, and t_a is the time when the simulation is aborted. The second term indicates the mean fitness over the simulated period. We check each criterion only in some time

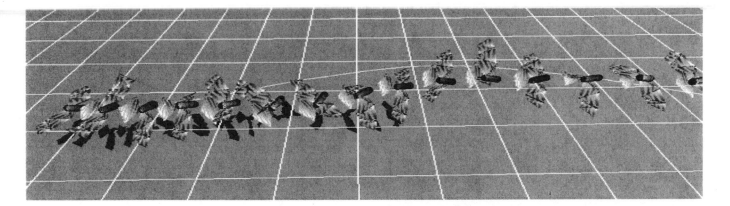

Figure 18: Sequential image of flying creature.(sampled at each 30 time steps)

Parameter	Value or Range
number of wing segment	4~10
cylinder length	0.2~0.8m
cylinder Radius	wings : 0.05m central body : 0.1~0.4m
ψ	initially : 10° ~ 70° during simulation : ± initial ψ
θ	initially : ± 0° during simulation : ± 45°
controller duration	0.2~0.6 sec
density of cylinder	1000kg/m^3
ρ	1.21kg/m^3
k_a, k_b, k_c	20.0, 1.0, 2.0
k_1, k_2, k_3	1.0, 3.0, 2.0

Table 1: Some values used in the simulations.

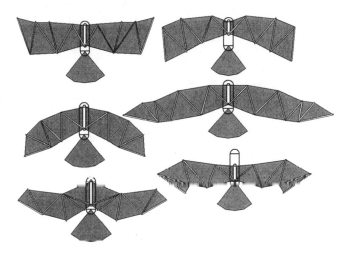

Figure 16: Various evolved wings.

Discussion and Future Work

We have described a system which generates virtual flying creatures and their path-following behaviors by artificial evolution. Several interesting creatures have evolved that exhibit organic and physically plausible flapping flight motions with various wing shapes. Additionally, we showed that the extremely difficult problem of evolving flying creatures capable of maneuvering through the air can be successfully addressed by a two-stage evolution process with robust sinusoidal controllers and a neural network. A new subspace of virtual flying creatures has been discovered in the hyperspace of possible creatures.

However, in our design, the wing structure is closer to that of a hang glider than that of a living creature, because of its non-flexible patagia. The flexibility of feathers or patagia is an important factors in the flight of real winged animals. The evolution of highly complicated shapes such as feathered wings is an extremely challenging problem. Flexible patagia would be a good start to enhancing this work, but

the computational cost of simulating the aerodynamics of deformable objects would need to be addressed for the time consuming evolution process. Since a creature's neural network is optimized only for a given trajectory, the sufficient training strategy or other more effective training algorithms need to be developed to produce creatures capable of following an arbitrary path. Going a step further, there is a need for research which investigates the way to evolve higher-level behaviors, such as diving toward a prey on the ground or soaring by sensing an ascending current of atmosphere. We expect that this method could also be applied to creatures in other environments (land, water) more easily with less running time, because of their comparatively stable conditions compared to creatures in the air. Another direction for future work might be to make an entire digital nature with a complete environment (land, water, sky) including evolved plants. As computers become more powerful, we believe that this sort of research will be a new frontier, which will give infinite diversity to the virtual world in the near future. Some movie clips of the work reported in this paper

Figure 17: Creatures in motion.

are available at http://www.melkzedek.com/efc.htm.

References

Sims, K. 1994. Evolving Virtual Creatures, *ACM Computer Graphics (SIGGRAPH '94)*, pp. 15–22.

Sims, K. 1994. Evolving 3D Morphology and Behavior by Competition. R. Brooks, & P. Maes, (eds.) *Artificial Life IV: Proceedings of the Fourth International Workshop on the Synthesis and Simulation of Living Systems*, pp. 28-39, MIT Press.

Ventrella, J. 1998. Attractiveness vs. Efficiency: How Mate Preference Affects Locomotion in the Evolution of Artificial Swimming Organisms. Adami, C. et al. (eds.) *Artificial Life VI: Proceedings of the Sixth International Conference on Artificial Life.* pp. 178-186: MIT Press.

Komosinski, M and Ulatowski, S. 1999. Framsticks: Towards a Simulation of a Nature-Like World, Creatures and Evolution. Floreano, D. et al. (eds.) *Advances in Artificial Life: Proceedings of the Fifth European Conference on Artificial Life*, pp. 261-265: Springer Verlag.

Ray, T. S. 2000. Aesthetically Evolved Virtual Pets. Maley, C.C. & Boudreau, E. (eds.) *Artificial Life VII Workshop Proceedings.* pp. 158-161.

Lipson, H. and Pollack, J. B. 2000. Automatic Design and Manufacture of Robotic Lifeforms. *Nature 406,* pp.974-978.

Shim, Y. S. and Kim, C. H. 2003. Generating Flying Creature using Body-Brain Co-Evolution. in *ACM SIGGRAPH / Eurographics Symposium on Computer Animation,* pp.276-285.

Reynolds, C. W. 1987. Flocks, Herds, and Schools: A Distributed Behavioral Model. in *Computer Graphics, 21(4) (SIGGRAPH '87 Conference Proceedings),* pages 25-34.

Ramakrishnananda, B. and Wong, K. C. 1999. Animating bird flight using aerodynamics. *The Visual Computer.* 15(10) pp. 494-508.

Wu, X. C. and Popović, Z. 2003. Realistic Modeling of Bird Flight Animations. in *Proceeding of SIGGRAPH '87 ,* pages 888-895.

Smith, R. 1998. Intelligent Motion Control with an Artificial Cerebellum. *PhD Thesis, Dept of Electrical and Electronic Engineering, University of Auckland,* New Zealand. (Available online including ODE engine at http://opende.sourceforge.net)

Anitescu, M. and Potra, F. A. 1997. Formulating rigid multibodydynamics with contact and friction as solvable linear complementarity problems, *Nonlinear Dynamics 14,* 231.247.

Stewart, D. E. and Trinkle, J. C. 1996. An implicit time-stepping scheme for rigid-body dynamics with inelastic collisions and Coulomb friction, *International J. Numerical Methods in Engineering 39,* 2673-2691.

Withers, P. C. 1981. An aerodynamic analysis of bird wings as fixed aerofoils. *Journal of Experimental Biology 90,* 143.162.

Wejchert, J. and Haumann, D. 1991. Animation aerodynamics. *SIGGRAPH '91 Proceedings.* volume 25, number 4.

Mueller, T. J. 2001. Fixed and flapping wing aerodynamics for micro air vehicle applications. *Progress in Astronautics,* volume 195.

Fox, R. W. and McDonald, A. T. 1976. Introduction to fluid mechanics. fifth edition, Wiley.

Rivals, I., Personnaz, L., Dreyfus, G., and Canas, D. 1993. Real-time Control of an Autonomous Vehicle : A Neural Network Approach to the Path Following Problem. in *5th International Conference on Neural Networks and their Applications (NeuroNimes93),*

Behavioural Categorisation: Behaviour makes up for bad vision

Emmet Spier[1]

[1]Centre for Computational Neuroscience and Robotics
Department of Informatics, Sussex University
emmet@sussex.ac.uk

Abstract

The performance of a mobile robot with a vision system is assessed in an everyday object categorisation task. The ability of the robot to arrive at a specified object, *behavioural categorisation*, is compared to the moment to moment results from the computations of its vision system, here called perceptual classification. It is found that the mobile robot using the vision system is significantly more accurate at behavioural categorisation than the underlying performance of the visual system's perceptual categorisation. This result is discussed as supporting the hypothesis that embodied systems using real time algorithms find that 'fast, cheap' visual systems are sufficient for their needs.

Introduction

Animals use vision, along with the rest of their senses, in the control of their behaviour. The creation of internal states that correlate with the presence of objects in the world - the goal of most of computer vision research, and for which there is enticing neuroscientific data (Logothetis et al. 1995) - is not the purpose of vision in the animal. Rather, animals use these products of their visual processing in order to achieve the tasks they set upon. Thus an assessment of the performance of a particular artificial vision system should be judged in terms of the overall performance of the complete system incorporating a vision capacity. Certainly for many industrial classification tasks working with controlled positioning and lighting the recognition ability of the visual system will be the main arbiter of performance. However, as Ballard (1991) observed under the label of *active vision*, a visual system need not rely on computational algorithms alone in order to function; moving the camera can make the difficult easy (e.g. Borotschnig et al., 2000).

When a vision system is integrated as part of the control system of a mobile robot then the internal states of the system are irrelevant to how we judge the performance of the robot. Harvey, Husbands and Cliff (1994) used artificial evolution to discover a network controller for a visually guided robot that can behaviourally distinguish between a triangle and a square. The controller needed to use only two pixels from its visual field to discriminate between a triangle and a square. The discrimination ability of the two-pixel-vision robot was derived from its ability to move, and the use of this self-controlled visual information in concert with its control system to aid its decision. The performance of the visual system is inseparable from the robot's behaviour. Brooks (1991) called such a robot *situated* and Clark (1997) describes the research program as 'embodied cognitive science' to emphasise that animals and robots possess a body whose controlled behaviour will permit their (artificial) brains to solve the 'agent's' problems in a more interactive manner, often reducing the task's complexity and therefore requiring a 'cheaper brain'. More recent embodied mobile robot experiments have also provided further evidence that control architectures combining 'minimal pixel' vision and behaviour can achieve proportionately impressive results (e.g. Scheier, Pfeifer and Kunyioshi, 1988; Marocco and Floreano, 2002). Mobile robots exploiting high resolution sensors in the control of their behaviour have followed two possible strategies, either to make strong assumptions about the visual scene (e.g. the robot Polly, Horswill, 1993, recognised people if they waggled their foot), or to mark the objects of interest in the scene with easy to detect features (e.g. the robot Nomad, Krichmar and Edelman, 2002, uses objects marked with horizontal and vertical lines and an edge detecting filter for discrimination).

This conference paper provides a preliminary investigation of the effect of integrating a mobile robot with a visual system that uses a high resolution image and possesses the acuity and capacity to discriminate between arbitrary everyday objects. We believe that this is the first behaviour-based robot with this capacity. In the next section the perceptual categorisation model incorporated into the robot's control system is described followed by a description of the robot and its experimental environment. The results of an experiment that compares the performance of the visual system alone (*perceptual categorisation*) with the performance of the robot incorporating the visual system (*behavioural categorisation*) are provided and discussed.

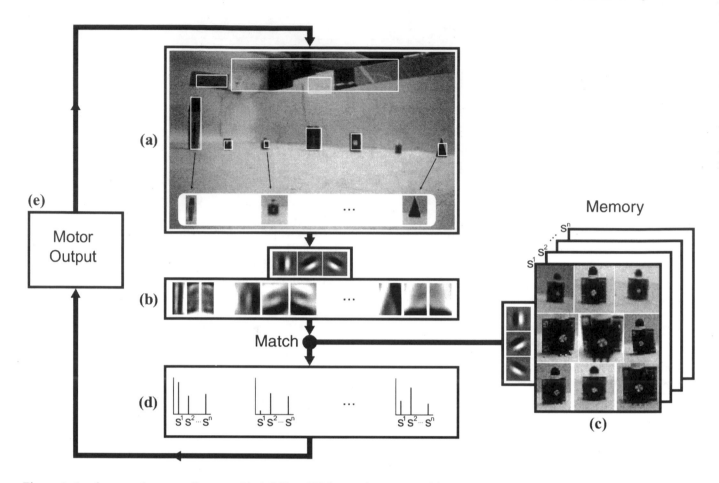

Figure 1: Implemented system diagram. (a) A 760×570 image is segmented into rectangular sections I_k scaled to a maximum dimension of 40 pixels, (b) the I_k are convolved with size 21×21 Gabor filters at angles of π, $\frac{\pi}{3}$ and $\frac{2\pi}{3}$, (c) A memory of stored classes S^i of snapshots (the complete set used for the light sensor class shown here) are compared with the results of stage (b) providing a set of measures of similarity (d) between the blobs I_k and the classes S^i. (e) Motor commands are sent to the robot according to whether the closest I_k to target S^* is to the left or right of the centre of the 760×570 image.

The Perceptual Categorisation Model

An image processing method for a mobile robot would experience the same three dimensional object at various degrees of rotation, perspective, illumination, scale and translation. A traditional (though, in general, neither solved nor automatic) approach to this problem is to attempt to build a 'CAD' style three dimensional model of the input image (after Marr, 1982). For over a decade an increasingly productive alternative approach has been to calculate similarity measures of the input against a class of multiple snapshot-like views. This method was first described using radial basis functions with vertex data input (Poggio and Edelman, 1990) and later using receptive field input (Edelman and Duvdevani-Bar, 1997), though other methods of image approximation can also be used, for instance eigenvectors (Murase and Nayar, 1995). Such 'view' methods provide invariance to rotation and perspective, and a simple version exploiting receptive fields that offer a degree of illumination

invariance is described below. Further, invariance to scale and translation can be achieved through enabling the visual system with the capacity for (virtual) saccades towards every identified visual blob (group of pixels). A crude method of blob segmentation (in general a challenging and completely unsolved issue) was used that thresholded the pixel values of an image scene possessing a highly contrasting background.

Figure 1 is a schematic of the complete classification system. An input image is first segmented into rectangular blobs (a) using both (i) the result of thresholding the distance between each input image pixel and the input image's average colour, and (ii) the result of an adaptive thresholding of the distance between each input pixel and that pixel's local average colour. The first method works well with uniform illumination (a reasonable approximation when close to objects, or within a carefully lit room) and the second provides a good level of robustness to varying light conditions across the input image when objects are of a similar size (the threshold adjusted to maintain a minimum and maximum blob pixel size). Monochrome rectangular cut outs from the input image are made containing each identified contiguous pixel area in the threshold images (the virtual saccades, figure 1 has identified nine) then scaled to a common size. A resulting cut out image (I_k) offers the visual system a reasonable invariance to object translation (from the saccade) and changes of size (with the loss of the ability to discern large from small) within the input image.

The activity of the visual system's low level receptive fields to each I_k can now be computed (b). The products of the convolution of an input image with a collection of Gabor functions of different phases is seen as good first approximation to the receptive fields found in the primary visual cortex (V1), sensitivity to local edges of various orientations (Marcelja, 1980). A Gabor function convolution also effectively removes the input image's non-zero bias (mean illumination) from the resultant receptive field activity. Figure 1b shows the results of each I_k under three convolutions using Gabor functions with rotations of 0, $\frac{\pi}{3}$ and $\frac{2\pi}{3}$, for formal purposes we describe the collection of convolutions of I_k as a real valued vector $g(I_k)$. Howell and Buxton (1998) found this to be a good representation for image matching and unlike eigenspace methods (Murase and Nayar, 1995) it does not require the recalculation of the eigenspace when new objects are learnt.

Each object class S^i is a collection of stored I_k snapshots S^i_j, figure 1c. The selection of snapshots within any S^i is under human control, their automatic recruitment needing to be addressed separately from here. The similarity measure between an I_k and S^i is a distance metric defined as

$$d^i(I_k) = \operatorname{argmin}_j[\| \, g(I_k) - g(S^i_j) \, \|^2 \, \alpha(I_k, S^i_j)]$$

where $\alpha(I_k, S^i_j)$ is a scale factor which is equals 1 when the aspect ratios of I_k and S^i_j are the same and increases in proportion to their difference. (An alternative method (Edelman and Duvdevani-Bar, 1997) is to use a radial basis function with centres at each $g(S^i_j)$ but here we are only interested in the best match.) Figure 1d shows a sketch of the d^i calculated for each of the I_k identified in (a). These d^i constitute the instantaneous categorisation assessments of the visual system; the smallest d^i for each I_k identifying the best fit, the *perceptual categorisation* of that area of the visual space. Using the best fit is equivalent to mutual inhibition between the S^i class responses and controls for the possibility that within any particular visual scene some I_k generally responds with smaller d^i values than other I_k.

Robot Design and Experimental Apparatus

Figure 2 shows the robot used for the reported experiment. The top section comprises a 24MHz 68332 microcontroller mounted in a Lego compatible box. Attached to this box, from left to right, was a wireless video transmitter, a wire-

Figure 2: The robot used, 16cm×12cm×15cm excluding aerials, weighing 800g and supported by two independently driven wheels and a coaster attached front centre. The maximum speed used for this experiment was 2.5cms^{-1}.

less serial transceiver (Abacom Technologies), a 40 character liquid crystal display and a fixed focus CCD video camera with a 62° field of view. The bottom section comprised of two Lego motor, motor controller and gear-chain sets and a front whisker touch sensor responsive to a pressure of 1μN. The Ni-MH rechargeable cell (seen inserted below the serial transceiver) contained sufficient energy to power all attached devices for one hour.

The robot microcontroller handled the motor control and touch sensor monitoring locally with all other computation carried out on a remote workstation (Pentium 4, 1.7GHz) connected via two radio linked channels. The video channel comprised a standard video surveillance broadcast device receiving its signal from the robot's video camera and broadcasting to a remote monitor unit which provided a composite video signal output that was connected to a video capture card (FlashBus MV Lite, Integral Technologies) installed in the workstation. The capture card was programmed to transfer both a monochrome and colour version of the video signal by direct memory access to the workstation's memory at 24 frames per second without significant processor interaction. The robot command channel ran wirelessly over an error corrected 38400 baud serial connection exchanging sensor data from the robot and commands from the workstation 30 times per second. With nine classes comprising a total of 66 snapshots and a captured frame containing ten segmented

Figure 3: The objects, clockwise from left: blue marker pen, rubber, light sensor, wheel, battery, IR sensor, triangle, pencil sharpener, Lego right angle.

blobs the perceptual model implemented on the workstation would run at around 14Hz.

The experimental arena comprised a floor area of 90cm×120cm covered with matt yellow paper rising to a height of 20cm at the edges. To the right of the arena was an external window providing strong side illumination during the day (top centre in Figure 1a) and otherwise lit from a normal office tubular fluorescent light fixed to the ceiling. Figure 3 shows the nine objects used for the experiment, the criteria for selection of these objects was that they should have a width between 1cm and 4cm and neither be yellow nor possess a monochrome intensity close to the yellow arena covering's own intensity.

Experiment and Results

The performance of the robot and its perceptual model was assessed in an object categorisation task. The robot control system as specified in figure 1 would move forward, left or right depending on the whether the blob I_k best matching target class S^*, that moment's particular perceptual categorisation, was in the centre, left or right of the input image. If there was no blob whose best match was class S^* for ten perceptual cycles then alternating and increasing ballistic turns to the left or right would be made as part of a sweeping search strategy.

The criteria for the robot's *behavioural categorisation* of an object was defined to be the single object in the robot's field of view when the robot was sufficiently close to the object that only it could be seen in the image.

Each trial an object was selected as the target class S^* and specified to the control system. The control system was pre-loaded with the nine S^i for the objects shown in figure 3, the particular S^i_j chosen manually during training trials as a sparse sampling that provided good performance. The nine objects were arranged 9cm apart in a randomised order tracing an arc of a circle with radius 50cm. The robot was placed 50cm from the objects and arranged so that the centre

Object	Trials	Mean Accuracy %	Arrival %
blue marker pen	11	67.92	90.91
IR sensor	15	59.65	93.33
Lego right angle	11	89.29	100.00
light sensor	12	72.28	91.67
pencil sharpener	13	63.90	84.62
rubber	10	95.40	100.00
triangle	12	72.14	100.00
wheel	7	77.86	100.00
battery	7	85.16	100.00

Table 1: Summary data of the robot's performance at the perceptual categorisation (mean accuracy %) and behavioural categorisation (arrival %) of each object.

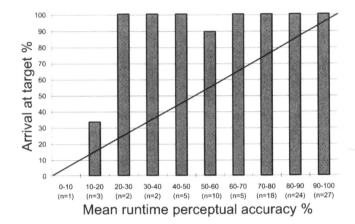

Figure 4: A histogram showing the behavioural categorisation accuracy of the robot for varying perceptual categorisation accuracy (grouped into 10% bins, lower bound inclusive, with the number of trials noted). The continuous line marks the situation where behavioural categorisation performance would be equal to perceptual categorisation performance. Data above the line indicates that behavioural categorisation is more accurate than perceptual categorisation.

of its view field was to the left or right of the target object. The trial was then initiated and terminated when the robot had made a behavioural categorisation.

A complete log of the perceptual system's input and computations during the trial was stored. This permitted, during post trial analysis, a judgement to be made concerning the accuracy of each cycle of perceptual categorisation made by the control system. Every cycle's frame was tagged with a label 'correct' or 'incorrect' according to a human judgment that either the system correctly identified S^i, or the system identified an object judged not be to S^i or did not identify any object as S^i.

Table 1 shows a summary of the trials carried out to measure the robot's ability to recognise each of the objects tested (over 30000 perceptual cycles). Here we note that while

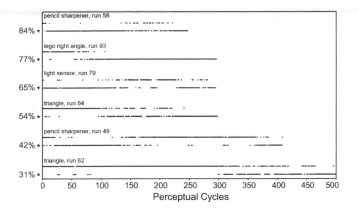

Figure 5: The perceptual categorisation accuracy during specific trials. The row marked with a '∗' records cycles when the perceptual categorisation was 'correct' and its sibling records the 'incorrect', the specific percentage accuracy recorded on the left of the figure. The trials were randomly selected from their 10% bins.

the mean perceptual categorisation accuracy (mean accuracy %) for each object is mediocre the behavioural categorisation accuracy (arrival %) is much improved. A better understanding of the performance of the system can be gained from consideration of figure 4. This histogram considers the behavioural categorisation accuracy of the robot for sets of trials grouped together with a similar perceptual accuracy. When a particular histogram bar is higher than the diagonal line shown in the figure then the behavioural categorisation accuracy is better than the perceptual categorisation accuracy; that is, the robot arrives at the target object S∗ more reliably than would be expected from consideration only of the performance of the perceptual model.

Figure 4 indicates that for trials with overall perceptual categorisation accuracy of 60% (rather average although comparable to state of the art systems using an equivalent number of views per class, Murase and Nayar, 1995) the robot successfully achieves 100% accurate behavioural categorisation. As the data collected is categorical it is only possible to use non-parametric statistical tests. A Mann-Whitney U test was carried out within each bin to test the hypothesis that the robot arrival success was different from the perceptual accuracy. For bins having a mean perceptual accuracy greater 20% the arrival success was significantly different ($p < 0.001$). It should be noted that the number of trials in the bins recording perceptual categorisation accuracy below 40% are too few to make firm statements however the trend of significantly better behavioural categorisation than perceptual categorisation is continued.

Discussion

What was it that the embodied robot system provided that enabled it to so significantly improve upon the performance

of the perceptual categorisation system? Figure 5 discounts some of the possible trivial explanations, that the segmentation or classification aspects of the perceptual classification system are distance sensitive. If the segmentation system only worked within a particular distance range (nearby for instance where further away the 'objects' would not be identified in the visual scene to be categorised) then it would be expected that during every trial the perceptual categorisation accuracy would have a fixed period of error, and therefore the mean value of perceptual accuracy would be reduced. A similar argument can be made for the perceptual classification system being accurate only within a certain range. However, as can be seen from the 84% run there does not appear to be a consistent period of error, and from consideration of the other runs this judgement appears to remain, each run (of a selected successively lower performance) having its own characteristic periods of error.

One possible explanation, relying on the embodiment of the robot, can be based upon the assumption that when the perceptual system makes an error, this error can be broken down into two types. An error could be systematic or random. A systematic error would occur when the perceptual system incorrectly categorised an object outside the class S∗ as being part of that class because the object's I_k matched a particular view within S∗. A random error would occur when through noise in the segmentation algorithm, blurring through movement of the robot, periods when the actual I_k of the target object falls in between two views in S∗, and other sources meant that the target object was not matched correctly by the perceptual system. If a significant proportion of the errors were random errors (and this needs to the tested by a future experiment), these errors would be expected to be distributed uniformly to the left and the right of the robot and over a course of a trial they would cancel out. The consequence of this cancelling out of random errors would be that the *effective* perceptual accuracy, as a result of being part of a control system performing behavioural categorisation, would be higher than the recorded perceptual accuracy. A further prediction can be made under this assumption, it would be expected that the behavioural recognition accuracy of the robot would decrease if, because of the arrangement of objects, the proportion of systematic errors increased.

Conclusion

It is functionally advantageous for an agent to have as accurate as possible classification ability towards relevant objects in its environment. However, an agent's actual ability, as with ever other, constitutes a trade-off between the competing demands of other systems and constraints on the agent's resources be they, amongst others, computational power, heat dissipation, glucose or tissue capacitance. This work provides a demonstration that an agent through its behaviour can reduce the demand it makes upon one part of its nervous system while possibly increasing the demand on another. This flexibility offers a space in which brain or control system design can be optimised for its resource use, thus permitting 'cheaper' solutions than that which would be available if only a disembodied system was used.

The experiment presented provides a demonstration of how agent's which behave in real time do not necessarily require the products of sensory information processing to be highly accuracy. Any particular momentary error the robot makes can be corrected in the next and successive perceptual cycles. Spier and McFarland (1997) showed that sequences of simple reflex decisions can offer better performance than a more complex static calculation. Here we see a relatively computationally cheap vision system (six vector distance calculations per object per image and naturally parallelisable) can provide near perfect performance when situated by a mobile robot.

Acknowledgements

The author would like to thank Bill Bigge for his design and technical support of the robot hardware, Ian Macinnes for the robot development environment, Edgar Bermudez Contreras for help with the post trial analysis, the anonymous reviewers for their comments, and appreciates conversations about this work with John Anderson, Chrisantha Fernando, John Howell and Andrew Philippides. This work has been supported in part by a grant from the Nuffield Foundation (NAL/00669/G).

References

Ballard, D. 1991. Animate Vision. Artificial Intelligence 48:57–86.

Borotschnig, H., Paletta, L., Prantl, M. and Pinz, A. 2000. Appearance-based active object recognition. Image and Vision Computing 18:715–727.

Brooks, R. A. 1991. Intelligence Without Reason. In Proceedings of the 12th international conference on artifical intelligence (IJCAI–91), 569–595.

Clark 1997. Being There. Putting brain, body and world together again. MIT Press.

Edelman, S. and Duvdevani–Bar, S. 1997. A model of visual recognition and categorization. Phil. Trans. R. Soc. Lond. B 352:1191–1202.

Harvey, I., Husbands, P. and Cliff, D. 1994. Seeing the Light: Artificial Evolution, Real Vision. In Proceedings of the Third Conference on the Simulation of Adaptive Behaviour., 392–401. MIT Press.

Horswill, I. 1993. Polly: A Vision-Based Artificial Agent. In Proceedings of the 11th National Conference on Artificial Intelligence (AAAI-93).

Howell, A. J. and Buxton, H. 1998. Learning Identity with Radial Basis function Networks. Neurocomputing 20:15–34.

Krichmar, J. L. and Edelman, G. M. 2002. Machine Psychology: Autonomous Behavior, Perceptual Categorization and Conditioning in a Brain-Based Device. Cerebral Cortex 12:818-830.

Logothetis, N.–K., Pauls, J. and Poggio, T. 1995. Shape recognition in the inferior temporal cortex of monkeys. Current Biology 5:552–563.

Marcelja, S. 1980. Mathematical description of the responses of simple cortical cells. Journal of the Optical Society of America 70:1297–1300.

Marocco, D. and Floreano, D. 2002. Active Vision and Feature Selection in Evolutionary Behavioural Systems. In Proceedings of the Seventh Conference on the Simulation of Adaptive Behaviour (SAB2002), Edinburgh, UK. MIT Press.

Marr, D. 1982. Vision. W.H. Freeman and Company, New York.

Murase, H. and Nayar, S. K. 1995. Visual Larning and Recogntion of 3–D Objects from Appearance. International Journal of Computer Vision 14:5–24.

Poggio, T. and Edelman, S. 1990. A network that learns to recognize three-dimensional objects. Nature 343:263–266.

Scheier, C., Pfeifer, R. and Kunyioshi, Y. 1998. Embedded Neural Networks: Exploiting Constraints. Neural Networks 11:1551–1569.

Spier, E. and McFarland, D. 1997. Possibly Optimal Decision Making under Self-sufficiency and Autonomy. Journal of theoretical biology 189:317–331.

The Evolution of Control and Adaptation in a 3D Powered Passive Dynamic Walker

Eric D. Vaughan[1], Ezequiel Di Paolo[1], Inman R. Harvey[1]

[1]Centre for Computational Neuroscience and Robotics,
University of Sussex,
Brighton, BN1 9QH
{e.vaughan, ezequiel, inmanh}@sussex.ac.uk

Abstract

Humans demonstrate speed, efficiency, and adaptability when traveling over rugged terrain. Bipedal robots modeled on biological designs could replace or assist people working in difficult environments. However, current research into humanoid robots has not produced practical machines. This paper explores the use of evolutionary robotics to evolve a simulation of a ten-degree of freedom bipedal robot. This machine demonstrates many of the properties of human locomotion. By using passive dynamics and compliant tendons it conserves energy while walking on a flat surface. Its speed and gait can be dynamically adjusted and it is capable of adapting to discrepancies in both its environment and its bodies' construction.

Introduction

The development of bipedal machines could allow robots to replace or assist humans in dangerous occupations such as firefighting, bomb disposal, and reconnaissance. Traditionally these kinds of tasks involve rugged environments such as forested, mountainous, and urban terrain, which are challenging for wheeled and tracked vehicles. While people demonstrate dynamic, efficient, and adaptable locomotion in these environments, bipedal robots have demonstrated few of these qualities. Some of these issues may be a result of the differences between biological approaches and technological ones. People have analog nervous systems that are highly parallel, while modern technology is based on deterministic digital computers. Humans are reactive and dynamic, while traditional artificial intelligence is based on linear ideas like: sense, plan, and act (Brooks, 1991). To develop machines with similar properties to humans, it is reasonable to assume they might need to have similar designs. One of the reasons human locomotion is so efficient is that it leverages passive dynamics to reduce energy consumption and uses the elastic nature of tendons to store and release energy. When electrodes are placed in the leg muscles of humans they show almost no activity in the swing leg during walking, except at the beginning and the end of the swing phase (Basmajian, 1976). This is because muscles initiate each step and then allow the stepping leg to swing passively past the stance leg. To determine how much energy can be attributed to tendons, Biewener and Blickhan (1988) recorded the amount of force horses feet exerted when they ran on force plates. They estimated that during galloping the forces released by the tendons recoil contributed to up to 40% of the positive work. Human locomotion is not only efficient but it can also dynamically adapt to changes in environment such as the ruggedness of terrain and even to anomalies in the body due to injury or disabilities. This is accomplished through pulsating collections of neurons called central pattern generators (CPG) in the spine. A CPG is used to generate different gaits that are later modulated by sensors in the muscles and tendons (Grillner, 2003). Through sensor feedback and plasticity the human gait can adapt to both external and internal influences.

This paper explores an approach to bipedal walking that is more closely based on biological designs. Through the use of evolutionary robotics techniques we evolve a simulated bipedal machine that is dynamic, efficient, and adaptable. In our model the body and control system form a single dynamic system whose basin of attraction is walking. Neural networks are used for control, passive dynamics for efficiency, and a CPG to initiate and regulate the walking gait. The result is a simulated machine with ten-degrees of freedom that embodies many of the characteristics of humanoid walking. Its gait and speed can be controlled dynamically, it has a passive swing-leg, and it can adapt to both external and internal disturbances.

Previous work and background

Evolutionary robotics is the use of biologically inspired techniques such as artificial neural networks and genetic algorithms to evolve the morphology and control systems of robots. We see these as dynamic systems whose basin of attraction is the performance of a specific behavior. When a machine enters a situation not encountered during evolution, the attractor will often pull the machine back into stability (Harvey et al., 1996). In 2003 we used evolutionary robotics techniques to evolve both two and three-dimensional passive dynamic walkers. These machines were powered on a flat surface using only their ankles and simple neural networks. Experiments revealed that passive dynamic walkers could have multiple knees as well as having a natural robustness to external noise (Vaughan, 2003).

In 1998 Honda developed a humanoid android P3 (Hirai et al., 1998) and later a smaller machine Asimo. These machines were bipedal and had 12 degrees of freedom in the legs. While they demonstrated the ability to ascend stairs and adapt to subtle slope changes they did not make use of tendons or passive dynamics. This resulted in unnatural walking that was both slow and inefficient.

At MIT a bipedal robot simulation M2 was created with 12 degrees of freedom (Pratt and Pratt, 1999). It had passive leg swing and used actuators that mimicked tendons and muscles. Its control system was composed of a series of hand written dynamic control algorithms. A genetic algorithm was used to carefully tune the machines parameters. When constructed physically this machine was never observed to walk. This may have been the result of discrepancies between the simulation and the physical robot. In our model we aim to resolve this by demonstrating the ability to adapt dynamically to anomalies in the body.

McGeer (McGeer, 1990) designed and simulated a two-dimensional bipedal passive dynamic walker (PDW) with knee joints and curved feet. By carefully selecting the leg mass, leg length, and foot size this robot was able to walk down a four-degree slope with no motors and no control system (Figure 1). Endo (Endo et al., 2002) attached neural oscillators to the joints of a simulated two-dimensional biped, which successfully walked on a flat surface. Bongard (Bongard and Paul, 2001) used evolutionary robotics techniques to evolve the body and control system of a simulated bipedal walker. Their machine had six degrees of freedom and spherical feet. Collins (Collins et al., 2001) physically built a three-dimensional PDW that walked a three-degree slope. The estimated amount of potential energy used by their walker was only three watts.

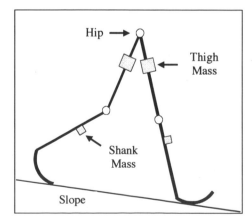

Figure 1: A passive dynamic walker

The simulation

The body of our simulated machine had ten-degrees of freedom: two at each hip, one at each knee, and two at each ankle (Figure 2).

The physics of the body were simulated using the open

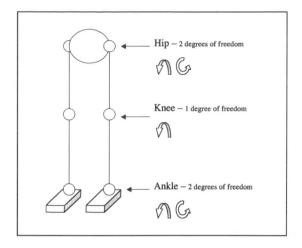

Figure 2: Degrees of freedom in the body

dynamics engine (ODE) physics simulator (Smith, 2003). Weights and measures were computed in meters and kilograms and gravity was set to earths constant of 9.81. The body on average was one meter tall and had 17 parameters (Figure 3): M_w is the mass of waist, M_t is the mass of thigh, M_s is the mass of shank, M_f is the mass of foot, L is the length of a leg segment, Y_t is the offset of the thigh mass on the y-axis, Y_s it the offset of shank mass on y-axis. X_t is the offset of the thigh mass on the x-axis, X_s is the offset of shank mass on x-axis, L_f is the length of foot, A_x is the ankle spring/damper around x-axis, K_x is the knee spring/damper around x-axis, H_x is the hip spring/damper around the x-axis, W is the radius of the waist, B_y is the starting angle of hips joint around y-axis, H_y is the spring/damper of hip around y-axis, A_y is the ankle spring and damper around the y-axis. Parameter ranges were selected based on observations of the human body. The mass of the foot was restricted to be less than that of the shank, the mass of the shank was less than that of the thigh, etc. All parameters were encoded in the genome.

The PDWs explored by McGeer (McGeer, 1990) had curved rigid feet. However, humans have relatively flat feet with ankles. This allows them to increase their traction and stability by keeping more of their foot surface on the ground. The ankle acts as a lever to inject energy in the gait as well as storing and releasing energy through compliant tendons. The PDW developed by Collins (Collins et al., 2001) using curved feet tended to pivot on each step decreasing stability. Kuo (Kuo, 1999) found a similar problem when exploring the lateral stability of a three dimensional straight-legged passive dynamic walker. The addition of roll motion created

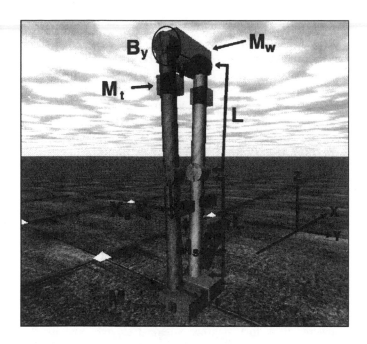

Figure 3: Illustration of body parameters

unstable modes in the periodic gait causing it to fall to one side or the other. We addressed this issue by using ankles and flat feet to increase traction. *(Figure 4)*.

In the human leg, muscle spindles and golgi tendon or-

Figure 4: Springs and dampers in the hips and ankles

gans are used to sense the angle and relative forces placed on them (Alexander, 2002). Through this mechanism they can dynamically adjust their movement. To simulate compliant tendons each joint was composed of a linear actuator that was attached to a spring/damper similar to the series elastic actuators found in MITs M2 (Pratt and Pratt, 1999). Angle sensors on the joint and deflection sensors on the spring/damper were used to acquire feedback *(Figure 4)*.

Almost all life on earth is symmetric, either radial (starfish) or bilateral (insects, mammals, and reptiles) (Miller and Levine, 1998). Even the human brain is symmetrically divided between left and right hemispheres. To explore whether independent neural networks without synaptic coupling can coordinate to produce walking, we used two

symmetric networks and attached one to each leg *(Figure 5)*.

The walking gaits in frogs are generated in the spinal column and then transformed into force patterns that direct their limbs to an equilibrium point in space (Bizzi et al., 1995). In the Lamprey the spinal column is composed of segments each with local touch sensors that modulate its rhythmic swim gaits (Grillner, 2003). In our model we created a virtual spinal cord of neurons that run down each leg. The cord was segmented into three sections one for each joint *(Figure 6)*. Each segment contained two hidden neurons for each degree of freedom that were connected to local sensors for detecting the angle and forces applied to the joint. To power them, a symmetric central pattern generator was attached that sent alternating pulses of opposing signs to each network. There were no axons between the two networks so their only method of communication and interaction was through the bodys actuators and sensors.

Humans have an inner ear with three semicircular canals. They allow people to detect orientation as well as acceleration changes. To mimic this mechanism, a group of three simulated gyroscopes and accelerometers around the x, y, and z-axes were attached to the neural network. To reduce the search space the wiring between sensors and motor neurons was not fully connected. MIT's M2 (Pratt and Pratt, 1999) robot placed a strict separation between the control systems for lateral motion (x-axis) and forward motion (y-axis) and demonstrated natural dynamic walking. In our model we have made a similar design decision. In general sensors around or along one axis were connected to the corresponding motor neurons for the same axises. To give the networks symmetric behavior the sign of the inputs and outputs along the x-axis were inverted in one of the networks. To ensure the default behavior of the ankles and hips was to return to a zero degree angle, negative connections were placed directly between some sensors and motor neurons: specifically between the hips rotating around the y axis and the ankles rotating around the x axis. *(Figure 6)*

Feed-forward continuous time neural networks (CTNN)

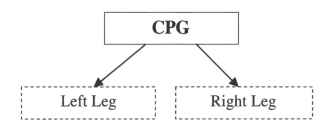

Figure 5: Symmetrical neural networks. The CPG sends pulses to two identical networks one for the left leg and one for the right.

were used to add power to the machine. Unlike traditional neural networks, a CTNN uses time constants to allow neu-

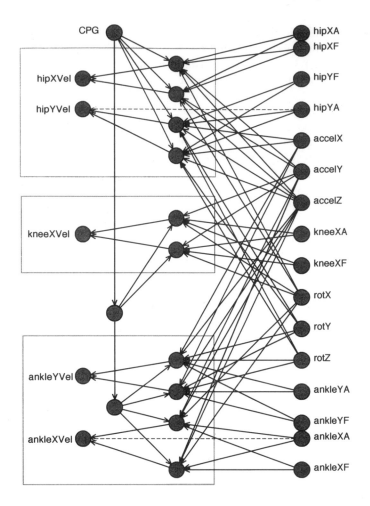

Figure 6: Detail of neural network. Segments are enclosed in boxes. Outputs are on the left and inputs are on the right. 'X', 'Y', or 'Z' are rotation around an axis. Accelerometers 'accel' are a special case where 'X','Y','Z' is acceleration along the axis. 'A' is an angle sensor and 'F' is a force sensor (i.e deflection of spring). 'Vel' is the desired velocity of the output motor. Dotted lines indicate a negative weight and solid lines indicate either a positive or negative weight. All weights were encoded in the genome.

rons to activate in real time and out of phase with each other. For a detailed analysis of this kind of network refer to (Beer, 1996). The state of a single neuron was computed by the following equation:

$$\tau_i \dot{y}_i = -y_i + \left[\sum_{j=1}^{N} w_{ji} \sigma(g_j(y_j)) \right] + I_i + \Omega \quad (1)$$

Where y is the state of each neuron, τ is a time constant, w is the weight of an incoming connection, σ is the sigmoid activation function tanh(), g is the gain, I is an external input, Ω is a small amount of Gaussian noise in the range of [-0.0001, 0.0001]. The state of each neuron was integrated with a time step of 0.2 using the Euler method. In our model neurons were encoded in the genome with τ and g while axons were encoded with real values in the range of [-5, +5].

Biases were omitted.

Four islands (Whitley et al., 1999) of a geographically distributed genetic algorithm (Husbands, 1994) were used each with a population of 50 individuals. The mutation rate was set to 0.5 and then lowered slowly during evolution. Crossover was random. This kind of evolutionary algorithm was used as it has proved previously effective in this context but we do not discount other algorithms being equally effective.

Walking

As demonstrated by Basmajian (1976), the human body takes advantage of passive dynamics during walking. A good question to ask is whether a CPG can initiate walking in a PDW. To find out, the body and control system of a PDW was evolved. The machine was placed on a four-degree incline and the CPG was turned on. Upon completion of a single step the network was completely disconnected from the machine and the ankle actuator was powered back to 0 degrees where it stayed fixed for the duration of the simulation. The fitness function was:

$$f = d \left(\frac{1}{1+t} \right) \left(\frac{1}{1+x} \right) \left(\frac{1}{1+z} \right) \left(\frac{1}{1+r} \right) \left(\frac{1}{1+y} \right) \quad (2)$$

Where: f is the fitness, d is distance travelled, t is the torque used, x is rotation of hip around the x-axis, z is the acceleration of the hip along the z dimension, r is rotation of feet around the z-axis, y is the rotation of the hip around the y-axis. t, x, z, r, and y were averages taken over the entire evaluation time. This fitness function was chosen because it selects for machines that walk as far and straight as possible without explicitly specifying how they move their legs. This allows their leg trajectories to emerge from the dynamics of their bodies rather than from the observations of a human gait. The result was a machine that passively walked down a four-degree slope *(Figure 7)*.

Central pattern generators in animals often dynamically change their rhythm. This can be seen when animals move from one gait to another or wish to increase or decrease their speed. In order to achieve this, the neural networks in our system must be responsive to the pulses of the CPG. If the network in figure 6 is used without any changes it is possible for the machine to walk by ignoring the CPG and setting up a feedback loop with its sensors. This is called a reflexive pattern generator (Beer, 1995) and has been successfully used to power passive dynamic walkers on a flat surface when little or no control of the gait is required (Vaughan, 2003). What is needed is to initially evolve a network with an attractor that is sensitive to the CPG and then allow the sensors to modulate its activity. To do this a subsumption approach was taken. Popularized by Brooks (Brooks, 1991) the idea is to first build and test a simple system and then add additional systems on top. To implement this in our model, first the network was evolved without sensor feedback and

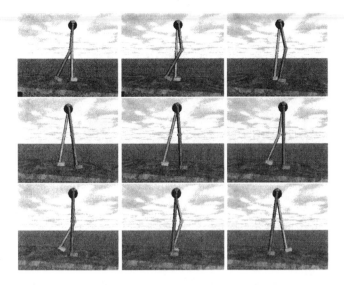

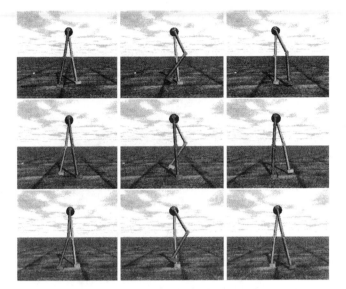

Figure 7: The gait of the passive dynamic walker un-powered on a four-degree slope.

Figure 8: Gait of powered walker on a flat surface.

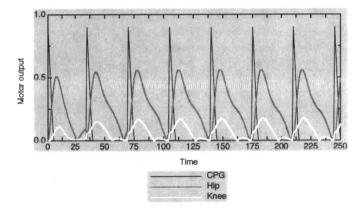

Figure 9: Graph of hip and knee motor outputs on the swing leg while walking on a flat surface. The graph of the knee motor indicates passive swing toward the the end of the swing phase.

then later sensors were reconnected to ensure the system was sensitive to the CPG. The fitness function was modified by multiplying it by the additional factor $\frac{1}{1+v}$ where v is the difference between the powered and passive machines average velocity. The timing of foot strike was recorded from the passive walker and the CPG was updated to pulse with the same timing. The machine was placed on a four-degree slope and the connections to all sensors except those with negative feedback were removed from the network. An oscillating rhythm was applied and the machine was evolved to walk powered down the inclined platform. Once the machine walked for more than ten steps the axons were reconnected to their sensors with very small weights and the population was evolved for an additional number of generations. Over hundreds of generations the slope was incrementally lowered from a four-degree slope to a flat surface. The result was a dynamically stable machine that was not observed to fall even after thousands of steps *(Figure 8)*.

The passive dynamic nature of the machine was preserved. Observation of the motor neurons revealed an increased activity at the start of a step and then a decrease in activity in the swing leg. This is illustrated in *(Figure 9)*. At the start of a pulse both the hip and knee motors are activated. For the first half of the swing phase the knee is extended but on the second half it is nearly turned off to allow it to passively swing past the stance leg.

Robustness

Animals are very adaptive to environmental changes. If the ruggedness of the ground changes suddenly they can dynamically adjust their gait to regain equilibrium. If they are injured or are born with physical abnormalities their body and gait adapts. They do this through the plasticity of their mo-

tor neurons as well as feedback loops between muscles and sensors. To determine the extent evolution can shape the body and control system to cope with noise, the same machine was evolved for an additional 100 generations. On each evaluation small force vectors along the x, y, and z-axis in a Gaussian distribution were applied. The result was a machine that developed dynamic mechanisms to adapt to noise. When pushed too far to one side, the machine was observed to adjust its foot placement by stepping inward to regain balance *(Figure 10)*.

A second experiment was to try to determine the machines ability to adapt to internal noise. The previous experiment was repeated except this time the random forces were replaced by mistakes in the body's construction. Each time the machine was built, errors were introduced to all body

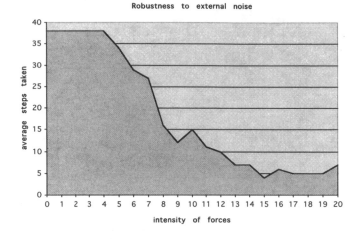

Figure 10: The y-axis is the average number of steps taken over 10 trials and the x-axis the magnitude of forces applied in a Gaussian distribution. Steps were capped at 38.

parameters. The result was a machine that even when built incorrectly could still walk in simulation *(Figures 11 and 12)*.

This result is very important because it demonstrates that

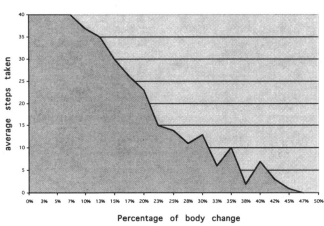

Figure 11: Robustness to internal noise. The y-axis is the average number of steps taken over 20 trials and the x-axis is the amount of noise in construction taken as a percent. (i.e. 10 would be a 10% random change in the overall size of a parameter). Steps were capped at 40.

a control system based on feedback can adapt to changes in the body. It also provides a possible solution to a long-standing problem with robotic simulations. When a simulated control system is transferred to a physical robot, it often fails because it did not take into account small differences between the simulation and reality. Our control system on the other hand appears to compensate by using feedback with its sensors and can adjust to changes dynamically. Even with a 20% error in all body parameters this machine

Figure 12: Two different bodies that walk with the same control system.

still manages to take on average 25 steps. This adaptability may increase its likelihood of making the transfer to a physical robot.

The walking attractor intentionally evolved before the sensors were reconnected continued to be responsive to the CPG. When the time between the CPGs pulses was reduced or increased the machine slowly adjusted its gait to line back up with the new pulse timing *(Figure 13)*. This not only demonstrates that the machines attractor can adapt to situations not experienced during evolution, but it also provides a possible mechanism for controlling the machines speed or changing its gait.

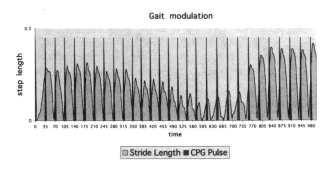

Figure 13: After the first 8 steps the CPG pulse rate is decreased slightly causing the machine to take smaller steps. On step 16 the rate is increased causing the machine to take larger steps.

Conclusion

Humans demonstrate dynamic, efficient, and adaptable locomotion. This is accomplished by the combination of passive dynamics, sensor feedback, and central pattern generators in the spine. These mechanisms are highly parallel and quite different from traditional artificial intelligence approaches. By applying techniques based closer on biological principles such as evolutionary robotics we have demonstrated that it is possible to build dynamic, efficient, and adaptable robots. Instead of a control system based on a linear program a parallel attractor was evolved to keep the machine in a walking gait. Control was decentralized into two symmetrical networks that demonstrated coordinated walking despite the fact there was no direct communication between them. Passive dynamic leg swing was observed indicating that physical machines based on this model could be energy efficient. Through manipulation of the CPGs rhythms the speed of these machines could be adjusted dynamically providing a form of external control. In the future this mechanism may also be used to change the gait between walking and running. The machines attractor was shown to dynamically adapt to both external and internal environmental changes. This is an interesting result since the CTNNs of our model do not store information through weight changes, as many conventional artificial neural networks do. Instead it had to rely entirely on the feedback between its sensors and actuators. This adaptability may provide a mechanism for transferring simulated control systems to physical robots.

This technique is very powerful and we are currently using it to explore more complex bipedal machines with torsos and spines. Some of these simulated machines have up to 25 degrees of freedom and have demonstrated the ability to dynamically run. We are now beginning to build a physical android based on this model and hope to discover further insights into how to use these methods to develop practical bipedal machines. Videos of our simulated machines can be found at (www.droidlogic.com).

References

Alexander, R. (2002). *Principles of Animal Locomotion*. Princeton University Press, New Jersey.

Basmajian, J., V. (1976). The human bicycle. *Biomechanics*, 3-A.

Beer, Randal, D. (1995). A dynamical systems perspective on agent-environment interaction. *Artificial Intelligence*, 72:173-215.

Beer, Randal, D. (1996). Toward the evolution of dynamical neural networks for minimally cognitive behavior. In Maes, P., Mataric, M., Meyer, J., Pollack, J., and Wilson, S., editors, *From animals to animats 4: Proceedings of the Fourth International Conference on Simulation of Adaptive Behavior*, pages 421-429. MIT Press.

Bizzi, E., Giszter, S. F., Loeb, E., Mussaivaldi, F. A., and Saltiel, P. (1995). Modular organization of motor behavior in the frog's spinal-cord. *Trends In Neurosciences*, 18:442-446.

Bongard, J. C. and Paul, C. (2001). Making evolution an offer it can't refuse: Morphology and the extradimensional bypass. In Keleman, J. and Sosik, P., editors, *Proceedings of the Sixth European Conference on Artificial Life, Prague, CZ*, pages 401-412.

Brooks, R. A. (1991). Intelligence without representation. *Artificial Intelligence*, 47:139-159.

Collins, S. H., Wisse, M., and Ruina, A. (2001). Passive dynamic walking robot with two legs and knees. *The International Journal of Robotics Research*, 20(7):607-615.

Endo, I., Yamasaki, F., Maeno, T., and Kitano, H. (2002). A method for co-evolving morphology and walking patterns of biped humanoid robot. In *Proceedings of the IEEE Conference on Robotics and Automation*.

Grillner, S. (2003). The motor infrastructure: From ion channels to neuronal networks. *Nature Reviews Neuroscience*, 4(7):573-586.

Harvey, I., Husbands, P., Cliff, D., Thompson, A., and Jakobi, N. (1996). Evolutionary robotics: The sussex approach. *Robotics and Autonomous Systems*, 20:205-224.

Hirai, K., Hirose, M., Haikawa, Y., and Takenaka, T. (1998). The development of the honda humanoid robot. In *IEEE Int. Conf. Robotics and Automation*, pages 1321-1326.

Husbands, P. (1994). Distributed coevolutionary genetic algorithms for multi-criteria and multi-constraint optimisation. In Fogarty, T., editor, *Evolutionary Computing, AISB Workshop Selected Papers*, volume 865 (LNCS), pages 150-165. Springer-Verlag.

Kuo, A. D. (1999). Stabilization of lateral motion in passive dynamic walking. *International Journal of Robotics Research*, 18(9):917-930.

McGeer, T. (1990). Passive walking with knees. In *Proceedings of the IEEE Conference on Robotics and Automation*, volume 2, pages 1640-1645.

Miller and Levine (1998). *Biology*, chapter 26, page 662. Prentice Hall.

Pratt, J. and Pratt, G. (1999). Exploiting natural dynamics in the control of a 3d bipedal walking simulation. In *Proceedings of the International Conference on Climbing and Walking Robots (CLAWAR99), Portsmouth, UK*.

Smith, R. (2003). The open dynamics engine user guide. http://opende.sourceforge.net/.

Vaughan, E. (2003). Evolution of 3 dimensional bipedal walking with hips and ankles. Msc thesis, Dept. of Informatics, University of Sussex. (http://www.droidlogic.com/sussex/papers.html).

Whitley, D., Rana, S., and Heckendorn, R. (1999). The island model genetic algorithm: On reparability, population size and convergence. *Journal of Computing and Information Technology*, 7:33-47.

Evolving Imitating Agents and the Emergence of a Neural Mirror System

Elhanan Borenstein[1] and Eytan Ruppin[1,2]

[1]School of Computer Science, Tel Aviv University, Tel-Aviv 69978, Israel
[2]School of Medicine, Tel Aviv University, Tel-Aviv 69978, Israel
borens@post.tau.ac.il

Abstract

Imitation is a highly complex cognitive process, employing vision, perception, representation, memory and motor control. The underlying mechanisms that give rise to imitative behavior have attracted a lot of attention in recent years and have been the subject of research in various disciplines, from neuroscience to animal behavior and human psychology. In particular, studies in monkeys and humans have discovered a neural mirror system that demonstrates an internal correlation between the representations of perceptual and motor functionalities. In contradistinction to previous engineering-based approaches, we focus on the evolutionary origins of imitation and present a novel framework for studying the *emergence* of imitative behavior. We successfully develop evolutionary adaptive autonomous agents that spontaneously demonstrate imitative learning, facilitating a comprehensive study of the emerging underlying neural mechanisms. Interestingly, some of these agents are found to embody a neural "mirror" device analogous to those identified in biological systems. Further analysis of these agents' networks reveals complex dynamics, combining innate perceptual-motor coupling with acquired context-action associations, to accomplish the required task.

Introduction

Imitation is an effective and robust way to learn new traits by utilizing the knowledge already possessed by others. The past twenty years have seen a renewed interest in imitation in various fields of research such as developmental psychology, experimental studies of adult social cognition, and most important, neurophysiology and neuropsychology (Prinz and Meltzoff, 2002). Research in this last field had led to the exciting discovery of *mirror neurons*. These neurons, found in the ventral premotor cortex (area F5) in monkeys, discharge both when the monkey performs an action and when it observes another individual making a similar action (Gallese et al., 1996; Rizzolatti et al., 2001). An analogous mechanism, whereby cortical motor regions are activated during movement observations was also demonstrated in humans using TMS, MEG, EEG and fMRI (e.g. Iacoboni et al., 1999). Imitation of motor skills requires the capacity to match between the visual perception of a demonstrator's action and the execution of a motor command. The neural mirror system, demonstrating an internal correlation between the representations of perceptual and motor functionalities, may form one of the underlying mechanisms of imitative ability.

Learning by imitation has already been applied by researchers in the fields of artificial intelligence and robotics in various experiments. Hayes and Demiris (1994) presented a model of imitative learning to develop a robot controller. Billard and Dautenhahn (1999) studied the benefits of social interactions and imitative behavior for grounding and use of communication in autonomous robotic agents. Borenstein and Ruppin (2003) employed learning by imitation to enhance the evolutionary process of autonomous agents. For an up-to-date introduction to work on imitation in both animals and artifacts see the cross-disciplinary collection (Dautenhahn and Nehaniv, 2002b). Furthermore, some researchers, motivated by the recent discovery of a neural mirror system, have implemented various models for imitative learning, embodying neurophysiologically inspired mechanisms. Billard (2000) presented a model of a biologically inspired connectionist architecture for learning motor skills by imitation. The architecture was validated through a mechanical simulation of two humanoid avatars, learning several types of movements sequences. Oztop and Arbib (2002), focusing on the grasp-related mirror system, argued that mirror neurons first evolved to provide visual feedback on one's own "handstate" and were later generalized to understanding the actions of others. They have conducted a range of simulation experiments, based on a schema design implementation of that system, providing both a high-level view of the mirror system and interesting predictions for future neurophysiological testing. Other researchers (Marom et al., 2002) claimed that the mirror system structure can be acquired during life through interaction with the physical or social environment and demonstrated models whereby perceptual and motor associations are built up from experience during a learning phase.

The studies cited above, however, assume that the agents' basic ability and incentive to imitate are innate, explicitly introducing the underlying functionality, structure or dynamics of the imitation mechanism into the experimental system.

In contrast to this engineering-based approach, we wish to study the neuronal mechanisms and processes underlying imitation from an evolutionary standpoint, and to demonstrate how imitative learning *per se* can spontaneously **emerge** and prevail. Clearly, acknowledging the evolutionary origins of imitation and examining the emerging (rather than engineered) imitative learning device can shed new light on the common fundamental principles that give rise to imitative behavior.

In this study, we thus set out to pursue two objectives: Acknowledging the significance of embodied imitation, **we first present a novel experimental framework for evolving context-based imitative learning in evolutionary adaptive autonomous agents** (Ruppin, 2002; Floreano and Urzelai, 2000). We demonstrate the emergence of imitating agents that embody a simple, yet biologically plausible mechanism of imitative behavior. **We then turn to systematically analyze the structure and dynamics of the resulting neurocontrollers**. This analysis surprisingly reveals neural devices analogous to those found in biological systems, including clear examples of internal coupling between observed and executed actions. Further analysis of the network adaptation dynamics demonstrates the innate nature of these internal links with direct bearing on one of the key questions in imitation theory, concerning the ontogeny of mirror neurons (Prinz and Meltzoff, 2002; Hurford, 2003). We conclude with a discussion of the implications of our findings for imitation theory and a description of future work.

Context-Based Imitation

Learning by imitation, like any cognitive process, must be considered an intrinsically embodied process, wherein the interaction between the neural system, the body and the environment cannot be ignored (Keijzer, 2002; Dautenhahn and Nehaniv, 2002a). In particular, every action, either observed or performed, occurs within a certain *context*. A context can represent the time or place in which the action is made, various properties of the environment, or the state of the individual performing the action. Clearly, there is no sense in learning a novel behavior by imitating another's actions if you do not know the context in which these actions are made – a certain action can be extremely beneficial in one context, but have no effect (or even be deleterious) in a different context. We hence use the term *context-based imitation* in the sense of being able to reproduce another's observed action whenever the context in which the action was originally observed, recurs.[1] For example, an infant observ-

---
[1] Animal behavior and human psychology literature introduces a wide range of definitions of imitation, focusing on what can constitute true imitation vs. other forms of social learning (Billard and Dautenhahn, 1999; Zentall, 2001). Our definition addresses the importance of the observed action's *context* for a successful *future* behavior.

ing his parents may learn by imitation to pick up the phone (*action*) whenever the phone is ringing (*context*).

Context-based imitation can thus be conceived as constructing a set of associations (or a mapping) from contexts to actions, based on observations of a demonstrator performing different actions within various contexts. These associations should comply with those that govern the demonstrator's behavior, and should be learned (memorized) so that each context stimulates the production of the proper motor action even when the demonstrator is no longer visible. Such a learning scheme can be seen as an imitation-based analogue of a partially observable hidden Markov model of classical operant learning. It should be noted however, that "*action*" is an abstract notion, and in reality, an imitating individual (agent) should also be capable of matching a *visual perception* of the demonstrator's action with the corresponding *motor command* that activates this action. The key objective of this study is to gain a comprehensive understanding of the mechanisms that govern such context-based imitative learning and in particular to examine the nature of the associations between visual perception, motor control and contexts that are being formed in the process. To address these questions, we employ an experimental setup that embodies context-based imitation within an evolutionary framework.

The Experimental Setup
The Environment

The agents in our simulation inhabit a world that can be in one of several *world states* $\{s_1, s_2, \ldots, s_n\}$. In each time step, the world state is randomly selected from $\{s_1, s_2, \ldots, s_n\}$ with a uniform distribution. These states can represent for example the presence of certain food items or the size of an observed object and hence form the *context* in which actions are observed and performed. An additional set, $\{a_1, a_2, \ldots, a_m\}$, represents the repertoire of motor *actions* that can be performed by the agent or by the demonstrator. Within the simulations described below, both n and m are set to 4. A *state-action mapping* is also defined, assigning a certain action as the proper action for each world state s_i. Regularly performing the proper action assigned to the current state of the world is deemed a successful behavior and confers a positive fitness. It is assumed that the environment is also inhabited by a demonstrator (teacher), successfully performing the proper action in each time step. However, the world state and demonstrator are not visible in every time step and can be seen with probabilities 0.6 and 0.2 respectively. **Furthermore, the above mapping, from world states to actions, is randomly selected anew in the beginning of each agent's run in the world.** The motivation for this state-action mapping shuffle is twofold. First, it prevents such a mapping from becoming genetically determined. To demonstrate a successful behavior, agents must *learn* the proper mapping by observing the demonstrator, promot-

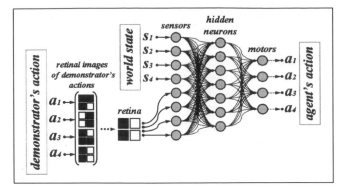

Figure 1: The agent's sensorimotor system and neurocontroller. The sensory input is binary and includes the current world state and a retinal "image" of the demonstrator's action (when visible). The retinal image for each possible demonstrator's action and a retinal input example for action a_4 are illustrated. The motor output determines which actions are executed by the agent. The network synapses are adaptive and their connection strength may change during life according to the specified learning rules.

ing an imitation based mechanism to emerge. Second, it represents a scenario of a changing environment, wherein novel world states appear over time (new food sources, other species, etc.), making prior state-action mappings obsolete.

The Agent

Figure 1 illustrates the structure of the agent's sensorimotor system and neurocontroller. The agent's sensory input in each time step includes the current world state (if visible) and a 4-cell retinal "image" of the demonstrator's action (if visible). The retinal image is determined according to a predefined mapping from actions to retinal binary patterns which remains fixed throughout the simulation. Each of the agent's output neurons represents a motor command, determining which actions (if any) will be executed by the agent.

Each agent embodies a simple feed-forward neural network as a neurocontroller. These networks however are *adaptive*, whereby the genotype of each individual encodes not only the initial synaptic weights but also a *Hebbian learning rule* and *learning rate* for each synapse. In particular, each synapse in the network, (i, j), connecting neuron j to neuron i, is encoded by 4 genes, defining the following properties: (i) w_{ij}^0 - the initial connection strength of the synapse (real value in the range [0,1]); (ii) s_{ij} - the connection sign (1 or -1); (iii) η_{ij} - the learning rate (real value in the range [0,1]); and (iv) Δw_{ij} - the learning rule applied to this synapse. Δw_{ij} encodes 1 of 5 learning (modification) rules: no learning, plain Hebb rule, postsynaptic rule, presynaptic rule and covariance rule. Each synaptic weight w_{ij} is initialized with w_{ij}^0 at the beginning of the agent's life and is updated after every time step (a sensory-motor cycle) ac-

cording to $w_{ij}^t = w_{ij}^{t-1} + \eta_{ij}\Delta w_{ij}$. For a detailed description of the adaptation dynamics see Floreano and Urzelai (2000). The network topology is static throughout the process and for the purpose of our simulation was set to 8-7-4 (i.e., 8 input neurons, a hidden layer with 7 neurons, and 4 output neurons), with an additional threshold unit in each layer. Such *evolutionary adaptive autonomous agents*, inspired by those presented in Todd and Miller (1991) and Floreano and Urzelai (2000), demonstrate a learning process that is supervised only indirectly, through natural selection.

The Evolutionary Process

A haploid population of the agents described above evolve to successfully behave in the environment. Each agent lives in the world for 500 time steps. Fitness is evaluated according the the agent's success in performing the proper action assigned to the current world state (i.e. activating only the appropriate motor neuron), according to the state-action mapping, in each time step. An agent should perform an action only if the world state is visible and regardless of the demonstrator's visibility. We use the Mean-Square Error (MSE) measure to calculate the distance between the agent's motor output and the desired output, averaged over the agent's life. The agent performance during the first 100 time steps is not evaluated (infancy phase). Fitness value is then calculated as $(1-Error)$ and averaged over 20 trial runs in the world.

The initial population is composed of 200 individuals, each assigned a randomly selected genome (encoding the initial connection weights, learning rules and learning rates). Each new generation is created by randomly selecting the best agents from the previous generation according to their fitness, and allowing them to reproduce. During reproduction, 2% of the genes are mutated. The genomes of the top 20% of individuals are copied to the next generation without mutation.

Results

Figure 2 portrays the fitness of the evolving agents across evolution. As evident, the evolved agents successfully master the behavioral task, regularly executing the proper action is each world state. Obviously, given the way the task is designed, this would not have been possible in the absence of an emerged imitation-based learning strategy.

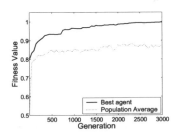

Figure 2: The fitness of the best agent in the population and the population average fitness as a function of generation.

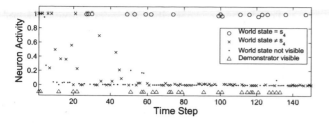

Figure 3: The activation level of one motor neuron (m_2) during the first 150 time steps. The different shapes indicate whether the world state was s_4 and whether it was visible. The triangles at the bottom further represent time steps in which the demonstrator was visible.

Having successfully evolved imitating agents, we turned to examine the structure, dynamics and neural mechanisms that these agents embody. In the rest of this section, we analyze one such successful agent – the best agent in the last generation of a specific evolutionary simulation run. Other successful agents, from various simulation runs, were analyzed and demonstrated similar phenomena (not shown here). Direct evidence of the agent's successful imitative behavior and the resulting learning dynamics are demonstrated in Figure 3, depicting the activity of one of the motor neurons (m_2) in different states of the world. In this specific simulation run, the state-action mapping was arbitrarily set so that a_2 is the proper action in world state s_4 and not in any other state. In the beginning of its life, the agent activates motor m_2, and therefore performs action a_2, whenever the world state is visible. However, after only a few demonstrations of the appropriate behavior, the proper state-action mapping is learned and this motor is activated only when the world state is s_4, as expected.

Furthermore, examining the network hidden layer reveals an interesting phenomenon with regard to the internal representation of actions. As stated above, to support imitative learning, wherein associations from contexts to motor commands should be inferred from observations of the demonstrator's actions, an agent should be capable of matching the visual perception of an observed action to the motor command that generates the corresponding action. Figure 4, depicting the activation level of 3 hidden neurons, attests to the emergence of such inherent perceptual-motor coupling. Apparently, various neurons in the hidden layer are active both when the agent performs a certain action and when it observes the demonstrator making a similar action, **forming internal mirror neurons analogous to those found in biological systems**.[2] Such mirror neurons were found in most

of the agents that evolved in our simulation environment. However, typically, not all actions in the repertoire were associated with a corresponding mirror neuron, and there have been cases where successful agents did not seem to incorporate any identifiable mirror neurons. An additional set of intervention experiments, wherein hidden neurons are externally activated (stimulated) or inactivated (lesioned), was performed (not detailed here). These experiments demonstrated that even actions that could not be associated with a fully localized representation (i.e. a mirror neuron) may still be represented in the hidden layer through a distributed neuronal configuration. These findings are further discussed in the following section.

We finally turn to examine the ontogenic, developmental aspects of the resulting neurocontroller. Our main objective is to identify which components in the neural mechanism are innate and which are acquired during the agent's life. We first determine which synapses play a significant role in the *learning* process. Clearly, variation in the synapse strength during life or the genetically coded learning rate are not appropriate indicators as they cannot differentiate between learning processes that genuinely adapt the agent to the world and unrelated self-organization processes. We thus measure the variance in the connection strength *at the end of the agent's life across 1000 simulation runs*. A low

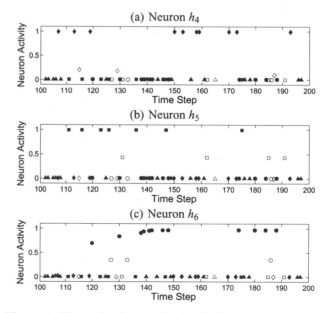

Figure 4: The activation level of 3 hidden neurons (h_4, h_5 and h_6) during time steps 100-200 with an indication of the executed or observed action. Circles, squares, diamonds and triangles represent actions a_1, a_2, a_3, a_4 respectively. A filled shape indicates that the action was executed by the agent (stimulated by a visible world state), while an empty shape indicates that the action was only observed but not executed. Time steps where actions are both observed and executed are not drawn.

---

[2]As seen in Figure 4, the activation level of mirror neurons during action observation is typically lower than the activation level during action execution. An analogous phenomenon can also be detected in neuronal recording data in the literature, and should be further investigated. However, in our simulation, the relatively small number of hidden neurons may account for this phenomenon, forcing mirror neurons to participate also in motor excitation.

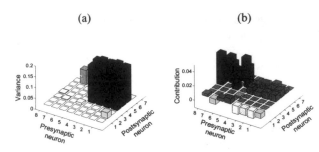

Figure 5: An illustration of the connection strength variance (a) and the overall contribution (b) of the synapses connecting the input layer to the hidden layer.

variance value indicates that the synapse dynamics are independent of the world characteristics (e.g. the state-action mapping), and thus cannot contribute to the learning process that adapt the agent to the world. As demonstrated in Figure 5a, this measure highlights the acquired nature of the synapses connecting the world state neurons (input neurons 1-4), with the mirror neurons we have identified (hidden neurons 4-6). **Clearly, the acquired state-action associations are induced by these synapses**. The markedly lower variance values in other synapses from this layer and in synapses connecting hidden layer neurons to motor neurons (not illustrated here), suggest that these synapses do not play an important part in the learning process. To measure the overall importance of each synapse to the agent's behavior, we have utilized the Multi-perturbation Shapley value Analysis (MSA), a rigorous way to determine the contributions of system elements (Keinan et al., 2004). The resulting contribution of each synapse connecting the input layer to the hidden layer is illustrated in Figure 5b. Evidently, the synapses that have been identified above as participating in the learning process, possess a non-negligible contribution value. However, the most important synapses are among those connecting the retinal neurons (input neurons 5-8), representing the observed action, with the mirror neurons (hidden neurons 4-6). **These connections manifest the strong innate associations between the visual perception of observed actions and the internal representation of these actions, developed during the evolutionary process.**

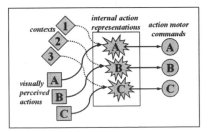

Figure 6: A simple model of context-based imitation. Solid arrows represents innate associations, while dashed arrows represents associations that are acquired during the agent's life via Hebbian learning.

Based on the findings described above, a simple model of the mechanism that evolved in our settings to support imitative behavior can be inferred (Figure 6). Notably, the required perceptual-motor coupling was not explicitly engineered into the agents, but rather emerged through evolution as an *innate* property. Furthermore, to support an effective mechanism of imitation, visually perceived actions are linked to the corresponding motor commands via fully localized internal elements, representing each action, in the form of mirror neurons. The acquired context-action stimuli can then be constructed through a simple mechanism of Hebbian learning without external supervision or reinforcement signals. This model can account for simple, low-level forms of imitation that exhibit a clear example of innate perceptual-motor link such as infant facial imitation (Meltzoff, 1996).

Discussion

This study presents an experimental framework for studying the emergence and dynamics of imitation in evolutionary autonomous agents. This framework provides a fully accessible, yet biologically plausible, distilled model for imitation and can serve as a vehicle to study the mechanisms that underlie imitation in biological systems. Our confidence in this framework is motivated by two observations: First, being an evolutionary emerging mechanism, rather than an engineered one, we believe it is likely to share the same fundamental principles driving natural systems. Second, our analysis of the resulting mechanism reveals phenomena analogous to those found biological neural mechanisms.

The model presented in this paper addresses the very essence of questions concerning the mechanism underlying imitative behavior. It successfully demonstrates how the required associations between perceived actions, motor commands and contexts can be constructed within a hybrid adaptation process, combining evolution and lifetime learning. In particular, addressing the ontogeny of mirror neurons, an issue which is currently in the center of imitation theory research, our model offers a simple schema for the origins and dynamics of the neural mirror system.

The mirror neurons that emerged in our simulation also provide interesting insights to the ongoing debate about the role of representation in embedded systems. Our model, promoting the use of observed actions of *"others"* for learning proper motor actions of *"self"*, suggests a hypothesis for the origins of internal representation. However, the emergence of fully localized internal representations was not absolute and various distributed representations were also demonstrated, supporting Cliff and Noble's (1997) call for an operational definition of representation. The evolution of such distributed representations also confirms that the emergence of mirror neurons within our experimental setup is not trivial and we thus wish to use this model to further determine the physical and social environmental conditions that promote the emergence of localized mirror neurons.

The framework presented in this paper can be further enhanced to simulate a more realistic scenario of social learning. In particular, we wish to examine how an extension of the agent's sensory input, and a complex social environment inhabited by demonstrators with varying levels of success, affect the resulting imitation strategy. Questions concerning the dependencies between observed and executed actions and the formation of mirror neurons are especially of great interest: How will the representation of actions that cannot be executed by the observer (e.g. due to different embodiment) differ from those of imitated actions? How will a hierarchical repertoire of actions affect the emerging representation? Can emerging mirror neurons help predict the actions of others (Ramnani and Miall, 2004)? We hope that further extensions of this basic model will allow us to obtain testable predictions regarding imitative behavior in humans and primates, and shed new light on some of the key issues concerning perception, internal representation and cognition.

References

Billard, A. (2000). Learning motor skills by imitation: a biologically inspired robotic model. *Cybernetics & Systems*, 32:155–193.

Billard, A. and Dautenhahn, K. (1999). Experiments in learning by imitation: grounding and use of communication in robotic agents. *Adaptive Behavior*, 7(3/4):411–434.

Borenstein, E. and Ruppin, E. (2003). Enhancing autonomous agents evolution with learning by imitation. *Journal of Artificial Intelligence and the Simulation of Behavior*, 1(4):335–347.

Cliff, D. and Noble, J. (1997). Knowledge-based vision and simple visual machines. *Philosophical Transactions of the Royal Society: Biological Sciences*, 352:1165–1175.

Dautenhahn, K. and Nehaniv, C. (2002a). The agent-based perspective on imitation. In Dautenhahn, K. and Nehaniv, C., editors, *Imitation in Animals and Artifacts*. The MIT Press.

Dautenhahn, K. and Nehaniv, C. (2002b). *Imitation in Animals and Artifacts*. MIT Press, Cambridge, Mass., USA.

Floreano, D. and Urzelai, J. (2000). Evolutionary robots with on-line self-organization and behavioral fitness. *Neural Networks*, 13:431–443.

Gallese, V., Fadiga, L., Fogassi, L., and Rizzolatti, G. (1996). Action recognition in the premotor cortex. *Brain*, 119:593–609.

Hayes, G. and Demiris, J. (1994). A robot controller using learning by imitation. In *Proceedings of the 2nd International Symposium on Intelligent Robotic Systems*.

Hurford, J. (2003). Language beyond our grasp: what mirror neurons can, and cannot, do for language evolution.

In Kimbrough Oller, U. G. and Plunkett, K., editors, *Evolution of Communication Systems: A Comparative Approach*. MIT Press, Cambridge, MA.

Iacoboni, M., Woods, R., Brass, M., Bekkering, H., Mazziotta, J., and Rizzolatti, G. (1999). Cortical mechanisms of human imitation. *Science*, 286:2526–2528.

Keijzer, F. (2002). Representation in dynamical and embodied cognition. *Cognitive Systems Research*, 3:275–288.

Keinan, A., Sandbank, B., Hilgetag, C., Meilijson, I., and Ruppin, E. (2004). Fair attribution of functional contribution in artificial and biological networks. *Neural Computation*, to appear.

Marom, Y., Maistros, G., and Hayes, G. (2002). Toward a mirror system for the development of socially-mediated skills. In *Proceedings Second International Workshop on Epigenetic Robotics: Modeling Cognitive Development in Robotic Systems*, volume 94, Edinburgh, Scotland.

Meltzoff, A. (1996). The human infant as imitative generalist: a 20-year progress report on infant imitation with implications for comparative psychology. In Hayes, C. and Galef, B., editors, *Social Learning in Animals; The Roots of Culture*, New York Academic Press.

Oztop, E. and Arbib, M. (2002). Schema design and implementation of the grasp-related mirror neuron system. *Biological Cybernetics*, 87:116–140.

Prinz, W. and Meltzoff, A. (2002). *The imitative mind: Development, evolution and brain bases*. Cambridge University Press.

Ramnani, N. and Miall, R. (2004). A system in the human brain for predicting the actions of others. *Nature Neuroscience*, 7:85–90.

Rizzolatti, G., Fogassi, L., and Gallese, V. (2001). Neurophysiological mechanisms underlying the understanding and imitation of action. *Nature Reviews Neuroscience*, 2:661–670.

Ruppin, E. (2002). Evolutionary autonomous agents: A neuroscience perspective. *Nature Reviews Neuroscience*, 3(2):132–141.

Todd, P. and Miller, G. (1991). Exploring adaptive agency II: Simulating the evolution of associative learning. In Meyer, J. and Wilson, S., editors, *From animals to animats: Proceedings of the First International Conference on Simulation of Adaptive Behavior*, pages 306–315, Cambridge, MA. MIT Press.

Zentall, T. (2001). Imitation in animals: evidence, function, and mechanisms. *Cybernetics and Systems*, 32(1-2):53–96.

A Comparison of Population Learning and Cultural Learning in Artificial Life Societies

Dara Curran and Colm O'Riordan

Department of Information Technology
National University of Ireland, Galway

dara.curran@nuigalway.ie
colmor@it.geminga.nuigalway.ie

Abstract

This paper examines the effect of the addition of cultural learning to a population of agents. Experiments are undertaken using an artificial life simulator capable of simulating population learning (through genetic algorithms) and lifetime learning (through the use of neural networks). To simulate cultural learning, the exchange of information through non-genetic means, a group of highly fit agents is selected at each generation to function as teachers which are assigned a number of pupils to instruct. Cultural exchanges occur through a hidden layer of an agent's neural network known as the verbal layer. Through the use of back–propagation, a pupil agent imitates the teacher's behaviour and overall population fitness is increased. We show that the addition of cultural learning is of great benefit to the population and that in addition, cultural learning causes the population to converge on a fixed lexicon describing its environment.

Introduction

The two primary evolutionary forces in nature can be described as population learning and lifetime learning. The first involves genetic inheritance and its role in the behaviour of living things. Populations of creatures evolve through the processes of genetic recombination and mutation. Over long periods, genetic encodings emerge which produce phenotypic traits suitable for a particular environment, such as webbed feet or enhanced vision, giving individuals competitive advantage over others. As generations progress, useful traits will be passed down to successive populations resulting in increases in overall fitness.

Lifetime learning represents each individual's ability to interact and learn from its environment. Creatures which are capable of correctly interpreting novel situations will be more likely to survive the un–predictability of many habitats. Furthermore, creatures which are capable of memory will recall previous errors in similar situations, greatly reducing the risk of repeated damaging behaviour.

Culture is a means of exchange of information through non–genetic means. Examples of such exchanges are language, symbols and artifacts. Means of cultural exchanges between organisms may be considered a subset of lifetime learning, known as cultural learning. Generally, these exchanges allow more experienced organisms to impart some domain knowledge to less informed organisms.

These learning types can be simulated in computer systems using genetic algorithms and neural networks respectively. Genetic algorithms represent potential problem solutions as genetic codes which are then evaluated for fitness. Pairs of codes are selected in proportion to their fitness and are combined together to produce offspring. These offspring become part of the next generation and the process is repeated. Genetic algorithms have been shown to be useful in a vast variety of problem domains.

Neural networks are simplified mathematical models of nervous systems, inspired by the neurons and synapses of living creatures. Neural networks function by reading input patterns, feeding these pattern values through a succession of weighted synapses between neurons, and finally displaying output values at specified output neurons. By comparing a network's output pattern and the desired output pattern for a given input, a measure of error can be obtained. This error is then used to alter the weighting value of synapses that connect neurons in the network. One such algorithm for performing this weighting adjustment is known as error back propagation. Through a series of training iterations, the overall error of a network is reduced, improving its performance.

The combination of genetic algorithms and neural networks provides a framework for evolutionary experimentation in popular domains such as language evolution, neural network design optimization and games.

This paper examines the benefits of employing cultural evolution in populations of reactive agents. Experiments are conducted where agents are placed in an environment where they must distinguish between food and poison bit–patterns in order to achieve high levels of fitness. Cultural evolution is introduced using the teacher–pupil model and its performance is compared to that of population learning alone.

We show that the addition of cultural learning is of great benefit to the population and that the population converges to a common lexicon describing its environment.

The remainder of the paper is organised as follows: the next section gives some background information on related work. The third section presents the artificial life simulator employed in the experiments. Each of its components are discussed: genetic algorithm, neural network architecture and the encoding used to generate valid genetic codes from neural network structures. In the fourth section, we detail the results and in the last section, conclusions are presented and future directions are outlined.

Related Work

A number of learning models can be identified from observation in nature. These can roughly be classified into two distinct groups: population and life-time learning. In this paper we consider another form of lifetime learning, cultural learning.

Population Learning

Population learning refers to the process whereby a population of organisms evolves, or learns, by genetic means through a Darwinian process of iterated selection and reproduction of fit individuals. In this model, the learning process is strictly confined to each organism's genetic material: the organism itself does not contribute to its survival through any learning or adaptation process.

Lifetime Learning

By contrast, there exist species in nature that are capable of learning or adapting to environmental changes and novel situations at an individual level. Such learning, known as life-time learning, still employs population learning to a degree, but further enhances the population's fitness through its adaptability and resistance to change. Another phenomenon related to life-time learning, first reported by Baldwin(Baldwin, 1896), occurs when certain behaviour first evolved through life-time learning becomes imprinted onto an individual's genetic material through the evolutionary processes of crossover and mutation. This individual is born with an innate knowledge of such behaviour and, unlike the rest of the populations, does not require time to acquire it through life-time learning. As a result, the individual's fitness will generally be higher than that of the population and the genetic mutation should become more widespread as the individual is repeatedly selected for reproduction.

Research has shown that the addition of life-time learning to a population of agents is capable of achieving much higher levels of population fitness than population learning alone(Curran and O'Riordan, 2003c; Curran and O'Riordan, 2003a). Furthermore, population learning alone is not well suited to changing environments(Curran and O'Riordan, 2003b).

Cultural Learning
Culture can be succinctly described as a process of information transfer within a population that occurs without the use of genetic material. Culture can take many forms such as language, signals or artifactual materials. Such information exchange occurs during the lifetime of individuals in a population and can greatly enhance the behaviour of such species. Because these exchanges occur during an individual's lifetime, cultural learning can be considered a subset of lifetime learning.

An approach known as synthetic ethology(MacLennan and Burghardt, 1993; Steels, 1997; Kirby and Hurford, 1997; Smith, tted) argues that the study of language is too difficult to perform in real world situations and that more meaningful results could be produced by modelling organisms and their environment in an artificial manner. Artificial intelligence systems can create tightly controlled environments where the behaviour of artificial organisms can be readily observed and modified. Using genetic algorithms, the evolutionary approach inspired by Darwinian evolution, and the computing capacity of neural networks, artificial intelligence researchers have been able to achieve very interesting results.

In particular, experiments conducted by Hutchins and Hazlehurst(Hutchins and Hazlehurst, 1995) simulate cultural evolution through the use of a hidden layer within an individual neural network in the population. This in effect simulates the presence of a Language Acquisition Device (LAD), the physiological component of the brain necessary for language development, whose existence was first suggested by Chomsky(Chomsky, 1976). The hidden layer acts as a verbal input/output layer and performs the task of feature extraction used to distinguish different physical inputs. It is responsible for both the perception and production of signals for the agent.

A number of approaches were considered for the implementation of cultural learning including fixed lexicons(Yanco and Stein, 1993; Cangelosi and Parisi, 1996), indexed memory(Spector and Luke, 1996), cultural artifacts(Hutchins and Hazlehurst, 1991; Cangelosi, 1999) and signal–situation tables(MacLennan and Burghardt, 1993). The approach chosen was the increasingly popular teacher/pupil scenario(Billard and Hayes, 1997; Denaro and Parisi, 1996; Cangelosi and Parisi, 1996) where a number of highly fit agents are selected from the population to act as teachers for the next generation of agents, labelled pupils. Pupils learn from teachers by observing the teacher's verbal output and attempting to mimic it using their own verbal apparatus. As a result of these interactions, a lexicon of symbols evolves to describe situations within the population's environment.

Experiment setup

The experiments outlined in the next sections are designed to simulate an environment where agents may encounter food or poison bit patterns. Agents that correctly identify these patterns are awarded higher levels of fitness than agents who mistake food for poison.

For this set of experiments, food and poison bit patterns are 5-bit patterns representing the 5-bit parity problem, where food is assigned the value 1 and poison the value 0. The next section outlines the artificial life simulator that is employed in these experiments.

Simulator

The architecture of the simulator is based on a hierarchical model (figure 1). Data propagates from the simulator's interface down to the simulator's lowest level. The neural network and genetic algorithm layers generate results which are then fed back up to the simulator's highest level.

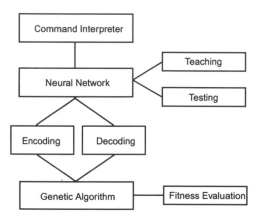

Figure 1: *Simulator Architecture*

Command Interpreter Layer

The command interpreter is used to receive input from users in order to set all variables used by the simulator. The interpreter supports the use of scripts, allowing users to store parameter information for any number of experiments. A parsing algorithm is used to break down a user's commands and to pass appropriate instructions to the relevant components of the simulator.

Genetic Algorithm Layer

The genetic algorithm layer is responsible for the creation of successive generations of neural networks. The layer is responsible for the simulation of population learning within the simulator. The layer accepts data from the encoding layer as a genetic algorithm data structure. This structure is then manipulated to generate the next generation using the selection, crossover and mutation operators.

Selection The selection process examines each individual in the population and must determine which individuals to place into the intermediate population. The simulator genetic algorithm layer uses the network's error and accuracy as a fitness measure to make this evaluation.

Once the fitness value is computed for each individual in the population, linear based fitness ranking is used to normalize fitness values evenly over the entire population. This is done to overcome the potential for stagnation across the population gene pool. Traditional proportional fitness assignment may produce a population with very similar fitness values, thereby deteriorating the selection process and possibly resulting in a loss of diversity stemming from a small number of individuals being allowed to reproduce many times. Ranking introduces a uniform scaling across the population's fitness so that fitness values are evenly spread and making the selection process more successful. In addition, fitness ranking provides a simple means of controlling selective pressure, the probability that the best individual is selected compared to the average probability of selection of all individuals(Goldberg, 1989).

The total number of individual networks in the population N_{ind} is sorted such that each individual occupies a ranking position *pos* where *pos*=1 represents the least fit individual and *pos*=N_{ind} the most fit individual. For selective pressure SP in [1.0,2.0], a network's fitness value $Fitness_{net}$ is calculated as:

$$Fitness_{net} = 2 - SP + 2(SP-1)(pos-1)/(N_{ind}-1)$$

Roulette wheel selection is then applied to the population to select individuals for the intermediate population. The number of times that an individual may be copied to the intermediate population is a parameter which can be set using the command interpreter.

Crossover As a result of the chosen encoding scheme, crossover may not operate at the bit level as this could result in the generation of invalid gene codes. Therefore, the crossover points are restricted to specific intervals — only whole node or link values may be crossed over. Two–point crossover is employed in this implementation.

Mutation The mutation operator introduces additional noise into the genetic algorithm process thereby allowing potentially useful and unexplored regions of problem space to be probed. The mutation operator usually functions by making alterations on the gene code itself, most typically by altering specific values randomly selected from the entire gene code. In this implementation, weight mutation is employed. The operator takes a weight value and modifies it according to a random percentage in the range [-200%,200%].

Encoding and Decoding Schemes

Before the encoding and decoding layers can perform their respective tasks, it is necessary to arrive at a suitable encoding scheme. Many schemes were considered in preparation of these experiments, prioritising flexibility, scalability, difficulty and efficiency. The scheme chosen is based on Marker Based Encoding which allows any number of nodes and interconnecting links for each network giving a large number of possible neural network permutations.

Marker based encoding represents neural network elements (nodes and links) in a binary string. Each element is separated by a marker to allow the decoding mechanism to distinguish between the different types of element and therefore deduce interconnections(Kitano, 1990; Miller et al., 1989). A gene code produced using this scheme is treated as a circular entity. Thus, the code parsing mechanism reading the end of the gene code will begin reading the start of the gene code once the end is reached until all available information is correctly retrieved.

Encoding and Decoding Layers

To perform genetic algorithm tasks, the neural network structures must be converted into gene codes on which the genetic algorithm will perform its operations. Conversely, once the genetic algorithm has performed its tasks, the genetic code structure must be converted back to a neural network architecture. The encoding and decoding layers must therefore accept both neural network data structures and genetic algorithm data structures to function correctly.

Neural Network Layer

The neural network layer is responsible for all functions carried out by the neural networks in the simulator's population. The neural network layer accepts a population of neural network data structures and performs a number of functions.

Simulating Cultural Evolution

In order to perform experiments related to cultural evolution, it was necessary to adapt the existing simulator architecture developed by Curran and O'Riordan(Curran and O'Riordan, 2003c) to allow agents to communicate with one another. This was implemented using an extended version of the approach adopted by Hutchins and Hazlehurst. The last hidden layer of each agent's neural network functions as a verbal input/output layer (figure 2).

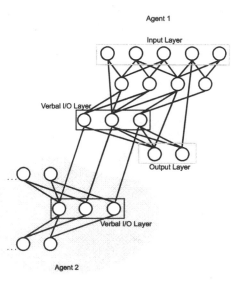

Figure 2: *Agent Communication Architecture*

At the end of each generation, a percentage of the population's fittest networks are selected and are allowed to become teachers for the next generation. The teaching process takes place as follows: a teacher is stochastically assigned n pupils from the population where $n = \frac{N_{pop}}{N_{teachers}}$, where N_{pop} is the population size and $N_{teachers}$ is the number of teachers. Each pupil follows the teacher in its environment and observes the teacher's verbal output as it encounters what it believes to be food or poison bit patterns. The pupil then attempts to emulate its teacher's verbal output using backpropagation. Once the teaching process has been completed, the teacher networks die and new teachers are selected from the new generation.

Unlike previous implementations, the number of verbal input/output nodes is not fixed and is allowed to evolve with the population, making the system more adaptable to potential changes in environment. In addition, this method does not make any assumptions as to the number of verbal nodes (and thus the complexity of the emerging lexicon) that is required to effectively communicate.

Experiments

The purpose of this set of experiments is to identify to what degree the addition of cultural learning aids a population of agents to attain high levels of fitness. We compare the performance of a population of 250 agents endowed with the ability to communicate using the teacher–pupil model to that of a population that uses purely genetic means to advance its fitness levels. The parameters for the cultural learning setting are chosen as follows: the teacher ratio, that is the percentage of the population that is chosen as teachers, is set at 0.1. The number of teaching cycles, the exposure each pupil has to the teachings of a particular teacher is set at 5 cycles. These parameters where chosen following a series of preliminary experiments. The results presented below are averaged from twenty experiment runs over 250 generations.

Results

It is clear from this experiment result illustrated in figure 3 that cultural learning is indeed enhancing the population's fitness and its ability to correctly distinguish between food and poison patterns. Figures 5, 6 and 7 show the evolution of cultural exchanges for one of the food patterns during one of the experiments. The illustration shows the verbal output for the population at generation 1, 125 and 250. These results show that the population's cultural exchanges are converging to specific verbal output patterns and that the population is generating a lexicon of patterns to represent instances of food and poison patterns.

The verbal layer of an agent's neural network is responsible for the cultural exchanges that take place during the agent's lifetime. In order for a cultural exchange to be successful, a teacher's verbal layer must be compatible to that of its pupils. This means that a pupil must have at least the

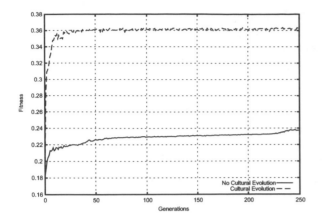

Figure 3: *Cultural Evolution*

same number of verbal nodes in its verbal layer as its teacher. If this does not happen, some information being transmitted from the teacher nodes will be lost.

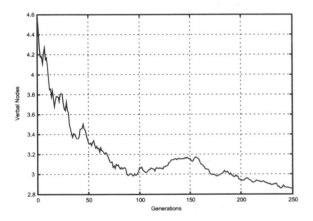

Figure 4: *Verbal Node Evolution*

The evolution of the number of verbal nodes can be seen in figure 4 showing that the number of verbal nodes in the population has converged. This result is intuitive, because to successfully and consistently communicate with one another, it is necessary that all parties in a cultural exchange have a similar number of verbal outputs. Teachers with a larger number of verbal outputs than pupils would effectively be attempting to communicate information that the pupils cannot perceive, let alone replicate.

Conclusion

These experiments have shown that the addition of cultural learning to a population of agents can greatly improve their overall fitness. This is particularly important when compared to supervised life–time or population learning approaches where problem solutions must be known in advance in order to correctly train populations. By selecting teachers from the populations, we allow the population to

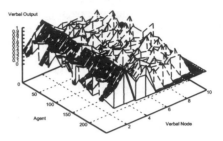

Figure 5: *Verbal Output at Generation 1*

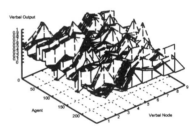

Figure 6: *Verbal Output at Generation 125*

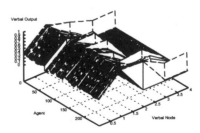

Figure 7: *Verbal Output at Generation 250*

self–supervise its progress, without the need for problem knowledge or human supervision.

Future work will concentrate on employing the system on problems which would be impossible to solve using traditional life–time learning approaches and will investigate the effect of cultural evolution in changing environments.

Acknowledgements

This research is funded by the Irish Research Council for Science, Engineering and Technology.

References

Baldwin, J. (1896). A new factor in evolution. In *American Naturalist 30*, pages 441–451.

Billard, A. and Hayes, G. (1997). Learning to communicate through imitation in autonomous robots. In *7th International Conference on Artificial Neural Networks*, pages 763–738.

Cangelosi, A. (1999). Evolution of communication using combination of grounded symbols in populations of neural networks. In *Proceedings of IJCNN99 International Joint Conference on Neural Networks (vol. 6)*, pages 4365–4368, Washington, DC. IEEE Press.

Cangelosi, A. and Parisi, D. (1996). The emergence of a language in an evolving population of neural networks. *Technical Report NSAL–96004, National Research Council, Rome.*

Chomsky, N. (1976). On the nature of language. In *Origins and evolution of language and speech*, pages 46–57. Annals of the New York Academy of Science, New York. Vol 280.

Curran, D. and O'Riordan, C. (2003a). Artificial life simulation using marker based encoding. In *Proceedings of the 2003 International Conference on Artificial Intelligence (IC-AI 2003)*, volume II, pages 665–668, Las Vegas, Nevada, USA.

Curran, D. and O'Riordan, C. (2003b). Lifetime learning in multi-agent systems: Robustness in changing environments. In *Proceedings of the 14th Irish Conference on Artificial Intelligence and Cognitive Science (AICS 2003)*, pages 46–50, Dublin, Ireland.

Curran, D. and O'Riordan, C. (2003c). On the design of an artificial life simulator. In Palade, V., Howlett, R. J., and Jain, L. C., editors, *Proceedings of the Seventh International Conference on Knowledge-Based Intelligent Information & Engineering Systems (KES 2003)*, University of Oxford, United Kingdom.

Denaro, D. and Parisi, D. (1996). Cultural evolution in a population of neural networks. In *M.Marinaro and R.Tagliaferri (eds), Neural Nets Wirn-96.New York: Springer*, pages 100–111.

Goldberg, D. E. (1989). *Genetic Algorithms in Search, Optimization and Machine Learning*. Reading, MA, Addison-Wesley.

Hutchins, E. and Hazlehurst, B. (1991). Learning in the cultural process. In *Artificial Life II, ed. C. Langton et al.* MIT Press.

Hutchins, E. and Hazlehurst, B. (1995). How to invent a lexicon: The development of shared symbols in interaction. In Gilbert, N. and Conte, R., editors, *Artificial Societies: The Computer Simulation of Social Life*, pages 157–189. UCL Press: London.

Kirby, S. and Hurford, J. (1997). Learning, culture and evolution in the origin of linguistic constraints. In *4th European Conference on Artificial Life*, pages 493–502. MIT Press, Cambridge, MA.

Kitano, H. (1990). Designing neural networks using genetic algorithm with graph generation system. In *Complex Systems, 4, 461-476.*

MacLennan, B. and Burghardt, G. (1993). Synthetic ethology and the evolution of cooperative communication. In *Adaptive Behavior 2(2)*, pages 161–188.

Miller, G. F., Todd, P. M., and Hedge, S. U. (1989). Designing neural networks using genetic algorithms. In *Proceedings of the Third International Conference on Genetic Algorithms and Their Applications*, pages 379–384.

Smith, K. (submitted). The cultural evolution of communication in a population of neural networks. *Connection Science.*

Spector, L. and Luke, S. (1996). Culture enhances the evolvability of cognition. In *Cognitive Science (CogSci) 1996 Conference Proceedings.*

Steels, L. (1997). The synthetic modeling of language origins. In *Evolution of Communication*, pages 1–34.

Yanco, H. and Stein, L. (1993). An adaptive communication protocol for cooperating mobile robots.

A Computational Framework to Simulate the Coevolution of Language and Social Structure

Tao Gong*, Jinyun Ke, James W. Minett and William S-Y. Wang

Language Engineering Laboratory, City University of Hong Kong, Hong Kong

*50005488@student.cityu.edu.hk

Abstract

In this paper, a multi-agent computational model is proposed to simulate the coevolution of social structure and compositional protolanguage from a holistic signaling system through iterative interactions within a heterogeneous population. We implement an indirect meaning transference based on both linguistic and nonlinguistic information in communications, together with a feedback without direct meaning check. The emergent social structure, triggered by two locally selective strategies, *friendship* and *popularity*, has small-world characteristics. The influence of these selective strategies on the emergent language and the emergent social structure are discussed.

1. Introduction

Recently, computational modeling of language evolution has grown rapidly, as exemplified by many anthologies and reviews (Standish et al 2003; Cangelosi and Parisi 2001; Wagner et al. 2003). Many computational models, based on evolutionary or artificial life theories, have been reported, such as the neural network models (e.g., Munroe and Cangelosi 2002), the vocabulary coherence model (e.g., Ke et al. 2002), and the Iterative Learning Framework (ILM) (Kirby 2002; Smith et al. 2003). These "emergent" models (according to Schoenemann 1999) share several assumptions related to language development. However, there are still several limitations.

First, most of them assume *direct meaning transference* (excluding Munroe and Cangelosi (2002)) in the interactions among agents, i.e., the intended meanings encoded in linguistic utterances and sent by speakers are always accurately available to listeners. However, it is obvious that expression and interpretation are independent in speakers' and listeners' minds, and that there are at least no *direct* connections among them. Other channels, such as pointing while talking or primitive feedback, can only provide a certain degree of confirmation. Interpretation is a complex process requiring linguistic and nonlinguistic information. It is unrealistic to assume direct meaning transference.

Second, these models either fail to model syntax (e.g., Ke et al. 2002), build in the syntactic features (e.g., Munroe and Cangelosi 2002), or else do not adopt a coevolutionary view of the emergence of syntax and the lexicon

(e.g., Smith et al. 2003). However, syntax in language is likely to have become conventionalized through language use, rather than as the result of an innate, grammar-specific module (Schoenemann 1999). The syntax is assumed to have emerged because of a pre-adapted cognitive capacity reflected in other cognitive processes, i.e. the sequencing ability, which can be attested in other primates and pre-language infants (Christiansen and Ellefson 2002). The emergence of the lexicon and the convergence of syntax should be interwoven, i.e. they should coevolve.

Third, these models often use random interactions, which disregard the influence of social structure. Although sociological research has studied structures that have emerged based on stable or global factors, very little research has touched upon the emergence of structure based on the evolution of language. Mutual understanding based on the evolving language can be a factor to trigger social structure and so is worth studying. Recent developments in complex networks (Newman 2003) provide an efficient methodology to study it.

Fourth, most current models are based on homogeneous populations. However, sociolinguists have shown there to be dramatic variations in the speech community and various dichotomies in the learning styles of children (Shore 1995). Heterogeneity of natural characteristics and linguistic behaviors among agents should therefore be considered in the computational models that are adopted.

Addressing these limitations and based on the "emergent" theory of Wray (2002), we present a computational model which uses an indirect meaning transference and simulates the coevolution of lexicon and syntax (simple word order) during the transition from a holistic signaling system to a compositional language. Based on mutual understanding of the evolving language and two locally selective strategies, this model also simulates the emergence of social structure during the emergence of language. In Section 2, we describe the model. Results and discussions are presented in Section 3. Finally, we draw some conclusions and point out some future directions in Section 4.

2. Description of the model

This model is basically a linguistic communication game among independent agents in a population, focusing only on horizontal transmission. Agents express and interpret two types of meanings: "predicate<agent>", such as "run<tiger>", and "predicate<agent, patient>", such as "chase<tiger, wolf>" or "eat<tiger, meat>". Nonlinguistic information (*Cues*) is used to assist the meaning interpretation, especially meanings like "chase<tiger, wolf>" — without cues, it is not clear who is chasing whom, especially in the early stages of the language evolution. Cues, pragmatic meanings describing environmental events, are integrated meanings all with the same strength, e.g., "fight<dog, cat>" (0.5); 0.5 is the strength. Cue Reliability (*CR*) manipulates the probability that the intended meaning is contained in one of the cues.

This model uses a rule-based system to represent the language. Linguistic rules includes *lexical rules* (meaning-utterance mappings), such as holistic, phrase and word rules, and *word order rules* which cover all possible sequences to regulate utterances for expressing integrated meanings with two or three meaning constituents, such as "agent first, predicate last, patient medial" (denoted by SVO for simplicity). Rule strengths, numerically indicating the frequency of successful use of the rules, can be adjusted by self-organizing strategies in rule competition. Agents start with a holistic signaling system (sharing a set of common holistic rules) and no dominant word order (all word order rules have the same strengths). Through iterative communications, a common set of rules shared by all agents indicates the convergence of the language.

A two-scale storage system, inspired by a *Classifier System* based model (Holland 2001), is used to handle lexical rules. It includes a buffer (storing "previous experiences" — meaning-utterance mappings (M-U mappings) obtained in previous communications) and a rule list (storing rules generalized from M-U mappings in the buffer when it is full). Rules in the rule list are used to express integrated meanings and interpret utterances, together with nonlinguistic information, in future communications.

There are two mechanisms for agents to acquire new rules: *random creation* in meaning expression (as in Kirby et al.'s model), i.e., with a certain probability, speakers create lexical rules (holistic or compositional) to help their production of integrated meanings, and *rule generalization*, i.e., a flexible detection of *recurrent patterns* (recurrent constituents in meanings and recurrent syllables in utterances among two M-U mappings) without syntax or location restriction. Rule generalization occurs when the buffer is full. Figure 1 shows some rule generalization examples. By extracting recurrent patterns as new compositional rules, some holistic signals are decomposed.

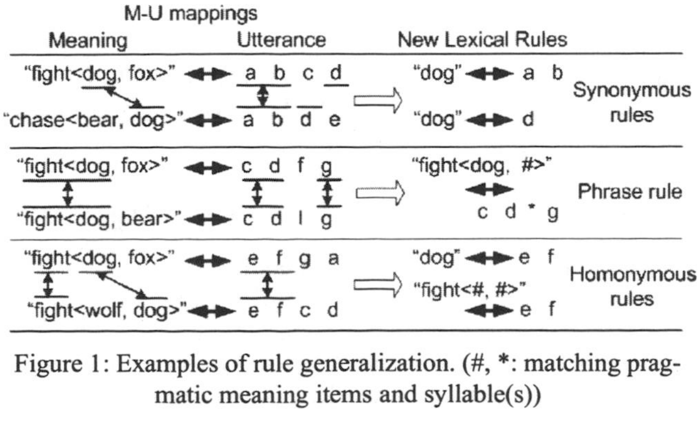

Synonymous and homonymous rules emerge inevitably during the execution of these two mechanisms because there is no clear access to other agent's language, and flexible rule generalization does not consider the existent rules. Due to the limited size of storage, the lack of context (meanings expressed in communications are independent of each other) and unreliable cues (otherwise, it would still be *direct meaning transference*), homonym avoidance, in which a form that is used "successfully" is reinforced and others which it competed are weakened, is built in. As for synonyms, agents randomly learn one form from a set of synonymous rules based on the *Principle of Contrast* (Clark 1987).

Heterogeneity means different agent can have different buffer and rule list sizes and different linguistic abilities for random creation and rule generalization.

Communications in this model are *concurrent*; during each time step, many communications between different pairs of agents happen simultaneously. An indirect meaning transference is implemented in communication. Communication proceeds as follows (summarized in Figure 2). First, the speaker selects a meaning to express. Based on his current linguistic rules, the speaker encodes the selected meaning into the utterance of his winning rules which have the highest combined rule strength, CS_{speak}, calculated from the formula

$$CS_{speak} = \frac{\text{Str}(\text{combinable activated rules})}{+ \text{Str}(\text{applicable word order rules})} \quad (1)$$

The utterance, built up accordingly, is transferred to the listener, who attempts to interpret the utterance. The listener sometimes also receives cues from the environment. Interpretation involves a more complex process of rule competition, considering not only linguistic but also

Figure 1: Examples of rule generalization. (#, *: matching pragmatic meaning items and syllable(s))

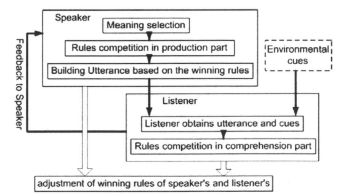

Figure 2: Indirect meaning transference.

nonlinguistic information, in the listener's mind, based on the combined rule strength CS_{listen} of his linguistic rules:

$$CS_{listen} = LangWeight \left\{ \begin{array}{l} Str(\text{combinable activated rules}) \\ + Str(\text{applicable word order rules}) \end{array} \right\} \qquad (2)$$
$$+ EnvWeight \left\{ Str(\text{Environmental Cues}) \right\}$$

The listener interprets the meaning based on his winning rules. If the combined rule strength of the listener's winning rules exceeds a certain threshold, a positive feedback is sent to the speaker indicating the listener's confidence in the interpretation. Otherwise, a negative feedback is sent, meaning that the listener was either unable to infer a meaning or else was not confident of inferring the intended meaning. Finally, based on this feedback rather than on a direct meaning check, both the speaker and the listener adjust their own rules, increasing the strengths of the winning ones and decreasing those of the losing ones if the feedback is positive. During the whole process of communication, expression and interpretation are independent and the interpretation is based on the interaction of linguistic and nonlinguistic information.

Mutual understanding based on the evolving language can influence the possibility of future communication between these two agents. A fully-connected weighted network is used to indicate the social relationships among members; see Figure 3. The connection weight, adjusted in both successful and failed communications, indicates the cumulative probability of successful communication between the two agents. Once the connection weight exceeds a threshold, a permanent edge is built. However, such permanent connections can still be broken after many failed communications. Agents permanently connected to each other are linguistic "friends", and have a higher chance to understand each other. The number of permanent edges of one agent indicates his linguistic "popularity", i.e., his propensity to communicate successfully with other agents. Friendship and popularity are local factors focusing on individual agents.

To enhance the realism of the model, we introduce a *local-view* assumption similar to Li and Chen's paper (2003), i.e., in one communication, one agent can only view several agents (local-view) instead of all group members and communicate with some of them.

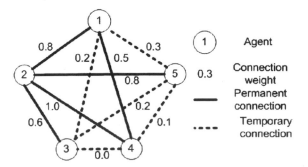

Figure 3: Social network used in this model.

We run two types of simulation. **Simulation 1:** Each generation, each agent selects the agents in his local-view, those to whom he is permanently connected having a higher chance to be chosen. Agents communicate with a subset of agents in their local-view, preferring to communicate with agents having higher popularity. A new generation begins after all agents have executed this process. **Simulation 2:** Each generation, agents randomly select the agents in their local-view and randomly attempt to communicate with some of them.

In this model, heterogeneities, such as different buffer or rule list size, different random creation and generalization rates (simulated by the Gaussian distribution), and different mechanisms to acquire new rules (simulated by random assignment), are allowed to make the model more realistic.

Finally, several major factors are used to study the performance: a) the *understanding rate* (*UR*), defined by

$$UR = \frac{\sum_{i,j} (\text{number of understandable meanings between agent } i, j)}{(\text{number of all possible pairs of } i, j)} \qquad (3)$$

indicates the average number of meanings understandable by every pair of agents in the population based on linguistic information only — it tests the real representation ability of the acquired language, considering not only the expressivity of meanings that might not happen in immediate environment, but also the understandability of these expressions (Displacement) (Hockett 1960); b) the *degree distribution* (P_k) indicates the distribution of the number of permanent connections (degree) versus the number of agents having such degree; c) the *number of sub-clusters* of connected agents, indicates the divergence within the population. Other parameters, such as the *rule expressivity* (*RE*), *convergence time* (*CT*) (the number of rounds of communication by which the highest *UR* is reached), the *average degree* (*AD*), the *clustering coefficient* (*C*), and the *average shortest path length* (*L*) are also considered. The following simulations' conditions are: 50 agents, 500 generations, *RC*=0.8. Buffer size=35±5, Rulelist size=45±5. Random creation rate=0.5±0.2, Rule generalization rate=0.5±0.2

3. Results and discussions

3.1 Coevolution of lexicon and simple syntax

Coevolution of the lexicon and syntax is simulated in this model, as shown in Figure 4. Figure 4(a) shows the *RE* of both holistic rules and compositional rules; the decrease of the former and the increase of the latter show the transition from initially holistic signals to a compositional language. The *UR* in Figure 4(a) shows the convergence to a common lexicon. The *UR* undergoes an S-shaped evolution, matching the result of Ke et al.'s model (2002). The *RE* of compositional rules used in combination increases rapidly, but the use of compositional rules may cause some meanings understandable when expressed by holistic rules to be misunderstood, causing the *UR* to briefly drop slightly. However, the recurrence of these compositional rules in successive communications allows them to win the competition with the holistic rules, which finally makes possible the emergence of a common lexicon.

Figures 4(b–c) show the convergence of the dominant word order from all possible sequential order rules; the curves trace the average strength of each of the eight order rules. Mutual understanding requires not only common

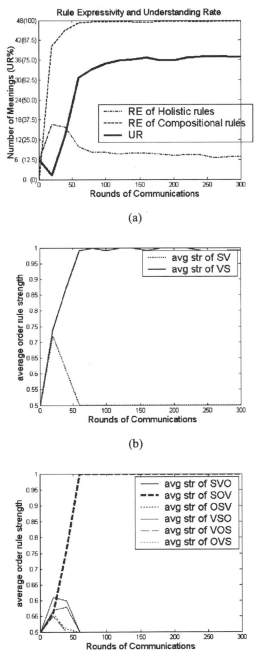

(a)

(b)

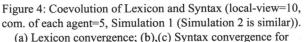

(c)

Figure 4: Coevolution of Lexicon and Syntax (local-view=10, com. of each agent=5, Simulation 1 (Simulation 2 is similar)).
(a) Lexicon convergence; (b),(c) Syntax convergence for "predicate<agent>" and "predicate<agent, patient>" meanings.

lexical rules but also a shared syntax to combine compositional rules. Two dominant word orders emerge from the initial state of no syntax, one for each of the two meaning types. There is no prior bias conferred to any particular word order; each is initially equally likely. Finally, combining Figures 4(a–c), we observe the coevolution of lexicon and syntax: the use of compositional rules triggers the convergence of syntax, which in turn boosts the convergence of the lexicon; the sharp increase of the *UR* and the dominant order rules' strengths are almost *synchronized*.

3.2 Emergence of social structure

During the emergence of language, the selective strategies trigger a global social structure based on the mutual understanding of the evolving language. The *AD* and *C*, also following an S-curve, trace the emergence of the social structure (see Figures 5(a–b)). Due to the restriction of the selective strategies, the *AD* and *C* of Scenario 1 is smaller. Besides, these strategies have their own influences. For example, the local, "self-centered" strategy of friendship can trigger an earlier increase of *AD* and *C* compared with Simulation 2. It also triggers an earlier emergence of sub-clusters (see Figure 5(c)). The high *C* and low *L* indicate that the emergent social structures of both simulations have small-world (Watts 1999) characteristics. However, their structures are different due to the influence of the friendship and popularity strategies (see Figure 5(d)). In Simulation 2, a network that is almost fully-connected emerges, with most agents having the same, high degree. However, in Simulation 1, the degree distribution is more uniform. Although almost all members can understand each other, the degrees of some agents do not increase much. This is because friendship tends to restrict agents to communicate with other agents belonging to their local-view and popularity only triggers a local convergence within the local-view. Agents within the local-view might have intensive connections with one another, but they do not connect to outsiders frequently. This local centralization prevents the degrees of some agents from greatly increasing.

On the other hand, different local-view sizes in the selective strategies can influence the emergent social structure. With the increase of the local-view size, the influence of friendship is gradually reduced, which breaks down the local convergence. Then, the degree of every agent increases gradually. This can be seen in Figure 6(a).

As for the language that emerges, with the increase of the local-view size, the centralization is more global; there seems to be an optimal local-view size for peak *UR* (Figure 6(b)) — either too much "democracy" or too much "dictatorship" cannot achieve the best *UR*. Actually, centralization around some agent(s) has two effects. First, popular agents connect many unpopular agents, like a network hub. Centralization around them can increase the chances for unpopular agents to exchange information, and thereby accelerate the convergence of linguistic rules. On the other hand, effective information transference between two agents (say, Agent1 and Agent2) requires direct

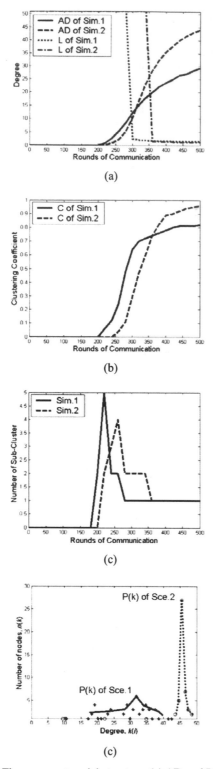

much, so that the information received by Agent2 via the popular agent does not change much from the original information sent by Agent1). However, with the increase of global centralization, other agents have higher chances to contact the popular agent and influence his rules. This makes the popular agent unstable, i.e., although the input information is the same, the output information differs greatly from time to time. This greatly affects the information transference and the convergence of linguistic rules between Agent1 and Agent2 through the popular agent. To compromise these two contradictory factors, the optimum performance occurs at an intermediate level of centralization.

Finally, the structure of Simulation 1 is triggered by the social strategies which are based on the evolving language. The evolution of the language has its influence on the final result; these social strategies, if based on a non-evolving language, can trigger a local-world (Li and Chen 2003) (see Figure 6(c)) or, if local-view is the whole group, a scale-free (Barabási 1999) structure. However, with an evolving language, this model has no such structures. This shows that when using language-related factors to trigger structure, one should consider the evolution of language.

4. Conclusions and future directions

Considering the evolutionary point of view and real communication situations, coevolution and indirect meaning transference are more realistic. Mutual understanding of the evolving language triggers a social structure showing small-world characteristics during the emergence of language. Without considering the evolving property of the language, results are different, which shows the influence of the evolving property on the emergent social structure. This model imports the study of emergent social structure based on evolving factors using artificial life modeling, an appropriate tool to simulate and study the influences of these evolving factors.

Several future directions are promising. First, the current model can be "situated" in an artificial world, and the Genetic Algorithm (GA) (Holland 1975) used to evaluate the fitness, both with and without language. Second, other types of communication, such as "one speaker, multiple listeners", need to be simulated to further study the language emergence in a more realistic situation. Third, in the social structure aspect, it is worth comparing the structures triggered by linguistic communication with those triggered by other nonlinguistic factors. Besides, when an agent chooses other agents to communication, he should consider not only linguistic-related factors such as "friendship", "popularity", but also other factors, such as kinship, economic status.

Figure 5: The emergent social structure. (a) AD and L of the two simulations; (b) C of the two simulations; (c) Sub-Clusters; (d) P_k.

connection or connection through a "stable" intermediary (say, a popular agent, whose internal rules do not change

Acknowledgments

This work is supported by City University of Hong Hong grants No. 9010001-570, 9040781-570. The authors offer thanks to Profs. J. Holloand, G. R. Chen and T. Lee for useful discussions and suggestions.

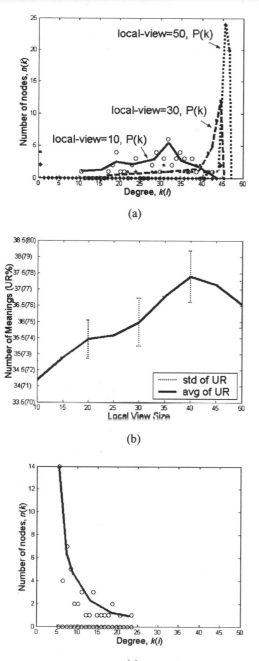

(a)

(b)

(c)

Figure 6: Local-view size effects. (a) P_k; (b) UR; (c) P_k of local world structure (based on Li and Chen (2003)'s model)

References

Barabási, A.L. 1999. Mean-field Theory for Scale-free. Physica A 2(72): 172–182.

Cangelosi, A., and Parisi, D. eds. 2001. Simulating the Evolution of Language. London: Springer Verlag.

Christiansen, M. H., and Ellefson, M. H. 2002. Linguistic Adaptation without Linguistic Constraints: The Role of Sequential Learning in Language Evolution. In Wray, A. ed.

The Transition to Language, 335–358. Oxford: Oxford University Press.

Clark, E. 1987. The principle of contrast: A constraint on language acquisition. In MacWhinney, B. ed. Mechanisms of language acquisition, 1–33. Hillsdale, NJ: Lawrence Erlbaum Assoc.

Hockett, C. F. 1960. The Origin of Speech. Scientific American 203: 88–96.

Holland, J. H. 1975. Adaptation in Natural and Artificial Systems. Michigan: University of Michigan Press.

Holland, J. H. 2001. Exploring the Evolution of Complexity in Signaling Networks. Complexity 7(2): 34–45.

Ke, J.; Minett, J. W.; Au, C. P.; and Wang, W. S-Y. 2002. Self-organization and selection in the emergence of vocabulary. Complexity 7(3):41–54.

Li, X. and Chen, G. R. 2003. A Local-World Evolving Network Model. Physica A 328(1-2):274–286.

Newman, M. E. J. 2003. The Structure and Function of Complex Networks. SIAM Review 45.

Munroe, S. and Cangelosi, A. 2002. Learning and the evolution of language: the role of culture variation and learning cost in the Baldwin Effect. Artificial Life 8(4):311–340.

Schoenemann, P. T. 1999. Syntax as an Emergent Characteristic of the Evolution of Semantic Complexity. Minds and Machines 9:309–346.

Shore, C. M. 1995. Individual Differences in Language Development. Thousand Oaks, CA. Sage Pub.

Smith, K.; Kirby, S.; and Brighton, S. 2003. Iterative learning: A framework for the emergence of language, Artificial Life 9:371–386.

Standish, R. K.; Bedau, M. A.; and Abbass, H. A. eds. 2003. Artificial life VIII: proceedings of the eighth International Conference on Artificial Life. Cambridge, MA: MIT Press.

Kirby, S. (2002). Learning, Bottlenecks and the Evolution of Recursive Syntax. In: Briscoe, T. ed. *Linguistic Evolution through Language Acquisition: Formal and Computational Models*. Cambridge, MA, Cambridge University Press. pp. 173–205.

Wagner, K.; Reggia, J. A.; Uriagereka, J.; and Wilkinson, G. S. 2003. Progress in the simulation of emergent communication and language. Adaptive Behavior 11(1):37–69.

Watts, D. J. 1999. Small Worlds. Princeton, NJ: Princeton University Press.

Wray, A. 2002. Formulaic Language and the Lexicon. New York: Cambridge University Press.

Boom and Bust:
Environmental Variability Favors the Emergence of Communication

Patrick Grim and Trina Kokalis

Group for Logic & Formal Semantics, Dept. of Philosophy, SUNY at Stony Brook
pgrim@notes.cc.sunysb.edu

Abstract

Environmental variability has been proposed as an important mechanism in behavioral psychology, in ecology and evolution, and in cultural anthropology. Here we demonstrate its importance in simulational studies as well. In earlier work we have shown the emergence of communication in a spatialized environment of wandering food sources and predators, using a variety of mechanisms for strategy change: imitation (Grim, Kokalis, Tafti & Kilb 2000), localized genetic algorithm (Grim, Kokalis, Tafti & Kilb 2001), and partial training of neural nets on the behavior of successful neighbors (Grim, St. Denis & Kokalis 2002). Here we focus on environmental variability, comparing results for all of these mechanisms in a range of different environments: (a) environments with constant resources, (b) environments with random resources around the same mean, and (c) sine-wave variable environments with cycles of 'boom and bust'. Communication, it turns out, is strongly favored by environmental variability on the pattern of 'boom and bust'.

Introduction

Sometimes the same idea appears persistently across a range of very different disciplines. Environmental variability seems to be such an idea.

In behavioral psychology, environmental variability has long been established as an important factor in operant conditioning. Intermittent schedules of reinforcement prove far more effective than constant reinforcement, with variable-ratio schedules producing the highest number of responses per time period and establishing behavior most resistant to extinction (Honig & Staddon 1966, Nye 1992).

In ecology and evolution, rates of environmental fluctuation have been proposed as a major factor in inter-species dynamics (Chesson and Huntly 1997). It has recently been proposed that Pleistocene climatic fluctuations are responsible for the evolution of larger brained mammals in general and higher primates in particular, with suggested links to social learning (Opdyke 1995, Potts 1996, Boyd & Richerson 2000).

In cultural anthropology, variable environments appear to play a major role in the transition from foraging cultures to incipient agriculture (Reynolds 1986).

It is tempting to think that these appeals to environmental variability may have something in common. Perhaps there is some central mechanism of variability and selection which, in different forms, is responsible for the way that individuals learn, the way that species evolve, and the way that cultures develop. The results we offer here indicate that an important role for environmental variation shows up even in simple computer simulations.

The Basic Model

We work throughout with an initially randomized 64 x 64 two-dimensional cellular automata array of 4,096 individuals carrying different behavioral strategies (Figure 1). All action and reproduction are local: individuals interact only with their eight immediate neighbors.

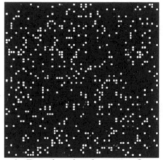

Figure 1 Randomized array of strategies

Our individuals alter their behavior in terms of what is happening immediately around them, but they do not move. In the simple models offered here, it is food sources that move, migrating in a random walk across the array. If a food source lands on an individual with its mouth open, that individual 'feeds' and gains points. Our food sources are not consumed, however; like a cloud of plankton or a school of fish, they continue their random walk, offering nourishment for the next individual down the line.

On any given round, an individual's strategy may dictate that it opens its mouth or does not, where mouth-opening carries a particular cost in energy. Its strategy also dictates whether or not it makes a sound on that round, heard by itself and its immediate neighbors. Sound-making also carries a cost in energy.

For even these simple individuals in this simple environment, there are forms of behavior that would seem to qualify as elementary forms of signaling or communication. Imagine a community of individuals that share the following strategy:

They make a sound when they are successfully fed.
They react to hearing a sound from their neighbors by opening their mouths.

When an individual in such a community feeds, it makes a sound. Its immediate neighbors, which share the same strategy, open their mouths in response. Since the food source continues its random walk, it will then fall on an open mouth on the next round. The result, in a community sharing such a strategy, is a chain reaction in which the food source is successfully exploited on each round (Figure 2). We term individuals with such a strategy 'Communicators'.

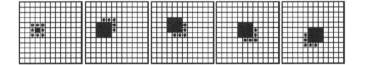

Figure 2 Migration of a food source across an array of Communicators. Gray shading indicates an open mouth, * the range of a sound made.

In previous work, using both this model and a more complex variation incorporating predators, we have shown that simple forms of communication can emerge from initially randomized arrays in synchronous updating using any of several mechanisms for strategy change. In our earliest studies we use strategy change by simple imitation. After 100 centuries of gain and loss, each cell surveys its immediate neighbors in order to see if any had garnered a higher score. If so, it adopts the strategy of its most successful neighbor (Grim, Kokalis, Tafti, and Kilb 2000). In later studies we use strategy change by local genetic algorithm. Here the strategy of a less successful cell is replaced with a hybrid , created by genetic algorithm from its strategy and that of its most successful neighbor (Grim, Kokalis, Tafti, and Kilb 2001). Most recently, we have instantiated strategies in the weights of simple neural nets, and have used strategy change by partial training on the behavior of more successful neighbors (Grim, St. Denis, and Kokalis 2002). Using any of these mechanisms, we have been able to show that communities of Communicators will emerge and grow. Figure 3 shows a typical emergence of two forms of Communicators—here with signals for both food and predators—in an array of randomized neural nets over 300 generations.

In this background research, however, we use an environment of constant resources: although our food sources migrate in a random walk across the array, the total number of food sources remains constant from generation to generation. Here we focus instead on the role of a *variable* environment. Is change in the environment a factor of importance in the emergence of communication? Does the *pattern* of change matter? The

results that follow indicate that a variable environment does indeed have a major impact on the emergence of communication, even in computer simulations as simple as those explored here. The pattern of variability, it turns out, is also crucial.

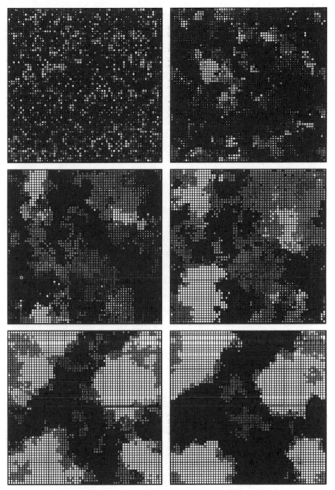

Figure 3 Emergence of two dialects of Communicators, shown in solid black and white, in a randomized array of neural nets with partial training on successful neighbors. (Grim, St. Denis, and Kokalis 2002)

Comparison Agents: Imitators, Localized Genetic Algorithms, and Neural Nets

Here, for the sake of simplicity, our environments contain wandering food sources but no predators. The behavioral repertoire of our individuals is similarly limited: they can open their mouths or not, and can make a single sound heard by their immediate neighbors or can remain silent. Mouth opening carries an energy cost of .95 points, with an energy cost of .05 points for sounding.

We code the behavior of these simple individuals in terms of binary four-tuples $<f, \sim f, s, \sim s>$. Variable f dictates whether an individual makes a sound or not when it is fed, $\sim f$ whether it makes a sound when it is not fed, s

dictates whether it opens its mouth when it hears a sound from itself or an immediate neighbor, and ~*s* whether it opens its mouth when it hears no such sound.

This gives us only sixteen possible strategies, of which four are of particular note. Those cells that carry strategy <1,0,1,0> are our 'Communicators'. They make a sound when fed, and open their mouths when they hear a sound. A hypothetical community of Communicators will therefore behave as illustrated in Figure 2. Strategy <0,0,1,0> is a 'free rider'; it opens its mouth when it hears a sound, benefiting from a signal from a Communicator neighbor, but does not signal reciprocally when fed. The null strategy <0,0,0,0> does nothing—it never opens its mouth and never makes a sound—and so pays no energy costs. 'All Eat' <0,0,1,1> keeps its mouth open constantly, harvesting any food that comes by but never making a sound.

It should be noted that we use 'imperfect' worlds throughout. All cells follow their programmed strategies subject to a 5% measure of error. Nowak and Sigmund (1990) argue that a measure of stochastic 'noise' makes for a more realistic model of cooperation. In previous work we have outlined its importance for the emergence of communication as well (Grim, Kokalis, Tafti and Kilb, 2000).

Although our sample space of behaviors is the same across our studies, those behaviors are instantiated in different ways—as coded behaviors or as operating neural nets. This allows us to compare strategy change by imitation, by localized genetic algorithm, and by local training of neural nets side by side.

In one series of runs our individuals carry behaviors coded as series of binary digits, and follow an imitation algorithm for strategy change. After 100 rounds of food gathering, point gain and energy loss, each cell surveys its immediate neighbors and sees if any has garnered a higher score. If so, it adopts the neighbor's strategy in place of its own.

In a second series of runs we use the same coding for behaviors, but employ a localized genetic algorithm for strategy change. After 100 rounds, should a cell have a more successful neighbor, its strategy is replaced with a genetic algorithm hybrid formed from its current strategy and that of its most successful neighbor. We use two-point crossover, choosing one of the offspring at random to replace the parent. It should be noted that ours is a *localized* genetic algorithm. All genetic recombination is local: cells with locally successful neighbors change their strategies to local hybrid recombinations.

In a third series of runs we generate the same sample space of behaviors using very simple neural nets (Figure 4). For simplicity, we use bipolar inputs with weights and biases 'chunked' at one-unit intervals between -3.5 and +3.5. If the sum at the output node exceeds a threshold of 0, the output is treated as +1, and the individual opens its mouth, for example. If less than or equal to 0, the output is treated as -1, and the individual keeps its mouth closed.

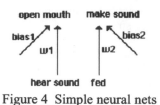

Figure 4 Simple neural nets

For our neural nets, strategy change is by partial training on successful neighbors. With bipolar coding and within the limits of our value scale, using 'target' for the neighbor's output, we can calculate the delta rule as simply $w_{new} = w_{old} + (target \times input)$ and $bias_{new} = bias_{old} + target$.

Though their behavior ranges are identical, our agents instantiate three very different forms of updating mechanisms. What we want to compare is their behavior across a range of different environments.

Environmental Variability and the Emergence of Communication

1. Communication in a Constant Environment

Our constant environment contains exactly 50 food sources each generation, moving in a random walk as before. We use the gain allotted for successful feeding as an independent variable: tests are run with gains from 1 to 140 points for each successful feeding. We plot what strategy an array evolves to—Communicators or otherwise—and in what number of generations.

Figure 5 shows new results across our three modes of strategy change. Runs are to 1500 generations, with the height of each bar indicating how many generations were required to fixation on a single strategy. Should no single strategy occupy the entire array by 1500 generations, the bar tops out, showing the strategy dominant in the array at that point.

It is immediately obvious, and somewhat surprising, how large the window for communication is in each of these cases. Communicators dominate the array from the

Imitation in a Constant Environment

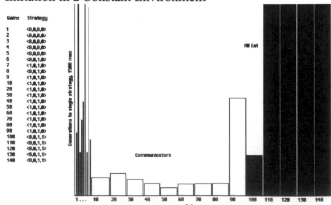

Localized Genetic Algorithm in a Constant Environment

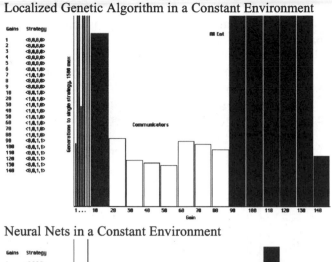

Neural Nets in a Constant Environment

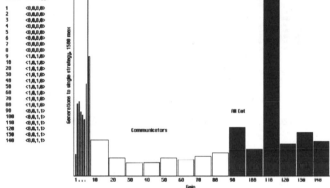

Figure 5   Conquest by All Eat at gains of 90 and above using strategy change by imitation, by localized genetic algorithm, and by partial neural net training on successful neighbors.

case in which each successful feeding is worth 10 points to the case in which it is worth 9 times as much.   But it is also clear   that communication has an upper terminus: above a gain of 100 points it is a strategy of All Eat proves dominant.

2. Communication in a Random Environment

In a second series of studies we assigned a random number of food sources between 0 and 100 each generation, randomly placed and moving.   The average number of food sources remained at 50, but the particular number any generation might be anywhere between 0 and 100. The amount of gain allotted for successful feeding was again our independent variable: tests were run with gains for each successful feeding from 1 to 140 points for each successful feeding.   Figure 6 shows results in a random environment for strategy change by imitation, localized genetic algorithm, and neural nets.

With any of our mechanisms of strategy change, it turns out, results in a randomized environment show at most a slight gain in the upper limit for Communicators.   In all

cases All Eat continues to prove dominant above a gain of 90 or 100.

Imitation in a Random Environment

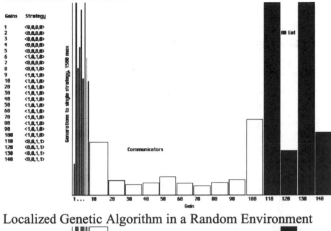

Localized Genetic Algorithm in a Random Environment

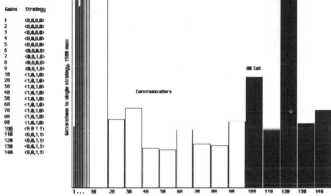

Neural Nets in a Random Environment

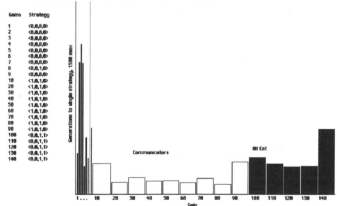

Figure 6   Window for communication in an environment of randomized food sources with a mean of 50.   Conquest by All Eat in each case at the upper end.

3. Strong Emergence of Communication in a Sine-wave Variable Environment

An environment with a random number of food sources produces much the same effects as one with a constant number of food sources. But what if we use an environment which, though variable, shows greater regularity in the variability of food resources? What if there is a cycle of 'boom and bust', for example—will this make a difference in the emergence of communication?

The decision to test environments with 'boom and bust' cycles still leaves a great deal of latitude, since patterns of 'boom and bust' may vary greatly. We conceived of different patterns in terms of different intervals marked out on a regular sine wave oscillating between 0 and 100. With values of that wave taken at intervals of 2 (Sin+2), we get one pattern of numbers for our food sources. With values taken at intervals of 3 (Sin+3), we get a different series (Figure 7).

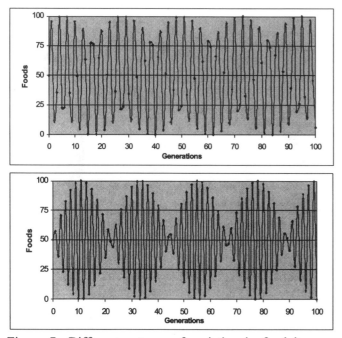

Figure 7 Different patterns of variation in food items: [sin(x) + 1] * 50 for different incremental series x_0, x_1, ... x_n. In the Sin+2 series at top, $x_{n+1} = x_n + 2$. In the sin+3 series below, $x_{n+1} = x_n + 3$.

What impact does a sine-wave variable environment have on the emergence of communication? Figure 10 shows emergence of communication in an environment changing on the pattern of sin+2 for each of our three mechanisms of strategy change.

The surprising result is that a variable environment allows conquest by our Communicators all the way up. Unlike constant and random environments, increased gains in a variable environment on the pattern of sin+2 do not favor All Eat at any point within the scope of the graph. We have tried larger gains for successful feeding up to 500, beyond the scope of the graph; it is still the Communicators that succeed.

We have found the result to be sensitive to patterns of variability, of course, but it is by no means confined to the pattern of sin+2. Resources following the pattern of sin+3 show similar results.

Across all of our modes of strategy change, then, sine-wave variable environments show a dramatic widening of the window of gain values in which Communicators appear and flourish. Although the average number of food

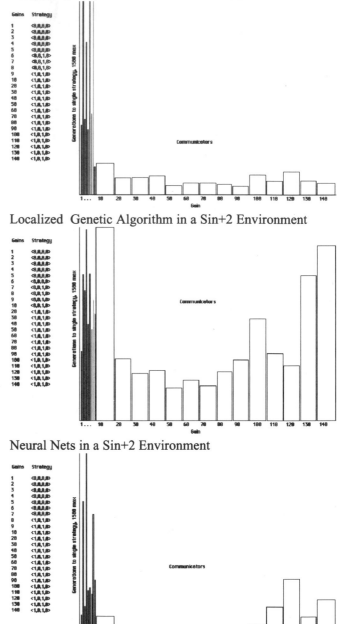

Figure 10. Triumph of Communicators at all gains above 10 in a 'boom and bust' environment on the pattern of sin+2, for all forms of strategy change.

sources remains the same as in our constant and randomly variable environments, an environment of 'boom and bust' strongly favors the emergence of communication. Although we have focused on simpler studies here, more complicated environments involving both food sources and predators show a similar effect.

Conclusion

In earlier studies we found that communities of Communicators can emerge from an initially randomized array of strategies in an environment of wandering food sources and predators. Communication can emerge, moreover, using any of three different mechanisms of strategy change: imitation of successful neighbors, localized genetic algorithm with most successful neighbors, and partial neural net training on the behavior of most successful neighbors (Grim, Kokalis, Tafti, & Kilb 2000, 2001; Grim, St. Denis, and K okalis 2002).

Here our attempt has been to expand those studies to questions of environmental variation: is communication about resources more favored in an environment in which the level of resources are variable than in which they are constant?

For an environment with randomly variable resources, the answer is 'no'. Random variation shows much the same effect as constant resources with the same average. In an environment with sine-wave variable resources, on the other hand—an environment of 'boom and bust' resource cycles—the answer is clearly 'yes'. It is thus not merely variability but the particular pattern of variability that is of importance; communicative strategies are much more strongly favored in sine wave variable environments. That effect holds whether the mechanism of strategy change at issue is one of imitation, localized genetic algorithm, or partial training on neural nets.

Environmental variability has been appealed to as an important explanatory factor in a range of different disciplines. In ecology, environmental fluctuation has been seen as playing an important role in species diversity (Chesson and Huntly, 1997). In cultural anthropology, cycles of boom and bust have been linked to the growth of agriculture (Reynolds 1986). Pleistocene climatic fluctuations have recently been proposed as instrumental in the evolution of lager brained mammals and higher primates, with speculative links to social learning (Potts 1996, Opdyke 1995, Boyd & Richerson 2000). The impact of environmental variability on individual learning is perhaps most developed in decades of careful work on schedules of reinforcement (Nye 1992). We take it as a suggestive fact, worthy of further investigation, that a clear impact of one form of environmental variation is evident even in simulations as simple as those we have outlined here.

References

Boyd, R., and Richerson, P. 2000. Built for Speed: Pleistocene Climate Variation and the Origin of Human Culture. In Tonneau, F. and Thompson N. eds. Perspectives in Ethology, vol. 13: Evolution, Culture, and Behavior. New York: Kluwer Academic/Plenum, pp. 1-45.

Chesson, P. L. and N. Huntly, 1997. The Roles of Harsh and Fluctuating Conditions in the Dynamics of Ecological Communities. American Naturalist 150: 519-553.

Grim, P., Kokalis, T., Tafti, A., & Kilb, N. 2000. Evolution of Communication in Perfect and Imperfect Worlds. World Futures: The Journal of General Evolution 56: 179-197.

Grim, P., Kokalis, T., Tafti, A., & Kilb, N. 2001. Evolution of Communication with a Spatialized Genetic Algorithm. Evolution of Communication 3: 105-134.

Grim, P., St. Denis, P., & Kokalis, T. 2002. Learning to Communicate: The Emergence of Signaling in Spatialized Arrays of Neural Nets. Adaptive Behavior 10: 45-70.

Honig, W. and Staddon, J. eds. 1977. Handbook of Operant Behavior. Englewood Cliffs, NJ: Prentice Hall

Nowak, M., and Sigmund, K. 1990. The Evolution of Stochastic Strategies in the Prisoner's Dilemma. Acta Applicandae Mathematicae 20: 247-265.

Nye, R. 1992. The Legacy of B. F. Skinner: Concepts and Perspectives, Controversies and Misunderstandings. Pacific Grove, CA: Brooks/Cole.

Opdyke, N. 1995. Mammalian Migration and Climate over the Past Seven Million Years. In Vrba, E., Denton, G., Partridge, C., and Burckle, L. eds. Paleoclimate and evolution with emphasis on human origins. New Haven, CT: Yale Univ. Press. Pp. 8-23.

Potts, R. 1996. Humanity's Descent: The Consequences of Ecological Instability. New York, William Morrow.

Reynolds, R. G. 1986. An Adaptive Computer Model for the Evolution of Plant Collecting and Early Agriculture in the Eastern Valley of Oaxaca. In Flannery, K. ed. Guil☐ Naquitz: Archaic Foraging and Early Agriculture in Oaxaca, Mexico. New York: Academic Press, pp. 439-500.

Expressing and Understanding Desires in Language Games

Michael Klein[1,2], Hans Kamp[2], Guenther Palm[3], and Kenji Doya[1,4]

[1]ATR Computational Neuroscience Laboratories
[2]Institute for Natural Language Processing, Stuttgart University
[3]Department of Neural Information Processing, Ulm University
[4]Crest, Japan Science and Technology Agency

Abstract

We speak because we want to get certain things accomplished. In this study we present the simulation of a multiagent language game which takes this into account. In contrast to previous language game studies, our agents use reinforcement learning to learn a function assigning a value to every state of the game. This value, that tells the agent how desirable the state is, is used along with a forward model to select actions. The agent can select verbal and non-verbal actions, depending on whether speaking or manipulating the world *directly* is more likely to bring about the change which the agent desires. On top of these capabilities, we used two rule-based agents to train a language learner. The learner trains a forward model of context-dependent utterance effects, which he then uses to express his desires and understand the desires of other players.

Introduction

Verbal communication normally serves some purpose (Wittgenstein, 1953; Austin, 1961). We speak because we want to get certain things accomplished. And often our concern is to get our addressees to do something for us, or to assist or counsel us in connection with our own actions. It is in such a setting of purposeful language use that language is acquired. As soon as a child begins to make use of language, it does so, most of the time, in the hope of getting its environment to comply with its wants and desires. And when it finds out that langauge can serve this purpose quite well, it will be all the more motivated to improve its linguistic skills further. In particular, this will impel it to learn the linguistic conventions that determine which expressions correspond to which states of affairs - conventions that are observed by those with whom the child interacts and whom provide the inputs to its acquisition process. Once these conventions have been mastered, the child can choose the utterances that express the states of affairs it desires in accordance with these conventions, with the effect that the one or ones it addresses will tend to understand what it wants and do what is needed to bring the wanted state about. Therefore, it seems reasonable enough to assume that the interplay between what linguistic expressions *mean* and the purposes to which language is put is an important, and probably crucial factor to the way in which human languages are learned. So it seems natural to try and see if the essential features of such a learning situation, in which the correlation between the meaning of language and the extralinguistic purposes of its use is of paramount importance, can be modelled in a somehow testable manner. Such a model should place the language-acquiring agent together with other agents who already possess the linguistic skills he still has to acquire in a common environment, and should explore the conditions that are needed for the agent to acquire the lingusitic knowledge through interaction with the agents which have mastered them already.

Surely, however, linguistic conventions are not the only things that children learn. They must also learn what states are desirable. Not just because they are agreeable in themselves, but also because of the subsequent rewards they promise.

In this study we build a computational model of language acquisition, in which two agents with a rule-based language module train one language learner. Using supervised learning, the learner trains a *forward model* which predicts the (context-dependend) effects of utterances. To train this model the learner observes the communication of the other agents. He can then use this model (i) to find the right utterance to express his own desires, and also (ii) to understand other agents by mapping speakers' utterances on the state they desire. We used a game environment, because it allows the agents to form their own desires. Agents learned a *value function* that can assign a value to every states telling the agent how desirable it is. Along with the forward model this function is used to select actions. It gives the agent the freedom to speak or not speak. In other words, the agent needs to decide for himself whether speaking or manipulating the world *directly* is the best action in a situation. This and the *effect-oriented* approach to language are the main differences between our study and previous studies in language games (Steels, 1996; Steels, 2001).

Language Production

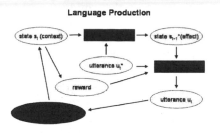

Figure 1: A simplified graph of how utterances (and other actions) are selected in our model: The forward model predicts the context-depended utterance effects and the value function determines whether the utterance is selected on the basis of how desirable the effect is.

Theoretical Framework

In order that an agent can express desires[1], he must have desires that he can express. That is, the agents of our model need to have the possibility of representing desires, and that in a form that is independent of the language that they use to express them and that the learning agent is to acquire. The form we have chosen is that of representing desires as *valued states of the world*. For every state of the world, the agent has a positive or negative numerical value expressing how much it desires this state. A function mapping every state on such a value is called a value - function (equation 2). Such a function is an estimation of how good it is for an agent to be in a given state. The notion of how good is defined in terms of future rewards that can be expected, or, to be precise in terms of the *expected return* (Sutton and Barto, 1998). The expected return is the sum of discounted rewards, which can be expected in and after a certain state (equation 1) The γ-parameter is the discount factor, which determines the value of future rewards. Given this definition of the expected return, the value function is defined in equation 2.

$$R_t = \sum_{k=0}^{\infty} \gamma^k r_{t+k+1} \qquad (1)$$

$$V^{\pi}(s) = E_{\pi}\{\sum_{k=0}^{\infty} \gamma^k r_{t+k+1}\} \qquad (2)$$

Reinforcement learning is a suitable method of generating such a function during interaction with the environment.

These methods allow us to determine the *desired state* of every agent in every state: It is the state with the highest

---

[1] We use the term *desire*, instead of *goal*, as we regard *desire* as the broader term. Desires can sometime have the form of a goal state, but in general can be a set of states instead of a goal point.

value. But not every state can be reached from a particular other state. Therefore, we need a *desire hierarchy* with more desired and less desired states. The value function gives us exactly what we need.

We propose a theory of communication and language acquisition, with the the following five essential points:

(i) An agent experiences utterances used by others speakers to have effects on the observable world. These effects of utterances are dependent on the context.

(ii) These effects on the observable world are indirect and are achieved by a more direct effect on unobservable states, such as the mental states of the addressees.

(iii) Linguistics events, i.e. experienced context-dependent effects of utterances are used to train a function mapping context configurations and utterances to effects.

Such a function, which predicts a sensation s_{t+1}^* based on the state s_t (context) and the action u_t (utterance) has been be called a *forward model* (Jordan and Rummelhart, 1992).

$$s_{t+1}^* = F(s_t, u_t^*) \qquad (3)$$

(iv) A speaker uses a certain expression, because he desires the effects he expects the expression to produce in the present context (according to his experience). By using his forward model and his value function, he chooses the action which will lead to the state of the world which he desires most.

This output function (or utterance function) maps the observed state and the desired observation into an utterance (equation 4).

$$u_t = argmax_u V(F(s_t, u_t^*)) \qquad (4)$$

(v) An addressee understands an utterance by using his forward model applied to the context and the utterance with which he was addressed. He thereby maps the utterance on the expressed desire or the intention of the speaker. (This is based on the plausible assumption, that human speakers share a core model of context-depended utterance effects.)

In other words, an addressee understands what the speaker is trying to achieve by using his *predictor*.

From this theoretical framework, we derive the following two hypotheses: (i) In a game environment, where only certain accomplishments are rewarded, agents equipped with value function (trained with reinforcement learning), a (rule-based) forward model, and a set of verbal and non-verbal actions can learn to behave in an optimal way, employing language and other actions whenever appropriate. (ii) In such an environment, an agent equipped with a optimal value

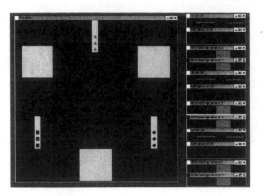

Figure 2: This shows the initial game state. The long yellow standing rectangles are the trees, each holding 3 pieces of food. The grey squares are the agents. They have the capacity of storing 5 pieces of each food type. The bar on the right displays scores and utterances. The green bars show, which agents cooperated with which other agents in their last move.

function and a rule-based model for non-verbal actions, can learn to use language to achieve his goals by expressing his desires and to understand the desires of other agents.

The Game

We test our hypotheses about language acquisition and communication in a multi-agent simulation. In this simulation, *food* grows in certain intervals in *trees*. In the present work, we use three trees growing three types of food. Every tree can hold maximally 5 pieces of food, and 3 pieces of food grow simultaneously, once the amount of food in the game is below a certain threshold.

There are three agents in the game. Every agent can store 5 pieces of each food type. Always after a certain time interval one piece of food gets *digested*, i.e., it disappears. This is to guarantee, that the agents need to act and cannot rest, after they have gained a sufficient amount of food items. However, they do not *starve* if they have no food for a number of time steps, but they get a low reward.

Agents can perform one of the following actions:

- harvest tree (take down all the food)

- give one piece of food to another agent

- ask another agent for a type of food

- do nothing

Generally, the agents take turns. However, when an agent asks another agent for a type of food, the normal order pauses for one time step, as then it is the turn of the addressee to give (or not to give) the desired object to the speaker. In our simulation, agents give objects away, when they are asked to. We do not address the emergence of cooperation, but assume a cooperative attitude from the very

start. Of course, we agree, that children must learn to become competent social agents, e.g. to abide by the social conventions of certain forms of give and take, and to understand how the use of language to get others to do things for you fits in with social interactions generally. However, a small, language learning child is not expected to be on equal footing with those from whom it learns, and certainly is not expected to give as much as it gets. In fact, children would have considerable difficulties to master language if e.g. their parents would not behave in a beneficial way.

The goal of the agents in the games is to have one piece of each food types at every time step. Therefore, the reward function was designed in the following way: Each agent gets a reward at every time step. If an agent has at least one item of every food type, it gets a reward of +3, otherwise it gets −1 for every food type which is missing in his store at that time step.

The agents in the game are simulated independent of each other. Every agent observes the relevant features of the environment at every time step. To detect state changes correlating with utterances, an agent needs to memorize past states and utterances. Therefore, every agent has his own short term memory, storing the complete observable game state (including the utterances) for a constant number m of time steps. Further, every agents is equipped with long term memory in the form of weights between neural units storing the value function and the forward model. The agents interact with the world and other agents only by their perception, actions and utterances.

An utterance of an agent is defined by its content (i.e. which word is used), its speaker (the agent), and its addressee. An agent can only address one of the other agents, never both of them, but the third party can observe every utterance that is made. This important, because in our setting language is learned by observing the context-dependent effects of the utterances of other agents. Who an agent talks to and what he says, is up to him. The content of an utterance is defined by the *vocabulary* of the agents in the game. In the present study, it consists of the three words *triangle*, *square*, and *diamond*. The content of an utterance can only be one word. An agent can use only one utterance at every time step.

With respect to their linguistic capabilities, agents can either be a *teacher* or a *learner*. Teacher-agents use a rule-based dialogue system to produce and understand utterances. Teachers in this game have no interest in the learner-agents learning of language. They do not perform any actions or utterance with the goal to teach language to the learners. Instead, learner-agents can learn language by observing the language use of the teacher-agents.

To chose their actions (or utterances), the agents predict the outcome of the action in the present context with a forward-model. The forward model is rule-based for *no action*, harvesting trees, donating objects, and for the verbal

Figure 3: This shows an arbitrary state during the early stages of training. The last action of agent 1 was to ask agent 2 for the square. Obviously, this is not the best move. A better move would be to harvest the *square tree*, as with this action, the agent would get 3 squares instead of one. The agent has not correctly estimated the values of these two states and, therefore, has chosen the wrong action.

actions of the teacher-agents.

These outcomes are evaluated with the value function and the action (verbal or non-verbal) which will bring about the state with the highest value is chosen.

If a verbal action is selected and the addressed agent is a teacher, then the addressed agent will give the desired object to the speaker. If the addressed agent is a language learner, this agent applies its *forward model* to the utterance and the game state. With this model, he can estimate what kind of change the speaker desires, i.e. he computes the intention from the utterances and the context. In other words, the language learner understands the utterance, because he *wonders* what effect, according to its own experience, such an utterance has in the present context. Using the present state of the game and the estimation of the desired state of the speaker, the addressee then uses a rule-based algorithm to computed which action would bring this desired state of the game about.

Learning Algorithms

The value function maps states of the game to real numbers. A state s is a 6-tuple of three $m \times n$ binary matrices (one for each agent) and three n-dimensional binary vectors (one for each tree). In all simulations described in this paper $m = 3$ and $n = 5$. At every point in time t, every agent can perceive the complete state $s(t)$.

$$s = <A_1, A_2, A_3, T_1, T_2, T_3> \qquad (5)$$
$$A_i = \mathbf{A}_{mn} \qquad (6)$$
$$T_i = \mathbf{t}_n \qquad (7)$$

The value function is implemented as a neural network with one neuron for every binary value of the vectors and the matrices of s. The output of the network is the linear combination of the weighted binary inputs. To train the value function, we used TD(0) reinforcement learning (Sutton, 1988) as described in equation 8. The term given in 9 is the so-called TD-error, giving distance and direction to the correct prediction and determine the weight changes.

$$V(s_t) \leftarrow V(s_t) + \alpha[r_{t+1} + \gamma V(s_{t+1}) - V(s_t)] \qquad (8)$$
$$r_{t+1} + \gamma V(s_{t+1}) - V(s_t) \qquad (9)$$

Because of the huge number of possible states, we used a neural network function approximation of the value function. As exploration mechanism we used a *softmax* method.

The linguistic capabilities of the language learner in the game are represented by a forward model. This model learns the context-dependent consequences of utterances. The context of each utterance is the game-state, as described above. The forward-model is implemented by a single-layer perceptron, mapping utterances and game-states onto game-states. We used supervised learning to train this forward model (equations 10, 11, and 12).

$$e_k = y_k^* - y_k \qquad (10)$$
$$\delta w_{ik} = \alpha e_k x_i \qquad (11)$$
$$w_{ik} \leftarrow w_{ik} + \delta w_{ik} \qquad (12)$$

Note that the same forward model can be trained by observing the effects of other agent's utterance as well as the effect of the agents own utterances (i.e. in our approach these two types of predictions are not distinguished, which of course is a considerable simplification).

Results

In the first stage of training, only the value function is trained. The value function enables the agents to estimate which states are desirable. At this stage, agents use a hard-coded forward-model to compute the outcome of possible action. They chose the action which brings about the outcome with the highest value. At the very early stages of the training, agents very often selected *no action* or senseless actions (such as donating objects to other players without being asked). Sensible, but suboptimal actions, as described in figure 3, did occur in the intermediate stage of training. After training, no more suboptimal action could be detected. Agents performed extremely well (approximately at the level of a human player or even better). Agents used language if appropriate, harvested trees whenever possible and optimal. They did not choose no action any longer, as usually some action or request would improve the state of agent. The value function gave the appropriate desire-hierarchy. We tested the performance of the model with

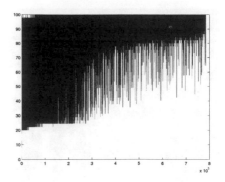

Figure 5: This shows an episode during language learning. The language learner (agent 2) asks agent 1 for the square, although agent 1 does not have one. Although the agent has a fully trained value function telling him which states are desirable, the forward model is not yet capable of predicting effects of the agent's utterances correctly. Therefore, the language learner is not able to understand the the context conditions and the normal *effects* his utterance has.

gamma = 0.9 / 0.7 / 0.5 / 0.3 / 0.1 (figure 4). The model showed stable performance for all values. The TD-error decreased faster with smaller *gamma*, while the overall performance was better with larger gamma. Closed to optimal performance could already be observed after about a million time steps.

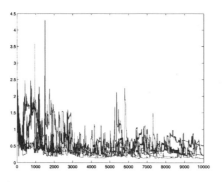

Figure 4: This is the development of the TD-error with γ = 0.3. The figure shows the development for 5 runs and 10^7 time steps (x is time in ksteps)

In the second stage of training, the forward-model of a language learning agent was trained. He was placed in an environment with two *teachers*. The initial language capabilities of the learner could best be described by *random utterances* and no understanding. Random utterances means that although the learner is a fully trained game player (i.e. he knows which state should be achieved), he has no idea which utterances are likely to bring these states about. This leads to situations, where e.g. he wants an object of one

Figure 6: The prediction error of language learning changes over time. This graph shows the percentage of correct prediction through approximately $5 * 10^6$ learning episodes

agent, but addresses the other agent, asking for a different object. Figure 5 illustrates another example where the language learner asks an agent for an object, which the addressed agent does not even have. This is because initially and in the early stages of training the learner has no or no accurate representation of the context condition for successful use of an utterance and also does not know its conventional effect. The language understanding of the learner faces the same problem. In early stages of training he usually shows no reaction upon being asked for a food item, because he wrongly maps the utterance to desired states of the speaker, that are unachievable from the current state or, in rare cases, he shows wrong reactions. While the training progresses, however, reactions are more and more in accordance with the utterances he is addressed with. Also the learner's utterances are more and more suitable with the situation and the desire until after training no difference between teacher and learner can be detected. As can be seen figure 6, the prediction error decreased very fast to a level close to 0.

Discussion

From the theoretical framework of human communication and language learning sketched above, we derived and tested two hypotheses: (i) In a game environment, where only certain accomplishments are rewarded, agents equipped with value function (trained with reinforcement learning), a (rule-based) forward model, and a set of verbal and non-verbal actions can learn to behave in an optimal way, employing language and other actions whenever appropriate. (ii) In such an environment, an agent equipped with a optimal value function and a rule-based model for non-verbal actions, can learn to use language to achieve his goals by expressing his desires and to understand the desires of other agents. We were able to confirm both hypotheses in our setting.

The aim of this simulation was to test our ideas in an environment which resembles the environment of humans. Of course, our simulation can only be a first approximation. A considerable simplification was our decision to enable our agents to observe the complete environment at all times (including the object possession of other agents). However, such a setting is justified and necessary at this stage, as in early language acquisition the effects of utterances need to be observable by the learning children.

In general, it remains difficult to compare the behavior and architecture of our agents with the human neural system and human verbal behavior. While the computational model of reinforcement learning we used is generally considered an appropriate model for human learning, so far no behavioral human data which is directly comparably to our task has been obtained. Although our system generally allows a human subject to play the game, it is not suitable for children in the early stages of language development, and it is clearly in these early stages, where children learn the nature of language use. Even if such a detailed comparison between the behavior of the model and human infants is not possible at present, our theoretical ideas are well in accord with the ideas of psycholinguists on language acquisition (Tomasello, 2000) and we look forward to give a more detailed comparison, when suitable behavioral data has been obtained.

Concerning the relation of our neural architecture to the real human brain we have refrained from claiming to model certain brain regions. However, there is considerable evidence that the basal ganglia is involved in reinforcement learning, while the cerebellum performs supervised learning and acts as a forward model in certain tasks (Doya, 1999). These two learning types are the ones involved in our computational model. Of course, it would be wrong to assume, that these two brain structures handle language acquisition alone, but they need to interact with the cerebral cortex. The cortex would be involved in tasks we did not model in this study, but which are nevertheless at the core of language. The unsupervised learning of the cortex plays a major role in the acquisition of word forms and their association with concepts (Klein and Billard, 2001) and is necessary to form concepts in the first place (Klein and Kamp, 2002). Word forms have been taken as given in this study, while the idea of concepts has been totally ignored. However, in real language acquisition concepts play a important role and an extension of our model able to generalize and learn more abstract feature of the predicted effects of utterances is very likely to make language learning faster and more efficient. The exact relation between the structures of our model and real brain structures has to be worked out in more details. When we have better hypothesis, we would like to evaluate them with functional brain imaging methods.

Finally, we would like to mention, that, although the language in our present study is restricted to single-word requests, the framework is can be extended to multi-word utterances and especially different kinds of speech acts, such as questions and the sharing of information. The next step

in our research is to test this considerably more complex linguistic challenge.

Acknowledgments

This work was supported by the German Academic Exchange Service (DAAD), the German Research Foundation (DFG), and the National Institute of Information and Communications Technology of Japan (NICT).

References

Austin, J. L. (1961). *Philosophical Papers*. Oxford University Press.

Doya, K. (1999). What are the computations of the cerebellum, the basal ganglia and the cerebral cortex? *Neural Networks*, 12:961–974.

Jordan, M. and Rummelhart, D. E. (1992). Forward models: Supervised learning with a distal teacher. *Cognitive Science*, 16:307–354.

Klein, M. and Billard, A. (2001). Words in the cerebral cortex - predicting fmri-data. In *Proceedings of the 8th Joint symposium on neural computation - The brain as a dynamical system, San Diego*.

Klein, M. and Kamp, H. (2002). Individuals and predication - a neurosemantic perspective. In Katz, G., Reinhard, S., and Reuter, P., editors, *Sinn und Bedeutung 6, Proceedings of the sixth meeting of the Gesellschaft fuer Semantik, Osnabrueck, Germany, October 2001*.

Steels, L. (1996). Perceptually grounded meaning creation. In Tokoro, M., editor, *Proceedings of the International Conference on Multi Agent Systems*, pages 338–344. AAAI Press.

Steels, L. (2001). Language games for autonomous robots. *IEEE Intelligent systems*, pages 16–22.

Sutton, R. S. (1988). Learning to predict by the methods of temporal differences. *Machine Learning*, 3:9–44.

Sutton, R. S. and Barto, A. G. (1998). *Reinforcement Learning - An Introduction*. MIT Press.

Tomasello, M. (2000). First steps towards a usage-based theory of language acquisition. *Cognitive Linguistics*, 11:61–82.

Wittgenstein, L. (1953). *Philosophical Investigations*. Blackwell.

The Evolution of Affect-Related Displays, Recognition and Related Strategies

Robert Lowe, Lola Cañamero, Chrystopher L. Nehaniv and Daniel Polani
Adaptive Systems Research Group
School of Computer Science, University of Hertfordshire
College Lane, Hatfield Herts AL10 9AB, United Kingdom
{R.Lowe, L.Canamero, C.L.Nehaniv, D.Polani}@herts.ac.uk

Abstract

This paper presents an ecologically motivated, bottom-up approach to investigating the evolution of expression, perception and related behaviour of affective internal states that complements game-theoretic studies of the evolutionary success of animal display. Our results show that the perception of displays related to affect greatly influences both the types of display produced and also the survival prospects of agents. Relative to agents that do not perceive rival agent internal state, affect perceivers prosper if the initial environment in which they reside provides numerous opportunities for interaction with other agents and resources. Conversely, where the initial environment with sparse resources does not allow for regular interaction, ability to perceive affect is not as facilitatory to survival. Furthermore, the agents evolve particular display strategies distorting the expression of affect and greatly influencing the proportion of affect perceiving to non-affect perceiving agents over evolutionary time.

1 Introduction

Expression of internal and external attributes as a possible precursor to communication has been the subject of much ethological and biological research. The animal kingdom provides us with numerous functional examples of animals using stereotyped displays to encourage observers to focus on particular attributes. Stotting behaviour in gazelles, for example, has been explained [11] as the sending of a message to a would-be predator that it should not waste its energy in attempting to catch such a vigorous gazelle. Cephalopods may use finely modified patterned visual displays in courtship rituals, for defence from predators (e.g. via crypsis), for predation and in antagonistic contests [14]. Cephalopod colour patterns are directed by neuronal and physiological change which allows for the subtle and speedy deviations in skin colour, patterning and shading that are crucial for their survival and reproduction.

Theoretical biology in the form of game theoretic models has similarly tended to focus on the functional behavioural component of communication where the pay-offs for using certain antagonistic behavioural strategies associated with displays are proposed. Examples include the "Hawk vs Dove" model of Maynard Smith and Price cited in [11] and the Enquist-Hurd contest pay-off matrix [12, 16, 13]. In the case of the latter model, pay-offs are calculated for animals producing distorted displays of actual strength which perceiving agents are assumed to interpret as honest. Applications of game theoretic models to communicative expression have generally been limited to field observations (e.g. red deer stags and the handicap principle [7, 8]) and dynamic simulation environments [17]. In the latter research, the applicability and limitations of the Enquist-Hurd game theoretic model were investigated; however, studies of this type are unavoidably constrained by the need to trade off empirical validity with faithful adherence to an often rigid and simplistic model.

The evolution of communication and its precursors in behavioural interactions [23, 22, 19] resonates with the area of ethology. Of the four essential aspects Tinbergen [23] defines, game theoretic models have concentrated on 'function' whereas 'mechanism', 'ontogeny' and 'phylogeny' have been largely neglected. Noble [20], for example, argues that function – the adaptive value of a given behaviour – can be studied independently from mechanism – the underlying physiological processes that give rise to physical behaviours. However, since affect refers to internal states such as motivations, drives and emotions, affective state may be modelled mechanistically and hypothesised to be of significant relevance to communication and signalling – and consequently to function. Indeed, affect related expression – an example of co-dependence between mechanism and function – has stimulated interest in psychologists [21] since the discoveries of Charles Darwin [9] and is an area of growing research emphasis among modellers of Artificial Intelligence [4, 10, 5]. The importance of affect to signalling and display is stated by Hauser [15], who suggests that "in a majority of species, affective states are responsible for the production of communicative signals".

While research into affective robots [10, 5] exploits the functional communicative value of affective expression, research with regard to group dynamics has been largely confined to psychological and ethological study. The evolutionary origin and development of such expression – which refers to Tinbergen's 'phylogeny' – has been the subject of investigation since Darwin; however, attempts to replicate the emergence and evolution of expressive and perceptive strategies over evolutionary time are lacking.

The current paper reports the initial findings of research into the use and distortion of affect expression on agent performance in a dynamic, 'season-based' environment. The research assesses the performance of *affect perceiver* agents versus *generic perceiver* agents, blind to displays of affect. *Both* types always *express* affect. Evolving expression of internal state was analysed in order to gauge the emergence and development of display strategies. Two hypotheses were made: 1) agent mortality will be influenced by ability to perceive affect in other agents, 2) displays of affect will evolve in ways that influence the proportion of surviving affect perceivers and generic perceivers. The layout of the paper is as follows: Section 2 describes the dynamic simulation environment used, the agent behavioural repertoire and action selection architecture governing behaviour. Section 3 details the performance of the two types of agent over evolutionary time with respect to behaviour and viability. Section 4 offers concluding remarks.

2 Agent Architecture and Environment
2.1 Environment Configuration
Our 2-D toroidal, continuously spaced, discretely timed environment consists of food resources and competing, reproducing, asexual agents. Agents perceive and display colours that carry information about internal state. Internal state is determined by the interactions among a behaviour-based architecture, an internal physiology and a set of motivational states, similar in spirit to [6, 2]. Following the common distinction used in ethology [18], the behavioural repertoire of agents includes both *consummatory* (goal-achieving) and *appetitive* (goal-seeking) behaviours. Agents' consummatory behaviours include *eat*, *hit* and *escape* while appetitive behaviours include *look for food* and *look to attack*.

2.2 Action Selection Architecture
Agents' decision making is governed by a "voting based" action-selection architecture, following [1, 3]. The main elements of the architecture are survival-related: homeostatic physiological variables, motivational states and behaviours.

These aspects of the architecture serve to link rival agent affect to perceiving agent affect through the influence of perceived affective state on perceiver motivational state and consequent displays and behaviour.

Agents have two physiological variables, 'energy' E and 'social stimulation' S that they need to maintain within viable bounds and that decrement by default, i.e. in the absence of any external stimuli. Motivations are urges to action that combine drives to correct physiological errors (deviations from the homeostatic ideal physiological value) and the perceived value of external stimuli: rival agents or food.

Rival agents whose perceived affective state indicates that they represent a threat will strongly influence the perceiving agent's motivational states. Consummatory and appetitive behaviours are carried out as a result of a solitary 'winner' behaviour being selected owing to a combination of the likelihood of them satisfying to the greatest extent all the motivations that contribute to its selection and the extent to which they correct associated physiological errors. Consummatory behaviours *eat* and *hit* help agents satisfy motivations to increase E and S, respectively, and to avoid death through exhaustive depletion of these two variables. An *escape* behaviour serves to allow agents to avoid a potentially life-threatening excess of S. Appetitive behaviours allow agents to continue their search for food or other agents.

The physiological variables are abstractions and not meant to be (strictly) biologically plausible. However, the use of these variables serves to create the sort of conflicting pressures appropriate to the emergence and evolution of signalling strategies. This owes to the need to satisfy motivations via behaviours that drive the physiological variables towards their homeostatic ideals. In the current simulation environment agents are thus required to trade off the utilisation of two resources, food and other agents, since neither resource serves to correct errors in both physiological variables.

Scarcity of resources and a fast metabolism ensure a need for agents to compete in order to survive and reproduce at the end of each generation. Being able to perceive the affective state of another agent, in this respect, may provide a competitive advantage since a rival agent who is in S excess will not be motivated to attack and therefore would not represent a survival threat since a further increase in S could lead to its death. The affect perceiving agent is then more able to select the pay-off optimising behaviour in a given interaction.

The external rival agent stimulus is calculated by affect perceivers using an objective measure of rival agent depiction of E and S values via colour saturation and hue respectively. Generic perceivers, on the other hand, attribute a fixed value to all rival agents for both E and S. This evolves over generational time to a potentially more adaptive value.

2.3 Expression and Recognition
Agents have two sensors: 1) 'stalker' detector, and 2) vision. In the case of 1) agents are able to detect the presence of unseen rival agents in close proximity. On detection of such rival agents, the perceiving agent will abruptly turn around and evaluate the threat or resource value of the rival using a forward-facing discretized visual field. Affect perceiving agents view colour hue and saturation, which in generation 1 of each simulation run faithfully reflect the physiological variables S and E, respectively. The use of colour, while not being an attempt at faithful colour modelling in biological agents, relates at an abstract level to cephalopod display

behaviour. In both cases, the physiological component of affect influences visual expression and its perception. Additionally, agents can potentially flash displays at rival agents when they are looking for food or to attack.

2.4 Evolving Displays

Displays represent deviations from the faithful 1:1 mapping of internal physiological state to colour configuration. This occurs over evolutionary time as a consequence of a 100 percent rate of small mutations of the mapping of the two physiological variables to dimensions of colour. There are a total of four agent displays from which one is selected in the presence of a stimulus. The two display-inducing appetitive behaviours (looking for food, looking to attack) each have a pair of weights pertinent to the types of stimuli present: agent, or agent and food. Display values are derived via a calculation relating to the direction of deviation from the faithful mapping (i.e. exaggerated or inhibited depiction of S value). Only the display of S value and not E can be affected and thus distorted by mutation. E (expressed faithfully as colour *saturation*), influences visual search for resources rather than targeted agent search. Displays are calculated via use of the following equation which determines the extent to which an agent can exaggerate or inhibit depicted S value:

$$hue = \begin{cases} W_1 > W_2 & : \quad S + (1 - S) * (W_1 - W_2) \\ W_1 \le W_2 & : \quad S - S * (W_2 - W_1) \end{cases}$$

where *hue* corresponds to the updated colour hue, and W_1, W_2 are weight values for exaggerated and inhibited depictions of physiological values, respectively. High evolved values of W_1 relative to W_2 and vice versa could lead to what observers might interpret as 'bluff' and 'trojan' strategies respectively [16]. A 'bluffer' has its hue value exaggerated relative to the actual S value. Conversely a 'trojan strategist' has a reduced hue depiction of S. The more the display value deviates from zero, the less accurate is the depiction of affective state. In general, it might be expected that a 'bluff' display would be adaptive to agents looking to avoid conflict whilst a 'trojan' display would enable aggressive agents to attack unsuspecting agents more easily.

3 Experiments, Results and Analysis

3.1 Experimental Set-up

Simulations are run over 30 generations of 10000 time steps on a JBuilder Applet. The artificial life simulation environment is initialised in every run with one of two possible food resource configurations: 1) distribution clustered with quantity relatively abundant (*summer*), 2) distribution scattered with quantity relatively scarce (*winter*). Each simulation run consists of 30 generations and half way through each generation, resource configuration changes from 1) to 2) or vice-versa to increase environmental dynamism. Out of a total of

100 runs, half are initialised with configuration 1) and half with 2) to control for order effects. The first generation of every simulation run is initialised with a fixed population of 80 agents (40 affect perceivers and 40 generic perceivers). Such a population size allows for a sufficient number of survivors at the end of each generation to minimise the potential effects of drift convergence. Surviving agents contribute an equal number of asexually produced offspring to the next generation of agents. Inherited continuous parameters in [0,1] for generic perceived stimulus (colour) values, agent type (i.e. affect perceiver or generic perceiver) and display values for different stimulus configurations are all subject to mutation.

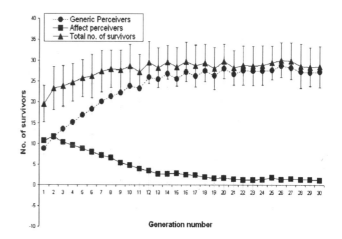

Figure 1: Graph of affect-perceiver, generic-perceiver and all agent average survivor rate over 30 generations in winter-first condition (average of 50 runs).

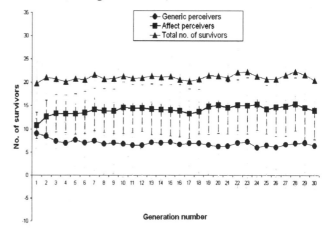

Figure 2: Graph of affect-perceiver, generic-perceiver and all agent average survivor rate over 30 generations in summer-first condition (average of 50 runs).

3.2 Results Summary

Fig.1 and Fig.2 depict the average numbers of agent survivors of affect perceiver and generic perceiver type and the applicable standard deviation error bars over all simulation runs. Fig.2 shows that affect perceiving agents, on average, outperform the generic perceivers when the first season encountered is summer. Fig.1 indicates that affect perceiving agents perform considerably worse than generic perceiving agents when winter precedes summer. While the standard deviation error bars in Fig.2 overlap the fact that the summer-first results showed affect perceivers outperforming generic perceivers on 40/50 occasions suggests that there is a strong bias towards affect perceiver survival success. Winter-first results showed generic perceivers outperforming the affect perceivers in 50/50 runs. Therefore, ability to perceive affect greatly influences survival prospects in accordance with *hypothesis 1*.

These results suggest that the adaptive value of affect perception is dependent on the order of seasons. Since the effects on survival of the first season influence survival prospects in the following season, it might also be extrapolated that the value of affect perception is environment dependent, i.e. dependent on the number and distribution of resources. Interestingly, in both conditions affect perceivers tend to outperform, on average, the generic perceivers in the first two generations. In the summer-first condition the affect perceiver survivor mean equals 10.72 compared to 9.04 for generic perceivers after the first generation. In the winter first condition affect perceiver mean equals 10.74 and generic perceiver mean equals 8.8. Agents with indiscriminate, generic perception of rival agents are seemingly disadvantaged by a lack of affect perception in initial generations. Generic perceivers may benefit, however, from evolving an agent stimulus value that is, averaged over agent interactions, more conducive to adaptive behaviour. Affect perceiving agents, with their more circumspect evaluation of external stimuli, at least in early generations, out-perform generic perceivers with respect to survival over the two seasons.

3.3 Evolution of Generic Perceivers

The tendency for generic perceivers to proliferate over time in the winter-first condition may owe to the fact that, with a lesser prospect of encountering agents with which to socially interact (owing to an absence of food clusters around which agents gather, as in the summer condition), there is a greater need to 'take the risk' and attack opportunistically or generically. Attributing generic stimulus values of excess S (i.e. S values above the 0.5 homeostatic level) to rival agents increases the likelihood that such generic perceivers will attack owing to the hard-wired nature of the agent action-selection architecture. Note that agents are only motivated to 'hit' if their own S values are in deficit and therefore when in S excess they do not represent a threat. In the winter-first condition, generic perceivers evolve, over time, agent

stimulus values conducive to 'hitting'. By generation 30, the S value, generically attributed, evolves to 0.7 with standard deviation just 0.04. The mean attributed E value of 0.3 (with standard deviation a very low 0.1) is that which the agents attribute to all other agents indiscriminately. The low standard deviations indicate a strong pressure over the 50 simulation runs for agents with generic perception to converge to these stimulus values attributed generically to rival agents. A perception of low E in the rival agent might serve to focus the agent on attacking rather than searching for food since higher perceived E values are associated with less specific and more generic searching (via an increase in visual field size with the same discretization) which does not facilitate 'homing in' on rival agents. This suggests that in such a harsh environment it may not be useful to attempt to discern rival agent intentions but rather to focus unreservedly on the most vital behaviours according to given motivations. The less reactive behaviour of affect perceiving agents may be more appropriate in an environment that allows for numerous interactions and opportunities to obtain food resources but may be less productive in harsher environments where opportunities to find stimuli are reduced. In the summer-first condition in 10/50 cases generic perceivers dominate and evolve a generically attributed social stimulation value of 0.24 ($\sigma = 0.12$). This tendency for generic perceivers to attribute a high degree of threat to rival agents in the summer condition could owe to the higher frequency of agent interactions allowing for more opportunity to 'hit' or be hit.

3.4 Physiological Determinants of Death

From a physiological perspective, agents could die due to S excess, deficit or alternatively E deficit. In the summer-first condition, it was found that on average generic perceivers were more likely to die of S deficit and less likely to die of S excess than affect perceivers. In the winter-first condition, the opposite was found true. E deficit deaths were significantly higher in the winter-first condition and more likely to be suffered by affect perceiving agents in this case. In the summer-first condition E deficit deaths were rare for both types of perceivers.

3.5 Social Interaction

Agents in the summer-first condition generally attacked more, presumably as a consequence of being clustered around the densely distributed, numerous resources. Respective total hits (averaged over all runs) 45.9 ($\sigma = 5.32$) and 40.63 ($\sigma = 5.69$) for affect perceivers and generic perceivers after the summer of generation 1 were registered in this condition. This is considerably more than the 28.84 ($\sigma = 5.33$) and 21.56 ($\sigma = 4.31$) by the equivalent agents in the winter season of the first generation of the winter-first condition. The legacy of these contrasting hitting totals served to maintain a considerable difference between the two conditions. By the end of the first generation the respective average hitting totals stood at 61.16 ($\sigma = 6.55$) and 56.38 ($\sigma = 6.92$) compared to 42.88 ($\sigma = 6.89$) and 37.68 ($\sigma = 7.13$).

3.6 Emergence and Evolution of Display Strategies

Displays can be used to manipulate the behaviour of rival agents. To give an example, if an agent is perceived as being in S deficit and looking to attack then an affect perceiving agent in S excess is likely to escape on perceiving the threat. A 'trojan' display in this case might allow an agent to mask its true intentions since an evolved set of displays that leads to an increase of S display to 0.5 or above i.e. beyond the homeostatic, mid-point of the internal S value will fool affect perceivers into 'believing' that the threat is non-existent. This occurs since the rival would be perceived as being in S excess. This is important given that agent survival is contingent upon being able to hit other agents to avoid death from S deficit and being able to avoid being hit in order not to die from S excess.

A number of strategies were seen to emerge among individual agents and groups of agents for the various values associated with the presence of an agent or the presence of food and an agent. The focus of the following analysis was on the summer-first condition owing to the fact that generic perceiver agents dominated the winter-first condition whereas use of displays should be associated with agents able to perceive such displays.

Fig.3 and Fig.4 show an interesting example of possible emergent display strategies being used by agents in order to facilitate survival prospects. In this particular run, the overall number of survivors (35) at the last generation was the highest of all runs in the summer-first condition, and therefore represents an interesting case study. Survivors in this run were predominantly generic perceivers (33) as opposed to affect perceivers (2). In Fig.3 agents show, on average, a general inclination towards 'bluffing' when looking for food while both a rival agent and food resource is present. The effect momentarily drops around generation 25 but then increases rapidly up to the 30th generation. In general this might be a useful strategy to adopt if the displaying agent is not looking to attack since a depicted exaggerated threat will be more likely to 'frighten' off rivals. In Fig.4 agents show an increasing tendency to adopt an increasingly strong 'trojan' strategy while looking to attack in the presence of an agent. This again may be adaptive since an agent looking to attack could derive benefit from lulling the rival into a false sense of security through a display of non-aggression.

In order for such displays to be adaptive there would need to be agents capable of interpreting the displays i.e. affect perceiver agents. At an early stage in this run we see generic perceiver agents dominating the survivor rates and at a similar stage 'trojan' displays begin to evolve. It might be ex-

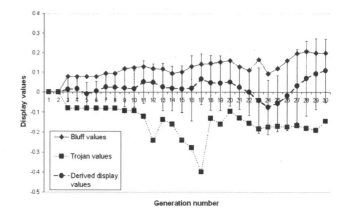

Figure 3: Graph showing average 'bluff'(*W1*), 'trojan'(-*W2*) and derived final display values when agent perceives agent and food and is currently *looking for food*, over 30 generations

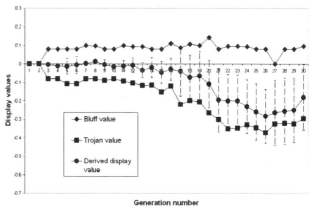

Figure 4: Graph showing average 'bluff'(*W1*), 'trojan'(-*W2*) and derived final display values when agent perceives agent only and is currently *looking to attack*, over 30 generations

pected that 'trojan' displays would evolve prior to the onset of generic perceiver agent domination and that such displays would be a cause of such survival success. Indeed, from generations 9 until 20 (during which time 'trojan' displays begin to evolve) affect perceiver agent survivors remain between 3 and 5. During the same period the ratio of 'trojan' to 'bluff' agents changes from 1:1 to 13:5. Could this strong effect just be an artefact of random drift? The highly evolved 'trojan' displays by generation 30 suggest otherwise. It is possible that such 'trojan' and 'bluff' strategies are effective even when affect perceivers are in a sufficiently large minority. Using 'trojan' or 'bluff' displays in the presence of such 'gullible' agents could entail a significant survival advantage and create evolutionary pressure to select for such a strategy. The fact that generic stimulus values did not stabilise at an early stage in the simulation run, together with the increasing evolutionary emphasis on 'trojan' display evolution from generation 9 suggest that the number of affect perceivers extant in the environment in the early stages of the simulation run were sufficiently high from generation 9 to create an evolutionary pressure towards use of a 'trojan' display strategy.

A number of other examples of possible strategies emerging and evolving in summer-first simulation runs in which generic perceiver agents begin to dominate over evolutionary time occur. For example, a symmetric 'bluff' strategy to that mentioned above is in evidence in two runs. In other simulation runs, however, other apparent strategies emerge that appear to be highly contingent and context dependent. Over the 50 simulation runs, there was no consistent direction in which displays emerged and evolved but numerous instances of highly evolved 'trojan' or 'bluff' displays in different runs is indicative of evolutionary pressure that may have an adaptive end in such environments. These findings therefore support *hypothesis 2*.

4 Concluding Remarks

A major benefit of using the game theoretic approach to studying signalling and behavioural interactions is that the rules established are simple, comprehensible and intuitively appealing; however, such theoretical models measure these benefits in a very abstract evolutionary setting. An ecological approach to the assessment of (pre)communicative behaviour, its underlying mechanisms and its connection to interaction over evolutionary time is always susceptible to problems of complexity. Reducing environments to their componental parts runs the risk that experiments and findings will be trivial but even slight increments to such complexity can lead to an exponential increase in analytical requirements. Therefore, a disciplined, systematic incremental approach that identifies the factors influencing the evolutionary trends and adaptive value of displaying and perceiving affective expression is necessary. The results of our experiments showed that perception of affect and use of displays can greatly influence survival prospects. Understanding the emergence and rationale of the rules that allow to categorise the ways in which such perception and expression are used and evolved is, hence, the aim of our approach.

Acknowledgements

We would like to thank Alexander Klyubin and Lars Olsson for providing excellent technical support. Robert Lowe is supported by a scholarship from the University of Hertfordshire.

References

[1] O Avila-García and L Cañamero. A comparison of behavior selection architectures using viability indicators. In *Intl. Wksp. Biologically-Inspired Robotics*, pages 86–93, 2002.

[2] O Avila-García, L Cañamero, and R te Boekhorst. Analyzing the performance of winner-take-all and voting-based action selection policies within the two-resource problem. In *Advances in Artificial Life (Proc. European Conf. Artificial Life - ECAL '03), Springer LNAI vol. 2801*, pages 733–742, 2003.

[3] O Avila-García, E Hafner, and L Cañamero. Relating behavior selection architectures to environmental complexity. In *From Animals to Animat (Proc. SAB'02)*, 2002.

[4] R Aylett and L Cañamero, editors. *Animating Expressive Characters for Social Interactions (Proc. AISB'02 Symp.)*. Imperial College, London, UK, April 4-5, 2002.

[5] C.L Breazeal. *Designing Sociable Robots*. MIT Press, 2002.

[6] L Cañamero. Modelling motivations and emotions as a basis for intelligent behavior. In *Proc. First Intl. Conf. on Autonomous Agents*, pages 148–155, 1997.

[7] T.H Clutton-Brock, F.E Guinness, and S.D Albon. *Red deer. Behaviour and ecology of two sexes*. Edinburgh University Press, 1982.

[8] T.H Clutton-Brock and N McIntyre. *Red deer*. Colin Baxter Photography Ltd, Moray, 1999.

[9] C Darwin. *The Expression of Emotions in Man and Animals*. Julian Friedmann Publ. London, second edition, 1979 [1872].

[10] K Dautenhahn, A.H Bond, L Cañamero, and B Edmonds, editors. *Socially Intelligent Agents*. Kluwer, 2002.

[11] R Dawkins. *The Selfish Gene*. Oxford University Press, second edition, 1989.

[12] M Enquist. Communication during aggressive interaction with particular reference to variation in choice of behaviour. *Animal Behaviour*, 33:1152–1161, 1985.

[13] M Enquist and P.L Hurd. Conventional signalling in aggressive interactions: the importance of temporal structure. *Journal of Theoretical Biology*, 192:197–211, 1998.

[14] R.T Hanlon and J.B Messenger. *Cephalopod Behaviour*. Cambridge University Press, 1996.

[15] M. D. Hauser. *The Evolution of Communication*. MIT Press, 2000.

[16] P.L Hurd. Is signalling of fighting ability costlier for weaker individuals? *J. Theoretical Biology*, 184:83–88, 1997.

[17] R Lowe and D Polani. Preventing bluff agent invasions in honest societies. In *Advances in Artificial Life (Proc. ECAL'03), Springer LNAI vol. 2801*, pages 118–127, 2003.

[18] D McFarland. Opportunity versus goals in robots, animals and people. In H.L Roitblat and J.-A Meyer, editors, *Comparative Approaches to Cognitive Science*, pages 415–433. MIT Press, 1993.

[19] C.L Nehaniv. Meaning for observers and agents. In *IEEE International Symposium on Intelligent Control/Intelligent Systems and Semiotics*, pages 435–440, 1999.

[20] J Noble. *The Evolution of Animal Communication Systems: Questions of Function Examined Through Simulations*. PhD thesis, COGS, University of Sussex, 1998.

[21] J.A Russell and J.M Fernandez-Dols. *The Psychology of Facial Expression*. Cambridge University Press, 1997.

[22] W.J Smith. *The Behavior of Communicating: An Ethological Approach*. Harvard, 1977.

[23] N. Tinbergen. On aims and methods of ethology. *Zeitschrift für Tierpsychologie*, 20:410–433, 1963.

Language, Altruism and Docility: How Cultural Learning Can Favour Language Evolution

Marco Mirolli[1,2], Domenico Parisi[1]

[1] Institute of Cognitive Sciences and Technologies, National Research Council, 15 Viale Marx, 00137 Rome, Italy
[2] Department of Philosophy and Social Sciences, University of Siena, 47 Via Roma, 53100 Siena, Italy
mirolli2@unisi.it

Abstract

Human language serves a number of different functions, one of the most prominent being communicating about relevant features of the environment. From the point of view of the speaker, if the communicated information is advantageous for the hearer but not for the speaker, this is an altruistic use of language, and, as such, it requires an explanation of its evolution. Simon 1990 proposed an explanation of altruism in humans based on the genetically inherited 'docility' of our species. In this paper we present artificial life simulations that apply Simon's ideas to the problem of the emergence of the altruistic use of language described above. From the point of view of evolutionary theory, the present work represents the first attempt to test Simon's 'docility' theory of altruism with agent-based computer simulations. From the point of view of language evolution, our simulations give an original explanation of (the altruistic aspect of) human language based on one of its most peculiar characteristic, namely, the fact that it is culturally transmitted.

Introduction

Among the many functions carried out by human language, one of the most important is its use to inform another individual about some significant feature of the environment. It is an open problem whether this use of language was the principal function for which human language evolved (Bickerton 2002) or language evolution started for more social reasons such as facilitating social interaction and social coordination (Knight et al. 2000). In any case, the use of language for communicating about the environment posits a problem for an adaptationist account of language evolution. In fact, if what is communicated about the environment is useful for the hearer but the act of communicating has no advantage for the speaker, sending appropriate messages about the environment is an altruistic behavior on the part of the speaker.

In this section we describe various evolutionary explanations of altruism, then we briefly review agent-based simulations that have been used to solve the problem of altruistic communication, and finally we describe the rationale of the present work. In the next section we describe our simulations and their results and, finally, in the last section we discuss these results and make some conclusive remarks.

Evolutionary explanations of altruism

To solve the puzzle of the presence of altruistic behaviors in the animal kingdom evolutionary theories have usually adopted one (or more) of the following four kinds of arguments: reciprocity, kin selection, group selection, and cultural selection.

Arguments based on the concept of reciprocity state that if individuals interact repeatedly with each other, then altruistic behavior can evolve because the individuals can adapt their strategies according to the results of previous interactions (Axelrod and Hamilton 1981; Trivers 1971).

Kin selection theory (Hamilton 1964) constitutes the most unchallenged explanation for the evolution of altruism (but see Henrich 2003 and Queller 1992 for some interesting discussions). According to this theory, altruistic behavior can evolve if the product between the benefit the behavior gives to the receiver (B) and the coefficient of kin-relatedness between the emitter and the receiver (r) is greater than the cost for the emitter (C), as stated by Hamilton's Rule: $-C + rB > 0$, that is, $rB > C$.

Another classical - but more controversial - mechanism which has been used for the explanation of the evolution of altruistic behaviors is group selection. According to group selection theory, if a population is divided into groups competing with each other, the total number of altruists in the population can increase even though the number of altruists inside each group is bound to decrease: the reason is that groups with small percentages of altruists will tend to disappear in favour of groups with a larger number of altruists (Sober and Wilson 1998).

Finally, some form of cultural evolution (Cavalli-Sforza and Feldman 1981; Boyd and Richardson 1985) has been suggested as a possible explanation of altruistic behavior (Richerson et al. 2002; Simon 1990). In particular, Simon's explanation of altruism runs as follows: if cultural learning is advantageous for individuals, genetic selection will favour docile individuals, that is, individuals that tend to learn from others[1]. Therefore, altruistic behavior can

---

[1] This is not in contradiction with the fact that cultural learning is unfrequent in the animal kingdom apart from the human species. The reason is that cultural learning requires a number of social, cognitive and neural pre-adaptations which happened to be present in early hominids but not in any other species.

emerge given that its cost is lower than the benefit of being docile and that individuals are not able to distinguish between selfish and altruistic behaviors.

Previous Simulations

In this section we review very briefly some simulations that have focused on the problem we are discussing in this paper, that is, the altruistic character of the use of language for informing conspecifics about some feature of the environment if the information benefits the hearer but not the speaker.

As far as we know, no simulations have been done for testing the plausibility of an account of language evolution which relies on reciprocal altruism. Simulations by Ackley and Littman 1994 and Oliphant 1996 have explored the possibility that kin selection could have played a role in the evolution of language, but Di Paolo 1999 has criticized this work for its improper use of kin-selection. Noble et al. 2001 explore some other adaptive factors that may be relevant for the emergence of a simple signalling system, including group selection, sexual selection, and the handicap principle (Zahavi 1975).

In the present work we test yet another possible mechanism for the evolutionary explanation of the altruistic use of human language which we are dealing with, a mechanism which relies on one of the most peculiar characteristic of human language, that is, the fact that human language is culturally transmitted.

Docility and the evolution of language

In Simon's model the conditions for the evolution of an altruistic behavior are the following:

1) there is some advantage d in being disposed to learn from others, i.e., in being docile;
2) organisms are not able to evaluate the contribution of each behavior to their own fitness;
3) the advantage d of being docile is greater than the cost c of the altruistic behavior.

With his argument Simon intended to show how the presence of altruism among human beings should be considered differently from the presence of altruism in other species in that cultural learning plays a crucial role in humans and this makes possible the emergence of altruism with a mechanism peculiar to our own species. As the use of language that we are dealing with is altruistic and as human language is the only communicative system in the animal kingdom to be culturally transmitted, Simon's explanation of altruism seems to be applicable to language evolution.

The present work represents the first attempt (to our knowledge) to test Simon's docility theory with agent-based computer simulations and, at the same time, gives an original explanation for the evolution of (the altruistic aspect of) human language based on its learned character.

Simulations

In order to test the soundness of Simon's theory with respect to the evolution of human language, we ran two simulations based on a simplification of the simulative scenario used by Cangelosi and Parisi 1998 and discussed in Parisi 1997. First, we describe the simulative set-up common to both simulations, then we indicate what makes the two simulations different, and finally we describe the main results of the simulations.

The simulative set-up

The population is a succession of 500 generations of 100 individuals each. In each generation the 20 individuals which have the most energy at the end of life are selected for reproduction. Each individual generates 5 offspring and the $20 \times 5 = 100$ offspring constitute the next generation[2]. All individuals have the same network architecture but connection weights can vary among individuals (see below). Each individual lives in its own copy of the environment which is a linear succession of 11 cells. At the beginning of each 'epoch' an individual is placed in the first cell while the last cell contains a 'mushroom'.

There are 210 edible mushrooms and 210 poisonous ones, each different from all the others. The perceptual properties of the mushrooms are represented as variations from two prototype sequences of ten bits ($+1$ and -1) each, one for the edible mushrooms and one for the poisonous mushrooms. The 210 edible mushrooms are the 210 bipolar patterns of ten bits which have four $+1$s (that is, all and only the patterns that differ by 4 bits from the prototype which has all -1s). The perceptual properties of poisonous mushrooms are the 210 bipolar patterns which have six $+1$ (those that differ by 4 bits from the prototype which has all $+1$s). If the individual enters the mushroom cell located at the end, the individual eats the mushroom. If the mushroom is one of the 10 edible mushrooms, the individual's energy is increased by 30 energy units. However, if the mushroom is poisonous, its energy is decreased by 5 units. Furthermore, an individual's energy is decreased by 1 unit for each step from the initial to the final cell. This is the fitness formula:

$f(x) = 30 \times$ number of edible mushrooms eaten $- 5 \times$ number of poisonous mushrooms eaten $-$ number of steps.

As the life of each individual last 420 'epochs', one for each possible mushroom, the maximum possible fitness will be:

---

[2] The results of our simulations are robust with respect to changes in most of the simulation's parameters, including: selection algorithm (range-based vs. proportional to fitness); use vs. non-use of cross-over; probability of mutations; type of mutation (substitution of a weight vs. gradual change); presence vs. absence of limits to weights' values; learning rate and momentum values (for the simulation with learning).

FMax = 30 × 210 (edible mushrooms eaten) − 5 × 0 (poisonous mushrooms eaten) − 210 × 10 (steps necessary for eating a mushroom) = 6300 − 2100 = 4200

The organism's behavior is controlled by a neural network with 12 input units, 2 hidden units, and 3 output units. Ten input units are 'visual' units. If the organism is sufficiently near the mushroom, that is, it is in the last but one cell of the corridor, the perceptual properties of the mushroom are encoded in this set of input units; otherwise, the activation of all the visual units is set to zero. The other two input units are the linguistic ones: their activity depends on the linguistic signal which is produced by another individual randomly chosen from the rest of the population, the speaker. All the input units are fully connected with the two hidden units which in turn are fully connected with all three output units. The activation of the first output unit is thresholded to a value of either 1, in which case the organism moves one step forward, or − 1, in which case the organism stays still. The continuous activations of the other two output units, in the interval [− 1; 1], constitutes the signal produced which is copied, when the organism acts as a speaker, in the two linguistic input units of the hearer.

We ran two simulations: in the first one, the genome of our organisms contains their connection weights, while in the second one, the genome is constituted by a single gene, the docility gene, which stands for the disposition of the organisms to learn from their parents.

The two conditions: genetic selection and cultural selection through docility

In the first simulation – the "genetic simulation" – the values of the connection weights of the initial generation are randomly chosen in the range [− 0.5; 0.5], but the connection weights of each successive generation are inherited from parents to offspring with each weight having a probability of 2% to have its value changed by a random number in the interval [− 0.5; 0.5].

In the second simulation – the "docility simulation" - the connection weights of all individuals are always random at birth and they are not inherited from parents. Instead, the genome of these organisms is constituted by one only gene, encoded as an integer number, which stands for an individual's 'docility', i.e., the number of learning trials for that particular individual. In the first generation each individual is assigned a random value in the interval [0; 200] for this gene and this value is genetically transmitted with a 2% probability of being changed by adding or subtracting a random number in the range [− 100; 100]. In any case, docility is forced to stay in the interval [0; 500].

The life of organisms in this second simulation is divided into two periods: infancy and adulthood. During infancy, the organism is supposed to follow its parent and learn from it how to behave in different situations. In short, the docility gene determines the number of back-propagation learning cycles to which the infant exposes itself: the learning is imitative in that the teaching input of the back-propagation algorithm comes from the output of the infant's parent. Since there are three kinds of situations to which organisms are exposed during their life, there are also three different learning conditions: 1) 'comprehension learning', 2) 'decision learning' and 3) 'naming learning'.

Comprehension learning corresponds to the situation in which the tested organism is distant from the mushroom and has to decide whether to move or not to move according only to the signal it receives from another organism; decision learning corresponds to the situation in which the organism is near the mushroom and so its decision whether to move or not to move depends on both the visual input and the linguistic input; finally, naming learning corresponds to the situation in which the organism acts as a speaker: it receives only the perceptual properties of a mushroom and has to produce a linguistic signal.

In short, for each learning cycle determined by one's docility, this is what happens:

1) one of the three learning situations is randomly chosen together with one of the 420 mushrooms;
2) the appropriate input is given both to the learner and to its parent;
3) both the organism's output and its parent's output are calculated;
4) the output of the parent is given to the child as teaching input (a random value chosen in the interval +0.25/-0.25 is added to the teaching input[3]);
5) finally, the child's connection weights are changed according to the back-propagation algorithm (with a learning rate of 0.3 and a momentum of 0.8).

After infancy, individuals start adult life, which is identical to that of the genetic simulation.

Results

We describe the average results of 10 replications of both simulations using three measures: language quality, average fitness, and average value of the docility gene (in the pictures, all values are normalized with respect to their maximum possible value).

This is the way we calculate language quality. A linguistic signal is constituted by a vector of two continuous numbers in the range [-1; 1] (the vector of activation of the linguistic output units of an organism). So, a signal can be considered as a point in a bi-dimensional Cartesian space. Let's call E and P the set of points ('clouds') which represent the signals produced by the organisms of one generation in presence of all edible and poisonous mushrooms, respectively. A good language is one in which the two clouds E and P are 1) small (all

---

[3] The results are robust even with respect to the quantity of noise added, provided that this quantity is adequate for cultural evolution: if there is no noise, there is no room for improvement in behavioral capacity; on the other hand, if there is too much noise, good behaviors cannot be preserved. All noise values between 0.1 and 0.4 produce the same qualitative results.

mushrooms belonging to the same category are named in similar ways) and 2) distant from each other (mushrooms of different categories are named in different ways). A measure for 1) is the mean distance of the points of a cloud from its geometric center; a measure for 2) is the distance between the two centers. We normalize those two values in the range [0; 1] so that the maximum quality obtainable (1) is achieved when each cloud is a single point, in opposite corners of the space. In order to plot a single value for each simulation, we measure the language quality as the product of the two normalized values.[4]

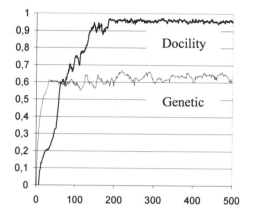

Figure 1: Average language quality of the genetic (thin line) and docility (thick line) simulations as a function of generations.

As it turns out, the language quality of the genetic simulation (fig. 1, thin line) is very bad: in this condition a good communication system does not emerge. The reason is the altruistic character of the type of communication we are dealing with here. Since producing appropriate messages (good speaking) gives advantages only to hearers but not to speakers, egoists (bad speakers) are selected against altruists (good speakers); hence the very low quality of language. As a result, the average fitness of the organisms of this simulation is sub-optimal. In fact, the presence of a good communication system is a necessary condition for optimal behavior. Without being told by others an organism cannot know which kind of mushroom is present at the end of the corridor and has to waste energy to go and check by itself. We can also calculate the maximal possible fitness for organisms which are not helped by language at all:

---

[4] The reason for using the product instead of the mean is that the product gives high values only if both values are high, and this seems quite appropriate. For example, the language quality of a communication system in which both clouds are collapsed in the same point (all mushrooms are named in exactly the same way) is certainly 0: this is in fact the value this system would reach using the product of the two measures while by using the mean the same system would reach the inappropriate value of 0.5.

FMaxWithoutLanguage = FMax − 210 (poisonous mushrooms) × 9 (steps to be made in order to see the perceptual properties of the mushroom) = 4200 − 1890 = 2310

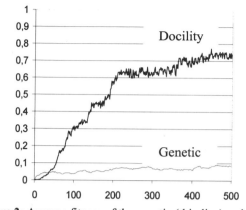

Figure 2: Average fitness of the genetic (thin line) and docility (thick line) simulations as a function of generations.

If we normalize this quantity, we get 2310 / 4200 = 0.55, that is just a little lower than the average fitness reached by the genetic simulation, which fluctuates in between 0.6 and 0.65 (fig. 2, thin line).

The situation of the docility simulation is very different. In this case, there is a strong selective pressure for the evolution of docility, since organisms that do not learn culturally are bound to behave randomly. As a result, the value of the docility gene increases constantly until it reaches almost its maximum value (fig. 3). To this increase in docility corresponds a parallel increase in the quality of the language produced by those organisms, which reaches the quite high value of about 0.75 (fig. 1, thick line). As it turns out, the correlation between docility and language quality is very high: 0,988. As a result, the organisms of this simulation can exploit all the advantages given by a good communication system and consequently their average fitness reaches almost the maximal possible value (fig. 2, thick line).

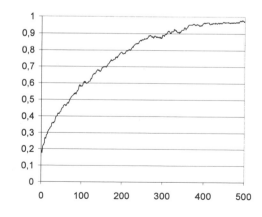

Figure 3: Average value of the docility gene of organisms of the docility simulation as a function of generations.

Discussion and Conclusions

There are a number of possible solutions to the problem of the evolution of altruism which rely on reciprocity, kin selection, group selection or cultural evolution, and we think that more than one of these factors may have played a role in the evolution of the use of language which is taken into consideration in our simulations.

Pedone and Parisi 1997 have suggested that the crucial factor for the evolution of an altruistic behavior is the similarity of behavior between interacting individuals. Henrich 2003 presents a generalization of Hamilton's Rule which substantiates this suggestion: what really matters for the evolution of altruism is the probability for an altruist to encounter another one. Consequently, all the theoretical solutions to the problem of altruism consist in finding plausible mechanisms for maintaining this probability high. Our docility simulation strongly confirms Simon's theory according to which docility can be such a mechanism. In fact, if docility evolves due to the egoistic advantages it confers to organisms, then cultural transmission can guarantee the similarity of behaviors necessary for altruism to emerge.

Turning specifically to the evolution of human language, our simulations suggest that the use of language for informing others about the environment (unless it benefits both speaker and hearer in coordinating their behavior, e.g., in group hunting) might have emerged relatively lately during hominid evolution, namely, after hominids had become docile and cultural evolution had started.

In fact, from the point of view of the informer, this use is an altruistic behavior, present in the human species, which needs an evolutionary explanation. The results of the simulation in which behaviors are genetically inherited (through the neural networks' connection weights) confirm the theoretical prediction that in such a condition a good communication system does not emerge because of its altruistic character. Consider also that in our simulations communication is in a sense hardwired: our organisms are *forced* to produce signals for others. So, we are not actually dealing with the emergence of communication as such, but with the emergence of a *good* system for communicating about the environment. This is justified by the assumption that hominids' proto-language may have initially served other, more 'social', functions, such as strengthening social relationships (Dunbar 1996). This use of language does not pose the theoretical problems of altruism, since it is not altruistic.

So, this is the evolutionary scenario we are suggesting. First, a very complex social structure created the basis for the evolution of the first kind of hominids' communication system which in turn favoured the development of a still more complex social structure. Second, this more complex social structure constituted one of the preconditions for the evolution of cultural learning. Finally, (proto-)language started to be used also for informing others about the environment. Our simulations show that the homogenisation of behavior induced by cultural transmission could have favoured the emergence of this kind of altruistic use of language.

Acknowledgments. The research presented in this paper is supported in the framework of the European Science Foundation EUROCORES programme "The Origin of Man, Language and Languages"

References

Ackley, D. H.; and Littman, M. L. 1994. Altruism in the Evolution of Communication. In R. A. Brooks and P. Maes eds. Artificial Life IV: Proceedings of the International Workshop on the Synthesis and Simulation of Living Systems, Cambridge: MIT Press, 40-48

Axelrod, R.; and Hamilton, W. D. 1981. The Evolution of Cooperation. Science 211: 1390-1396

Bickerton, D. 2002. Foraging Versus Social Intelligence in the Evolution of Protolanguage. In A. Wray ed. The Transition to Language. Oxford: Oxford University Press

Boyd, R.; and Richerson, P. J. 1985. Culture and the Evolutionary Process. Chicago: University of Chicago Press

Cangelosi, A.; and Parisi, D. 1998. The emergence of a 'language' in an evolving population of neural networks. Connection Science 10(2): 83-97

Cavalli-Sforza, L. L.; and Feldman, M. W. 1981. Cultural transmission and evolution: a quantitative approach. Monographs in Population Biology 16, Princeton: Princeton University Press

Di Paolo, E. A. 1999. A little more than kind and less than kin: the unwarranted use of kin selection in spatial models of communication. In D. Floreano, J.D. Nicoud, and F. Mondada eds. *Advances in Artificial life Proc. ECAL'99*, LNAI 1674, Lausanne: Springer-Verlag, 504-513

Dunbar, R. I. M. 1996. Grooming, gossip and the evolution of language. London: Faber and Faber

Hamilton, W. D. 1964. Genetic evolution of social behavior. Journal of Theoretical Biology 7(1): 1-52

Henrich, J. 2003. Cultural group selection, coevolutionary processes and large-scale cooperation. Journal of Economic Behavior and Organization

Knight, C.; Studdert-Kennedy, M.; and Hurford, J. eds. 2000. The evolutionary emergence of language: social function and the origins of linguistic form. Cambridge, Mass.: Cambridge University Press

Knudsen, T. 2003. Simon's selection theory: why docility evolves to breed successful altruism. Journal of Economic Psychology 24: 229-244

Noble, J.; Di Paolo, E. A.; and Bullock, S. 2001. Adaptive factors in the evolution of signalling Systems. In A. Cangelosi and D. Parisi eds. Simulating the evolution of language. London: Springer-Verlag, 53-78

Oliphant, M. 1996: The dilemma of Saussurean communication. Biosystems 37(1-2): 31-38

Parisi, D. 1997. An Artificial Life approach to language. Brain and Language 59: 121-146

Pedone, R.; and Parisi, D. 1997. In what kinds of social groups can altruistic behavior evolve?. In R. Conte, R. Hegselmann and P. Terno eds. Simulating social phenomena. Berlin: Springer-Verlang, 195-201

Queller, D. C. 1992. A general model for kin selection. Evolution 46: 376-380

Richerson, P.; Boyd, R.; and Henrich, J. 2003. The cultural evolution of cooperation. In P. Hammerstein ed. Genetic and cultural evolution of cooperation. Cambridge, Mass.: MIT Press

Simon, H. A. 1990. A mechanism for social selection and successful altruism. Science 250: 1665-1668

Sober, E. R.; and Wilson, D. S. 1998. Unto others: The evolution and psychology of unselfish behavior. Cambridge, Mass.: Harvard University Press

Trivers, R. L. 1971. The evolution of reciprocal altruism. Quarterly Review of Biology 46(1): 35-57

Zahavi, A. 1975. Mate selection. A selection for a handicap. Journal of Theoretical Biology 53: 205-214

Evolution of Plastic Sensory-motor Coupling and Dynamic Categorization

Gentaro Morimoto and Takashi Ikegami

Graduate School of Arts and Sciences
The University of Tokyo
3-8-1 Komaba, Tokyo 153-8902, Japan
{genta, ikeg}@sacral.c.u-tokyo.ac.jp

Abstract

We study the dynamic categorization ability of an autonomous agent that distinguishes rectangular and triangular objects. The objects are distributed on a two-dimensional space and the agent is equipped with a recurrent neural network that controls its navigation dynamics. As the agent moves through the environment, it develops neural states which, while not symbolic representations of rectangles or triangles, allow it to distinguish these objects. As a result, it decides to avoid triangles and remain for longer periods of time at rectangles. A significant characteristic of the network is its plasticity, which enables the agent to switch from one navigation mode to another. Diversity of this switching behavior will be discussed.

Introduction

Gibson reports blind touch experiments with a cookie cutter where a subject can tell the shape of a cutter when he moves it by himself (Gibson, 1962). O'Regan stresses the importance of active vision and argues perception as mastering sensory-motor coupling (O'Regan and Noë, 2001).

The significance of Gibson's perception theory that deals with perception associated with action is that it highlights the difference between memory and experience. For example, to know a paper crane having seen one in a photograph is qualitatively different from making a paper crane yourself. The difference is due to the fact that our representation of "paper crane" isn't a simple, static labeling. One's experience of constructing a paper crane, a non-trivial task, involves a number of complex perceptive experiences which are organized to give a representation of a paper crane. In other words, a mental representation is a combination of diverse somatosensory experiences. For real-world situations, one imagines that such representations are not static and algorithmic but rather more dynamic in nature.

In ECAL95, Pfeifer and Scheier reported a study on evolutionary robots that could perceive the size of an object by their bodily movements (Scheier and Pfeifer, 1995). They prepared two sizes of cylinder, one of larger diameter than the other. Robots were small vehicles with a single arm that could grasp only the smaller cylinders. By training the internal network of the robots, robots could eventually categorize the large and small cylinders. When a robot successfully grasped an object, it assumes it is small, otherwise it thinks it is large. As a result, the robot neglects a cylinder it can't grasp and spends more time with smaller ones. It looks as if a robot could obtain a representation of sizes of cylinder. This example clearly shows that categorization is established by his embodied active perception. Recently, there have been reported many such examples of dynamic categorization (Tani and Nolfi, 1998; Marocco and Floreano, 2002; Nolfi and Marocco, 2002).

The simulation results presented here provide another example of such dynamic categorization. In contrast to existing models, our model exhibits a dynamic repertoire of seemingly purposeful, lifelike behaviors whilst categorizing objects. In particular, while agents are trained to ignore triangles and linger at rectangles, they are also required to fill in the area inside rectangles. When doing this, our agents show a behavior that is distinct to that of finite state machines coupled with random noise, because long-term correlations with previous object interactions may be observed.

Model Description

The basic idea of our model is inspired by the active vision system of (Kato and Floreano, 2001). The differences are:

1) Categorization is not a direct task but a required feature to perform the task.

2) Objects and sensory area are rotatable. Therefore direction independent categorization should be required.

3) Focusing or leaping feature of sensory area is not considered here. The size of sensory area is fixed and moves continuously in the space.

4) No noise in the environment. Agent should use his internal dynamics to move in a non-deterministic way.

Field and Task

The field is a two-dimensional discrete lattice space of 100×100 points. Each point is in one of two states, empty(0) or occupied(1). An agent is situated in a point in the field and

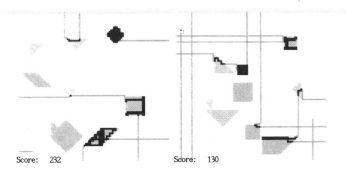

Figure 1: Examples of the field conditions and trajectories of an evolved agent.

has a direction of the heading. He receives sensory inputs from the point he stays and the 8 neighboring points. The positions of sensory inputs can be distinguished relative to his direction. Therefore 9 bits of information in total can be used to decide the next movement. On every discrete time step the agent changes the position to a neighboring point and change the direction of the heading according to the motor output. Motor output can be one of the 3 directions, namely, straight, left, and right.

If an agent crosses a boundary of the field, he appears from random position of the opposite boundary with the same direction of the heading. [1]

There are some objects classified as rectangles or triangles in the field. This classification is determined by the global arrangement of the occupied points. Using upright and slanted edges, we designed 2 types of triangles and 4 types of rectangles. Possible shapes of objects are enumerated in Fig. 4–7 in the latter section. The size of the objects are distributed between 5 and 15 units each side.

The task upon agents is to fill more points in rectangles and less in triangles. This task implicitly requires the categorization of objects. Because they can see only 3×3 points in the field simultaneously, the categorization is required to be dynamic.

Fig. 1 shows examples of an object arrangement in the field and a trajectory of an evolved agent. The internal structure of the agent is explained in the next subsection.

Network Architecture

An agent is equipped with a recurrent neural network to decide the next movement from the current states. Recurrent

---

[1]The reason why we used such a strange boundary condition is not essential but practical. Evolved agents tend to go straight when no objects are seen. Therefore, if the boundaries are precisely periodic, he often fail to find any object and get no score in experiments. Of course if we assume the boundary condition to be periodic, more intelligent agents that can explore the field more broadly might evolve. But in our simulations no such agents evolved. That's possibly because of the cheapness of their internal network structures.

neural network is interpreted as a mapping of internal variables depending on the inputs as parameters in terms of dynamical systems. Activations of context neurons represents the internal variables which can be used to keep memory such as "I've turned to left at the corner 2 time steps before".

Fig. 2 depicts the internal network structure. The activations of neurons are updated internally as follows:

$$y_i(t) = g\left(\Sigma_j w_{ij} y'_j(t-1) + b_i)\right), \tag{1}$$

$$g(x) = \left(1 + \exp(-\beta x)\right)^{-1}, \tag{2}$$

where $y'_i(t)$ is the actual activation of the ith neuron at the time t, $y_i(t)$ is the value generated by the internal dynamics. w_{ij} is the weight of connection from the jth neuron to the ith neuron, and b_i is the bias of the ith neuron. The summation is taken over all neurons which have a connection into the ith neuron. $g(x)$ is the sigmoid function. As a result, the activations take the value from $(0, 1)$. β is the nonlinearity coefficient and 1.0 in this paper.

Actual activations of input neurons are modified as follows:

$$y'_i(t) = (1 - \mu)s_i(t) + \mu y_i(t), \tag{3}$$

where s_i is the raw sensory input and take one of two values, namely, 0(empty) or 1(occupied). For context and output neurons $y'(t) = y(t)$. The meaning of Eq.3 is that the actual activations of input neurons are modified by the value which is generated from the internal dynamics. In this paper, we used $\mu = 0.3$. Therefore the activation of an input neuron takes the value from $(0, 0.3)$ or $(0.7, 1.0)$, depending on the state of the corresponding position in the field. When the value from internal dynamics is close to the sensory inputs, actual activations of input neurons get distinguishable clearly.

In addition, we introduced the plasticity of weights. Weights from input neurons and context neurons into input neurons change during the interaction with the environment so that input neurons can play a role like a prediction. These weight are updated according to the following difference:

$$\Delta w_{ij}(t) = \eta \left(s_i(t) - y_i(t)\right) y'_j(t-1), \tag{4}$$

where $\eta (= 0.01)$ is the learning rate. By introducing the plasticity in this way, the value y_i corresponding to the input neurons, which is generated from the internal dynamics, has the tendency to get closer to s_i in general.

Genetic Algorithm

To get the agents that have relevant sensory-moter coupling for dynamic categorization, the genetic algorithm is used.

Weights and biases of neural networks are binary encoded with 8-bit strings. In this model every genotype of individual has 8×195 bits of length. Gene strings are decoded to

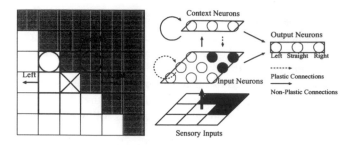

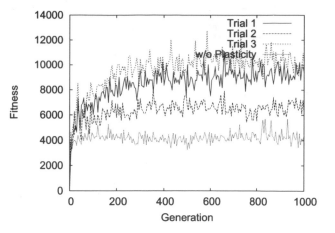

Figure 2: An agent situated in the environment and his neural network architecture: 9 sensory inputs are connected to corresponding input neurons. 3 context neurons are used to keep short term memory in unit cubic space. Colors of the circles in front of the agent show the value of context neurons and indicate his internal state. Movements are decided by the most activated neuron of the 3 output neurons related to moving Straight, Left and Right. Connections into input neurons are plastic and updated according to sensory inputs.

Figure 3: Fitness of the best individual from each generation. Trial 1,2,3 are the cases with neural plasticity. The last line shows the case without plasticity. Without plasticity, fitness saturate to lower value. The difference of the saturated value in the cases with plasticity depends on the dynamic repertoire of the movements discussed later by examples.

real values in the range $[-4, 4]$. In the case of plastic connections, values decoded from gene strings are interpreted as the initial values of the weights.

The number of population is 100. In each generation the performance of every agent is evaluated in 100 patterns of object arrangement randomly generated. The movement is simulated from the random initial position and direction for 1000 time steps. Weight values of plastic connections are reset to the initial values when the agent is put in the new environment. The score of the agent is determined by the number of footprints on rectangles minus the number of footprints on triangles. Footprints on the same position are not counted again. Scores in 100 arrangements are summed up to the fitness of the agent.

To create the next generation, bottom 30 individuals are killed and top 30 individuals are duplicated. In addition, 20 pairs of individuals exchange part of their genes by one point crossover and 50 individuals suffer point mutation. The mutation rate is 5 % for every bit.

Experimental Results

Fig. 3 shows the increase of the fitness of the best individual from each generation. Those grow faster in early generations and saturate in about 1000 generations. In the case without plasticity, fitness saturates to lower value than the cases with plasticity. Because the volume of genotype space are equal, we can say that this kind of plasticity has a good effect in this categorization task.

Fig. 4–7 are the collection of moving patterns of agents. They are the best individuals from the 1000th generation of 3 GA trials with plasticity and 1 trial without plasticity corresponding to Fig. 3. In general, the reaction to an object depends on the entrance point and direction. Some typical

entrance patterns to the first object the agent meet in the field are enumerated. The dark line shows the trajectory of the agent. In some figures the position of the agent is on the boundary. It means that the agent has left from the object. More explanations are found in each figure caption.

Fig. 8 shows the internal distinction between objects by the best individual in the trial 1.

Discussion

Although agents' predictive ability due to neural plasticity is not itself a component of agent fitness, it improves the fitness indirectly. While interacting with objects, agents form expectations based on the continuity of the object and their movement. The difference between these expectations and actual sensory input changes the internal dynamics and gives rise to the variety of behavior observed. Such relatively long time-scale dynamics (compared to the internal dynamics) should be a common feature of perceptual experience. We claim that the coupling between the short time scales of sensory-motor interaction and long time scales of adaptation and learning mechanisms plays an important role in cognitive processes. With this in mind, we can discuss the relationship between learning process and evolutionary process of adaptive systems.

The previous section does not show all the agent movement patterns. Agents may change their pattern of movement depending on what they previously interacted with. To solve the shape discrimination problem algorithmically, this kind of instability might be harmful. But our evolved agents can solve the problem reliable in spite of their intrinsic instability, indeed appearing as if they are able to autonomously

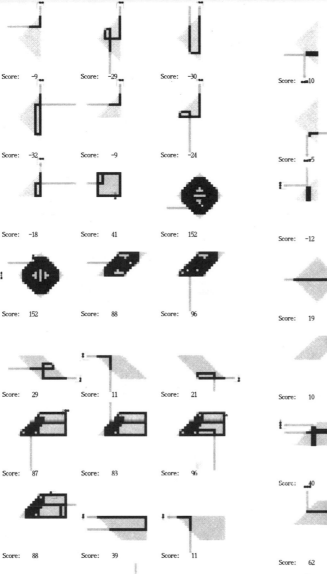

Figure 4: The best individual of the 1000th generation in the trial 1: Showing the movement patterns when the first object is found. This agent tries to distinguish the shape of object by moving along the edges of the object. After he categorized the object as "a rectangle", he tries to fill broad area near slanted edges by the rolling movement. He eventually enters to periodic movement and never leaves "a rectangle". The switching mechanism of this agent from the viewpoint of internal dynamics is shown and discussed later.

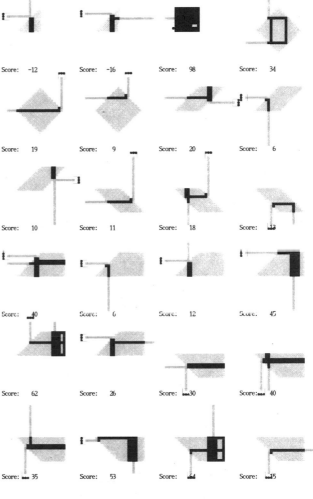

Figure 5: The best individual of the 1000th generation in the trial 2: Showing the movement patterns when the first object was found. This agent tries to fill between upright edges by turning into the object from random position. If the object is a upright square, he stays on the object forever. But if the object is a trapezoid, he leaves the object from a slanted edge. To realize the random movement while filling rectangles, chaotic or quasi-periodic internal dynamics is used.

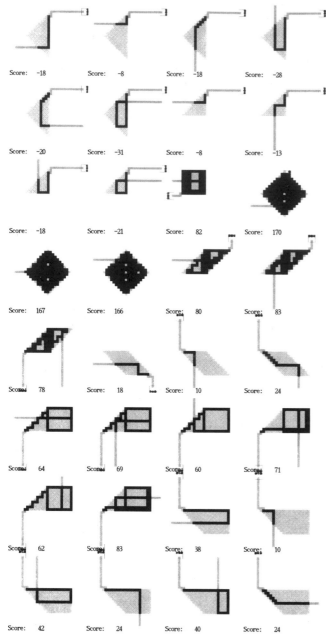

Figure 6: The best individual of the 1000th generation in the trial 3: Showing the movement patterns when the first object was found. This agent takes more flexible strategy compared to the other 2 agents before. Fitness is slightly higher than others. He leaves upright squares, parallelograms, and trapezoids after staying for a while. In addition, He can fill almost all the points of slanted squares. He moves as if wondering from one object to others and stay relatively short time around triangles and long time around rectangles. Actually, the trajectories in Fig. 1 is the movements of this agent. It can also be seen that after leaving an object, his internal states differ depending on the object.

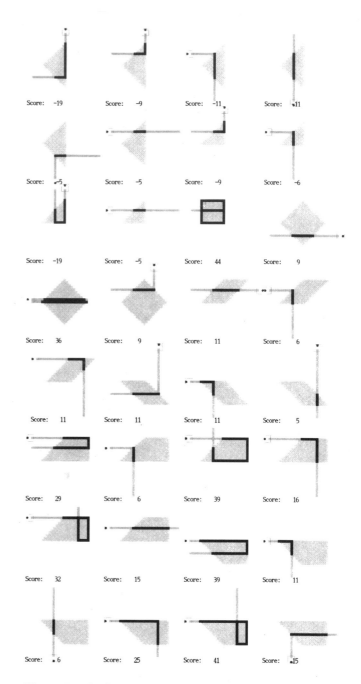

Figure 7: The best individual of the 1000th generation in the trial without plasticity: Showing the movement patterns when the first object was found. Without plasticity, evolved agent shows quite simple pattern of movement which can be realized without internal states or dynamics. That seems not only because it's hard to switch the movement by the internal dynamics. It may be harmful to have an unstable internal dynamics and switch the movement without memorizing long-term correlation.

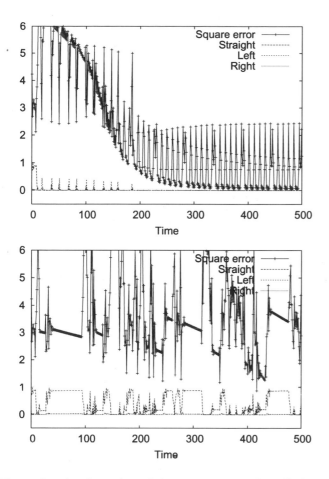

Figure 8: The dynamics of the square error of prediction and the activations of output neurons. Square error of prediction is defined as $\Sigma(s_i - y_i)^2$ by the difference between the raw sensory inputs and the generated values from internal dynamics. Although it is hard to read the actual movement from the activities of output neurons, the change of the pattern of activities shows the change of the pattern of movements. The agent is the best one from the trial 1 which shows the clear difference between interactions with a rectangle and a triangle. The top figure shows the dynamics while interacting with the slanted square. Square error decreases smoothly. The movement have changed from traveling along the edges of the square to rolling to fill the broad area along the edges around 200 time steps. The bottom one shows the dynamics while interacting with many triangles. In this environment the agent repeats entering and leaving triangles. Square error doesn't decrease smoothly. Internal dynamics is perturbed while moving around a triangle, but the switching to filling movement doesn't occur.

decide their movements, and at times even seem bored. In Kato and Floreano's active vision system, the environment is noisy and the agent tries to stabilize its movement to discriminate objects successfully. This is the main difference between our approach and theirs.

The agents presented here cannot discriminate objects perfectly; different agents have difference preferences. We feel that this is a positive result attributable to the autonomous and embodied characteristics of living systems. In particular, exploring behavior should be investigated as one way of understanding the difference between living systems and machines. In the narrow context here, generation of radial categories (Lakoff, 1987) with dynamics having both stable and unstable directions is a possible candidate for lifelike categorization.

Acknowledgments

This work is partially supported by Grant-in aid (No. 09640454) from the Ministry of Education, Science, Sports and Culture and from The 21st Century COE (Center of Excellence) program (Research Center for Integrated Science) of the Ministry of Education, Culture, Sports, Science, and Technology, Japan.

References

Gibson, J. (1962). Observations on active touch. *Psychological review*, 69:477–491.

Kato, T. and Floreano, D. (2001). An evolutionary active-vision system. In *Proceedings of Congress on Evolutionary Computation*. IEEE-Press.

Lakoff, G. (1987). *Women, fire, and dangerous things*. The University of Chicago Press.

Marocco, D. and Floreano, D. (2002). Active vision and feature selection in evolutionary behavioral systems. In *Proc. of the 7th Intl. Conf. on SAB*, pages 247–255. MIT Press.

Nolfi, S. and Marocco, D. (2002). Active perception: A sensorimotor account of object categorization. In *Proc. of the 7th Intl. Conf. on SAB*, pages 266–271. MIT Press.

O'Regan, J. K. and Noë, A. (2001). A sensorimotor account of vision and visual consciousness. *Behavior Brain Science*, 24:939–1011.

Scheier, C. and Pfeifer, R. (1995). Classification as sensory-motor coordination: A case study on autonomous agents. In *Proc. of the 3rd ECAL*, pages 657–667. Springer.

Tani, J. and Nolfi, S. (1998). Learning to perceive the world as articulated: an approach for hierarchical learning in sensory-motor systems. In *Proc. of the 5th Intl. Conf. on SAB*, pages 270–279. MIT Press.

Song Grammars as Complex Sexual Displays

Kazutoshi Sasahara and Takashi Ikegami
Department of General Systems Studies, Graduate School of
Arts and Sciences, University of Tokyo,
3-8-1, Komaba, Meguro-ku, Tokyo 153-8902, Japan
{sasahara, ikeg}@sacral.c.u-tokyo.ac.jp

Abstract

We study the complex evolution of song grammars of the Bengalese finch. Their mating songs have the remarkable feature that they are described by finite-state automata. (Honda and Okanoya, 1999) In addition, it has been experimentally confirmed that complex songs are preferred by females and that the Bengalese finch's song is more complex than that of its ancestors. (Okanoya, 2002) These facts suggest that complex grammar-like systems may have evolved as a result of sexual selection. In order to explore this hypothesis, we study the communication between male and female finches, modeling their co-evolution by asymmetric finite-state automata. By introducing a complexity measure for song grammars, we study the effect of females' preferences. We observe that a gradual transition from lower complexity to higher complexity grammars is associated with the changing of male birds' courting strategy.

Introduction

Language is where the phonemes of a vocal communication system are arranged in a complex, one-dimensional manner with precise articulation. Whether or not this language contains grammar is a main difference distinguishing human and animal communication (Hauser et al., 2002). Interestingly, such functional features of human language are very similar to the courtship songs of songbirds and whales and different to the vocal signals of chimpanzees (Aitchison, 2000; Michael and Allison, 2000). This suggests that grammatical language behavior may have evolved independently in different species. Therefore, we believe that the study of language-like behavior is one way to understand the origins and evolution of language.

In recent years, experiments studying the grammar of one particular songbird, the Bengalese finch (*Lonchura striata var. domestica*), have been carried out and the following facts were established: (Honda and Okanoya, 1999; Hosino and Okanoya, 2000; Okanoya, 2002; Okanoya, 2003)

- The courtship song of the male Bengalese finch consists of a combination of chunks, each of which is a sequence of sound elements. Unlike alarm calls and threats, these may be recursively constructed by finite-state automata.

- More complex songs are preferred by females over monotonous ones. They promote the reproductive behavior of females, for example, the frequency of nest-making and mating poses etc.

- The Bengalese finch is a domesticated species of White-Backed Munia (*Lonchura striata*). After domestication, the song of the Bengalese finch has become much more complex than that of the White-Backed Munia.

In light of these facts, it has been hypothesized that males with complex song grammars have been chosen by females and that song grammars have evolved as a result of sexual selection (Okanoya, 2002). To explore this hypothesis, we study the co-evolution of males' song grammars and females' preferences by a synthetic approach which represents birds as asymmetric finite-state automata (FAs).

A significant aspect of our modeling is the way in which female birds gauge the complexity of the songs they hear. While complex songs generated by a FA enhance the reproductive behavior of females, the frequency of this reproductive behavior tends to be relatively low when the females are listening to monotonous or random songs (Okanoya, 2003). This suggests that the female birds can innately discern grammatical features such as recursive arrangement of song elements, and therefore distinguish interesting songs from monotonous or random ones. We therefore assume that female birds may be sensitive for arrangement of chunks and may have innate preferences for phrasing and rhythm in a song, so that they can gauge songs according to her preferences. As a results, the song grammars of males may have become complex due to the diversity of females' preferences.

To model this process, we constructed artificial birds as asymmetric FAs, one type of which is used only for song generation, the other type being used only for listening. We then introduced the following communication interaction between males and females. The female interjects in synchrony with the male song (by wagging her tail or chirping softly, for example), measuring how many interjections succeed according to her preferences before she evaluates her satisfaction with the song. [1] In this model, we call such a interaction *"song-interjection"* communication.

With this model, we demonstrate co-evolution of male song grammars and female preferences.

Model

Here the co-evolution of male song grammars and female preferences is modeled as a communication game.

Male and female birds

The song grammar of a male bird is expressed as a finite-state automaton (FA) with an output as follows:

$$G = (Q, \Sigma, \Delta, \delta, \lambda, q_0), \qquad (1)$$

where Q is a finite set of states, q_0 is an initial state, Σ is a finite set of input symbols, Δ is a finite set of output symbols, δ is a state transition function, $Q \times \Sigma \to Q$, λ is an output function and $Q \times \Sigma \to \Delta$ (Hopcroft and Ullman, 1979). In this model, $\Delta = \{blank, A, B, ..., J\}$, where each letter represents a song chunk and *"blank"* represents a silent interval between chunks. The bunch of combinatorial chunks between blanks expresses a phrase and the whole output sequence expresses a courtship song. A male bird arranges the chunks in accordance with his song grammar.

On the other hand, the preference of the female bird is expressed by a FA with an input:

$$P = (Q, \Sigma, \delta, q_0, F), \qquad (2)$$

where Q, Σ, δ, and q_0 are the same as above, and F is a set of accepting states, which is a subset of Q. This expresses a female's preference for the phrasing, rhythm, and arrangement in a courtship song. She changes her internal state by listening to the song and interjects if she is in an accepting state. Examples of male and female FAs are shown in Fig.1. Note that every node in a female's FA has one transition for every possible input, which is completely different from males' FAs.

Communication and co-evolution

Given the number of male and female birds, N^{male}, N^{female}, in the initial state, communication occurs as follows.

Each male bird attracts a female at random and sings a song for length L_{song} according to his grammar, G, where L_{song} denotes the length of the courtship song. Since a male's song must be a signal for tempting a female, *"novelty"* is an important factor in a courtship song (Miller and Todd, 1993; Werner and Todd, 1997). So, we assume that females may pay attention to the novelty of males' song as follows:

---

[1]The female Zebra finch has been observed to chirp in synchrony with the male song. However, no such observation has yet been made for the female Bengalese finch.

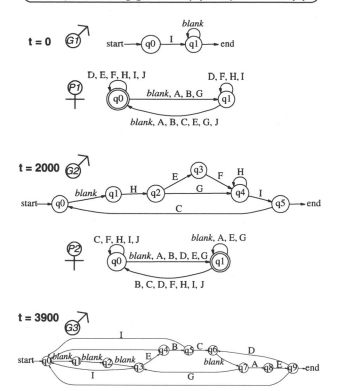

Figure 1: Examples of song grammars (G) and preferences (P): $G1$, $G2$ and $G3$ shows male song grammars and $P1$, $P2$ show female preferences. $G1/P1$ and $G2/P2$ are well-suited pairs that yielded good communication at t = 0 and t − 2000, respectively. $G3$ shows a grammar more complex than $G1$ or $G2$. The double circle represents a accepting state.

(i) There must be at least one chunk in the song which is not *blank*.

(ii) The female must make at least one mistake in interjecting. This is because a song to which a female can interject perfectly (i.e. which is perfectly predictable by her FA) is boring for her.

Unless these conditions are satisfied, the birds involved are not eligible candidates for mating. Listening to a novel song that fulfills the above conditions, a female bird interjects to the song in accordance with her preference, P and evaluates her contentment. Each male bird can sing for a length up to L_{song}^{max} and can sing to several females within this length, where L_{song}^{max} denotes the maximum length he can sing. For example, if a male bird has $L_{song} = 10$ and $L_{song}^{max} = 50$, he can attract five females. Each time step, every male bird behaves in the above way.

On the other hand, the female interjects to all songs that she hears and that she is satisfied with according to her preference P and the above novelty criteria (i) and (ii). In this

Examples of communication

t = 0

G1(N^{node}=2,L$_{song}$=10,LI=0.67) vs. P1(N^{node}=2)

```
I__I_I_I_I
_#__##__##
```

score = 0.38

t = 2000

G2(N^{node}=6,L$_{song}$=13,LI=0.67) vs. P2(N^{node}=2)

```
_HEFIC_HGI_HE
#_#___#_#_#_#
```

score = 0.43

t = 3900

G3(N^{node}=10,L$_{song}$=148,LI=0.63) vs. P3(N^{node}= 8)

```
__I__EC__EBI__EBI__I__EBC_GEC__I__I__I__EBC_GEC__EC__EBI...
###_###__###___###___###_###___##__###_###_###_###___##__###_###___...
```

score = 0.74

t = 2000

G4(N^{node}=9,L$_{song}$=28,LI=0.64) vs. P4(N^{node}=2)

```
GEGEJC_IGEJC_EGEGIC_IGIC_IJC
##___#_#___#__##_##_#_##_#_#
```

score = 0.0

Figure 2: Examples of communication: $G1$ vs. $P1$, $G2$ vs. $P2$ and $G3$ vs. $P3$ lead to suitable communications. $G4$ vs. $P4$ shows an example of poor communication.

model, the male can't discern the female's contentment for his song; only the female herself knows how much she likes a particular song. Some examples of communication are illustrated in Fig.2.

After a pair of birds communicate, their communication is assigned a score calculated as follows:

$$S = \frac{1}{3}\left\{ \frac{1}{N_{interj}^{th}} min(N_{interj}^{succ}, N_{interj}^{th}) + \frac{N_{interj}^{succ}}{N_{interj}^{all}} + \frac{N_{chunk}}{L_{song}} \right\}, \quad (3)$$

where $0 \leq S \leq 1$. The first term denotes the evaluation of the number of successful interjections. The score is proportional to the number of successful interjections N_{interj}^{succ} below the threshold N_{interj}^{th}. If $N_{interj}^{succ} \geq N_{interj}^{th}$, the female bird's evaluation is saturated and the first term becomes 1. The second term denotes the success rate of interjection, that is, the ratio between the total number (N_{interj}^{all}) and successful number (N_{interj}^{succ}) of interjections. The third term denotes the fraction of non-empty chunks in a song. In other words, in total communication score (3) considers the evaluation of both quantity and quality of interjection, and the richness of song elements. In this way, females prefer longer songs

to which they may relatively easily interject, based on their preferences with the novelty criteria (i) and (ii).

According to the communication scores, females select the males with the highest score as their mating partners. Assuming that they produce offspring in proportion to their communication score, the number of offspring is calculated as $C_{offs} \cdot S$. Then, their offspring's genders are randomly assigned and they are added into the system as new child birds.

Since child birds study songs from their fathers or may have similar song preferences to their mothers as a result of their upbringing, their characters become similar to those of their parents. Therefore, in our model child birds inherit the FAs of their parents, changed according to the following genetic mutation operations:

(a) **Arrow Mutation**: Change the transitions of the FA with the number of nodes remaining fixed.

(b) **Node Mutation**: Change the number of nodes (± 1) and then add or remove arrows as required.

(c) **Random Mutation**: A new FA is made at random.

These (a)-(c) express the possible inaccuracy in child birds inheriting their parents' characteristics, song grammars G and preferences P. In particular, the accuracy of inheritance is highest in (a) and (c) represents complete failure to inherit any characteristics.

In addition, the following mutation is performed in the male bird population only:

(d) **Song Mutation**: Change L_{song} (± 5), and change L_{song}^{max} (± 2)

In this artificial ecosystem, each bird has a life time T_{life}, after which they are removed from the system. In order to limit the maximum number of birds, some birds are removed due to a fixed ecological capacity of $C_{echo}(N^{male} + N^{female})$.

In our simulations, these procedures are iterated over time.

Simulation Results

Here we describe the typical results of this artificial evolution.

The parameters of our simulations were as follows. The initial populations of males and females were 100 respectively. Every male bird had a FA constructed randomly with $N_{node} = 2$, $L_{song} = 10$ and $L_{song}^{max} = 50$. The maximum length of song was 500. Meanwhile, every female bird also had a randomly constructed FA with $N_{node} = 2$. Two examples of initial FAs are shown in the top of Fig.1. Other significant parameters were $N_{interj}^{th} = 100$, $C_{offs} = 3.5$, $T_{life} = 5$ and $C_{echo} = 0.3$.

Population Dynamics

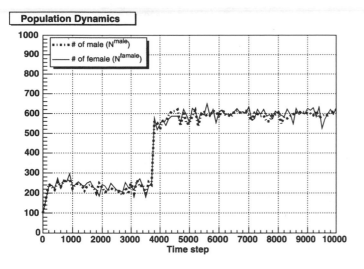

Figure 3: Population dynamics: Step-like evolution is observed. After the critical period around t = 3700, the number of birds rapidly increases.

Features of communication

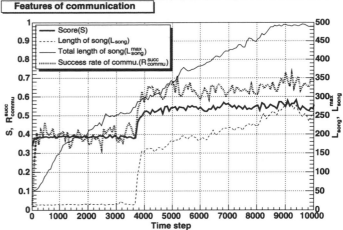

Figure 4: Features of communication: Shortly after the critical period t = 3700, a trend in preferring longer songs emerged. Subsequently, communication scores and successful communication rates increased.

Co-evolution

The population as a function of time is shown in Fig.3. We see a step-like evolution in which the population increased rapidly at about $t = 3,700$ and then remained almost constant.[2] After this period, we can see a change of strategy in the courtship behavior of both males and females. In Fig.4, we can see a rapid increase in the length of males' songs, L_{song}. Before this period, the male birds sang relatively short songs, even if the maximum song length was much higher. They could find partners successfully by singing to many females. This is clear from the N_{court} in Fig.5., which represents the average court count a female bird. Before the critical period at $t = 3700$, the communication scores were low in Fig.4., but N_{court} was very high in Fig.5.

On the other hand, shortly after that period, a trend in preferring longer songs emerged. Once such a trend appeared in the system, the character of males and females was drastically changed as may be seen in Fig.4 and Fig.5. In particular, Fig.5 shows that the male birds began to sing longer and more complex songs as the number of FA nodes increased. Subsequently, the female birds had a tendency to become sensitive to the arrangement of song chunks and also had FAs with increasing numbers of nodes. It is because of larger FAs with many nodes that females could distinguish the order of complex male songs.

The male birds never sang for the maximum length L_{song}^{max} shown in Fig.4. This results from a kind of dilemma. The

male birds wanted the females to listen to their songs to get high scores, but if their affinity was bad, the cost of failure would be more serious. We can say that the male birds evolved a survival strategy as a result of co-evolution, avoiding the risky behavior of singing to only one female a song of length close to L_{song}^{max}.

On the whole, the communication evolved to become quite successful judging from the change in average communication score and the success rate of communication which represents the frequency non-zero scores in Fig.4. [3]

In summary, the emergence of male birds which could sing novel and longer songs triggered the co-evolution of song grammars and female preferences.

Complexity of song grammars

For the following discussion, we require a measurement of song complexity. We can define LI, the linearity of a song grammar as

$$LI \equiv N_{node}/N_{arrow}, \qquad (4)$$

where N_{arrow} is the number of arrows leaving a node. If $N_{node} = N$, this value ranges between $1/N \leq LI \leq 1$ as N_{arrow} varies from N^2 to N. More complex FAs have lower values of LI. However, when the number of nodes is low, this index isn't suitable for measuring the complexity of song grammars. For example, the song grammars $G1$ and $G2$ of Fig.1 have the same value of $LI(= 0.67)$, but the song generated by $G2$ is clearly much more complex as may be seen in Fig.2.

---

[2]Whether or not we see such a stepwise change depends on both the parameters N_{interj}^{th} and C_{offs} that affect the number of offspring. The step-like evolution is observed in wide region of these parameters, provided they are not too big or too small (i.e. the production rate is not too high or too low).

---

[3]If a male's song doesn't fulfill the novelty criteria (i) and (ii), the male gets no score. Moreover, if a target female cannot interject properly at all, the male is also unable to get a score.

In the initial state, only simple song grammars like $G1$ exist in the system. In such a case, even grammars that could only generate *blank* and chunks in turn could get a score. As the evolution proceeds, the song grammars became more complex. In Fig.1, more complex song grammars $G2$ and $G3$ are shown. $G2$ has a feature typical of more complex songs. That is, it has a node, q_4 which may be reached by two different paths, $q_1 \rightarrow q_2 \rightarrow q_3 \rightarrow q_4$ and $q_1 \rightarrow q_2 \rightarrow q_4$. Song grammars which include such branches can arrange non-deterministic chunks, thereby avoiding perfect interjection, which is a prerequisite for song novelty (ii). However, if LI is less than 0.5, song grammars have more than 2 branches per node and it becomes difficult for female birds to successfully interject.

The evolution of song grammar linearity is shown in Fig.5. In this picture, we see that LI decreases gradually from 0.6 to 0.4 over time. LI never reaches 0 or 1. Such song grammars are not too simple and not too stochastic. This suggests that the song grammars have to be understandable by female birds. Therefore, we may conclude that the novel song-interjection communication is itself a driving force of the evolution of song grammars while at the same time placing an upper bound on their complexity.

In addition, we can confirm in Fig.5 that before the critical period at $t = 3700$ where the male birds sang only short songs, LI was almost constant. On the other hand, shortly after this period, the number of females' nodes increased rapidly and LI started to decrease, i.e. the song grammars became complex. This suggests that the female birds require longer songs before they can evaluate them.

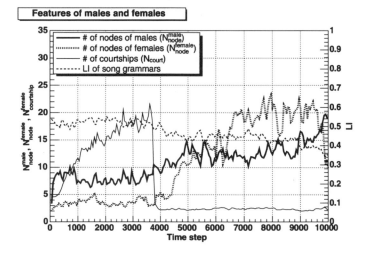

Figure 5: Features of males and females: The number of times the male birds sang to females N_{court} and the number of nodes of females N_{node}^{female} drastically changed at the critical time t = 3700. On the other hand, the number of nodes of males N_{node}^{male} and the linearity of song grammars LI changed gradually over time.

Discussion

It has been suggested by Okanoya that the mating song of the Bengalese finch may have evolved as a result of female preference. In the case of male peacock's sexual displays, the females' preferences are said to be the shapes and hole patterns in the males' plumage (Zahavi and Zahavi, 1997). We wonder, what kind of characteristics do female Bengalese finches prefer? If they prefer complex song grammars, how could these have evolved?

In order to explore these issues, we have modeled the co-evolution of male and female birds by asymmetric FAs. As a result of our simulations, we confirmed that song grammars could evolve to become complex via relatively brief, "*novel song-interjection*" communications. In this system, successful communication and song novelty are key concepts. Our results suggest that the driving force behind the evolution of grammar-like systems could be communication itself, not a diverse external environment (Suzuki and Kaneko, 1994; Hashimoto and Ikegami, 1996; Steels, 1999; Komarova et al., 2001; Sasahara and Ikegami, 2003).

The process of song grammar evolution observed in our model may be thought of as a "*runaway process*" (Zahavi and Zahavi, 1997). That is, singing a complex song doesn't assist a male bird in his survival. However, if this character, which is preferred by females, can be inherited by his offspring, complex songs could have an advantage in the conservation of his genes.

In addition, by introducing a complexity measure for song grammars, we studied the effect of females' preferences. We found that the linearity of song grammars LI gradually decreased from 0.6 to 0.4. This suggests that the song grammars may have evolved as a result of the cognitive ability of the female birds, in agreement with Okanoya's hypothesis.

Acknowledgments

This work was partially supported by Grant-in aid (No. 09640454) from the Ministry of Education, Science, Sports and Culture and from The 21st Century COE (Center of Excellence) program (Research Center for Integrated Science) of the Ministry of Education, Culture, Sports, Science and Technology, Japan. KS would like to thank Okanoya for his helpful suggestions.

References

Aitchison, J. (2000). *The Seeds of Speech: Language Origin and Evolution.* Cambridge University Press.

Hashimoto, T. and Ikegami, T. (1996). Emergence of net-grammar in communicationg agents. *Biosystems*, 38:1–14.

Hauser, M., Chomsky, N., and Fitch, W. (2002). The faculty of language: What is it, who has it, and how did it evolve? *Science*, 298:1569–1589.

Honda, E. and Okanoya, K. (1999). Acoustical and syntactical comparisons between songs of the white-backed munia(lonchura striata) and its domesticated strain, the bengalese finch(lonchura striata var. domestica). *Zoological Science*, 16:319–326.

Hopcroft, J. and Ullman, J. (1979). *Introduction to Automata Theory, Languages and Computation*. Addison Wesley.

Hosino, T. and Okanoya, K. (2000). Lesion of a higher-order song control nucleus disrupts phrase-level complexity in bengalese finches'. *NeuroReport*, 11:2091–2095.

Komarova, N., Niyogi, P., and Nowak, M. (2001). The evolutionary dynamics of grammar acquisition. *Journal of Theoretical Biology*, 209:43–59.

Michael, S. and Allison, J. (2000). What songbirds teach us about learning? *Nature*, 417:351–358.

Miller, G. and Todd, P. (1993). Evolutionary wanderlust:sexual selection with directional mate preference. In Jean-Arcady, M., Herbert, L., and Stewart, W., editors, *Proceedings of the Second International Conference on Simulation of Adaptive Behavior*, pages 21–30. MIT Press.

Okanoya, K. (2002). Sexual display as a syntactical vehicle. In W., A., editor, *The Transition to Language*, pages 46–63. Oxford University Press.

Okanoya, K. (2003). *Kotorino Uta kara Hito no Kotoba he (in Japnaese)*. Iwanami Syoten.

Sasahara, K. and Ikegami, I. (2003). Coevolution of birdsong grammar without imitation. In Wolfgang, B., Thomas, C., Peter, D., Jan, T., and Jens, Z., editors, *Advances in Artificial Life, 7th European Conference on Artificial Life*, pages 482–490. Springer.

Steels, L. (1999). *The Talking Heads Experiment. Volume 1. Words and Meanings*. Antwerpen.

Suzuki, J. and Kaneko, K. (1994). Imitation game. *Physica D*, 75:328–342.

Werner, G. and Todd, P. (1997). Too many love songs: Sexual selection and the evolution of communication. In W., A., editor, *Fourth European Conference on Artificial Life*, pages 434–443. MIT Press.

Zahavi, A. and Zahavi, A. (1997). *The Handicap Principle*. University Press.

Analogies between Genome and Language Evolution

Luc Steels[1,2]
[1] Sony CSL - Paris - 6 Rue Amyot, 75005 Paris
[2] University of Brussels (VUB AI Lab)
E-mail: steels@arti.vub.ac.be

Abstract

The paper develops an analogy between genomic evolution and language evolution, as it has been observed in the historical change of languages through time. The analogy suggests a reconceptualisation of evolution as a process that makes implicit meanings or functions explicit.

Keywords: Language evolution. Evolution of communication. Cultural evolution and learning.

Introduction

In November 1998, a small workshop involving linguists, biologists, and Artificial life researchers was held in Paris with the goal of exploring analogies between language evolution and genomic evolution.[1] As one might expect, the discussion was both enormously stimulating but also very inconclusive. It brought out great gaps between the fields, partly caused by a lack of clear theories, particularly for language evolution. Analogies are very risky. Nevertheless, they play an important role in scientific discovery (such as the analogy between planets circling around the sun and electrons orbiting the nucleus). In the best case, they lead to a conceptual revision, both of the source of the analogy and its target. The analogy between language evolution and species evolution was first proposed by Darwin, who was strongly influenced by August Schleicher, Ernst Haeckel, and other 19th century linguists who viewed language as a living system (Richards, 1987). More recently, the syntactic structure of genomes is being described using the same formalisms as used in

---

[1]Participants included F. Cambien, P. Hogeweg, T. Ikegami, D. Krakauer, T. Kuteva, L. Steels, E. Száthmary, B. Vittori, and G. Weisbuch

linguistics, and genomic evolution is being modeled in terms of changes to formal grammars (see e.g. (Dassow, 1996)). With the background of better Artificial life models of language evolution (as surveyed for example in (Cangelosi and Parisi, 2001) and (Steels, 2003)) and better knowledge of the functions of the genome in development and genomic evolution, I develop in this paper another kind of analogy between language systems and genomes, emphasising meaning or function.

The paper is intended to be a discussion paper at a conceptual level. The main purpose is to formulate constraints and issues that must be addressed in models of language or genomic evolution. At the same time, the paper provides background and justification for computational and robotic experiments discussed in other more technical papers ((Steels and Kaplan, 2000), (Steels, 2003), (Steels, 2004)). There are two major points: First I emphasise the benefit of looking at the whole system (form, meaning, and effect in the case of language; genes, biochemical function, and structure/behavior in the case of genomes), as opposed to only focusing on the evolution of syntax. Second I will emphasize that both language evolution and genomic evolution are concerned with making certain meanings/functions explicit which were implicit before, or vice-versa. The big issue is then: how we can understand the mechanisms underlying this process and how we can synthesise them in artificial systems.

Evolution of language and languages

A distinction must first of all be made between the evolution of language, in the sense of the origins of language, and the evolution of languages throughout human history. This is analogous to the distinction between the origins of life itself and the subsequent evolution of living organisms over millions of years. Investigations into the evolution of language focus on finding developmental histories of how different brain areas could have become recruited for language and what factors might have caused verbal behavior to become such an important part of human activity. The work of Deacon (Deacon, 1997) is representative for this research challenge.

Investigations into the evolution of languages take the form of empirical investigations surveying the actual change in language, for example from Latin to French, Spanish, or Italian (see e.g. (Hopper and Traugott, 2003), (Heine and Hnnemeyer, 1991)), and theoretical investigations trying to identify and/or simulate the cognitive processes that give rise to these changes (see e.g. (Heine, 1997), (Steels, 2004)). It is generally assumed that language change cannot be based on genetic evolution because (1) it is very fast compared to genetic evolution, and (2) a person born in one linguistic community can learn the language of another community quite easily, even though the earlier one starts the better.

There are possibly very strong relations between the original evolution of language and the subsequent evolution of languages, in line with the uniformitarian hypothesis (adopted by Lyell in geology and Darwin in biology): The same processes that have molded languages throughout history must also have been playing a role in the genesis of language *ab initio*, and indeed they have been observed when a lexical language (pidgin) evolves into a creole (DeGraff, 1999). There also appears to be obvious connections between language learning and language change, in the sense that the cognitive operators which have been hypothesised as driving language change, are highly relevant to the ones underlying the socio-cultural learning of language (Heine, 1997). This paper focuses only on the evolution of languages without exploring these additional ramifications.

Defining Language Evolution

In order to characterise language evolution more precisely, I am going to take a functionalist point of view, which means that language is primarily seen as a vehicle for communication, and so its origins and evolution fit within the general process of evolving communication systems. Communication is here defined as the process whereby one agent (the speaker) deliberately influences the behavior of another agent (the listener) using (conventionalised) signs. Language therefore involves three aspects: forms, meanings and effects.

- The *forms* of language are sounds, words, word order patterns, intonation, stress, etc. They are the observable building blocks with which utterances are made.

- The *meanings* are what is expressed by utterances. Meanings are here defined as distinctions relevant to the agent-environment interaction. For example, the distinction between red, green, and orange traffic lights is relevant for deciding whether to cross the street or not. Meanings are assumed to be coded as information states so that they can play a role in semantic processes, instantiated as transformations over information states.

- The *effects* of an utterance are the behaviors carried out by the listener as a result of the meanings deduced from the form of an utterance. The most basic effect of language is to draw attention to an object or event in the world but many other effects are possible.

Consider a scene where two people are walking towards a bus stop. One looks behind and suddenly shouts "the bus", after which both start to run. The forms here are the words "the" and "bus" put in a particular order. The meanings include (1) a specific class of autonomously moving objects (buses), and (2) an indication (using the word "the") that there is a unique bus being expected in the present context. The effect of this utterance is to draw the attention of the listener to the fact that a bus is approaching and to take immediate action to catch it.

There are four important properties of human natural languages which are of crucial importance for the present discussion:

(1) Typically the form of the utterance only gives a hint about expected behavior. It influences behaviors which might already be going on anyway, without fully causing or determining them. In the example above, the speaker did not say that the listener should start running or whether a bus was approaching, she just said "the bus". The participants were already walking towards the bus and shared the context and goals. Natural language is therefore not a code in the sense of Shannon, which simply translates information from one form into another (Sperber and Wilson, 1987). Part of the meaning must be reconstructed based on the shared situation, common ground, joint attention, inference, etc.

This is why it is extraordinarily difficult for computers and robots to parse and interpret human language and it raises doubts whether information theory is a good framework for studying human natural language and language evolution.

(2) The relation between form, meaning, and effect is very indirect and multi-layered. Several words and grammatical constructions often collaborate in a non-modular way to constrain the possible meanings of the utterance, and there is a multi-layered hierarchical structure with certain words and constructions having a purely regulatory effect on the meanings and effects of others. For example, the word "back" has many meanings: a body part ("my back hurts"), a spatial area ("in the back of the car"), a temporal relation ("back in the good old days"), an adverbial particle ("I will be back"). The syntactic context and semantic expectations help the listener to pick out effortlessly the intended meaning. The influence by the meanings of an utterance on action selection (the 'illocutionary force') is even more determined by the context. For example, whether the utterance "the bus" evokes running or not depends entirely on what is happening in the present situation.

(3) There are important differences between languages in what meanings they make explicit, either in the lexicon or in the grammar. For example, European languages typically express tense and aspect (present/past/future, progressive, perfective/imperfective) through morphology and grammatical constructions. Compare: "I will write a letter" (future) with "I was writing a letter" (past imperfective). In Chinese, tense is not explicitly expressed grammatically but must be circumscribed indirectly, or inferred from the context, even though aspect is made explicit (for example with the particle -le for perfective aspect).

Thus the following sentence is unclear whether the washing was in the past or the present.

```
Akiu xi-zhe na jian dayi.
Akiu wash-prg that clothing-cl coat
Akiu is/was washing that coat.
```

This example illustrates also that Chinese, similar to most African languages, makes a distinction between different classes of nouns which are expressed through classifiers like "jian". English weakly uses a distinction between male/female/neuter, but otherwise does not express the distinctions implied by the Chinese classifiers at all.

(4) Finally, there is substantial evidence from all the world's languages and over all periods of recorded history, that profound changes take place, both in what meanings are made explicit and in how they are made explicit. For example, many languages (like Latin, Old-Germanic, Chinese, Polish) do not have a separate syntactic class of articles (like "the", "a", etc.) to express determination (definiteness/indefiniteness with respect to present context, quantification, etc.). It has been shown that a grammatical system for determiners may evolve in a language, as indeed it did in most languages that evolved from Latin (French, Italian) or from Old-Germanic (German, Dutch, Danish), typically by changing the form and function of demonstrative pronouns, like "that" => "the" or "ille" (Latin) => "le" (French).

The process by which new grammatical subsystems arise is generally known as grammaticalisation (Hopper and Traugott, 2003) and is discussed further below. It has also been shown that certain grammatical systems may disappear, at which point their function is totally lost or it is taken over by another system which develops often in competition with the first. A well known example is the case system in old English (consisting of morphological affixes or inflections to make the role of the referent of a noun phrase in an event explicit, as in German der/dem/den/des). The case system of early Old English was similar to that of Latin or (old-)German in complexity but largely disappeared by the advent of Middle English. This meant that other means had to be found in order to express these roles, leading to a tightening of word order and the extended use of prepositions. These grammaticalisation phenomena are precisely the processes that any theory of language evolution must explain.

In conclusion, we can view language evolution as follows:

> Language evolution is the process whereby meanings which were implicit, become explicit or vice-versa.

The Analogy with Genomic Evolution

An organism's genome and its role in the development and functioning of an organism is of course in many (if not most) respects very different from an utterance or set of utterances. Nevertheless we can view the genome as a kind of communication which influences

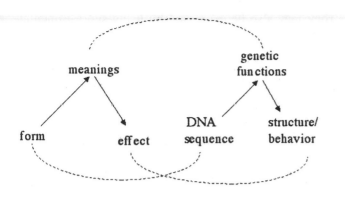

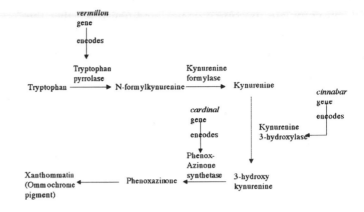

Figure 1: An analogy is suggested between the form/meaning/effect of language and the DNA/gene function/structure relation in development of organisms.

Figure 2: Example of a simple pathway for the synthesis of color pigments in the *Drosophilia* eye. Most biochemical pathways are much more complex, forming networks rather than linear chains.

how the organism is to develop, maintain itself, and behave. Just as in the case of language and in line with functional genomics, we will adopt a functionalist viewpoint, looking at the whole process from genomes to behaving organisms. The cores of this system are the biochemical pathways that determine the development of cells, tissues, and organs, and their structural maintenance and functioning. For example, the synthesis of ommochrome pigments in the *Drosophilia* eye is based on the 'tryptophan degradation pathway' schematically shown in figure 2 ((Wilkins, 2002) p. 104). Each step in a pathway synthesises molecular substrates, regulated by enzymes acting as catalysts. The function of genes is to act as such regulators, co-determining whether a transition takes place or not. Some of them have meta-functions, regulating the activities of other genes, or repairing gene copying. But other factors may intervene in the success of a transition as well, for example, certain substrates or catalysts might have to be provided by the environment or maybe byproducts of earlier biochemical transitions.

We can identify three aspects to genome-steered development, analogous to the form, meaning, and effect of language (see figure 1).

- An organism's genetic material, the DNA sequence, is similar to the form aspect of an utterance.

- The functions of genes in establishing transitions in biochemical pathways play the same role as the meanings of utterances.

- The effects of gene function (in specific contexts) are the behaviors and structures of the cells, tissues, and organs that allow the organism to function in a particular way.

Given this analogy, we can see that the genomic system has some of the same properties as natural language based communication systems: (1) Genes influence the transitions in biochemical pathways but are not necessarily the sole cause or controller of a transition. Substances available in the environment and even environmental stimuli processed by specific sensors, such as a pheromones, may co-determine whether a morphogenetic pathway unfolds. (2) The relation between the genome, the functions of the genes in orchestrating particular biochemical pathways, and the resulting structure and behavior of the organism is very indirect and multi-layered. For example, many genes are concerned with setting up the context for others, forming gene regulation networks with activators, inhibitors, pleotropic regulators, etc. Just as words and grammatical constructions are polysemous and ambiguous, the same gene product may be used multiple times in different cell or tissue types and at different times. (3) Differences between species (and even among members of the same species) concern what transitions in biochemical pathways are explicitly regulated or influenced by genes or not, in other words whether a certain biological function is genetically determined. In the most obvious case, this regulation establishes whether the biochemical pathway can complete its course or not, and hence whether the organism has certain structures or behaviors. But it can also be a matter of variation in the probability or speed of certain transitions. (4) There are profound changes in whether certain biochemical transitions are mediated by gene-encoded enzymes or not. Ernst Mayr already pointed out that morphological variation can not only occur by genetic variation, typically in highly canalised systems, but that there is also non-genetic variation, for example due to environmentally induced plastic responses (Mayr, 1963).

Based on these observations, it makes sense to view genomic evolution as analogous to language evolution:

> Genomic evolution is the process whereby transitions in biochemical pathways which where were implicit, become (genetically) explicit, or viceversa.

Empirical Data on Language Evolution

A lot is known from empirical observations by historical linguists how language evolution takes place ((DeGraff, 1999), (Heine and Hnnemeyer, 1991)), even though there are today hardly any good theoretical models of the causal mechanisms that underly them. Basically, the following five phenomena have been identified.

1. *Lexicalisation* Lexicalisation is the process by which a word becomes associated with a (new) meaning. The activities of human beings are in constant flux and new behaviors come up all the time, for example driven by the development of new technologies. Hence new distinctions become relevant and they may lead to a subsequent need to express them in verbal interactions. There may also be a desire to express existing meanings in new ways in order to 'keep the listener on her toes'. In principle, any word can be associated with any meaning because there is an arbitrary relation between form and meaning in language. But usually the word form is not constructed *de novo*, but rather an existing word whose meaning has some analogical or metaphorical relationship to the new meaning, is recruited. A listener must be able to guess the (new) meaning of a word and if the word has already a meaning which is close that becomes more likely to succeed. For example, the color word "orange" is adopted by metonymy from the orange fruit.

2. *Syntacticisation* New meanings (or existing meanings) are not necessarily expressed by new word forms.

They may also be expressed by using suprasegmental form characteristics, such as the ordering of words, intonation, stress on certain words or syllables, etc. Moreover words do not occur in isolation but are linked to each other in syntactic contexts which determine part of the interpretation process. For example, in the utterance "the oldest girl sent a letter to her father" we know that "oldest" and "girl" cooperate to identify a particular referent, that "her" probably refers to the same referent, that "her" and "father" cooperate to refer to another person, etc. Syntactic structures imply that words or word groups become members of particular syntactic categories (such as adjectives, nouns, noun phrases, etc.) which constrain their combination and thus their interpretation.

3. *Grammaticalisation* Grammaticalisation is the process whereby a word or syntactic structure shifts to carry a grammatical function, in the sense that it looses some of its original lexical meaning (a process known as semantic bleaching) to express a more abstract meaning or gain a new one, and it looses some of its original syntactic properties (syntactic bleaching) (Hopper and Traugott, 2003). In the process of grammaticalisation, a new syntactic category may emerge or the recruited word or syntactic construction may become assigned to another syntactic category, which often implies a new syntactic context, e.g. different word order, ability to engage in morphological variation, etc.

For example, a verb of volition ("will") has become a future tense auxiliary in English (as in "It will rain tomorrow") (Bybee and W.Pagliuca, 1994). The verb "will" was originally a main verb with the meaning of "want" (it still is in Dutch), but it became recruited to express the more abstract sense 'future' and shifted syntactic category to auxiliary. This implied a specific position in the sentence (for example, before the subject in interrogative sentences, as in "Will it rain tomorrow?") and loss of the ability to have direct objects (one can no longer say "I will a book" in the sense of "I want a book").

4. *Cliticisation and Affixation* Once a lexical form has become grammatical it has a strong tendency to loose some of its original phonological structure and become first a clitic, that is a highly simplified word form which must occur next to its host (such as "ll" in "I'll see him tomorrow"). In a further process of phonetic erosion and simplification, the clitic may gradually become a morphological affix that is an inherent part of

the word, such as "-ed" in "walk-ed".

5. *Deletion* A final step is that the clitic or affix becomes so weak that it is no longer clearly audible and looses its function altogether. In that case the construction may progressively disappear and the cycle repeats itself, beginning with a new lexical item that is recruited to serve the same purpose. Often some debris of earlier evolutions is left behind and then later becomes available for recruitment to new functions.

These steps in language evolution (usually called grammaticalisation chains or 'clines') do not occur in any kind of predictable time frame and often there is competition between newer forms and existing ones. For example, for the expression of past tense in English, a system of form variation ("do" vs. "did") still co-exists with the use of a morphological affix ("walk" vs. "walked") and an auxiliary ("I wrote him" vs. "I did write him"). Nevertheless there is considerable regularity in evolutionary paths. For example, the recruitment of a verb of volition for expressing future is found in totally diverse language families (Bybee and W.Pagliuca, 1994) There is also a consensus that this type of evolution is basically uni-directional: from lexicalisation to grammaticalisation and cliticisation or affixation.

Historical linguists usually take a global view when tracking historical changes in language. But language does not exist as an abstract entity separate from the use of language in situated interactions between speakers and listeners. It is rather like a species, as Darwin already suggested. Each language user has his or her own private knowledge of the language, which may differ considerably from others depending on the history of interactions of the individual and the network structure of the population. It it the cumulative microchanges made by speakers and listeners that cause the global evolution in the language, which can usually only be observed in hindsight.

This raises two crucial questions (1) What are the social and physical behaviors and cognitive operations that individual language users carry out so that their net effect gives rise to the evolutionary phenomena observed and (2) what are the forces that drive the grammatical pathways forward?

The first question can only be adequately addressed against the background of theories of the cognitive processes of language understanding and production, for which we have today better and better (computational) models (Levelt, 1989). Next, we have to add to the normal operation of these processes the cognitive mechanisms that language users employ to perform 'language engineering': recruit existing words and syntactic structures for new meanings, to guess meanings of unknown elements, to push a lexical item towards a more grammatical item, etc. Heine (Heine, 1997) is one of the few diachronic linguists who have attempted to circumscribe these language molding operations but much work remains to operationalise them and embed the required mechanisms in a global theory of language use.

With respect to the second question, many historical linguists implicitly adopt the following hypothesis (see e.g. (Hopper and Traugott, 2003)): The goal of a communication between two participants (speaker and listener) is to reach communicative success with minimal effort. Making additional meanings explicit may help because it restricts the possible set of interpretations. The speaker also forces the listener to make similar distinctions in a particular situation and hence making meanings explicit helps to coordinate them in a population. On the other hand, there is a cost associated with expressing meaning, in terms of memory, processing, and learning. Hence languages try to find a balance between expressive power and effort. There is no unique solution for this multiple constraint satisfaction problem and choices are necessarily made, either based on historical accidents which continue to propagate or because certain meanings are more important to the culture in which the language developed.

Parallels with Genomic Evolution

The mechanisms underlying genomic evolution are progressively being unraveled thanks to the growing availability of complete genomes and techniques for tracking the function of individual genes in the development of cells and cellular structures. Earlier views based on the notion that genetic material consists of a set of individual, well-defined genes, each with a discrete function, have been abandoned in the face of the discovery of complex gene-regulatory networks. By implication, genetic evolution, which used to be thought of in terms of the cumulative stochastic operations over genes (mutation, duplication, recombination), is now viewed as a much richer process that can be understood in terms of the recruitment and co-optation of genes and gene regulatory networks for new functions (Carroll, 2001). In this sense, genomic evolution is based on 'genetic engineering' processes that are analogous to the 'language engineering' activities underlying language evolution. For example, Radman (Radman, 1999) and colleagues have shown that genetic mutations and re-combinations are not purely stochastic events but may partly take place under genomic regulatory control, induced by environmental stress such as increased errors in DNA-copying (SOS response). The environment in other words triggers the need for exploring ways in which certain functions become explicitly encoded for. Similar to lexicalisation, an existing gene (already used in another pathway) may become inserted in a pathway and thus leads to a new function that then undergoes selection (Wilkins, 2002).

Conclusion

One of the goals of Artificial Life is to synthesise in artificial systems evolutionary phenomena such as the ones that are discussed in this paper. It is clear that this is going to be extraordinarily difficult in the case of language evolution (and even more difficult for 'realistic' genomic evolution). Verbal communication engages all areas of cognition and is situated within concrete settings experienced through embodied interactions. It is impossible to capture all that in computational models. The use of robots (as advocated in (Steels, 2001)) already introduces the real-world interaction and the embodiment, but brings of course additional complexities to set up and carry out experiments.

The analogy between language evolution and genomic evolution has a heuristic value for developing frameworks and artificial life experiments, both in the domain of language evolution and in the domain of genomic evolution. When we view evolution as 'making explicit meanings or functions which were implicit before' or on the contrary 'eliminating explicit expression when no longer needed', then interesting parallels and similar questions start to appear, such as: how may intermediary levels appear, how can basic structures (such as intermediary control genes in the genome or basic sentence patterns in language) be conserved, despite constant change, etc. We need to understand the 'genetic engineering' that organisms use to adapt themselves to environmental conditions, just as a theory of language evolution requires mapping out the 'language engineering' that language users engage in to adapt the language to their needs. Of course there are also tremendous differences between language and genome evolution but that should not prevent us from exploiting

the analogies.

Acknowledgement

This reseach was conducted at the Sony Computer Science Laboratory. Additional funding came from the CNRS OHLL project, the ESF OMLL project, and the EU FET program on ECAgents. I thank C.L. Nehaniv for many interesting comments in his review of the paper.

References

Bybee, J. L. Perkins, R. and W.Pagliuca (1994). *The Evolution of Grammar. Tense, Aspect, and Modality in the Languages of the World.* Chicago, University of Chicago Press.

Cangelosi, A. and Parisi, D. (2001). *Simulating the Evolution of Language.* Berlin: Springer-Verlag.

Carroll, S., J. G. S. W. (2001). *From DNA to Diversity. Molecular Genetics and The Evolution of Animal Design.* London, Blackwell Science.

Dassow, J., V. M. (1996). Evolutionary grammars: A grammatical model for genome evolution. In Dassow, e., editor, *German Conference on Bioinformatics*, pages 199–209. Berlin, Springer-Verlag.

Deacon, T. (1997). *The Symbolic Species; The Co-evolution of Language and Brain.* New York, W. Norton and Company.

DeGraff, M. (1999). *Language Creation and Language Change. Creolization, Diachrony, and Development.* Cambridge Ma, The MIT Press.

Heine, B. (1997). *Cognitive Foundations of Grammar.* Oxford. Oxford University Press.

Heine, B., U. C. and Hnnemeyer, F. (1991). *Grammaticalization: A Conceptual Framework.* Chicago. The University of Chicago Press.

Hopper, P. and Traugott, E. (2003). *Grammaticalization.* Cambridge.Cambridge University Press, Cambridge.

Levelt, W. (1989). *Speaking: From intention to articulation.* Cambridge.Cambridge University Press, Cambridge.

Mayr, E. (1963). *Animal Species and Evolution.* Cambridge Ma. Harvard University Press.

Radman, M. (1999). Mutation: enzymes of evolutionary change. *Nature*, 401:866–869.

Richards, R. (1987). *Darwin and the emergence of evolutionary theories of mind and behavior.* Chicago, The University of Chicago Press, Chicago.

Sperber, D. and Wilson, D. (1987). *Relevance, communication and cognition.* Oxford, Basil Blackwell.

Steels, L. (2001). Language games for autonomous robots. *IEEE Intelligent Systems*, September/October:16–22.

Steels, L. (2003). Evolving grounded communication for robots. *Trends in Cognitive Science*, 7,7:308–312.

Steels, L. (2004). Self-organising grammars in embodied situated language games. In Hurford, e., editor, *Proceedings Evolution of Language V*, page 40. Leipzig, MPI.

Steels, L. and Kaplan, F. (2000). Collective learning and semiotic dynamics. In Foreano, D., N. J.-D. and Mondada, F., editors, *Advances in Artificial Life (ECAL 99)*, pages 679–688. Berlin, Springer-Verlag.

Wilkins, A. (2002). *The Evolution of Developmental Pathways.* Sunderland Ma, Sinauer Assoc.

The Effects of Learning on the Evolution of Saussurean Communication

Edgar E. Vallejo[1,2] and **Charles E. Taylor**[2]

[1]Tecnológico de Monterrey, Campus Estado de México, Atizapán de Zaragoza 52926, México
[2]University of California Los Angeles, Los Angeles, CA 90025, USA

Abstract

This paper presents a computational framework for studying the influence of learning on the evolution of communication. In our model, an evolving population of learning agents is engaged in pairwise comunicative interactions. Simulation results show the genetic assimilation of trasmission behaviors as a consequence of saussurean learning.

Introduction

The evolution of communication is an excellent domain for studying fundamental questions of artificial life research. Previous work by Ackley and Littman (2), Arita and Taylor (3), Di Paolo (5), Hashimoto and Ikegami (8), and Steels (18), among others, have shown that we are able to explore important issues such as emergence, self-organization and cultural evolution within this framework.

The aim of this work is to study the influence of learning on the evolution of communication. We believe this is an instrumental aspect for understanding the origin and evolution of communication systems with the complexity of human languages.

In nature, learning seem to influence both the association of signaling behaviors to appropriate external referents and the development of responses to different signal types (10). However, there are several examples of innate underpinnings in transmission behaviors. For example, young vervet monkeys seem predisposed to alarm calling, dividing up the universe of predators into different ill-defined categories, leaving social experience to sharpeb the boundaries of exclusive predator categories and its corresponding association to particular call types (4). Moreover, there is indication of innate predispositions in human languages (16). On the other hand, there is a growing conviction that learning plays a substantial role in the development of how a receiver responds on hearing a signal (10).

A fundamental question that arises in the context of animal communication is whether a comunication system is likely to be symbolic. According to Marler, for an animal communication system to qualify as symbolic, information about external referents has to be both encoded by signalers and decoded by receivers (10). This definition is consistent with the nature of the saussurean linguistic sign (17).

In his seminal paper, Hurford proposed a model for studying the evolution of saussurean communication (9). In this model, an agent consists of two probabilistic matrices that provide a framework for semiotic interactions. Computational simulations were conducted to investigate the evolutionary potentials of different learning strategies. Further studies by Oliphant (14), Oliphant and Batali (15) and Nowak et al (13) have contributed to elucidate the fundamental properties of this model.

These studies have been largely conducted within the framework of Lamarkian evolution. In most cases, learned communicative behaviors are written back to the genetic description of agents and thus transmitted to offspring during reproduction. We believe that Darwinian evolution provides a more convenient framework for studying the evolution of communication. In their influential work, Hinton and Nowlan proposed a computational framework for conducting studies on the effects of learning on evolution in Darwinian evolution (7) (1). This holds much promise. For example, Turkel has shown that this computational framework can be used for exploring how the capacity of human language could have evolved via natural selection (19).

In this work, we conducted computational studies about the influence of learning on the evolution of communication. Experimental results show the genetic assimilation of transmission behaviors. This results are consistent with dominant theories on the evolution of communication (6).

The model

Agent architecture

The formal definition of the agent architecture presented below is based on considerations of the models proposed by Hurford (9) and Oliphant (14).

Agent Let $O = \{o_1, \ldots, o_n\}$ be a finite set of n objects and $S = \{s_1, \ldots, s_m\}$ be a finite set of m signals. A *learning communicative agent* is a triple (δ, ϕ, σ), where

1. $\delta : O \rightarrow S \cup \{s_\#\}$ is the transmission function, where $s_\#$ is the undetermined signal,

2. $\phi : S \rightarrow O \cup \{o_\#\}$ is the reception function, where $o_\#$ is the undetermined object, and

3. σ is the learning strategy.

Communication An agent $A_1 = (\delta_1, \phi_1, \sigma_1)$ comunicates to an agent $A_2 = (\delta_2, \phi_2, \sigma_2)$ as follows. Initially, A_1 perceives an object o_i and produces a signal s_j according to the mapping described by the transmission function δ_1, such that $\delta_1(o_i) = s_j$. Once A_1 produces the signal s_j, the agent A_2 interprets the signal s_j as the object o_k according to the mapping described by the reception function ϕ_2, such that $\phi_2(s_j) = o_k$. A communication event from A_1 to A_2 is successful if the following conditions are satisfied

1. $\delta_1(o_i) = s_j$,

2. $\phi_2(s_j) = o_k$, and

3. $o_i = o_k$

Homonymy An agent $A = (\delta, \phi, \sigma)$ is said to possess homonymy if there exist a signal $s_j \neq s_\#$ and a pair of objects o_i and o_k that satisfy the following conditions

1. $\delta(o_i) = s_j$,

2. $\delta(o_k) = s_j$, and

3. $o_i \neq o_k$

Synonymy An agent $A = (\delta, \phi, \sigma)$ is said to possess synonymy if there exist an object $o_i \neq o_\#$ and a pair of signals s_j and s_k that satisfy the following conditions

1. $\phi(s_j) = o_i$,

2. $\phi(s_k) = o_i$, and

3. $s_j \neq s_k$

Innate transmission Let $A = (\delta, \phi, \sigma)$ be an agent. A transmission from A for a given object o_i is said to be innate if $\delta(o_i) \neq s_\#$ and is said to be subject to learning if $\delta(o_i) = s_\#$.

Innate reception Let $A = (\delta, \phi, \sigma)$ be an agent. A reception of A for a given signal s_j is said to be innate if $\phi(s_j) \neq o_\#$ and is said to be subject to learning if $\phi(s_j) = o_\#$.

Learning In our model, both transmission and reception behaviors are partially learned. Before a communication event from A_1 to A_2 takes place, A_1 replaces the undetermined signals in δ_1 with signals in S using the learning strategy σ_1. Similarly, A_1 replaces the undetermined objects in ϕ_1 using the learning strategy σ_1. Agent A_2 proceeds similarly.

A fundamental aspect of our model is that learning is performed for communication purposes and does not permanently modify the actual description of an agent.

Learning strategies

We consider three different learning strategies: imitator, calculator and saussurean, after (9). In addition, we introduce a fourth learning strategy: random learner.

Imitator An imitator agent replaces the undetermined signals in his transmission function by the corresponding signals in the transmission function of another agent. Similarly, he replaces the undetermined objects in his reception function by the corresponding objects in the reception function of another agent.

Formally, an agent $A_1 = (\delta_1, \phi_1, \sigma_1)$ imitates an agent $A_2 = (\delta_2, \phi_2, \sigma_2)$ as follows.

1. $\delta_1(o_i)$ is set to $\delta_2(o_i)$ if $\delta_1(o_i) = s_\#$, $\delta_2(o_i) \neq s_\#$, for $i = 1, \ldots, n$, and

2. $\phi_1(s_j)$ is set to $\phi_2(s_j)$ if $\phi_1(s_j) = o_\#$, $\phi_2(s_j) \neq o_\#$, for $j = 1, \ldots, m$.

Calculator A calculator agent replaces the undetermined signals in his transmission function in such a way that a communication event to another agent would be successful. Similarly, he replaces the undetermined objects in his reception function in such a way that the communication event from another agent would be successful.

Formally, an agent $A_1 = (\delta_1, \phi_1, \sigma_1)$ calculates an agent $A_2 = (\delta_2, \phi_2, \sigma_2)$ as follows.

1. $\delta_1(o_i)$ is set to s_k if $\phi_2(s_k) = o_i$, $\delta_1(o_i) = s_\#$, for $i = 1, \ldots, n$, and

strategy	code
imitator	0
calculator	1
saussurean	2
random	3

Table 1: Codification of learning strategies

o_i	$\delta_1(o_i)$		s_j	$\phi_1(s_j)$
1	b		a	$o_\#$
2	$s_\#$		b	3
3	a		c	2

Table 2: Transmission and reception functions

2. $\phi_1(s_j)$ is set to o_k if $\delta_2(o_k) = s_j$, $\phi_1(s_j) = o_\#$, for $j = 1, \ldots, m$.

Saussurean A saussurean agent replaces the undetermined signals in his transmission function by the corresponding signals in the transmission function of another agent. Similarly, he replaces the undetermined objects in his reception function in such a way that a communication event to himself would be successful.

Formally, an agent $A_1 = (\delta_1, \phi_1, \sigma_1)$ is saussurean with respect to and agent $A_2 = (\delta_2, \phi_2, \sigma_2)$ as follows.

1. $\delta_1(o_i)$ is set to $\delta_2(o_i)$ if $\delta_1(o_i) = s_\#$, $\delta_2(o_i) \neq s_\#$, for $i = 1, \ldots, n$, and

2. $\phi_1(s_j)$ is set to o_k if $\delta_1(o_k) = s_j$, $\phi_1(s_j) = o_\#$, for $j = 1, \ldots, m$.

Random A random agent can assume any of the strategies described above. Before a communication event takes place, a random agent randomly selects one of the following strategies: imitator, calculator or saussurean.

For convenience, we consider a codification of learning strategies as shown in table 1.

Evolution of communication

In our model, a population of learning communicative agents are intended to evolve successful communication at the population level. We use genetic algorithms for this purpose. The design decisions presented below are based on considerations of the performance of genetic algorithms in practical applications (12).

Genome representation A learning communicative agent $A = (\delta, \phi, \sigma)$ is represented linearly as follows

$$A = (\delta(o_1), \ldots, \delta(o_n), \phi(s_1), \ldots, \phi(s_m), \sigma)$$

For example, consider a set of objects $O = \{1, 2, 3\}$, a set of signals $S = \{a, b, c\}$ and a calculator agent $A_1 = (\delta_1, \phi_1, \sigma_1)$. In addition, consider the functions δ_1 and ϕ_2 described in table 2.

The linear representation of agent A_1 is

$$(b, s_\#, a, o_\#, 3, 2, 1)$$

Genetic operators Agents produce a new offspring by means of genetic operators. Fitness proportional selection, one-point recombination and point mutation operate on the linear representation of agents described above.

Fitness function Fitness was defined as the communicative accuracy of learning communicative agents. The communicative accuracy is the ability of an agent to successfully communicate with a collection of other agents.

Let P be a finite population of agents, A be an agent in P, and $Q \subseteq P$ be a non empty collection of agents. The communicative accuracy of A with respect to Q given the set of objects $O = \{o_1, \ldots, o_n\}$ and the set of signals $S = \{s_1, \ldots, s_m\}$, $C(A, Q, O)$, is defined as

$$C(A, Q, O) = \frac{\sum_{o_i \in O} \sum_{A_k \in Q} c(A, A_k, o_i) + c(A_k, A, o_i)}{|Q|}$$

where $c(A, A_k, o_i) = 1$ if the communication event from A to A_k is successful given the object o_i, and 0 otherwise; $|Q|$ is the cardinality of Q. $c(A_k, A, o_i)$ is defined similarly.

Experiments and results

Experiments were conducted to investigate whether a population of learning communicative agents is likely to arrive to successful communication at the population level. In addition, we validated the evolutionary performance of competing learning strategies. Most importantly, we were interested in exploring the effects of learning on the genetic description of an evolving population of learning communicative agents. The simulation procedure is described in table 3.

Several simulations were conducted using different combinations of parameter values as shown in table 4. The following were the major results:

1. Create an initial random population P of agents

2. **Do until** number generations is met

   (a) **For each** individual $A_i = (\delta_i, \phi_i, \sigma_i) \in P$ **do**

      i. Perform the learning process of A_i according to the learning strategy σ_i with respect to a random agent $A_h \in P$

      ii. Select a random subpopulation of agents $Q \subseteq P$

      iii. Perform the learning process for all $A_j \in Q$ according to the learning strategy of A_j with respect to a random agent $A_k \in P$

      iv. Measure the communicative accuracy of A_i with respect to Q, $C(A_i, Q, O)$, given the set of objects O

      End for

   (b) Select two individuals $A_{mother} \in P$ and $A_{father} \in P$ for reproduction using fitness proportional selection

   (c) Produce an offspring A_{new} from A_{mother} and A_{father} using one-point recombination and point mutation

   (d) Select a random individual $A_{old} \in P$

   (e) Replace A_{old} by A_{new}

   End do

Table 3: Simulation procedure

Parameter	Value
generations	2000–3000
population P	128–512
subpopulation Q	4–16
signals	4–8
meanings	4–8
crossover probability	0.6–0.7
mutation probability	0.001–0.01

Table 4: Parameters for simulations

1. Agents arrived at highly successful communication at the population level. However, simulations showed that there exists a threshold condition on the number of interactions (size of set Q) required to achieve accuracy in communication. Figure 1 shows the results of the simulations for different number of interactions. Agents reached local minima in communication accuracy as the number of interactions is reduced.

2. Local minima were produced mostly by the presence of homonymy – i.e. when the same signal is used to describe two or more different objects. Synonymy – when two or more signals are interpreted as the same object, did not appear consistently.

3. In pairwise strategy contests, populations of calculators dominated a populations of imitators. However, populations of saussureans outcompeted both calculators and imitators.

4. In all strategy contests: imitators, calculators, saussureans and randoms, experimental results showed the superiority of the saussurean learning strategy. In most cases, saussureans took over the entire population. Only rarely, did imitators or calculators dominate such populations. Random learners never became dominant. Figure 2 shows the frequency of strategies in the population for a typical simulation where all strategies were initially present.

5. The undetermined signal trait disappeared in the population. All the signals produced for every object were genetically assimilated in the transmission gene segment of the genome for all agents. The frequency of undetermined signals in the population as evolution proceeded is shown in figure 3.

6. The undetermined objects trait prevailed in the population. Objects interpreted for every signal were not genetically assimilated in the reception gene segment of the genome. The frequency of undetermined objects in the population as evolution proceeded is shown in figure 3.

7. Most simulations considered a set of 4 objects and a set of 4 signals. Both homonymy and synonymy began to appear more consistently when both the number of objects and signals was increased. However, homonymy continued to be more common.

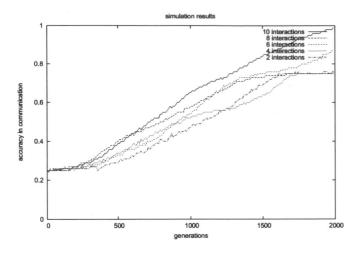

Figure 1: Threshold condition on interactions

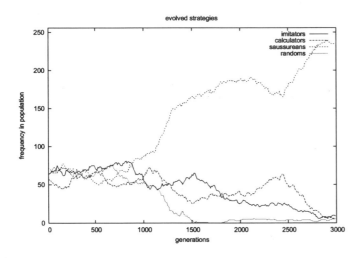

Figure 2: Evolved strategies

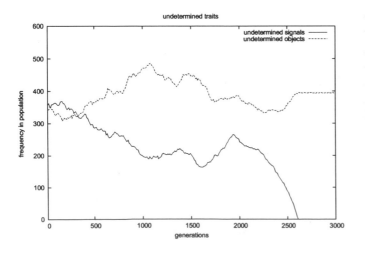

Figure 3: Genetically assimilated traits

Discussion

The overall experimental results indicate that given a sufficient number of interactions, a population of learning communicative agents is capable of arriving at highly successful communication. Surprisingly, transmission behaviors became innate as a consequence of the evolution saussurean learning. On the other hand, reception behaviors prevailed as characteristics that are subject to learning. Our results are consistent with several hypotheses formulated for the evolution of communication (10).

Why transmission was genetically assimilated

In general, experimental results showed that there is an evolutionary preference for saussurean agents. Saussurean learning implements a tight coupling between transmission and reception behaviors.

Why did signaling behaviors become innate? First, imitation of transmission in a static environment provides the opportunity for genetic assimilation of signaling behaviors. Second, a tight coupling between reception and innate transmission provide the oportunity for successful learning in reception behaviors.

Why saussureans prevailed

From the evolutionary perspective, how good is saussurean learning for evolving communication?. Could an alternative communicative strategy invade a population of saussurean learners?.

Maynard-Smith(11) has demostrated that game theory can be used as a framework to explain the evolution of most phenotypic traits in situation in which fitness of a trait depends on what others are doing. He has also provided the notion of evolutionary stable strategy (ESS). An ESS is a phenotype such that, if almost all individuals have that phenotype, no alternative phenotype can invade the population.

In our model, experimental results showed that saussurean learning is likely to be an ESS. However, further studies are required to formally demostrate what types of learning are evolutionary stable strategies.

In this study, we did not consider the cost of producing a signal. Previous studies suggest that honesty of communication become an issue when cost is considered (2).

Acknowledgements

This work was supported by the UC MEXUS-CONACYT, the UCLA Center for Embedded Sensor Networks, the Defense Advance Research Projects (DARPA), administered by the Army Research Office under Emergent Surveillance Plexus MURI Award No. DAAD19-01-1-0361, and DARPA MURI administered by the US Air Force No. F49620-01-1-0361. Any opinions, findings and conclusions or recommendations expressed in this publication are those of the authors and do not necessarily reflect the views of the sponsoring agencies.

References

[1] Ackley, D.H. and Littman, M.L. 1992. Interactions between learning and evolution. In C.G. Langton, C.E. Taylor, J.D. Farmer and S. Rasmussen.1992. *Artificial Life II*. Addison Wesley.

[2] Ackley D,H. and Littman, M.L. 1994. Altruism in the Evolution of Communication. In R.A. Brooks, P. Maes, (eds), *Artificial Life IV. Proceedings of the Fourth International Conference on Artificial Life*, The MIT Press, pp.40–48.

[3] Arita, T. and Taylor, C.E. 1996. A simple model for the evolution of communication. In: L.J. Fogel, P.J. Angeline, T. Back, (eds), *Evolutionary Programming V*. The MIT Press, pp.405–410.

[4] Cheney, D. and Seyfarth, R.M. 1990. *How Monkeys See the World*. University of Chicago Press.

[5] Di Paolo, E. A. 1998. An investigation into the evolution of communication. *Adaptive Behavior*, 6(2):285–324.

[6] Hauser, M. 1996. *The Evolution of Communication*. The MIT Press.

[7] Hinton, G.E. and Nowlan, S.J. 1987. How learning can guide evolution. *Complex Systems*1, pp.495–502.

[8] Hashimoto, T. and Ikegami, T. 1996. Emergence of net-grammar in communicating agents, *BioSystems* 38:1–14.

[9] Hurford, J. R. 1989. Biological evolution of the saussurean sign as a component of the language acquisition device. *Lingua* 77, pp.187–222.

[10] Marler, P. 1998. Animal communication and human language. In N.G. Jablonski and L.C . Aiello (eds.) *The Origin and Diversification of Language*. University of California Press.

[11] Maynard-Smith J (1982), *Evolution and the Theory of Games*. Cambridge University Press.

[12] Mitchell, M. 1996. *An introduction to genetic algorithms*. The MIT Press.

[13] Nowak M.A., Plotkin, J.B. and Krakauer, D. 1999. The evolutionary language game. *Journal of Theoretical Biology*, 200(2):147–162.

[14] Oliphant, M. 1996. The dilemma of saussurean communication. *BioSystems*, 37(1-2), pp.31–38.

[15] Oliphant, M. and Batali, J. 1997. Learning and the emergence of coordinated communication. *Center for Research on Language Newsletter*, 11(1).

[16] Pinker, S. 1994. *The Language Instinct*. William Morrow and Company.

[17] de Saussure, F. 1986. *Course in general linguistics*. Open Court.

[18] Steels, L. 1996. Self-organizing vocabularies. In: C. Langton and T. Shimohara (eds), *Artificial Life V. Proceedings of the Fifth International Conference on Artificial Life*, The MIT Press, pp.179–184.

[19] Turkel, W.J. 2002. The learning guided evolution of natural language. In T. Briscoe (ed.) *Linguistic Evolution through Language Acquisition*. Cambridge University Press, pp.235–254.

Minimum cost and the emergence of the Zipf-Mandelbrot law

Paul Vogt[1,2]

[1]Language Evolution and Computation Research Unit, University of Edinburgh, U.K.
[2]Induction of Linguistic Knowledge / Computational Linguistics, Tilburg University, The Netherlands
paulv@ling.ed.ac.uk

Abstract

This paper illustrates how the Zipf-Mandelbrot law can emerge in language as a result of minimising the cost of categorising sensory images. The categorisation is based on the discrimination game in which sensory stimuli are categorised at different hierarchical layers of increasing density. The discrimination game is embedded in a variant of the language game model, called the selfish game, which in turn is embedded in the framework of iterated learning. The results indicate that a tendency to communicate in general terms, which is less costly, can contribute to the emergence of the Zipf-Mandelbrot law.

Introduction

One of the most sound universal tendencies observed in human languages is that when words are ranked according to their occurrence frequency in a descending order, the frequency f is inversely proportional to its rank k according to $f = Ck^{-B}$, where $B \approx 1$ and C is a constant, see Figure 1. This finding was discovered by G. K. Zipf (Zipf, 1949), and has since been called *Zipf's law*. Besides its observation in linguistics, Zipf's law has also been observed in economy, physics, biology, demography, social sciences etc. (Günther et al., 1996).

Zipf explained his finding in terms of least effort (Zipf, 1949). He assumed that speakers want to minimise articulatory effort, thus minimising the length of an utterance, which tends to promote ambiguity in language. On the other hand, hearers want to have optimal clarity to interpret the meaning of an utterance unambiguously with the least effort. Fulfilling the needs of both agents leads to a trade-off, which Zipf called the *principle of least effort* (Zipf, 1949). Although the observation of Zipf's law in real linguistic data is sound, it has only recently been shown empirically in an alife model that the principle of least effort indeed leads to a Zipfian distribution (Ferrer i Cancho and Solé, 2003).

In 1953, Mandelbrot derived a more general expression of Zipf's law, which explains small differences between Zipf's law and real linguistic data, notably for the first few ranks (Mandelbrot, 1953). According to the *Zipf-Mandelbrot law* the frequency f of a word is related to its rank k as $f =$

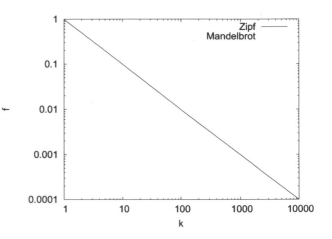

Figure 1: This plot shows both Zipf's law and the Zipf-Mandelbrot law. Mandelbrot's formula was drawn with parameters $C = 10$, $V = 10$ and $B = 1.35$ (these parameters are illustrative and not justified). The graph shows the frequency distribution f as a function of the rank k.

$C(k+V)^{-B}$, where C, V and B are constants. When $V = 0$ and $B \approx 1$, this expression equals Zipf's law. Figure 1 shows both laws plotted on the usual log-log scale. It appears that Mandelbrot's equation fits linguistic data better than Zipf's equation (Mandelbrot, 1953). The derivation of Mandelbrot's formula was based on minimising the articulatory cost in terms of word length (Mandelbrot, 1953).[1]

In addition to Zipf's least effort and Mandelbrot's minimum cost explanations, many other explanations of Zipf's law have been proposed that, too, focus on the relation between word length and its frequency. For example, it has been shown that randomly generated texts exhibit Zipf's law (Li, 1992), and so does the frequency distribution with which monkeys press the keys of a typewriter (Miller, 1957). This was explained by noting that generating shorter se-

---

[1] I sometimes refer to Zipf's law when I refer to the phenomenon that ranked words occur with are frequency distribution described by a hyperbola, irrespective whether expressed by Zipf's or Mandelbrot's equation. Where relevant, the distinction will be made.

quences is more statistically likely than generating long ones (Li, 1992). Explanations that do not explain the emergence of Zipf's law in terms of articulatory effort focus on frequency effects. More frequently used words are more likely to be selected in communication (Günther et al., 1996). This effect has been shown in a simulation where speakers select words based on occurrences in the preceding discourse (Tullo and Hurford, 2003).

This paper investigates the emergence of a Zipfian distribution in language as the result of a minimum cost principle (Mandelbrot, 1953), based on Steels' language game model (Steels et al., 2002). However, in contrast to Mandelbrot's derivation, the cost will not be minimised by optimising the word length, but rather by trying to minimise computational costs at the cognitive level of categorisation. As common in language game models, categorisation will be done using *discrimination games* (Steels, 1996). A recent discovery (Vogt, 2004) revealed that Zipfian distributions of word frequencies had emerged in robotic studies based on the language games, such as reported in (Vogt, 2000). This paper studies the hypothesis, suggested in (Vogt, 2004), that the emergence of Zipf's law may be explained by a tendency to use general categories in communication as a principle of minimum cost.

The next section outlines the model with which the study was done. Then the results are presented, which are discussed in the subsequent section. The final section provides conclusions.

The model

The study was done using the simulation toolkit THSim (Vogt, 2003b), which mimics aspects of the Talking Heads experiment (Steels et al., 2002).[2] THSim implements a number of different language games that can be incorporated by a population of agents. In the current study, the population plays *selfish games* – independently developed by Smith and Vogt (Smith, 2003; Vogt, 2000) – where hearers guess the reference from utterances produced by speakers, and learning is achieved by cross-situational statistical learning (Vogt and Smith, 2004). By engaging in selfish games, agents develop a repertoire of categories, which form the meanings of word-forms the agents develop as part of the selfish game.

The exact details of the selfish games are irrelevant for the purposes of this paper, similar results have been observed with – on language games based – guessing games (unpublished) and observational games (Vogt, 2004).[3] Figure 2 illustrates the working of the selfish games. In a selfish game, two agents – a speaker and a hearer – are selected from the population. Both agents look at a context

---

[2]The THSim toolkit containing the code of the present study is available at http://www.ling.ed.ac.uk/~paulv/thsim.html.

[3]Consult, e.g., (Vogt, 2000; Vogt, 2003b) for a description of these language game models.

Figure 2: Two situations of selfish games illustrate the working of cross-situational learning. When a learner first hears the word "wateva" in a context with squares and a circle (situation 1), and later in a situation with a circle, a triangle and a polygon (situation 2), then she can induce – based on co-occurrences – the knowledge that "wateva" refers to the circle.

(or situation) that contains a number of coloured geometrical shapes. The speaker selects one shape as the topic, tries to categorise this topic and produces an utterance to convey its reference. If the speaker has no way to express a particular category (or meaning), she invents a new word-form. When the hearer receives an utterance, she tries to interpret the utterance by guessing which shape the speaker intends to convey. The hearer categorises all shapes and searches the association between the utterance and category that best matches the utterance in the given situation. This selection is based on maximising the probability $P(m|w)$ that given an utterance w, it means meaning m, provided the reference of the meaning is in the context. These probabilities are estimated according to word-meaning occurrences in previous situations. This learning mechanism has been called *cross-situational statistical learning* (Vogt and Smith, 2004) and works on the same principle as the cross-situational learning model introduced by Siskind (Siskind, 1996).

The selfish games are embedded in a cultural evolution where the language originates and is transmitted from one generation to the next culturally, i.e., the agents' morphologies remain the same throughout the course of evolution. At the start of each agent's lifetime, her linguistic knowledge is non-existent; this develops ontogenetically. The evolution is modelled using the iterated learning model (Kirby, 2002), which implements the population dynamics through iterating large sequences of selfish games played by the population. In each iteration, where a given number of selfish games are played, the population contains adult and learner agents. The adults are assumed to have mastered the language, which the learners learn by acting as hearers in selfish games with an adult as speaker. At the end of each iteration, the adults 'die' and are replaced by the learners, and new

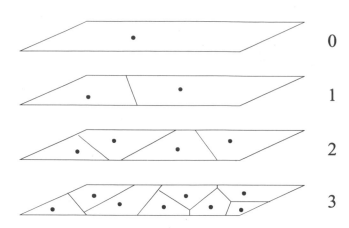

Figure 3: The hierarchical layering of categories using a prototype representation. Each layer l accepts up to n^l categories, which form Voronoi segments in the n-dimensional conceptual space. In the example here $n = 2$.

learners enter the population. This continues, in principle, indefinitely. The remainder of this section explains the categorisation process of the discrimination game in more detail.

The discrimination game

The agents categorise and form meanings using the discrimination game model proposed by Luc Steels (Steels, 1996). The aim of an agent playing a discrimination game is to categorise an object (the *topic*) such that it distinguishes the topic from all other objects in the context. In the original implementation (Steels, 1996), categories were represented by combinations of nodes in a binary tree. In the current implementation, categories are represented by prototypes that are points in an n-dimensional *conceptual space* (Gärdenfors, 2000), where the dimensions are *quality dimensions* which can be measured by feature detectors. For the present study, the conceptual space is 6-dimensional and is spanned by the R, G and B channels of the RGB colour space, a shape feature[4] and the x and y coordinates.

The distribution of prototypes leads to a segmentation of the space into Voronoi areas formed by the regions nearest to the prototypes; these Voronoi segments constitute the categories. A hierarchical layering of prototypes allows the emergence of a *taxonomic hierarchy* of categories (Rosch, 1978), see Fig. 3. In this taxonomy, categories on layers of low density are more general than those that are on layers of high density. The density $D(l)$ of layer l is given by $D(l) \leq n^l$, where n is the dimension of the conceptual space $S(l)$. The cost of categorisation is proportional to the time required to find a category that distinguishes the topic from the rest of the context. In the layered model, the agents can minimise categorisation cost by searching the different layers from the least dense layer to the more dense ones until a

---

[4]The shape feature is a value proportional to the shape's area divided by the area of its smallest bounding box (Vogt, 2003b).

distinction can be made – i.e., the agents try to find the most general categories to be used in the communication act.

When an agent participates in the selfish game, she looks at the context C, which contains a number of objects ($|C| = 4$). The objects are selected randomly with a *uniform distribution* from a set of 10 different shapes, they are combined with an arbitrarily selected colour from a set of 11 colours and are placed at an arbitrarily selected point on the 2-dimensional display. For each object $o_i \in C$, the agent extracts a 6-dimensional feature vector $\mathbf{f}_i$ describing the objects in terms of its quality dimensions as mentioned above.

The speaker of the selfish game now selects a topic $o_t \in C$ and plays a discrimination game starting at layer $l = 1$ and continues at the next layer until $l = l_{max}$ or until the speaker found a *distinctive category*, see below. (l_{max} is the final layer that has at least one category for increasing values of l.) Given this category, the speaker tries to produce an utterance by searching its lexicon for a matching word-meaning association. If no such association is found, the speaker continues at the next layer until $l = l_{max}$. When the hearer receives an utterance, she plays a discrimination game for each object $o_i \in C$ starting at layer $l = 1$ and tries to interpret the utterance at layer l. This continues until the hearer interprets the utterance or until $l = l_{max}$. Note that an utterance is interpreted when the hearer found an association in its vocabulary that unambiguously identifies the utterance with a meaning (read *distinctive category*) that is consistent with the result of the discrimination games at layer l. This does not necessarily means that the interpretation is correct, which is only the case if the identified object $o = o_t$.

The discrimination game works as follows:

1. The feature vectors $\mathbf{f}_i$ of all objects $o_i \in C$ are categorised using the 1-nearest neighbourhood search (Cover and Hart, 1967) applied to the conceptual space at layer l. This results for each vector $\mathbf{f}_i$ in a category c_i, represented by the prototype $\mathbf{c}_i$ nearest to $\mathbf{f}_i$.

2. The agent then verifies whether the topic's category c_t distinguishes the topic from all other objects in the context. This holds when there is a *distinctive category* (or meaning) $m_t = c_t$ for which $\neg \exists o_j \in C \backslash \{o_t\} : c_j = c_t$.

3. If there does not exists such a meaning, then add a new category c – for which the topic's feature vector $\mathbf{f}_t$ is taken as an exemplar (i.e., $\mathbf{c} = \mathbf{f}_t$) – to the first hierarchical layer l that has space (i.e., $D(l) < n^l$) and return with failure.

4. Otherwise the category's prototype $\mathbf{c}_t$ is moved toward $\mathbf{f}_t$ such that it becomes the centre of mass of the feature vectors it distinctively categorised and return the distinctive category m_t. Note that if two categories become closer than within a given threshold, they are merged.

Note that as the hearer has to guess the topic, she considers all objects $o_i \in C$ as a potential topic and therefore plays a discrimination game for each potential topic at the given layer. This yields a distinctive category set M, which contains the meanings of those objects that have successfully been distinguished.

As the complexity of the search in layer l is of order $o(D(l))$, the computational cost increases exponentially at lower levels in the taxonomic hierarchy (i.e., increasing values of l).

Results

In order to test the hypothesis that minimising the categorisation costs can lead to the emergence of Zipf's law, an experiment with 2 different conditions was carried out.

Condition 1 No hierarchical layering of conceptual spaces was used (there was only one layer available to each agent – and the density of the conceptual space was limited by the agents' memory sizes, which limits were never reached).

Condition 2 The hierarchical layering as described above was present.

For both conditions 10 trials of the simulation were run for 10 iterations of 100,000 selfish games. The population in each each iteration contained a total of 10 agents (5 adults and 5 learners). All speakers were selected from the adult population, but in order to prevent the emergence of many different languages, only 90% of the hearers were learners, the others were adults, consult (Vogt, 2003a) for a discussion.

Figure 4 summarises the most important results of the experiment. The two top figures and the leftmost figure on the bottom row show the ranked frequency distributions of 9 randomly selected agents throughout their lifetime (from each generation 1 agent) plotted on a log-log scale. In addition, these figures show the approximated curve of the Zipf-Mandelbrot equation with parameter settings as specified shortly.

The top left graph shows the results of condition 1, i.e., the run without hierarchical layering of categories. Clearly, the frequency distribution does not reveal the Zipf-Mandelbrot law $f = C(k + V)^{-B}$ with $B \approx 1$. This plot can be approximated by Mandelbrot's equation with $V \approx 1400$ and $B \approx 4$, which means that relatively many high ranked words occur almost equiprobable (V is high), while the occurrence frequencies of the remaining words drop faster than the Zipf-Mandelbrot law with $B \approx 1$ would predict.[5]

The top right graph of Fig. 4 shows the frequency of occurrences of word-meaning associations emerging under condition 2. This plot shows a curve similar to the Zipf-Mandelbrot curve (Fig. 1), which after a first small period transfers in an almost straight line with a slope near

-1 ($V \approx 3$ and $B \approx 1.2$), which is typical for the Zipf-Mandelbrot law observed in natural languages. This figure shows the frequency distribution of *word-meaning* associations rather than the distribution of *word* occurrences. When the ranked frequency distribution of word occurrences is plotted against the rank k, an approximation of the Zipf-Mandelbrot law emerges with a value of $V \approx 80$ and $B \approx 5$, see Fig. 4 (bottom left).

In the condition 2 simulation, the categories of word-meanings that occupy conceptual spaces of higher density (i.e., higher values of l) have lower frequencies, which confirms the hypothesis that reducing the computational cost of categorisation can lead to the emergence of the Zipf-Mandelbrot law, see Fig. 4 (bottom right).

Not shown in the graphs are the communicative success of the two experiments. Communicative success measures the average number of successful selfish games over some window of time (a selfish game is successful if the hearer guessed the right topic). The simulation of condition 1 leads to an average communicative success of 69.3 ± 0.3 % over the final 10,000 games, whereas condition 2 yields an average of 51.2 ± 0.3 %. Note that the communicative success is averaged over the final 10,000 games of the simulation and averaged over all 10 trials.

Discussion

Minimising the computational costs of categorisation by trying to generalise the categorisation as much as possible does, indeed, lead to the emergence of the Zipf-Mandelbrot law (Mandelbrot, 1953). Moreover, Mandelbrot's equation matches the results better than Zipf's original equation. That the emergence of the Zipf-Mandelbrot law is caused by minimising the cost of categorisation through the hierarchical layering of conceptual spaces is prominently visible in the bottom right graph of Fig. 4, as the categories of lower frequency tend to occupy conceptual spaces of higher density.

The Mandelbrot-Zipf law emerges only for the ranked frequency distributions of the occurrence of *word-meaning associations*, rather than those of *words*, because as an artifact of the language game model, words tend to have multiple associations with different meanings at different layers. Nevertheless, these different meanings refer to the same objects in different situations. In human language, however, meanings at different hierarchical layers tend to have different words – e.g., animal, dog and spaniel. Assuming we can better model a one-to-one bias in word-meaning associations, see, e.g., (Smith, ming), thus specifying different taxonomies with more specific words, I decided to look at word-meanings as atomic elements rather than at words alone.

The agents (both as speaker and as hearer) do not optimise the effectiveness of their language use, but rather minimise the computational cost of categorisation. This aspect is a prominent reason why the communicative success of condi-

---

[5]The parameter values given are rough estimations; they are not obtained from statistical analysis, but the solid lines in the plots of Fig. 4 show the corresponding curve.

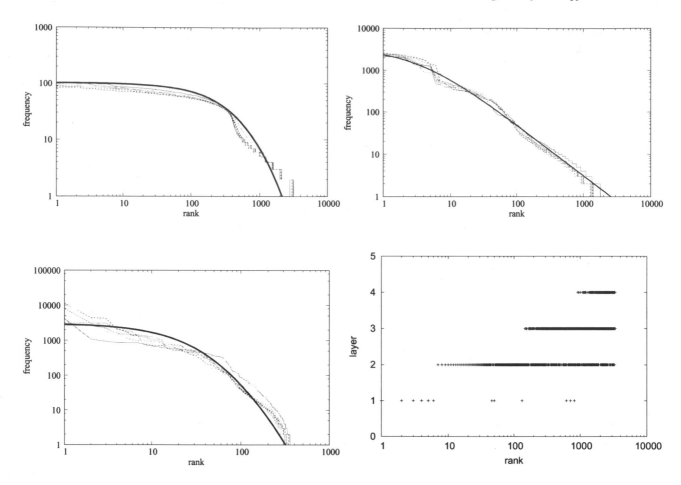

Figure 4: **Top left:** The frequency distribution of the occurrences of *word-meaning associations* as a function of the rank in condition 1. **Top right:** The frequency distribution of the occurrences of *word-meaning associations* in condition 2. **Bottom left:** The frequency distribution of *word* occurrences in condition 2. **Bottom right:** The hierarchical layers *l* occupied by the categories of rank *k* relating to one of the agents in condition 2 from the previous two graphs. The first three graphs additionally shows the Mandelbrot equation imposed with parameter settings as given in the text (thicker solid lines).

tion 2 stays well behind the success of condition 1, where the agents do search the entire lexicon for associations, thus including those that allow more effective communication. Hence, minimising the cost of categorisation while not optimising communicative success causes a trade-off between effort and understandability, which was the basis of Zipf's *principle of least effort* (Zipf, 1949). It is likely that Zipf's law will not emerge when the agents optimise for effectiveness, which – of course – is costly, when a hierarchical taxonomy of categories is maintained. However, the current level of success is too low to be acceptable. Future work should investigate how the success of the communication can increase without increasing the cost too much.

Conclusions

This paper shows that the Zipf-Mandelbrot law can emerge as a result of minimising the cost of categorising sensory stimuli. The emergence is striking, because no aspect of a Zipfian distribution was put in the model: not in the distribution of objects the agents categorise, nor in the distribution of categories in the hierarchically layered conceptual spaces.

To investigate the validity of the general hypothesis that minimising the cost of categorisation does lead to the emergence of Zipf's law in human language, it would be interesting to investigate to what extent the shorter words used in real languages are indeed the more general terms. If that is the case, it should also be investigated to what extent the more general categories are indeed cognitively less costly to categorise, such as appears to be the case for basic level categories as opposed to their superordinate and subordinate categories (Rosch, 1978).

It is important to stress that I do not claim that minimising cost of categorisation is *the* mechanism for the observation of the Zipf-Mandelbrot law in natural languages; there are too many possible explanations around to make such a hard claim (Günther et al., 1996). Furthermore, the most frequently used words are function words, such as 'the' and 'a', which carry no meaning in the sense used here (although one might argue that these words have very general meanings, or at least are applicable in a very general way).

However, humans undoubtedly try to minimise the cognitive effort to categorise sensorimotor events. It is therefore plausible that the tendency to use generalised categories is a bias that – *in addition to other biases*, such as reducing articulatory effort (Mandelbrot, 1953; Zipf, 1949) and other frequency related approaches (Günther et al., 1996; Tullo and Hurford, 2003) – yields the emergence of the Zipf-Mandelbrot law.

Acknowledgements

This work was supported by a VENI grant provided by the Netherlands Organisation for Scientific Research (NWO). Many thanks go to Martin Reynaert and Andrew Smith and the four anonymous reviewers for their useful suggestions and comments on this work.

References

Cover, T. M. and Hart, P. E. (1967). Nearest neighbour pattern classification. *Institute of Electrical and Electronics Engineers Transactions on Information Theory*, 13.21–27.

Ferrer i Cancho, R. and Solé, R. V. (2003). Least effort and the origins of scaling in human language. *Proceedings of the National Academy of Sciences*, 100(3):788–791.

Gärdenfors, P. (2000). *Conceptual Spaces*. Bradford Books, MIT Press.

Günther, R., Levitin, L., Schapiro, B., and Wagner, P. (1996). Zipf's law and the effect of ranking on probability distributions. *International Journal of Theoretical Physics*, 35(2):395–417.

Kirby, S. (2002). Learning, bottlenecks and the evolution of recursive syntax. In Briscoe, T., editor, *Linguistic Evolution through Language Acquisition: Formal and Computational Models*. Cambridge University Press.

Li, W. (1992). Random texts exhibit Zipf's law like word frequency distributions. *IEEE Transactions on Information Theory*, 38(6):1842–1845.

Mandelbrot, B. B. (1953). An information theory of the statistical structure of language. In Jackson, W., editor, *Communication Theory*, pages 503–512, New York. Academic Press.

Miller, G. A. (1957). Some effects of intermittent silence. *American Journal of Psychology*, 70:311–314.

Rosch, E. (1978). Principles of categorization. In Rosch, E. and Lloyd, B. B., editors, *Cognition and Categorization*. Lawrence Erlbaum Ass.

Siskind, J. M. (1996). A computational study of cross-situational techniques for learning word-to-meaning mappings. *Cognition*, 61:39–91.

Smith, A. D. M. (2003). Intelligent meaning creation in a clumpy world helps communication. *Artificial Life*, 9(2):559–574.

Smith, K. (forthcoming). The evolution of vocabulary. *Journal of Theoretical Biology*.

Steels, L. (1996). Perceptually grounded meaning creation. In Tokoro, M., editor, *Proceedings of the International Conference on Multi-Agent Systems*, Menlo Park Ca. AAAI Press.

Steels, L., Kaplan, F., McIntyre, A., and Van Looveren, J. (2002). Crucial factors in the origins of word-meaning. In Wray, A., editor, *The Transition to Language*, Oxford, UK. Oxford University Press.

Tullo, C. and Hurford, J. R. (2003). Modelling Zipfian distributions in language. In Kirby, S., editor, *Language Evolution and Computation, Proceedings of the workshop at ESSLLI*.

Vogt, P. (2000). Bootstrapping grounded symbols by minimal autonomous robots. *Evolution of Communication*, 4(1):89–118.

Vogt, P. (2003a). Grounded lexicon formation without explicit meaning transfer: who's talking to who? In Banzhaf, W., Christaller, T., Dittrich, P., Kim, J. T., and Ziegler, J., editors, *Advances in Artificial Life - Proceedings of the 7th European Conference on Artificial Life (ECAL)*. Springer Verlag Berlin, Heidelberg.

Vogt, P. (2003b). THSim v3.2: The Talking Heads simulation tool. In Banzhaf, W., Christaller, T., Dittrich, P., Kim, J. T., and Ziegler, J., editors, *Advances in Artificial Life - Proceedings of the 7th European Conference on Artificial Life (ECAL)*. Springer Verlag Berlin, Heidelberg.

Vogt, P. (2004). Generalisation as a bias toward the emergence of Zipf's law. In *Proceedings of Evolang 5*.

Vogt, P. and Smith, A. D. M. (2004). Quantifying lexicon acquisition under uncertainty. In Lenaerts, T., Nowe, A., and Steenhout, K., editors, *Proceedings of Benelearn 2004*.

Zipf, G. K. (1949). *Human Behaviour and the Principle of Least Effort: An Introduction to Human Ecology*. Addison-Wesley, Cambridge, MA.

Modeling Multicellular and Tumorous Existence with Genetic Cellular Automata

Armand Bankhead III[1], Nancy Magnuson[2] and Robert B. Heckendorn[1]

[1]Bioinformatics and Computational Biology; University of Idaho, Moscow, ID 38843
[2]School of Molecular Biosciences; Washington State University, Pullman, Washington 99164
bank2192@uidaho.edu, heckendo@uidaho.edu

Abstract

We model a population of cells using cellular automata with genetically-based rules. As in actual multicellular systems, each cell's state is based on environment and genetics. With imperfect reproduction and the accumulation of random mutations, tumorous behavior naturally emerges in a stochastic manner. We validate our model by reproducing results used to confirm other published models. We also demonstrate that our model exhibits both homeostatic and tumorous behavior using metrics based on clinical biopsy diagnosic techniques.

Introduction

Multicellular organisms are composed of cells that co-exist despite resource limitations (ie. glucose, oxygen). Unlike unicellular organisms, such as bacteria or eubacteria, multicellular organisms possess proto-oncogenes that activate or inhibit reproduction to benefit the organism (Heath, 2001). When these proto-oncogenes are mutated, cells reproduce uncontrollably and tumorous behavior emerges to benefit individual cells and defy the organism.

As tissue cells are damaged or intentionally sloughed, reproduction is required (Tannock and Hill, 1998). Cellular reproduction within the tissue of an organism is somatic and semi-conservative (Fairbanks and Andersen, 1999). This process is imperfect; both parent and offspring cell genomes experience random mutations. As the population is replenished, mutations may accumulate to cause genetic dysfunction. Mutational damage to genomic areas controlling reproduction may result in tumorous behavior.

We define multicelluar existence as a population of cells that co-exist using reproductive control. We define tumurous existence as a population of cells that reproduce to maximize population size and resource consumption. In this paper we present a genetic cellular automata abstraction capable of modeling multicellular and tumorous existence.

Cellular automata (CA) are a logical choice for modeling both population types. Just as a biological cell reacts to a local environment using genetic expression, cellular automata react to a local neighborhood using a set of rules. Because multicellular and tumorous existence are the direct result of genetic functionality, we believe these CA rules should reflect the genetic mechanisms involved.

The next section discusses other CA models used to describe tumor growth. The following section introduces genetic CA (GCA) and highlights important features of the design. Next, using results from other published CA models, we show that GCA can evolve similar tumorous behavior. Then we use biopsy-based metrics to classify multicellular and tumorous existence to illustrate the validity of our model. Finally, we discuss results and future work.

Previous Work

CA's have been applied to model avascular tumorous behavior (Moreira and Deutsch, 2002).

A model produced by Qi *et al.* (1993) uses a two dimensional CA to illustrate critical aspects of Gompertzian population growth curves of cancerous cells. Probability-based rules for proliferation and death are parameterized by experimental data. Nutrient resources are dependent on the size of the population–as the population increases less nutrients are available. Simulated population growth curves are then compared to Gompertzian growth curves for validation.

Work by Kansal *et al.* (2000) uses a three dimensional CA to model brain tumor growth. Rules are parameterized by experimental data such that layers of proliferative, non-proliferative, or necrotic cells form based on location relative to the center of the tumor. These layers result from pressure and nutrient constraints on the tumor growth. Rules are deterministic and based on a single set of experimental data. The model is validated via experimentally derived *in vitro* growth curves and Gompertzian growth curves. *Of course*, the layering effects defined by the model's rules are exhibited.

More recently, a two dimensional CA model presented by Dormann and Deutsch (2002) uses a hybrid lattice-gas approach to account for varying channels for migration and chemical diffusion. Rules are also probability-based and derived from experimentally derived differential equations. Cellular rules include behaviors such as mitosis (reproduction), apoptosis (suicide) and necrosis (death). Nutrients are explicitly accounted for using a lattice-gas diffusion process, although the diffusion process is assisted by expanding the local neighborhood of each cell to three layers. The proliferative, non-proliferative, and necrotic layering effect is also used to validate this model.

Although these models are insightful to specific data sets, they do not shed light on the underlying general cause of tumorigenesis. All of the models mentioned above simulate a contrived starting population of tumor cells. Tumor development is not contrived, but is the result of genetic dysfunction (Tannock and Hill, 1998). Normal multicellular populations evolve this behavior contrary to the organism. A more complete tumorigenesis model should encompass both multicellular and tumorous existence.

Genetic Cellular Automata Design and Implementation

Cellular Automata Framework

Our model is similar to other CA tumor models in several regards. We model space as a 100x100 lattice of boxes; cell enitities may exist within these boxes. Each cell has a discrete state defined by its binary genome; cell rules are probabilistically determined using this binary genome. Using these rules, cells react to a local Moore neighborhood of eight boxes. Also, time is observed as discrete cycles and all cells on the lattice are updated synchronously.

GCA are atypical to CA tumor models. Nutrients diffuse across the lattice using a stochastic averaging technique. Signals from a cell's environment and interacting genes are interpreted by genes that control the rules of the cell. As a result, biological behaviors, such as reproduction and apoptosis, are controlled by genetic rules. Upon reproduction, cells copy their genome with some mutation rate; the resulting parent and offspring are dissimilar. More than one cell may exist within each box of the lattice. These atypical properties are described below in further detail.

Nutrient Diffusion

All possible forms of eukaryotic cell diet (i.e. glucose, oxygen, etc.) are reduced to one generic nutrient. In biological tissue populations, nutrients are typically delivered by nearby blood vessels and diffuse through adjacent cells. We represent these nutrient sources on the outer edge of the lattice. Nutrient source boxes produce enough nutrients to sustain a lattice population with an average density of 5 cells per box.

Diffusion is emulated by stochastically updating Moore neighborhoods throughout the lattice. The update involves feeding cells within the box and then mixing the nutrient concentration evenly in the neighborhood of boxes. The rate of diffusion may be adjusted by varying the number of updates over the lattice.

This nutrient diffusion process allows our model to simulate nutrient deficiencies and circumvent using a static diffu-

sion representation specific to a particular experimental data set.

Genes Produce Genetic Signals

Cell rules are based on genetic signals. Similar to biological genes, genetic functionality is dependent on the fidelity of a gene; a mutated gene is less likely to be expressed properly. Each gene is implemented as a 64-digit binary string divided into three biologically-based (Fairbanks and Andersen, 1999) regions shown in Figure 1.

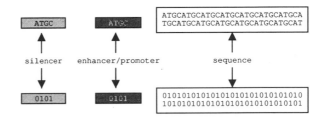

Figure 1: Schematic of a bit string (bottom) used to represent biological genes (top). non-mutated bits = 0; mutated bits = 1.

- Silencer: Negative control. When damaged the gene is over-expressed.

- Enhancer Promoter: Positive control. When damaged the gene is under-expressed.

- Sequence: Coding region. When damaged the gene product expressed does not function.

Eq. 1 shows genetic functionality (F_g) is related to the number of non-mutated bits of the gene.

$$F_g = \frac{1}{F_s} * F_{ep} * F_s \tag{1}$$

The functionality of the silencer (F_s), enhancer/promoter (F_{ep}), and sequence (F_s) are the percentages of non-mutated bits within each region.

Finally, the genetic signal or probability (P_g) of a gene's proper expression is

$$P_g = Sig_g(c_g, s_g, F_g) = \frac{1}{1 + exp(-s_g(F_g - c_g))}, \tag{2}$$

where Sig_g is a sigmoid function parameterized by slope (s_g) and centering (c_g) for realistic behavior.

Genetic Networks

The actual regulatory genetic networks that control eukaryotic cell behavior are currently incomplete for many multicellular organisms. Integrating the complete genetic networks involved would also be computationally unfeasible. Abstractions are necessary.

Using a coarse-grained approach we focus on basic environment-to-gene and gene-to-gene interactions by collapsing major genetic pathways into genetic classes. For simplicity, we refer to these genetic classes as genes and represent these genes using the bit string design presented in the previous section. Figure 2 shows the collapsed genetic networks used for each cell while Table 1 summarizes these relationships.

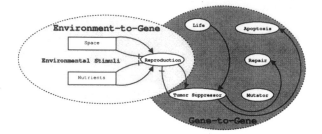

Figure 2: Collapsed genetic network indicating relationships between environment and genes. Environment-to-Gene interactions are shown to the left; Gene-to-Gene interactions are shown to the right. Environmental signals are labeled in rectangles; genetic signals are labeled in ovals. Arrows indicate gene activation; crossed connections indicate inhibition.

		Input Signals				
		Envioronment		Gene		
		Nutrients	Space	Life	Repro.	Tumor Sup.
Genes	Life					
	Reproction	+/-	+/-			-
	Tumor Suppressor			+	+	
	Apoptosis					+
	Repair					+
	Mutator					

Table 1: Genetic Network Summary. '+' means positive control (activation); '-' means negative control (inhibition).

Cellular Automata Rules

Each cell's rules are dictated by the genetic relationships described in the previous section. Rule behavior is probabilistic and based on the summation of environmental and genetic input signals of relevant genes shown in Table 2.

For example, a cell's overall mutation rate is based on a mutation rate (10^{-6}) that does not cause premature mutational meltdown. Repair signals lower the mutation rate (Fairbanks and Andersen, 1999) whereas mutator signals increase the mutation rate (Kuby, 1997). If the repair gene becomes damaged the mutation rate may increase whereas if the mutator gene becomes damaged the mutation rate may decrease.

With rules based on multicellular genetics, we capture general multicellular behavior. Instead of cells reproducing to fill the lattice to a maximum density (10 cells per box),

		Genes					
		Life	Repro.	Tumor Sup.	Apop.	Repair	Mutator
Rules	Reproduction		X				
	Mutation Rate					X	X
	Genetic Death	X					
	Suicide				X		

Table 2: Genetic Rules. 'X' indicates a direct relationship between cell rules and genes.

the cells are genetically designed to reproduce to an acceptable density (3 cells per box). As individuals accumulate mutations, genetic rules may ignore proper environmental signalling–just as genetic dysfunction leads to tumorous existence in organisms.

Algorithmic Description

Initially a normal cell (with no mutations) is placed randomly on the lattice. The cell reproduces itself and clonal reproduction continues until an appropriate density of cells is present within each box. Cell densities are kept relatively homogenous as a natural consequence of the GCA rules. After the lattice has reached population homeostasis, random cell removal (.0005 chance a cell will be removed) encourages reproduction.

A Comparison to Previous Work

Several of the CA models mentioned above verify their work using comparisons to experimentally derived features. Some of these tumorous features are discussed and evidence is presented to show that our model evolves similar behavior without directly designing the model to do so.

Gompertzian Growth

Typically unicellular populations grow exponentially in the presence of excess nutrients. As mentioned above, exponential growth is prevented in multicellular organisms through proto-oncogenes; cells are genetically designed to reproduce when appropriate to the organism. A population of tumorous cells typically displays exponential growth until space or nutrient limitations are met (Tannock and Hill, 1998).

Models presented by Qi *et al.* and Kansal use Gompertzian growth to illustrate validity. Gompertzian growth curves are used to describe growing populations in the presence of resource limitations. Typical Gompertzian curves are initially exponential and dampen to a stationary phase with little growth (Qi et al., 1993) (Kansal et al., 2000).

$$V = V_0 = \frac{A}{B} - exp(-Bt)) \qquad (3)$$

The above equation utilizes several parameters to fit the growth curve to some experimentally derived data set. Figure 4 shows growth curves derived using this equation with the parameters A = 0.997, B = .0787, V_0 = 0.000103.

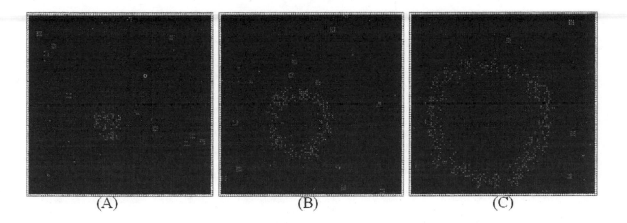

Figure 3: Cells dividing are light gray. (A) Shows initial tumor growth. (B) Proliferative (light gray) and non-proliferative (dark gray) layers are visible. (C) Necrotic center is colored black. Note random background growth due to random cell removal.

After a tumorous population of cells emerges, nutrient and space limitations inhibit growth. The result is a population growth curve that is implicitly Gompertzian (Figure 4).

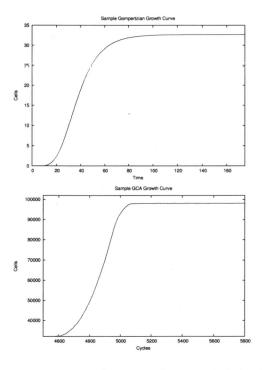

Figure 4: Comparison of Gompertzian growth derived from experimental data (Steel, 1977) (top) and a sample GCA growth curve (bottom).

Tumor Layering

As a result of nutrient and space limitations (Tannock and Hill, 1998), biological tumor masses often manifest three distinct layers:

1. Proliferative: Tumorous cells on the exterior of the tumor actively reproducing.

2. Non-Proliferative: Quiescent tumor cells without enough nutrients and space to reproduce.

3. Necrotic: Dead tumor cells.

Models by Kansal *et al.* (2000), Dormann and Deutsch (2002) use this layering phenomena to confirm their models. Such behavior emerges from our GCA model due to nutrients unable to support the dense tumor cell population. As a result, Figure 3 shows cells at the center of the tumor mass starve and may become non-proliferative or necrotic.

Tumor data (Kansal et al., 2000) also suggests that the outer proliferative layer is fixed in depth. If the density of the tumor is homogenous, nutrients would only be able to penetrate a fixed depth into the tumor mass. Our data support this feature: The outer circumference of the proliferative layer is a steady distance from the non-proliferative layer of the tumor.

A Clinical Approach to Classifying Multicellular and Tumorous Growth

To diagnostically confirm cancer, physicians perform clinical biopsies (AJCC, 2002). A sample of suspect tissue is removed and used for labratory testing. These tests are used to classify the tissue as being normal or tumorous.

We validate our GCA's ability to express both multicellular and tumorous existance using metrics which parallel those used by physicians for cancer diagnosis.

Nodule Formation

Tumorous nodules are the result of abnormally high cellular density (Tannock and Hill, 1998). By the definition of

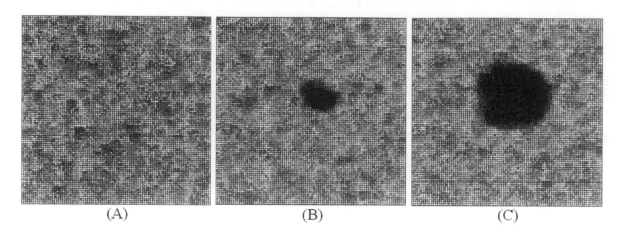

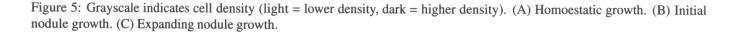

Figure 5: Grayscale indicates cell density (light = lower density, dark = higher density). (A) Homoestatic growth. (B) Initial nodule growth. (C) Expanding nodule growth.

cancer, this high cellular density is the result of abnormal reproduction. Tumorous nodules are often used as the initial indication of a tumorous population.

Figure 5(A) shows a homeostatic population of cells that exist at a moderate density throughout the lattice. Figures 5(B) and 5(C) show a tumorous population emerging (in darker coloring) via exponential growth to take over the lattice. The tumorous population is surrounded by a homeostatic population of cells–this contrast in density results in visible nodules.

Genetic Instability

Chromosomal annealing tests are used to observe major genomic inconsistencies of tumorous cells (Tannock and Hill, 1998). The DNA from a normally behaving cell of an organism may be compared to the genome of a tumorous cell to observe a ratio of similarity. With increased mutation and reproductive rates, tumor cells often genetically diverge from the organism's genome. We simulate chromosomal annealing by analyzing the average number of mutations present for cells at each box of the lattice. This test takes into account the magnitude of mutations.

Unlike actual biological genomic experiments, the GCA model accounts for **all** mutations that have acrued in the population since all cell genomes are accessible for analysis. Results shown in Figure 6 are based on the entire population, not samples that are used to statistically infer the entire population.

Growth Curves

After removing a biopsy tissue sample, cells may be grown in a culture for examination (AJCC, 2002). Features of Gompertzian growth are used above to validate tumorous growth curves. We use these characteristics to classify both

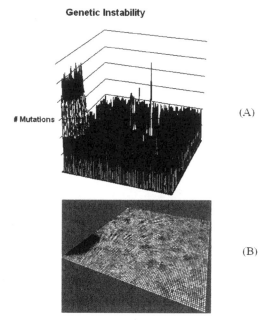

Figure 6: (A) Plot of relative mutations. (B) Corresponding grayscale density lattice.

normal and tumorous cells. Figure 7 shows growth curves resulting from several GCA runs.

The observed stochastic nature that tumorous populations emerge at different simulated time points is notable. This result is encouraging because tumorous growths occur stochastically throughout the lifetime of organisms.

Conclusion

We have designed a CA model with these unique features:

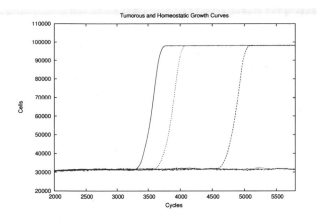

Figure 7: Growth curves generated using six GCA runs. Three runs remain homeostatic at 30,000 cells. Three runs stochastically grow in a Gompertzian manner to 100,000 cells.

- Genetically based rules
- Tumorous behavior evolves stochastically
- Non-tumorous and tumorous states are modeled

We validate our GCA model by presenting features used to validate previous CA models. We also present data based on metrics used to classify biospy tissue samples to verify that GCA can exhibit both multicellular and tumorous existence. Although this model is abstract, it reflects the basic genetic mechanisms that cause tumor growth.

GCA models may be extended to provide a framework for the addition of established genetic networks. Such genetic simulations may help the understanding of genetic dysfunction pathways and possibly provide treatment strategies for tumorigenesis. Also, unlike biological analyses, GCA models allow for the observation of the entire population–instead of drawing statistical relevance from a population sample. The GCA model presented evolves behaviors produced by other CA models that are specific to experimentally derived data sets. The behaviors evolved are the result of generalized genetic mechanisms that are known to result in tumorous behavior.

It is well known that cancer is an evolutionary disease. The task of treating a disease that evolves to the host's defenses and the physician's medications is a difficult one. It is our hope that this model will be used to simulate useful experiments and treatment strategies.

Future Work

We hope to expand the genetic networks represented by our GCA to involve established pathways derived from experimental data. Also, a three dimensional GCA is being considered that would allow a more realistic comparison to experimentally derived *in vitro* and *in vivo* data sets–since these data are produced in three dimensions. Finally, we hope to use this model to generate hypotheses about genetic dysfunction pathways that lead to tumorigenesis.

Acknowledgements

This work is funded by NIH Grant #P20 RR16448-01. Experiments were simulated on the University of Idaho Beowulf Cluster Supercomputer funded by Intel Corporation and NSF Grant #EPS80935.

References

AJCC (2002). *AJCC Cancer Staging Manual*. American Cancer Society.

Fairbanks, D. J. and Andersen, R. W. (1999). *Genetics: The Continuity of Life*. Brooks/Cole Publishing Company.

Heath, J. K. (2001). *Principles of Cell Proliferation*. Iowa State University Press.

Kansal, A., Torquoato, S., Harsh IV, G., Chiocca, E., and Deisboeck, T. (2000). Simulated brain tumor growth dynamics using a three-dimensional cellular automaton. *Journal of Theoretical Biology*, 203:367–382.

Kuby, J. (1997). *Immunology*. W.H. Freeman and Company.

Moreira, J. and Deutsch, A. (2002). Cellular automaton models of tumor development: A critical review. *Advances in Complex Systems*, 5:247–267.

Qi, A.-S., Zheng, X., Du, C.-Y., and Bao-Sheng, A. (1993). A cellular automaton model of cancerous growth. *Journal of Theoretical Biology*, 161:1–12.

Steel, G. G. (1977). *Growth Kinetics of Tumors*.

Tannock, I. F. and Hill, R. P. (1998). *The Basic Science of Oncology*. Mcgraw-Hill.

Whatever Emerges should be Intrinsically Useful

Hugues Bersini

IRIDIA-CP 194/6
Université Libre de Bruxelles
50, av. Franklin Roosevelt – 1050 Bruxelles
Belgium
bersini@ulb.ac.be

Abstract

The practical work presented in this paper uses a GA to evolve a cellular automata (CA) implementation of a binary numbers adder. One very useful way to compress the enormous search space and eventually find an optimal CA consists in adopting a macro-coding of the states and the rule table. It is further discussed how this work illustrates and defends our favorite position in the currently vivid epistemological debate around the notion of "emergence". This position is Crutchfield's "intrinsic emergence" one, in which to say that a macro-property is emergent requires that this "property" supplies some mechanical and non-human observer with additional functionality.

Introduction and Intrinsic Emergence

For many years now, cellular automata (CA) [16] have been the favorite computational platform to experiment and illustrate emergent phenomena. It is far from surprising that many authors have relied on their CA experimentation to quest for formal definitions of the nature of "emergence" and to practically validate them [3][10][14]. This paper is following a similar trend by fully adopting the practice of CA. On the whole, all authors interested in the rationalization of emergence converge to the fact that at least two levels of observation are required: A first one in which the micro-states and micro behavioral rules are specified and implemented, and a second one, which by only depending upon the underlying micro-characteristics, exhibits interesting macro-phenomena. They are obtained by unfolding in space and time the micro-rules though the micro-states, most of the time in a non-decomposable way (see [10] for an attempt to formalize this non-decomposability).

An observer, so far always human, is necessary to instantiate this second and more abstract level of observation and to spot, follow and trace these interesting and new phenomena. This characterization of emergence has turned out to be quite common [2][3][8][10][13] and

could be symbolized by the little UML class diagram of figure 1 showing the three basic actors: the parts, the whole, allowing to iterate the parts in space and time, and the human observer. Nevertheless, this paper considers that such a classical characterization, though including necessary ingredients (i.e. the two levels of observation and the abstraction in space and time of the second with respect to the first), is far from sufficient, severely limited and incomplete on one essential aspect: the identity and the role of this second level observer. The problem is not so much that "the whole is more than the sum of its parts" but rather who is responsible for observing that "whole". For CA, Neural Networks, other computer simulations of networks and whatever computational source of emergence, the observer is generally accepted to be human. However, this "anthropomorphisation" of the phenomenon of emergence is antagonistic to any scientific practice that, in principle, aims at not leaving subjectivism a leg to stand on. Basically, if the formalization of emergence demands the intervention of a human observer, even worse to be "psychologically surprised" [14], its intrusion in the vocabulary of physics is compromised right off the bat.

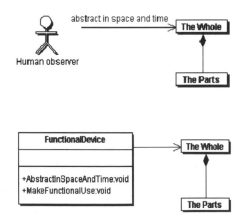

Fig. 1. The "Intrinsic Emergence": from the human observer to the functional device.

To our knowledge, such a limitation has been faced and removed mainly by two authors [5][6][7] who, in their writing, have answered this preoccupation by supplying the characterization of emergence with a key ingredient. Like the updated UML class diagram shows, a "functional device" must substitute the human observer that, for whatever utility or performance reasons, will fine-tune its observation of any macro-phenomena produced by the system. Cariani [5] claims that, for a phenomenon to be said emergent, devices need to be built, able to find new observable, autonomously relative to us, whose selection and tuning must rely on performance measures. However the most convincing reply to this limitation and which provides the main guidelines for the work to be described in this paper is the Crutchfield's definition of "intrinsic emergence" [6]:

"... *Pattern formation is insufficient to capture the essential aspect of the emergence of coordinated behaviour and global information processing... At some basic level though, pattern formation must play a role... What is distinctive about intrinsic emergence is that the patterns formed confer additional functionality which supports global information processing... During intrinsic emergence there is an increase in intrinsic computational capability, which can be capitalized on and so lends additional functionality.*"

For instance, the game of life "glider" [12][3][10] should not be characterized as "emergent" unless some functional device able to observe the CA and to aggregate its cells in space and in time (the glider covers 5 cells and is a period-four phenomenon) will make a specific use of it. Emergence is a bottom-up phenomenon but like in any top-down and hierarchical construction of complex systems, some external entity needs to make sense and use of whatever emerges at any level of the construction. Following trends initiated by Packard, Crutchfield, Mitchell and others [1][7][11], the work described in this paper consists in using an evolutionary algorithm to discover a CA able to perform an engineering task, in our case, the binary addition of numbers coded on n bits. It is shown that the discovery of such an efficient CA is very painful by adopting the natural micro-coding of the states and the rules.

An important improvement and acceleration of this discovery is allowed by adopting a simplified macro or abstract characterization of the states and the rules of the CA. This new and one level up "macro observation", intrinsically emerges, since autonomously tuned by the system itself on the basis of performance measures. Once a new macro-way of observing the system is adopted, new phenomena emerge since being detectable only by this new observing device. Allowing part of the evolutionary process to take place by adopting such an emergent observable accelerates in a consequential way the discovery of a quasi-optimal CA. This is why and only

why, according to the definition of "intrinsic emergence", this observable can be labeled as "emergent", since it is intrinsically selected by the system to improve its performance and needs no external human observer.

Apart from casting some fresh light on the characterization of emergence, any engineering of distributed computation should often be able to reproduce such a practice: find a way of observing the system which helps the optimizing of its structure and behaviour. The next section will describe the task to be achieved by the CA. The third one will describe how a simple evolutionary algorithm can search for a satisfactory non-uniform CA, finally followed by the two last sections discussing and experimenting how a useful way of observing the CA can emerge in order to boost the discovery of this satisfactory solution.

The CA Task: Binary Addition

Binary addition is an interesting task to be performed by CA for essentially two reasons. First, it is much more reminiscent of what real digital circuits perform in computers than "density classification" or "synchronization" [1][7][11][15], so that the developments presented here could more easily be transposed for the automatic discovery of useable logical circuits. Moreover, the computer digital circuitry for binary additions is built by growing up in functional abstraction i.e. big circuits (able to add big numbers) are composed of small circuits. Basically the little circuits, optimally shaped for adding very small binary numbers, are kept as the basic blocks for building larger circuits. This fact will be important to help at better understanding the type of observables we want our system to be able to "emerge".

The CA used to perform this task is a two dimensional CA (this dimensionality was favored to the one-dimensional case by analogy with digital circuitry and for reasons that will become obvious later on), with periodic boundaries and characterized by the classical 8-cells Moore neighborhood. At time 0, like indicated in fig.2, two n bits binary numbers are installed in the first line of the CA (the first number on the extreme left, the second number on the extreme right, in the figure, numbers are coded on 5 bits). All other cell initial states are tuned to 0. Time step after time step, the CA then updates all its cells in a synchronous way, by complying with the rule table associating with each of the 2^8 possible neighborhood configurations the next value of the state (so that the rule table here has 2^8 entries with value "0" or "1" for each). The goal of the CA global computation is, like indicated in fig.2, after a given number of time steps, taken here to be equal to the vertical size of the CA (the number of lines), to obtain, still in the first line and on its extreme left, the n+1 bits composing the result of the addition of the two numbers.

Fig.2. The CA Task: Binary addition of two five bits binary numbers. Initially, the numbers are shown at the extreme left and extreme right of the first line of the first CA (a 25x25 periodic CA). The two numbers are 10001 and 00111. All the other cells are set to 0. After 25 time steps, the correct sum: 011000 should appear at the extreme left of the first line of the second CA.

In the software, all parameters, the size of the CA, the size of the numbers to add, the number of time steps to reach the result, can easily be changed. However, here for sake of clarity, we will maintain one set of number, 25 for the size of a square CA (so that 25 five time steps are computed in order to get the answer) and 5 for the size of the two numbers to add by the same CA. Finally, this addition must be performed for a certain number of couples of binary numbers, say 5 again, like, for instance given below:

```
    N1      N2     N1+N2
   01110   11110   101100
   00111   10101   011100
   11100   00110   100010
   11001   01111   101000
   10111   11110   110101
```

This way of doing makes CA very similar to classical addition logical circuits were the numbers to add are injected as input on the top of the circuit, then cascade down through a sequence of logical gates to finally give the result of the addition at the bottom of this circuit. It must be clear that this procedure does not guarantee at all obtaining a universal adder but just a specific one, which works well for these n numbers. However, the bigger this n the more likely you'll make the adder universal.

What Kind of GA Evolves the CA

The kind of GA and how well it allows optimizing the rule table is not a central topic here and an infinite number of possible evolutionary algorithms could be applied. The rule table to be optimized is composed of 2^8 bits so that the search must take place in this huge binary space (2^{2^8} possibilities). A population of 20 individuals is being evolved. After computing the fitness of every individual, the best one is kept in the subsequent population (the "elitist version") and the best half of the population is

selected on which to apply the mutation and the crossover mechanisms responsible for generating the subsequent population. Following a series of mutation and crossover upon the 10 best individuals, a new population of 20 individuals is generated. More precisely, the first individual of the new population is the same as the first individual of the previous one. Then the second and the third individuals of the new population are obtained by randomly mutating one bit of two individuals arbitrarily taken among the 10 best individuals of the previous population. Since a large preference is given to the crossover mechanism, the 17 remaining individuals are obtained by applying the simplex crossover on three individuals.

The simplex crossover first introduced in [4], and which has been shown to improve on the two parents classical one, is described below.

Simplex Crossover:
1) Take three parents and rank then by decreasing fitness: P1>P2>P3
2) For each bit, if parent 1 and parent 2 agree on the value of the bit, the offspring will have the same value for this bit. If parent 1 and parent 2 disagree on the value of the bit, take the reverse of the value of this bit given by parent 3.
3) The idea is that both parent 1 and 2, the good ones, attempt at shaping the offspring similar to them while parent 3, the worst of all three, and as soon as it can say something, will force the offspring to be very different from him.

For instance:

```
P1:  1001100110
P2:  1011110100
P3:  0001101111
-----------------------------------
Offspring:  1011110100
```

In our case and provided the individuals in the population are ranked by decreasing fitness, the three parents are selected so that the first resides in the first third of the 10 best, the second in the second third and the third in the last third. Again, we are not so much interesting here in how well this GA performs. This is one very simple instance to implement and much better versions could be imagined. It largely suffices to test and validate the main idea behind this work. The fitness is computed by summing, over all five couple of numbers, the Hamming distance between the desired result (the correct sum) and the one obtained after 25 time steps of the CA run. For all possible rule tables and for 5 couples of 5 bit numbers, the fitness value is comprised between 0, the best fitness, and 30, the worst one (i.e. 5*6).

After many attempts, the performances obtained were very poor, giving an average fitness around 6.5. In substance, after thousands of iterations, no CA could be found able to sum these 5 binary numbers. Like done by other authors before [15], a first recovery decision consisted in allowing the CA to become non-uniform i.e. to allow distinct rule tables to characterize the update mechanism of different cells. Then a cell could be characterized by one rule table and its neighboring cell by a distinct one. While this increase in the degrees of freedom should naturally improve the computational capacity of the CA, the original motivation was to be able to replicate the heterogeneity of computer logical circuits (composed of "AND", "NOR", "XOR", etc. gates). Consequently, an additional attribute has to be associated with every cell i.e. which one of the rule tables it complies with. The way the rule tables were distributed among the cells was random and the figure below shows the increase in performance obtained by allowing the number of distinct rule tables to vary from 1 to 5 (the parameter "v" for "variety"). But, still no CA could be found with fitness better than 6 after 500 iterations of the GA search.

The new individual to optimize would now comprise $v*2^8$ bits increasing by many order of magnitude the size of the search space. Indeed, the same figure shows that, although a small increase of v allows an improvement in the fitness, in order to obtain this improvement and to discover better individuals, the number of GA iterations must equally increase in a drastic way (500 GA iterations in the first case and 2000 in the second one). So, while increasing the heterogeneity increases in proportion the chance to find a better CA (for instance with v=5, the average best fitness obtained after 2000 GA iterations is 5,4), the computational time must also be consequently increased to allow this discovery. A very practical problem turns out to be how to accelerate the discovery of an optimal CA in a fundamental way.

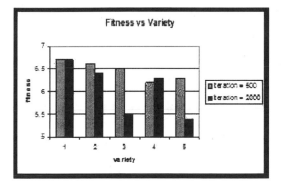

Fig.3. The fitness as a function of the variety of the rule tables for 500 and 2000 GA iterations

Discovery of a Useful Way to Observe CA

Facing this huge search space, among the many ways to reduce its size, the original one proposed in this paper requires to modify the way CA is observed and to move one level up in abstraction to look at the states and the rules. This new observable makes any cell to be characterized by only x out of the 8 states composing its neighborhood and to mask the (8-x) others. For instance, x could be taken to be 3, like indicated in fig.4. If only 3 neighbors are taken into account, many cells, whatever distinct precise neighborhood they have, turn out to be in identical state. This is very reminiscent of the "don't care" symbol of Holland's definition of GA schema. Instead of "10010000", for instance, the state of a cell will become "#0####00". A formidable collapse in the dimension of the search space follows from this abstract observation which, in presence of 5 don't care (we will keep with this value in the following although a bigger or smaller one is equally possible), boils the coding of the rule table down to 2^3*v (v still being the variety factor).

Fig. 4. Here the mask "02367" has been applied

The way a cell is updated now complies with this reduced coding of the rule table and 2^5 distinct neighborhood (the masked "don't care cells") will give the same next state. The (8-x) hidden neighboring cells will be called the "mask" and C_8^5 masks are possible. A mask will be defined, for instance, as "02367" i.e. the five hidden neighboring cells (this is the case shown in fig.4). The first experiments we tried consisted in seeing whether such a reduced search space, although skipping the fine tuning probably necessary at the discovery of an optimal CA, degraded or not the performance obtained so far. The GA was exactly the same apart from tackling a consequently smaller binary space and the 5-cells masks were randomly generated. The experimental results were quite surprising and supported the initiative since, despite this degradation in characterizing the cell state, the reduction of the search space would largely compensate for it. In average, the fitness, following a same number of GA iterations, was slightly improved. For instance, for the "variety" equal to 5, one point of fitness could be gained, around 5 instead of 6.

Emergence of a Useful way to Observe CA

At this stage of the work, and although this masking sounds as a promising proposal, several remarks can be made. First, though slightly improved, the fitness is still unsatisfactory and an average of five is the best we can reach. Second, it is clear that by limiting the observation to such a small part of the cell state, it's hard to imagine how we could reach one very best individual the coding of which should demand for a fine tuning. Last but not least, with respect to the problematic of emergence, this masking can't be said to emerge since it is manually imposed by the human user.

One way to answer all these remarks is by the following new set of experiments that we indeed performed. First, let's define the emergent "observable" to be the "mask" given the best fitness after a certain number of random trials. In such a way, this masking will be intrinsically selected by the system itself with no need for human intervention. So there should be an initial set of trials to reveal this promising mask. The right mask will emerge out of this initial sequence of random trials. Afterwards, both the mask and the population of the best rules obtained with this mask should be memorized so as to release a new set of simulations in which the complete coding of the cell states and the rules will be re-established. The new algorithm works as follows:

1) For "s" trials, generate one random mask and compute the best fitness after "y" GA iterations
2) Out of these s trials, memorize the best mask and the best rules set for this mask
3) Re-establish the complete coding and generate the initial rule set from the rule set memorized after the "masking" phase. To do so, the $v*2^8$ bits of the "unmask" version bit will be obtained by taking the value of the corresponding $v*2^3$ bits of the memorized "masked" version, so that $v*2^5$ bits will share the same initial value. It is easy to understand that this procedure guarantees that the best fitness of this new unmasked rules set is equal to the best fitness of the previous masked one. So an elitist evolutionary algorithm can only improve on the best fitness of the unmasked version.
4) compute the best fitness after "z" GA iterations.

In order to verify in the most "honest" way the benefit allowed by this algorithm, results need to be compared with a "unmasked" version of the algorithm allowing z+y*s iterations. In the table below, one can read three output of the algorithms for z=1500, y=300 and s=5. First, for comparative purpose, a simple run of the "unmasked" algorithm is done for 3000 GA iteration, then 5 trials are made of the "masked" version for 300 GA iterations, to be concluded by 1500 GA iterations of the re-establish "unmasked" version, with the best mask extracted from the precedent trials. In such a way, the unmasked version and the "emergent masked" version will be fairly compared following a same number of 3000 GA iterations. In the table below the three outputs are quite representative of the many run we did of this same algorithm.

Focusing on the first output (i.e. the first column), the best fitness attained by the unmasked version after 1500 GA iterations was 5. Then 5 trials of 300 GA iterations were run with 5 different masks. The best fitness, 2, was obtained with the fifth mask: "01457". A final release of 1500 GA iterations was performed by re-establishing the

complete coding of the states and the rules, which eventually lead to the optimal rule table. Here the right way to observe the CA turned out to be by exploiting the mask "01457", an observable that was really selected by the system itself. Following this last run, a final perfect adder was obtained for the 5 numbers (fitness = 0). In general, this algorithm leads to the best final fitness we could ever have, with an average of 1.5 after the 5 masks testing and the final 500 iterations instead of 6 by simply using the original unmasked coding during 3000 iterations. A considerable improvement results from the use of this emerging observable of the CA. Results shown in the 2nd and the 3rd columns are also interesting since, despite the mediocre results obtained after 500 iterations with mask "01456" in the first case and "02356" in the second, the following "masked" version considerably improves the final fitness (2 and 1). So in all cases, it seems that the use of the masks constrain the GA search space in a very profitable way.

No masking: fitness = 5	No masking: fitness = 7	No masking: fitness = 7
With masking: 01347 fitness = 7	With masking: 01456 fitness = 7	With masking: 12357 fitness = 7
With masking: 01245 fitness = 4	With masking: 23567 fitness = 10	With masking: 02356 fitness = 6
With masking: 24567 fitness = 6	With masking: 12467 fitness = 9	With masking: 12356 fitness = 8
With masking: 01234 fitness = 6	With masking: 12457 fitness = 10	With masking: 01237 fitness = 10
With masking: 01457 fitness = 2	With masking: 12357 fitness = 9	With masking: 01346 fitness = 7
No masking: 01457 fitness = 0	No masking: 01456 fitness = 2	No masking: 02356 fitness = 1

Conclusions and Related Work

It is necessary here to spend some lines on the closely connected and very influential work of Crutchfield and Mitchell [6][7] who also evolve CA both in an engineering and in an epistemological perspective. Regarding the epistemological impact of their work, a new way of observing the CA in terms of "regular domains", "particles", and "particle interactions" is proposed. To some extent, this new observable can be said to emerge since it provides the system with a better understanding of how the CA performs. Now this is somewhat different from our proposal where this new macro way to observe the CA behaviour is not aiming at a better comprehension of how it performs but straightforwardly at a better performance. Roughly said, the adoption of this new sophisticated and high-level semantics based on "particles and particle interactions" allow to understand why the CA can classify or synchronize but does not make it synchronize or classify better (indeed the coding and general framework they used was enough to find the optimal CA), whereas the "emergent abstract observable" discussed in this paper aims at improving (here by simply accelerating the search) how the CA can achieve addition. Both works illustrate the necessary condition for a phenomenon to "intrinsic emerge" since, in both cases, additional functionally is provided to the system. Simply, the nature of this functionality turns out to be different, oriented towards a better understanding on the one hand and towards a better performance on the other hand. In contrast with more dynamical phenomena like the game of life "glider", this very small increment in abstraction provided by superposing the mask on the CA is a very elementary structural form of emergence, but what is more relevant here is the way it autonomously appears rather than its definitive nature. Besides, but much more work should be done, by observing the CA dynamics when adding the two initial numbers, it clearly appears that some kind of particle is crossing the space to encounter others of the same kind, and that the adoption of this "mechanical" high level semantics could, here also, clarify the rules of addition evolved by the GA. It would have been even more convincing if, instead of the mask, it is these new reading of CA in terms of particles and particle encounters that would have been autonomously selected to accelerate the discovery of the optimal adder.

Adopting a more engineering perspective, this manner of reducing and constraining the search space by adopting a more coarse grained observation of the system to optimize could be effective for a lot of combinatorial and real optimization applications. Take for instance the Traveling Salesman Problem, a way of grouping the cities in some kind of region and first optimize the path among the regions, then keep the best "regionalization" and, from this new solution, try to optimize the real one, is definitely something to try when the number of cities prohibit any acceptable solution by just staying at the lowest level. Also, this technique is similar to some form of automatic feature selection which is a rather common attitude when having to optimize real functions with many dimensions. An obvious next thing to try will be to put the mask under genetic control to let a population of masks to evolve as part of the whole GA's evolutionary process.

Finally, the idea that adopting some "distorted" way of observing the reality can make the observer more adapted in his environment than by trying too match all details will not surprise researchers interested in the study of perception both correct and illusory. Living systems calibrate their perceptual apparatus not to perfectly match the external reality but to extract what is needed for a better life. We don't see the world as it is but as it fits. The work presented here resonates with this generally accepted view of perceptive systems: Here, the evolving system does not observe the CA with all possible details but just in a way which allows rushing towards the most adapted one.

Whenever a functional perspective is adopted in biological systems, Darwinism is just around the corner so that, like in the practical work presented in this paper, an emergent observable should be selected in a biological system on account of the adaptive value it provides this system with.

References

1. Andre, D., Forrest H. Bennet and J.R. Koza. 1996. Discovery by genetic programming of a cellular automata rule that is better than any known rule for the majority classification problem. In J.R. Koza, D.E. Goldberg, D.B. Fogel and R.L Riolo (eds) – Genetic Programming 1996: Proceedings of the first conference, pp. 3-11, MA. The MIT Press.
2. Baas, N.A. 1994. Emergence, hierarchies, and hyperstructures. In C.G. Langton (Ed.) Artificial Life III. Sante Fe Studies in the Sciences of Complexity, Proc. Vol. XVII. (pp. 515-537). CA: Addison-Wesley.
3. Bedeau, M.A. 1997. Weak emergence. In J. Tomberlin (Ed.) Philosophical perspectives: Mind, causation and world, Vol. 11 (pp. 375-399), MA: Blackwell.
4. Bersini, H. and G. Seront (1992): "In search of a good optimization-evolution crossover" - In Parallel Problem Solving from Nature, 2 - Männer and Manderick (Eds.) - pp. 479 - 488.
5. Cariani, P. 1997. Emergence of new signal-primitives in neural networks. Intellectica, 2, pp. 95-143.
6. Crutchfield, J.P. 1994. Is Anything Ever New? Considering Emergence. In Integrative Themes. G. Cowan, D. Pines and D. Melzner (eds.) Santa Fe Institute Studies in the Sciences of Complexity XIX. Addison-Wesley, Reading , MA.
7. Crutchfield, J.P. and M. Mitchell. 1995. The evolution of emergent computation. Proceedings of the National Academy of Science, 23(92):103. van Leeuwen, J. (ed.): Computer Science Today. Recent Trends and Developments. Lecture Notes in Computer Science, Vol. 1000. Springer-Verlag, Berlin Heidelberg New York.
8. Emmeche, C., S. Koppe and F. Stjernfelt. 1997. Explaining emergence: Towards an ontology of levels. Journal for General Philosophy of Science, 28:83-119.
9. Holland, J.H. 1992. Adaptation in Natural and Artificial System. 2nd edition. Cambridge, Mass. The MIT Press.
10. Kubik, A. 2003. Toward a formalization of Emergence. In Artificial Life 9: pp. 41-65. MIT Press.
11. Packard, N.H. 1988. Adaptation toward the edge of chaos. In J.A.S. Kelso, A.J. Mandell, M.F. Shlesinger (eds.) Dynamic Patterns in Complex Systems, pp. 293-301. Singapore World Scientific.
12. Poundstone, W. 1985 The Recursive Universe. Chicago: Contemporary Books.
13. Rasmussen, S., Baas, N.A., Mayer, B., Nilson, M., & Olegen, M.W. 2002. Ansatz for dynamical hierarchies. Artificial Life, 7, pp. 367-374.
14. Ronald, E.A., Sipper, M. & Capcarrère, M.S. 1999. Design, observation, surprise! A test of emergence. Artificial Life, 5, pp. 225-239.
15. Sipper, M. 1998. Computing with cellular automata: Three cases for nonuniformity. Physical Review E, 57(3).
16. Wolfram, S. 1994. Cellular Automata and Complexity. Addison-Wesley, Reading, MA.

Shaping collective behavior: an exploratory design approach

Pablo Funes[1], Belinda Orme[1] and Eric Bonabeau[1]

[1]Icosystem Corporation, 10 Fawcett Street, Cambridge, MA 02138, USA
{pablo, belinda, eric}@icosystem.com

Abstract

In order to fulfill the true promise of decentralized, self-organizing intelligence, a major design problem has to be overcome. Designing the individual-level rules of behavior and interaction that will produce a desired collective pattern in a group of human or non-human agents is difficult because the group's aggregate-level behavior may not be easy to predict or infer from the individuals' rules. While the forward mapping from micro-rules to macro-behavior in self-organizing systems can be reconstructed using computational modeling techniques such as agent-based modeling, the inverse problem of finding micro-rules that produce interesting macro-behavior poses significant challenges, all the more as what constitutes "interesting" macro-behavior may not be known ahead of time. An exploratory design method is described in this paper. It relies on interactive evolution. We show how it can be used to discover new, "interesting" patterns of collective behavior when one does not know in advance what the system is capable of doing, a generic situation in the design of collective intelligent systems.

Introduction

Designing the individual-level rules of behavior and interaction that will produce a desired collective pattern in a group of human or non-human agents is difficult because the group's aggregate-level behavior may not be easy to predict or infer from the individuals' rules. For example, the aggregate-level properties of traffic jams (Helbing *et al.*, 2000), a crowd evacuating a public space (Still, 1993) or the stock market (Palmer *et al.*, 1994) cannot easily be derived from knowing the rules of behavior and interaction of drivers, people or investors. Agent-based modeling (ABM) (Reynolds, 1987; Epstein & Axtell, 1996; Axelrod, 1997; Bonabeau, 2002), or micro-simulation, is often the only way to capture the emergent properties resulting from the behavior and interactions of the group's constituent units or agents, particularly but not only when the individual-level rules are discrete. While ABM is useful in producing aggregate-level patterns from individual-level rules, finding the appropriate rules still requires manual search and tinkering when (1) the collective-level patterns may be difficult to formalize into a mathematical detector and therefore the evaluation of a solution cannot be automated, and/or (2) the collective-level patterns made possible by the individual-level rules are not even known ahead of time. If the desired collective-level pattern is known and its detection can be automated, then traditional search and optimization algorithms can be utilized. This paper focuses on cases where (2) is true (but of course, (2) implies (1)). We show that in such cases a technique originally developed to generate "interesting" images and pieces of art (Dawkins, 1987; Sims, 1991, 1992, 1993) can be used to design the individual-level rules of behavior and interaction to produce "interesting" collective-level patterns.

The technique (see Takagi, 2001 for a review) is a directed search evolutionary algorithm which requires human input to evaluate the fitness of a collective-level pattern (here, the fitness might be how close the collective-level pattern is to the desired pattern, or how interesting the pattern is) and uses common evolutionary operators such as mutation and crossover (Goldberg, 1989; Forrest, 1993) to breed the individual-level rules that produced the fittest collective-level patterns. Using a simple example of a human game that can be played in small groups (Bonabeau, 2002; Funes *et al.*, 2003), we show that this approach is particularly powerful as an exploratory design technique, when the aggregate-level capabilities of the system are not known. Interactive evolutionary computation (IEC), as this technique is known, combines computational search with human evaluation (Takagi, 2001).

Some of the individual-level rules discovered using IEC are presented together with their corresponding striking, unexpected patterns. In itself, the discovery of the rules is important as it shows that it is possible to design simple rules to produce robust (genetic robustness being a by-product of the evolutionary method: to be discovered, a fitness peak has to be reasonably stable under a number mutations), collective-level patterns; in addition, the IEC technique used to discover these rules is very generic and makes it possible to systematically discover novel phenomena in self-organizing systems (Bonabeau *et al.*, 2000).

A Simple Game

To illustrate the approach, a game of aggressors and defenders is used with a group of N players. Two rules can be used by the players:

- Rule #1: Pick two people A and B in the group. Then start moving in such a way that B (your defender) is between yourself and A (your aggressor).

- Rule #2: Pick two people *A* and *B* in the group. Then start moving in such a way that you are between *A* (the aggressor) and *B* (the defendee).

If for example everyone in the group follows Rule #1 and picks *A* and *B* randomly, the resulting collective-level pattern is chaotic, sometimes room-filling motion across the room, often constrained by the walls, which leads to some wall following. If on the other hand everyone in the group follows Rule #2 and picks *A* and *B* randomly, the resulting collective-level pattern is the rapid formation of a cluster, that is, the whole group collapses onto a single cluster. These two simple versions of the game can easily be played with a group of 8 and more (our experience includes playing the game with up to 400 people) and produce aggregate-level patterns that cannot easily be predicted from an examination of the individual rules. For example, in both versions there is exactly the same number of aggressors, defenders, defendees. While it is difficult to predict the outcome of even the simplest versions of the game, Anderson (2003) has established elements of a mathematical proof for the two situations described above (all participants following the same rule and picking their *A*'s and *B*'s randomly). A complementary approach consists of simulating the mapping from micro-rules to macro-behavior through agent-based modeling (ABM) (Epstein & Axtell, 1996; Bonabeau, 2002). Figure 1 sketches how Rule #1 and Rule #2 are represented in the model.

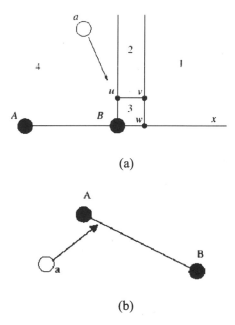

(a)

(b)

Figure 1: Definition of Micro-rules #1 (a) and #2 (b) in the agent-based model. (a). The goal for agent *a* is to have *B* protecting it from *A*. Depending on the relative positions of *A* and *B*, it proceeds as follows: in region 1 *a* moves toward the closest point in line *x*. In region 2, *a* moves toward point *v*. In region 3, *a* moves toward point *w*. In region 4, *a* moves toward point *u*. Points *u*,

v and *w* are defined as follows: let *r* be the radius of one agent, then the distance between *u* and *B*, *u* and *v*, *v* and *w* is equal to 2*r*. The rule is satisfied, and the agent stands still, anywhere over the line *x* and to the right of *w*. Symmetrical regions (not shown) exist below *x*. (b) A similar approach was used to define the defender rule. *a* tries to satisfy the defender rule "place yourself between person *A* and person *B*" by moving toward the closest point to its current position on the line from *A* to *B*.

Based on the angle of motion θ_i calculated as indicated in Figure 1, the position A_i, represented as a complex number, of agent *i* is updated:

$$A_i \leftarrow A_i + \sigma e^{i\theta_i}, \qquad (1)$$

where σ represents speed, unless there is a collision with a wall or another agent, in which case σ is decreased so as to avoid the collision. Discrete time steps are used in the simulation, with $\sigma = \delta$ step^{-1}, where δ is the distance covered in one time step; here $\delta = 2r$. Figure 2 illustrates the patterns observed in the two simple cases described above when simulated using agent-based modeling.

In some situations other than the two simple ones described above, we discovered by playing the game live that the aggregate-level pattern could be quite different from the above-mentioned patterns. For example, if by chance everyone picks the same person as aggressor or defender of defendee, the resulting aggregate-level pattern is very different. In other situations the relationship graph has several disconnected components (although the probability of observing more than one component if participants pick A and B tends uniformly randomly toward 0 as N^3 (Anderson, 2003)), leading to the formation of several clusters.

After these phenomena were observed *in vivo*, the corresponding micro-rules were simulated using ABM and could be re-created *in silico*, proving the predictive power of ABM. The observation of these unexpected collective patterns of behavior prompted us to ask the following question: would it be possible to design social networks (characterized by relationship graphs: who interacts with whom and what is the nature of the interaction?), instead of creating random ones, to produce interesting aggregate-level patterns, with no *a priori* knowledge of what interesting patterns this system could produce? More precisely, every individual is characterized by the following set of rules or properties:

- Does the individual follow Rule #1 or Rule #2?
- Who is *A*?
- Who is *B*?

The size of the space of relationship graphs (which we will also call rule space) grows fast with the number of individuals *N*. While rule space is large, it is likely that most resulting collective patterns will either be random-looking or reducible to one of the two basic patterns described above (chaotic behavior or clustering).

However we do not know what such a system can or cannot do, that is, we do not know what kind of collective-level patterns to expect other than chaotic motion and clustering. In Section 3 we describe a method to search for the relationship graphs, or individual-level rules, that will produce the most interesting collective-level patterns with only a vague notion of what interesting means: neither random nor reducible to one of the two basic patterns.

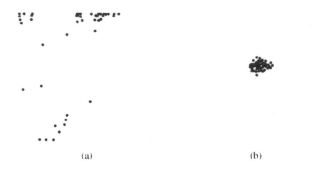

(a)   (b)

Figure 2: Micro-rules #1 (a) and #2 (b) lead to two dramatically different collective behavior in reality and in the predictive agent-based simulation.

Interactive evolution

The IEC search method works as follows:
1. A small initial population of relationship graphs is generated.
2. The resulting collective-level dynamical patterns are generated using an agent-based simulation and presented to a human observer.
3. The observer selects the collective-level patterns that are the most interesting –the fittest individuals in the population according to whatever set of objective and subjective criteria the observer may be using.
4. A new population (new generation) of relationship graphs is generated by applying mutation and crossover operators to the relationship graphs that correspond to the previous generation's fittest collective-level patterns (Goldberg, 1989; Forrest, 1993). In addition to the offspring and mutated versions of those graphs, randomly generated graphs are injected into the population to ensure diversity.
5. The new population is then simulated and the resulting collective-level patterns presented to the observer, and so forth.
6. This procedure is iterated until interesting patterns emerge from the search.

The user interface, which is a critical component of the method as it is based on visualizing the solution that the observer evaluates solutions, is shown in Figure 3. Obviously this method can only work if the population size is kept small and if interesting patterns emerge after a reasonably small number of generations.

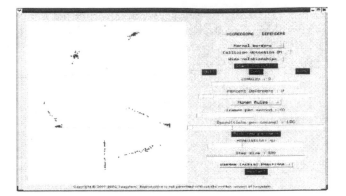

Figure 3: User interface with six playgrounds. Clicking on the playground that displays the preferred behavior results in the remaining ones becoming mutants or recombinations of the selected one.

Details of the evolutionary algorithm
The notation we use to pertain to a given generation is the subscript notation such that game i in generation n and $n+1$ is given by i_n and i_{n+1} respectively.

Selection of the fittest game and elitism
The game designated as the fittest by the user in generation n, j_n, is selected for preservation in generation $n+1$. The remaining games in generation $n+1$, i_{n+1}, will consist of agents derived from either a recombination or a mutation of some the agents in the original game, j_n. This implies mutation and recombination occur with a probability of 0.5. For example if population size is 9 games, one will be identical to the fittest game selected in the previous generation; four will have some agents in their population altered by recombination; and four will have some agents in their population altered by mutation.

Recombination.
If the rules for game i_{n+1} are to be created from a recombination of the rules describing agents in game j_n, then a uniformly random fraction drawn between 0% and 10% of the total number of agents in game i_{n+1} are derived in the following manner: a typical agent selected to be changed is the target agent, p; another source agent, q, is chosen randomly from the population to provide the rules to replace some but not all of those in the target p. The rules which can be substituted from q into p are
1. Person A,
2. Person B,
3. Rule of the game to be played (#1 or #2).

Each agent in the game has its own set of these three features. In recombination, while the source agent's features remain unchanged, some (not all) of the target agent's features can be replaced by those from the source agent. In other words, q's Person A (and/or B) may also become p's Person A (and/or B), and/or q's Rule may also become p's Rule. The number of features of the genotype of p which are substituted for those in q is chosen uniformly randomly (one, two or three features are substituted with probability 1/3 each).

Mutation

If the rules for game i_{n+1} are to be created from a mutation of the rules describing agents in game j_n, then a uniformly random fraction drawn between 0% and 10% of the total number of agents in game i_{n+1} are derived in the following manner. To mutate an agent's genotype, one of three possible mutations can occur with probability 1/3. The mutated feature is either

1. Person *A*,
2. or Person *B*,
3. or Rule of the game to be played (#1 or #2).

If Person *A* or Person *B* is mutated, a random agent different from the agent being mutated is selected to be the new Person *A* or Person *B*. If the Rule is mutated, it becomes either Rule #1 or Rule #2 with probability ½ each.

Results

A number of patterns discovered using the IEC technique described above are presented in this section. Figure 4 shows several such patterns.

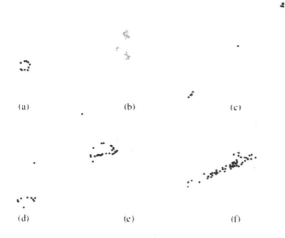

Figure 4: A few examples of evolved behaviors. (a) Circle: agents chase each other around in a circle. (b) Juggle: two blobs fuse and reemerge and sometimes toss a smaller blob at each other. (c) Corner-middle: two groups of agents go to opposite corners while one stays in the middle. (d) Pursuer-evaders: an agent follows a larger group that slows down, is reached by the pursuer, then escapes again. (e) Chinese streamer: a D shape that moves around. (f) Somersault: a thick line that makes a 360 degree turn, then stops, then turns back in the opposite direction.

Figure 5 shows snapshots of the spatio-temporal dynamics of one of the patterns, the Chinese streamer (CS). Figure 6 shows the social network structure (who

interacts with whom) of a typical CS pattern. The network has a highly specific structure that makes the CS pattern possible.

Figure 5: "Chinese streamer" pattern. From a random initial placement, a pattern quickly emerges (a-d) and starts turning, stabilizing in a shape with a handle and trailing ribbon which rotates smoothly. The direction of rotation can be clockwise or counterclockwise (as here), presumably depending on the initial positions.

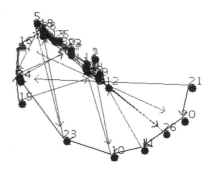

Figure 6: Social network structure of the CS pattern.

The CS pattern is particularly interesting in that it is unexpected (it was impossible to predict that the system could display this type of behavior under the right relationship graph), robust to mutations used to evolve it (a by-product of evolutionary search) and totally insensitive to initial conditions (when the appropriate relationship graph is in place, the Chinese streamer always forms regardless of the initial positions of the participants; initial conditions influence the rotation of the pattern, which can be clockwise or counter-clockwise depending on the details of the participants' locations). It also happens to be robust to noise in the behavior of the participants. The noise-adding procedure that was used is described in Figure 7. We found that below a noise level of $\nu=2$, the Chinese streamer still forms. A noise level of $\nu=2$ is quite large since the random vector (E) added to the initial displacement vector (D) is up to twice as large as the initial vector.

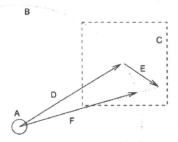

Figure 7: Altering an agent's behavior by adding random noise. Agent A is moving at small, discrete displacements of length δ, which is the radius of circle B. D is the intended movement vector as computed by the rules. E is a random vector uniformly chosen in the square C whose side's length is $v\delta$, where v is called the noise level. The vector obtained from adding E and D is normalized to obtain F, the final displacement vector.

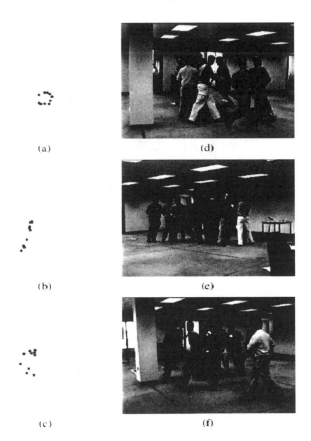

Figure 8: Three rules designed for swarms of 10 agents, evolved using the IEC interface (a-c) and then given to a group of people (d-f). Rule "circle" (a,d) makes all agents run around in a circle; rule "align" (b,e) made them form a straight line and rule "Chinese streamer" resulted in a central cluster with a tail or "ribbon" circling behind.

Lastly, Figure 8 illustrates the ultimate real-world test of the approach: when the rules and interaction graph discovered by applying IEC to the dynamic output of the agent-based simulation are given to humans, the predicted patterns do emerge. Obviously, a reliable agent-based model is necessary to get real-world validation.

5 Discussion

We have illustrated with a simple game example how to tackle the issue of designing decentralized, self-organizing systems with "interesting" properties.

Using this approach requires a shift in mindset from the traditional top-down design approach: here, design is *exploratory* because the aggregate-level capabilities of the system are not known ahead of time. Although it might be possible to create a fitness function *a posteriori* that will make the evolutionary algorithm discover, say, the Chinese streamer automatically (without human intervention), defining such a fitness function requires that the user know the existence of the Chinese streamer as a potential aggregate-level pattern. The exploratory design approach outlined and illustrated in this paper enables the user to navigate the space of possible aggregate-level behaviors without *a priori* knowledge of what is possible or what to look for.

Some of the patterns discovered in the simple game system were extremely surprising and could not be foreseen by any human engineer. Obviously the patterns discovered may sometimes be difficult to understand and need to reverse-engineered. For example the Chinese streamer pattern has been reverse-engineered based on its social network structure (Figure 6) and we now understand its mechanisms.

This approach to designing self-organizing systems has a wide range of applications, from collective robotics to distributed control. One example: radio-frequency tags, known as RFIDs. Although RFIDs have recently become very popular, most users intend to use them with a centralized mindset without knowing what a self-organizing collection of RFIDs might be able to do collectively. Our exploratory design approach enables an open-minded search for the hidden capabilities of such a system.

There are limitations to the approach. First, it requires the complete pre-specification of the micro-rules, that is, one needs to know in advance the complete space of possible microscopic behaviors. Second, it requires a mapping from micro-rules to aggregate-level behavior, such as bottom-up simulation. Third, it requires a way of visualizing the aggregate behavior. Fourth, it can only work with small populations and a small number of generations because of user fatigue. Lastly, it is based on the premise that, although the user may not know ahead of time what aggregate-level behaviors may emerge, she will recognize and interesting pattern when she sees one.

References

Anderson, C. 2003. Linking micro- to macro-level behavior in the aggressor-defender-stalker game. Pages 7-14 in: *Proceedings of the Second International Workshop on the Mathematics and Algorithms of Social Insects* (T. Balch & C. Anderson, eds.), Georgia Tech.

Axelrod, R. 1997. *The Complexity of Cooperation: Agent-Based Models of Competition and Collaboration.* Princeton University Press, Princeton, NJ.

Bonabeau, E. 2002. Agent-based modeling: methods and techniques for simulating human systems. *Proc. Nat. Acad. Sci. USA* **99**: 7280-7287

Bonabeau, E., Guérin, S., Snyers, D., Kuntz, P., Theraulaz, G. & Cogne, F. 2000. Complex three-dimensional architectures grown by simple agents: an exploration with a genetic algorithm. *BioSystems* **56**: 13-32.

Dawkins. R. 1987. *The Blind Watchmaker.* W. W. Norton, New York.

Epstein J. M., Axtell R. L. 1996. *Growing artificial societies: social science from the bottom up.* MIT Press, Cambridge, MA.

Forrest, S. 1993. Genetic algorithms: Principles of adaptation applied to computation. *Science* **261**: 872-878.

Funes, P., Orme, B. & Bonabeau, E. 2003. Evolving emergent group behaviors for simpe humans agents. Pages 76–89 in: *Proceedings of the 7th European Conference on Arti.cial Life (ECAL 2003)* (P. Dittrich, J.T. Kim, eds.), Dortmund, 14–17 September, 2003.

Goldberg, D. E. 1989. *Genetic Algorithms in Search, Optimization and Machine Learning,* Addison-Wesley Longman Publishing.

Helbing, D., Farkas, I. & Vicsek, T. 2000. Simulating dynamical features of escape panic. *Nature* **407**: 487-490.

Palmer, R. G., Arthur, W. B., Holland, J. H., Le Baron, B. & Tayler, P. 1994. Artificial economic life: a simple model of a stockmarket, *Physica D* **75**: 264-274.

Reynolds, C. 1987. Flocks, herds, and schools: a distributed behavioral model. *Computer Graphics* **21**: 25-34.

Sims, K. 1991. Artificial evolution for computer graphics. *Computer Graphics* **25**: 319-328.

Sims, K. 1992. Interactive evolution of dynamical systems. Pages 171-178 in: *Towards a Practice of Autonomous Systems: Proceedings of the First European Conference on Artificial Life* (F. J. Varela & P. Bourgine, eds.), MIT Press, Cambridge, MA.

Sims, K. 1993. Interactive evolution of equations for procedural models. *Vis. Comput.* **9**: 446-476.

Still, K. G. 1993. New computer system can predict human behaviour response to building fires. *Fire* **84**: 40-41.

Takagi, H. 2001. Interactive evolutionary computation: fusion of the capabilities of EC optimization and human evaluation. *Proc. IEEE* **89**: 1275-1296.

Updating Schemes in Random Boolean Networks: Do They Really Matter?

Carlos Gershenson

Centrum Leo Apostel, Vrije Universiteit Brussel. Krijgskundestraat 33 B-1160 Brussel, Belgium

cgershen@vub.ac.be   http://homepages.vub.ac.be/~cgershen

Abstract

In this paper we try to end the debate concerning the suitability of different updating schemes in random Boolean networks (RBNs). We quantify for the first time loose attractors in asyncrhonous RBNs, which allows us to analyze the *complexity reduction* related to different updating schemes. We also report that all updating schemes yield very similar critical stability values, meaning that the "edge of chaos" does not depend much on the updating scheme. After discussion, we conclude that synchonous RBNs are justifiable theoretical models of biological networks.

Introduction

Random Boolean Networks (RBNs) have been useful tools for modelling and studying the functional and computational requirements and possibilities of life (Kauffman, 1993). They are simple and general. Their advantage is that one does not need to assume any previous functionality. Exploring different parameters of the network, such as number of nodes and connections, one can find regions in the parameter space where computation, such as the one required by life, is very probable. This region is characterized by being stable enough to keep information, but flexible enough to transmit, manipulate and transform it. Not too frozen and ordered, but not too variable and chaotic. That is why it has been referred to as the *edge of chaos*.

There has been a debate on wether RBNs should have a synchronous or asynchronous, deterministic or non-deterministic, updating scheme (Harvey and Bossomaier, 1997; DiPaolo, 2001; Gershenson, 2002; Gershenson et al., 2003). We attempt to close this debate in this paper motivated by the results presented within.

In the next section, we present the background of RBNs and different updating schemes. Then, we show results of experiments which include the quantification of loose attractors, and also present results related to the stability of different RBNs. The main discussion follows from these results. Concluding remarks close the paper.

Background

RBNs are a generalization of cellular automata (CA), where the functionality of each node is not restricted to its neighbourhood. The state (zero or one) of the n nodes of a RBN depends on the states of k other nodes connecting to each. Which nodes affect which (the connectivity) is initially generated at random. The way nodes affect each other (the functionality) is determined by logic lookup tables, which are also initially randomly generated. We can bias the connectivity to achieve different topologies, such as scale-free (Aldana, 2003). We can bias the functionality as well, and this will result in different network properties (Derrida and Pomeau, 1986). The RBNs we study have homogeneous topology and no functional bias, as this is the standard in the literature.

We have proposed a classification of RBNs according to their updating scheme (Gershenson, 2002). We have seen that the updating changes considerably the properties of the same networks, such as number of attractors and attractor lengths, and also affects drastically the shapes of the basins of attraction. The change of updating scheme also affects drastically the behaviour of models based on RBNs or CA (Bersini and Detours, 1994).

Classical RBNs (**CRBNs**) (Kauffman, 1969) have synchronous updating: all nodes at time $t + 1$ take into account nodes at time t for their updating. Since the dynamics are deterministic, and the state space is finite, sooner or later a state will be repeated, and the network will have reached an *attractor*. If it consists of only one state, it will be a *point* attractor. Otherwise, it will be a *cycle* attractor. CRBNs have been widely studied (Wuensche, 1997; Aldana-González et al., 2003).

Asynchronous RBNs (**ARBNs**) (Harvey and Bossomaier, 1997) have asynchronous and non-deterministic updating. A node is randomly chosen and the network updated. There are point attractors, but no cycle attractors due to the non-determinism. However, not all states are revisited. We can identify as *loose* attractors the regions of the state space which capture the dynamics of the network. Until now, loose attractors had been ignored in statistical studies, including ours.

Generalized Asynchronous RBNs (**GARBNs**) (Gershenson, 2002) are similar to ARBNs but semi-synchronous: each time step they select randomly which nodes to update synchronously. Therefore, at a time step some of the nodes can be updated, only one, or even all of them.

Deterministic Asynchronous RBNs (**DARBNs**) (Gershenson, 2002) introduce two parameters per node, p and q, which allow asynchrony in a deterministic fashion. A node is updated if the modulus of p over time equals q. The set of all p's and q's can be seen as the *context* of the network(Gershenson et al., 2003). The context is initially randomly generated, where p's can take integer values between one and $maxP$, and q's between zero and $maxP - 1$. DARBNs have point and cycle attractors.

Deterministic Generalized Asynchronous RBNs (**DGARBNs**) (Gershenson, 2002) are the semi-synchronous coutnerpart of DARBNs: they update synchronously all nodes which fulfill the condition $p \bmod t == q$.

Mixed-context RBNs (**MxRBNs**) (Gershenson et al., 2003) are non-deterministic in a particular way: They are DGARBNs with m "*pure*" contexts (sets of p's and q's), and each P time steps, one context is chosen randomly.

We have recently studied the sensitivity to initial conditions of different types of RBNs (Gershenson, 2004), and we found out that the updating scheme almost does not affect the phase transition between "ordered" and "chaotic" regimes of the networks.

Experiments

For our experiments we used a software laboratory we have been developing for this purpose. It is an open source project, available at http://rbn.sourceforge.net.

Loose attractors

An attractor can be seen as a part of the state space which a dynamical system has a high probability of reaching[1]. In deterministic systems this can be precisely defined, since from one state the dynamics will follow to only one another state. However, in non-deterministic systems several states can follow from one state. The state transitions can be analysed carefully for specific systems, but this becomes intractable for large families of networks. Therefore, we can approximate with simulations loose attractors in RBNs as the part of the state space which is reached after some time.

Here we consider for the first time in statistical analysis the existence of loose attractors. We believe that their study is very important, since they reflect the complexity reduction which non-deterministic RBNs can achieve. The algorithm we devised to find them is as follows: We first let run the

---

[1]In a system with random state transitions, a loose "attractor" would be equal to the whole state space. As we will expose, this is not at all the case of non-deterministic RBNs: the set of their "preferred" states is significantly smaller than the state space.

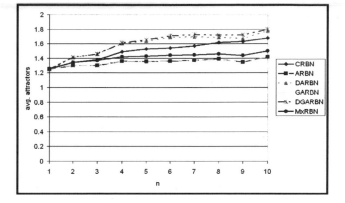

Figure 1: Average number of attractors for k=1

network from an initial state for it to reach an attractor (we used more than 10000 time steps). Then, if the state is already in an attractor we have previously found, we begin again with a different initial state. If not, we create a new attractor composed of that state, and the states which we obtain following the dynamic of the network are added to the attractor (omitting states which were already visited). This search for states in an attractor lasts until a maximum search period is reached (we used 3000 iterations). After this, another initial state is chosen and the algorithm is run again. After all the possible initial states of a network, or a certain number, have been explored, there might be some overlapping attractors, since the state of a loose attractor might be missed at first by the algorithm, but then it would construct a redundant attractor over it. Therefore, a trimming process takes place, in which repeated states are sought in different attractors, and if found, the attractors are merged, removing redundant states.

The algorithm is general enough so that any type of attractor (point, cycle, loose) can be detected with it, though it is redundant and computationally expensive. The trimming of the attractors is necessary only for non-deterministic networks (ARBN, GARBN, and MxRBN). Notice that for these networks the sequence of states in the found attractors does not indicate necessarily the actual possible state sequences. The algorithm cannot handle networks of $n > 20$ in a computer with 1Gb RAM. The best supercomputer now would not achieve $n = 30$. Probably the algorithm could be optimized. Also, an analytical solution would be very helpful to explore loose attractors further, but this is not an easy task.

We can appreciate some of the results of our experiments in Figures 1-6 for $k = 1$ and $k = 3$. The averages are of one thousand networks, exploring at most two thousand initial states. We used $maxP = 7$ for contextual networks (DARBNs, DGARBNs, and MxRBNs). MxRBNs have $m = 2$ pure contexts, randomly chosen at each $P = 100$ time steps. The figures sum point, cycle, and loose attractors.

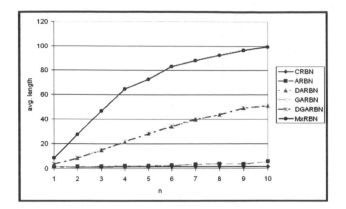

Figure 2: Average length of attractors for k=1

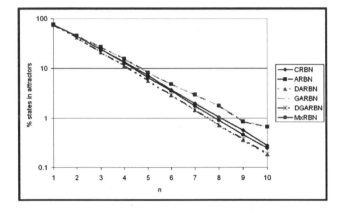

Figure 3: Average percentage of states in attractors for k=1

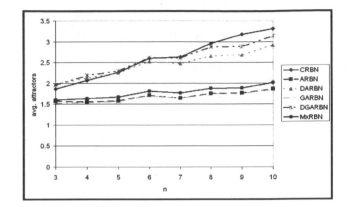

Figure 4: Average number of attractors for k=3

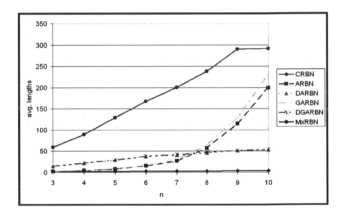

Figure 5: Average length of attractors for k=3

The full data, including standard deviations, and more graphics ($k = 2$) with better resolution are available at http://homepages.vub.ac.be/~cgershen/rbn.

Edge of chaos

We have recently studied the sensitivity to initial conditions of different types of RBNs (Gershenson, 2004). We first created randomly an initial state A, and flip one node to have another initial state B. We run each initial state in the network for ten thousand time steps, obtaining states A' and B'. Then we compare the normalized Hamming distance (1) of the final states with the one of the initial states to obtain a parameter δ (2).

$$H(A,B) = \frac{1}{n} \sum_{i}^{n} |a_i - b_i| \qquad (1)$$

$$\delta = H_{t \to \infty} - H_{t=0} \qquad (2)$$

If δ is negative, it means that the Hamming distance was reduced. Since the initial distance is minimal ($\frac{1}{n}$), a negative δ indicates that both initial states tend to the same attractor. This implies that the network is stable, in an ordered phase.

A positive δ indicates that the dynamics for very similar initial states diverge. This is a common characteristic of chaos in dynamical systems.

Since the initial states are chosen randomly, the comparison we make is equivalent to see B as a perturbed version of A, and observe if the perturbation affects the dynamics.

To compare the regimes of different types of RBN, we created *NN* number of networks (200), and evaluated for each *NS* number of states (200) for all six types of RBN.

We can observe the averages of δ for networks with $n = 5$ in Figure 7. The error bars indicate the standard deviations. We can see that all networks have an average phase transition from ordered to chaotic for values of k between one and three (although the standard deviations indicate us that there can very well be chaotic networks for $k = 2$ and ordered for $k > 2$). They have all a similar "edge of chaos".

Further results and analysis on the phase transitions of RBNs, including statistics for networks of different sizes, can be found in (Gershenson, 2004).

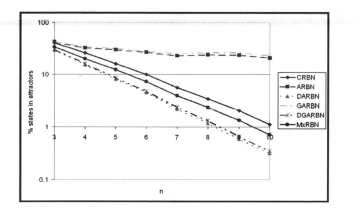

Figure 6: Average percentage of states in attractors for k=3

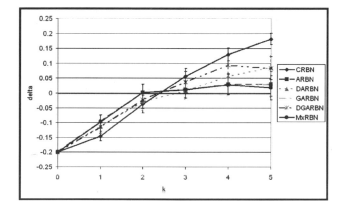

Figure 7: Sensitivities to initial conditions for n=5

Discussion

We should say that all the presented data related to attractors have *very* high standard deviations (this is why these are not shown in the figures). This diversity is due to the fact that some networks have only one attractor, others several. Some attractors are of length one (point), while few ones are very long. This makes some networks to have very few states in their attractors and others not so few. This high variance seems to be common in these type of statistical studies (Mitchell et al., 1993).

Nevertheless, one thing we can deduce from the first experiments, is that all network types can perform *complexity reduction*. This means that compared to all the possible states of a network, only few ones are favoured: those in attractors. This property has also been dubbed as "order for free" (Kauffman, 1993). This is more noticeable as k decreases: fewer connections allow more order, and more order implies more complexity reduction[2]. Even when there is an exponential increase of attractor length (states in an attractor) for ARBNs and GARBNs, there is an even higher

increase in complexity reduction, since the percentage of states in attractors reduces exponentially with n for low values of k (see Figure 3). ARBNs and GARBNs have also less, but larger attractors, i.e. loose attractors include much more states. We can see that the dynamics of non-deterministic RBNs with low k are very different from random transitions in a state space, because the percentage of states in loose attractors is considerably small (independently of standard deviations). We should notice that MxRBNs, even when non-deterministic, reduce much more complexity than ARBNs and GARBNs, since they have a constrained set of possible attractors, determined by their pure contexts. This tells us that there are many ways in which a network can be constrained in order to have complexity reduction. Other ways include topological or functional bias. It is worth mentioning that this analysis could not have been made without the inclusion of loose attractors in the statistics.

Another property independent of the updating scheme we have found is that networks have almost the same phase transition from an ordered to a chaotic phase. This is, they have very similar "edge of chaos". This region is interesting, because it is precisely where computation and life can take place (Langton, 1990; Fernández and Solé, 2003). It is flexible and stable enough in order to allow the storage, transmission, and processing of information. Thus, in principle, any updating scheme can be good for modelling life and computation. Moreover, due to the high variances found, evolution has a myriad of parameters to play with in order to find suitable networks. Not only the size and connectivity are important. Also the functionality, topology, updating scheme, and other criteria can be modified to reach an "edge of chaos". The precise numbers for this do not matter much for theoretical studies, but we know that there is a much higher probability to find it for values $1 < k < 3$ (Gershenson, 2004)[3] when there are no functional nor topological biases. We speak about probabilities, but nature selects the convenient configurations of parameters, making them more probable if they are useful in a particular context. This changes the probability space and the constraints of systems. Therefore, in some contexts we can expect to find networks which reached their "edge of chaos" via functional bias, others through topological bias, others through number of connections. But in general, nature would fiddle with all of these parameters in order to find viable networks.

However, it seems that determinism, or at least quasi-determinism, will be favoured by evolution, since RBNs with these types of updating schemes can perform more complexity reduction, as seen in Figure 6. They can have the similar stability than non-deterministic RBNs at the "edge of chaos", but they will need, in general, less nodes to per-

---

[2]But too much complexity reduction implies no adaptability nor evolvability. That is why we find life at the "edge of chaos".

[3]In analytical studies n has not been taken into account, but we have found with simulations that the size of the networks does matter, since larger nets seem to have indirect "interferences" which decrease the precise critical value of k as n increases.

form the same functions. This can be thought in the following way: a non-deterministic network can perform computations and complexity reduction, but in order to cope with the non-determinism, a lot of redundancy is required. And excessive redundancy costs. Contextual RBNs can perform even more complexity reduction because they "throw" information into their context (but they are more complex). We can assume that there are many constraints in nature which prevent "full" non-determinism (where any node can be updated indiscriminately), so in this case, plain ARBNs and GARBNs would not be good models of natural networks. Still, they can be constrained through artificial evolution to show rhythmic behaviour (DiPaolo, 2001). We can also assume other constraints, such as delays (Klemm and Bornholdt, 2003), or limited non-determinism, such as the one of MxRBNs.

Entering into the main debate concerning the use of synchronous RBNs as realistic models, we have found that the main difference of ARBNs with classical RBNs was due to non-determinism, not to asynchronicity (Gershenson, 2002). And the determinism of natural networks has just been justified: deterministic or quasi-deterministic networks are more efficient, especially in the case of large networks. There is not a big difference between the properties of DARBNs and CRBNs, and moreover, they can be mapped into one another (Gershenson, 2002). Therefore, for *simplicity*, CRBNs seem to be justifiable models of real networks, if we are interested in theoretical studies of the possibilities of RBNs, such as the ones carried out by Kauffman[4]. This is because even when contextual RBNs can perform even more complexity reduction[5] than CRBNs, they are harder to study. CRBNs give similar results than DARBNs and DGARBNs, and we should not be interested in the *precise* numbers we obtain. As we have seen, there is a huge variance in RBNs, and different parameters, such as biases or constraints, can change the precise numbers considerably. Another reason for justifying CRBNs is that if we are not assuming any functionally, how could we assume some updating period for a DARBN? Real networks are not fully synchronous, but they are also not fully boolean, nor with homogeneous connectivity. We believe that the synchronous assumption is justifiable for *theoretical* studies, especially compared to plain ARBNs. But if we are interested in modelling a particular network, then the type of synchronicity should be that which resembles more the one of the particular system modelled.

---

[4]Kauffman used RBNs to explain how there could be so few cell types with so many genes. Indeed, there is a complexity reduction towards few cell types (attractors). But the precise calculation related to the number of genes (roughly known in 1993) and the precise number of expected cell types (attractors) seems more like numerology, since there is a huge variance in RBNs, and real regulatory networks do have topological and functional biases.

[5]This is a bit tricky, because contextual RNBs have much more possible states, since these include the actual phase of the context (Gershenson et al., 2003).

Then we should model a suitable updating scheme, but also a suitable functionality and topology. DGARBNs seem to be a good alternative for this. MxRBNs are also promising, since even when they are non-deterministic, they have enough limits so that they can perform much better complexity reduction than plain ARBNs or GARBNs. Other methods already mentioned are also worth exploring.

The main lesson from the presented data is that there is always a critical region, and that nature *thrives* (selects) towards it, since it is of selective advantage. But in general, it does not matter which updating scheme is being used, since all schemes *have* this region. We can conclude, as it has been stated by others, that life is very probable in our universe (Kauffman, 1993), and almost inevitable in a planet like ours. This contributes to the understanding of the general conditions for life. But this understanding generates further questions. What about environments which do not allow the exploration, selection, or retention of life? They might be too ordered (frozen) or too chaotic (boiling). How abundant are they? Which are the paths from them to a life-supportive environment? Which are the paths *out* of a life-supportive environment? Which are the mechanisms used by an environment to *maintain* or *propagate* its ability to promote life? These are questions which eagerly await exploration.

Conclusions

In the XIXth century, many Latin American countries tried to develop with "order and progress" (It was the lemma of Mexican president Porfirio Díaz, and the Brazilian flag bears the inscription "Ordem E Progresso"). In order to have both, a careful balance is required: too much order does not allow changes, thus progress. Too much progress can destabilize the order. The evolution of life requires the same "order and progress" principle: order to retain acquired characteristics, progress to explore new possibilities. This is the "edge of chaos". In this paper, we have defended that in random Boolean networks, many parameters influence the precise location of this region, but it exists, and evolution can find it.

Even when interactions in real systems may be non-deterministic, the responses can be at a higher level deterministic, or close to deterministic. It is convenient to have determinism, because of computational reasons: information can be manipulated much easier and with less requirements. We can see that for the same networks, deterministic updating offers much more complexity reduction, therefore, this should be favoured by evolution. We can assume that nature can find cyclic or quasi-cyclic behaviours, varying different parameters, because it did, presumably more than once.

The main criticism to CRBNs was that the synchronicity was not justified (Harvey and Bossomaier, 1997). We believe that synchronicity can be justified with our results, since we have seen that synchronous networks are able to compute and to reduce complexity better than asynchronous non-deterministic ones. Therefore, it is expected, and observed, that synchronicity will evolve in living systems. How could this happen, and the precise mechanisms by which asynchronous components can synchronize, are other questions, very interesting ones, and people have been already studying them (Rholfshagen and DiPaolo, 2004; Strogatz, 2003).

Acknowledgments

I thank Jan Broekaert, Hughes Bersini, and two anonymous referees for useful comments. This work was funded in part by the Consejo Nacional de Ciencia y Tecnología (CONACyT) of Mexico.

References

Aldana, M. (2003). Boolean dynamics of networks with scale-free topology. *Physica D*, 185(1).

Aldana-González, M., Coppersmith, S., and Kadanoff, L. P. (2003). Boolean dynamics with random couplings. In Kaplan, E., Marsden, J. E., and Sreenivasan, K. R., editors, *Perspectives and Problems in Nonlinear Science. A Celebratory Volume in Honor of Lawrence Sirovich*. Springer Applied Mathematical Sciences Series.

Bersini, H. and Detours, V. (1994). Asynchrony induces stability in cellular automata based models. In *Proceedings of the IVth Conference on Artificial Life*, pages 382–387. MIT Press.

Derrida, B. and Pomeau, Y. (1986). Random networks of automata: A simple annealed approximation. *Europhys. Lett.*, 1(2):45–49.

DiPaolo, E. A. (2001). Rhythmic and non-rhythmic attractors in asynchronous random boolean networks. *Biosystems*, 59(3):185–195.

Fernández, P. and Solé, R. (2003). The role of computation in complex regulatory networks. Technical Report 03-10-055, Santa Fe Institute.

Gershenson, C. (2002). Classification of random boolean networks. In Standish, R. K., Bedau, M. A., and Abbass, H. A., editors, *Artificial Life VIII: Proceedings of the Eight International Conference on Artificial Life*, pages 1–8. MIT Press.

Gershenson, C. (2004). Phase transitions in random boolean networks with different updating schemes. *submitted*.

Gershenson, C., Broekaert, J., and Aerts, D. (2003). Contextual random boolean networks. In Banzhaf, W., Christaller, T., Dittrich, P., Kim, J. T., and Ziegler, J., editors, *Advances in Artificial Life, 7th European Conference, ECAL 2003 LNAI 2801*, pages 615–624. Springer-Verlag.

Harvey, I. and Bossomaier, T. (1997). Time out of joint: Attractors in asynchronous random boolean networks. In Husbands, P. and Harvey, I., editors, *Proceedings of the Fourth European Conference on Artificial Life (ECAL97)*, pages 67–75. MIT Press.

Kauffman, S. A. (1969). Metabolic stability and epigenesis in randomly constructed genetic nets. *Journal of Theoretical Biology*, 22:437–467.

Kauffman, S. A. (1993). *The Origins of Order*. Oxford University Press.

Klemm, K. and Bornholdt, S. (2003). Robust gene regulation: Deterministic dynamics from asynchronous networks with delay. q-bio/0309013.

Langton, C. (1990). Computation at the edge of chaos: Phase transitions and emergent computation. *Physica D*, 42:12–37.

Mitchell, M., Crutchfield, J. P., and Hraber, P. T. (1993). Dynamics, computation, and the "edge of chaos": A re-examination. Technical Report 93-06-040, Santa Fe Institute.

Rholfshagen, P. and DiPaolo, E. A. (2004). The circular topology of rhythm in random asynchronous boolean networks. *BioSystems*, 73(2):141 152.

Strogatz, S. (2003). *Sync: The Emerging Science of Spontaneous Order*. Hyperion.

Wuensche, A. (1997). *Attractor Basins of Discrete Networks*. PhD thesis, University of Sussex.

Reducing Prejudice: A Spatialized Game-Theoretic Model for the Contact Hypothesis

Patrick Grim[1], Evan Selinger[2], William Braynen[1] , Robert Rosenberger [1], Randy Au [1], Nancy Louie [1] and John Connolly[1]

[1]Group for Logic and Formal Semantics, Dept. of Philosophy, SUNY at Stony Brook, Stony Brook NY 11794-3750
[2]Dept. of Philosophy, Rochester Institute of Technology, Rochester, NY 14623-5604
pgrim@notes.cc.sunysb.edu

Abstract

There are many social psychological theories regarding the nature of prejudice, but only one major theory of prejudice reduction: under the right circumstances, prejudice between groups will be reduced with increased contact. On the one hand, the contact hypothesis has a range of empirical support and has been a major force in social change. On the other hand, there are practical and ethical obstacles to any large-scale controlled test of the hypothesis in which relevant variables can be manipulated. Here we construct a spatialized model that tests the core hypothesis in a large array of game-theoretic agents. Robust results show that prejudicial strategies flourish in a segregated environment but are eliminated in an integrated environment. We take this to offer a new kind of support for the contact hypothesis. The model also suggests a deeper game-theoretic explanation for some of the social phenomena at issue.

Introduction

There are a number of social psychological theories on the nature of prejudice (Adorno 1950, Campbell 1965, Tajfel and Turner 1986), but only one major theory of prejudice reduction: the contact hypothesis. According to the contact hypothesis, prejudice against members of one group by members of another will be reduced with increased social contact between members of the groups (Allport 1954). The hypothesis is simple and accords with common sense; it is understandable that it underlies a number of social policies, its most famous association being the desegregation of U.S. public schools (Patchen 1982, Schofield and Sagar 1977, Stephan 1978). Social psychological support for the contact hypothesis comes from laboratory, field, and survey methods (Cook 1985, Desforges et al. 1991, Robinson 1980, Sigelman and Welsh 1993, Stephan and Rosenfield 1978, Wilner et al. 1955). As with most large-scale social psychological hypotheses, however, there are practical and ethical obstacles to conducting controlled tests in which relevant variables can be manipulated. Those obstacles also impede the search for more fundamental explanation: if increased contact decreases prejudice, precisely how does it do so? As Pettigrew (1998) notes, the contact hypothesis itself does not address process. The attempts that have been made to understand mechanism, moreover, appeal to complex psychological processes of conceptual re-organization and the social dynamics of acquaintance and friendship (Brewer and Miller 1984, Gaertner et al. 1993, Pettigrew 1997).

We have found both a new type of confirmation for the contact hypothesis and hints toward deeper explanation in a game-theoretic simulation (Axelrod 1984, Epstein and Axtell 1996, Gilbert and Conte 1995, Schelling 1996).

A Minimal Model for Social Prejudice

Any model regarding prejudice in general must be capable of representing at least two different groups. In order to study prejudicial behaviors, as opposed to non-prejudicial, there has to be some range of behaviors that in some cases depend upon the group-identification of agent and recipient. Additionally, since prejudice has significant social effects, advantages and disadvantages can be expected to accrue depending on the behaviors that agents take and behaviors that are taken toward them. If prejudice is represented within the parameters of the contact hypothesis, moreover, changes in prejudicial behavior have to be analyzed with reference to circumstances of (a) contact and (b) lack of contact between members of at least two different groups.

These conditions dictate a minimal model using: (i) distinct groups, (ii) behaviors which may or may not be differentiated by actor and recipient groups, (iii) consequent advantages and disadvantages of those behaviors, (iv) some mechanism for updating patterns of behavior, and (v) conditions of greater and lesser contact between members of the groups. We think of the spatialized game-theoretic model used here as perhaps the simplest possible model of this form; little is built in beyond the minimal factors required for any model of prejudice adequate to the parameters of the contact hypothesis. An attempt at understanding ethnocentrism, with some points of contact with this model, appears as Axelrod 1997.

Agents are instantiated as cells in a 2-dimensional cellular automata array (Gilbert and Troitzsch 1999, Gutowitz 1990). Each cell interacts with only its 8 immediate neighbors—those cells touching it on sides and diagonals. Each cell is also of one of two colors—green or red—identifying its group. We can thus construct different conditions of contact by using arrays with different configurations of the two colors. Integrated contact can be modeled by randomizing the array by color, for example.

Segregation can be modeled by dividing the array into distinct color groups. This satisfies minimal conditions (i) and (v) above.

For the interaction between agents we have each cell play 200 rounds in an iterated Prisoner's Dilemma game with each of its 8 neighbors. We use the standard matrix. Each player gains 3 points for joint cooperation and 1 point for joint defection. Should one player defect and the other cooperate, the defector gets 5 points and the cooperator gets 0 (Table 1). The advantages and disadvantages of interaction in condition (iii) are reflected in each cell's total score. Here again we have constructed our model as simply as possible, using the standard *e. coli* game-theoretic model for conflict and cooperation, familiar from over 20 years of simulation research [3, 4].

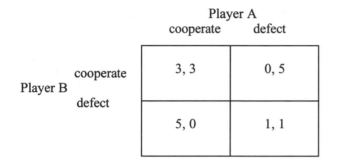

Table 1. Standard Prisoner's Dilemma matrix, left gains to Player B.

We take as a basis just the 8 reactive strategies in an iterated Prisoner's Dilemma: strategies with behaviors on a given round determined only by behavior of the opponent on the previous round. These are shown in Table 2 using 0 for defect, 1 for cooperate, and a coding <i,c,d> to indicate a strategy's initial move *i*, its response *c* to cooperation by the opponent on the previous round , and its response *d* to defection by the opponent on the previous round.

<0,0,0> All-Defect
<0,0,1> Suspicious Perverse
<0,1,0> Suspicious Tit for Tat
<0,1,1> D-then-All-Cooperate
<1,0,0> C-then-All-Defect
<1,0,1> Perverse
<1,1,0> Tit for Tat
<1,1,1> All-Cooperate

Table 2. The 8 reactive strategies in an iterated Prisoner's Dilemma.

Suppose we start with a randomized cellular automata array of these 8 strategies. After 200 rounds of play with each of its neighbors, our cells total their scores. If a cell has a neighbor with a higher score, it adopts the strategy of its highest-scoring neighbor. In the case of a tie between higher-scoring neighbors, the strategy of one is

chosen at random. This gives us a simple mechanism, well explored in the literature (Kennedy et al. 2001, Nowak and May 1993a, Nowak and May 1993b), as our updating schema for condition (iv). If we start with a randomized array of these 8 strategies, it is well known that dominance goes first to All-Defect and C-then-All-Defect, but that Tit for Tat (TFT) then grows in clusters and eventually conquers the entire array: a vindication for the robustness of TFT in a spatialized environment (Grim 1995, Grim 1996, Luna and Stefansson 2000).

Each of these 8 simple strategies is 'color-blind': each reacts to its opponent's previous play, but without regard to color. In order to meet condition (ii) in modeling prejudicial behavior, we add a single strategy PTFT ('Prejudicial Tit for Tat'). PTFT plays TFT with an opponent of its own color, but plays All-Defect against any opponent of the other color (Grim et al. 1998, Poundstone 1992).

In summary, we work with a 64 x 64 toroidal or 'wrap-around' array of 4096 cells, each of which has a background color of red or green. Different arrangements of those colors allow for different test conditions regarding the contact hypothesis. Each cell plays 200 rounds of an iterated Prisoner's Dilemma strategy with its 8 neighbors, following one of 8 'color-blind' strategies or a 'color-sensitive' strategy PTFT. After 200 rounds the gains and losses are totaled for each cell. If a cell has a higher-scoring neighbor, it adopts that strategy that has been most successful in its immediate neighborhood. Strategies are changed, but never colors, and strategy-updating is synchronous across the array. With a new configuration of strategies, we begin a new round of local play.

This is our minimal model for the conditions of the contact hypothesis. There are no complications of genetic algorithms or learning in neural nets, our agents do not construct any internal representations and indeed have no psychological depth at all. In Allport's original presentation, the contact hypothesis is qualified by a set of conditions that have been further elaborated and debated in the literature since: in order to reduce prejudice the contact at issue must be carried out by participants of equal status, who share common goals, participate in inter-group cooperation, and receive the support of authorities (1954). These complications are also largely missing in our model. While equal status for our cells is assured, cells operate in terms of purely individual gains and losses rather than common goals. Although there may be cooperation between individuals, there is nothing to model 'intergroup cooperation.' Since none of our cells represent authoritative figures, our model does not instantiate any kind of authoritative support.

Simulational Confirmation for the Contact Hypothesis

Despite the simplicity of the model, and despite the absence of the additional Allport conditions, our simulation robustly and persistently generates the phenomena predicted by the contact hypothesis. This suggests that the basic principles of contact networks and advantage, modeled in spatialialized game theory, may be sufficient to explain at least some aspects of the dynamics of real prejudice that have been noted in the social psychological literature.

First, consider an array that is carefully segregated in terms of background color. The array, divided in half down the middle, consists of green individuals on the one side and red individuals on the other (Figure 1).

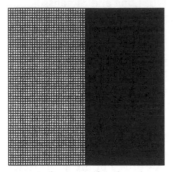

Figure 1. A segregated array of red and green (shown here as shades of gray)

Over this array we layer a randomized distribution of strategies. A red cell might thus instantiate any of our 8 'color-blind' reactive strategies <0,0,0>, <0,0,1>, ..., <1,1,1>, or might instead instantiate the color-sensitive strategy PTFT, representable as <1,1,0>/<0,0,0>. From that initial randomized array of strategies we follow the updating algorithm outlined: after playing 200 rounds of the iterated Prisoner's Dilemma with each of its immediate neighbors, each cell surveys the success of its neighbors. If any has proven more successful, the cell copies the strategy of its most successful neighbor.

With a segregated background, the array converges within approximately 12 generations to a mixture of TFT and PTFT. The 'prejudicial' strategy, in other words, proves successful in occupying roughly 50% of the final array. In different runs, starting from different initial randomizations, either TFT or PTFT may show a slight dominance. The development of a typical array is shown in Figures 2 and 3.

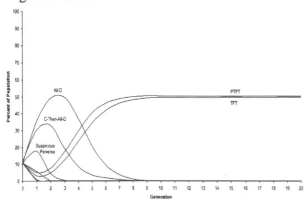

Figure 2. Percentages of the population for 9 strategies in an array segregated by color. 20 generations shown.

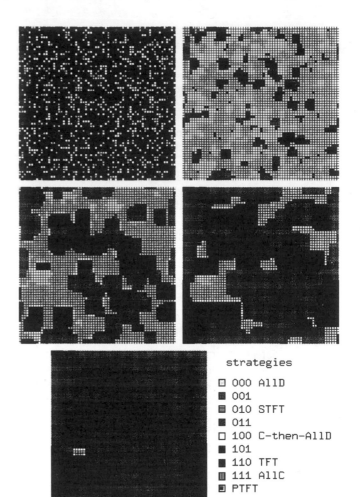

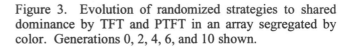

strategies
- ☐ 000 AllD
- ■ 001
- ▦ 010 STFT
- ■ 011
- ☐ 100 C-then-AllD
- ■ 101
- ■ 110 TFT
- ▥ 111 AllC
- ▤ PTFT

Figure 3. Evolution of randomized strategies to shared dominance by TFT and PTFT in an array segregated by color. Generations 0, 2, 4, 6, and 10 shown.

The claim of the contact hypothesis is that increased contact between groups will reduce prejudice. We therefore introduce a second array, with randomized background color (Figure 4).

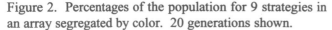

Figure 4. An integrated array of red and green (shown as shades of gray)

We overlay this integrated array with an initial randomization of our 9 strategies, as before, and repeat the simulation. Within 20 generations, the array shows a nearly complete conquest by TFT. Except for lone individuals or very small clusters, PTFT has been eliminated (Figures 5 and 6).

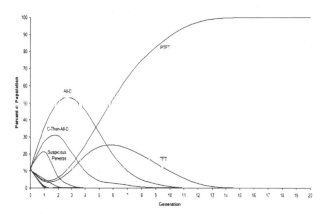

Figure 5. Percentages of the population for 9 strategies in an array randomized by color. 20 generations shown.

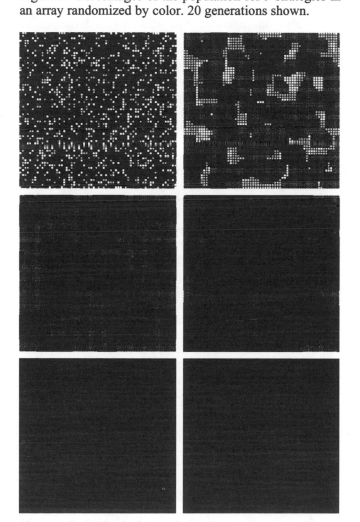

Figure 6. Evolution of randomized strategies to dominance by TFT (shown in black) in an integrated color array. Generations 0, 4, 8, 12, 16, and 20 shown.

Social Identity Theory and a Stronger Result

We take this result to be a strong simulational instantiation of the basic phenomena predicted by the contact hypothesis. The result can be further strengthened by introducing a modeling factor borrowed from another theory of the nature of prejudice.

Social identity theory posits that much of one's identity is informed by the groups to which one belongs, and by the positive or negative perceptions of those groups. People are strongly motivated to develop a positive social identity; positive attitudes towards their own group and prejudice against others is one effect (Tajfel and Turner 1986). In our model PTFT is the only strategy that makes a distinction as to color. In order to model an additional value for 'social identification' we might then add a single point to PTFT cells when they are playing with neighbors that share the same color. A green PTFT playing a green All-C, for example, will be awarded 601 points instead of 600 points for 200 rounds of the iterated prisoner's dilemma.

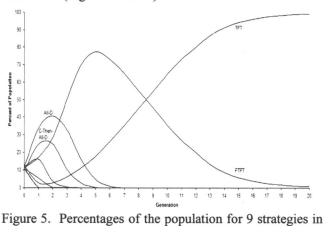

Figure 7. Percentages of the population for 9 strategies in a segregated array, with one extra 'social identification' point for PTFT playing a cell of its own color.

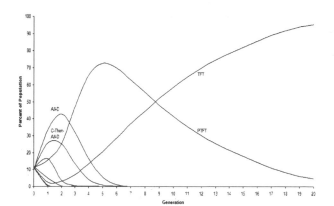

Figure 8. Percentages of the population for 9 strategies in an array randomized by color, with one extra 'social identification' point for PTFT playing a cell of its own color.

With this extra 'social identification' point for PTFT, the result across all runs is that the array shown in Figure 1 now goes entirely to the prejudicial strategy PTFT (Figure 7). The array in which green and red strategies are mixed at random, on the other hand, uniformly goes to an array occupied almost entirely by 'color-blind' TFT (Figure 8).

Eliminating established prejudice

Our models begin with a randomized array of strategies. In an environment mixed as to background color, the model shows evolution to dominance by TFT. In an environment segregated with regard to background color, the model shows evolution to co-dominance between TFT and PTFT in the simpler case, and evolution to full dominance by PTFT with a single additional 'social identification' point.

The contact hypothesis, however, is a hypothesis about prejudice reduction. What our models most directly show, it might be objected, is not reduction of established prejudice but the effect of contact in discouraging the spread of prejudice.

Many criticisms of model-building simply call for better models. In this case we can address the objection directly by starting not with a randomized array of strategies but with scattered territories of TFT and PTFT such as those shown in the final frame of Figure 2. What if we begin with this distribution of just these strategies, but with a mixed color background? Will established prejudice be eliminated?

The answer is 'yes'. Figure 9 shows evolution from such a distribution to clear dominance by TFT in a mixed environment. A similar shift to dominance by TFT can be shown if we start with an array dominated by PTFT except for very small patches of TFT, and give an additional 'social identification' point to PTFT when it plays its own color. Against a mixed color background PTFT is still progressively eliminated.

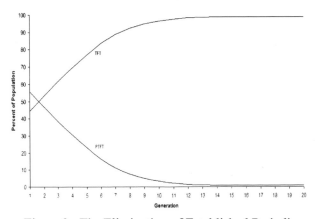

Figure 9. The Elimination of Established Prejudice: Triumph by TFT from scattered territories of PTFT and TFT in an array randomized by color.

Conclusion

Our attempt has been to construct a minimal model of prejudice adequate to the basic parameters of the contact hypothesis. The computationally interesting fact is that phenomena of precisely the sort the contact hypothesis would predict are evident in even this minimal model. In this respect, our results offer a model-confirmation of the contact hypothesis.

What our results further suggest is that patterns of individual advantage in different contact networks—— captured by game-theoretical payoffs in a spatialized cellular automata—may be sufficient to explain why the contact hypothesis holds. Previous attempts to explain reduction of prejudice have appealed to complex psychological mechanisms of conceptual re-organization and the social dynamics of acquaintance and friendship (Brewer and Miller 1984; Gaertner et al.; Pettigrew 1997). Spatialized game-theoretic considerations of advantage and imitation seem to offer a simpler and deeper explanation for at least some of the phenomena at issue.

As aids to intuition and theoretical development, models such as this one may prove useful regarding other sociological and social psychological hypotheses that are difficult to test under strictly controlled conditions: hypotheses regarding deterrence and the death penalty, harm and pornography, or trickle-down economics.

It must be admitted that the methodological use of artificial societies is still at an early stage of development. Like both animal experimentation on the one hand and economic modeling on the other, simulational sociology has major limitations. Questions regarding the realism of a model and thus its generalizability to real societies inevitably remain. In this case, the matter is complicated by the fact that our model is so simple as to abstract away from many of the Allport provisos: the condition requiring equal status is satisfied, for example, but qualifications regarding common goals and intergroup cooperation play no role. Our results thus provide grounds for questioning whether these are in fact necessary for the basic phenomena of the contact hypothesis. Further research may be able to establish whether these conditions might play a more complicated role. Intergroup cooperation may not be strictly required for contact to reduce prejudice, for example, although certain types of intergroup conflict may work against the prejudice-reducing effect of contact.

References

Adorno, T. W., Frenkel-Brunswick, E., Levinson, D. J., and Sanford, R. N. 1950. The authoritarian personality. New York: Harper.

Allport, G. W. 1954. The nature of prejudice. Cambridge, MA: Addison-Wesley.

Axelrod, R. 1984. The evolution of cooperation. New York: Harper.

Axelrod, R. 1997. The dissemination of culture: A model with local convergence and global polarization. Journal of Conflict Resolution 42(2): 203-226.

Brewer, M. B., Miller, N. 1984. Beyond the contact hypothesis: Theoretical perspectives on desegregation. In Miller, N., and Brewer, M. B. eds. The psychology of desegregation. New York: Academic Press, p. 281-302.

Campbell, D. T. 1965. Ethnocentric and other altruistic motives. In Levine, D. ed. Nebraska Symposium on Motivation. Lincoln, NE: Univ. of Nebraska Press, p. 283-311.

Cook, S. W. 1985. Experimenting on social issues: The case of school desegregation. American Psychologist 40: 452-460.

Desforges, D. M., Lord, C. G., Ramsey, S. L., Manson, J. A., Van Leeuwen, M. D., West, S. C., & Lepper, M. P. 1991. Effects of structured cooperative contact on changing negative attitudes toward stigmatized social groups. Journal of Personality and Social Psychology 60: 531-44.

Epstein, J. M., & Axtell, R. 1996. Growing artificial societies—Social science from the bottom up. Cambridge, MA: MIT Press.

Gaertner, S. L., Dovidio, J. F., Anastasio, P. A., Bachevan, B. A., Rust, M. C., 1993. The common group ingroup identity model: Recategorization and the reduction of intergroup bias. In Stroebe, W., & Hewstone, M. eds. European review of psychology 4, Chichester: John Wiley, p. 1-26.

Gilbert, N., and Conte, R. eds. 1995. Artificial societies: The computer simulation of social life. London: UCL Press.

Gilbert, N., and Troitzsch, K. G. 1999. Simulation for the social scientist. Philadelphia: Open Univ. Press.

Grim, P. 1995. Greater generosity in the spatialized Prisoner's Dilemma. Journal of Theoretical Biology 173: 353-359.

Grim, P. 1996. Spatialization and greater generosity in the stochastic Prisoner's Dilemma. BioSystems 37: 3-17.

Grim, P., Mar, G., & St. Denis, P. 1998. The philosophical computer. Cambridge, MA: MIT Press.

Gutowitz, H. ed. 1990. Cellular automata: Theory and experiment. New York: North-Holland.

Kennedy, J., Eberhart, R., and Shi, Y. 2001. Swarm intelligence. Menlo Park, CA: Morgan Kaufmann.

Luna, F., and Stefansson, B. 2000. Economic simulations in Swarm: Agent-based modeling and object oriented programming. Dordrecht: Kluwer.

Nowak, M. and May, R. 1993a. Evolutionary games and spatial chaos. Nature 359, 826-829.

Nowak, M. and May, R. 1993b. The spatial dimensions of evolution. Int. Journal of Bifurcation Chaos 3: 35-78.

Patchen, M. 1982. Black-white contact in schools: Its social and academic effects. West Lafayette, IN: Perdue Univ. Press.

Pettigrew, T. F. 1998. Intergroup contact theory. Annual Review of Psychology 49: 65-85.

Pettigrew, T. F. 1997. The affective component of prejudice: Empirical support for the new view. In Tuch, S. A. and Martin, J. K. eds. Racial attitudes in the 1990s: Continuity and Change. Westport, CT: Praeger, p. 76-90.

Poundstone, W. 1992. Prisoner's dilemma. New York: Doubleday.

Robinson, J. L. Jr. 1980. Physical distance and racial attitudes: a further examination of the contact hypothesis. Phylon 41: 325-332.

Schelling, T. C. 1996. Micromotives and macrobehavior. New York: Norton.

Schofield, J. W., and Sagar, H. A. 1977. Peer interaction patterns in an integrated middle school. Sociometry 40: 130-138.

Sigelman, L., Welsh, S. 1993. The contact hypothesis revisited: Black-white interaction and positive racial attitudes. Social Forces 71: 781-795.

Stephan, W. G. 1978. School desegregation: An evaluation of predictions made in Brown vs. Board of Education. Psychological Bulletin 85: 217-238.

Stephan, W. G. and Rosenfield, D. 1978. Effects of desegregation on racial attitudes. Journal of Personality and Social Psychology 36: 795-804.

Tajfel, H. and Turner, J. 1986. The social identity theory of intergroup behavior. In Wotchel, S., and Austin, W. G. eds. Psychology of intergroup relations. Chicago: Nelson, p. 7-24.

Wilner, D. M., Walkley, R., and Cook, S. W. 1955. Human relations in interracial housing: A study of the contact hypothesis. Minneapolis: Univ. of Minnesota Press.

Quasi-Stable States in the Iterated-Prisoner's Dilemma

Philip T. Mueller

Fittest Bits, Niwot, CO
phil.mueller@fittestbits.com

Abstract

This paper describes the states of a heterogenous population of agents playing the Iterated-Prisoner's Dilemma. The interactions of the agents are governed by five interaction processes which range from highly localized interactions to complete mixing; while the evolution of the agents is governed by five adaptive processes which range from local processes to global processes. For certain combinations of interaction processes, adaptive processes and control parameters, the populations alternate between periods of cooperation and defection while spending relatively little time in between. In addition, even at high rates of mutation, the population does not degenerate into random play.

Introduction

The Prisoner's Dilemma is a two player simultaneous non-zero sum game. Each player can chose to cooperate or defect with the payoffs for the various combinations shown in Table 1.

Payoff table (player1, player2)		Player 2	
		cooperate	defect
Player 1	cooperate	R=3, R=3	S=0, T=5
	defect	T=5, S=0	P=1, P=1

Table 1: Prisoner's Dilemma payoff table. T is the temptation to defect; S is the suckers payoff; R is the reward for mutual cooperation; and P is the punishment for mutual defection.

The dilemma occurs when $T > R > P > S$ and $2R > T + S$. For those constraints, the two players are collectively better off if they both cooperate, but individually will do at least as well as the other player if they defect.

When two players play the Prisoner's Dilemma multiple times, it is called the Iterated-Prisoner's Dilemma (IPD). The IPD has been used extensively over the past 40 years to study cooperation and conflict in economic, social, biological and political settings (Axelrod, 1984; Boyd and Lorberbaum, 1987; Poundstone, 1992; Nowak and Sigmund, 1995; Hoffman and Waring, 1996; O'Riordan, 2001). In addition, the IPD has been used to study the evolution of strategies (Axelrod, 1997). Changing the relative payoffs has also been studied (Angeline, 1994; Yoshida et al., 1998; Delahaye and Mathieu, 1996).

Cyclical and irregularly oscillating populations have been studied (Nowak and Sigmund, 1989; Nowak and Sigmund, 1993). These experiments were based on populations containing only a few unique strategies and at each time step the proportion of each strategy was adjusted based on that strategy's performance. However, in the experiments here, the strategies themselves evolve.

In addition, the experiments in this paper focus on errors in the adaptive process; while others have focused on errors in the play of the IPD (Molander, 1985; Bendor, 1993; Wu and Axelrod, 1995).

The interaction processes of these experiments are derived from the fact that most human interactions are based on location, social labeling or organizational roles that make some interactions more likely than others. In addition, the adaptive processes of these experiments are derived from the fact that humans learn rapidly from comparisons between their own performance and the performance of others. The variation in interaction processes and adaptive processes from local to global allow us to see how these parameters affect the stability of the population. As we shall see, the combination of local interactions with global adaptations leads to unstable behavior.

In this paper, we are interested in the dynamics of these instabilities and not the cases which have a single evolutionarily stable strategy.

Experiments

Each experiment consists of a population of 256 agents implementing a stochastic strategy for playing the IPD, an interaction process determining how opponents are chosen, and an adaptive process determining how the agents evolve over time. There are five interaction processes which move from fixed localized interactions to complete mixing in small steps; and five adaptive processes, one global, two local and two control.

The stochastic strategy for playing the IPD is made up of a 3-tuple of real (64-bit floating point) values (i, p, q) (Nowak and Sigmund, 1992). The value i is the probability of cooperating on the first move; p is the probability of cooperating if the other player cooperated on the previous move; and q is the probability of defecting if the other player defected on the previous move. Thus, TIT-FOR-TAT (TFT) (Axelrod, 1984) would be represented by (1.0, 1.0, 0.0); always defect (ALLD) by (0.0, 0.0, 0.0); and always cooperate (ALLC) by (1.0, 1.0, 1.0). These three values are the "genes" that will be acted on by the adaptive processes described below.

The initial strategies are uniformly distributed across the (p,q) space, with one agent at $i = p = \{0, 1/16, 2/16, ..., 15/16\} \times q = \{0, 1/16, 2/16, ..., 15/16\}$. The agents are randomly placed on a 2 dimensional world grid.

The interaction processes are described in (Cohen et al., 1999) and are summarized here:

- 2DK – 2 Dimensional grid (16 × 16 torus) with agents keeping the same position each time step and playing the same four neighbors (North-South-East-West) each time step.

- FRNE – Fixed random network of symmetrical neighbors. A random network of four paired neighbors is selected at time step 0 and kept for the entire run.

- FRN – Fixed random network of asymmetrical neighbors. Each agent picks four other agents at random to play at time step 0 and plays the same neighbors for the entire run.

- 2DS – 2 Dimensional grid (16 × 16 torus) with the agent's position shuffled on each time step and playing the four neighbors (North-South-East-West) each time step.

- RWR – Each agent picks four other agents at random (with replacement) to play at each time step.

Adaptive processes determine how the agents evolve over time. Two of the adaptive processes (1FGA and BMGAS) are described in (Cohen et al., 1999) and summarized here, while three others (MBMGAS, R1FGA and RBMGAS) are described here:

- 1FGA – For each agent, select another agent at random, if it played better, copy its strategy (with errors). There are two sources of errors, the comparison of the randomly chosen agent with the current agent and possible mutation when copying.

- BMGAS – For each agent, if the best agent it played was better, copy its strategy (with errors). There are two sources of errors, the comparison of the best agent with the current agent and possible mutation when copying.

- MBMGAS – Same as BMGAS except with errors while selecting the best agent played. That is, there are three sources of errors, selecting the best agent played, the comparison of the (so called) best agent with the current agent and possible mutation when copying.

- R1FGA – For each agent, select another agent at random, then select randomly between the two agents. This is a control for 1FGA.

- RBMGAS – For each agent, select randomly amoung the agent and the agents it played. This is a control for BMGAS and MBMGAS.

Each agent plays the IPD at each time step and each game consists of four moves. Which opponents are chosen and the total number of games played by each agent depends on the interaction process for that experiment. However, each agent will play at least four games each time step. The agent's final score is the average of its score across all of the games it played that time step. After all of the agents have played their games, the agent's strategies are synchronously updated based on the adaptive process for that experiment.

The non-control adaptive processes all involve choosing the best agent based on score and then copying that agent's strategy. When choosing the best agent, errors are made based on the selection error rate; when copying the agent's strategy, errors are made based on the mutation rate and the mutation spread. The selection error rate is the probability of choosing the wrong agent at each comparison; the mutation rate is the probability of independently mutating each of the genes; and the mutation spread is the standard deviation of the normal random number (with mean 0) which is added to mutate a gene. If after mutation a gene is greater than 1.0, it is set to 1.0 and if it is less than 0.0, it is set to 0.0.

For the control adaptive processes, instead of selecting the best agent based on score, the "best" agent is selected at random. Copying the selected agent's strategy is still subject to the same errors as in the non-control cases.

A history consists of 3000 time steps; and a complete run of an experiment consists of 1000 histories. The primary figure of merit, the average score for the population at each time step, gives a measure of the amount of cooperation in the population. To avoid biases relating to startup, data is only collected for the last 1500 time steps of each history. To obtain two numbers to describe the results of an experiment, the average and standard deviation of the population average score is computed for the last 1500 time steps for each history. Then the averages and standard deviations are averaged across the histories. However, we shall see that these two numbers do not tell the whole story.

The parameters varied in these experiments are the selection error rate, the mutation rate and the mutation spread.

Interaction Process	Adaptive Process	Mean Score (sd) (Cohen et al., 1999)	Mean Score (sd)	Mean Score Diff. (%)
2DK	1FGA	2.025 (0.069)	2.032 (0.064)	0.3
	BMGAS	2.554 (0.009)	2.555 (0.007)	0.0
	MBMGAS		2.502 (0.008)	
	R1FGA		2.247 (0.020)	
	RBMGAS		2.224 (0.019)	
FRNE	1FGA	2.035 (0.089)	2.032 (0.065)	−0.1
	BMGAS	2.572 (0.007)	2.572 (0.007)	0.0
	MBMGAS		2.526 (0.008)	
	R1FGA		2.247 (0.020)	
	RBMGAS		2.230 (0.020)	
FRN	1FGA	1.884 (0.120)	1.906 (0.093)	1.1
	BMGAS	2.476 (0.026)	2.420 (0.015)	−2.3
	MBMGAS		2.383 (0.016)	
	R1FGA		2.247 (0.020)	
	RBGMAS		2.239 (0.024)	
2DS	1FGA	1.484 (0.086)	1.495 (0.072)	0.7
	BMGAS	1.089 (0.003)	1.089 (0.004)	0.0
	MBMGAS		1.095 (0.004)	
	R1FGA		2.244 (0.020)	
	RBMGAS		2.245 (0.021)	
RWR	1FGA	1.502 (0.109)	1.487 (0.090)	-1.0
	BMGAS	1.098 (0.036)	1.103 (0.020)	0.5
	MBMGAS		1.110 (0.016)	
	R1FGA		2.246 (0.020)	
	RBGMAS		2.244 (0.022)	

Table 2: Comparison of results.

Results

As shown in Table 2, this study verifies the results in (Cohen et al., 1999) for the ten experiments that match. The selection error rate was 0.1, the mutation rate was 0.1 and the mutation spread was 0.4 for all of the experiments listed in the table.

Figure 1 is a histogram of average population score values for a single combination of interaction process and adaptive process (2DK and 1FGA). Each line shows the histogram with a selection error between 0.00 and 0.25 and with a mutation rate of 0.10 and a mutation spread of 0.20. The histogram for 2DK with R1FGA is also shown for reference.

We see that there is a relatively smooth transition from a cooperating population when the selection error is 0.00 to a non-cooperating population when the selection error is 0.25. When the selection error is near the middle of the range (0.10 and 0.15) the population alternates between cooperative and highly non-cooperative states, while spending relatively little time in between those two states.

From an evolutionary point of view, there are two stable states in the fitness landscape at about average score 1.25 and 2.3, and there is sufficient "energy" in the selection error rate, mutation rate and mutation spread to allow the popula-

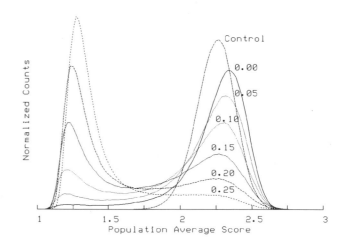

Figure 1: Score histograms for 2DK/1FGA with selection error 0.00-0.25, mutation rate 0.10 and mutation spread 0.20 and the control score histogram 2DK/R1FGA.

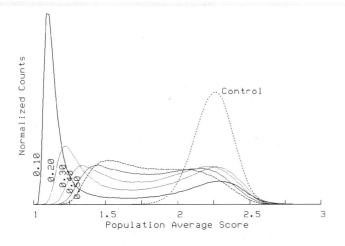

Figure 2: Score histograms for 2DK/1FGA with selection error 0.15, mutation rate 0.10 and mutation spread 0.10-0.50 and the control score histogram 2DK/R1FGA.

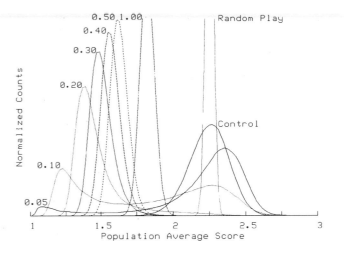

Figure 3: Score histograms for 2DK/1FGA with selection error 0.15, mutation rate 0.05-1.0 and mutation spread 0.20, the control score histogram 2DK/R1FGA and the score histogram for random play.

tion to jump between these two states. In addition, in Figure 2, we see that for larger values of the mutation spread, there is so much "energy" that all states between an average score of 1.5 and 2.5 become equally probable.

Figure 3 is a set of histograms of average population scores for the 2DK/1FGA experiment with varying values of mutation rate. For comparison, the histogram for random play is included.

The adaptive process used in the random play experiment sets each "gene" of each player's strategy to a random number uniformly distributed between 0 and 1, inclusive, at each time step.

For this set of experiments, adding "energy" to the system by increasing the mutation rate changes the fitness landscape from having two quasi-stable states to a single highly stable state. The stability of the state is inferred from the steepness of the distribution.

The mutation algorithm used in these experiments tends to scatter the values of the "genes" in the interval [0,1] with a bias to the boundary values. In this case, ALLD performs better than the other strategies. So, the selection pressure of the adaptive process will tend to favor the ALLD strategy which is what we see from the position of the histogram peaks in the figure.

Figure 4 shows histograms corresponding to Figure 1 for all of the experiments. That is, the average score histograms for all of the combinations of interaction processes and adaptive processes with a selection error rate between 0.00 and 0.25, a mutation rate of 0.10 and a mutation spread of 0.20. Each column is a different interactive process and each row is a different adaptive process.

Quasi-stable states only appear for the adaptive process 1FGA, that is, only for the global adaptive process. The two

states are well defined for the interaction processes that have fixed local interactions, but the cooperate state is practically non-existent for the two interaction processes with a high degree of mixing. None of the BMGAS or MBMGAS cases have quasi-stable states. Thus, it appears that it is the global nature of 1FGA that gives rise to the two quasi-stable states.

Conclusion

One way to look at this data is to consider the surface $S: \mathbf{R}^3 \rightarrow \mathbf{R}^2$, where the domain is the three control parameters: selection error, mutation rate and mutation spread and the range is the population average score and the normalized population count. For these experiments, the range and domain were restricted to $[0, 0.25] \times [0, 1] \times [0, 0.5] \rightarrow [1, 3] \times [0, 1]$. Figures 1–3 are orthogonal slices of this surface.

From the figures, we can see that the quasi-stable states are weakly dependent on the selection error but strongly dependent on the mutation rate and mutation spread. In fact, the quasi-stable states only occur in the region bounded by $[0.05, 0.20] \times [0.05, 0.20] \times [0.1, 0.3]$ which is only about 4% of the entire domain.

These experiments show that a stable cooperative population can suddenly and rapidly become a stable non-cooperative population (and vice versa). In addition, this happens when the interactions between the agents is localized and fixed and when the agents have access to global information about the best strategy, and then use that information erroneously. That description bears an eerie resemblance to modern society, where most people interact mostly with nearby people, yet are informed of the actions of others via mass media.

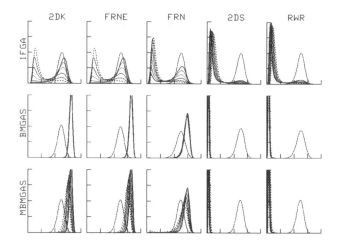

Figure 4: Score histograms for all experiments with selection error 0.00-0.25, mutation rate 0.10 and mutation spread 0.20 and their control score histograms.

Further Research

The experiments in this paper leave many questions unanswered. For example, what causes the population to transition from one quasi-stable state to the other? One approach to answering this question is to look at the distribution of agents in (p,q) space for different population average scores. That is, to look at the three dimensional histogram of population average score, p and q.

Other questions include: How does the population size affect the fitness landscape? Will smaller populations increase the stability of the two states reducing the time spent between them? Will larger populations take longer to transition from one state to the other thus blurring the distinctions between the states?

What about other interaction processes such as tag-mediated selection (Riolo, 1996; Cohen et al., 1999). Under what conditions, if any, does that system exhibit quasi-stable states?

Another area of interest is to investigate the effects of "teaching" the agents by biasing them toward a particular strategy. For example, biasing the agents toward ALLC by adding a small positive random offset to i, p and q after the mutation step of the adaptive process. The question is how do various biases affect the fitness landscape?

Acknowledgements

I would like to thank Chris Platt for the many conversations on this research and Mary Mueller for helping track down some of the references and for her encouragement. I would also like to thank all of the reviewers for their comments.

References

Angeline, P. J. (1994). An alternate interpretation of the iterated prisoner's dilemma and the evolution of non-mutual cooperation. *Proceedings 4th Artificial Life Conference*, pages 353–358.

Axelrod, R. (1984). *The Evolution of Cooperation.* Basic Books, New York.

Axelrod, R. (1997). *The Complexity of Cooperation. Agent-Based Models of Competition and Collaboration.* Princeton University Press, Princeton, New Jersey.

Bendor, J. (1993). Uncertainty and the evolution of cooperation. *Journal of Conflict Resolution*, 37(4):709–734.

Boyd, R. and Lorberbaum, J. P. (1987). No pure strategy is evolutionarily stable in the repeated prisoner's dilemma game. *Nature*, 327:58–59.

Cohen, M. D., Riolo, R. L., and Axelrod, R. (1999). The emergence of social organization in the prisoner's dilemma: How context-preservation and other factors promote cooperation. Sante Fe Institute Working Paper 99-01-002, Sante Fe Institute, Sante Fe, NM.

Delahaye, J.-P. and Mathieu, P. (1996). Random strategies in a two levels iterated prisoner's dilemma : How to avoid conflicts. *Proceedings of the ECAI 96 Workshop: Modelling Conflicts in AI*, pages 68–72.

Hoffman, R. and Waring, N. (1996). The localisation of interaction and learning in the repeated prisoner's dilemma. Sante Fe Institute Working Paper 96-08-064, Sante Fe Institute, Sante Fe, NM.

Molander, P. (1985). The optimal level of generosity in a selfish, uncertain environment. *Journal of Conflict Resolution*, 29(4):611–618.

Nowak, M. and Sigmund, K. (1989). Oscillations in the evolution of reciprocity. *Journal of Theoretical Biology*, (137):21–26.

Nowak, M. and Sigmund, K. (1992). Tit for tat in heterogeneous populations. *Nature*, 355:250–253.

Nowak, M. and Sigmund, K. (1993). Chaos and the evolution of cooperation. *Proceedings of National Academy of Science, USA*, 90:5091–5094.

Nowak, M. and Sigmund, K. (1995). Invasion dynamics of the finitely repeated prisoner's dilemma. *Games and Economic Behavoir*, pages 364–390.

O'Riordan, C. (2001). Iterated prisoner's dilemma: A review. Technical Report NUIG-IT-260601, National University of Ireland, Galway, Ireland.

Poundstone, W. (1992). *Prisoner's Dilemma*. Doubleday, New York.

Riolo, R. L. (1996). The effects of tag-mediated selection of partners in evolving populations playing the iterated prisoner's dilemma. Sante Fe Institute Working Paper 97-02-016, Sante Fe Institute, Sante Fe, NM.

Wu, J. and Axelrod, R. (1995). How to cope with noise in the iterated prisoner's dilemma. *Journal of Conflict Resolution*, 39(1):183–189.

Yoshida, S., Inuzuka, N., Naing, T. T., Seki, H., and Itoh, H. (1998). A game-theoretic solution of conflicts among competitive agents. *Lecture Notes in Computer Science*, 1441:193–205.

Evolving Memory: Logical Tasks for Cellular Automata

Luis Mateus Rocha

Modeling, Algorithms, and Informatics Group
Los Alamos National Laboratory, MS B256
Los Alamos, NM 87545, USA
e-mail: rocha@lanl.gov, URL: http://www.c3.lanl.gov/~rocha

Abstract

We present novel experiments in the evolution of Cellular Automata (CA) to solve nontrivial tasks. Using a genetic algorithm, we evolved CA rules that can solve non-trivial logical tasks related to the density task (or majority classification problem) commonly used in the literature. We present the particle catalogs of the new rules following the computational mechanics framework. We know from Crutchfield et al (2002) that particle computation in CA is a process of information processing and integration. Here, we discuss the type of memory that emerges from the evolving CA experiments for storing and manipulating information. In particular, we contrast this type of evolved memory with the type of memory we are familiar with in Computer Science, and also with the type of biological memory instantiated by DNA. A novel CA rule obtained from our own experiments is used to elucidate the type of memory that one-dimensional CA can attain.

1. Background

An important question for both Cognitive Science and Artificial Life is that of the origin of symbols from the dynamic interaction of many components. By symbols we mean memory structures which can be used to store and manipulate information used to produce and re-produce some behavior. There are currently two main camps in Cognitive Science and Artificial Life with very distinct approaches to the concepts of symbols, representations and even information: the representationalist and dynamicist camps. The first regards information as the most important feature of Life and Cognition, emphasizing genotype/phenotype relations (e.g. Langton, 1989) and internal representations of the environment (e.g. Pinker, 2002), respectively. The second, in its radical form, regards information as an unnecessary concept to explain Life and Cognition. Instead, explanations based solely on dynamical systems theory are preferred (e.g. [Beer, 1995]).

This feud has been discussed in detail in (Rocha and Hordijk, 2004) where we emphasized that both of these camps, while choosing to work either with symbols or equations of dynamics, fail to approach the study of the origin of memory, symbols, representations, information, and the like, from dynamics. As also detailed (Ibid) the biological organization clearly uses the genotype as a type of memory which can be accessed very much like Random Access Memory in a computer. Thus, studying the origin of memory from a dynamic milieu should be a fundamental goal of Artificial Life.

Indeed, we proposed that using the known living organization as a guideline, artificial life can become the ideal laboratory to study the problems of origin of memory (Ibid), as well as the relative advantages of alternative forms of implementing memory in evolving systems (Rocha, 2001). This research program follows directly from previous work on complex systems, where Mitchell (1998) and Rocha (1998, 2000) have proposed a set of experiments with Cellular Automata as paradigmatic examples of the process of emergence of representations from a dynamical substrate.

2. Evolving Cellular Automata

2.1 Nontrivial Tasks

One-dimensional cellular automata (CA) consist of a one-dimensional *lattice* of N identical cells, each a state-determined automaton with k possible states; here $k=2$. Let $s_i(t)$ denote the state of cell i at time t, with $s_i \in \{0,1\}$. Each cell is "connected" to $2r$ other cells which we think of as its neighborhood of *radius r*. Usually, periodic boundary conditions are employed, i.e., cells 1 and N are each other's neighbor. In homogeneous CA, each cell's automaton is defined by the same update rule ϕ which takes as input the cell's neighborhood state, $\mu_i = (s_{i-r}(t),..., s_i(t),..., s_{i+r}(t))$, and outputs the new state of the cell at time $t+1$: $s_i(t+1) = \phi(\mu_i)$.

The initial conditions for a CA are defined by a particular *initial configuration* (IC) of (typically random) cell states. In discrete time steps, all the cells subsequently update their state synchronously according to the update rule ϕ. This update rule can be represented by a *lookup table* with one entry for each of the 2^{2r+1} possible neighborhood configurations μ, and their corresponding output values for $s(t+1)$. Here we use CA rules with $r=3$, thus the lookup table contains 128 entries: there are 2^{128} such rules.

Das et al (1994) such CA rules using genetic algorithms (GA) to solve several non-trivial computational tasks, such as the *density classification* task (a.k.a majority classification problem). Each CA rule is encoded in the GA as a 128 bit string, where each bit encodes the outcome of each entry in the rule's lookup table. The goal of the density task is to find a CA that decides whether or not the IC contains a majority of 1s (i.e., has high density). Let ρ_0 denote the density of 1s in the IC. If $\rho_0 > 1/2$, then within M time steps the CA should reach the fixed-point attractor configuration of all 1s (i.e., all cells in state 1 for all subsequent iterations); otherwise, within M time steps it should reach the fixed-point configuration of all 0s. Since the CA cells have access only to local interactions (with other cells within radius r), this task requires the CA to propagate information across the lattice in order to achieve global coordination. In this sense, this is a nontrivial task.

The unbiased performance $\mathcal{P}_{N,I}(\phi)$ of a CA rule ϕ on a given task is defined as the fraction of I randomly generated ICs for which ϕ reaches the desired behavior within M time steps on a lattice of length N. Here, we employ $N = 149$, $M = 2N$ and $I = 10^5$.

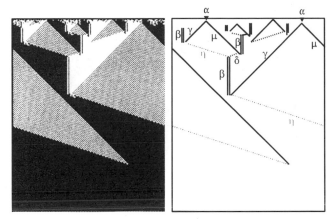

Figure 1: (a) Space-time diagram for ϕ_{DMC} given a random IC with a majority of dark cells. The rule correctly classifies the IC in 141 iterations. (b) Space-time diagram with regular domains filtered out, depicting particles and their interactions after the initial transient is removed.

Figure 1.a shows the space-time diagram of one of the CA rules evolved by Das et al (1994): ϕ_{DMC}. This rule is defined by a 128-bit string as discussed above, in hexadecimal: 0504058705000F77037755837BFFB77F. The lattice is started with a random IC (0 is denoted by white, 1 by dark). Each row in the space-time diagram shows the CA lattice at a particular time step t; time increases down the page.

2.1 Particle Interactions

The large regular, relatively stable regions in the space-time diagram are called *regular domains*. Examples in figure 1.a are the all white, all dark, and checkerboard (alternating white and dark) regions. Crutchfield et al (2002) refer to the boundaries between domains as *particles*. Domains and particles were defined formally in the *computational mechanics* framework (Hanson and Crutchfield, 1992), which provides a way of suppressing (by way of filtering out) the domains in a space-time diagram, making the particles more explicit. An example of this filtering process for ϕ_{DMC} is shown in figure 1.b.

Particles are localized patterns that behave according to certain rules. For example, they have a certain constant velocity at which they move through the lattice. Velocity is defined as the number of cells the particle moves at each iteration of the CA; it is positive if the movement is to the right of the lattice, and negative to the left. Particles also interact with one another according to deterministic rules. These rules and the velocities of particles are referred to as a *particle catalog* for a given CA. Typically, such a catalog is based on a small number of particles, α, β, δ, γ, η, and μ, and a small number or rules such as: $\beta + \gamma \to \eta$, meaning that when particles β and γ collide, the η particle results. The two-particle interaction catalog for ϕ_{DMC} is shown on table I.

Table I: Catalog of regular domains, particles and particle interactions for rule ϕ_{DMC}

Regular Domains	$\Lambda^0 = \{0+\}$, $\Lambda^1 = \{1+\}$, $\Lambda^2 = \{(01)+\}$	
Particles (velocities)	$\alpha \sim \Lambda^0 \Lambda^1 (-)$, $\beta \sim \Lambda^1 \Lambda^0 (0)$, $\gamma \sim \Lambda^0 \Lambda^2 (-1)$, $\mu \sim \Lambda^2 \Lambda^1 (1)$, $\delta \sim \Lambda^2 \Lambda^0 (-3)$, $\eta \sim \Lambda^1 \Lambda^2 (3)$	
Observed Interactions	decay	$\alpha \to \gamma + \mu$
	react	$\beta + \gamma \to \eta$, $\mu + \beta \to \delta$, $\eta + \delta \to \beta$

Particles transfer information about properties of local regions across the lattice to distant sites. Crutchfield, et al (2002) defend that particle collisions are the loci of information processing and result in either the creation of new information in the form of other particles or in annihilation.

3. Emergent Memory in Evolving Automata

Most CA rules evolved tackle the density task by block-expansion, that is, by expanding large neighborhoods of either "1" or "0" states in the initial configuration. But, unlike rule ϕ_{DMC}, they lack the ability to integrate local information to produce an accurate global result. Indeed, the performance of block-expansion rules is quite inferior to ϕ_{DMC}, which grants an obvious evolutionary advantage to latter.

Furthermore, whereas the ϕ_{DMC} rule maintains a similar level of performance for larger lattices, block expansion rules performs very close to random guessing. Thus, the CA rule ϕ_{DMC} is indeed capable of effectively integrating information from local areas of large lattices, whereas block-expansion rules are not (Crutchfield and Mitchell, 1995).

3.1 Memory and Communication

The CA space-time domains, being regions that are "space- and time-translation invariant" (Crutchfield et al ,2002, page 17) can be seen as *memory* structures. Each domain is defined by a cyclic repetition of strings (words) from its regular language (the 0's and 1's of the CA) in space and time. Unless otherwise perturbed, these domains retain their cycles in space and time. For instance, for the CA rule ϕ_{DMC} (see figure 1) we observe the three domains specified in Table I. Λ^0 and Λ^1 refer to the two desired outcomes for the density task, while Λ^2 refers to an intermediate domain used in the process of integrating lattice information and producing the final outcome.

Indeed, the introduction of intermediate domains in CA with intricate particle systems, is their key difference from block-expansion rules, which simply propagate the final outcome domains Λ^0 and Λ^1. Here we define CA with intricate particle systems, as those that employ at least one intermediate domain.

Domains interact by one taking over the other or by establishing an inalterable border. In either case, their interaction defines the particles described in section 2. In the first case, we obtain particles (e.g. μ and γ in figure 1.b) which propagate in the direction of the receding domain, at greater or lesser velocity, while in the second case we obtain a particle (e.g. β in figure 1.b), with zero velocity, which maintains the same lattice position in time, creating a vertical line in the space-time diagram.

The CA with intricate particle systems use the intermediate domains as *memory* stores for intermediate results, and the particles to *communicate* these results across the lattice. Furthermore, the particle interaction rules are used to *integrate* the information stored in the various intervening domains to ultimately produce a final homogeneous lattice state. The inclusion of an additional memory state in ϕ_{DMC}, establishes a more effective means to solve the density task in a distributed manner.

3.2 Building up Memory: Logical Tasks

The role of domains as emergent memory structures used for distributed information processing via the particle interaction scheme can be further appreciated as we notice that memory can be built upon in order to solve more complicated tasks. Rocha (1998,2000), conducted some additional experiments to evolve CA's which solve more than one task. The goal was the evolution of CA rules with radius 3 to solve both the density task and some related, but more complicated, logical tasks (Ibid).

From a machine learning perspective, the idea of evolving CA rules to solve more than one task, especially tasks that at times depend on conflicting demands as discussed below, is rather odd. But the idea of these experiments was to see if one could evolve CA rules that can use the evolved particle system as a more flexible computational system. Such motivation has been previously discussed in detail (Ibid). Here we present a novel analysis of the particle systems evolved for these tasks.

To implement logical tasks the CA lattice is functionally divided in two halves (the center cell is not used): A and B

(figure 2.a). Here we describe results for the logical AND task only, which depends on the density value of the A and B lattice halves. Each half is interpreted as a separate logical variable in traditional logical operations. A variable is "1" if there is a majority of "1" cells in its respective lattice half, and "0" otherwise. Notice that since the boundary conditions of the lattice are periodic, this lattice has two boundaries between the two variables (halves) A and B (figure 2.b). The cells on the neighborhood of these boundaries compute their values from cells in both halves. However, since we are looking for global integration across the lattice, the local errors at the boundaries are not too relevant, especially as lattices grow in size.

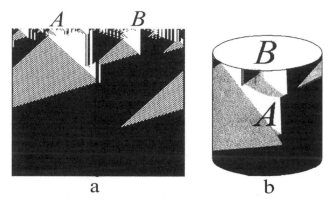

Figure 2: Implementation of logical tasks in one-dimensional CA. (a) The lattice is divided into two halves A and B, each interpreted as a separate logical variable whose value is "1" if it contains a majority of cells in state "1", and "0" otherwise. (b) The space-time lattice is periodic.

The AND task is thus related to the density. It differs from the density task for some of the cases when the two halves of the lattice have opposing densities (see details below). The gist of the logical tasks is that they should ideally perform the density task in each half, and then integrate the results appropriately.

Several rules were evolved with a GA, with elite selection, whose initial population of 100 individuals was composed of 20 individuals encoding some of the best rules evolved so far for the density task, including ϕ_{DMC}, ϕ_{ABK}, as well as a rule evolved by Juillé and Pollack (1998), and others, plus 80 randomly generated individuals. The fitness function used in this GA was calculated by presenting each rule with 100 different IC's, 50 to be analyzed by the density task, and the other 50 by the AND task. The 50 IC's presented to the density task had their density of "1's" uniformly distributed over the unit interval (just as the experiments of Das et al (2004)). The 50 IC's presented to the AND task were similarly biased to a uniform distribution of lattices where for 50% of lattices the density of both halves was "1", and for the other 50% the density of at least one of the halves A or B was "0". More specifically, in 50% of the lattices the density of "1's" in both halves is computed from a uniform distribution on [0.5, 1], and for another 50% of the lattices, the density of "1's" in at least one half (A or B) is computed from a uniform distribution on [0, 0.5]. In this last case, 1/3 of the time both halves have majority density "0", and 2/3 of the time only one of the halves has majority density "0". Notice that if we were to use an unbiased distribution of lattices, only 25% of the time would the case of both halves having density "1" be generated, thus making rules that always tend to "0" too favorable in the evolutionary process.

The AND task can only differ from the density task when the density of both halves is distinct, and the density of "1's" in the majority "1" half is greater than the density of "0's" in the majority "0" half. Without loss of generality, assume that the density of half A is predominantly 0, this means that the density of "0's" in half A, $d_A(0)$ is uniformly distributed in [0.5, 1]. Assume also that the density of half B is predominantly 1, this means that density of "1's" in half B, $d_B(1)$ is also uniformly distributed in [0.5, 1]. In this setting, the density task conflicts with the AND task only when $d_A(0) < d_B(1)$, which happens on average half the time each lattice half has opposed densities.

In our experiments, to compute fitness, 50% of the lattices were presented to the density task, and 50% to the AND task. The density of "1's" of lattices presented to the density task is uniformly distributed in the unit interval. Thus, the value of density used to generate each half of these lattices is equal and also uniformly distributed in the unit interval. Only when this value of density is in the near neighborhood of 0.5, would we find lattice halves with opposing densities – and only in half of those would the tasks conflict. Therefore, for the 50% of lattices presented to the density task, seldom will we find lattices where the density task should lead to a different result than the AND task.

Regarding the 50% of lattices presented to AND task to compute fitness, half are biased to produce both lattice halves with majority density "1". Of the remaining half, only 2/3 produce lattices with halves of opposing densities, that is a total of 1/3 of the lattices presented to the AND task. Since only in half of the latter do the rules conflict, in our experiment, only 1/6 of the lattices presented to the AND task conflict with the density task. In the case of our fitness function, this means on average 8.3 lattices.

Even though the tasks conflict in only a small fraction of lattices, the unbiased performance on the AND task of the best rules previously evolved for the density task was much smaller than that of the new CA rules evolved in our experiments. This means that solving the density task alone was not the best strategy found by the genetic algorithm – neither was solving the AND task alone. Indeed, several CA rules were evolved that can perform very well simultaneously on the density task and on the AND task. Notice that to calculate the unbiased performance, IC's are randomly produced with independent values for each cell with probability 0.5. This means that the density of "1's" in the IC's used to calculate performance of evolved rules tends to

be around 0.5, where we find greater conflict between the two tasks. (performance details in Rocha ,1998, 2000, Rocha and Hordijk, 2004).

3.3 The Evolved CA Rules

The significance of having rules that can perform well on more than one task was discussed in (Rocha ,2000). What we want to highlight here is the manner in which evolved CA particle systems dealt with the different requirements for information integration across the lattice demanded by the AND task. Because the logical tasks divide the lattice into two halves, we expected evolved CA rules to create additional domains and particles which would behave more like static, local memory stores, whose information could be accessed at a latter time as needed.

Table II: Catalog of regular domains, particles and particle interactions for rule ϕ_{AND}

Regular Domains	$\Lambda^0 = \{0+\}$, $\Lambda^1 = \{1+\}$, $\Lambda^2 = \{(01)+\}$, $\Lambda^3 = \{(110)+\} \vee \{(001)+\}$	
Particles (velocities)	$\alpha \sim \Lambda^1 \Lambda^0$ (–) , $\beta \sim \Lambda^0 \Lambda^1$ (0), $\beta` \sim \Lambda^0 \Lambda^3$ (0), $\beta`` \sim \Lambda^3 \Lambda^1$ (0), $\gamma \sim \Lambda^1 \Lambda^2$ (-1), $\delta \sim \Lambda^2 \Lambda^3$ (-3), $\epsilon \sim \Lambda^1 \Lambda^3$ (3), $\eta \sim \Lambda^0 \Lambda^2$ (3), $\mu \sim \Lambda^2 \Lambda^0$ (1), $\nu \sim \Lambda^3 \Lambda^0$ (-3), **Note:** The domain combinations $\Lambda^2 \Lambda^1$, and $\Lambda^3 \Lambda^2$ were not observed as stable boundaries or particles.	
Observed Interactions	decay	$\alpha \rightarrow \gamma + \mu$
	react	$\beta + \gamma \rightarrow \eta$, $\beta`+ \gamma \rightarrow \nu + \eta$, $\mu + \beta \rightarrow \delta + \beta``$, $\mu + \beta` \rightarrow \delta$, $\eta + \delta \rightarrow \beta`$, $\gamma + \delta \rightarrow \epsilon$, $\epsilon + \nu \rightarrow \gamma + \mu$
	annihilate	$\beta` + \nu \rightarrow \Lambda^0$, $\eta + \mu \rightarrow \Lambda^0$, $\epsilon + \beta`` \rightarrow \Lambda^1$

Indeed this is what we observed in the best CA rule for the AND task, ϕ_{AND} (005F1053405F045F005FFD5F005DFF5F) – performance details in (Rocha and Hordijk, 2004). The strategy of this rule builds on rule ϕ_{DMC} by creating an additional intermediate domain, which keeps local lattice information without expanding. The particle catalog of ϕ_{AND} is detailed in Table II. Figures 4 shows a space-time diagram for this rule, with particle interaction schematics.

The most striking feature of the particle catalog of rule ϕ_{AND} is the existence of several particles with zero velocity. These are particles which remain in the same position in the lattice until other particles collide with them. Whereas rule ϕ_{DMC} had only one particle with zero velocity (β), rule ϕ_{AND} produces three such particles (β, $\beta`$, and $\beta``$). We named all these particles β, to highlight the similarity of their behavior with particle β of rule ϕ_{DMC}.

Both particles $\beta`$ and $\beta``$ exist due to the fourth domain Λ^3 introduced by rule ϕ_{AND}. This domain does not expand into final domains Λ^0 and Λ^1, so the respective particles with these domains have zero velocity. It only expands into intermediate domain Λ^2 with particle δ. We note that Λ^3 typically exists as $\{(110)+\}$ but it can also exist as $\{(001)+\}$. We consider these patterns to be the same domain because they behave in exactly the same manner in terms of particle interactions, and are in effect interchangeable.

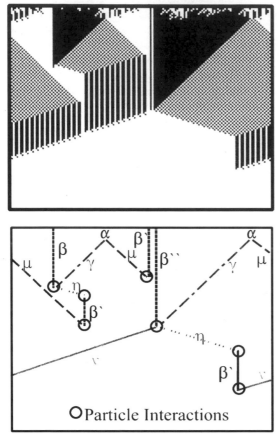

Figure 3: Space-time diagram and respective particle interactions for ϕ_{AND} given a random IC, leading to an all "0" lattice.

Domain Λ^3 functions as a static intermediate memory store. In rule ϕ_{DMC}, without Λ^3, when the particles involving domain Λ^2 collide with others, the result is always one of the final domains Λ^0 or Λ^1, while in rule ϕ_{AND} some collisions result in the additional intermediate domain Λ^3. This way, domain Λ^3, contained by static particles $\beta`$ or $\beta``$, preserves an intermediate result without spreading it into neighbor domains. The intermediate result can later be integrated with particles from other lattice regions: a collision with particle ν results in the all "0" domain Λ^0, and a collision with particle ϵ results in the all "1" domain Λ^1.

The existence of the fourth domain and its static particles is particularly useful for the logical tasks ("AND" in this case). Because two arbitrary halves are defined, the task encourages the evolution of rules that can "hold" intermediate results in one part of the lattice to be integrated with those

from another part. Indeed, the logical task can be better executed when a more *static type of memory* is produced to hold intermediate results, which in this case is implemented by domain Λ^3 and its static particles $\beta`$ and $\beta``$.

4. The Nature of Evolved Memory

We can think of the set of particle interaction rules that emerge in the evolving CA experiments, as a process that maps between the random initial state of the CA lattice (IC), into a final desired state for the task. Crutchfield et al (2002) regard this process as a computation that produces a final outcome from the IC input. As we detail below, we do not see this process as a computation, but the individual particles are certainly the elements in the space-time behavior of the CA which communicate information across the lattice: the loci of information processing (Ibid). Therefore, the collection of particle interactions in space-time, is a dynamic process of integration of the information carried by each of the individual particles into a final domain.

The type of memory the CA domains implement is quite different from the concept of random-access memory (RAM) we are familiar with in universal computers, and also quite distinct from the inert type of memory that DNA grants living organisms (Rocha and Hordijk, 2004).

RAM is a type of memory that can be accessed at any time, and whose value is independent of dynamics. This is the same as saying that the value of the memory is the same independently of the rate of access to it. When a computer stores the value of a variable in a memory store (e.g. the tape in a universal Turing machine), that value remains unchanged when accessed, and the speed of the computer to access the memory and perform computations also does not change it. Similarly, a computation is a process of integrating memory in store, with algebraic and logical operations. But speed of the computer does not change the value of the computation: 2+2=4 in any computer.

Clearly this does not happen with the domains and particles of the evolved CA. Particles have a velocity, and the resulting domains of all particle interactions in space-time depend on when the particles meet each other. If the particles start from different locations in the lattice, even preserving lattice density, they may collide differently and produce a different outcome for the tasks we studied here. It is as if 2+2=4, only when 2 and 2 meet at the right time. This is why we do not see the process of particle interactions as a full-blown computation, but rather a process of information integration based on dynamic memory, rather than RAM.

It was because of this issue that we created the logical tasks. In this case, the evolved CA came as close as possible to creating static (RAM-like) memory stores. Indeed, the fourth domain Λ^3 created by rule ϕ_{AND}, is a domain that preserves its memory without spreading it into the final outcome domains Λ^0 and Λ^1. In a sense, it keeps its memory until it is accessed. The several particles β created by this rule have zero velocity, therefore they preserve the same information until a particle of non-zero velocity collides with them. In this sense, the domain functions more like a traditional memory store.

However, the information stored is still not separated from the dynamics. The domains are not rate-independent like RAM nor inert in the sense that DNA is (Rocha and Hordijk, 2004). It is by virtue of their dynamics, the way their particles collide, that information is expressed. Conversely, in DNA or RAM, information is read out by "third-party" machinery, without destroying or reacting with the memory. So while the β particles of the evolved CA were able to create static memory stores, these are still reactive with and destroyed by the embedding dynamics.

This point is obvious when we notice that while processes such as the transcription of mRNA from DNA and RNA Editing work on genetic memory without access to its content (the encoded proteins), our evolved CA cannot manipulate their particles without access to their content. Particle reactions are simply domain interactions. In this sense, information carriers and content are inseparable. This way, we can say that domains and particles do not function as inert memory stores to be manipulated without access to content.

Does this mean that we cannot witness the emergence of a type of memory more like RAM (and genotypes) computationally? Our stumbling block was in obtaining a means to manipulate memory without recourse to its content. This has been a recurrent stumbling block in Artificial Life. For instance, Langton (1986) proposed a self-reproduction scheme in CA in which the separation between genotype (memory) and phenotype (content) was blurred. This lack of separation was actually seen as a worthwhile model for studying Artificial Life, with a generalized concept of genotype/phenotype mappings (Langton, 1989). But as it was clear for theoretical biologists looking at Artificial Life, a strict separation between genotype and phenotype is the key feature of life-as-we-know-it and a necessary condition for open-ended evolution (Pattee, 1995) (Rocha, 2001). Thus, the study of the emergence of a strict separation between genotype and phenotype, between memory and content, from a purely dynamic milieu should still be the number one goal of Artificial Life.

We submit that the dynamics produced by one-dimensional CA may be too simple to achieve what we desire to model. Indeed, homogeneous CA as a model of material dynamics, our artificial chemistry, is rather poor. In Biology, the genotype/phenotype mapping is based on the existence of two basic, distinct types of material (chemical) structures: DNA/RNA and aminoacid chains. Both are quite different: DNA is remarkably unreactive, or biochemically inert, whereas aminoacid chains are incredibly rich biochemical machines. In contrast, our one-dimensional homogenous CA compute the same exact update rule in each cell.

It seems reasonable that in order to evolve a system in which more reactive structures use non-reactive structures as information stores, we need to work with more heterogeneous dynamical systems where different populations of artificial "chemistry" structures interact.

References

Crutchfield,J.P., Mitchell,M., (1995). "The evolution of emergent computation". *PNAS*. **92**, 10742-10746.

Crutchfield,J.P., Mitchell,M., Das,R., (2002). "The Evolutionary Design of Collective Computation in Cellular Automata". In: *Evolutionary Dynamics: Exploring the Interplay of Selection, Neutrality, Accident, and Function*. Crutchfield,J.P., Schuster,P.K. (Eds.). Oxford University Press, pp. 361-412.

Das,R., Mitchell,M., Crutchfield,J.P., (1994). "A genetic algorithm discovers particle-based computation in cellular automata". In: *Parallel Problem Solving from Nature - PPSN III*. Davidor,Y., Schwefel,H.-P., Manner,R. (Eds.), Springer-Verlag, pp. 344-353.

Hanson,J.E., Crutchfield,J.P., (1992). "The attractor-basin portrait of a cellular automaton". *Journal of Statistical Physics*. **66** (5/6), 1415-1462.

Juillé,H., Pollack,J.B., (1998). "Coevolving the "ideal" trainer: application to the discovery of cellular automata rules.". In: *Genetic Programming Conference (GP-98)*, . Koza,J.R., et al (Eds.), Morgan Kaufmann Publishers.

Langton,C.G., (1986). "Studying artificial life with cellular automata". *Physica D*. **22** (1-3), 120-149.

Langton,C.G., (1989). "Artificial Life". In: *Artificial Life*. Langton,C. (Ed.). Addison-Wesley, pp. 1-47.

Mitchell,M., (1998). "A complex-systems perspective on the "computation vs. dynamics" debate in cognitive science". In: *Proc. 20th Conf. of the Cog. Sci. Society*. Gernsbacher,M.A., Derry,S.J. (Eds.), pp. 710-715.

Pattee,H.H., (1995). "Artificial Life needs a real Epistemology". *Lecture Notes in Artificial Intelligence*. **929**, pp. 23-38.

Pinker,S., (2002). *The Blank Slate: The Modern Denial of Human Nature*. Penguin.

Rocha,L.M., (1998). "Syntactic autonomy". In: *Joint Conference on the Science and Technology of Intelligent Systems ISIC/CIRA/ISAS* IEEE Press, pp. 706-711.

Rocha,L.M., (2000). "Syntactic autonomy : Why there is no autonomy without symbols and how self-organizing systems might evolve them". *Annals of the New York Academy of Sciences*. **901**, 207-223.

Rocha,L.M., (2001). "Evolution with material symbol systems". *Biosystems*. **60** (1-3), 95-121.

Rocha,L.M., Hordijk,W., (2004). "Material Representations: From the Genetic Code to the Evolution of Cellular Automata". *Artificial Life*. In Press.

Van Gelder,T., Port,R., (1995). "It's about time: an overview of the dynamical approach to cognition". In: *Mind as Motion: Explorations in the Dynamics of Cognition*. Port,R., Van Gelder,T. (Eds.). MIT Press, pp. 1-43.

Complex Genetic Evolution of Self-Replicating Loops

Chris Salzberg[1,2], Antony Antony[3] and Hiroki Sayama[1]

[1]Department of Human Communication, University of Electro-Communications, Tokyo 182-8585, Japan
[2]Graduate School of Arts and Sciences, University of Tokyo, Tokyo 153-8904, Japan
[3]Section Computational Science, Universiteit van Amsterdam, 1098 SJ Amsterdam, the Netherlands
chris@cx.hc.uec.ac.jp   antony@phenome.org   sayama@hc.uec.ac.jp

Abstract

It is generally believed that self-replication models constructed on cellular automata have quite limited evolutionary dynamics in both diversity and adaptive behavior. Contrary to this view, we show that complex genetic diversification and adaptation processes may occur in self-replicating loop populations. Applying newly developed tools for detailed genetic identification and genealogy tracing to evoloop populations, we uncovered a genotypic permutation space that expands combinatorially with replicator size. Within this space populations demonstrate broad behavioral diversity and non-trivial genetic adaptation, maximizing colony density while enhancing sustainability against other species. We also found a set of non-mutable subsequences enabling genetic operations that alter fitness differentials and promote long-term evolutionary exploration. These results reveal the amazing potential of cellular automata to re-create complex genetic evolution of self-replicators in a simple, deterministic framework.

Introduction

Since von Neumann's seminal work on self-reproducing automata (von Neumann 1966), models of artificial self-replicators based on cellular automata (CA) have formed one of the mainstreams in Artificial Life (Langton 1984; Reggia et al. 1993; Sipper 1998). Recent developments indicate that simple CA systems can reproduce natural selection processes occuring on different self-replicating structures (Sayama 1999). Their evolutionary dynamics, however, are generally believed to be quite limited in both diversity and adaptive behavior (Sayama 1999; McMullin 2000; Suzuki et al. 2003). Previous results point to a seemingly well-defined fitness landscape in which optimization converges to a single global maximum: homogeneous populations dominated by a single species of the smallest size and shortest replication time.

Contrary to these earlier observations, here we show that complex genetic diversification and adaptation processes may occur in such simple CA. We investigate a system of evolving self-replicating loops (evoloops) (Sayama 1999) in which replication, variation and natural selection emerge solely from local rules. Applying newly developed tools capable of sophisticated genetic identification and genealogy tracing to evoloop populations (Salzberg 2003; Salzberg, in press), we uncovered a genotypic permutation space that expands combinatorially with replicator size. Within this space populations demonstrate broad behavioral diversity and non-trivial genetic adaptation, maximizing colony density while enhancing sustainability in the presence of other competing species. Such adaptation was observed even within species of the same size, thought to be of equal fitness in previous treatment. Intriguing genetic features were also found that may parallel issues in molecular genetics, including the discovery of non-mutable subsequences enabling genetic operations that alter relative fitness differentials. Simulations with such "genetically modified organisms" demonstrate continuously changing, long-lasting evolutionary behavior. These results reveal the amazing potential of CA to re-create complex genetic evolution of self-replicators in a simple, deterministic framework.

Model

The evoloop (Sayama 1999) we investigate is a deterministic nine-state 2D CA model with von Neumann neighborhoods, designed after Langton's self-replicating loop (Langton 1984). An evoloop individual contains an identifiable modular structure describing the shape of offspring (genotype) and an external structure of its own body (phenotype). The former is a sequence of moving signal states (genes) and the latter is a looped sheath of square or rectangular shape, with an arm thrust outward [Fig. 1(a)]. A viable gene sequence contains several '7' states for straight growth of the arm and a pair of consecutive '4' states to control left turning of the arm. In a process of self-replication, cyclic propagation of signal states coordinates the external arm to create a new structural entity. The growing arm is guided through three successive turns and eventually meets its own root, causing tip and root to bond together to form a new, separate loop [Fig. 1(b)]. The truncated arm then retracts, completing the self-replication process.

Loops are destroyed by the appearance and propagation of the dissolver state '8' through contiguous loop structures. Triggered by local configurations non-integral to the normal self-replication cycle, this process of structural dissolution typically arises from shortage of space due to overcrowding and exhibits highly complex dynamics. Its spread is af-

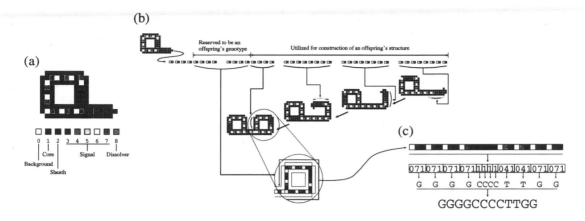

Figure 1: (a) An evoloop individual. (b) Self-replication of an evoloop. Gene sequence is utilized five times during replication, first to construct the umbilical cord (omitted in the figure) and four times to construct the offspring loop. Following loop closure, the truncated arm retracts towards the parent loop and both loops commence the next replication cycle. (c) Labeling scheme of gene sequence of an evoloop. Starting from the bonding location, the mapping transforms '071' triplets to G's, '041' triplets to T's, and '0' states to C's.

fected by minor variations in gene sequence permutation and spacing, often producing leftover sheath cells that form a static, reactive environment. Collisions of loop sheath structures during replication often lead to a change in the gene sequence of offspring. Resting solely on the local interactions of states on a CA grid, such "mutation" events result in an emergent process of variation and natural selection that collectively shapes the path of evolution.

This distinctly bottom-up feature of the evoloop system distinguishes it from other well-studied artificial evolutionary systems of computer programs (Yedid and Bell 2002; Lenski et al. 2003); in these systems a central system manager tracks living organisms and applies probabilistic mutations to their genomes so that explicit control is partially possible. In contrast, the evoloop system supplies no universal structural cues; everything down to the separation between replicator and its environment has to be specified by the system observer. The challenge of analyzing such a potentially "messy" system — coupled with the belief that its evolution always converges homogeneous populations of smallest-sized, fastest-replicating species — has left the detail of its dynamics practically untouched to date.

Methods

We attempt a complete analysis by structurally identifying every birth and death event at the highest level of detail (Salzberg 2003; Salzberg, in press). Unique local configurations are used as markers for detecting such events: the appearance of an umbilical cord dissolver (state '6') for birth detection and the disappearance of an inner sheath (state '2') for death detection. The detection mechanism was embedded in simulator software as an event-driven function requiring almost no additional computational overhead.

At birth, the detection mechanism extracts information about evolutionary identity of the newborn loop, i.e. a genotype corresponding to the configuration of genes in its gene sequence traced counter-clockwise starting at the location of

the umbilical cord dissolver [Fig. 1(b)] and a phenotype describing the size (length and width) of its sheath structure. A pair of genotype and phenotype describes a species. To write a gene sequence, we represent a triplet '071' that describes a gene for straight growth by G, a triplet '041' that describes a gene for left turning by T, and a single core state '1' that fills in the sheath by C. For example, the gene sequence of the newborn in Fig. 1(b) is written as GGGGCCCCTTGG [Fig. 1(c)].

Each different species observed during a run is assigned a unique integer label. As a run progresses, a database is compiled containing the mapping between species labels and their evolutionary identities (gene sequence and loop size). Each newborn loop appearing in the CA space is first checked with all the species registered in the database; if its identity is not matched, it is assigned a new label and added to the database. Then we record each such birth event with the labels of both parent and offspring with a time stamp indicating the moment of loop closure. From this level of detail, entire genealogical histories may be reconstructed and every evolutionary transition precisely pinpointed. This new analysis scheme has enabled us to discover a richness of evolutionary phenomena in the evoloop system that were largely overlooked in earlier studies.

Results

Genetic and behavioral diversity

In the birth event records compiled during our simulation runs, self-replicating species have the same labels for both parent and offspring and are hence easily identified. We collected gene sequences of all self-replicating species and discovered, to our surprise, a far larger and more diverse set than expected beforehand. From this set we extracted the following constraints imposed to the sequences for successful replication: (1) the sequence must include the same number of G's as the size of its phenotype, (2) the sequence must

Size	Number of species	Size	Number of species
4	15	12	646,646
5	56	13	2,496,144
6	210	14	9,657,700
7	792	15	37,442,160
8	3,003	16	145,422,675
9	11,440	17	565,722,720
10	43,758	18	2,203,961,430
11	167,960	19	8,597,496,600

Table 1: Number of different self-replicating loop species for a given loop size. We estimate this number by calculating the number of possible gene arrangements in a fixed-length sequence within the constraints for self-replication described in text. For a loop of size n, this estimate amounts to $_{2n-2}C_{n-2}$ different species (Salzberg 2003). Loops of size 3 or less cannot self-replicate as there is insufficient space in their genome to fit the required G and T genes.

include a pair of T's, (3) the two T's must have no intervening G between them, and (4) the trailing T must be immediately followed by G. Within these constraints, permutation of G's, T's and C's amounts to a set of viable genotypes whose number grows combinatorially with loop size. We analytically derived an estimate of this number to be $_{2n-2}C_{n-2}$ where n is the loop size (Salzberg 2003), listed in Table 1 for sizes from 4 to 19. By size 18, this figure already amounts to over two billion different viable self-replicators. Such huge genetic diversity has been totally ignored in the earlier classification based on loop size only.

Each genotype in this large possibility space may have quite different behavioral patterns. We carried out exhaustive simulation runs up to size-9 loops to make sure that all the permutated genotypes counted in the above estimate are actually self-replicating. Figures 2 and 3 show the results for size-4 and size-6 loops, demonstrating a striking behavioral diversity within the same-sized loops that were considered as a single species in earlier treatment. Note that these patterns — though seen at larger scales than individual loop bodies — are solely dictated by their gene sequences through their non-trivial interactions via transition rules, and thus should be considered as an "extended phenotype" (Dawkins 1990) of each species that can also be subject to natural selection.

Genetic adaptation

The rapid convergence toward smallest self-replicators, quite commonly found in artificial evolutionary models including evoloops (Sayama 1999; Yedid and Bell 2002), tells that the replication time is clearly one of the key quantities being optimized through evolution. However, the huge genotypic permutation space presented in the previous section implies that there may be more room for fine tuning in genetic adaptation, even among the same-sized loops that basically share the same replication time. Whether such microevolution occurs in the evoloop system has remained unresolved to date. To answer this question, we focus on two characteristic quantities for each species and evaluate how they evolve in actual simulation runs.

The first quantity we choose is the sustainability of each species in the presence of other competing species. We characterize this by a relative population ratio of that species after a given period of time in competition with another species, each of which starts from one ancestor. If the given time period is not too long, this ratio captures a shapshot of the population composition under gradual dominance by one over the other, which quantitatively indicates the competitive strength and evolutionary stability of the species against the competitor. Computing an average of such ratios with all the possible competitors would give a mean *survival rate* of that species in the melee of various other species in the "wild". To actually compute this rate, however, one has to restrict the competitor candidates in a practical number. We thus limit ourselves to size-4 species only, assuming that their possible competitors are also of size 4 due to the natural selection favoring shortest replication time. We carried out a round robin among all the fifteen size-4 species and used the results to obtain the mean survival rate for each species, which is shown in Fig. 4. It is clearly seen that there are significant differences of sustainability within the same-sized species, even of the smallest size-4 ones. We note that two species (1 and 15) show particularly low sustainability due to their evolutionary instability; they quickly evolved into other species in most cases.

The second quantity being measured is the colony density of each species. We characterize this by a quadratic coefficient of a parabola[1] fitted by the least-squares method to the population growth curve of that species in an infinite domain. Specifically, we fit a parabola $p(t) = at^2 + bt + 1$ to the population curve and used a as a characteristic quantity of colony growth, which we call *colony density index*. This quantity can be easily measured and defined to each species for its own. It depends, however, on the choice of time range of data point sampling for fitting from the population growth curve. We have tested 0–1500, 0–2000, 0–3000 and 0–5000 updates for the sampling time range. The results with 0–2000 are shown in Fig. 5, reflecting a diversity of growth patterns illustrated in Fig. 2.

These two quantities are found to positively correlate with each other (Fig. 6). Their correlation coefficient varies with different time ranges used for the measurement of colony density index (0.420 with time range 0–1500, 0.674 with 0–2000, 0.423 with 0–3000, and 0.274 with 0–5000) and is highest when the range 0–2000 is chosen. This implies that the sustainability of a population is determined by natural selection acting at a time scale around 2000 updates in the evoloop system. This can be understood in that time scales shorter than this would produce no significant difference in colony structure and time scales longer than this would not be relevant for selection since such a large colony would rarely appear in actual evolutionary processes.

Interestingly, the above two quantities both increase during evolution of loops *in vivo*. Figure 7 shows an exam-

---

[1]Note that a population of evoloops grow parabolically, not exponentially, due to the geometric constraint of the 2D space.

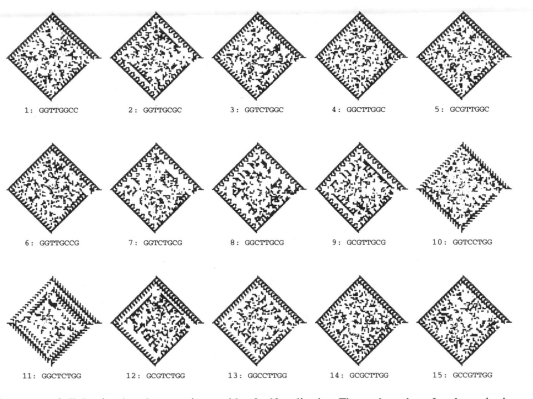

1: GGTTGGCC 2: GGTTGCGC 3: GGTCTGGC 4: GGCTTGGC 5: GCGTTGGC

6: GGTTGCCG 7: GGTCTGCG 8: GGCTTGCG 9: GCGTTGCG 10: GGTCCTGG

11: GGCTCTGG 12: GCGTCTGG 13: GGCCTTGG 14: GCGCTTGG 15: GCCGTTGG

Figure 2: Growth patterns of all the size-4 evoloop species capable of self-replication. The total number of such species is $_{2\times4-2}C_{4-2} = 15$. Each snapshot is taken after 5000 updates starting from one ancestral loop. An integer label is attached to each species, which will be used in the following figures.

Figure 3: Growth patterns of all the size-6 evoloop species capable of self-replication. The total number of such species is $_{2\times6-2}C_{6-2} = 210$. Each snapshot is taken after 5000 updates starting from one ancestral loop. Empty areas indicate unsuccessful species that can self-replicate just once in their lifetime so that there is always only one individual alive in the space. More results for different sized loops can be found at http://complex.hc.uec.ac.jp/loops/.

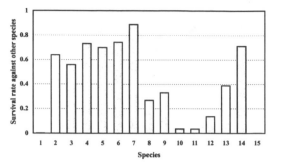

Figure 4: Mean survival rate of size-4 species obtained from the results of the round robin among all the fifteen size-4 species. Relative population ratios are measured after 100000 updates in each competition. The space used is of 1000×1000 grid with opponents placed at opposite ends of the periodic space. We also ran another set of experiments on a 500×500 grid, confirming these results.

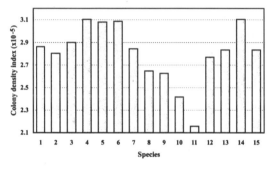

Figure 5: Colony density index of size-4 species. Sample data points for parabola fitting are taken between 0 and 2000 updates at intervals of 10 from the population growth curve of each species.

ple of such processes starting from a size-8 ancestral loop. The evolutionary transition of dominant species in this run is mapped onto Fig. 6; the population moves diagonally in the plot to optimize both quantities. This result gives a clear-cut answer to the question we posed above: there *is* microevolution taking place in the evoloop system, even among the same-sized loops with the same replication time. Natural selection not only favors short replication time but also increases colony density of loops and enhances sustainability against other species through non-trivial genetic adaptation.

Non-mutable subsequences

Moreover, from extensive simulation results, we recently discovered empirically that any subsequence of the form G{C}T{C}TG, where {C} represents any number of C's, will always survive mutations leading to other self-replicating species. Such non-mutable subsequences are a non-trivial outcome of dynamic properties of the evoloop's CA rules and have yet to be rigorously explained. Their existence implies that the genetic state space is partitioned into distinct groups of self-replicating species, each possessing the same conserved subsequence, between which no connecting evolutionary path exists. Each group enforces a minimum loop size for which exact self-replication is possible;

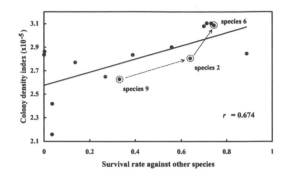

Figure 6: Correlation between survival rate and colony density index plotted in Figs. 4 and 5, respectively. The dashed arrows represent the actual evolutionary transition of dominant species seen in the experiment in Fig. 7.

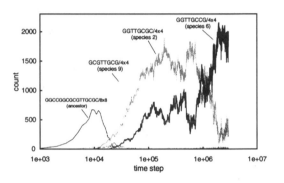

Figure 7: Example of evolution of loops *in vivo* starting from a size-8 ancestor with gene sequence GGCCGGCGCGTTGCGC. Time axis is log-scaled since the selection becomes slow as time proceeds. The transition of dominant species is mapped onto Fig. 6. The space used is of 1000×1000 grid. The same experiments were also performed on size 800×800 and 1200×1200 grids to verify the robustness of the observed dynamics to boundary conditions.

shorter gene sequences cannot contain both the conserved subsequence and the sufficient number of G genes required for exact self-replication.

We use this property to configure "genetically modified organisms", species which cannot evolve below a given minimum threshold size. Experimenting with this threshold enabled us to reduce the size-based fitness differential normally leading to strong competitive exclusion. Figure 8 shows evolutionary dynamics starting from a size-15 "GMO" evoloop injected with the subsequence GCCCCCTCCCCCCCCTG, enforcing a minimum size of 15 on all viable descendants. Although size-based fitness alone favors this minimal-sized species, the fitness differential in this case is relatively weak and gives way to other, emergent behavioral characteristics. As a result, the system fluctuates between dominant species, demonstrating continuous, long-lasting evolutionary behavior, covering over six million iterations and ending with incidental extinction. The progression of major species appearing in this exploration process is shown in Fig. 9. Interestingly, this progression seems to show the presence of some general pattern in the

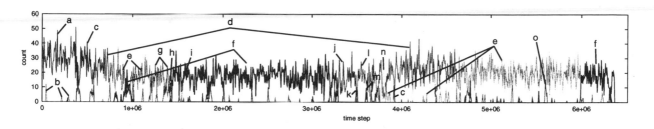

Figure 8: Evolutionary dynamics of "GMO" evoloops with subsequence GCCCCCTCCCCCCCCTG injected to set minimal viable loop size to 15. The space used is of 401x401 grid. Of the total 7106 species observed during this run (including 58 different self-replicating species), only exact self-replicators with populations exceeding 10 individuals are plotted. The same experiments were also performed on size 399x399 and 400x400 grids to verify the robustness of observed dynamics to boundary conditions.

genetic modification process. In this and other experiments, C states are preferentially inserted between G genes alongside the conserved subsequence, producing a general evolutionary tendency towards larger species. This trend is at least weakly reversible, as evidenced by the emergence of certain species with added C states in the middle of their sequence (species n and o) and by the re-appearance of certain smaller species (e.g. species d and e).

Conclusion

The complexity and diversity of CA dynamics has been well known to many for long. Still, it is quite surprising, especially to researchers well-acquainted with the capabilities and practical limitations of CA, that a system so simple can produce such genetic and behavioral diversity of self-replicators and their complex genetic evolution as an emergent property solely arising out of local transition rules. Our findings manifest the importance of developing sophisticated observation and interpretation techniques to capture the full richness of evolutionary phenomena emerging at multiple scales within the system, which has long been underestimated compared to model construction in self-replication studies.

Acknowledgments

C.S. acknowledges financial support by grants from the International Information Science Foundation, the Netherlands Organization for International Cooperation in Higher Education (Nuffic), and the VSB Funds.

References

Dawkins, R. 1990. *The Extended Phenotype.* Oxford, UK: Oxford University Press.

Langton, C. G. 1984. Self-reproduction in cellular automata. *Physica D* 10:135–144.

Lenski, R. E., Ofria, C., Pennock, R. T. and Adami, C. 2003. The evolutionary origin of complex features. *Nature* 423, 139–144.

McMullin, B. 2000. John von Neumann and the evolutionary growth of complexity: Looking backward, looking forward... *Artificial Life* 6:347–361.

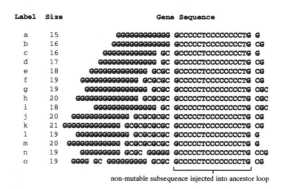

Label	Size	Gene Sequence

Figure 9: List of gene sequences of self-replicating species that appeared in Fig. 8, shown in the order of their first appearance in large numbers. Sequences a – m all have C genes added between G genes along either side of the subsequence. Species n and o, which appear only briefly in large numbers, have C genes injected at positions further away from the subsequence.

Reggia, J. A., Armentrout, S. L., Chou, H. H. and Peng, Y. 1993. Simple systems that exhibit self-directed replication. *Science* 259:1282–1287.

Salzberg, C. 2003. *Emergent Evolutionary Dynamics of Self-Reproducing Cellular Automata.* M.Sc. Thesis. Section Computational Science, Universiteit van Amsterdam.

Salzberg, C., Antony, A. and Sayama, H. Visualizing evolutionary dynamics of self-replicators: A graph-based approach. *Artificial Life*, in press.

Sayama, H. 1999. A new structurally dissolvable self-reproducing loop evolving in a simple cellular automata space. *Artificial Life* 5:343–365.

Sipper, M. 1998. Fifty Years of Research on Self-Replication: An Overview. *Artificial Life* 4:237–257.

Suzuki, H., Ono, N. and Yuta, K. 2003. Several necessary conditions for the evolution of complex forms of life in an artificial environment. *Artificial Life* 9:153–174.

Yedid, G. and Bell, G. 2002. Macroevolution simulated with autonomously replicating computer programs. *Nature* 420:810–812.

von Neumann, J. 1966. *Theory of Self-Reproducing Automata.* Urbana, IL: University of Illinois Press.

Redrawing the Boundary between Organism and Environment

Tim Taylor

Institute of Perception, Action and Behaviour, University of Edinburgh
JCMB, The King's Buildings, Mayfield Road, Edinburgh EH9 3JZ, U.K.
tim.taylor@ed.ac.uk

Supplementary material available at http://homepages.inf.ed.ac.uk/timt/papers/rboe/

Abstract

In this position paper, I argue that a fruitful, and as yet largely unexplored, avenue for artificial life research lies in modelling organisms (specifically, phenotypes) and environment as a single dynamical system. From this perspective, the origin and evolution of life is the progressive control of the dynamical system at a local level by constraints which are represented on an organism's genome. Such an approach shifts the focus of artificial life models away from the design of individuals, towards the *interaction* of an individual with its dynamic environment. It also blurs the boundary between organism and environment; the most important modelling distinction is no longer between an organism's body and its external environment, but rather between the genome (which is treated as an essentially symbolic structure) and phenotype-plus-environment combined. An evolutionary cellular automata system, called EvoCA, is introduced as a tool to explore these ideas. To demonstrate how this approach differs from traditional studies, two example applications of EvoCA are presented. One concerns sensor and effector evolution; the other concerns the origin of life, and in particular the evolution of genome-regulated self-stabilising dynamics. Advantages of the new approach are summarised, and some potential criticisms are considered. The paper concludes with a discussion of some implications of this shift in perspective.

Introduction

One of the key challenges facing artificial life researchers, as well as biologists, is to explain the origin of living organisms from a non-living environment (Bedau et al., 2000; Maynard Smith, 1986). Furthermore, in order to build artificial evolutionary systems, we would like to know how to produce highly evolvable systems, in which agents can control and exploit their environment in unlimited and increasingly complex ways.

Most ALife work on the evolution of life has employed a strong representational distinction between living organisms and their environment. Examples include Tierra (Ray, 1991) and PolyWorld (Yaeger, 1994). In Tierra, for instance, individuals are computer programs with associated instruction pointers, registers, stacks, etc. Interactions between an individual and its environment can only be achieved in a limited number of predefined ways, such as by the allocation of memory in order to reproduce (an interaction with the abiotic environment), or by reading machine instructions from a neighbouring program (an interaction with the biotic environment). In these systems the environment is often modelled as a rather inert medium, the only significant role of which is just to provide a "place" in which organisms exist. For further discussion of this topic, see (Taylor, 2001).

Even in work where no such distinction exists between organisms and environment, individuals, and the dynamical laws of the environment, are carefully crafted to achieve a particular type of behaviour. Examples of this type include von Neumann's self-reproducing automata (von Neumann, 1966), simulations of autopoietic systems (Varela et al., 1974; McMullin and Varela, 1997), and Holland's α-Universes (Holland, 1976).

Neither of these approaches—using a strong representational distinction between living and non-living entities, or carefully crafting the "laws of physics" of the world for a particular purpose—can provide much insight of how life first originated from a non-living environment which, presumably, was not specifically designed to support it.

Much of this work is characterised by an emphasis on the computational capacities of the organisms. The Tierran language, for example, is computationally universal, but this does not mean that Tierran programs can interact with their environment in unlimited ways. Accompanying this perspective has been an (over-)emphasis on the process of self-reproduction, often to the exclusion of other important issues, such as the properties of the environment, and the representational relationship between organisms and environment. If nothing else, the poor evolvability of these systems demonstrates that the processes of self-reproduction with heritable mutation and selection, by themselves, are insufficient to explain the evolutionary origin of complexity.

Howard Pattee, a physicist by training, has devoted much of his career to the question of the origin of life (Rocha, 2001). His particular perspective is the issue of how semiotics (i.e. symbol systems, such as genomes, and their associated semantics in the context of an organism) can originate from a purely physical environment.

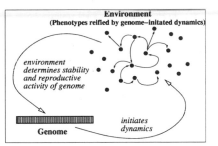

Figure 1: Semantic Closure: Closing the loop between genotype, phenotype and environment

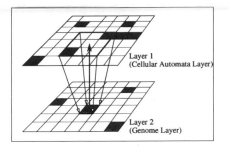

Figure 2: Schematic illustration of EvoCA's design. This specific example shows EvoCA-B. See text for details.

Pattee argues that the distinction between the material and symbolic aspects of living organisms, seen as an example of the more general epistemological distinction between laws and initial conditions, is a defining feature of life, and also a necessary condition for open-ended evolution (Pattee, 1995a; Pattee, 1995b). He explains the relationship between the two as follows:

> Writing symbols is a time-dependent dynamic activity that leaves time-independent structure or record. ... Symbols are read when these structures re-enter the dynamics of laws as constraints. Any highly evolved formal symbol system may be viewed as a particularly versatile collection of initial conditions or constraints, often stored in a memory, producing significant or functional behavior that is usefully described by locally selected rules rather than physical laws. ... [A]ll symbol systems must have material embodiments that obey physical laws. But for the reasons just stated, the lawful description of symbols, even though correct in all details, can reveal no significance. (Pattee, 1995b)

The symbols recorded on the genome ultimately acquire semantics in an organism in the context of the survival value of the dynamics that they initiate (i.e. natural selection of phenotypes). It is this autonomous structure-function self-referent organisation that is entailed in Pattee's term "semantic closure" (Figure 1).

This perspective, then, sees organisms as entities whose phenotypes are embedded within an environment viewed as a dynamical system, and whose genotypes interact with the environment by specifying constraints[1] upon its dynamics, thereby generating the phenotypes. That is, the abiotic environment has its own dynamics and self-organisational properties; genotypes act to "sculpt" these pre-existing dynamics by supplying constraints. From this point of view, the most important distinction is not between organisms and their abiotic environment, but rather between the environment as a

whole (including organism phenotypes) and organism genotypes. It is the relatively time-independent genotypes, by supplying local constraints to the dynamics of the environment, that reify phenotypes as distinct entities within the environment.

The EvoCA System

A simulation platform, called EvoCA, was designed to explore this dynamical systems view of organisms and environments.[2] EvoCA is built upon a cellular automaton (CA) system. Cellular automata were chosen because they are fairly simple, discrete time and space dynamical systems, whose behaviour has been extensively studied.

The system consists of two layers (see Figure 2). Layer 1 is the environment, modelled as a standard CA. The system can be adapted to use any kind of CA; the examples described here use a 2D environment, with two possible states per cell, and the standard Game of Life update rules. Layer 2 is a discrete grid of the same dimensions as Layer 1. Each grid position in Layer 2 can contain zero or one genomes, which will be described later. The action of a genome is centred upon the corresponding grid position in Layer 1. Layer 2 has no dynamics as such (i.e. genomes are relatively inert structures).

An evolutionary algorithm is used to evolve populations of genomes, with selection based upon the dynamics that a genome produces.[3] Two distinct versions of EvoCA exist, which employ different kinds of evolutionary algorithm: EvoCA-A (abiotic selection) uses a standard generational genetic algorithm, while EvoCA-B (biotic selection) uses a natural selection algorithm. The details of genome design and action are also different in each system. The two systems are described in the following sections, and typical results from each are presented in order to demonstrate particular aspects of their behaviour. As these results are included

---

[1]Throughout this paper the general term 'constraint' is used to cover initial conditions, constraints and boundary conditions. For further discussion of these concepts, see (Pattee, 1995a).

[2]The source code is included in the supplementary material.

[3]Various authors have experimented with evolving the transition function of a CA to achieve a particular task, e.g. (Crutchfield and Mitchell, 1995). This is fundamentally different to the current approach as it entails evolving the "laws of physics" of the environment rather than constraints to control a given set of laws.

solely for the purposes of illustrating certain features and consequences of the general modelling approach being advocated, full details of the experiments are omitted. These details are included in the supplementary materials.

EvoCA-A

Genomes   A genome in EvoCA-A comprises a variable length list of genes. Two types of gene are available: timed and conditional. Both types specify a particular target cell in Layer 1 (whose position is defined relative to that of the genome) and a target state for that cell. A maximum radius is defined for each dimension of the CA to confine the position of the target cell relative to the genome.

Each gene additionally specifies a precondition that must be satisfied in order for it to activated. Timed and conditional genes have different types of preconditions. Timed genes specify a time (i.e. a specific iteration of the CA) at which they act. At the specified iteration, the gene sets the state of the target cell to the target state. Conditional genes specify a watch cell and watch state. The watch cell specification is confined to the set of cells that are direct neighbours of the target cell. Whenever the specified watch cell is in the specified watch state, the conditional gene is triggered, setting the state of the target cell to the target state.

Every gene in the genome is checked at each iteration of the CA to see whether it should be activated for that iteration. Whenever any gene is activated, its action overrides the normal CA transition function for the target cell for that particular iteration.

The Evolutionary Algorithm   A standard generational genetic algorithm is used to evolve a population of individuals. Each individual is evaluated in isolation, and placed in the same grid position in Layer 2. The iteration count of the CA is reset to zero at the start of each evaluation. All cells are initially set to the quiescent state, except those which have non-quiescent states specified by timed genes acting at time zero, or those that are influenced by external signals (described later). The CA is then allowed to run for a given number of iterations, with the genes of the genome setting specific cell states when they become active.

The fitness function depends upon the particular task in hand; an example is given in the next section. In addition to one-point crossover and gene mutation, a number of other genetic operators are also available: gene insertion (a random gene is inserted into an existing genome); gene deletion (an existing gene is deleted from a genome); gene reversal (the order of a sequence of genes between two selected points in the genome is reversed); and gene duplication (a sequence of genes between two selected points in the genome is duplicated at the end of the genome). A limit on the maximum allowable genome length is defined.

Example Application: Sensor and Effector Evolution
The following example demonstrates the application of the

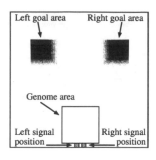

Figure 3: Goal pattern for EvoCA-A experiments

approach to the topic of sensor and effector evolution. The system was set up as shown in Figure 3, using a 2D toroidal CA of size 75x75 cells. To evaluate a genome, it was placed in the centre of the genome area shown. The maximum radius of gene action is shown by the boundary of this area (i.e. genes could directly set the state of only those cells within the genome area). In addition, two goal areas were defined, along with two signal positions, as shown. Note that the signal positions adjoin, but do not overlap, the genome area.

The task that the organisms were set is as follows. Each genome was tested under three conditions: left signal, right signal, and no signal. For the left signal condition, the cells in the 2x2 left signal position (Figure 3) were initially set to state 1 (the non-quiescent state). In this condition, the task of the organism was to produce activity in the left goal area. Any activity in this goal area over the course of the evaluation was rewarded, with the darker shaded cells in Figure 3 being rewarded more than the lighter cells in the goal area. However, any activity in the right goal area was penalised.

For the right signal condition, the opposite signal positions and goal areas were used in the fitness calculation. For the no signal condition, any activity in either goal area was penalised. The scores from the three conditions were combined to produce a final fitness value for the organism. Full details are given in the supplementary materials.

Results   In all experiments the system was able to evolve organisms that could perform the task well (i.e. they initiated activity in the appropriate goal area, and no activity in the other goal). In some runs, the best evolved individual would produce a "glider" (a moving pattern of activation) that travelled from the genome area to the appropriate goal area. In other runs, more extensive and complex patterns of activation were observed; an example is shown in Figure 4. The implications of these results are discussed later.

EvoCA-B

EvoCA-B has modifications to allow natural (biotic) selection of organisms, rather than the artificial selection method (the genetic algorithm) used in EvoCA-A. In EvoCA-B, the whole population of genomes exists in Layer 2 concurrently, and the survival and reproduction of each genome is deter-

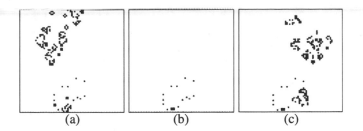

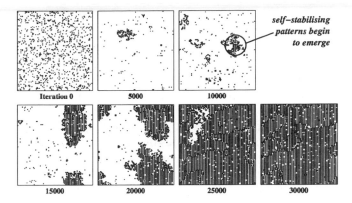

Figure 4: EvoCA-A: Sensor and effector evolution. Snapshots of the activation of the CA towards the end of the evaluation period for each of the three signal conditions: (a) Left signal (b) No signal (c) Right signal. For reference, the goal areas, genome area and signal areas are shown in gray.

Figure 5: EvoCA-B: Evolution of self-stabilising dynamics

mined by its local environment in Layer 1 (which can be influenced by the genome—and by other nearby genomes). A schematic illustration of the design is shown in Figure 2.

Genomes For practical implementational reasons, genomes in EvoCA-B are somewhat simplified. A genome can only set the state of a *single* target cell in Layer 1 corresponding to the genome's position in Layer 2, rather than being able to set states of target cells over an extended area as is the case in EvoCA-A. The genome consists of conditional genes only (no timed genes are used). The condition section of a gene specifies a combination of the current state of the associated cell, plus the *number* of neighbouring cells that are currently in a non-quiescent state. So, for 2 states and 8 neighbours, there are $2\times(8+1)=18$ possible conditions. If a gene's condition is met, then the gene sets the state of the cell as before. In addition, a gene also encodes a direction for the genome to move on the Layer 2 grid if the gene's condition is met (which could be to one of the directly neighbouring cells, or to stay put). Thus the genomes are acting in a manner reminiscent of Turing machines, reading and writing "data" from Layer 1 and conditionally moving position at each iteration.[4]

The Evolutionary Algorithm Each possible local configuration (i.e. the configuration of the states of a cell and its 8 immediate neighbours) has a genome reproduction probability associated with it. Similarly, each local configuration has associated with it a genome death probability and a genome reproduction-fidelity probability. (All these probabilities are hard-wired and constant.) For a genome at a given moment in time, its probability of reproduction (and, if it does reproduce, its probability of producing a faithful copy with no mutations), and its probability of death—i.e. all of the factors determining its evolutionary success (Dawkins, 1976)—are therefore wholly determined by the local environment in

which it finds itself. Thus, some local environments will be particularly conducive to the success of a genome, conferring upon it high stability and fecundity, others will be fairly neutral, while others still will be harmful.

In this way, there is a natural selection pressure for genomes that generate local conditions that promote their own survival and reproduction. Genomes can influence their local environments through the action of their genes, but only to a degree—the environment is subject to perturbations from the action of other nearby organisms. Selection pressure for self-generating and self-maintaining dynamics, i.e. autopoiesis (Varela et al., 1974), is therefore an inherent feature of the model.

Example Application: Natural Selection for Genome-Regulated Self-Stabilising Dynamics Typical results for the system as described are shown in Figure 5. These were from a run on a 100x100 grid, seeded with 100 randomly generated genomes placed at random positions. For the first 10000 iterations, fairly random dynamics are observed in Layer 1, as the genomes initiate dynamics in the environment. However, at around iteration 10000, a more dynamically stable pattern begins to appear in a portion of the CA. The pattern, identified by vertical stripes, is not completely static, but is able to regenerate itself under perturbations from the rest of the environment. Over the next 20000 iterations, this pattern spreads to fill the whole space as organisms that possess this phenotype out-compete their rivals. (A movie is available in the supplementary materials.) These dynamics demonstrate the natural selection of genomes that generate dynamics that promote their own survival—in other words, genome-regulated, self-stabilising dynamics.

Discussion

Advantages of the Approach

The results just described illustrate a number of features of the modelling approach being advocated.

Let's consider the EvoCA-A results first. Success in this

---

[4]This mechanism for genome movement was included to compensate for the fact that a genome can only directly affect a single cell at any iteration; at least it can now move from one iteration to the next, and can therefore affect different cells at different times.

task requires an individual to be sensitive to the signal's presence and location. It also requires that the individual exploits the environmental dynamics for long-range communication, to activate the cells in the goal area. Even in this simple situation, it could be argued that the organisms have evolved sensory and effector apparatus. To say that an individual is responding to a signal rather than just following the "laws of physics" (i.e. the normal CA update rules)—which of course it still is—is justified because the evolutionary selection process has introduced semantics to the signals, from the perspective of the individual, as indications of the task to be performed. During evolution, the successful organisms were selected precisely because they behaved *as if* the state of the cells in the signal position area was a signal, and responded in the appropriate way. Evolution has therefore introduced the potential for a new level of description of the system, where it is more informative to describe the action of an organism in terms of local rules (e.g. organism *A* responds to signal *B* by producing action *C*) rather than in terms of the universal laws of physics (cf. the quote from Pattee in the Introduction section).

Similarly, even though individual genes act by setting the state of single cells in the CA (there is no "glider gene," for example), genomes are able to produce complex actions such as the production of gliders and other patterns of spreading activation. This is because the genes are interacting with the pre-existing dynamics of the environment, by setting initial conditions for those dynamics. Again, these actions acquire semantics from the perspective of the organism through the process of selection during evolution.

The environment in this example is very simple. We could imagine environments with many more possible states per cell, and with much richer dynamics. In such environments, even if an organism's genes could still only directly respond to and activate a limited number of states, we could nevertheless imagine the organism being able to deal with a much wider variety of states, indirectly, by harnessing the environmental dynamics (e.g. by setting up a "chain reaction" to eventually achieve the desired result). In this way, organisms could potentially evolve to exploit almost any property of the environment, even if their genes were still able to perform only a limited subset of actions at the lowest level. Any property or process so incorporated can be expected to be retained if it promotes the evolutionary success of the organism. From this perspective, the evolutionary acquisition of new sensory or effector capabilities is not the problem that it is with other approaches (Dautenhahn et al., 2001).

Another feature of the approach is that phenotypes and abiotic environment are represented as a single system. In cases where we allow multiple organisms to coexist in the environment (e.g. EvoCA-B), organisms are therefore part of the environment experienced by other organisms. This introduces the possibility of rich co-evolutionary dynamics and high evolvability (Waddington, 1969; Odling-Smee

et al., 2003). Evolvability is also increased by the fact that there is no pre-defined specification of the organisation of a phenotype, so this is free to evolve over time.

The approach shifts the focus of the "problem" of evolvability away from the process of self-reproduction (which is taken for granted in appropriate environmental conditions), towards the issue of organism interactions (both organism–organism and organism–environment). It emphasises the view of organisms as self-generating and self-organising organisations, rather than self-reproducing automata.

Potential Criticisms

No modelling approach is perfect, and there are many potential criticisms that could be levelled at EvoCA and at the ideas that it embodies. Some of these are now considered.

Genetic System is Immutable   In EvoCA, the perspective of a genome as a source of constraints for a dynamical system is taken to the extreme; genomes play *no* part in the dynamics of the system other than to specify constraints (i.e. they have no material embodiment). This is largely for practical, rather than theoretical, reasons, and means that the design of the system can be kept very simple. The design may be compared to an artificial chemistry; the main difference is this separation of representation of genetic material from the rest of the system. This simplification is not without consequences. It means that an external mechanism is required for interpreting genomes as constraints (this happens at each iteration of the CA), and for writing genomes, with noise, at reproduction (this is performed by the evolutionary algorithm). Another consequence is that the genes are restricted to specifying constraints in a predefined way—in the particular design of EvoCA they are defined to map to the lowest level of the CA dynamics by constraining a specific cell to be in a specific state at a specific time. These restrictions all arise because genomes in EvoCA do not participate in the dynamics of the system at all, except through supplying constraints. This design decision is justified because of the perspective of genomes taken here—that the *fundamental* role of the genome is to supply constraints to the dynamical environment. It should also be noted that it is conceivable that a more complex genotype–phenotype mapping could evolve on top of the given system.

Where are the Organisms?   The examples presented, particular for EvoCA-B, are open to the criticism that an organism's phenotype lacks individuality (i.e. there is no recognisable boundary between it and the rest of the environment). The approach takes the view that an organism's phenotype is the set of genome-initiated dynamics in the environment. If it is advantageous for an organism to have a distinct membrane defining its boundary, then this is something we might expect to see evolving. However, particularly if we are modelling the origin of life, we should not assume such a distinction *a priori*. Also note that the question of whether such a membrane evolves is a question of the *properties of the environment*—is this existence of such a structure possible within the given environment?

The Limitations of Computational Models It could be argued that computational models such as EvoCA are unsuitable for the purpose of studying open-ended evolution, because of their digital nature. Each cell can only exist in one of a small number of discrete states, and therefore the number of states of the system as a whole is limited. A simple counter-argument is that, as the size of the system under consideration grows, the number of possible states soon becomes astronomically large. Furthermore, the complexity of the environment can always be increased, for example by using a multi-layer CA with different dynamics in each layer, or by allowing real-numbered states. More pertinently, we are also interested in dynamics and cycles of states, rather than the state of the system at a single instance—emergent behaviour can of course arise in the dynamics of discrete dynamical systems as well as analogue ones. The crucial point, however, is that the process of evolution, as demonstrated earlier, can endow states with semantic significance from an organism's perspective, at which point it becomes appropriate to describe the system at the level of local rules—the shape of which will depend on the system's specific history—rather than in terms of the underlying laws of physics. More sophisticated arguments have been put forward as to why purely digital devices cannot self-complexify, e.g. (Cariani, 1989). Cariani accepts that "the absence of gradualist pathways [in digital devices] would not preclude evolution entirely" but suggests that "[t]he important point is that purely computational devices do not construct or modify their primitives, and this does foreclose the possibility of fundamental novelty" (*ibid*, p.111). While these arguments hold for the system as a whole (i.e., in the case of EvoCA, the "laws of physics" of the environment), we can still address, given this limitation, how organisms evolve *within* the environment (with the corresponding emergence of semantics already described), starting from very simple forms to progressively control and exploit more and more of the environment's properties. We are not looking for fundamental novelty in the environment itself, but in the way in which organisms interact with it.

Implications

A corollary of this approach is that we can increase the complexity of evolved organisms—while still assuming only a simple set of mechanisms for the evolution of genomes—by increasing the complexity of the abiotic environment. From this perspective, some of the most important research questions are: What features must the environment possess to enable open-ended evolution? What features must it possess to enable the evolution of organisms that we might reasonable regard as *living*? Indeed, how are these two sets of fea-

tures related? (Is one a subset of the other? Are they identical?) We can address these questions not only at the level of features that have long been argued as being necessary for life (e.g. the requirement of a water-like substance in a liquid phase, or the possibility of semi-permeable membranes), but also at the level of fundamental physical properties such as the conservation of matter, energy flow, entropy increase, etc. How critical are each of these features for allowing the possibility of open-ended evolution and the emergence of life? Of course, I am not arguing that this approach should replace traditional ones; rather, it is complementary to them. By taking a different perspective, it highlights the significance of some different questions to consider in our attempt to understand life-as-it-is and life-as-it-could-be.

References

Bedau, M. A., McCaskill, J. S., Packard, N. H., Rasmussen, S., Adami, C., Green, D. G., Ikegami, T., Kaneko, K., and Ray, T. S. (2000). Open problems in artificial life. *Artificial Life*, 6(4):363–376.

Cariani, P. (1989). *On the Design of Devices with Emergent Semantic Functions*. PhD thesis, State University of New York at Binghampton.

Crutchfield, J. P. and Mitchell, M. (1995). The evolution of emergent computation. *Proceedings of the National Academy of Sciences U.S.A.*, 92(23):100742–100746.

Dautenhahn, K., Polani, D., and Uthmann, T. (2001). Guest editors' introduction: Special issue on sensor evolution. *Artificial Life*, 7(2):95–98.

Dawkins, R. (1976). *The Selfish Gene*. Oxford University Press.

Holland, J. H. (1976). Studies of the spontaneous emergence of self-replicating systems using cellular automata and formal grammars. In Lindenmayer, A. and Rozenberg, G., editors, *Automata, Languages, Development*, pages 385–404. North-Holland, New York.

Maynard Smith, J. (1986). *The Problems of Biology*. Oxford University Press.

McMullin, B. and Varela, F. J. (1997). Rediscovering computational autopoiesis. In Husbands, P. and Harvey, I., editors, *Fourth European Conference on Artificial Life*, pages 38–47. MIT Press/Bradford Books.

Odling-Smee, F., Laland, K., and Feldman, M. (2003). *Niche Construction: The Neglected Process in Evolution*. Princeton University Press.

Pattee, H. (1995a). Artificial life needs a real epistemology. In Morán, F., Moreno, A., Merelo, J., and Chacón, P., editors, *Advances in Artificial Life: Third European Conference on Artificial Life*, LNAI, pages 23–38. Springer.

Pattee, H. (1995b). Evolving self-reference: Matter, symbols, and semantic closure. *Communication and Cognition—Artificial Intelligence*, 12(1–2):9–28.

Ray, T. S. (1991). An approach to the synthesis of life. In Langton, C., Taylor, C., Farmer, J., and Rasmussen, S., editors, *Artificial Life II*, volume X of *SFI Studies in the Sciences of Complexity*, pages 371–408. Addison-Wesley.

Rocha, L. M., editor (2001). *The Physics and Evolution of Symbols and Codes: Reflections on the Work of Howard Pattee*. Special issue of *BioSystems* 60(1-3).

Taylor, T. (2001). Creativity in evolution: Individuals, interactions and environments. In Bentley, P. J. and Corne, D. W., editors, *Creative Evolutionary Systems*, chapter 1, pages 79–108. Morgan Kaufman.

Varela, F. J., Maturana, H. R., and Uribe, R. (1974). Autopoiesis: The organization of living systems, its characterization and a model. *BioSystems*, 5:187–196.

von Neumann, J. (1966). *The Theory of Self-Reproducing Automata*. University of Illinois Press, Urbana, Ill.

Waddington, C. (1969). Paradigm for an evolutionary process. In Waddington, C., editor, *Towards a Theoretical Biology*, volume 2, pages 106–128. Edinburgh University Press.

Yaeger, L. (1994). Computational genetics, physiology, metabolism, neural systems, learning, vision and behavior or poly-world: Life in a new context. In Langton, C., editor, *Proceedings of Artificial Life III*, pages 263–298. Addison-Wesley.

Imitation and Inequity in Avoiding the Tragedy of the Commons

Terry Van Belle and **David H. Ackley**
Department of Computer Science
University of New Mexico
Albuquerque, NM, USA

Abstract

We present a tragedy of the commons model in which individuals have no *a priori* knowledge of the immediate consequences of their actions: Each agent chooses actions using a simple neural network, which it gradually modifies to more closely imitate those of its wealthier neighbors. For a small commons size, the model leads neither to the tragedy of complete resource exhaustion nor to complete cooperation, but instead to the emergence of polarized 'economic classes' of poor and altruistic agents living amongst rich and greedy ones. The tragedy does emerge with larger commons sizes; we found that adding a degree of enforced local sharing among neighbors staves off tragedy there, and once again the economic stratification emerges. Though simple, the model displays a surprising range of dynamic behaviors at multiple temporal and spatial scales, as two fundamentally conflicting 'ideologies' war for control of agent behavior.

The Tragedy of the Commons

The Tragedy of the Commons (Hardin, 1968), like its sister model, the Prisoner's Dilemma (Axelrod and Dion, 1988), is an abstraction of social interactions in which the long-term common good of a group of agents is in conflict with each agent's short-term individual interests. In the classical example, as long as the village's common pasture is not overgrazed, it can provide ample feed to the villagers' livestock, but each individual livestock owner can gain advantage—if only temporarily—by grazing more animals there.

Because of its importance in everyday life, the tragedy has been extensively studied. Many solutions to the problem have been proposed, using both formal models (Axtell et al., 2000; Axelrod, 1997; Riolo et al., 2001; Axtell, 2003), and real-world examples (Milinski et al., 2002; Ostrom, 1990).

It is typically presumed that each agent in a commons model is inherently aware of the relationship between its action and its short-term marginal return—that putting another cow in the common pasture will increase that cow owner's share—and the model assumes that each agent makes (fully or boundedly) rational decisions on that basis. By contrast, we consider a commons model in which each agent chooses its actions based on a simple three weight neural network that maps from agent wealth and local resources to a decision about what fraction of the available resource to consume. Depending on the weights the neural network can produce a significant range of behavioral strategies; we investigate what happens when agents begin with random neural network weights, but then over time modify themselves by imitating the neural networks of the richest people in their local neighborhoods. For some parameter values, we observe neither the tragedy outcome of universal defection nor the 'utopian' outcome of unanimous cooperation, but instead the emergence of polarized economic classes and the exploitation of the obliging poor by the greedy rich.

In such cases, a quasi-stable balance is achieved by an extreme disparity in resource usage, with some in the population maximizing their take, and the remainder minimizing their take to compensate. For fully independent and rational agents, such a situation is unstable because the exploited can increase their resources by being just as greedy as their exploiters. However, in a system in which agents (willingly or under coercion) adopt the behavioral strategies employed by the rich, a system of economic differentiation and exploitation can emerge and spread. We find that in a variety of circumstances in which the tragedy of the commons is avoided, such endemic social inequities often arise and persist.

The Model

The world consists of a two-dimensional 64×64 grid. Each grid square permanently contains exactly one agent. The neighborhood of an agent consists of the 8 immediate squares, or fewer if the agent is on an edge of the world.

Each square initially contains one unit of some spontaneously growing *resource* that is valuable to the agents, but the squares are grouped into many disjoint rectangular *commons*, and the resources are pooled within each commons. For example, if the commons are 2 squares high and 1 square wide, then the 4,096 squares of the grid would be grouped into 2,048 commons, each which contains 2 villagers who have 2 units of resource initially at their disposal. The neighborhood of an agent, however, is *not* restricted by the boundary of the commons, allowing information to flow between commons via imitation. See Figure 1.

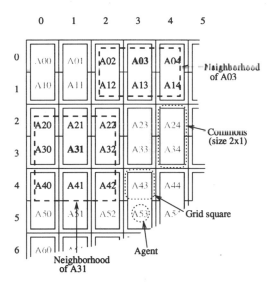

Figure 1: The upper-left corner of a world with 2 × 1 commons. The neighborhoods of agents A_{31} and A_{03} are shown with dashed lines; note that only A_{21} is in the same commons as A_{31}.

The model is run for some specified number of timesteps. During a timestep, the following phases occur effectively simultaneously in all commons: First, the agents may take a certain amount from the currently available resource in their commons. Then agents pay 'taxes', then agents imitate a richer neighbor (if any), then their neural network weights change randomly a bit, and then the commons resource grows.

Take Each agent is controlled by a neural net (Rumelhart and McClelland, 1986) with one output and three inputs: A *bias* input B that is always set to 1, a *resource* input R providing the current average resources per square in the agent's commons, and a *wealth* input W specifying the agent's current holdings in terms of accumulated resource. Each input is scaled by a modifiable link weight (x_B, x_R, x_W), which is initialized to a uniform random value from the range $[-1, +1]$ at the start of a model run. The output of the neuron represents the desired *take* of agent A_i in the current circumstances, and is determined by a sigmoid of the net scaled input:

$$t_i = \frac{1}{1 + e^{-(Bx_B + Rx_R + Wx_W)}}$$

What agent A_i actually gets, g_i, is determined by its desired take t_i, and the current amount of resource R in the commons:

$$g_i = \frac{t_i}{n}R$$

where n is the number of agents sharing the commons. R is then reduced by the sum of the g_i's.

This commons model is inherently 'kinder' than it might be, in the sense that an agent can never take more than a $\frac{1}{n}$th share of the available resources on any timestep. The essential tragic opportunity remains, however, because if everyone in a commons takes their full share, then the resource will be fully exhausted on that turn. Only if at least some agents choose to exercise self-restraint, and take less than their full share, will significant resources be left over to grow in the future.

Tax Next, each agent's wealth W_i is reduced by a combination of a fixed term (parameter τ_{fixed}) to cover the cost of existing, and a proportional 'spoilage' (parameter τ_{spoil}) term:

$$W_i' = (1 - \tau_{spoil}) \cdot \max(W_i - \tau_{fixed}, 0)$$

Imitate Next, each agent A_i considers imitating one of its neighbors and computes an *imitation weight change* (vector $\Delta \mathbf{x}_i$). If all the neighbors A_j have the same or less wealth ($\forall j \ W_i > W_j$), then $\Delta \mathbf{x}_i = \mathbf{0}$. Otherwise, A_i selects randomly from set K of richer neighbors based on how much richer ($\Delta W_j = W_j - W_i$) they are, picking neighbor N_j with probability

$$prob(N_j) = \frac{\Delta W_j}{\sum_{k \in K} \Delta W_k}$$

Given the selected neighbor A_j, the imitation weight change is:

$$\Delta \mathbf{x}_i = \alpha(\mathbf{x}_j - \mathbf{x}_i)$$

with imitation rate parameter α.

Noise   Whether or not imitation occurs, a noise vector $N(\mathbf{0}, \beta)$ is drawn from a random Gaussian vector distribution, with mean $\mathbf{0}$ and standard deviation specified by parameter β.

Update   Once all agents have performed the previous phases, then finally each agent A_i's weights are updated, based on the (possibly zero) $\Delta \mathbf{x}_i$ and the Gaussian noise:

$$\mathbf{x}'_i = \mathbf{x}_i + \Delta \mathbf{x}_i + N(\mathbf{0}, \beta)$$

Grow   In the final timestep phase, in each commons the pooled resource R grows according to a logistic equation, increasing exponentially when R is small, with growth tapering off as the resource per square approaches maximum resource parameter R_{max}. In addition, minimum resource parameter $R_{min} > 0$ gives each commons some chance to recover from poor management:

$$R' = \max(nR_{min}, \ R + \gamma R\big(1 - \frac{R}{nR_{max}}\big))$$

Experimental Results

Although this is a new model, and many experiments remain to be performed, we have already gained substantial experience exploring the model. Here, first, we discuss results in the degenerate case of a size 1×1 'commons', then focus primarily on data and observations from the smallest non-trivial commons, of size 2×1, and finally touch on results from larger commons. Except for the commons size and timestep limits, model parameter values were held constant for all experiments reported in this paper, and are given in Table 1.

1×1 Commons: Private Ownership

To calibrate our understanding of the model, we tested its behavior with the world divided into 1×1 'commons', so the results of each agent's choices directly affect only itself. Though tapping separate resources, the agents' behaviors are still coupled via the imitation process. Any time an agent overuses its resource, its wealth drops and it will start to

Parameter	Value	Parameter	Value
World width	64	τ_{fixed}	0.1
World height	64	τ_{spoil}	0.1
R_{min}	0.01	α	0.1
R_{max}	10	β	0.1
γ	1.25		

Table 1: Experimental parameter values. See text.

imitate its more temperate neighbors; a similar reaction occurs if it underuses its resource. Without any resource sharing, there is no possibility of cooperation or defection.

Figure 2 displays typical behavior of this model: The average resource per square quickly climbs to around 5 units—just about half of the $R_{max} = 10$ saturation value. The average take settles down to about 0.3, and the average wealth slowly rises past 15. In contrast to this seemingly sedate high-level view of the model evolution, the top of Figure 3 shows a close-up view of the 'take' of a sample agent within the grid. With 1×1 commons, agents develop a 'farming' strategy that involves choosing a very low take for one or a few timesteps, depleting their wealth in the process, until the resource reaches a threshold size, at which point they 'harvest' the sizable bounty, and the cycle begins again.

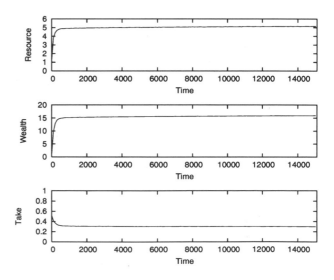

Figure 2: Average common resource levels, agent wealth and take, for 1×1 commons, averaged across 50 runs.

This is not the only sensible strategy in a 1×1 commons, and we have also shown that agents choosing a constant take at an optimal level do significantly better than these emergent farmers,

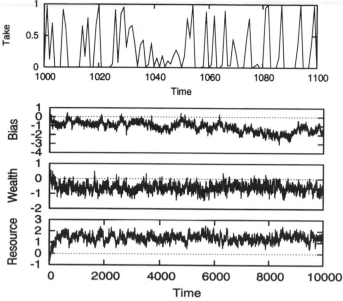

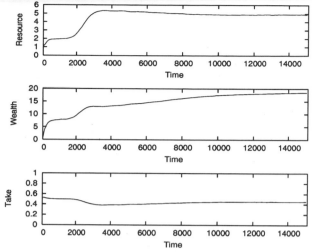

Figure 4: Average common resource levels, agent wealth, and take for 2×1 commons, averaged across 50 runs.

Figure 3: *(Top graph)* A sequence of takes from timesteps 1000–1100, for a sample agent using the 'farming' strategy in a 1×1 commons. *(Bottom graphs)* The evolution of the weights of a sample agent in a 1×1 commons.

but the farming strategy can be implemented by the neural network controller without requiring precisely chosen weights, as the bottom of Figure 3 suggests: A negative bias tends to shut down the agent take except when the resource multiplied by the resource weight is large enough to overcome it.

2×1 Commons: Emergent Inequity

Of course a 1×1 'commons' isn't a commons at all in any significant sense. The 2×1 commons, with two agents sharing each pooled resource, displays far more complex and interesting behaviors. Figure 4 displays overall behavior of this model, averaged over 50 runs varying only the random number seed.[1] The inflection points in Figure 4 suggest temporal phenomena effects at multiple scales, and indeed this seems to be the case.

Consider Figure 5, which presents a snapshot from a 2×1 commons run captured at time 50,000, showing the distribution of agent takes (left panel) and wealth (right panel), with darker grays indicating larger values. We see that most of the vertically oriented 2×1 commons have polarized so that they contain a single rich, greedy agent and a single poor, altruistic one. Figure 6 reveals the essential

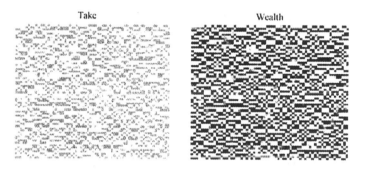

Figure 5: Distributions of t_i (left pane) and W_i (right pane) in a world of 2×1 commons. Most commons consist of a rich, greedy agent above or below a poor, altruistic one. See text.

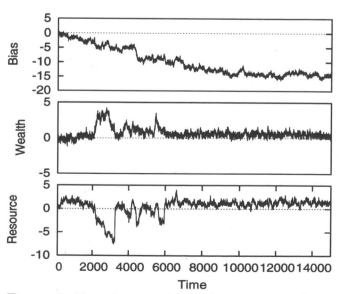

Figure 6: Neural network weights vs. time for a sample agent in a 2×1 commons.

---

[1] A 95% confidence interval around the final values in Figures 2 and 4 is ± 0.025 for Resources, ± 0.05 for Wealth, and ± 0.002 for Take.

trick at work—a negative bias weight encourages taking little, while a positive wealth weight allows agents that are already wealthy to choose a large take. When poor agents imitate rich agents possessing this network, they eventually embrace this "rich get richer" philosophy, even though following it specifically prevents them from realizing the benefits that flow to the rich agent employing the same strategy.

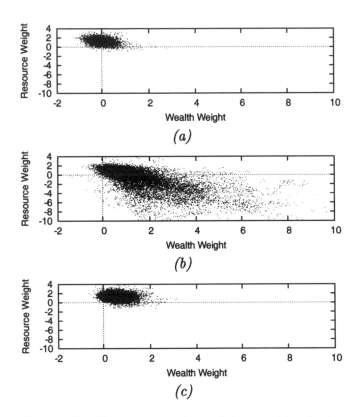

Figure 7: Resource and wealth weights during timesteps 1–2000 *(a)*, 2001–10,000 *(b)*, and 10,001–18,000 *(c)*, for a sample agent in a 2×1 commons. Aggregated across 20 runs. Some points in *(b)* are clipped beyond (10, -10).

There is another unexpected aspect of this strategy visible in Figure 6: Particularly in the era around timesteps 2000–6000, the resource weight is frequently actually *negative*, meaning that other things being equal, the more resource is available, the *less* of it you should take. Of course other things aren't equal; for the rich the positive wealth weight is more than enough to overcome this pathological aversion to resources and keep on taking. Over the 20 runs that we have currently examined, it appears this "big lie" strategy eventually gives way to an alternative "rich get richer" strategy employing a large negative bias weight without requiring a negative resource weight. The weight distributions

in Figure 7 suggest the progression through three distinct phases is quite robust.

Larger Commons: Avoiding Tragedy With Enforced Sharing

Moving beyond the 2×1 commons, we found that similar polarizations can occur in 3×1 and 4×1 commons, though the class emergence outcome occurs less frequently. In the basic model, with larger or fatter commons such as 2×2 and 4×4, the usual outcome is the tragedy, ending with all agent takes near 1 while their gets languish near 0.

In an attempt to stave off tragedy in larger commons sizes without moving to a completely centralized approach, we added a neighborhood sharing mechanism to the model: Between the **Take** and **Tax** phases, each agent A_i shares a fraction (parameter η) of its wealth W_i evenly and simultaneously with its n_i neighbors:

$$W_i' = W_i - \eta W_i + \frac{1}{n_i} \sum_{j=1}^{n_i} \eta W_j$$

Setting η to a non-zero value can stabilize the 2×2 and 4×4 cases, though it takes more sharing to stabilize the larger commons: Sharing as little as 10% ($\eta = 0.1$) can stabilize the 2×2 case, while at least $\eta = 0.6$ is necessary for 4×4 commons.

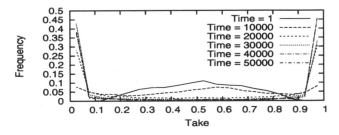

Figure 8: Histogram of the takes for agents in a world with 4×4 commons, $\eta = 0.6$.

Again, however, once the tragedy has been averted, the result is social inequity. Figure 8 displays the progressive loss of the middle class in a 4×4 commons model, and Figure 9 reveals the characteristic banded patterns associated with exploitation. Each 4×4 commons tends to organize itself into roughly half selfish and half altruistic agents, and the selfish agents tend to cluster together along the borders between commons, so that they are close to as many rich, selfish neighbors as possible. This arrangement allows them to share relatively little of their resources with nearby

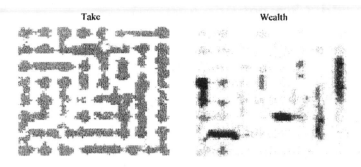

Figure 9: Distributions of t_i (left pane) and W_i (right pane) in a run with 4×4 resources and $\eta = 0.6$.

poor, and mostly to wash it back and forth between agents that are in a similar situation to themselves.

Discussion and Conclusion

A key aspect of our model is that the imitation process does not focus on the overt behavior of the agent imitated, but on that agent's entire programming. If poor agents imitate only the rich's 'take the max' actions, the tragedy of the commons promptly ensues. Stable patterns of social inequity can emerge because the poorer agents buy into an entire "philosophy," even if that philosophy then dictates they behave differently than those they look to for guidance.

Even when the internal details of an agent's decision-making processes are largely unobservable, machine learning techniques (Rumelhart and Mc-Clelland, 1986) may be used for imitation if the inputs associated with the agent's selected actions are available. In a general sense, things like story plots and morality plays are also ways of conveying input-output associations to the receptive; it is easy to speculate on ways that computational dependencies that encourage social inequity may be embedded in some of society's most familiar ideas.

It is important to note that although the exploitation patterns we observed are often quite stable, with an agent remaining rich or poor for many hundreds of consecutive timesteps, such situations are not frozen for all time. Perhaps because of the inescapable **Noise** step in the algorithm, the patterns drift like sand dunes over the course of thousands of timesteps, while maintaining the overall distribution of rich and poor. Sometimes a rich agent *is* brought down and a poor one *does* rise. By contrast, once a model goes firmly for global Hobbesian defection, we have never seen any sustained reemergence of resources and wealth.

The ubiquity of social inequity in these quite simple models suggests that perhaps something fundamental is at work. While it seems a clearly honorable goal to seek methods that avoid the tragedy of the commons with evenhanded fairness and justice, if one is stuck with mechanisms that allow inequities to emerge and persist, perhaps a most immediate goal is to ensure that the inequities remain mobile.

Acknowledgments

The research presented in this paper benefitted greatly from discussion and contributions by the members of the 2003 UNM Artificial Life Research Group. It was supported in part by DARPA contract F30602-00-2-0584, and in part by NSF contract ANI 9986555.

References

[1] Axelrod, R. (1997). *The Complexity of Cooperation: Agent-Based Models of Competition and Collaboration.* Princeton University Press, Princeton, NJ.

[2] Axelrod, R. and Dion, D. (1988). The further evolution of cooperation. *Science*, 242:1385–1390.

[3] Axtell, R. (2003). The emergence of institutions of self-governance on the commons. Brookings Institution Working Paper, Center on Social and Economic Dynamics.

[4] Axtell, R., Epstein, J. M., and Young, H. P. (2000). The emergence of classes in a multi-agent bargaining model. Working Paper 9, Center on Social and Economic Dynamics.

[5] Hardin, G. (1968). The tragedy of the commons. *Science*, 162:1243–1248.

[6] Milinski, M., Semmann, D., and Krambeck, H.-J. (2002). Reputation helps solve the 'tragedy of the commons'. *Nature*, 415:424–426.

[7] Ostrom, E. (1990). *Governing the Commons: The Evolution of Institutions for Collective Action.* Cambridge University Press, New York.

[8] Riolo, R. L., Cohen, M. D., and Axelrod, R. (2001). Evolution of cooperation without reciprocity. *Nature*, 414:441–443.

[9] Rumelhart, D. E. and McClelland, J. L. (1986). *Parallel Distributed Processing, Volume 1: Foundations.* MIT Press, Cambridge, MA.

The Quantum Coreworld: Competition and Cooperation in an Artificial Ecology

Alexander (Sasha) Wait[1]

[1] Harvard Medical School, Department of Genetics, 77 Avenue Louis Pasteur, Boston, MA 02115, USA.
await@genetics.med.harvard.edu

Abstract

Evolving systems, in principle, can exploit any tool in their genetic repertoire. This work specifies a computational chemistry consistent with the rules of Quantum Mechanics and a distributed artificial ecology that permits intervention by interested participants. A preliminary demonstration of two programs that use Quantum operations to authenticate one another is described. Competitors cannot impersonate the cooperating Quantum users without themselves using Quantum mechanics. Some limitations of this example are discussed. Readers can also visit the Quantum Coreworld ecology on the Internet at http://science.fiction.org.

Introduction

Does the underlying physics of a living system change its properties in a qualitative way? Could a different physics permit entirely new types of life? The aim of the Quantum Coreworld project is to engineer, or discover, a toy life-form with a different underlying physics. The success of such an organism—at exploiting available resources before competitors or at cooperating with genetically identical friends—must depend on its use of Quantum operations. If this endeavor required delicate control of large Quantum systems, there would be no way to get started with current technology. As it happens, however, interesting Quantum operations can be simulated on ordinary digital computers.

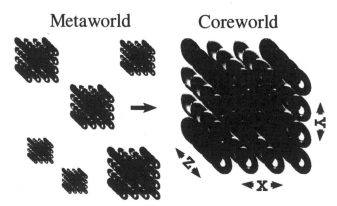

Figure 1. A Quantum Coreworld in the Metaworld.

The Quantum Coreworld was built by integrating "libquantum" (Butscher and Weimer 2003) with "pMars" (Ma et al. 1995), the *de facto* standard for the game Corewar (Dewdney 1984). The whole ecology—referred to as the Metaworld in Figure 1—combines elements from Avida (Adami and Brown 1994, Adami 1998, Adami and Lenski 2002), Tierra and Network Tierra (Ray 1992, 2000, 2004) and ongoing Internet Corewar tournaments (Dewdney 1987, Pihlaja 2004). Its name is inspired by the Coreworld, the first artificial life simulator based on Corewar (Rasmussen et al. 1990). This article provides an overview of the new world; more details can be found at: http://genetics.med.harvard.edu/~await/alife9.

Quantum strangeness is particularly obvious when unexpected correlations show up in spatially separated Quantum systems. To give organisms a better chance of exhibiting this strangeness, the world consists of isolated films of locally interacting compartments or cores. Films interact by periodic exchanges, random "bubbles", along vertical columns of cores. In Figure 1, the cores are donuts arranged into films along the Z axis. An energy source—energy is called privilege in the Quantum Coreworld—warms the world from behind, into the page, so that some cores are better suited to life than others. Cores along the Z axis furthest from the reader will have more privilege added by the energy source. Along the X and Y axes the greatest likelihood that privilege will be added is in the center, an intermediate likelihood at the center edges and the lowest likelihood at the corners. As time passes cores are cleared, or "washed away", so that life remains mostly in the middle of the world. This process is nearly the inverse of adding new privilege, so the cores at the corners are most likely to be cleared, center edges have an intermediate likelihood and cores at the center the lowest likelihood.

Each Coreworld in the Metaworld consists of a fixed number of molecules. These molecules are the basic constituent parts of cores and organisms, and operate according to the rules of the computational chemistry. Important global parameters—such as the dimensions of the world in units of cores—are referred to as the climate of the Coreworld. Shifts in the climate and opportunities for outside intervention are measured in epochs. The world's evolution is completely determined given a description of interventions (if any), the climate, and every molecule. At the beginning of every epoch, a Quantum Coreworld simulator starts with this description and produces a new one that describes the world at the end of the epoch. The simulator also reports the temporal and spatial distribution of genotypes in the world.

The Chemistry

Corewar, Avida, and Tierra are part of the well known family of artificial chemistries called assembler automata (Dittrich 2000). The Corewar language, Redcode, was last standardized in 1994; this standard has sustained ongoing tournaments typically run using the pMars simulator. Tutorials for that standard are readily available (Karonen 2004). Molecules in the Quantum Coreworld are described with an extended Redcode language that supports the notion of privilege and Quantum operations. This language also supports a more biological—2004 style—biotic execution model. A biotic molecule is active in the 2004 style model if it has at least one unit of privilege. Traditional 1994 standard programs can still run without change as a type of active abiotic environment.

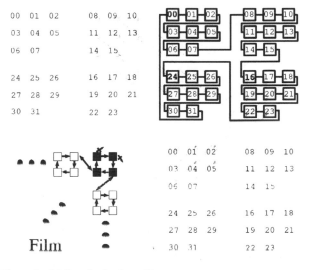

Figure 2. Molecules in three Cores

A typical arrangement of molecules in a core is shown in Figure 2. The number of molecules per core—referred to as core-size—is equal to the total number of molecules in the world divided by the total number of cores. There are three cores in the figure, core-size 32, corresponding to the upper right corner of the Coreworld in Figure 1. The world does not wrap around, so there is nothing beyond the edges of the world. From the core in the upper right corner of Figure 2 the accessible molecules in its neighbors are illustrated by lines connecting molecules across cores. The inset in Figure 2 represents this relationship in another way by dividing cores into four quadrants. In any case, molecules can interact with any other molecule in the same core, and with exactly one quarter of the molecules in each neighboring core. The numbering of molecules indicates their absolute position in the core. The peculiar arrangement allows the entire world to be displayed on a standard 1600x1200 display with one pixel per molecule.

Every molecule in the Coreworld has one: instruction field, modifier field, A mode, A field, B mode, B field, privilege field, qubit field, and tag. The instruction field determines the type of thing that each molecule can do.

The available instructions are listed in Table 1. The modifier specifies what field(s) an instruction uses. Modifiers are listed in Table 2. The A mode determines how the A-field is to be interpreted and the B mode determines how the B-field is to be interpreted. Valid modes are listed in Table 3. Both the A-field and B-field are integers between 0 and core-size - 1. Privilege is a positive integer. Its value can be unlimited in principle, but it cannot be greater than the total privilege in the world, which is itself bounded. The tag is a cryptographic hash of the intervention bringing that molecule into the world. Molecules with biotic tags execute under the 2004 style model—requiring privilege greater than zero to be active—while molecules with abiotic tags execute under the 1994 style model and use more traditional Corewar task-queues. Abiotic molecules, by definition, are not part of organisms, and are removed from the world at the end of each epoch.

DAT	terminate process (1994) do not advance privilege (2004)
MOV	move from A to B (1994) use privilege if necessary (2004)
ADD	add A to B, result in B
SUB	subtract A from B, result in B
MUL	multiply A by B, result in B
DIV	divide B by A, result in B if A is not zero, else DAT
MOD	divide B by A, remainder in B if A is not zero, else DAT
JMP	execute at A
JMZ	execute at A if B is zero
JMN	execute at A if B is not zero
DJN	decrement B, if B is not zero, execute at A
IJN	increment B, if B is not zero, execute at A (2004 only)
SLT	skip if A is less than B
SEQ CMP	skip if A is equal to B
SNE	skip if A is not equal to B
NOP	no operation
SPL	new abiotic task (1994 only)
QOP	Quantum operation on B, specified by A (2004 only)
QCN	Quantum not of target B controlled by A (2004 only)

Table 1. Valid instructions. Note: Instruction behavior can depend on execution style: biotic (2004) or abiotic (1994).

.A	Instr. read and write A-fields
.B	Instr. read and write B-fields
.AB	Instr. read A-field of A-instr. and B-field of B-instr. and write B-field
.BA	Instr. read B-field of A-instr. and A-field of B-instr. and write A-field
.F	Instr. read both A&B fields of A&B instr. and write to both A&B fields (A to A and B to B).
.X	Instr. read both A&B fields of A&B instr. and write to both A&B fields (A to B and B to A).
.I	Instr. read and write Instr., Modifier, Modes, A & B fields
.P	Instr. read and write privilege
.Q	Instr. read and write qubit
.D*	B pointer refers to dual core
.E*	Use privilege, execute faster

Table 2. Valid modifiers. Note (*) indicates modifier can be used in combination with other modifiers.

#	immediate
$	direct
@	indirect using B-field
<	predecrement indirect using B-field
>	postincrement indirect using B-field
*	indirect using A-field
{	predecrement indirect using A-field
}	postincrement indirect using A-field

Table 3. Valid addressing modes.

Several departures from 1994 Redcode can be seen in the above Tables. IJN, Increment Jump if Not zero, is analogous to DJN, Decrement Jump if Not zero. It has been discussed, but not added to Redcode in the past. In combination with the new .Q modifier—which specifies that the instruction deals primarily with the qubit field—IJN has a non-quantum application as a "concurrency primitive" (atomic test and set). It also plays a role in adding privilege, essentially spawning a new task, when used in combination with the new modifier .P, which specifies that the instruction deals primarily with the privilege field. The QOP and QCN instructions are one qubit and two qubit operations, respectively, and discussed in the example of a Quantum organism. The .D modifier specifies that the B-pointer refers to the dual core (recall Figure 2), and the .E modifier uses privilege in exchange

for a greater likelihood of execution; the .E modifier is the only permanent way privilege is removed from the world.

A consideration in developing this chemistry was the hope that community participation would improve the chances of discovering a toy Quantum life-form. The seemingly arbitrary decision to support both 1994 and 2004 styles of Redcode is one concrete step in this direction; this allows hundreds of existing Redcode programs to be tested in the Quantum Coreworld as interventions that specify an abiotic environment. In the spirit of other scientific free software projects, such as TeXmacs (van der Hoeven 2003), I believe this openness is a natural way to do science.

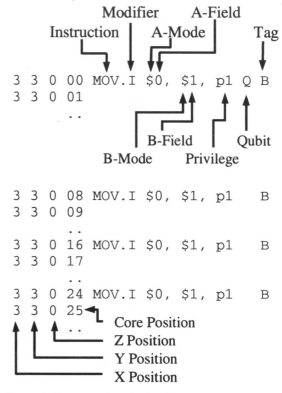

Figure 3. Four organisms in the Coreworld

Organisms and Genotypes

With the chemistry out of the way, it is possible to define organisms and genotypes in the Coreworld. In general, an individual organism is a contiguous sequence of biotic molecules; the tag of the molecule must specify it came from a "biotic" intervention. A simple one-molecule organism is shown in Figure 3. This organism is called an "Imp", and it circles around the core until it is cleared. If at least one of the molecules in an individual organism has privilege greater than zero, the organism is alive. An organism's genotype, in the Coreworld, has one genetic element for every molecule in the individual. A cryptographic hash of the encoding of these genetic elements is the unique genotype; see Deutsch (2002) for the implementation. The genetic portion of a molecule is the instruction, modifier, A mode, B mode and tag. These fields are all copied by a MOV.I, along with the A-field and B-field. MOV.I requires one unit of privilege; the free privilege is adjusted, as necessary, so that privilege is conserved. In the case of the Imp, it is reduced to privilege zero, and then after one unit of privilege is moved to the next instruction, it has privilege negative one. If this occurs, the instruction with negative privilege is cleared, which is why the Imp in the biotic execution model does not leave a trail of Imps in core; the molecules are consumed to provide the necessary privilege . From time to time, the Imps in Figure 3 can catch up to each other, briefly creating a larger two-molecule organism. The Coreworld gives this larger organism its own genotype, and tracks the number of cycles that these genotypes—and all genotypes—occur core-by-core.

Epochs

At the beginning of a new epoch, the simulator chooses one climate parameter—X, Y, Z or M—and perturbs it by adding or subtracting one. If the parameter is already at the limit of its range, the direction of change is forced. These parameters are defined so that the X dimension is equal to 2^X. Y and Z are defined in a similar way. M is defined so that the multiplier that determines the allocation of new privilege is equal to 2^M-1. By convention, X,Y, and Z vary from 2 to 4 and M from 1 to 5. The total number of molecules in each world is fixed to be 2^{20}, while the X,Y, and Z dimensions of the world can vary between values of 2^2, 2^3, or 2^4 cores. Thus, core-size varies between the powers of two from 2^8 to 2^{14}. Molecules are adjusted as necessary to accommodate a new core-size; this adjustment benefits from restricting core-sizes to powers of two. Since the X,Y, and Z dimensions are part of a Coreworld's climate, and remain constant throughout a given epoch, core-size remains constant throughout an epoch.

Interventions, as already mentioned, can occur at the beginning of an epoch. Whether or not an intervention occurs, pairs of films in the Coreworld are collapsed by picking individuals from each film, one at a time, and randomly placing them in a single film—possibly in nearby cores, but in the same location. Privilege lost in this process and throughout the past epoch—except for privilege consumed by the .E modifier—is referred to as free privilege. At the beginning of each epoch, the free privilege from the previous epoch is distributed in the world; this is done randomly or by repeatedly stamping out the molecules of intervention(s).

The stochastic behavior of the Quantum Coreworld is determined by the state of its pseudo-random-number generator (perturbed by interventions). A cryptographic quality generator is used—see Jenkins (1996) for the implementation—the state of the generator must be saved at the end of each epoch.

A Quantum Redcode example

The programs used to illustrate Quantum operations in this article are reminiscent of a bacterial replicator and a temperate bacteriaphage that imparts a beneficial capability to its host. These programs use a Quantum Hadamard operation as an integral part of their survival strategy. Quantum bits and Quantum operations are defined according to the standard conventions in Quantum computation (Nielsen and Chang 2000). When the host encounters a molecule from an unknown phage—or another replicator—it eventually executes the following sequence of Redcode instructions (from "rep-qphage.red"):

```
      ...   found something ...

modulebverify  QOP   #1,                @attack
               DJN.Q >attack,
oldploc        JMP   moduleverify,      #0

      ...   not friend ...

moduleverify   DJN   modulebverify,   #confidence

      ...   friend ...
```

These instructions, or molecules, were specified earlier in Table 1. As mentioned, molecules consist of a segment of "code", two segments of "data" and—added in the Quantum Coreworld—a Quantum bit or qubit. Privilege is not used in this example. Since the execution style is closer to standard 1994 Redcode, two co-operating programs can gain a benefit—effectively faster execution —over a single program. An analogous, Hadamard-based advantage is possible in the new style execution model, but not completed at the time of writing.

A qubit, $|\psi\rangle$, is defined $|\psi\rangle = \alpha |0\rangle + \beta |1\rangle$ where α and β are complex numbers. The states $|0\rangle$ and $|1\rangle$ are known as computational basis states. The probability that a qubit is in state $|0\rangle$, when measured, is $|\alpha|^2$ and the probability that it is in state $|1\rangle$, again when measured, is $|\beta|^2$. In the Quantum Coreworld—and physicists believe in the real world—there is no way to access α and β directly. In the code above, the DJN.Q operation performs a measurement. DJN then clears the qubit to $|0\rangle$ (setting $\alpha=1$ and $\beta=0$). We need only one more Quantum operation to understand the code. QOP #1 [address] specifies that the qubit at address should be modified by Quantum operation #1. In the Quantum Coreworld, QOP #1 is a Hadamard. The operation turns a $|0\rangle$ into $(|0\rangle+|1\rangle)/\sqrt2$ and a $|1\rangle$ into $(|0\rangle-|1\rangle)/\sqrt2$; conceptually, these are halfway between $|0\rangle$ and $|1\rangle$ (measuring either state gives a $|0\rangle$ or $|1\rangle$ with equal probability.) That is it!

If a pattern of #confidence (ten in the example) qubits are initialized to a one, and then disguised with a Hadamard, the pattern of ones can only be identified by another Quantum program. Measurements of these qubits will return a 0 or 1 with equal probability if a Hadamard is not applied first. Other Quantum operations are available, and it is suspected that some sequences of such operations are too difficult to simulate—due to time and space constraints—on any digital computer. For example, such operations can permit the factoring of large integers that no known technique can tackle. (Shor 1994)

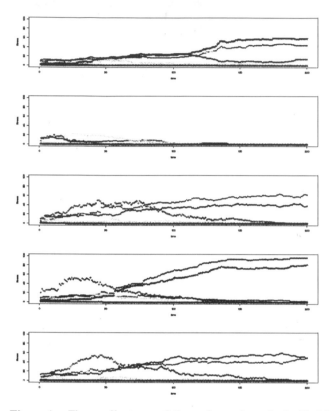

Figure 4— Five replicators and three viruses in a single 10x10 film in the Quantum Coreworld. Each panel is a separate run. The horizontal axis is time, and the vertical access is fitness, which is defined, for purposes of this example only, as the number of compartments in which a program is active.

Testing the example

In Figure 4, replicators and viruses were placed at random on a ten-by-ten film in the Quantum Coreworld. The results of five runs are shown. Two of the programs are identical, except for their ability to authenticate a potentially helpful—but possibly harmful—virus. The non-quantum virus is present in the panel second from the bottom. The Quantum virus is active in the top panel. In this experimental set-up, all viruses are mixed with all programs. If we added a second or third film, we could put the viruses on one film and either the classic or Quantum host on the other. If the authentication works, then the Quantum organism will have no detrimental effect when it is in the presence of one of the other viruses, but the non-quantum organism will be harmed. Experimental setup is

summarized in Table 4, and complete results are available in the supplementary information.

Film 1 (all virus)	Film 2	Film 3
10x10film (all)	no film	no film
5x5film	no film	no film
no-quantum	quantum	no film
quantum	no-quantum	no film
no-quantum	quantum	"quick" (yellow)
quantum	no-quantum	"quick" (yellow)
no-quantum	quantum	"unlock" (cyan)
quantum	no-quantum	"unlock" (cyan)
no-quantum	quantum	"retry" (purple)
quantum	no-quantum	"retry" (purple)

Table 4. Set-up for experiment to test Quantum advantage.

A simple classifier was used to evaluate each run. If a program is active in 23/25 compartments on each film then it "wins"; if it occupies no compartments on any film it "loses" and otherwise it draws. The results are **47 wins, 22 draws, and 31 losses** for the Qunatum organism; and **31 wins, 21 draws, and 48 losses** for the classical organism. Repetition produced comparable results. **Quantum wins!**

Future Work

A small but significant benefit can be obtained in the Quantum Coreworld by using the Quantum Hadamard operation for authenticating a friend. If an "unfriendly" Quantum virus existed, this would neutralize much of the benefit of Quantum authentication. These programs, in any case, are not very biological, since they predate the use of privilege as described here. This does not preclude the existence of more truly Quantum life-forms in the Coreworld. One possible way to increase the chances of finding such life-forms could be to study communication in the Coreworld, since the spatial structure and more biological execution model of this article should give organisms more to communicate about. Perhaps organisms could benefit from using Quantum Cryptography (Bennett and Brassard 1984) to obtain unconditional security for communication between cores.

As currently devised, the Quantum Coreworld can evolve slowly by mutation and Natural selection, as well as through participant-engineered interventions. If—or when—an interesting Quantum organism is discovered, a next step is to evolve it *de novo*. After seeding the Metaworld with a minimal replicating ecology, we can restrict interventions to organisms written by genetic algorithms. This direction of research attempts to answer the question: "From what starting point could Quantum Mechanical life evolve?" as opposed to the question being currently asked: "Could Quantum Mechanical life differ from life?"

On one hand, the most intriguing—and the most far-fetched—outcome of this research is the possibility that toy Quantum life-forms in the Quantum Coreworld could help us to recognize such behavior in the real world. On the other hand, the most practical benefit from studying Quantum Artificial Life might be a greater understanding of biological possibilities with more standard physics. A variety of experiments exploring phenomena in evolutionary theory are under consideration; examining multi-level selection, frequency-dependent selection, and mechanisms of diversity maintenance are possibilities.

The operations permitted in a classical world and a Quantum world—especially the operations we can model on a digital computer—are only subtly different. The Quantum Coreworld is a specific model in which these differences can be examined; this exploration can lead to a better understanding of the world as it is.

Acknowledgments

This project could never have been started in earnest without the support of my graduate advisers in Computer Science, Gilles Brassard and Claude Crépeau. It could not have been completed without the support of my current graduate adviser in Biophysics, George Church.

References

Adami. C. 1998. *Introduction to Artificial Life.* Santa Clara: TELOS Springer-Verlag.

Adami, C. and Brown, T. 1994. Evolutionary Learning in the 2D Artificial Life System 'Avida'. In Brooks, R. and Maes, P. editors, *Artificial Life IV*, 377-381. Cambridge, MA: MIT Press.

Adami, C. and Lenski, M. 2002. Avida. Software on the web at: http://sourceforge.net/projects/avida

Bennett C.H. and Brassard G. 1984. An update on Qunatum cryptography. In *Advances in Cryptology – proceedings of Crypto'84*, 475-480. Springer Verlag.

Butscher, B. and Weimer, H. 2003. Simulation eines quantencomputers, unpublished (in German.) Software on the web at: http://www.enyo.de/libquantum/

Deutsch, L.P. 2002. RFC1321-based (RSA-Free) MD5 library. Software on the web at: http://sourceforge.net/project/showfiles.php?group_id=42360

Dewdney, A.K. 1984. In the game called core war hostile programs engage in a battle of bits. *Scientific American*, 250:14–22.

Dewdney, A.K. 1987. A program called MICE nibbles its way to victory at the first Core War tournament, Computer Recreations, *Scientific American*, 14-20.

Dittrich, P.; Ziegler, J.; and Banzhaf, W. 2000. Artificial Chemistries - a Review. *Artificial Life*, 7:225-275.

van der Hoeven, J. 2003. Why Freedom is Important for Scientists, part of GNU TeXmacs. On the web at: http://www.texmacs.org/tmweb/about/philosophy.en.html

Jenkins, R.J. 1996. Isaac. In Fast Software Encryption, third international workshop, 41-49. Lecture Notes in Computer Science, 1039, Springer. Software on the web at: http://burtleburtle.net/bob/rand/isaacafa.html

Karonen, I. 2004. Beginner's guide to Redcode. On the web: http://vyznev.net/corewar/guide.html

Ma, A.; Sieben, N.; Strack, S.; and Wangsaw, M. 1995. PMars. Software on the web at: http://www.koth.org/pmars

Nielsen, M.A. and Chuang, I.L. 2000. *Quantum Computation and Quantum Information*. Cambridge University Press.

Pihlaja, M.J. 2004. New KOTH server. Usenet post in REC.GAMES.COREWAR on the web at: http://groups.google.com/groups?th=1f82856630ed460d

Ray, T.S. 1992. An approach to the synthesis of life. In C. G. Langton, C. Taylor, J.D. Farmer, and S. Rasmussen, editors, *Artificial Life II*, 371–408, Redwood City, CA, Addison-Wesley.

Ray, T. S. 2000. Evolution of Complexity: Tissue Differentiation in Network Tierra. ATR Journal, 40:12-13

Ray, T.S. 2004. Tierra. Software on the web at: http://www.isd.atr.co.jp/~ray/tierra/source/index.html

Rasmussen, S.; Knudsen, C.; Feldberg, R; and Hindsholm, M. 1990. The Coreworld: Emergence and evolution of cooperative structures in a computational chemistry. *Physica D*, 42:111–134.

Shor, P. 1994. Algorithms for Quantum computation: Discrete logarithms and factoring. In *35th Annual Symposium on Foundations of Computer Science*, 124-134, IEEE.

Self-reproduction by glider collisions: the beehive rule

Andrew Wuensche
Discrete Dynamics Lab, Santa Fe, NM 87506
andyw@cybermesa.com, www.ddlab.com

Abstract

We present a 3-value cellular automaton which supports self-reproduction by glider collisions. The complex dynamics emerge spontaneously in both 2d and 3d according to the 6-neighbor, k-totalistic, "beehive" rule; the 2d dynamics on a hexagonal lattice is examined in detail. We show how analogous complex rules can be found, firstly by mutating a complex rule to produce a family of related complex rules, and secondly by classifying rule-space by input-entropy variance. A variety of complex rules opens up the possibility of seeking a common thread to distinguish those few rules from the rest: an underlying principle of self-organization?

Introduction

Structure emerging by local interactions, self-reproduction and evolution; these themes are central to understanding natural processes. A system's complexity, according to this approach, relates to the number of levels on which it can be usefully described (Wuensche, 1994; Wuensche, 1999).

The simplest artificial systems able to capture the essence of these dynamical processes are cellular automata (CA), where "cells" connected on a regular lattice synchronously update their color by a logical function of their neighbor's colors. Just a tiny proportion of possible logics (complex rules) allow higher levels of description, greater complexity, to emerge from randomness.

In a movie of successive patterns on the lattice, recognizable sub-patterns emerge; mobile structures (gliders [1]) interact, aggregate, make glider-guns, and gliders self-reproduce or self-destruct by colliding.

In discrete CA everything can be precisely specified: rules, connections, dynamics. So for a given complex rule it should be possible to find causal links between the underlying "physics" and the ascending levels of emergent structure. We can also ask if there is a common thread that distinguishes those few rules that support complex dynamics from the vast majority that do not: an underlying princi-

---

[1] "Gliders" and other terminology is taken from John Conway's famous Game-of-Life (Conway, 1982). Gliders can also be regarded as particles or waves

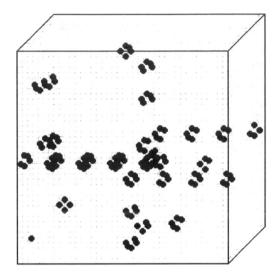

Figure 1: A snapshot of the beehive rule running on a 3d (40x40x10) lattice. The $k = 6$ neighborhood is shown in figure 3. The complex dynamics includes the spontaneous emergence of gliders, self-reproduction by glider collisions and glider guns, analogous to the 2d case. Gliders move in the direction of their red heads. Read this figure as if looking down into a shallow box.

ple of self-organization? That investigation would require a good sample of complex rules, which is now accessible.

Of the variety of complex behaviors in CA, self-reproduction (or self-replication) is perhaps the most prized (von Neumann, 1966; Conway, 1982; Langton, 1984) and provided the early motivation for ALife - but there have been few further examples of non-trivial self-reproduction until recent work by Antonio Lafusa (Bilott et al., 2003). He has been searching among the multi-value k-totalistic rules by genetic algorithm, using a fitness function of high input-entropy variance (Wuensche, 1999) and related measures.

We have also looked at multi-value k-totalistic rule-space, but for much smaller lookup tables than Lafusa's. We have limited both the value-range v (range of colors) and the neighborhood k to keep our look-up tables short, and make it easier to understand how specific entries relate to gliders. We have results for $v = 3$ and $k = 4$ to $k = 9$, but we will mainly describe results for $k = 6$ on a hexagonal 2d lattice.

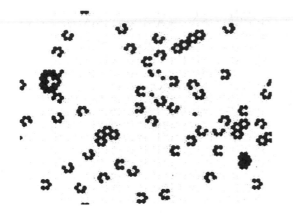

Figure 2: A snapshot of the beehive rule running on a 2d hexagonal lattice. Gliders move in the direction of their red heads.

Complex rules are easily found in these smaller rule-spaces by the classification methods in (Wuensche, 1999). Small lookup tables also make it easier to study mutations. It turns out that a large proportion of 1-value mutations are quasi-neutral; they make little difference to the complex dynamics. Some mutations result in modified but equally interesting complex dynamics. So mutations create families of related complex rules. Of course, there are also sensitive positions in the lookup table were a mutation completely disrupts the complex dynamics.

This paper outlines the ideas and methods. The "beehive" rule, which supports spontaneous self-reproduction by glider collisions in both 2d and 3d, is examined in detail for 2d, and other examples are presented. Further details and results can be found at www.ddlab.com (Wuensche, 2004).

k-totalistic rules

We will consider a subclass of CA rules, the k-totalistic rules (Adamatzky, 1994; Bilott et al., 2003)[2], where a cell's update depends only on the frequency of values (colors) in its neighborhood, not their position (figure 4). Because of this, the dynamics conserve symmetry; whatever happens in one direction or reflection can also happen in all others. k-totalistic lookup tables (kcode) are much smaller than the general case, $G = v^k$. The size L of the kcode is given by $L = (v+k-1)!/(k!(v-1)!)$. For $[v,k] = [3,6]$, $L = 28$, as opposed to $G = 729$. For greater $[v,k]$, L increases rapidly. If complex behavior can indeed be found for small $[v,k]$, it is of course worthwhile to think small and deal with short kcode.

The beehive rule

The beehive rule is a multi-value k-totalistic rule with $[v,k] = [3,6]$. The rule created the snapshots in figures 1 and 2, and spontaneously self-organizes a basic glider which becomes

---

[2]Thanks to Antonio Lafusa for introducing this class of rules to us. There is a prior attribution to (Adamatzky, 1994), and his identical class "ATOT".

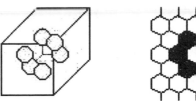

Figure 3: The $k = 6$ neighborhoods of 3d, and 2d hex, CA.

```
kcode = 0022000220022001122200021210
              kcode index
          /  totals: 2s+1s+0s=k=6
         /  /            kcode
basic   /  /         /
glider /  /         /   mutations
------- / 2_1_0     /   2___1___0
background->  0:  0 0 6  -> 0    o    c   -
     head+->  1:  0 1 5  -> 1    0    -    0
              2:  0 2 4  -> 2    -   Sg   cg
              3:  0 3 3  -> 1 -+ G    -    G
       out4   4:  0 4 2  -> 2 -+ -    G    G
       out3   5:  0 5 1  -> 0 -+ G    G    -
       out1   6:  0 6 0  -> 0 -+ G    G    -
     side2->  7:  1 0 5  -> 0    c    c    -
     side1->  8:  1 1 4  -> 2    -    c    c
      side1+  9:  1 2 3  -> 2    -   cg    G
             10:  1 3 2  -> 2 -+ -    G    G
       out2  11:  1 4 1  -> 1 -+ G    -    G
       tail  12:  1 5 0  -> 1 -+ G    -    G
      head-> 13:  2 0 4  -> 0    c    c    -
             14:  2 1 3  -> 0   Gs    c    -
             15:  2 2 2  -> 2        gc   gc
             16:  2 3 1  -> 2 -+ -    G    G
             17:  2 4 0  -> 0 -+ G    G    -
             18:  3 0 3  -> 0    g    c    -
             19:  3 1 2  -> 2    -    c   cg
             20:  3 2 1  -> 2    -   cg   Gd
             21:  3 3 0  -> 0 -+ G    G    -
             22:  4 0 2  -> 0    G    c    -
  center->   23:  4 1 1  -> 0    g   cg    -
             24:  4 2 0  -> 2    -   cg    G
             25:  5 0 1  -> 2    -   cg    G
             26:  5 1 0  -> 0    g   gc    -
             27:  6 0 0  -> 0    G   Gd    -
```

key to mutations:
quasi-neutral G=25/56, wildcards -+ 10/28
G/g=gliders, G=same/similar dynamics,
g=weak/different, S=spirals, d=dense,
s=sparse, c=chaos, o=order, 0=all 0s

Figure 4: The lookup table (kcode) of the k-totalistic beehive rule, showing its construction. This also shows the entries that make the basic glider in figure 5, and the consequences of all 56 possible 1-value mutations, 25 of which are quasi-neutral.

the predominant structure in both a cubic 3d and hexagonal 2d lattice, with neighbors as in figure 3; the cell itself is not included in its neighborhood.

The complex dynamics includes self-reproduction by glider collisions (figure 6), and polymer-like gliders and glider-guns (figures 6 and 7), but no permanently static patterns. We chose the beehive rule for closer scrutiny because its self-reproduction is especially clear in a live simulation.

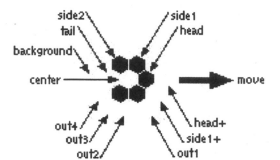

Figure 5: The basic $k = 6$ 2d glider moves in the direction of its red head. Each cell that forms the glider and its surroundings must blink to the correct color at the next time-step according to the kcode. 12 cells are indicated which cover all cases because of symmetries. The cells are controlled by 6 kcode entries in figure 4; mutation of these disrupts the dynamics (except "center").

Self-reproduction by glider collisions

We will look in some detail at the 2d dynamics, firstly the outcomes of all possible, non-equivalent, types of collisions between pairs of basic gliders, bearing in mind that different direction on the hex lattice, and reflections, are equivalent. Self-destruction, survival, conservation and self-reproduction all occur, depending on the exact point and direction of impact, summarized in the table below. Of the 21 collision types (8 head-on and 13 angular), 4 lead to self-reproduction, where 2 gliders release either 4, 5, or 6 after an interaction phase of several time-steps. Figure 6 shows some examples.

		gliders		
	type	no	before	after
self-destruction:.	2->0	10	20	0
one-survivor:.....	2->1	4	8	4
conservation:.....	2->2	3	6	6
self-reproduction:	2->4	1	2	4
	2->5	1	2	5
	2->6	2	4	12
		------------		
	totals	21	42	31

The glider before/after ratio is 31/42, so if collision types were equiprobable, and ignoring other interactions, we would expect a high population density of gliders to decrease over time; though this is observed in the long run, other structures and interactions make the dynamics more complex. Gliders can crash into the transient patterns following collisions. An isolated red cell, from collision debris, explodes to make 6 new gliders, so outside perturbations, noise, would tend to repopulate the space with gliders; the dynamics in general is robust to noise. Polymer-like gliders made up of sub-units, also emerge.

Most notably, there are a variety of glider-guns[3] that eject from 1 to 4 glider streams in different directions.

Some examples of all of the above are given in figure 7. These processes combine with self-reproduction to produce an extremely complex hive of activity.

---

[3]Strictly speaking these are a cross between "glider-guns" and "puffer trains" (Conway, 1982)

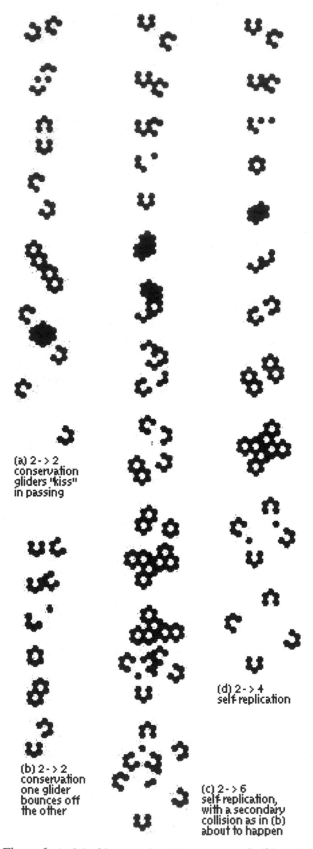

(a) 2->2 conservation gliders "kiss" in passing

(b) 2->2 conservation one glider bounces off the other

(c) 2->6 self replication, with a secondary collision as in (b) about to happen

(d) 2->4 self replication

Figure 6: 4 of the 21 types of collisions between 2 gliders (time-steps from the top). Conservation: (a) and (b). Self-reproduction: (c) and (d). For all collisions types see (Wuensche, 2004).

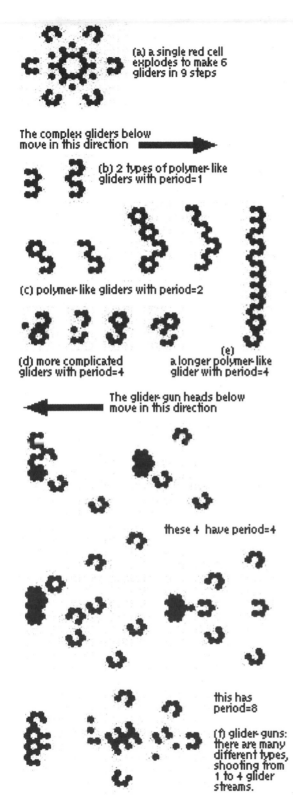

Figure 7: (a) an exploding red cell makes 6 new gliders. (b,c,e) polymer-like gliders made from subunits. (d) a glider that is also present in (e) as a subunit, (e) a longer polymer-like glider made of subunits from (b) and (d). (f) 5 examples of the various types of glider-gun, which shoot from 1 to 4 glider streams. For more examples see (Wuensche, 2004)

Figure 8: The result of a 1-value mutation to the beehive rule, at index 2 (the output 2 is changed to 1). Glider activity is gradually overwhelmed by spirals. A snapshot on a 2d (60×60) hex lattice

Mutations

The consequences of all possible 1-value mutations to the beehive rule are tabulated in figure 4, and snapshots of all can be found at (Wuensche, 2004).

The lookup table has 28 entries, and each can be changed from its present value to two alternatives, giving 56 possible minimal (1-value) mutations. The results of this experiment (Wuensche, 2004) show that for 10 of the entries, changing to either alternative (20 mutations) is quasi-neutral; it appears not to make much difference to the dynamics; experiment confirms that these 10 entries can actually be wildcards. A further 5 mutations elsewhere, to just one value, are also quasi-neutral, making 25/56. Multiple mutations in these neutral regions needs examining.

On the other hand, mutations to any of the 6 sensitive entries that maintain the basic glider destroy the dynamics - with one exception - a mutation at index 23 (the glider's center) which sets the glider's tail at the next time-step. This mutation closes the glider's tail (a black cell, value 2), but otherwise conserves complex dynamics.

Another interesting mutation is at index 2, which causes glider activity to be gradually overwhelmed by spirals, as shown in figure 8.

The beehive kcode is set out below, indicating these mutations, the 10 wildcards (+), and the 6 glider entries (ˆ),

```
index  23                          2
       |                           |
     002200+220++200+++220++++210
       ˆ                ˆ    ˆˆ    ˆˆ
```

It would be possible then, to explore the family of related rules by gradually mutating away from the beehive rule, and entering into the network of related complex rules in rules-space.

Finding complex rules

To find new complex rules from scratch (without mutating old ones), and in particular rules that support gliders, we use

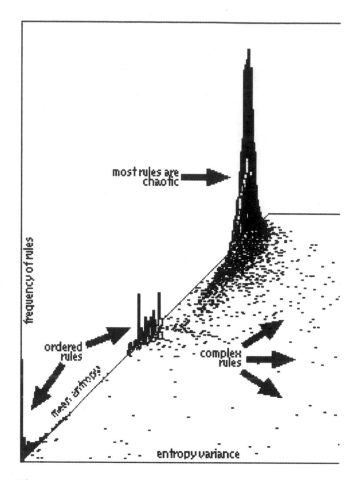

Figure 9: About 15800 $[v,k] = [3,6]$ k-totalistic rules classified by input-entropy variance.

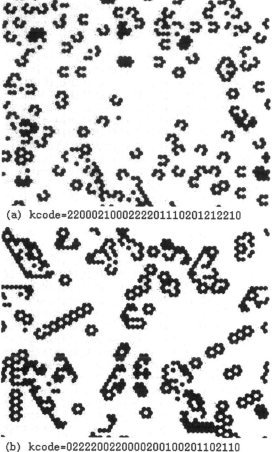

(a) kcode=22000210002222011102012122 10

(b) kcode=0222200220000200100201102110

Figure 10: Complex rules on a 60×60 hexagonal lattice.

the method for automatically classifying 1d rule-space by input-entropy variance (Wuensche, 1999), but which applies equally well to k-totalistic rules, and to 2d and 3d.

We track how frequently the different entries in the kcode (as in figure 4) are actually looked up, once the CA has settled into its typical behavior. The Shannon entropy of this frequency distribution, the input-entropy S, at time-step t, for one time-step (w=1), is given by $S^t = -\sum_{i=0}^{L-1} \left(\frac{Q_i^t}{n} \times log \left(\frac{Q_i^t}{n} \right) \right)$, where Q_i^t is the lookup frequency of neighborhood i at time t, L is the kcode size, and n is the size of the CA. In practice the measures are smoothed by being averaged over a moving window of $w = 10$ time-steps. The measures are started only after 200 time-steps, and are then taken for a further 300 time-steps. The 2d CA 100×100 is run from a sample of 5 random initial states. The sizes of these parameters can be varied, of course.

Average measures are recorded for (a) entropy variance (or standard deviation), and (b) the mean entropy. This is repeated for a sample of randomly chosen rules. The sample is then sorted by both (a) and (b), and data plotted as in figure 9, The plot classifies rule-space between chaos, order and complexity. Individual rules can be selected by various

methods, including directly from the plot, to check their behaviors.

The basic argument is that if the entropy continues to vary in settled dynamics, moving both up and down, then some kind of self-organizing collective behavior must be unfolding. This might include competing zones of order and chaos, or two differnt types of chaos, as well as glider dynamics.

In the case of the beehive rule and other glider rules, at any given moment there may be a bias in the dynamics towards a preponderance of gliders (low entropy) or post-collision transient patterns (high entropy). The lattice (or a patch undergoing the analysis) must not be too large in relation to the scale of possible emergent structures, otherwise the effects would cancel out. By contrast, stable/high entropy indicates chaos (most rules); stable/low entropy indicates order - in both cases the entropy variance is low.

Other complex rules

In figures 10 and 11, we show 4 examples of $[v,k] = [3,6]$ complex rules, found independently by the input-entropy variance method (more can be seen at (Wuensche, 2004), also for $k = 7,8,9$). The basic beehive glider is sometimes present, but we also see different gliders and complex struc-

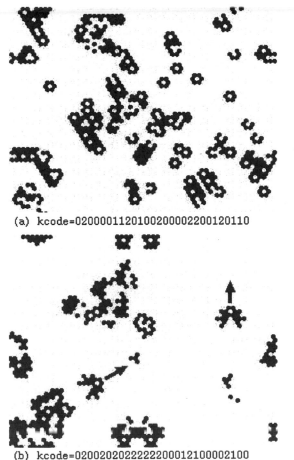

(a) kcode=02000011201002000002200120110

(b) kcode=0200202022222200012100002100

Figure 11: Complex rules on a 60 × 60 hexagonal lattice.
(b) note 2 large slow moving gliders (period=3), their motion is
indicated by arrows

exceptional; a 2 at index 23 closes the basic glider's tail, and
a 2 at index 26 keeps it closed (a black cell, value 2). There
is also a correlation with the frequencies of values.

If a common thread or bias in kcodes can be identified among these and other complex rules, which distinguishes them from the vast majority of rules-space, then this
could become the basis for an underlying principle of self-
organization in k-totalistic cellular automata.

Discussion

There is a network of complex rules in k-totalistic rule-
space, connected by mutations, where large scale collective
behaviors emerge spontaneously. The complex dynamics includes self-reproduction by glider collisions, polymer-like
gliders, glider guns, and possibly other structures and interactions. This implies higher levels of description beyond the
underlying "physics", the kcode. The levels could conceiv-
ably unfold without limit given sufficient time and space; the
number of these emergent levels is our qualitative measure
of complexity.

Some questions arise; what is the mechanism of self-
reproduction? how do glider-guns self-assemble? are these
systems computation universal? how does complexity scale
with greater v or k? how do the various complex rules relate? is there an underlying principle of self-organization?
and what is it?

Discrete Dynamics Lab

The software used to research and produce this paper was multi-
value *DDLab*, in which the dynamics can be seen live, and the
rules are provided. It is available at www.ddlab.com.

References

Adamatzky, A. (1994). *Identification of Cellular Automata*. Taylor
and Francis.

Bilott, E., Lafusa, A., and Pantano, P. (2003). Is self-replication
an embedded characteristic of the artificial/living matter? In
Standish and Bedau, editors, *Artificial Life VIII*, pages 38–48.
MIT Press.

Conway, J. (1982). *What is Life?*, chapter 25 in "Winning
ways for your mathematical plays", Vol.2, by Berlekamp,E,
J.H.Conway and R.Guy. Academic Press, New York.

Langton, C. (1984). Self-reproduction in cellular autonata. *Physica
D 10*, 10:135–144.

von Neumann, J. (1966). *Theory of Self-Reproducing Automata*.
Univ. of Illinois Press. edited and completed by A.W.Burks
from 1949 lectures.

Wuensche, A. (1994). Complexity in one-d cellular automata.
Santa Fe Institute working paper 94-04-025.

Wuensche, A. (1999). Classifying cellular automata automatically.
COMPLEXITY, 4/no.3:47–66.

Wuensche, A. (2004). www.ddlab.com (follow the links to
"self-reproduction" and "dd-life").

tures, which we have not yet examined in detail. The example in figure 11(b) has a remarkably complex glider and
glider-gun.

In these examples, the kcode has been transformed with a
value-swapping algorithm to an equivalent kcode, but with
the colors (values) made to correspond with the beehive rule,
where the background value is 0 (green), the leading head of
gliders is 1 (red). This allows the different kcode tables to be
compared to look for common biases. Below we compare
the kcodes of our 4 examples with the beehive rule. The
wildcards(+) and glider entries (ˆ) are indicated.

```
26 23                       frequency of values
 |  | +   ++   +++   ++++          2__1__0
00220002200220011222000021210     11   4 13
      ^        ^        ^
22000210002222011102012122210 - 10a 11   7 10
02222002200020010020110201102110 - 10b  9   5 14
02000011201002000002200120110 - 11a  6   6 16
0200202022222200012100002100 - 11b 11   3 14
 |          |     ||       ||
 2          4     34       34 - matches
```

We can see that there is a high correlation with glider entries, except for index 23 which we have already noted is

The value of death in evolution: A lesson from Daisyworld

Matthew Bardeen

Department of Informatics, University of Sussex, Brighton, BN1 9RH
mdb20@sussex.ac.uk

Abstract

The first Daisyworld model showed how simple interactions and varying growth rates of daisies could lead to a type of global homeostasis. This idea has been extended to include variable mortality rates as a form of natural selection. This paper shows that this extension can increase the temperature regulation and persistence of Daisyworld systems, even with higher overall mortality rates.

Introduction

Death is the natural endpoint of all living things, and while it is tempting to think of death as an inherently random process, this is not always the case. Environmental factors, such as lack of food and temperature variations, increase the chance of death in many organisms. This idea is one of the central tenants of Darwin's theory of natural selection. Organisms which are less fit for their environment are replaced by those which are more fit.

Standard theory is that the fitness of an organism is determined by how many breeding offspring it bears and how many of those live to breeding maturity. This fitness is really a combination of two factors – fecundity and survival. For example, Leatherback sea turtles lay many hundreds of eggs per season but only about 1 to 3 individuals per 100 eggs survive to breed. Compare this with Asian elephants, which give birth to one calf every 3 to 4 years with about 80% of the calves surviving to a breeding age. However, there is another force, environmental modification, hidden within the survival factor. It has been known for some time that organisms modify their environment during their lifetimes and these modifications often result in a more favorable environment for their offspring.

In 1978 Andrew Watson and James Lovelock introduced the Daisyworld model as a way of lending credence to the Gaia hypothesis (Watson and Lovelock, 1983). This hypothesis suggests that organisms can not only change their local environment, but the global environment as well. Daisyworld is a toy-world, intended as a proof of concept, rather than a model of a real physical system. The idea is simple –

localized interactions can affect global dynamics and generate homeostatic behavior. The model consists of a "planet", heated by the sun and populated by black and white daisies. The black daisies have a lower albedo (reflectiveness) than the white daisies, thus they absorb a greater amount of solar radiation and raise the local temperature. White and black daisies have albedos of 0.75 and 0.25 respectively, and the surface of the planet has a neutral albedo of 0.5. The growth rate of the daisies is linked to the local temperature, which is directly influenced by albedo. This difference in growth rate causes the area covered by black and white daisies area to vary, thus warming and cooling parts of the planet. This creates a homeostatic response to external forces, such as increasing incoming solar radiation (insolation), and keeps the temperature of the planet relatively constant.

One major criticism of the Daisyworld model when it was introduced was that it did not model natural selection. Since this time the model has undergone been numerous extensions and reformulations that have addressed this concern to some extent (Saunders, 1994; Stöcker, 1995; Lenton and Oijen, 2002; Harding, 1999; Weber, 2001; Lenton, 1998; von Bloh et al., 1997). However, all of these subsequent models have concerned themselves only with varying the fecundity of the daisies and not varying their survival rates, assuming a constant mortality rate. For natural selection to work efficiently in this context, there must be some differential selection pressure on organisms – a point which will be addressed in more detail later in the paper. For now, the relevance of a few of the Daisyworld models to natural selection will be reviewed.

Saunders suggests a scenario where evolution happens without natural selection – organisms mold their environment to their own needs, rather than being molded to the environment (Saunders, 1994). He shows if the daisies adapt their birth rate to local conditions, they cover slightly more area, at the cost of a decreased range in planetary temperature regulation.

As a counterpoint, Stöcker takes the stance that the daisy's growth rate is more a matter of physics than genetics, and assumes that the mutations affect daisy albedos rather than birth rates. Based on this he shows that the mutation of albedos increases the temperature regulation range as well as the stability of the system (Stöcker, 1995). Other studies of mutation in the Daisyworld model confirm these findings (Lenton and Oijen, 2002; von Bloh et al., 1997). In particular, von Bloh attributes this increase to the fact that mutation allows higher albedos than the original "white" daisies.

Of particular interest is von Bloh's formulation and analysis of a 2D cellular automata version of the Daisyworld, which will be used as the basis for these simulations (von Bloh et al., 1997). In this paper, his model is extended with temperature-dependent mortality rates. The next section describes the original model and the extension in detail. This extended model is then compared to von Bloh's original model to determine the effect of variable mortality rates on the homeostatic response of the system. The results are analyzed and related back to the concepts in the introduction.

Model Description

As mentioned previously, this paper extends von Bloh's 2D Daisyworld model (henceforth referred to as the VB model) (von Bloh et al., 1997). While the implementation differs from the original Watson-Lovelock (WL) model, the salient features remain the same. There are white and black daisies on a "planetary sheet" that have slightly different growth rates and thus exhibit the same homeostatic response as the WL model.

The basic equations governing the effective temperature of the original WL model are

$$\sigma_B(T_e)^4 = S(1-A), \tag{1}$$
$$d\alpha_b/dt = \alpha_b(x\beta(T) - \gamma), \tag{2}$$
$$d\alpha_w/dt = \alpha_w(x\beta(T) - \gamma), \tag{3}$$

where σ_B is the Stefan-Boltzmann constant, S is the luminosity of Daisyworld's sun, and A is the albedo of the planet. The fraction of total planetary area covered by the black and white daisies is α_b and α_w is respectively. The fractional area of the uncovered ground is x, $\beta(T)$ is the birth rate of the daisies with respect to local temperature, and γ is a constant mortality rate.

Interestingly, if we set the growth rate of the daisies to a constant rate and vary the mortality rate with temperature, the system still exhibits homeostasis. Reformulating the general growth rate equation by swapping the terms for the mortality and growth rates yields

$$d\alpha/dt = \alpha(x\gamma - (1 - \beta(T))). \tag{4}$$

This is functionally equivalent to the original equation and, with the correct values, exhibits a similar homeostatic response. Restating the equations in this provides way to

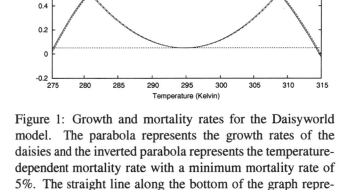

Figure 1: Growth and mortality rates for the Daisyworld model. The parabola represents the growth rates of the daisies and the inverted parabola represents the temperature-dependent mortality rate with a minimum mortality rate of 5%. The straight line along the bottom of the graph represents a constant mortality rate of 5%.

model the idea of survival in a Daisyworld context. This paper will build on this premise and link both mortality and growth rates to the temperature such that

$$d\alpha/dt = \alpha(x\beta(T) - \gamma(T)). \tag{5}$$

Returning to the von Bloh's model, this 2D version of the Daisyworld model includes heat diffusion between cells as well as the effect of the daisies on the local temperature. The equation used is

$$C\frac{\partial T(x,y,t)}{\partial t} = D_{Tw}(\frac{\partial^2}{\partial x^2} + \frac{\partial^2}{\partial y^2})T(x,y,t)$$
$$-\sigma_B T(x,y,t)^4 + S(1 - A(x,y,t)), \tag{6}$$

where D_T is the heat diffusion constant and $A(x,y,)$ is the space/time distribution of albedo.

Growth patterns for the daisies are generated using a cellular automata (CA) model (von Bloh et al., 1997). If a cell is empty, then there is a chance that a daisy in a neighboring cell will produce offspring in the empty cell. This chance is based upon the temperature of that cell and is shown graphically in Figure 1 and given by

$$\beta(T) = \frac{4}{(T_{max} - T_{min})^2}(T - T_{min})(T_{max} - T) \tag{7}$$

where T_{max} and T_{min} are the maximum and minimum temperatures at which the daisies can grow and T is the current temperature of the cell. Heat diffusion between the cells allows the temperature of uncovered cells to be affected by neighboring covered cells, making the probability of colonization higher than otherwise possible. The daisy albedos

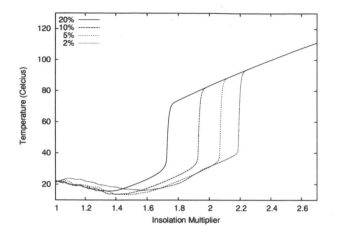

Figure 2: Average temperature of the constant mortality rate model with no mutation ($r = 0$) and mortality rates of 2%, 5%, 10%, and 20%.

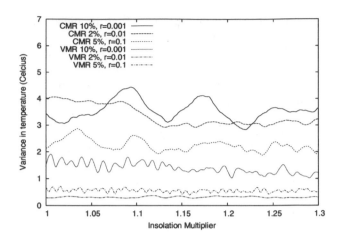

Figure 3: Variance in average temperature over a range of parameter values, detail from insolation multiplier 1 to 1.3.

vary from 0 to 1 inclusive and are evolved at birth with mutation rate r by adding a random value $\mathcal{R} \in [-r, r], \overline{\mathcal{R}} = 0$ to the parent albedo.

If a cell is covered by a daisy, then the probability that it dies is λ. The model was tested using two different methods of determining the mortality rate of the daisies. In the first method, denoted as CMR (constant mortality rate), λ is held constant regardless of temperature. This was the method used in von Bloh's formulation. With the second, denoted as VMR (variable mortality rate), λ was varied with respect to temperature. The curves used (Figure 1) were obtained by

$$\gamma(T) = 1 - \frac{4\rho}{(T_{max} - T_{min})^2}(T - T_{min})(T_{max} - T) \quad (8)$$

where $\rho \in [0, 1]$ and serves set the minimum mortality rate. If ρ is large, then the minimum mortality rate will be low. This is just a reformulation of the birth rate equation and serves as a means of introducing a temperature-dependent mortality rate to the system.

Results

All of the experiments were performed on a 50 by 50 square lattice and consisted of 50 simulation runs for any given parameter value. In each, 10% of the the planet surface area was populated randomly by approximately equal numbers of black and white daisies. The temperature of the planet was uniformly set to 26° C. Each simulation was allowed to reach an equilibrium temperature and then the solar output was increased linearly. The transition from a cold dead planet to an equilibrium state is not considered, only the transition from this equilibrium state to planetary "death" from increasing temperatures.

Figure 2 shows the model response with no mutation and a variety of different mortality rates after averaging the 50 runs. The system performs the best when mortality rates are low, as is expected, maintaining a temperature of 40 ° C when the insolation is 2.2 times the initial values. The runs with a mortality rate of 20% fare the worst – only regulating temperature to 1.7 times the initial insolation.

Figures 4 and 5 show the results of the model with a mutation rate r of 0.01. With minimum mortality rates less than 5% the VMR model persists for significantly longer than the CMR model, even though the mortality rates are higher (Figure 1). At minimum mortality rates higher than 5% the CMR models persist for longer.

However, this does not tell the whole story. Looking at the variance of the simulation runs (Figure 3) shows that the temperatures in the VMR models are much less variable than that of the CMR models across all mortality and mutation rates. This means the VMR models remain closer to the optimum growth temperature over a wide variety of daisy configurations.

To determine the effects of mutation on these models, the simulations were run with varying mutation rates. As can be seen by Figure 6, without mutation the VMR models are less persistent than the corresponding CMR models. At higher mutation rates, the situation is reversed. All the VMR models perform far better than the corresponding CMR models, as shown by Figure 7. The VMR models with a 10% minimum mortality rate regulate far better than the CMR models with only a 2% mortality rate.

Discussion and Conclusion

These results are rather surprising. How could a higher mortality rate yield increased persistence, and why does this effect not hold true all the time?

To answer the first question, consider the CMR model

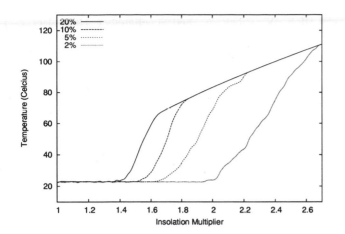

Figure 4: Average temperature of the VMR model with mutation rate 0.01 and minimum mortality rates of 2%, 5%, 10%, and 20%.

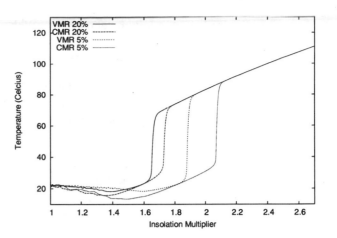

Figure 6: Comparison of average temperatures between the VMR and CMR models with no mutation.

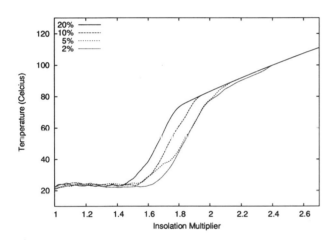

Figure 5: Average temperature of the CMR model with mutation rate 0.01 and minimum mortality rates of 2%, 5%, 10%, and 20%.

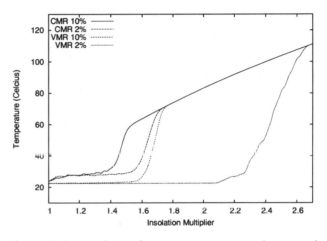

Figure 7: Comparison of average temperatures between the VMR and CMR models with a mutation rate of 0.1.

where daisies are killed randomly. Daisies well suited to their local environments could be killed while those that are not are allowed to persist. This could create local instabilities in temperature and cause the global system to be less stable. However with the variable mortality rate, those daisies that are unsuitable for their local environment are more likely to die while those that are suited remain. This effect is most noticible when when mutation rates are high. As the potential difference between parent and offspring increases, the chance for a unsuitable offspring being born also increases. Under the VMR model, these offspring die off fairly quickly and open up the cells that they inhabited for regrowth, while under the CMR model they could persist and serve to destabilize the neighborhood.

The higher rate of mortality outside the optimum temper-

ature under the VMR model yields a higher selection pressure. This provides important feedback to daisy patches and allows a patch to tune albedos to the optimum values for a given local temperature. However, it also means less variation in the albedos – a fact that helps in some case and hinders in others. In an unperturbed system, this high selection pressure means that the variance in daisy albedo across the planet will tend to some minimum value, determined mainly by the albedo mutation rate. While fine for unchanging systems, a low variance in albedos means that the daisies must mutate in order to persist when faced with increasing insolation. With a high variance in albedos, it is far more likely that there could be a daisy suitable for the increasing temperature. Systems with high mutation rates have a double advantage – a greater rate of change in response to the envrionment and more inherent variation.

As the variance in daisy albedo is higher in the CMR model due to the lower selective pressure, this allows the systems to persist for longer at low mutation rates. With higher high mutation rates, the higher selective pressure of the VMR model weeds out the unsuitable daisies quicker and leads to those systems persisting for longer.

To see how this works, imagine a patch of light daisies surrounded by uncovered land. As the temperature increases in the VMR model, the daisies on the outside of the patch die more often than those in the middle. With low mutation rates, they stand to be replaced by daisies very similar to themselves – thus the replacement daisies stand a similar chance of dying. Eventually the combined effect of lower birth rates and higher death rates becomes too much and, unable to replace daisies fast enough, the patch dies from the outside-in. With high mutation rates, there is more variation in albedos and a higher chance of a daisy being born which is capable of surviving the increasing temperatures. This creates a small scale feedback loop where the daisy patch can replenish its borders at a higher rate.

Local interactions of these types play a large part in the dynamics of any system – especially in small systems. The model world used in this paper is incredibly small. For comparison, von Bloh used a CA lattice of 1600 cells a side in his simulations (von Bloh et al., 1997). Larger worlds result in less variation in global temperature, especially in the CMR models, and they tend to persist for longer, though more work needs to be done to fully detail this relationship.

Natural selection and the environment provides pressure on organisms. Gaia theory suggests organisms provide pressure on their environment. Combining the two gives surprising results – even with much higher overall mortality rates, natural selection can increase temperature regulation and persistence within Daisyworld. Whether this effect extends to real world systems remains to be seen.

Acknowledgments

I would like to thank my supervisor, Dr. Emmet Spier, for his guidance and assurance.

References

Harding, S. (1999). Food web complexity enhances community stability and climate regulation in a geophysiological model. *Tellus*, 51B:815–829.

Lenton, T. (1998). Gaia and natural selection. *Nature*, 394:439–447.

Lenton, T. and Oijen, M. (2002). Gaia as a complex adaptive system. *Phil. Trans. R. Soc. Lond. B*, 357:683–695.

Saunders, P. (1994). Evolution without natural selection: Further implications of the daisyworld parable. *J. theor. Biol.*, 166:365–373.

Stöcker, S. (1995). Regarding mutations in daisyworld models. *J. theor. Biol.*, 175:495–501.

von Bloh, W., Block, A., and Schellnhuber, H.-J. (1997). Self stabilization of the biosphere under global change: a tutorial geophysiological approach. *Tellus*, 49B:249–262.

Watson, A. and Lovelock, J. (1983). Biological homeostasis of the global environment: the parable of daisyworld. *Tellus*, 35B:286–289.

Weber, S. (2001). On homeostasis in daisyworld. *Climatic Change*, 48:465–485.

The Flexible Balance of Evolutionary Novelty and Memory in the Face of Environmental Catastrophes

Andrew Buchanan, Mark Triant and Mark A. Bedau
Reed College, 3203 SE Woodstock Blvd., Portland OR 97202 USA
bedau@reed.edu

Abstract

We study the effects of environmental catastrophes on the evolution of a population of sensory-motor agents with individually evolving mutation rates, and compare these effects in a variety of control systems. The evolution of mutation rates must balance (i) the need for evolutionary "novelty," which pushes mutation rates up, and (ii) the need for evolutionary "memory," which pushes mutation rates down. We observe that an environmental catastrophe initially shifts the balance toward evolutionary novelty and causes mutation rates to evolve upwards. Then, as the population adapts to the new environment, the balance shifts back toward evolutionary memory and the mutation rate falls. These observations support the hypothesis that second-order evolution of the mutation maintains a flexibly shifting balance between evolutionary novelty and memory.

Introduction

The evolution of life on Earth has repeatedly been forced to adapt to environmental catastrophes (Raup, 1991), and evolution through such catastrophes has been suggested as a mechanism behind adaptive radiation (Raup, 1986). As artificial evolving systems increasingly interact with unpredictable environments, their robustness in the face of environmental catastrophes will also become increasingly critical. Some (e.g., (Stanley, 1973; Stanley, 1990; Sepkoski, 1997)) account for diversity trends in the fossil record by appealing to the tendency of ecologically unspecialized taxa to better survive drastic environmental change. This may likewise figure into an explanation of a taxon's resilience in the face of environmental catastrophe. In any case, understanding the nature and source of evolutionary robustness in the face of environmental catastrophes is important for understanding both natural and artificial evolution.

The ability of a system to adapt to environmental catastrophes is related to its evolvability, i.e., the capacity for evolution to create new adaptations. Evolvability has received considerable attention recently in both the biological and evolutionary computation communities. There is general agreement that evolvability is crucial for understanding the origin of complex adaptations (Conrad, 1982; Wagner

and Altenberg, 1996) as well as the process of open-ended evolution (Taylor, 1999; McMullin, 2000), a central open problem in artificial life (Bedau et al., 2000). This paper treats the issue of how evolvability evolves and specifically focuses on the evolution of evolvability in a finite population adapting to a dynamic endogenous fitness function in a spatial environment.

A system's evolvability depends on its ability to produce adaptive phenotypic variation, and this hinges on both the extent to which the system's phenotype space contains adaptive variation and the ability of evolutionary search to locate it while avoiding maladaptive traps. Two main factors control the effectiveness of the evolutionary search process: the way in which genetic operators traverse genotype space, and the way that genotypes are phenotypically expressed (the genotype-phenotype mapping). For evolutionary search to explore a suitable variety of viable evolutionary pathways, genetic operators must generate sufficient amounts of the right kind of genetic *novelty*. At the same time, since evolutionary adaptations are built through successive improvements, genetic *memory* is required for the evolutionary process to retain incremental improvements.

Genetic operators like mutation rate simultaneously affect the demands for both evolutionary novelty and memory, so evolvability requires genetic operators to balance these competing demands successfully. (In evolutionary computation this principle is known as the tension between "exploration" and "exploitation" (Holland, 1975).) Furthermore, the appropriate balance between evolutionary novelty and memory can continually change as evolution progresses, so evolvability requires genetic mechanisms to adjust the balance flexibly. Thus, understanding evolvability involves understanding how the balance between evolutionary novelty and memory shifts during the course of evolution, and what general mechanisms allow this balance to flexibly shift. One way to address these questions is to let second-order evolution control the genetic mechanisms that structure first-order evolution, enabling us to study the evolution of evolvability. There are a variety of such genetic mechanisms, but perhaps the simplest is the mutation rate. This paper addresses the evolution of evolvability in this context.

Optimal and evolving mutation rates are discussed in both the biological and evolutionary computation literatures (Kimura, 1960; Eiben et al., 1999); further references are cited elsewhere (Bedau and Packard, 2003). But it remains unclear what, if anything, such results might reveal about the evolution of mutation rates in a finite spatial population with many loci and many alleles per locus, with heavy neutral evolution, and when the context of evolution continually changes and thus the fitness function unpredictably varies— the context of the present study. The discussion of evolving mutation rates in the evolutionary computation literature mainly focuses on the issue of automated control of evolution, in the context of efforts to minimize the time required to solve function optimization problems. But this work typically presumes that evolution is driven by a fixed and externally-specified fitness function, whereas the theoretical issues that interest us concern evolution with implicitly-specified fitness functions that continually change in the course of evolution.

An earlier treatment of these issues provided preliminary evidence that second-order evolution of mutation rates allows a flexible response to exogenously shifting demands for evolutionary novelty and memory (Bedau and Packard, 2003). The present paper systematically explores this earlier result, in order to determine how robust the process is and to identify the mechanisms behind it. In particular, we study the robustness of this process by observing hundreds of catastrophes caused by transitions to a variety of different kinds of environments. In addition, we isolate the mechanisms behind what we observe by comparing these results with a variety of control systems.

Our experiments with environmental catastrophes also have the potential to address a further issue: whether and how evolving through a series of environmental catastrophes increases a population's ability to adapt to novel environments in the future. Earlier work has shown that autonomous agents with the ability to learn during their lifetime adapt to changing environments (Menczer and Belew, 1996), choose from a variety of environmental niches (Walker, 1999), and evolve increased genomic and neural complexity when subjected to noisier environments (Seth, 1998). Environmental catastrophes have also been shown to affect the diversity dynamics of evolving communicative systems (Green et al., 2000). Because our system includes second-order evolution, we can look at similar issues in a simpler and more fundamental context, and thus perhaps explain what happens in more complex settings.

The Model

Our evolutionary system is composed of many agents that could be called organisms, on analogy with biological systems. It has been used in various studies of the evolution of sensory-motor functionality; see the references in earlier

work (Bedau and Packard, 2003). The system consists of many agents that exist together in an environment, in this case a toroidal lattice. The lattice has a binary field defined on it, $\mathcal{E}(x)$, which may be interpreted as a resource field. The pattern of the resource field is static and resources are not depleted by agents. Thus, agents are constantly gathering resources and expending them through metabolism and reproduction, but the resource pattern in the environment remains fixed. The exception to this are the periodic environmental catastrophes, as described below.

The resource patterns in our system consist of three qualitatively distinct types. The first, B, maps a square onto the lattice. The absolute amount of resources available to the agents across different simulations in B is fixed, but the location is chosen at random. The second, C, maps resources onto the environment in numerous pairs of equivalent, mutually bisecting, perpendicular line segments. The placement of each pair of line segments in C is chosen at random independently of other pairs leading to partial overlap. The absolute amount of resources availble is thus an inverse function of the degree of overlap in that particular instantiation of C. The third resource pattern, R, maps individual resource sites onto the environment at random. As with C, the sites are chosen independently of one another leading to a degree of overlap and variable absolute amounts of food. However, in all but improbable cases of extreme overlap, the total resources, $\Sigma\mathcal{E}$, are such that $\Sigma\mathcal{E}_R > \Sigma\mathcal{E}_C > \Sigma\mathcal{E}_B$.

In analogy with biological systems, the dynamics of the population as a whole are comprised of all the birth-life-death cycles of the agents. Births occur when agents accumulate enough resources to reproduce (see below), deaths occur when agents run out of resources, and the lives of agents consist of their interactions the environment. We label each agent with the index i, let I^t be the set of agents existing at t. Time is discrete. One unit of time is marked by each agent acting based on their local environment.

During its lifetime, each agent extracts information from the environment by sensing and reacting to the local resource field. There is a discrete set of different possible sensory states, $s \in S = \{s_1, ..., s_{N_S}\}$. The agents exchange no information with each other directly (although this would be an easy generalization). Each agent (labeled with the index i) has certain information associated with it: (i) a current location, x_i^t, (ii) a current sensory state, s_i^t, (iii) a current reservoir of resources, E_i^t, and (iv) a sensory-motor map, φ_i, that yields a behavior given sensory information as input. In this model, an agent's behavior is a vector, $\vec{b}$, denoting the agent's movement in the environment, $x_i^t \rightarrow x_i^{t+1} = x_i^t + \vec{b}_i^t$ where $\vec{b}_i^t = \varphi_i(s_i^t)$. Every $\vec{b}_i^t$ is a member of a discrete set of different possible behaviors, $\vec{b}_i^t \in B = \{\vec{b}_1, ..., \vec{b}_{N_B}\}$. (Another easy generalization would be to include other kinds of possible behaviors.) Apart from reproduction, the agent's supply of resources may be changed in one of two manners: an augmentation from extracting resources at the agent's new location, or a constant-sized universal reduction, $E_i^t \rightarrow E_i^{t+1} = E_i^t + \alpha(\mathcal{E}(x_i^{t+1})) - \beta$, where $\alpha(z)$ is a resource extraction function and β is a constant metabolic cost of surviving.

The sensory-motor map, φ_i, operating on a sensory input, s_i^t, has a particularly simple form because the sensory input is discrete. Since $s_i^t \in \{s_1, \ldots, s_{N_S}\}$, we may identify the function φ_i with its graph, a set of N_S behavior values, $\{\varphi_{is}\}$. Pursuing the biological analogy, we will consider the sensory-motor strategy elements $\{\varphi_{is}\}$ as i's *genome*, with N_S loci, and each particular element $\vec{b} = \varphi_{is}$ as the trait (or allele) at the sth locus of i's genome.

Reproduction occurs when an agent's resource supply exceeds a threshold, E_r. The parent splits its resources with its child, and the child inherits its parent's strategy elements $\{\varphi_{is}\}$, apart from changes introduced by mutation. The genome of each agent i contains a special gene, μ_i, which controls the rate at which i's strategy elements mutate when i reproduces, i.e., the probability that a strategy element of i's children is chosen (with equal probability) from the set of possible behaviors, B. We also introduce a *meta-mutation* rate parameter, μ_μ—the probability that i's children's mutation-rate gene is chosen (with equal probability) from the interval $[\mu_i - \varepsilon, \mu_i + \varepsilon]$. The value of μ_μ is fixed during the course of a given simulation, and the value of μ_i is fixed during the course of i's lifetime (as is φ_i).

This model provides a simple setting for empirical study of the evolution of evolvability. Agents' immediate environments produce sensory states that then trigger actions by means of the agents' sensory-motor maps. Since the agents' survival and reproduction depends directly on their ability to continually find resources in their environment, the implicit fitness functions in this model are constantly buffeted by the contingencies of natural selection and, thus, unpredictably change. This first-order evolution is structured by the sensory-motor maps actually compared and tested by natural selection. One especially simple mechanism that regulates the variety of maps available for evolutionary exploration is the mutation rate; the higher the mutation rates, the greater the variety. Thus, by allowing mutation-rate genes to evolve, we can study second-order evolution of evolvability.

Methods

We collected data from the model in an experimental situation and three kinds of controls. In the experimental condition, the agents' mutation rates were allowed to evolve and the environment cycled through a sequence of qualitatively different patterns, B, C and R, at a regular interval. The effect of these environmental catastrophes on the agents in this phase and all subsequent phases was measured in terms of population and average mutation rate across the entire population. The Same-Environment (SE) control was identical to the experimental situation except that the catas-

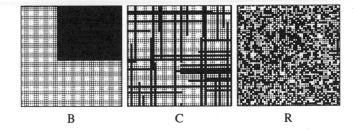

B   C   R

Figure 1: Top-down pictures of a part of the three environments B, C, and *R*.

trophes did not involve a qualitative change in environment; rather, whichever pattern was in place was redistributed so that local environments were changed, but the overall environment was qualitatively identical. In the Population-Only (PO) control, the environmental catastrophe was replaced by an artificial population catastrophe. Ninety percent of the population was selected at random and killed, mimicking the effect that environmental catastrophes had on the population. We conducted 80 trials in the experimental situation with counterbalanced sequences, and these data were averaged for analysis. In addition, we generated a total of 48 runs in the SE control and 29 in the PO control.

The size of the torroidal lattice was held fixed at 141×141 sites and all trials had 120,000 iterations with catastrophes every 10,000 iterations. When the mutation rate was allowed to evolve, the initial mutation rates μ were set to 0.3 (30% of the loci mutated each reproduction event). In order to prompt quick evolution of mutation rates, μ_μ was set to 0.5 ε to 0.1 (e.g., 50% chance of the mutation rate mutating up or down by as much as 0.1). The ratio of metabolic cost to the resource intake rate and to the reproduction threshhold was 10:11:50 in all conditions. Initial populations always consisted of 1,000 agents. Reproduction was always asexual and always produced exactly one new agent. Population and mutation rate data were sampled every 30 model updates. Resource pattern B was 45×45 sites for a total of 2,025 resource sites. Pattern C consisted of 70 pairs of lines with 101 total resource sites for a total of 7,070 resource sites less any overlapping sites. 12,500 resource sites were distributed in the R pattern (less some overlap). See Figure 1.

Results

Figure 2 shows the population level response to the following regular sequence of environmental catastrophes: R, B, C, R, B, etc. The results from 80 runs were averaged, and the data before the first catastrophe was discarded. The catastrophes occur in such quick succession that the population level never reaches equilibrium in an environment, thereby enabling us to study the population's immediate response to catastrophe. The catastrophes have a dramatic impact on population level, truncating the population to less than 10%

env: R B C R B C R B C

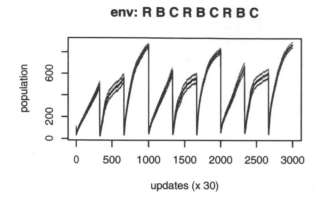

env: R B C R B C R B C

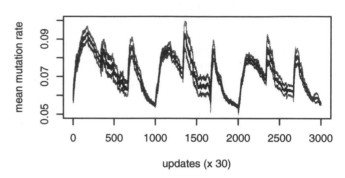

Figure 2: Dynamics of the population level (dark) plus and minus the standard error (light), averaged across 80 runs. Every 10,000 model updates the environment suffers an environmental catastrophe.

Figure 3: Dynamics of the mean mutation rate (dark) plus and minus the standard error (light), averaged across the same runs as in Figure 2. Every 10,000 model updates the environment suffers an environmental catastrophe.

of its pre-catastrophe size, but the population recovers as it adapts to its new environment. By comparison, the population bottleneck in the SE control is never even as small as 50% of the population size (data not shown). Evidently, when the population experiences a new instance of the same kind of environment in the SE control, most agents can successfully use the same genes in the new environment. This implies that after an environmental catastrophe in the experimental condition, the population cannot rely on genes it already has but must readapt. Even if the population experienced that same kind of environment earlier in its history, adaptive behavioral patterns are no longer in genetic memory; it remembers little if anything about how to behave in that environment since it has most recently been in two very different kinds of environments.

When the population adapts to a given environment and never experiences environmental catastrophes, the mutation rate evolves down to an equilibrium value (data not shown). Figure 3 shows how the mutation rate responds to a series of environmental catastrophes. After each kind of catastrophe, the mean mutation rate rises significantly, and the population level starts to recover from the catastrophe. While the population level continues to rise, after some time the mean mutation rate then starts to fall and is still falling when the next catastrophe strikes. This starts the same mean mutation rate dynamics again.

Figure 4 shows the mean mutation rate recovery from the three different kinds of environmental catastrophes. Note that each kind of environment produces a characteristic mutation rate dynamic, and that the R dynamic differs significantly from the characteristic B and C dynamics, which are similar but distinguishable. Comparing the mutation rate dynamic in the PO control (light) shows that the mutation rate response to environmental catastrophes is not solely a

result of population bottleneck the catastrophes cause. Furthermore, the mutation rate dynamic in the SE control (data not shown) is like that in the PO control. Thus, the mutation rate response to environmental catastrophe is not caused by any type of environmental disruption. Rather, it is caused by experiencing a new kind of environment, even if that same kind of environment had been experienced at some point in the past.

Figure 5 shows how the population level responds to different kinds of environmental catastrophes. Note that different kinds of environment produce characteristically different kinds of population dynamics. Also note that in the R environment the population is adapting more quickly to later R catastrophes (the thick line has lower slope than the thin line). We see a different but related effect in the B environment (higher maximum population levels after experiencing more catastrophes).

Discussion

The mutation rate dynamics seen in Figure 3 fit the hypothesis that second-order evolution can fluidly balance evolutionary memory and novelty. The low population level immediately after an environmental catastrophe indicates that the population is not well adapted to its new environment, so the balance shifts toward evolutionary novelty. The observed rise in mutation rate achieves this. As the population becomes adapted to its new environment, the population level climbs, and the balance shifts back toward evolutionary memory. The observed fall in mutation rate brings this about.

What is the mechanism by which the mutation rate adapts? Having a higher or lower mutation rate does not affect an individual's chances of surviving or reproducing; its benefit would be felt only by an individual's offspring. If an

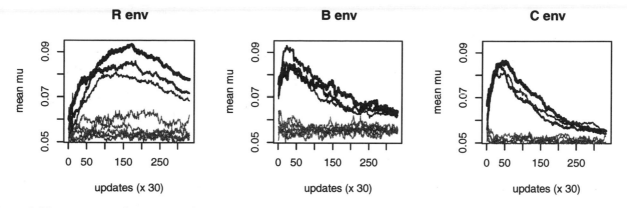

Figure 4: The response of mean mutation rate (dark) in three different kinds of environmental catastrophes, averaging data from 80 runs. The three different curves in a given environment correspond to the first (thick dark line), second (medium dark line), and third (thin dark line) time that kind of environment was encountered in the series of catastrophic environmental changes. This is compared with the mean mutation rate response in the same environment to 6 catastrophes in the PO control (light), averaging 11 runs in the *C* environment, 11 in the *B* environment, and 7 in the *R* environment.

individual has the appropriate mutation rate in a given context, it is more likely that the individual's offspring will survive and flourish. So, if an individual's fitness is measured also by the survivorship and fecundity of its offspring, then having a contextually appropriate mutation rate increases an individual's fitness. Just as "inclusive fitness" extends an individual's fitness over space through kin (Hamilton 1964), an individual's fitness also can be extended over time through a lineage. Thus, the mechanism behind the adaption of mutation rates can be viewed as a kind of group selection.

The different characteristic mutation rate dynamics elicited by different kinds of environmental catastrophes shown in Figure 4 further indicate the fluidity of this balancing process. When and how the balance shifts depends on the nature of the environment and the extent to which the population has already adapted to it.

We would explain the characteristic mutation rate dynamics in the different environments by reference to their different local niche structure. The balance remains shifted towards evolutionary novelty (higher mutation rates) longer in *R* than in *B* and *C* because *R* contains a greater number of different local niches that call for different local behavioral strategies, and, thus, adaptation takes more time. By contrast, it is relatively easy to find the local strategies needed in *B* and *C* so the balance quickly tips back toward evolutionary memory. In addition, we hypothesize that the faster shift back toward memory in *B* than in *C* is a result of resources in *B* occurring in one contiguous space, while the resources in *C* are distributed across space, so it takes the agents longer to physically explore all the niches.

The local niche structure in the different environments also explains the characteristic population dynamics seen in Figure 5. The population in *R* continues to adapt to new local niches throughout its time in the *R* environment, so the population is continually rising. In both the *B* and *C* environments the population eventually starts to saturate the available niches, and the population starts to level off.

The fact that the population adapts to new *R* environments better after it has experienced more *R* catastrophes is intriguing (left panel of Figure 5). This provides evidence for the hypothesis that experiencing environmental catastrophes increases a population's ability to adapt to new environments, a hypothesis suggested by the fossil record (Raup, 1986). The way in which the final population level in *R* and *B* environments increases with the number of catastrophes experienced tends to further corroborate this hypothesis.

Topics for future work include getting experimental evidence about the robustness of the flexible balance of evolutionary memory and novelty by studying a wider range of models subject to a wider range of environmental catastrophes, and constructing a quantitative theory of the flexible balance of evolutionary memory and novelty that enables the mutation dynamics to be predicted, and that explains the connections of this phenomenon with other theories about mutation rate such as that concerning quasispecies (Eigen and Schuster, 2001; Eigen et al., 1988).

Acknowledgements

Thanks to Mike Raven and Albyn Jones for help with code, to Jeffrey Fletcher and John Huss for helpful discussion, and to the ALife IX referees for helpful comments.

References

Bedau, M. A., McCaskill, J. S., Packard, N. H., Rasmussen, S., Adami, C., Green, D. G., Ikegami, T., Kaneko, K., and Ray, T. S. (2000). Open problems in artificial life. *Artificial Life*, 6:363–376.

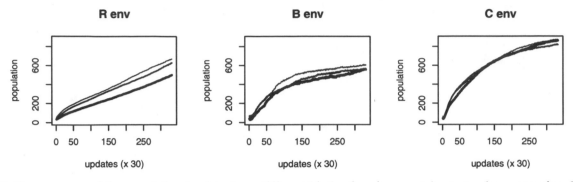

Figure 5: The response of the population level to three different kinds of environmental catastrophes, averaging data from 80 runs. The three curves in a given environment show the response to the first (thick line), second (medium line), and third (thin line) environmental catastrophe of a given kind.

Bedau, M. A. and Packard, N. H. (2003). Evolution of evolvability via adaptation of mutation rates. *Biosystems*, 69:143–162.

Conrad, M. (1982). Natural selection and the evolution of neutralism. *Biosystems*, 15:83–85.

Eiben, A. E., Hinterding, R., and Michalewicz, Z. (1999). Parameter control in evolutionary algorithms. *IEEE Transactions on Evolutionary Computation*, 3:124–141.

Eigen, M., McCaskill, J., and Schuster, P. (1988). Molecular quasispecies. *Journal of Physical Chemistry*, 92:6881–6891.

Eigen, M. and Schuster, P. (2001). *The hypercycle: a principle of natural self-organization*. Springer-Verlag.

Green, D. G., Newth, D., and Kirley, M. G. (2000). Connectivity and catastrophe: Towards a general theory of evolution. In Bedau, M. A., McCaskill, J. S., Packard, N. H., and Rasmussen, S., editors, *Artificial Life VII*, page 153. Cambridge, MA: MIT Press.

Hamilton, W. D. (1964). The genetical evolution of social behavior. *Journal of Theoretical Biology*, 7:1–52.

Holland, J. H. (1975). *Adaptation in Natural and Artificial Systems: An Introductory Analysis with Applications to Biology, Control, and Artificial Intelligence*. University of Michigan Press, Ann Arbor. (Second Edition: MIT Press, 1992.).

Kimura, M. (1960). Optimum mutation rate and degree of dominance as determined by the principle of minimum genetic load. *Journal of Genetics*, 57:21–34.

McMullin, B. (2000). John von neumann and the evolutionary growth of complexity: looking backwards, looking forwards. *Artificial Life*, 6:347–361.

Menczer, F. and Belew, R. K. (1996). From complex environments to complex behaviors. *Adaptive Behavior*, 4:317–363.

Raup, D. M. (1986). Biological extinction in earth history. *Science*, 231:1528–1533.

Raup, D. M. (1991). *Extinction: bad genes or bad luck?* Norton.

Sepkoski, J. J., J. (1997). A kinetic model of phanerozoic taxonomic diversity ii: early phanerozoic families and multiple equilibria. *Paleobiology*, 5:222–251.

Seth, A. K. (1998). The evolution of complexity and the value of variability. In Adami, C., Belew, R. K., Kitano, H., and Taylor, C. E., editors, *Artificial Life VI*, page 209. Cambridge, MA: MIT Press.

Stanley, S. M. (1973). An explanation of cope's rule. *Evolution*, 27:1–26.

Stanley, S. M. (1990). Delayed recovery and the spacing of major extinctions. *Paleobiology*, 16:401–414.

Taylor, T. (1999). On self-reproduction and evolvability. In Floreano, D., Nicoud, J.-D., and Mondada, F., editors, *Advances in Artificial Life*, pages 94–103. Springer-Verlag.

Wagner, G. P. and Altenberg, L. (1996). Complex adaptations and the evolution of evolvability. *Evolution*, 50:967–976.

Walker, R. (1999). Niche selection and the evolution of complex behavior in a changing environment – a simulation. *Artificial Life*, 5:271–289.

Kin-Selection: The Rise and Fall of Kin-Cheaters

Sherri Goings*, Jeff Clune*†, Charles Ofria and Robert T. Pennock

Michigan State University, East Lansing, USA

*Both authors contributed equally to this paper
†Article Correspondence: jclune@msu.edu*

Abstract

We demonstrate the existence of altruism via kin selection in artificial life and explore its nuances. We do so in the Avida system through a setup that is based on the behavior of colicinogenic bacteria: Organisms can kill unrelated organisms in a given radius but must kill themselves to do so. Initially, we confirm results found in the bacterial world: Digital organisms do sacrifice themselves for their kin—an extreme example of altruism—and do so more often in structured environments, where kin are always nearby, than in well-mixed environments, where the location of kin is stochastically determined. Having shown that helping one's kin is advantageous, we turn our attention to investigating the efficacy and implications of the strategies of kin-cheaters, those who receive help from kin but do not return it. Contrary to the expectations of current theory, we find that kin-cheaters outcompete kin-altruists. Our results cause us to question the stability of strategies that involve altruism between kin. Knowing that kin-altruism persists in biological systems, however, we search for, and find, conditions that allow kin-based altruism to persist in evolving systems despite the presence of kin-cheaters.

Introduction

At first glance, the persistence of altruistic behavior (defined as helping other organisms at a net cost to the acting organism) is puzzling from an evolutionary perspective. However, in nature we frequently see acts that seem altruistic. Many theories exist that help explain such apparent acts of altruism. They usually demonstrate that the acting entity, when properly identified, actually receives a net gain by the apparently altruistic behavior. A prominent theory is kin selection, where an individual suffers or risks a net cost to help its kin (Hamilton 1963). The theory of kin selection recognizes the gene as the acting agent and argues that a certain altruistic gene (e.g. gene A) is actually helping out copies of itself by causing an individual that possesses it to act altruistically toward that individual's kin (since the individual's kin is likely to have gene A) (Dawkins 1976). In this paper we demonstrate the existence of kin selection in the Avida digital evolution system and investigate an aspect of the theory that is often unexamined; whether kin that cheat on their relatives are favored by natural selection. We accomplish this by comparing organisms that are altruistic

towards close relatives (kin-altruists) and those that are only altruistic towards identical copies of themselves (clone-altruists). We refer to these clone-altruists as 'kin-cheaters' to emphasize that they are not altruistic towards non-identical kin even though such non-identical kin are altruistic towards them. Most theorists have overlooked the potential success of this type of cheater. We investigate whether the assumption generally made by theorists that kin-cheaters should not be selected for is a valid one.

It is often helpful to look at extreme cases when investigating a theory. The extreme act of altruism is giving one's life for another. For this reason we chose to study the evolution of this behavior as our means of studying kin selection in artificial life. A further reason is because kin-selection driven by an organism sacrificing its life has been well studied in the bacterial world (Chao and Levin 1981). Checking our results against these findings enables us to confirm that the computational system we are using to investigate evolutionary phenomena is behaving similarly to the biological systems to which we wish to extrapolate our findings.

A small percentage of colicinogenic bacteria will produce a toxin until they explode, releasing the toxin into the surrounding area. This toxin is harmful to those that are not immune to it. Since colicinogenic bacteria are immune to this toxin, those harmed by it are typically non-kin. This strategy has been likened to that of 'suicide bombers' (Lenski and Velicer 2000). Chao and Levin found that this trait is more likely to be beneficial in structured environments, where one's kin are next to one, versus well-mixed environments, such as a liquid culture, where resources and spatial location are randomized. They attribute the greater benefit in structured environments to the increased likelihood that the extra resources provided by killing non-kin will be received by kin due to their frequent proximity, thus differentially (and "selfishly") helping those likely to share one's genes.

We modified the Avida digital evolution system to facilitate the study of kin-altruism. To do this we added to Avida the element that makes colicinogenic bacteria ideal for studying kin-altruism: the ability of organisms to kill a number of non-kin in their surroundings by sacrificing their lives. Initially, we find that this strategy is overwhelmingly selected for. We then replicate some of the findings of Chao and Levin: sacrificing one's life to aid one's kin is more effective in structured environments than in well-mixed (randomized) environments. We further investigate whether the presence of kin-cheaters can prevent kin-based altruism from being a stable evolutionary strategy, and show that

1) Kin-altruists thrive in the absence of kin-cheaters.
2) Once kin-cheaters arrive on the scene, they outcompete kin-altruists, raising the question of how the altruism predicted by kin-selection theory persists in nature.
3) Limiting factors on the destructive power of kin-cheaters exist, allowing kin-based altruism to persist.

Methods

All of the following experiments use the Avida digital evolution system, a virtual environment in which digital 'organisms' evolve through random mutation and natural selection (Ofria and Wilke 2004). At the start of each 'run' of Avida (one execution of the software program), a virtual world is seeded with a digital organism that can self-replicate. Each digital organism has a sequence of instructions considered to be its genome. Self-replication involves copying this genome and then dividing into two child organisms. The copy process is imperfect, however, so each instruction has a chance of mutating to any random instruction when copied. These organisms quickly fill up the virtual environment and compete for a limiting resource: SIPS (Single Instruction Processing units). These are the basic unit of energy available to an organism. Since organisms need this energy to execute their genomes, and thereby replicate, those that earn more SIPS (by performing tasks) or use less (via efficiency) will tend to be selected for naturally. As Daniel Dennett says, "evolution will occur whenever and wherever three conditions are met: replication, variation (mutation), and differential fitness (competition)" (Dennett 2002). The Avida system includes these conditions and is thus a tractable system that we can use to investigate the general properties of any evolving system (Lenski et al. 2003).

We modified the system by adding an instruction called `explode` to the list of instructions an organism is able to execute. An organism that executes this instruction will probabilistically "explode", killing itself and emitting virtual toxins. The percent probability that executing the `explode` instruction will cause an organism to explode varies from organism to organism depending on its genotype (see detailed methods). For example, one organism may explode 90% of the time it executes `explode`, whereas another may do so only 2% of the time. A probabilistic approach is necessary for otherwise all or none of a lineage would explode. If the instruction does not cause the organism to explode, it has no effect except to use one of the organism's allotted SIPS. If the organism does explode, it kills itself and all non-kin organisms within a given explosion radius. The kin/non-kin distinctions are made based on the genetic (Hamming)

distance between the exploding organism and those inside this radius. Whether an exploding organism considers another its kin depends on whether the Hamming distance (number of genomic differences) between the organisms is less than or equal to the exploding organism's Hamming distance threshold (HDT). This Hamming distance threshold can be different for different organisms and is set before the run, unless otherwise specified. The radius parameter gives the distance an exploding organism can propel its fatal toxins when exploding. The radius was set to two for all experiments except for baseline runs, meaning an exploding organism affects the 24 organisms surrounding it. This parameter is also set at the start of each run and does not change during the course of a run. A population size of 3,600 organisms and a genomic mutation rate of .2 were used for all runs unless otherwise noted.

Experiments and Results

Evolution of Kin-Altruism

Once we had modified the system to allow the `explode` instruction to mutate in, we were interested in seeing if it would be selected for. We set a global HDT of 0, so that an exploding organism kills all organisms with non-identical genomes in the blast radius of 2. We performed twenty runs seeded with the default organism that does not have the `explode` instruction in its genome and thus never explodes. In every run (20/20) the number of explosions went up substantially (fig. 1). In these structured runs, when organisms replicate, their children are placed into neighboring cells (killing the organism that previously existed in that cell). The killing of non-kin by an exploding organism thus eliminates organisms that had the potential to kill the exploding organism's kin (if any such kin were nearby). This benefit conferred on an exploding organism's kin was large enough to outweigh the cost of death, as evidenced by the strong selection for this behavior in our initial twenty runs. In order to assure that this selection was due to the benefit provided to kin, we performed two different baseline tests that eliminated this benefit. In the first set of baseline runs we set the radius to zero so that an exploding organism killed only itself. In the second set of baseline runs the exploding organism killed every organism in a radius of two, kin and non-kin alike. In both of these cases, the frequency of explosions did not rise above the minimal level that is to be expected by the high rate at which the `explode` instruction mutates in. The clear difference between the experiment and the control runs can be seen in figure 1.

In studying colicinogenic bacteria, Chao and Levin found that sacrificing one's life to benefit one's kin is a more effective strategy in structured environments, where kin remain near one another, than in well-mixed environments, where the locations of kin and resources are continuously randomized (Chao and Levin 1981). They reason that the benefits of sacrificing one's life—more available resources—are more likely to be shared by one's kin in structured environments than in well-mixed environments. The same reasoning should apply to our setup: The benefit of sacrificing one's life—eliminating those that can kill one's kin—will only be of benefit if one's kin are nearby when one explodes. Chao and Levin's experiments focused on whether small colonies of colicinogenic bacteria were more likely to *invade* non-colicinogenic bacteria in these two different environments. We took a slightly different approach, looking at whether strategies involving sacrificing one's life were more likely to *evolve* in these disparate environments. Having already shown that such strategies evolve in structured environments (see above), we did a set of 20 runs in a well-mixed environment that were otherwise identical to the first set. Our findings support Chao and Levin's results and explanation: the strategy of sacrificing oneself for one's kin did not evolve in any (0/20) of our well-mixed runs (fig. 1).

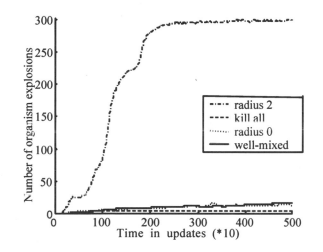

Figure 1: Number of organism explosions per 100 updates averaged over all 20 runs. The "...." line represents one base-line experiment in which an exploding organism kills only itself. The "----" line signifies the other base-line run in which an exploding organism kills itself and all surrounding organisms, regardless of kinship, within a radius of 2. The ".-.-" line represents the first experiment in which an exploding organism kills itself and all surrounding non-kin within a radius of 2. These three treatments all occurred in a structured environment. The "—" line represents the second experiment: It is identical to the first except that it occurs in a well-mixed environment.

While this does not show that such behavior cannot evolve in well-mixed environments (for it would surely evolve if the blast radius were drastically increased) these results do show that it is much less likely for the strategy to evolve in well-mixed environments, all else being equal[1]. Note that this also reinforces our conclusion above that the reason this strategy is evolving is due to the benefits it confers on kin. It should be noted that, as far as we know, organisms in natural systems are unable to determine the exact genetic distance between themselves and other organisms. Instead, they tend to use whatever reliable indicators of kinship are available (e.g. spatial proximity, smell, physical similarity, etc.), and rely on the fact that close kinship and genetic similarity are typically correlated. While this difference in information may mean that natural organisms will be less able to implement beneficial strategies effectively, we do not believe this difference will substantively change the underlying strategies themselves. Therefore it should not preclude us from extrapolating our findings from the digital organisms in our setup to organisms in nature.

The Problem of Kin-Cheaters

Having shown that it is sometimes advantageous to give one's life to aid one's kin, we turned our attention to the problem of kin-cheaters (organisms that receive benefits if a relative sacrifices its life but will kill this relative if they explode). To analyze the effect of these cheaters we started by assessing what occurs in their absence. We wanted to test if, in the absence of kin-cheaters, it is advantageous to be less than maximally discriminating about who one considers to be kin. To translate the question into the language of sexual species, is it more beneficial to be maximally discriminatory, and thus only help identical twins, or should one be less discriminatory, and thus help brothers, nieces, cousins, second cousins, etc.?

Efficacy of Kin-Altruism: Organisms in the original runs (described above) can be considered maximally discriminatory because they killed anyone with a non-identical genotype when they exploded. To test whether less discriminatory strategies would invade, we seeded 50 runs with two groups of organisms; one maximally discriminatory and the other non-maximally discriminatory. Logistically, organisms from these two groups differed only in two respects. The first difference was the Hamming distance threshold (HDT) setting in their explode instruction (see methods). The maximally discriminatory organisms had HDTs set to 0 (HDT.0) and thus killed all non-identical genomes. The less discriminatory organisms were set to HDT.5, and thus killed all creatures with a genetic distance of 6 or greater (i.e. more than 5 of the 100 instructions in an affected organism's genotype differed from the exploding organism). The second difference was the changing of some of the neutral nop-x instructions to effectively neutral nop-a instructions, to create a genetic distance of 12 between organisms of the two groups, ensuring that each

---

[1] Chao and Levin found that this strategy is less likely to invade in well-mixed environments, where it only invades if it begins above a certain frequency, versus in structured environments, where it can invade no matter how small its initial frequency (Chao and Levin 1981).

type considered the other to be non-kin at the outset. 95% of the initial population consisted of maximally discriminatory organisms and 5% were less discriminatory. In all 50 runs the more altruistic lineage, which began as a small minority, invaded and went to fixation. From this we conclude that less discriminatory (more altruistic) strategies are more beneficial in the absence of kin-cheaters than maximally discriminatory (selfish) strategies.

Efficacy of Kin-Cheating: Where one finds altruism, one typically finds cheating. With regard to altruism amongst kin, however, evolutionary theorists largely ignore the existence of kin-cheaters. While the strategy of faking that one is indeed related to a kin-altruist is often discussed, as in the case of the cuckoo, rarely does the literature on kin selection probe the idea that a true relative may not be reciprocating the generosity of its brethren. The main reason for this seems to be the persuasive argument that a gene is better off helping copies of itself in another organism and thus would be worse off by cheating (see, for example, Dawkins 1982). It has been argued, however, that the cheating phenomenon should be no less prevalent for genes than individuals: any gene that receives help from its kin but does not reciprocate should do better off compared to its altruist relatives (Sober and Wilson 1999). We used our setup to evaluate these competing theories by testing the efficacy of a kin-cheating strategy.

In our setup, kin-cheaters are those who, if they explode, will kill less discriminating relatives but will not be killed by such relatives (because the less discriminating relatives consider them kin). It should be noted that we are stretching the concept of 'cheater' a bit. Cheaters in our setup are altruistic towards some organisms (namely identical copies of themselves), but can be considered cheaters because, when interacting with more altruistic relatives, they receive a benefit but do not return it. To test whether such kin-cheaters would invade kin-altruists, we again ran 50 runs seeded with two groups of organisms; one maximally discriminatory (HDT.0) and one less discriminatory (HDT.5). This time, however, the initial genetic (Hamming) distance between the organisms of the two groups was one. This means that organisms in the less discriminatory group are altruists that consider cheaters kin—and thus do not kill them when exploding—but the maximally discriminatory organisms are kin-cheaters that kill altruists when exploding. The population was seeded with 95% kin-altruists and 5% kin-cheaters. In 45 out of the 50 runs the kin-cheater lineage went to fixation. Whether the two are related, then, makes a significant difference (p-value < .001 using Fisher's exact test). Figure 2 represents two of these runs; one where the kin-cheater went to fixation and one where it did not.

Conditions that Enable the Success of Kin-Altruists in the Presence of Kin-Cheaters: This result is interesting for two reasons. Initially, it shows that in most cases kin-cheaters eliminate kin-altruists. This result is consistent with Sober and Wilson's argument that kin-cheaters are at least a phenomenon that kin selection theory needs to account for. The widely held assumption that organisms

that cheat on their kin will be worse off, all else being equal, seems incorrect. Also interesting is the fact that kin-altruists win 10% of the time (5/50). How is the destructive power of kin-cheaters countered, here or in nature where we see kin altruism all the time? Sober and Wilson propose one theory, a group selectionist account, that requires kin-groups that do not have cheating members to outperform those that do. This cannot be the explanation for why kin-cheaters failed to fixate in these five runs, however, since our setup does not include the differential survival of multiple groups. Thus, whether or not one accepts Sober and Wilson's explanation of one force that may mitigate kin-cheaters, there appears to be at least one more force at work that manifests itself in our setup.

We watched a few of the runs in which altruists win and ran a few more in which the organisms were allowed to change their Hamming distance threshold via mutation

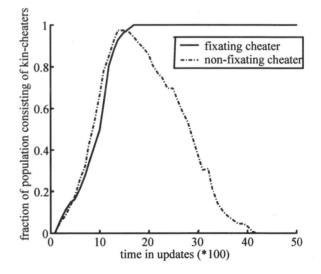

Figure 2: Two sample runs seeded with 95% kin-altruistic organisms (HDT.5) and 5% kin-cheaters (HDT.0) with a genetic (Hamming) distance of 1 between the two groups. In one of the displayed runs (solid line) the kin-cheater quickly fixates. This occurred in 45/50 runs. In the other run (dotted line) the kin-cheater quickly expands in the population, coming very close to fixation, but is stopped at the last moment by the kin-altruist, which has mutated away until it no longer considers the kin-cheater to be its relative, and thus is no longer altruistic towards it. This phenomenon occurred in 5/50 runs.

during the course of the run. We observed two independent, non-concurring phenomena that prevent cheaters from going to fixation in the population. The first cause occurs when a group unrelated to the cheaters evolves an altruistic strategy of self-sacrifice with a HDT greater than 0. This group outcompetes the cheaters because, as shown in our earlier experiment, a less discriminatory group of organisms will outcompete an unrelated group of maximally discriminatory organisms.

The second cause is if the diversity within the kin-altruist group reaches a point at which one section of the group is not willing to be altruistic toward the cheater because the genetic distance between the organisms in that section of the group and the cheater is greater than 5 (fig. 2). These two causes boil down to the same principle: The cheater will not go to fixation if it encounters a group before it fixates that is more altruistic than it but is not altruistic *towards* it (i.e. does not consider it to be kin). We hypothesize that there are a number of different environmental conditions that will increase the frequency of such an encounter occurring, including larger population sizes, decreased fitness of the cheaters (they will fixate less quickly), higher mutation rates, and the degree of altruism (how great a Hamming distance threshold) in kin-altruist groups (the higher the Hamming distance threshold the longer it takes to reach sufficient diversity).[2] We tested one of the predictions of this hypothesis by varying the population size. Kin-cheaters had gone to fixation in 45 of the 50 runs with populations of 3,600 organisms. We performed 50 more runs with populations of 10,000 organisms. Lending strong support to the hypothesis, cheaters fixate in only 2 of the 50 runs (p-value < .001 using Fisher's exact test). Figure 3 summarizes these results.

	More Altruistic (Less Discriminatory)	More Selfish (More Discriminatory)
Unrelated	50/50	0/50
Related	5/50	45/50
Related Large Population	48/50	2/50

Figure 3: Summary of three experiments where two different groups of organisms competed. The data report the number of runs out of 50 in which that group went to fixation, eliminating the other group entirely. More selfish organisms kill all but their closest kin (HDT.0). More altruistic organisms are less likely to kill distant relations (HDT.5). When the two groups of organisms are unrelated, the more altruistic—or less discriminatory—group wins. When the two groups are related, the more selfish—or maximally discriminatory—variant is a 'kin-cheater.' Kin-cheaters will outcompete their more altruistic relatives under certain conditions but are less likely to do so in larger populations.

---

[2] This last force will rarely be significant in anything but the smallest of groups or at very low mutation rates, as group diversity will reach the Hamming distance threshold rather quickly otherwise.

The two types of events that check the ability of kin-cheaters to fixate are both interesting for different reasons. If the first type were the only mitigating force, where kin-cheaters are successful at outcompeting all of their relatives but lose when they encounter a non-related band of more altruistic organisms, kin-altruism would be individually unstable even though it could be a persistent feature within the population. In other words, every time kin-altruism evolves, its level of discrimination will continuously be reigned in by more discriminatory offspring until it ends up as maximally discriminatory. Under these harsh conditions, non-maximally discriminatory altruism could only be persistent in the population if such altruism evolved with a high enough frequency that there were always groups that had not yet transitioned to a maximally discriminatory strategy. While theoretically possible, such a setup in the natural world is unlikely and does not describe the more successful versions of kin-altruism we see in biological systems.

The second class of events relates more to the world we live in. In these situations, a cheater evolves within a group of related altruists and spreads until it encounters the border at which those that are related to it are no longer willing to give it preferential treatment. An analogy could be drawn to the human cultural practice of treating distant relations equal to strangers. One may find an open palm when seeking a loan from brothers, cousins, and even second cousins. At some genetic distance, however, the fact that relatives are treated on par with strangers means that one must secure a loan on merit instead of nepotism. This practice, whether conscious or not, limits the size of the area any given kin-cheater can exploit.

Conclusions

We evolved kin selection in the Avida digital evolution system using a setup similar in respects to colicinogenic bacteria. Organisms could sacrifice themselves and, in the process, kill non-kin greater than a certain genetic (Hamming) distance away. We further investigated the question as to whether kin-cheaters would outcompete their kin-altruist relatives. We show that without cheaters a more altruistic (less discriminatory) strategy outcompetes a more selfish (more discriminatory) strategy. With kin-cheaters, however, more discriminatory strategies are selected for. Unchecked, and assuming that cheaters will eventually arise, this tendency of more discriminatory strategies to outcompete less discriminatory strategies should prevent the ability of anything but maximally discriminatory types of altruism from persisting. These implications, the possibility of which have been overlooked by most theorists, make necessary an explanation for how altruism amongst kin persists in nature. We observe one candidate explanation: The possibility that kin-cheaters will run into altruists they cannot exploit serves as a check on their tendency to eliminate altruism towards distant kin. We further show that many conditions exist that make this possibility quite likely, providing one potential way to explain how non-maximally discriminatory altruism amongst kin endures.

Detailed Methods

All of the experiments were performed using version 2.0 beta7 of the Avida software. The default organism used in our experiments consisted of a 15-instruction long copy loop that performs self-replication and 85 nop-x instructions that perform no function when executed. The length of the organism is fixed at 100. An Avida organism consists of its "genome" (sequence of instructions), 2 stacks (only one of which is active at any given time), and 3 registers (A, B, and C). Unless otherwise noted, the default instruction set was used with the addition of the `explode` instruction. This set contains instructions for self-replicating, performing logic operations, and manipulating numbers in the stacks and registers. The added `explode` instruction, when executed, inputs the number in the organism's register A by default, although if it is followed by a nop-B or nop-C it will input the number in the specified register (B or C) instead. The instruction then mods this number by 100 to get the percent chance that the organism will explode. This percentage is then compared to a random number and, if the percentage is greater than or equal to the random number, the organism explodes (otherwise it continues executing its code).

The organism also inputs the number on top of the currently active stack and sets the Hamming distance threshold (HDT) parameter to this value. The Hamming distance between two organisms is determined by comparing their genomes site by site and totaling the number of differences found. The Hamming distance between two organisms is therefore the genetic difference between those organisms. An exploding organism with a HDT of 0 will kill all non-identical organisms; with a HDT of 1 it will kill any organism with more than one non-identical site, etc. All experiments use organisms with fixed lengths so no alignment is necessary before the determination of Hamming distances. In the first set of experiments (those researching the evolution of self-sacrifice) the HDT was fixed at 0 for all organisms. In the second set of experiments (the competitions), the HDT is set at 5 for one group of organisms and 0 for the other for the duration of the run. The units of the radius parameter are cells on the environment grid, so an organism exploding with a radius of 2 will affect the 24 organisms that are no more than 2 cells away. In the base run in which the radius is set to 0 the organism kills only itself. The mutation rate used is .002 per instruction copied, which equals .2 mutations per genome, unless otherwise noted. The population consists of 3,600 organisms placed on a 60x60 grid, unless otherwise noted. Two birth methods were used: well-mixed (a.k.a. randomized, mass-action, mass-habitat, liquid, well-shaken, etc.) and neighborhood (structured). In well-mixed environments a new organism is placed in a cell randomly chosen from the entire population. In neighborhood environments a new organism is placed in a cell randomly chosen from the 8 cells in the 3x3 square surrounding the parent organism. In both methods the organism currently occupying the chosen cell is killed.

Acknowledgments

We would like to thank Richard Lenski and Dusan Misevic for their insightful suggestions. We also thank all the members of the digital evolution group at Michigan State University. This work was supported, in part, by an NSF grant (DEB-9981397) to Richard Lenski and colleagues and by the College of Arts and Letters Dean's Recruitment Fellowship at Michigan State University.

References

Adami, C. (1998). *Introduction to Artificial Life*. Santa Clara: TELOS Springer-Verlag.

Adami, C., Seki, R., and Yirdaw, R. (1998). Critical exponent of species-size distribution in evolution. In Adami, C., Belew, R., Kitano, H., and Taylor, C. editors, *Proceedings of "Artificial Life VI,"* p. 221-227. Cambridge, MA: MIT Press.

Chao, L. and Levin, B. R. (1981). Structured habitats and the evolution of anticompetitor toxins in bacteria. *Proc. Natl. Acad. Sci. USA* 78:6324-6328.

Dawkins, R. (1982). *The Extended Phenotype*. Oxford and San Francisco: Freeman.

Dawkins, R. (1976). *The Selfish Gene*. Oxford: Oxford University Press.

Dennett, D. (2002). The new replicators. Pp. E83-92 in Pagel, M., ed. *Encyclopedia of Evolution*. New York: Oxford University Press.

Hamilton, W.D. (1963). The evolution of altruistic behavior. *American Naturalist* 97:354-356.

Lenski, R. E., Ofria, C., Pennock, R. T., and Adami, C. (2003). The evolutionary origin of complex features. *Nature* 423:139-144.

Lenski, R. E. and Velicer, G. J. (2000). Games microbes play. *Selection* 1:51-57.

Ofria, C. and Wilke, C. O. (2004). Avida: a software platform for research in computational evolutionary biology, *Artificial Life* 10:191-229.

Sober, E. and Wilson, D. S. (1999). *Unto Others*. Cambridge, MA: Harvard University Press.

Homeostasis and Rein Control: From Daisyworld to Active Perception

Inman Harvey

Evolutionary and Adaptive Systems Group
Centre for Computational Neuroscience and Robotics
Centre for the Study of Evolution
COGS/Informatics, School of Science and Technology, University of Sussex, BN1 9QH, UK
inmanh@cogs.susx.ac.uk

Abstract

Homeostasis refers to the ability of organisms to maintain vital properties, such as body temperature, within a zone of viability, or of comfort, and the Gaia Hypothesis proposes that the Earth with its biota acts as a homeostatic whole. The Daisyworld model was proposed as one possible mechanism for providing this homeostatic regulation. Here a new and much simplified version of this model is presented, demonstrating that the combination of any 'Hat function' with any feedback, positive or negative, can lead to homeostasis through 'Rein Control'. This principle is so general that it can be extended to other domains such as active perception, here demonstrated in a simulated robot.

Introduction

Living organisms have many physiological variables that must be maintained within upper and lower bounds for continued survival. Typically there are regulatory mechanisms that maintain these variables within these bounds even in the face of substantial environmental perturbations. These homeostatic systems can take many forms. For mammals such as humans that need to maintain their body temperature within fairly tight limits, the variety of mechanisms includes physiological, reflex behaviour such as shivering and sweating, and more considered behaviour such as moving towards or away from heat.

Such regulatory mechanisms are important for the survival of the organism, and often appear to be complex, subtle, and so crafted as to provide near-optimal conditions for the organism; their origins and maintenance are usually attributed to the power of selection over many generations of Darwinian evolution. So controversy was inevitable when Lovelock (1972) proposed the Gaia Hypothesis that the Earth with its biota acts as a homeostatic geo-physiological system that regulates global properties, e.g. temperature at the Earth's surface, within a range that provides viable living conditions for the biota; for a review see (Lenton 1998). No immediate explanation was available for the origin or cause of any such regulation, as clearly the Earth as a whole had not evolved through successive generations of selection within a population of Earths. Further, as species within the biota, and individuals within each species, were competing with each other, it was difficult to see how selection at the individual level could favour behaviour that led to global cooperation in regulating a global variable. Individual acts of behaviour typically have a cost to the individual; yet the net effect from just one individual on global temperature is insignificant. So selection at the individual level will favour those profligates who do not care to do their bit towards global homeostasis. These arguments appeared initially to give sound evolutionary reasons why the global regulation proposed in the Gaia Hypothesis could not be maintained, or indeed even have arisen in the first place.

Daisyworld

The Gaia Hypothesis was originally restricted to the claim (based on observation) that such global regulation existed, without any theory or mechanism to explain how it might happen; it initially met with much skepticism. The 'parable of Daisyworld' was then proposed as a possible mechanism (Lovelock 1983, Watson and Lovelock 1983).

Whereas the surface temperature of a lifeless planet would have changed dramatically with the increase in luminosity of the sun over geological timescales, observation indicates that on our Earth it has remained remarkably constant around temperatures suitable for life. Daisyworld is a deliberately simplified model of an imaginary planet with just two species of daisies, black and white, that demonstrates how this could happen. The growth rate of the daisies depends only on their local temperature, but in turn the daisies modify this because of differences in the way they absorb radiation; black daisies have low albedo (or reflectivity) and heat up easily, whereas white daisies with higher albedo tend to reflect the sun's radiation back.

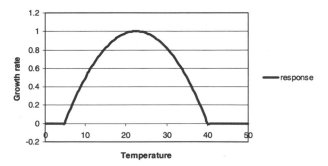

Figure 1.Hat function: daisy response to temperature

The dependency of growth rates on temperature is assumed to be a Hat-shaped function as in figure 1. The Daisyworld model demonstrates how these feedbacks via the environment, both positive for black and negative for white daisies, result in regulation of the planetary temperature. Watson and Lovelock (1983) demonstrate that with their particular parameter values and equations, the resulting close-coupled system regulates the temperature to within the viability zone (here 5^0 to 40^0C) over a far wider range of solar luminosities than would have been the case in the absence of any daisies.

Comprehensibility of Daisyworld

Daisyworld is intentionally a simplified model; temperature is taken as just one example of an essential variable that can be regulated, and the lessons from the Daisyworld parable are meant to have far wider scope. One point made early on is that the precise form of the Hat function is unimportant, provided it has the general peaked character around an optimum temperature, with the brim of the hat representing here a zero growth rate outside the viability range. Nevertheless, the use of the Stefan-Boltzmann law to calculate absorbed and emitted radiation, and use of the equations governing the comparative and indeed competitive growth rates of the different daisies, means that analysis of this system is not a trivial problem.

I have therefore adopted the strategy of radically simplifying the usual Daisyworld model, to see how much can be left out whilst still retaining the homeostatic regulation. In doing so, firstly it becomes clearer how crucial is the difference in local temperatures between black and white daisies, something often obscured by the conventional graphs shown; secondly it becomes much easier to visualize the very simple underlying feedback interactions; thirdly it becomes plainer just how much can be generalized from this one example to other domains.

Organisation of Paper

In the following sections I shall start by describing the cut-down version of Daisyworld. Visualisations of the conditions for steady state will be shown. Then results obtained through computer simulations integrating the equations to a steady state will be given. The conditions and parameter values will be manipulated to see just how far they can change whilst maintaining robustness.

I shall draw some very general conclusions, and to demonstrate their generality apply them to a very different domain of active perception. Here a simulated robot is supplied with oriented light-sensors that display a similar Hat function response to a light source. Feedback directly coupled to this response will change the orientation of the light-sensor in (a random choice of) either positive or negative direction. Collectively the coupling between many such individual light-sensors determines the global orientation of the robot. The result is homeostasis in the sense that collectively the system acts so as to maintain, as far as possible, the light-sensors oriented to the light so as

to stay within their sensitive regions; in other words, phototaxis despite the random nature of the feedback.

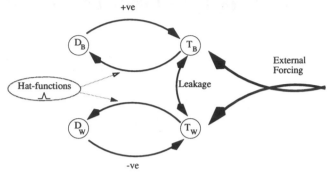

Figure 2. Interactions in the cut-down model: a black daisybed above, separate white one below. Both receive external forcing from the sun, and the only interaction between them is by 'leakage' or heat conductance.

Cut-down Daisyworld

The simplifications are twofold: firstly, the model is idealized into a simpler form with fewer interactions; secondly, the Hat function and the putative underlying laws of radiation and heat conductance are simplified into linear or piecewise linear form.

The black and white daisies can be modeled as growing on separate daisybeds, in other words not competing for space. The interactions and feedbacks are then limited to those shown in figure 2. The Hat function can be replaced by a piecewise linear function of similar general form, which I call a 'Witch's Hat' function. T_B and T_W are the average temperatures of each daisybed, D_B and D_W are the proportional coverage of each daisybed by black and white daisies respectively, as determined by the Witch's Hat function. The temperature T of each bed (taken as uniform within the bed) is determined by a combination of factors: external forcing by the sun, feedback (positive or negative) proportional to D (the coverage of daisies in that bed), and a 'leakage' factor whereby some (parameterized) proportion of the heat flows from the hotter to the cooler daisybed. Unlike the original model, there is no direct interaction between D_B and D_W.

A Single Daisybed

Initially we can simplify still further by just looking at the behaviour of a single daisybed; consider one half only of figure 2. T is the temperature of a bed with albedo α, S the temperature of the Sun, and deep space is at zero temperature. In a simplified, distorted version of physics, heat flow into the bed from the Sun is $(1 - \alpha)(S - T)$, and out of the bed into deep space is $(T - 0)$, i.e. T.

D is now the quantity of daisies (rather than growth rate), which varies according to a Hat function of the local temperature T: $D = H(T)$. In the simplest version, where we assume there is feedback linearly proportional to D to raise (or lower) the local temperature T, this feedback is uD: for black daisies u is a positive feedback to increase the temperature, for white daisies u is negative.

The rate of change of flowerbed temperature is

$$\frac{dT}{dt}=(1-\alpha)(S-T)-T+uH(T)$$

Equilibrium is when the rate of change is zero:

$$0=(1-\alpha)S-(2-\alpha)T+u.H(T)$$

$$H(T)=\frac{(2-\alpha)T-(1-\alpha)S}{u}$$

For fixed S, u, α, this is linear in T with zero value when

$$T=\frac{(1-\alpha)S}{(2-\alpha)}$$

The line has slope $(2-\alpha)/u$. The equilibrium points are where this straight line crosses the Hat function.

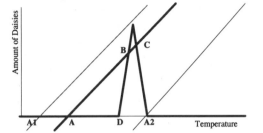

Figure 3. Slope meets Witch's Hat function at A, B, and C.

In figure 3 the heavy sloping line has a positive slope (black daisies for positive feedback) and crosses H(T) at A, B, and C. A implies no daisies, B proves to be an unstable equilibrium, but C is a stable non-zero equilibrium. For different amounts of external forcing from changing luminosity of the sun, the sloping line shifts, in parallel fashion, along the temperature axis. There will be a stable equilibrium point C, with a positive quantity of daisies, for any such line between the lighter sloping lines through A1 and A2 in the figure; these indicate the limits for intersecting the RHS of the Hat. This implies that the range of viability that allows some daisies to survive extends all the way from A1 to A2, rather than the limited range D to A2 available if there is no feedback. The slope of the line reflects the degree of feedback, with a vertical line corresponding to u=0. The stronger the feedback, the larger u is, the further away the slope is from vertical; and hence the further away to the left the viability range is extended. A1 lies at a distance $u/(2-\alpha)$ to the left of the central optimum temperature of the Witch's Hat, if one takes the maximum height of that hat to be scaled to 1.0. If the line has negative slope (white daisies for negative feedback) then the mirror image case holds, and the range of viability is extended out to the right instead. So regardless of the sign of the feedback, the range of viability is extended.

Rein Control

In a thoroughly mixed metaphor, in figure 3 the line AC can be thought of as a rein pulling the zone of viability towards A, rather than its default left limit of D. Clynes (1969) put forward the notion of *Rein Control*, in commenting that biological systems typically have (at least) two channels for sensing and regulating variables: one (or more) in one direction from the norm, another in the other direction. This notion has received relatively little currency, although it is taken up in recent work drawing ideas from Daisyworld theory and applying them in modified form to physiological control (Saunders et al. 1998). The rein metaphor is appropriate as a rein can only pull, not push. Hence for control in both directions we need a further feedback loop, as in the following extension to the simulation; we need both reins.

Two Daisybeds

For the simulation, we assume 2 daisybeds whose bare ground is grey with albedo 0.5. One bed can support only black daisies with a lower albedo (typical value used 0.0); the other can only support white daisies (typical albedo 1.0). For each bed, the average albedo depends on the proportion of cover by black daisies ($0 \le D_B \le 1$) or white daisies ($0 \le D_W \le 1$). The consequent temperatures, assumed to be uniform across each bed, are T_B and T_W. These temperatures are then potentially modified by heat transfer between the beds from the hotter black one to the white. This transfer is parameterised by a factor $0 \le L \le 1$. When L=0, no heat transfer takes place, but if L=1 then the beds each have their temperature modified to the mid-temperature $(T_B + T_W)/2$; for intermediate values of L the temperatures are scaled linearly between these two extreme cases.

To find through computation any non-zero stable equilibrium point for fixed values of S and L, we initialise D_B and D_W to 0.5, and then iterate this loop:

1. Calculate albedo for each bed from D_B, D_W.
2. Calculate T_B and T_W from S and these albedos.
3. Adjust these temperatures by the between-bed heat transfer, or 'leakage', parameterised by L.
4. Use the Hat function to calculate $D'_B = H(T_B)$ and $D'_W = H(T_W)$
5. Adjust D_B and D_W a small proportion of the way towards these new values D'_B and D'_W by:

   $$D \Leftarrow (1-\delta)D + \delta D' \text{ for a small value of } \delta.$$

6. Go back to 1.

δ should be chosen small enough to ensure that the values change smoothly over successive iterations of this loop, and then the loop must be repeated sufficiently many times until the changes in values at each iteration are vanishingly small. In practice it was found, for the range of parameters used here, that $\delta = 0.0001$ and 200000 iterations of the loop made further changes in the variables invisible at the level of double precision floating point numbers.

Results are shown in graphical form for various values of the parameter L. In each case the resulting equilibria are shown across the full range of external forcing by the sun, as it varies from excessively cool to excessively hot.

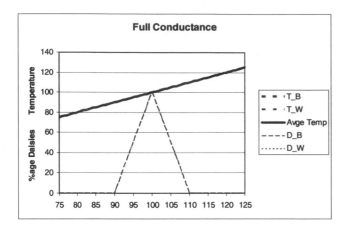

Figure 4. Conductance or Leakage L=1. The superimposed lines for D_B and D_W indicate the Witch's Hat function, with a viability zone between 90 and 110 on the lower scale. Vertical axis indicates temperatures (T_B, T_W and average, here all superimposed) and also percentage of Daisies in each bed ((D_B, D_W, here also superimposed).

Maximum heat conductance. If L=1, the temperatures of black and white beds are identical. Since they follow the same Hat function, there is always the same number of black daisies in one bed as white daisies in the other. Hence the overall average effect is that of grey, albedo 0.5. In other words, there is no net feedback, and (regardless of how many daisies there are) the temperature is the same as if there were none. Maximum conductance means minimum, or zero, homeostasis. See figure 4, where the horizontal axis indicates the sun's output, scaled according to the corresponding temperature of a lifeless planet; here the temperatures T_B and T_W are the same as this.

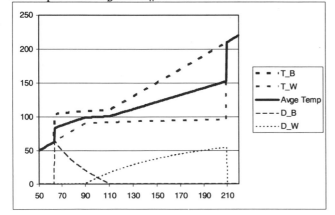

Figure 5. L=0. Daisybeds are independent, and only in the range 90-110 on horizontal axis do both daisies coexist.

Minimum conductance. When L=0, the two daisybeds are completely unconnected, and behave as if they were separate planets each regulating itself; the black bed extends its viability only towards lower sun temperatures, the white bed only towards higher. See figure 5.

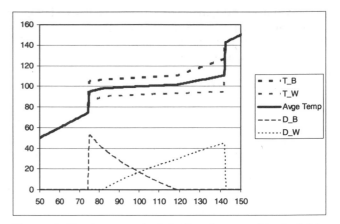

Figure 6. Conductance L=0.5. Coexistence of both daisies between 82 and 118.5

Intermediate conductance. Figure 6 shows the more general picture, where although the extension of black daisies left, white daisies right, is not as far as in figure 6, the range of coexistence of both daisies is greater. Note that at all times that either type of daisy is viable, the black daisies are hotter than their optimal temperature of 100^0, and the white daisies are cooler than this.

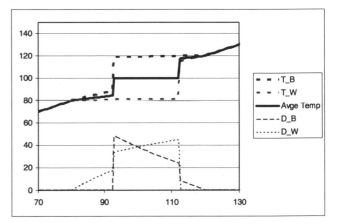

Figure 7. The daisies are given different, narrower, Hat functions: black's from 115-120, white's from 80-85. There is coexistence of both daisies between 93 and 112.

Moving the Hat functions. Figure 7 shows that there is still homeostasis when the viability zones (Hat functions) of the daisies are shifted relative to each other. Note from the figures in the caption, black is here shifted towards the hotter end, white towards the cooler end.

Moving to Active Perception

We have seen above how the simple yet powerful combination of a Hat function with a feedback loop (either positive or negative) produces *Rein Control*, and in the context of Daisyworld, homeostasis; the Hat function directly relates to the idea of a zone of viability, and Rein Control tends to regulate a system to stay within it. The cut-down Daisyworld has reproduced the basic homeostatic results of the usual version; though the simpler equations used means that it does not reproduce the phenomenon whereby the average planetary temperature actually decreases slightly as solar luminosity increases.

Such a powerful principle can be extended to other domains, and here it is demonstrated with active perception in a simple simulated 2-dimensional robot. Despite the very different domain, the underlying principles are identical.

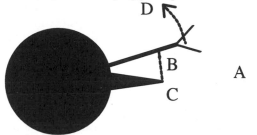

Figure 8. Circular agent can only rotate about its centre; orientation is indicated by nose C, currently facing East. One tentacle is shown, with sensor at end A.

One sensory tentacle is shown in the plan view of figure 8. For fans of *Doctor Who*, think in terms of a *Dalek*. The tentacle rotates around the centre of the robot, and has a sensory angle of acceptance as indicated at the end A. If a light source passes across this receptive field, the sensor response is given by a Witch's Hat function, with maximum response when the tentacle points directly at the source. This response produces a torque D on the tentacle A, in this case shown as left or counter-clockwise. D is counteracted by a restraining spring B attached to the nose C. Two parameters, specific to this tentacle, modulate the torque response D and the spring modulus B.

Now consider 100 such tentacles, each with different randomly chosen angles of acceptance, different directions and torque parameters for D, different spring constants for B. As a light source passes in front of them, each tentacle will respond independently. But collectively they will be held together via the springs attached to the nose, some pulling in one direction, some in the other. In the absence of any light, all these tentacles will be drawn together over the nose, but in the presence of any stimulation they will rotate apart independently in different directions, restrained only by the springs. The resultant balance of these spring forces on the nose will rotate the robot as a whole in one direction or the other, about its centre.

The translation from Daisyworld is direct as far as the underlying equations go, even though conceptually it is quite a leap. Each tentacle corresponds to a daisy species; on average half will provide feedback in one direction and half the other. The Hat function on sensor response corresponds to daisy dependence on temperature. The springs correspond to the leakage or conductance of heat between daisybeds, and the resulting direction of the nose corresponds to the average planetary temperature. The dynamics of the motion of each tentacle, and consequent

rotation of the robot, are calculated in simulation with similar liberties and simplifications used for modeling the physics as were used in modeling Daisyworld, whilst respecting the general principles. The end result is that, despite the random parameterization of all the tentacles, this robot efficiently performs phototaxis.

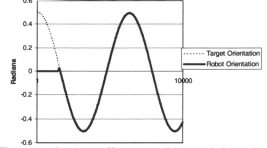

Figure 9. Time runs horizontally across this graph from the left, and the heavy line indicates the orientation of the nose in radians. The lighter line indicates a light source passing in wide, sinusoidal fashion across the front of the robot.

As indicated in figure 9, the robot will immediately pick up on a passing light source, and track it so accurately that thereafter the plots coincide. Further testing shows that this behaviour is exceptionally robust to changes in the allowed ranges for the randomly chosen parameters. The maximum angle of acceptance can be allowed to vary over 3 orders of magnitude, out as far as 3 radians (or nearly 180°) each side. The upper limit on torque parameters and spring constants can be allowed to vary over more than 2 orders of magnitude; phototaxis is still reliable.

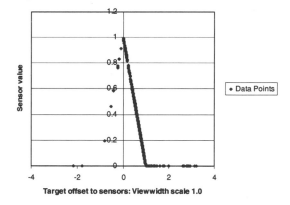

Figure 10. Samples from all the left-moving sensors whilst the robot is performing phototaxis. Each data point indicates on the vertical scale the sensor response, calibrated to a maximum of 1.0, and on the horizontal scale the angular offset of target source to tentacle direction. Although different tentacles have different angles of acceptance, this offset is here rescaled so as make all the individual Hat functions coincide on this graph.

Analysis. Although each tentacle can move independently, and each response to sensory stimulus is in a random direction, their collective coupling means that almost all the tentacles will stay approximately oriented towards the light source nearly all the time. While unqualified teleological language is just as inappropriate here as it is in Daisyworld, we can carefully say: "Although this robotic system only functions this way as a whole, through multiple feedbacks, it can seem to a casual observer **as if** the tentacles are trying to maintain their sensory stimulation; just as in Daisyworld it might seem **as if** the daisies are trying to regulate the temperature so as to stay within the viability zone."

Figure 10 shows samples of sensory inputs from all the counterclockwise or left-moving sensors during a run. The outline of the Witch's Hat function is clearly visible, with almost all the data points on the right-hand slope, which is where Rein Control is acting for regulation in that direction. The data from right-moving sensors is the mirror image, giving the second of the pair of metaphorical reins.

Discussion

Many people have been mystified as to how homeostatic regulation is achieved in the Daisyworld model. How could such regulation have arisen, since surely it requires some care in setting up the feedback structure and the parameters? An evolutionary origin appears unrealistic. Indeed any ongoing system that includes biota and *seems* to require global collaboration *seems* susceptible to exploitation by evolution of sub-groups towards cheating.

The cut-down version of Daisyworld presented here makes several useful pedagogical points. The analysis of a single daisybed (see figure 3) shows that extension of the range of viability in one direction arises from the simple interaction of a feedback of any sign (the sloping heavy line) and a Hat function. The feedback need not be linear, though normally it should be monotonic; the Hat function can be anything to suit your millinery tastes, although the Witch's Hat seems near-ideal from a mathematical stance. Simple feedbacks are universal in natural systems, and Hat functions are also widespread; the zone of viability associated with any homeostasis automatically implies a Hat function, and so does the typical response of any active sensor. So no special design process needs to be postulated for the basic phenomenon illustrated here with a single daisybed. We should observe such systems everywhere.

The idea of Rein Control (Clynes 1969) deserves wider currency. This phenomenon, like a rein, can only 'pull' and not 'push'. So for homeostatic regulation in both directions we need feedbacks in both directions. In the active perception example above, the 100 simulated sensor feedbacks in random directions would average out at roughly half in each direction, so any system that includes Hat functions and many arbitrary feedback loops is likely to incorporate both reins of Rein Control.

In cut-down Daisyworld, the interaction between black and white daisybeds is limited to 'leakage' and its role is made clear. Too much coupling means the opposing homeostatic tendencies will nullify each other, whereas being uncoupled would imply, in effect, separate planets. So some intermediate loose coupling between the different systems is essential, but no further global organisation is needed. In this simple version we can observe directly the

(uniform) temperature of the black bed, and likewise that of the white bed. By plotting these separately for didactic purposes, it is easier to appreciate the importance of the difference between these temperatures, something obscured in much of the previous literature where typically only average planetary temperatures have been displayed.

The original Daisyworld model has extra layers of complexity on top of this cut-down version, so it is of interest to see what is common to both. Homeostatic regulation is already apparent in the simpler version, but the phenomenon whereby average planetary temperature can actually decrease slightly as solar luminosity increases is only seen in the more complex version.

The plots shown in the various figures indicate that the black daisies are almost always living in a hotter climate than their optimum temperature. They are mostly on the right slope of the Witch's Hat; and vice versa for white daisies. Figure 10 shows the equivalent for the perception example. What if daisy evolution allowed either species to modify their metabolism, and so 'shift their viability zone' (subject to underlying physical constraints) in the direction of the climate they actually experience? Then the black Hat function would shift to the right, and the white one to the left, as shown in figure 7.

The principles shown here are very simple and of wide applicability. They do not require an evolutionary origin or explanation, but may be quite compatible with evolution.

Acknowledgment This work was motivated and stimulated by the participants in the series of EPSRC-funded workshops *Daisyworld and Beyond*, organised by Tim Lenton and myself in Sussex and Edinburgh, 2001-03.

References

Clynes, M. 1969. Cybernetic implications of rein control in perceptual and conceptual organization. Ann. NY Acad. Sci. 156: 629-670

Lenton, T. M. 1998. Gaia and natural selection. Nature 394: 439-447

Lovelock, J. E. 1972. Gaia as seen through the atmosphere. Atmos. Environ. 6: 579-580.

Lovelock, J. E. 1983. Gaia as seen through the atmosphere. In: P. Westbroek and E. W. d. Jong. Biomineralization and Biological Metal Accumulation. Dordrecht: D. Reidel Publishing Company, 15-25.

Saunders, P. T., Koeslag, J. H., and Wessels, J. A. 1998. Integral rein control in physiology. J. Th. Biol. 194: 163-173.

Watson, A. J. and Lovelock, J. E. 1983. Biological homeostasis of the global environment: the parable of Daisyworld. Tellus 35B: 284 -289.

Measuring Biological Complexity in Digital Organisms

Wei Huang, Charles Ofria, and Eric Torng

Department of Computer Science and Engineering
Michigan State University, East Lansing, MI 48824
Article Correspondence: ofria@msu.edu

Abstract

We define biological complexity as the genetic information that an organism has about its environment. We have significantly improved methods to measure complexity based on Shannon Information Theory and the principle of mutation-selection balance from population genetics. The previous method of Adami et al. was a population-based measure; it examined the information content of all genomes corresponding to the same phenotype. This population-based method had inherent limitations to its ability to approximate complexity: it requires a full population that must be at equilibrium, genomes must be fixed-length, and the environment must have only a single niche. Our new method overcomes these difficulties because it is genome-based rather than population-based. We approximate the total information in a genome as the sum of the information at each locus. The information content of a position is calculated by testing all of the possible mutations at that position and measuring the expected frequencies of potential genes in the mutation-selection equilibrium state. We discuss how this method reveals the way information is embedded in the organism during the evolutionary process.

Introduction

Our goal is to understand the relationship between biological complexity and evolution. For example, does a detrimental mutation always imply a decrease in complexity? Does a beneficial mutation always increase complexity? How does new complexity arise? Is it always an uphill climb? The answer to these questions depends on the nature of the complexity definition that you use. Many definitions have been proposed in the past, but almost all of them have serious flaws, and few use a rigorously mathematical approach.

We first review some previously used complexity measures and their limitations. We then discuss the concept of physical complexity developed by Adami and Cerf (1) and refined by Adami, Ofria, and Collier (2). To facilitate this discussion, we briefly review Shannon Information Theory and describe the Avida digital life platform. Finally, we discuss our new approach, its advantages, and initial results.

Complexity Measures

The most used metric for the complexity of a sequence is KCS (Kolmogorov-Chaitin-Solomonoff) complexity, defined as the size of shortest algorithm that can generate that sequence. This definition works in many intuitive cases, but has serious problems: some apparently complex structures can be coded in short programs such as fractals and cellular automata (3), while a long sequence with no pattern, and no meaning (effectively random) needs a long program to generate it; one that just lists the entire sequence. Thus, the KCS definition fails to be a useful measure of biological complexity.

A related complexity definition is logical depth (4). Bennett defines the logical depth of a sequence as the running time of the shortest program that computes it. Thus, it overcomes some of the problems with KCS complexity because more complex structures may take a while to generate, but there are still problems when it comes to random sequences with no meaning behind them.

A count of the number of "parts" in an organism is perhaps the simplest definition of complexity, as suggested by Hinegardner and Engelberg (5). This, of course, depends on what we recognize as parts. Hinegardner and Engelberg suggest that at root, organisms are composed of molecules, but they do not take the differences in the complexity of those molecules into account. This definition may provide a useful approximation of complexity, but it neglects any complexity inherent in gene regulation or other interactions.

In 2000 Adami and Cerf (1) developed physical complexity as a method to compute the complexity of symbolic strings. One of the authors (Ofria) then worked with Adami (2) to translate this concept to study the biological evolution of complexity.

Conceptually, the physical complexity of an organism is the amount of information that is stored in its genome about its environment. A genome stores information that is expressed into the functional capabilities of the organism in a given environment. Thus, the physical complexity of a genome or organism should mirror its functional capabilities (phenotype). They relate phenotype and physical complexity by building on some basic concepts from Shannon Information Theory. We review these concepts now.

Shannon Information Theory

The field of Information Theory uses quantitative mathematics to formally define measures of disorder and uncertainty, which are used, in turn, to define the information content of a message as the reduction of uncertainty attributed to that message. Originally, information theory was designed for telecommunications to maximize information transmittal over a noisy channel. In evolutionary biology, we can consider the replication of a genome from parent to child as a channel that genetic information is passed through. Mutations are the noise in this channel, and the quality of the resulting message will determine if a mutation is detrimental, neutral, or even beneficial.

Information theory defines *uncertainty* (also called entropy) as the number of bits we expect to need to fully specify a situation, given a set of probabilities. Uncertainty is maximized when all probabilities are equal—we have no idea what the outcome will be. Uncertainty is defined as:

$$H = -\sum p_i log_2 p_i \qquad (1)$$

In the context of information transmission, we are primarily concerned with how much information the output symbol of a channel tells us about the input symbol of the channel. If the output symbol is random, it thus provides no information about the input symbol. If, however, we have an error-free channel, then the output symbol provides complete information about the input symbol.

More formally, for a given channel, let X represent the input symbol, Y the output symbol, and m the possible symbols that can go through this channel. $H(X)$ is our base uncertainty of the input without knowing the output Y. We define our uncertainty about X after receiving Y as:

$$H(X|Y) = \sum_{y=1}^{m} \sum_{x=1}^{m} -p(y)p(x|y) \log p(x|y) \qquad (2)$$

Finally, we define the *information* (also called mutual information or mutual entropy) that goes through this channel with each symbol as the difference between these two uncertainties.

$$I(X:Y) = H(X) - H(X|Y) \qquad (3)$$

Living organisms are a special case when we study their replication as an information transmission process. The information contained within a genome determines how the organism behaves; in particular, it determines whether or not the organism can replicate. This becomes a self-reinforcing process since if required information is destroyed no further copies can be made, while if unimportant positions in the genome are mutated, this has no bearing on further replication. Thus, outside of an adaptive event, only changes in the non-informative portions of the genome will persist over time.

Population-Based Complexity

Adami et al. use Shannon Information Theory relate phenotype to physical complexity. If we know nothing about the organism, then we have maximal uncertainty about its genome (any genome is possible). On the other hand, if we know the organism's phenotype, we have less uncertainty about what the organism's genome is (only a small fraction of possible genomes corresponds to any specific phenotype). The difference between these uncertainties represents the information stored in the genome about its environment, and thus its physical complexity.

Unfortunately, it is difficult to define the entropy or uncertainty of a genome given its phenotype. To approximate this uncertainty, Adami proposed that most encodings of a phenotype would be similar to each other – typically only differing by neutral mutations. Given this assumption, a large population with the same phenotype should contain the distribution of genomes needed to calculate physical complexity. Unfortunately, it is difficult to get a large enough population, so Adami showed that in most cases, it is sufficient to calculate the entropy of a population of genomes site by site. If there is no epistasis (non-linear interactions between genome positions), this will give us the same result.

To illustrate this technique, suppose we wish to approximate the physical complexity of DNA-based organisms. We first need a population in an equilibrium state that have identical phenotypes. Without any information about the genome, we must assume that each of the four nucleotides has equal probability of occurring at any site i leading to a maximum site entropy of $H_{\max} = 2$. Let the frequencies for each nucleotide at site i within the actual population be $p_C(i)$, $p_G(i)$, $p_A(i)$, $p_T(i)$. The population entropy of this site is then

$$H_i = - \sum_{j}^{C,G,A,T} p_j(i) \log p_j(i) \qquad (4)$$

The information content at site i would then be:

$$I(i) = H_{\max} - H_i = 2 - H_i \qquad (5)$$

Finally, the physical complexity for this phenotype is approximated by applying this equation to each site and summing them together.

The Avida Platform

To apply this population-based physical complexity, we must have a population of genomes with the same phenotype. Adami, Ofria and Collier used the Avida Platform (8) to generate populations of evolved digital organisms. The Avida software maintains a population of self-replicating computer programs (similar to computer viruses) that evolve subject to natural selection in a complex environment. The phenotype of an organism corresponds to the set of actions it can perform and related information such as timing. Organisms receive energy for performing specific computations. The fitness of the organism is then its total energy intake divided by the energy required to produce an offspring. The genome of an organism is composed of a Turing-complete programming language; that is they can perform any computable mathematical function—no explicit limitations are imposed on what can be evolved. Indeed, we have witnessed a wide variety of unexpected and seemingly clever adaptations arise through evolution in Avida.

Measurement Limitations

In their previous study, Adami et al. used the Avida platform to examine the evolution of physical complexity in digital organisms. Avida was setup in single-niche, mass-action mode. The single-niche aspect means that the organisms are in direct competition against each other and the species with the highest fitness phenotype will dominate. The fact that the population is mass-action means that there is no local structure so if a higher fitness species evolves it can to take over rapidly due to an exponential growth rate. Finally, they forced all organisms to have the same length genomes (100 sites) so that sequence alignment would be unnecessary. During each experiment, they calculated the frequency of each instruction at each site by counting the number of organisms with that instruction.

This prior, population-based technique allowed for good estimates of physical complexity over time in many instances, but suffers from a number of limitations. First, the technique only produces accurate measurements if the population has reached an equilibrium state. For example, if a beneficial mutation causes a new species (and thus new phenotype) to take over a population, all otherwise neutral sites in the genome hitchhike to fixation, and it will take time before equilibrium is reached where most genotypes of the new phenotype are represented. It is often the case that a new beneficial mutation will arise before this equilibrium preventing us from determining the true complexity of that phenotype. We can see these effects in Figure 1, the upper line of which displays the physical complexity over time using the population-based technique for a typical Avida experiment. Notice that each time complexity increases, it overshoots its mark and then gradually comes down again, typically to a higher resting level than it started.

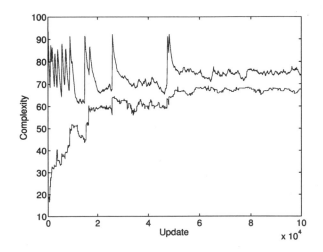

Figure 1: A comparison between the population-based method of calculating physical complexity (upper line) and our new method (lower line). The new method is applied to the lineage of the most abundant organism at the end of the experiment.

The second problem with population-based physical complexity is the constraints that must be placed due to computational concerns. The population size must be small (typically 10,000 or fewer) and the length of their genomes is fixed to prevent alignment problems. The finite population size limits the possible range of genotypes compounding the population diversity limitation noted earlier. If there are too many neutral sites, it is unlikely for them all to be represented in such a limited population even at equilibrium. The fixed length genome puts an inherent cap on complexity growth (there is only so much "blank tape" to write information into) and precludes many powerful forms of mutations such as gene duplications.

A final problem is ensuring that all organisms possess the same phenotype. This can only be achieved by using a single niche environment, and even with a single niche, a single phenotype only occurs at equilibrium. This limits the the range of interesting experiments that can be studied using population-based physical complexity, and excludes the possibility of studying ecosystem complexity.

Our Approach

Here, we demonstrate a new method for calculating physical complexity that has the ability to transcend the limitations listed above. We calculate the complexity of a single genome by examining its local mutational landscape by finding the fitness of all possible single-step mutants and using the principles of mutation-selection balance to approximate an infinite landscape. Since we calculate the complexity one organism at a time, we never have to worry about overshooting the correct complexity due to a biased sample, we do not impose any genome size limitations, and we can allow the environment to vary as long as we always test an organism using the state of the environment during its lifetime. In essence, we shift the complexity measure from the phenotype level to the genotype level.

This method does, however, suffer from limitations of its own. First, we only consider single-step mutants. In the future we plan to refine this method by examining multiple sites at once in an attempt to decipher epistatic interactions and improve our complexity measure. Second, a significant amount of extra processing power is required to generate all possible single point mutations from a genome and to test the fitness of each. In a computer this may be feasible, but in a natural system it is

nearly impossible given our current technology. We must therefore limit ourselves to applying this physical complexity measurement technique to computational systems for the moment. This is not as severe a problem as it may seem since the main goal of quantifying complexity is to study its origin. In the natural world, evolution progresses too slowly to see significant changes to species in time spans shorter than centuries, so experimental macro-evolution already has this restriction placed on it. Furthermore, as we improve our techniques for calculating complexity in the digital world, we can use this to determine the quality of other complexity approximation algorithms that can more easily be applied to natural systems.

The Technique

As our first step in calculating the physical complexity of Avida genomes, we developed a test environment that organisms can be inserted into. The test environment is initialized to the exact same conditions as the environment that the population is evolving in, but only one organism is tested at a time. That organism is processed until either it gives birth and we can measure its fitness, or else it dies of old age (indicating a zero fitness).

As in the previous method, we calculate the complexity of the whole genome by summing the complexities of the individual sites in that genome. To determine the complexity of site i, we start by mutating this site to all other possible states and then use the test environment to calculate the fitness of each. In the case of Avida, there are 26 instructions in the genetic alphabet, so we need to generate 25 new genomes to represent each possible mutation at site i. We then run each of the resulting 26 genomes through the test environment to determine the fitness of each. With these fitnesses and a mutation rate, we can predict the abundance of each instruction at this site were a population at equilibrium.

Intuitively, it is clear that if a genome has equal fitness no matter which instruction is at site i, then we would expect all possible instructions to appear with about equal frequency. Further, this would translate to a maximal entropy for that site, and thus a zero complexity. On the other hand, if only the original instruction has a non-zero fitness, then we expect that instruction to dominate in an equilibrium population (the others would persist at a small frequency due to detrimental mutations creating them.) In this case, the population would have a low entropy at this genomic position, and it would contribute maximally to complexity. It is slightly more complicated to calculate the expected abundance at sites with mixed fitness levels; our techniques are discussed below.

We show a sample sub-sequence from a genome in Table 1 where a single site is mutated throughout. Its original state was 'm', but all others are tested as well and their fitness recorded.

Sequence	Fitness
...akapbkawbjbo**a**cpbnaqblafpq...	0
...akapbkawbjbo**b**cpbnaqblafpq...	6.46734
...akapbkawbjbo**c**cpbnaqblafpq...	0
...akapbkawbjbo**d**cpbnaqblafpq...	5.94
...	...
...akapbkawbjbo**m**cpbnaqblafpq...	6.46734
...	...
...akapbkawbjbo**z**cpbnaqblafpq...	3.23367

Table 1: Samples for genome sequences with all single-site mutations and the resulting fitness of each. The site being changed is marked in bold.

We use the mutation-selection balance principle from population genetics to take the fitness values and determine the portion of the population that we expect each genotype to fill at equilibrium. Fisher, Haldane, and Wright, pioneers of population genetics, developed mathematical models quantifying the relative importance of selection and mutation in maintaining genetic variation. We simplify and specialize these equations to Avida, which has populations that are asexual, haploid, and have overlapping generations, and where we only consider the possibility of site i mutating, since we are not considering interactions between sites.

Let p_j denote the percentage of the population occupied by genotype j at equilibrium, ω_j the fitness of genotype j, D the alphabet size (in our case $D = 26$), and μ the per-site mutation rate. Furthermore, we assume all mutations are equally probable. The average fitness is defined by

$$\bar{\omega} = \sum_{k=1}^{D} p_k \omega_k \qquad (6)$$

At equilibrium, the following equation must hold:

$$p_j = (p_j \omega_j / \bar{\omega})(1 - \mu) + \sum_{k=1}^{D} (p_k \omega_k / \bar{\omega}) \mu(1/D) \qquad (7)$$

In this equation, $p_j \omega_j / \bar{\omega}$ is the relative replication rate of genotype j, and $1 - \mu$ is the probability that genotype j replicates without mutation at site i. These two factors are multiplied together to give us the rate of perfect replication within genotype j. For the second part of the equation, $\mu(1/D)$ is the probability that any genotype (including j) mutates to genotype j. We then multiply this by the relative replication rate for each genotype to determine the rate that each genotype mutates to genotype j. These two factors summed together represent the rate at which genotype j enters the population. Since all organisms leave the population with equal probability, at equilibrium, the rate at which genotype j enters the population must be the same as p_j, the percentage of the population occupied by genotype j at equilibrium.

We use equation 6 to simplify equation 7:

$$p_j = (p_j \omega_j / \bar{\omega})(1 - \mu) + \mu(1/D) \qquad (8)$$

To determine the final abundance of each of the 26 genotypes at equilibrium, we generate the 26 equations and solve them. This will always provide us with a unique solution that will predict the abundance of each possible instruction at this site, were we at equilibrium in an infinite population—exactly what we need. We can then calculate the physical complexity of this site and repeat this process for each other site in the genome, summing them up to determine the physical complexity of the genome as a whole.

Experiments and Results

Our first experiments test how accurately our models predict the abundance of single-step mutants. We initiate Avida experiments with a population size of 3600 where only a single site is allowed to mutate. Given our instruction set of 26, there can only be 26 possible genotypes in the population. We then compare our predicted abundances to the observed abundances once an equilibrium is reached as shown in Figure 2. We have performed over 30 such comparisons, and all performed similarly well.

Our second set of experiments highlight the improved accuracy of our new method when a population is not at equilibrium compared with the previous population-based method for calculating complexity. In order to be able to calculate the population-based complexity, we perform Avida experiments with large (3600), single-niche populations with fixed-length genomes. According to the Natural Maxwell's Demon proposed by Adami et

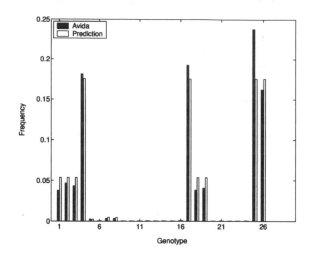

Figure 2: Mutation-selection balance at site 53 of a genome. White bars are mathematical predictions of instruction frequency, black bars are experimental results.

al. (2), complexity should increase over time in such populations.

Within each experiment, we first identify the lineage of the most abundant genotype at the end of the experiment. At each Avida update, we then compute both the population-based complexity (using all genotypes alive at that update) and our single-step mutant complexity of the current genotype on the isolated lineage. Figure 1 contains a plot of both complexity measures at each Avida update for one Avida experiment.

As we can see from Figure 1, when the population is not at equilibrium, the population-based complexity is inaccurate for many of the updates as it suffers from hitch-hiking effects. On the other hand, the proposed growth in complexity over time (with minor fluctuations) is clearly visible in the single-step mutant complexity (lower line in Figure 1). The minor fluctuations in complexity are expected; there are occasional decreases due to detrimental mutations, drift, or (occasionally) evolution of a more compressed way to code for a phenotypic trait. A detrimental mutation can be an important step on the way to significant fitness improvements(9). Some fluctuations may also be caused by inaccurate approximations of the actual complexity due to the fact that we do not yet account for epistasis. At equilibrium, both methods provide a qualitatively similar result, though the population-based complexity measure is always higher than the single-step mutant complexity.

Our third set of experiments highlight the improved accuracy of our new method for calculating complexity when a population is at equilibrium compared with the previous population-based method. We perform 30 sets of Avida experiments (identical to those performed in our second set of experiments) with three different population sizes: 900, 3600, and 14,400, and we focus on the comparison between the computed complexities once the population reaches a final equilibrium. As noted in Figure 1, the population-based complexity is higher than the single-step mutant complexity at equilibrium. The mean difference between complexity

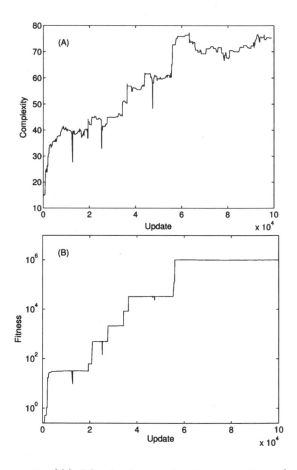

Figure 3: (A) Physical complexity over time for a lineage from an Avida experiment where genome length is allowed to change. (B) Fitness over time for the same lineage from the same experiment.

measures at equilibrium in the size 900 population experiments was 21.35 bits. The mean difference between complexity measures at equilibrium in the size 3600 population experiments was 10.04 bits. Finally, the mean difference between complexity measures at equilibrium in the size 14,400 population experiments was a nearly negligible 1.61 bits.

Clearly, as population size increases, these measures become dramatically closer. This can be easily explained as our new method approximates an infinite population, and is thus better reflected by larger population sizes.

Our final experiments demonstrate the increased flexibility of our new method for calculating physical complexity. In particular, we show that it can be applied to variable length genomes. We perform a sample Avida experiment with a population size of 10,000 organisms where we allow the size of the genome to change. Figure 3A displays our new physical complexity measure of a lineage from this experiment. There are many clear jumps in complexity over time. Figure 3B shows how these complexity jumps correlate with fitness increases. This suggests that we are accurately reflecting the true complexity as at each fitness jump, more information about the environment is encoded into the genome. Downward spikes occasionally occur in both graphs due to detrimental mutations that briefly exist along the lineage.

Discussion and Future Work

Does complexity always have an increasing trend? This is an age-old question. Since Darwinian evolution is a unique and long-term procedure in nature, it is nearly impossible to get a conclusive result from the natural world. In Avida, we are able to perform many experiments and observe macro-evolutionary dynamics to study how this process works. The data we have collected thus far has concurred that complexity does seem to always increase over time, in accordance to the Natural Maxwell's Demon proposed by Adami et. al (2). These experiments, as well as the theory that led to them, assume a single-niche environment. If the environment is in any way unstable this law of increasing complexity breaks down. If, for example, a resource is no longer present in the environment, genomic information about that resource is wasted and is no longer reinforced by selection—it should decay over time or be replaced by more pertinent information.

Initial experiments in environments with multiple, limited resources show that many species can easily co-exist in an Avida population and form primitive eco-systems (10). While theory does not dictate that populations in such naturally fluctuating environments must increase in complexity, we have observed a much more rapid fitness increase in these populations. These new techniques will allow us to examine both the complexity of individuals within an ecosystem, and examine the information shared between organisms as a first step in calculating the complexity of the ecosystem as a whole. It would be impossible to show that any single species must always gain in complexity, but there is much more that may be said about the ecology as a whole.

Acknowledgments

We would like to thank Chris Adami, Richard Lenski, Thomas Schmidt and the Avida Group for helpful discussions. This work was supported by NSF grants EIA-0219229 and DEB-9981397.

References

[1] Chris Adami and Nicholas Cerf. Physical complexity of symbolic sequences. *Physica D*, 137:62–69, 2000.

[2] Chris Adami, Charles Ofria, and Travis C. Collier. Evolution of biological complexity. *Proc. Nat. Acad. Sci*, 97:4463–4468, 2000.

[3] Ben Goertzel. *The Evolving Mind*. Routledge, June 1993.

[4] Charles H. Bennett. Logical depth and physical complexity. In R Herken, editor, *The Universal Turing Machine, A Half-Century Survey*, pages 227–257. Oxford University Press, Oxford, 1988.

[5] R Hinegardner and J Engelberg. Biological complexity. *Journal of Theoretical Biology*, 104:7–20, 1983.

[6] John T. Bonner. *The Evolution of Complexity*. Princeton University Press, 1988.

[7] Daniel W. McShea. Metazoan complexity and evolution: Is there a trend? pages 477–492.

[8] C. Ofria and C. Wilke. Avida: A software platform for research in computational evolutionary biology. *Artificial Life*, 10:191–229, 2004.

[9] Richard E. Lenski, Charles Ofria, Robert T. Pennock, and Christoph Adami. The evolutionary origin of complex features. pages 139–144, May 2003.

[10] T. Cooper and C. Ofria. Evolution of stable ecosystems in populations of digital organisms. In MA Bedau RK Standish and HA Abbass, editors, *Eighth International Conference on Artificial Life*, volume 119, pages 227–232, Boston, MA, 2002. MIT Press.

The Role of Nearly Neutral Mutations
in the Evolution of Dynamical Neural Networks

Eduardo Izquierdo-Torres

Department of Informatics
University of Sussex, BN1 9QH. UK.
e.j.izquierdo-torres@sussex.ac.uk

Abstract

The evolution of continuous time recurrent neural networks is increasingly being employed to evolve nervous systems for autonomous agents. Nonetheless, the picture of populations engaged in hill-climbing rugged fitness landscapes poses a problem of becoming trapped on a local hilltop. Developments in evolutionary theory and molecular biology have pointed to the importance of selective neutrality. The neutral theory claims that the great majority of evolutionary changes are caused not by Darwinian selection but by random drift of selectively neutral or nearly neutral mutants. However, with a few exceptions neutrality has generally been ignored in artificial evolution. This paper addresses the distribution of fitness effects of new mutations when evolving dynamical systems and provides evidence of an improved evolutionary search process when incorporating nearly-neutral drift. This is one of the most fundamental problems in artificial evolution, because it lies at the heart of maintaining a constant-innovative property.

Introduction

Continuous time recurrent neural networks (CTRNNs) are increasingly being employed as simple model nervous systems for robot controllers. For this, the approach has been an evolutionary one, evolving increasingly complex neural controllers (Beer & Gallagher, 1992; Harvey, Husbands, Cliff, Thompson & Jakobi, 1997). However, in attempting to address the features of the fitness landscapes as an abstraction of the dynamics of evolution a picture has become ingrained of a landscape as a rugged terrain. This poses the problem of populations becoming trapped on a local hilltop (Barnett, 1997). Yet, developments in evolutionary theory and molecular biology point towards the importance of selective neutrality (Kimura, 1983). Kimura introduces this phenomenon to question the preeminence of selection as the sole mediator of the dynamics of biological evolution. More recently, the identification of neutral networks – connected networks of genotypes mapping to common phenotypes (and therefore equal fitness) in RNA secondary structure folding (Schuster, Fontana, Stadler & Hofacker 1994) and protein structure (Babajide, Hofacker, Sippl & Stadler 1997) has revived the interest of selective neutrality.

Our interest stems from the growing evidence that such neutrality and indeed neutral networks may be a feature of fitness landscapes which arise in evolving neural controllers for autonomous agents (Cliff, Husbands & Harvey, 1993; Smith, Husbands & O'Shea, 2001; Smith, Phillippides, Husbands & O'Shea, 2002), and from the evidence that the dynamics of evolutionary processes on such landscapes is qualitatively different from the traditional rugged landscapes (Harvey & Thompson, 1996; Barnett, 1997, 1998, 2001).

Within the last few years there has been an increasing amount of research into the possible benefits and applications of neutral networks in evolution. However, nearly all of this work has been primarily concerned with discrete alphabet genotypes. Conversely, the most common approaches to evolving dynamical neural networks as neural controllers for autonomous agents have been based on real-value encodings. It will be argued herein that when dealing with real-valued landscapes emphasis needs to be made on nearly-neutral mutations (as opposed to exactly neutral), given that in real-valued landscapes it is unlikely for two points to have the same fitness.

This has led us to ask: What proportions of mutations are deleterious, neutral, and advantageous? What is the strength of selection that acts on non-neutral mutations? In other words, what is the distribution of fitness effects of new mutations? This question is central to our understanding of near neutrality and the maintenance of a constant-innovative property at both the genotypic and phenotypic level. Unfortunately, even in biology relatively little is known about the distribution of fitness effects, despite its importance. Recently, analysis of mutation accumulation experiments have shown that the distribution of fitness effects for deleterious mutations is highly leptokurtic (Piganeau & Eyre-Walker, 2003); that is with a few mutations having large effects, and the vast majority having mild effects. A similar distribution of effects is obtained when evolving dynamical neural networks, placing great importance on neutral drift during its evolution. In this paper we consider the usefulness of such neutrality.

Neutrality in Artificial Evolution

Neutral theory was first introduced by Kimura (1983) as genetic change without selection pressure in evolution. The importance of this theory is that genetic code can be constantly altered with no fitness disadvantage. Thus, neutrality provides an explanation for a crucial aspect of natural evolution – the constant-innovative property.

In the abstraction of a search space as a fitness landscape, a neutral network has been proposed in the form of a plateau connecting two peaks (Barnett, 1997; Smith, Husbands & O'Shea, 2001). This leads to a population being able to achieve any possible phenotype, given enough time.

Barnett (1997) and Shackletton et al (2000) have applied neutralist ideas to binary search spaces, namely, adding tunable neutrality to Kauffman's NK landscapes and evaluating various redundant mappings. Many benefits have been supported by these studies in which drift allows for the discovery of better adapted phenotypes. Smith (2001; 2002) has suggested benefits in evolvability and the speed of evolution and Jakobi (1995) has suggested benefits in producing increasingly complex behaviors.

Currently, little or no research has been carried out on real-valued search spaces and converting the genotype to a binary representation requires certain level of quantization. Zendric-Ahsmore (2003) explores several different encoding schemes when evolving a fully connected recurrent neural network. In his work, exploitation of neutrality is studied via redundant mappings and achieved only when binary dimensions are added to the real-valued search space. In the case of evolving hardware (Harvey & Thompson, 1993), results encourage the application of potentially useful junk genotype.

This work takes a different approach to exploiting neutrality based on the hypothesis that when evolving dynamical neural networks as model nervous systems neutrality comes for free. The intuitive reason behind this idea is the implicit redundant genotype-phenotype mapping generated by the infinite possibilities of real-valued genotypes and the finite number of significantly different phase portraits.

Finally, in the rarer cases when artificial evolution practitioners consider neutrality, often neutral and non-neutral mutations are treated as 'all or nothing' with respect to selection. However, when exploring neutrality in real-value search spaces exact neutrality will be practically impossible. For this reason, in this work we consider mutations whose behavior is influenced by both selection and random drift.

Experimental Design

In order to measure fitness effects from mutations and to take advantage of nearly neutral networks in the evolution of CTRNNs, we experiment with the commonly used real-value encoding on a very simple optimization problem using a slightly modified version of a population of 1+1 evolutionary technique.

Real-Valued Direct Encoding CTRNN

The property of neutrality in the dynamics of evolution is greatly determined by the encoding scheme and the genotype-phenotype mapping of the model being evolved.

Our model represents the common encoding and mapping scheme used in the evolutionary robotics literature, where the genotype is real-valued and maps almost directly to the CTRNN. The genotype is a vector of n^2 real numbers between [0, 1], where n is the number of neurons in the network. These values are mapped to [-5, 5] to represent the connection weights w_{ji} in the CTRNN, which mediate the interaction between the neurons, positive (excitatory), negative (inhibitory). This model encodes a fully connected real-valued CTRNN. For simplicity, the biases in the transfer functions and time-constant parameters were kept fixed, and are thus not included in the genotype. This model allows for no structural neutrality, that is, there is no redundancy in the genotype   CTRNN-architecture mapping, on the basis that any mutation, as small as it may be, will cause a change in the strength of the connections of the network. However, will mutations inevitably affect the overall behavior of the resulting dynamical system? Small changes in the parameters of the system may not change its attractors and overall behavior significantly. Therefore, it is likely that the genotype   phenotype mapping is redundant (considering the phenotype as the long-term behavior of the system). As has been discussed, when evolving CTRNNs various levels of phenotype can be characterized: phenotype as architecture of the CTRNN and phenotype as the long-term behavior of the resulting dynamical system. For the reasons herein described we will treat the phenotype as the latter.

The Task

Long-term behaviors of dynamical systems are of particular interest to agent-environment interaction (Beer, 1995). For this reason, the optimization problem – evolution of a logic gate by a CTRNN – is intended to represent a "generic" scenario where, from a dynamical systems perspective, a successful solution depends essentially on the phase portrait of the system; that is, a successful solution depends on the attractors – the long term behavior of the system. This problem has been adapted from a very similar problem tackled by Barnett (2002).

The activity of each 'neuron' in the CTRNN is given by the general form (see for detail Beer 1995a):

$$y'_i = \frac{1}{\tau_i}\left(-y_i + \sum_{j=1}^{N} w_{ij}\sigma(y_i+\theta_j)+I_i\right) \tag{1}$$

where y_i is the state of neuron i; y_i' denotes the time rate of change of this state (i.e. dy/dt); τ_i is its time-constant ($\tau > 0$); I_i represents an external (sensory) input; w_{ji} is the strength of the connection from unit j to unit i and $\sigma(x)$ is a sigmoid:

$$\sigma(x) = \tanh(x) \tag{2}$$

with bias term (threshold) θ_j. In simulation, node activations are calculated forward through time by straight-forward time-slicing using Euler integration.

The experiments were performed for two logic gates: AND and XOR. The CTRNN has to solve the logic gate as follows: two nodes (node 1 and 2) are designated as inputs, and the sigmoided output of a third node as output (hyperbolic tangent so it is guaranteed to lie between -1 and 1). A network trial consists of four runs of the network, corresponding to the particular logic task (see table 1).

Inputs were chosen as *low*=-0.5, *high*=+0.5 and target outputs were chosen as: *low*=- , *high*=+ with =0.6. Node activations are initialized for each of the four runs to zero, with the exception of the output node which is initialized with -0.2. The off-centre initial activation was found to aid finding solutions in the very similar problem attempted by Barnett (2002).

Input 1	Input 2	AND output	XOR output
High	*High*	*High*	*Low*
High	*Low*	*Low*	*High*
Low	*High*	*Low*	*High*
Low	*Low*	*Low*	*Low*

Table 1. AND and XOR Logic

All time-constants are set to 5 and the step size for Euler integration is 0.02. During a run the appropriate (constant) inputs are fed continuously to the input nodes. Networks are run for a stabilization period T_s=500 time steps to allow for the network to fall into an attractor state, followed by and evaluation period of a further T_e=500 time steps; the idea is that a network be evaluated on its steady state network. Networks receive a fitness value f based on the mean distance between output and target (i.e. distance integrated over the evaluation period) averaged over the four runs; specifically:

$$f = \left[1 + \frac{k}{4(1+\lambda)}\sum_{r=1}^{4}\Delta^{(r)}\right]^{-1} \quad (3)$$

with mean distance between target and output for the r-th run defined by:

$$\Delta^{(r)} = \frac{1}{Te}\sum_{i=Ts}^{Ts+Te}\left|\sigma(x_3^{(r)}{}_i) - \lambda^{(r)}\right| \quad (4)$$

where $x_3^{(r)}{}_i$ is the activation of the output node and $^{(r)}$ the appropriate target for the r-th run. The constant "stretch factor" k may be deployed to control selection pressure. We used k=9 for our experiments. Note that maximum possible fitness is 1 and minimum possible fitness is $1/(1+k)$=0.1 since the maximum mean distance $^{(r)}$ for each of the four runs is 1+ .

Artificial Evolution: *Netcrawler*

The evolutionary technique used consists of a very simple population of 1+1 with a random mutation strategy, or netcrawler (Barnett, 2000). This technique has been originally designed as an optimum way to drift along fitness landscapes of high neutrality. It is here used as an intuitive measure for the usefulness of the neutrality taking place in the dynamics of evolving dynamical systems.

The original netcrawler consists of a hill-climber which is initially dropped into weight space according to a uniformly distributed random vector between [0, 1]. The individual is then mutated, and replaced if the mutated

individual is better or equally fit than the original. Given that the search space is real-valued it is unlikely that an equally fit individual be found. Therefore, in order to explore random drift, we experiment and compare slight variations to this technique. The variation consists in allowing individuals with certain slightly deleterious mutational loads to replace its ancestor. As the range of fitness is from 0.1 to 1, a 1% nearly-neutral netcrawler will be one were the current individual may be replaced by an individual who is at maximum 0.009 worst fit. A walk in a nearly-neutral network is considered when a slightly neutral replacement occurs. We conduct experiments with 0% (original netcrawler), 1%, 5%, 10%, 15% and 20% nearly neutral netcrawlers. The mutation operator used is a one locus per genotype variation taken from a uniformly random distribution in [0, 1].

Neutrality Measurements

Despite the lack of an exact measurement for neutrality, Barnett has introduced several indicators which provide a meaningful idea of the presence of neutrality in fitness landscapes of discrete binary spaces. This work makes use of some of these measurements with certain adaptations for the real-valued scenarios. Due to space constraints only a brief account of the measurements will be provided here (for more detail see Barnett, 1997).

Neutral and Nearly Neutral Mutations: a mutation is neutral if the fitness of a genotype g and the fitness of a mutated version g' are equal or nearly equal.

The *Diffusion Coefficient* is an estimate of the velocity at which the population is moving in genotype space, as given by the squared Euclidean distance of the centroid of the population in adjacent generations. For our experiments, the centroid of the population is the genotype of the netcrawler as such. The squared Euclidean distance between centroids is defined by:

$$d_t = \sum_{i=1}^{D}\left(c_i^t - c_i^{t-h}\right) \quad (5)$$

where d_t is the distance traveled from generation t-k to t, D is the size of the genome or the dimensionality of the genotypic search space, c_i^t represents the value of the centroid along the i-th axis at time t.

The *Lagged Diffusion Coefficient* is an estimate of the distance traveled by the population's centroid in a certain time lapse. This is given by a lagged squared Euclidean distance between the centroid of the population at generation i and k generations prior, as defined by equation 5. In our experiments a time lag of k=500 was used.

Results

Results taken from several evolutionary runs using the experimental design described above are given in this section. The role of nearly-neutral mutations are interpreted and discussed with attention paid to its possible usefulness.

Firstly, the distribution of effects caused in fitness by one-locus mutations for 120 evolutionary runs using the netcrawler to evolve the 3-neuron CTRNNs for the duration of 3000 generations was studied - a total of 360,000 mutations considered (see figure 1). From this we obtained that 78.8% were deleterious, 17.9% were advantageous and 3.3% were very close to exactly neutral[1].

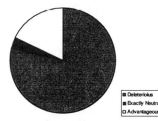

Figure 1. Proportion of fitness effects of a one-locus mutation: advantageous (white), deleterious (gray), and neutral (black).

Unsurprisingly, deleterious mutations are most common, given that in the face of random change fitness will most likely deteriorate. Despite the high proportion of deleterious mutations, the distribution of fitness effects of mutations shows that a great part of them cause only very small effects (figure 2). In fact, half the mutations cause an effect of less than 15% in fitness.

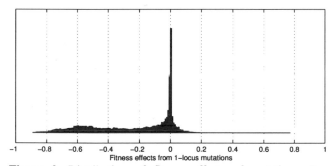

Figure 2. Distribution of fitness effects of mutations. Few mutations have large effects while most mutations have only slight effects on fitness.

As can be seen from the distribution, most changes in the real-valued genotype cause only slight changes in the fitness – as the fitness is given by the long-term behavior of the resulting dynamical system. Then this result hinges towards the potential for implicit redundancy in the genotype-phenotype mapping when evolving CTRNNs. So, what would happen if selection could not perceive a certain degree of too slight changes in the phenotype? Would the population engage in useful genetic drift? Or would individuals accumulate small deleterious mutations?

The neutral theory of molecular evolution has proven that slightly deleterious mutations are not necessarily part of adaptive evolution but can actually become part of random drift. To study the role of nearly-neutral mutations in the dynamics of evolution we allowed certain variations of effects on fitness not to be perceptible to selection. We

---

[1] The main factor causing the 'exactly' neutral effect is the limit of precision with which was experimented. Differences smaller than 10^{-5} are unperceived.

experimented and compared evolutionary processes with various ranges of nearly-neutral mutations for 20 runs each, starting at random using the netcrawler for 3000 generations on both problems: AND and XOR.

The mean fitness of the evolutionary runs using no-near neutrality, 1% and 5% near-neutrality are presented in Figure 3. As depicted, when slightly deleterious mutations are not perceived by the selection forces an improved evolutionary process is obtained. This implies that random drift plays a significant role in the evolutionary adaptation of the CTRNN.

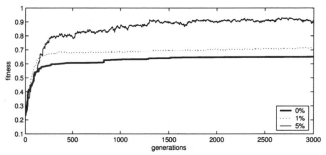

Figure 3. Comparison of averages fitness during evolution of original netcrawler solving the logic AND task against the modified versions of the algorithm which permit nearly neutral mutations.

However, increments in the percentage of selectively neutral mutations will cause an undesired evolutionary effect. This is due to the gradual accumulation of slightly deleterious mutations which in the long run has a detrimental effect on the fitness of the individuals. Figure 4 shows the mean fitness of evolutionary runs using 10%, 15% and 20% nearly neutral mutations.

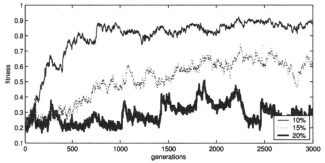

Figure 4. Average fitness of netcrawler solving logic AND task when bigger ranges of nearly neutral mutations are allowed.

At 10% the evolutionary process maintains a high enough fitness, regardless of certain temporary fitness losses. However, when increased to 15% the process becomes highly erratic. Finally with 20% the process becomes completely unsustainable. Although these effects are particular to this problem, we can generalize the effect of slightly neutral mutations as requiring a sort of balance, where a little enables useful drifting, and too much generates undesired deleterious accumulations. These results place great importance in the interaction and balance to be maintained between random genetic drift and positive selection.

The importance of nearly neutral mutations is further exemplified by its potential for vanishing the picture of local optima in the abstraction of fitness landscapes. Whereas conventional evolutionary techniques assume that all or nearly all of the variation depends on the initial random population and recombination operators - and hence evolution will cease when this initial variation is exhausted; for the netcrawler the reverse is true, mutation is considered the driving force behind evolution (no recombination at all). For this approach, if the fitness landscape comprises local optima there is a high probability that the netcrawler will get stuck on it. Nevertheless, when nearly-neutral mutations are considered, the population is never stuck, but constantly moves through genotype space; when it is not climbing a fitness slope it will be drifting along a 'nearly' neutral network.

The picture of individuals engaged in useful drifting along nearly-neutral networks is supported when comparing the lagged diffusion coefficient among the different ranges of nearly-neutral mutation effects. Figure 5 shows the average movement of the netcrawler in genotype space when no nearly-neutral mutations are considered, with 5% and 20% ranges for neutrality from 20 evolutionary runs.

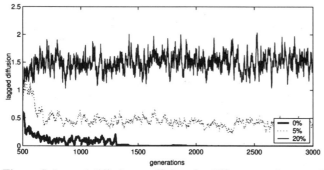

Figure 5. Lagged diffusion coefficient for different nearly-neutral ranges of mutations.

As can be appreciated, the exactly neutral netcrawler (0%) rapidly ceases to move or moves relatively very little after the first thousand generations, thus showing potential for local entrapment. The 5% nearly neutral keeps moving enough to reach higher fitness without accumulating too many deleterious mutations, thus improving the evolutionary search process. On the other extreme, the 20% nearly neutral shows too much random drift, which in turn doesn't allow it to accumulate advantageous mutations (as shown by its poor evolutionary run in figure 4). Figure 6 depicts the drifting through nearly-neutral networks for a particular run using the 1% nearly neutral netcrawler. The dark dots represent the periods of selectively neutral drift and the thin vertical lines represent jumps across networks of higher fitness.

Figure 7 shows the average of the fitness of the last 500 generations on the experiments shown previously for each of the different ranges of nearly-neutral mutations. The actual data is represented in circles and the line is a simple

spline interpolant fitting of the data. An analogous result is shown for the evolutionary runs solving the XOR problem. Results for both problems indicate certain enhancement from the use of small amounts of drifting.

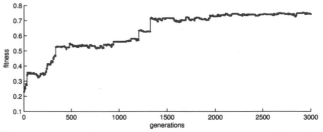

Figure 6. Evolutionary run from the 1% near neutrality netcrawler drifting through percolating nearly-neutral networks.

An important point that is made here is that big ranges of permissible drift results in individuals accumulating too many deleterious mutations. There lies a potentially dangerous use of neutrality which needs be taken into account when evolving CTRNNs, despite 'some' random drift actually improving the evolutionary search.

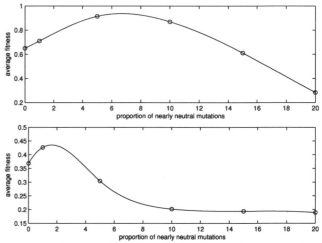

Figure 7. Average fitness on the last generations of evolved CTRNNs solving the logic AND (top) and XOR (bottom) problem using different ranges of nearly-neutral mutations.

Discussion

Although selection is the main force behind evolution, with this paper we wish to put emphasis on an approach to artificial evolution that takes into account other factors as well. In particular, we wish to point out that the approach to evolving dynamical neural networks may be greatly enhanced by random genetic drift. For this reason several issues have been addressed. Firstly, this paper has sketched the distribution of fitness effects of new mutations when evolving CTRNNs in real-valued search spaces. The results show that a big proportion of mutations cause little effect on fitness. Our results agree to a certain extent with recent theoretical work from molecular biology which states that the distribution of mutational effects on fitness is leptokurtic. This has led us to explore the limits of nearly neutral mutations and how they affect this evolutionary process. From neutral theory, we know that small changes in fitness (less than $1/K$ where K is the size of the population) do not cause a selectively effective advantage or disadvantage. For simplicity we have experimented with an extreme evolutionary case where the population is of size 1+1. Thus, unless explicitly stated any change in the population will fix, no matter how small or in fact neutral. This has allowed us to explore the ranges of selectively neutral effects from new mutations. We are, however, very careful of extrapolating our conclusions to (larger) populations. Our results show that ranges of effects of near neutrality from 1% to 10% actually improve the evolutionary process for the tasks treated here. We observed that the process of drifting through phenotypes of nearly-neutral fitness is reminiscent of biological neutral networks. Additionally, we observe that if the range of near neutrality is too big then accumulation of deleterious mutations increases and the population actually deteriorates significantly. The most important generalization drawn from this study is not the ranges which worked better for these problems but the fact that genetic drift plays an important role in the evolutionary search of dynamical systems. We put forward the hypothesis that this is due to an implicit redundant genotype   phenotype mapping. Future work should be directed towards determining the exact relation between the genotype and the behavior when evolving dynamical systems.

Conclusions

In this paper we have done the following:
1. Emphasized the importance of nearly neutral networks and random genetic drift in artificial evolution.
2. Pointed out the implicit redundancy in the genotype-phenotype mapping when evolving dynamical neural network.
3. Considered the possibility that, for the case of evolving real-valued CTRNNs, useful neutrality comes for free.
4. Analyzed the importance of the distribution of fitness effects from mutations in the dynamics of evolution.
5. Highlighted the use of mutation as the driving force behind evolution in landscapes featuring neutrality.

All of these issues deserve further development. They have been presented here in full knowledge of this fact. Nonetheless, if this paper raises awareness of the importance of near-neutrality in the dynamics of artificial evolution of dynamical systems then its objectives will have been accomplished.

Acknowledgements

The author thanks Inman Harvey for insightful discussions. The author is supported by Programme Al an, the European Union Programme of High Level Scholarships for Latin America, identification number E03M12883VE.

References

Babajide, A., Hofacker, I., Sippl, M., and Stadler, P.F. 1997. Neutral networks in protein space: A computational study based on knowledge-based potentials of mean force. Folding Design, 2, 261-269.

Barnett, L. 1997. Tangled webs: Evolutionary dynamics on fitness landscapes with neutrality. M.Sc. diss., University of Sussex. UK.

Barnett, L. 2000. Netcrawling – optimal evolutionary search with neutral networks. In Proceedings of the 2001 Congress on Evolutionary Computation.

Barnett, L. 2002. Explorations in evolutionary visualisations. In Artificial life VIII Workshops. The 8th International Conference on the Simulation and Synthesis of Living Systems.

Beer, R.D., and Gallagher, J.G. 1992. Evolving dynamical neural networks for adaptive behavior. Adaptive Behavior 1:92-122.

Beer, R.D. 1995a. On the dynamics of small continous-time recurrent neural networks. Adaptive Behavior 3(4):475-511.

Beer, R.D. 1995b. A dynamical system perspective on agent-environment interaction. In Artificial Intelligence 72. 173-215.

Cliff, D., Husbands, P., & Harvey, I. 1993. Evolving visually guided robots. In Meyer, J.-A., Roitblat, H., and Wilson, S. (Eds.), From Animals to Animats 2: Proc. of the Sec. Intl. Conf. on Simulation of Adaptive Behaviour, (SAB92). 374-383.

Harvey, I. and Thompson, A. 1996. Through the labyrinth evolution finds a way: A silicon ridge. In Proc. 1st Internatl. Conf. Evol. Sys.: From Biology to Hardware (ICES 96). Springer-Verlag.

Harvey, I., Husbands, P., Cliff, D., Thompson, A. and Jakobi, N. 1997. Evolutionary robotics: the Sussex approach. In Robotics and Autonomous Systems, v. 20: 205-224.

Jakobi, N. 1995. Facing the facts: necessary requirements for the artificial evolution of complex behaviour. Cognitive Science Research Paper 422. University of Sussex. UK.

Kimura, M. 1983. The Neutral Theory of Molecular Evolution. Cambridge, UK: Cambridge University Press.

Piganeau, G. and Eyre-Walker A. 2003. Estimating the distribution of fitness effects from DNA sequence data: implications for the molecular clock. Proc Natl Acad Sci U S A; 100(18):10335-40.

Shackleton, M., Shipman, R. and Ebner M. 2000. An investigation of redundant genotype-phenotype mappings and their role in evolutionary search. In Proceedings of the International Congress on Evolutionary Computation. IEEE Press. pp:493-500.

Shuster, P., Fontana, W., Stadler, P.F., and Hofacker, I. 1994. From sequences to shapes and back: A case study in RNA secondary structures. Proc. Roy. Soc (London) B, 255:279-284.

Smith, T., Husbands, P. and O'Shea, M. 2001. Neutral networks and evolvability with complex genotype-phenotype mapping. European Conference on Artificial Life: ECAL2001. 272-281. Springer

Smith, T., Philippides, A, Husbands, P. and O'Shea, M. 2002. Neutrality and ruggedness in robot landscapes. Congress on Evolutionary Computation: CEC2002. pp. 1348-1353. IEEE Press.

Zendric-Ashmore, L.A. 2003. An investigation into the benefits of neutrality within real-valued search landscapes. MSc. diss. University of Sussex. UK.

Sustained Evolution from Changing Interaction

George Kampis[1] and Laszlo Gulyas[2]

[1]Department of History and Philosophy of Science, Eötvös University, Budapest
[2]Computer and Automation Research Institute, Hungarian Academy of Science
gk@hps.elte.hu

Abstract

We develop and analyze an agent-based simulation model the purpose of which is to achieve sustained evolution under sympatric conditions. Evolution is understood in the form of speciation, i.e. the emergence of reproductively isolated and functionally distinct populations. In the model, reproduction is expressed as a function of phenotype to phenotype interaction. A population with a given set of phenotypic interactions tends to adaptive stasis, whereas if the interaction space changes, so that new dimensions are added in the course of the process, new species can emerge. We show that this behavior is stable and leads to a persistent production of new species.

Introduction

Sustained evolution is a notorious problem of evolutionary simulations (Bedau et al. 2000, Brooks 2001, Standish 2002, Holland 2003) as well as of fundamental theory (Stenseth and Maynard Smith 1984). In particular, species emergence is difficult to achieve without the artificial introduction of reproductive barriers (such as allopatric conditions).

Motivated in part by niche construction (Laland et al. 2000), exaptation (Gould and Vrba 1982), and studies of causality (Kampis 2003), we develop a model of species emergence based on phenotypic interactions. The key ingredient of the model is the updating of selection forces that arise from phenotype to phenotype interaction change, implemented here as a dimensionality increase of the interaction space.

Theory

Evolutionary simulations tend to 'run out of steam' as a result of the very adaptation process defined by the selection forces. Adaptation is akin to optimization: when optima are approached, little room is left for further development. In order to proceed towards open-ended, free evolution, the task of the evolution process needs to be redefined 'on the fly'. This recognition is present in various forms in biological evolution theory, e.g. niche construction and related concepts (Lewontin 1983, Odling-Smee et al. 2003). The idea explored in this paper comes from the study of natural causation (Kampis 2002a,b). Causal interactions are 'fat', i.e. they occur in parallel, and invoke many simultaneous levels and variables. A real-world evolution process of genetic 'replicators' is always built upon the underlying phenotype-to-phenotype interaction of

individuals as 'interactors'. If the process is modeled at the interactor level, effects of 'fat' causal interactions should be taken into account. These effects can induce new evolutionary change from the relational flexibility of 'fat' phenotype (Kampis and Gulyas 2003).

The basic concept here is that of *interaction space*. The interaction space defines how phenotypes 'see' each other, and is itself an evolutionary product. As a consequence, phenotypes are less rigidly defined than genotypes. This is easily understood in a simple example such as sexual selection. What is a relevant variable when pairing two organisms is, in the one individual, a matter of the other individual's interacting variables. Mating occurs, roughly speaking, if the two sets of variables fit, e.g. when the female prefers male antlers (variable 1) and the male possesses such antlers (variable 2). Then, a fit, or match, is possible, and reproduction will occur, which propagates both the genotypes and the phenotypes. Such a match is "groundless", however – it is based on nothing but the relational properties of the two phenotypes.

This characteristic of sexual selection is, we suggest, a suitable metaphor for more general ecological and evolutionary interactions. The idea is that interaction space can radically change with little or no genetic component. Once changed, however, it has dramatic effects on differential survival, leading to major genetic change.

The current simulation we present is confined to the case of sexual selection. To test the theory, we demonstrate that in a sexual selection-based species, evolution can transform and eventually split a population if individuals with a new phenotype which redefines interactions (such as a female that prefers body size instead of antlers) are produced.

It is important to understand how interaction space works. With the introduction of 'dissenter' individuals having a new interaction component, silent phenotype traits of every other individual (i.e. their body size, in the given example) become suddenly highlighted, and become part of a changed ecological interaction space. Every individual is effected by such a transition, as they are all potential mating partners of the dissenter(s); some preferred, some not. Not affected by prior selection, such phenotypic properties can vary more widely along the new dimension. As a result, the old sexual selection pressure becomes supplemented by a new one, arising spontaneously and endogenously, within a fully sympatric population. Next, new sexual selection can lead to the development of a new best match, and to a novel, sexually reproducing, stable sub-population, reproductively isolated from the original.

The Model

We study sexually reproducing gender-less (i.e. 'snail-like') agents, hereafter called organisms or individuals. Reproductive success depends on a similarity metric which introduces sexual selection. Genotypes and phenotypes are assumed to coincide; however, the model's behavior is based on a phenotype interpretation. Individuals are represented as parameter vectors in phenotype space. In the current version of the simulation, the only internally (i.e. in the course of a run) adjustable parameters are the phenotype vectors. The evolutionary process is an outcome of the pairing of phenotypes for reproduction. Any two organisms will reproduce if they ever meet and if they are sufficiently similar. Depending on the degree of similarity, more or less offspring will be produced. This selection force is variable (any individual can become the 'center' of a species) yet well defined (once a center is formed, distant individuals are selected against). A species is a dynamically maintained cluster of interbreeding individuals around an emergent center, characterized by the property vector of its dominant types.

Interaction change

Interaction change is represented in the model as the change of the dimensionality of the phenotype vector at the birth of a new 'dissenter' agent. The transformation of dimensionality introduced in one individual globally redefines interaction space, as it is instantaneously 'felt' across the whole population, as reflected in the global nature of the concept of interaction space. For simplicity, we will only study effects of dimension increase. This corresponds to the situation where the introduction of a new interaction, such as when paying attention to a newly defined sexual trait in mating, does not imply the neglect of the previously favored traits.

Once this general framework is specified, the key question of the model is to allocate new phenotype values to the new interaction dimension. In accordance with the theoretical considerations mentioned, we allow new values to be 'stretched' along the new dimension, which increases variance. For this so-called 'new dimension' procedure, we implemented two different, arbitrary methods. We studied type-independent and type-based methods. The latter assign identical values to identical phenotypes, the former don't.

The evolution engine

Experiments were conducted using a basic evolution engine, the task of which was to maintain a stable population before we develop new species. The evolution engine was written using an agent-based approach, in the REPAST environment developed by the University of Chicago (Repast 2003). The engine defines the usual evolutionary background, complete with reproduction, mutation, crossing over, aging, and death. It provides a 'sympatric' (i.e. panmictic) environment without a spatial component. With this engine, we simulate a partial artificial ecology that has one resource, namely, energy. Each organism has an equal chance to 'eat' in every time-step. Note that this already introduces an implicit competition for energy, which leads to density-dependent effects: genotypes represented with a higher number of individuals obtain a larger share. Reproduction is implemented similarly. More abundant genotypes have more offspring on the average, even if reproduction rates are equal. This yields a higher-than-linear growth, i.e. 'hypercompetition', a phenomenon typical for many evolution models.

The evolution engine, left alone, maintains a single species under a very wide range of parameters. We applied conditions where a constant energy influx constrains the maximum average number of individuals. That is, new species can only emerge at the indirect cost of others, as the overall number of the individuals cannot increase on the average. Together with the density-dependent effects (such as hypercompetition), this places a slight limitation on the species production dynamics of the presented model. At the same time, maintaining a meaningfully stable minimal 'base line' ecosystem was considered a key to the study of species emergence.

Realization details

The model consists of a population of agents and a non-spatial environment shared by them. Each agent has its phenotype represented as a variable length vector of integers from the interval $[V_{min}, V_{max}]$. (The length of the vector is always identical for all agents.) The agent's phenotypic traits remain fixed during the lifetime of the agent. In addition to their phenotype, agents only have the minimum of properties: age and accumulated energy.

The evolution engine uses a quasi-parallel activation regime. Each agent gets to act once per every time step in a dynamically randomized order. The agents' activity consists of three steps: energy intake, energy consumption, and reproduction. The agent first seeks E_{in} units of energy from the shared environment, and, depending on the amount available, it receives e_{in} units (possibly 0). The efficiency of energy intake decays with age:

$$e_{accumulated} = e_{accumulated} + e_{in} \cdot (E_{discounting})^{age},$$

where $0 < E_{discounting} < 1$. Next, the agent consumes $E_{consumption}$ units of its accumulated energy. If it does not have this amount available, the agent dies. Surviving agents attempt to reproduce with probability $P_{encounter}$ (i.e. every time step produces part of an entire new generation). Updating the shared environment completes the iteration. This means the addition of $E_{increase}$ units to the energy pool.

In reproduction an active agent picks a random mate from a list of potential partners, individuals similar enough to bear an offspring with the given agent. Similarity is measured by Euclidean distance between the agents' phenotype vectors. An advantage of this metric is that it is dimension independent. It allows for sexual selection to occur in the same way between any selected phenotype pairs of arbitrary dimensionality. Given two parents, and similarity d defined as above, the number of

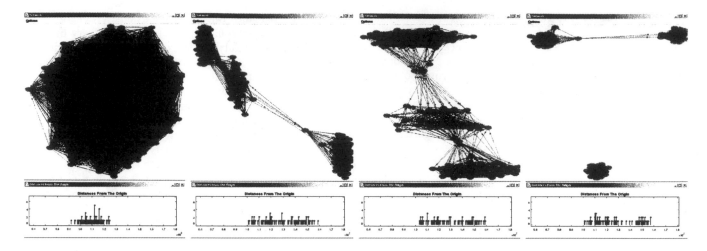

Figure 1. Segregation of a species in a hand-controlled experiment (t=75,90,114). Complete separation occurs at t=117. A stable species splits as a response to the artificial introduction of a new interaction dimension at t=75. (Old phenotype space was 5 dimensional.)

offspring is $M_{const} + (M_{limit} - d) \cdot M_{slope}$ (for $d > M_{limit}$). New agents inherit the active parent's phenotype, except for mutation and crossing over that occur with probabilities $P_{mutation}$ and $P_{crossing}$, respectively, per gene. Mutation shifts the value of a gene by a random value in $[-V_{mutation}, +V_{mutation}]$. If the mutated value falls outside the interval $[V_{min}, V_{max}]$, the offspring is dropped.

A new phenotype dimension is introduced with probability P_{change} per offspring. When this occurs, a new component is added to the agents' phenotype vector. The particular value an agent receives depends on the used 'stretch' method. The type-based method calculates the value from the agent's old phenotype. For simplicity, only the value v of a single dimension (the last one) is used:

$$v_{new} = V_{min} + (v \cdot V_{stretch}) \bmod (V_{max} - V_{min} + 1),$$

where $V_{stretch}$ is a positive parameter. The type-independent method selects a uniform random value from $[V_{min}, V_{max}]$.

Experiments

We conducted several elementary experiments to test how interaction change can affect evolution. All experiments were done on populations with several hundred or a thousand individuals. To minimize complications with hyperspace, in the present, test-of-principles model we kept dimensionality low. In a typical run we started with phenotypes of 5 dimensions. During a run the number has increased to about 50 over a few thousand time steps.

Hand-Controlled Experiments

We first present a hand-controlled experiment for interaction change (with $P_{change} = 0$). After the population converges to a stable species (Fig.2.left), a new phenotype dimension is added by hand. On the example of a transition from an 5- to a 6-dimensional phenotype, the

Figure demonstrates how the operation changes the distribution of phenotypes without changing the (old) phenotype values. Using the event as 'fuel', the relaxation process of the evolution engine tends to produce one or more new stable species, as detailed on Fig. 1. The networks shown here are results of a Fruchterman – Reingold (1991) algorithm, or spring model (FR). Nodes are organisms, edges show mutual reproduction ability; an absent edge means reproductive isolation between individuals (i.e. connected components are species). Distances in FR are not proportional and the placement is arbitrary. Fig. 1. (bottom part) shows the same time steps represented in one-dimensional true Euclidean distance, measured from the origin of the phenotype space. Height represents the number of individuals at a given distance. We see that a well-defined morphological cluster gets extended because of the introduction of the higher dimension, and it finally splits, in terms of the one-dimensional distance parameter.

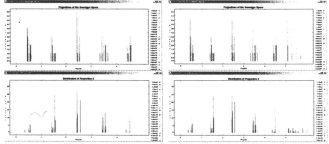

Fig. 2. Distribution of phenotype vectors before (left) and after (right) interaction change in a sample run. Note the role of the new dimension. Both the type-based (top) and type-independent method (bottom) are shown.

In several hand-controlled experiments we invariably found that a species at adaptive stasis (i.e. where there is no other change except drift) destabilizes and tends to yield offspring species upon the artificial introduction of a new interaction dimension (using either method). Simultaneous FR and phenotype distribution plots (not shown) reveal that the source of this behavior is that different types of new reproduction events occur that rapidly change the species composition and facilitate separation.

Autonomous Experiments

In a series of autonomous experiments the system was run for 6-12,000 steps (or "generations") with various nonzero values of P_{change} that introduces new interactions at uneven intervals. Qualitatively, we experienced similar behavior as in the hand-controlled case. With the subsequent change of interaction dimension, a series of speciation event occurs that together with the basic behavior of the evolution engine produces a complicated dynamics of speciation and extinction. Animations from data show that the process is guided by the interaction change events.

Morphological (and functional) separation of emergent species is visualized using a standard ordination method in a sample run in Fig. 3. The phenotype space is 5 and 50 dimensional at the left and the right hand, respectively. Figures come from the same autonomous run.

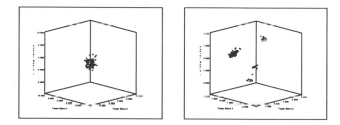

Figure 3. Emergence of species in n-space, shown by correlation-based 3-factor analysis (using SPSS). Scale is arbitrary & varies.

Theoretically, it is possible that several factors other than interaction change (such as the growing mutational load in higher dimensions) are also responsible for the experienced behavior in autonomous runs. However, hand-controlled experiments and convergence of the basic engine indicate that the former is a sufficient condition.

Discussion will be left to sensitivity analysis (below).

Artificial Phenotypes and Interaction Change

Our primary motivation is to demonstrate the evolutionary effects of interaction change in abstract form. Yet there is a temptation to seek biological meaning in our procedures. A complete discussion of the issue is left to another paper.

The type-based method assumes a simultaneous switch in the entire population at the time step when a new interaction for a single organism is introduced. This is unrealistic but less artificial than it seems. Phenocopies, bistability of selected traits, and environmentally induced epigenetic effects can equally produce similar transitions. The chosen method, at the same time, conserves the

number of different phenotypes during interaction change. This limits arbitrariness in the model by constraining the newly introduced variability in the population.

The type-independent method spreads the species in hyperspace, giving more chance for distant regions to be occupied by organisms that may find new sexual matches among themselves. Biologically, this corresponds to the case where despite the homogenization of overt traits due to selection, a significant hidden variability remains there in the unused phenotype traits, to become expressed later.

As clarified by the analysis below, the behavior of the model was insensitive to the choice between the two assignment methods. In future models, an obvious task would be to study the effect of different biologically meaningful assignments, down to single point mutations. A further objective is to allow emergent species to drop some of the old interaction dimensions, further structuring global interaction space. Finally, allowing more complex phenotype-based interactions (such as predation) would yield ecosystems with emergent producers and consumers that avoid sharing the same resources as here.

Sensitivity Analysis

To test the robustness of our results, we conducted extensive experiments with various parameters. We varied each parameter in a wide range, while keeping the rest at its default value. These studies yield results consistent with the experiments above. The more detailed analysis, and discussion of the mechanisms of species evolution (to rule out trivial effects) are left to a Technical Report (Gulyas 2004) and to a subsequent paper, respectively.

V_{min}	0	M_{limit}	15
V_{max}	100	M_{slope}	0
$P_{encounter}$	0.1	M_{const}	1
$P_{crossing}$	0.2	$E_{consumption}$	5
$P_{mutation}$	0.1	E_{in}	10
$V_{mutation}$	2	$E_{discounting}$	0.9
		$E_{increase}$	1000

Table 1: Default parameter settings.

Experiments with the Evolution Engine

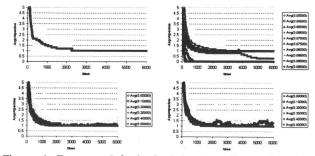

Figure 4. From top-left: basic behavior; effects of varying $P_{encounter}$ (in the range 0.05-0.095), $P_{mutation}$, and $P_{crossing}$. (10 different random seeds; 6,000 steps)

For these experiments we initially placed agents in a hypersphere of diameter 30 using a uniform random distribution. Under the parameter settings studied, sexual selection plus hypercompetition led to a fast convergence to one species from the about 25-30 in the hypersphere (see Gulyas 2003 and Fig.4.). The value of $P_{encounter}$ was tested in the interval [0.05, 0.095]. The evolution engine produces its basic behavior for $P_{encounter}$ values above 0.065. At lower probabilities the population gradually becomes extinct due a to lack of sufficient reproduction. To speed up the processes of interest, we introduced high probabilities for mutation and crossing over. Both were tested in the range 0.1-0.5. As demonstrated in Fig.4., basic convergence to one species is robust against these values. (The system trivially converges when any of the two parameters is set to 0.) The behavior of the evolution engine is robust across a wide range of $V_{mutation}$ values (i.e., 1-10) as well. The number of species drops down to a few (≥ 1) and stays there. Higher values yield more species, but the number of constant species is relatively low even for extreme values of the mutation parameter.

Similarity-related parameters were also tested: M_{limit} in the interval [0, 20], M_{slope} between 0 and 2.0, and M_{const} in the range of 0-10. Of these parameters, only the first has a pronounced effect. For $M_{limit} \geq 5$, the engine converges to a single species, while lower values increase the number of constant species, as the possibility of mating will be fairly low among agents with different phenotypes. At $M_{limit} = 0$ the population dies off.

Naturally, the system is more sensitive to the energy-related parameters. If $E_{consumption}$ (tested in the interval [1, 10]) is above 6 (with $E_{in} = 10$), the population goes extinct. Otherwise, the dynamics settles down to a single species. Similarly, $E_{in}>9$ yields a single species, while other values (tested in the range 5-25) kill off the population. The empirical limit for $E_{discounting}$ is 0.875. For values below that, agents get old too soon to survive. For other values in [0.8, 0.975], the engine produces a single species. (The speed of convergence is roughly inversely related to the value of the tested parameter.)

With $E_{in}=10$, an agent population of size ≥100 needs at least 1000 units of energy input per time step. $E_{increase}$ was tested in the interval [900, 2000]. This produces a single species. Higher values yield longer-lived agents, and slow down convergence.

The Evolution of Species

In the evolution engine we experience fast convergence to a single species. However, as shown on Fig.5., the gradual introduction of new phenotype dimensions at $P_{change} \neq 0$ results in a growing number of reproductively isolated groups, i.e. species. As the energy-system can only support a certain number of individuals, there is a natural upper limit to this. Apart from this, interaction change as simulated in this model yields a persistent evolution of new species.

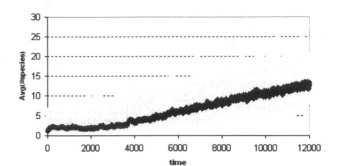

Figure 5. The evolution of species. Graphs show the average number of species (connected components of FR networks) versus time, using default settings (in 10 runs). Error bars show minimum and maximum numbers.

P_{change}	0.0075
'Stretch' method	Type-based / Type-independent
$V_{stretch}$ (type-based only)	2

Table 2: Default values for the interaction change parameters.

There are three parameters related to interaction change, as summarized in Table 2. In a limited test, we experimented with changes imposed on these, one by one. Changing P_{change} between 0.0005 and 0.001 yields results similar to those shown already. The number of species undergoes a monotonic growth. The curve gets steeper for higher values of the parameter, with the number of produced species falling between 5-15 after 12,000 iteration, and increasing with the parameter value.

The value of $V_{stretch}$ was varied between 1 and 20. As shown on Fig.5., the results are similar to the default behavior, although higher parameter values yield higher variance. Finally, the type-independent 'stretch' method also reproduces the qualitative hand-controlled results, although again with a much greater variance (Fig.6.).

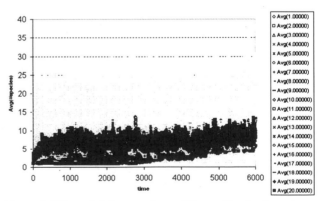

Figure 5. Effect of different values of $V_{stretch}$. Graphs show the average number of species versus time (10 runs), with default settings. Error bars show minimum and maximum values.

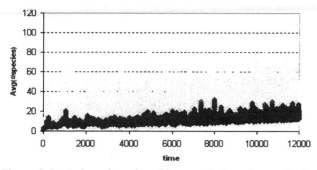

Figure 6. Evolution of species with a type-independent method. The average number of species (10 runs) versus time using default settings. Error bars show minimum and maximum values.

Conclusions

In this paper we presented a first simulation report of an interactions-based evolutionary model (kept in development for more than a year) The model uses similarity-based sexual selection to approach open ended evolution. We demonstrated that changing phenotype-to-phenotype interaction (adding new dimensions to interaction space) can repeatedly split species by the production of new selection constraints. We hypothesize that sustained ecological evolution proceeds by similar mechanisms. Prospects for further studies have been briefly outlined.

The current version of the simulation is downloadable from the site http://hps.elte.hu/~gk/EvoTech/.

Acknowledgments

Research reported here was supported by several organizations, which is gratefully acknowledged. Part of the work was carried out during the first author's stay at JAIST, Japan, as Fujitsu Associate Professor of Complex Systems, School of Knowledge Science. Computer simulations were done on the BeoWulf cluster of the Center for Complex Systems Studies, Physics Department, Kalamazoo College, MI (during the first author's stay as Guest Scholar), as well as on the SUN E10K, E15K system of Hungary's NIIF Supercomputing Center.

References

Bedau, M.A.; McCaskill, J.S.; Packard N.H.; Rasmussen, S.; Adami, C.; Green, D.G.; Ikegami, T; Kaneko, K.; and Ray, T. S. 2000. Open Problems in Artificial Life. *Artificial Life*, 6:363—376.

Brooks, R.A. 2001. The Relationship Between Matter and Life. *Nature*, 409:409—411.

Fruchterman, T.M.J., and Reingold, E.M. 1991. Graph Drawing by Force-directed Placement, *Software-Practice and Experience*, 21(11):1129—1164.

Gould, S.J., and Vrba, E.S. 1982. Exaptation - A Missing Term in the Science of Form. *Paleobiology*, 8:4—15.

Gulyas, L. 2004. Summary of EvoTech-V v5.0.8, Technical Report, Eötvös University. http://hps.elte.hu/~gk/EvoTech/Summary.pdf

Holland, J.H. 2003 Challenges for Agent-based Modeling. http://www.nd.edu/~swarm03/Keynote/keynote.html

Kampis, G. 2002a. Towards an Evolutionary Technology (in Japanese, in press). English version: http://www.jaist.ac.jp/~g-kampis/EvoTech/Towards.html

Kampis, G. 2002b. A Causal Model of Evolution, In *Proceedigs of. SEAL 02* (4th Asia-Pacific Conference on Simulated Evolution And Learning),Singapore, 836—840.

Kampis, G. 2003. Causal Depth and the Modal View of Causality. Presented at the IUHPS/DLMPS World Conference, Oviedo, Spain, August 8-13. http://www.jaist.ac.jp/~g-kampis/Oviedo/CausalDepth.html

Kampis, G., and Gulyas. L. 2003. Causal Structures in Embodied Systems, *ERCIM News*, No 53:23—25.

Laland, K.N.; Odling-Smee, J.; and Feldman, M.W. 2000. Niche Construction, Biological Evolution and Cultural Change, *Behavioral and Brain Sciences*, 23(1):131—146.

Lewontin, R. 1983. The Organism as the Subject and Object of Evolution. *Scientia*, 118:65—82.

Odling-Smee, J.; Laland, K.N.; and Feldman, M.W. 2003. Niche Construction: The Neglected Process in Evolution, Princeton, NJ: Princeton UP.

Repast (2003): http://repast.sourceforge.net/

Standish, R.K. 2002. Prospects for Open-Ended Evolution in Artificial Life. In *Proceedings of the 7th AROB conference*, Beppu, Japan, 455.

Stenseth N. C., and Maynard Smith, J. 1984. Coevolution in Ecosystems: Red Queen Evolution or Stasis? *Evolution*, 38:870—880.

This document was last changed on 14 April 2004.

See How She Runs: Towards Visualising Artificial Red Queen Evolution

James A. R. Marshall[1] and Simon Tokumine[2]

[1]Centre for Behavioural Biology, Department of Computer Science, University of Bristol, BS8 1UB, UK.
[2]Department of Computing, Imperial College London, SW7 2AZ, UK.
marshall@cs.bris.ac.uk

Abstract

Since Darwin originally proposed that one of the primary drives in evolution towards increased complexity might be biotic competition, the search for examples of this phenomenon has been a major objective for students of evolution. The Red Queen hypothesis proposes that this competition is never-ending; as an organism or species improves in some manner relative to its competitors, these competitors are also improving. This results in an evolutionary "arms race" in which neither side maintains its advantage for long. Examples of the Red Queen effect from the natural world can be found, but it is perhaps in artificial evolutionary systems that there is the greatest scope for gaining a deeper understanding of the nature of evolutionary change under such circumstances, whether gradualistic or punctuational for example. Here we present a model of co-adaptation and introduce a visualisation method for tracking changes in the fitness landscape and the population's location on it. This enables the Red Queen's endless run within the model to be visualised.

Introduction

It was Darwin that first proposed, with his metaphor of the wedge, that competition between species could drive evolution to create ever more complex adaptations [5]. Since then naturalists and biologists have found many examples of this in the real world, such as brood parasitism [8] to name but one. More recently this idea has been refined and dubbed the "Red Queen" effect [17], or "evolutionary arms race" [6], as adaptations of a species or organism provide it with no net benefit because its competitors are continually keeping up with adaptations of their own. Like the character in Lewis Carroll's "Through the Looking Glass", or nuclear-armed superpowers, species and organisms have to run continually just to maintain their position.

It is perhaps in artificial life models of evolution that the greatest potential for gaining deeper understanding of the Red Queen effect lies. As well as confirming that the effect can occur in principle, it is of interest to consider the different evolutionary modes with which it might operate. For example, does Red Queen evolution typically proceed in a gradualistic manner, as traditional Darwinism would expect, or in punctuated bursts of evolutionary activity between intervening periods of stasis [7]? Similarly, can populations be observed to descend fitness gradients, as expected by Wright's "shifting balance theory" [18]?

Earlier approaches to visualising artificial evolutionary or co-evolutionary systems have included measurement of co-evolutionary progress in populations [3], observation of the dynamics of co-evolution in ecosystems [10], and visualising fitness landscapes as sampled by populations [1, 9]. In this paper we take an alternative approach and visualise the Red Queen in an artificial life model using Wright's [18] concept of the fitness landscape, animating this and the population's position on it as both change.

The Model

The model underlying the visualisation[1] introduced in this paper is that presented by Marshall & Rowe [12, 13, 14] and only the relevant details are very briefly discussed here. The model was designed to study the evolution of cooperation. Interaction within the model takes the form of an iterated 2 x 2 game, typically the Iterated Prisoner's Dilemma (IPD). Individuals have both a memory of their interactions with other individuals and a strategy that specifies the current action in the game from two choices, A or B, based on the action choices of both participants on their last interaction together. The model is an evolutionary one, with selection pressure applied on payoffs from interactions through the imposition of a per time step living cost, so individuals die and are replaced by the offspring of the individuals in the population that score higher in their interactions. Individuals are also terminated with a random death probability at each time step, to prevent successful individuals becoming "immortal". Reproduction is sexual and uses crossover and point mutation operators to combine two parents' chromosomes into an offspring chromosome. The strategy chromosome encoding follows Mar & St Denis [11], but with a binary alphabet, to specify a two-dimensional strategy. Thus five loci are used (one for each potential interaction history, one for the first interaction), each with two alleles (A or B).

Visualisation Method

The inspiration for our visualisation method is Wright's [18] fitness landscape. Wright posed the question "if the entire field of possible gene combinations be graded with respect to adaptive value under a particular set of

---

[1] The source code for the model and visualisation tool is downloadable from
ftp://ftp.swarm.org/pub/swarm/apps/objc/contrib/EPD-2.1.2-2.1.1.tar.gz

conditions, what would be its nature?" (*ibid*). Wright presented a few examples of visualising simple genospaces, from two to five paired allelomorphs (see figure 1), and noted the high dimensionality of even these simple examples, with another dimension also necessary to represent adaptive value or fitness.

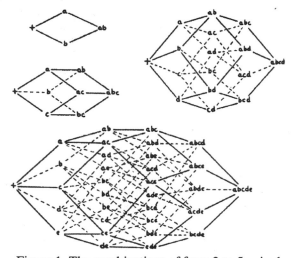

Figure 1. The combinations of from 2 to 5 paired allelomorphs. Reproduced from Wright [18].

To convey his concept of the fitness landscape, Wright thus presented a hypothetical example in which a high-dimensional genospace was compressed into a two-dimensional plane, and drew contour lines to delineate areas of different fitness (see figure 2), much like a cartographer uses contour lines to indicate altitude on a map.

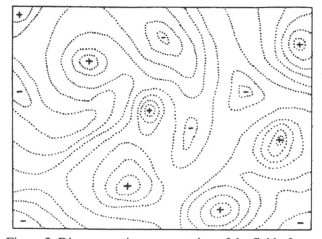

Figure 2. Diagrammatic representation of the field of gene combinations in two dimensions instead of many thousands. Dotted lines represent contours with respect to adaptiveness. Reproduced from Wright [18].

Genospace Compression

The technique presented in this paper therefore aims to compress the high-dimensional genospace of the Marshall & Rowe model into a representation of sufficiently low dimensionality to present meaningfully on paper or a computer screen (i.e. in two dimensions). A key requirement of this compression, however, is that it maintains the relationships of points in the genospace and points in the compressed representation, i.e. points which are genetically close in genospace (limited genetic difference, or Hamming distance), should be plotted close to each other in the representation. A means of representing the adaptive value of each point in the compressed genospace is also required.

Although the genetic representation used in the Marshall & Rowe model has 5 loci per chromosome and is therefore five-dimensional, as there are only 2 alleles per locus there are only 32 different possible genotypes. The Search Space Matrix technique, described by Collins [4], provides an elegant means of compressing such a genospace into two dimensions, while minimising the Hamming distance between neighbouring points in the compressed representation. The result of applying the Search Space Matrix technique to the genospace of the Marshall & Rowe model is presented below in figure 3.

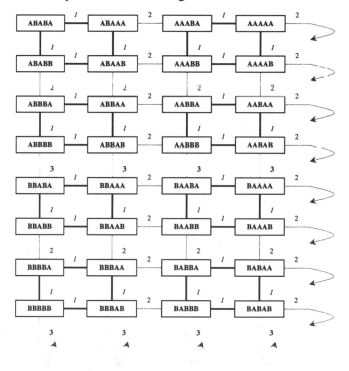

Figure 3. Search Space Matrix representation of genospace in the Marshall & Rowe model. Cells are points in genospace, lines connect neighbouring points and indicate Hamming distance between them. As the Search Space Matrix is toroidal, Hamming distances at the bottom and right edges indicate genetic distance to the opposite edges of the matrix.

Fitness Evaluation

In order to produce a fitness landscape, it is not enough to simply represent the genospace; the adaptive value of each point in that genospace must also be represented. With a two-dimensional representation such as that produced by the Search Space Matrix technique the obvious way to represent fitness is using the third dimension, namely height. However a means of determining the adaptive value of a given genotype must be specified.

In most traditional applications of evolutionary algorithms the objective is to optimise a static problem, therefore the fitness landscape is fixed and the fitness of any individual solution can be evaluated in isolation. However in co-evolutionary or co-adaptive systems such as ours, fitness is a function of the co-evolving species, or of the population as a whole. Here we specify how we calculate fitness in the Marshall & Rowe model, using expected utility from game theory.

The approach to calculating fitness is simple. Each point in the model's genospace represents a unique strategy for an iterated 2 x 2 game. At each model timestep, the frequencies of each of the n strategies are logged from the model in the row matrix

$$[\mathbf{F}] = (f_1, f_2, ..., f_n). \qquad (1)$$

The value of a strategy against another, $V(I|J)$, represents the score of strategy I when interacting with strategy J, e.g. for the IPD $V(\text{ALL D}|\text{TFT}) = T + wP/(1 - w)$ where ALL D is the strategy of unconditional defection, TFT is the well known strategy Tit-for-Tat, T is the temptation payoff in the Prisoner's Dilemma, P is the punishment payoff in the Prisoner's Dilemma, and w is the "shadow of the future" or probability of continued interaction from one iteration to the next in the game [2]. The payoff of any one strategy S_i in the model against all the n other strategies, given a known shadow of the future w, is thus specified by the matrix

$$\mathbf{O} = \begin{pmatrix} V(S_1|S_1) & V(S_2|S_1) & \cdots & V(S_n|S_1) \\ V(S_1|S_2) & \ddots & & \\ \vdots & & & \vdots \\ V(S_1|S_n) & & \cdots & V(S_n|S_n) \end{pmatrix}. \qquad (2)$$

Then the expected utility of each strategy in a certain population is given by the row matrix

$$[\mathbf{E}] = [\mathbf{F}] \times \mathbf{O}. \qquad (3)$$

It is easy to normalise the values in this matrix using the maximum and minimum values in O, in order to simplify the calculation of the appropriate height values in the visualisation. The value of w used to calculate O is set to reflect the average interaction length typically observed in the model with the parameters under investigation. Note that this calculation of expected utility assumes a panmictic population, because the frequencies of strategies as experienced by each individual strategy are identical, whereas the Marshall & Rowe model implements a spatial population. However the model can be parameterised to be equivalent to a panmictic population, and that is the approach taken here.

Population Distribution

Having determined a means of representing the genospace, and the adaptive value of each point within it, the last remaining component of our visualisation technique is the visualisation of the areas of genospace that the population occupies. In our three-dimensional visualisation the x and y-axes are used to represent the genospace, and the z-axis is used to represent the adaptive value of points in the genospace. The obvious means to represent the population's distribution in the genospace is therefore colouring of the landscape representation, so each point in the genospace is assigned a value from the row matrix [F] representing the frequency of that genotype in the population. This value is translated to a colour from a colour range and used to colour the landscape in the visualisation.

Results

Figure 4 below presents a sample visualisation from a simulation run of the Marshall & Rowe model using the Apology game[2].

Apology, the payoff matrix for which is shown below, was used in this instance as it generates a more interesting fitness landscape than the Prisoner's Dilemma and consequently more interesting evolutionary dynamics.

	A	B
A	1,1	5,3
B	3,5	0,0

Table 1. Apology payoff matrix (payoffs to row player shown first).

The main difference with the Prisoner's Dilemma is that Apology reserves the highest payoffs for players that make different stage-game choices, while players co-ordinating on one outcome do less well. An evolving population of Apology players should therefore be expected to be especially susceptible to changes in the frequency of different strategies, and hence provide good conditions for observation of the Red Queen effect.

---

[2] The model was run with the following parameters: horizontal environment size in cells = 1, vertical environment size in cells = 1, initial individual energy = 15, energy living cost = 3, crossover rate = 0.1, mutation rate = 0.01, maximum population size = 50, initial population size = 25, death probability $d = 0.05$, interaction length per encounter = 3. Apology payoffs were as follows: AA = 1, AB = 5, BA = 3, BB = 0. For the visualisation the "shadow of the future" was set to $w \approx 0.7$.

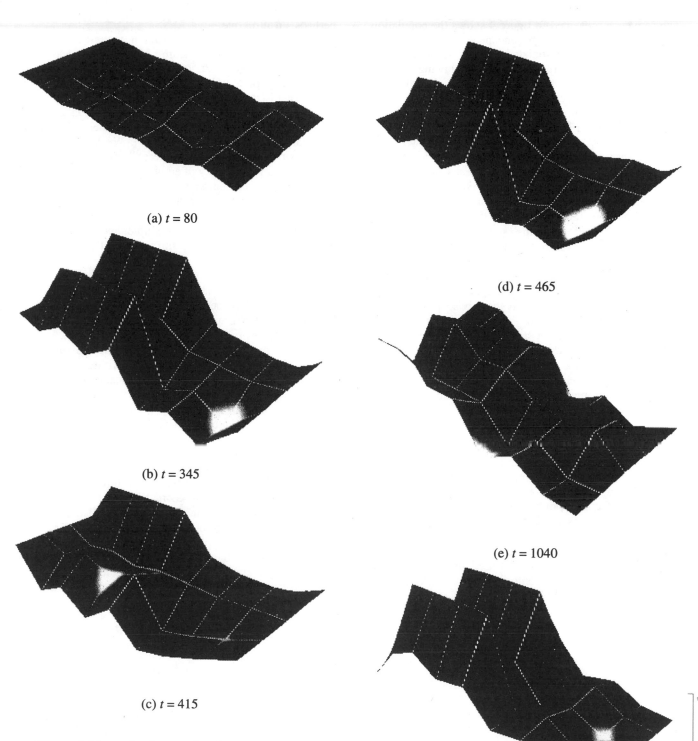

(a) *t* = 80

(b) *t* = 345

(c) *t* = 415

(d) *t* = 465

(e) *t* = 1040

(f) *t* = 1210

Figure 4. Fitness landscape visualisations for one
simulation. Peaks in the landscape indicate high-fitness
genotypes. Gridlines are used to indicate the relief of the
landscape. Colour represents the frequency distribution of
the population in the genospace (white = 1, black = 0). The
origin (top left) of the SSM is the point BBBBB.

The sequence presented in figure 4 shows the population in the model moving through genospace over a period of simulated time, and the associated changes in the fitness of different points in that genospace. Following several hundred timesteps of what might best be described as genetic drift over a largely uniform but constantly changing fitness landscape (figure 4a), a putative instance of the Red Queen Effect is seen in figures 4b to 4d. In figure 4b the population mostly occupies a fitness valley, yet two obvious fitness peaks exist. Attempting to occupy one of these fitness peaks (4c), that peak becomes depressed by the change in the population's constituency while the fitness valley the population began at is simultaneously elevated. The population abandons the depressed fitness peak for its newly elevated fitness valley, only to achieve, by this shift, a reversion to the former shape of the landscape (4d). This dynamic tension between occupying the fitness peak and remaining in the fitness valley continues for several hundred timesteps, followed by more drift, at the conclusion of which the landscape adopts an entirely new form on which the population finds itself concentrated once again in a fitness valley (4e). In the course of trying to adapt to this new landscape, the population restores the fitness landscape to a state very similar to that first seen in (4b), on which it finds itself in a very similar position to that it previously occupied (4f).

Discussion

While the evolutionary dynamics illustrated by the sample results presented above are interesting, they pose more questions than they answer. Even a more thorough investigation of these results alone, such as explaining the evolutionary dynamics from a game theoretic perspective, is beyond the scope of this paper, concerned as it is with visualisation. However, these results do shed a little light on some questions that have exercised biologists for the past decades or even centuries. For example, the observation of what might be characterised as genetic drift in the model addresses an issue that Wright and even Darwin himself were concerned with, the exclusivity or otherwise of natural selection as a mechanism of evolution. In particular the comparatively uniform fitness landscape of figure 4a could represent a temporary change of the landscape into a "neutral network" [15] through which the population is able to percolate, although with non-adaptive consequences in this case (as the population becomes stuck in a fitness valley in figure 4b). Similarly the question of the mode of evolution, whether gradual or punctuational, finds some answers in the model through the observation of prolonged periods of apparent stasis followed by rapid shifts in the population's genetic constitution. Of course the observation of any such phenomena within a model does not settle these questions, and detailed investigation is required armed additionally with suitable metrics, such as those developed by Cliff & Miller [3]. However a visualisation approach similar to that described here provides an intuitive entry-point for the problem and

perhaps a fruitful one.

We believe that the visualisation technique we have presented here differs from earlier approaches by combining the representation of the fitness landscape with that of the population's position on it. Earlier approaches have visualised only the fitness landscape (by population sampling) [1], or have separately visualised the population's fitness distribution and the population's trajectory through the genospace [9]. We believe that the preliminary results presented here demonstrate both the practicality and the utility of combining both these elements into a single representation, as Wright himself proposed doing [18]. We hope that such visualisation techniques may be usefully applied to other co-evolutionary systems, artificial immune systems for example, and that others will find it interesting to explore further the ideas presented in this paper.

For application to other co-evolutionary systems, several points will have to be taken into account. Most obviously, the target model presented here has a sufficiently limited genospace that it is practical to exhaustively evaluate the adaptive value of every point in it given a certain population distribution. In more complex evolutionary systems this will not be the case, and measures will have to be taken to sample the fitness landscape and build an approximate representation of it (this is indeed the approach taken in [1] and [9]). Also, the genospace compression technique used here, Search Space Matrix, has some limitations, particularly the existence of neighbouring points of high genetic difference (from figure 3 it can be seen that there are two "Hamming cliffs" where neighbouring points are different at three alleles). One approach would be to have geometric space between points in the visualisation related to Hamming distance in the genospace representation. Alternatively, it would be helpful to investigate the application of different compression techniques such as Sammon mapping [16]. In the future we hope to investigate these issues through the visualisation of progressively more complex co-evolutionary systems.

Acknowledgements

Thanks to Tim Kovacs and members of the Ant Lab at the University of Bristol for helpful comments and suggestions.

References

1. Anastasoff, S. J. 2000. A 'Fitness Landscaping' Comparison of Evolved Robot Control Systems. In Proceedings of the Seventh International Conference on Artificial Life, 273-281. Cambridge, Mass.: MIT Press.
2. Axelrod, R. 1984. The Evolution of Cooperation. New York: Basic Books.
3. Cliff, D. and Miller G. F. 1995. Tracking the Red Queen: Measurement of Adaptive Progress in Co-Evolutionary Simulations. In Advances in Artificial Life:

Proceedings of the Third European Conference on Artificial Life (ECAL95). Lecture Notes in Artificial Intelligence 929, 200-218. Berlin: Springer Verlag.

4. Collins, T. D. 1997. Using Software Visualisation Technology to Help Evolutionary Algorithm Users Validate Their Solutions. In Proceedings of the Seventh International Conference on Genetic Algorithms (ICGA'97), 307-314. San Francisco, Calif.: Morgan Kaufmann.

5. Darwin, C. R. 1859. On the Origin of Species by Means of Natural Selection or the Preservation of Favoured Races in the Struggle for Life. London: John Murray.

6. Dawkins, R. and Krebs, J. R. 1979. Arms Races Between and Within Species. Proceedings of the Royal Society of London B 205: 489-511.

7. Eldredge, N. and Gould, S. J. 1972. Punctuated Equilibria: an Alternative to Phyletic Gradualism. In Models in Paleobiology, 82-115. San Francisco, Calif.: Freeman.

8. Hamilton, W. J. and Orians, G. H. 1965. Evolution of Brood Parasitism in Altrical Birds. Condor 67: 361-382.

9. Harvey, I. and Thompson, A. 1997. Through the Labyrinth Evolution Finds a Way: A Silicon Ridge. In Proceedings of the First International Conference on Evolvable Systems. Lecture Notes in Computer Science 1259, 406-422. Berlin: Springer Verlag.

10. Kauffman, S. A. and Johnsen, S. 1991. Coevolution to the edge of chaos: coupled fitness landscapes, poised states, and coevolutionary avalanches. Journal of Theoretical Biology 149: 467.

11. Mar, G. and St Denis, P. 1994. Chaos in Cooperation: Continuous-Valued Prisoner's Dilemmas in Infinite-Valued Logic. International Journal of Bifurcation and Chaos 4(4): 943-958.

12. Marshall, J. A. R. and Rowe, J. E. 2000. Investigating the Mechanisms Underlying Cooperation in Viscous Population Multi-Agent Systems. In Proceedings of the Seventh International Conference on Artificial Life, 348-352. Cambridge, Mass.: MIT Press.

13. Marshall, J. A. R. and Rowe, J. E. 2003a. Kin Selection May Inhibit the Evolution of Reciprocation. Journal of Theoretical Biology 222(3): 331-335.

14. Marshall, J. A. R. and Rowe, J. E. 2003b. Viscous Populations and Their Support for Reciprocal Cooperation. Artificial Life 9(3): 327-334.

15. Maynard Smith, J. 1970. Natural selection and the concept of a protein space. Nature 255: 563-564.

16. Sammon, J. W. 1969. A Nonlinear Mapping for Data Structure Analysis. IEEE Transactions on Computers C 18(5): 401-408.

17. van Valen, L. 1973. A New Evolutionary Law. Evolutionary Theory 1: 1-30.

18. Wright, S. 1932. The Roles of Mutation, Inbreeding, Crossbreeding and Selection in Evolution. In Proceedings of the Sixth International Congress on Genetics, 356-366. Ithaca, New York: Brooklyn Botanic Garden.

Sexual reproduction and Muller's ratchet in digital organisms

Dusan Misevic[1], Richard E. Lenski[1,2] and Charles Ofria[3]

[1]Department of Zoology, [2]Department of Microbiology & Molecular Genetics, [3]Department of Computer Science & Engineering, Michigan State University, East Lansing, MI 48824
Corresponding author: Dusan Misevic (dule@alife.org)

Abstract

The evolution of sexual reproduction has long been a major problem in biology. According to one theory, sex opposes the fitness-destroying process of Muller's ratchet, which occurs by the stochastic loss of high-fitness genotypes in small populations. Sex opposes the ratchet by allowing genotypes with different deleterious mutations to produce mutation-free offspring. We used the Avida digital-evolution software to investigate sex in relation to Muller's ratchet. Populations of digital organisms mutated, competed, and evolved in a complex environment. Populations were either asexual or sexual; in the latter case, parental genomes recombined to produce offspring. We also varied genomic mutation rates and population sizes, which at extreme values often caused mutational meltdowns and population extinctions. Our results demonstrate that sex is advantageous for population survival under some conditions. However, differences in extinction probabilities were usually small, occurred over a narrow range of mutation rates and population sizes, and the advantage of sex for population survival required many generations. Also, the mean fitness of surviving asexual populations was often greater than in sexual populations. This last result indicates the nccd for work that compares the statistical distribution of mutational effects and epistatic interactions in asexual and sexual populations.

Introduction

The mixing of genomes via sexual reproduction and the resulting recombination of genetic variants are widespread and important to life on Earth (Maynard Smith; 1978 Bell 1982). For over a century, biologists have sought to explain the evolutionary origin and maintenance of sex (Weismann 1889; Ghiselin 1988; West et al. 1999). While there have been important advances, there are still more questions than answers, making this research area both interesting and active (Rice 2002).

There are many potential costs associated with sexual reproduction, including time and energy spent in searching for a suitable mate and in the act of mating (Lewis 1983). Moreover, at the genetic level there often exists a two-fold cost of sex, which is variously called the cost of meiosis or the cost of males. This cost alone implies that asexual populations should grow twice as fast as sexual ones, all else being equal (Maynard Smith 1971). This two-fold cost presents a major hurdle that must be overcome if the evolution of sex is to be understood. Many theories have been proposed to explain why sex arose and how it persists

despite these costs, but most have not been adequately tested (Kondrashov 1993) and none are broadly supported by those tests that have been performed to date (West et al. 1999). One of the main theories is attributed to the famous geneticist H. J. Muller, and it proposes that sex is beneficial in opposing what is now referred to as Muller's ratchet.

Muller's ratchet is a stochastic process that leads to the loss of genetic information from asexual populations via the loss of those high-fitness genotypes that are free of any deleterious mutations (Muller 1964; Felsenstein 1974). The ratchet depends on both mutation and drift, where drift refers to changes in gene frequencies caused by the random sampling of a finite number of genes from the previous generation. Drift is especially important, and selection is relatively weak, in small populations that are most sensitive to sampling effects. By chance, the most fit genotypic class can be lost from a small population, especially if the genomic mutation rate is high. Once this class is lost from an asexual population, it cannot be recovered owing to the low probability of beneficial mutations in a small population. The metaphorical ratchet has thus advanced one notch and its action is irreversible; each subsequent loss of the most fit genotypic class advances the ratchet another notch. By contrast, in a sexual population the most fit class can potentially be reconstructed, even after it has been lost, via recombination between two genotypes that carry different mutations. The harmful effects of Muller's ratchet are thus opposed by sex.

Small populations subject to the ratchet may even go extinct if the resulting fitness decay causes the birth rate to drop below the death rate. This feedback can produce a vicious cycle, in which declining fitness leads to a drop in population size, which speeds up the ratchet causing further fitness loss. This cycle has been described as a mutational meltdown (Lynch et al. 1993, 1995). In this study, we use population survival as one metric to compare the effect of Muller's ratchet in sexual and asexual populations.

Several experiments have shown the fitness-destroying effect of Muller's ratchet in diverse microorganisms including viruses (Chao 1990; Duarte et al. 1992), bacteria (Andersson and Hughes 1996), and protozoa (Bell 1989). At least one study with viruses further showed that genetic recombination could oppose the ratchet (Chao et al. 1997). Research in evolutionary computation has also examined the ratchet, with an emphasis on designing strategies to eliminate its adverse effects from applied optimization algorithms (Nowak and Schuster 1989; Prügel-Bennett 1996; Zitzler et al. 2000; Laumanns et al. 2001). However, questions about the operation of Muller's ratchet and its interaction with reproductive mode are difficult to answer using biological systems (Kondrashov 1982; Maynard Smith 1988). In this paper, we therefore use digital organisms to examine the effects of Muller's ratchet over a range of population sizes and mutation rates, and we test whether sexual reproduction can substantially impede the ratchet's harmful effects.

Methods

We use digital evolution software called Avida to study Muller's ratchet and its interaction with reproductive mode. Avida maintains and monitors experimental populations of digital organisms, which are self-replicating computer programs written in a customized assembler-like language (Adami 1998; Ofria and Wilke 2004). Avida has been used for several other studies of evolutionary dynamics and outcomes (Lenski et al. 1999; Wilke et al. 2001; Lenski et al. 2003). Digital organisms in Avida evolve on rugged fitness landscapes characterized by diverse epistatic interactions, they have intricate genotype-phenotype maps that emerge from a complex developmental program, and they can exhibit quasi-species dynamics (Lenski et al. 1999; Adami et al. 2000; Wilke et al. 2001; Lenski et al. 2003). They approach the level of complexity of organic viruses, making results obtained with Avida biologically relevant and of general interest (Adami 2002; Wilke and Adami 2002). In Avida, genomes have 26 possible instructions at each position. All organisms descend from an ancestral program used to seed a population. Organisms execute the programs encoded by their genomes, including commands that enable them to copy and divide their genomes. The `copy` instruction duplicates a single instruction. During this duplication process, the instruction has a probability of being miscopied and changed to a different instruction in the offspring's genome; mutations from one instruction to any other are equally likely. In this study, we held the genome length constant by setting rates of insertion and deletion mutations to zero. The genomic mutation rate, U, equals the mutation probability per instruction copied multiplied by the genome length. The value of U is controlled by the investigator and was varied in our experiments.

Each digital organism occupies a cell in a rectangular lattice. The size of the lattice sets the maximum population size, which was also varied in our experiments. After a `divide` instruction is executed, the genome is split into two; the duplicated genome (the offspring), is placed into a random cell in the lattice, which kills the organism that previously occupied that position. Although death is random, the danger of being overwritten provides a selective advantage to organisms that replicate faster. Also, if any organism has not reproduced after executing its instructions an average of 15 times each, it dies and is removed from the population. If all the individuals in a

population fail to reproduce within this allotted time, then the population has become extinct. In this study, organisms could accelerate the execution of their genomic instructions, and thus their reproduction, by performing certain logic functions (Lenski et al. 2003). If an organism performs one of these functions, then it receives some corresponding resource that provides energy and accelerates execution of its genomic program. Aside from differences in their ability to perform logic functions, all organisms would execute their genomes at the same rate. Even in that case, fitness can vary among organisms depending on their relative gestation time (number of executed instructions necessary to produce an offspring). An organism's expected fitness equals the product of the baseline energy available to all organisms (made proportional to genome length to eliminate selection on genome size per se) and bonuses received for performing logic operations, divided by the gestation time. Organisms do not have access to, and cannot manipulate, their expected fitness. Realized fitness is affected, however, by population structure and interactions among organisms.

In this study, we introduce a new command to Avida that causes the digital organisms to reproduce sexually. We use this variant command to compare evolution in asexual and sexual populations. When executed, the `divide-sex` command separates a copied genome from its parent, but it does not immediately place that new genome into the population. Instead, the new genome goes into a separate location called the *birth chamber*. If the chamber is empty, the new genome remains there until a second genome arrives. When two genomes are present, they recombine and then both resulting offspring are placed at random into the population. [Notice that this mechanism for sexual reproduction does not involve the two-fold cost of sex, although we could have introduced such a cost by placing only one of the two recombinant offspring in the population. It is likely that the most primitive biological forms of sex did not have to overcome this two-fold cost (Maynard Smith 1978), and so we began this research by placing both recombinants in the population. Even so, as we show below, the conditions favoring sexual reproduction with respect to Muller's ratchet are fairly narrow.] Recombination occurs by taking a single continuous region (with two random endpoints) from one genome and swapping it with the corresponding region from the other genome. Genomes are circular and fixed in length; genomic positions are defined by distance from the first command executed and direction of execution. The initial speed of execution of an offspring's genome is set to the weighted average of its two parents, with weights based on the proportion that each parent contributed to the offspring's genome. Under asexual reproduction, the initial speed is inherited from the sole parent.

We performed the evolution experiments with Avida in two stages. Briefly, the *first stage* used large populations in order to evolve digital organisms that were well adapted to their environment. For the *second stage*, these organisms were moved into much smaller populations to examine the effects of Muller's ratchet. Maximum population size was identical in all cases during the first stage, but this size was varied in the second stage. Mutation rates were also varied across runs, but the rate was held constant in both stages of a given lineage. Similarly, reproductive mode (asexual or sexual) varied across runs, but this mode was held constant during both stages of any lineage. Further details on the two evolutionary stages are provided below.

First evolutionary stage: All runs started with a hand-written ancestor, which had a genome of 100 instructions. The ancestor was capable of self-replication, but it could not perform any logic functions. Ten replicate experiments were run with each of five genomic mutation rates (U = 0.1, 0.3, 1, 3, 10) and with reproduction being either strictly asexual or sexual for the population. Thus, there were 100 runs of the first stage. Replicates differ only in the random number seed, which then affects all the stochastic events during the run, such as mutations and offspring placement. The maximum population size (N) was 3600 organisms for all runs in the first stage. Genome length was held constant in all runs. Nine different resources could be obtained by digital organisms that evolved the ability to perform logic functions; these resources were available in infinite supply. Experiments ran for 100,000 updates, where an update is an arbitrary

unit of time in Avida corresponding to the execution of 30 instructions, on average, per individual organism. In these first-stage runs, one generation required roughly 10 updates; the exact value depends on the number of instructions needed to produce an offspring, which often changes during evolution. At U = 0.1 and 0.3, the experiments ran for an additional 500,000 updates in order to compensate for the slower adaptation at these lower mutation rates; this extension ensured there were genotypes that could use all nine resources in each first-stage treatment. During each run, we recorded the numbers of organisms using each resource as well as the mean and highest fitness in the population. At the end of each run, we saved the most fit genotype (provided it was able to use all nine resources) for use in the second evolutionary stage.

Second evolutionary stage: These runs used the pool of well-adapted genotypes from the first evolutionary stage as starting material to investigate the effect of Muller's ratchet on small populations. Each small population had the same mutation rate and same reproductive mode as its first-stage progenitor. For each of the five mutation rates, one sexual and one asexual organism were randomly chosen from the pool of genotypes saved at the end of the first-stage runs. Each of these ten genotypes (also referred to as proximate ancestors) was then used to start 100 replicate experiments

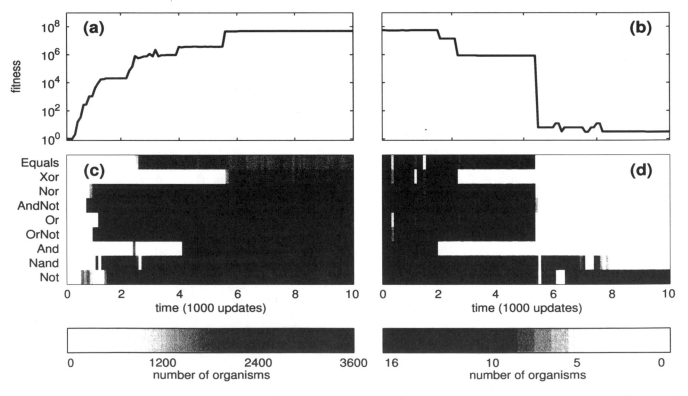

Figure 1. Trajectories for maximum fitness and resource use during evolution in large and small asexual populations. (a) Maximum fitness in a large population (N = 3600) during the initial 10,000 updates with genomic mutation rate U = 0.3; **(b)** Maximum fitness in a small population (N=16) that began with the most fit genotype from (a), and which continued at the same mutation rate; **(c)** Number of organisms performing each of the nine rewarded logic functions, indicated by shading intensity (scale below), in the same large population as in (a); **(d)** Number of organisms performing these logic functions in the same small population as in (b). Note the different scales in (c) and (d).

Reproduction: asexual, sexual		Genomic mutation rate, U				
		0.1	0.3	1	3	10
Population size, N	4	75, 80	**8, 41**	0, 1	0, 0	0, 0
	8	98, 100	92, 98	**60, 87**	0, 3	0, 0
	16	100, 100	100, 100	98, 99	**67, 95**	0, 0
	32	100, 100	100, 100	100, 100	97, 99	**18, 70**
	64	100, 100	100, 100	100, 100	99, 100	98, 99
	128	100, 100	100, 100	100, 100	100, 100	100, 100

Table 1. Survival of asexual and sexual populations when population size is small and Muller's ratchet operates. The two numbers in each cell show the number (out of 100 runs) of surviving asexual and sexual populations, respectively, for each combination of genomic mutation rate and maximum population size. Paired values are shown in **bold** when they were significantly different (see text for details).

at each of six small population sizes (N = 4, 8, 16, 32, 64, or 128), for a total of 6000 second-stage runs. All parameters other than population size were identical to those in the first-stage experiments. The second-stage runs lasted for 500,000 updates, during which we again recorded the number of organisms using each resource, as well as the mean and highest fitness. We also recorded whether the population had gone extinct by the end of the run.

Results and Discussion

Adaptation and decay: In the first stage of the evolution experiments, several populations at each mutation rate produced highly adapted genotypes that could perform all nine logic functions and thereby obtain the corresponding resources. Figure 1a shows the maximum-fitness trajectory over the first 10,000 updates for a representative first-stage population that evolved with U = 0.3 while reproducing asexually. Figure 1b shows the corresponding trajectory for a second-stage population founded by the most fit genotype from the first stage, but with the maximum population size now reduced from 3,600 to 16 organisms. The step-like changes in maximum fitness are typical of the experiments. The steps reflect, in large measure, the adaptive gains and maladaptive losses of logic functions that occurred in the large and small populations, respectively (Figs. 1c and 1d).

Population survival and extinction: In order to test if sexual reproduction could substantially impede Muller's ratchet, we compared the number of sexual and asexual populations that survived to the end of the second stage. Recall that mutation accumulation by Muller's ratchet can cause individuals to fail to reproduce and die, leading to a decline in population size which, if severe, might cause a mutational meltdown and eventual extinction. In total, we compared the fate of asexual and sexual populations under 30 different combinations of mutation rate and population size (Table 1). For 13 combinations with relatively large population sizes, low mutation rates, or both, all 100 asexual and all 100 sexual populations survived to the end of the experiment. In 4 combinations subject to both high

mutation rates and small population sizes, all 200 populations went extinct, regardless of their reproductive mode. In the remaining 13 combinations of mutation rate and population size, the number of surviving sexual populations was greater than the number of surviving asexual populations. For 4 of these combinations, the difference was significant based on Fisher's exact test (two-tailed p < 0.05) with a Bonferroni correction to adjust for performing 30 tests (Sokal and Rohlf 1995). Also, the likelihood that all 13 cases in which there was a difference would, by chance, trend in the same direction is very small (binomial test, p < 0.001). The data on population survival and extinction therefore support the hypothesis that sexual reproduction can slow the advance of Muller's ratchet and prevent mutational meltdown. On the other hand, most of the differences in extinction probabilities are fairly small, they depend on the particular parameter values for mutation rate and population size, and the survival advantage to sex requires thousands of generations to be manifest.

Mean fitness of surviving organisms: While the data on population survival are consistent with the hypothesis that sex is beneficial in opposing Muller's ratchet, the mean fitness values of survivors suggest a more complicated picture. Owing to the large number of experiments (100 populations for each of 60 combinations of population size, mutation rate, and reproductive mode), we cannot present all of the fitness data. However, Figure 2 shows the most important patterns. All of the populations in this figure evolved with genomic mutation rates set to 0.3; the three panels show data obtained for population sizes of 4, 16, and 64. Fitness values are expressed relative to the proximate ancestor, and were transformed owing to their tremendous range. At the lowest population size (Fig. 2a), surviving sexual populations had slightly higher mean fitness values than did their surviving asexual counterparts, although this difference was not significant (two-tailed t-test, p > 0.5). For both reproductive modes, the final mean fitness values were very low relative to the ancestors. The situation was more complicated, however, at somewhat larger population sizes (Figs. 2b and 2c). As expected, the

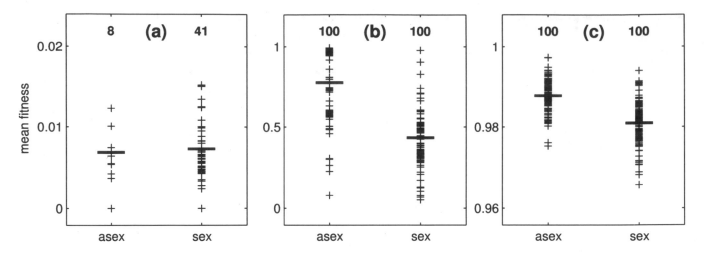

Figure 2. Distributions of mean fitness in asexual and sexual populations that survived Muller's ratchet. Panels **(a)**, **(b)**, and **(c)** correspond to maximum population sizes of 4, 16, and 64, respectively. All populations shown here evolved with a 0.3 genomic mutation rate. Asexual and sexual populations are shown at the left and right, respectively, in each panel. Each + symbol shows the mean fitness of one surviving population; the horizontal bar shows the mean value across the surviving populations for each treatment. The numbers of surviving populations in each treatment are shown along the top (see Table 1). Owing to the tremendous range of fitness values within and between treatments, mean fitness, W, is transformed as $\log(W+1)/\log(W_0)$, where W_0 is the fitness of the proximate ancestor. Note the changes in fitness scale between panels.

mean fitness under both reproductive modes was much higher at N = 16 than at N = 4, and mean fitness was higher still with N = 64. Unexpectedly, however, the asexual populations had higher mean fitness than did the sexual populations at these larger population sizes (two-tailed t-test, both p < 0.0001). It appears that sexually reproducing populations, while better able to survive Muller's ratchet in very small populations, may accumulate more harmful mutations than asexual populations at somewhat larger population sizes. Data obtained from other combinations of population size and mutation rate gave similar results.

Distribution of mutational effects: As a first effort to understand one factor that might have contributed to lower mean fitness of sexual than asexual populations, we subjected the proximate ancestors to an in-depth mutational analysis. We constructed all 2500 one-step mutants (25 alterative instructions at each of 100 genomic sites) for the first-stage sexual and asexual genotypes that were ancestral to the populations shown in Figure 2. Table 2 shows the fraction of one-step mutations that were lethal, deleterious

	Lethal	Deleterious	Neutral	Beneficial
Sexual	0.2420	0.6392	0.1088	0.0100
Asexual	0.4236	0.5052	0.0708	0.0004

Table 2. Distribution of one-step mutational effects on fitness in sexual and asexual ancestors of some second-stage populations. The proportion of the 2500 different mutations are shown for one sexual genotype and one asexual genotype that evolved during the first stage with U = 0.3, and which then served as the ancestors for the second-stage evolution shown in Figure 2.

(non-lethal), neutral, and beneficial. The asexual genotype had a substantially higher proportion of mutations that were lethal, while the sexual type had a correspondingly higher proportion of deleterious but non-lethal mutations. Those populations derived from this asexual genotype would have had a higher risk of extinction, especially in the smallest populations, as a consequence of the higher fraction of lethal mutations. But surviving asexual populations might also have been purged of their most deleterious mutations, leaving these survivors with higher fitness than those from the sexual populations. The generality of these differences as a function of reproductive mode remains to be seen, as do such other factors as the extent and form of epistatic interactions between mutations (Lenski et al. 1999). But these preliminary data do suggest that prior evolution under the different reproductive modes can influence subsequent evolution. In other words, there are multiple interacting and dynamical feedbacks that shape evolving genomes, and they will complicate efforts to discern the various forces responsible for the origin and maintenance of sexual reproduction (Lenski 1999).

Summary and future directions: Sexual reproduction has several disadvantages relative to asexual reproduction, which begs the question of why sex is common in nature. Many potential advantages of sex have been hypothesized, including that sexual reproduction opposes the maladaptive effect of Muller's ratchet in small populations. We used the Avida software to perform evolution experiments with digital organisms that would test this hypothesis. Our results demonstrate the effect of Muller's ratchet in small populations. At high mutation rates and in very small populations, the ratchet often led to mutational meltdowns caused by the vicious cycle of mutation accumulation and population decline. Sexual populations survived this effect significantly better than asexual populations, but only over a fairly narrow range of parameter values. Opposing this advantage, surviving organisms in asexual populations unexpectedly had higher mean fitness than those in sexual populations at some other parameter values. This last result points toward the need for systematic analyses of the effect of reproductive mode on genetic architecture, including the distribution of mutational effects on fitness as well as the extent and form of epistatic interactions among mutations. Avida is well-suited for such analyses, which we intend to pursue in our future work on the evolution of sex.

Acknowledgments

We thank Chris Adami as well as members of the Ofria and Lenski research groups for helpful discussions and valuable suggestions. This research was supported by an NSF grant (DEB-9981397) to R.E.L. and C.O., and by a fellowship from the MSU Center for Biological Modeling to D.M.

References

Adami C (1998) *Introduction to Artificial Life.* Springer, New York

Adami C (2002) Ab initio modeling of ecosystems with artificial life. *Natural Resource Modeling* **15**:133-145

Adami C, Ofria C, Collier TC (2000) Evolution of biological complexity. *Proceedings of the National Academy of Sciences, USA* **97**: 4463-4468

Andersson DI, Hughes D (1996) Muller's ratchet decreases fitness of a DNA-based microbe. *Proceedings of the National Academy of Sciences, USA* **93**:906-907

Bell G (1982) *The Masterpiece of Nature.* Univ. California Press, Berkeley

Bell G (1989) *Sex and Death in Protozoa.* Cambridge Univ. Press, Cambridge

Chao L (1990) Fitness of RNA virus decreased by Muller's ratchet. *Nature* **348**:454-455

Chao L, Tran TT, Tran TT (1997) The advantage of sex in the RNA virus φ6. *Genetics* **147**:953-959

Duarte E, Clarke D, Moya A, Domingo E, Holland J (1992) Rapid fitness losses in mammalian RNA virus clones due to Muller's ratchet. *Proceedings of the National Academy of Sciences, USA* **89**:6015-6019

Felsenstein J (1974) The evolutionary advantage of recombination. *Genetics* **78**:737-756

Ghiselin MT (1988) The evolution of sex: A history of competing points of view. In: Michod RE, Levin BR (eds) *The Evolution of Sex.* Sinauer, Sunderland, Mass., pp 7-23

Kondrashov AS (1982) Selection against harmful mutations in large sexual and asexual populations. *Genetical Research* **40**:325-332

Kondrashov AS (1993) Classification of hypotheses on the advantage of amphimixis. *Journal of Heredity* **84**:372-387

Laumanns M, Zitzler E & Lothar T (2001) On the effects of archiving, elitism, and density based selection in evolutionary multi-objective optimization. In Zitzler E, Deb K, Thiele L, Coello Coello CA, Come D (eds) *Evolutionary Multi-Criterion Optimization: First International Conference Proceedings.* Springer-Verlag Heidelberg, pp 181-195

Lenski RE (1999) A distinction between the origin and maintenance of sex. *Journal of Evolutionary Biology* **12**:1034-1035

Lenski RE, Ofria C, Collier TC, Adami C (1999) Genome complexity, robustness and genetic interactions in digital organisms. *Nature* **400**:661-664

Lenski RE, Ofria C, Pennock RT, Adami C (2003) The evolutionary origin of complex features. *Nature* **423**:139-144

Lewis WM (1983) Interruption of synthesis as a cost of sex in small organisms. *American Naturalist* **121**:825-834

Lynch M, Bürger R, Butcher D, Gabriel W (1993) The mutational meltdown in asexual populations. *Journal of Heredity* **84**:339-344

Lynch M, Conery J, Bürger R (1995) Mutational meltdowns in sexual populations. *Evolution* **49**:1067-1080

Maynard Smith J (1971) What use is sex? *Journal of Theoretical Biology* **30**:319-335

Maynard Smith J (1978) *The Evolution of Sex.* Cambridge Univ. Press, Cambridge.

Maynard Smith J (1988) The evolution of recombination. In: Michod RE, Levin BR (eds) *The Evolution of Sex.* Sinauer, Sunderland, Mass., pp 106-125

Muller HJ (1964) The relation of recombination to mutational advance. *Mutation Research* **1**:2-9

Nowak M, Schuster P (1989) Error thresholds of replication in finite populations, mutation frequencies and the onset of Muller's ratchet. *Journal of Theoretical Biology* **137**:375-395

Ofria C, Wilke C (2004) Avida: A software platform for research in computational evolutionary biology. *Artificial Life* **10**:191-229

Prügel-Bennett A (1997) Modeling evolving populations. *Journal of Theoretical Biology* **185**:81-95

Rice WR (2002) Experimental tests of the adaptive significance of sexual recombination. *Nature Reviews Genetics* **3**:241-251

Sokal RR, Rohlf FJ (1995) *Biometry.* Freeman, New York

Weismann A (1889) *Essays upon Heredity and Kindred Biological Problems.* Clarendon Press, Oxford

West SA, Lively CM, Read AF (1999) A pluralistic approach to the evolution of sex and recombination. *Journal of Evolutionary Biology* **12**:1003-1012

Wilke CO, Wang J, Ofria C, Lenski RE, Adami C (2001) Evolution of digital organisms at high mutation rates leads to survival of the flattest. *Nature* **412**:331-333

Wilke CO, Adami C (2002) The biology of digital organisms. *Trends in Ecology & Evolution* **17**:528-532

Zitzler E, Deb K, Thiele L (2000) Comparison of multiobjective evolutionary algorithms: Empirical results. *Evolutionary Computation* **8**:173-195

Chaotic Population Dynamics and the Evolution of Aging
Proposing a Demographic Theory of Senescence

Joshua Mitteldorf[1]

[1]Temple University , Ambler, PA 19002
josh@mathforum.org

Abstract

According to accepted evolutionary theories, aging has evolved as a side-effect of strong selection pressure for early fertility, despite the fact that it has no adaptive value of its own. I have argued elsewhere that recent experimental results make these theories untenable, and that there is now a broad array of evidence indicating that aging has evolved as an adaptation, selected for its own sake. To explain nature's preference for aging is a substantial theoretical challenge. The classical Weismann hypothesis, "making room for the young," fails because the benefit to the population accrues in the form of enhancement to the rate of increase of population average fitness, while the cost affects individual fitness directly and efficiently. In multi-level selection models, the aging genes are lost before their benefit can accumulate. I propose here that aging has evolved based on a different benefit: its contribution to demographic homeostasis. I argue that population dynamics are inherently chaotic, and that the stable ecosystems that we commonly observe in nature are a highly evolved phenomenon. Natural selection for population homeostasis is far more efficient than selection for rate of evolution because chaotic population dynamics can be lethal on a time scale of just a few generations, while enhanced rate of evolution takes far longer to affect population mean fitness. My thesis is that aging can evolve based on its ability to damp population fluctuations. For illustration, I offer an individual-based computational model that reproduces chaotic population dynamics with a delayed-feedback logistic equation. Genes for aging emerge handily.

Introduction

Two classic dilemmas

The problem of altruism is an area where computational modeling has made a decisive contribution to evolutionary theory. Historically, there have been two classic cases of extreme altruism, where fundamental understanding has been elusive, and substantial controversy remains. These are reproductive restraint and the evolution of aging. They constitute "extreme" altruism in the sense that the cost to the individual is high and direct, while the benefit to the population is far too diffuse to be accounted for by kin selection. They are open questions because theory has robustly predicted the impossibility of effects that field biologists claim are commonplace.

Population self-regulation. Do individuals temper their reproductive potential in order to stabilize population swings and avoid exhausting food species on which they depend? Field biologists have collected extensive documentation of this effect (Wynne-Edwards 1962;

Nudds 1987; Kolenosky 1972). Yet prevailing theoretical arguments dismiss the possibility. Williams (1966) argued persuasively that altruistic populations would be susceptible to invasion by selfish individuals. This reasoning is widely accepted to this day. But recently, computer modelers and complexity theorists have demonstrated mechanisms by which reproductive restraint might evolve (Pepper & Smuts 2001; Rand *et al* 1995; Haraguchi & Sasaki 2000; Rauch *et al* 2003; Pels *et al* 2002). All these models have confirmed a thesis that had been meticulously demonstrated in an early monograph by Gilpin (1975), which remains widely unappreciated. Because of the communication barrier that insulates the biological community from the complexity community, the prevailing wisdom in the former has not yet adapted.

My own experience (Mitteldorf *et al* 2002) as well as these published accounts has convinced me that it is not difficult to model the evolution of reproductive restraint in an ecological context.

Evolution of Aging is an area where a theoretical consensus has prevailed, but a broad array of experimental data has emerged in contradiction to that theory.

Once again, the theory is founded on the primacy of individual selection. Since aging detracts unambiguously from individual fitness, it must have evolved as a side-effect, via *pleiotropy*, or genetic linkage. The pleiotropic theories come in two flavors, both invoking tradeoffs between longevity and fertility. The original flavor (Williams 1957) is based on direct genetic tradeoffs. The more recent flavor is the Disposable Soma theory (Kirkwood 1977), based on metabolic tradeoffs, apportioning a scarce resource (food energy) between demands of reproduction and the cellular repair and maintenance that is the basis for longevity.

I have recently amassed evidence against these theories, and made the case that substantial experimental evidence compels us to consider an adaptive theory, in which aging has been selected for its own sake (Mitteldorf 2004). There are four major lines of evidence (as well as several lesser lines):

- Laboratory animals bred for longevity fail to show depressed fertility. (Leroi *et al* 1994) This is direct evidence against the hypothesis of tradeoffs.

- Some mechanisms of aging appear to be conserved over vast stretches of evolutionary time. (Guarente & Kenyon 2000) Maintenance of these mechanisms in the genome implies that the mechanism has served a purpose.

- In caloric restriction experiments, animals evince the

ability simultaneously to forestall aging and increase stress resistance and immune function, even while under dietary stress. (Weindruch & Walford 1986) Why should the body withhold its best efforts to forestall aging when it is unstressed and well-fed?

- Genes have been discovered in wild populations of mice, worms and flies that appear to have no other function than to hasten the progress of senescence. When such genes are knocked out or artificially disabled, experimental animals live longer, and without apparent cost. (Migliaccio *et al* 1999; Dillin *et al* 2002; Lin *et al* 1998) Why has natural selection failed to eliminate such genes?

The need for a new theory of aging

These observations demand an adaptive theory of senescence. But whereas modeling the evolution of population regulation has shown itself to be suprisingly straightforward, no comparable model for the evolution of aging has emerged. What makes modeling the evolution of senescence so difficult?

Since the pioneering work of Weismann (1889), proposed mechanisms for the evolution of aging as an adaptation have been based on a benefit to the group described as "making room for the young." When this effect is fully elaborated (as Weismann never did), it takes the following form:

Mortality rates for young organisms in the wild are elevated by competition with adult conspecifics, which are generally larger and stronger, (and sometimes smarter and more experienced). Aging drains the population of strong adults, so that more of the young can advance to maturity. Therefore, a population that knows aging has a higher turnover rate, and will adapt more nimbly to changing environmental conditions. In addition, aging tempers the advantage of the more fit over the less fit, enhancing population diversity, which also contributes to the adaptability of the population, and enhances the rate of evolutionary change.

The cost of aging is forgone reproductive opportunity. Models can be constructed in which the population-level advantage is pitted against the individual cost, and these models invariably fail. Genes for aging reliably diminish in frequency, and are quickly extinguished from the population. This is my experience, in five years of trying more than a dozen distinct ideas; and it is attested by the fact that there are no published computational models that evolve aging (while there are several that evolve reproductive restraint.)

The reason that aging is not able to get a toehold in the population is that its costs act directly and quickly against individual fitness, while its benefits accrue on a much longer time scale. The costs of aging affect the reproductive success of any individual that carries an aging gene; but the benefits accrue on an evolutionary timescale, and are spread widely over the evolving population. Genes for aging must persist and dominate a population while

- First, the population grows gradually more diverse

- Then the greater diversity leads to better gene combinations
- These gene combinations grow in prevalence and the population as a whole becomes more competitive, relative to neighboring groups
- Finally, this change in fitness must prove decisive in group-wise competition that drives competing populations to extinction.

For all this time, the aging population must be protected from invasion by freeloaders that do not carry the aging gene, and would rapidly take over the population if they had the chance.

I have found it impossible to construct a plausible model with these characteristics, and I now understand why.

How population dynamics changes the picture

The above argument is predicated on the assumption that aging carries a high cost to the individual, because aging must evolve in a context where there is intense competition for individuals reproductive rate. But enforced reproductive restraint can change that context. If Gilpin and his successors are correct, individual reproductive rate is not maximized. It cannot be maximized, because high reproductive rates lead to chaotic population dynamics that are fatal to the group.

In a context where individual reproductive rate is not maximized, it becomes plausible to evolve senescence as an adaptation. At the least, a *weak hypothesis* would be that the cost of aging can be offset 100%, by pairing higher aging with higher fertility. (This hypothesis gives new meaning to fertility/longevity tradeoffs, and stands pleiotropy on its head.) At best (*strong hypothesis*) aging can make its own unique and valuable contribution to taming the dragon of chaos.

Logistic Population Dynamics

Dynamics of the time-delayed logistic equation

The logistic equation is the oldest and simplest dynamic model of a population limited by finite resources.

$$\frac{1}{x}\frac{dx}{dt} = b(1 - x/K)$$

where $(1/x)\ dx/dt$ is the logarithmic population growth rate, b is the maximal growth rate in the absence of intraspecific competition, and K is the steady state population level. For x that is small compared to K, the solution exhibits exponential growth; and for x $\gg x_{ss}$, the solution declines exponentially. It is well-known that populations governed by the logistic equation are extremely well-behaved: x approaches K asymptotically from either above or below, without overshooting (Abrams 2000).

But the logistic equation is equally prominent in another context entirely: as a difference equation, it is frequently used to study dynamic chaos. The behavior of the logistic equation with finite time increments may be either smooth or chaotic, depending on the size of Δt. For small Δt

(compared to the timescale $1/b$), the behavior is very much like the differential equation; for larger Δt, there are cycles in which x overshoots x_{ss}, and if Δt is increased further, the behavior undergoes a transition to dynamic chaos, such that x jumps wildly about from one time step to the next (Bar-Yam 1997).

Fig 1 shows solutions to the logistic difference equation, starting with very small x, for Δt=1, 2, 2.5 and 2.99. As Δt approaches 3 from below, dynamics become increasingly chaotic. For Δt>3, solutions tend to negative infinity exponentially fast.

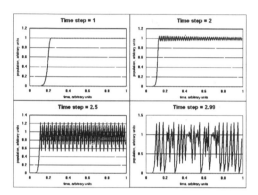

Fig 1. Logistic dynamics, for four values of Δt. The series illustrates the approach to chaos at Δt=3.

Of course, natural population systems cannot afford this kind of dynamic; they would soon fluctuate to extinction. But from a theoretical perspective, the behavior with finite Δt may be a more plausible model for the behavior of real populations than the smooth differential version. Hence it is reasonable to hypothesize that nature has maintained strong selection pressure at the group level to keep population dynamics out of the chaotic regime.

The reason that we find a predominance of stable ecosystems in nature is not that population dynamics are inherently stable, but that demographic homeostasis has been the target of intense selection.

Ways to stabilize population dynamics

If we believe that stable population dynamics is a major target of natural selection, what kinds of adaptations may nature be expected to have found? All such adaptations have a direct, negative impact on individual reproductive value. In this sense, there is a substantial tradeoff between individual and group levels. But while r (reproductive value) must needs be lowered, it is not necessary to lower K (steady-state population level) in order to stabilize dynamics; on the contrary, higher numbers are generally associated with less volatility and less risk of extinction.

(Benton and Grant (2000) make a compelling case that K is a meaningful measure of fitness while r is not; but invocation of r as a proxy for fitness is deeply ingrained in the culture of population genetics. Many of the evolutionary models invoked in the ALife community

implicitly reward r.)

We may assume that nature has found the way to population stability that has the least cost, measured by some combination of group and individual fitness. But it is not easy for us to gauge the cost of nature's various options. Birth rates have a direct impact on the effective Δt. In fact, the parameter that determines approach to chaos is the product of the birth rate and the time delay. Lowering the birth rate b offers the fringe benefit of lowered resource requirements, shrinking the environmental footprint.

Lower birth rate decreases r proportionately. The impact on K is less clear. If you think of K as the carrying capacity of a niche, it may seem to be independent of b; but if you imagine the death rate to be a function determined by the set of all environmental and crowding conditions, you may equally well conclude that K is directly proportional to b. It is probably safe to say that reality is somewhere in between. There are two lesser reasons, however, that nature may choose to keep birth rates high. One is population diversity, and the impact on population adaptability in the face of a changing environment. The second is the insurance against accidents that larger numbers provides.

Raising the death rate is *not* a path to demographic stability. In the logistic equation, the only death term is proportional to crowding, and raising the death rate is equivalent to lowering K. Thus decreased K leads to the same population dynamics in a smaller population – not at all a winning proposition. You can, of course, add an "accidental death" term that is independent of population density. This is equivalent to lowering b; in fact, the b that appears in the logistic equation is really the net population growth rate, or difference between birth and accidental death rates. Increasing the accidental death rate rather than decreasing the birth rate avoids the two "lesser reasons" but also misses the resource savings described above.

Senescence operates differently from accidental death, and has a special advantage. When population density is low and expanding freely, the death rate from competitive forces is low, so many individuals live to an age where senescence matters. So in times of population expansion, aging effectively tempers the growth rate. But in tightly competitive times, when the population is high and contracting, starvation and resource scarcity will prevent most individuals from attaining an age at which senescence becomes a life-limiting condition. So as a force for demographic stability, aging has the potential to damp population growth when it is too fast, but then to get out of the way when populations are shrinking and its action would be counterproductive.

In the real world, we find an additional adaptation associated with senescence that further improves its utility as a stabilizer of population dynamics. Most animals are able to lessen the effects of senescence in response to caloric deprivation. This is the "Caloric Restriction effect" (CR), that has been observed in the laboratory for a wide range of species, from yeasts to primates (Weindruch &

Walford 1986; Masoro 2002). The CR adaptation further enhances the effectiveness of senescence for avoiding unstable dynamics. The death rate from senescence is at its lowest under the conditions of starvation that attend the contracting phase of the population cycle. Aging does not reduce *r*, and arguably has minimal impact on *K*.

Thus, if nature has sought a least-cost path to demographic homeostasis, it is understandable that she has recruited senescence as a useful device.

Description of the Model

To illustrate the operation of chaotic population dynamics, and the opportunity to evolve aging, I have implemented a multi-level selection model. An asexual population is arrayed on an n*n grid of sites, (opposite edges identified to avoid boundary effects). A small rate of migration links each site with its four von Neumann neighbors. Within each site is a variable population of individuals (I call them "orgs") that reproduce clonally, with a birth rate that determines the probability of reproduction in each time step. The probability per time step of individual death is modeled as the sum of two terms: One term derives from aging, and increases exponentially with the age of the org after maturity (a Gompertz (1825) function). The other term derives from crowding, and is proportional to the number of orgs sharing the site.

The crowding variable that determines the death rate at each site is measured with a time delay. In other words, the death term in the logistic equation is proportional to the site population a number of timesteps in the past. The time delay is crucial to the model, because it makes the difference between a logistic equation that approaches steady state smoothly and a population dynamic that is powerfully unstable.

The delay feature mimics an attribute of real ecosystems. Population growth can have a momentum that continues even after resources on which the population depends have been depleted.

Principal parameters of the model

grid size: length and width of the population grid (typically 16*16)

maturity: the age before which an org is not yet able to reproduce, and senescence has not yet become effective. (typically the reciprocal birth rate)

delay: population levels feed back to death rates, after this delay (typically 50 timesteps, corresponding to a "chaos parameter" of 2.25)

K=steady-state population: the denominator in the term that invokes death from crowding; the number of orgs at each site to which the system would relax with stable dynamics. (typically 100)

migration: the rate at which orgs from one site diffuse to a random neighbor site. (typically 3E-5)

mutation rate: the probability that an aging gene will randomly change its value during reproduction. (typically 3E-2)

The model supports an option for **smart aging**, emulating the CR adaptation. When this option is invoked, aging is muted in response to (delayed) crowding.

Heuristic dynamics of the model

Within each site, selection rewards the orgs with genes to create the most offspring, i.e., high birth rates and low rates of aging. But high individual birth rates and low aging rates are a recipe for chaotic population dynamics, leading to extinction. When the system is allowed to evolve globally, a steady-state is established in which individual growth parameters skirt the edge of chaos. Sites in which individual selection has taken the growth too high are constantly blinking out of existence, then re-seeded by migrating orgs from sites where the growth rate is lower (on average).

Experiments with the model

Three kinds of experiments were programmed:

1. Population dynamics were calibrated with non-evolving orgs. In the absence of aging, the maximum sustainable birth rates were recorded.

2. The grid was populated with orgs that carry a gene that determines their individual rate of aging. The aging gene is permitted to mutate, and its evolved distribution was subsequently measured. Evolved rate of aging was plotted as a function of programmed birth rate.

3. Orgs were given two evolving genes, one of which determines individual birth rate and the other individual rate of aging. Birth rate and aging rate were permitted to evolve independently.

Since population dynamics can be stabilized either by moderating the birth rate or by increasing the rate of aging, this latter option allowed for direct competition between these two solutions to the chaos problem.

Model Results

1. Demonstration of population dynamics

In this calibration experiment, **delay** was fixed and different (constant) values of **b** were initialized on the grid, with no aging and no evolution. Values of **b** that were too high led to global extinction. The maximum viable **b** was found to be inversely proportional to **delay** for a given **K** parameter. Small **K** per site increased the risk of extinction, leading to lower values of maximum viable **b**. But this effect saturated for large values of **K**. Unlike the drift-and-dominate effects described in Wright's (1931) shifting balance theory, the effect of chaotic population dynamics is not limited to small sub-population sizes.

The straight lines in this plot are an indication that the chaos parameter, (**time delay * birth rate**), is the operational determinant of the limits of growth. In deterministic models, dynamics becomes chaotic as this parameter approaches 3. In the results below, (from a stochastic model) the slopes of the lines range from 1.8 for the smallest population per site to 2.5 for the largest.

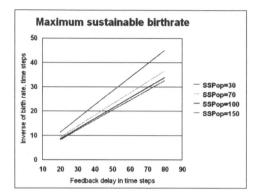

Fig 2. The chaos parameter (time delay * birth rate) determines stability of population dynamics. Higher **K** (steady-state population per site) permits a slightly higher chaos parameter.

2. Evolution of aging when birth rate is fixed

Rate of aging was programmed by an individual gene that was permitted to evolve through 10^6 time steps, attaining a steady state. The terminal value depends on the fixed value of *b*: the higher the birth rate, the more aging is required to stabilize population dynamics.

In Fig 3, a family of three curves is plotted for different values of the migration parameter. The migration parameter is important because a very high value of migration is equivalent to a single panmictic population, while a very low value corresponds to n^2 independent panmictic populations.

The migration parameter programs the relative importance of group selection vs. individual selection. This is the "rescue effect": if migration is high, then extinct sites are quickly repopulated from neighboring sites, while low values of migration impose a higher cost for extinction. In the present results, lower values of migration are associated with slightly higher evolved rates of aging. (For the lowest value of migration and highest values of *b*, extreme volatility extinguished the global population before a stable level of aging could be established.)

All deaths in the model are either caused by crowding or by aging. The y axis in Fig 3 corresponds to the proportion of all deaths attributable to aging, a dimensionless quantity that might be crudely compared to the impact of aging in nature. The proportion of aging deaths ranges up to 35% in the model results. Measuring the corresponding parameter for real ecosystems is a field biologist's nightmare, and data are scarce. Bonduriansky and Brassil (2002) estimate that the fitness cost to antler flies in the wild is about 20%.

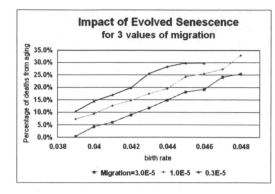

Fig 3. Birth rate is fixed, and rate of aging is allowed to evolve for 10^6 time steps. Larger fixed birth rates necessitate higher rates of aging for stability. The three series correspond to different values of the migration parameter.

3. Simultaneous evolution of aging and birth rate

In these runs, each individual carried two genes, determining its rate of aging and its birth rate, respectively. In principle, the chaos problem could be solved either with a high birth rate and a high aging rate, or a low birth rate and low aging rate.

Fig 4 below is a scatter plot placing each of ~10^4 individuals on a map according to their evolved birth rate (x axis) and aging rate (y axis). Always a compromise was reached, where both moderated birth rate and substantial aging contribute to demographic stability.

Two separate runs are superimposed in the scatterplot. In one of the runs, aging was programmed to be "smart" in the sense of the CR response. In the smart aging run, the effect of aging was moderated during times of high (delayed) population density when, presumably, animals would be stressed by tight food supplies. The figure shows how smart aging shifts the center of the distribution toward higher values of aging, and correspondingly higher values of *b*.

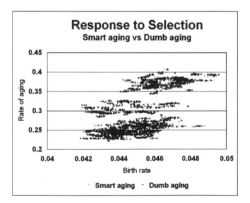

Fig 4. Scatter plot of evolved individuals by birth rate (x) and aging rate (y). With "smart aging" (CR response) turned on, aging is preferred over low birth rate as a solution to the chaos problem.

Summary and conclusions

Experimental results demand an adaptive theory of aging. There has been a century of ideas on the subject, but no quantitative models in which senescence evolves as an adaptation have emerged. Considerations of demographic homeostasis and chaotic population dynamics may be able to bridge this gap, and provide a plausible mechanism by which senescence may be selected.

It would be dishonest to minimize the radical revision in thinking that I am proposing. The idea that natural selection may not be optimizing individual reproductive value casts a shadow on a great body of evolutionary theory. This theory enjoys wide acceptance, although much of its foundation is untested and therefore vulnerable, because complex real-world ecosystems so seldom provide opportunities for clean tests of simple hypotheses.

Perhaps 95% of evolutionary biologists think that individual selection is all the selection that there is. 5% have argued bravely that group selection can sometimes overcome the more direct effect of individual selection.

In this landscape of opinion, I am staking out a position on the far left wing, claiming that evolution not just of populations but of entire ecosystems is the rule rather than the exception. The imperative for demographic homeostasis is a primary agent of natural selection, rivaling reproductive potential in its import.

References

Abrams, P.A. 2000. The evolution of predator-prey interactions: Theory and evidence. *Annual Review of Ecology and Systematics.* 31:79-105.

Bar Yam Y. 1997. *Dynamics of complex systems.* Reading, Mass. Addison-Wesley.

Benton T.G. & Grant A. 2000. Evolutionary fitness in ecology: Comparing measures of fitness in stochastic environments. *Evol. Ecol. Res,* 2:769-789

Bonduriansky R and Brassil C.E. (2002) Senescence: rapid and costly ageing in wild male flies. *Nature* 420:377

Gilpin, M.L. 1975. *Group Selection in Predator-Prey Communities.* Princeton University Press, Princeton, NJ.

Gompertz B. 1825. On the nature of the function expressive of the law of human mortality and on a new mode of determining life contingencies. *Philosophical Transactions of the Royal Society of London* **A** 115**:**513-585.

Guarente, L.; and Kenyon, C. 2000. Genetic pathways that regulate ageing in model organisms. *Nature* 408:255-262

Haraguchi, Y.; and Sasaki, A. 2000. The evolution of parasite virulence and transmission rate in a spatially structured population. *Journal of Theoretical Biology* 203:85-96.

Kirkwood T. B. L. 1977. Evolution of aging. *Nature.* 270:301–304.

Kolenosky, G. B. 1972. Wolf predation on wintering deer in east-central Ontario. J. Wildl. Manage. 36:357-368

Leroi,\ A. M., Chippindale A. K. and Rose M. R. 1994. Long-term laboratory evolution of a genetic life-history tradeoff in *Drosophila melanogaster. Evolution.* 48:1244-1257.

Lin Y J, Seroude L and Benzer, S. 1998. Extended life-span and stress resistance in the drosophila Maynard Smith, J. 1964. Group selection and kin selection. *Nature* 20:1145-1147.

Maynard Smith, J. 1976. Group selection. *Quarterly Review of Biology* 51:277-283.

Migliaccio E., Giorgio M, Mele S, Pelicci G, Reboldi P, and Pandolfi P. 1999. The p66shc adaptor protein controls oxidative stress response and life span in mammals. *Nature* 402:309-313.

Mitteldorf J. Croll D and Ravela S C. 2002. Multilevel selection and the evolution of predatory restraint. Proceedings of the eighth international conference on artificial life. MIT Press. Cambridge, MA.

Mitteldorf, J. 2004. Aging selected for its own sake. *Evol. Ecol. Res.* Forthcoming.

Nudds, T. 1987. The Prudent Predator. In: Wild furbearer management and conservation in North America ed M. Novak, J. Baker, M. Obbard and B. Mallock.

Pels, B.; deRoos, A.M.; and Sabelis, M.W. 2002. Evolutionary dynamics of prey exploitation in a metapopulation of predators. *American Naturalist* 159:172-189.

Pepper, J.W.; and Smutts, B.B. 2001. *Agent-based modeling of multilevel selection of feeding restraint as a case study* pp. 57-68 *in* W.C. Pitt editor Swarmfest 2000, Proceedings of the 4th Annual Swarm User Group Conference. Natural Resources and Environmental Issues, Volume XIII, S.J. and Jessie E. Quinney Natural Resources Library, Logan UT.

Rand D A, Keeling M, and Wilson H.B. 1995. Invasion, stability and evolution to criticality in spatially extended, artificial host-pathogen ecologies. *Proc Royal Society of London B, Biological Sciences* **259**:55-63

Rauch, E. M.; Sayama, H.; and Bar-Yam, Y. 2003. Dynamics and genealogy of strains in spatially extended host-pathogen models. *J Theor Biol.* **221**:655-664.

Weindruch R. & Walford, R. 1986. *The Retardation of Aging and Disease by Dietary Restriction,* Springfield, IL: Thomas

Weismann A. 1889. *Essays upon heredity and kindred biological problems,* Clarendon Press. London.

Williams, GC. 1957. Pleiotropy, natural selection, and the evolution of senescence. *Evolution* **11**:11-21

_____ 1966. *Adaptation and Natural Selection: A Critique of Some Current Evolutionary Thought.* Princeton University Press, Princeton, NJ.

Wright, S. 1931. Evolution in mendelian populations. *Genetics* 16:97-159.

Wynne-Edwards, V.C. 1962. *Animal Dispersion in Relation to Social Behaviour.* Oliver and Boyd, Edinburgh, UK.

The Role of Non-Genetic Change in the Heritability, Variation, and Response to Selection of Artificially Selected Ecosystems

Alexandra Penn and Inman Harvey
Centre for Computational Neuroscience and Robotics, University of Sussex, Brighton. BN1 9QG
alexp@cogs.susx.ac.uk

Abstract

A response to selection on the level of the ecosystem has been demonstrated in artificial selection experiments, and poses interesting challenges to concepts of heritability, variation and phenotype in biological systems. We use ecosystems modeled as Lotka-Volterra competition systems, and subject to an ecosystem-level selection process, to illustrate and discuss the potential, and possible mechanisms, for ecosystem-level evolution without genetic change of the component species. A limited positive response to selection is demonstrated by the selection of alternative stable ecosystem states.

Introduction

The ability of whole ecosystems to respond to selection has recently been demonstrated in artificial selection experiments, (Swenson et al., 2000b; Swenson et al., 2000a). However the concept of ecosystem selection remains controversial; artificial selection experiments demonstrate only the phenomenon of ecosystem-level evolution and not the mechanism by which it occurs (Johnson and Boerlijst, 2002). A common question concerns the fact that an ecosystem has no genome and hence should not be able to respond to selection. In fact the requirements for a response to natural selection are general properties, not restricted to any particular level of the biological hierarchy, or indeed systems with their own genomes in the conventional sense (Lewontin, 1970). In "The Units of Selection"(1970), Lewontin set out three fundamental conditions which a population of units must satisfy in order to respond to natural selection:

1. Phenotypic variation amongst units.

2. Fitness consequences of phenotypic variation.

3. Heritability of fitness.

If ecosystems were to act as units of selection then they would need to fulfil all these requirements, including the implicit assumption of the individuality or separateness of units. For communities "in the wild" this would present significant problems. (The possibility of self-organised spatial structures allowing selection between subcommunities has been addressed elsewhere (Johnson and Boerlijst, 2002).) In the artificial selection experiments however, two of the properties are imposed by the experimenters. A population of ecosystems is provided by the experimental setup, and the fitness consequences of the measured "ecosystem-level" phenotypic variation are imposed. Selected ecosystems are reproduced by sampling the medium of the "parent" ecosystem (eg soil, pond water), and inoculating new sterile medium. This still leaves two important criteria to be satisfied. In order to respond to selection the ecosystems must still vary phenotypically one from the other, and this variation must be at least partially heritable. The concepts of phenotype, variation and heritability at the ecosystem level are far from straightforward, and deserve further exploration. In particular, Wilson and Swenson have suggested that when whole ecosystems are subject to artificial selection on an ecosystem-level property, the response to selection could be due to change in species composition of the community only, without species genetic change (Wilson and Swenson, 2003). In previous work, (Penn, 2003), we showed that ecosystems with simple dynamics responded to ecosystem-level selection if genetic change was allowed. Here we wish to investigate the possibility of evolution without genetic change in a simple model. This paper will explore the ideas and implications of variability, heritability and phenotype in self-contained experimental ecosystems (microcosms), and discuss the feasibility of evolution at the ecosystem level without genetic change in the constituent species.

Evolution at the Ecosystem Level

What is an ecosystem "phenotype"?

The term ecosystem phenotype refers to the measured property of the experimental microcosm that is selected for. As examples of ecosystem phenotypes, Swenson and Wilson chose above-ground plant biomass of Aribidopsis thaliana for a soil ecosystem, and pH and rate of biodegradation of 3-chloroaniline, for aquatic, microbial ecosystems. Their choices of phenotype were based on their definition of an ecosystem as the "interactions of the species with each other and their physical environment". Because the traits pH and biodegradation of chloroaniline are properties of the physical environment and the trait of plant biomass "is likely to be mediated through effects on the physical environment", they claim that their experiments qualify as ecosystem-level selection.

We may consider ecosystem-level properties as those that cannot easily be attributed to one species in the assemblage. Rather they will be consequences of the complex interactions of the many species present, (as well as the particular environmental conditions which they are subjected to.) If we consider that it is possible to represent the ecosystem as some sort of state vector of species composition, species numbers and genotypes, then the particular measurable phenotype in which we are interested would be some sort of function on this vector (given for the moment a constant environment). From this perspective, stable phenotypes would correspond to attractors in the ecosystem's state space [1]. A further complicating factor is the possible form of this "phenotype function" with respect to the underlying dynamics and ecosystem state, which will have a significant effect on the response to selection. Phenotype could be an additive or non-additive function of ecosystem state. Many to one mappings from ecosystem state to phenotype value are probable, leading to areas of selective neutrality.

Variation

In order for selection to act there must be variation between the "phenotypes" of the ecosystems in a population. This can, broadly speaking, be generated in two ways: by variation between offspring and parent ecosystems caused by the sampling process; and by variation arising during the ecosystems' development stage during an experimental "ecosystem generation". Variation in either of the stages could be heritable or non-heritable, because in ecosystem reproduction there is no distinction between germ and soma. The heritability of any variation arising depends on its reliability of transmission not only via the sampling procedure, but also after being subject to another "generation" of ecosystem development.

According to Swenson and Wilson (Swenson et al., 2000b; Swenson et al., 2000a), ecosystems are complex systems, sensitively dependent on initial conditions, and it is this that can potentially give rise to wide phenotypic variation from small variations due to sampling error when ecosystems are reproduced. This position implicitly assumes that the important sources of variation in ecosystem phenotype are endogenously generated, that is, have at their root natural variability of the population dynamics. Such behaviour has been shown in both models and experimental

microcosms with chaotic dynamics, (Schefer et al., 2003), in which small initial differences in numbers of each of several species competing for resources give rise to different stable outcomes after a period of transient chaos.

However, there are many other potential sources of variation which could alter the state of the ecosystem phenotype when it is measured. Amongst others: sampling error on community composition when ecosystems are reproduced; heterogeneous spatial distribution of species within the microcosm potentially magnifying sampling error; environmental stochasticity; and species' own effects on the abiotic environment. In addition, given sufficient diversity within a population of ecosystems, "sexual" recombination, i.e. mixing, of ecosystems could perform a similar role. The pertinent question about all of this variation from the perspective of selection is, is it heritable?

Heritability

In order for evolution to occur we require partially heritable phenotypic variation. The phenotype in question is whatever function of the community composition that we define, measured at the end of an ecosystem generation. This could be an instantaneous or cumulative value. Heritability of the phenotype will depend on the reproducibility of an ecosystem's state at time T from the process of taking a sample of that system and using it to seed an offspring system.

To try and express this in a more formal fashion, if we imagine the ecosystem state at the end of a generation, E, as a vector of species numbers, then heritability requires that the combination of 2 vector operators, sampling and development (ecosystem dynamics plus additional sources of variation), on E produces the same end state E. (For ease of description, and to suit the purposes of this paper, we ignore intra-species genetic variation and change.) Variation requires that either the sampling or development operator acts so as to produce E' rather than E. The response to selection depends on the balance between these two outcomes with given sampling and development operators.

So, for an ecosystem phenotype to be heritable, it must be robust to the operations of sampling and development. This implies that heritability depends upon the existence of attractors in the ecosystems' state space, and evolvability on the existence of multiple stable attractors. In order for a sustained response to ecosystem selection to be possible, a network of attractors of varying stability is required. The attractors must be reachable via the variation incurred during the sampling and development stages. The heritability of a given phenotype then, will depend on the nature and extent of the basin of attraction of the ecosystem's state at the end of the generation. It is interesting to note that, unlike selection at the level of the organism, ecosystem selection must search for phenotypes that are not only fit, but also stable.

[1] Even limit cycles could produce a stable ecosystem phenotype if the cycle length was short with respect to the length of the ecosystem generation, and the selected trait was a cumulative rather than an instantaneous function of the ecosystem state.

Evolution without genetic change

Wilson and Swenson, (Wilson and Swenson, 2003), note that evolution at the phenotypic level of the community could theoretically occur either through genetic changes in the constituent species, changes in the species composition of the community, or both. That is, that evolution could occur without changes in the genes of the species present. They emphasize that changes in species composition or population numbers of species in a community literally are changes in gene frequency at the level of the community. In the context of the ideas of heritability and variation discussed above, evolution of a community without genetic change would be possible if that community possessed multiple stable states. Sampling error during reproduction, or noise or chaotic dynamics during development could be enough to knock the community into different basins of attraction, and hence allow evolution to occur.

Model

In order to investigate the possibility of this mode of evolution occurring, we use a simple, yet widely-used, model of basic ecological dynamics, and subject it to an ecosystem selection procedure. No genetic change within species is allowed. As an initial simple approximation, the within-ecosystem dynamics are modeled using the generalised Lotka-Volterra competition equations (MacArthur, 1972), these equations potentially have multiple stable equilibria and so are suitable for our purposes. We make the assumption that, as under laboratory conditions, ecosystems are closed after sampling. That is, unlike the model of Ikegami and Hasimoto, (Ikegami and Hasimoto, 2002), species that have been eliminated from the population cannot reappear.

$$N_{i,t+1} = N_{i,t}\left[1 + \frac{R_i}{K_i}\left(K_i - \sum_{j=1}^{S} N_{j,t}\alpha_{ij}\right)\right]; \quad (1)$$

Where $N_{i,t+1}$ is the density of species i at the next time step, S is the total number of species, K_i is the carrying capacity of species i, R_i is the rate of increase of species i, and α is a matrix of interaction coefficients representing the per capita effect of each species on every other (α_{ij} is the per capita effect of species j on species i). Source ecosystems for each selection run were randomly initialised with K_i's set at uniform random in the range 100:1000, R_i's at uniform random 1.5:2.5, and each α_{ij} set randomly on a skewed distribution in the range 0:2, unless $i = j$ and then $\alpha_{ij} = 1$. The α_{ij}s were drawn from the distribution $\alpha_{max}x^{1.5}$, where x is randomly chosen from 0:1. This gives a weakly skewed interaction matrix with many weakly interacting species and fewer strongly interacting ones. Note that although all direct interactions are competitive ($a_{ij} > 0$), indirect effects may give rise to mutualisms or commensualisms. The initial populations for the source ecosystems are set at uniform random in the range 0:100

Ecosystem reproduction involves taking fixed-size samples from a selected parent ecosystem (eg samples of soil or water containing the microbial ecosystem (Swenson et al., 2000b)) and using them to inoculate or "seed" offspring ecosystems. Reproduction is asexual, that is, samples from different parent ecosystems are not mixed before being used to inoculate new ecosystems. In real ecosystem selection experiments, the process of sampling can introduce variation between offspring both in species genetic composition, and initial species population sizes. In this model only variation in population size is considered. The initial population size for each species in an offspring ecosystem is calculated on the assumption that a sample contains individuals chosen at random from the parent ecosystem, thus the expected frequency of a species in a sample is equal to its frequency within the sampled ecosystem. Since species population sizes are continuous variables, sampling was modeled using the standard Gaussian approximation to a binomial distribution. Thus, N_i, the size of the species in the new sample, was generated at random from a Gaussian distribution with mean, Bp_i, and standard deviation, $\sqrt{Bp_i(1-p_i)}$, where p_i is the frequency of the species in the parent ecosystem, and $B = 100$ and $B = 10$ were the mean sample sizes.

Each ecosystem was run for 50 time steps which constituted an ecosystem generation. During this "developmental stage", developmental noise was added to the dynamics by multiplying each of the $N_j\alpha_{ij}$ interaction terms by a number drawn from a uniform random distribution in the range 0.5:1.5.

Selection is for the maximization of a random linear function of the population sizes, normalized to the range 0:1.

$$F = \sum_{i=1}^{S} N_i\beta_i; \quad (2)$$

Where the β_i are randomly generated coefficients in the range -1:1. This is an appropriate ecosystem-level fitness function because, as is likely to be the case with macroscopic properties of real ecosystems, the optimal strategy for each species to achieve maximum ecosystem fitness is context dependent. That is, dependent on the dynamics and population levels of the other species within the community.

Results

With both the larger ($B = 100$, smaller sampling error,) and smaller ($B = 10$, larger sampling error), sample size we see a small positive response to directed selection, and a slightly negative or no response to random selection. Figures 1 and 2, show the mean response to directed and random selection respectively, of 30 randomly generated ecosystem populations, $B = 10$. Each population is created from a different randomly initialised source ecosystem with different interaction coefficients, carrying capacities and growth rates, and 20 species. Figures 3 and 5 show the final species population

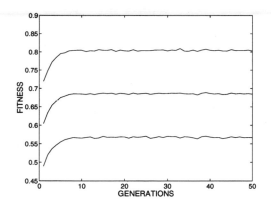

Figure 1: Directed Selection: Mean fitness (+ and - Std Dev) of 30 ecosystem populations, $B = 10$, over 50 generations.

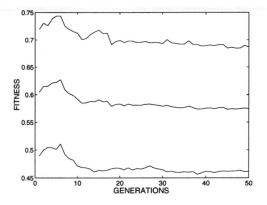

Figure 2: Random Selection: Mean fitness (+ and - Std Dev) of 30 ecosystem populations, $B = 10$, over 50 generations.

states for the best ecosystem in population 1 over 50 generations, under directed and random selection respectively. Note that this is in fact the same set of ecosystem parameters in both cases, effectively the same "ecosystem" under a different selective regime. We can see that two different stable attractors have been reached. Although not shown, in the case $B = 100$, the same "ecosystem" also falls into two different stable attractors under selection and random selection, both of which have a different species composition to either of those seen when $B = 10$. These dynamics are typical of the majority of our randomly generated ecosystems. Figures 4 and 6 show the corresponding fitness values over the course of 50 generations for the ecosystem populations in figures 3 and 5. Best, worst, mean and upper and lower quartiles are shown to give an indication of the diversity within the population. It is evident that under both selection and random selection the fitnesses undergo a period of change and then settle to a stable value with the diversity of the population reduced. This is particularly noticeable in the case of the randomly selected population (fig.6). Here fitness does in fact increase, but to a low stable mean value, and only after a prolonged period of change caused by the dynamics of a 2-species transient (fig.5). When one species is eliminated from the ecosystem, it has reached a stable point and the fitness jumps to a new steady value. In figure 4, directed selection, we see the same dynamic of fitness change during a transient period (fig.3), then settling to a stable, higher, value when a stable attractor in the ecosystem state space has been reached. In this case the attractor consists of 2 species. Variation in their dynamics is simply caused by the added developmental noise. If it is removed it can be seen that the ecosystem has reached a fixed point.

Discussion

In the results above we see a combination of two types of variation, "instantaneous variation", and variation caused by movement to a different basin of attraction. "Instanta-

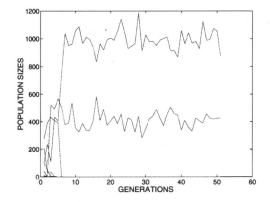

Figure 3: Final species population states for the best ecosystem in population 1 over 50 generations, directed selection.

neous variation" is the phenotypic variation of ecosystems in a population which may be following different trajectories but are ultimately destined for the same attractor. Stochastic sampling from a point on a particular trajectory will still tend to lead to the same attractor if the basins of attraction are broad and attractors widely spaced in state space. If species are close to an alternative equilibrium point/basin of attraction then variation due to sampling error or developmental stochasticity could allow the community access to a new stable state corresponding to a different phenotypic value, and potentially a higher fitness. In this way the community could increase its fitness. Even in systems subject to random rather than directed selection this effect could lead to change in fitness and more stable community compositions via a "ratchet" effect. The key issue for ecosystem evolvability is the reachability of those different basins of attraction. Our simple models show that fitter attractors can be selected at an early stage before the community reaches equilibrium. However, once these attractors are reached the perturbations due to sampling error and developmental noise are not large enough to allow a progression to attractors cor-

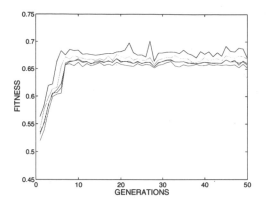

Figure 4: Fitness, best, mean, worst, and upper and lower quartiles, for ecosystem population 1 over 50 generations, directed selection.

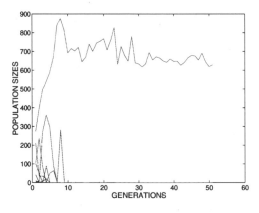

Figure 5: Final species population states for the best ecosystem in population 1 over 50 generations, random selection.

responding to higher fitness. Thus, the ecosystems' evolvability with these dynamics is very limited. The only option to increase fitness is by attractor switching in the early non-equilibrial stages, when we assume that the population sizes of the species within each ecosystem may be close to many different attractor boundaries.

Properties of Lotka-Volterra competition equations

In an N species Lotka-Volterra competition system there are 2^N possible stable equilibria corresponding to presence or extinction of each of the species. In any given system the number of realisable attractors will in fact be much less as many of the non-zero species values at equilibrium will correspond to negative numbers and hence not be allowed. If we assume that the probability of any non-zero valued species in an attractor being positive or negative is 0.5, then the expected number of equilibria with only positive or zero species sizes will be $(1 + 0.5)^{10} - 1 = 56$. However, not all of these potential equilibria will be stable. Limit cycles and chaotic dynamics are also possible although the latter

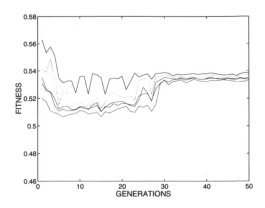

Figure 6: Fitness, best, mean, worst, and upper and lower quartiles, for ecosystem population 1 over 50 generations, random selection.

appear to be restricted to a fairly narrow parameter range (Vano et al., 2004). In theory, though, we should expect enough attractors to be present in a randomly generated set of ecosystem parameters to be able to achieve a response to selection.

One limitation of Lotka-Volterra dynamics is that the only possible change in attractor with asexual reproduction is via loss of species through competitive exclusion. This of course means that once we have reached a particular stable attractor with fewer species than the starting number, the attractors available to us are very much reduced in number, no matter how many might potentially exist. Additional noise, from environmental variation etc, could keep the populations out of equilibrium for longer, maintaining more species in the population, and allowing more attractors to be reached. In addition, more skewed interaction dynamics, with many more weakly than strongly competing species, and sparser connections, could both allow the maintenance of more species in the population and hence more potential for evolution. This sort of interaction distribution is often seen in natural ecosystems. Interestingly, we found no response to selection of ecosystems in which the interaction coefficients were generated from a uniform random distribution. The effects of the form of the interaction matrix on ecosystem evolvability deserve to be further investigated. Other options exist, however, given enough population diversity, sexual reproduction (mixing) could offer the opportunity to gain species and shift attractor that way.

Are there multiple stable states in nature?

Potentially, in real ecosystem dynamics the possibility exists of alternate stable states which do not involve complete competitive exclusion, but stability of different proportions of the same sets of species. Dynamics such as these would allow a much more flexible response to selection. The existence of multiple stable equilibria in natural populations is still uncertain although several potential examples have been observed in marine ecosystems, both in competitive and predator-prey systems (Sutherland, 1981; Barkai and McQuaid, 1988). Both types of outcomes are seen, competitive exclusion and persistence at different proportional population sizes of the same species. In general however, the long time-series data needed to draw conclusions about the reality of these phenomena are not available. An additional cause of the existence of multiple stable states in ecosystems are environmental feedbacks, in which organisms affect their environment and alter it to their own preference. Such dynamics may play an important part in the ecosystem selection process in real ecosystems.

Conclusions

We have demonstrated a limited response to selection without genetic change in simple model ecosystems modeled as Lotka-Volterra competition systems. The response to selection depended on the skewed distribution of species interaction coefficients. Ecosystems were able to move between different attractors corresponding to phenotypes of varying fitness during the early stages of evolutionary runs. However, once stable attractors had been reached, the variation due to developmental noise and sampling error was not sufficient to allow movement to new attractors. It seems that evolution of whole ecosystems without genetic change is possible in principle. A requirement for such evolution is not only the potential for multiple stable (or locally stable) states in a particular ecosystem, but also the reachability of those states. In our model, evolution was limited by the competitive exclusion dynamics, which severely curtailed the number of available attractors once low species numbers were reached. In other more complex and realistic ecosystem dynamics this might not be the case. Our model is based on the restricted case of artificial, rather than natural, selection of ecosystems. Hence conclusions can only be drawn about the response to selection under a restricted set of circumstances, in which several of the requirements for a response to selection are enforced by the experimenter. This is an interesting and potentially useful topic in its own right. However, it is hoped that the results may ultimately be pertinent in considering the possibility of selection of communities outside the laboratory.

Acknowledgements

Thanks to all at the CCNR for support and discussion. A.Penn is supported by a BT Exact CASE studentship.

References

Barkai, A. and McQuaid, C. (1988). Predator-prey reversal in a marine benthic ecosystem. *Science*, 242:62–64.

Ikegami, T. and Hasimoto, K. (2002). Dynamical systems approach to higher-level heritability. *J. Biol. Phys.*, 28(4):799–804.

Johnson, C. and Boerlijst, M. (2002). Selection at the level of the community: the importance of spatial structure. *Trends Ecol Evol*, 17:83.

Lewontin, R. (1970). The units of selection. *Annu.Rev.Ecol.Syst.*, 1:1–18.

MacArthur, R. (1972). *Geographical Ecology*. Harper and Row.

Penn, A. (2003). Modelling artificial ecosystem selection:a preliminary investigation. In Banzhaf, W., Christaller, T., Dittrich, P., Kim, J., and Ziegler, J., editors, *Advances in Artificial Life, 7th European Conference, ECAL 2003, Dortmund, Germany, September 14-17, 2003, Proceedings*, volume 2801 of *Lecture Notes in Computer Science*, pages 659–666.

Schefer, M., Rinaldi, S., Huisman, J., and Weissing, F. (2003). Why plankton communities have no equilibrium: solutions to the paradox. *Hydrobiologia*, 491:9–18.

Sutherland, J. (1981). The fouling community at beaufort, north carolina:a study in stability. *Am. Nat.*, 118:499–319.

Swenson, W., Arendt, J., and Wilson, D. (2000a). Artificial selection of microbial ecosystems for 3-chloroaniline biodegradation. *Environ. Mirobiol.*, 2:9365.

Swenson, W., Wilson, D., and Elias, R. (2000b). Artificial ecosystem selection. *PNAS*, 97:9110.

Vano, J. A., Wildenberg, J. C., Anderson, M. B., Noel, J. K., and Sprott, J. C. (2004). Chaos in low-dimensional lotka-volterra models of competition. *submitted to Physics Letters A*.

Wilson, D. and Swenson, W. (2003). Communtiy genetics and community selection. *Ecology*, 84:586.

Ecolab, Webworld and self-organisation

Russell K. Standish

School of Mathematics, University of New South Wales

r.standish@unsw.edu.au, http://parallel.hpc.unsw.edu.au/rks

Abstract

Ecolab and Webworld are both models of evolution produced by adding evolution to ecological equations. They differ primarily in the form of the ecological equations. Both models are self-organised to a state where extinctions balance speciations. However, Ecolab shows evidence of this self-organised state being critical, whereas Webworld does not. This paper examines the self-organised states of these two models and suggest the likely cause of the difference. Also the lifetime distribution for a mean field version of Ecolab is computed, showing that the fat tail of the distribution is due to coevolutionary adaption of the species.

Introduction

In models of evolving ecologies, a "drip feed" of mutated species are added to a simulation of ecological dynamics. As new species are incorporated into the ecology, they create new links in the food web, perturbing the system dynamics. When enough links are added, feedback loops will form, and the simulated ecology will suffer a mass extinction. Over time, the system *self organises* to a state where the introduction of new species will be balanced by extinctions, and the system diversity fluctuates around some mean value.

But what is this state that the system self organises to? The first suggestion was a *critical* state (Bak and Sneppen, 1993), characterised by long range influences of a species extinction over others in the food web. The original model of Bak and Sneppen used to illustrate this idea is no more than a cartoon. The interactions between species in this model had no relation to biological interactions. The first attempt to use some real biologically inspired dynamics was probably Ecolab (Standish, 1994), which employed the well known Lotka-Volterra equations, for which a quite a bit of theoretical information is available. This model clearly self organises to a state where speciation is balanced by extinction of average (Standish, 1999), although a variation of the model (incorporating a mechanism of specialisation) produces unbounded growth in diversity (speciation exceeding extinction) (Standish, 2002).

So is this state a critical state? One problem is that criticality in self-organised systems is only achieved in the limit

of zero driving rate — in this case zero mutation rate. Sole *et al.* (Solé et al., 2002) prefer the term *self-organised instability*. Whilst I am sympathetic to this notion, I would also like to point out that *stability* is very precise term in dynamical systems theory, referring to the behaviour of the linearised system around an equilibrium point. Unstable ecosystems do not have to fall apart — the classical Lotka-Volterra (Maynard Smith, 1974) limit cycle is a case in point. Rather the notion of an ecosystem persisting in time without falling apart is captured by *permanence*, for which a few modest results are known for Lotka-Volterra systems (Law and Blackford, 1992). So perhaps self-organised impermanence would be a more accurate description.

Self organised critical systems are characterised by a power law distribution of extinction avalanches, and also a power law distribution of lifetimes. Traditionally, the presence of power law signatures in a self-organising system is taken as evidence of self-organised criticality. Newman (1997) developed another toy evolutionary model that exhibited power law spectra, with neither self-organisation nor criticality in sight. However, when the artificial constant diversity restriction is lifted in the obvious way, self-organisation reappears (Standish, 1999), and the model can also be understood as a mean field approximation of coevolutionary system that potentially admits critical behaviour.

Ecolab demonstrates power law spectra of lifetimes (Standish, 1999), with an exponent of -1. However, it has proven very difficult to measure the distribution of extinctions, as extinction avalanches overlap in Ecolab due to the finite rate of speciation. Conversely, studies of a similar model called Webworld claim an absence of any power law signatures (Drossel et al., 2001). I have implemented the Webworld model using the *Ecolab* (Standish,) simulation system. I was similarly unable to see evidence of power law signatures, and propose a possible explanation.

In this paper, I show that the Fourier transform of the diversity time series is related to the lifetime distribution. Furthermore, in the limiting case of infinitesimal speciation, this transform is the distribution of extinction avalanches (extinction frequency).

Ecolab model

We start with a generalised form of the Lotka-Volterra equation

$$\dot{n}_i = r_i n_i - n_i \sum_j \beta_{ij} n_j. \qquad (1)$$

Here n_i is the population of species i, r_i is the difference between reproduction and death and β_{ij} is the interaction between species i and j.

Periodically, each species i generates a number of mutant species, proportional to $n_i r_i \mu_i$, where μ_i is the mutation rate for species i. For each mutant species, the parameters r_i, β_{ij}, and μ_i are mutated from the parent species according to additive or multiplicative processes — the exact details aren't important here, but are described in (Standish, 1994).

One crucial property that is preserved by the mutation operator is *boundedness* (Standish, 2000). Boundedness ensures that population sizes in eq (1) can never exceed a particular limit.

It turns out that a necessary condition for permanence in eq (1) is that the matrix β has positive determinant (Law and Blackford, 1992). The determinant can be written as a sum

$$\det|\beta| = \sum_{\mathbf{p} \in \text{perm}(1...,n)} (-1)^{s(p)} \beta_{1p_1} \beta_{2p_2} \cdots \beta_{np_n} \qquad (2)$$

where $\text{perm}(1...,n)$ is the set of permutations of the numbers $1...,n$ and $s(p)$ is the number of swaps involved in the permutation.

All diagonal terms of β must be positive to ensure boundedness of eq (1).

Now consider permutations with one swap ($i \to j, j \to i$. If the terms β_{ij} and β_{ji} are of opposite sign (predator-prey case), then the contribution to the determinant is positive. However, if the terms have the same sign, (eg +ve, the mutual competition case, an increase in n_i causes n_j to decrease, which reduces competition on n_i, reinforcing the original change) then it describes a positive feedback loop between species i and j.

Likewise, it can be seen that the term $T = (-1)^{s(p)} \beta_{1p_1} \beta_{2p_2} \cdots \beta_{np_n}$ describes an $s(p)$ feedback loop through the ecosystem, which is a negative feedback loop if $T > 0$, and a positive feedback loop if $T < 0$. The necessary condition for permanence can be interpreted as saying that negative feedback loops must dominate over positive feedback loops for the ecosystem to be permanent.

As species are added to the system through speciation, new links are added to the foodweb at random. The chance of a positive feedback loop forming increases dramatically as the foodweb approaches its percolation threshold (Green and Klomp, 1999). Once this happens, an extinction avalanche is almost certain. The twin pressures of speciation and extinction through impermanence oppose each other leading to a state where the food web lies on its percolation threshold. Newth *et al.* (2002) examined the scaling

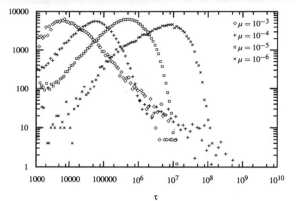

Figure 1: Lifetime distributions for different values of the maximum mutation rate μ in Ecolab. Histograms have been scaled so that the peaks are comparable in size.

structure of the Ecolab model, and observed the critical behaviour here. This is the strongest evidence yet that Ecolab self-organises to a critical state.

Plots of the lifetime distribution for several different values of the maximum mutation rate (mutation rates in Ecolab are allowed to vary, but can never exceed the maximum value) are shown in figure 1. These can be compared with other published data, such as (Standish, 1999). At higher mutation rates, the distributions exhibit a power law tail with exponent -1. As the mutation rate is turned down, the power law tail disappears, leaving a lognormal distribution. It is unclear whether the power law has disappeared altogether, or whether with the collection of more data it will be resolved out of the noise at the base of the graph.

Relationship between diversity time series and lifetime distribution

When a species becomes extinct, it may trigger secondary extinctions in other species, in a chain of extinctions known as an *extinction avalanche*. In the Bak-Sneppen model, these avalanches follow a power law distribution in avalanche size with exponent -1. However, it only becomes meaningful to discuss avalanche size in the limit of infinitesimal mutation rate, as otherwise the extinction avalanches overlap each other. In the Ecolab case, speciation occurs continuously, as do the resulting extinctions. More interesting is to discuss the frequency spectrum of extinctions, obtained by Fourier transforming the extinction time series. Diversity (number of species in the ecosystem at any point in time) is simply the difference between the speciation and extinction time series — in the infinitesimal speciation limit, the diversity spectrum is identical to the extinction spectrum.

The diversity time series can be written as a sum over speciation events s_j and associated lifetimes τ_j:

$$D(t) = \sum_j \Theta(t - s_j) - \Theta(t - s_j - \tau_j), \qquad (3)$$

where

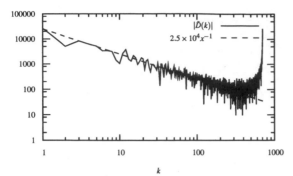

Figure 2: Fourier transform of a typical diversity time series in Ecolab showing hyperbolic behaviour.

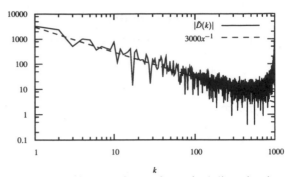

Figure 3: Fourier transform of a typical diversity time series in Webworld showing hyperbolic behaviour.

$$\Theta(x) = \begin{cases} 0 & x < 0 \\ \frac{1}{2} & x = 0 \\ 1 & x > 0 \end{cases}$$

is the usual Heaviside step function. Taking the Fourier transform of this (and ignoring constant factors):

$$\tilde{D}(k) = \sum_j \exp(iks_j) \frac{1 - exp(ik\tau_j)}{ik}. \tag{4}$$

Now we assume that speciation events occur every timestep (ie $s_j = j$), and that the lifetimes τ_j are sampled from a normalised lifetime distribution $T(\tau)$. Integrating over this distribution yields:

$$\tilde{D}(k) = \sum_j \exp(ikj) \frac{1 - \tilde{T}(k)}{ik} = \frac{1 - \tilde{T}(k)}{(1 - e^{ik})ik}. \tag{5}$$

Now, if $\lim_{\tau \to 0} T(\tau) < \infty$, then $\tilde{T}(k) \to 0$ as $k \to \infty$. So we can predict that asymptotically,

$$|\tilde{D}(k)| \sim k^{-1} \text{ as } k \to \infty \tag{6}$$

As $k \to 0$, $1 - e^{ik} \sim -ik$; $1 - \tilde{T}(k) \sim \langle \tau \rangle ik$, where $\langle \tau \rangle$ is the mean of $T(\tau)$. Even though a power law lifetime distribution would lead to an infinite $\langle \tau \rangle$, and any experiment, there is an upper cutoff to the lifetimes observed, which would reflect a finite $\langle \tau \rangle$. Therefore also $\tilde{D}(k) \sim k^{-1}$ as $k \to 0$. Figures 2 and 3 show $\tilde{D}(k)$ for typical Ecolab and Webworld runs respectively, and both data sets demonstrate this hyperbolic law. The power law observed in the time series spectra is uninteresting, as it is a general feature of all such stochastic processes. Adami (1998) makes a similar point in his book — that power laws in the time series spectra are necessary, but not sufficient for self-organised criticality.

Webworld

The Webworld model was introduced by Caldarelli *et al.* (1998), with some modifications described in Drossel *al.* (2001). The model implemented here is taken verbatim from the latter paper, so I will give only a brief synopsis of the

model here. The source code is available as part of the *Ecℓab*[1] software suite.

Webworld has a population dynamics which is a generalisation of the Lotka Volterra dynamics (eq 1) used in Ecolab:

$$\dot{n}_i = -n_i + \lambda n_i \sum_j g_{ij}(t) - \sum_j n_j g_{ji}(t). \tag{7}$$

This equation is called a *functional response* equation. λ is a model parameter called *ecological efficiency*, and usually taken to be $\lambda = 0.1$. $n_0 = R/\lambda$ is a special species, called the *environment*. By choosing $g_{ij}(t)n_i - g_{ji}(t)n_j = \beta_{ij}n_in_j$, $\forall i, j > 0$ and $g_{i0} = 1$, equation (1) is recovered. However, unlike Ecolab, Webworld tracks resources, and so n_i is perhaps better interpreted as the amount of biomass represented by species i than a population size.

In Webworld, the *functional response* term g_{ij} is given by

$$g_{ij}(t) = \frac{S_{ij} f_{ij}(t) n_j(t)}{b n_j(t) + \sum_k \alpha_{ki} S_{kj} f_{kj}(t) n_k(t)} \tag{8}$$

where the *efforts* f_{ij} are given recursively:

$$f_{ij}(t) = \frac{g_{ij}(t)}{\sum_k g_{ik}(t)} \tag{9}$$

Drossel *et al.* (2001) show that allowing species to vary the amount of effort in this way is an *evolutionary stable strategy*. The $\alpha_{ij} \le 1$ terms above represent that different species do not compete as strongly as members of the same species ($\alpha_{ii} = 1$, $\forall i$).

$$\alpha_{ij} = c + (1 - c)q_{ij} \tag{10}$$

where $0 < c < 1$ is a *competition* parameter that strongly influences the final steady state diversity of the model. The precise definition of the interaction terms S_{ij} and q_{ij} is very interesting, but not germane to the argument here.

In (Drossel et al., 2001), the equations are evolved in time until the ecosystem reaches equilibrium, or until a large period of time has elapsed before another species is added to

---

[1] http://parallel.hpc.unsw.edu.au/ecolab

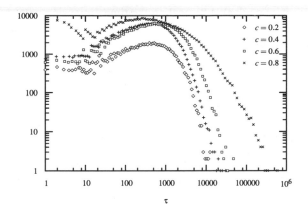

Figure 4: Lifetime distributions for different values of c for Webworld.

the system. This is to mimic the "infinitesimal speciation rate" mentioned in relation the self-organised criticality. In this study, a single species is added periodically every 20 time units, which was chosen empirically as being sufficiently rare for the ecosystem to stabilise between speciations.

Figure 3 shows the spectrum of the diversity time series for a run with $c = 0.4$. Figure 4 shows a lifetime distribution from these runs, illustrating that the best functional form is a lognormal distribution of lifetimes. Its important to note that lognormal distributions are often confused with power law distributions (Mitzenmacher, 2003), however are distinctly different in that lognormal distributions have a mean $\langle \tau \rangle$, whereas power law distributions do not. However, as mentioned in the section describing the Ecolab model, very low speciation rates were used in this model, and it is possible that a power law tail is hidden within the noise at tail of the graph.

Quite another explanation for the absence of critical behaviour in Webworld comes from considering the dynamics of the effort coefficients f_{ij}. These are iterated within a timestep to determine evolutionary stable values — this is a model of how predators exhibiting high phenotypic plasticity might adapt to variations in food sources. It would seem to be a less plausible model of less complex organisms that might not be so choosy about their food source. Whatever the biological realism of this process, the f_{ij} coefficients describe the effective foodweb, which tends to be quite sparse and without many loops (Quince et al., 2002). Could it be that this process prevents the percolation threshold of the foodweb from being reached?

Mean Field Ecolab model

In (Standish, 1999), I introduced a mean field[2] version of the Ecolab model (which I dubbed Ecolab--). That model is a simple multiplicative process, which is related by log-

---

[2]Some people use *mean field* to mean *panmictic*. The Ecolab model described in previously is already panmictic. By mean field, I mean that each species experiences a stochastic force that is the average of the interspecific interactions in the full model.

arithms to the standard isotropic 1D random walk process. The lifetime distribution is known as the *first-passage time* distribution in this subject, and is known to exhibit a $\tau^{-3/2}$ tail (Redner, 2001). Note that this is different from, but still compatible with, the upper bound of τ^{-1} derived in (Standish, 1999).

However, this model has a lognormal limiting distribution of population sizes n of the form:

$$p(n,t) = \frac{1}{n\sqrt{t}} \exp\left(-\frac{(\ln n - rt)^2}{4t}\right) \quad (11)$$

This distribution does not satisfy boundedness.

In order to introduce boundedness, we need to reintroduce the quadratic term into the mean field model:

$$\dot{n} = n(r - \beta n + \gamma), \quad (12)$$

where γ is an uncorrelated stochastic variable, with zero mean.

Taking logarithms $\xi = \ln n$ and applying the Ito transformation formula (Karlin and Taylor, 1981, p. 372, e.g.), eq (12) can be written as a Langevin equation:

$$\dot{\xi} = (r - \frac{1}{2} - \beta e^\xi + \gamma). \quad (13)$$

The extra term of $\frac{1}{2}$ comes from the effect of change of variables on the stochastic term γ. Langevin equations can be converted into a Fokker-Planck equation describing the probability distribution $w(x,t)$ that ξ has the value x at time t (Risken, 1984):

$$\frac{\partial w}{\partial t} = \frac{\partial^2 w}{\partial t^2} - \frac{\partial}{\partial t}(r - \frac{1}{2} - \beta e^x)w \quad (14)$$

Taking the Laplace transform of equation (14) yields a second order homogeneous ordinary linear differential equation:

$$\frac{\partial}{\partial x}\left(w'(x,s) - (r - \frac{1}{2} - \beta e^x)w\right) + sw = 0. \quad (15)$$

The full time dependent equation doesn't appear to be amenable to analytic treatment, however the time independent equation ($s = 0$) can be reduced to a 1st order ODE. Let

$$y(x) = \exp\left((r - \frac{1}{2})x - \beta e^x\right) \quad (16)$$

$$y'(x) = \left(r - \frac{1}{2} - \beta e^x\right)y(x) = g(x)y(x), \quad (17)$$

and write $w_0(x) = w(x,0) = y(x)v(x)$. Substitute this into equation (15), and one obtains:

$$\frac{d}{dx}(yv') = 0$$

$$v(x) = A_0 \int^x \frac{dx'}{y(x')} + A_1$$

$$= A_0(-\beta)^{r-\frac{1}{2}}\Gamma(\frac{1}{2} - r, -\beta e^x) + A_1 \quad (18)$$

where $\Gamma(a,x)$ is an incomplete gamma function (Abramowitz and Stegun, 1965, 6.5.3), and A_j are constants of integration. Substituting this into the expression for $p(n,0)$ yields:

$$p(n,0) = \frac{1}{n}w(\ln n, 0) =$$

$$n^{r-\frac{3}{2}}e^{-\beta n}\left(A_1 + A_0(-\beta)^{-\frac{1}{2}}\Gamma(\frac{1}{2} - r, -\beta n)\right) \quad (19)$$

From the series $\Gamma(a,x) \sim \Gamma(a) - \frac{x^a}{a} + \frac{x^{a+1}}{a+1}\cdots$ (Gradsteyn and Ryzhik, 1980, 8.354), one can see:

$$p(n,0) \quad \sim \quad n^{r-\frac{3}{2}}e^{-\beta n}\left(A_0(-\beta)^{r-\frac{1}{2}}\Gamma(\frac{1}{2} - r) + A_1\right) +$$

$$\frac{A_0}{n(r-\frac{1}{2})}e^{-\beta n} \quad (20)$$

which is normalisable if and only if $r > \frac{1}{2}$ and $A_0 = 0$. We may therefore set $w_0(x) = y(x)$.

The asymptotic behaviour at large times translates into the the small s regime. We can compute $w_1(x) = \left.\frac{\partial w(x,s)}{\partial s}\right|_{s=0}$ by differentiating eq (15) with respect to s.

$$\frac{d}{dx}(w_1' - (r - \frac{1}{2} - \beta e^x)w_1) + w_0 = 0 \quad (21)$$

to which the solution is:

$$w_1(x) = w_0(x)\left(A_1 + \int^x \frac{1}{w_0}(x')\int^{x'}w_0(x'')dx''\right). \quad (22)$$

The innermost integral can be evaluated, the answer being (Abramowitz and Stegun, 1965, 6.5.2)

$$\int^x w_0(x')dx' = \beta^{-r+\frac{1}{2}}\gamma(r - \frac{1}{2}, \beta e^x) \quad (23)$$

with γ being another of the incomplete gamma functions. This can be represented as a series (Gradsteyn and Ryzhik, 1980):

$$\int^x w_0(x')dx' = \beta^{-r-\frac{1}{2}}\sum_{n=0}^{\infty}\frac{(-1)^n(\beta e^x)^{r-\frac{1}{2}+n}}{n!(r-\frac{1}{2}+n)} \quad (24)$$

Performing the integral on each term of the series yields:

$$w_1(x) = w_0(x) \times \quad (25)$$

$$\left(A_1 + \exp(\beta e^x)\sum_{n=1}^{\infty}\sum_{j=0}^{n-1}\frac{(-1)^j\beta^j e^{jx}}{j!n(r-\frac{1}{2}+n)} + E_1(\beta e^x)\right)$$

where $E_1(x)$ is the exponential integral (Abramowitz and Stegun, 1965, 5.1.1).

Interestingly, for the special case $r - \frac{1}{2} \in \mathbb{Z}^+$, the result can be expressed as a finite series. It might seem that r can

be chosen to any value by scaling the time dimension without loss of generality, however that is not the case, as the timescale is already set by the variance of the stochastic term γ in eq. (12).

Considering the special case $r = \frac{3}{2}$, and making use of the identity $\gamma(1,x) = (1 - e^{-x})$ (Gradsteyn and Ryzhik, 1980, 8.352), we have:

$$w_1(x) = w_0(x)(A_1 - \beta^{-1}e^{-x} - \Gamma(-1, \beta e^x)) \quad (26)$$

To compute the asymptotic form of the lifetime distribution, we make use of *first passage theory* (Redner, 2001). The first passage probability $F(n,t|n_0)$ of the population having value n, given a starting value n_0 at $t = 0$ is related to $p(n,t|n_0)$:

$$p(n,t) = \delta(t)\delta(n - n_0) + \int_0^t F(n, t - t'|n_0)p(n_0, t|n_0)dt' \quad (27)$$

Taking the Laplace transform, and rearranging gives us

$$\tilde{F}(n,s) = p(n,s)/p(n_0,s) \quad n_0 \neq n \quad (28)$$

(as we're not interested in the $n = n_0$ case). For concreteness, let $n_0 = 10$ and $n = 1$, as is taken in the case of Ecolab experiments computing the lifetime distribution (Fig. 1). The asymptotic form can be computed directly from $F(n,s)$:

$$\tilde{F}(n, 1/\tau) = \int^{\infty}e^{-\tau t}F(n,t)dt \sim \int^{\tau}F(n,t)dt \quad (29)$$

$$F(n,\tau) \sim \frac{\partial}{\partial\tau}\tilde{F}(n, 1/\tau) \quad (30)$$

$$= -\frac{p(n_0,0)\left.\frac{\partial p(n,s)}{\partial s}\right|_{s=0} - p(n,0)\left.\frac{\partial p(n_0,s)}{\partial s}\right|_{s=0}}{\tau^2 p(n_0,0)^2}$$

Unless the numerator vanishes, the long time tail will obey a τ^{-2} power law. Substituting equations (16) and (26) yields for the case $r = \frac{3}{2}$:

$$F(n,\tau) \sim \frac{e^{2\beta n_0}}{\tau^{-2}}\left(\frac{1}{\beta n_0} - \frac{1}{\beta n} + \Gamma(-1, n_0) - \Gamma(-1, n)\right) \quad (31)$$

Since $\Gamma(-1, n) > \Gamma(-1, n_0)$ for $n < n_0$, this derivative term is negative. It seems unlikely to vanish for any value of $r > 0$, however this will need to be checked numerically.

This result is interesting. The mean field model can be considered as a neutral model, in the sense of the neutral shadow models proposed by Bedau and Packard (Bedau et al., 1998). An observed excess of lifetimes over the mean field case (in Fig 1 a τ^{-1} distribution is observed) would represent coadaption by the species in the ecosystem.

Conclusion

In this paper, I consider the question of self-organised criticality in a couple of evolutionary ecology models (Ecolab and Webworld). In spite of their similarity, only Ecolab appears to self-organise to criticality, whereas Webworld's self-organised state appears to be noncritical, in agreement with Webworld's creator's statements.

Whilst it is possible that experiments have not been run long enough to observe critical behaviour, a more likely explanation is organismal plasticity in Webworld prevents long range interdependence of species in the foodweb from building up.

A mean free approximation to the Ecolab model is solved, and the lifetime distribution from this model is expected to have a τ^{-2} asymptotic behaviour. The fact that the real Ecolab model appears to have a τ^{-1} asymptotic behaviour hints at adaption occurring within that system.

Finally, the spectral density of the diversity time series (which is related to the distribution of extinction avalanches) is expected to have a $1/f$ behaviour, regardless of the underlying process, so this should not be taken as evidence of self-organised criticality.

Acknowledgements

I would like to thank the *Australian Centre for Advanced Computing and Communications* (ac3) for computer time use in these simulations. I would also like to thank Ben Goldys for helpful comments on the manuscript.

References

Abramowitz, M. and Stegun, I. A. (1965). *Handbook of Mathematical Functions*. Dover, New York.

Adami, C. (1998). *Introduction to Artificial Life*. Springer.

Bak, P. and Sneppen, K. (1993). Puntuated equilibrium and criticality in a simple model of evolution. *Phys. Rev. Lett.*, 71:4083.

Bedau, M. A., Snyder, E., and Packard, N. H. (1998). A classification of long-term evolutionary dynamics. In Adami, C., Belew, R., Kitano, H., and Taylor, C., editors, *Artificial Life VI*, pages 228–237, Cambridge, Mass. MIT Press.

Caldarelli, G., Higgs, P. G., and McKane, A. J. (1998). Modelling coevolution in multispecies communities. *J. Theor. Biol.*, 193:345–358.

Drossel, B., Higgs, P. G., and McKane, A. J. (2001). The influence of predator-prey population dynamics on the long-term evolution of food web structure. *J. Theor. Biol.*, 208:91–107.

Gradsteyn, I. S. and Ryzhik, I. (1980). *Table of Integrals, Series and Products*. Academic, New York.

Green, D. G. and Klomp, N. I. (1999). Environmental informatics – a new paradigm for coping with complexity in nature. *Complexity International*, 6.

Karlin, S. and Taylor, H. M. (1981). *A Second Course in Stochastic Processes*. Academic Press, New York.

Law, R. and Blackford, J. C. (1992). Self-assembling food webs: A global viewpoint of coexistence of species in lotka-volterra communitites. *Ecology*, 73:567–578.

Maynard Smith, J. (1974). *Models in Ecology*. Cambridge University Press, London.

Mitzenmacher, M. (2003). A brief history of generative models for power law and lognormal distributions. *Internet Mathematics*, 1. to appear. See http://www.eecs.harvard.edu/~michaelm/.

Newman, M. E. J. (1997). A model of mass extinction. *J. Theo. Bio.*, 189:235–252.

Newth, D., Lawrence, J., and Green, D. G. (2002). Emergent organization in dynaic networks. In Namatame, A., Green, D. G., Aruka, Y., and Sato, H., editors, *Proceedings Complex Systems '02: Complexity with Agent Based Modeling*, pages 229–237. Chuo University.

Quince, C., Higgs, P. G., and McKane, A. J. (2002). Food web structure and the evolution of ecological communities. In Laessig, M. and Valleriani, A., editors, *Biological Evolution and Statistical Physics*, volume 585 of *Lecture Notes in Physics*, page 281. Springer, Berlin.

Redner, S. (2001). *A Guide to First-Passage Processes*. Cambridge UP, Cambridge.

Risken, H. (1984). *The Fokker Planck Equation*. Springer, Berlin.

Solé, R. V., Alonso, D., and McKane, A. (2002). Self-organised instability in complex ecosystems. *Phil. Trans. Royal Soc. B*, 357:667–681.

Standish, R. K. Ecolab documentation. Available at http://parallel.acsu.unsw.edu.au/rks/ecolab.

Standish, R. K. (1994). Population models with random embryologies as a paradigm for evolution. *Complexity International*, 2.

Standish, R. K. (1999). Statistics of certain models of evolution. *Phys. Rev. E*, 59:1545–1550.

Standish, R. K. (2000). The role of innovation within economics. In Barnett, W., Chiarella, C., Keen, S., Marks, R., and Schnabl, H., editors, *Commerce, Complexity and Evolution*, volume 11 of *International Symposia in Economic Theory and Econometrics*, pages 61–79. Cambridge UP.

Standish, R. K. (2002). Diversity evolution. In Standish, R., Abbass, H., and Bedau, M., editors, *Artificial Life VIII*, pages 131–137, Cambridge, Mass. MIT Press.

Tierra's missing neutrality: case solved.

Russell K. Standish

School of Mathematics, University of New South Wales,Sydney, 2052,Australia
R.Standish@unsw.edu.au, http://parallel.hpc.unsw.edu.au/rks

Abstract

The concept of neutral evolutionary networks being a significant factor in evolutionary dynamics was first proposed by Huynen *et al.* about 7 years ago. In one sense, the principle is easy to state — because most mutations to an organism are deleterious, one would expect that neutral mutations that don't affect the phenotype will have disproportionately greater representation amongst successor organisms than one would expect if each mutation was equally likely.

So it was with great surprise that I noted neutral mutations being very rare in a visualisation of phylogenetic trees generated in *Tierra*, since I already knew that there was a significant amount of neutrality in the Tierra genotype-phenotype map.

It turns out that competition for resources between host and parasite inhibits neutral evolution.

Keywords: Tierra, neutral evolution, genotype-phenotype map, Vienna RNA package

Introduction

The influence of *neutral networks* in evolutionary processes was first elucidated by Peter Schuster's group in Vienna in 1996 (Huynen et al., 1996; Reidys et al., 1997). Put simply, two *genotypes* are considered *neutrally equivalent* if they map to the same *phenotype*. A *neutral network* is a set of genotypes connected by this neutrality relationship on links with Hamming distance 1 (i.e. each link of the network corresponds to a mutation at a single site of the genome). It should be noted that this definition is subtly different from that employed in Kimura's *neutral evolution theory* (Kimura, 1983), as in that theory, neutrality is defined as equivalence of fitness values, a notion that is ill-defined in coevolutionary systems. However as phenotypically equivalent organisms are neutral in Kimura's sense when a fitness function exists, much of neutral theory can be carried over into discussion of phenotypic neutrality.

Schuster's group noted that evolution tended to proceed by diffusion along these neutral networks, punctuated occasionally by rapid changes to phenotypes as an adaptive feature is discovered. The similarity of these dynamics with the theory of Punctuated Equilibria (Eldridge, 1985) was noted by Barnett (1998). It was also noted that if a *giant network* existed that came within a hop or two of every possible genotype, evolution will be particularly efficient at discovering solutions, since only a few non-neutral mutations are needed to reach the optimum solution.

Most work on neutrality in evolution uses the genotype-phenotype mapping defined by folding of RNA (Schuster et al., 1994). This mapping is implemented in the open source Vienna RNA package[1], so is a convenient and well-known testbed for ideas of neutrality in evolution.

Also in 1996, I developed a definition of the genotype-phenotype mapping for Tierra, which was first published in 1997 (Standish, 1997). I noticed the strong presence of neutrality in this mapping at that time, which was later exploited to develop a measure of complexity of the Tierran organism (Standish, 1999; Standish, 2003). In 2002, I started a programme to visualise Tierra's phylogenetic trees and neutral networks (Standish and Galloway, 2002) in order to "discover the unexpected". Two key findings came out of this: the first being that Tierra's genebanker[2] data did not provide clean phylogenetic trees, but had loops, and consisted of many discontinuous pieces. This later turned out to be due to Tierra's habit of reusing genotype labels if those genotypes were not saved in the genebanker database. This might happen if the population count of that genotype failed to cross a threshold. This is all very well, except that a reference to that genotype exists in the parent field of successor genotypes. The second big surprise was the paucity of neutral mutations in the phylogenetic tree. We expect most mutations to an organism to be deleterious, and so expect that neutral mutations will have disproportionately greater representation amongst successor genotypes than one would expect if each mutation was equally likely.

Neutrality in Tierra

Tierra (Ray, 1991) is a well known artificial life system in which small self-replicating computer programs are ex-

---

[1] http://www.tbi.univie.ac.at/~ivo/RNA

[2] The *genebanker* is a database in which Tierra stores the genotypes that arise during evolution.

ecuted in a specially constructed simulator. These computer programs (called digital organisms, or sometimes "critters") undergo mutation, and radically novel behaviour is discovered, such as *parasitism* and *hyperparasitism*.

It is clear what the genotype is in Tierra, it is just the listing of the program code of the organism. The phenotype is a more diffuse thing, however. It is the resultant effect of running the computer program, in all possible environments. Christoph Adami defined this notion of phenotype for a similar artificial life system called *Avida* (Adami, 1998). In Avida, things are particularly simple, in that organisms either reproduce themselves at a fixed replication rate, or don't as the case may be, and optionally perform range of arithmetic operations on special registers (defined by the experimenter).

In Tierra, organisms do interact with each other via a template matching mechanism. For example, with a branching instruction like `jmpo`, if there is a sequence of `nop0` and `nop1` instructions (which are no-operations) following the branch, this sequence of 1s and 0s is used as a template for determining where to branch to. In this case the CPU will search outwards through memory for a complementary sequence of `nop0`s and `nop1`s. If the nearest complementary sequence happens to lie in the code of a different organism, the organisms interact.

To precisely determine the phenotype of a Tierran organism, one would need to execute the soup containing the organism and all possible combinations of other genotypes. Whilst this is a finite task, it is clearly astronomically difficult. One means of approximation is to consider just interaction of pairs of genotypes (called a tournament). Most Tierran organisms interact pairwise — very few triple or higher order interactions exist. Similarly, rather than running tournaments with all possible genotypes, we can approximate matters by using the genotypes stored in a genebanker database after a Tierra run. In practice, it turns out that various measures, such as the number of neutral neighbours, or the total complexity of an organism are fairly robust with respect to the exact set of organism used for the tournaments.

So the procedure is to pit pairwise all organisms in the genebanker against themselves, and record the outcome in a table (there is a small number of possible outcomes, which is detailed in (Standish, 1997)). A row of this table is a phenotypic signature for the genotype labeling that row. We can then eliminate those genotypes with identical signatures in favour of one canonical genotype. This list of unique phenotypes can be used to define pragmatically a test for neutrality of two different genotypes, that may have generated by mutation from genotypes recorded in the genebanker. Pit each organism against the list of unique phenotypes, and if the signatures match, we have neutrality. The source code for

this experiment is available from the author's website.[3]

Tierra has three different modes of mutation:

Cosmic Ray A site of the soup is randomly chosen and mutated;

Copy Data is mutated during the copy operation;

Flaw Instructions occasionally produce erroneous results

Furthermore, in the case of cosmic ray and copy mutations, a certain proportion of mutations involve bit flips, rather than opcodes being substituted uniformly. This proportion is set as a parameter in the soup_in file (`MutBitProp`) — in these experiments, this parameter is set to zero.

In order to study the issue of whether neutrality is greater or less than expected in Tierra, I generated three datasets with each of the 3 modes of mutation operating in isolation. The sizes of each data set was 69,139, 87,003 and 198,982 genotypes respectively, generated over a time period of about 1000 million executed instructions. Genebanker's threshold was set to zero, so all genotypes were captured. This led to a proper phylogenetic tree. After performing a neutrality analysis, a set of 83, 86 and 158 unique phenotypes was extracted as the test set for the tournaments.

Since the neighbourhood size increases exponentially with neighbourhood diameter, I restrict analysis to single site, or point mutations. In each data set, around 7% of these genotypes were created by a mutation at a single site and were neutrally equivalent to its parent. For each of these, I compute the number of neutral neighbours n_i existing in the 1 hop neighbourhood of the parent genotype i. The 1 hop neighbourhood size is 32^{ℓ_i}, where ℓ_i is the length of the genome. For a given parent i, the ratio

$$r_i = \frac{v_i 32^{\ell_i}}{o_i n_i} \tag{1}$$

gives the proportion of neutral links actually followed relative to the number of neutral links available (*neutrality excess*), where v_i is the number of neutrally equivalent offspring, and o_i the total number of offspring and n_i the size of the 1 hop neutral neighbourhood. Fig. 1 shows the running average of this quantity over these transitions, with the genotypes numbered in size order.

Since all daughter genotypes are recorded, no selection is operating. In this case, one would expect that the proportion of neutral variants seen should be identical to the proportion of neutral variants within the 1 hop neighbourhood, and hence the neutrality excess should be identical to 1. However, in the case of instruction and cosmic ray flaws, not every daughter genome will make it into the genebanker. In the case of instruction flaws, it is rather unpredictable what the effect is. In the case of cosmic ray mutations, 50% of

---

[3]http://parallel.hpc.unsw.edu.au/getaegisdist.cgi/getsource/eco-tierra.3, version 3.D3

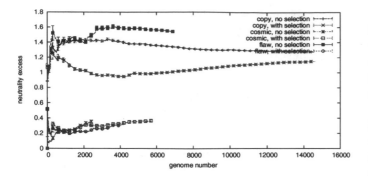

Figure 1: Running average of neutrality excess ($\langle r_i \rangle$). Genomes are ordered according to size, and neutrality excess is averaged over all genomes to the left of that data point. Three different datasets are analysed, with each of the three modes of mutation turned on. Then the datasets are further filtered to only include offspring whose maximum population count is greater than 1, i.e. selection is operating.

time one would expect the parent to be mutated, rather than the daughter. In the case of a mutation affecting a crucial gene of a parent genotype, the organism may not be able to reproduce at all, thus favouring neutral mutations. Only copy mutations should affect all sites of the genome equally, leading to a neutrality excess equal to one. The measured value, however is about 1.3, substantially greater than one. The reason for this is not known at this point in time.

The datasets were further subsetted to include just those transitions whose daughter organism successfully reproduced, i.e. with a maximum population count greater than 1. The neutrality excess in this case is substantially less than 1, so something in Tierran evolution is favouring non-neutral evolution.

Competition Effects

Consider a single species ecosystem with logistic dynamics:

$$\dot{x} = rx(1 - x/K), \qquad (2)$$

where x is the population size, r the net reproductive rate and K the carrying capacity. A phenotypically equivalent genotype attempting to invade this ecosystem will have the following dynamics:

$$\dot{x}' = rx'(1 - x/K) \approx 0, \qquad (3)$$

(x' being the population size of the invading genotype) as $x \approx K$ at equilibrium. So there is a substantial likelihood that the neutral variant fails to invade the ecosystem.

This argument is of course an extreme case. Stochastic effects due to finite population sizes will increase the chances of a neutral variant invading the ecosystem, however the point still remains that the neutral variant is not on an equal footing as the incumbent.

In Tierra, however, there is an age structure in the population, with organisms being placed in a *reaper queue*, from which the oldest organisms are selected when death is required. This fact alone implies that neutral variants of self-replicating organisms will successfully replicate, and hence cannot be responsible for the neutrality deficiency.

However, consider a Tierran ecosystem consisting of hosts and parasites, where the parasite require the presence of a host organism within a certain distance of itself in the soup, in order for the parasite to replicate. Since parasitic organisms replicate faster than the hosts (due to their smaller program lengths), they tend to displace host organisms until there are not enough hosts to go around. At which point, the parasite's fecundity drops. At equilibrium, the effective reproductive rates of host and parasite are equal.

A neutral variant will therefore be quite likely to not have a suitable host in its neighbourhood to allow it to replicate. Consequently, neutral evolution is suppressed amongst parasites. In the next section I will test this idea by setting up an artificial host-parasite coevolutionary system, using the well known RNA genotype-phenotype map.

Vienna RNA Folding Experiments

It is quite well known that evolution using the RNA folding map (Schuster et al., 1994) exhibits a great deal of neutrality, at least for a standard genetic algorithm optimising a well defined fitness function. Until now, evolutionary systems based on the RNA map exhibit the unsurprising result of neutrality excess defined by eq (1) being greater than or equal to 1. I now present results of an RNA map experiment that demonstrates neutrality supression ($r_i < 1$), based on the resource competition explanation elabortaed earlier. We need two types of organism (host and parasite) competing for a fixed space that can support $N = 100$ organisms. Parasites can only reproduce if they are situated next to a host (neighbourhood size $\nu = 2$), but reproduce twice as fast as the host type.

Once an organism has reproduced, it replaces the least fit organism. Fitness is determined by how close the parasitic phenotype is to any hosts in the neighbourhood of the parasites, and decreases in a similar way with the similarity of the parasites in the neighbourhood for host organisms:

$$F_{\mathrm{h}} = 1 - \frac{1}{\nu \ell} \sum_{i \in \mathcal{P}_\nu} d(i, h) \qquad (4)$$

$$F_{\mathrm{p}} = \frac{\rho}{\nu \ell} \sum_{i \in \mathcal{H}_\nu} d(i, p) \qquad (5)$$

where h and p are host and parasite genotypes respectively, $\mathcal{H}_\nu$ and $\mathcal{P}_\nu$ the set of hosts and parasites respectively within the neighbourhood of size ν of p and h respectively. $d(i, j)$ is the string edit distance between the phenotypes[4], and ℓ

---

[4]The *string edit* distance is related to the Hamming distance

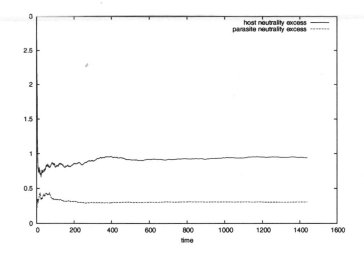
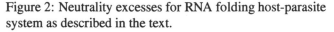

Figure 2: Neutrality excesses for RNA folding host-parasite system as described in the text.

is the gene length (set equal to 20 for all organisms in this experiment).

The factor ρ adjusts the relative dominance of parasites over hosts. Set it too low, and hosts will eliminate the parasites by virtue of replacing then when replicating. Set it too high, and hosts will only be competing with themselves. In this experiment, a value of $\rho = 3.1$ was found to give intermediate behaviour.

An alternative version of this experiment where organisms were selected at random for death, rather than according to a fitness relationship showed similar dynamics, although the neighbourhood size ν needed to be increased to 4 to allow a stable population of parasites to persist.

Figure 2 shows the neutrality excess for this experiment. The model will consistently produce a neutrality deficiency for the parasites over a broad range of model parameters. If the parameter ρ is set too high, the hosts will compete strongly with themselves, suppressing neutrality in the host population also.

Source code for this experiment is available from the author's website. [5]

Conclusion

The suppression of neutrality in Tierran evolution is a real effect. An explanation couched in terms of host parasite competition was found, and a model was constructed using

---

(no. of base pairs that differ between two strings), but allows for gaps in the strings. Given a set of edit operations (eg insertionsq and deletions) and edit costs, the edit distance is given by the minimum sum of the costs along an edit path converting one object into the other. Please consult the Vienna RNA package documentation for a precise definition of string edit distance

[5]http://parallel.hpc.unsw.edu.au/getaegisdist.cgi/getsource/rnafold/, version D1

the well-known RNA folding map that illustrated this explanation.

This finding is potentially important. It has been argued that neutral diffusion is an important feature of evolutionary processes allowing efficient search of phenotype space. The sort of competition effects seen here to impede neutral diffusion are characteristic of climax ecosystems. This would imply that disturbed ecosystems will have greater evolvability than climax systems. This "brake" on neutral diffusion being released during times of environmental stress could provide an alternative explanation for the patterns of adaptive radiation seen after mass extinction events.

Acknowledgments

I would like to thank the *Australian Centre for Advanced Computing and Communications* for a grant of computing time used in this project.

References

Adami, C. (1998). *Introduction to Artificial Life*. Springer.

Barnett, L. (1998). Ruggedness and neutrality — the *NKp* family of fitness landscapes. In Adami, C., Belew, R., Kitano, H., and Taylor, C., editors, *Artificial Life VI*, pages 18–27, Cambridge, Mass. MIT Press.

Eldridge, N. (1985). *Time Frames — The Rethinking of Darwinian Evolution and the Theory of Punctuated Equilibria*. Simon and Schuster, New York.

Huynen, M., Stadler, P. F., and Fontana, W. (1996). Smoothness within ruggedness: The role of neutrality in adaption. *Proc. Nat. Acad. Sci. USA*, 93:397.

Kimura, M. (1983). *The Neutral Theory of Molecular Evolution*. Cambridge UP, Cambridge.

Ray, T. (1991). An approach to the synthesis of life. In Langton, C. G., Taylor, C., Farmer, J. D., and Rasmussen, S., editors, *Artificial Life II*, page 371. Addison-Wesley, Reading, Mass.

Reidys, C., Kopp, S., and Schuster, P. (1997). Evolutionary optimization of biopolymers and sequence structure maps. In Langton, C. and Shimohara, K., editors, *Artificial Life V*, page 379, Cambridge, Mass. MIT Press.

Schuster, P., Fontana, W., Stadler, P. F., and Hofacker, I. L. (1994). From sequences to shapes and back: A case study in RNA secondary structures. *Proc. Royal Soc. London B*, 255:279–284.

Standish, R. K. (1997). Embryology in Tierra: A study of a genotype to phenotype map. *Complexity International*, 4.

Standish, R. K. (1999). Some techniques for the measurement of complexity in Tierra. In Floreano, D., Nicoud, J.-D., and Mondada, F., editors, *Advances in Artificial Life: 5th European Conference, ECAL 99*, volume 1674 of *Lecture Notes in Computer Science*, page 104, Berlin. Springer.

Standish, R. K. (2003). Open-ended artificial evolution. *International Journal of Computational Intelligence and Applications*, 3:167.

Standish, R. K. and Galloway, J. (2002). Visualising Tierra's tree of life using Netmap. In Bilotta, E. et al., editors, *ALife VIII Workshop proceedings*, page 171. http://alife8.alife.org.

Drastic Changes in Roles of Learning in the Course of Evolution

Reiji Suzuki and Takaya Arita
Graduate School of Information Science, Nagoya University
Furo-cho, Chikusa-ku, Nagoya 464-8601, Japan
{reiji, arita}@is.nagoya-u.ac.jp

Abstract

An interaction between evolution and learning called the Baldwin effect is known as the two-step evolutionary scenario caused by the balances between benefit and cost of learning in general. However, little is still known about dynamic evolutions on these balances in complex environments. Our purpose is to give a new insight into the benefit and cost of learning by focusing on the quantitative evolution of phenotypic plasticity under the assumption of epistatic interactions. For this purpose, we have constructed an evolutionary model of quantitative traits by using an extended version of Kauffman's NK fitness landscape. Phenotypic plasticity is introduced into our model, in which whether each phenotype is plastic or not is genetically defined and plastic phenotypes can be adjusted by learning. The simulation results have clearly shown that the drastic changes in roles of learning cause the three-step evolution of the Baldwin effect (Suzuki and Arita, 2003) and also cause the evolution of the genetic robustness against mutations. We also conceptualize four different roles of learning by using a hill-climbing image of a population on a fitness landscape.

Introduction

The Baldwin effect is known as one of the interactions between evolution and learning, which suggests that individual lifetime learning can influence the course of evolution without the Lamarckian mechanism (Baldwin, 1896). This effect explains these interactions by paying attention to balances between benefit and cost of learning through the following two steps (Turney et al., 1996). In the first step, lifetime learning gives individual agents chances to change their phenotypes. If the learned traits are useful for agents and make their fitness increase, they will spread in the next population. In the second step, if the environment is sufficiently stable, the evolutionary path finds innate traits that can replace learned traits, because of the cost of learning. This step is known as genetic assimilation (Waddington, 1942). Through these steps, learning can guide the genetic acquisition of learned traits without the Lamarckian mechanism in general.

Hinton and Nowlan conducted the first computational experiment of the Baldwin Effect (Hinton and Nowlan, 1987).

They assumed an extremely simplified version of a network connection model. The essential point of this study is that they introduced the quantitative evolution of phenotypic plasticity into their model, in other words, they allowed a population to adjust how much it depends on these two adaptive mechanisms through evolution. They revealed the existence of the Baldwin effect by showing the increase and subsequent decrease in the phenotypic plasticity. However, the learning mechanism in their model was too simple on the ground that its benefit was approximately proportional to the number of plastic phenotypes and the cost of learning was explicitly introduced. Thus, further investigations were necessary so as to understand this effect in more realistic situations. Since then, this effect has been discussed in various contexts such as the evolution of the strategies for iterated Prisoner's Dilemma (Arita and Suzuki, 2000), the evolution of developmental mechanisms (Downing, 2004) and so on.

The effects of epistasis are of interest in this field because epistatic interactions among loci are ubiquitous in modern genetics and evolutionary biology. For instance, Mayley conducted an evolutionary experiment using the Kauffman's NK fitness landscape (Mayley, 1996). He pointed out that there should be a neighborhood correlation between genotype and phenotype space to guarantee a genetic assimilation to occur. Bull also discussed the evolution on NK landscapes using a different learning mechanism, and then concluded that whether the learning can increase the fitness or not depends on the ruggedness of the landscape, the probability of learning, and the number of learning iterations (Bull, 1999). However, all phenotypes were plastic and the quantitative evolution of phenotypic plasticity was not introduced into their models. In this sense, the two steps of the Baldwin effect were not clearly discussed in these models when compared with Hinton and Nowlan's model.

Our purpose is to give a new insight into the dynamic evolution of the benefit and cost of learning in complex environments. Especially, we focus on the effects of epistatic interactions on the quantitative evolution of phenotypic plasticity. As a first approach, we have investigated the quantitative evolution of phenotypic plasticity in a neural network as a

Table 1: Three-step evolution of the Baldwin effect. The arrows represents the transitions of indices in each step as follows: "steady ($\rightarrow$)", "increasing ($\nearrow$)", "slightly increasing ($\rightarrow$)" and "decreasing ($\searrow$)".

step	1st	2nd	3rd
lifetime fitness	$\nearrow$	$\nearrow$	$\rightarrow$
innate fitness	$\rightarrow$	$\rightarrow$	$\nearrow$
phenotypic plasticity	$\nearrow$	$\searrow$	$\rightarrow$
phenotypic variation	$\rightarrow$	$\nearrow$	$\searrow$
the standard interpretation	1st	1st and 2nd	2nd

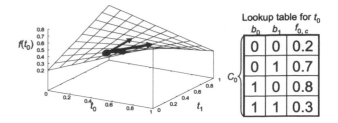

Figure 1: The example of the interpolated fitness of the trait t_0 for $N=2$ and $K=1$.

realistic situation (Suzuki and Arita, 2003). The transitions of the phenotypic plasticity and the phenotypic variation revealed that the evolutionary scenario consists of three steps unlike the standard interpretation of the Baldwin effect.

The next approach, discussed in this paper, is to clarify the dynamic changes in roles of learning through the course of evolution by paying attention to the effects of epistasis and to genetic robustness against mutations. For this purpose, we have constructed an evolutionary model based on Kauffman's NK fitness landscape (Kauffman, 1993) in which we can explicitly adjust the degree of epistasis. We discuss the evolution of quantitative traits by extending the fitness evaluation of the NK model. We introduced the phenotypic plasticity into our model, in which whether each phenotype is plastic or not is genetically defined and the plastic phenotype can be adjusted by a simple learning process. By conducting experiments with various degree of epistasis, we show that the drastic changes in roles of learning cause the three-step evolution of the Baldwin effect and then the evolution of the genetic robustness against mutations.

Three-step Evolution of the Baldwin Effect

We investigated the evolution of connection weights in a neural network as a situation where there are epistatic interactions among loci (Suzuki and Arita, 2003). It was observed that the evolutionary scenario consists of three steps by focusing on the transitions of four indices as shown in Table 1. The *lifetime fitness* represents the actual fitness after learning in the population and the *innate fitness* is the potential fitness before learning based on initial phenotypes. The *phenotypic plasticity* represents the proportion of plastic phenotypes in the population. The *phenotypic variation* is the absolute difference in phenotypic values between before and after learning among plastic phenotypes.

The first step, that is the increase in both lifetime fitness and phenotypic plasticity, was simply caused by the benefit of learning. It is noteworthy that the second step has both properties of the first and second step in the standard interpretation of the Baldwin effect. The decrease in the phenotypic plasticity corresponds to the second step in the standard interpretation of the Baldwin effect in the sense the

increased fitness by learning becomes dependent on fewer plastic phenotypes. At the same time, the increase in phenotypic variation means that the population becomes strongly dependent on the remaining plastic phenotypes. Thus, we can also say that the population was still in the first step in this point of view. This phenomenon is supposed to be due to the implicit cost of learning caused by the epistatic interactions among plastic phenotypes through the learning processes. The third step corresponds to the second step in the standard interpretation because the genetic assimilation occurred in the remaining plastic phenotypes.

Model
NK Landscape with Real Valued Traits

We have constructed an evolutionary model based on Kauffman's NK fitness landscape (Kauffman, 1993), so as to discuss the evolution of phenotypic plasticity in quantitative traits with / without epistatic interactions among loci. There are P individuals in a population and each individual has N traits of which initial phenotypes are determined by genes in a N-length chromosome GI. Each gene represents the quantitative trait t_i ($i=0, \cdots, N-1$) which consists of a real value within the range $[0.0, 1.0]$. We adopt NK fitness landscapes for evaluation of fitness because we can explicitly adjust the degree of epistasis by using the parameter K. It represents the number of other traits that affect the fitness contribution of each trait. However, the standard NK fitness landscape only assumes the binary traits ("0" or "1"). Then we extended the definition of the fitness evaluation so as to deal with the fitness contributions of quantitative traits.

Each trait t_i has epistatic interactions among other K traits $t_{i+j \bmod N}$ ($j=1, \cdots, K$). For each t_i, we prepare a lookup table which defines its fitness corresponding to all possible (2^{K+1}) combinations of interacting traits when these phenotypes consist of only binary values ("0" or "1"). The value of each fitness in the lookup table is randomly set within the range $[0.0, 1.0]$. These tables are similar to those of the standard NK landscape.

The fitness for quantitative trait is defined as the linearly interpolated value among the fitness for binary combinations of interacting phenotypes using the following equation:

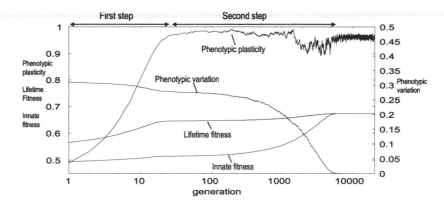

Figure 2: Evolutionary dynamics of fitness and phenotypic plasticity for K=0.

$$f(t_i) = \sum_{c \in C_i} [f_{i,c} \cdot \prod_{j=i}^{i+K} \{(1.0 - b_{j \bmod N}) $$
$$\cdot (1.0 - t_{j \bmod N}) + b_{j \bmod N} \cdot t_{j \bmod N}\}], \qquad (1)$$

where $f(t_i)$ is the fitness of the trait t_i, C_i denotes the all possible 2^{K+1} combinations of binary traits, $f_{i,c}$ is the fitness of t_i when the combination of traits is c. b_j represents the j th binary phenotype in c. Figure 1 shows an example of the interpolation of the fitness of quantitative traits for $N = 2$ and $K = 1$. The table on the right side represents the lookup table which determines the fitness of t_0 corresponding to four binary combinations of t_0 and t_1. The left figure shows the interpolated fitness of t_0 generated by the right table and the equation (1). The individual fitness is regarded as the average fitness over all traits. Note that if we assume only binary phenotypic values, this model is equivalent to the standard NK fitness landscape.

Learning

Each agent has another N-length chromosome GP which decides whether the corresponding phenotype of GI is plastic ("1") or not ("0"). Each trait whose corresponding bit in GP equals to "1" is adjusted by repeating the following procedure L times. First, for each plastic trait t_i, we calculate the difference in t_i between time t and $t+1$ (Δt_i) using the following equation:

$$\Delta t_i = \begin{cases} -\beta(F_0 - F_c) & \text{if } max(F_0, Fc, F_1) = F_0, \\ \beta(F_1 - F_c) & \text{if } max(F_0, Fc, F_1) = F_1, \\ 0 & \text{otherwise,} \end{cases} \qquad (2)$$

where F_c represents the individual fitness of the current combinations of traits and F_0 is the individual fitness when t_i is set to 0, F_1 is the individual fitness when t_i is set to 1. Next, we actually adjust all values of the plastic traits by adding

Δt_i at the same time. This process means that the individual gradually adjust its own plastic phenotypes toward fitter extreme phenotypic value ("0.0" or "1.0") in proportion to the increase in the fitness. The gray arrows in Figure 1 show examples of Δt_0 and Δt_1. The black arrow corresponds to the resultant direction and distance of learning process.

Evolution

After all individuals have finished their learning processes, the population in the next generation is generated by a simple genetic operation as follows: First, the worst individual's chromosomes (GI and GP) are replaced by copies of the best individual's. Then, every gene for all individuals is mutated with a probability p_m. A mutation in GW adds a randomly generated value within the range $[-d, d]$ to the current value and a mutation in GP flips the current binary value. If a mutated phenotypic value in GI exceeds the domain of the phenotypic space, another mutation is operated on the original value again. We adopted these procedures so as to observe the gradual transitions of four indices explained previously.

Experiments

Experiments without Epistasis (K=0)

We have conducted evolutionary experiments using the following parameters: P=20, N=15, K=0 or 4, L=5, β=10.0, p_m=0.003 and d=0.03. The initial population was generated on condition that initial values in GI were taken at random within the range $[0, 1]$ and the proportion of "1" in GP for each individual was uniformly distributed also within the range $[0, 1]$.

First, we have conducted the experiments without epistatic interactions among loci. Figure 2 shows the course of evolution over 20000 generations with K=0. The results shown are averages over 10 trials. The horizontal axis represents the generation in logarithmic scale. The lines represent the four indices that we have explained above. Specifically, the *lifetime fitness* denotes the average actual fitness

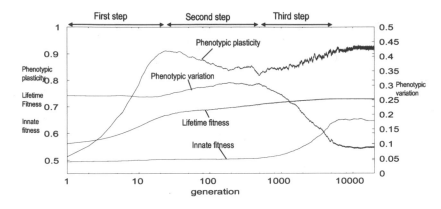

Figure 3: Evolutionary dynamics of fitness and phenotypic plasticity for $K=4$.

among all individuals calculated after the learning process, and the *innate fitness* is the average potential fitness calculated before the learning process using initial phenotypic values. The *phenotypic plasticity* represents the average proportion of "1" in all GPs and the *phenotypic variation* is the average absolute difference between the initial value and the resultant value adjusted by the learning process among all plastic phenotypes.

As shown in the transitions of these indices, the evolutionary process basically consists of the standard two-step evolution of the Baldwin effect. From the initial population, we observe an increase in both lifetime fitness and phenotypic plasticity while the innate fitness remained steady. The phenotypic plasticity rapidly rose and exceeded 0.97 at around the 28th generation. This means that more plastic individuals could obtain higher fitness and could occupy the population due to the benefit of learning.

Next, the innate fitness slowly increased and the phenotypic variation gradually decreased until around the 6000th generation. We can regard that the genetic assimilation occurred on the learned traits because the initial phenotypic values were getting closer to resultant phenotypic values after learning. The main reason for this phenomenon is due to the limitation in the number of iterated learning processes (L). Besides, as the innate fitness got closer to the lifetime fitness, the phenotypic plasticity was slightly decreased until around the 2300th generation. It is due to the genetic drift because there is no explicit cost of learning when $K = 0$.

However, in contrast with the evolutionary scenario in Table 1, the phenotypic plasticity increased again and kept high even after the genetic assimilation had completely finished, despite the fact that the learning did not increased the fitness of the population at all. Thus, another role of learning must occur after the Baldwin effect. This will be discussed later in detail.

Experiments with Epistasis ($K=4$)

Figure 3 shows the course of evolution for $K=4$. From the initial population, we observe approximately the same transitions as those for $K=0$ during the first step, but the peak value of the phenotypic plasticity, 0.91 (at around the 24th generation), was relatively smaller than that for $K=0$.

However, a clearly different scenario caused by epistatic interactions among loci was observed further on. While the lifetime fitness still slowly increased, the phenotypic plasticity gradually decreased to about 0.82 and then the phenotypic variation increased until around the 500th generation. This phenomenon corresponds to the second step in the three-step evolution of the Baldwin effect, in which the benefit and cost of learning worked together as previously described. The cost of learning is considered to bring about the decrease in the phenotypic plasticity. A contribution of each phenotypic value to the individual's fitness strongly depends on the other phenotypic values when there are epistatic interactions. Similarly, the learning in a plastic phenotype also affects the learning processes of the other plastic phenotypes. However, when we calculate Δt_i for each plastic trait t_i respectively, we do not consider any changes in the other plastic traits. Thus, the learning in too many plastic phenotypes does not always yield an effective increase in the whole fitness. This is the implicit cost of learning caused by the epistatic interactions among loci. In fact, when we maximized the plasticity of individuals in this step, their lifetime fitness tended to be smaller than that of the original individuals. At the same time, the benefit of learning is reflected in the steady transition of the innate fitness and increase in the phenotypic variation, because these transitions mean that the lifetime fitness increased by learning was getting more strongly dependent on the remaining plastic phenotypes.

Finally, the innate fitness eventually began to increase, however in contrast, the phenotypic variation decreased. Thus, the genetic assimilation occurred in the remaining plastic phenotypes because these initial phenotypic val-

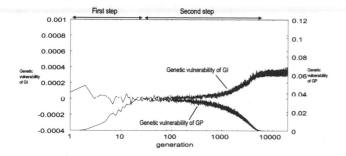

Figure 4: The evolution of the genetic vulnerability for *K*=0.

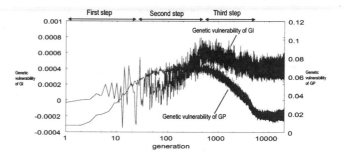

Figure 5: The evolution of the genetic vulnerability for *K*=4.

ues were getting closer to resultant phenotypic values after learning. In comparison with *K*=0, the innate fitness converged to around 0.64 and the genetic assimilation did not occurred completely. This step approximately corresponds to the third step in our three-step evolution of the Baldwin effect, except that the phenotypic plasticity gradually increased again to high values as observed for *K*=0.

Evolution of Genetic Robustness

As discussed in the previous section, we found that the three-step evolution of the Baldwin effect emerges when there are epistatic interactions among loci. However, it is still open to question why the phenotypic plasticity increased again through the last step in both cases of *K*=0 and *K*=4. Here, we focus on another different role of learning, that is, the genetic robustness against mutations (DeVisser, 2003).

Instead of measuring the genetic robustness, we measured the genetic vulnerability in view of its adaptive property based on the following procedures: First, for every individual in each generation, we generated a copy of the individual. Then, we conducted the mutational operations on its randomly selected genes in *GI* (or *GP*) for 5 times. We defined the genetic vulnerability of *GI* (or *GP*) as the average difference between the average lifetime fitness in the original population and in the mutated individuals. Thus, the genetic vulnerability becomes smaller when the genetic robustness gets larger.

Figure 4 and 5 shows the evolution of the genetic vulnerability of *GI* and *GP* in the same experiments as Figure 2 and Figure 3 respectively. As a whole, genetic vulnerabilities of *GI* and *GP* for *K*=4 are larger than those for *K*=0. We see that the genetic vulnerability of *GP* increased and subsequently decreased in both cases. It should be noticed that its peak exists between the second and third step when *K*=4. This implies that the population became strongly dependant on learning despite the decrease in the phenotypic plasticity in the second step.

When *K*=0, the genetic vulnerability of *GI* increased and converged to around 0.0003 along with the increase in the innate fitness. This is because that as the initial phenotypic

values get closer to the optimal values, mutations tend to become detrimental for maintaining adaptive traits. However, owing to the recovering adaptive phenotypes by learning with the large phenotypic plasticity, the increase in the genetic vulnerability kept relatively small in comparison with a non-plastic population. Actually, when we eliminated the plasticity of all traits of all individuals in the last generation, the genetic vulnerability of *GI* became much larger than that of the original population.

When *K*=4, we observe a peak of the genetic vulnerability of *GI* between the second and third step of the Baldwin effect. The increase in the second step implies that the initial values of phenotypes became more important factors for the learning processes in the other plastic phenotypes due to the epistatic effects than the previous step. Its gradual decrease through the third step was accompanied by the increase in the phenotypic plasticity which was restrained by the implicit cost of learning in the previous step. Thus, the increase in phenotypic plasticity was caused by the selective pressure for the evolution of the genetic robustness against mutations on *GI*. The reason why the implicit cost of learning vanished through the third step is due to the fact that epistatic effects among plastic phenotypes got smaller as the initial phenotypes approached to the learned phenotypes through the genetic assimilation in this step.

Conclusion

In the literature of Darwinian evolution, effects of non-genetic factors on genetic evolution (such as Waddington's genetic assimilation or the Baldwin effect) had not been treated as important mechanisms of possible evolutionary change for a long time (Weber and Depew, 2003), while these effects have been investigated based on theoretical or constructive approaches in the field of artificial life or complex systems for more than a decade. However, recent progresses in the molecular and developmental biology have experimentally demonstrated that these mechanisms actually exist and play important roles for genetic evolution in many aspects. They have also revealed that the complex genetic regulations are fundamental basis for developmen-

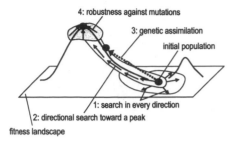

Figure 6: The roles of learning on fitness landscape.

tal process or plasticity of traits. Thus, it is now essential to investigate into evolutionary models based on theoretical or constructive approaches with epistatic effects in conjunction with experimental biology in order to understand these mechanisms in real environments.

As the new first approach, we have discussed the quantitative evolution of phenotypic plasticity based on an extended version of the NK fitness landscape. By conducting the evolutionary experiments with various degree of epistasis, we have found that a three-step evolution of the Baldwin effect emerges when the degree of epistasis is relatively large. It also turned out that the phenotypic plasticity brings about the genetic robustness against mutations.

In conclusion, what needs to be emphasized is that the drastic changes in roles of learning emerged through the course of evolution in order, and then each role was the main selective pressure that guided the complex evolution of phenotypic plasticity. Here, we conceptualize this phenomenon by using a hill-climbing image of a population on a fitness landscape. Figure 6 shows an example of a fitness landscape which consists of all possible phenotypic configurations. Let us assume that the initial population existed on the black filled circle on the right hand. The gray region around it represents the potential area where the current population can reach through learning.

Our experiments suggest that the role of learning changes as follows: 1) The learning in many phenotypes allows the population to search adaptive phenotypes in every direction on the phenotypic space owing to the benefit of learning. 2) The learning in less phenotypes enables the population to get to more adaptive phenotypic configurations by transforming the shape of the potential area due to the implicit cost of learning. This corresponds to the decrease in the phenotypic plasticity and increase in the phenotypic variation. 3) If the potential area reaches a maximum phenotypic configuration, the learning guides the genetic configuration to approach the maximum because of the cost of learning resulted from the limit of the learning ability. This phenomenon corresponds to genetic assimilation. 4) When the genetic configuration completely reaches the maximum, the learning in every direction prevents mutations from dropping down the popula-

tion from the optimum. This state continues until the population looses its stability due to some kind of internal or external factors.

In this scenario, the Baldwin effect corresponds to 1)-2), and 2) is clearly observed when there exists implicit cost of learning such as epistatic interactions. We believe that these detailed investigations into the dynamic evolution of the benefit and cost of learning can help further understanding of the phenotypic plasticity in real biological systems.

References

Arita, T. and Suzuki, R. (2000). Interactions between learning and evolution: The outstanding strategy generated by the baldwin effect. In Bedau, M., McCaskill, J., Packard, N., and Rasmussen, S., editors, *Artificial Life VII*, pages 196–205. Cambridge, MA: MIT Press.

Baldwin, J. M. (1896). A new factor in evolution. *American Naturalist*, 30:441–451.

Bull, L. (1999). On the baldwin effect. *Artificial Life*, 5(3):465–480.

DeVisser, J. A. G. M. e. a. (2003). Perspective: Evolution and detection of genetic robustness. *Evolution*, 57(9):1959–1972.

Downing, K. L. (2004). Development and the baldwin effect. *Artificial Life*, 10(1):39–63.

Hinton, G. E. and Nowlan, S. J. (1987). How learning can guide evolution. *Complex Systems*, 1:495–502.

Kauffman, S. (1993). *The Origins of Order: Self-Organization and Selection in Evolution*. New York: Oxford University Press.

Mayley, G. (1996). Landscapes, learning costs and genetic assimilation. *Evolutionary Computation*, 4(3):213–234.

Suzuki, R. and Arita, T. (2003). The baldwin effect revisited: Three steps characterized by the quantitative evolution of phenotypic plasticity. In Banzhaf, W., Christaller, T., Dittrich, P., Kim, J. T., and Ziegler, J., editors, *Seventh European Conference on Artificial Life*, pages 395–404. Springer.

Turney, P., Whitley, D., and Anderson, R. W. (1996). Evolution, learning, and instinct: 100 years of the baldwin effect. *Evolutionary Computation*, 4(3):4–8.

Waddington, C. H. (1942). Canalization of development and the inheritance of acquired characters. *Nature*, 150:563–565.

Weber, B. H. and Depew, D. J., editors (2003). *Evolution and Learning - The Baldwin Effect Reconsidered -*. Cambridge, MA: MIT Press.

Niche Construction and the Evolution of Complexity

Tim Taylor

Institute of Perception, Action and Behaviour, University of Edinburgh
JCMB, The King's Buildings, Mayfield Road, Edinburgh EH9 3JZ, U.K.
tim.taylor@ed.ac.uk

Abstract

An individual-based model of the process of niche construction is presented, whereby organisms disturb the environment experienced by their neighbours. This disturbance in local conditions creates a niche that potentially could be filled by another species (which would then create still more niches and so on). The model is unique in allowing the complexity of the organisms—measured by the number of genes they possess in order to be well adapted to their local environment—to evolve over time, and is therefore the first model with which it is possible to study the contribution of niche construction to the evolution of organism complexity. Results of experiments demonstrate that the process of niche construction does indeed introduce an active drive for organisms with more genes. This is the first explicit example of a model which possesses an intrinsic drive for the evolution of complexity.

Introduction

For more than half a century, engineers, computer scientists and biologists have tried to recreate the dynamics of biological evolution in synthetic systems, ranging from computer simulations to evolving RNA molecules *in vitro*. It is a surprising fact that none of these attempts has succeeded in producing ongoing evolutionary dynamics in which innovations continue to arise, or where the complexity of the system, given any suitable measure, can be said to increase in an unbounded fashion.

The artificial life platform Avida has recently been used to study the evolution of complex features (Lenski et al., 2003), but the observed increase in the complexity of organisms arose only because a series of nine progressively more complex reward functions was added to the environment. After the organisms had evolved solutions to these functions, the system reached a fairly stable end point.

Studies of evolution *in vitro*, such as Spiegelman's and Orgel's experiments with evolving RNA sequences using a viral enzyme (Orgel, 1979), have also demonstrated the need for a better theoretical understanding of these issues. Reflecting on Orgel's results, Maynard Smith comments: "In this simple and well-defined system, natural selection does not lead to continuing change, still less to anything that could be recognized as an increase in complexity: it leads to a stable and rather simple end point. This raises the following simple, and I think unanswered, question: *What features must be present in a system if it is to lead to indefinitely continuing evolutionary change?*" (Maynard Smith, 1988, p.221, emphasis added)

The theoretical biology literature provides some suitable starting points for answering Maynard Smith's question. Waddington recognised the need to develop a framework that described the logical structure of open-ended evolution, and published a suggestion of what such a framework might look like (Waddington, 1969). He went as far as to call this characterisation a new paradigm under which biological evolution should be studied.

The overall scenario is summarised as follows: "The complete paradigm must therefore include the following items. A genetic system whose items (Qs) are not mere information, but are algorithms or programs which produce phenotypes (Q^*s). There must be a mechanism for producing an indefinite variety of new Q'^*s, some of which must act in a radical way which can be described as 'rewriting the program.' There must also be an indefinite number of environments, and this is assured by the fact that the evolving phenotypes are components of environments for their own or other species. Further, some at least of the species in the evolving biosystem must have means of dispersal, passive or active, which will bring them into contact with the new environments (under these circumstances, other species may have the new environments brought to them). These environments will not only exert selective pressure on the phenotypes, but will also act as items in programs, modifying the epigenetic processes with which the Qs become worked out into [Q^*s]." (Waddington, 1969, p.120)[1]

This demonstrates that Waddington was fundamentally interested in situations where organisms constitute part of the environment experienced by other organisms. This leads to *niche construction*, whereby the presence of one species may introduce new niches in which other species can flour-

---

[1] The final symbol in the original text is printed as $Q's$, but this appears to be a typographical error.

ish (which in turn may introduce yet more niches and so on). Waddington predicted that such a system would produce a continually increasing number of species, and continually increasing phenotype complexity (*ibid.* p.119).[2]

Little work has been devoted to exploring this proposal, presumably because of the difficulties in capturing it fully with an analytical model. One of the motivations behind the work described here was to build a individual-based model that complied with Waddington's paradigm. This could then be used to test his predictions about the evolution of complexity and diversity.

In addition to the work with Avida already mentioned, a small number of other relevant models have recently appeared in the literature. Walker described a model specifically based upon Waddington's ideas of niche selection (but not niche *construction*) (Walker, 1999). However, in this case the potential niches were predefined and were *not* dependent on the presence of particular species (i.e. inter-organism interactions were not modelled). This was a serious flaw, and the results showed that it generally did not lead to the evolution of complex organisms.

Odling-Smee and colleagues have studied the evolutionary consequences of niche construction using more traditional population genetic models (Laland et al., 1999; Odling-Smee et al., 2003). They have shown that niche construction is a potent evolutionary force and can lead to unusual dynamics. In the current context, though, a disadvantage of such models is that they are only capable of describing the change in frequencies of a fixed number of genes over time. They are unable to model the introduction of new genes during the course of evolution, and are therefore not ideal for studying questions relating to possible increases in community diversity and organism complexity.

A different approach has been taken by Christensen and colleagues, who present a model in which individuals are described as vectors in a genotype space, and interactions between species are described by a pre-generated interaction matrix (Christensen et al., 2002). Their model produced dynamics in which relatively stable communities of species (which they refer to as quasievolutionary stable strategies, or q-ESS) exist for extended periods, separated by short transition periods of hectic reorganization. Furthermore, the average duration of the q-ESS periods increased slowly over time, in agreement to analysis of the fossil record. Despite the many attractive features of this model, it does not include a concept of organism complexity, and therefore cannot be used to address questions regarding how this evolves.

Finally, Pachepsky and colleagues described a spatially-explicit, individual-based model in which mutualistic relationships could evolve between organisms (Pachepsky et al.,

2002). They found that the possibility of such relationships lead to increased community diversity and stability. In agreement to the results of (Christensen et al., 2002), the composition of the communities tended to show periods of stability separated by short transition periods leading to new relatively stable states. However, the model was again not designed to address the evolution of organism complexity. It was also limited by the fact that mutualistic relationships could only arise between pairs of organisms rather than more extended ecological webs.

In the following sections, a new model is described which has been specifically designed to study the possible consequences of niche construction for the evolution of community diversity and organism complexity. The model overcomes many of the shortcomings of previous work in addressing these questions.

Description of the Model[3]

Before presenting a formal description of the model, it is worth giving a general overview of its design. It is an individual-based model in which the local environment experienced by an organism is expressed as an arbitrary vector of real numbers. An organism's behaviour, and the degree to which it is of adaptive value in the local environment, is abstractly modelled by requiring the organism to define a mathematical function which should closely match this "environment vector".[4] Reproductive success is proportional to the closeness of match over a defined subset of the vector. The crucial part of the model is that the presence of an organism causes a perturbation in the environment vector experienced by all organisms in the local neighbourhood, thereby producing an environmental niche that potentially could be filled by a different species. As organisms can evolve arbitrarily complicated functions in order to match their environment, the model can therefore be used to study the role of niche construction in the evolution of organism complexity.

The formal definition of the model is as follows. Space is represented as a discrete two dimensional grid with wrapped boundaries in both dimensions (so the topology is toroidal). Each grid position, or *patch*, is denoted P_{ij}, where i and j are its spatial coordinates. A patch can contain zero or more *organisms*, up to a number limited by the local density parameter T_M. Each patch provides a local environment, which is experienced by all organisms located within it. This is represented as a vector $\mathbf{E}_{ij} = (E_{ij}^1, E_{ij}^2, \ldots, E_{ij}^{L^E})$, where E_{ij}^k is a real number in the range $[0, 1]$. These elements E_{ij}^k can be interpreted as observables or attributes of the physical environment (e.g. temperature, humidity, levels of a nutrient, etc.). The local environment $\mathbf{E}_{ij}$ can be influenced by the

---

[2]The sentence about "rewriting the program" shows that Waddington was also concerned with the problem of how fundamentally novel phenotypic traits can arise during evolution, which is another topic of major importance to artificial life research.

[3]The source code of the model is available at http://homepages.inf.ed.ac.uk/timt/papers/ncec/

[4]This method of representing an environment, and the way in which organisms generate functions to match the environment, were inspired by and adapted from (Rocha, 2001).

presence of organisms in a local neighbourhood of patches P_{xy}, where $x \in \{i - R_N, i + R_N\}, y \in \{j - R_N, j + R_N\}$, and R_N is the neighbourhood radius (a parameter of the model). The details of an organism's influence on the environment are described later. In the absence of any organisms in the local neighbourhood, $\mathbf{E}_{ij} = \mathbf{E}^A$, which represents the virgin abiotic environment.

An organism O_n is defined by the triplet $(S_n, \mathbf{V}_n, D_n)$ which can be interpreted as, respectively, its sensitivity to particular environmental attributes, its preferred operating values for those attributes, and the manner in which it alters the environment through its behaviour. A schematic diagram of the structure of an organism is shown in Figure 1, and of a local collection of patches in Figure 2.

S_n is a pair (B_n, L_n) with $1 \leq (B_n, L_n) \leq L^E$. This specifies that the organism is sensitive to all environmental attributes in the range $[E^{B_n}, E^{B_n + L_N})$ (indices on $\mathbf{E}$ wrap around such that $E^{L^E + 1} = E^1$ and so on). This information is used in the calculation of an organism's fitness in a given environment, to be described later.

$\mathbf{V}_n$ (the second element of O_n) is a vector of *genes*, $\mathbf{V}_n = (G_n^1, G_n^2, \ldots, G_n^{C_n})$, of variable length $|\mathbf{V}_n| = C_n \geq 1$. The general idea is that the elements of $\mathbf{V}_n$ represent transformation operations that can be applied sequentially to a default "preferences vector" $\mathbf{P}^D$ (of length L^E equal to that of an environment vector $\mathbf{E}$) in order to generate a new vector $\mathbf{P}_n$ representing the organism's preferred operating values for each of the attributes of the environment. The larger an organism's vector of genes $\mathbf{V}_n$, the more operations are applied to $\mathbf{P}^D$, and hence the more complicated the form of $\mathbf{P}_n$ can become. The definition of a gene, and the way in which a vector of genes $\mathbf{V}_n$ is used to generate a vector $\mathbf{P}_n$ which can then be compared to the local environment vector $\mathbf{E}_{ij}$, follows the method described in (Rocha, 2001). For reasons of space, only a general outline of the method will be described here; for a full description, the reader is referred to Rocha's paper. A gene G_x is defined by the quintuple $(F_x, p, s, r, \odot_x)$. F_x is an element of $\mathcal{F}$, a small set of fuzzy set shapes each defined over a generic interval $[0, L^F]$. Similarly, $\odot_x$ is an element of O, a small set of fuzzy set operations. The action of a gene G_x on a given preferences vector $\mathbf{P}_y$ is to apply the fuzzy set shape F_x to $\mathbf{P}_y$ using the fuzzy set operation $\odot_x$. The elements p, s and r of G_x provide further details of the action, such as specifying the subset of elements of $\mathbf{P}_y$ on which F_x is to be applied. See (Rocha, 2001) for details. For a given set of genes, the final form of $\mathbf{P}_n$ is sensitive to the order in which the genes are applied, so even if we consider only organisms with a given, small number of genes, a large variety of different $\mathbf{P}_n$ is possible.

D_n (the final element of O_n) is a quadruple of the form (F_n, p, s, r), the elements of which have the same meaning as the corresponding elements of a gene. D_n determines the way organism O_n disturbs the local environment vector $\mathbf{E}_{ij}$. Specifically, the environment vector at a given patch P_{ij}

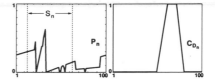

Figure 1: Structure of an organism. S_n specifies the environment attributes to which the organism is sensitive, and $\mathbf{P}_n$ specifies its preferred values for those attributes. $\mathbf{C}_{D_n}$ specifies the way in which the organism affects the local environment experienced by itself and other organisms.

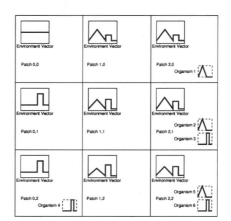

Figure 2: A local collection of patches. In this example the local neighbourhood radius R_N is 1, and all unshown patches neighbouring this local collection are assumed to contain no organisms. For each organism present, its environmental disturbance pattern $\mathbf{C}_{D_n}$ is shown. Patch 0,0 displays the virgin abiotic environment because there are no organisms in its neighbourhood. The environment in all other patches has been modified by the presence of organisms.

at time t, $\mathbf{E}_{ij}^t = \frac{1}{|\mathcal{N}_{ij}| + 1}\left(\mathbf{E}^A + \sum_{n \in \mathcal{N}_{ij}} \mathbf{C}_{D_n}\right)$, where $\mathcal{N}_{ij}$ is the set of organisms currently residing in all patches within the local neighbourhood radius R_N, and $\mathbf{C}_{D_n}$ is the fuzzy set shape defined by D_n, given by F_n appropriately modified by p, s and r. In experiments where the model flag ENV_DIST_RESTRICTED=true, s is clamped at zero for all organisms, which limits the number of elements of $\mathbf{C}_{D_n}$ which can take on nonzero values—that is, an organism can only disturb a limited subset of environmental attributes. This definition of $\mathbf{E}_{ij}^t$ was adopted because it is additive with respect to the members of $\mathcal{N}_{ij}$, so a given collection of organisms will always produce the same disturbance regardless of the order in which their effects are considered.

The fitness $f_n(i, j)$ of an organism O_n in patch P_{ij} is defined as $f_n(i, j) = 1 - \frac{1}{L_n}\sum_{x=B_n}^{x=B_n+L_n-1}|\mathbf{E}_{ij}(x) - \mathbf{P}_n(x)|$ (where B_n and L_n are the components of S_n, as described earlier). The probability p_n^R of organism O_n reproducing at a given iteration of the simulation is given by $p_n^R = c^R f_n(i, j)$, where c^R is a parameter of the model. Each organism also has a probability of death, p_n^D, associated with it. In some experiments this is kept constant, $p_n^D = c^D$, where c^D is a model parameter, and in others it is set to $p_n^D = c^D(1 - f_n(i,j))$.

Time in the model proceeds in discrete iterations. The top level algorithm is shown in Figure 3. The procedure `initialise` sets the environment vector of each patch to be $\mathbf{E}^A$, randomly generates a number of organisms (determined by the initial local density parameter T_0), each of size $|\mathbf{V}_n| = 1$, and places them in randomly chosen patches. `updateEnvironment` updates $\mathbf{E}_{ij}$ for patch P_{ij} according to the organisms currently present within the local neighbourhood. `updateOrganisms` recalculates the fitness f_n and reproduction probability p_n^R (and, if appropriate, the probability of death p_n^D) of each organism within a patch, according to the current form of $\mathbf{E}_{ij}$. `killOrganismsStochastic` considers each organism O_n within a patch in turn, and kills it with probability p_n^D. Similarly, `reproduceOrganisms` makes a copy of each organism with probability p_n^R and places it in a randomly chosen patch within a dispersal radius R_D. When an organism reproduces, mutations may be introduced into the offspring: a randomly generated new gene is inserted into $\mathbf{V}_n$ with probability p_+^M; a randomly selected gene in $\mathbf{V}_n$ is deleted with probability p_-^M (provided $|\mathbf{V}_n| > 1$); for each gene in $\mathbf{V}_n$, with probability $p_g^M/|\mathbf{V}_n|$, a randomly chosen element of the gene is mutated; similarly, with probability p_d^M, a randomly chosen element of D_n is mutated; and, with probability p_s^M, the value of the elements of S_n are independently altered by either -1, 0 or $+1$. Finally, `checkOvercrowding` considers, for each patch, the mean number of organisms present in each patch within the local neighbourhood. If this exceeds the maximum local density threshold T_M, more organisms are killed from patches in the local neighbourhood (stochastically, according to their p_n^D values) until the density returns below T_M.

Experiments

Five sets of experiments were conducted in order to answer the question: *Does niche construction introduce an intrinsic drive for the evolution of organism complexity?* All experiments used the standard model configuration unless otherwise stated.[5] Five experiments were conducted in each set, which had identical configurations apart from the seed supplied to the random number generator at run initialisation.

Set I was the standard trial set, using the default configuration unaltered. Set II was the main control set, in which ENV_DIST_ENABLED=false (i.e. the presence of organisms did not affect a patch's environment vector, so no niche

[5]This was: ENV_SIZE=50x50, NUM_ITERATIONS=5000, $R_N = 1$, $R_D = 2$, $T_0 = 0.1$, $T_M = 10$, $L^E = 100$. $\mathbf{E}^A = \mathbf{P}^D = (0.5, 0.5, \ldots, 0.5)$, $p_+^M = p_-^M = p_g^M = p_d^M = p_s^M = 0.05$, $c^R = 0.5$, $c^D = 0.1$, $L_n = 20$, DEATH_PROB_POLICY=constant, ENV_DIST_ENABLED=true, ENV_DIST_RESTRICTED=true.

```
initialise

iteration = 0

while (iteration < NUM_ITERATIONS)
{
  if (ENV_DISTURBANCE_ENABLED)
    for each patch P; do
      updateEnvironment(P)
    end
  endif

  for each patch P; do
    updateOrganisms(P)
    killOrganismsStochastic(P)
    reproduceOrganisms(P)
  end

  for each patch P; do
    checkOvercrowding(P)
  end

  iteration = iteration + 1
}
```

Figure 3: The top level algorithm

construction could occur). Set III was similar to Set II, but the abiotic environment $\mathbf{E}^A$ was more complex (see inset of Figure 4(f)) than the homogeneous version used in other experiments. This was designed to see to what extent any difference in dynamics between Sets I and II was due to the more complex environments present in Set I *per se*, rather than to the continuous process of niche construction. In Set IV, ENV_DIST_RESTRICTED=false, so organisms could affect an unlimited number of environmental attributes. Sustained evolutionary trends in this case were expected to be less likely, as ecologies were always open to the threat of a new species being introduced that had a devastating affect on the environment. Finally, in Set V, an organism's probability of death, $p_n^D = c^D(1 - f_n(i,j))$ [with $c^D = 0.5$], rather than being constant. Thus fitter organisms should tend to live longer as well as reproduce more frequently, effectively increasing the selection pressure for well-adapted organisms.

Results and Analysis

In all sets, the results obtained from each experiment of the set were qualitatively similar. In the following discussion, typical results from each set are highlighted.

The main results from Set I are shown in Figure 4(a)-(d). Figure 4(a) shows that the mean number of genes $|\mathbf{V}_n|$ of organisms in the population steadily grew over time, starting from 1 and reaching approximately 14–15 genes after 5000 iterations. The mean age of organisms in these runs (not shown) stayed fairly steady at about 3 iterations per organism, so 5000 iterations represents approximately 1667 generations. The overall population size (not shown) settled to a level of around 21500 individuals within the first 100 iterations, and stayed at that level for the rest of the run. The diversity of different species, where a species is

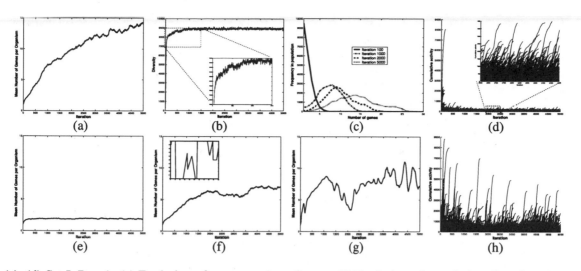

Figure 4: (a)–(d) Set I, Run 1: (a) Evolution of mean number of genes (b) Evolution of population diversity (c) Evolution of gene number frequencies (d) Species activity waves, (e) Set II, Run 1: Evolution of mean number of genes (f) Set III, Run 1: Evolution of mean number of genes (inset: $\mathbf{E}^A$ used for Set III), (g)–(h) Set V, Run 4: (g) Evolution of mean number of genes (h) Species activity waves

defined as a collection of individuals that share exactly the same $(S_n, \mathbf{V}_n, D_n)$, also settled to a fairly constant value (of around 9000 species), although it only settled at this value after around 1000 iterations (Figure 4(b)). The inset of this figure shows that even after the population level had settled down after 100 iterations, the diversity of the system gradually increased for the next 900 iterations. However, the size of this increase was not large. Furthermore, if we look at the diversity of *significant* species (not shown), arbitrarily defined at a given iteration as those species that had 10 or more individuals present in the population, then the reverse trend is seen—the diversity of these species is around 300 at iteration 1, and gradually falls to around 200 by iteration 1000, after which time it remains fairly constant.

Figure 4(c) shows the distribution of genome sizes ($|\mathbf{V}_n|$) in the population at various points during the run. This demonstrates that the move to longer genomes is an *active* trend (McShea, 1994)—that is, it is not an artifact due to the presence of a lower limit of genome size, but indeed the whole population is moving towards higher genome sizes. Figure 4(d) shows an activity wave diagram of significant species throughout the run. Each line represents a single species, and is defined by the cumulative count of individuals of the species over time (Bedau and Brown, 1999). The inset of this figure demonstrates that many significant species coexist at any given time, and multiple species often tend to die out at around the same time (shown by the cessation of multiple activity waves)—although there is always a base level of active species that prevents the population from collapsing when such extinctions occur.

The change in genome size over time for a run from the control group in Set II is shown in Figure 4(e). In this case,

the mean size remains at around 1.8 genes for the duration of the run. In addition, the frequency distribution of genome sizes (not shown) does not change over time, but remains in a state similar to the plot for iteration 100 in Figure 4(c) for the duration of the run. These results demonstrate that there is no drive, either active or passive, for increased genome size in the situation where organisms do not cause disturbances in their local environment. This indicates that the active trend observed in Set I was indeed caused by niche construction.

In Set III, organisms again did not cause environmental disturbances, but the abiotic environment $\mathbf{E}^A$ was more complex. In this situation, an increase in genome size was observed over time (Figure 4(f)), but this tended to tail off after reaching a level of around 7 genes. This shows that complex environments introduce selection pressure for complex organisms (i.e. with a large number of genes). However, in this case, the environmental complexity has been provided extrinsically—this is similar to the situation in (Lenski et al., 2003). In contrast, in Set I, we started with a simple environment, but the system displayed intrinsic dynamics for generating complex organisms (and environments).

The Set IV experiments involved a situation where organisms could again cause disturbances to the environment, but this time the breadth of the disturbance was unlimited. As expected, this tended to slow down the pace of evolution; the mean number of genes (not shown) did rise over the course of the runs, but more slowly than in Set I. In some runs it increased steadily, reaching a value of approximately 8 genes after 5000 iterations, whereas in other runs it seemed to peak at around 7 genes after around 3000 iterations and remain around that level of the rest of the run.

Perhaps the most surprising results, and the most variation between runs in a set, came from Set V, which had a similar configuration to Set I except that probability of death of an organism p_n^D was not constant, but rather inversely proportional to fitness. In these runs, the mean number of genes oscillated wildly throughout the run. An example is shown in Figure 4(g). The population diversity (not shown) was also somewhat lower then in Set I, in the region 6000–7000, and this also oscillated more than was observed in the other sets. However, the diversity of *significant* species was higher than in Set I, in the range 300–350 throughout the course of the run. Species activity waves were much higher than in other sets (Figure 4(h)). These results for diversity and activity waves are in line with the expectation that the new method of calculating probability of death would increase selection pressure—so the better adapted species tended to last for longer. The behaviour of the mean genome size still remains a little puzzling. One possibility is that because the population tends to comprise a higher proportion of highly adapted organisms that have been around for longer (which is confirmed by higher mean ages and mean probabilities of reproduction in this set), when one of these species finally does become extinct, a larger number of other species are likely to have depended upon it, and so the greater the impact on the stability of the ecosystem. Further experiments are underway to investigate this issue.

Discussion

Theoretical biologists have described situations in which there may exist drives for ongoing evolution rather than stasis, e.g. (Dawkins and Krebs, 1979; Van Valen, 1973), but the model presented here is the first to explicitly demonstrate a system with an intrinsic drive for the evolution of complexity. Organism-induced perturbations in the environment can be either beneficial or harmful to other organisms in the model. It therefore also provides a very general representation of symbiotic relationships.

It should be emphasized that, when an organism reproduces, the probability of mutation for each gene in $\mathbf{V}_n$ is defined as $p_g^M/|\mathbf{V}_n|$. So the probability of a genome being copied with no mutations is roughly $(1 - p_g^M)$, regardless of length. If the mutation rate per gene was not normalized by $|\mathbf{V}_n|$, the model would soon reach the error threshold beyond which it is impossible to reliably transmit genomes to the next generation (Eigen and Schuster, 1977)—and this has indeed been verified experimentally. That is, the organisms are able to continually increase in complexity only because we are assuming they have found a way of reliably replicating long genomes, thereby avoiding the error threshold.

In terms of complying with Waddington's paradigm (see Introduction), the model still falls short. It successfully captures the concept of niche construction, but misses features such as the ability to fundamentally rewrite the genotype–phenotype mapping, and modelling epigenetic processes.

However, the results have demonstrated that niche construction by itself is sufficient to introduce an intrinsic drive for the evolution of organism complexity.

The question of how long this trend might continue in the model remains unanswered—in a couple of runs in Set I there were signs that the increase in genome length was starting to tail off towards the end of the run. Further work will investigate this, as well as factors that could prevent (or at least delay) any tailing off (such as allowing organisms to evolve different degrees of sensitivity to environmental attributes (i.e. allowing L_n in S_n to evolve), or even allowing the length of $\mathbf{E}_{ij}$ to grow to simulate the evolution of fundamentally new types of behaviour). More work is also required to analyse the occupancy of different niches in the model, their stability, duration and evolution.

Finally, the environment described in the model is very abstract. A challenge for artificial life researchers is to understand how to build real (or artificial) environments which possess the same capacity for niche construction. The main challenge here is to build systems in which organisms constitute significant components of the environment experienced by other organisms. Some suggestions for progress in this direction are offered in (Taylor, 2004).

References

Bedau, M. A. and Brown, C. T. (1999). Visualizing evolutionary activity of genotypes. *Artificial Life*, 5(1):17–35.

Christensen, K., di Collobiano, S., Hall, M., and Jensen, H. (2002). Tangled nature: A model of evolutionary ecology. *Journal of Theoretical Biology*, 216:73–84.

Dawkins, R. and Krebs, J. (1979). Arms races between and within species. *Proceedings of the Royal Society London*, B 205:489–511.

Eigen, M. and Schuster, P. (1977). The hypercycle: A principle of natural self-organization. *Die Naturwissenschaften*, 64(11):541–565.

Laland, K., Odling-Smee, F., and Feldman, M. (1999). Evolutionary consequences of niche construction and their implications for ecology. *Proceedings of the National Academy of Sciences USA*, 96:10242–10247.

Lenski, R. E., Ofria, C., Pennock, R. T., and Adami, C. (2003). The evolutionary origin of complex features. *Nature*, 423:139–144.

Maynard Smith, J. (1988). Evolutionary progress and levels of selection. In Nitecki, M., editor, *Evolutionary Progress*, pages 219–230. Univ. Chicago Press.

McShea, D. W. (1994). Mechanisms of large-scale evolutionary trends. *Evolution*, 48(6):1747–1763.

Odling-Smee, F., Laland, K., and Feldman, M. (2003). *Niche Construction: The Neglected Process in Evolution*. Princeton University Press.

Orgel, L. (1979). Selection in vitro. *Proceedings of the Royal Society, London*, B, 205:435–442.

Pachepsky, E., Taylor, T., and Jones, S. (2002). Mutualism promotes diversity and stability in a simple artificial ecosystem. *Artificial Life*, 8(1):5–24.

Rocha, L. (2001). Evolution with material symbol systems. *Biosystems*, 60:95–121.

Taylor, T. (2004). Redrawing the boundary between organism and environment. In *Proceedings of Artificial Life IX*. MIT Press.

Van Valen, L. (1973). A new evolutionary law. *Evolutionary Theory*, 1:1–30.

Waddington, C. (1969). Paradigm for an evolutionary process. In Waddington, C., editor, *Towards a Theoretical Biology*, volume 2, pages 106–128. Edinburgh University Press.

Walker, R. (1999). Niche selection and the evolution of complex behavior in a changing environment—a simulation. *Artificial Life*, 5(3):271–289.

A Model for Exploring Genetic Control of Artificial Amoebae

Barry Drennan[1] and Randall D. Beer[1,2]

[1]Department of Electrical Engineering and Computer Science
[2]Department of Biology
Case Western Reserve University
btd@po.cwru.edu, rxb9@po.cwru.edu

Abstract

We develop a computer simulation of several cellular processes seen in amoebae, including the production and regulation of proteins via a genome; the production, release, and destruction of diffusible chemicals; and regulated chemotaxis through a lattice environment facilitated by the interactions of proteins and diffusible chemicals. We also test this model by adapting biological situations to this model to evaluate its ability to model genetic networks and genetically regulated chemotaxis. The model will be used to simulate evolution in artificial amoebae to produce behavior seen in biological organisms such as *Dictyostelium discoideum*.

Introduction

The amoeba *Dictyostelium discoideum*, or soil-dwelling slime mold, is of particular interest to developmental biologists due to the unusual cooperative behavior it demonstrates. When its supply of food runs low, individual *Dictyostelium* amoebae collaborate in order to move to a new location where more food may be found. This process involves intercellular signaling (Söderbom and Loomis 1998) and cell differentiation (Brown and Firtel 1999), which are critical tools in the development of multicellular organisms.

Under normal circumstances, *Dictyostelium* exists as numerous individual amoebae. When food becomes scarce, these cells begin emitting pulses of a chemical signal, cyclic AMP (cAMP). Neighboring cells are attracted to this signal, and they exhibit chemotaxis along the upward gradient of cAMP. The cells thus aggregate into a small multicellular "slug" which subsequently propagates as one unit for some distance. During this time, the cells differentiate, forming two types of cells: a large number which will eventually become spores; and a small number that instead form a stalk to better disperse the spores, and then die.

Of interest to us is a model which can simulate these and other similar behaviors in the context of a genetic and protein framework. Savill and Hogeweg (1997) developed a model of *Dictyostelium* which exhibited its life cycle behaviors of aggregation, motility, and stalk formation. However, this model based its behavior on fixed numeric constants controlling various factors such as cell adhesivity and the release of cAMP, rather than placing them under genetic control.

Dellaert and Beer (1994) developed a model for evolving morphology of an artificial multicellular organism on a grid, using Boolean networks for genetic regulation. Goel and Thompson devised models of bacteriophage self-assembly and operation (1988a), and protein synthesis (1988b) based on movable finite automata. The E-Cell project (Takahashi *et al.* 2003, Tomita *et al.* 1999) provides an open framework for modeling of cellular and biochemical processes, in an effort to create a highly detailed simulation of an entire cell.

A combination of some of these concepts is used here, with enhancements made to biological realism by modeling genetic regulation and protein production. In addition, various behaviors, such as chemical production, chemoattraction, and membrane channels are placed under the domain of these proteins rather than governing those behaviors with predefined constants. Our goal is to develop a model that can simulate arbitrary cell behavior, on the level of cell motility and intercellular signaling, and that, by providing a genome as an interface to these behaviors, permits the artificial evolution of these behaviors. We will ultimately use the behaviors observed in *Dictyostelium* as test cases for examining the evolutionary trajectory by which these behaviors are generated in the model.

Methods

The model developed herein combines simulations of the basic biological principles of cells, genes, and proteins (Kimball 2003) acting within a latticed environment with a genetic algorithm (GA) to evolve behavior of these cells. The model can also be used separately from the GA to test manually crafted genomes or to examine the behavior of previously evolved genomes.

Cells

The representation of a cell in this model is inspired by amoebae and other single-cell organisms which do not have a nominally-fixed shape. In other words, there are no structural features providing rigidity to the cell. Instead, a cell is free to expand, move, and contract throughout the environment, obtaining any contiguous shape at any time through means of minimizing its calculated free energy.

A simulated cell occupies any number of lattice points, or cellular automata (von Neumann 1966), in the three-dimensional cellular automaton (CA) grid. During each time step, a number of automata are selected for evaluation, to determine whether the occupation of that automaton by a new cell will create a lower free energy condition (Glazier and Graner 1992, 1993; Savill and Hogeweg 1997; Marée and Hogeweg 1999, 2002). For example, a cell has a nominal volume and surface area, and free energy is minimized by occupying or releasing automata to maintain that perimeter. Another factor affecting free energy is the satisfaction of membrane-bound receptors which are attractive or repulsive to other membrane-bound receptors or to diffusible chemicals.

Contrary to eukaryotes such as amoebae, however, the cells represented here do not have discrete organelles. Genetic material, proteins, and diffusible chemicals are considered to be available everywhere within the cell. While the inclusion of intracellular transport mechanisms may prove important for the evolution of more complex organisms, this is beyond the scope of this project.

Cells can also produce, destroy, absorb, or release any of a number of chemicals which diffuse throughout the environment. These diffusible agents can thus function as chemical messengers between cells.

Genome

In biological cells, DNA is the encoding chemical of genetic information, consisting of the four bases adenine, guanine, cytosine, and thymine (A, G, C, and T). Each base is paired with a corresponding base (A-T, C-G). In this simulation, however, the gene-to-protein translation process does not correspond to the biological process. We therefore use different "names" for our four bases - W, X, Y, and Z.

Genetic strings in DNA are represented by the two matching halves of the DNA: the sense and antisense strands. The presence of two different coding strands could be important to the process of evolution; however, implementing both strands only serves to place constraints on the evolution process, in the event that both the sense and antisense strands had overlapping portions of genetic material which were translated into proteins (mutation of one base might thus alter two proteins). Thus, only one strand is implemented in this system.

Eukaryotes (organisms with discrete organelles) often have sections of noncoding DNA, called introns, interspersed with the sections of DNA which actually code for proteins. Prokaryotes, however, do not make use of introns, and as the process by which introns are removed during transcription is unclear, introns are not implemented in this system. In other words, every gene is contiguous on the genetic strand.

Transcription and Translation

Transcription is the process whereby DNA is used as a blueprint for forming messenger RNA (mRNA), which is then transported to ribosomes for translation into proteins. The mRNA strand is the base-pair complement of the DNA strand. As intracellular transport mechanisms are not modeled in this system, the process of transcription retains only one significant feature of biological transcription: the regulatory effect of promoters and operons (we use the scheme seen in prokaryotes). Promoters indicate the binding site of RNA polymerase, an enzyme which binds to DNA and begins the transcription process. Frequently, the activity of RNA polymerase is enhanced by the binding of a catabolite activator protein (CAP) to a region nearby on the DNA strand. Operons are regions of DNA located just downstream of the promoter which permit the binding of proteins to inhibit transcription of a particular gene. Through this process, the inhibitive property of operons inhibits any enhancements provided by CAP. The model developed here includes a region of noncoding DNA which indicates the start of transcription for each gene, and may also include regions which allow proteins to bind to the DNA for enhancing or inhibitive effects.

In the model, mRNA is not actually generated, and the effect of producing a protein through transcription and translation occurs all at once. That is, the intermediate stages of gene-to-protein translation are not tabulated, but rather, once a gene is selected for representation, the protein is subsequently produced.

Translation produces proteins from mRNA blueprints based on an encoding scheme which divides the mRNA strand into pieces three base-pairs long, called codons. Each codon encodes a separate amino acid using tRNA, which includes an antisense codon at one end and the corresponding amino acid at the other. The process for production of tRNA is genetically encoded in an organism, but in this model, sufficient tRNA is assumed to be present in this model without explicitly encoding a synthesis mechanism. In the model developed here, translation occurs virtually simultaneously with transcription; once a gene is selected for transcription, a protein is translated from the gene and is introduced into the cell.

One difference introduced here is that one strand of mRNA is capable of being a translation instruction for only one protein. However, mRNA manipulation strategies add yet another layer of complexity, and are thus beyond the scope of this model.

Amino Acids

Each codon represents one amino acid for a protein encoded by a gene. In biological organisms, there are 20 common amino acids translated from mRNA. As the chemical basis for protein behavior is not modeled accurately, there is no need to be faithful to the number of amino acids present in real organisms, and so we use 16. The amino acids, numbered 0-9 and A-F, have functions determined in part by their locations within the protein.

The representation indicates with what amino acid or DNA sequences a protein can bind, and by what diffusible agents a protein is affected. In addition, one amino acid shares a duty indicating the start of a protein for transcription/translation purposes, and another flags the end of a "field" in the protein's functional representation.

Proteins

Proteins in this model have their function determined through the interpretation of certain fields (domains) in the protein. The lead domain of a protein indicates its function according to the encoding shown in Table 2; the lead domain will also always start with the amino acid 6 as this amino acid encodes the start of a protein transcription sequence. An amino acid 9 indicates the stop of the lead domain (and the stop of every subsequent domain), while the amino acid following 9 indicates the function of the next domain.

This representation of proteins is not chemically accurate, by any means. However, one might find some small justification for such a method by considering the existence of domains within real proteins. Each domain frequently has a separate function from the other domains of the protein, whether it is to embed a protein in the membrane of a cell, or to bind to another protein or other chemical, etc. Thus, the portions of proteins modeled here are referred to as domains as well. Additionally, the head end of some mRNA contains a short sequence which indicates the destination of the final protein; other small pieces of mRNA known as signal recognition particles detect these sequences and help to direct the protein to the transport mechanism which will bring it to its final destination where it can fill its particular designed role.

Proteins can serve one of several different duties in the model. They can serve as enzymes for the production or destruction of a diffusible agent; channels for the passage of a diffusible agent through the cell membrane; chemoreceptors for adjusting the free energy calculations used in determining chemotaxis; or ligands for binding to other proteins or to DNA in order to change that protein's function or inhibit that DNA's transcription (see Table 1). Thse functions are determined by the lead amino acid of the active domain of a protein, and the active domain is selected by the binding of a diffusible agent or another protein to a selector domain of the protein.

In real biological systems, protein folding occurs as a protein is synthesized. Either on its own or with the guidance of certain enzymes, a protein assumes a folded state where amino acids within the protein form hydrogen bonds with other amino acids in that same protein. The net result is a more compact protein which has a much lower free energy than the newly-synthesized, unfolded protein. The surface of the protein determines its function through its folded geometry; the folded geometry is determined by the sequence of proteins (which was determined by genetic information in DNA).

Protein folding is unfortunately not modeled here. While the process of protein folding largely determines the function of a protein, and provides a mechanism for the introduction of mutations which do not immediately affect protein function, and thus permits the evolution of accumulated genetic traits, that process is also NP-complete, even in a 3-d lattice model (Berger and Leighton 1998). Instead, we model protein function solely by the amino acid sequence, as described above.

While all of these mechanisms together represent a fraction of the processes seen in biological cells, they represent a "cut" through those processes to provide the most basic interactions needed for the behaviors to be studied (chemotaxis, multicellular interaction, or survival).

Genetic Algorithm

The model also includes a facility for evolving genomes via genetic algorithm. Supported are one method of crossover (two-point) and four mutation methods (single base mutation, block insertion, block deletion, and block copy). Fitness functions can be defined to describe an arbitrary desired behavior. This code is also being modified to parallelize evaluations among several computers.

Results and Discussion

The success of a model depends upon its ability to simulate the modeled phenomenon. In order to verify this model's fidelity, we have devised a number of test situations, two of which are described here.

Three-state Genetic Oscillator / "Repressilator"

A genetic oscillator can be formed by linking the repressors of three genes in a loop (Elowitz and Leibler 2000), as shown in Figure 1. The proteins generated by

Amino acid	Domain function	Codons	DNA Bind
0	None	wxw wxx wxy	none
1	Unused	wyw wyx wyy wyz	w
2	Chem Producer	wzw wzx wzy wzz	x
3	Activator	xww xwx xwy xwz wxz	wx
4	Chem Channel	xxw xxx xxy xxz	y
5	Actin	xyw xyx xyy xyz wwy	wy
6	Start Protein	www wwx	xy
7	Chemical Pump	xzw xzx xzy xzz	z
8	Chem Selector	yww ywx ywy ywz	z
9	End Domain	zzw zzx	none
A	Chem Attractor	yxw yxx yxy yxz wwz	xz
B	DNA Operon	yyw yyx yyy yyz	y
C	Chem Repeller	yzw yzx yzy yzz zyz	yz
D	Protein Attractor	zww zwx zwy zwz	x
E	Protein Selector	zxw zxx zxy zxz	w
F	Protein Repeller	zyw zyx zyy	wz
--	End Protein	zzy zzz	----

Table 1. Shown are the model's 16 amino acids, their functions when interpreted as domain function markers, the codons which represent them, and the DNA bases to which they can bind.

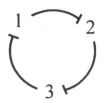

Figure 1. This diagram shows the general layout for the three-state repressilator. The links represent inhibition of the production of one protein by the previous protein in the sequence.

each gene serve as the repressors for the next gene in the sequence. As the concentration of that protein increases, its inhibitory effect increases as well, shutting down the next gene in the series. As the concentration of that protein decreases due to the inactivation of its gene, its inhibitory effect on the next gene decreases, and transcription of that gene begins anew. The result is a three-state oscillation of the concentrations of the three proteins where each state oscillates at a 120-degree phase difference from the other two. The behavior of the three-state "repressilator" is well-described, and forms a useful basis case for evaluating the fidelity of this model.

We recreate the repressilator within the model using the artificial genome shown in Figure 2. Each gene produces a protein which binds to the operon site of the next gene, thus preventing it from being transcribed. For example, the sequence in the first gene, "xxx yww wzw wzw xxx yww," creates a protein with amino acid sequence "482248," which can bind to the operon "yzxxyz" in the second gene.

The cell model, including the genome and the protein production facilities, is evaluated without the environment to obtain the effects of this genome. Figure 3a, a graph of the protein concentrations within the cell over time, indicates that the three protein concentrations in the model do in fact oscillate at about a 120-degree phase difference.

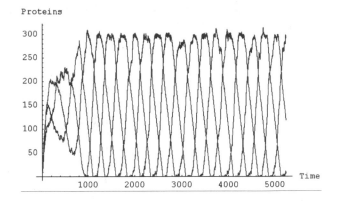

Figure 3. Shown are the model's resulting concentrations over time of the three proteins involved in generating the three-state "repressilator."

Genetically Regulated Chemotaxis

In order to exercise the environmental simulation and evaluate the ability of cells to signal each other, we devise a test setup where one cell is alternately attracted to two chemical signals produced by two other cells. In order to maintain state information regarding whether the active cell is attracted to the first or second chemical (C1 or C2, respectively), a bistable genetic switch is implemented. Such a switch has been implemented in living *E. coli* by Gardner *et al.* (2000). The network used here is an extension of that concept. A schematic of the genome implementing the active cell is shown in Figure 4.

The bistable genetic switch works through mutual inhibition of two genes. Under normal circumstances, the system falls toward a stable state where only one of these two genes is transcribed, since the resulting proteins inhibit transcription of the other gene. This can be counteracted by applying an external influence to overcome the inhibition and cause the other of the two genes to dominate.

In this system, the two mutually-inhibitive genes are *modeC1inh* and *modeC2inh*. They are named so because when the cell is in the mode where it is attracted to chemical C1, the gene *modeC1inh* is being transcribed. When a large amount of C1 or C2 enters the cell, the protein produced by *switchC2* or *switchC1*, respectively, is activated, enhancing the transcription of *modeC2inh* or *modeC1inh*. If the gene being enhanced is being inhibited by its counterpart currently, the enhancement causes transcription of the gene to occur anyway, which causes increasing inhibition of the counterpart gene. The continued enhancement leads the enhanced gene to become the dominating gene, and transcription of its counterpart falls to zero. In other words, the cell switches between modes "C1" and "C2".

	Operons	Function (DNA binder)		DNA binding sequence		
wzwz	xxyyzz	www	yyy	xxx yww wzw wzw xxx yww	zzx zzz
wzwz	yzxxyz	www	yyy	wzw xxx yww wzw xxx yww	zzw zzz
wzwz	xyzxyz	www	yyy	wzw wzw xxx xxx yww yww	zzx zzz

Figure 2. The genome for the three-state repressilator is shown here. Each gene is shown in its own row, spaced to show functional units such as codons and operons. All three genes carry one domain, marked by the "yyy" codon, which indicates that the domain binds to DNA when active. The remainder of the domain, shown in the three boxed sequences to the right, specifies the operon sequence to which the resulting protein can bind.

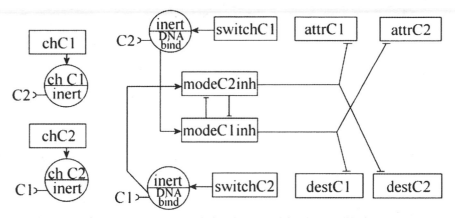

Figure 4. Shown is a diagram of the genetic network implemented in the oscillating attractor. Genes are shown in rectangles, and their proteins are denoted by circles. Above the bar in a circle is the primary function, and below is the secondary function when the protein is activated by the chemical noted adjacent to that region. At the left, the two genes *chC1* and *chC2* code for channel proteins that pass C1 and C2 through the cell membrane, but are inactivated in the presence of C2 and C1, respectively. In the center are the genes *modeC1inh* and *modeC2inh*. Their mutual inhibition permits only one of these proteins to be produced at one time. They also inhibit the production of proteins which metabolize the two chemicals inside the cell (*destC1* and *destC2*) and cause chemoattraction to chemicals outside the cell (*attrC1* and *attrC2*). The proteins generated by *switchC1* and *switchC2* cause a state switch to occur when the concentration of C2 or C1 increases, by enhancing the transcription of *modeC1inh* and *modeC2inh*.

In addition, the *modeC1inh* and *modeC2inh* genes affect the attraction of the cell to chemicals C1 and C2 outside the cell. They do this by inhibiting attractor genes. In mode C1, the cell inhibits the *attrC2* gene, and thus is not attracted to C2; likewise, the *attrC1* gene is inhibited in mode C2. These genes also inhibit metabolism genes (*destC1* and *destC2*) which eliminate the two chemicals C1 and C2. In mode C1, the metabolism of chemical C1 is inhibited, since the cell requires a large enough concentration of C1 in order to switch to mode C2; likewise, *destC2* is inhibited in mode C2.

Finally, the channels permitting C1 and C2 to flow through the cell membrane are directly influenced by the concentration of C1 and C2 inside the cell. A large concentration of C1 causes the C2 channels to close. This generally occurs in mode C1, where the cell is attracted to the C1 chemical. Once the cell switches modes – C1 also triggers the switch to mode C2 – the cell moves toward the chemical C2 and begins metabolizing C1. When the concentration of C1 drops, the C2 channels open and eventually cause the C1 channels to close.

The experimental setup also includes two other cells placed in the environment. Their genomes are much simpler – each of these passive cells produces one of the two chemicals, and has membrane channels to let that chemical pass out of the cell. They are otherwise not responsive to the environment.

Initial conditions place the active cell close to the passive cell producing C1 (Figure 5a, i.). At this point, the stable state which the cell chooses is still unaffected by the tiny concentrations of C1 and C2 in the cell; in this case, the cell randomly chooses mode C1, and is thus attracted to passive cell 1.

The active cell follows the gradient of C1 upwards, thus leading it to push passive cell 1 out of the way. In this way, it "follows" passive cell 1 around the environment (5a, ii.) until the concentration of C1 reaches a critical level (5b, iii.). Once this occurs, the cell switches modes (5c, iv.) and becomes attracted to C2. It then moves to passive cell 2 (5a, iv.) and pushes it around the environment until the mode switches back to C1. This process repeats indefinitely, though the frequency is in part determined by the amount of chemical in the environment.

Conclusions

The model developed here is able to reproduce some of the basic notions of engineered genetic networks, such as three-state oscillations and bistable switching. We plan next to apply a genetic algorithm to this model to evolve behaviors observed in *Dictyostelium*, such as aggregation, migration, and stalk formation, as well as the transitions between these behaviors. Ultimately, as computing power improves, this model could also be used to evolve open-ended simulations where survival is the only measure of fitness.

Acknowledgments

This work was made possible in part by the NSF IGERT program in Neuromechanics at Case Western Reserve University and NSF grant EIA-0130773.

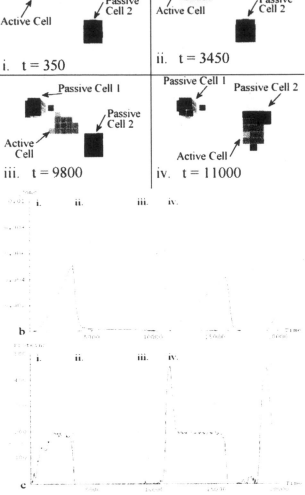

Figure 5. **a.** Four snapshots of the genetically-regulated chemotaxis test are shown here. **b.** Concentrations of C1 (solid) and C2 (gray). **c.** Amounts of proteins *modeC1inh* (solid) and *modeC2inh* (gray). In **b** and **c**, the dotted lines i-iv correspond to the times of the four snapshots in **a.**

References

Berger, B. and Leighton, T. (1998) Protein folding in the hydrophobic-hydrophilic (HP) model is NP-complete. Proceedings of the Second Annual International Conference on Computational Molecular Biology 30-39.

Brown, J. and Firtel, R. (1999) Regulation of cell-fate determination in *Dictyostelium*. Developmental Biology 216:426–441.

Dellaert, F. and Beer, R. (1994) Toward an evolvable model of development for autonomous agent synthesis. Proceedings of Artificial Life IV 246-257.

Elowitz, M.B. and Leibler, S.A. (2000) Synthetic gene oscillatory network of transcriptional regulators. Nature 403:335-338.

Gartner, T.S., Cantor, C.R., and Collins, J.J. (2000) Construction of a genetic toggle switch in Escherichia coli. Nature 403:339-342.

Glazier, J. and Graner, F. (1993) Simulation of the differential adhesion driven rearrangement of biological cells. Physical Review E 47(3):2128-2154.

Goel, N.S. and Thompson, R.L. (1988b) Movable Finite Automata models for biological systems II: Protein Biosynthesis. Journal of Theoretical Biology 134:9-49.

Graner, F., and Glazier, J. (1992) Simulation of biological cell sorting using a two-dimensional extended Potts model. Physical Review Letters 69(13):2013-2016.

Kimball, John W. (2003) Kimball's Biology Pages. http://biology-pages.info/

Marée, A.F.M., Panfilov, A.V., and Hogeweg, P. (1999) Migration and thermotaxis of *Dictyostelium discoideum* slugs, a model study. Journal of Theoretical Biology 199:297-309.

Marée, A.F.M. and Hogeweg, P. (2002) Modelling *Dictyostelium discoideum* morphogenesis: the culmination. Bulletin of Mathematical Biol. 64:327-353.

Savill, N. and Hogeweg, P. (1997) Modelling morphogenesis: from single cells to crawling slugs. Journal of Theoretical Biology 184:229-235.

Söderbom, F. and Loomis, W. (1998) Cell–cell signaling during *Dictyostelium* development. Trends in Microbiology 6:402-406.

Takahashi, K., Ishikawa, N., *et al.* (2003) E-Cell 2: Multi-platform E-Cell simulation system. Bioinformatics 19:1727-1729.

Thompson, R.L. and Goel, N.S. (1988a) Movable Finite Automata models for biological systems I: Bacteriophage assembly and operation. Journal of Theoretical Biology 131:351-385.

Tomita, M., Hashimoto, K., *et al.* (1999) E-Cell: software environment for whole cell simulation. Bioinformatics 15:72-84.

von Neumann, J., and Burks, A.W. (ed.) (1966) Theory of Self-Reproducing Automata. University of Illinois Press.

Asymmetric cell division and its integration with other developmental processes for artificial evolutionary systems

Peter Eggenberger Hotz

Artificial Intelligence Laboratory, Department of Information Technology, University of Zurich
Winterthurerstrasse 190, CH-8057, Zurich, Switzerland
Emergent Communication Mechanisms Project, Information Sciences Division, ATR International,2-2-2, Hikaridai
Seika-cho, Soraku-gun, Kyoto 619-0288, Japan
eggen@ifi.unizh.ch

Abstract

Artificial evolutionary techniques are more and more coupled with mechanisms abstracted from developmental biology. Artificial cells endowed with genetic regulatory networks were used to evolve and develop simulated creatures. This paper reports on the evolution of a simple moving creature using developmental mechanisms such as asymmetric cell division, genetic regulation and cell adhesion and physical interactions between cells. Surprisingly, artificial creatures were evolved able to move using only genetic regulatory networks without the need to employ neural controllers.

Introduction

The last years have seen an increased interest in combining evolutionary algorithm with concepts borrowed from developmental biology. As simple direct encoding schemes, where each primitive of the phenotype is represented by a single parameter (gene), no longer work for complex evolutionary tasks, new concepts have to be found to tackle such tasks successfully. The question arises on which level of abstraction one should start with developmental evolution. In our case the cellular level was chosen: the task to evolve cellular mechanisms is way to complex to be approachable with the current simulation technology. Higher level approaches have also their drawbacks, putting a simulator for the non-trivial task to choose among a plethora of possible structures and mechanisms. As the cell has only a limited set of behaviors (Wolpert, 1998), it is much easier to use those as a guideline to implement an interesting artificial evolutionary system. Still, processes and structures have to be chosen and implemented. The method of investigation to choose among the cellular mechanisms is to analyze the possible value of biological concepts for artificial evolution, to implement the chosen concepts, to investigate the possible advantages of a given concept in simulation and to understand why such a mechanism is useful. In this paper asymmetric cell division (ACD)was combined with other developmental mechanisms in an artificial evolutionary system and their use was investigated by evolving simple behaving creatures.

Although other authors (Kaneko, 1992; Kitano, 1994; Gruau and Whitley, 1993; Cangelosi et al., 1994; Fleis-cher and Barr, 1992; Vaario and Shimohara, 1997; Kodjabachian and Meyer, 1998) propose to combine developmental processes with evolutionary algorithm, I discuss here only those models that tried to evolve genetic regulatory networks to control the developmental processes. Dellaert and Beer (Dellaert, 1995) proposed a model based on Boolean networks to evolve autonomous agents. Two different models of neurogenesis were developed: one simple and one complex. With the complex model, the authors succeeded in designing an autonomous agent by hand-coding a genome. (Eggenberger, 1997) proposed the use of genetic regulatory networks coupled with developmental processes to use in the field of artificial evolution and was able to evolve simple shapes and simple neural networks. (Bongard, 2002) combined a physical simulator with a genetic regulatory network able to simulate simple organisms such a box pusher. (Eggenberger, 2003b) combined genetic regulatory networks with physical processes during morphogenesis and was able to evolve processes mimicking invagination. (Kumar and Bentley, 2003) also discuss asymmetric cell division.

Methods

Cell fates during development are controlled by two different mechanism: a) cytoplasmic determinants distributed asymmetrically to the cells by ACD and b) inductive signals released from neighboring cells. The exclusive use of either mechanism defines two extremes tactics: mosaic and regulative development. The former development is controlled entirely by cytoplasmic factors. Each cells undergoes autonomous differentiation, so that its fate is determined by lineage independent of its position. Regulative development is controlled entirely by inductive interactions. Each cell is specified conditionally, according to its interactions with other cells. Cell fate is determined by position, irrespective of lineage. Most species use a combination of both mechanisms in their developmental program. ACD allows to control genetically the cell division plane and therefore to control the positioning of the cells individually and precisely. If each cell division is controlled by one gene, this approach is quite similar to a direct encoding scheme, where a parameter specifies an entity of the phenotype. During cell division cytoplasmic factors are asymmetrically distributed coupling cell division with cell differentiation. This coupling allows artificial evolution to explore mechanisms, which are able to position differentiated cells at precise positions by controlling precisely the cell division. To be useful ACD has to be integrated with other developmental processes. That this is possible, simple moving creatures were evolved in which the shape and the physical interactions among cells are specified by the ACD, the movements were triggered by signalling molecules; all these mechanisms were controlled and integrated by ligand-receptor interactions and genetic regulatory networks.

Gene Regulation

In the current evolutionary simulator the genome consists of two sets of genes: the structural genes encoding cellular functions and the cis-regulators represent switches able to turn on or off a structural gene.

Two classes of structural genes were implemented: in one class artificial substances are produced simulating a simple artificial chemistry in the cells, in the other class a structural gene represents a cellular function such as cell division or cell adhesion. These functions are not evolved but implemented as programs which are called in case the corresponding structural gene was activated. The properties of each structural gene are encoded by seven parameters. The first parameter determines to which class the structural gene belongs and which function this gene will have if it is activated. It is up to the designer to decide how many members the classes of structural genes are made available to the artificial evolutionary system (AES). The second parameter is used to calculate the probability of an interaction with partner molecules. This affinity parameter aff determines which molecule (signalling molecule or axonal receptor) interacts with which partner (regulatory unit or receptor). To each molecule in the simulation a real valued number and a function is assigned, which calculates an artificial binding affinity between the molecules. The third parameter determines the sign of the effect e_{ij}, i.e. inhibitory or excitatory. The fourth parameter specifies the threshold ϑ. This parameter determines how high the sum of the products of all the affinities a_{ij} between the signalling molecules and the regulators times their concentrations has to be in order to turn on or off the associated structural gene. The fifth parameter designates the decay d_i rate for the product. The sixth parameter is used to store the affinity parameter α The seventh parameter is used as the diffusion parameter D_1

$$G_i(c_{sm_0}, .., c_{sm_m}) = \frac{1}{2}(1.0 + \tanh(\beta y(c_{sm_0}, .., c_{sm_m}))) \quad (1)$$

$$y(x) = 2.0x - 1.0 \quad (2)$$

$$x = \sum_{j=0} (\Theta(a_{ij} * c_{sm_j} - \vartheta_j)) \quad (3)$$

$$\Theta(x) = \left\{ \begin{array}{ll} 1.0 & : \quad if \ x > 0 \\ 0.0 & : \quad otherwise \end{array} \right. \quad (4)$$

Where:

- G_i is the activity of the i-th gene (see equation 1).

- $tanh(x)$ is the hyperbolic tangent.

- a_{ij} affinity to encode the effect between the regulatory unit i and the signalling molecule j (also referred to as transcription factors if they regulate the gene activity).

- $y(x)$ is just a scaling function allowing to have gene activities between 0.0 and 1.0 (see equation 2).

- c_{SM_j} concentration of the signalling molecules j

- ϑ_j is a threshold value controlling the influence of a signalling molecule on a gene (see equation 4).

- $\Theta(x)$ is a normalizing function to make sure that the impact on a structural gene is between 0 and 1.

- β is a parameter affecting the steepness of Θ

One or several regulatory units control a structural gene. Regulatory units are switches that control the activity of the structural gene. Active regulatory units influence the activity of the structural gene, but only an activated structural gene is able to a response such as cell migration or the production of a receptor. Two parameters are assigned to every regulatory unit: an affinity aff_{RU} and a threshold. The affinity aff_{RU} has the same use as the affinity parameter in a structural gene. A signalling molecule is defined by the parameters encoded in the structural gene. Both affinity parameters are used to calculate the probability for an interaction between a regulator and a signalling molecule. Both factors are variables of the affinity function $aff_{Tot} = f_{aff}(aff_T, aff_S)$ and its value will influence the probability of a gene's activation. The threshold defines the limit of the minimal impact able to activate a gene: the product of the affinity aff_{Tot} and the concentration of the signalling molecule has to exceed the threshold's value.

Whether a given gene at position (i, j, k) in a cell on the grid will be activated depends on the affinity and concentration of all the signalling molecules at that position. All these influences are summed up and if this sum exceeds the gene's threshold the gene will be activated or inhibited according to the sign of the effect. All these parameters are varied by the evolutionary strategy and used to explore the interaction space for useful developmental processes able to solve the designer defined tasks. The gene activity of the i-th gene depends on parameters of the structural gene and its regulatory units.

The activation of a gene leads to two types of responses: Either a simulated molecule is produced or a function (implemented as a procedure) is executed. The link between the activation of a gene and its response depends on the first of the seven parameters of a structural gene. The following responses were implemented for the experiments in this paper:

1. Production of chemical substances

   (a) a signalling molecule is produced to communicate between the cells.

   (b) a cell adhesion molecule (CAM) is produced to connect the current cell to another one.

   (c) receptors are produced for signalling molecules.

2. The activation of a gene calls a predefined function of the following types:

   (a) asymmetric cell division (ACD)

   (b) random cell division

Asymmetric Cell Division

Cell differentiation, the process which gives rise to cells with different subsets of active genes, is coupled with cell division. During development there are cells which divide asymmetrically by segregating protein determinants into only one of the daughter cells. ACD is interesting because it provides a mechanism for placing specific cells at precise positions in a developing organism (Horvitz and Herskowitz, 1992). Recently a general concept of how the cell division planes are oriented is emerging: In a first step, an axis of polarity is established in the mother cells and coordinated with the surrounding cells or even the body plan. Along this established axis cell determinants are localized asymmetrically and as the mitotic spindle is oriented along this axis, the cell determinants are also distributed asymmetrically to the two daughter cells(Knoblich, 2001).

There are two types of asymmetric cell division: One is intrinsic (see Figure 1,the other is extrinsic (see Figure 2). ACD allows to control the cell's behavior in detail. It corresponds in many aspects to a direct encoding scheme allowing to specify in detail the cells and it is possible to evolve recursive developmental schemes. Some authors in artificial evolution contrast recursive and developmental approaches, but in fact, developmental processes can also be recursive as shown in (Eggenberger, 2003a). A cell contains asymmetrically distributed cytoplasmic factors, which by the following cell division may be distributed in different amounts to the two daughter cells and therefore influence their fates in two different directions. In order to simulate ACD a new gene class, the asymmetric cell division gene class, was introduced. Two kinds regulate two angles specifying the cell division plane depending on their activities. By regulating these genes, for instance by diluting regulating factors inside the cell, continuous changes of positions can be implemented as illustrated in Figure 3. Additonally, the proposed

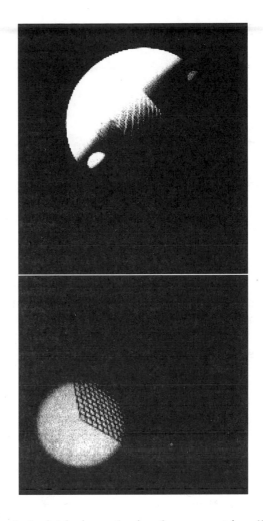

Figure 1: An intrinsic mechanism for asymmetric cell division. Intrinsic determinants are molecules that are expressed in the mother cell (on the left) and then forwarded during mitosis to the two daughter cells. In this example, two different determinants (green and red spheres) are forwarded to the two daughter cells, which will determine two different fates for the two cells.

artificial genome is able to control concurrently its own gene activities, ACD, cell adhesion and physics. This allows the AES to mimic morphogenetic processes in order to create artificial creatures. The idea to include physical processes such as diffusion and cellular interactions into the AES is to exploit these to use the genetic information parsimoniously. The genetic information of shaping or behavior can be reduced, because the intercellular communication allows to change not only single cells, but whole groups of them to perform a given function (reducing the number of parameters means in general a reduction of the search space and an increase in evolvability). If at a given position a morphogen diffuses, it can possibly change the cellular interactions, which will lead to different cellular behaviors and in

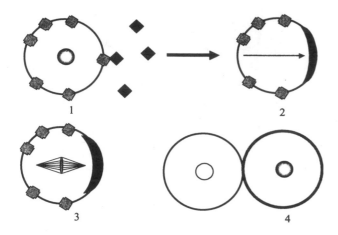

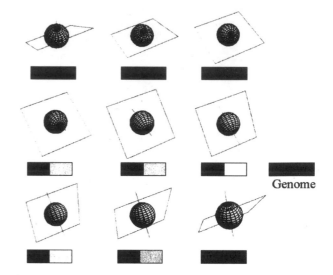

Figure 2: An extrinsic mechanism for asymmetric cell division. An external signal (red blocks) are binding to a membrane bound receptor (1). The excited ligand-receptor complex allows the binding of a protein to the membrane(2). This protein determines in turn the cell division axis (3) and the protein will be only distributed in one cell (4).

Figure 3: Cell Lineage. The activity of two genes control the cell division plane and will determine the future position of the two daughter cells. Illustrated is an example in which the left gene is inactive and the right gene has an increasing activity (symbolized by colors (blue = no activity, red = highest activity))and the effect of each gene activity pattern on the cell division plane is shown above each genome.

the end to different shapes. Another noteworthy property of the proposed approach is that it allows for continuously changing development. The artificial morphogenetic processes can be regulated in such a way that a cell performs a task quite similar, but slightly different than its neighboring cells.

In order to show the possible advantages of this concept, a simple simulator was written allowing to simulate a large number of cells interacting by viscoelastic elements consisting of passive springs and active dashpots. Each cell was connected to its six nearest neighbors. Movements of the cells, the production of pressure etc. will displace the cells in the direction of the applied force, but will also create a contracting force which restricts the movements. The link to the genome was established by assuming that the amount of produced cell adhesion molecules was proportional, for instance, to the spring constant allowing the AES to explore different mechanical interaction patterns between the cells. To each cells a set of differential equation is assigned, which are solved by an ordinary differential equation solver using a fourth-order Runge-Kutta method. Small changes in the genome result in small changes in the shape due to concurrent effects of the cells behavior and their physical properties. The reduction of parameters for the formation of certain shapes (obviously, for random shapes no reduction is possible) is due to the physics taking care of the positioning of cells, which react on tensions and pressures. Again, a physical process takes care of the shaping, but the genes can influence and change the outcome by changing the interaction forces between the cells (for more details see (Eggenberger, 2003b)).

Evolution Strategy

A (2,16)-ES was used to perform the evolutionary runs. Usually a moving solution was found after having tested 100-300 generations. The initial mutation rate Dn was set to 0.3.

Results

The result (illustrated with Figure (5)shows that it is possible to evolve moving creatures based on the above developmental mechanisms. ACD was the mechanism enabling to specify precisely the positions of the cells as well as their physical interactions between the cells where it was necessary. First, ACD determined a T-shaped creature for which a fitness function was defined. The length of the creature was determined by diluting a regulatory factor with each cell division until the impact of the factor on the cell division gene become insufficient and the cell division process in this axis stopped. The cells growing in the perpendicular direction to the first axis used a similar mechanism: A second factor controlled a second cell division gene determining a second cell division plane perpendicular to the first one; this factor was also diluted during cell division which resulted in a cease of cell growth along this second axis.

The final shape was also dependent on the cell adhesion molecules a cell expressed. If two cells had their cell adhesion molecule gene on, they produced a link between each other. The link was weaker or stronger depending on the ac-

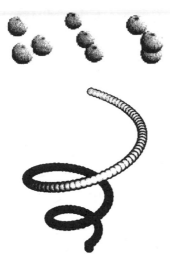

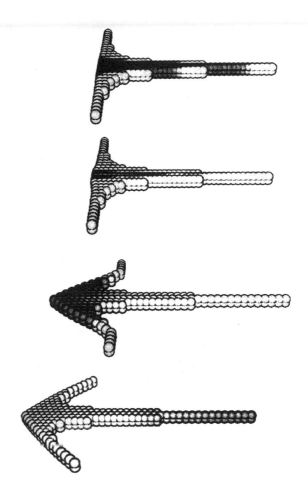

Figure 4: Simulation of asymmetric cell division. In order to simulate asymmetric cell division, a specific type of structural gene was defined representing two parameters determining the cell division plane (upper part of the Figure illustrating the effect of different planes on cell division). In the lower part of the figure an example is illustrated: By diluting regulatory factors controlling the activity of these special structural genes it is possible to change continuously the cell division plane of each cell division allowing to build spirals.

tivation state of the cell adhesion molecule gene. Another cell adhesion molecule was used to get friction on a simulated plane on which the creature was moving. By rhythmically changing the concentration of the adhesion between the cells and the surface the evolutionary system was able to produce a moving creature based on the genetic regulatory network without the need to use a neural controller. Note that ACD specified a T-structure, which due to the physical interactions changed to an arrow-like structure.

Figure 5: Dynamic shape change during movement. An artificial creature was first specified in a T-Shape, which during the movement of the creature is dynamically reshaped to an arrow-like structure. Note that surprisingly no neural network is needed to control the movement, because the genetic regulatory network is able to control the movements by rhythmically emitting signalling molecules, which change the adhesion properties between the cells as well as the adhesion between the cells and the ground!

Discussion

In order to control precisely the positioning of cells to build functioning structures, ACD implemented as proposed above showed very useful. This paper discusses the mechanisms of internal and external ACD, which allows to distribute artificial cell factors asymmetrically to the cells with concurrent control of the cell division plane allowing to get differentiated cells during the development of a cellular structure. The advantage is that where is no longer a need for symmetry breaking external sources to get the development started and that a evolutionary algorithm endowed with the mechanism of ACD is able to control single cells precisely, if this is needed. In this respect ACD is similar to direct encoding schemes, because factors can control p.e. the positioning of cells very precisely. This leads to the insight that a combination of direct and indirect encodings is probably the best way to go, each having its advantages for differ-

ent problems. As mosaic and regulative development are both used in the developmental program of most organisms, the combination of ACD with developmental processes allowing inductive signalling will expand the power of evolutionary algorithms. In addition, as ACD is also based on a genetic regulatory network, it is easy to combine ACD with other developmental processes such as cell adhesion as was shown with the example of dynamic shaping. Three points are noteworthy: First, movement does not need neural networks to control it, the genetic regulatory networks is able to control movement by rhythmically changing the adhesion to the environment. Second, the genetic specification of a shape may be altered during its use as in biology for instance with bones, where the shape of the bones depends heavily on their use. Third, artificial evolutionary systems endowed by developmental mechanisms controlled by ligand-receptor interactions and genetic regulatory networks are easy to extend and can solve different problems such as morphogenesis, neurogenesis or learning tasks for robots (see (Eggenberger et al., 2002; Ishiguro et al., 2003; Eggenberger, 2003a)).

Acknowledgement

This work was sponsored by the EU-Project HYDRA (IST-2001-33060).

References

Bongard, J. C. (2002). Evolving modular genetic regulatory networks. In *Proceedings of The IEEE 2002 Congress on Evolutionary Computation (CEC2002)*, pages 1872–1877.

Brooks, R. and Maes, P., editors (1994). *Artificial Life IV: Proceedings of the Workshop on Artificial Life*, Cambridge, MA. MIT Press. Workshop held at the MIT.

Cangelosi, A., Parisi, D., and Nolfi, S. (1994). Cell division and migration in a 'genotype' for neural networks. *Network*, 5:497–515.

Dellaert, F. (1995). *TOWARD A BIOLOGICALLY DEFENSIBLE MODEL OF DEVELOPMENT*. PhD thesis, Case Western Reserve University.

Eggenberger, P. (1997). Evolving morphologies of simulated 3d organisms based on differential gene expression. In (Husbands and Harvey, 1997).

Eggenberger, P. (2003a). Exploring regenerative mechanisms found in flatworms by artificial evolutionary techniques using genetic regulatory networks. In *To appear in : Proceedings of the Congress of Evolutionary Computation (CEC 2004), Canberra*.

Eggenberger, P. (2003b). Genome-physics interaction as a new concept to reduce the number of genetic parameters in artificial evolution. In *To appear in : Proceedings of the Congress of Evolutionary Computation (CEC 2004), Canberra*.

Eggenberger, P., Gomez, G., and Pfeifer, R. (2002). Evolving the morphology of a neural network for controlling a foveating retina - and its test on a real robot. In *Proceedings of The Eighth International Symposium on Artificial Life*, pages 243–251.

Fleischer, K. and Barr, A. H. (1992). A simulation testbed for the study of multicellular development: The multiple mechanisms of morphogenesis. In (Langton, 1994), pages 389–416.

Gerstner, W., Germond, A., Hasler, M., and (Eds.), J.-D. N., editors (1997). *Seventh International Conference of Artificial Neural Networks (ICANN'97)*, Cambridge, MA. Sprinter.

Gruau, F. and Whitley, D. (1993). The cellular developmental of neural networks: the interaction of learning and evolution. Technical Report 93-04, Laboratoire de l'Informatique du Parallélisme, Ecole Normale Supérieure de Lyon, France.

Horvitz, H. R. and Herskowitz, I. (1992). Mechanisms of asymmetric cell division: Two bs or not two bs, that is the question. *Cell*, 68:237–255.

Husbands, P. and Harvey, I., editors (1997). *Fourth European Conference of Artificial Life*, Cambridge, MA. MIT Press.

Ishiguro, A., Fujii, A., and Hotz, P. E. (2003). Neuromodulated control of bipedal locomotion using a polymorphic cpg circuit. *Adaptive Behavior*, 11(1):7–17.

Kaneko, K. (1992). Overview of coupled map lattices. *Chaos*, 2(3):279–282.

Kitano, H. (1994). Evolution of metabolism for morphogenesis. In Brooks, R. A. and Maes, P., editors, (Brooks and Maes, 1994), pages 49–58.

Knoblich, J. A. (2001). Asymmetric cell division during animal development. *Nature Reviews*, 2:11–20.

Kodjabachian, J. and Meyer, J.-A. (1998). Evolution and development of neural controllers for locomotion, gradient-following, and obstacle-avoidance in artificial insects. *IEEE Transactions on Neural Networks*, 9(5):796–812.

Kumar, S. and Bentley, P. J. (2003). *On Growth, Form and Computers*. Elsevier Academic Press.

Langton, C. G., editor (1994). *Artificial Life III: Proceedings of the Workshop on Artificial Life*, Reading, MA. Addison-Wesley. Workshop held June, 1992 in Santa Fe, New Mexico.

Vaario, J. and Shimohara, K. (1997). Synthesis of developmental and evolutionary modeling of adaptive autonomous agents. In (Gerstner et al., 1997), pages 721–726.

Wolpert, L. (1998). *Principles of Development*. Oxford University Press.

A Functional Model of Cell Genome

Alessandro Fontana[1] and Walter Steven Fraccaro[2]

[1]IEEE member, Milan (Italy)
[2]University of Milano-Bicocca, Milan (Italy)
alessandro.fontana@ieee.org

Abstract

This paper is concerned with a model of the cell genome called Artificial Genome, that tries to model some aspects of the cell cycle, in particular those related to gene expression, cell differentiation and cell growth. The functioning of the model during interphase and mitosis is explained in detail through an example, that shows how the four functional categories of the Artificial Genome (Functions, Code, Data and Buffer) interact to determine the phenotype. The capacity of the model of generating phenotypical patterns, represented as 2-dimensional shapes, is explored through a simulation, that evolves in 9 cycles a cell to become a small face made up of 132 cells. Finally some parlallels between the Artificial Genome and the natural one are discussed.

Introduction: the Genome

The discovery of the structure and replication mechanism of the DNA, performed by J. Watson and F. Crick in 1953, is undoubtedly one of the most important achievements of the 20th century. Another major milestone, achieved in the 21st century thanks to the simultaneous efforts of the international Human Genome Consortium and of the private company Celera Genomics, has been the sequencing of the human genome. Other species's genomes (including Drosophila Melanogaster, C. Elegans, mouse, etc.) are also currently available.

Having at disposal the whole base sequence, the biologists are now faced with the challenge of interpreting the meaning and functioning of the sequence, that is understanding how the code is utilized to produce what is called the "phenotype".

Some known facts about the genome follow. The genome is made up of chromosomes (23 pairs in human), that are very long molecules of deoxyribonucleic acid (DNA), composed of basic elements called Nucleotides, whose distinctive parts are four sub-elements called Bases: Adenine, Cytosine, Guanine, Thymine. Triplets of these bases (codons) are *transcribed* into RNA and then *translated* into amino-acids, the building-blocks of proteins, which in the end determine the cell's (chemical) behaviour and function.

Some *unknown* facts about the genome follow. One of the biggest surprises coming from the sequencing of the DNA of man and other species is that the genome, which can be very large (3 GB in human), seems almost empty, meaning that the overwhelming majority of it is made up of non-coding sequences (like for example "TTAGGG", which is repeated several times), never transcribed into RNA and never translated into proteins. The coding part represents only 2% of the whole base sequence; the remaining 98% of the sequence, that apparently has no function, has been labeled "junk DNA". However, recent experiments have given evidence that "junk DNA" may be involved in regulating the activity of the coding part.

According to the replication mechanism proposed by Watson and Crick, when a cell duplicates each of the two daughter cells inherits one of the two single-stranded helices of the mother cell's DNA, on which the DNA-polymerase enzyme rebuilds the complementary chain, thus yielding a perfect double-stranded copy of the parental DNA; as a result all the cells of an organism have the same DNA. A major question that arises is then how can different cells (e.g. a nervous cell and a heart cell) of the organism behave differently, to perform different functions. This is achieved by selectively activate in each cell only a subset of genes by means of several biochemical mechanisms (e.g. the genes that don't have to be activated are "silenced" through the methylation of the corresponding genome segments, which inhibits their transcription). However, the question how a cell knows which parts of the DNA have to be silenced and which not, is still unanswered.

A strictly related question is how can the zygote, through a series of duplications, grow into a fully developed organism (embryogenesis). Also this process seems to imply a series of selective activations and de-activations of different groups of genes, whose control mechanism is still largely unknown.

It is the aim of this paper to present a functional model of the cell genome that, without going into biochemical details, tries to model some key aspects of cell dynamics, like gene expression, differentiation and morphogenesis. The remainder of the paper is organized as follows. In the following three sections the model of the genome is introduced and its functioning during interphase and mitosis is explained in detail through an example. In the subsequent two sections a simulation is shown that demonstrates the potential of the model and some parallels between the artificial genome and the natural one are outlined. The final section draws the conclusions.

Artificial Genome

The information content of the Artificial Genome is divided in three categories:

- Functions
- Code
- Data

The Functions hold information embedded in intrinsic mechanisms, not expressed explicitly (that is as a sequence of quaternary digits 0–1–2–3 called bases) and cannot be modified during the organism's life: they can be compared to a computer hardware.

The Code holds information that is expressed explicitly but again is not (normally) modified during the organism's life: it can be compared to the computer software that contains instructions (code). Modifications occurring to the Code are called *Mutations*.

The Data contain information that is expressed explicitly and is normally modified during the organism's life: they can be compared to the computer software that contains data.

There is also a fourth category, called Buffer, which contains information that is used temporarily during the cell cycle and then discarded. The Buffer does not need to be inherited or duplicated and therefore is not part of the Artificial Genome.

We note that, according to the said definitions, the Artificial Genome is not the same for all the cells of the organism, that will have the same Functions and Code but, in general, different Data (and Buffer).

The overall information content of the Artificial Genome is made up of the following units:

Unit Name	Symbol	Category
Gene Activator	GAF	Funct.
Gene Controller	GCF	Funct.
Gene Transcriptor	GTF	Funct.
Cell Duplicator	CDF	Funct.
Gene Activation Code	GAC	Code
Gene Control Code	GCC	Code
Gene Transcription Code	GTC	Code
Duplication Plan Code	DPC	Code
Cell Type	CT	Data
Development Stage	DS	Data
Gene Number	GNB	Buffer
Exon Address	EXA	Buffer
RNA	RNA	Buffer

As already said, the last three units are not strictly part of the Artificial Genome.

We will now analyze in detail all these units and we will explain how they interact to determine the cell cycle events, but first let us introduce the concept of Artificial Phenotype.

The Artificial Phenotype, that is the phenotype of the Artificial Genome, is represented as a drawing on a bi-dimensional grid, where each grid cell represents an artificial cell and the colors represent the result of the cell's gene expression. The picture below gives an example of Artificial Phenotype.

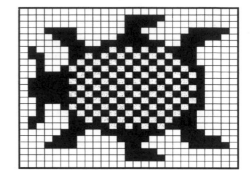

Figure 1: Example of Artificial Phenotype.

Let us now get back to the cell cycle, which is divided in two phases: interphase and mitosis.

Cell Cycle: Interphase

During the interphase the genes are activated and expressed and the phenotype is modified accordingly. Let us assume that at the beginning of the interphase the organism has the following phenotype (two cells expressing no genes –color white–, numbers inside represent Cell Types):

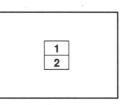

Figure 2: Phenotype at the beginning of interphase.

In the interphase, three functions are executed in sequence (Inputs > Outputs indicated in the parentheses)

Gene Activator	(GAC,CT,DS	> Gene #)
Gene Controller	(GCC,Gene #	> Exon Addr.)
Gene Transcriptor	(GTC,Exon Addr.	> RNA)

The whole process is triggered by the couple of variables (CT,DS) and ends with a modification of the phenotype.

Gene Activation

The function Gene Activator reads the couple (CT,DS) and GAC, and writes Gene Number. Let us have a closer look at these units.

The data-type number Development Stage (DS) is simply a counter that is incremented at each mitosis (with the exception of the "idle" case, explained later). It is assumed, for simplicity's sake, that all cell cycle events are synchronous and have no duration.

Beside the Development Stage, another data-type number is assigned to each cell, representing its "type": the Cell Type (CT). The same type can be assigned to more than one cell, so types can be interpreted as a set of categories into which the whole of the cells is divided, conceptually corresponding to tissues or organs. Cell Types can vary with time, allowing to model aspects related to the development.

As far as GAC is concerned, it is basically a matrix whose rows represent the different Cell Types of the organism (ranging from 0 –zygote– to n) and whose columns represent the Development Stages (ranging from 0 to m). In the following picture an example of GAC & DPC combined is reported.

For each (CT,DS) couple, four digits are given, two for GAC (grey background, used for interphase) and two for DPC (white background, used for mitosis). The GAC indicates the gene(s) activated for each (CT,DS) couple (only one, for simplicity).

CELL TYPES	DEVELOPMENT STAGES							
	0		1		2		3	
	GAC	DPC	GAC	DPC	GAC	DPC	GAC	DPC
0	1 0	1 0						
1			1 0	1 3				
2			1 1	0 2				
3					1 0	1 2		
4					1 1	1 2		
5					1 2	1 2		
6					1 3	1 2		
7							1 0	1 3
8							1 0	1 3
9							1 2	1 3
10							1 2	1 3
11							1 3	1 3
12							1 3	1 3
13							3 0	1 3
14							1 2	1 3
15								
	INTERPH.	MITOSIS	INTERPH.	MITOSIS	INTERPH.	MITOSIS	INTERPH.	MITOSIS

Figure 3: GAC (Gene Activation Code) and DPC (Duplication Plan Code).

The function Gene Activator gets the couple of values (CT,DS) and reads the digits written in the GAC table at these coordinates. If we assume DS = CT = 1 the read digits are [1 0]. This array, written into Gene Number, indicates the gene to be activated.

Gene Control

The function Gene Controller reads Gene Number and GCC and writes Exon Address. Let us have a look at these units. A picture of GCC is reported below.

GENE #		EXON ADDRESS #1				EXON ADDRESS #2				EXON ADDRESS #3			
0	0	1	0	0	0	1	0	0	0	1	0	0	0
0	1	1	0	1	0	1	0	0	0	0	0	0	0
0	2	0	0	0	0	1	0	3	0	1	0	3	3
0	3	1	0	0	0	1	0	0	0	1	0	0	0
1	0	1	1	1	1	1	2	0	1	1	2	1	3
1	1	1	1	3	1	1	1	2	3	1	2	0	3
1	2	1	1	2	0	0	1	3	2	1	3	3	1
1	3	0	3	0	2	1	3	1	2	1	3	1	3
2	0	1	1	1	2	1	2	2	2	1	3	3	3
2	1	0	0	0	3	0	0	3	3	0	1	3	3
2	2	1	0	0	0	0	0	1	0	1	0	2	2
2	3	0	0	0	0	1	0	2	2	1	0	0	0
3	0	1	0	1	0	1	0	2	2	1	0	1	0
3	1	1	0	0	0	1	0	0	0	1	0	0	0
3	2	1	0	0	0	0	0	0	0	0	0	0	0
3	3	1	0	0	0	0	0	0	0	1	0	0	0

Figure 4: GCC (Gene Control Code).

The GCC contains the code necessary to sort out the exons, that is the portions of the coding sequence that will be transcribed. The table has 16 rows, corresponding to the total number of genes (in this example). Each row is divided into n blocks (three in our case, grey cell indicating block start), each of which represents the address of an exon. The array returned by the function Gene Activator ([1 0]) is looked up in the two left columns of the GCC table. The following digits of the matching row are then used to build up the exon addresses. For each block, the first digit (1) tells whether the exon is to be transcribed or not, that is whether the following three digits are to be read or discarded. In our case the addresses of the three exons are [1 1 1], [2 0 1], [2 1 3]. This array is written into Exon Address.

Gene Transcription

The function Gene Transcriptor reads Exon Address and GTC and writes RNA, putting together the exons. A picture of GTC is reported below.

0	1	2	0	0	1	2	3	0	1	2	3	0	1	2	0
0	1	2	3	3	0	3	3	1	1	2	1	0	2	2	3
0	2	0	1	0	1	1	0	0	1	0	3	0	1	2	3
0	1	2	3	0	1	3	1	0	1	2	3	0	2	2	2

Figure 5: GTC (Gene Transcription Code).

GTC represents the "coding" portion of the genome, the one that is actually transcribed into RNA. The exon addresses returned by Gene Controller tell which cells must be read in the GTC table to build-up the RNA sequence. The first digit of an exon address indicates the row of GTC, the other two digits the position in the row. In our example, the read sequence is [0 2 0]: this value is written into RNA. Finally, RNA is used to modify the phenotype, thus completing the expression of the gene. The mapping between the RNA values and the relevant phenotypical effects is reported in the Color Table below.

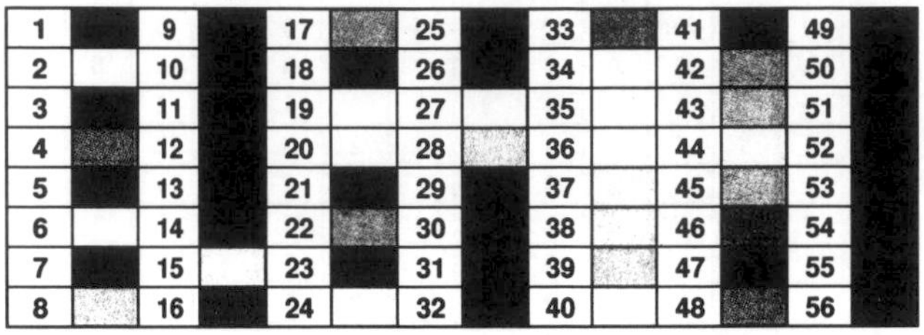

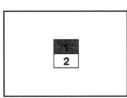

Figure 6: The Color Table maps RNA values to the corresponding phenotypical effects, represented as colors.

In our example, the corresponding base 10 number of the RNA [0 2 0] (16), is looked up in the Color Table and the relevant Cell Type (1) is colored with the corresponding color (dark grey). The phenotype becomes:

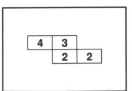

Figure 7: Phenotype at the end of interphase.

This procedure is carried out for all the cells. When all the cells have been processed, the interphase is over. The work-flow of the interphase is summarized in the following diagram.

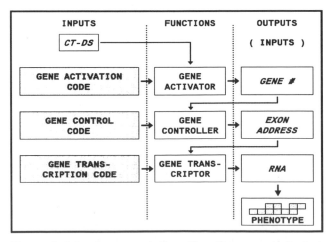

Figure 8: Interphase work-flow (Functions are indicated, Code is highlighted in grey, Data and Buffer in italic).

Cell Cycle: Mitosis

During mitosis the cell duplicates, giving origin to two daughter cells; this operation is executed by the function Cell Duplicator, which reads the couple (CT,DS) and DPC and modifies the phenotype.

The Duplication Plan Code contains the scheme for the duplication of the whole organism, that tells each Cell Type in each Development Stage if and how it must duplicate. This information is contained in two digits: the mitosis type and the displacement type.

The mitotis type has four possible values: Grow, Differentiation, Aging and Idle. If the value is "Differentiation", the two daughter cells are assigned two new Cell Types, starting from the first "free" (not in use by any cell) Cell Type. If the value is "Grow" the two daughter cells inherit the Cell Type of the mother cell. In both cases DS is incremented by one for the two daughter cells. If the value is "Aging" the DS is incremented for the cell but no duplication occurs. If the value is "Idle" nothing happens. The displacement type has also four possible values: down, up, right, left. It indicates the position of the second daughter cell, while the first remains at the same position of the mother cell.

A summary of mitosis and displacement types and the relevant coding is reported in the following table.

Mitosis Type	Coding	Displacement Type	Coding
Grow	0	2. daught. cell down	0
Differentiation	1	2. daught. cell up	1
Maintain (aging)	2	2. daught. cell right	2
Idle cycle	3	2. daught. cell left	3

A picture of GAC & DPC combined is reported in Figure 3. DPC is made up by the digits associated to the mitosis period (white background), representing the mitosis type and the displacement type.

In our example, at the coordinates (CT = 1, DS = 1), we have [1 3]. The first digit (1) indicates that during this mitosis a differentiation is performed; the second digit (3) indicates that the displacement type is "2. daughter cell left". As a result, two new Cell Types, 3 and 4, are created. At the coordinates (CT = 2, DS = 1) we have [0 2], meaning that no new Cell Type is created and the second daughter cell is placed to the right. The final phenotype is the following (no genes are expressed during mitosis: cells are all white):

Figure 9: Phenotype at the end of mitosis.

Simulation

A simulation has been carried-out, where the following code (here reported in a compacted form)

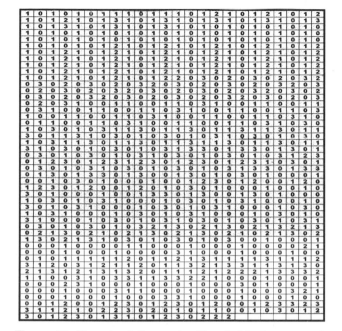

Figure 10: The simulation code. The 1. block represents GAC-DPC (grey), the 2. GCC (white), the 3. GTC (grey).

was used to develop the zygote to become in 9 cycles a small face made up of 132 cells.

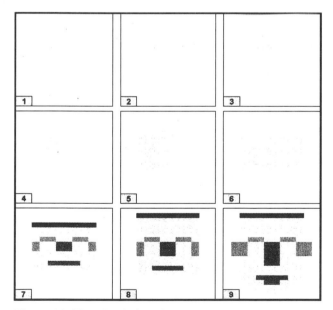

Figure 11: The simulation phenotype.

Some statistics relevant to the simulation:

- Total genome size: 1145 bases
- GAC and DPC %   75%
- GCC %   20%
- GTC ("coding" portion) %:   5%
- Number of cells: 132

In this simulation the vast majority of the sequence is represented by the Duplication Plan Code, the Gene Activation Code and the Gene Control Code, while the "coding" part, represented by the Gene Transcription Code, accounts for just 5% of the total.

Discussion: What the Proposed Model Can Tell on the Nature of the Genome

Hereafter some observations on the proposed model.

In the proposed model the cascade of events that occur during interphase and mitosis is triggered by the values taken by CT and DS, that tell the cell who it is and what it has to do during the different phases of the cell cycle. The (CT,DS) couple represents therefore the primary source of information for both differentiation and morphogenesis.

The gene expression mechanism may look complicated (as the natural one, that makes us of exons, introns, three types of RNA, does), but it has the advantage of allowing a more flexible means of modulating the genes. Without the addressing mechanism provided by GCC, modifying the genes would necessarily require changing some digits in the coding sequence GTC. With such mechanism this can also be achieved by changing the exon addresses, thus modifying the combination of the picked exons. This gives more possibilities to a (either artificial or natural) genetic algorithm to explore and exploit new regions of the search space. Furthermore, the addressing mechanism offers an explanation for the phenomenon of gene overlapping, by which a DNA portion can participate to more genes.

The first part of the simulation (until DS no. 7) has been constructed using only the mitosis type "differentiation", generating a genotypical variety without a corresponding phenotypical variety (all the cells expressed the same gene - color light grey). The choice of this development scheme was motivated by the necessity to "mark" the different parts of the organism in order to be able, at a later stage of development, to selectively trigger the expression of different genes in those parts. We define the "cell diversity ratio" as the number of cell types divided by the total number of cells:

$$CDR = \text{number of cells types} / \text{number of cells}$$

This parameter can be taken as a measure of the "diversity" of the organism: if it equals unity, it means that the organism's cells are all (genotypically) different; if it is less than 1, it means that more cells share the same cell type. CDR is equal to 1 in the first DS's of the simulation, then progressively decreases as more type "grow" mitoses are been used. The similarity of the face at DS = 7 with a child's face and the face at DS = 9 with a more adult face could be a clue that something similar occurs also in nature.

The proposed framework can be used to model genetic diseases. An interesting artificial genetic disease is induced by mutations in the (CT,DS) couple, that make the cell "jump" onto another point of the GAC-DPC matrix, causing the re-activation of embryonal development mechanisms and the expression of the relevant genes. The presence in (some) colon-cancer cell clones of the Carcinoembryonic Antigen (CEA - normally produced by embryo cells) could be a clue that something similar occurs also in nature.

In the model the "non-coding" part of the genome is made up mostly by the GAC-DPC matrix. The dimension of such matrix can be increased either by increasing the number of cell types (a measure of the "complexity" of the organism) or by increasing the number of development stages (a measure of life duration). This could explain why certain plants have very large genomes, even larger than humans: they live longer.

The proposed coding scheme can of course be optimized through compression techniques, e.g. storing only the differences between one DS and the next one: this would allow the storage of more information with the same genome size.

The described model can be enriched by introducing a description of inter-cell dynamics, that is modeling the chemical signals that cells exchange to self-organize as a population (most of the models present in literature are actually concerned primarily on inter-cell dynamics).

We said that, when a mitosis of type "differentiation" is performed, the first "free" cell types (i.e. not used by any cell) are assigned to the daughter cells, which seems to imply a kind of central control that has no biological grounding. An alternative, biologically more plausible method consists in representing cell types with binary sequences; the cell types of the daughter cells are then generated by appending at the end of the mother cell's sequence a "0" for the first daughter cell and a "1" for the second. With such a method, the cell type holds the cell's "duplication history" starting from the zygote onwards.

Not necessarily the proposed model implies a violation of the dogma by which all cells have the same DNA, if the natural counterpart of the (CT,DS) couple, which holds the part of the Artificial Genome that varies between cells of the same organism, is realized through epigenetic mechanisms.

Conclusion

A model of the genome has been proposed, that tries to reproduce some aspects of the cell cycle, with particular reference to gene expression, cell differentiation and morphogenesis. The model is a rather high-level one, nevertheless it seems to hold the potential to model real biological phenomena, even though there are for sure cell biology aspects it cannot explain: further efforts are needed to customize the model for a better adherence to these aspects. A final remark: our work follows the philosophy of the artificial life community, to try to reproduce life as it could be, with the hope it leads us to a better understanding of life as it is.

References

De Garis, H. 1999. Artificial Embriology and Cellular Differentiation. Evolutionary Design by Computers: 281-295.

Eggenberger, P. 1997. Evolving Morphologies of Simulated 3d Organisms Based on Differential Gene Expression. In Proc. of the 4. European Conference on Artificial Life. Cambridge, MA: MIT Press.

Furusawa, C., and Kaneko, K. 2000. Complex Organization in Multicellularity as a Necessity in Evolution. Artificial Life 6(4): 265-281.

Gibbs, W. W. 2003. The Unseen Genome: Beyond DNA. Scientific American, 2003-12.

Hood, L., and Galas, D. 2003. The Digital Code of DNA. Nature. 2003 Jan 23; 421(6921): 444-448.

Kitano, H. 1995. A Simple Model of Neurogenesis and Cell Differentiation Based on Evolutionary Large-Scale Chaos. Artificial Life, 2(1):79-99.

Moore, M. J. 1996. When the Junk isn't Junk. Nature 379: 402-403.

Olovnikov, A. M. 1996. The Molecular Mechanism of Morphogenesis: a Theory of Location DNA. Biochemistry, Vol. 61.

Prodan, L., Tempesti, G., Mange, D., and Stauffer, A. 2002. Embryonics: Electronic Stem Cells. In Proc. of the 8. International Conference on Artificial Life, 101-105. Sidney, Australia: MIT Press.

Reil, T. 1999. Dynamics of Gene Expression in an Artificial Genome – Implications for Biological and Artificial Ontogeny. In Proc. of the 5. European Conference on Artificial Life, 457-466. New York, NY: Springer Verlag.

Asynchronous Dynamics of an Artificial Genetic Regulatory Network

Jennifer Hallinan[1] and Janet Wiles[2]

[1]ARC Centre for Bioinformatics, Institute for Molecular Biosciences and School of ITEE The University of Queensland, AUSTRALIA
[2]School of ITEE and School of Psychology, The University of Queensland, AUSTRALIA
j.hallinan@imb.uq.edu.au

Abstract

The synchrony / asynchrony dichotomy prevalent in models of genetic regulatory networks can be replaced by a parameter, s, which is the probability of a node being updated in a single time step. Here we apply the idea of such parameterized synchrony to study the dynamics of the genetic regulatory network extracted from an artificial genome model. We find that the relationship between degree of synchrony and the number of limit cycles is not linear. The number and length of limit cycles peaks at intermediate values of s. The proportion of state space explored and the length of transient trajectories also follows this pattern. The richer behavior found at intermediate values of the synchrony parameter is much more characteristic of biological systems than either full synchrony or complete asynchrony.

Introduction

A characteristic property of biological systems is the robustness of their dynamic behaviors despite the inherent stochasticity of the underlying molecular interactions. A current issue in the design of genetic regulatory networks is how such characteristic dynamics can best be modeled while acknowledging the underlying stochasticity of the network components.

Genetic regulatory networks developed using Reil's Artificial Genome (AG) model (Reil, 1999) reflect the behavior of real genetic regulatory networks, implementing and extending random boolean networks by grounding their design in sequence-level models of DNA. They exhibit complex dynamic behavior, ranging from rapid "freezing" at a point attractor in state space, through limit cycles of varying length and complexity, to apparently chaotic dynamics, depending on the connectivity and degree of inhibition of the network. However, these models suffer from several important deviations from biological plausibility. Perhaps the most important of these deviations pertains to the gene update rules.

In the standard AG, as in the classical Random Boolean Network (RBN), all genes are updated simultaneously, on the basis of their inputs from the previous time step. While computationally convenient, such synchronous updating does not occur in biological GRNs; factors such as mRNA and protein synthesis, degradation and transport times mean that the system is replete with delays of varying amounts, and genes are activated or inhibited in a fundamentally asynchronous manner. Unfortunately, with Boolean GRN models, the implementation of asynchronous gene update, whether deterministic or not deterministic, completely alters the network dynamics. Unless connectivity and inhibition levels are very high, most networks quickly move to a point attractor in state space and remain frozen there. The synchrony simplification is therefore widely accepted, largely because synchronous Boolean networks do exhibit dynamic behavior similar to that of biological cells.

The existence of complex dynamic behavior in network models is of interest because biological cells are assumed to function on the "edge of chaos", in the regime between totally frozen and chaotic dynamics. This region is characterized by the presence of limit cycle attractors with wide basins of attraction. Such attractors are widely assumed, following Kauffman (1993), to be models of cell types—each cell type is an attractor in gene expression phase space.

The biological implausibility of synchronous updating is widely recognized, and the effects of synchrony have been examined in a variety of models, including globally coupled logistic maps (Abramson & Zamette, 1998), Conway's Game of Life (Blok & Bergersen, 1999), cellular automata (Schönfisch & de Roos, 1999) and random Boolean networks (Harvey & Bossomaier, 1997; Di Paolo, 2001).

Limit cycles *per se* do not generally exist in an asynchronously updated networks, although networks with such properties can be specifically handcrafted (Nehaniv, 2002). However, several authors have demonstrated the existence of pseudo-periodic "loose" attractors in asynchronously updated networks (Harvey & Bossomaier, 1997; Di Paolo, 2001). Hallinan and Wiles (2004) have demonstrated that networks based upon the AG model can be evolved to exhibit such loose attractors, even when asynchronously updated.

Synchrony / asynchrony is often thought of as a binary condition. An alternative is to consider degree of synchrony as a tunable parameter, implemented as the probability of each node being updated at each time step. Under such an update scheme a network with synchrony 1.0 is the standard synchronous network, while one with synchrony $1/n$, where n is the number of nodes in the network, is equivalent to the usual conception of an asynchronous network.

In this study we examine the effects of modifying the degree of synchrony upon the dynamics of AG-generated networks.

Methods

Our model is based on the Artificial Genome model (AG) developed by Reil (1999). A genome is generated at random, using equal proportions of *b* "bases". There are four bases in real DNA – adenine (A), thymine (T), guanine (G) and cytosine (C), so we used four bases, designated 0, 1, 2 and 3. The genome is then searched for instances of a gene marker string of length *l* (we used 0101) analogous to the TATA box to which biological transcription factors bind. The following *g* bases are then designated a gene. The region between the end of a gene and the beginning of the next 0101 marker string becomes the promoter region for the downstream gene.

Each gene is "translated" into a gene product by incrementing each base by 1. A gene with the sequence 012130 will therefore result in a product with the sequence 123201. All of the promoter regions in the genome are searched for matches with each gene product; if a match to the product of gene **A** is found in the promoter region of gene **B**, we say that gene **A** controls gene **B**. This control may be either excitatory—**A** promotes the transcription of **B**—or inhibitory. In this way a genetic regulatory network is constructed from the randomly generated genome (Figure 1).

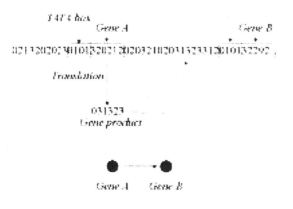

Figure 1. Reil's artificial genome model of a genetic regulatory network.

Ten random genomes and their corresponding networks were generated, with the parameters shown in Table 1. These parameter values were selected because previous experiments indicated that they produce networks which, when updated synchronously, show limit cycle dynamics, but when updated asynchronously rapidly collapse to a point attractor (Hallinan & Wiles, 2004).

Table 1. Parameters used for network generation

Parameter	Value
Chromosome length	14000
Number of bases	4
Gene Marker	0101
Gene length	6
Proportion of inhibitory links	0.4
Maximum number of timesteps	1000

The resulting networks had an average of 57.9 nodes and 172.3 links, giving an average connectivity of 2.97.

Our model differs from that of Reil (1999) in two major ways. One is the manner in which inhibitory links are implemented. In the original model a link was deemed inhibitory if its last base has a particular value. Using this approach a network can have a degree of inhibition of 0, 0.25, 0.50 or 1.0, and the links emanating from a given gene will always be either inhibitory or excitatory, no matter which to gene it links. We designate individual links as inhibitory with a specific probability as they are formed. This scheme allows much finer-grained inhibition, and is more biologically plausible in that it allows a single gene product to participate in some reactions as an inhibitor and in others as an activator.

The other difference, as discussed above, is our update scheme. Instead of synchronous updating we use a synchrony parameter, *s*, which represents the probability of a node being updated at any time step. This approach has been applied to the Game of Life, which shows characteristic phase transitions (Blok and Bergerson, 1999).

Each network was run *n* times, where *n* is the number of nodes in the network, each time with a different initial node activated, for synchrony values ranging from 0.1 to 1.0. With a synchrony value of 1.0 each node has a chance of 1.0 of being updated at each timestep, making it a synchronous updating scheme. In this way much of the state space of each network is explored in a systematic manner, although the stochastic element in the update scheme means that the entire state space is almost certainly not fully explored.

Each network was run for 1000 timesteps and a record kept of the states visited. This state list was then used to construct a state transition diagram for each network run, from which statistics pertaining to the number and length of limit cycles encountered could be compiled.

Results

Typical state transition graphs for networks with low synchrony (0.3) and high synchrony (0.9) are shown in Figure 2.

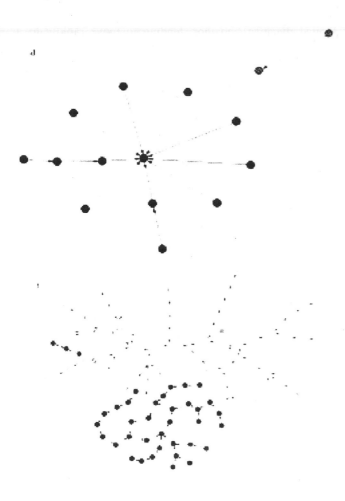

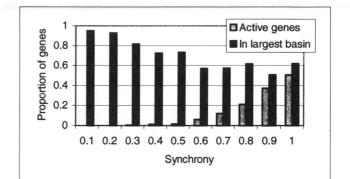

Figure 3. Change in network dynamics with synchrony as evidenced by the proportion of genes active per time step and proportion of states in the largest basin of attraction.

As expected, networks with low synchrony displayed fewer and shorter limit cycles that those with high synchrony. The relationship between synchrony and limit cycle behavior is not, however, linear. For most measures of dynamic behavior a peak occurs a synchrony rate of about 0.9 (Figure 4).

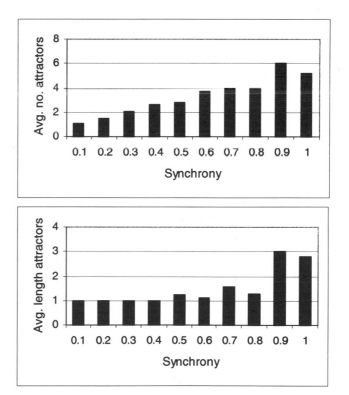

Figure 2. State transition diagrams. a. Network with synchrony 0.3 has two basins of attraction, each leading to a single point attractor and the longest transient in the network is 3. b. Same network with synchrony 0.9 has four basins of attraction, three of which have limit cycle attractors and the longest transient is >15.

The networks with high synchrony visit many more states, and settle to longer attractors than the networks with lower synchrony. This pattern is common to all of the networks; the network dynamics change with the degree of synchrony.

Despite the higher number of basins of attraction in the more synchronous networks, the proportion of all genes in the largest basin of attraction decreases relatively slowly with increasing synchrony (from 0.95 to 0.61), whereas the proportion of genes actually active in each time step increases sharply. In networks with very low synchrony very few genes are active per time step (Figure 3).

Figure 4. Average length and number of attractors in networks with different levels of synchrony.

Limit cycle length varied from 1 to 14 states, with most limit cycles less than five states long. The longest limit cycles were found, once again, at a synchrony of 0.9 (Figure 5).

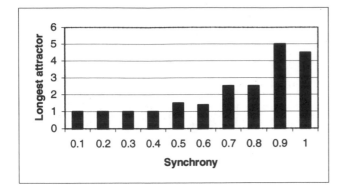

Figure 5. Average length of limit cycles for networks with different levels of synchrony

Although the number of long limit cycles increases with increasing synchrony, all networks have a significant proportion of point attractors (Figure 6).

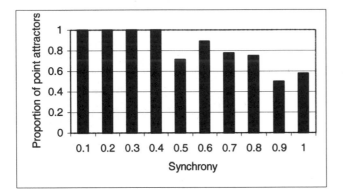

Figure 6. Proportion of attractors which are point attractors, for different levels of synchrony

Discussion

An issue in the design of artificial life simulations is the identification of which features of biological systems are most significant for robust dynamics and what kinds of abstractions provide informative analogues. Computational models of genetic regulatory networks are increasingly taking into account the role of stochasticity, not as an irrelevant detail, but as a core property of the system.

The observation that networks updated asynchronously display limited dynamic behaviour, collapsing rapidly to a stable state, is of concern if these networks are to be used as models of genetic regulatory networks. In previous work we have shown that networks can be evolved to display rich dynamic behaviour under asynchronous updating (Hallinan & Wiles, 2004). The current work builds upon this finding be exploring in more detail the nature of asynchrony in order to make the simplified Boolean model more biologically relevant.

We implemented the degree of asynchrony as a parameter, s. Networks with s of below about 0.5 have very limited dynamic behavior, tending to move rapidly to a single point attractor from any point in the state space. In contrast, networks with s between 0.5 and 1.0 display a range of interesting dynamic behaviors, with the number and length of fuzzy limit cycles increasing with s to peak at about 0.9. Interestingly, fully synchronous networks, with an s of 1.0, have fewer and shorter limit cycles, on average, than networks with a s of 0.9. We are currently testing whether this holds in larger systems, examining factors such as the number of active n nodes, and the way in which basins of attraction change with changes in parameter values.

All of the networks in our study were run repeatedly, with a single node active at timestep 1, a different node each time. This means that each network explored its state space n times (where n is the number of nodes in the network) per value of s, and the order of exploration was the same each time. Although the entire state space of any network is not fully explored, because of the stochasticity in the update rule, this protocol means that the number of trajectories used for state space exploration is equivalent each time, and the state transition diagrams reflect the different dynamics of the networks under different update schemes. As s increases, the state spaces of the networks tend to have more basins of attraction and longer limit cycles. The basins of attraction are also larger, as reflected by the longer transients evident in networks of higher s (Figure 2).

Although networks with higher s have more long limit cycles, all networks have a significant proportion of point attractors. Once again, the proportion of attractors that were point attractors was lowest at an s of 0.9 (0.5 compared with 0.57 for $s = 1.0$).

Genetic regulatory networks are characterized by robust temporal dynamics. Kauffman (1993) hypothesizes that an attractor in gene expression space represents a cell type. Models of genetic regulatory networks such as the artificial genome, display limit cycle behavior with large basins of attraction. A large basin of attraction for a limit cycle is necessary for robustness; it implies that a small perturbation in gene expression will leave the network in the same basin of attraction, eventually to return to the same attractor. Cell type is therefore stable, as observed in biological systems.

The observation that both limit cycle length and size of basin of attraction reach a maximum at an s of less than 1.0 demonstrates that biologically plausible dynamic behavior in a genetic regulatory network is not dependant upon synchronous node updating. Although the asynchrony parameter, s, is still an extreme simplification of the multiplicity of variable temporal delays induced by differential rates of transcription, translation, degradation and other cellular processes, it illustrates that stochasticity, feedback and delay are fundamental to network robustness in both biological and computational networks.

References

Abramson, G. & Zanette, D. H. 1998. Globally coupled maps with asynchronous updating. Physical Review e 58(4): 4454 - 4460.

Blok, H. J. & Bergersen, B. 1999. Synchronous versus asynchronous updating in the "game of Life". Physical Review e 59(4): 3876 - 3879.

Di Paolo, E. A. 2001. Rhythmic and non-rhythmic attractors in asynchronous random Boolean networks. Bio systems 59: 185 - 195.

Hallinan, J. & Wiles, J. 2004. Evolving genetic regulatory networks using an artificial genome. In Chen, Y.-P. (ed.) 2nd Asia-Pacific Bioinformatics Conference (APBC2004), Dunedin, New Zealand. Conferences in Research and Practice in Information Technology, Vol. 29.Australian Computer Society, Inc.

Harvey, I. & Bossomaier, T. 1997. Time out of joint: Attractors in random Boolean networks. In P. Husbands & I. Harvey (eds.) Proceedings of the Fourth European Conference on Artificial Life (ECAL97). Cambridge, MA: MIT Press: 67 - 75.

Jeong, H., Tombor, B., Albert, R., Oltvai, Z. N. & Barabasi, A.-L. 2000. The large-scale organization of metabolic networks. Nature 407: 651 - 654.

Kauffman, S. A. 1993. The Origins of Order: Self Organization and Selection in Evolution. New York: Oxford University Press.

Nehaniv, C. L. 2002. Evolution in Asynchronous Cellular Automata, In R.K. Standish, M.A. Bedau, and H.A. Abbass (eds.) Proceedings of the Eighth International Conference on Artificial Life, MIT Press, pp. 65-78.

Reil, T. 1999. Dynamics of gene expression in an artificial genome: Implications for biological and artificial ontogeny. In Floreano, D., Mondada, F. & Nicoud, J. D. (eds.) Proceedings of the 5th European Conference on Artificial Life. New York, NY: Springer Verlag: 457 - 466.

Schonfisch, B. & de Roos, A. 1999. Synchronous and asynchronous updating in cellular automata. Bio systems 51: 123 - 143.

Sole, R. & Pastor-Santorros, R. 2002. Complex networks in genomics and proteomics. Santa Fe Institute Working Paper 02-06-026.

Wagner, A. & Fell, D. 2001. The small world inside large metabolic networks. Proceedings of the Royal Society of London, Series B 268: 1803 - 1810.

Small World and Scale–Free Network Topologies in an Artificial Regulatory Network Model

P. Dwight Kuo and Wolfgang Banzhaf
Memorial University of Newfoundland, St. John's, NL, Canada A1B 3X5
{kuo,banzhaf}@cs.mun.ca
http://www.cs.mun.ca/~{kuo,banzhaf}
Tel: (709) 737-8652 Fax: (709) 737-2009

Abstract

Small world and scale–free network topologies commonly exist in natural and artificial systems. Many mechanisms for producing these topologies have been presented in the literature. We present an artificial regulatory network model generated by a duplication / divergence process on a randomly generated genetic string and show that networks with small world and scale–free topologies can be produced with some regularity.

Introduction

Recently, there has been significant interest in small world and scale–free network topologies and potential methods or processes which may generate them (Romualdo et al., 2003; Valverde et al., 2002; Barabasi et al., 2001; Barabasi and Albert, 1999). In the majority of these contributions, the mechanisms for generating such topologies are based on preferential attachment (Romualdo et al., 2003; Valverde et al., 2002; Barabasi and Albert, 1999). In this contribution we work within the framework of a model of an artificial regulatory network first presented by Banzhaf (Banzhaf, 2003a; Banzhaf, 2003b) generated by a duplication / divergence process similar to that presented in (Romualdo et al., 2003). However, their model operates directly on the nodes and edges of the model, bypassing any genetic–type representation of the network.

Here, we show that scale–free and small world network topologies appear with some regularity in the gene–protein network interaction diagram generated by parameterizing the networks by the degree of matching between genes and proteins and discuss possible implications. Duplication and divergence are performed directly on the genetic–string, not on the actual nodes and edges of the interaction network.

It has also been shown that this model can reproduce phenomena found in natural genetic regulatory networks such as heterochrony (Banzhaf, 2003a). As such, this model can relate changes in the timing and intensity of gene expression to tiny pattern changes on bit strings which could possibly provide the algorithmic " missing link " between genotypes subject to constant evolutionary changes and the remarkably stable phenotypes found in the real world.

Background

Regulatory Networks

Regulatory networks are an important new research area in biology (Bower and Boulouri, 2001; Davidson, 2001; Kitano, 2001). With the realization that in higher organisms only a tiny fraction of DNA is translated into proteins, the question of what the rest of the DNA is actually doing becomes all the more pressing. Regulation appears to be a very reasonable answer for a functional role for unexpressed DNA. According to Neidthardt et al. (Neidhardt, 1996), 88% of the genome of the bacterium *E. Coli* is expressed with 11% suspected to contain regulatory information (also see Thomas (Thomas, 1999)).

In addition, it has been recognized that the DNA information controlling gene expression is the key to understanding differences between species and thus to evolution (Hood and Galas, 2003).

The three major genetic mechanisms, all tied to regulation (Davidson, 2001) which allow such a variety of reactions of living organisms to the pressure for survival are:

1. Interactions between the products of genes

2. Shifts in the timing of gene expression (heterochrony)

3. Shifts in the location of gene expression (spatial patterning)

These mechanisms allow nature to set up and control the mechanisms of evolution, development and physiology. Since many evolutionary effects can be traced back to their regulatory causes, regulatory networks mediate between development and evolution thus unfolding the patterns and shapes of organism morphology and behaviour (Davidson, 2001; Banzhaf, 2003b).

Studying models of regulatory networks can help us understand some of these mechanisms providing lessons for biology and in the area of artificial evolution.

Scale–Free Network Topologies

It has been found that a high degree of self–organization may characterize the large–scale properties of complex networks

(Barabasi and Albert, 1999). Many researchers have shown that the probability $P(k)$ that the number of nodes connected to k (vertex degree) other nodes in a network decays as a power law, following: $P(k) \sim k^{-\gamma}$ in systems as diverse as the internet (Faloutsos et al., 1999), protein interaction networks (Wuchty, 2001), the electrical power grid of the western United States of America (Watts, 2003), the neuronal network of the worm *Caenorhabditis Elegans* (Watts, 2003), and the network of citations of scientific papers (Barabasi et al., 2002).

It has thus been suggested that scale–free networks emerge in the context of a dynamic network with the addition of new vertices connecting preferentially to vertices which are highly connected in the network (Barabasi and Albert, 1999), as well as through explicit optimization (Valverde et al., 2002).

Small World Network Topologies

Small world graphs can be defined as any graph with n vertices and average vertex degree k that exhibits $L \approx L_{random}(n,k) \sim \frac{\ln(n)}{\ln(k)}$, and $C \gg C_{random} \sim \frac{k}{n}$ for $n \gg k \gg \ln(n) \gg 1$ (Watts, 2003). C is referred to as the clustering coefficient (if vertex v has k_v neighbors, $C = \frac{2}{n}\sum_{v=1}^{n}\left(\frac{k_v(k_v-1)}{2}\right)$) of the network while L is the characteristic path–length of the network (average number of links connecting two nodes). L_{random} and C_{random} refer respectively to the characteristic path–length and clustering coefficient for a completely random graph with the same k and n.

Like scale–free network topologies, the small world topology has also been noted in many networks (including those with scale–free topology) such as the electrical power grid of the western United States of America (Watts, 2003), the neuronal network of the worm *Caenorhabditis Elegans* (Watts, 2003), and the network of film actors who have acted in the same films (Watts, 2003).

Artificial Regulatory Network Model

Our artificial regulatory network (ARN) model is based on work by Banzhaf (Banzhaf, 2003a; Banzhaf, 2003b). In this model, the ARN consists of a genome represented by a bit string with direction (i.e. 5' → 3' in DNA) and mobile "proteins" which are equipped with bit patterns for interactions with the genome. The proteins are able to wander about in order to interact with the genome, notably at "regulatory" sites located upstream (3' → 5' direction) from genes. Attachment to these sites inhibits or activates the production of the corresponding protein thereby demonstrating the mechanisms of activation and inhibition.

Creation of the genome commences with the generation of a random 32–bit string. This string is then used in a series of whole length duplications similar to those found in nature (Wolfe and Shields, 1997) followed by mutations in

order to generate a genome of length L_G. A "promotor" bit sequence of 8–bits was then arbitrarily selected to be "01010101". In a genome generated by randomly choosing "0" s and "1" s, this one–byte pattern can be expected to appear with probability $2^{-8} = 0.39\%$. Since the promotor pattern itself is repetitive, overlapping promotors or periodic extensions of the pattern are not allowed, i.e. a bit sequence of "0101010101" (10–bits) is detected as a single promotor site starting at the first bit. The promotor signals the beginning of a gene on the bit string which is analogous to an open reading frame (ORF) on DNA - a long sequence of DNA that contains no "stop" codon and therefore encodes all or part of a protein. This gene is set to a fixed length of $l_{gene} = 5$ 32–bit integers which results in an expressed bit pattern of 160 bits for each gene. Therefore, genes can thus be created by complete duplications of previously created genes, mutation, and / or combinations of the end and starting sequences of the genome during duplication.

Immediately upstream from the promotor exist two 32–bit segments which represent the enhancer and inhibitor sites. As previously mentioned, attachment of proteins to these sites results in changes to protein production for the corresponding genes. In this model, we assume only one regulatory site for the expression and one site for the suppression of protein production. This is a radical simplification since natural genomes may have 5–10 regulatory sites that may even be occupied by complexes of proteins (Davidson, 2001; Banzhaf, 2003b).

The model presented here completely disregards processes such as transcription, and neglects elements such as introns, RNA–like mobile elements and translation procedures resulting in a different alphabet for proteins. This last mechanism is replaced as follows: Each protein is a 32–bit sequence which results from a many–to–one mapping of its corresponding gene which contains five 32–bit integers. The protein sequence is created by performing the majority rule on each bit position of these five integers so as to arrive at a 32–bit protein. Ties (not possible with an odd number for l_{gene}) for a given bit position are resolved by chance.

These proteins may then be examined to see how they may "match" with the genome. This comparison is implemented by using the XOR operation which returns a "1" if both inputted bits are complementary. In this scheme, the degree of match between the genome and the protein bit patterns is specified by the number of bits set to "1" during an XOR operation. In general it can be expected that a Gaussian distribution results from measuring the match between proteins and bit sequences in the random genome (Banzhaf, 2003b).

If we make the simplifying assumption that the occupation of two regulatory sites per gene modulates the expression of the corresponding protein, we may deduce an interaction network comprising the different genes and proteins which can be parameterized by strength of match.

By examining the interaction networks at different matching strengths (thresholds) we may obtain different network topologies for the same connected network components. An example is shown in Figs. 1 and 2. Each node in the diagram represents a gene found in the genome along with its corresponding protein forming a gene–protein pair. Edges in the diagram represent some form of influence of one gene's protein on another gene. For the diagrams presented, a random genome was created by the previously mentioned duplication and mutation procedure with the network interaction diagrams being created at thresholds of 21 and 22.

It must be stressed that although the actual genome has not changed, by simply changing the threshold parameter, we have obtained a different network topology. It may be noted by the more astute reader that the diagrams in Figs. 1 and 2 possess different numbers of genes and proteins. This is due to the fact that only connected gene–protein pairs are displayed in the diagrams. Should a change in the parameterized threshold lead to the creation of an isolated node, it is deleted from the diagram. Also note that only the largest network of interactions is displayed.

It is possible to have multiple clusters of gene–protein interactions that are not interconnected. This is likely to occur as the threshold level is increased. As connections between gene–protein pairs are lost due to the threshold, each cluster of gene–protein pairs begins to become isolated from the others. This often occurs abruptly indicating a phase transition between sparse and full network connectivity.

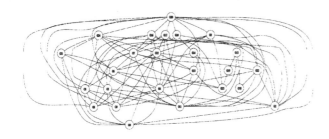

Figure 1: Gene-protein interaction network for a random genome at a threshold of 21 bits.

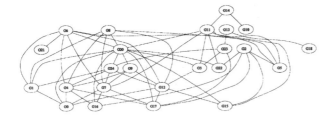

Figure 2: Gene-protein interaction network for a random genome at a threshold of 22 bits.

Results

At mutation rates of 1% and 5%, 200 genomes were generated by 12 duplication events per genome leading to individual genomes of length $L_G = 131072$. From these genomes, the number of genes were then determined based on the number of promotor patterns present. The distribution of the number of genes present in the genome of size L_G is shown in Figs. 3 & 4.

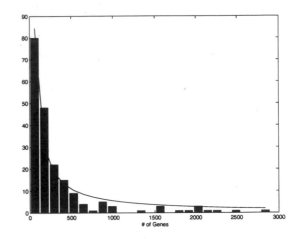

Figure 3: Histogram of the number of genes in each genome (200 genomes) fitted to a power law: $P(g) \sim g^{-\gamma}$ for a mutation rate of 1.0%. γ was calculated to be 0.9779.

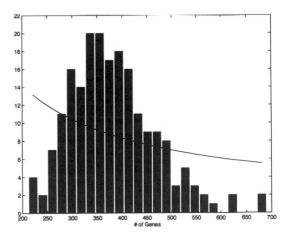

Figure 4: Histogram of the number of genes in each genome (200 genomes) fitted to a power law: $P(g) \sim g^{-\gamma}$ for a mutation rate of 5.0%.

It can be observed that the distribution of the number of genes in Fig. 3 follows a power–law distribution. As well, if we turn the mutation rate lower to 0.1% (results not shown) the distribution of the number of genes again shows a scale–free like distribution.

However, in Fig. 4 the apparent distribution is disrupted. This is attributed to the higher rate of mutation. At such a mutation rate, the rewiring of the network becomes so prevalent that it begins to disrupt the duplication of nodes leading to a randomly connected network. For an 8 bit promotor, the probability that it remains intact after one duplication event is only 66% at a mutation rate of 5%. Therefore, it can be expected that many of the genes copied during the duplication process will be subsequently destroyed in later duplication steps. However, there will also be other genes which arise from this higher mutation rate. But, these new genes will also be easily destroyed via mutation. Genomes which start with very large numbers of genes are disrupted early on in the duplication process by mutation, while those with few genes obtain additional genes through mutation.

To test this explanation, we created genomes of length L_G completely at random without the use of duplication / divergence. The distribution of these completely randomly generated networks are shown in Fig. 5. As can be seen, this distribution is quite similar to that generated in Fig. 4 lending additional support to the hypothesis that at 5% mutation the network topology becomes randomized. Therefore, we may use the distribution of the number of genes in networks generated by duplication / divergence as an estimate of the effect of mutation rate on the network as compared to randomly generated genomes.

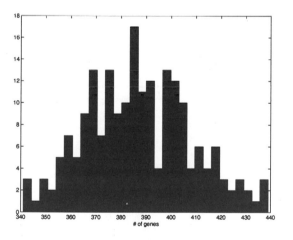

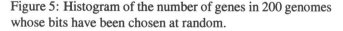

Figure 5: Histogram of the number of genes in 200 genomes whose bits have been chosen at random.

In general, the duplication process, despite being performed directly on the genetic string can be considered to be similar to the mechanism of preferential attachment.

Consider the duplication process on a string which contains multiple genes while neglecting the effects of mutation. For the case of this argument, we also assume that no additional genes are created from a duplication event by joining the end and beginning of one genome string. We start with a network of 5 gene–protein pairs connected as shown on the left side of Fig. 6 and proceeding through a single duplication event generating the network shown on the right side.

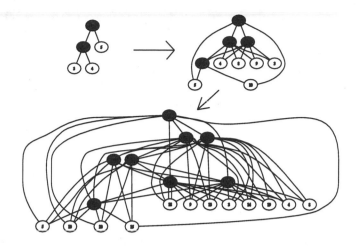

Figure 6: An example of the effect of two duplication events. Highly connected (shaded) nodes become even more higly connected (preferential attachment). Each node represents a gene / protein pair; each edge represents an interaction between gene / protein pairs.

It can be seen that the more highly connected nodes on the left, nodes 1 and 2 and their copies 6 and 7 (shown in gray), become even more highly connected after a single duplication event. This can again be seen in the third diagram which shows the result of another duplication event. As the number of duplication events increases, the difference in the number of connections between highly connected nodes and less connected nodes increases. This can be thought of as a form of preferential attachment since nodes that are already highly connected will become even more so after subsequent duplication events. Preferential attachment has been shown to be a mechanism which can generate scale–free networks (Barabasi and Albert, 1999; Romualdo et al., 2003).

However, this neglects the mechanism of mutation. Mutation may be thought of as an operator which reorganizes the network. If mutations should occur on a gene, this may either change the gene–protein pair's binding site, or the generated protein thus reorganizing a portion of the network. The other possibilities are that mutations may either disrupt the promotor pattern in effect deleting a gene–protein pair from the network, or create a new gene-protein pair by creating a new promotor site.

With these considerations in mind, we may then examine the networks generated by these genomes to see whether their topologies may be considered scale–free and / or small world.

The network of gene–protein interactions was parameterized by the threshold value leading to a maximum of 32 possible networks for each genome. The histograms of the probability of being connected with k components were fitted to the equation $\alpha k^{-\gamma}$ for each threshold value using the sum of least squares method. The threshold value which produced a γ value closest to 2.5 was kept. It has been found that a large number of networks which have displayed scale–free behaviour exhibit values of $2 < \gamma \leq 3$ (Goh et al., 2002).

Values for the parameter γ characterizing scale–free networks were also calculated for each of the genomes and are shown in Figs. 7 & 8.

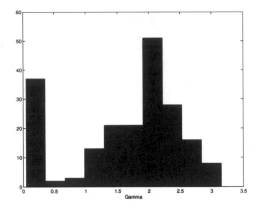

Figure 7: Distribution of values of γ for the best fit of $P(k) \sim k^{-\gamma}$ with a mutation rate of 1.0%.

It can be seen that there exist many genomes created at random which may be considered to satisfy the definition of a scale–free network. In Fig. 7 there is a large number of networks whose coefficient γ is close to zero. This can be attributed to the fact that since the mutation rate is low, the probability of discovering new promotor patterns through subsequent duplication / divergence steps is not high. Therefore, if there were few promotors in the initial starting string, there will often be few genes in the overall genome. With a small number of genes, the scale–free coefficient γ will often be of small magnitude.

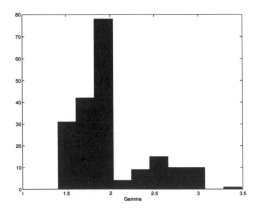

Figure 8: Distribution of values of γ for the best fit of $P(k) \sim k^{-\gamma}$ with a mutation rate of 5.0%.

For each network the clustering coefficient, C, and the characteristic path length, L, were calculated and compared

to the corresponding metric for a randomly connected network of the same size and vertex degree distribution. The threshold value that produced a network with the smallest absolute difference $| L - L_{random} |$ that also satisfied $C \gg C_{random}$ were taken to be those most characteristic of the small world network topology. The additional constraint that $L > 1.3$ was also enforced so as to try to exclude graphs that were close to being fully connected. The distributions for

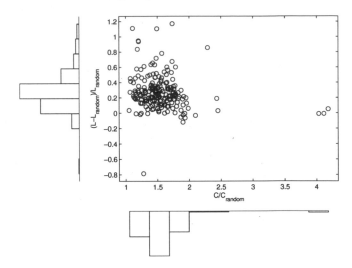

Figure 9: Scatter plot and histograms of values of $\frac{C}{C_{random}}$ and $\frac{L_{random} - L}{L_{random}}$ for each of the randomly generated genomes (200 genomes) with a mutation rate of 1.0%.

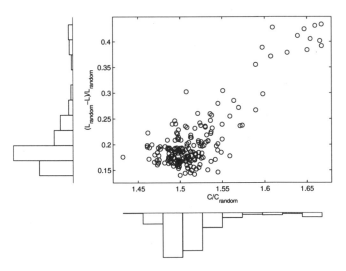

Figure 10: Scatter plot and histograms of values of $\frac{C}{C_{random}}$ and $\frac{L_{random} - L}{L_{random}}$ for each of the randomly generated genomes (200 genomes) with a mutation rate of 5.0%.

the clustering coefficient and the characteristic path length obtained from the 200 genomes for both rates of mutation are shown in Figs. 9 & 10.

From these figures, it can be seen that in the majority of genomes, there exists a threshold at which the interaction network approaches or satisfies the definition of a small world network topology.

Conclusions

A model of an artificial regulatory network model has been presented. The construction of such a network using a simple whole genome duplication process directly on a genetic-string representation of the genome produces a network construction scheme similar to preferential attachment. The addition of a mutation operator introduces a kind of rewiring of the network topology by changing activation / inhibition sites, creating / destroying gene–protein pairs and changing the configuration of proteins which the genes code for. Examining networks generated in this way by varying the threshold at which genes and proteins may interact shows that many of these regulatory networks display the charateristics of small world and scale–free network topologies with some regularity.

Note that we have assumed that duplication proceeds by duplicating the whole genome which occurs relatively rarely in nature (Wolfe and Shields, 1997; Nadeau and Sankoff, 1997). Future work may include investigating the effects of shorter length duplication events on regulatory network topologies.

Acknowledgements

The authors would like to kindly thank François Képès of Atelier de Génomique Cognitive, CNRS for helpful discussions and suggestions.

References

Banzhaf, W. (2003a). Artificial regulatory networks and genetic programming. In Riolo, R. L. and Worzel, B., editors, *Genetic Programming Theory and Practice*, chapter 4, pages 43–62. Kluwer.

Banzhaf, W. (2003b). On the dynamics of an artificial regulatory network. In Banzhaf, W., Christaller, T., Dittrich, P., Kim, J. T., and Ziegler, J., editors, *Advances in Artificial Life – Proceedings of the 7th European Conference on Artificial Life (ECAL)*, volume 2801 of *Lecture Notes in Artificial Intelligence*, pages 217–227. Springer Verlag Berlin, Heidelberg.

Barabasi, A. L. and Albert, R. (1999). Emergence of scaling in random networks. *Science*, 286:509–512.

Barabasi, A. L., Erzsebet, R., and Vicsek, T. (2001). Deterministic scale-free networks. *Physica A*, 299:559–564.

Barabasi, A. L., Jeong, H., Ravasz, R., Neda, Z., Vicsek, T., and Schubert, A. (2002). Evolution of the social network of scientific collaborations. *Physica A*, 311:590–614.

Bower, J. and Boulouri, H., editors (2001). *Computational Modeling of Genetic and Biochemical Networks*. MIT Press, Cambridge, MA.

Davidson, E. H. (2001). *Genomic Regulatory Systems*. Academic Press, San Diego, CA.

Faloutsos, M., Faloutsos, P., and Faloutsos, C. (1999). On power-law relationships of the internet topology. In *SIGCOMM*, pages 251–262.

Goh, K. I., Oh, E., Jeong, H., Kahng, B., and Kim, D. (2002). Classification of scale–free networks. *Proceedings of the National Academy of Sciences, USA*, 99(20):12583–8.

Hood, L. and Galas, D. (2003). The digital code of DNA. *Nature*, 421:444–448.

Kitano, H., editor (2001). *Foundations of Systems Biology*. MIT Press, Cambridge, MA.

Nadeau, J. and Sankoff, D. (1997). Comparable rates of gene loss and functional divergence after genome duplications early in verebrate evolution. *Genetics*, 147:1259–1266.

Neidhardt, F. C. (1996). *Escherichia Coli and Salmonella typhimurium*. ASM Press, Washington, DC.

Romualdo, P., Smith, E., and Solé, R. (2003). Evolving protein interaction networks through gene duplication. *Journal of Theoretical Biology*, 222:199–210.

Thomas, G. H. (1999). Completing the e.coli proteome: a database of gene products characterised since completion of the genome sequence. *Bioinformatics*, 7:860–861.

Valverde, S., Ferrer Cancho, R., and Solé, R. (2002). Scale–free networks from optimal design. *Europhysics Letters*, 60:512–517.

Watts, D. (2003). *Small Worlds: The Dynamics of Networks between Order and Randomness*. Princeton, NJ: Princeton University Press.

Wolfe, K. and Shields, D. (1997). Molecular evidence for an ancient duplication of the entire yeast genome. *Nature*, 387:708–713.

Wuchty, S. (2001). Scale-free behavior in protein domain networks. *Molecular Biology & Evolution*, 18(9):1694–1702.

Inertia of Chemotactic Motion as an Emergent Property in a Model of an Eukaryotic Cell

Shin I. Nishimura* and Masaki Sasai[†]

Department of Complex Systems Science, Graduate School of Information Science, Nagoya University

Nagoya, Japan 464-8601

Abstract

Chemotaxis is widely seen in many biological systems. Among them amoebic cells from unicellular slime molds to immune cells are believed to directly sense chemical gradients. Here, we construct a model of amoebic cell by taking account of the chemical kinetics as well as a cellular body. The model is composed of discrete grids and a set of rules which define chemical and motional events on each grid. The model can explain the observed features of the cellular locomotion. We find that the simulated cell tends to keep the direction of motion, which reminds us of "inertia" of motion in Newtonian dynamics. The averaged motion of amoebic cells approximately obeys an "underdamped" equation of motion for a short time scale. "Inertia" of chemotactic motion is an emergent property of the system where motion and the signal processing are strongly coupled to each other.

Introduction

Amoebic cells in animal species have important roles. For example, a phagocyte, a defender of an animal body, is an amoebic cell that migrates and kills external microbes by phagocytosis (Roitt et al., 1998). In general, an amoebic cell does not have a definite shape, but cells that have characteristic shapes also possess the amoebic property more and less. For instance, a neuron in its developmental phase elongates an axon, which can be thought of as an amoebic motion.

It is believed that an eukaryotic cell can directly sense the small difference of chemical concentration between two points at the cellular membrane (Pollard, 2003). This ability of amoebic cells is called "chemotaxis". When external microbes invade tissues, immune cells migrate from vessels to the tissues driven by the chemotaxis: They direct their courses along the gradient of chemoattractants produced from the invaded tissues (Springer, 1994; Katanaev, 2001).

Recent studies have revealed detailed biochemical mechanisms how actins and other molecules control the chemotactic motion (Pollard, 2003; Pollard and Borisy, 2003; Pollard et al., 2000; Iijima et al., 2002). An actin monomer in

*shin@sasai.human.nagoya-u.ac.jp

[†]sasai@info.human.nagoya-u.ac.jp

the cytoplasm binds an ATP and a profilin. External signals indirectly activate the Arp2/3 complex which binds to an ATP bound actin. This Arp2/3-ATP-actin complex initiates the actin polymerization. The polymerization is terminated when a caping protein binds to the actin polymer. ATP-actin subunits in thus formed actin polymer are changed into ADP-actin by phosphate dissociation and ADP-actin subunits are severed from the actin polymer. Force acting on the cellular membrane depends on the balance between the actin polymerization and depolymerization.

We introduce a model that treats actin filaments and some chemicals as well as the cellular membrane. By using this model, we study in this paper whether there exists a simple rule to which an eukaryotic cell should be subject.

Model

Our model consists of discrete two-dimensional grids on which concentrations of relevant molecules are defined. A cell is defined on the grids as a domain. We adopt hexagonal grids for convenience. A grid is either external or internal of the cellular domain. When the grid is in the cellular domain, three real numbers are defined on the grid, which indicate concentrations of activator, inhibitor and actin filaments.

We give four rules in order to move the cell: Chemical Kinetics, Diffusion, Actin filaments extending the cellular domain and Keeping the cell. The following paragraphs explain those rules.

(1) **Chemical Kinetics**: Both activator and inhibitor are produced by the stimulation of the external signal (Levchenko and Lglesias, 2002). The activator enhances polymerization of actins, whereas the inhibitor suppresses the polymerization. First, this rule selects a grid in the cellular domain randomly. If concentrations of activator, inhibitor and actin filaments at the selected grid j are expressed as A_j, I_j and F_j, respectively, those variables are updated to A'_j, I'_j and G'_j obeying the following equations:

$$A'_j = A_j + \alpha S_j - k_\alpha A_j \qquad (1)$$

$$I'_j = I_j + \beta S_j - k_\beta I_j \qquad (2)$$

$$F_j' = \begin{cases} \gamma - k_f F_j & (\frac{A}{I} > h) \\ -k_f F_j & (\text{otherwise}) \end{cases}, \quad (3)$$

where α, β, γ, k_α, k_β, k_f and h are constants. S_j indicates the concentration of chemoattractants or the strength of the external signal at the jth grid. Grids at the border of the cellular domain are regarded as the cellular membrane. We say that a grid in the cellular domain is in the membrane if at least one of the six nearest grids is external. S_j is set to zero if the jth grid is in the cellular domain but not in the membrane. The functional form of S_j represents chemical gradients.

(2) **Diffusion**: Only the inhibitor diffuses into the whole cytoplasm (Levchenko and Lglesias, 2002). This rule selects a grid from the whole cellular domain. At the selected jth grid and its nearest cellular lth grid, I_j and I_l are redistributed obeying the following equations:

$$I_j' = I_j - DI_j \quad (4)$$

$$I_l' = I_l + \frac{DI_j}{n}, \quad (5)$$

where D is the diffusion constant. n is the number of the nearest cellular grids. D should be smaller than 1 by definition.

(3) **Actin filaments extending the cellular domain**: The rule randomly selects a grid from the membrane. When F_j at the selected jth grid in the membrane reaches the threshold F_{th}, an external grid in the six nearest grids of the jth grid is turned into a cellular grid. When there are two or more than two external grids around the jth grid, a grid is randomly selected. If this grid is referred to as l, $F_l = F_j/2$ and other variables are set to zero. G_j' equals to $F_j/2$ by definition, where the prime indicates the value at the next time step.

(4) **Keeping the cell**: We also give a rule to prevent a cell from breaking into pieces. The cellular volume is kept and the cellular surface length is constrained to be as small as possible. This rule randomly selects a grid from the membrane. Then the rule decides either to remove the grid or to add a new cellular grid around the grid. This rule checks the cellular "tension" by calculating energy of tension as:

$$E = (V - V_0)^2 + cl^2, \quad (6)$$

where V and l are the cellular volume and length of the membrane and V_0 and c are constants. When E' denotes the energy after either removing or adding a cellular grid, we define the probability P as follows:

$$P_e = \exp\left(-\frac{E' - E}{kT}\right), \quad (7)$$

where kT is a constant. We generate a random number between 0 and 1 and then compare the number with P_e. If the number is smaller than P_e, we "undo" the event of removing/adding. From the definitions of P_e and E, the volume of

the cell tends to be V_0 and the length of the membrane becomes as small as possible. Note that if removing is chosen, the values of A, I and F in the removed grid are added into the nearest cellular grid.

We also give the "master" rule that randomly selects one of the above rules. Each rule has the probability of selection. The probabilities of selection for rules from (1) to (4) are written as P_1, P_2, P_3 and P_4. $P_1 + P_2 + P_3 + P_4$ should equal to 1. After the master rule selected one of the four rules, the selected rule is executed. We iterate this process several millions times.

Results

First, we try to clarify what our cell does in a simple linear gradient. We select a set of parameters to make the cell go up the gradient as follows: $\alpha = 1.0$, $\beta = 0.1$, $k_\alpha = 0.9$, $k_\beta = 0.02$, $\gamma = 4.0$, $k_f = 0.99$, $D = 0.45$, $h = 10.0$, $F_{th} = 1.0$, $P_1 = 0.0419$, $P_2 = 0.03$, $P_3 = 0.03$, $P_4 = 0.898$, $V_0 = 900$, $c = 1.2$ and $kT = 100$. The initial diameter of the cell equals to 30 grids. Although we have not yet exhaustively tried different parameter sets, we expect that the cell behaviors are robust against the parameter change. We let $S_j = -y + \text{constant}$ which indicate a linear gradient.

Figure 1 shows snapshots of our simulation, in which a cell clearly moves downward according to the gradient. First, actin filaments increases in all membrane grids, then actin filaments remains in only the "head" of the cell. Figure 2 clarifies that the cell does not move in the completely same direction as the gradient direction but motions of the cell have large fluctuations with regard to the x-direction.

Jeon et al. have succeeded in making a new device that can maintain complex but static gradients (Jeon et al., 2002). By using this device, they analyzed the locomotion of neutrophils in several patterns of gradient. They tested "Hill" gradient, for example, in which concentration initially increases then decreases at a middle point along a certain direction. Neutrophils starting form a bottom climbed up the hill, and they continued to proceed without stopping at the top until they finally returned from the downhill side. The cell of our model shows the same behavior at least qualitatively. In Figure 3 the simulated cells move beyond the top of the hill, where the gradient is drawn in the left side of the figure.

We also analyze "Flat & Drop" gradient. This gradient starts with a flat pattern. From a certain point, the concentration of chemicals drops gradually. Initially, the cell is "pushed" by an additional event defined as follows: First a membrane grid is randomly selected. If the vertical position y of the selected grid is larger than l, the grid is removed. This event has an effect as if to push the cell by a plank. l is set to 7.5, about a half of the cellular radius. Figure 4 depicts trajectories of the cells pushed. After pushed, cells move down in the decreasing y direction even though there is no gradient. After passing the line $y = -50$, the cell moves

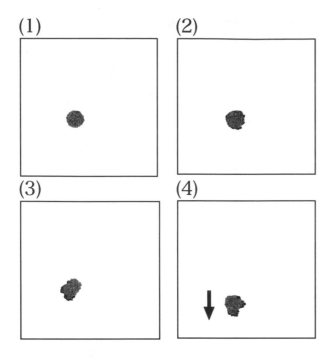

Figure 1: Snapshots of the cell moving. The gray collared area is the region of cellular grids. Dark part of the cell indicates positions where F_j is larger than $F_{th}/2.0$. S increases in the downward direction.

against an upward gradient until they finally return.

Cases of Hill and Flat & Drop gradients imply that the present model cells have a tendency to keep their motions. Figure 4 shows that cells continue to move even without gradient. The cells do not stop soon in both cases. This property of the simulated cells has some resemblance to "inertia" in Newtonian dynamics. Cells move as if balls move in the potential which has the reversed pattern of the chemical gradient. In the case of Hill gradient, for example, cells passing the top of the gradient look like balls going up an uphill by inertia. In the case of Flat & Drop gradient, cells moving the flat part look like balls rolling on a flat floor.

The cell should turn around a source point of chemoattractants if it has an inertia-like property. We define "Central" gradient as follows: Strength of signals S equals to $C - u|\vec{r} - \vec{r}_0|^2$, where C and u are constant. All gradient vectors point to the center $\vec{r}_0$. This gradient is an analogy of a linear spring that binds a material point to a fixed point. In Figure 5 we can see that a fluctuating trajectory indeed turns around a center (using $C = 4000$, $u = 1$ and $\vec{r}_0 = (70,0)$). Here, the cell was initially pushed down by calling the event procedure defined in the Flat & Drop gradient study, during a short period of starting steps. Although not all the trajectories starting with different random seeds showed the circular orbit as shown in Figure 5, most of them did not move directly toward the center but turned around the center for a while, showing the existence of the inertia effect working at

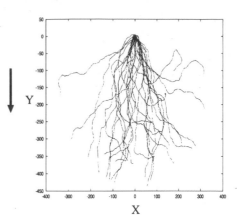

Figure 2: Trajectories of centers of cells in Linear gradient. Each trajectory starts with a different random seed but the same initial condition.

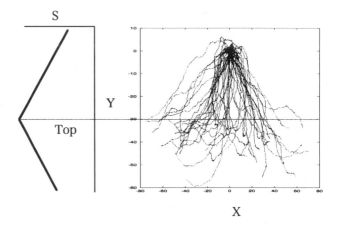

Figure 3: Motions of cells in Hill gradient. The shape of the gradient is drawn in the left.

least for a short time scale.

Why does the cell have an inertia-like property? Although the definite answer is not yet known, the distribution of Inhibitor I_j gives a clue to the answer. Figure 6 indicates I_j in a cell moving linearly in a flat field without gradient. I_j tends to be concentrated on the rear of the cell. This heterogeneous distribution of Inhibitor is kept until the cell stops moving. We speculate that such heterogeneity is maintained by the following mechanism: While moving, the head of the cell is always extended, whereas the rear is always retracted. Inhibitor gradually gathers into the rear if this extending-retracting processes are slightly faster than diffusion of Inhibitor. The large concentration of Inhibitor thus accumulated suppresses increase in actin filaments F_j, which should decrease the probability that F_j at the rear exceeds the threshold. In this way motion of the cell maintains the bias

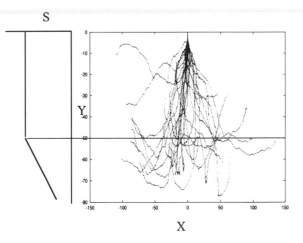

Figure 4: Behavior of cells in Flat & Drop gradient. The pattern of gradient is changed below the line $y = -50$ as is drawn in the left figure. In the area of $y < -50$ the direction of the gradient is upward. The cells move against the gradient until they finally turn to the gradient.

of Inhibitor which further stabilizes the cell motion. The cell does not stop immediately once the cell starts to move. When the gradient is opposite to the direction of cellular motion, we easily conjecture that the opposite gradient does not extinguish the bias at once.

Discussion

An inertia-like property of amoebic cells has been observed in experiments though the term "inertia" has not been used. For example, Verkhovsky et al. observed that the polarized cellular fragments undergo locomotion even without the gradient of chemoattractant (Verkhovsky et al., 1999). Those fragments sometimes stayed still but stared to undergo locomotion when a mechanical stimulus was applied. They showed that the actin-myosin II bundle is formed at one edge of the fragments associating with the locomotion. In general, cellular locomotion in a uniformed concentration is called "chemokinesis". Cells in a uniformed concentration can show long straight runs although their directions are random relative to one another in a short time scale (Wilkinson, 1998). From studies of Jeon et al., neutrophils were shown to move without changing directions immediately after the change of interleukin-8 gradient (Jeon et al., 2002).

What type of equation does the cellular motion with "inertia" should obey? Figure 7 depicts the averaged y-position of cells which move in Linear gradient as a function of time as well as a fitting curve. In this figure initial 1400 steps are shown. Data are same as those used in Figure 2 and the average was taken over all cells. Note that the total steps executed in Figure 2 are about 6000 steps. The fitting curve

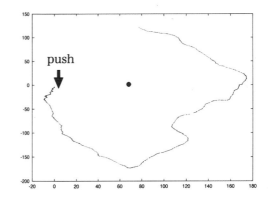

Figure 5: A trajectory in "Central" gradient. The cell starts from the point (0,0). The center is set to (70,0). The cell is pushed down during a short period of starting steps. The cell then turns around the center counterclockwise.

is a solution of the following underdamped equation:

$$\frac{d^2\vec{r}}{dt^2} = -p\frac{d\vec{r}}{dt} + q\vec{\nabla}S(\vec{r}) \qquad (8)$$

where $\vec{r}$ indicates the mean position of the cell, $S(\vec{r})$ is the chemical concentration at $\vec{r}$ and p and q are unknown functions but approximately regarded as constants. This equation is a Newtonian, underdamped equation. Since the solution curve do not fit the averaged y-position of cells after about 2000 steps, Equation (8) may not hold for the long time scale. An equation that should hold for the longer time scale might be more complex.

Does inertia of cells have any biological advantage? We do not have any clear answer for this question yet. A naive idea is that cells can avoid local maxima of concentration by inertia to find out the global maximum. Another idea is that when amoebic cells catch foods or enemies by phagocytosis, they should be moving around for a moment to prevent from missing foods or enemies. Anyway, this is an open question to be investigated.

Conclusion

We succeeded in making a model that explains amoebic chemotaxis. The cellular locomotion and the chemical processes are strongly coupled to each other, which stabilizes the "inertia"-like motion of the cell. The cell in our model moves in a way similar to the motion in Newtonian underdamped dynamics.

Acknowledgments

This work was supported by ACT-JST project of Japan Science and Technology Corporation.

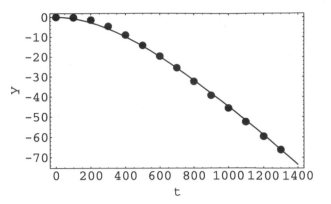

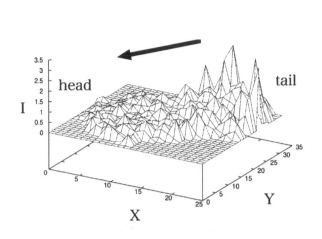

Figure 6: Distribution of Inhibitor I_j. X and Y axes indicate spatial coordinates of grids. Z axis indicates the value of I_j.

Figure 7: Horizontal and vertical axes indicate time and the y-position, respectively. Dots are sampled points of center positions averaged over all cells examined in Linear gradient. A solid curve indicates the function $y(t) = p^{-1}\{\exp(-qt) + qt - 1\}$ (which is the solution of $\ddot{y} = -p\dot{y} + q$, where $p = 0.002$, $q = -0.025$, $y(0) = 0$ and $\dot{y}(0) = 0$) that approximately fits the sampled points.

References

Iijima, M., Huang, Y. E., and Devreotes, P. (2002). Temporal and spatial regulation of chemotaxis. *Dev. Cell*, 3:469–478.

Jeon, N. L., Baskaran, H., Dertinger, S. K. W., Whitesides, G. M., Water, L. V. D., and Toner, M. (2002). Neutrohil chemotaxis in linear and complex gradients of interleukin-8 formed in a microfabricated device. *Nat. Biotechnol.*, 20:826–830.

Katanaev, V. L. (2001). Signal transduction in neutrophil chemotaxis. *Biochemistry (Moscow)*, 66(4):351–368.

Levchenko, A. and Lglesias, P. A. (2002). Models of eukaryotic gradient sensing: Application to chemotaxis of ameobae and neutrophils. *Biophys. J.*, 82:50–63.

Pollard, T. D. (2003). The cytoskeletion, cellular motility and the reductionist agenda. *Nature*, 422:741–745.

Pollard, T. D., Blanchoin, L., and Mullins, R. D. (2000). Molecular mechanisms controlling actin filament dynamics in nonmuscle cells. *Annu. Rev. Biophys. Biomol. Struct.*, 29:545–76.

Pollard, T. D. and Borisy, G. G. (2003). Cellular motility driven by assembly and disassembly of actin filaments. *Cell*, 112:453–465.

Roitt, I., Brostoff, J., and Male, D. (1998). *Immunology*. Mosby International Ltd.

Springer, T. A. (1994). Traffic signals for lymphocyte recirculation and leukocyte emigration: The multistep paradigm. *Cell*, 76:301–314.

Verkhovsky, A. B., Svitkina, T. M., and Borisky, G. G. (1999). Self-polarization and directional motility of cytoplasm. *Curr. Biol.*, 9:11–20.

Wilkinson, P. C. (1998). Assays of leukocyte locomotion and chemotaxis. *J. Immunol. Meth.*, 216:139–153.

Phenotypic Variability in Canalized Developmental Systems

Sean T. Psujek[1] and Randall D. Beer[1,2]

[1]Department of Biology
[2]Department of Electrical Engineering and Computer Science
Case Western Reserve University, Cleveland, Ohio, USA 44106
stp4@case.edu, rxb9@case.edu

Abstract

The evolution of nervous systems has been relatively conservative despite a large diversity of behavior. Novel behaviors appear to be produced during evolution by changing synaptic connectivity (Edwards and Palka 1991, Shaw and Meinertzhagen 1986, and Nishikawa et al. 1992). Due to a process of canalization, in which the development of organisms becomes buffered against genetic and environmental changes, phenotypic variation is reduced (Waddington 1942). Therefore, we explore phenotypic variability in simulations of neural development in which the developmental pathways have become canalized. Variation is determined under point mutations and gene knockouts. Although a genetic regulatory network utilizing activation and controlling a single process of neural development is evolvable and can become canalized, large genetic perturbations do not result in increased phenotypic variability as expected from related work (Bergman and Siegal 2003).

Introduction

Over evolutionary time, the behavior of organisms has varied considerably despite the relatively conservative evolution of nervous systems (Nishikawa 1997). For example, insect nervous systems have a similar cellular composition yet the class shows remarkable diversity in behavior (Edwards and Palka 1991). Instead of cell type alterations, connectivity changes in nervous systems appear to play a large role in generating new behaviors in invertebrates (Edwards and Palka 1991, Shaw and Meinertzhagen 1986) as well as vertebrates (Nishikawa et al. 1992).

Connectivity changes occur as modifications in neural development; yet, developmental systems are canalized, that is, they are buffered against genetic and environmental perturbations and show decreased phenotypic variation (Waddington 1942). Phenotypic variability can be regained through large genetic and environmental perturbations which decanalize the developmental system (Rutherford and Lindquist 1998, Bergman and Siegal 2003). Do the synaptic changes that lead to novel behaviors occur as a result of small scale genetic variation (e.g. point mutations) or are they due to large genetic perturbations such as gene knockouts?

To answer our motivating question, we seek to create a model system that demonstrates canalization as well as a release from this buffering. Recently, another simulation study has demonstrated decanalization in genetic regulatory networks (GRN) through gene knockouts (Bergman and Siegal 2003). The phenotype in their study was the pattern of gene expression at equilibrium. We use the connectivity matrix of the neural network that results from the pattern of gene expression as our measure of phenotype. We aspire to have the behavior of an artificial organism as the phenotype upon which selection will be based.

In this paper, we introduce our model of neural development and demonstrate that it can reproduce the connectivity pattern of biological neural networks. We also analyze the evolvability of the model and illustrate canalization in evolved genetic regulatory networks. Using canalized and non-canalized developmental systems, we explore phenotypic variation produced under two genetic perturbations. Using this model, we hope to elucidate the mechanisms in developmental systems underlying phenotypic variability.

Target Identification Model of Neural Development

There are numerous processes (e.g. cell migration, cell death, axon pathfinding, target cell identification, synaptogenesis) involved in neural development. Even though these processes all work together to create the final pattern of connectivity, neurites must make a final selection for synaptogenesis among nearby cells (Spencer et al. 2000). Therefore, we focused on the process of target cell identification to generate neural circuits.

A generalized interpretation of Sperry's chemoaffinity hypothesis is used to determine neural specificity (Sperry 1963). In our model, cell surface proteins identify the pre-synaptic and post-synaptic cell and determine the type of connection. Neurons will make synaptic connections only if the surface proteins of the cells are complementary. The post-synaptic protein determines whether the connection is inhibitory or excitatory. Dellaert and Beer (1996) and Fleischer (1995) used a similar mechanism to identify target neurons although both models incorporated the guidance of axon growth cones as an additional developmental process. In our model, neurites are not guided and are assumed to have spread throughout the two-dimensional space in which the nervous system is grown.

Using only one process of neural development allows us to add additional complexity to the model in the direction of other neural developmental processes, regulatory mechanisms, or genetic operators (e.g. gene duplication/deletion) based on our findings.

Our interpretation of Sperry's ideas is supported by experiments in cultured neurons from the snail, *Lymnaea stagnalis* (Syed et al. 1990, Magoski and Bulloch 1998). After removing the neurons that form the respiratory central pattern generator (CPG) from the snail and culturing the cells in a brain-conditioned medium, the neurons extended neurites in all directions and formed the appropriate synaptic connections to recreate the CPG.

To simulate this process, our model of neural development proceeded through four steps. In the first step, cells were placed by the user at specific positions in a planar environment. These positions did not change during development. Following the placement of cells, a growth period occurred in which a regulatory gene network involving intra- and extra-cellular signaling proteins controlled the expression of synapse-determining surface proteins. At the end of the growth period (the length of which is arbitrarily determined by the experimenter), a list was produced of the surface proteins expressed by each cell. Finally, this list was converted by a function into the connectivity matrix for the network. The connectivity matrix indicated which neurons were connected and whether the connections were inhibitory or excitatory.

Figure 1 shows the final step of development in the formation of three synaptic connections from the pattern of expressed surface proteins. After the growth period, each cell has two surface proteins expressed represented in the figure by squares, triangles, and circles. Similar proteins have the same shape. The filled shapes are pre-synaptic proteins and open shapes are post-synaptic. A neural connection is formed between cells in which a complementary pair is expressed. The function ignores complementary surface proteins expressed in the same cell (i.e. no self-connections are formed).

Surface proteins encoded in the genome are expressed by intra- and extra-cellular signaling proteins during the growth period of development. At each time step during growth, the developmental algorithm checks each cell's intra- and extra-cellular environments for signaling proteins causing transcription events. Expressed surface proteins are then bound in the cell's membrane and do not degrade over the remainder of the growth period. If new signaling factors are expressed, they go into a new cell environment list which replaces the old one after each time step. In effect, signaling proteins act for only one time step before being degraded.

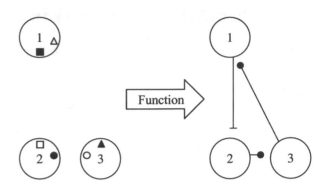

Fig. 1. Expressed surface proteins are used by a function to create the synaptic connections. Each shape in the left group of cells represents a unique surface protein. Complementary protein pairs are designated by filled and open shapes. In the right group of cells, excitatory connections are indicated by short bars, and inhibitory connections are indicated by filled circles.

Figure 2 shows a gene template of our model which encodes the various proteins and allows for gene regulation. We used two bases represented as '0' or '1', instead of the four used in DNA (adenine, guanine, cytosine, thymine), to create the functional unit of a gene. Alleles, therefore, are genes that express different proteins. There are four regions of an individual gene: activation, response range, function, and protein identifier.

Gene regions:

Activation	Response Range	Function	Protein Identifier
xxxx	xxxx	xxxxx	xxxxxx

Fig. 2. Gene template. Each 'x' represents the location of a single base. There are only two bases in our model, '1' and '0'.

The activation and response regions act as a gene's promoter by determining if a protein causes expression of a gene. An intra-cellular signaling protein need only match the activation region of a gene to cause its expression. An extra-cellular protein must not only match the activation region but must also have been expressed close enough to another cell for a particular gene to respond. This mimics a concentration gradient. The response region encodes distance and is used to determine if the distance of the origin of the extra-cellular factor is too far away for the gene to respond. This encoded distance is mutable, allowing a gene, over evolution, to respond to a chemical further away or to prevent expression if the chemical is released too close.

The functional region encodes the protein's role in development. Proteins in the model can act as signaling chemicals within and between cells or as cell surface proteins that determine synaptic connections. Intra- and extra-cellular signaling proteins result in transcription of genes. Cell surface proteins could be of three forms: a pre-synaptic, an inhibitory post-synaptic, or an excitatory post-synaptic chemical tag. Mutations in this region can result in the function of a protein being altered, for example, turning a pre-synaptic tag into an extra-cellular signaling protein, but mutations will never result in a functionless protein.

Last in the gene sequence is the protein identifier region. The bit sequence here uniquely describes the protein expressed; there was no translation component to this model. Converting this bit sequence from base 2 to base 10 gave the protein's numerical identifier used in the examples below.

Mutations occurring in the two bits at the end of the functional and protein identifier regions do not affect the respective region's function/property and thereby add stability to the developmental system.

Transcriptional regulation of the genes is accomplished by the signaling proteins. In Figure 3, a signaling protein indicated by an oval binds to an equivalent sequence on the activation region of *Gene 1* resulting in the expression of another signaling protein, '0001'. It, in turn, activates *Gene 2* to produce a cell surface protein, '0110'.

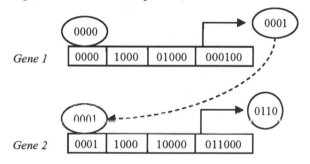

Fig. 3. Transcriptional regulation. Signaling proteins (ovals) act as transcription factors by binding to the activation region of the gene initiating the production of other signaling proteins and surface proteins (circle). Solid arrows indicate transcription of a gene's protein coding region.

Figure 4 shows a genome that creates a three-connection neural network in three cells. It will be used to demonstrate how the developmental process results in a neural network. The binding of signaling proteins to activation regions is shown by shading the protein and the bound region. '0000' is the symmetry-breaking protein introduced during the first time step. It results in transcription of genes 1, 2, and 3. The last column indicates the function and identifier of the encoded protein. For example, the bit sequence in the functional region of *Gene 2*, '1100', (minus the last two bits) denotes that the protein is an inhibitory post-synaptic tag. The protein identifier, '0010', converted to base 10 is '2'. For ease of reading, we use the English description of the protein instead of its binary sequence.

The expression pattern of this genome in three neurons is shown in Figure 5. A protein was introduced only into *Neuron 1* at the beginning of the developmental period to break the symmetry between the developing cells. The arrows indicate expression of genes with the resultant proteins at the head of the arrow. Extra1 and Extra2 are

proteins that can leave the cell and induce expression events in nearby cells (they do not result in gene expression in the cell from which they were expressed). Note that Extra1 does not result in the transcription of Post0_Exc, Pre1, and Extra2 in *Neuron 3* due to the response region of those genes. Extra1 was expressed too far away for the genes corresponding to those proteins in *Neuron 3* to respond to the extra-cellular signaling protein. The neuronal circuit for this expression pattern is shown in Figure 1 on the right.

	Act.	Resp.	Fctn.	Id.	
1		1000	10000	000000	Pre(-Synaptic)0
2		1000	11000	001000	Post(-Synaptic)2_Inh
3		1000	01000	000100	Extra(-cellular)1
4	0001	1000	10000	000100	Pre(-Synaptic)1
5	0001	1000	10100	000000	Post(-Synaptic)0_Exc
6	0001	1000	01000	001000	Extra(-cellular)2
7	0010	0110	10000	001000	Pre(-Synaptic)2
8	0010	0110	11000	000100	Post(-Synaptic)1_Inh

Fig. 4: Hand-designed genome for a three-connection circuit in three neurons. Shaded regions represent the signaling proteins and the activation regions to which they bind (ignoring the last two bits of the protein identifier region). The darkest transcription regions are those which are bound by the symmetry breaking protein (0000) in the first stage of development. _Inh represents an inhibitory protein, and _Exc represents an excitatory protein.

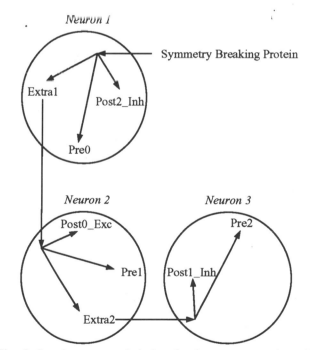

Fig. 5. Proteins expressed during development to produce the three-connection network with three neurons. Arrows indicate the cascade of expression events. Distances between cells shown above are representative of the differences in cell spacing used in development.

To demonstrate that this model is capable of reproducing the connectivity of biological neural networks, we sought to replicate the three-interneuron *Lymnaea* respiratory CPG (Syed et al. 1990) with an additional connection from interneuron RPeD1 to a fourth interneuron, VD2/3 (Magoski and Bulloch 1998). This target network was reproduced using 15 genes. The result of development from this new genome is show in Figure 6.[1]

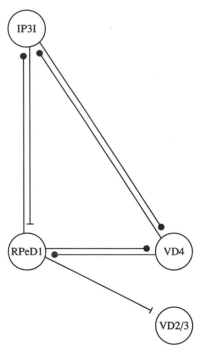

Fig. 6. The circuit architecture and cell spacing for the *Lymnaea* respiratory network. Neuron IP3I controls expiration and VD4 controls inspiration by opening and closing, respectively, the snail's pneumostome. Excitatory connections are indicated by short bars. Inhibitory connections are indicated by filled circles.

Characterization of Single-bit Mutations

We explored the effects on phenotype of single-bit mutations in the genome that produced three-connections. For our genome of eight genes of nineteen bits each, we performed all possible one hundred fifty-two single-bit mutations and found nineteen unique architectures resulted. The majority of the mutations (fifty-nine) resulting in the original circuit. Only ten mutations resulted in no synaptic connections being formed. Ten architectures occurred only once from a point mutation. Figure 7 shows the original three-connection circuit and all of the unique architectures resulting from the point mutations.

Mutations had the effect of removing or adding synaptic connections or performing a combination of both resulting in high variability from this genome. This analysis

provides some evidence for the evolvability of the system with the caveat that it was based on a hand-designed genome. It also indicates that small scale genetic variation, i.e. single-bit mutations, can produce large phenotypic variation, the number of unique networks produced.

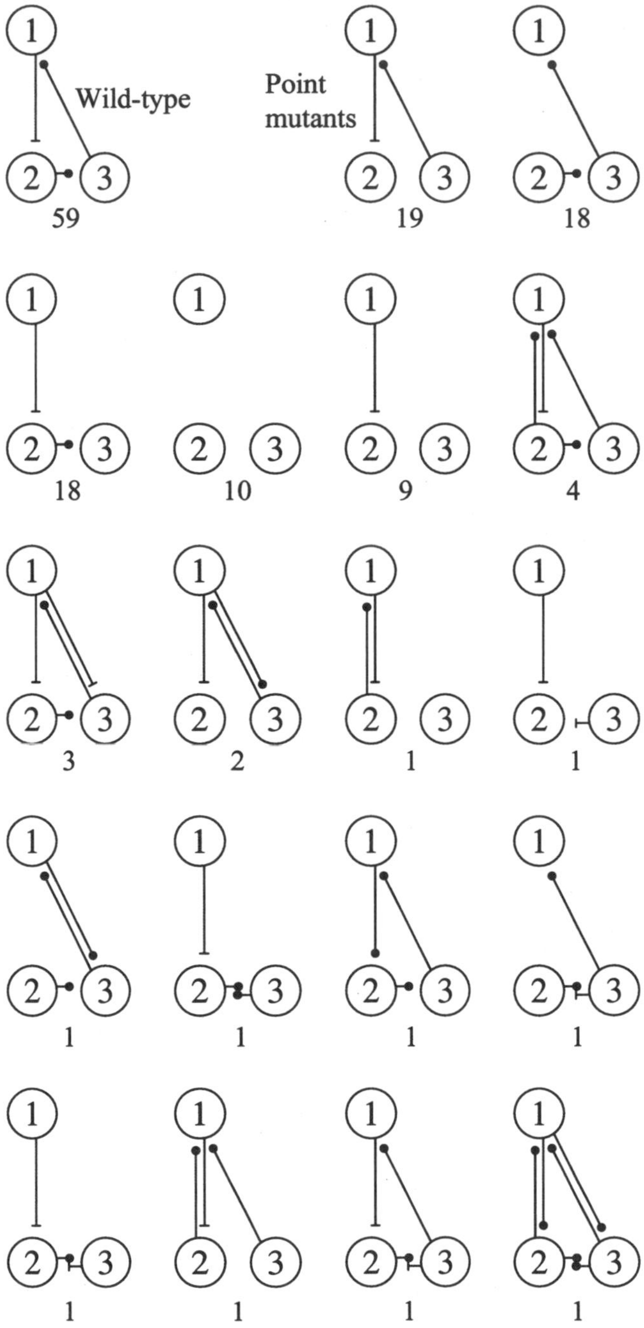

Fig. 7. The circuit architecture and cell spacing for a three-connection network and all unique architectures produced through all possible single-bit mutations. The number of mutations resulting in the network is shown below each circuit.

---

[1] RPeD1 to IP3I is actually a mixed inhibitory-excitatory connection but whose action is mainly inhibitory.

Canalization of Development after Evolution

We performed fifty genetic algorithm (GA) runs from random initial genomes. As the hand-designed genome was composed of eight genes, we decided to use ten genes for the GA runs to allow for more possibilities of cell-cell signaling and expression of surface proteins.

We used a mutation rate that averaged one mutation per genome. No crossover or an elitist fraction was implemented. Individuals were haploid and reproduction was asexual.

As a measure of fitness, we decided to use the ratio of actual to maximum number of connections. This fitness function was chosen for its simplicity during the initial testing of evolving neural networks using the developmental model. Ignoring self-connections the maximum number of connections in a given n-neuron network is n^2-n. The ratio was entered into a parabolic function that returns a value from zero to one with a maximum value of one at a ratio of ½. Selection was fitness proportionate.

Forty-one of the GA runs were able to reach the maximum fitness in three hundred generations. The other nine runs never produced synaptic connections. In those runs, the random initial genomes and subsequent mutations never resulted in the minimum expression of a pre-synaptic, a post-synaptic, and an extra-cellular signaling protein required to form at least one synaptic connection.

Phenotypic Variation of Canalized Runs

If there is an implicit selective pressure for developmental robustness, it would be strongly evident in the sub-population of maximum fitness individuals; otherwise, the explicit selective pressure from our fitness measure would be stronger. Therefore, we calculated the average number of neutral mutations (the number of mutations that leave the circuit architecture unchanged) across all genomes producing maximum fitness circuits in each generation. In Figure 8, we show the increase in this sub-population's robustness over evolution for one GA run.

To determine if there exists a selective pressure for this pattern of increasing robustness to mutations, we performed the neutral mutation count on all of the forty-one GA runs that produced genomes resulting in maximum fitness. There is a highly significant difference in the number of neutral mutations in the best genome at the generation when the GA run first reaches a maximum fitness (and continues unbroken until the last generation) and at the last generation (P=0.000685, one-tailed Fisher Sign test).

To explore phenotypic variability, we determined the number of unique networks produced through all point mutations for the best individual at the beginning of the maximum fitness continuation and for the best individual at the last generation for all forty-one runs. The results of

this analysis are show in Table 1. These data indicate that buffering of development against genetic perturbations also decreased the phenotypic variation.

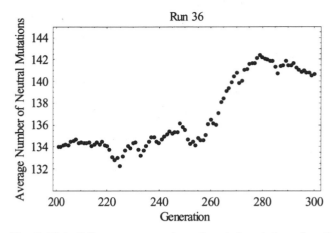

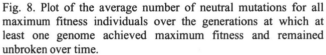

Fig. 8. Plot of the average number of neutral mutations for all maximum fitness individuals over the generations at which at least one genome achieved maximum fitness and remained unbroken over time.

Table 1. *Phenotypic variability in canalized and non-canalized GA simulations*

		Variation		
GA simulations	N	Decreased	Increased	Unchanged
Canalized	27	19	4	4
Non-canalized	14	1	11	2

N, number of runs

Phenotypic Variation after Gene Knockouts

Bergman and Siegal (2003) show an increase in phenotypic variability after gene knockouts in their regulatory gene networks. For the canalized runs in our simulation, we deleted each gene in turn from the best genome at the last generation and then determined the phenotypic variation. All but one of the twenty-seven canalized runs remained unchanged or decreased in variation for the gene knockouts. The unique run increased its phenotypic variation by one network after knocking out a gene.

Conclusion

Our work demonstrated that a process of target cell identification encoded into a genetic regulatory network can generate the synaptic connectivity of a known biological neural network and produce wide variation in neural network architectures through single-bit mutations. Regulatory networks provide combinatorial expression of developmental controlling factors leading to large phenotypic variation, the substrate upon which selection acts (Carroll et al. 2001).

But developmental pathways can become buffered against genetic and environmental disturbances thereby reducing phenotypic variation. In our simulations, after the maximum fitness had been obtained, genomic changes occurred that increased a GRN's ability to withstand mutations yet decreased the number of different networks produced through low genetic variation.

Phenotypic variability could not be regained through larger genetic perturbations in contrast to Bergman and Siegal's work (2003). This suggests a lack of regulatory complexity. Increased complexity can be achieved through two common regulatory motifs: gene repression (Hanna-Rose and Hansen 1996) or multi-component promoters (Lee et al. 2002), which were present in Bergman and Siegal's model. We plan to add gene repression next to the model and continue our investigation of phenotypic variability. Gene repression may allow for a release from canalization and greater phenotypic variation after gene knockouts.

Eventually, instead of using a topological fitness measure, we will implement behavioral fitness measures to explore the phenotypic variability necessary for the origin of behavioral novelty through connectivity changes.

Acknowledgements

We thank Jennifer Talley and four anonymous reviewers for useful comments. This research was made possible in part by a NSF IGERT fellowship in Neuromechanics at Case Western Reserve University and by NSF grant EIA-0130773.

References

Bergman, A. and Siegal, M.L. (2003) Evolutionary capacitance as a general feature of complex gene networks. *Nature* 424:549-552.

Carroll, S.B., Grenier, J.K., and Weatherbee, S.D. (2001) From DNA to Diversity. Walsworth Publishing Company.

Dellaert, F. and Beer, R.D. (1996) A developmental model for the evolution of complete autonomous agents. In P. Maes, M. Mataric, J. Meyer, J. Pollack and S. Wilson (eds.), *From animals to animats 4: Proceedings of the Fourth International Conference on Simulation of Adaptive Behavior*. MIT Press.

Edwards, J.S., and Palka, J. (1991) Insect neural evolution – a fugue or an opera? *Semin. Neurosci.,* 3:391-398.

Fleischer, K. (1995) A multiple-mechanism developmental model for defining self-organizing geometric structures. Ph.D. diss., Dept. of Computer Science, California Institute of Technology.

Hanna-Rose, W., and Hansen, U. (1996) Active repression mechanisms of eukaryotic transcription repressors. *Trends Genet.* 12:229-234.

Lee, T.I., et al. (2002) Transcriptional regulatory networks in *Saccharomyces cervisiae. Science* 298:799-804.

Magoski, N.S. and Bulloch, A.G.M. (1998) Trophic and contact conditions modulate synapse formation between identified neurons. *J. Neurophysiol.* 79:3279-3283.

Nishikawa, K.C. (1997) Emergence of novel functions during brain evolution. *Bioscience* 47:341-354.

Nishikawa, K.C., Anderson, C.W., Deban, S.M., and O'Reilly, J.C. (1992) The evolution of neural circuits controlling feeding behavior in frogs. *Brain Behav. Evol.* 40:125-140.

Rutherford, S.L. and Lindquist, S. (1998) Hsp90 as a capacitor for morphological evolution. *Nature* 396:336-342.

Shaw, S.R. and Meinertzhagen, I.A. (1986) Evolutionary progression at synaptic connections made by identified homologous neurons. *Proc. Natl. Acad. Sci. USA* 83:7961-7965.

Spencer, G.E., Lukowiak, K., and Syed, N.I. (2000) Transmitter-receptor interactions between growth cones of identified *Lymnaea* neurons determine target cell selection *in vitro. J. Neurosci.* 20:8077-8086.

Sperry, R.W. (1963) Chemoaffinity in the orderly growth of nerve fiber patterns and connections. *Proc. Natl. Acad. Sci. USA.* 50:703-710.

Syed, N.I., Bulloch, A.G.M., and Lukowiak, K. (1990) In vitro reconstruction of the respiratory central pattern generator of the mollusk *Lymnaea. Science.* 250:282-285.

Waddington, C.H. (1942) Canalization of development and the inheritance of acquired characters. *Nature* 150:563-565.

Self-repairing and Mobility of a Simple Cell Model

Keisuke Suzuki and Takashi Ikegami

General Systems Sciences,
The Graduate School of Arts and Sciences,
The University of Tokyo,
3-8-1 Komaba, Tokyo, 153-8902, Japan

Abstract

The evolution of a mobile cell system is studied here. While a primary definition of life is self-reproduction, mobility is also a major characteristic. Typically, most alife models assume cell mobility a priori. Here, using a development of a cell model due to Varela and McMullin, we show the emergence of cell mobility. A balance between cell repair and mobility is demonstrated. The introduction of a new functional particle is a critical part of the present model.

Introduction

Self-reproduction, one of the primary characteristics of living systems, has been studied both experimentally and theoretically for more than 40 years. Initially, von Neumann showed the existence of a universal constructor in a 2-dimensional cellular automaton with 29 cell states (Neumann, 1966). More recently, Langton, Tempesti, Morita and Imai, Sayama, Suzuki and Ikegami have demonstrated the dynamics of self-replicators with special rule sets (Langton, 1984; Tempesti, 1995; Morita and Imai, 1997; Sayama, 1999; Suzuki and Ikegami, 2003). Although in these models self-replication occurs successfully, the question of how the underlying physical chemistry could support such specific symbolic rule sets has not received much attention. Recently, a series of studies by Rasmussen, McCaskill, Ono and Ikegami and others have indeed shown that simple catalytic networks can develop cellular membranes and self-reproduction processes(Mayer et al., 1997; Breyer et al., 1998; Ono and Ikegami, 1999; Ono and Ikegami, 2000). Self-reproduction in 3-dimensional space has also been demonstrated by extending Ono's model(Madina et al., 2003).

What has not been achieved so far is self-mobility; a unit that acquires mobility in some physical space. In this paper, we will study the origin of mobility in a cell system based on a simple discrete cell model, the so-called SCL model which was introduced by Varela(Varela et al., 1974) and reformulated by McMullin(McMullin and Varela, 1997). Varela used the model as a practical example of autopoiesis, a principle of biological autonomy common to all living things.

An autopoietic system is organized as a network of processes that composes the system. Each process recreates the other processes which together comprise a coherent unit in some physical domain. Indeed, Varela claims that autopoiesis is a necessary and sufficient condition for what we call life (Varela, 1979). A cell that generates its own boundary to enclose a catalyst which in turn sustains the boundary is a simple example of autopoietic organization.

However, we believe that a new principle is required to understand the dynamics and evolution of living systems. Previous cellular models of autopoiesis stressed the notion of autonomous self-maintenance. Here, we stress the notion of motility by extending the SCL model. In the original model, components of the cell are used for maintaining the cell boundary. Here we also employ them for cell movement. We introduce a new kind of membrane component, a functional LINK particle. We use the term functional as these particles work both to repair membranes and move cells.

The Model

We use a modification of the "Substrate-Catalyst-Link" (SCL) model originally introduced by Valera to explain the idea of autopoiesis. Before introducing our modifications, we first explain how the original model works.

The SCL model consists of three kinds of particles which reside on a square lattice. They are called "SUBSTRATE" "CATALYST" and "LINK" particles. They all diffuse freely in the space at given constant rates. If two SUBSTRATE particles are in the Moore neighborhood of a CATALYST particle, they disappear and a single LINK particle is formed in place of one of them. LINK particles neighboring each other are bonded. Each LINK particle is allowed to bond with at most two other LINK particles, so that the bonding tends to form a linear chain. A quasi-stable structure is a closed loop of bonds. SUBSTRATE particles can diffuse freely through the loop, while CATALYST particles cannot.

Each LINK particle, whether bonded or not, decays into two SUBSTRATE particles at a certain rate. A loop of LINK particles may be regarded as a proto-membrane, as the loop is maintained by the constant production of LINK particles via a CATALYST. Thus, we may refer to such a loop enclosing a catalyst as a cell unit.

In order to add mobil

ity to the model, first we introduce some restrictions in the bonding of free LINKs. In the original bonding rule, the catalyst's movement leaves some fragments of bonded LINKs, which disturb the maintenance of valid membrane structure.

One rule already introduced in the original model was that a free or single bonded LINK particle cannot bond with other LINKs when double bonded LINK exists in the vicinity. New restrictions we introduce here are:

RULE A Bonded Links may decay to two Substrate as long as there does not exist free LINK particles in either of the bonded LINK particles's Moore neightborhoods.

RULE B A bond between two LINK particles break in the Moore neighborhood of a CATALYST.

RULE C A bond between two LINK particles break when free LINK particles exist on *both* halves of the Moor neighbors as divided by the bond.

Rule A, B are to assist cell movement as will be discussed below and Rule C is introduced to suppress entanglement of bonded loops.

The main feature of our new SCL model is the introduction of a new LINK particle, which we call "functional LINK" particles (Fig. 1). As this functional LINK particle can actively push a catalyst when it is in the catalyst's neighborhood, catalysts can move. Also, functional LINKs in cell membranes let substrate particles pass through the membrane. The other, normal LINK particles do not. These new functional LINK particles are produced when a catalyst is surrounded by a sufficient number of substrates. Namely, we have:

RULE D Two substrate particles in the Moore neighborhood of a catalyst may form a functional LINK particle if the total number of substrates in the region is above a given threshold V. Otherwise, normal LINK particles are produced.

Considering the Moore neighborhood of a catalyst particle, we compute the number functional LINKs in the North, South, East and West regions. Each side contains at most 3 particles. With these numbers, we compute the following probability for the direction (North, South, East and West) in which the catalyst will move.

$$P_k(i) = \alpha \cdot \frac{\exp(\beta\, N_k(i))}{\displaystyle\sum_{all\ l} \exp(\beta\, N_l(i))} \qquad (1)$$

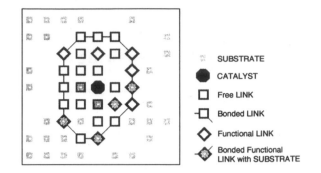

Figure 1: An example of cell formation in our model. A catalyst, surrounded by free LINK particles, is enclosed by the membrane (i.e. a closed loop of two kinds of bonded LINK particles). SUBSTRATE particles can only cross the membrane where functional LINK particles exist. Inside the membrane, SUBSTRATE particles can occupy the same sites as LINK particles.

This is the probability of the catalyst at the i-th site moving in the k-th direction. N_k is the number of the functional LINKs in the opposite side of the k-th direction. If another particle except the bonded LINK already exists in the catalyst's destination, we check to see if there is room for the other particle to move in the same direction, to make room for the catalyst. If there is room, both the catalyst and the other particle move. We say that the catalyst "pushes" the other particle. However, if there is not room for the other particle to move, the catalyst and the other particle exchange positions. This behavior is already a feature of Varela's original model. We take $\alpha = 0.05$, $\beta = 3.0$ in the following simulations.

In contrast to the original model, substrate particles can only cross a boundary of bonded LINK particles at the functional LINK particles. The more that a boundary is composed of functional LINK particles, the more substrates may cross the boundary. Therefore, where and how many functional LINK particles are embedded in the boundary determines where and how many substrate particles may enter or leave the cell.

Observation

There are some unique phenomena observed in our model. First, we describe these before analyzing them. In our simulations, we use the fixed parameters shown in Table 1. In addition, we allocate sufficient grid space for the movement of the cell.

Movement of the Cell

As Varela's model shows, simply establishing a self-sustaining unit just repairs itself, but does not change otherwise. However, in the original model, if any cell movement occurs, the boundary is easily damaged and as a result,

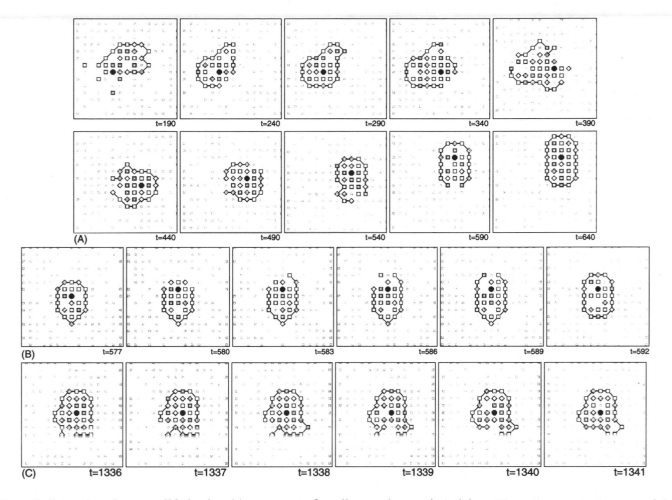

Figure 2: Examples of proto-cell behavior: (a) movement of a cell over a long period of time. The cell moves to the upper right corner while repairing its boundary. (b) The cell motion process in detail. Within 15 steps, the cell as a whole is moved in the upwards direction. (c) Repairing of the cell boundary with free LINK particles. Due to Rule A, bonded-links release LINK particles frequently and holes are created in the membrane. These holes are rapidly restored by the neighboring LINK particles.

whole-cell movement cannot be observed. But in our model, the stability of the boundary seems much stronger because Rule C suppresses a formation of irregular structures of bonded LINKs such as a spiral structure, with which bonded LINKs does not enclose CATALYST. The cell is able to move around, continuously repairing its boundary(Fig. 2a). Indeed, the motion of a cell requires continuous repairing of the boundary. This happens as follows:

1. A catalyst moves into the vicinity of a boundary.

2. Due to the Rule B, a boundary particle in the vicinity of a catalyst disintegrates with a relatively high rate.

3. A substrate (often a functional LINK) will be inserted as a new component of the boundary.

4. Other parts of the boundary will be naturally replaced by LINK particles due to Rule A.

By both continually recreating the part of the boundary close to the catalyst and repairing the part of the boundary further away, a cell can gradually move as a whole(Fig. 2b). The addition of the new functional LINK particle is crucial to this ability. Fig. 3 shows the spatial trail of a cell's movement by tracing the position of its CATALYST for 10,000 time steps.

Such a stable dynamics is sustained by the result of a autopoietic organization. Bonded LINKs are frequently disconnected by Rule A when no free LINK particles exist in the vicinity. This event leads to the formation of holes in the middle of a membrane. However, these holes are immediately repaired by free LINKs near the hole. This suppresses undesirable irregular formation of bonded-LINKs. This repairing process is illustrated in Fig. 2c. Generally, a cell maitains the number of each LINK particles for hundreds of time steps due to the rapid repairing processes. But sometimes, a cell decays and reforms with very different compo-

Parameter	Values
the initial substrate density	0.6
the decay rate of LINK particles	0.003
the diffusion rate of CATALYST	0.01
the diffusion rate of LINK	0.3
the diffusion rate of SUBSTRATE	1.0
free-link-bond-restriction (Rule A)	0.5
opposite-side-link-bond-restriction (Rule B)	1.0
decay-with-catalyst (Rule C)	1.0
The threshold of functional LINK production V	5

Table 1: The parameters used in our simulations

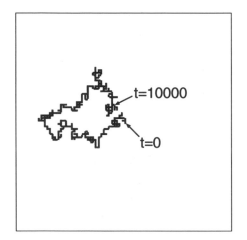

Figure 3: A spatial trail of a cell's catalyst after 10,000 steps on a 72x72 cell space.

nents.

Dynamic Processes

A functional LINK particle is generated from two substrates by Rule D only when a catalyst is surrounded by enough substrates. Since the substrates can only cross the cell boundary where the functional LINK particles are present, a cell in a poor environment (i.e. one with a low density of substrate particles) can only generate normal LINK particles. Since normal LINK particles cannot penetrate the boundary, the production of LINK particles is significantly suppressed. As a result, few repairing processes occur and the cell unit diminishes. Therefore, there exists a critical density of substrates to create a mobile cell structure (see Fig.4). We found that the critical density is around 0.5, where a membrane begins to last for significantly longer periods of time than with a substrate density of 0.4. It is worth noting that the mobility of a catalyst is accelerated by the formation of a membrane. The maximum speed of catalyst mobility is found when the substrate density is around 0.7. This shows that mobility requires a synchronous repairing process. As repairing and mobility processes are comple-

mentary to each other, a perfect repairing process suppresses mobility. Therefore in rich environments above density 0.7, the mobility is again significantly suppressed.

When a cell has a sufficient number of functional LINKs in its boundary, more substrates are accumulated inside the cell. When enough substrates are found around a catalyst, more functional LINK particles are generated. Indeed, we observe that the fraction of functional LINK particles in a cell increases as time proceeds(see Fig. 5). If a cell has a boundary composed only of functional LINK particles, the catalyst constantly generates excess functional LINK particles. Therefore any breakage of the boundary is immediately repaired by the functional LINK particles. However, if there are many ppfunctional LINK particles surrounding a catalyst, there is no free space and so the cell can only move when these particles decay. Such a cell unit is the most stable one. However, a cell typically breaks up before reaching this state.

In order to investigate the disorder of cell movement, we compared the deviation of the CATALYST movement in three cases: the model with only normal LINK particles (like the original model), the model with functional LINK particles but where SUBSTRATEs may pass through the nomal LINKs (that is, functional LINKs act only as "motors"), and the complete model with functional LINKs as already described above (Fig. 6). Cell units with fully functional LINK particles ("motor" and "sensor") show deviate significantly from random walking after around 100 time steps, while the lifetime of a cell is several hundred time steps. Therefore, it can be said that a cell can move directionally for a certain period of time until a reconfiguration of internal LINK particles occurs.

The Functional LINK Particle as a "Sensor"

Functional LINK particles on the membrane act as protosensors of a cell system, since substrates can across the membrane only via the functional LINKs. The substrates that come across the membrane will compose either normal or functional LINK particles. A part of functional LINK stays around the catalyst to push it. Therefore, functional LINK gives a way of coupling inputs and motor outputs. External substrates are detected by the functional LINK on the boundary. The absorbed substrate will compose LINK particle as described. Then the final displacement of a catalyst is determined by the cooperative effect of functional LINK particles.

This cooperative effect has some time delays. In Fig.7, we computed the time correlation between inputs of SUBSTRATE with displacement of CATALYST $< S(t)D(t + \tau) >$ and repairing a membrane $< S(t)R(t+\tau) >$. Here, $S(t)$ is the number of substrate through the membrane. $D(t)$ is a moving average with width 40 of the moving events of the CATALYST. Similarly, $R(t)$ is a width-40 moving average of the number of discrete double bonding events which only

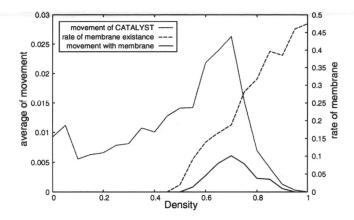

Figure 4: Average values of the displacement of the CATALYST, membrane existence time and displacement of the CATALYST when a membrane exists are computed for a range of initial substrate densities. The lower substrate densities cannot support membrane formation. A mobile cell only appears when the density is around 0.6-0.9. It is worth noting that the speed of movement is increased by membrane formation.

occur repairing a membrane.

Comparing with the two peaks, we estimate that the substrate input is used as a repairing output around 5 time steps and as a motor output around 10 time steps. As the latter correlation lasts longer, it suggests that some functional LINK particles are accumulated a few dozen steps in a cell being recruited for repairing or pushing. Combining the present result, we can determine some primitive form of diffusive coupling between the inputs and motor output.

Discussion

Varela's original cell model demonstrates cells that keep their identity whilst maintaining their own boundaries. Varela noted, however, that living autonomous systems must do more than this; they should not only keep their identity but at the same time couple with their environment(Varela, 1979). In his model, a coupling between the cell system and its environment was not shown explicitly. The present model is a first realization of such coupling, with the membrane as an input channel and movement of whole cell units.

Shifting our focus from self-reproduction to self-motility, we presented some necessary conditions for acquiring motility in a self-sustainable cell system. Such a system has two complementary aspects: movement and repair. By introducing a new type of LINK particle, the functional LINK particle, we can treat sensors and active movement of a cell system in a same level.

We consider it is very effective for discussing a origin of sensor-motor-coupling. Generally, sensory-motor coupling is discussed in artificial neural system with navigation.

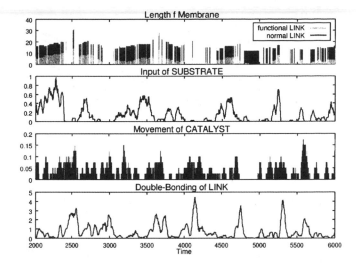

Figure 5: Time evolution of the length of membranes, the rate of substrate input, the catalyst displacement and the rate of double bonding events. The length of membranes is defined whenever the bonded LINK particles form a closed loop. The percentage of functional LINKs of the membrane increases during each cell lifetime. The substrate input is detected only when membrane is closed. The value increases when the membrane lives long enough. The movement of the catalyst may be computed regardless of the state of membrane formation. The double-bonding event corresponds to repairing of membrane formation. This occurs when the membrane diminishes.

There are sensory inputs and motor outputs connected to a separate controller. A primitive life system could not have such sophisticated neural connections. Indeed, insects and protozoa use internal chemical networks to make sensor-motor couplings. In our model, functional LINK particles bifurcate into sensors on a cell boundary and engines giving a catalyst motility. The coupling between these is still ambiguous, however some adequate evolutionary pressure may strengthen this coupling to make it more reliable. An extension of our model in this direction will be reported elsewhere.

Acknowledgments

This work is partially supported by Grant-in aid (No. 09640454) from the Ministry of Education, Science, Sports and Culture and from The 21st Century COE (Center of Excellence) program(Research Center for Integrated Science) of the Ministry of Education, Culture, Sports, Science, and Technology, Japan.

References

Breyer, J., Ackermann, J., and McCaskill, J. S. (1998). Evolving reaction-diffusion ecosystems with sefl-

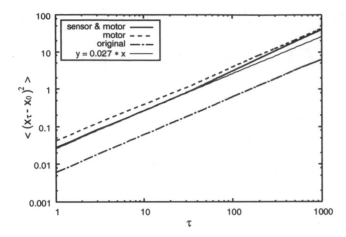

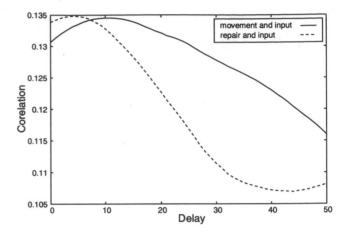

Figure 6: deviations of the displacement of the CATALYST of three cases: the original model with three bonding restriction(no functional LINK), the model with functinal LINK as only a "motor" not "sensor"(SUBSTRATE can penetrate normal LINK), and the model with functional LINK as both "motor" and "sensor"(full role of functional LINK). a thin line is a $y = 0.027x$. only the model with full role of functional LINK shows a declination from a random walk from 100 to 1000 time steps.

Figure 7: Two Time correlation functions are computed. The first one is between the displacement of catalyst and substrate flows through the membrane (solid line). The second one is between the repairing events and the substrate flows (segmented line).

assembling structures in thin films. *Artificial Life*, 4:25.

Langton, C. G. (1984). Self-reproduction in cellular automata. *Physica D*, 10:135.

Madina, D., Ono, N., and Ikegami, T. (2003). Cellular evolution in a 3d lattice artifical chemistry. In Banzhaf, W., Christaller, T., Dttrich, P., and Kim, J. T., editors, *Proceedings of the 7th European Conference on Artificial Life*, page 59. Dortmund, Springer.

Margulis, L. and Fester, R. (1991). *Symbiosis as a Source of Evolutionary Innovation Speciation and Morphogenesis*. MIT Press.

Mayer, B., Köhler, G., and Rasmussen, S. (1997). Simulation and dynamics of entropy-driven, molecular self-assembly processes. *Physical Review*, 55(4):4489.

McMullin, B. and Varela, F. R. (1997). Rediscovering computational autopoieses. In Husband, P. and Harvey., I., editors, *Proceedings of the 4th European Conference on Artificial Life*, page 38. Cambridge, MA: MIT Press/Bradford Books.

Morita, K. and Imai, K. (1997). A simple self-reproducing cellular automaton with shape-encoding mechanism. In Langton, C. G. and Shimohara, K., editors, *Artificial Life V : ,Proceedings of the 5th International Workshop on the Synthesis and Simulation of Living Systems*, page 489. MIT Press.

Neumann, J. V. (1966). *Theory of Self-Reproducing Automata*. Illinois: University of Illinois Press Urbana.

Ono, N. and Ikegami, T. (1999). Model of self-replicating cell capable of self-maintenance. In Floreano, D., Nicoud, J. D., and Mondada, F., editors, *Proceedings of the 5th European Conference on Artificial Life*, page 399. Switzerland, Springer, Lausanne.

Ono, N. and Ikegami, T. (2000). Self-maintenance and self-reproduction in an abstract cell model. *J.Theor. Biol.*, 206:243.

Sayama, H. (1999). A new structurally dissolvable self-reproducing loop evolving in a simple cellular automata space. *Artificial Life*, 5(4):343.

Suzuki, K. and Ikegami, T. (2003). Interaction based evolution of self-replicating loop structures. In Banzhaf, W., Christaller, T., Dttrich, P., and Kim, J. T., editors, *Proceedings of the 7th European Conference on Artificial Life*, page 89. Dortmund, Springer.

Tempesti, G. (1995). A new self-reproducing cellular automaton capable of construction and computatio. In et al., F. M., editor, *Advances in Artificial Life*, page 555. LNAI-929, Springer-Verlag.

Varela, F. R. (1979). *Principles of Biological Autonomy*. New York: North Holland.

Varela, F. R., Maturana, H. R., and Uribe, R. (1974). Autopoiesis: The organization of living systems, its characterization and a model. *BioSystems*, 5:187.

Evaluating an Evolutionary Approach for Reconstructing Gene Regulatory Networks

Dion J. Whitehead[1†], Andre Skusa[1†] and Paul J. Kennedy[2]

[1] NRW Graduate School in Bioinformatics and Genome Research, Center of Biotechnology (CeBiTec), University of Bielefeld, Postfach 10 01 31, D-33501 Bielefeld, GERMANY
[2] Faculty of Information Technology, University of Technology, Sydney, PO Box 123, Broadway, NSW 2007, AUSTRALIA
[†] Both authors contributed equally
andre.skusa@cebitec.uni-bielefeld.de

Abstract

Reconstructing networks from (partial) incomplete data is a general problem in biology. We use an evolutionary approach in an artificial network creation and reconstruction framework to investigate limitations of gene expression network inference from simulated microarray data. For this, the simulated dynamics of the evolved networks are optimized to fit the target dynamics. Evolving networks with similar dynamics is not as difficult as comparing the resulting network topologies to the original network to be reconstructed.

Introduction

A common problem in biology is the reconstruction of networks from incomplete data, for example biochemical networks (Arkin et al., 1997). An important example of this problem is the inference of gene regulatory networks (GRNs) from time series of gene expression data. GRNs may be modelled using a variety of formalisms of differing complexities ranging from directed graphs, boolean networks, differential equations and Bayesian networks (De Jong, 2002).

Time series gene expression data from high throughput assays (eg. cDNA microarray experiments (Baldi and Hatfield, 2002)) may be used to recover the original GRNs that generated the data. This is called a *reverse problem*. The reverse problem has been approached in many different ways, depending on the formalism used to model the GRN. For example, (Liang et al., 1998) infer GRNs modelled using random boolean networks (RBN); (Bower and Bolouri, 2000) used hierarchical clustering on a boolean model; (Kikuchi et al., 2003) find weights for systems of differential equations using genetic algorithms (GA) and (Friedman et al., 2000) infer Bayesian network representations of a GRN.

Microarray experiments (Parmigiani et al., 2003) are a recent cellular biology technology able to measure the level of expression of thousands of genes in cells at one instant. There are two kinds of experiments: cDNA microarray experiments that compare the relative gene expression levels of two samples on the same chip, and oligonucleotide microarrays (Affymetrix) that use one chip per sample.

Microarray data is difficult to use for modelling networks of gene expression. There are challenges processing the data itself, partly due to the difficulty of comparing data from different samples and partly because of the large number of genes (tens of thousands) compared to the number of repetitions of experiments (tens), which is often referred to as the high dimensionality of microarray data. See (Huber et al., 2003) for a recent model of the error associated with microarray experiments and (Yang et al., 2000) for details on normalising data within and between experiments. However, the biggest problem using microarray data to build models of gene expression networks is not the noise of the sample, nor the intrinsic noise of the biological phenomena. It is the fact that microarrays only measure one part of the "real" network, that is, the transcription stage of gene regulation, ignoring RNA regulation, post–translational modification, localization signals, phosphorylation, etc. Noise starts to look unimportant against all the missing data.

A challenge when inferring GRNs is that it is difficult to assess the effectiveness of algorithms because the actual GRNs are mostly unknown. Some researchers approach this challenge by comparing the induced GRNs with known GRNs for the same data. See, for example, (Wahde and Hertz, 2000).

This difficulty has led some researchers, eg. (Repsilber and Kim, 2003), (P. Mendes and Ye, 2003), to test proposed methods with artificially generated, but plausible GRNs. Microarray data is simulated for the artificial GRN and the proposed solution to the reverse problem is compared with the known artificial GRN.

This contribution follows a similar approach but looks at a slightly different question: what are the similarities and differences between artificial GRNs that generate the same gene expression data time series.

Our methodology consists of four stages (illustrated in Fig. 1): (i) create an artificial GRN (called the *target GRN*) at random including marked mRNA and gene nodes, (ii) simulate microarray data for the target GRN, (iii) evolve artificial GRNs which produce simulated microarray data matching the simulated microarray data for the target GRN, and (iv) compare the evolved GRNs with the target GRN.

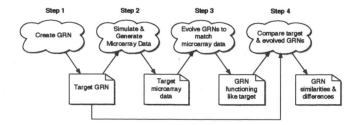

Figure 1: Methodology used in this paper

Network Model

GRNs consist of many different types of reactions, but with current technologies only a subset of these reactions can be directly measured. In the case of microarray experiments, only mRNA levels can be directly measured. Crucial measurements that are not made are the actual concentration of the protein for which the mRNA is translated, the resulting variant if different exon combinations are possible, post-translational modification, protein localization, its current state (for example bound by an inhibitor protein) and many others, as eg. the recent discovery of the impact of small RNA strands (*microRNA*) on all levels of genomic regulation (Ambros, 2001). The result is a network with many unknown nodes and edges (Fig. 2). This is the basis for our inquiry: assuming it is *possible* for a network to be reconstructed from experimental data, what affect do invisible nodes have on the reconstructed networks?

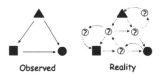

Figure 2: Example of incomplete measurement. Dark nodes are mRNA levels measured by microarray technologies. Unknown nodes are the unmeasured entities.

In most models of GRNs the nodes represent equivalent entities ie. gene products. The GRNs we use are different in that nodes represent all of the entities that regulate protein production (eg. genes, mRNA, modified protein states etc.) rather than simply gene products. This generalisation is motivated by the fact that gene regulation also occurs at places other than just promoter sites.

In our model, based on microarray experiments, nodes are assigned to one of two categories: *visible* ie. experimentally measured or *invisible* ie. not measured. Visible nodes represent entities that may be detected in the simulated microarray experiment (ie. mRNA molecules). Conversely, invisible nodes represent molecules undetectable by the simulated microarray. Both kinds of nodes, however, contribute towards the network dynamics.

Method Overview

Step 1: Create Random Gene Regulatory Networks

One possibility of randomly creating artificial regulatory networks (ARNs) is to generate a bit string and mimic the process of gene regulation occurring from this genome (Banzhaf, 2003). The network then emerges from the interaction of the components. In contrast to this we generate the network topology directly. Though it is trivial to create networks at random it is not as straightforward to create networks at random which should "make sense", ie. they should exhibit dynamic behavior that is arbitrary, but reasonable in the domain they are created for. In our case, reasonable GRNs are those which do not stop expression at an early time step or where activity is confined to only a small subset of the whole network. Therefore, we applied a method to prune and modify random networks to produce interesting dynamics:

First, a directed random network is created by applying the Erdös-Renyi approach (Erdös and Renyi, 1960) to directed graphs: for a number of nodes each possible edge is tested and drawn if a random number is lower than a given probability p. Each edge is assigned a uniform random weight in the range of $[-1, 1]$. The node which reaches the most other nodes is determined and set to the *starting node* for the simulation (for the concept of a starting node see also *Step 2* in this section). All nodes unreachable by the first node are connected at random to the reachable nodes (also chosen at random).

All nodes in the network are now reachable from the first node in respect to the edge directions, but not necessarily in respect to the reaction rates, because negative reaction rates might inhibit the expression of complete subnetworks "behind" a node. Thus, a second pruning is performed considering the assigned edge weights (which serve later as reaction rates) and all nodes which are not reachable in respect to negative weights are simply deleted from the network. At the end, the nodes are randomly assigned as visible or invisible.

Step 2: Simulate Microarray Data

In this step we simulate microarray data for the target GRN. Simulation of microarray data has been accomplished by other researchers in a variety of ways and with different motivations. Repsilber and Kim (Repsilber and Kim, 2003) start with an artificial GRN and simulate the network dynamics for a number of time steps to build a time series. They then change the initial conditions of the GRN and simulate another time series. A simple model combines the two sets of expression values to give a time series of log ratios. Cui et al (Cui et al., 2003) take a different approach. They use a more complicated model of the error implicit in microarray experiments but generate the raw expression values randomly from statistically plausible probability density functions.

Our approach is more similar to that of Kim and Repsilber. We chose to implement a relatively simple and widespread formalism: ordinary differential equations (De Jong, 2002), based on the so-called Hill function, a sigmoidal function that agrees with experimental data. That the function is non-linear, is important. Even though linear functions are simpler, most biochemical interactions are non-linear, which has important consequences in terms of network dynamics. A degradation term is added to model gradual decay.

$$\frac{dg_i}{dt} = h^+(g_i) - \gamma_i g_i \tag{1}$$

where g_i is the expression level of the ith gene and i ranges from 1 to N for N total genes. $\gamma_i = 0.3$ is the degradation term and $h^+(g_i)$ is the Hill function given by

$$h(g_i) = \frac{x(g_i)^3}{x(g_i)^3 + K} \tag{2}$$

with

$$x(g_i) = \sum_{j=1}^{N} W_{ji} g_j \tag{3}$$

where W is the matrix of interactions and $K = 1$ the threshold for regulatory influence on g_i. One can imagine the rows representing the nodes, and the columns represented the interaction, such that w_{ij} represents the interaction from node i to node j. To allow for repression of genes, then if $x(g_i)$ is negative then $h^+(g_i)$ is replaced by $h^-(g_i) = 1 - h^+(g_i)$ and K changes sign.

Using Euler's method and a fixed step size, the artificial GRN is simulated using numerical integration of the differential equations (ie. equation 1) for a small number of time steps. At the initial time step, the first node (*starting node*) is given a positive value (2 in our simulations) and the value of all other nodes are set to 0. The network is then simulated for s time steps with node values recorded for each time step to form a time series of network activity (s is chosen such that the steady state starts right after that). In an approach inspired by oligonucleotide microarray experiments (eg. Affymetrix experiments), a single gene expression value is generated for each measured node. At this stage, sampling error and other forms of noise are not added to the simulated data. The result of this step is a matrix of values with a row for each measured node and a column for each time step.

Step 3: Evolve Gene Regulatory Networks

Next, GRNs are evolved using an evolutionary computation algorithm to discover networks that produce similar output to the target GRN. To use an evolutionary approach to discover networks, it must be possible to encode each candidate GRN on an artificial genome. In *Step 2* above, we describe how GRNs are represented as a square matrix W of interaction weights. This matrix completely defines the behaviour of a particular GRN from given initial conditions. For this reason, we simply encode each GRN as its matrix W. The top leftmost $n \times n$ sub–matrix (where n is the size of the target matrix) is constrained such that these nodes may not be lost by evolution. These are the visible nodes. Wiring between these nodes, however, is adaptable by evolution. Also, the first node (ie. at W_{11}) has a hard–wired feedback to itself to simplify the evolution task. All other weights, however, are set by evolution.

The artificial evolution begins with a population of randomly created networks 10% larger than the target matrix with weights drawn uniformly randomly from the range [-1, 1]. However, evolution may increase or decrease the number of genes (rows/columns) in the matrix. The evolution protocol consists of a set of network mutations with an associated probability of occurrence at each time step, for example edges may switch nodes or be deleted, weights may be changed by a small amount, nodes can be duplicated. For the experiments in this contribution, we typically used population sizes of 1000 individuals and run lengths up to 6000 generations.

Each member of the population is scored by a fitness function that quantifies how close the output from the GRN corresponding to the genome is to the output of the target GRN. So, for each genome, the GRN is simulated using equations (1), (2) and (3) with the same initial conditions and constant parameters as for the target GRN. Only expression values for the first n genes (ie. the upper leftmost $n \times n$ submatrix matching the target W matrix) are recorded. The distance of the resulting time series of expression values for each of the n genes to the target values is computed using a simple squared error approach:

$$f = \sum_{i=1}^{n} \sum_{t=1}^{T} (g_{it} - y_{it})^2 \tag{4}$$

where f is the fitness of the genome. Evolution seeks to minimise the value f. f sums over each of the n visible genes y_{it} indexed by gene with i and by time with t for each of the T time steps. The value g_{it} is the target gene expression value of gene i at time t.

An elitist strategy is used to manage the population. After calculating the fitness of the population, the weakest 40% of the population is deleted and replaced with the fittest 40%. Genetic operators are then applied to the population with (small) fixed probability. Genetic operators include: node duplication, node deletion, edge creation, edge deletion, edge mutation (ie. changing the weight value) and edge duplication. Genetic operators do not modify the feedback edge to the "initial" node ie. W_{11} nor delete any of the hard–coded visible nodes in the top leftmost $n \times n$ submatrix. Fitness is computed for members of the new population and evolution continues until networks with suitably small fitness are designed or the maximum number of generations is reached.

Step 4: Compare Gene Regulatory Networks

No sophisticated comparison methods were applied in this paper because the evolved networks were much larger and densely connected than expected. Applying and developing meaningful comparison methods is the most important next step. Here we "manually" compared networks only in parts.

Experiments and Results

The goal for the experiments was to evolve networks similar to a given target network regarding the behavior, ie. the expression time series. We chose two target networks for evolution from the randomly created networks, each with 20 nodes. One had a simple sequential connection topology with resulting simple dynamics. The other had a more complicated topology including negative feedback. Both networks contain only visible nodes (see section *Network Model*). However, the evolution protocol allowed the addition and deletion of invisible nodes. By manually comparing the time series of networks of size 20 we concluded that a distance smaller than or equal to approx. 1.0 indicated very similar network dynamics. The same mutation probabilities were used in both experiments. Several parameter settings have been tested. Different simulation lengths (40 and 20 time steps respectively) were used for the two networks to capture the significant dynamics before the steady state is reached. To simplify the plots some significant nodes were selected for drawing the expression level time series.[1]

Simple Sequential Network

The first target we chose can be seen in Fig. 3. It contains mostly sequentially connected genes and therefore the resulting expression time series consists mainly of sequentially increasing expression levels for each gene reaching a final steady state (Fig. 4, top). The main characteristics of this network is the division of the nodes into two groups: one in which the nodes are immediately expressed, and the another where the nodes are expressed around 20 time steps later at much lower expression levels. This reflects the structure of the network, which can be divided into mainly two groups of genes activated separately by the starting gene but then do not subsequently interact. A resulting time series for one of the evolved networks can be found in the bottom of figure 4. The distance to the target network for this was approx. 0.74, and it contains 64 nodes and 826 edges. Comparing the two time series shows that the main characteristics described previously are in the evolved network. Due to the much higher number of nodes and edges in the resulting network, a comparison of the network topology is not trivial and is deferred to future work. The simplicity of the target network topology suggests that a large number of possible topologies could exhibit the target dynamics.

---

[1]Parameter descriptions and data available on request.

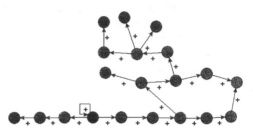

Figure 3: Topology of the simple sequential network. Starting node marked black. The two parts of the network (left and right sides) are arranged according to the two main activity patterns in resulting dynamics (Fig. 4).

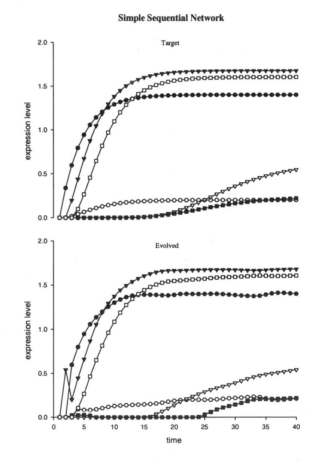

Figure 4: Dynamics of the simple sequential network. Top: Target network behavior. Bottom: Evolved network behavior. Upper and lower expression levels refer to right and left sides in network topology in Fig. 3. Absolute differences between time series points have mean 0.013 and standard deviation 0.0272.

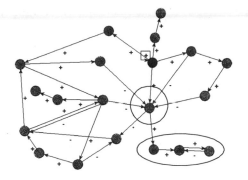

Figure 5: Topology of the network with negative feedback. Starting node is black. Single circled node is crucial to the dynamics and refers to ○ curves in Fig. 7. Circled lower nodes are a motif preserved in evolved network (Fig. 8).

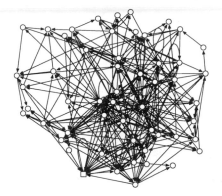

Figure 6: Example of evolved network from the negative feedback network. Network dynamics shown in Fig. 7.

Network with Negative Feedback

The other network chosen for analysis is shown in Fig. 5 and the matching time series at the top of Fig. 7. This network is more complex than the previous due to negative feedback structures leading to delayed inhibition. Dynamics are driven by a node (see Fig. 5) activated by the starting node and inhibited later by all other connecting nodes. The time series shows this node (○ in Fig. 7) initially increasing (by the starting node) then being inhibited to zero, with subsequent nodes also decaying to zero. The time series from one of the evolved networks (Fig. 6) demonstrates that more complicated dynamics can also be approximated (Fig. 7, bottom). The distance between networks is approx. 1.0 and the evolved network consists of 55 nodes and 298 edges. The tendency of networks to grow under evolution is still present, although growth is less than the previous network. This may have been caused by the more restricted space for possible solutions. Comparison of network topologies is non–trivial. Manual examination shows that one substructure in the target network is preserved in the evolved network. The three preserved nodes in the target motif are circled in Fig. 5 and the matching subgraph in the evolved network is shown in Fig. 8. In both cases expression levels of the middle node □ increase after the central node ○ decreases from other influences. Edges in the evolved network motif (Fig. 8) not contributing to these dynamics are not shown. The remaining structure differs only in the position of negative feedback (moved into a self-loop of node □). So, the evolved network preserves not only the overall dynamics, but also some parts of the original topology. Further detailed analysis and development of similarity detection tools is deferred to the future.

Conclusions and Future Work

Although target and evolved networks have similar dynamics, comparison of topologies is nontrivial. Until structural

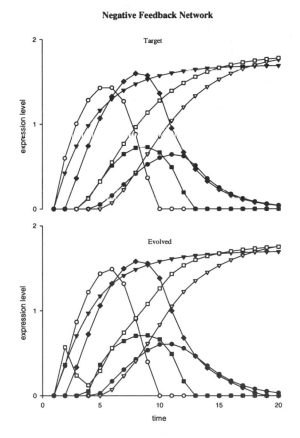

Figure 7: Negative feedback network dynamics. Top: Target. Bottom: Evolved. Delayed inhibition (○ line) belonging to single marked node in Fig. 8 is crucial for network dynamics. Other levels change once the ○ expression decreases (eg. □ or △). Absolute differences between time series points have mean 0.024 and standard deviation 0.0555.

Figure 8: Preserved motif. Edges not contributing to node dynamics removed. Differs by inhibiting edge (self–loop).

(in addition to daynamic) similarity can be measured, results of evolutionary approaches to GRN reconstruction appear unreliable. However, the preserved substructure in the negative feedback network shows future directions. The increase of nodes and edges compared to target networks is a challenge. This may be ameliorated by an evolutionary parsimony constraint, but initial tests showed that this complicated the objective function. Producing a desired behavior by adding nodes is probably easier than initially connecting nodes with optimal reaction rates: nature explores parameter space by gene duplication then modification. Future steps will also complete the stages of our methodology (eg. adjusting the ratio of visible to invisible nodes (ie. the extent of data incompleteness) allows us to tune detection of the artificial GRN by the microarray and adding noise to expression values creates more "realistic" artificial data). Also, other established GRN reconstruction methods will be used.

Acknowledgements

We would like to thank Jacob Köhler for helpful comments. PJK would like to thank Klaus Prank and the NRW Graduate School in Bioinformatics and Genome Research in Bielefeld for hosting his visit and AS and DJW would like to thank the same institution for financial support.

References

Ambros, V. (2001). microRNAs: tiny regulators with great potential. *Cell*, 107:823–826.

Arkin, A., Shen, P. D., and Ross, J. (1997). A test case of correlation metric construction of a reaction pathway from measurements. *Science*, 277(5330):1275–1279.

Baldi, P. and Hatfield, G. W. (2002). *DNA microarrays and gene expression*. Cambridge University Press, Cambridge, UK.

Banzhaf, W. (2003). On the dynamics of an artificial regulatory network. In Banzhaf, W., Christaller, T., Dittrich, P., Kim, J. T., and Ziegler, J., editors, *Proc. ECAL 2003, LNAI 2801*, pages 217–227. Springer–Verlag.

Bower, J. M. and Bolouri, H. (2000). *Computational Modeling of Genetic and Biochemical Networks*. MIT Press, Cambridge, MA.

Cui, X., Kerr, M. K., and Churchill, G. A. (2003). Data transformations for cDNA microarray data. *Statistical Applications in Genetics and Molecular Biology*, 2(1).

De Jong, H. (2002). Modeling and simulation of genetic regulatory systems: A literature review. *Journal of Computational Biology*, 9(1):67–103.

Erdös, P. and Renyi, A. (1960). On the evolution of random graphs. *Publ. Math. Inst. Hung. Acad. Sci.*, 5:17–61.

Friedman, N., Linial, M., Nachman, I., and Pe'er, D. (2000). Using bayesian networks to analyze expression data. In *Proc. Conf. on Research in Computational Molec. Biology*, pages 127–135, New York. ACM Press.

Huber, W., von Heydebreck, A., Sueltmann, H., Poustka, A., and Vingron, M. (2003). Parameter estimation for the calibration and variance stabilization of microarray data. *Statistical Applications in Genetics and Molecular Biology*, 2(1).

Kikuchi, S., Tominaga, D., Arita, M., Takahashi, K., and Tomita, M. (2003). Dynamic modeling of genetic networks using genetic algorithm and S–system. *Bioinformatics*, 19(5):643–650.

Liang, S., Fuhrman, S., and Somogyi, R. (1998). REVEAL, a general reverse engineering algorithm for inference of genetic network architectures. In *Proc. Pacific Symp. on Biocomputing 3 1998*, pages 18–29.

P. Mendes, W. S. and Ye, K. (2003). Artificial gene networks for objective comparison of analysis algorithms. *Bioinformatics*, 19 Suppl. 2:ii122–ii129.

Parmigiani, G., Garrett, E. S., et al. (2003). The analysis of gene expression data: An overview of methods and software. In Parmigiani, G., Garrett, E. S., Irizarry, R. A., and Zeger, S. L., editors, *The analysis of gene expression data*, pages 1–45, Heidelberg. Springer–Verlag.

Repsilber, D. and Kim, J. T. (2003). Developing and testing methods for microarray data analysis using an artificial life framework. In Banzhaf, W., Christaller, T., Dittrich, P., Kim, J. T., and Ziegler, J., editors, *Proc. ECAL 2003*, pages 686–695. Springer–Verlag.

Wahde, M. and Hertz, J. (2000). Coarse–grained reverse engineering of genetic regulatory networks. *Biosystems*, 55:129–136.

Yang, Y., Buckley, M. J., Dudoit, S., and Speed, T. P. (2000). Comparison of methods for image analysis on cDNA microarray data. Technical report 584, Department of Statistics, University of California, Berkeley.

Bonding as an Emergent Phenomenon in an Abstract Artificial Chemistry

Dominique Chu[1] and Rune Vabø[2]

[1]School of Computer Science, University of Birmingham, UK
[2]Institute for Marine Research, Bergen, Norway
hsvdg@uib.no

Abstract

In this article we describe an implementation of a very simple toy artificial chemistry originally proposed as a "Gedanken" model by Groß and McMullin. They predicted that this model would have a number of interesting properties including emergent bonding, and top-down constraints on bond-stability. We found that not all of the original claims could actually be verified.

The Problem of Novelty in Artificial Life Systems

In a contribution to the Artificial Life 8 conference Groß and McMullin (Groß and McMullin, 2003) discussed the problem of novelty emerging in Artificial Life particularly focusing on agent-based models (ABM) (Casti, 1997; Holland, 1995). While acknowledging the usefulness of ABMs in general, they pointed out that ABMs are not particularly good at creating what they call perpetual novelty. As one of the main limitations for the emergence of novelty they identified what they called "meta-rules," that is hard-coded procedures that define the scheduling of events in the model, the order in which agents are called, mutated, resources are supplied to the environment and so on. While necessary for the simulation those meta-rules are "hard-coded" and as such significantly constrain the possible behaviors displayed by the model.

Groß and McMullin discussed several ways to avoid the limitations of meta-rules. One possibility is to evolve/modify the meta-rules themselves at run time. Unfortunately this only looks like a solution, as this would require the specification of meta-meta-rules that specify in which way the meta-rules are to be changed. Given the insight that meta-rules are indispensable, Groß and McMullin decided that it would be better to think about ways to minimize the limitations posed by meta-rules, rather than to attempt to avoid them altogether.

In their article Groß and McMullin focused the discussion on the emergence of novelty in agent-based artificial chemistries (Dittrich et al., 2001). In such artificial chemistries the behavior of particles is determined in a bottom-up fashion, meaning that the rules of behavior function at the level of the individual agent. While this is a plausible way to implement artificial chemistries, it is not entirely unproblematic from the point of view of emerging novelty. To see this consider the usual way to represent bonding in such models. A common way to do this is to implement an explicit bonding rule that specifies under which conditions in the simulation bonds are formed and break. Groß and McMullin see inherent limitations in this method to define bonding.

> The "new" high level particle formed as a result of this kind of bonding is little more than a collection of its primitive components plus a few motion constraints which have been preconceived and explicitly pre-programmed. In this sense bonds—their formation behavior, rupture—are completely pre-specified[...]. [...]This seems distinctively unlike the formation of new molecular species in nature and it also restricts the potential for novelty creation in the model. (Groß and McMullin, 2003)

Groß and McMullin call this approach to bonding "bottom-up," because whether or not a bond will be formed is determined exclusively by the inner and outer states of the particles that participate in the bonding. This is opposed to their suggested "top-down" approach to bonding, where the bond is an emergent result of the environmental conditions of the particles, while there are no explicit rules for bonding given. Top-down bonding, they conjecture, will lead to artificial chemistries with a much higher potential for emergent novelty.

In their paper they describe a simple "Gedanken" artificial chemistry as an example for a system that has top-down constraints on bonds. Admitting that this system is probably too simple to display perpetual novelty, they list a few key-features even this very simple model will possess:

- **Emergent Bonding:** Particles can form bonds with one another, although no explicit rules for bonding is specified.

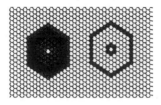

Figure 1: The 2 types of particles. Different gray-scales indicate the field-strength. The right side shows an *A*-particle, on the left is a *B*-particle. The particles themselves are not visible in this picture, but located in the center of the fields.

- **Variety:** The model supports a variety of compound configurations (isomers).

- **Configuration Dependence (Top Down Constraint):** Which particles of a compound can bond with yet another particle depends on specifics of the compound itself.

In this article we report some experimental results obtained from an implementation of this "Gedanken" artificial chemistry. Particularly, we are trying to establish whether or not the phenomena predicted by Groß and McMullin can indeed be observed.

Description of the Model

The model is implemented in Java using the Repast (Repast, 2004) simulation libraries. Every particle type is characterized by the particular "force-field" it induces in its neighborhood (see figure 1 for an example of an agent type). The details of this field are defined by the user at run-time and remain fixed for each particle. In principle the model allows an arbitrary amount of different particle types to be defined at run time; for all simulations reported here we consider only 2 types. Each agent has 6 momentum components (one component for each possible direction in a hexagonal grid; see below) and 2 variables describing its position. Position and momentum variables are updated at each time step. The momentum components can take a maximal value of 100 and a minimal value of 0.

The environment of the model is a 2d hexagonal discrete grid (with periodic boundary conditions). Each grid site is occupied by at most one particle at a given time. Every site is associated with a particular value (either zero, or a negative or positive integer) describing the field induced by an agent. If the force fields of agents overlap on one grid site, then the net value of the grid site is the sum of all contributions.

The movement rules of the model are highly schematic and no attempt is made to be physically realistic or even plausible: At each time step particles either stay where they are or move to a neighboring site (so maximum speed is one cell per time-step). In order to determine the direction of movement, agents check the field values at all 6 neighboring sites. If agents move in a direction d, then their d

momentum-component is updated by adding the difference between the field values after and before movement to the d momentum component. For example, if an agent had a momentum of one in the north direction, his current site has a field value of -2 and the new site has a value of -1, then the updated component of the north component of the momentum is 2.

At every time step agents attempt to perform a movement that maximizes their momentum in the following way: In order to determine the direction of movement agents check the field values at all neighboring sites and calculate the momentum they would have after moving to this site. The direction chosen for the actual movement is the direction in which the momentum component *after* movement is maximal. If several momentum components are equal, a random one is chosen among the maximal ones. If all momentum components (before *or* after movement) are zero, no movement is performed. Particles do not feel their own force field.

An example might clarify this movement rule. Consider a particle that is on a site with a force-field of 2. Its north-component of the momentum is 5 while all other momentum components are 1. All neighboring field values surrounding it are 0 except for the north field value, which has a value of -1. Thus if the particle choose the northern direction its momentum component after the movement will be 5-3=2. In all other movement directions the momentum after (and before) the movement will be 1. Since the north-direction will be maximal after the movement into the north direction, the north dirction is chosen for movement.

Agents are subject to some random kicks which change (with a user defined probability) the value of a randomly selected momentum component by ± 1. In absence of any fields, the agents thus perform a random walk over the grid driven by the random kicks.

Emergent Bonding

The phenomena that can be observed in the model strongly depend on the choices of the force fields of the agents. We experimented with a number of values, but here we will report only results obtained with agents of the following type: Agent $A = (-40, -35, -30, 15, 15, -30)$ and $B = (-20, 0, 0, 0, 0, -20)$, where the numbers in parenthesis denote the field values an agent induces in its first, second, third, ... neighborhood; the field values of those agents are depicted in figure 1. (At the moment the model supports up to 8 neighborhoods, but in the simulations reported here we will only use 6). The qualitatively results do not depend on the particle types as long as the particle types do support stable bonding; there are of course also particle types that show no interesting behavior (for example particles with no field). As anticipated by Groß and McMullin, the system does support emergent bonding between particles. Bonding between *A*-particles can be achieved by placing the agents in each others fourth neighborhood (the fifth

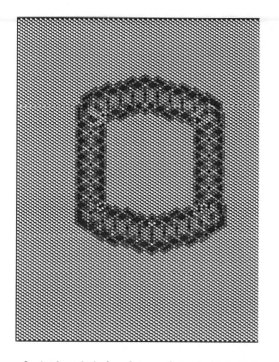

Figure 2: A closed chain of *A*-particles in its initial state.

one also works). The particles are then trapped between two "repulsive" fields effectively preventing them from coming too close to each other and from leaving each others' neighborhood. The same arrangement, but with more flexibility for relative movement is achieved by placing *B*-particles into each others second, third, fourth or fifth neighborhood. We also found many other particle types that support bonding. A necessary condition for emergent bonding is that the particles have an inner and outer, strongly repulsive field and—sandwiched between those—a neutral or positive field.

Both *A* and *B*-particles can form strong and stable bonds with one another; note that due to the random kicks the particles receive even the most stable configurations might rupture.

Chains

Both types of particles can not only engage in pairwise bonding, but can also be arranged into linear chains of the form $\cdots - A - A - A - A - A - A - \cdots$. For all practical purposes those chains can be made arbitrarily long. In the case of *A*-particles chains can be made as follows: Place two *A* particles into the fourth neighborhood on opposite sides of another *A*-particles to make an initial short chain. The chain can then be extended by adding *A* particles on either side of the initial chain. Simulations show that those chains have very good stability properties and can even be arranged into closed chains that typically remain bonded over very long times. Figure 2 shows an example of such a closed chain.

While chains (closed or open) are very stable, isomers might have very different stability properties. As an exam-

ple consider the linear chain of 12 *A*-particles. Like all linear chains of *A*-particles this configuration is stable. One possible isomer of this chain is a cross-shaped configuration, consisting of a chain of length 6 and 2 side chains of size 3 attached to the main chain between particle 3 and 4. We simulated this configuration 10 times and found that only in 2 of those runs the cross remained intact during the first 200 time steps. In the eight remaining runs one or several of the side chains broke away. Attaching only one side chain (such that the overall configuration looks like a "T") improves the stability. Again we ran 10 simulations for the T-configuration and 10 for the inverted T-configuration (those cases are actually equivalent). In our experiments we observed that the configuration remained stable in 10 out of 10 runs and in 8 out of 10 runs respectively.

A likely reason for the instability of the cross configuration is that particles 3 and 4 in the main chain each have to bond to 4 other particles. This interpretation is suggested by separate experiments with only five particles that are arranged in a linear chain of 3 particles and one particle attached to each side in the middle of the chain. In this configuration the central particle has to bond to 4 other particles which does not seem to be stable. We conclude that *A*-particles can form up to three bonds with other *A*-particles. If a particle has more than three bonds, then all bonds tend to be unstable, which leads to the disintegration of the compound.

The strength of the bonds depend on the context: Pairs of bonded *B* particles in isolation are extremely stable configurations; so are chains of *A*-particles. We performed 10 simulations with an *A*-particle chain of length 6 and (spatially separated) a pair of bonded *B*-particles. As expected both did not disintegrate in 10 out of 10 runs. However, if the bonded *B* particles are both placed into the 4-neighborhood of the middle particle of the *A* chain, then the compound did not remain stable. This configuration dissolved in 10 out of 10 simulation runs. It is particularly interesting to note that in all those cases the bonds between the *B* particles broke. Thus while in isolation inter-*B* bonds are extremely stable, within a compound of *A* particles those bonds are extremely unstable.

Encapsulation

While performing our experiments we discovered another property that quite naturally emerges from certain configurations in the models, however without having been predicted by Groß and McMullin: Encapsulation of particles into larger wholes. Loosely, by encapsulation we mean that a compound of interacting particles starts to behave as a whole. There are two aspects to this phenomenon.

Firstly, the dynamics of compounds of particles can be understood to operate on two time-scales. Structures of bonded particles (as for example chains or closed chains) change their internal configuration on a fast time scale. On a

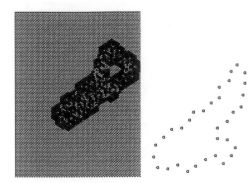

Figure 3: Two rings can collide with one another, without dissolving. Note that the model has periodic boundary conditions, so a part of the left chain is on the right side of the picture.

Figure 4: Chains tend to fold onto themselves. To the left we see a folded chain. To the right another folded chain without the grid or the fields represented. Those and similar chains have life times of the order of hundred-thousand time-steps.

much slower time-scale the compound as a whole will drift through the environment; this overall movement is of course a result of the individual particles' motion, but at the same time imposes a constraint on it. This is best explained by an example: Consider a closed chain of A-particles. The chain exists because the constituent particles are trapped between the repellent force-fields of their neighbors. The individual particles see themselves surrounded by repulsive fields; consequently, the number of available degrees of freedom available to them is reduced. While free particles can move in any direction, the bonded ones are restricted to stay within the attractive fields of their neighbors. If a particle is bonded then its possible degrees of freedom will very much depend on where its neighbors move; at the same time, the neighbors' movement is itself limited by their neighbors' movement. In this sense the local movement of one particle in a compound is influenced by the global state of the compound as a whole.

Secondly, compounds remain intact even if they collide with other objects (i.e. other compounds or particles; see figure 3). Again this can be illustrated by a closed A-particle chain. If we put two or more chains in the environment, then those chains might collide with one another, however without being destroyed. The chains will not exchange particles with one another and particles of the chain will not form bonds with one another (which they would do in the free state). This second aspect of encapsulation is not only important when two chains interact, but also when various parts of the chain interact with one another. In simulations chains tend to fold onto themselves (see figure 4), bringing different parts of the chain into contact. Chains are only stable structures because the various parts of the chain do not react with one another when brought into contact.

In the particular case of A-particle chains (open and closed) the encapsulation property is a consequence of the outer negative field. Because of the particular choices of the fields of A-particles, properly bonded chains will be lined by a strongly repulsive field at each side of the chain. This outer field is particularly strong because it is a result of the overlay of the repulsive fields of the neighbors in the chain. This functions as an efficient barrier against most incoming particles and also allows chains to fold onto themselves without being dissolved by themselves.

Discussion

In the present model, there was no explicit rule for bonding, yet a simple chemistry with characteristic bonding properties (i.e. arity and stability of bonds) did emerge. It is instructive to contrast bonding in the present model with artificial chemistries that rely on explicitly defined bonding rules, such as McMullin's SCL model (B. McMullin, 1997). In the SCL model (as in most similar artificial chemistries) establishing bonds between particles is a rather involved process requiring a degree of coordination between particles: At every time step for each particle the following things need to be checked:

- Check whether the particle is already engaging in the maximum number of allowed bonds.

- Check whether or not a suitable bonding partner is in the neighborhood

- Establish communication with potential bonding partners in order to agree on whether or not to form a bond.

- Finally, keep a record of all bonding partners in order to

  - ...coordinate motion
  - ...communicate possible breakage of the bond.

This contrasts well with the present model where all an agent needs to do in order to calculate its own movement is to look at the force fields in its immediate neighborhood. The particles themselves do not "know" whether or not they are bonded. In fact there is no explicit notion of bonding in the model. Yet still this model shows emergent "chemical" properties similar to the bonding of particles in the SCL model.

We could also verify that the model supports isomeric variants of compounds. Particles can not only form chains, but a variety of different configurations. When it comes to top-down constraints, the evidence is more ambiguous. In order to show that there are clear top-down constraints, it would be necessary to show that, all local conditions equal, an effect or interaction is only possible given some global configurations but not others.

One could interpret the loss of stability of the cross-shaped configurations relative to the "T" shaped configurations as a top-down constraint in this sense. However, closer examination shows that this difference in stability is a result of the fact that A-particles do not support more than 3 stable bonds, thus a local effect. The next best evidence is that of the disintegration of the, otherwise stable, $B - B$ chain, when placed into the neighborhood of a 6 long A-particle chain. In order to really prove the point of the configuration dependence of the bond stability, it would have been necessary to show a configuration in which the $B - B$ compound is stable despite being in the immediate neighborhood of a A-particle chain. We could not find such a configuration. We therefore conclude that the case for top-down constraints is at best weak.

A very interesting property that is supported by the model, but has not been predicted by Groß and McMullin, is encapsulation: Some compounds naturally form a unit in that they can interact with other particles or compounds (for example by colliding) yet retain their structural integrity. It should be noted that the property of encapsulation *per se* is very common in artificial chemistries. As an example one might think of the membrane in McMullin's SCL model (B. Mc-Mullin, 1997; McMullin and Groß, 2001)); this membrane is a structure that is very similar to the closed chain in the present model. However, the crucial difference is that in the SCL model the encapsulation property is a direct result of the bonding rules of the membrane particles. Membrane particles can bond with at most 2 other particles; once they have two active bonds, as all particles in a closed chain, they will not engage in any further interactions with other particles. In the SCL-model encapsulation was thus explicitly programmed into the system, while in the present model encapsulation is an emergent result of the way in which particles interact.

Conclusion

Groß and McMullin conjectured that their proposed artificial chemistry would have emergent bonding, support isomers and demonstrate top-down constraints. We could readily verify the first two properties, but found only weak evidence for the third one. However, we also found an effect that has not been predicted by Groß and McMullin, namely emergent encapsulation.

We think that the modeling strategy proposed by Groß and McMullin is a fruitful one and can be a useful complement to existing artificial chemistries that operate on a higher level. Particularly, top-down constraints would indeed be an interesting property to have in artificial chemistries. The modification of local properties by non-local configurations seems to be an important property of real chemistry, particularly in the realm of bio-chemical processes. It seems, though, that the artificial chemistry proposed by Groß and McMullin does not display top-down constraints. Whether or not a simple extension to the model can overcome this limitation is unclear at present.

For future extensions of the model it would be desirable to have an even richer phenomenology without substantially increasing the complexity of the basic rules of the model. Particularly it would be desirable to have emergent catalysts. In the current model we have only observed rudimentary forms of catalytic activity when two parts of a molecule push two unbounded particles together. However, this is a very primitive and unreliable effect. Extensions will most likely require continuous space/higher dimensions. Work in this direction is currently being undertaken.

References

B. McMullin (1997). Scl: An artificial chemistry in swarm. Dublin City University, School of Electronic Engineering, Technical Report: bmcm9702 and Santa Fe Institute Working Paper: 97-01-002.

Casti, J. (1997). *Would-Be Worlds*. John Wiley & Sons, New York.

Dittrich, P., Ziegler, J., and Banzhaf, W. (2001). Artificial Chemistries—A Review. *Artificial Life*, 7:225–275.

Groß, D. and McMullin, B. (2003). The creation of novelty in artificial chemistries. In *Proceedings of the eighth international conference on artificial life*, pages 400–409. MIT Press.

Holland, J. (1995). *Hidden Order*. Addison-Weseley, Reading.

McMullin, B. and Groß, D. (2001). Towards the implementation of evolving autopoietic artificial agents. In *Proceedings of the 6th European Conference on Advances in Artificial Life*, pages 440–443. Springer-Verlag.

Repast (2004). http://repast.sourceforge.net/.

Evolution of Robust Developmental Neural Networks

Alan N. Hampton[1] and Christoph Adami[1,2]

[1]Digital Life Laboratory 136-93, California Institute of Technology, Pasadena, CA 91125
[2]Jet Propulsion Laboratory 126-347, California Institute of Technology, Pasadena, CA 91109
adami@caltech.edu

Abstract

We present the first evolved solutions to a computational task within the *N*euronal *Org*anism *Ev*olution model (*Norgev*) of artificial neural network development. These networks display a remarkable robustness to external noise sources, and can regrow to functionality when severely damaged. In this framework, we evolved a doubling of network functionality (double-NAND circuit). The network structure of these evolved solutions does not follow the logic of human coding, and instead more resembles the decentralized dendritic connection pattern of more biological networks such as the *C. elegans* brain.

Introduction

The complexity of mammalian brains, and the animal behaviors they elicit, continue to amaze and baffle us. Through neurobiology, we have an almost complete understanding of how a single neuron works, to the point that simulations of a few connected neurons can be carried out with high precision. However, human designed neural networks have not fulfilled the promise of emulating these animal behaviors.

The problem of designing the neural network *structure* can be generalized to the problem of designing complex computer programs because, in a sense, an artificial neural network is just a representation of an underlying computer program. Computer scientists have made substantial progress in this area, and routinely create increasingly complicated codes. However, it is a common experience that when these programs are confronted with unexpected situations or data, they stall and literally stop in their tracks. This is quite different from what happens in biological systems, where adequate reactions occur even in the rarest and most uncommon circumstances, as well as in noisy and incompletely known environments. It is for this property that some researchers have embraced evolution as a tool for arriving at robust computational systems.

Darwinian evolution not only created systems that can withstand small changes in their external conditions and survive, but has also enforced *functional modularity* to enhance a species' evolvability (Kirschner and Gerhart, 1998) and

long-term survival. This modularity is one of the key features that is responsible for the evolved system's robustness: one part may fail, but the rest will continue to work. Functional modularity is also associated with component re-use and developmental evolution (Koza et al., 2003).

The idea of evolving neural networks is not new (Kitano, 1990; Koza and Rice, 1991), but has often been limited to just adapting the network's structure and weights with a bias to specific models (e.g., feed-forward) and using *homogeneous* neuron functions. Less constrained models have been proposed (Belew, 1993; Eggenberger, 1997; Gruau, 1995; Nolfi and Parisi, 1995), most of which encompass some sort of implicit genomic encoding. In particular, developmental systems built on artificial chemistries (reviewed in Dittrich et al. 2001) represent the least constrained models for structural and functional growth, and thus offer the possibility of creating modular complex structures. Astor and Adami (2000) introduced the **Norgev** (*N*euronal *Org*anism *Ev*olution) model, which not only allows for the evolution of the developmental mechanism responsible for the *growth* of the neural tissue or artificial brain, but also has no *a priori* model for how the neuron computes or learns. This allows neural systems to be created that have the potential of evolving developmental robustness as found in nature. In this paper, we present evolved neural networks using the Norgev model, with inherent robustness and self-repair capabilities.

Description of Norgev

Norgev is, at heart, a simulation of an artificial wet chemistry capable of complex computation and gene regulation. The model defines the tissue substrate as a two-dimensional hexagonal grid on which proteins can diffuse through discrete stepped diffusion equations. On these hexagons, neural cells can exist, and carry out actions such as the production of proteins, the creation of new cells, the growth of axons, etc. Proteins produced by the cell can be external (diffusible), internal (confined within the cell and undiffusible) or neurotransmitters (which are injected through connected axons when the neuron is excited). Cells also produce a constant rate of cell-tag proteins, which identify them to other cells and diffuse across the substrate.

Each neural cell carries a genome which encodes its behavior. Genomes consist of genes which can be viewed as a genetic program that can either be executed (expressed) or not, depending on a gene *condition* (see Fig. 1). A gene condition is a combination of several condition *atoms*, whose values in turn depend on local concentrations of proteins. The gene condition can be viewed as the upstream regulatory region of the genetic program it is attached to, while the atoms can be seen as different binding modules within the regulatory region. Each gene is initially active (activation level $\theta = 1$) and then each condition atom acts one after another on θ, modifying it in the $[0,1]$ range, or totally suppressing it ($\theta = 0$). Table 1 shows all the possible condition atoms and how they act on the gene expression level θ passed on to them.

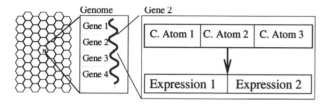

Figure 1: Neural cells are placed into an hexagonal grid and then start executing their genome, which consists of a series of conditions followed by a series of expression actions.

Cond. Atom	Evaluation value θ
SUP[*CTPx*]	$= \theta$, if cell **is** of type *CTPx*
NSUP[*CTPx*]	$= 0$, if cell **is not** of type *CTPx*
ANY[*PTx*]	$= \theta$, if $[PTx] \neq 0$
NNY[*PTx*]	$= \theta$, if $[PTx] = 0$
ADD[*PTx*]	$= R_0^1(\theta + [PTx])$
SUB[*PTx*]	$= R_0^1(\theta - [PTx])$
MUL[*PTx*]	$= \theta * [PTx]$
AND[*PTx*]	$= min(\theta, [PTx])$
OR[*PTx*]	$= max(\theta, [PTx])$
NAND[*PTx*]	$= 1 - AND[PTx]$
NOR[*PTx*]	$= 1 - OR[PTx]$
NOC[*PTx*]	$= \theta$, the neutral condition

Table 1: Repressive and evaluative condition atoms: The SUP and NSUP conditions evaluate the cell-type of the cell in which they are being executed. On the other hand, ANY and NNY repress the gene under the influence of *any* type of protein (internal, external, cell-type or neurotransmitter), where '[*PTx*]' stands for the concentration of protein *PTx*. The neutral condition is special and acts as a silent place holder. $R_0^1()$ saturates the activation into the $[0,1]$ range.

Once a gene activation value θ has been reached, each of the gene's expression atoms are executed. Expression atoms can carry out simple actions such as producing a specific protein, or they can emulate complex actions such as cell division and axon growth. Table 2 contains a complete list

of expression atoms used in Norgev. A more complete description of the Norgev model and its evolution operators (mutation and crossover) can be found in (Astor and Adami, 2000).

Expr. Atom	Action description
PRD[*XY*]	produces substrate *XY*
SPL[*CTPx*]	divide. offspring of type *CTPx*
GRA[*XY*]	grow axon following *XY* gradient
GDR[*XY*]	grow dendrite following *XY* gradient
EXT	excitory stimulus *XY*
INH	inhibitory stimulus *XY*
MOD+[*NTx*]	increase connection weights
MOD-[*NTx*]	decrease connection weights
RLX[*NTx*]	relax weights
DFN[*NTx*]	define cells neurotransmitter
NOP	null action, neutrality

Table 2: Expression atoms. Each is influenced by θ in a different way. For PRD it states the production quantity; for SPL, GRA and GDR the probability of execution; for EXT and INH the stimulus amount; for MOD+, MOD- and RLX the increase, decrease and multiply factor; and for DFN and NOP, θ has no influence.

We know that in cellular biology, gene activation leads to the production of a specific protein that subsequently has a function of its own, ranging from enzymatic catalysis to the docking at other gene regulatory sites. In this model, the most basic expression element is the production of proteins (local or externally diffusible) through the PRD[*PTx*] atom. These can then interact and modulate the activation of other genes in the genome. In this sense, it can be argued that they are only regulatory proteins. However, at least abstractly, genes in this model need not only represent genes in biological cells but can also represent the logic behind enzyme interaction and their products. Thus, Norgev's genome encodes a dynamical system that represents low level biological DNA processes, as well as higher level enzymatic processes including long-range interaction through diffusible substances like hormones. However, the objective is not to create a complete simulation of an artificial biochemistry, and thus other expression atoms are defined that represent more complex actions, actions that in real cells would need a whole battery of orchestrated protein interactions to be accomplished.

Organism example

The best way to understand the model is probably to sit down and create by hand a functional organism. Here we will present a handwritten organism (Fig. 2) and explain how it develops into a fully connected neural network that computes a NAND logical function on its two inputs and sends the result to its output.

The organism, which we named *Stochastic*, relies on the random nature of the underlying chemical world to form its

1. SUP(*cpt*) ANY(*cpt*)   ⇒ SPL(*acpt0*)
2. SUP(*acpt0*) ADD(*apt0*) SUB(*cpt*)   ⇒ SPL(*acpt3*)
3. SUP(*acpt0*) ADD(*spt0*) SUB(*spt1*)   ⇒ SPL(*acpt1*)
4. SUP(*acpt0*) ADD(*spt1*) SUB(*spt0*)   ⇒ SPL(*acpt2*)
5. SUP(*acpt0*) ADD(*cpt*)   ⇒ SPL(*acpt0*)
6. SUP(*acpt1*) ANY(*spt0*)   ⇒ GDR(*spt0*) DFN(*NT1*)
7. SUP(*acpt2*) ANY(*spt1*)   ⇒ GDR(*spt1*) DFN(*NT2*)
8. SUP(*acpt3*) ANY(*apt0*)   ⇒ GDR(*acpt1*) GDR(*acpt2*) GRA(*apt0*)
9. SUP(*acpt3*) ADD(*NT1*) NAND(*NT2*) ⇒ EXT0
10. ANY(*eNT*)   ⇒ EXT0

Figure 2: Genome of *Stochastic*

tissue structure. When an organism is first created, a tissue seed (type *CPT*) is placed in the center of the hexagonal grid, two sensor cells on the left of the grid and an actuator cell on the right. These then diffuse their marker proteins *CPT*, *SPT0*, *SPT1* and *APT0* respectively. In the first time step, only the first gene (Fig. 2) is active in the tissue seed and all the rest are suppressed. This gene will always be active and step after step will split off cells of type *ACPT0* until all the surrounding hexagons are occupied by these cells. After that, the seed does not execute any further function other than secrete its own cell type protein *CPT*. The new cells will, in turn, also split off more cells of type *ACPT0* (gene 5), and so make the tissue grow larger and larger (time=4 in Fig. 3). In a sense, these cells provide a cellular support for further development of the actual network, and could thus be called *glial*-type cells, in analogy to the supportive function glial cells have in real brains. These glial cells can split off three different types of neurons. If the signal from the actuator *APT0* is greater than the signal from the tissue seed *CPT*, then a neuron of type *ACPT3* will split off with probability $p > 0$ (gene 2). On the other hand, if the external protein signal of sensor *SPT0* is strong compared to the external protein of sensor *SPT1*, then instead a neuron of type *ACPT1* will split off with $p > 0$ (gene 3). Last of all, if the signal *SPT1* is greater than *SPT0*, then it is more likely that a neuron of type *ACPT2* will split off (gene 4). This is all that these glial cells of type *ACPT0* do: split off more glial cells, or any of three differentiated neuron types depending on how close they are to the sensors or the actuators.

These three cell types (*ACPT1*, *ACPT2* and *ACPT3*), will then form the actual neural network that will do all the processing. Through gene 6, cells of type *ACPT1* will grow a dendrite towards sensor *SPT0* and define their default neurotransmitter as *NT1*. In the same way, cells of type *ACPT2* will have gene 7 active and will grow a dendrite towards sensor *SPT1* and define their neurotransmitter as *NT2*. Last of all, gene 8 is active in cells of type *ACTP3*, and will direct the growth of dendrites towards cells of type *ACPT1* and *ACPT2* and an axon towards the actuator *APT0*. In the end, each sensor *SPT0* and *SPT1* is connected to every neuronal cell *ACPT1* and *ACPT2*, and all the *ACPT3* neuronal cells are connected to the actuator *APT0* (time=120 in Fig. 3). However, which and how many *ACPT1* and *ACPT2* neurons

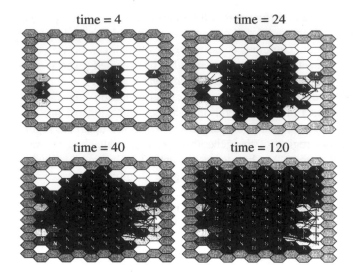

Figure 3: Successive stages in the developmental growth of the *Stochastic* neural tissue.

connect to which and how many of the *ACPT3* neurons relies on stochastic axonal growth, preferably connecting neurons that are nearer on the hexagonal grid. Moreover, all neurons end up connected after the axonal growth process has finished, forming a fully functional NAND implementation.

We still need to understand how the neurons actually process the signals passing through them. This is mediated through genes 9 and 10. Neurons *ACPT1* and *ACPT2* act as *relays* of the sensor signals through gene 10. That is, whenever they receive any neurotransmitter of type *eNT* (default sensor neurotransmitter) they will become excited and inject their gene-defined neurotransmitters through their axons. Neuronal cells of type *ACPT3* will then compute the NAND evaluative action on the amount of neurotransmitters *NT1* and *NT2* injected into their cell bodies and activate accordingly (gene 9). Their activity causes the default neurotransmitter to be injected into the actuator, thus finalizing the simulated input-output NAND computation.

Robustness of *Stochastic*

While *Stochastic*'s neural tissue will always look different every time it is grown because of the stochastic nature of neuronal splitting, it always forms a processing network that correctly computed the NAND function. This confers some robustness to the phenotype of the network in spite of the stochastic, but genetically directed, growth process.

However, the developmental process is far more robust than that. For example, we can manually kill (remove) neurons of a fully developed tissue and have a similar functional (but somewhat scarred) tissue grow back. Fig 4 shows an example where we even removed the tissue seed *CPT*, which has an important role in the organisms development (without its external signal, glial cells of type *ACPT0* do not prolif-

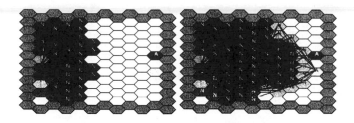

Figure 4: Robustness of *Stochastic* under cell death. Half the neural tissue from Fig. 3 was removed (left). After 80 time steps a different, but functional tissue arises (right).

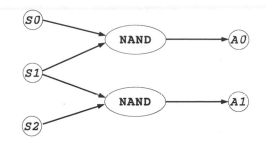

Figure 5: Evolution objective: to *double* the functionality of the original organism.

erate). While the morphology of the self-repaired tissue has changed, it still computes the NAND function. More than anything, this observation helps illustrate the potential capabilities of developmental processes in artificial chemistries to create robust information processing neural tissues even under the breakdown of part of their structure. Note that the self-repair property of *Stochastic* was not evolved (or even hand-coded), but rather emerged as a property of the developmental process. Naturally, these robustness traits can be augmented and exploited under suitable evolutionary pressures.

Evolution of organisms in Norgev

Here, we present the evolutionary capabilities of Norgev, that is, how its genetic structure and chemistry model allow for the evolution of developmental neural networks that solve pre-specified tasks. In the previous section we presented the *Stochastic* organism, which grew into a neural tissue that computed a NAND function on its inputs. Our goal was to study how difficult it would be to *double* the tissue's functionality and compute a double NAND on three inputs, and send the result to two outputs (Fig. 5). Because one of the mutational operators used in the Genetic Algorithm is *gene doubling* (see Astor and Adami, 2000), we surmised that there was an easy route through duplication and subsequent differentiation. Because of the universality of NAND, showing that more complex tissues can evolve from *Stochastic* suggests that arbitrary computational tissues can evolve in Norgev.

The input signal was applied for four time steps (the time for the input to pass through the tissue and reach the output), and then the output was evaluated by a reward function $R = 1 - \sqrt{\sum_i (y_i(x) - t_i(x))^2}$ where x is the input, y the tissue's output, and t the expected output. Organisms were then selected according to a fitness function given by the average reward over 400 time steps, and a small pressure for small genome sizes and neuron numbers. Mutation rates were high and evolution was mainly asexual. Details of the experiments will appear elsewhere (Hampton and Adami, in preparation).

We evolved organisms that obtained the double NAND

functionality in two separate runs on massively parallel cluster computers, over several weeks. The two solutions were very different in both structure and algorithm. The simplest, *Stochastic A*, evolved the fastest with the more straightforward morphology (Fig. 6). Its genome is short (Fig. 7) when compared to evolved organisms in other runs, but is substantially more difficult to understand compared to its ancestor.

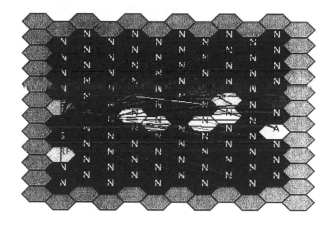

Figure 6: *Stochastic A* neural tissue expressing 6 different cell types. Most of the axonal connections that spread out from the central sensor are not utilized. Instead, the actual computation takes place in a compact area near the center.

After careful analysis of the genome, paired with an evaluation of the physical connections present in the neural tissue, we came to the conclusion that the organism had not reused *any* genomic material to double the NAND function, but had instead *completely* rewritten its code to implement a shorter and more efficient algorithm when compared to the ancestor we wrote. Let us embark once again in a quick step-by-step genome analysis. Gene 1 is active in the tissue seed, which then splits off a cell of type *ACPT0* and *APT2*. After this, the gene is forever shut off because of the repressive *NNY(apt2)* condition. Cell *ACPT0* then splits off cells of type *ACPT1*, *ACPT2* and *ACPT3* through gene 2. This gene is always active, and thus *ACPT0* cells are always in an

1.	MUL(*cpt*) NNY(*apt2*) SUB(*ep2*)	⇒	SPL(*acpt0*) SPL(*apt2*) GRA(*ep2*) DFN(*NT1*)
2.	SUP(*acpt0*) SUB(*spt3*) SUB(*ep2*)	⇒	SPL(*apt2*) SPL(*acpt1*) INH1 GRA(*acpt5*) MOD-(*NT1*) SPL(*acpt3*)
6.	SUP(*acpt1*) ANY(*spt0*) ADD(*cpt*) MUL(*NT1*) ADD(*acpt2*)	⇒	GDR(*spt0*) SPL(*acpt1*) GRA(*ep2*) GRA(*acpt5*) MOD-(*NT1*)
7.	SUP(*acpt2*) NAND(*spt1*) NSUP(*spt1*) ADD(*acpt0*)	⇒	GDR(*spt1*) DFN(*NT2*) GRA(*ep2*) GRA(*apt1*)
8.	SUP(*acpt3*) ANY(*apt0*)	⇒	GDR(*acpt1*) GDR(*acpt2*) GRA(*apt0*)
10.	NAND(*eNT*) OR(*ep2*)	⇒	EXT0 PRD(*ip0*)
11.	ANY(*acpt3*) NSUP(*acpt3*) MUL(*acpt1*) NAND(*NT2*) AND(*acpt0*) NNY(*rfp*)	⇒	DFN(eNT) INH1 MOD-(*NT1*) GRA(*apt0*)

Figure 7: Genome of evolved **Stochastic A** organism. Gene numbering is maintained from the ancestral genome, and gene 11 is a new gene which was randomly created. Gene atoms in light gray appear to be useless and are considered "junk".

inhibitive activation state (due to action atom *INH1*). Gene 6 makes *ACPT1* cells grow a dendrite to sensor *SPT0* and have same-type daughter cells. These are the cells that cover the whole substrate in Fig. 6. Gene 7 causes *ACPT2* cells to grow a dendrite towards sensor *SPT1*, an axon towards actuator *APT1* and define its neurotransmitter as *NT2*. Through gene 8, *ACPT3* cells grow a dendrite to sensor *SPT2*, a dendrite to cells *ACPT2*, and an axon to actuator *APT1*. Gene 11 is the most cryptic. This gene is only active in the first ~3 time steps of the organism's life, and effectively makes cells of type *ACTP0*, *ACPT1* (only the ones in the center, not all the rest) and *ACPT2* grow an axon towards the actuator *APT0*. Once the tissue has developed, gene 10 is used by all cells for processing sensory information (neurotransmitter *eNT*), on which it performs a NOT function.

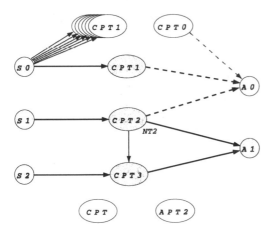

Figure 8: Effective neural circuit grown by **Stochastic A**. Dashed axonal connections grow due to gene 11, which is only active during the first moments of the organisms life. Axons and neurons that have no influence on the final computations are rendered in light gray.

The effective neural circuit is shown in Fig. 8. The result is processed in three time steps instead of the incorrectly postulated minimum of four time steps. This is due to an implicit OR function computed by the actuator cells that we did not anticipate, but which was discovered and exploited by the organism. The neural tissue is applying a NOT function at a relay of its inputs, and then an OR on the actuators

to arrive at the double NAND (Fig. 9). The resulting simplicity of the organism is apparent from the fact that only gene 10 is used for neural processing once the tissue has developed, and it thus has a structure more conducive to further function doubling.

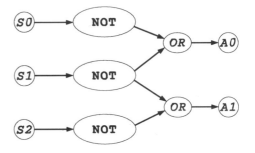

Figure 9: The computation carried out by **Stochastic A**.

Another organism that solved the problem was **Stochastic B**, which took considerably longer to evolve, and that turned out to be highly complex and difficult to understand. In Fig. 10, cellular structures can clearly be seen in which stripe-like patterns of two different neural types succeed one another. These stripes were different for each organism, and reflect a stochastic development. The axonal connections linking all these neurons are so interwoven that it is difficult to believe that this organism is actually acting on its inputs instead of undergoing some recurrent neuronal oscillation.

We were unable to describe the development and internal workings of this organism due to its complexity. However, a complete description is in principle always possible because of our access to all of the organism's internal state variables, and more importantly, to its genetic code: the source of its dynamics. Taking the first steps in that direction, we studied the neuronal activation under each of the eight possible input configurations (Fig. 10). We can clearly see neuronal activity that follows the striped pattern on the right-hand side of the tissue (for inputs of the form $x0x \rightarrow 11$). Remarkably, the left side of the tissue does not follow the same organization and thus we theorize that although they have the same cell type, they have differentiated internally even further depending on their position on the tissue. We came to the conclusion that this organism is not performing the same internal computation as **Stochastic A**. We can see this by inspect-

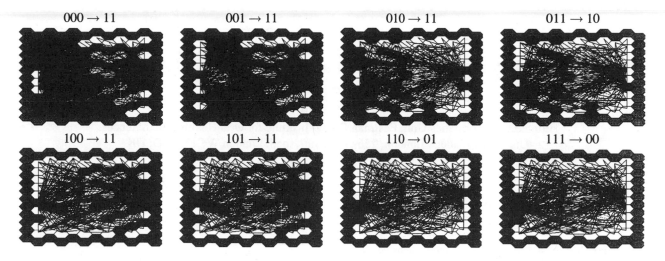

Figure 10: Neuronal activity of a *Stochastic B* neural tissue under the eight possible binary input combinations, where active neurons are shaded, and inactive neurons white. This activity only reflects neurotransmitters that will be injected by active neurons down their axonal branches.

ing input $110 \rightarrow 01$, and noticing that no tissue neurons are activated, and thus there is no neuron performing the NOT function on the last input.

Conclusions

Biology baffles us with the development of even seemingly simple organisms. We have yet to recreate insect neural brains that perform such feats as flight control. As an even simpler organism, the flatworm *C. elegans*, has a nervous system which consists of 302 neurons, highly interconnected in a specific (and mostly known) pattern, and 52 glial cells, but whose exact function we still do not understand. Within Norgev, we have shown that such structural biocomplexity can arise *in silico*, with dendritic connection patterns surprisingly similar to the seemingly random patterns seen in *C. elegans*. And we might have been baffled at the mechanism of development and function of our in silico neural tissue if it were not for our ability to probe every single neuron, study every neurotransmitter or developmental transcription factor, and isolate every part of the system to understand its behavior. Thus, we believe that evolving neural networks under a developmental paradigm is a promising avenue for the creation and understanding of robust and complex computational systems that, in the future, can serve as the nervous systems of autonomous robots and rovers.

Acknowledgements

Part of this work was carried out at the Jet Propulsion Laboratory, California Institute of Technology, supported by the Physical Sciences Division of the National Aeronautics and Space Administration's Office of Biological and Physical Research, and by the National Science Foundation under grant DEB-9981397. Evolution experiments were carried out on a OSX-based Apple computer cluster at JPL.

References

Astor, J. and Adami, C. (2000). A developmental model for the evolution of artificial neural networks. *Artificial Life*, 6:189 218.

Belew, R. R. (1993). Interposing an ontogenetic model between genetic algorithms and neural networks. In *NIPS5 ed J Cowan (San Mateo), CA: Morgan Kaufmann*.

Dittrich, P., Ziegler, J., and Banzhaf, W. (2001). Artificial chemistries–A review. *Artificial Life*, 7:225–275.

Eggenberger, P. (1997). Creation of neural networks based on developmental and evolutionary principles. In *Proc. ICANN'97, Lausanne, Switzerland, October 8-10, 1997*.

Gruau, F. (1995). Automatic definition of modular neural networks. *Adaptive Behaviour*, 3:151–183.

Kirschner, M. and Gerhart, J. (1998). Evolvability. *Proc. Natl. Acad. Sci. USA*, 95:8420–8427.

Kitano, K. (1990). Designing neural network using genetic algorithm with graph generation system. *Complex Systems*, 4:461–476.

Koza, J. R., Keane, M. A., and Streeter, M. J. (2003). The importance of reuse and development in evolvable hardware. In *5th NASA/DoD Workshop on Evolvable Hardware, Chicago, IL, USA*. IEEE Computer Society.

Koza, J. R. and Rice, J. P. (1991). Genetic generation of both the weights and architecture for a neural network. *IEEE Intl. Joint Conf. on Neural Networks*, 2:397–404.

Nolfi, S. and Parisi, D. (1995). Evolving artificial neural networks that develop in time. In *Advances in Artificial Life, Proceedings of the Third European Conference on Artificial Life*, pages 353–367. Springer.

A Functional Self-Reproducing Cell in a Two-Dimensional Artificial Chemistry

Tim J. Hutton

Biomedical Informatics Unit, Eastman Dental Institute for Oral Health Care Sciences,
University College London, 256 Grays Inn Road, London WC1X 8LD, UK
T.Hutton@eastman.ucl.ac.uk

Abstract

We show how it is possible to make a self-reproducing cell in
an artificial chemistry by surrounding a replicating molecule
with a semi-permeable membrane. The molecule can carry an
arbitrary amount of information, encoded in a material form
as a sequence of bases, as in DNA. The cells produce enzymes through a decoding of their base sequence, and these
enzymes trigger reactions essential to the cell's survival. Earlier work in a similar artificial chemistry showed that replicators free in solution could obtain no survival advantage from
producing enzymes; here we show that when surrounded by a
membrane the replicators *can* obtain an advantage. We show
that the cells reliably reproduce over many generations under
environmental pressure for resources. By creating cells in a
material-based artificial chemistry we hope that the system
might have the potential for open-ended, creative evolution.

Introduction

A deeper understanding of the requirements for creative,
open-ended evolution could be obtained if we were able to
recreate it for experimental purposes. This remains an important open problem in artificial life (Bedau et al., 2000) although information-carrying replicators have been created,
often specifically for the purpose, in many different media, including cellular automata, machine-code, artificial
chemistries and wet chemistry.

Open-ended evolution is defined as the unbounded appearance of adaptive activity, for which we have a test,
suggested in (Bedau et al., 1998) and refined in (Channon,
2002). The term creativity, on the other hand, goes beyond
this by requiring the appearance of innovative design solutions such as "mechanisms for sensing new aspects of [the]
environment and for interacting with it in new ways ... and
also for the very notion of individuality to change in radical ways (e.g. the evolution of multicellular organisms from
unicellular ones)." (Taylor, 2001) Undergoing an evolutionary transition (Maynard-Smith and Szathmáry, 1995) such
as the one mentioned would certainly be an example, no artificial system to date has demonstrated such evolutionary
creativity.

It should be noted that systems with a predefined fitness
function can give an insight into some aspects of evolution-

ary design creativity (Lenski et al., 2003) but are of limited use for understanding how to build systems that exhibit
open-ended creativity.

Taylor (Taylor, 1999; Taylor, 2001) discusses open-ended
evolutionary creativity in depth and argues that, for it to be
possible, the replicators must a) be fully embedded in their
arena of competition, b) have rich, unlimited interactions
between each other and with their environment, c) initially
replicate implicitly, rather than using some encoding of the
replication process, and d) be constructed entirely of 'material' components, allowing the possibility of different encodings of information. These suggestions give a possible
reason why creative, open-ended evolution has not yet been
demonstrated in an artificial system: no system to date satisfies all four requirements, with the theoretical exception
of (von Neumann, 1966). In this paper we give details of
a system that extends (Hutton, 2002) and discuss whether it
satisfies all of Taylor's requirements.

Background

As a computationally less-expensive alternative to molecular dynamics simulation, artificial chemistries (ACs) provide a way of modelling chemical processes abstractly. Recent spatially-explicit AC models include lipid micelles and
membranes (Madina et al., 2003) and template-replicating
molecules (Hutton, 2002; Smith et al., 2003).

In (Ono and Ikegami, 2002), self-reproducing cells were
demonstrated to have the capacity to evolve through the
selection of catalysts that affect the formation of the lipid
membranes. However, the evolutionary capacity in this system was very limited since little information could be inherited.

In (Hutton, 2002), a set of reactions in an AC was shown
that allowed molecules carrying any amount of information to replicate. These molecules qualify as units of evolution and unlimited hereditary replicators (Szathmáry and
Maynard-Smith, 1997) but failed to demonstrate any interesting behaviour because they could do nothing but replicate
- their phenotype was minimal.

In (Hutton, 2003a), a method for the replicating molecules

to produce reaction-specific *enzymes* was presented, allowing the molecules some way of controlling their chemical environment. However, it was found that in the absence of a surrounding membrane the enzymes could confer no evolutionary advantage to the molecule that produced them, since they would diffuse away. Here we extend this work by showing how it is possible to put each molecule inside a semi-permeable membrane, allowing the molecule to retain sole use of the enzymes it produces and thus potentially to derive from them some survival advantage to outweigh the extra replication time required to produce them.

To continue to replicate successfully, the molecule must cause the membrane to grow and divide, and ensure that a copy of the molecule is present in each daughter cell. We show how this can be achieved by connecting the molecule to the membrane temporarily, as is common in bacterial reproduction. Ideally the molecules themselves would engineer this kind of functionality, by producing the enzymes that trigger the required reactions but this is not yet done because of the large size of the resulting cell. We evaluate the potential of the cells to demonstrate the evolutionary growth of complexity (McMullin, 2000) and whether such a system is capable of evolutionary creativity.

System Description

Our AC extends that of (Hutton, 2002), where 'atoms' have a fixed *type* $\in \{a, b \ldots t\}$ and a variable *state* $\in \{0, 1, 2 \ldots\}$. The atoms move around a finite world and a set of reactions (Table 1) determine what happens when they bump into each other; bonds between them may be formed (to make 'molecules') or broken and their states can be changed. Bonded atoms are prevented from moving apart.

Previously a lattice-based physics was used for speed, however this can lead to inflexibility because some of the atoms in large molecules will be unable to move. One solution is to allow bonded atoms to move further apart on the lattice but then it becomes difficult to make useful membranes since the contents of a bonded loop of atoms will be able to escape through the gaps.

A continuous-space physics (see eg. (Smith et al., 2003)) has better flexibility properties (at greater computational cost) and allows us to make membranes that prevent their contents from escaping, by using a simple spring force for volume exclusion. Semi-permeability is achieved by turning off the volume exclusion force for unbonded atoms in state 0 which can therefore pass through membranes (and other structures) freely. A more realistic model of lipid molecules might avoid the need for such measures at still greater computational cost.

Reactions 1-24 permit the molecule to make a copy of itself and cause the membrane to divide with one copy in each half. Figure 1 shows the various stages in the process, with the atoms in state 0 shown as dots and other atoms as filled circles (for clarity). The starting point is a loop of

a8's, with a molecule such as e9-a1-b1-c1-d1-f1 inside, although importantly any sequence of a1's, b1's, c1's and d1's with an e9 at one end and an f1 at the other will replicate also. The molecule first attaches the 'e' end to the membrane, and the division ends with the molecules being released back into the daughter cells. Modern eukaryotic cells do not connect the strands of DNA to the membrane while duplicating but this is common in bacteria and appears to be a simpler way of achieving cell division.

	A	b_1	B	b_2	C	b_3	A'	b_1'	B'	b_2'	C'	b_3'
1	e9	✗	a8				10	✓	8			
2	e10	✗	e0	✗	a8	✓	4	✓	3	✓	8	✓
3	x3	✓	x4	✓	y1	✗	4	✓	4	✓	2	✓
4	x2	✗	x0				3	✓	5			
5	x3	✓	x5	✗	y4	✓	4	✓	3	✓	4	✗
6	f4	✓	f3	✗	a8	✗	7	✓	8	✓	10	✓
7	a10	✓	a8	✗	f8	✓	10	✓	11	✗	7	✗
8	x6	✓	y4				1	✓	7			
9	x7	✓	y7				6	✗	6			
10	e6	✓	a8				1	✓	8			
11	a8	✓	e1	✓	a12	✗	9	✗	1	✗	9	✓
12	a9	✓	a9	✓	a10	✗	10	✓	9	✗	14	✓
13	a14	✓	a10	✓	a8	✗	15	✗	14	✗	8	✓
14	a14	✓	e1	✗	a9	✓	8	✗	16	✗	8	✓
15	a15	✓	e1				8	✗	16			
16	e16	✗	f1				17	✗	17			
17	e17	✓	x1				17	✓	2			
18	e17	✓	x3	✓	y5	✗	17	✓	4	✓	3	✓
19	f17	✓	x4	✓	y3	✗	1	✓	6	✗	6	✓
20	x6	✓	e17	✓	x6	✗	18	✗	9	✓	1	✗
21	a10	✓	a11	✗	x1	✓	10	✓	12	✓	1	✓
22	x1	✓	y1	✗	a12	✓	1	✓	1	✓	13	✗
23	x1	✓	a13	✓	a10	✗	1	✗	11	✓	13	✓
24	x1	✓	y1	✓	a13	✓	1	✓	1	✓	10	✗
25	a8	✓	a7	✗	a8	✓	8	✓	8	✓	8	✗
26	zi	✓	y1				0	✗	j			
27	di	✓	y1				i	✗	18			
28	di	✗	xp	u	yq	✗	i	✗	r	v	s	✗

Table 1: The reactions used in our system. Reactions involve either two or three atoms, the column notation refers to the figure above. The types of the atoms are not given in the right-hand half of the table because they do not change. For example, the first reaction occurs when an e9 bumps into an a8 - the result is an e10 bonded to an a8. The symbols *x* and *y* are variables standing for any type a-f. Reactions 26-28 are for the production and application of enzymes and are explained in the text.

The membrane is pulled between the strands by reactions 21-24, in a sequence that passes a chain of atoms in state 1 along past a fixed point on the membrane. Such engineering solutions can also be seen in modern cells, for example the microtubules on the mitotic spindle which act like conveyor belts or bargepoles to spatially organise the chromosomes and other components.

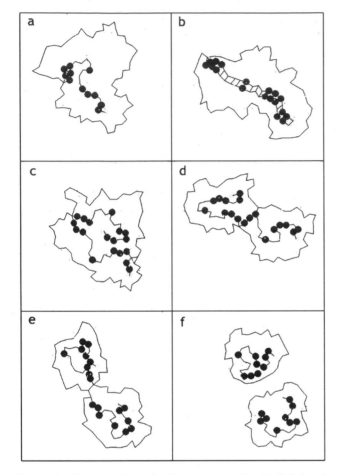

Figure 1: Six snapshots in the process of cell division in our artificial chemistry system. The information-carrying molecule first attaches itself to the membrane and duplicates itself (a, b). The membrane is pulled between the strands (c, d) before fusing at the top of the molecule (e) and dividing, releasing the molecules into the daughter cells (f) for the process to repeat.

There are several benefits to an information-carrying molecule of having a surrounding membrane, including a) the molecule is protected from any harmful external agents, and b) the molecule is able to control the chemical composition of its local environment. The second of these is achieved through reactions 26-28, which allow the production and application of enzymes.

The action of enzymes is implemented in reaction 28, which encodes any two-atom reaction involving states between 0 and 17 into a unique state i:

$$i = 2(2(6(6(18(18(18p+q)+r)+s) \\ +x)+y)+u)+v+18 \quad (1)$$

where x and y are encoded as $\{a = 0, \ldots f = 5\}$, and u and

v are 0 for unbonded and 1 for bonded. This formula is the most efficient way of encoding the values of p, q, r, s, x, y, u and v into a single number. For example, the enzyme d5731 codes for e0+a0→e2a3 and will trigger this reaction whenever the reactants bump into it.

Enzymes are produced through reactions 26 and 27. After each cycle of reproduction, the sequence of bases is turned from a non-reactive information sequence into an enzyme by a simple base-3 encoding given by reaction 26, where $i \geq 18$, z stands for any of the types a, b or c, and j is computed by:

$$j = 3(i-18) + \text{value}(z) + 18 \quad (2)$$

with $\text{value}(a) = 0$, $\text{value}(b) = 1$, $\text{value}(c) = 2$.

This reaction converts a special copy of the base sequence, produced after cell division, for example c18-a1-b1-d1, into a20-b1-d1, then b24-d1 and then d37. This base-3 'shift and add' procedure is a simple way of converting a string of a's, b's and c's into a single number.

Atoms in nature cannot store such amounts of information, the 'atoms' in our system would perhaps be better termed 'proteins'. The idea is that enzymes can be encoded as small objects (rather than large, complex three-dimensional molecules) because their core functionality (triggering a specific reaction) is simple. This is just one of the simplifications that we are using to make a computationally-amenable AC system that supports life.

Figure 2 shows two cells in different stages of producing their enzymes. Each base in turn is released in state 0, its information (its type) having been incorporated into the enzyme that is being produced. Reaction 27 allows more than one enzyme to be produced, for example by the sequence cbadbbaccd which would produce d37 and d134.

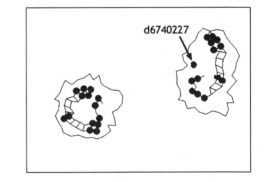

Figure 2: After cell division, the information-sequence is decoded into a reaction-specific enzyme. The sequence of bases is first duplicated (left) and then decoded into an enzyme (right). This particular enzyme (d6740227) causes the membrane to grow in length, helping the cell to continue to reproduce.

After division, the membranes of the daughter cells are approximately half the size of that in the original cell

(Fig. 1f), thus some mechanism for causing membranes to grow is necessary if reproduction is to continue. The cell achieves this by producing the necessary enzyme. In the cells shown the base sequence is `bbacaabaccbbbabd`, which is converted by repeated applications of reaction 26 into `d6740227` (Fig. 2), which triggers the reaction `a8+a0→a8a7`, adding extra atoms to the membrane. The atoms are incorporated into the membrane by reaction 25, making it bigger and allowing the cell to continue to reproduce. Cells without the correct base sequence to produce this enzyme or an equivalent one will run out of room inside the membrane and will cease to replicate.

Observations

Sometimes the cells depicted in Figs. 1 and 2 fail to divide evenly, leaving one daughter with insufficient room in the membrane to reproduce, this is one aspect of the design that could be improved, ideally by evolution. Also, there is no timing between the growth of the membrane and duplication of the information-sequence, this would also be desirable for the cell to reproduce more reliably. Even with these issues, the cells are capable of indefinite reproduction, as long as there are sufficient space and material resources.

To demonstrate this, we use the same experimental set-up as in (Hutton, 2002), where a fixed-size world is periodically 'flooded', with one half being cleared and replenished with raw material (unbonded atoms in state 0). The reproducing entities have until the next flood to repopulate the empty half, and if they cannot do so quickly enough their numbers will drop to zero. Figure 3 shows the world after around 13 floods of 200,000 timesteps each, surviving cells are visible in amongst the broken components of cells cut in half by the flood.

Figure 3: A screenshot at approximately 2,600,000 timesteps, with floods every 200,000 timesteps. Some surviving cells are visible, amongst broken bits where the flood has cut cells in half.

In Fig. 3 it can be seen that the information molecule can survive contact with the outside chemical environment, it can also continue to replicate as long as it encounters strips of membrane. One side-effect of the cell division process is that these loose molecules can even produce complete cells again by a process of invagination; by wrapping a length of membrane around one copy of the molecule.

To investigate if this process is accounting for the continued survival of the cells, we rerun the experiment and introduce caustic agents into the external environment. These agents break bonds between any atoms they encounter, with the exception of the membrane atoms, making survival inside a membrane the only option. To test the evolutionary stability of the cells we introduce a low probability mutation reaction (p=0.0000001 per atom per timestep) that either adds or deletes a base, as R15 in (Hutton, 2003b). In this experiment we have used a flood that rotates between the four quarters of the world, allowing a smaller area to be simulated. Figure 4 shows a screenshot after 1,136,000 timesteps (flood every 50,000 timesteps), surviving cells are visible in amongst many lengths of severed membrane.

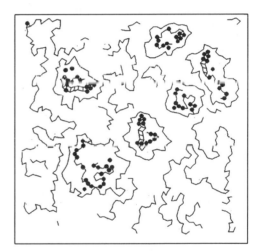

Figure 4: A screenshot after 1,136,000 timesteps, with a different quarter of the world being flooded every 50,000 timesteps. Six surviving cells are visible amongst the sections of membranes from cells cut in half by floods.

The need to keep the insides of cells separate from caustic agents in the external environment is seen in natural cells as well - if there were no barrier then the delicate machinery of reproduction would soon be perturbed. This is one of the main functions of a cell membrane.

By tracking the occurences of reaction 1 and reading off the string of bases connected to the 'e' atom we can observe which sequences are reproducing throughout the experiment. Figure 5 shows the reproduction rates for different sequences in a typical run, with the original sequence (`ebbacaabaccbbbabdf`) dominating, and mutants failing to reproduce for long. These results confirm that the cells are capable of repeated reproduction, even without timing between the growth of the membrane and the replication of the information-carrying molecule.

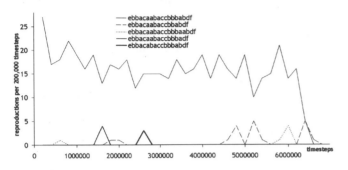

Figure 5: A plot of the reproduction rates for different cells in a typical run. The cell with the sequence necessary for membrane growth (`ebbacaabaccbbbbabdf`) is the only one that continues to survive. Several mutations appeared in this particular run, with bases either missing or added. None survived for very long since they did not produce the correct enzyme (or an equivalent one) that helps the membrane to grow in size. Due to the small size of the area being simulated, the weight of mutations caused an extinction event at 6,600,000 timesteps.

The selection pressure against mutated sequences (Fig. 5) shows that carrying a significant amount of information can be beneficial in an AC system of reproducing entities. In this system the information is useful because the enzyme it produces is retained for the exclusive use of the cell. This was not the case in previous work when the molecules didn't have a surrounding membrane, instead the molecules soon evolved to shed their bases, even if the enzymes were required for reproduction (Hutton, 2003a). In the current system the cell shown is the smallest possible (since the enzyme that causes the membrane to grow is essential to reproduction), however it is hoped that the cells might evolve the production of *additional* enzymes to help them reproduce more reliably, or faster.

Conclusions

We have demonstrated an artificial chemistry (AC) containing a simple form of cell. An arbitrary sequence of bases encodes information for producing enzymes that can assist in the cell's reproduction. The cell is functional in the sense that it is capable of doing things other than just reproduce, through the action of its enzymes. We have shown that the cell is capable of robust, repeated reproduction in a changing chemical environment. The genetic and membrane subsystems are tied together, each playing a role in the production of the other. This form of cooperation has been discussed at length in designs for synthetic cells in wet chemistry (Rasmussen et al., 2003) and in abstractions of the life processes in natural cells (Ganti, 1997).

Taylor's requirements for a system to exhibit open-ended, creative evolution (Taylor, 2001) seem to be satisfied in this system, with one exception. The cells are information-carrying reproducers, and are fully embedded in their environment. Currently the cells reproduce implicitly (the reproduction rules are given) although since they are encoded as material components the potential exists for reproduction to proceed in a different manner, perhaps by using a set of enzymes. However, the cells do not currently interact in a very rich way since they have no way of sensing each other's presence. This is potentially a problem for evolvability, since without rich interactions there is no drive towards higher sophistication. The cells as they are could compete in a passive way, by making their reproduction cycle more robust or rapid by producing additional enzymes but a great number of enzymes would be required to make structures for defence or aggression. For comparison, in natural cells the number of enzymes required just for growth and division has been estimated at between 2,000 and 5,000 (Murray and Hunt, 1993). One possibility for allowing interactions between cells is through cell signalling - if some enzymes could occasionally move through the membrane then simple messages might be usefully passed. Cell signalling is an important prerequisite for cell cooperation and multicellularity.

Many of the reactions in Table 1 could be encoded as enzymes and produced by the cells themselves. This would reduce the number of required reactions and increase the size of the smallest viable cell. While this would make the first replicator less likely to appear spontaneously it might improve the evolvability of the system. The three-way reactions could be replaced by a set of two-way reactions, although this would require additional states, and in theory a system with only reactions 26, 27 and 28 (or even less) could support self-reproducing cells.

The earliest proto-cells on Earth are likely to have been formed through encounters between lipid aggregates such as vesicles, and autocatalytic polymers such as RNA. Understanding how these entities could have become a single unit, with the components helping each other's survival, is an important goal in understanding the origin of life (Ganti, 1997; Szostak et al., 2001). This study is not directly aimed at this question but draws inspiration from speculations about early forms of life, and may help eliminate hypotheses, as long as the abstractions used in the simulation are justifiable in the context.

The current system is implemented in two dimensions to minimize the computational costs and make visualisation easier but there might be advantages to moving to a three-dimensional physics, due to the greater number of topological possibilities. The reaction rules for template-replication could remain the same in 3D but a different approach to membrane division would be required. Solutions seen in nature that could be explored include contractile rings (tightening a belt around a balloon) and septation (growing a sheet across the middle).

A more realistic simulation of lipids might reduce the complexity required to design a working AC cell, or rather transfer some of the design complexity into the environment. However, it seems that such an approach would necessarily incur a greater computational cost, and this is a major problem when trying to recreate an evolutionary system, since a fundamental tenet is that there must be many interacting entities.

The use of a semi-permeable membrane around an information-carrying replicator solves the problem encountered in (Hutton, 2003a) and discussed in (Szostak et al., 2001) and elsewhere, that a naked replicator can obtain no survival advantage from the production of enzymes. It is speculated that cells that additionally produce other enzymes might outcompete the ones presented here, and thus that the system might evolve to greater complexity. By following the principles of (Taylor, 2001) we hope that the system might eventually exhibit open-ended, creative evolution.

The source code for the implementation presented here is available at: `http://www.eastman.ucl.ac.uk/~thutton/Evolution/Squirm3/`

Acknowledgments

The author would like to thank Dan LaLiberte and Chrisantha Fernando for much interesting discussion and motivation. The anonymous reviewers were also very helpful in clarifying the important points of this paper.

References

Bedau, M., McCaskill, J., Packard, N., Rasmussen, S., Adami, C., Green, D., Ikegami, T., Kaneko, K., and Ray, T. (2000). Open problems in artificial life. *Artificial Life*, 6(4):363–376.

Bedau, M., Snyder, E., and Packard, N. (1998). A classification of long-term evolutionary dynamics. In Adami, C., Belew, R., Kitano, H., and Taylor, C., editors, *Proc. Artificial Life VI*, pages 228–237. MIT Press.

Channon, A. (2002). Improving and still passing the ALife test: Component-normalised activity statistics classify evolution in Geb as unbounded. In Standish, R., Bedau, M., and Abbass, H., editors, *Proc. Artificial Life VIII*, pages 173–181. MIT Press.

Ganti, T. (1997). Biogenesis itself. *J. Theor. Biol.*, 187(4):583–593.

Hutton, T. (2002). Evolvable self-replicating molecules in an artificial chemistry. *Artificial Life*, 8(4):341–356.

Hutton, T. (2003a). Information-replicating molecules with programmable enzymes. In *Proc. Sixth International Conference on Humans and Computers*, pages 170–175, University of Aizu, Aizu-Wakamatsu, Japan.

Hutton, T. (2003b). Simulating evolution's first steps. In Banzhaf, W., Christaller, T., Dittrich, P., Kim, J., and Ziegler, J., editors, *Proc. Seventh European Conference on Artificial Life*, pages 51–58, Dortmund, Germany.

Lenski, R., Ofria, C., Pennock, R., and Adami, C. (2003). The evolutionary origin of complex features. *Nature*, 423:139–144.

Madina, D., Ono, N., and Ikegami, T. (2003). Cellular evolution in a 3D lattice artificial chemistry. In Banzhaf, W., Christaller, T., Dittrich, P., Kim, J., and Ziegler, J., editors, *Proc. Seventh European Conference on Artificial Life*, pages 59–68, Dortmund, Germany.

Maynard-Smith, J. and Szathmáry, E. (1995). *The major transitions in evolution*. Morgan-Freeman.

McMullin, B. (2000). John von Neumann and the evolutionary growth of complexity: Looking backwards, looking forwards... *Artificial Life*, 6(4):347–361.

Murray, A. and Hunt, T. (1993). *The cell cycle: an introduction*. Oxford University Press.

Ono, N. and Ikegami, T. (2002). Selection of catalysts through cellular reproduction. In Standish, R., Bedau, M., and Abbass, H., editors, *Proc. Artificial Life VIII*, pages 57–64. MIT Press.

Rasmussen, S., Chen, L., Nilsson, M., and Abe, S. (2003). Bridging nonliving and living matter. *Artificial Life*, 9(3):269–216.

Smith, A., Turney, P., and Ewaschuk, R. (2003). Self-replicating machines in continuous space with virtual physics. *Artificial Life*, 9(1):21–40.

Szathmáry, E. and Maynard-Smith, J. (1997). From replicators to reproducers: the first major transitions leading to life. *Journal of Theoretical Biology*, 187:555–571.

Szostak, J., Bartel, D., and Luisi, P. (2001). Synthesizing life. *Nature*, 409:387–390.

Taylor, T. (1999). *From artificial evolution to artificial life*. PhD thesis, Division of Informatics, University of Edinburgh.

Taylor, T. (2001). Creativity in evolution: Individuals, interactions and environment. In Bentley, P. and Corne, D., editors, *Creative Evolutionary Systems*, chapter 1. Morgan Kaufman.

von Neumann, J. (1966). *Theory of self-reproducing automata*. Urbana: University of Illinois Press. Edited and completed by Arthur Burks.

Metabolic closure in *(M,R)* systems

Juan-Carlos Letelier[1], Jorge Soto-Andrade[1], Flavio Guíñez-Abarzúa[1],
Athel Cornish-Bowden[2] and María Luz Cárdenas[2]

[1]Facultad de Ciencias, Universidad de Chile, Santiago, Chile
[2]BIP, IBSM, CNRS, 31 chemin Joseph-Aiguier,
13402 Marseille Cedex 20, France
letelier@uchile.cl

Abstract

The work of Robert Rosen, related to metabolic networks called *(M,R)* systems is reviewed and clarified. We study the algebraic formulation of *(M,R)* systems particularly the mapping β, which encapsulates Rosen's solution to the problem of metabolic closure and infinite regress. We construct an arithmetical example of an *(M,R)* system and also an *(M,R)* system based on a three-step minimal metabolism.

Introduction

The theories of Robert Rosen pose a scientific enigma. The core of his theory, called here *Rosen's Central Result*, is that metabolism *bootstraps* itself without the help of external agents thus keeping cellular organization invariant in spite of continuous structural change (Rosen, 1958a; Rosen, 1959; Rosen, 1972; Rosen, 1991). In theory, Rosen's insights and results should have had a profound impact in theoretical biology, especially in the field of Artificial Life as he claimed that his theory was directly relevant to the problem of *fabrication* of living systems (Rosen, 1991). But, in spite of recents attempts to use Rosen's ideas in areas like bioinformatics (Wolkenhauer, 2002) and control theory (Casti, 2002), the impact and dissemination of Rosen's ideas have been extremely small. The lack of impact is partly due to the highly abstract nature of the central result and to the surprising fact that Rosen never gave biological or mathematical examples of his ideas. Because of this, and because it is such a special result regarding its biological as well mathematical aspects, we have found it necessary to revisit and clarify it as well as to connect it to other theoretical ideas concerning metabolic closure.

Overview of *(M,R)* Systems

Rosen's theory centers around a model of metabolic networks he called *(M,R)* systems. As the study of these systems is an essential first step to understand Rosen's result, here we give an outline of them. Initially an *(M,R)* system looks like a simple graph-theoretic view of metabolism, but this interpretation is misleading as *(M,R)* systems are endowed with a richer mathematical structure.

The *M* Components

In an *(M,R)* system every biochemical reaction is interpreted as a mapping. Thus the first reaction of glycolysis, catalyzed by the enzyme glucokinase:

$$\text{Glucose} + \text{ATP} \longrightarrow \text{Glucose 6-phosphate} + \text{ADP}$$

can be formalized as an operator M that transforms molecules a_1 and a_2 into b_1 and b_2:

$$a_1 + a_2 \xrightarrow{M} b_1 + b_2 \tag{1}$$

The catalyst M, acts as a mathematical mapping, transforming variables (a_1, a_2) into variables (b_1, b_2). As enzymes are not totally specific for the types of molecules that they transform, Rosen interpreted M as a mapping between two sets defined by Cartesian products, where sets A_i, B_i represent the admissible molecules that the enzyme can process.

$$
\begin{aligned}
M : A_1 \times A_2 &\longrightarrow B_1 \times B_2 \\
(a_1, a_2) &\mapsto M((a_1, a_2)) = (b_1, b_2)
\end{aligned}
\tag{2}
$$

Although Rosen did not mention it, this over-reaching formalization is extreme. An enzyme can be presented *in vitro* with artificially produced molecules that are accepted and processed as substrates because of their structural resemblance to the natural substrate, and it then appears that the set A_1 is "large". However, it is radically different *in vivo* because in the organism only one (or a few) acceptable substrates exist. For example, in some bacteria the enzyme glucokinase mentioned above will not accept any natural sugar substrate other than glucose.

This mathematical framework can be extended to take account of the network of thousands of biochemical reactions that constitute a living metabolism. We can interpret the overall metabolism M_{met} as the following transformation:

$$
\begin{aligned}
M_{met} : A = (A_1 \times \cdots \times A_p) &\rightarrow B = (B_1 \times \cdots \times B_q) \\
\mathbf{a} = (a_1, \ldots, a_p) &\mapsto \mathbf{b} = (b_1, \ldots, b_q)
\end{aligned}
\tag{3}
$$

In a very compact notation, the complete metabolism is the (huge) mapping f between the (huge) sets A and B.

$$f : A \longrightarrow B \qquad (4)$$
$$\mathbf{a} \longmapsto f(\mathbf{a}) = \mathbf{b}$$

As many metabolisms are theoretically possible between sets A and B, conceptually we define $\mathcal{M}$ as the set of all possible metabolisms between A and B. Does $\mathcal{M}$ have some sort of structure? After all, a metabolic network is much more than just a random collection of transformations between molecules. This is a crucial point that was never clarified by Rosen, who presented his arguments by assuming

$$f \in \mathcal{M} = Map(A, B) = \text{set of all mappings between sets } A \text{ and } B$$

We will see that this identification is too general, and that $\mathcal{M}$ must be a proper subset of $Map(A,B)$, consisting only of some selected, admissible mappings from A to B. Intuitively we might think of $\mathcal{M}$ as consisting of all mappings from A to B that preserve some sort of underlying, implicit, structure common to sets A and B as these two sets must be much alike (A represents the left-hand side of biochemical reactions and B the right-hand side). We develop this viewpoint in our arithmetical example in a section below.

The R Subsystems

Rosen's crucial insight concerns the long-term stability of a metabolic network. Because components M_i represent physical objects (enzymes) they must inevitably become degraded by a wide variety of processes. If the cell is to continue operating in a steady state, every M_i must be replaced as fast as it is degraded. Rosen posited that for every M_i there must exist a subsystem R_i, made of a subnetwork of biochemical reactions, that uses metabolites, from B, to replace M_i or, in Rosen's terminology, to "repair" it (the word *repair* was a confusing choice, and in this paper we use the term *replacement*).

This insight is essential for understanding the biological relevance of Rosen's work. In contrast to a man-made machine, a living organism is a self-made machine, and all of its components must be made and, when necessary, replaced within the system by components that are themselves products of the system, are themselves degraded, and must also be replaced, again by components within the system, which are likewise degraded, and need to be replaced, and so on indefinitely. In such a system the possibility of infinite regress is obvious, and Rosen's work can be interpreted as a search for a way to escape this, or in other words a search for *closure*.

The central result is an attempt to express how a system must be organized if it is to continue in operation indefinitely. The crux of the matter is thus to understand how the

R_i are replaced and how to avoid an infinite regress. But, as we have seen, the same wear-and-tear argument that was applied to M_i applies equally well to the R_i. It is possible, of course, but not elegant or useful, to invoke a second-level repairer to replace each R_i. But this "solution" just raises the new question of how to replace the second-level repairers, and is thus no solution at all. The central result is an intuition about the systemic nature (i.e. a property that depends on the connectivity of the network) of this maintenance or "replication" function[1]. Thus, in some (M,R) systems, the total network regenerates each R_i: these systems are the (M,R) *systems with organizational invariance*, and they constitute Rosen's model of living systems. Remarkably, the central result is a mathematical enunciation of this metabolic closure.

Algebraic Representation of *(M,R)* Systems

As we saw metabolism can be interpreted as a mapping f that transforms an instance $a \in A$ into an instance $b \in B$. But how can the collective action of subsystems R_i be represented as a mapping? The replacement mechanism is a procedure, denoted by Φ, that, starting with $b \in B$ as input, produces f according to:

$$\Phi(b) = f, \text{ with the condition } b = f(a) \text{ (for some } a \in A)$$

Because the net effect of Φ is to select from the relatively large set $\mathcal{M} \subset Map(A,B)$ the given metabolism f, we call Φ a *selector*. Thus as f represents *metabolism*, Φ represents *replacement*. As with f, Φ can also be represented by a mapping between the sets of metabolic configurations (B) and the set of possible metabolisms ($\mathcal{M}$). Again Rosen assumed the most general structure for the set of selectors $\mathcal{S}$.

$$\Phi \in \mathcal{S} = Map(B, \mathcal{M}) = Map(B, Map(A, B))$$

It is, however, trivial to find not only one but many mappings from one set to another that take a given value (f in this case) at a given point (b in this case), so that it is an essential feature of a sensible mathematical model of metabolism to ask for the set $\mathcal{S}$ of selectors to be a *proper* subset of $Map(B, \mathcal{M})$.

Then a (M,R) system has the following algebraic structure based on mappings $f \in \mathcal{M}, \Phi \in \mathcal{S}$, acting in synergy

$$A \xrightarrow{f} B \xrightarrow{\Phi} \mathcal{M} \subset Map(A,B)$$
$$a \longmapsto f(a) = b \longmapsto \Phi(b) = f$$

---

[1] *Replication* was another unfortunate choice of terms, evoking ideas of reproduction whereas the essential notion is *organizational invariance* (i.e. the network maintains its connectivity) under continuous structural change.

Now, in the full language of maps, we can rephrase the closure result sought by Rosen. How can the selector map Φ be produced by the network when the system is capable of organizational invariance, without implying infinite regress?

Rosen's Central Result in a Nutshell

Rosen's solution to avoid infinite regress, encapsulated in his central result, was to posit that, for a suitable b, for any metabolism $f \in \mathcal{M}$ there should exist *one and only one* selector $\Phi \in \mathcal{S}$ such that $\Phi(b) = f$. He called β the assignment $f \mapsto \Phi$. Thus, on purely formal grounds, Rosen's β is a mapping from $\mathcal{M}$ to $\mathcal{S}$ which is none other than the inverse mapping to the the mapping for evaluation at b, map $ev_b : \Phi \mapsto \Phi(b)$ from $\mathcal{S}$ to $\mathcal{M}$. Admittedly, the invertibility of an evaluation map ev_b is unusual, but Rosen, besides making this demanding hypothesis, never produced a clear-cut mathematical description or an algorithm to construct such an invertible evaluation and its inverse β. He only showed that it was logically possible (Rosen, 1959; Rosen, 1972; Rosen, 1991), and sometimes he admitted that the existence of β was mathematically difficult (Rosen, 2000, pages 261-265). The beauty of the concept of β is that Rosen introduced it as the inverse of the evaluation mapping ev_b, so that in some sense β simplifies to some b. This prompted Rosen to say that β was equivalent to an element $b \in B$, thus closing the loop and avoiding infinite regress.

The operation of an organizational invariant (M,R) system can therefore be viewed as (just) three mappings (f, Φ, β) acting in synergy

$$A \xrightarrow{f} B \xrightarrow{\Phi} \mathcal{M} \xrightarrow{\beta} \mathcal{S} \qquad (5)$$
$$f(a) = b$$
$$\Phi(b) = f$$
$$\beta(f) = \Phi, \text{ with } \beta = (ev_b)^{-1}$$

In this brief presentation, as in all Rosen's writings, the precise description of conditions that would entail the invertibility of the evaluation at b, and so the existence of β, are left open.

Also, Rosen's claim that the existence of β enables us to avoid infinite regress deserves further discussion. Indeed, we might legitimately ask whether a mapping $\gamma : \mathcal{S} \to H(\mathcal{M}, \mathcal{S})$ is not needed, which would reconstruct β as a product of the metabolism as well, i.e. $\gamma(\Phi) = \beta$. Following Rosen's insight, we could assume that since β is equivalent to b, then γ should be equivalent to something that produces b, namely f, since $f(a) = b$. In more precise terms then, a natural idea is to take γ to be the inverse of the evaluation at f, if possible. This requires that β must be the only mapping in $H(\mathcal{M}, \mathcal{S})$ which transforms f into Φ, in other words, the equation $\beta(f) = \Phi$, in β, has exactly one solution, namely $\beta = ev_b^{-1}$. However, natural as it may be, this property is, in general, not entailed by the assumption that ev_b is invertible, and needs to be stated as a supplementary assumption.

As β encapsulates the notion of metabolic closure we intend to clarify its nature in the next section.

(M,R) Systems with Organizational Invariance and the Notion of β

Consider the relations defining an organizationally invariant *(M,R)* system

$$A \xrightarrow{f} B \xrightarrow{\Phi} [\mathcal{M} = Map(A,B)] \xrightarrow{\beta} [\mathcal{S} = Map(B,\mathcal{M})]$$

Recall that β stands for the inverse of the evaluation map at b, denoted $ev_b : \Psi \mapsto \Psi(b)$, for any mapping $\Psi \in \mathcal{S}$. Now we can express the central result mathematically:

For β to exist, the functional equation in Φ, $\Phi(b) = f$ must have *one and only one* solution, where f and $b = f(a)$ are given.

We see then that the mapping $\Phi \in \mathcal{S}$ must have the quality of being completely determined by its value at a single element b. This is admittedly a rather unusual property for everyday mappings from one set to another, even for continuous mappings. Intuitively this property means that Φ somehow has a rigid behaviour.

To give an example of *rigidity*, think of a mapping Φ on the set $\mathbb{Z}$ of the integers into itself, which has the property of being *additive*, i.e. $\Phi(m+n) = \Phi(m) + \Phi(n)$. Then, if you know its value $\Phi(1)$ at 1, you can deduce its value at any positive integer n. Indeed, since $n = 1 + 1 + ... + 1$, n times, we must have $\Phi(n) = \Phi(1 + ... + 1) = \Phi(1) + ... + \Phi(1) = n\Phi(1)$. Moreover, since $\Phi(0) = \Phi(0+0) = \Phi(0) + \Phi(0)$, we see that $\Phi(0)$ must necessarily be 0. Then, since $0 = \Phi(0) = \Phi(1+(-1)) = \Phi(1) + \Phi(-1)$, we realize that $\Phi(-1)$ is forced to coincide with $-\Phi(1)$. It follows that for any negative integer $-n = (-1) + ... + (-1)$, n times, we must have $\Phi(-n) = \Phi((-1) + ... + (-1)) = \Phi(-1) + ... + \Phi(-1) = -n\Phi(1)$.

So the global behaviour of the additive mapping Φ is completely determined by its value $\Phi(1)$ at 1. We will develop this into an arithmetical example of a metabolism in the style of Rosen.

$\mathcal{M}$ Must be a Proper Subset of $Map(A,B)$: $\mathcal{M} = H(A,B) \subseteq Map(A,B)$

For the central result to hold, the mappings involved must be restricted *a priori* to a special type, i.e. f must belong to a strict subset of $Map(A,B)$, called here $H(A,B)$, and Φ must belong to strict subset of $Map(B, Map(A,B))$, called here $H(B, H(A,B))$. Note than in this notation $\mathcal{M} = H(A,B)$ and $\mathcal{S} = H(B, H(A,B))$.

Rosen's initial formulation: $H(A,B) = Map(A,B)$ and $H(B, H(A,B)) = Map(B, H(A,B))$, is definitely too general since there will be then many choices of Φ such that $\Phi(b) = f$. Of course, if the set $H(B, H(A,B))$ were too small, there might be no Φ such that $\Phi(b) = f$. Thus in order to work, Rosen's scheme requires a subtle balance in the size of the sets $H(A,B)$, $H(B,H(A,B))$, to achieve both existence and unicity for Φ. Intuitively, the set $H(A,B)$, for instance, must be strictly smaller than $Map(A,B)$ because it should consist only of those admissible mappings from A to B that preserve some sort of underlying or implicit structure on A and B. Thus an organizational invariant *(M,R)* system, should be represented as:

$$A \xrightarrow{f} B \xrightarrow{\Phi} H(A,B) \xrightarrow{\beta} H(B,H(A,B))$$

and can be interpreted as two coupled *(M,R)* systems acting in conjunction:

$$(M,R)_{internal} \quad : \quad A \xrightarrow{f} B \xrightarrow{\Phi} M$$
$$(M,R)_{external} \quad : \quad B \xrightarrow{\Phi} M \xrightarrow{\beta} S$$

The following properties must be emphasized:

- In these two coupled *(M,R)* systems Φ is the replacement function for one and the metabolic function for the second.

- Φ must be reconstructible from its value at a single point.

- Rosen's formulation, which appears to indicate $M = Map(A,B)$ and $S = Map(B, Map(A,B))$, is incorrect because M and S are too big.

Examples of *(M,R)* Systems

One problem in Rosen's work is the lack of examples of the theoretical notions (like Φ and β). To partially overcome this difficulty we introduce two examples of *(M,R)* systems. One is an arithmetical construction and the other uses a minimal and ideal metabolic network.

An Arithmetical Example of an *(M,R)* System

Let $A = B = \mathbb{Z}_{12}$, the integers $(mod\,12)$. So our metabolic states (a and b) are parameterized by the integers $(mod\,12)$. Since integers $(mod\,12)$ can be added, for example $9 + 7 = 16 = 4(mod\,12)$ our set of metabolic states is endowed with an additive structure. We posit that the mappings f representing metabolisms are to be additive mappings from A to B, i.e. from $\mathbb{Z}_{12}$ to itself. These mappings are necessarily of the form

$$f_c : n \mapsto f_c(n) = c \cdot n \;(mod\,12)$$

For example, the metabolism f_7 transforms the metabolic state $a = 5$ into the state $b = 11$, as $f_7(5) = 7 \cdot 5 = 11(mod\,12)$.

So, in terms of our previous notations, $M = H(A,A) = \{f_c \mid c \in A\}$.

Notice that $M = H(A,A)$ may be identified with A, via the one-to-one correspondence $f_c \longleftrightarrow c$. Under this identification the operation of addition of mappings in $H(A,A)$ corresponds to addition of numbers $(mod\,12)$ in A, i.e. $f_c + f_d = f_{c+d}$. Thus the first part of this (M,R) system is represented by:

$$A \xrightarrow{f_c} A \longrightarrow [M = H(A,A)]$$

We need to specify the set $S = H(A, M)$ of the selectors $\Phi : A \longrightarrow H(A,A) = M$. As we have essentially the same additive structure on A and M, as explained above, the mappings Φ in S must preserve this common additive structure on A and M, i. e. to satisfy

$$\Phi(c + d) = \Phi(c) + \Phi(d)$$

for all $c, d \in A$. It follows that any mapping Φ in S will be completely determined by its value at 1, which must be an $f \in M$. But we already know that any such f must be of the form f_k, for some $k \in A$. If we write for convenience $\Phi = \Phi_k$, we then have

$$\Phi_k(1) = f_k$$
$$\Phi_k(b) = b\Phi_k(1) = bf_k = f_{bk}$$

We consider three examples to illustrate the various possibilities for Φ.

Example 1. Let us choose $a = 4$, $f = f_5$. Then $b = f(a) = f_5(4) = 5 \cdot 4 = 20 = 8(mod\,12)$. Let us look for $\Phi = \Phi_k$ such that $\Phi_k(b) = f$, i.e. $\Phi_k(8) = f_5$. Since $\Phi_k(b) = f_{bk}$, this simplifies to $f_{8k} = f_5$, i.e. to the equation $8k = 5$, which has no solution in A.

Example 2. Let us choose $a = 5$, $f = f_3$. Then $b = f(a) = f_3(5) = 3 \cdot 5 = 15 = 3(mod\,12)$. Let us look for $\Phi = \Phi_k$ such that $\Phi_k(b) = f$, i.e. $\Phi_k(3) = f_3$. Since $\Phi_k(b) = f_{bk}$, this simplifies to $f_{3k} = f_3$, i.e. to the equation $3k = 3$ in A, which has solutions $k = 1, 5$ and $9(mod\,12)$. We see that in this case there are three mappings Φ (Φ_1, Φ_5 and Φ_9) such that $\Phi_k(b) = f$.

Example 3. Let us choose $a = 5$, $f = f_7$. Then $b = f(a) = f_7(5) = 7 \cdot 5 = 35 = 11(mod\,12)$. Let us look for $\Phi = \Phi_k$ such that $\Phi_k(b) = f$, i.e. $\Phi_k(11) = f_7$. Since $\Phi_k(b) = f_{bk}$, this simplifies to $f_{11k} = f_7$, i.e. to the equation $11k = 7$ in A, which has the unique solution $k = 5(mod\,12)$. Thus in this case, as an organization invariant (M,R) system demands, there is only one Φ (Φ_5) such that $\Phi(b) = f$. We see now that, for $b = 11$, in fact for every $f = f_c$, we can find a unique Φ such that $\Phi_k(11) = f_c$, where k is the only solution of the equation $11k = c$ in A, i. e. $k = 11c$. So we realize that for this happy choice of b, the core of Rosen's central result, i.e the invertibility of the evaluation at b, is fulfilled. Following Rosen's terminology, we then have $\beta(f_c) = \Phi_{11c}$, for all $f_c \in \mathcal{M}$.

A Metabolic Example of an *(M,R)* System

An example with a more biological flavor can be constructed from simplified rules representing idealized metabolisms. Consider the following three reactions, without specifying the nature of the catalysts M_1, M_2, and M_3, that represent a minimal metabolism

$$
\begin{aligned}
1) \quad & s+t \xrightarrow{M_1} st \\
2) \quad & s+u \xrightarrow{M_2} su \\
3) \quad & st+u \xrightarrow{M_3} stu
\end{aligned}
$$

These three equations specify a particular instance of a metabolism M between the sets $A = \{a\} = \{(s,t,u,st)\}$ and $B = \{b\} = \{(st,su,stu)\}$. Here the set $\mathcal{M}$ is simply *one* transformation f such that $f((s,t,u,st)) = (st,su,stu)$, i.e. $f(a) = b$.

To specify the corresponding R part, the subsystem of metabolic reactions producing each R_i must be specified. In this simplified network this specification is simply to identify one of the outputs $\{st,su,stu\}$ with one of the M_i; thus we are specifying the production mechanism by which a given R_i is continuously generated. A great number of assignments are possible, in total 3^3, but this number is decreased substantially by excluding autocatalytic assignments such as $M_1 = st$, or $M_3 = stu$, in which the product of a reaction catalyzes the same reaction that produces it. Although the point is arguable, and others may arrive at a different conclusion, we think that autocatalysis of this kind should be avoided in the theory of *(M,R)* systems as we are seeking systems in which the circularity is a property of the global connectivity in the entire network and not a property of a single reaction. This restriction requires, for example, the only valid assignment for $st + u \xrightarrow{M_3} stu$ to be $M_3 = su$, as su is neither a substrate nor a product of reaction 3. This kind of argument decreases the initial 27 possibilities to the following four valid assignments for Φ :

$$
\begin{aligned}
\Phi_1 : \quad & (M_1,M_2,M_3) \longrightarrow (stu,stu,su) \\
\Phi_2 : \quad & (M_1,M_2,M_3) \longrightarrow (stu,st,su) \\
\Phi_3 : \quad & (M_1,M_2,M_3) \longrightarrow (su,stu,su) \\
\Phi_4 : \quad & (M_1,M_2,M_3) \longrightarrow (su,st,su)
\end{aligned}
$$

Each one of these selectors generates a different *(M,R)* system, where the M part is similar:

$$
(M,R_1) = \begin{aligned} s+t &\xrightarrow{stu} st \\ s+u &\xrightarrow{stu} su \\ st+u &\xrightarrow{su} stu \end{aligned}
\qquad
(M,R_2) = \begin{aligned} s+t &\xrightarrow{stu} st \\ s+u &\xrightarrow{st} su \\ st+u &\xrightarrow{su} stu \end{aligned}
$$

$$
(M,R_3) = \begin{aligned} s+t &\xrightarrow{su} st \\ s+u &\xrightarrow{stu} su \\ st+u &\xrightarrow{su} stu \end{aligned}
\qquad
(M,R_4) = \begin{aligned} s+t &\xrightarrow{su} st \\ s+u &\xrightarrow{st} su \\ st+u &\xrightarrow{su} stu \end{aligned}
$$

Thus from the 27 choices for Φ that are theoretically compatible with this simple metabolism f we have discarded 23, leaving only four as valid assignments, and the set of selectors is reduced to $S = \{\Phi_1, \Phi_2, \Phi_3, \Phi_4\}$.

The procedure outlined here, which starts with the information provided by f and serves to define the set of possible selectors Φ is an explicit embodiment of the function β, which turns out here to be a multivalued function:

$$
\beta(f) = \{\Phi_1, \Phi_2, \Phi_3, \Phi_4\}
$$

The fact that $\beta(f)$ is not single-valued (as any honest function should be) shows that the condition of invertibility, which is the symptomatic property of *(M,R)* systems with organizational invariance, fails for this simple metabolic network. Thus although this metabolic network is an *(M,R)* system –and also an autocatalytic network (Kauffman, 1993)– it cannot be construed as an *(M,R)* system with organizational invariance because the rule to assign Φ starting from f gives more than one result. This example is also interesting as it shows that an autocatalytic set, such as the one represented by (M,R_1), is not necessarily an *(M,R)* system with organizational invariance. The two ideas, although related, are different in a fundamental way.

Discussion

The main objective of this paper has been to clarify some central aspects of Rosen's ideas. His central result refers to something most biologists will find extremely esoteric: an attempt to prove (from purely logical grounds) the necessity for a circular organization of metabolic networks. Furthermore the mathematical fact used to introduce this notion, the invertibility of certain evaluation maps, is unusual enough to make it very difficult to explain the context of the result even to mathematicians. Rosen himself never explained the mathematical context where his central result could hold true and provided neither mathematical nor biological examples.

As may be surmised, we have adopted the point of view that Rosen had a powerful intuition on the nature of metabolic networks and the necessary (but otherwise ignored) requirement of circularity. However, his intuition is far from being workable and ready to apply to current network analysis without major efforts, both to clarify the circumstances in which his central result applies, and to explain its meaning in biological terms. An intriguing possibility could be to combine Rosen's analysis with the notion of autopoietic systems, another theory that posits metabolic closure as the core of biological organization (Maturana and Varela, 1980; Letelier et al., 2003). This paper is intended as a step in the proper direction, as we have isolated from Rosen's extensive work what we think is its core, and we have clarified concepts like f, Φ, and β. We have explained the mathematical intuition behind the idea of organizational invariance as embodied in the operator β, a crucial concept as essentially it acts as a *generator* of the complete formal structure of an *(M,R)* system. In effect it is possible to reformulate the very definition of an organizational invariant *(M,R)* system as the kind of system where for some b the equation $\Phi(b) = f$ has exactly one solution Φ, for any given f, giving rise to the operator β, which sends any f to its associated Φ and implicitly generates the structure of the whole system.

Acknowledgements

It is a great pleasure to thank Diane Greenstein for her excellent editorial skills. Supported by Fondecyt 1030761 (JCL), Fondecyt 1040444 (JSA) and the CNRS (AC-B, MLC). We also thank J. Mpodozis and G. Marín for many discussions and for introducing a healthy degree of skepticism into our own discussions.

References

Casti, J. (2002). "Biologizing" control theory: How to make a control system come alive. *Complexity*, 7:10–13.

Kauffman, S. (1993). *The origins of order: Self organization and selection in evolution.* Oxford, Oxford University Press.

Letelier, J., Marin, G., and Mpodozis, J. (2003). Autopoietic and *(M,R)* systems. *Journal of Theoretical Biology*, 222(2):261–72.

Maturana, H. and Varela, F. (1980). *Autopoiesis and Cognition: The realization of the living.* Dordrecht, Reidel.

Rosen, R. (1958a). A relational theory of biological systems. *Bull. Math. Biophys.*, 20:245–341.

Rosen, R. (1958b). The representation of biological systems from standpoint of the theory of categories. *Bull. Math. Biophys.*, 20:317–341.

Rosen, R. (1959). A relational theory of biological systems II. *Bull. Math. Biophys.*, 21:109–128.

Rosen, R. (1972). Some relational cell models: The metabolism-repair system. In Rosen, R., editor, *Foundations of Mathematical Biology*. New York, Academic Press.

Rosen, R. (1991). *Life Itself.* Columbia University Press, New York.

Rosen, R. (2000). *Essays on Life Itself.* Columbia University Press, New York.

Varela, F., Maturana, H., and Uribe, R. (1974). Autopoiesis: The organization of living systems, its characterization and a model. *Biosystems*, 5(4):187–196.

Wolkenhauer, O. (2002). Mathematical modelling in the post-genome era: understanding genome expression and regulation—a system theoretic approach. *Biosystems*, 65:1–18.

Flows of information in spatially extended chemical dynamics

Kristian Lindgren[1], Anders Eriksson[1] and Karl-Erik Eriksson[2]

[1]Department of Physical Resource Theory, Chalmers University of Technology and Göteborg University,
SE-41296 Göteborg, Sweden

[2]Division of Engineering Sciences, Physics and Maths, Karlstad University, SE-65188 Karlstad, Sweden

frtkl@fy.chalmers.se

Abstract

A continuity equation for information is presented, involving flows in both space and scale in chemical pattern formation systems. The flows are connected to thermodynamic properties of the system. Information leaves the system (or is destroyed) at the smallest length scales, which corresponds to entropy production. Information enters the system at the largest length scales when the system is open to an inflow of Gibb's free energy. The continuity equation describes how information can be aggregated at different scales and positions during the pattern formation process. The formalism is applied to the Gray-Scott model, exhibiting self-replicating spots in a spatially extended system.

The aim of the present paper is to give a physically consistent picture on how spatial structure emerge in chemical systems. This picture should illustrate how aggregation of information in the form of spatial patterns can be connected to the physical constraints of the studied system posed by thermodynamics, in particular the driving forces in terms of flows of free energy. Our ambition is to contribute towards a theoretical framework, that may be used to find necessary conditions for the formation of higher-order structures from lower-level components and mechanisms – a framework to encompass both biological life and artificial life.

Pattern formation in chemical systems is an example of a process where the connection between information theory, statistical mechanics, and thermodynamics is useful for a description and characterisation of the dynamics. Patterns can be characterised by their information content - a characterisation that may include how information is distributed in the system, both with respect to position and scale of resolution (Eriksson and Lindgren, 1987). At the same time, the flows of Gibbs free energy that drives the pattern formation process, or maintains the spatial structures that have been formed, can be given an information-theoretic interpretation. By combining these information-theoretic perspectives with the general reaction-diffusion type of chemical dynamics, we derive a continuity equation for information in open chemical systems. Information flows in both space and scale; flows across system boundaries are due to inflow of Gibbs free energy, and at the microscopic level to entropy

production in chemical reactions and in the diffusion process. The results presented here is a generalisation of the equation derived earlier for closed systems (Eriksson et al., 1987).

The formalism is exemplified by an analysis of the Gray-Scott model (Gray and Scott, 1984), exhibiting self-reproducing "spots" in a spatially extended system (Lee et al., 1993; Pearson, 1993). In this model, two chemical components U and V react according to U + 2V → 3V, V → G, where G is not reacting with U or V.

The formalism is based on the information-theoretic concept of relative information or Kullback information (Kullback, 1959), defined by

$$K[P^{(0)};P] = \sum_i p_i \log \frac{p_i}{p_i^{(0)}} \geq 0,$$

where P and $P^{(0)}$ are normalised probability distributions. This measure quantifies the information one gains when one learns that the a priori distribution $P^{(0)}$ was not correct but that the system is described by P. An advantage with the use of this quantity as a basis for the information-theoretic analysis of the spatial structure in a chemical system is the strong connection to statistical mechanics and thermodynamics (Jaynes, 1957). This will make it possible to relate our information-theoretic analysis to thermodynamic properties of the system.

The exergy E, or available energy, for a system of volume V, characterised by pressure p, temperature T, and chemical potentials g_i (for M different components), in an environment characterised by the corresponding intensive variables (p_0, T_0, and g_{i0}), can be written (Reif, 1985)

$$E = S(T - T_0) - V(p - p_0) + \sum_{i=1}^{M} N_i(g_i - g_{i0}),$$

where S is the thermodynamic entropy of the system and N_i is the number of molecules of the different types. The total number of molecules is $N = \sum_i N_i$.

Assuming $T = T_0$ and $p = p_0$, and ideal gas expression for the chemical potential $g_i = k_B T_0(C + \ln c_i)$, with normalised

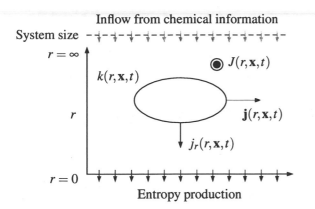

Figure 1: A schematic picture of the flows of information in a chemical pattern formation system. The pattern is characterised by an information density $k(r,\mathbf{x},t)$ distributed over spatial dimensions as well as over different length scales r. Information flows both in space and in scale, where the flow is destroyed when it gets down to the microscopic level. Here information disappears into microscopic degrees of freedom due to entropy production. Information enters the system at the very large scale due to a diffusion-controlled inflow of chemical information or Gibbs free energy. A pattern is formed when information that flows downwards in scale is aggregated at certain positions as described by the continuity equation.

concentrations $c_i = N_i/N$ and corresponding concentrations c_{i0} in the environment, we get a Kullback information expression for the exergy, or the Gibbs free energy, of the system

$$E = k_\mathrm{B} T_0 N \sum_{i=1}^{M} c_i \ln \frac{c_i}{c_{i0}} = k_\mathrm{B} T_0 N K[c_0; c]$$

or, in general, when spatially dependent concentrations are allowed,

$$E = k_\mathrm{B} T_0 n \int_V d\mathbf{x} \sum_{i=1}^{M} c_i(\mathbf{x}) \ln \frac{c_i(\mathbf{x})}{c_{i0}}$$
$$= k_\mathrm{B} T_0 n \int_V d\mathbf{x}\, K[c_0; c(\mathbf{x})] = k_\mathrm{B} T_0 n K,$$

where $n = N/V$. This relation between exergy E and information K allows us to connect the information-theoretic analysis to the thermodynamic properties of the system. The integrated Kullback information, K, between actual and equilibrium concentrations, is our starting point for the analysis of spatial structure. First, by using the average concentration the information can be decomposed into a structural part K_{struct} and a chemical part K_{chem}. The structural information measures deviation from a homogenous state, and the chemical information measures the deviation of average

concentrations $\bar{c}_i$ from chemical equilibrium:

$$K = \int_V d\mathbf{x} \sum_i c_i(\mathbf{x}) \ln \frac{c_i(\mathbf{x})}{\bar{c}_i} + V \sum_i \bar{c}_i \ln \frac{\bar{c}_i}{c_{i0}}$$
$$= K_{\mathrm{struct}} + K_{\mathrm{chem}}$$

In order to get a more detailed description, we continue with a decomposition of the structural information into contributions $k(r,\mathbf{x},t)$ from different positions $\mathbf{x}$ as well as from different length scales r. In order to do this, we introduce a resolution dependent concentration distribution. We define the concentration $\tilde{c}_i(r,\mathbf{x},t)$ at resolution r as

$$\tilde{c}_i(r,\mathbf{x},t) = \exp(\tfrac{1}{2} r^2 \nabla^2) c_i(\mathbf{x},t),$$

where $\exp(\tfrac{1}{2} r^2 \nabla^2)$ is the resolution operator in d dimensions, defined by the convolution with the kernel $(2\pi r^2)^{-d/2} \exp[-\mathbf{x}^2/(2r^2)]$. As can be seen in the Fourier space, this operator suppresses structures at length scales $\lesssim r$. At $r = 0$, the two concentrations coincide since $\tilde{c}_i(0,\mathbf{x},t) = c_i(\mathbf{x},t)$. If r is much larger than the system size, the concentration is approximately homogenous, $\tilde{c}_i(\infty,\mathbf{x},t) = \bar{c}_i(t)$.

This means that the structural information can be written

$$K_{\mathrm{struct}} = \int_0^{\infty} dr/r \int d\mathbf{x}\, k(r,\mathbf{x},t),$$

with a local information density at position $\mathbf{x}$ and scale (resolution) r,

$$k(r,\mathbf{x},t) = \sum_i \tilde{c}_i [r\nabla \ln \tilde{c}_i]^2 \geq 0.$$

In the following we present an information-theoretic description of how information is flowing in the system that connects to the thermodynamic loss of information due to entropy production. We will assume that the system evolves according to the reaction - diffusion equation,

$$\dot{c}_i = D_i \nabla^2 c_i + F_i(\mathbf{c}).$$

This will be formulated in a continuity equation for information density k, taking into account flows both in scale (r) and in space $(\mathbf{x})$, see Figure 1. There may also be sources or sinks due to the fact that we allow for an open system. By assuming that the following continuity equation holds, we can derive the terms for the flow in the scale direction $j_r(r,\mathbf{x},t)$, the spatial flow $\mathbf{j}(r,\mathbf{x},t)$ and the sources/sinks term $J(r,\mathbf{x},t)$,

$$\dot{k}(r,\mathbf{x},t) = r \frac{\partial}{\partial r} j_r - \nabla \cdot \mathbf{j} + J.$$

In order to properly define the flows, we also require (i) that the flows are rotation-free, (ii) that the system is spatially closed, i.e., the chemical flows across the system

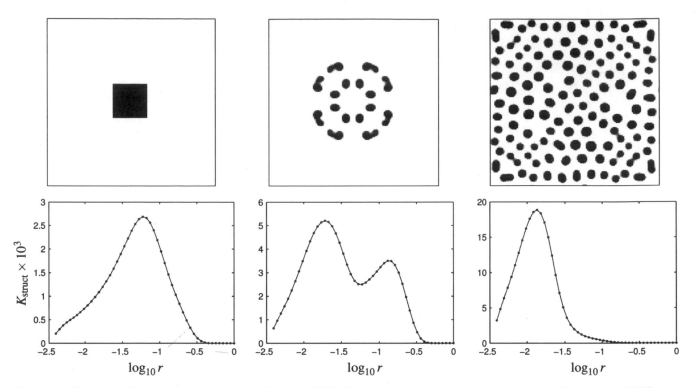

Figure 2: **Top row** The concentration of the chemical V in the system at three times; $t = 0$, $t = 1000$, and $t = 7000$ steps. White corresponds to zero concentration, and black corresponds to a concentration of one half. **Bottom row** The structural information, integrated over the system, as a function of the resolution r. The length of the system is 1, $D_u = 2D_v = 0.05$, $\hat{D} = 0.02$, $k = 0.058$.

boundaries are in a different direction (in which the system has no extension), (iii) that the information content of the inflow and outflow corresponds to the net flow of Gibbs free energy into the volume, and (iv) the destruction of information corresponds to the entropy production.

In the limit of $r \rightarrow \infty$ we cannot distinguish any spatial structure, but the chemical information K_{chem} is still present, unaffected by the resolution parameter. In a decomposition of the total information this part can therefore be considered as present at the $r \rightarrow \infty$ limit. The chemical information will be consumed by the chemical reactions and we should therefore expect that a proper definition of information flow shows how information will flow in the direction towards smaller length scales r. If the system gives rise to spatial structure, that should be captured in the continuity equation, resulting in a temporal accumulation of structural information.

The entropy production is determined by one term corresponding to the entropy produced due to diffusion in the system and one term given by the reactions that tend to even out the chemical non-equilibrium in the system. The entropy production certainly leads to a decay of the information in the system – decay of structural information as well as of chemical information.

It is reasonable to assume that information is leaving the system, through the thermodynamic entropy production, at the smallest length scales of the system, i.e., at $r = 0$. At this point information disappears from the macroscopic description of our system, and the information is spread out on microscopic degrees of freedom.

Therefore, we define the information flow j_r in the direction of smaller r, at the border $r = 0$, to be equal to the chemical entropy production. To define this flow for general resolution values r, we generalise by introducing the resolution operator into the expression for entropy production,

$$j_r(r, \mathbf{x}, t) = \sum_i D_i \frac{[\nabla \tilde{c}_i]^2}{\tilde{c}_i} + J_{\text{pot}}$$

$$J_{\text{pot}}(r, \mathbf{x}, t) = \left(\ln \frac{\tilde{c}_i}{c_{i0}} \right) \exp(\frac{r^2}{2} \nabla^2) F_i(\mathbf{c}(\mathbf{x}, t))$$

Now we use the continuity equation for the closed system, in which $J = 0$, to define the spatial flow by $\mathbf{j}(r, \mathbf{x}, t) = r^2 \nabla J_{\text{pot}}$.

In the case of an open system, the flow of chemical information through the system (typically an inflow of a substance acting as a fuel and an outflow of waste products) affects the pattern formation process in two opposite ways. If the flow across the system boundary is controlled by diffusion, the direct effect is a decrease of the spatial structure at all length scales, since such a flow tends to even out all

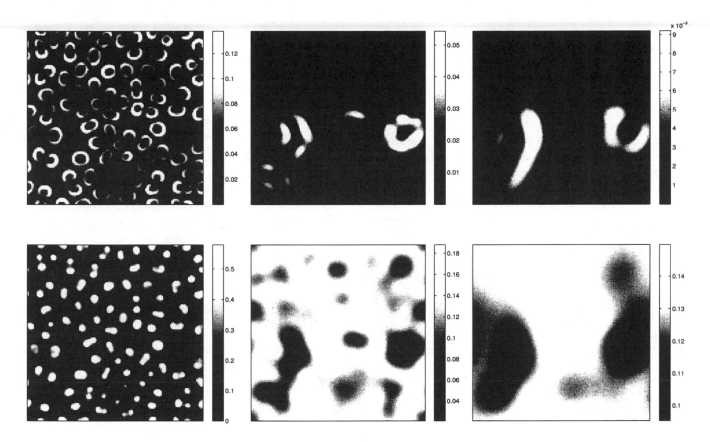

Figure 3: Top row The structural information in the system at $t = 10000$ steps, for three values of the resolution r; $r = 0.01$, $r = 0.05$, and $r = 0.1$. **Bottom row** The information flow $j_r(r, \mathbf{x}, t)$. The length of the system is 1, $D_u = 2D_v = 0.05$, $\hat{D} = 0.02$, $k = 0.058$.

spatial differences. The opposite effect is that the flow may lead to an increase in chemical information, i.e., the system is pushed away from equilibrium. This is of course what makes it possible to keep the driving information flow for the pattern formation processes in the system. The change in chemical information takes the following form

$$\dot{k}_{\text{chem}}(t) = -j_r(\infty, \mathbf{x}, t) + \sum_i \overline{X}_i \ln \frac{\overline{c}_i}{c_{i0}},$$

where $\overline{X}_i$ is the average net inflow of component i. Thus, in a stationary situation, the information flow towards smaller length scales, j_r, is balanced by the inflow of exergy (or Gibbs free energy) in units of $k_B T_0$, represented by the last term.

The negative effect from the diffusion over the system boundary is captured by the sink term J in the continuity equation. In a diffusion controlled flow, X_i is given by $\hat{D}_i(\hat{c}_i - c_i)$, where $\hat{D}_i > 0$ and $\hat{c}_i$ is the concentration outside the system, chosen so that $\hat{c}_i > c_i$ for inflow and $\hat{c}_i = 0$ for an outflow of i, respectively. This results in the following expression for J;

$$J(r, \mathbf{x}, t) = -\sum_i \hat{D}_i(c_i + \hat{c}_i) [r \nabla \ln \tilde{c}_i]^2 \leq 0,$$

which shows that J always is a sink term.

We apply the formalism to the pattern formation of the "self-replicating spots" system (Gray and Scott, 1984; Lee et al., 1993; Pearson, 1993), U + 2V → 3V and V → G, with the dynamics

$$\dot{c}_u = D_u \nabla^2 c_u - (c_u - k_{\text{back}} c_v) c_v^2 + \hat{D}(1 - c_u)$$
$$\dot{c}_v = D_v \nabla^2 c_v + (c_u - k_{\text{back}} c_v) c_v^2 - k c_v - \hat{D} c_v.$$

We have introduced a very slow back reaction ($k_{\text{back}} = 10^{-5}$) in order to get the relationship between equilibrium concentrations of U and V defined by the reactions. In Figure 2, the dynamics is illustrated starting from an initial state (left) with a square of high concentration of V. As the system evolves four concentration peaks (spots) emerges from the square, and these spots reproduce by growing and splitting until the system is filled with spots (middle and right). In the process, spots may disappear, which leaves space for other spots to reproduce. In the lower part of the figure, the decomposition of the information in the pattern with respect to scale is plotted for the three snapshots above (at time 0, 1000, and 7000, respectively). It is clear that the initial state has a longer characteristic length as detected by the information density. When the system produces the spots at the significantly shorter length scale, information is found both at the old length, now due to the size of the cluster (middle), and at the length scale of the spots. When the square distribution has been completely decomposed into spots, no information is left at the initial length scale.

In Figure 3, we show the information density over the system for three different length scales after long time (upper part), and the information flow in scale, j_r, for the same state (lower part). At low resolution, or large r (right), the information density is low and captures structures of longer lengths, while at finer resolution, small r (left), the information density is large and reflects the pattern of spots. Note that each spot is seen as a circle in the information density picture, since the information is sensitive to gradients in the pattern. The information flow in scale, , is close to homogenous for large r (right), but when information moves on to finer scales of resolution, the spatial flow **j** redistributes the flow so that a higher flow j_r is obtained at the concentration peaks. At finest resolution, $r = 0$, this flow leaves the system as entropy production, which is mainly located to the concentration peaks where the chemical activity is high.

In this paper we have presented an information-theoretic perspective on pattern formation in chemical systems. The continuity equation, with the corresponding information flows, connects to thermodynamic flows and thus to thermodynamics restrictions of the system. It is clear that the second law of thermodynamics leads to a destruction of information at the finest levels of resolution. Is there a second law of information-dynamics in thermo-dynamical systems, which makes information flow in the direction of smaller scales? Under what circumstances is this law valid? These are questions for further investigation.

In the Gray-Scott model (Gray and Scott, 1984), we detect an aggregation of information on macroscopic length scales. This illustrates how information theory can be used in order to determine whether spatial structure emerges on length scales much larger than those of the individual components and their interaction. It should be noted that there is also an information flow in the opposite direction (i.e. from micro scale to macro scales) due to noise, which govern the formation of spatial configurations. This flow is several orders of magnitude smaller, but may still control where in space the information from the free energy will be aggregated. The interplay between the driving and the controlling information flows will be investigated in a forthcoming paper.

Acknowledgements

This work is supported by the EU project *PACE* under FP6 contract 002035.

References

Eriksson, K.-E. and Lindgren, K. (1987). Structural information in self-organizing systems. *Physica Scripta*, 35:388–397.

Eriksson, K.-E., Lindgren, K., and Månsson, B. Å. (1987). *Structure, Context, Complexity, and Organization.* World Scientific, Singapore.

Gray, P. and Scott, S. K. (1984). Autocatalytic reactions in the isothermal, continuous stirred tank reactor – oscillations and instabilities in the system A+2B→3B, B→C. *Chem. Eng. Sci.*, 39(6):1087–1097.

Jaynes, E. T. (1957). Information theory and statistical mechanics. *Physical Review*, 106:620.

Kullback, S. (1959). *Information theory and statistics.* Wiley, New York.

Lee, K. J., McCormick, W. D., Ouyang, Q., and Swinney H. L. (1993). Pattern formation by interacting chemical fronts. *Science*, 261(5118):192–194.

Pearson, J. E. (1993). Complex patterns in a simple system. *Science*, 261(5118):189–192.

Reif, F. (1985). *Fundamentals of statistical and thermal physics.* McGraw - Hill.

Cellular Dynamics in a 3D Molecular Dynamics System with Chemistry

Duraid Madina[1] and Takashi Ikegami[2]

[1]The University of New South Wales, Sydney 2052, Australia
[2]The University of Tokyo, 3-8-1 Komaba, Tokyo 153-8902, Japan
duraid@unsw.edu.au

Abstract

We present a three-dimensional model of the formation of simple protocellular structures. The model is based on an earlier lattice artificial chemistry due to Ono and Ikegami which consisted of a primitive metabolic system built on an artificial chemistry. This model computed the interactions of simple amphiphilic molecules which organized into membrane-like structures. The current model, however, treats space as continuous rather than a lattice. Moreover, although forces between atoms are computed in a more realistic manner, an adaptive method of computing intermolecular forces allows for efficient computation.

Introduction

A number of simple reaction-diffusion systems have been shown to exhibit self-organizing and self-reproducing patterns, without requiring any detailed structures to support these processes(Turing, 1953; Pearson, 1993). However, these patterns do not permit any significant individuality and such systems seem too simple to be useful models of early pre-cellular structures. In an attempt to distinguish biological cells from simple, driven, dissipative devices, Varela and Maturana (Varela et al., 1974) highlighted one essential feature of living systems - "autopoiesis", the ability for a cell to produce and maintain its own boundary. In the spirit of their work, there have been more recent studies into the organization and maintenance of protocellular structures in computational models (McMullin and Varela, 1997; Breyer et al., 1998). In (Edwards et al., 1998) a simple model of the self-assembly of two-component lipids (similar to those considered here) is described. In contrast to these more abstract models, several realistic models of the molecular dynamics of amphiphilic lipid self-assembly have been studied, corresponding well to experiment (Marrink et al., 2000). Mayer and Rasmussen have studied the dynamics of micellar self-reproduction using a modified lattice gas technique (Mayer and Rasmussen, 2000). We note that the present work is inspired by some previous studies by Ono (see e.g. the third chapter of (Ono, 2001)). A primary distinction between these and the present work is that we place abstract models of protocellular chemistry in an accurate molecular

dynamics framework, comparable to the popular AMBER (Pearlman, 1995) and NAMD (Skeel, 1999) packages. In a previous attempt to bridge the realism gap between highly abstract and highly accurate models, we presented a three-dimensional version (Madina et al., 2003) of "Lattice Artificial Chemistry" (LAC) (Ono and Ikegami, 1999; Ono and Ikegami, 2001). LAC simulates both simple repulsive forces and chemical reactions between abstract molecules using a simple lattice method. In the 2D version of LAC, spontaneous emergence of cellular structures, maintenance and self-reproduction of cellular structures, and evolution of metabolic systems through cellular selection were observed. In the 3D version of LAC, a much richer variety of cellular morphologies was observed, while preserving many of the qualitative features of the 2D version. As a small step towards to studying the way in which cell membranes are not merely passive vehicles which delimit chemical networks, but instead more actively affect the chemical reactions which occur within them, we attempt to study more realistic membranes while maintaining an abstract, efficiently computable representation of cell metabolics. Our present model, therefore, allows for selected molecular dynamics (such as the inter-lipid interactions of a protocellular membrane) to be computed with near-arbitrary accuracy, while at the same time allowing metabolic or other intracellular reaction pathways to be implemented more abstractly.

The Model: A 3D Molecular Dynamics System with Chemistry

The present model is essentially a molecular dynamics system: various substances are represented as atoms, point particles with particular properties such as charge and mass, and which are bonded in particular ways. Atoms have continuously variable locations and velocities, and interact through various standard forces such as Coulomb (r^2) repulsion and Lennard-Jones (r^6 / r^{12}) attraction. The force field is given as:

$$F = \sum k_q \frac{q_i q_j}{r_{ij}} - \sum \left(k_{lj} \frac{A}{r^6} - \frac{B}{r^{12}} \right) + \sum k_{bond}(r_i - r_{rest})^2 + \sum k_{angle}(\theta_i - \theta_{rest})^2$$

Standard bond forces are also employed: bonds between

pairs of atoms are parameterised by a rest length and a force constant, while bonds between atom triplets employ a similar spring force which attempts to preserve bond angles. However, a key deviation of our model from standard molecular dynamics methods is that we also permit coarser, molecule/atom and molecule/molecule interactions in addition to ordinary atom/atom interactions, though these are defined in terms of the basic forces above. In particular, we additionally define a "hydrophobic" force which is a strong, Coulomb-like repulsive force between certain atoms tagged as hydrophobic (these comprise the hydrophobic half of our simplified lipids) and water molecules. We summarise the key differences between the present model and our previous work in the table below. In particular, a primary difference is that we no longer rely solely on repulsive diffusion forces. In addition, we no longer define an anisotropic potential function for pointlike membrane particles as in our previous work. In particular, we do not assign individual membrane particles an orientation. Instead, the present membrane *molecules* are free to move and change their orientation continuously. In other words, treat lipid particles and the remainder of the system equally. This allows the formation of more complex, self-organizing membrane bilayers. This is a key difference between the present work and the relatively unstructured membranes we have studied previously.

Our model contains a simple metabolic system as illustrated in Fig.1, below. We implement this as chemical reactions which spontaneously alter atoms and molecules. The following transitions may occur: Resource particles **X** are generally free to move through the system. If a pair of resource particles are seperated by a distance **r** which is similar to the natural (rest) bond length of lipid, they may "react" to form either an autocatalytic molecule (a bonded pair of autocatalytic particles **A** or a lipid molecule (a bonded pair of lipid particles **H** and **T**). The rate at at which these reactions may occur is proportional to both the distance **r** between resource particles and the local density of autocatalyst molecules. In this way, autocatalyst molecules catalyze both their own reproduction and the production of lipid (membrane) molecules. The rate of production of these molecules is bounded as this process consumes resource particles **X** which are in limited supply. Both autocatalytic and lipid molecules continually decay into nonbonded pairs of waste particles **Y**. Finally, the main driving force in the model is an externally imposed "recycling" which continuously transforms waste particles back into resource particles at a fixed rate.

Implementation

We use a somewhat unconventional implementation of molecular dynamics which we describe here. The dynamics of the system is evaluated over time which increases in relatively large, discrete intervals which are not generally of the

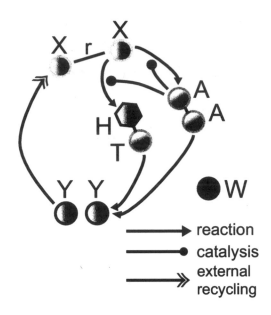

Figure 1: The simple metabolic system of the present model.

same size. This serves two main purposes. Firstly, it allows periods of relatively uninteresting dynamics to be processed more quickly. Periodically, the system is evaluated with both high and low and precision integrators. If the difference in the dynamics (specifically, total system energy) is greater than a certain threshold (typically around 0.1%), the results of the low precision integrations are discarded. Secondly, stepping through time in this manner allows the system to adaptively cope with the occasionally significant perturbations caused by the chemistry in our model, i.e. the instantaneous transformation of atom types and formation and removal of bonds between atoms. Note that while the reaction pathways presented here conserve the total number of atoms in the system, this need not be the case, and it is possible to explicity include flows of matter into or out of the system. We also use a multi-level integration scheme: strong bonding forces which operate on short timescales and are susceptible to ringing are integrated on timescales approximately one order of magnitude smaller than weaker, inter-molecular forces which contribute periodic "impulses" to the integration of forces in the system. Further, a "rip-up and try again" scheme is implemented such that localised increases in system energy above a certain threshold are assumed to be integration errors. When such errors are detected, the system state is rolled back and integration is reattempted using a shorter timestep. We note that this addition is primarily of use when our "chemistry", the instantaneous modification of atoms and/or bonds between them occurs. This leads to localised perturbations which would otherwise lead to fatal instabilities in the molecular dynamics. Finally, we employ a novel multigrid summation scheme where multi-molecule structures (primarily pairs and triplets of water molecules)

Feature	2D LAC	3D LAC	Present Model
Basic Element	Particle	Particle	Atom
Space	Hexagonal Lattice	Face-Centered Cubic Lattice	Continuous 3D
Time	Discrete, Constant	Discrete, Constant	Discrete, Variable
Forces	Repulsive	Repulsive	Attractive and Repulsive
Metabolism	Complex, Multiple Catalysts	Single-Catalyst	Single-Catalyst
Membrane Particles	Single	Single	Pair
Membrane Structure	Monolayer	Monolayer	Monolayer and Bilayer

Table 1: A summary of the key differences between the present work and its predecessors.

have their dynamics approximated by simpler proxy pseudo-molecules. These proxies may be described statically or determined dynamically; details of the algorithm by which the latter occurs will be described elsewhere (Madina, 2004). This effects of this summation scheme are generally restricted to regions of bulk water, waste and resource particles, away from membrane and catalytic activity. If used appropriately, the effect on the total system dynamics is negligble. Nevertheless, there is a price to be paid for the approximations made in our model; over very long timescales, results from our model can deviate substantially from physically correct values since small errors which would ordinarily lead to obvious errors in traditional molecular dynamics schemes may be suppressed by the approximation schemes we adopt. Having said that, since we are studying a driven system which artificially couples dynamics with very different timescales, such drift is fairly insignificant.

Results

Our model reproduces a number of well known phenomena of lipid aggregation, as well as the essential features of our metabolic system and its artificial chemistry. In addition to the spontaneous formation of spherical micells and vesicles, other structures such as bilayer sheets and tubes may form. Despite the constant transformation of certain lipid particles into waste particles which behave very differently, the structures which form can remain stable for long periods of time, provided that there are sufficient resource and catalytic particles present in the region. Moreover, the model exhibits a sensitive dependence on initial conditions. For the results presented here, 10000 molecules were placed randomly in a cubic volume with periodic boundary. Approximately 80% of the molecules are water, resource, catalysts and waste a further 5%, and the remainder are lipid molecules. Simulations were performed on Intel Pentium 4 systems which computed approximately one iteration per minute; basic protocell formation could be observed after a few hours of computation. One such randomly generated initial condition led to the formation of a tube structure as shown in Fig.2. We believe that this is simply due to the initial conditions having a distribution of lipids molecules that is denser is a roughly tube-like region, which seeds the formation of the tube later

Figure 3: Lipid molecules may spontaneously aggregate to form micelles and vesicles.

on. This tube becomes axis-aligned as a result of its seeing its mirror image in the periodic boundary.

Another randomly generated starting point, with the same parameters, instead led to the formation of the robust lipid protocell illustrated in Fig.3. A small number of lipids which gathered into a micelle are also visible in the upper right, though this relatively unviable structure ultimately decays.

By manually generating an initial condition consisting of a lipid bilayer in a water bath that is rich in resources and catalysts, a sheet of lipid molecules can be sustained for a long period of time, as shown in Fig.4.

The structures observed in this model exhibit a high degree of resistance to externally applied perturbations. For example, defects introduced into a spherical protocell (such as spontaneously removing a section of its membrane particles) are quickly repaired, providing that the defect is not so large that the contents of the protocell diffuse into the environment before it has a chance to recover. This resilience is due to the fact that our model is a driven system: whereas a defect introduced into a standard molecular dynamics simulation of a lipid membrane may quite easily destroy the structure, our model involves a constant external supply of

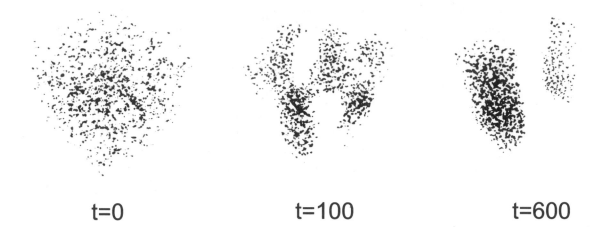

t=0   t=100   t=600

Figure 2: Random initial conditions may lead to stable structures such as tubes. Lipid molecules are nearby pairs of spots, while resources, catalysts are and water appear more faintly. Waste particles are not shown.

Figure 4: Bilayer sheets of lipid molecules are stable, provided there are sufficient resources in the surrounding water.

energy which is channeled by the metabolic system into the production of replacement membrane molecules and the repair of the membrane. There is an important relationship here: the metabolic system is only effective if the concentration of autocatalysts is sufficiently high, but the concentration of autocatalysts can only become sufficiently high if it is enclosed (at least partially) by membrane molecules.

Discussion

We have presented a model which combines molecular dynamics and a simple "chemistry" to support the construction of simple metabolic systems which are contained by simple lipid membranes. From random initial configurations, we observed the emergence of different, protocell-like struc-

tures. These structures can exhibit a significant degree of homeostasis despite residing in a continually driven, non-equilibrium system. Varela and others have stressed the importance of "autopoiesis" as an essential feature of living organisms: structures such as a protocells should have the ability to create and maintain their own boundary. It is difficult to envisage a straightfoward molecular dynamics simulation of an autopoietic structure: while Varela's model was defined on a two dimensional lattice, in the real world, autopoietic structures are highly complex. While molecular dynamics can yield realistic and often quite accurate biophysical phenomena such as lipid aggregation and protein folding, these are almost trivial compared to autopoiesis. Both computational and conceptual difficulties alike would plague any attempt to construct an autopoietic structure in a molecular dynamics system. While we trust that these difficulties will eventually be overcome, we suggest the approach we have taken here - to equip standard molecular dynamics with an abstract, oracle-like chemistry, may be a possible way to shortcut these difficulties. Computation is made much simpler as the long timescales associated with real biological reaction pathways may be avoided, while the problem of engineering (let alone understanding) a biologically realistic autopoetic structure is reduced to the vastly smaller problem of designing a simple set of reaction pathways that can support autopoiesis more trivially. We note that while our model is rather simplistic, it does not necessarily have to be so. For example, the model may be made more simple by replacing the accurate TIP5 water model (Mahoney and Jorgensen, 2000) with a simpler 3-point or single-point model as in previous work. Or it may be made more complex by replacing the simple two-atom lipid molecules with more realistic phospholipids. The primary barrier to applying the present methods to highly realistic protocellular modelling is implementing more detailed chemical pathways so that complex structures do not magically appear and disturb the system dynamics too greatly. Also, the increased computational cost of highly realistic models is also an important factor. Nevertheless, we feel that these difficulties are not insurmountable and that in the near future, it may be possible to synthesise a realistic protocell by adopting a hybrid molecular dynamics/chemistry model as presented here. Another advantage of the present model is that it permits cells to be somewhat more mobile than those in our previous work, where cells were fixed by rather strong potential forces, and each other. This is not to say that the cells we observe *do* move in any interesting sense - they simply repel each other and tend to relax into simple spatial configurations. Still, we wonder if it may be possible to add some source of cell motion to our model. Further, while only the simplest reaction pathways exist in the present model, we believe that it should be possible to implement more interesting chemistries which permits differentiation between cells, so that they may become a target of some simple evolution. This is one area of current work. Finally, we wonder if the range of membrane forms which may persist in the present model in spite of membrane decay may indicate that perhaps the shapes of the earliest proto-life forms may be significantly different from those of the living structures we observe today.

Acknowledgements

This work was partially supported by the COE project "complex systems theory of life", a Grant-in-aid (No. 13831003) from the Japanese Ministry of Education, Science, Sports and Culture and afiis project (Academic Frontier, Intelligent Information Science 2000-2004 Doshisha University). DM thanks TI for his help and generous support, Naoaki Ono for many helpful discussions, and Russell Standish of the UNSW HPCSU for his assistance.

References

Breyer, J., Ackermann, J., and McCaskill, J. (1998). Evolving reaction-diffusion ecosystems with self-assembling structures in thin films. *Artificial Life*, 4:25–40.

Edwards, L., Peng, Y., and Reggia, J. (1998). Computation models for the formation of protocell structures. *Artificial Life*, 4:61–77.

Madina, D. (2004). Logical dynamics in biological systems. Master's thesis, The University of New South Wales, Sydney 2052, Australia.

Madina, D., Ono, N., and Ikegami, T. (2003). Cellular evolution in a 3d lattice artificial chemistry. In Banzhaf, W., Christaller, T., Dittrich, P., Kim, J., and Ziegler, J., editors, *Proceedings of the 7th European Conference on Artificial Life (ECAL'03)*, pages 59–68. Springer, Berlin, Germany.

Mahoney, M. and Jorgensen, W. (2000). A five-site model for liquid water and the reproduction of the density anomaly by rigid, nonpolarizable potential functions. *Journal of Chemical Physics*, 112:8910–8922.

Marrink, S., Tieleman, D., and Mark, A. (2000). Molecular dynamics simulations of the kinetics of spontaneous micelle formation. *Journal of Physical Chemistry B*, 104:12165–12173.

Mayer, B. and Rasmussen, S. (2000). Dynamics and simulation of micellar self-reproduction. *Int. J. Mod. Phys. C*, 11:809–826.

McMullin, B. and Varela, F. (1997). Rediscovering computational autopoiesis. In Husbands, P. and Harvey, I., editors, *Proceedings of the 4th European Conference on Artificial Life*, pages 38–47. MIT Press, Brighton, UK.

Ono, N. (2001). *Artificial Chemistry: Computational Studies on the Emergence of Self-Reproducing Units*. PhD thesis, The University of Tokyo, Komaba 3-8-1, Meguro, Tokyo 153-8902, Japan.

Ono, N. and Ikegami, T. (1999). Model of self-replicating cell capable of self-maintenance. In Fkoreano, D., Nicoud, J., and Mondada, F., editors, *Proceedings of the 5th European Conference on Artificial Life (ECAL'99)*, pages 399–406. Springer, Lausanne, Switzerland.

Ono, N. and Ikegami, T. (2001). Computational studies on the emergence of self-reproducing units. In Kelemen, J. and Sosik, S., editors, *Proceedings of the 6th European Conference on Artificial Life (ECAL'01)*, pages 186–195. Springer, Prague, Czech Republic.

Pearlman, D. e. a. (1995). Amber, a computer program for applying molecular mechanics. *Computer Physics Communications*, 91:1–41.

Pearson, J. (1993). Complex patterns in a simple system. *Science*, 261:189–192.

Skeel, R. e. a. (1999). Namd2: Greater scalability for parallel molecular dynamics. *Journal of Computational Physics*, 151:283–312.

Turing, A. (1953). The chemical basis of morphogenesis. *Transactions of the Royal Society*, B237:37–72.

Varela, F., Maturana, H., and Uribe, R. (1974). Autopoiesis: The organization of living systems, its characterization and a model.,. *BioSystems*, 5:187–196.

Lipidia: An Artificial Chemistry of Self-Replicating Assemblies of Lipid-like Molecules

Barak Naveh[1], Moshe Sipper[1], Doron Lancet[2] and Barak Shenhav[2]

[1]Dept. of Computer Science, Ben-Gurion University, Beer-Sheva 84105, Israel

[2]Dept. of Molecular Genetics and the Crown Human Genome Center,
The Weizmann Institute of Science, Rehovot 76100, Israel

{barnav,sipper}@cs.bgu.ac.il, {doron.lancet,barak.shenhav}@weizmann.ac.il

Abstract

Lipidia is a new simulation system that is related to the "Lipid World" scenario for the origin of life. Lipidia allows for conducting experiments with a population of assemblies containing lipid-like molecules on a two dimensional grid. The dynamics of the assemblies is modelled using the Graded Autocatalysis Replication Domain (GARD) model. New experiments using a finite environment model with GARD were conducted with Lipidia. The experiments show that more self-replicating assembly species appear when using a model of finite environment than when using a model of infinite environment. In many species the number of individuals increases as well.

Introduction

The "RNA World" is possibly today's most popular theory for the origins of life (Gilbert, 1986; Joyce, 2002). Because RNA molecules can act as catalysts in addition to acting as templates, it is hypothesized they might have been able to do both: to store alphabet-based genetic information *and* to catalyze their own creation. Life, according to this theory, began when certain RNA molecules achieved the capability to replicate themselves. This scenario, despite its elegance, suffers from difficulties.

In an attempt to come up with a probable scenario, having observed that no known bio-molecule is capable of self replication in its naked form, it has been suggested that self replication might not have been achieved by a single molecule, but rather by a molecular ensemble (Kauffman, 1995). This work is based on "The Lipid World" scenario (Segre et al., 2001) which follows that line of thought. The scenario assumes that self-replication was initially achieved by non-covalent assemblies of lipid-like molecules that contained mutually catalytic sets (Segre et al., 2000). RNA according to this scenario, while possibly playing an important role, came later.

Why Lipids?

Lipid-like amphiphiles (molecules that have one end that "loves" water, and another end that "hates" water) are assumed to have been present in the primordial soup (Deamer, 1997; Luisi et al., 1999). They are known to be capable of self-organizing into higher-level structures (e.g. micelles and vesicles) (Luisi et al., 1999; Gompper and Schick, 1994; Tanford, 1978). Lipid vesicles have been "shown to be capable of enhancing the rates at which precursors are converted into vesicle-forming amphiphiles (Bachmann et al., 1992). In some settings, this leads to an auto-catalytic expansion of the molecular assemblies, a process resembling cell growth" (Segre and Lancet, 2000). Random fission process can cause occasional divisions. Altogether, we have assemblies of molecules that demonstrate a primitive form of growth and division, in a process that is, although noisy, capable of self-replication with a reasonable fidelity (Segre et al., 2000; Segre and Lancet, 2000). The inside of similes of such assemblies, namely lipid vesicles, is shielded from the surrounding environment and thus hypothesized to be capable of offering "hospitable conditions" under which RNA replication can be more likely. Once some coupling is formed between these two replication systems, an early cell could come into existence (Szostak et al., 2001).

Scope of Current Work

We use the *Graded Autocatalysis Replication Domain* (GARD) model (Shenhav et al., 2004; Segre et al., 1998) to quantitatively model and simulate the developmental process of non-covalent assemblies of lipid-like molecules. Previous studies using the GARD model have mostly examined such assemblies in a one-at-a-time fashion. The behavior of assembly populations has been largely unexplored (Segre et al., 2000). In this work we expand the model to a *population* of assemblies and obtain quantitative and qualitative results regarding its behavior. Also, previous studies assumed idealization of an infinite environment where the assembly's effect on the environment is negligible and "food" molecules are in infinite supply. In this work we introduce a finite environment to the model, which allows cross-interactions between assemblies via the environment. We compare the effect of finite environment vs. infinite environment.

Lipidia

This section introduces Lipidia's terminology and describes its objects and interactions.

Structure

Lipidia is based on a two dimensional interaction *grid*, as with cellular automata. Each square on the grid is called a *grid-location* (or *location* for short). For each location there is a defined *environment* containing a variety of *molecules*. Each location may contain zero or more *assemblies* of molecules. The location's environment is common for all assemblies contained within it. Molecules from the environment may *join* an assembly, and molecules from the assemblies may *leave* their assembly back to the environment. "Matter" on the grid is therefore preserved — no matter is ever lost or created[1].

Each grid location has eight neighboring locations, except the locations on the grid's edge, which border a surrounding *gutter*. The gutter is a special location that takes care of objects falling-off the grid. A few gutter policies can be applied. A common policy is to insert objects falling from one edge to the opposite edge, hence turning the grid into a toroid.

Initial Configuration

The environment is seeded with an arbitrary number of molecules, of N_G different *types*. We usually start with all grid locations empty of assemblies, and their environments uniformly seeded with a constant number e_0 of molecules for each of the N_G types.

Assembly Birth

Assemblies spontaneously come to existence at some constant low rate. We call this appearance *assembly birth*, and it can happen at any grid location. Ideally, birth rate should depend on the numbers and types of molecules available at each location. It is reasonable to assume that the rate of assembly creation will degrade as the environment material runs out. However, for simplicity we simulate a constant birth rate.

Assembly Growth

As the simulation progresses, molecules from the environment may join assemblies, and molecules from the assemblies may leave them and return back to the environment. Join and leave reactions establish the assembly growth; however, growth is also affected indirectly by other assemblies via the shared environment. The dynamics of assembly growth are governed by the GARD model (Segre et al., 2000) as follows.

The system contains a set of N_G types of molecules, and mutual catalysis can occur between any molecule pair.

---

[1] The finite environment model is replaceable with an infinite environment model of fixed concentrations.

The catalytic rate enhancement exerted by molecule type j on molecule type i is denoted by a matrix element β_{ij}. Values for β matrix are assigned in accordance with previously developed Receptor Affinity Distribution (RAD) model (Lancet et al., 1993). The basal reaction rates k_f and k_b (forward and backward) respectively specify spontaneous join and leave rates.

An assembly s is represented by an N_G-dimensional vector, where each component s_i denotes the number of molecules of the i-th type in the assembly. An environment e is represented similarly.

Join Reaction The join rate J_i of molecule type i in the environment e of an assembly s is given by:

$$J_i = k_f e_i (1 + \sum_{j=1}^{N_G} \beta_{ij} s_j) \qquad (1)$$

Therefore, J_i increases the higher the count, e_i, of molecules of type i in the environment. The spontaneous rate, k_f, is enhanced by a catalysis generated by molecules within the assembly: each molecule of type j contributes β_{ij} for that rate enhancement. Hence equation 1 above.

Leave Reaction The leave rate L_i of molecule type i from assembly s to its environment e is given by:

$$L_i = k_b s_i (1 + \sum_{j=1}^{N_G} \beta_{ij} s_j - \beta_{ii}) \qquad (2)$$

Therefore, L_i increases the higher the count, s_i, of molecules of type i within the assembly. The spontaneous rate, k_b, is enhanced by a catalysis generated by molecules within the assembly: each molecule of type j contributes β_{ij} for that rate enhancement, except of one molecule of type i that can not catalyze its own leave. Hence equation 2 above.

Assembly Division (Split)

When the number of molecules in an assembly reaches a certain value, denoted by N_0, the assembly is divided into two daughter assemblies. When this division takes place, every molecule in the original assembly is randomly joined to one of the daughter assemblies. The exact structure of the assembly is not modelled.

Assembly Diffusion

Each assembly diffuses at some low rate from its current location to a neighboring location on the grid. It may as well diffuse into the gutter and handled according to the gutter policy.

Environment Diffusion

As a result of diffusion, environments of two neighboring locations mix some percentage of their molecules at some low rate. Such diffusion may also occur into the gutter, then to be handled according to the gutter policy.

Population Control and Assembly Death

An assembly may divide into two daughter assemblies, which may divide further. In time we obtain an exponential explosion of assemblies. Following are means that limit population growth, some of which involve assembly *death*, whereby an assembly disassembles and return its molecules back to the environment.

Natural Death   An assembly undergoes spontaneous decomposition following a certain amount of time after its creation. This sort of death is natural to the model.

Finite Environment   A finite environment provides a natural means to limit population growth. It puts a bound on the total number of molecules, and therefore on the number of assemblies that can be created.

The Reaper   The above means are part of the model and therefore "natural". However, to avoid excessive computation we also need an "artificial" means. The reaper keeps the global assembly population on the grid below some bound. When the number of assemblies exceeds the bound, the reaper selects an assembly at random and "kills" it.

Scheduling of Random Events

All simulation reactions and behaviors mentioned above can be collectively called *events*. Many of these events are stochastic and occur at defined rates. The simulation schedules the events in a stochastic but fair method, which reflects their rates. The method is best visualized as a giant roulette wheel, where each event "owns" one roulette slot. Unlike true roulette, slots may vary in size making them more or less likely to occur. Thus, the size of the slot corresponds to the rate of the owning event.

Upon every cycle, the simulation engine gives the roulette wheel a spin to choose the next event. The event is activated and the state of the system is modified, possibly changing the rates of other events, whereby their corresponding slots become wider or narrower. The algorithm implementing the roulette wheel requires $O(log(n))$ time and $O(n)$ space for n events.

Attractors

This section discusses the important concept of *attractors* in the context of Lipidia; how they are defined and how to find them.

Composition Stability

The *normalized composition* (or *composition* for short) of an assembly s is given by:

$$\tilde{s} = s/\|s\|, \tag{3}$$

where $\|s\|$ is the norm of the vector s.

Due to mutual catalysis, we expect to find compositions that are *stable* over time. We say that a composition is stable if it remains *similar* along splits. This stability involves a quasi-stationary state, and should be distinguished from equilibrium-type stability. Thus, even though during assembly growth molecules can join or leave, when it splits (actually, just before it does) its composition is similar to its parent's composition (at the time of split).

Composition Similarity and Self Replication

We estimate the similarity of two assemblies s_p and s_q by using the scalar product of their compositions:

$$H(s_p, s_q) = \tilde{s}_p \cdot \tilde{s}_q \tag{4}$$

therefore, $H = 1$ denotes perfect similarity and $H = 0$ denotes perfect dissimilarity. Note that H only measures "how far" composition s_p is from s_q, but does not measure "how hard" it is, in terms of reactions, to get from composition s_p to composition s_q. When a parent assembly splits into two daughter assemblies that grow to have a similar composition as the parent, we say that the parent replicated.

Trajectories in Composition Space

A composition we measure at any time point, and specifically at the time of assembly split can be thought of as a point in the composition space, which includes all possible compositions. Each of the assembly's ancestors might have visited another point in the composition space. An assembly's lineage can be thought of as the *trajectory* along these points.

Attractors and Basins of Attraction

Each point in the composition space is theoretically reachable. However, due to mutual catalysis some points are more likely then others. The composition space may be perceived as a landscape, where low points represent compositions that are easier to get (in terms of reactions), while high points represent the opposite. The trajectories are therefore likely to "fall" into lower areas. If an area has a basin shape, each trajectory, having fallen in, will have a hard time escaping the basin: it will tend to keep falling back into the *attractor*, that is, the foot of the basin. Therefore, an assembly having a composition within an attractor will tend to be stable.

Attractors as Assembly Species

When we observe species in nature we see that individuals of the same species may vary from each other. However, they seem to be "trapped" within some "cloud of variations" that represents their species. We can therefore see our attractors as representing assembly types (or species). Assemblies of the same attractor will vary from each other, but will stay "trapped" within the attractor, in the same way as above[2].

---

[2]Differentiating cells in multicellular organisms have been similarly viewed as shifting away from each other while falling into

Finding Attractors

Finding attractors in the multidimensional composition space is not easy, and various clustering techniques can be employed. We used a simple algorithm that is by no means optimal. The algorithm has two components: *filter* and *clusterizer*. The filter checks that a new assembly s_n, upon split, is similar enough to its parent ($H(s_n, parent(s_n)) \geq T_{similar}$) and that such similarity has been maintained for the last T_{stable} splits. If it has, the clusterizer is invoked to decide to which cluster s_n belongs. The clusterizer holds a list of assemblies, $L = (s_1, s_2, ..., s_k)$, where assembly s_i represents the cluster i. When invoked, the clusterizer finds $s_m \in L$ such that $H(s_n, s_m)$ is maximal, that is, s_m that is most similar to s_n. If $H(s_n, s_m) < T_{similar}$ then the match is not considered good enough and the clusterizer adds s_n to L, thus creating a new cluster. The thresholds $T_{similar}$ and T_{stable} can be adjusted. The list L resulting from a simulation approximates the attractors.

Results

This section summarizes our preliminary results and observations, obtained from a series of experiments conducted using three types of simulation settings.

Type 1: Basic GARD

To establish reference results we first configured Lipidia to simulate basic GARD (Segre et al., 2000). It was achieved by: creating a 1×1 grid with only one location, setting the environment model to infinite, and setting the reaper to maintain a population of a single assembly.

Under these conditions some attractors were discovered, which means that self-replicating assemblies of various species were found. This result is consistent with previous works with GARD (Segre et al., 2000; Segre et al., 1998). It should be noted that in these settings, a very small number of attractors, sometimes a single one, tended to dominate and to attract most of the assemblies. Assemblies did occur in other attractors, but rarely (Fig. 1).

Type 2: Multi GARD with infinite environment

In the second type of experiments we kept the same conditions as in the first, except for setting the reaper to maintain a constant population of 16 assemblies, instead of one. Because the environment was infinite, the 16 assemblies did not have any effect on each other, and developed independently. The results obtained were therefore similar to the first type (Fig. 2). The overall simulated time extended to about $\frac{1}{16}$th of experiments of type 1. This was expected since the unchanged total of 15,000,000 reactions was "consumed" by 16 assemblies developing in parallel, instead of by one.

At time 50, about 30 attractors were discovered in the type-2 experiments, while at the same time about 18 were

stable cell types, that is, attractors (Kauffman, 1995).

discovered in type-1 experiments. Not surprisingly, the "parallel search" done by the 16 independent assemblies found more attractors than a single assembly did (for same amount of time). However, for the same total number of reactions, the type-1 experiments yielded a discovery of almost twice as many attractors as the type-2 experiments did. This "inefficiency" in type-2 experiments, in terms of number of reactions, may be attributed to the independence of the 16 assemblies: because there is no coupling between the assemblies, nothing prevents many of them of doing "the same". Such coupling was established in experiments of type 3 .

Type 3: Multi GARD with finite environment

In the third type of experiments we replaced the infinite environment model with a finite environment model. All other settings were kept as in the second type. As Fig. 3 shows, this change was significant.

At time 100, about 24 attractors were discovered in type-1 experiments, while at the same time about 130 were discovered in type-3 experiments: more than a five-fold increase in attractor discovery rate! The discovery was also more efficient in terms of reactions: for the same total number of reactions, experiments of type 3 yielded the discovery of almost three times as many attractors as experiments of type 1 did. It should also be noted that in experiments of type 3 assemblies have occurred more frequently in more attractors. Diversity has increased.

The increase of diversity is attributed to essential molecules coming in short supply in the finite environment. In the case of infinite environment, the composition that produces the fastest stably growing assemblies is quickly becoming dominant. Optimal assemblies can always be constructed to produce the strongest auto-catalysis possible by picking the most suitable molecules from the given molecular repertoire. This, however, is not possible with finite environment. Some optimal assemblies can surly be constructed but their very construction consumes the molecules essential for their own composition. Their count in the population is therefore limited and new niches are becoming available for compositions that take advantage of the remaining molecular repertoire.

Conclusions

We described Lipidia; a new simulation system that allows to conduct experiments with a population of lipid-like assemblies on a two dimensional grid, using finite and infinite environment models. We further described a series of experiments performed using Lipidia. Our results show that a finite environment produces more attractors (species), and faster, than an infinite environment. A finite environment allows more assemblies to occur in more attractors and in greater numbers. Thus, diversity increases.

The results might be considered surprising. One might think that having an infinite supply of resources, in the form

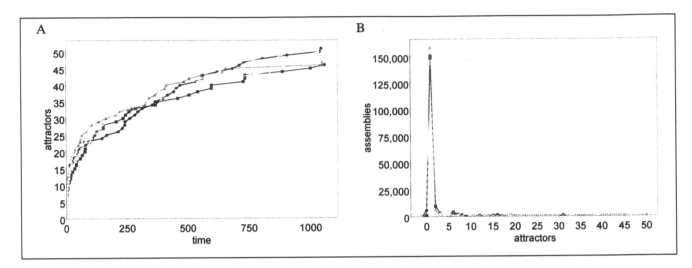

Figure 1: Results of Lipidia simulating basic GARD. (A) The number of attractors as they are discovered in time. (B) Attractor frequencies: the number of assemblies that occurred in each attractor throughout the entire simulation. Attractors are numbered sequentially as they are discovered, thus each attractor number identifies an attractor and also denotes the order of discovery. The graphs show that a very small number of attractors tend to dominate while others are rarer. Results were obtained using a grid with a single location and an infinite environment model. Population was limited to a single assembly and the environment included 100 molecules for each of the N_G types. The first assembly was seeded. Here $N_G = 100, k_f = 0.01, k_b = 0.00001$ and $N_0 = 80$. The rate enhancement factors β_{ij} were sampled from a log-normal distribution with mean $\mu = -6$ and standard deviation $\sigma = 4$, in accordance with RAD model (Lancet et al., 1993). The experiment was repeated 5 times, each simulated for 15,000,000 reactions. Each run is shown in a different shade.

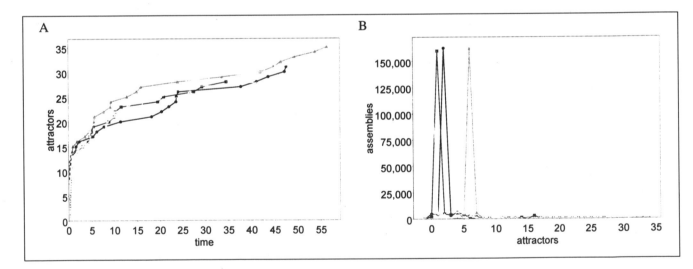

Figure 2: Results of Lipidia simulating 16 assemblies in an infinite environment. (A) The number of attractors as they are discovered in time. (B) Attractor frequencies: the number of assemblies that occurred in each attractor throughout the entire simulation. The graphs show that a single attractor tends to dominate while others are rarer. Attractor discovery rate is improved due to parallelism, but that parallelism is wasteful in terms of reactions: many of the 16 assemblies are "doing the same". Results were obtained using the same settings as in Fig. 1, except the population limit that was set to 16 assemblies, instead of one.

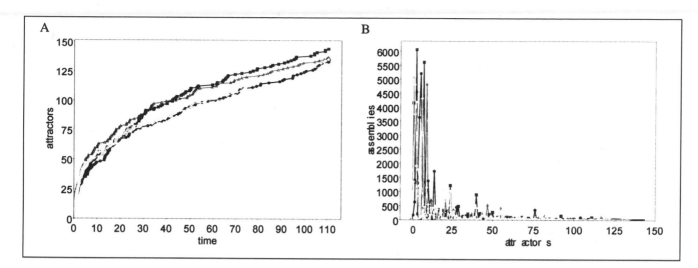

Figure 3: Results of Lipidia simulating 16 assemblies in finite environment. (A) The number of attractors as they are discovered in time. (B) Attractor frequencies: the number of assemblies that occurred in each attractor throughout the entire simulation. The graphs show that the introduction of a finite environment accelerates the discovery of attractors and relaxes the dominance of the single attractor, or the very few ones. Results were obtained using the same settings as in Fig. 2, except for the infinite environment model that was replaced with a finite environment model.

of "food" molecules, might help to "do more". According to our results, it only helps to "do more of the same". Diversity seems to spring when resources are limited. It is when resources for the "best solutions" run out that the race towards alternative solutions begins.

A complete description of Lipidia has been given, although some of its features have not yet been used and are due for future work. Lipidia is capable of simulating a few hundred assemblies within a few hundred grid locations. Future experiments could use these large scale capabilities to explore large grids containing large assembly populations. Diffusion, birth, and death could also be employed. Lipidia is implemented in Java and can run on many platforms. It is available online at: http://ool.weizmann.ac.il/lipidia.

References

Bachmann, P. A., Luisi, P. L., and Lang, J. (1992). Autocatalytic self-replicating micelles as models for prebiotic structures. *Nature*, 357:57–59.

Deamer, D. W. (1997). The first living systems: a bioenergetic perspective. *Microbiology and Molecular Biology Reviews*, 61(2):239–61.

Gilbert, W. (1986). The RNA world. *Nature*, 319:618.

Gompper, G. and Schick, M. (1994). *Self-assembling amphiphilic systems*. Academic Press, London.

Joyce, G. F. (2002). The antiquity of RNA-based evolution. *Nature*, 418(6894):214–21.

Kauffman, S. A. (1995). *At Home in the Universe: The Search for the Laws of Self-Organization and Complexity*. Oxford University Press.

Lancet, D., Sadovsky, E., and Seidemann, E. (1993). Probability model for molecular recognition in biological receptor repertoires: Significance to the olfactory system. *Proceedings of the National Academy of Sciences of the USA*, 90(8):3715–9.

Luisi, P. L., Walde, P., and Oberholzer, T. (1999). Lipid vesicles as possible intermediates in the origin of life. *Curr. Opin. in Colloid and Interface Science*, 4(1):33–39.

Segre, D., Ben-Eli, D., Deamer, D. W., and Lancet, D. (2001). The lipid world. *Origins of Life and Evolution of the Biosphere*, 31(1-2):119–45.

Segre, D., Ben-Eli, D., and Lancet, D. (2000). Compositional genomes: Prebiotic information transfer in mutually catalytic noncovalent assemblies. *Proceedings of the National Academy of Sciences of the USA*, 97(8):4112–7.

Segre, D. and Lancet, D. (2000). Composing life. *EMBO Reports*, 1(3):217–22.

Segre, D., Lancet, D., Kedem, O., and Pilpel, Y. (1998). Graded autocatalysis replication domain (GARD): Kinetic analysis of self-replication in mutually catalytic sets. *Origins of Life and Evolution of the Biosphere*, 28(4-6):501–14.

Shenhav, B., Kafri, R., and Lancet, D. (2004). Graded artificial chemistry in restricted boundaries. In *Proceedings of Artificial Life IX (this issue)*, Boston, MA. MIT Press.

Szostak, J. W., Bartel, D. P., and Luisi, P. L. (2001). Synthesizing life. *Nature*, 409:387–390.

Tanford, C. (1978). The hydrophobic effect and the organization of living matter. *Science*, 200(4345):1012–8.

Towards the Simulation of Reaction Networks in Astrochemistry

Pierre Philippe, David Weiss Solís, Tom Lenaerts and Hugues Bersini
IRIDIA - Université Libre de Bruxelles
CP 194/6
50, av. Franklin Roosevelt
1050 Bruxelles - Belgium
pphilipp@ulb.ac.be

Abstract

The aim of this paper is to apply artificial chemistry (AC) to the modelling of astrochemistry. More generally, we have attempted to construct a computational model of 'a' minimal real chemistry in an object oriented (OO) approach (UML, Java). To this end we map astrochemistry into the ACs theoretical framework thanks to the developed OO model. The OO approach has permitted us to separate the actual and logical representation of chemical structure. To perform this mapping we identify and simplify features from astrochemistry to tackle the issues of, representation and canonization, reaction rules, and meta-dynamics for a suitable level of abstraction. In the search for a chemical criterium for existence relations between objects we have used chemical thermodynamics. The result is an abstract, simplified model of a real chemical dynamical reaction network which conceptually captures real chemistry in general.

Introduction

Astrochemistry is the study of the chemicals found in outer space, usually in molecular gas clouds, and of their formation, interaction and destruction. For our purposes, the chemistry of interstellar molecule-bearing nebulae is the prime candidate in the search for a real-world archetype of a strongly constructive chemical network. The model subject was motivated by the following observations: primarily, the system behaves as a mixture of very diluted ideal gases, giving the possibility to restrict the molecules to a finite and computationally tractable set of chemical objects while still considering large variations in space and time. Secondly, the low size and diversity of interstellar molecules intrinsically eases the combinatorial explosion and the canonization problem. Thirdly, if we focus on a large enough portion of the cloud, the system behaves as a closed system which allows for simple thermodynamic treatment. Fourthly, the rate equations which in biological kinetics (and bioprocess engineering) give rise to non-polynomial non-linearities (Nicolis and Prigogine, 1989) are reduced to constants (Duley and Williams, 1984). On the other side computational approaches have been identified as the sole way to understand these inaccessible cosmic objects (Le Bourlot, 1997). As a

modelling framework we have chosen artificial chemistries (ACs) for their synthetic approach towards chemistry.

Artificial chemistries, in their most general form, have been defined as man-made systems which capture different features from real chemical systems (Dittrich et al., 2001). The ambitious goal put forward for these systems is to construct theoretical models which allow us to improve our understanding of the origin of evolutionary systems. To achieve this goal, real (bio)chemical systems were formally abstracted into a triplet $< S, R, A >$, with S referring to a collection of molecules, R referring to a collection of reaction rules and A identifying the reaction vessel or reactor. The intricate interaction between these components is used to study their complex dynamics in terms of organizational and functional properties. The focus of these systems is clearly on the results produced by A.

Although ACs provide interesting tools to formally study the kinetics and meta-dynamics of chemical systems, there is often a large gap with chemical reality. For instance, molecules captured in S are regularly represented as numbers or strings. Such a representation ignores the real complexity of molecular structures and their structure/function relationship. Moreover, the uniqueness of the representation is extremely relevant when the synthetic algorithm can construct different isomorphic forms of the same molecule (Schubert and Ugi, 1978). Furthermore, the reactions in R are often depicted as linguistic operations on molecules (Fontana, 1996). Yet in real chemistry this is not enough. There are additional energy considerations, apart from the structural ones that need to be taken into account, which determine the possible molecular structure. One of the main criticisms towards ACs is that they are models where one cannot quantify a distance between model and modelled system (Kauffman, 1995). This criticism however is not restricted to ACs but extends to biochemical or bioprocess models where parametric estimation is delicate. Hence, careful considerations are in order when drawing conclusions obtained from AC for real chemistry. In order to close the gap and to make well-founded observations and predictions, AC should move closer to real chemistry.

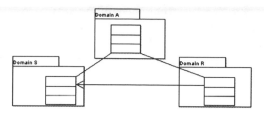

Figure 1: UML package diagram.

A solution can be found in that context. ACs do not stand on their own in their synthetic approach towards chemistry. In chemistry itself a rich research area called *computer chemistry* exists which has been addressing this topic already for a couple of decades (not to be mistaken with the area of computational chemistry) (Ugi et al., 1993). In this context both formal and semi-formal algorithms (which focus on S and R) have been constructed for the synthetic design of molecular structures (Dugundji et al., 1976; Ugi et al., 1993; Faulon and Sault, 2001).

These last two disciplines are closely related in their methodology but differ significantly in their goals. In order to integrate knowledge from both one requires a realistic case-study. The goal set for our research is double. First, we want to study the relation between reaction network topology and dynamics in a real chemical system. Second, we want to improve the dynamic perception of computer chemistry and pave the way towards full integration of AC and computer chemistry. To achieve this distant goal we address in this paper the issue of mapping the S, R, and A components into astrochemistry and making the necessary simplifying assumptions to effectively apply AC to astrochemistry, in a simple yet incremental approach.

In the following section we discuss the structure of our AC. Due to its roots both as a knowledge representation device in artificial intelligence research and as a simulation tool in the natural sciences (see SIMULA), object-oriented (OO) modelling is ideally suited for this task. Moreover, both in computer chemistry and AC this relevance has already been acknowledged (Bauerschmidt and Gasteiger, 1997; Bersini, 2000; Bersini, 1999). The resulting model allows us to reason about the logical structure of reactions and molecules in astrochemistry. Hence, in the next section, we develop a simple mathematical model to construct the actual algorithmic solution. In a modest attempt to close the loop, this procedure amounts to a subset of the original (too) complex problem that gave rise to ACs.

Object-Oriented Model

In this section we briefly discuss the different components of the triplet $< S, R, A >$. Fig. 1 shows the package diagram for a general AC. The object oriented approach permits to separate and develop independently the three parts of the model because each part can be encapsulated, i.e. each component's internal workings are invisible to the other components. Each component can be developed independently and each can rely on a particular level of abstraction, or algorithmic solution to test and improve the performance of the whole.

The component S is in charge of the molecular representation and canonization. The component R is responsible for defining the reactions, determining which reactions can take place, and the actual manipulation of molecules. The A component will take account of the kinetics and meta-dynamics of the system. Each part in the diagram can be specified in more detail.

Molecular Representation S It is customary for chemists to apply graphs (Lewis structures) to represent molecules and to add extra features to these structures to represent issues like stereo-chemistry. For instance, molecules can be represented, in their simplest form, by connection tables or matrices (Dugundji et al., 1976). These representations are not enough to allow for a complete expression of most molecular particularities. Therefore, extensions to these base representations or other representations have been suggested (Bauerschmidt and Gasteiger, 1997). These extensions were often the result of requirements posed on the structure by their users. Also a certain amount of flexibility is required from the actual molecular model since depending on the chemical problem, different levels of molecular detail are required. OO models allow for such a flexibility.

The novelty of the current OO model lies in the fact that we make a distinction between the actual representation of the molecule, for instance graph or matrix, and the logical components which constitute a molecule, for instance atoms and bonds. By making this distinction an OO chemistry can be constructed independent of the computational representation. This achievement is crucial since such a separation allows for a reasoning about a 'chemical world' without being bound to computational or representational issues. For current experiments we opted for an undirected graph representation which contains the principal chemical reactive elements in gas phase in its nodes.

Through the algorithm A, new and existing molecular types are constructed. Each of these molecules is represented by a directed graph. In such systems generated redundant structures should be identified or removed to ensure correct meta-dynamical results. Therefore, an algorithm has been designed which transforms constitutional similar graphs into a particular base-form. The algorithm is a simplification of general algorithms for canonical labelling and consists of three steps:
- Each atom in a molecule is assigned an arbitrary base value which is determined by its atomic number, its unbound electrons and the energy of the bonds.

-Afterward each atom is validated, i.e. it assigned a ranking using the base value. The higher the ranking value the higher the precedence of the atom in the molecule.

- Finally, the complete molecule is re-numerated according to the validated values.

As a result of this process, molecules can be differentiated in the system. At this point only structurally similar molecules can be detected, stereochemistry is not taken into account yet.

Class	Sub class	Reaction template	rate k
1	1.a	$A + \psi \rightarrow B^+ + e^-$	$10^{-17} s^{-1}$
	1.b	$A + \psi \rightarrow B^+ + C + e^-$	$10^{-19} s^{-1}$
2	2.a	$A^+ + B \rightarrow C^+ + D$	$10^{-9} \frac{cm^3}{s}$
3	3.a	$A + B \rightarrow C + D$	$10^{-11} \frac{cm^3}{s}$
4	4.a	$A^+ + e^- \rightarrow B$	$10^{-15} s^{-1}$
	4.b	$A^+ + e^- \rightarrow B + C$	$10^{-10} s^{-1}$
5	5.a	$A + h\nu \rightarrow B + C$	$10^{-11} s^{-1}$
	5.b	$H_2 + h\nu \rightarrow 2H$	$10^{-14} s^{-1}$
6	6.a	$2H \overset{grains}{\rightarrow} H_2$	$10^{-17} \frac{cm^3}{s}$

Table 1: Examples of classes of astrochemical reactions and their reaction rates (ψ= cosmic ray energy, $h\nu$ = electromagnetic radiation energy (photon) and e^- = electrons). Note that hydrogen reactions have special behavior within a given class of reaction.

Reactions R Traditionally each concrete reaction needs to be specified for the reaction generator to work. Yet general classes of reactions can be identified which capture the architecture of a reaction independent of the actual molecular elements. OO models allow for such a conceptualization. In Table 1 the different classes and subclasses are listed.

The functionality of a formal reaction is visualized in Fig. 2.

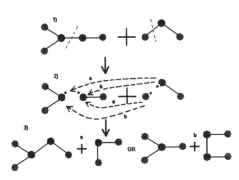

Figure 2: Reaction as conditional graph rewriting.

Due to our choice of undirected graphs for the molecular representation, a reaction corresponds to a graph rewrit-

ing rule (cfr. (Benkö et al., 2003)), although in our model this graph rewriting is conditional. No new atoms are introduced. Hence, as in real chemistry there is a conservation of mass (electrons are considered as non-massive particles). Formally, a reaction consists of two phases:

-A certain amount of bonds of the reactants are broken according to some chemically plausible criteria.

-The products are formed by trying all possible combinations of half-reactants.

As a consequence of this reaction scheme a combinatorial explosion in the number of possible combinations is possible. Especially in the case of larger molecules, this mechanism will lead to an inefficient method. The fact that hydrogen is in large excess (99 percent in molecular form), (cfr. mass action law) and acts as a chain terminator, already limits the combinatorial explosion. Note, also that the largest molecules in astrochemistry have only 13 atoms, composed of only a few elements. These astrochemical observations were one of the motivations to first investigate the applicability of AC.

Yet, using the described reaction mechanism, the number of different molecules still increases rapidly. Only 120 different molecules have been observed. The reaction graph (the number of reactions that have been identified as plausible) exceeds 4000 different reactions on a small set of atoms (including ionic and radical reactions). Astrochemical conditions allow us to define a general reaction template:

$$c_1 A + c_2 B \rightarrow c_3 C + c_4 D \qquad (1)$$

A, B, C and D are chemical species (ions radicals or neutrals) and the c_is are the stoichiometric coefficients; additionally:

- $c_i \in \{0,1\}$, and
- $\sum_{LHS} c_i - \sum_{RHS} c_i = \Delta N$.

That is only monomolecular and bimolecular reactions are allowed because the probability of a three body collision is nil in such extreme dilution conditions (Herbst, 2001). In theory most reactions can be decomposed in elementary reactions of the form of Eq. (1), representing a collision event, in interstellar chemistry, however, elementary reactions and reactions are confounded. Moreover, the classes of reactions derived from the general template reaction (1) and their reaction rates were adopted from the literature (Duley and Williams, 1984). The different reaction classes are summarized in Table 1.

Reactor Algorithm A Given both the molecules and their unique representation, and the reactions in combination with the suggested methods to reduce the combinatorial explosion, we can now address the reaction engine A.

We assume a mixture of molecular types present at certain densities. On this mixture A is executed producing the reaction network dynamics. The algorithm consists of the following steps;

- for every molecular species present in the sample, the algorithm tests if it can react on its own or if it can react with any of the other species present.

  - if a reaction is thermodynamically possible,
  - then examine whether the specific reaction already exists
    * if the reaction has already been added to the reaction list,
    * then reactant and product concentrations are updated
    * else a new reaction object is created and the products are generated.
      · if the products already exist
      · then the concentrations of reactants and products are updated,
      · else they are added to the list of species and the concentration of reactants are updated.

Although we already introduced some simplifications, the combinatorial explosion resulting from the examination of all possible species combinations can still be huge (Bersini, 2000; Bersini, 1999) if we consider the products as candidates to participate as reactants in further reactions. Mechanisms are required which can determine whether or not a particular reaction can take place. In the next section we for the first time introduce thermodynamics in a meaningful way in astrochemical reactions networks to define a criterion for the automated selection of relevant reactions. The mapping of the R and A components is done in some detail but the S component will not be discussed as there are no formal universal chemical rules to describe it.

Astrochemical Model

Reaction rules R   The graph rewriting mechanism acts as a generator of diversity and needs an objective, binary criterium that can discriminate between possible and impossible products. We argue that thermodynamics is a candidate to filter the combinatorial explosion faced by the R component. In interstellar chemistry, thermodynamics is reputed useless because despite the timescale involved, dilution is such that equilibrium is never attained. Therefore the whole time evolution of the species has to be solved to explain present observations. Furthermore, the set of molecules is given by astronomical observations and the only thing left to determine are the rate constants, this is refered in the literature as solving the problem kinetically (Herbst, 2001). We, in turn, are interested in the more general problem of automatically generating the species themselves from a minimal set of initial molecules or ideally, atoms, thus considering the constructivity of the system explicitly. We do not seek to use a thermodynamic criterion as a predictor of equilibrium concentrations, but rather as an energy constraint to differentiate probable from improbable molecules. In the remaining part of this section we will go in more detail into this

approach applying the principles of thermodynamics to the system at hand along with suitable approximations.

Because we consider a closed thermodynamic system at constant temperature and pressure, spontaneous processes (permitted reactions), are selected on the basis of the sign of the change in the Gibbs free energy function G:

$$\Delta G = \Delta H - T\Delta S \qquad (2)$$

where ΔH and ΔS are the enthalpy and the entropy change associated with a reaction.

Because we are dealing with an ideal gas, the work of expansion or contraction ($\kappa\Delta N$) must be taken into account:

$$\Delta H = \Delta U + \kappa\Delta N \qquad (3)$$

where κ is a constant depending on pressure and ΔN is the change in the number of molecules. In order to apply the suggested thermodynamic criterium, one must satisfy the hypotheses made, particularly the assumption that it is a closed system. We thus include as part of the single reaction system, electrons, cosmic rays and photons as non-massive entities which role is energetic. This leads to the definition of 4 classes of cosmic *events* as potential energy contributions:

- ψ cosmic ray energy,
- $h\nu$ electromagnetic radiation energy,
- F_i first ionization potential energy (we consider only singly ionized molecules or atoms), and
- E_{dp} energy transferred to an interstellar dust particle.

Some simplifying assumptions concerning ΔG can be made. Since the temperature is so low (10-50 K), the entropy term is negligible compared to the internal energy term which is independent of temperature. Also, the internal energy contributions from rotation, translation, and vibration are essentially negligible when compared to the potential energy stored in the chemical bonds E_b.

U is then approximated by the total potential energy E_p, contained in or transferred to a compound :

$$U \approx E_p = E_b + E_i + E_{dp} + \psi + h\nu \qquad (4)$$

Note that the possibility of cosmic events essential in order to have a constructive chemistry (accounting for E_b only would favor atoms over molecules). This equation lists only microscopic quantities and should be related to the macroscopic applicability domain of thermodynamics. The correct approach would consist in using statistical mechanics to determine the value of ΔG for a single reaction event, but for the first simulations, a rough approximation consists in defining a free energy variation per reaction event ΔG_r. In this approximation the standard conditions are not molar concentrations, but instead single particles that participate in the reaction:

$$\Delta G_r = \sum_{RHS} E_p - \sum_{LHS} E_p + \kappa\Delta N \qquad (5)$$

We can apply the criterium of Tab. 2 in astrochemistry with-

$\Delta G_r < 0$	Exergonic reaction	permitted
$\Delta G_r > 0$	Edergonic reaction	not permitted
$\Delta G_r = 0$	Equilibrium	irrelevant

Table 2: Selection rules, ΔG_r is the free energy change associated with a reaction.

out the fear that a thermodynamically favored product will be kinetically unfavored or vice-versa (see Fig. 3).

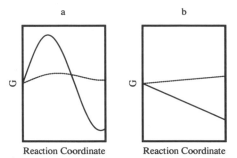

Figure 3: Schematically, starting with one educt on the left of the reaction coordinate axis two structurally possible products are considered a) in general, a thermodynamically unfavored reaction (the reaction leading to the product with the highest G value) can be kinetically favored: energy is available from the environment to pass through maximums along the reaction coordinate, the favored product is thus the one with the lowest maximum (dotted line path), b) at low temperature and pressure as in astrochemistry, if the system is correctly defined as a closed system, there is no possibility of energy exchange with the environment and the only reaction that takes place is the one resulting in a lower G value than the educt (solid line path).

Eq. (4) permits to define conditional reactions. If the bond potential energy change plus the contribution from $\kappa \Delta N$ is sufficient to account for a negative free energy change and if the reactants match reaction class 2 or 3, of Tab. 1 the reaction will exist unconditionally, otherwise, it is tested whether it can occur in the case of a cosmic event for all of the other reaction classes. Once the permitted reactions have been found, further selection is done kinetically, the reaction that can occur is the one that proceeds faster amongst the possible reactions. This does not mean that reactants are replaced by products in the vessel. This latter possibility will be managed by the reactor algorithm A that uses the rates of a reaction to determine if it was effectively reactive.

Reactor Dynamics A The value and thus the sign of ΔG in a macroscopic mixture will vary as a function of the compo-

sition and can be permitted at one time during the simulation and forbidden at another time. Given the macroscopic relation:

$$\Delta G = RT \ln \left(\frac{Q}{K_e} \right) \quad (6)$$

a reaction quotient:

$$Q = \frac{C^{c_3} D^{c_4}}{A^{c_1} B^{c_2}}$$

and a ΔG value computed from Eq. (5) the equilibrium constant K_e, that describes the fixed points of the reaction can be computed. Once it is known, any subsequent value of ΔG can be computed from a particular reaction quotient, which must be provided by the A component that has the information about the time evolution of the abundances.

This emphasizes again the importance of the synthetic approach of ACs where the whole problem solution is only accessible through the interaction of all of the components of the tuple.

Introducing the rate coefficient k, the reactor algorithm A uses the following standard differential chemical relations to describe a single reaction:

$$-\frac{dA}{dt} = -\frac{dB}{dt} = \frac{dC}{dt} = \frac{dD}{dt} = kA^{c_1} B^{c_2} \quad (7)$$

Tab. 1 lists the values of k for all reaction classes. It is important to remark that these rates do not have an arrhenius-like temperature dependence ($E_{act} \gg k_b T$) (Herbst, 2001) and are treated as constants.

To count the contributions from all reactions, reaction 1 must be recast as:

$$r : c_{r,1} x_{r,1} + c_{r,2} x_{r,2} \rightarrow c_{r,3} x_{r,3} + c_{r,4} x_{r,4}, r = 1, \ldots, \mathbb{R} \quad (8)$$

where the indices identify the position in the reaction of a compound for every reaction r, if each compound is also labelled in arbitrary order by a superscript and dropping the $c_{r,j}$ s then,

$$\frac{dx^i}{dt} = \sum_{\substack{r \in R \\ x^i \in RHS}} k_r x_{r,1} x_{r,2} - \sum_{\substack{r \in R \\ x^i \in LHS}} k_r x_{r,.} x^i, i = 1, \ldots, \mathbb{S} \quad (9)$$

with $x_{r,.}$ being x^i's partner in the reaction's LHS, describes the mathematical problem that has to be solved by explicit simulation of the molecules; or, for comparison, by standard ODE numerical integration routines. The numbers $\mathbb{S}$ and $\mathbb{R}$ describe the size of the list of species and reactions which grow over time. The whole problem amounts to the search of two object spaces that are dependent on the whole trajectory of the dynamic system.

Discussion

Our first goal was to build a realistic model and a new framework for AC developers. In fulfilling the goal set for this paper of making a computational model of a minimal actual chemistry, we have succeeded in constructing an OO model that captures chemistry in general in a realistic way. We have identified the *S*, *R* and *A* components in astrochemistry and have made the necessary abstractions to simplify astrochemistry in order to incorporate it in our AC OO framework. Along the way we have made two contributions: recognizing the importance of a separation between actual and logical representation of chemical structure, and using thermodynamics in astrochemistry in a meaningful way. We have also recognized computer chemistry as a source of knowledge for artificial chemists. At the same time as the conception of a theoretical model, we have developed a set of tools useful to achieve a complete AC. These tools are easily pluggable because they are based on the OO model, created for all type of AC. But the full integration of tools are still in progress and we will present the final software in a future work.

Moreover the objective for astrophysicists would be to predict results (new molecules and reactions) with the new approach of this model: indeed the study of dynamics and meta-dynamics interest greatly the searchers in this specific field. The need of computational model (difficulties to observe species, huge *ab initio* calculations) combined with this new approach go through a necessary simplification. It is a source of new ideas when explications of observations are out of reach (Le Bourlot, 1997). Our future work will focus on validation of the model against astrochemical data and results will be presented in our next paper.

Acknowledgements

The authors would like to acknowledge Jacques Reisse for his insight about the possibility of modelling interstellar chemistry with ACs.

References

Bauerschmidt, S. and Gasteiger, J. (1997). Overcoming the limitations of a connection table description: A universal representation of chemical species. *Journal of chemical information and computer science*, 37:705–714.

Benkö, G., Flamm, C., and Stadler, P. (2003). Artificial chemistry based on graph rewriting. In Banzhaf, W., Christaller, T., Dittrich, P., Kim, J. T., and Ziegler, J., editors, *ECAL*, volume 2801 of *Lecture Notes in Computer Science*, pages 10–19. Springer.

Bersini, H. (1999). Design Patterns for an Object-Oriented Computational Chemistry. In Floreano, D., Nicoud, J.-D., and Mondada, F., editors, *ECAL*, volume 1674 of *Lecture Notes in Computer Science*, pages 389–398. Springer.

Bersini, H. (2000). Reaction Mechanisms in the OO Chemistry. In Mark A. Bedau, John S. McCaskill, N. H. P. and Rasmussen, S., editors, *Proceedings of the 7th Internatinal Conference on Artificial Life*, Portland, Oregon. MIT Press.

Dittrich, P., Ziegler, J., and Banzhaf, W. (2001). Artificial Chemistries - A Review. *Artificial Life*, 7(3):225–275.

Dugundji, J., Gillespie, P., Marquarding, D., Ugi, I., and Ramirez, F. (1976). *Metric Spaces and Graphs Representing the Logical Structure of Chemistry*, chapter 6, pages 108–174. Academic Press.

Duley, W. and Williams, D. (1984). *Interstellar Chemistry*. Academic Press, 24-28 Oval Road, London NW1 7DX.

Faulon, J.-L. and Sault, A. (2001). Stochastic Generator of Chemical Structure: 3. Reaction Network Generation. *Journal of chemical information and computer science*, 41:894–908.

Fontana, W. (1996). *The future of science has begun: Approaches to Artificial Life and Artificial Intelligence*, volume 4, Digest On Organization, pages 23–40. Fondazione Carlo Erba.

Herbst, E. (2001). The chemistry of interstellar space. *Chem. Soc. Rev.*, (30):168–176.

Kauffman, S. (1995). *At Home in the Universe: The Search for Laws of Self-Organization and complexity*, chapter 3, pages 47–69. Oxford University Press, 198 Madison Avenue, New York, New York 10016.

Le Bourlot, J. (1997). *Images de la physique*, chapter La modelisation des nuages moleculaires interstellaires, pages 22–30. Département Sciences physiques et mathématiques du CNRS.

Nicolis, G. and Prigogine, I. (1989). *Exploring Complexity, An Introduction*. W. H. Freeman and Company, New York, fifth edition.

Schubert, W. and Ugi, I. (1978). Constitutional symmetry and unique descriptors of molecules. *American Chemical Society*, 100(1):225–275.

Ugi, I., Bauer, J., Bley, K., Dengler, A., Dietz, A., E.Fontain, Gruber, B., Herges, R., Knauer, M., Reitsam, K., and Stein, N. (1993). Computer-Assisted Solution of Chemical problems – The Historical Development and the Present State of the Art of a New Discipline of Chemistry. *Angew. Chem. Intl. Ed.*, (32):201–227.

Homochirality as Fixed Point of Prebiotic Chemistry

Raphaël PLASSON[1], Hugues BERSINI[2] and Auguste COMMEYRAS[1]

[1] OMEMF, CC17, Université Montpellier II, Place E. Bataillon, 34095 Montpellier Cedex 5, FRANCE
[2] IRIDIA, ULB, 50 avenue F. Roosevelt, CP194/6, 1050 Bruxelles, BELGIUM
rplasson@univ-montp2.fr

Abstract

The purpose of this work is to highlight the ability of a pre-biotic system of activation/polymerization/depolymerization (APD) of amino acids to have non-racemic fixed points, similarly to Frank's model, and thus to evolve spontaneously toward homochirality. Chemical kinetic simulations of reaction sets, from simple isolated polymerization systems to complete APD systems in presence of inversion reactions, are developed in order to understand the mechanisms of amplification of asymmetry. The results emphasize the emergence of autocatalysis thanks to the synergetic action between epimerization and APD reactions, allowing spontaneous symmetry breaking from racemic state. The APD system appears to be an original nonequilibrium chemical system model, as an extension of the Frank's model based on prebiotically relevant chemical reactions.

Introduction

One of the most fascinating characteristics of the molecular constituents of all living beings is a property called homochirality: these molecules are chiral — that is they are not identical to their mirror image, as a right hand compared to a left hand — and only one of either "right" or "left" form naturally exists. In order to describe the chirality of molecules, we will limit ourselves to the L/D nomenclature of Fischer. For example, the proteinic amino acids are L, while the ribose of ribonucleic acids is always D. This property is fundamental for molecular structures: a peptide characterized by the same sequence of amino acids as a given protein, but with a random distribution of D and L amino acids, will never adopt the three-dimensional conformation of the protein. Knowing that it is the conformation of the protein which determines its function, it is broadly accepted that homochirality of the molecular constituents of cells can be considered as a prerequisite for life (Avetisov and Goldanskii, 1991).

Homochirality is a very interesting phenomenon in the origin of life, since for symmetry reasons, prebiotic compounds are synthesized in an equally distributed L/D mixture (called a "racemic" mixture). As a matter of fact, this racemic situation should be quite attracting and stable in the long term, as the inversion of chiral centers tends to equilibrate L and D concentrations. How could the biomolecules have escaped from this attractor ? Two families of hypothesis have been proposed to explain the emergence of homochirality. The first one, exogenous, relies on the external origin of asymmetry (asymmetrical weak force, environmental asymmetry, ...) and will not be considered further in this paper. The second one, endogenous, relies on intrinsic properties of the chemical system (e.g. difference of affinity between L and D shaped compounds (stereoselectivity) or autocatalytic processes).

One of the most used models in the prebiotic field relies on the principle of amplification and accumulation of asymmetry. A series of experiments (Blair and Bonner, 1981; Hitz and Luisi, 2003) illustrates how a small difference in L and D concentrations can be multiplied in peptides by the stereoselective polymerization of amino acids derivatives. What is still unsatisfactory with this explanatory schema is that racemic and non-racemic remain stable situations, nothing is qualitatively new and the non-racemization of polymers just reflects the non-racemization of initial concentrations. Adopting the terminology of dynamical systems, this model is a conservative one in which any initial condition will lead to a different outcome. Such systems cannot produce homochirality in the long term (Bonner, 1999).

A more exciting model relies on autocatalytic reactions, in which each L or D form catalyzes its own formation. This model was first developed by Frank (Frank, 1953). From this perspective, Decker (Decker, 1974) described the minimal chemical open systems that can exist in several steady states. The formalism was extended by Kondepudi (Kondepudi and Nelson, 1984), and similar ones are currently developed (Iwamoto, 2003). All these models describe open-flow systems of autocatalytic synthesis of chiral compounds, according to the following scheme:

$$F \longrightarrow X_L \qquad F + X_L \longrightarrow 2X_L \qquad (1)$$
$$F \longrightarrow X_D \qquad F + X_D \longrightarrow 2X_D \qquad (2)$$
$$X_L + X_D \longrightarrow P \qquad (3)$$

In contrast with the precedent case, the initial racemic state becomes unstable, and a non-racemic state can spontaneously emerge as the attracting fixed point. This new dynamics is now dissipative, and every initial condition will lead to a non-racemic situation, favoring excess of either L or D.

Few experimental systems have shown such behavior of symmetry breaking: crystallization of sodium chlorate (Kondepudi et al., 1990), synthesis of cobalt complexes (Asakura et al., 2000), organometallic synthesis of pyrimidyl alkanol (Soai et al., 2001), etc. All these systems are based on direct autocatalytic reactions. Unfortunately, these experiments are far from the prebiotic field.

The simulations presented in this paper try to conserve the spirit of Frank's model, while based on prebiotic data and knowledge, in the field of peptides and amino acids reactivity (Taillades et al., 1999), for determining the ability of a prebiotic closed system to evolve toward homochirality, even in the absence of initial imbalance. A set of six types of chemical reactions are included in the simulation, which all together can lead to the autocatalytic effect giving rise to homochirality. The next section will describe the six types of reactions, and the third one will present and discuss results of simulations, where the reactions will be successively introduced. We will show that while only deactivation and polymerization is enough to obtain amplification of asymmetry, the whole set of reactions becomes necessary to obtain, like in Frank's model, a dissipative dynamics and an autocatalytic spontaneous generation of homochirality.

The Activation/Polymerization/Depolymerization Model

The activation/polymerization/depolymerization (APD) system involves three kinds of chemical species (X, X* and X_n, X being either L or D), and six kinds of reactions, summed up on Fig. 1.

Involved Species:

1. X represents the amino acids, that is the non-activated monomers, unable to polymerize by itself. X can be of configuration either L or D;

2. X* represents the activated monomers, which only in their active form can polymerize. They can be of configuration either L or D;

3. X_n represents the peptides, that is the polymer, composed of a sequence of L or D monomers.

Involved Reactions:

1. *Activation* corresponds to the conversion of X to X*:

$$X \xrightarrow{k_a} X^* \qquad (4)$$

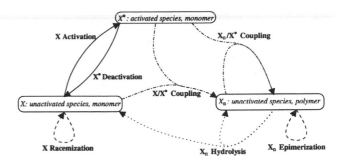

Figure 1: System of activation/polymerization/depolymerization of amino acids: a model of closed dissipative prebiotic system.

2. *Deactivation* corresponds to the conversion of X* to X:

$$X^* \xrightarrow{k_d} X \qquad (5)$$

3. *Polymerization* corresponds to the reaction of X* with either X or X_n, leading to the peptide elongation on the left side (by convention). For a given peptide, the kinetic constant $k_n^=$, between X* and the peptide's left side of same configuration (Eq. 7 and Eq. 8), is greater than the kinetic constant $k_n^{\neq}$, between X* and the peptide's left side of opposite configuration (Eq. 6 and Eq. 9).

$$L^* + DX_{n-1} \xrightarrow{k_n^{\neq}} LDX_{n-1} \qquad (6)$$

$$L^* + LX_{n-1} \xrightarrow{k_n^=} LLX_{n-1} \qquad (7)$$

$$D^* + DX_{n-1} \xrightarrow{k_n^=} DDX_{n-1} \qquad (8)$$

$$D^* + LX_{n-1} \xrightarrow{k_n^{\neq}} DLX_{n-1} \qquad (9)$$

4. *Peptide hydrolysis* corresponds to the conversion of one peptide to two shorter peptides or amino acids. Hydrolysis can take place between two residues of either same configurations (Eq. 10 and Eq. 13) or opposite configurations (Eq. 11 and Eq. 12). Here again, according to the difference in intensity of the values of the two constant rates, this reaction can be more or less stereoselective. Experimental data is lacking on this specific fact.

$$X_nLLX_m \xrightarrow{k_h^=} X_nL + LX_m \qquad (10)$$

$$X_nLDX_m \xrightarrow{k_h^{\neq}} X_nL + DX_m \qquad (11)$$

$$X_nDLX_m \xrightarrow{k_h^{\neq}} X_nD + LX_m \qquad (12)$$

$$X_nDDX_m \xrightarrow{k_h^=} X_nD + DX_m \qquad (13)$$

5. *Racemization* corresponds to the conversion between L and D.

$$L \underset{k_r}{\rightleftharpoons} D \qquad (14)$$

6. *Epimerization* corresponds to the conversion of one monomer inside the peptide chain.

$$\text{X}_n\text{LX}_m \overset{k_e}{\rightleftharpoons} \text{X}_n\text{DX}_m \qquad (15)$$

This model is based on real chemical reactions which are part of the Commeyras "primary pump" scenario about the prebiotic origin of peptides (Commeyras et al., 2002), but is compatible with other similar APD systems (Huber et al., 2003). Many of the kinetic rates used in the simulations were experimentally measured (Plasson et al., 2002), more particularly the rates concerning and supporting the stereoselectivity of polymerization. This scenario explains the appearance of peptide chains by a succession of activation reactions, taking place in dry phase, and polymerization and depolymerization reactions, possible in wet phase.

Chemical Kinetic Simulation Program

The program used for the kinetic simulations, Sa3, is based on numerical integration of a set of chemical equations, by a semi-implicit Runge-Kutta algorithm (Kaps and Rentrop, 1984). It was developed by J.-C. Micheau and D. Lavabre (*IMRCP, Université Paul Sabatier*, Toulouse, FRANCE).

As all considered reactions must be written, it was necessary to limit their number. In this model, the formation of peptides of polymerization degree greater than 4 were neglected. All stereoisomers of peptides up to the tetramers and the corresponding reactions were then considered.

Amplification of Non-Racemic Situations

Polymerization of Non-Racemic X*

In line with Blair and Bonner's experimental results, a first set of simulation shows that an initial excess of one form of X* will be amplified by its accumulation in peptides thanks to the stereoselectivity of the polymerization reactions.

In this first simulation, only the reactions of deactivation and polymerization were taken into account. In the initial conditions, only X* is present, with an excess of 10% of L*. The L excess in monomers is determined by $ee_1 = \frac{([\text{L}]+[\text{L}^*])-([\text{D}]+[\text{D}^*])}{([\text{L}]+[\text{L}^*])+([\text{D}]+[\text{D}^*])}$, and in homopeptides by $ee_n = \frac{[\text{L}_n]-[\text{D}_n]}{[\text{L}_n]+[\text{D}_n]}$ for $n \in [2,4]$.

Because of stereoselectivity of coupling, a majority of homochiral peptides is produced. Homochiral peptides are first generated with a much higher L excess than initially introduced: 20%, 29% and 38% for respectively ee_2, ee_3 and ee_4 (Fig. 2). These values decrease as polymerization go on, but when all L* are consumed, they still remain high: 10%, 14% and 25% for respectively ee_2, ee_3 and ee_4, while final excess of L is only 5%. Stereoselective polymerization allows concentration of the configuration in excess into homochiral polymers.

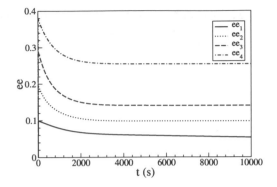

Figure 2: Evolution of ee_n during the polymerization of non-racemic X* ($ee^{\text{init}} = 0.1$) calculated by computer simulation using the Sa3 program (tolerance, 10^{-3}; maximal step, 1 s). Parameters: $k_d = 5 \cdot 10^{-4}$ s^{-1}; $k_1^= = 1.6 \cdot 10^{-2}$ s^{-1}.M^{-1}; $k_1^{\neq} = 1.1 \cdot 10^{-2}$ s^{-1}.M^{-1}; $k_2^= = 6 \cdot 10^{-2}$ s^{-1}.M^{-1}; $k_2^{\neq} = 1.3 \cdot 10^{-2}$ s^{-1}.M^{-1}; $k_3^= = 6 \cdot 10^{-2}$ s^{-1}.M^{-1}; $k_3^{\neq} = 1.3 \cdot 10^{-2}$ s^{-1}.M^{-1}; $[\text{L}^*]_0 = 1.1 \cdot 10^{-1}$ M; $[\text{D}^*]_0 = 9 \cdot 10^{-2}$ M.

Addition of Racemic X* to Non-Racemic Homochiral Dipeptide X$_2$

Only the reactions inverting L in D (and symmetric) can amplify the global excess of L or D (defined by $ee^{\text{tot}} = \frac{[\text{L}^{\text{tot}}]-[\text{D}^{\text{tot}}]}{[\text{L}^{\text{tot}}]+[\text{D}^{\text{tot}}]}$). Among these reactions are the racemization and the epimerizations.

In a new set of simulations, we add the racemization and modify the initial concentrations: an equal concentration of L* and D*, and an excess of homochiral dipeptide LL ($ee_{2,0} = 10\%$).

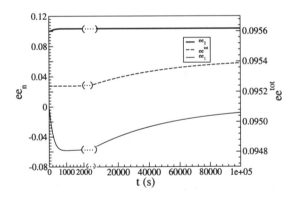

Figure 3: Evolution of ee_n and ee^{tot} during addition of racemic X* to non-racemic homochiral dipeptide ($ee_2^{\text{init}} = 0.1$). Same parameters as in Fig. 2, but initial conditions, $[\text{L}^*]_0 = 1 \cdot 10^{-2}$ M; $[\text{D}^*]_0 = 1 \cdot 10^{-2}$ M; $[\text{LL}]_0 = 1.1 \cdot 10^{-1}$ M; $[\text{DD}]_0 = 9 \cdot 10^{-2}$ M, and with addition of $k_r = 5 \cdot 10^{-5}$ s^{-1}.

As X* reacts on an excess of homochiral dipeptides, the main formed products are homochiral tripeptides coming from X*/X$_2$ coupling, and amino acids coming from X* hydrolysis. Two phenomena deserve some attention (Fig. 3).

At first, because of stereoselectivity, most of L* reacts on LL, and most of D* reacts on DD. As X* is initially racemic, about the same quantity of LL and DD are consumed. Thus, the absolute concentration difference between LL and DD remains roughly constant, while the total concentration of LL and DD decreases. This implies the increase of ee_2.

Secondly, as there are more LL than DD, L* is more quickly consumed than D*. Thus, during polymerization, concentration of D* is higher than the one of L^*, so that de-activation produces more D than L. At the completion of the reaction, an excess of D is obtained. Compensating for the D/L disequilibrium, the racemization increases consequently the total concentration $[L^{tot}]$, and decreases the total concentration $[D^{tot}]$: the global L excess increases.

While replicating the results of the previous simulation, namely the amplification of the L excess in the peptides, a new global effect occurs, thanks to the presence of racemization: the increase of the global L excess. This result shows that in some conditions, racemization may be able to enhance non-racemic situations rather than to tend to racemic state, via an "inversion of population" phenomenon. But still, we are in the presence of a simple amplification, with no qualitative change: the dynamics is still conservative.

Spontaneous Emergence of homochirality

Closed APD System

The preceding study has shown the ability of a simple system to amplify a first excess in either L or D. While keeping with classical prebiotic chemistry, would it be possible to obtain similar results to the Frank's Model: homochirality as a new fixed point of a now dissipative dynamics ?

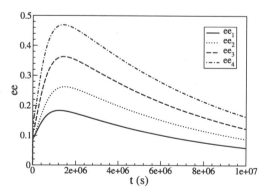

Figure 4: Evolution of ee_n in an APD system ($ee_2^{init} = 0.1$). Same parameters as in Fig. 3, but integration parameters, tolerance, 10^{-3}; maximal step, 10^4 s; and with addition of $k_a = 1 \cdot 10^{-3}$ s^{-1}; $k_h^= = 1 \cdot 10^{-5}$ s^{-1}; $k_h^{\neq} = 1 \cdot 10^{-6}$ s^{-1}.

Two new types of reactions are introduced: the activation one, possible if the system is no more isolated but energetically open, and the depolymerization one, allowing constant

regeneration of basic materials to be recombined. The addition of these two types of reactions introduces reactive loops in the system, which are well-known to promote dynamical irreversibility and to induce new stable configurations.

The experimental results (Fig. 4) show that departing from the same excess of LL, the system now converges to a single global fixed point, in which all compounds are racemic. Instead of being amplified, the dynamic evolution of the system re-establishes a perfect symmetrical situation and looses the memory of the initial condition. The system has turned out to be dissipative but with no self-amplification of the initial unbalanced presence of L or D. However, a transitory period still shows this amplification, even stronger than in the previous simulations. Depolymerization being much slower than polymerization, a rough intuitive explanation is that the system first behaves like in the absence of depolymerization.

Emergence of an Autocatalytic Set

So far, it has been shown that an APD system can temporarily amplify a first L or D excess, but no stable symmetry breaking can be observed, as the fixed point remains racemic. The addition of the last type of reaction, epimerization, is then introduced in the model in order to see how the symmetry breaking can be obtained. This new reaction type aims at favoring the appearance of homopeptides so as to, indeed, amplify an initial unbalanced situation. For simplicity, only one reaction of epimerization is introduced, applied exclusively on LDL and DLD tripeptides, all other epimerization reactions still being neglected.

$$ LDL \xrightarrow{k_e} LLL \qquad DLD \xrightarrow{k_e} DDD \qquad (16) $$

As Eq. 16 is considered with equal constant rates k_e for both reactions, the whole system remains symmetric. Supposing that homochiral peptides are more stable than heterochiral peptides, these reactions should be faster than the reverse reactions.

These epimerization reactions may have very interesting potential in an APD system, as shown Eq. 17 for LDL epimerization to LLL. Embedded in such a system, LDL comes from a multi-step activation/polymerization of two L and one D, and LLL is to be hydrolyzed into three L. This set of reactions can then be reduced to a conversion of D to L, consuming and regenerating two L, that is an autocatalytic inversion of D. In this scope, if this path can become more efficient than the non-catalyzed amino acid inversion, the whole racemic system may become unstable.

$$
\begin{array}{llll}
2\,L+D & \rightarrow & LDL \rightarrow LLL \rightarrow & 3\,L \\
2\,L+D & \xrightarrow{\hspace{4cm}} & & 2\,L+L
\end{array} \qquad (17)
$$

The simulation is done starting with racemic X*. As seen in Fig. 5, the system remains racemic for a long time, when

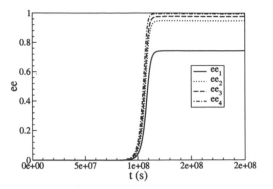

Figure 5: Evolution of ee_n in an APD system in presence of selective epimerization of peptides ($ee^{\text{init}} = 0$). Same parameters as in Fig. 4 but initial conditions, $[L^*]_0 = 1 \cdot 10^{-2}$ M; $[D^*]_0 = 1 \cdot 10^{-2}$ M; and with addition of $k_e = 1 \cdot 10^{-3}$ s^{-1}.

an exponential raising of L excess occurs, driving the concentrations to a non-racemic state, with high global L excess. This time, the racemic state is unstable, and the fixed points are non-racemic.

Several systems have been studied for different k_e values. For low values, the system always remains racemic (Fig. 6). For such cases, the autocatalytic inversion of amino acids is slower than the uncatalyzed inversion: the racemic state is stable. But up to a critical value of k_e, a bifurcation occurs, and strong L or D excesses are obtained. The system switches from a globally stable to a locally stable one, in which a certain memory of the initial condition is preserved. Whatever the amount of initial excess of L or D, the final homochiral situation will conserve the sign of the excess. For such cases, the autocatalytic inversion of amino acids becomes faster than the non-catalytic one. This bifurcation phenomenon turns out to be similar to the Frank's model.

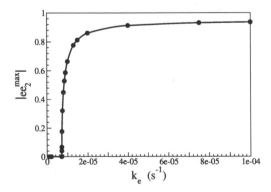

Figure 6: Bifurcation diagram of an APD system in presence of selective peptide's epimerization: variation of final $|ee_2|$ as a function of k_e. Same parameters as in Fig. 5 but k_e.

This critical value is still quite high compared with kinetic rates expected in mild conditions (Schroeder and Bada,

1976). However, this critical value is strongly dependent on a huge number of parameters that were not investigated, and effective epimerization rates may have been much higher in more drastic conditions (Huber et al., 2003).

So, while the basic racemization is opposed to the spontaneous emergence of homochirality, this new contextual form of it, favoring homopeptides, is the simplest autocatalytic effect which can lead to stable and attracting homochirality.

Conclusion

A first key distinction often misunderstood in biological and chemical literature needs to be clarified. The amplification of the initial excess obtained in the first simulations described in the paper is a different dynamical phenomenon from the autocatalytic rupture of symmetry obtained in the last simulation. In the last case, the dynamics is dissipative meaning that the final concentration values are independent of the initial values, and even a simple statistical fluctuation could cause homochirality. The previous simulations have been criticized as a possible explanation of homochirality, as they required a sufficient initial excess, and were wrongly generalized to the second set of simulations, in which the state of homochirality is qualitatively but not anymore quantitatively dependent on the initial conditions.

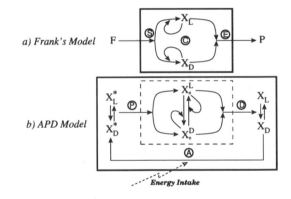

Figure 7: Spontaneous symmetry breaking in nonequilibrium chemical systems. (a) Frank's model: open-flow system, based on synthesis of chiral compounds from achiral compounds Ⓢ, autocatalysis of this synthesis Ⓒ, and elimination of the chiral compounds Ⓔ. (b) APD model: closed system, based on incoming energy flux through activation Ⓐ, complexification of the system with polymerization Ⓟ, and matter recycling with depolymerization Ⓓ.

While Frank's model is an excellent and very analytically detailed illustration of this rupture of symmetry, its prebiotical relevance can be questioned. Our model described in this paper tries to reconcile Frank's hypothesis on the origin of homochirality with the more likely chemical environment of the prebiotic world. It appeared that the polymerization/depolymerization/epimerization system presents implicit autocatalytic reactions, thanks to the synergy between all the reactions. This system is similar to the Frank's model, able to amplify asymmetry by consuming activated amino acids by polymerization, and generating non-reactive amino acids by depolymerization (Fig. 7).

The originality of the APD system is to embed this Frank-like open system in a closed system, by recycling materials thanks to energy intake, maintaining the nonequilibrium state. This model no more describes the synthesis of non-racemic chiral compounds from an achiral system, but the conversion of a preexistent racemic mixture of chiral compounds to a non-racemic mixture, and may constitute a "minimum connected, reflexively autocatalytic set" as defined by Steel (Steel, 2000). This system appears to be more robust than classical Frank-like models in a prebiotic perspective. The inversion reactions are no more disadvantages for long-term stability of the non-racemic states (Kondepudi and Asakura, 2001), as they on the contrary become the source of homochirality. The non-racemic state can thus spontaneously emerge, and become long-term stable.

Acknowledgments

This work was realized in collaboration between our respective laboratories within the framework of the COST D27 European Program.

Moreover, we have to thank Dominique Lavabre and Jean-Claude Micheau from IMRCP in Toulouse, FRANCE, for Oa3 program and their helpful advice about its use.

References

Asakura, K., Ikumo, A., Kurihara, K., Osanai, S., and Kondepudi, D. K. (2000). Random chiral asymmetry generation by chiral autocatalysis in a far-from-equilibrium reaction system. *J. Phys. Chem. A*, 104:2689–2694.

Avetisov, V. V. and Goldanskii, V. I. (1991). Homochirality and stereospecific activity: Evolutionary aspects. *Biosystems*, 25(3):141–149.

Blair, N. E. and Bonner, W. A. (1981). A model for the enantiomeric enrichment of polypeptides on the primitive earth. *Orig. Life*, 11(4):331–335.

Bonner, W. A. (1999). Chirality amplification - the accumulation principle revisited. *Orig. Life and Evol. Biosphere*, 29:615–623.

Commeyras, A., Collet, H., Boiteau, L., Taillades, J., Vandenabeele-Trambouze, O., Cottet, H., Biron, J.-P., Plasson, R., Mion, L., Lagrille, O., Martin, H., Selsis, F., and Dobrijevic, M. (2002). Prebiotic synthesis of sequential peptides on the hadean beach by a molecular engine working with nitrogen oxides as energy sources. *Polymer International*, 51:661–665.

Decker, P. (1974). The origin of molecular asymmetry through the amplification of "stochastic information"

(noise) in bioids, open systems wich can exist in several steady states. *J. Mol. Evol.*, 4:49–65.

Frank, F. C. (1953). Spontaneous asymmetic synthesis. *Biochem. Biophys. Acta*, 11:249.

Hitz, T. and Luisi, P. L. (2003). Chiral amplification of oligopeptides in the polymerization of α-amino acid N-carboxyanhydrides in water. *Helv. Chim. Acta*, 86(5):1423–1434.

Huber, C., Eisenreich, W., Hecht, S., and Wächterhäuser, G. (2003). A possible primordial peptide cycle. *Science*, 301:938–940.

Iwamoto, K. (2003). Spontaneous appearance of chirally asymmetric steady states in a reaction model including michaelis-menten type catalytic reactions. *Phys. Chem. Chem. Phys.*, 5:3616–3621.

Kaps, P. and Rentrop, P. (1984). Application of a variable-order semi-implicit Runge-Kutta method to chemical models. *Comp. Chem. Engineering*, 8:393–396.

Kondepudi, D. K. and Asakura, K. (2001). Chiral autocatalysis, spontaneous symmetry breaking, and stochastic behavior. *Acc. Chem. Res.*, 34:946–954.

Kondepudi, D. K., Kaufman, R., and Singh, N. (1990). Chiral symmetry breaking in sodium chlorate crystallisation. *Science*, 250:975–976.

Kondepudi, D. K. and Nelson, G. W. (1984). Chiral symmetry breaking states and their sensitivity in nonequilibrium chemical systems. *Physica*, 125A:465–496.

Plasson, R., Biron, J.-P., Cottet, H., Taillades, J., and Commeyras, A. (2002). Kinetic study of the polymerization of α-amino acid N-carboxyanhydrides in aqueous solution using capillary electrophoresis. *J. Chromatogr. A*, 952:239–248.

Schroeder, R. A. and Bada, J. L. (1976). A review of the geochemical applications of the amino acids racemization reaction. *Earth-Science Rev.*, 12:347–391.

Soai, K., Sato, I., and Shibata, T. (2001). Asymmetric Autocatalysis and the Origin of Chiral Homogeneity in Organic Compounds. *The Chemical Record*, 1:321–332.

Steel, M. (2000). The emergence of a self-catalysing structure in abstract origin-of-life models. *Appl. Math. Lett.*, 13:91–95.

Taillades, J., Collet, H., Garrel, L., Beuzelin, I., Boiteau, L., Choukroun, H., and Commeyras, A. (1999). N-carbamoyl amino acid solid-gas nitrosation by NO/NOₓ: A new route to oligopeptides via α-amino acid N-carboxyanhydride. Prebiotic implications. *J. Mol. Evol.*, 48:638–645.

An Evolvable Artificial Chemistry
Featuring Continuous Physics and Discrete Reactions

Thomas E. Portegys

Illinois State University
Campus Box 5150
Normal, Illinois, USA 61790-5150
portegys@ilstu.edu

Abstract

This paper describes an artificial chemistry featuring atoms and molecules moving and colliding in a continuous manner in a viscous fluid filling a 2D cellular space. Chemical reactions are mappings of discrete cellular configurations to parameterized actions on atoms. Actions allow atom creation and destruction, bonding and unbonding to make and break molecules, orientation, type change, and propulsion. Actions are easily added in this extensible model. An example involving a complex "foraging" reaction is provided as a demonstration of the capabilities of the framework. The reaction rules can be evolved by a genetic algorithm to exhibit a desired set of reactions. A portion of the foraging reaction was evolved to demonstrate this.

Introduction

This work was done in the spirit of the growing belief that artificial life requires a sufficiently rich physics and chemistry in which to develop (Bedau et al. 2000, Dittrich et al. 2001). The attempt here is to combine a simple continuous Newtonian particle physics with the well-known and readily computable cellular automaton (CA) model as a chemistry implementation. There are two additional purposes for using a CA: (1) the chemistry of real organic molecules, involving complex foldings and bondings, is computationally infeasible (Zagrovic et al. 2002); and (2) it serves the aim of artificial life research to discover underlying mechanisms necessary for the existence of living systems.

This project was primarily inspired by three recent efforts: Hutton's artificial chemistry Squirm3 (2002), Smith, Turney and Ewaschuk's JohnnyVon (2003), and my prior work on intercellular signaling (Portegys 2002). Prior to these, Dittrich et al. (2001) compared a wide range of artificial chemistry approaches, including assembler automata (Rasmussen et al. 1990, Ray 1992, Adami and Brown 1994), Ono and Ikegami's autocatalytic membrane formation (1999), and lattice molecular systems (McMullin and Varela 1997), in which the atoms comprising a molecule map discretely to cellular space.

Squirm3 is a lattice molecular system in which mobile molecules self-replicate using available atoms in a 2D

cellular space. Atoms have a fixed type and a variable state. Chemical reaction rules based on type, state and proximity determine the states and bonding status of atoms. Atoms move by jumping from cell to cell through simulated Brownian motion; however, atoms bound into molecules are largely immobile. Beginning with a soup of inert atoms exposed to state-disrupting cosmic rays, self-replication was shown to spontaneously occur.

In JohnnyVon, by contrast, T-shaped atoms called codons move through a continuous 2D space and interact exclusively through force fields; collisions do not occur. There are two types of codons, distinguished by their field "colors". Each codon is an automaton containing a set of rules governing its field strengths in response to signals represented by the proximity of other codon fields.

This project uses the intercellular signaling scheme put forth in my recent work addressing the problem of allowing cells in a CA to communicate without disturbing the state of intervening cells. In this project, signals are chiefly directed toward atoms residing within cells, rather than the cells themselves.

Description

This system is an artificial chemistry featuring atoms and molecules moving and colliding in a continuous manner in a viscous fluid filling a 2D cellular space. A molecule consists of a bound set of atoms. Chemical reactions are mappings of discrete cellular configurations to parameterized actions on atoms. Actions allow atom creation and destruction, bonding and unbonding to make and break molecules, orientation, type change, and propulsion. Time proceeds in discrete steps.

Atoms and molecules

Atoms are elementary particles possessing a type, mass, radius, charge, orientation, position, and velocity. A molecule is a set of atoms connected by bonds. The atoms within a molecule must form a connected set; there must be a bond path directly or indirectly connecting any pair of atoms. A bond may be of variable but limited length.

Forces resulting from elastic collisions and chemical reactions result in continuous velocity and position updates according to Newtonian physics. For the initial implementation, rotational momentum is not supported. The atoms within a molecule respond to forces as a single unit; so a collision of one of its atoms results in a force applied to the mass of the entire molecule. An atom is oriented in one of the eight compass directions. It may also be in a mirrored state to support symmetric reactions.

Space

The space is a 2D cellular grid filled with a viscous fluid that impedes the movement of atoms through it. In the current implementation, the diameter of an atom is equal to the dimensions of a cell.

Reactions

Reactions arise from the interaction of atoms. At any moment, the center of an atom resides within a unique cell. This cell forms the center of a 3x3 Moore neighborhood of cells. A reaction specifies a configuration of atom types residing within a neighborhood. Since atoms have an orientation, their cell neighborhoods are oriented accordingly, as shown in Figure 1.

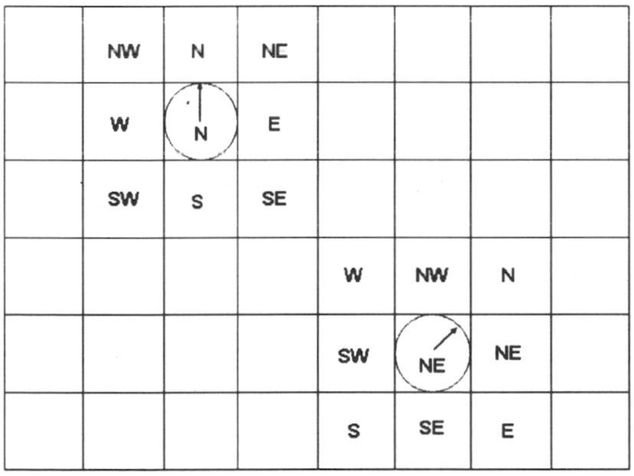

Figure 1: Oriented Cell Neighborhoods

A neighborhood cell value is one of {atom type, empty, occupied, ignore}. The neighborhood evaluation is a conjunction of all 9 value matches. An empty value specifies that a cell must not contain any atom. This could be used for inhibitory control of a reaction. The occupied value matches any atom type. The ignore value positively matches any cell condition.

A reaction consists of the following: {neighborhood, action, target, parameters}. Target is the location of the cell where the action is directed via a signal. This may reside outside of the neighborhood. Parameters apply to actions. These are: {atom type, orientation, strength, tendency, delay, duration}. The actions are:

- Create and bond atom. Parameters: type and orientation of created atom.
- Destroy atom. Parameter: target atom type.
- Bond/unbond acting atom to/from atom. Parameter: target atom type.
- Grapple atom: bond to and move atom to location in acting atom's neighborhood. Parameters: target atom type, orientation relative to neighborhood center cell.
- Orient atom relative to acting atom's orientation. For example, if acting atom is oriented east, and it orients an atom to its west, the atom will acquire a north orientation. Parameters: target atom type and orientation.
- Modify atom type. Parameter: target atom old and new types.
- Propel atom: apply a propulsion force to a specific atom (see below).

Propel atom action:
This is a reaction in which atoms apply propulsion forces to themselves and other atoms, which when combined with the viscous fluid medium gives the chemistry a more active and interesting nature, albeit at the expense of real world fidelity. More natural alternatives would be to employ charge forces or Brownian motion to achieve coarser and slower movements and reactions. The parameters for the propel action are: target atom type, force direction (orientation), speed (strength), tendency (relative probability), delay, and duration. Since the atoms within a molecule move as a unit, the propulsion forces are combined at a molecular level. Propulsion forces are accumulated with collision forces to update the positions and velocities of molecules.

Update cycle

1. Update positions and velocities of molecules based on accumulated forces.
2. Clear forces.
3. Compute new collision forces.
4. Match reactions to atom neighborhoods and distribute action signals to target cells.
5. Execute actions.

Results

This section presents a demonstration of the capabilities of the framework. Both Squirm3 and JohnnyVon solved the challenge of self-replication. For the sake of variety, a different problem is ventured here that I propose is of comparable complexity. The problem involves an orchestration of cooperating reactions that allows a particular molecule to randomly encounter and systematically destroy atoms of a particular type. The overall impression is one of 'foraging', a metaphor that will be used in the following description.

The molecule, called 'Maxwell' (for whimsical reasons) is shown in Figure 2.

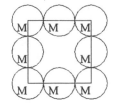

Figure 2: The Maxwell Molecule

This molecule is a ring of bonded atoms. Its atoms are labeled with 'M' in order to distinguish atoms 'belonging' to Maxwell from others. However, the corner atoms are of a different type than the side atoms. In addition, the orientations of all the atoms point directly away from the center, so for example the top center atom is oriented north, the top right atom northeast, etc. This allows the same reaction rules to work regardless of the molecule's relative position to external atoms. One set of reactions is structured to allow the Maxwell molecule to destroy 'food' atoms by "ingesting" them. In doing so, a variety of reaction types are brought into play.

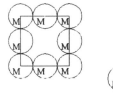

Figure 3: Beginning Ingestion Reaction

Figure 3 show the beginning of the ingestion reaction sequence as Maxwell approaches a food atom.

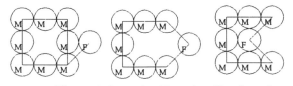

Figure 4: Ingestion Reaction Continued

Figure 4 show the next three steps. Proceeding from left to right: the lower right corner atom grapples the food atom, moving it counterclockwise over the side atom; the side atom has destroyed itself to create a pathway into the center, and the other nearby corner atom has bonded to the food atom; the food atom is grappled into the center.

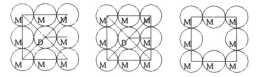

Figure 5: Ingestion Reaction Completion

The reaction completes in Figure 5 with the restoration of the side atom and rebonding of the molecule. The process is facilitated by a type change to the center food atom before it is destroyed. There are 21 reactions involved in the entire process.

Maxwell will systematically ingest food atoms it comes into contact with. The propel reaction type allows it to move about, or 'forage', in a space that is stocked with variable-sized patches of food atoms.

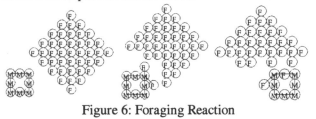

Figure 6: Foraging Reaction

Figure 6 shows Maxwell ingesting a food patch it has come into contact with. In the "programmed" version of foraging, it will move in a spiral counter-clockwise pattern around the perimeter of the patch, usually resulting in the complete ingestion of it. Figure 7 shows the finale of the process.

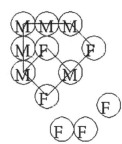

Figure 7: Foraging Reaction Finale

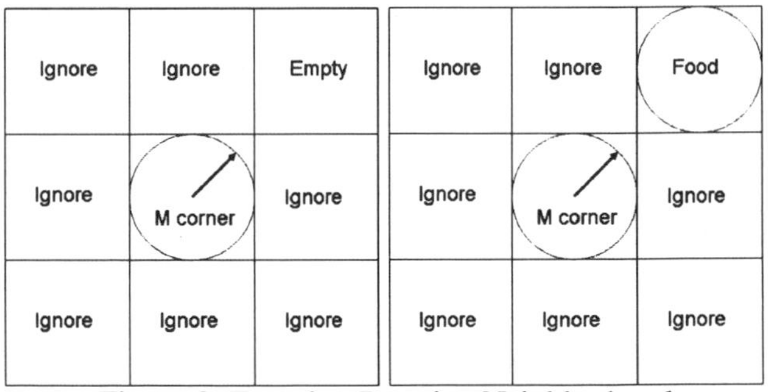

Ignore	Ignore	Empty	Ignore	Ignore	Food
Ignore	M corner	Ignore	Ignore	M corner	Ignore
Ignore	Ignore	Ignore	Ignore	Ignore	Ignore

Figure 8: Foraging Reaction Neighborhoods

Three propel reactions are involved in foraging. The left portion of Figure 8 shows the neighborhood of the first propel reaction that produces movement in free space (no food present). The neighborhood specifies a corner atom with an empty cell directly "north" of it. Table 1 shows the other attributes of this reaction. The direction and strength values translate to a fairly high rate of speed moving the entire molecule in a northeast direction. The tendency value indicates that, in the absence of other propel reactions, there is a 10% probability of this reaction occurring. This is important considering that this reaction applies to all four corners of the molecule. If all "fired" at every step, the propulsions would cancel and the molecule would not move. As it is, the result is a random Brownian-type motion with fairly long legs.

Attribute	Value
target type	corner
displacement	0,0
direction	north
strength	2.0
tendency	0.1
delay	0
duration	1

Table 1: First Foraging Reaction

The neighborhood in the right portion of Figure 8 applies to the second and third reactions. The second reaction attribute values are given in Table 2. This reaction moves a corner atom east when a food atom is north of it, which results in the molecule moving along the side of a food patch. The strength value causes it to move slowly, lest the food be knocked away by impact.

Attribute	Value
target type	corner
displacement	0,0
direction	east
strength	0.1
tendency	5.0
delay	0
duration	20

Table 2: Second Foraging Reaction

The tendency and duration of this reaction causes it to heavily outweigh and outlast the first reaction, so that an intermittent exposure to free space will not result in the molecule darting away from the food patch.

Attribute	Value
target type	corner
displacement	0,0
direction	north
strength	0.1
tendency	3.0
delay	20
duration	50

Table 3: Third Foraging Reaction

The third reaction is given in Table 3. The purpose of this reaction is to steer the molecule around food patch corners. The delay and duration values allow this reaction to remain in effect for a relatively long time after running off the edge of the patch. It will subsequently cause the atom to

move north, resulting in the corner atom counter-clockwise from it to "engage" the next edge using the second reaction.

Evolving the chemistry

In the previous discussion, the foraging reactions were programmed to perform suitably. Two additional experiments were done using a genetic algorithm to evolve the three reactions pertaining to foraging. For the evolution procedure, a population member consisted of a set of foraging reactions. The population size was twenty, with the ten fittest members selected out of each generation as measured by foraging success. Foraging success is defined by the number of food atoms ingested times a value of ten each. A set of eight randomly selected fit members were mutated, the nature of which depended on the experiment (see below). An additional two fit members were selected for mating. Mating involved randomly selecting one reaction from each of the three parental pairs of foraging reactions.

In the first experiment (hill-climb), the action, neighborhood, target type, and displacement attributes of the reactions were fixed at the programmed values, while the direction, strength, delay, and duration attributes were set to minimum values and subjected to a simple hill-climbing mutation procedure. The aim of this experiment was to "tune" the propulsion parameters.

In the second experiment (random), the reactions were initially rendered inert by clearing them to minimum values and setting the action to a null value. Mutation consisted of randomly setting the various attributes to values within set tolerances. The exception to this was that when the action was mutated, the rest of the attributes were again cleared.

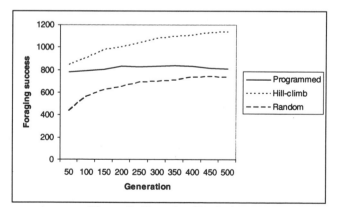

Figure 9: Foraging Evolution

Figure 9 shows the results of the experiments. The programmed (unevolved) version was run for the same number of generations and is shown as a control. The hill-climbing experiment proved to be an immediate success. Upon closer examination of the fittest members, it became obvious that the programmed foraging speed was too slow, especially the reactions controlling speed in the presence of food atoms. The random experiment quickly produced a functional set of reactions, but failed to improve dramatically after that. An inspection of the reactions of the fittest members revealed a basic propel reaction triggered by the presence of free space. Surprisingly, this was enough to approach the performance of the programmed version.

Conclusions

One of the aims of this project was to find a way to acquire some of the benefits of a continuous physical medium, such as that in which organic chemistry takes place, and also retain the computability benefits of a cellular automaton as a chemistry framework. The Maxwell molecule was designed to showcase some of the capabilities of this approach. The framework is easily extensible: new reaction types can be plugged in with few code changes.

This model has some obvious similarities and differences with Squirm3 and JohnnyVon; the former is CA based, and latter is based on continuous fields and movements. At this point it would be premature to attempt more definitive comparisons, since all are in early stages of development.

Besides further applications, such replicating molecules and semi-permeable membranes, there are a number of relatively easy enhancements that can be made to make the model more realistic, such as going to 3D (the physics code already supports this), adding rotation/angular momentum effects, and introducing charge as a source of attraction and repulsion force.

The open source C++ code is available at www.itk.ilstu.edu/faculty/portegys/research.html

References

Adami, C.; and Brown, C.T. 1994. Evolutionary learning in the 2D artificial life system avida. In Brooks, R.A. and Maes, P., editors, Artificial Life IV, Cambridge MA, MIT Press, 377-381.

Dittrich, P.; Ziegler, J.; and Banzhaf, W. 2001. Artificial chemistries – A review. Artificial Life 7:2245-275

Hutton, T. J. 2002. Evolvable Self-Replicating Molecules in an Artificial Chemistry. Artificial Life 8(4):341-356.

McMullin, B. and Varela, F.J. 1997. Rediscovering computational autopoiesis. In Husbands, P. and I. Harvey, editors, Fourth European Conference on Artifcial Life, pages 38-47, Cambridge, MA, MIT Press.

Ono, N. and Ikegami, T. 1999. Model of self-replicating cell capable of self-maintenance. In Floreano, D., J.-D. Nicoud, and F. Mondana, editors, Proc. Fifth European Conference on Artifcial Life (ECAL' 99), pages 399406, Berlin, Springer.

Portegys, T. E. 2002. An Abstraction of Intercellular Communication", in Standish, R.K., Bedau M.A. and Abbass, H.A., editors, Artificial Life VIII Proceedings, Cambridge, MA, MIT Press, 75-78.

Rasmussen, S., C. Knudsen, R. Feldberg, and M. Hindsholm. 1990. The coreworld: Emergence and evolution of cooperative structures in a computational chemistry. Physica D, 42:111-134.

Ray, T. S. An approach to the synthesis of life. In Langton, C. G., C. Taylor, J. D. Farmer, and S. Rasmussen, editors, Artifcial Life II, pages 371-408, Redwood City, CA, Addison-Wesley.

Smith, A.; Turney, P.; and Ewaschuk, R. 2003. Evolvable Self-Replicating Molecules in an Artificial Chemistry. Artificial Life 9(1):21-40.

Zagrovic, B.; Snow, C.D.; Shirts, M.R.; and Pande, V.S. 2002. Simulation of Folding of a Small Alpha-helical Protein in Atomistic Detail using Worldwidedistributed Computing. Journal of Molecular Biology.

The Role of RNA Editing in Dynamic Environments

Luis M. Rocha and Chien-Feng Huang

Modeling, Algorithms and Informatics Group (CCS-3)
Computer and Computational Sciences Division
Los Alamos National Laboratory, MS B256
Los Alamos, NM 87545, USA
rocha@lanl.gov and cfhuang@lanl.gov

Abstract

This paper presents a computational methodology based on Genetic Algorithms with Genotype Editing (GAE) for investigating the role of RNA editing in dynamic environments. This model is based on genotype editing characteristics that are gleaned from RNA editing processes as observed in several organisms. We have previously expanded the traditional Genetic Algorithm (GA) with artificial editing mechanisms (Rocha, 1995, 1997), and studied the benefits of including straightforward Genotype Editing in GA for several machine learning problems (Huang and Rocha, 2003, 2004). Here we show that genotype editing also provides a means for artificial agents with genotype/phenotype mappings descriptions to gain greater phenotypic plasticity. We simulate agents endowed with the ability to alter the edition of their genotype according to environmental context. This ability grants agents an adaptive advantage as genotype expression can become contextually regulated. The study of this genotype edition model in changing environments has shed some light into the evolutionary implications of RNA editing. We expect that our methodology will both facilitate determining the evolutionary role of RNA editing in biology, and advance the current state of research in Evolutionary Computation and Artificial Life.

1. RNA Editing

Evidence for the important role of non-protein coding RNA (ncRNA) in complex organisms (higher eukaryotes) has accumulated in recent years. "ncRNA dominates the genomic output of the higher organisms and has been shown to control chromosome architecture, mRNA turnover and the developmental timing of protein expression, and may also regulate transcription and alternative splicing." (Mattick, 2003, p 930).

RNA Editing (Benne, 1993; Bass, 2001), a process of post-transcriptional alteration of genetic information prior to translation, can be performed by ncRNA structures (though it can also be performed by proteins). The term initially referred to the insertion or deletion of particular bases (e.g. uridine), or some sort of base conversion. Basically, RNA Editing instantiates a non-inheritable stochastic alteration of genes, which is typically developmentally and/or environmentally regulated to produce appropriate phenotypical responses to different stages of development or states of the environment.

The most famous RNA editing system is that of the African Trypanosomes (Benne, 1993; Stuart, 1993). Its genetic material was found to possess strange sequence features such as genes without translational initiation and termination codons, frame shifted genes, etc. Furthermore, observation of mRNA's showed that many of them were significantly different from the genetic material from which they had been transcribed. These facts suggested that mRNA's were edited post-transcriptionally. It was later recognized that this editing was performed by guide RNA's (gRNA's) coded mostly by what was previously thought of as non-functional genetic material (Sturn and Simpson, 1990). In this particular genetic system, gRNA's operate by inserting, and sometimes deleting, uridines. To appreciate the effect of this edition let us consider Figure 1. The first example (Benne, 1993, p. 14) shows a massive uridine insertion (lowercase u's); the amino acid sequence that would be obtained prior to any edition is shown on top of the base sequence, and the amino acid sequence obtained after edition is shown in the gray box. The second example shows how, potentially, the insertion of a single uridine can change dramatically the amino acid sequence obtained; in this case, a termination codon is introduced.

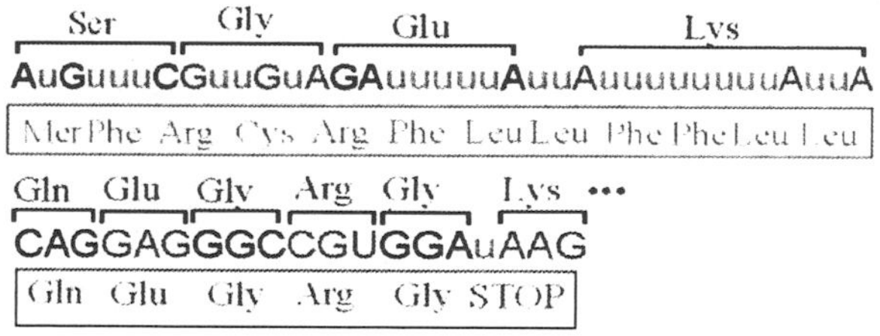

Figure 1. U-insertion in Trypanosomes' RNA

The importance of RNA Editing is thus unquestionable, since it has the power to dramatically alter gene expression: "cells with different mixes of (editing mechanisms) may edit a transcript from the same gene differently, thereby making different proteins from the same opened gene." (Pollack, 1994, P. 78). It is important to retain that, at least for certain RNA Editing mechanisms such as U-Insertion, a mRNA molecule can be more or less edited according to the concentrations of the editing operators it encounters. Thus, several different proteins coded by the same gene may coexist in an organism or even a cell, if all (or some) of the mRNA's obtained from the same gene, but edited differently, can be translated.

If the concentrations of editing operators can vary according to environmental contexts, different resulting phenotypes may be selected accordingly, and thus evolve a system which is able to respond to environmental changes without changes in the major part of its genetic information -- one genotype, different contexts, different phenotypes. Notice, however that what is inheritable, and subjected to variation, is the original non-edited genotype, which is ultimately selected and transmitted to the offspring of the organism (Rocha, 1995; 1997). This type of phenotypic plasticity may be precisely, for instance, what the Trypanosome parasites have achieved: control over gene expression during different parts of their complex life cycles.

The role of RNA editing in the development of more complex organisms has also been shown to be important. Lomeli et al. (1994) discovered that the extent of RNA editing affecting a type of receptor channel responsible for the mediation of excitatory postsynaptic currents in the central nervous system, increases in rat brain development. As a consequence, the kinetic aspects of these channels differ according to the time of their creation in the brain's developmental process. Another example is that the development of rats without a gene (ADAR1) known to be involved in RNA editing, terminates midterm (Wang et al., 2000). This showed that RNA Editing is more prevalent and important than previously thought. RNA editing processes have also been identified in mammalian brains (Simpson and Emerson, 1996). More recently, Hoopengardner et al. (2003) found that RNA editing plays a central role in nervous system function. Indeed, many edited sites alter conserved and functionally important amino acids, some of which may play a role in nervous system disorders such as epilepsy and Parkinson Disease.

2. Introducing Editing in Genetic Algorithms

Genetic Algorithms (GA) (Holland, 1975) have been used as computational models of natural evolutionary systems and as adaptive algorithms for solving optimization problems. GA operate on an evolving population of artificial organisms, or agents. Each agent is comprised of a genotype (encoding a solution to some problem) and a phenotype (the solution itself). Evolution occurs by iterated stochastic variation of genotypes, and subsequent selection of the best phenotypes in an environment – that is, according to how well the respective solution solves a problem (or fitness function). Table 1 depicts the process of a simple genetic algorithm.

The essence of GA lies on the separation of the description of a solution (the Genotype) from the solution itself (the Phenotype): variation is applied solely to the descriptions, while the respective solutions are evaluated,

and the whole selected according to this evaluation. Nonetheless, one important difference between evolutionary computation and biological organisms lies precisely on the relation between Genotype and Phenotype. In GA, typically, the relation between the two is linear and direct: one genotype produces a unique phenotype. In contrast, in biological organisms there exists a multitude of processes, taking place between the transcription of genes and their expression and subsequent development into a phenotype, responsible for the establishment of an uncertain, contextually regulated relation, between Genotype and Phenotype.

Table 1. Mechanism of a simple GA

1. Randomly generate an initial population of l n-bit agents, each defined by a genotype string of symbols from {0, 1}.
2. Evaluate each agent's (phenotype) fitness.
3. Repeat until l offspring agents have been created.
   a. select a pair of parent agents for mating;
   b. apply crossover operator to genotype string;
   c. apply mutation operator to genotype string.
4. Replace the current population with the new population.
5. Go to Step 2 until terminating condition.

In other words, the same genotype does not always produce the same phenotype; rather, many phenotypes can be produced from one genotype depending on states of the environment. One of the biological processes responsible for such phenotypic plasticity is RNA Editing.

In analogy with the process of RNA Editing, Rocha (1995; 1997) proposed an expanded GA with stochastic edition of genotypes (chromosomes), prior to translation into phenotypes. Here we present novel experiments to show how this GA with Genotype Editing can be successfully used to model the environmentally-regulated control of gene expression achieved by RNA Editing in real organisms.

Genotype Editing (Rocha, 1995; Huang and Rocha, 2003, 2004) is implemented by a set of editors with different editing functions, such as insertion or deletion of symbols in the original genotypes. Before genotypes can be translated into the space of phenotype solutions, they must "pass" through successive layers of editors, present in different concentrations. In each generation, each genotype encounters an editor in its layer with probability (given by the concentrations). If an editor matches some subsequence of the genotype when they encounter each other, the editor's function is applied and the genotype is edited. The detailed implementation of the simplest GA with Edition (GAE) is described in the following:

The GAE model consists of a family of r m-bit strings, denoted as $(E_1, E_2, ..., E_r)$, which is used as the set of editors for the genotypes of the agents in a GA population.

The length of the editor strings is assumed much smaller than that of the genotypes: $m \ll n$, usually an order of magnitude. An editor E_j is said to match a substring, of size m, of a genotype string, S, at position k if $e_i = s_{k+i}$, $i=1,2, \ldots, m$, $1 \le k \le n-m$, where e_i and s_i denote the i-th bit value of E_j and S, respectively. For each editor E_j, there exists an associated editing function F_j that specifies how a particular editor edits genotypes: when the editor matches a portion of a genotype string, a number of bits are inserted into or deleted from the genotype string.

For instance, if the editing function of editor E_j is to add one randomly generated allele at s_{k+m+1} when E_j matches S at position k, then all alleles of S from position $k+m+1$ to $n-1$ are shifted one position to the right (the allele at position n is removed). Analogously, if the editing function of editor E_j is to delete an allele, this editor will instead delete the allele at s_{k+m+1} when E_j matches S at position k. All the alleles after position $k+m+1$ are shifted in the inverse direction (one randomly generated allele is then assigned at position n).

Finally, let the concentration of the editor family be defined by $(v_1, v_2, \ldots, v_r)$. This means that the concentration of editor E_j is denoted by v_j: the probability that S encounters E_j. With these settings, the algorithm for the GA with genotype editing is essentially the same as the regular GA, except that step 2 in Table 1 is now redefined as:

"For each agent in the GA population, apply each editor E_j with probability v_j (i.e., concentration). If E_j matches the agent's genotype string S, then edit S with editing function F_j and evaluate the resulting agent's fitness."

It is important to notice that the "post-transcriptional" edition of genotypes is not a process akin to mutation, because editions are not inheritable. Just like in biological systems, it is the unedited genotype that is reproduced. One can also note that Genotype Editing is not a process akin to the Baldwin effect as studied by, e.g., Hinton and Nowlan (Hinton and Nowlan, 1987). The phenotypes of our agents with genotype edition do not change (or learn) ontogenetically. In Hinton and Nowlan's experiments, the environment is defined by a very difficult ("needle in a haystack") fitness function, which can be made more amenable to evolutionary search by endowing the phenotypes to "learn" ontogenetically. Eventually, they observed, this learning allows genetic variation to discover, and genetically encode fit individuals. In contrast, genotype edition does not grant agents more "ontogenetic learning time", it simply changes inherited genetic information ontogenetically but the phenotype, once produced, is fixed. Also, as we show below, it is advantageous in environments very amenable to evolution, such as Royal Road functions (the opposite of "needle in a haystack") (Huang and Rocha, 2003, 2004).

It is also important to retain that just like an mRNA molecule may be edited in different degrees according to the concentrations of editing operators it encounters, in the GAE the same genotype string may be edited differently because editor concentration is a stochastic parameter that specifies the probability of a given editor encountering a chromosome. Thus, if a genotype string is repeated in the population, it may actually produce different solutions (or phenotypes). This is akin to what happens with RNA editing in biological organisms where, at the same time, several different proteins coded by the same gene may coexist.

In (Huang and Rocha, 2003, 2004), we have conducted a systematic study of the GAE in several static environments to investigate if there are any evolutionary advantages of genotype editing, even without control of environmental changes. We demonstrated that genotype editing can improve the GA's search performance by suppressing the effects of hitchhiking. We have also showed that editing frequency plays a critical role in the evolutionary advantage provided by the editors -- only a moderate degree of editing processes facilitates the exploration of the search space. Therefore, one needs to choose proper editor parameters to avoid over or under-editions in order to develop more robust GAs. Here, we extend our study of the GAE to dynamic problems by linking concentrations of editors to environmental states (or contexts) – thus allowing editor concentrations to serve as a control switch for environmental changes.

3. Evolution in Dynamic Environments

How rapid is evolutionary change, and what determines the rates, patterns, and causes of change, or lack thereof? Answers to these questions can tell us much about the evolutionary process. The study of evolutionary rate in the context of GA usually involves defining performance measures that embody the idea of rate of adaptation, so that its change over time can be monitored for investigation.

In this paper, two evolutionary measures, the maximum fitness and the population fitness at each generation, are employed.[1] To understand how Genotype Editing works in the GAE model, we employ a testbed, the small Royal Road $S1$ (Huang and Rocha, 2003) due to its simplicity for tracing evolutionary advancement.

Table 2 illustrates the schematic of the small Royal Road function $S1$. This function involves a set of schemata $S = (s_1, \ldots, s_8)$ and the fitness of a bit (genotype) string x is defined as

$$F(x) = \sum_{s_i \in S} c_i \sigma_{s_i}(x),$$

where each c_i is a value assigned to the schema s_i as defined in the table; $\sigma_{s_i}(x)$ is defined as 1 if x is an instance of s_i and 0 otherwise. In this function, the fitness of the global optimum string (40 1's) is $10*8 = 80$.

---

[1] The maximum fitness is the fitness of the best individual in the current population; the population fitness here is defined as the value obtained by averaging the fitness of all the individuals in the current population.

Table 2. Small royal road function $S1$

$$s_1 = 11111\times : c_1 = 10$$
$$s_2 = \times\times\times\times\times11111\times : c_2 = 10$$
$$s_3 = \times\times\times\times\times\times\times\times\times11111\times : c_3 = 10$$
$$s_4 = \times\times\times\times\times\times\times\times\times\times\times\times\times11111\times\times\times\times\times\times\times\times\times\times\times\times\times\times\times\times\times : c_4 = 10$$
$$s_5 = \times\times\times\times\times\times\times\times\times\times\times\times\times\times\times\times\times11111\times\times\times\times\times\times\times\times\times\times\times\times\times : c_5 = 10$$
$$s_6 = \times11111\times\times\times\times\times\times\times\times\times : c_6 = 10$$
$$s_7 = \times11111\times\times\times\times\times : c_7 = 10$$
$$s_8 = \times11111 : c_8 = 10$$

As a step towards the study of linking editors' concentrations with environmental contexts, we introduce another testbed (fitness landscape) in which each schema is comprised of all 0's and the other parameters remain the same as used in $S1$. The fitness landscapes consisting of schemata of all 1's and all 0's are called **L1** and **L0**, respectively. These two testbeds are maximally different in the configurations of their fitness landscapes. By oscillating these two landscapes, we are able to investigate the effects of drastic environmental changes.

The GAE experiments conducted in this section are based on a binary tournament selection, one-point crossover and mutation rates of 0.7 and 0.005, respectively; population size is 40 for each of 50 GAE runs. A family of 5 editors, C1, was randomly generated, with editor length selected in the range of 2 to 4 bits (see (Huang and Rocha, 2003, 2004) for a set of guidelines for parameter choices of the editors). Table 3 shows the corresponding parameters generated for each editor in family C1: length, alleles, concentration and editing function. For example, editor 3 is a bit-string of length 4 (0101); its concentration, or the probability that a genotype string will encounter this editor is 0.7302; its editing function is to delete 1 bit, meaning that this editor deletes 1 genotype string allele at the position following the genotype substring that matches the editor's string.

Figure 2.a and 2.b display the averaged maximum fitness and averaged population fitness, respectively, for several GAs and the GAEs on static environments L0 and L1.[2] In the figure, L0 (GA) and L1 (GA) denote the results obtained for the traditional GA on landscapes L0 and L1, respectively. L1C1 (GAE) denotes a GAE with the family of editors C1 shown in Table 3, applied to the L1 landscape. L0C1 (GAE) denotes a GAE with the same family of editors C1 applied to the L0 landscape.

One can see that the family of editors C1 facilitates the population's adaptation on L1 with respect to the maximum fitness and population fitness, in comparison with the traditional GA without edition on the same

---

[2] The value of the averaged maximum fitness measure is calculated by averaging the fitness of the best individuals at each generation for all 50 runs, where the vertical bars overlaying the measure curves represent the 95-percent confidence intervals. This applies to all the results obtained for the measures employed in this paper.

landscape. However, C1 is by no means beneficial for the GAE on landscape L0.

Table 3. Parameters of the five editors

	Editor 1	Editor 2	Editor 3	Editor 4	Editor 5
Length	4	4	4	2	4
Alleles	1110	0011	0101	00	0111
Concentration	0.0635	0.0476	0.7302	0.2857	0.3175
Editing Fun.	Delete 4 bits	Add 3 bits	Delete 1 bit	Delete 3 bits	Delete 2 bits

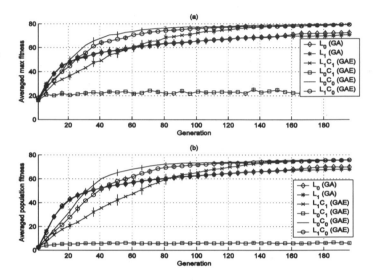

Figure 2. Evolutionary measures on static landscapes

To enhance the performance of the GAE population on L0, we produced another editor family, C0, whose only difference from C1 is a new set of editor concentrations, {0.31, 0.062, 0.989, 0.002, 0.05}, with all other editor parameters remaining the same as in Table 3. The results in Figure 2 show that the GAE with C0 now performs much better on L0 than with C1. We also notice that the L1C1 and L0C0 GAE clearly outperform the GA without edition on L1 and L0 respectively.

Consider now a dynamic environment which oscillates periodically between the landscapes L1 and L0. This oscillation models an environment with recurring dramatic changes in conditions. We know that some biological organisms, namely parasites that go through dramatic environmental changes, use the edition of mRNA molecules to their advantage, by associating the process of edition to environmental context. The ability to associate changes in the environment with internal parameters such as concentrations of editing agents, is one of the mechanisms that can be used to (contextually) regulate gene expression (Mattick, 2003) with potential adaptive advantages (Rocha, 1995).

Figure 3 depicts our modeling of this process with the oscillation of landscapes L1 and L0, at every 100 generations. Four scenarios are tested:

1. **L1L0.** Landscapes oscillate without genotype edition. The population evolves solely according to the traditional GA.
2. **L1C1L0C1.** Landscapes oscillate with genotype edition, but edition is always implemented with family C1.
3. **L1C0L0C0.** Same as above but with family C0.
4. **L1C1L0C0.** Landscapes oscillate with edition, but the family of editors changes with the environment: family C1 operates when landscape L1 is in place, and C0 operates with L0.

The dramatic oscillation of environments is very hard for scenario 1 that uses only the traditional GA (L1L0). The first time the environment changes, the population is forced to evolve new solutions for L0 from a population already evolved for L1 and, before it is able to produce good individuals in L0, the environment changes again. Subsequent oscillations produce the same result, and the population never reaches a good solution.

In scenario 2, when editor family C1 is used on both environments (L1C1L0C1) we observe that the population behaves very well on landscape L1 but poorly on L0. This is an improvement over L1L0, but worse than the other two scenarios.

The results for scenario 4 (L1C1L0C0) show that the association of editor concentrations to environmental contexts (i.e., the association of L1 with C1, and L0 with C0) is indeed beneficial, as the population of agents is capable of evolving very good solutions in both environments.

However, we also notice that in scenario 3 (L1C0L0C0), which uses solely editor family C0, the population is capable of producing very good individuals on both oscillating environments. This means that family C0 is good at editing genotypes in both landscapes. The results obtained for scenarios 3 and 4 thus show that genotype editing can lead to evolutionary advantages in two distinct ways to cope with dynamic environments: (1) by employing editors which can produce genotypes encoding good solutions in both landscapes (scenario3: L1C0L0C0), or (2) by changing the concentrations of editors when the environment changes (scenario 4: L1C1L0C0).

We do notice, however, that scenario 4 provides a quicker response immediately after the environment changes from L0 to L1. In figure 3, we can see that when this change occurs at generations 200 and 400, the averaged maximum fitness of L1C0L0C0 (scenario3) suffers a larger setback than that of L1C1L0C0 (scenario 4). In scenario3, the population needs to completely re-adapt to the new environment, whereas in scenario 4 the population contains some individuals that are very fit in L1, but, after edition, are somewhat fit in L0. Thus, whereas the average maximum fitness (for 50 runs of each scenario) at generations 200 and 400 is very close to 0 for scenario 3, it is about 15 for scenario 4. These values are

clearly significant given the 95-percentile confidence intervals computed and depicted in the figure.

Furthermore, a microscopic inspection shows that in the case of scenario 4, at generation 199, the chromosome of one individual of fitness 60 is defined by substring $\{0,1,0,1,0\}$ at the position of schema $S7$. When the landscape oscillates from L0 to L1 at generation 200, this individual undergoes some edition which results in these alleles being altered to $\{1,1,1,1,1\}$. This individual thus acquires a fitness amount of 10 from building block $S7$. This situation is relatively typical in scenario 4; yet in scenario 3, since more individuals converged to genotype strings of all 0's at generation 199, it is more difficult for the agents to acquire corresponding building blocks at generation 200 simply by genotype edition. All this means that under scenario 4, the GAE evolves genotypes which produce fair solutions in both landscapes, but which are edited differently accordingly: genuine phenotypic plasticity. Therefore, the same genotypes may exist in both landscapes, whereas in the case of scenario 3 the constant editor family, C0 seems to facilitate the evolution of new good genotypes after the landscape changes.

4. Conclusion and Future Work

This paper presents our computational methodology using Genetic Algorithms with Genotype Editing for investigating the role of RNA editing in dynamic environments. Based on several genetic editing characteristics that are gleaned from the RNA editing system, we show that the incorporation of editing mechanisms indeed provides a means for artificial agents with genotypes to gain greater phenotypic plasticity, and /or a mechanism to generate novel fit individuals when a population is faced with dramatic environmental changes. By linking changes in the environment with internal parameters such as concentrations of editors, the artificial agents can use genotype edition to their advantage, as gene expression can become contextually regulated, such ability thus gives organisms an adaptive advantage. In a nutshell, the results obtained have provided the following insights:

There are two strategies for artificial agents with genotype edition to produce phenotypic plasticity to cope with environmental changes: (1) by using different families of editors for different environmental demands, or (2) by employing a single family of editors that allows the evolutionary process to cope well with a changing environment.

We have thus far studied the association of editor families with different concentrations to environmental changes. The work here presented details simulations with two specific editor families. Based on such anecdotal evidence, our results simply show that Genotype Editing may provide evolutionary advantages in oscillating environments. In future work, we intend to allow the family of editors and the genotypes of agents to co-evolve, so that the artificial agents can discover proper editor

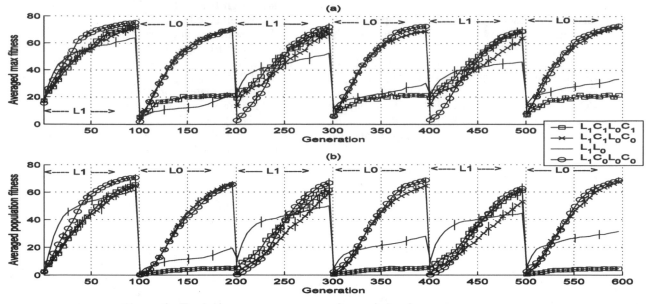

Figure 3. Evolutionary measures on dynamic landscapes

concentrations to adapt to changing environments. Since there are several internal editor parameters involved in an editing system, such as the size of the editor family, editor length and editor functions, in addition to the investigation of editor concentrations, our future work is also going to study the effects of associating other parameters with external environments. Since the length of oscillation period is expected to be another critical parameter that will affect how well the GAE's population adapts to changing environments, we will also study the effects of oscillation periods. With a systematic study on these editor parameters, our hope is to gain a deeper understanding of the role of RNA Editing in nature and also to design robust evolutionary computation algorithms for complex, dynamic real-world tasks, as we have done in Huang and Rocha, (2003; 2004) for non-changing environments.

References

Bass, B. L. (Ed.) (2001). RNA Editing. Frontiers in Molecular Biology Series. Oxford University Press.

Benne, R. (Ed.) (1993). RNA Editing: The Alteration of Protein Coding Sequences of RNA. Ellis Horwood.

Hinton, G. E. and Nowlan, S. J. (1987). "How learning can guide evolution." Complex Systems. Vol. 1, pp. 495-502.

Holland, J. H. (1975). Adaptation in Natural and Artificial Systems. University of Michigan Press.

Hoopengardner, B., Bhalla, T., Staber, C., and Reenan, R. (2003). "Nervous System Targets of RNA Editing Identified by Comparative Genomics." Science 301: 832-836.

Huang, C.-F. and Rocha, L. M. (2003). "Exploration of RNA Editing and Design of Robust Genetic Algorithms." Proc. 2003 Congress on Evolutionary Computation, IEEE Press, pp. 2799-2806.

Huang, C.-F. and Rocha, L. M. (2004). "A Systematic Study of Genetic Algorithms with Genotype Editing." *GECCO 2004*. In press.

Lomeli, H. et al. (1994). "Control of Kinetic Properties of AMPA Receptor Channels by RNA Editing." Science, 266: 1709-1713.

Mattick, J. S. (2003). "Challenging the Dogma: the Hidden Layer of Non-protein-coding RNAs in Complex Organisms." BioEssays. 25: 930-939.

Pollack, R. (1994). Signs of Life: The Language and Meanings of DNA. Houghton Mifflin.

Rocha, Luis M. (1995). "Contextual Genetic Algorithms: Evolving Developmental Rules." Advances in Artificial Life. Springer Verlag, pp. 368-382.

Rocha, Luis M. (1997). Evidence Sets and Contextual Genetic Algorithms: Exploring Uncertainty, Context and Embodiment in Cognitive and biological Systems. PhD. Dissertation. State University of New York at Binghamton. Science.

Simpson, L. and Emerson, R. B. (1996). "RNA Editing." Annual Review of Neuroscience, 19: 27-52.

Stuart, K. (1993). "RNA Editing in Mitochondria of African Trypanosomes." In: RNA Editing : The Alteration of Protein Coding Sequences of RNA. Benne, R. (Ed.). Ellis Horwood Publishers. pp. 26-52.

Sturm, N. R. and Simpson, L. (1990). "Kinetoplast DNA Minicircles Encode Guide RNA's for Editing of Cytochrome Oxidase Subunit III mRNA." Cell, 61: 879-884.

Wang, Q., Khillan, J., Gadue, P., and Nishikura, K. (2000). "Requirement of the RNA Editing Deaminase ADAR1 Gene for Embryonic Erythropoiesis." Science, 290 (5497): 1765-1768.

A Tangled Hierarchy of Graph-Constructing Graphs

Chris Salzberg[1,2], Hiroki Sayama[1] and Takashi Ikegami[2]

[1] Department of Human Communication, University of Electro-Communications, Tokyo 182-8585, Japan
[2] Graduate School of Arts and Sciences, University of Tokyo, Tokyo 153-8904, Japan

Abstract

The traditional construction paradigm of machine and tape is reformulated in a functionally homogeneous space of directed graph structures. Hierarchy-based roles, normally appointed to actors in a construction process, are dissolved and replaced by symmetric, level-less *engagement*. The separation between static (information carrying) and active (information processing) structures, imposed by mandate of the rules or physics in earlier theoretical models, results instead purely from graph topology. While encompassing traditional machine-tape paradigms as a special case, the formalism is shown to incorporate a wider class of construction relations. Exploiting its flexibility, a representation of a Turing machine is demonstrated, establishing computation universality. The concept of a "Tangled Construction Hierarchy" is introduced.

Introduction

The formalistic study of machine construction has its roots in the self-reproducing automata theory of mathematician John von Neumann (von Neumann, 1966). Inspired by the extreme "complication" of real living organisms, von Neumann sought to realize the emergence of complexity-increasing evolution in a functionally homogeneous medium governed only by local rules. With Stan Ulam he invented a formulation for this purpose, now known as *cellular automata* (CA) and widely adopted for the modeling of self-replication (Sipper, 1998), in which the outputs and inputs of a construction process are fundamentally made of the same "stuff". Embedded in this architecture von Neumann instantiated his automata theory in the form of a complicated universal construction machine capable of self-reproduction. By introducing static self-description (tape) and division of labour (translation/transcription), he thus demonstrated a loophole in the construction paradox stating that "a machine tool is more complicated than the elements that can be made with it" (von Neumann, 1966, p.79). Decades earlier, Turing (Turing, 1936) had initiated the study of *computing* machines on the basis of a similar machine-tape paradigm; von Neumann imported it to the realm of *constructing* machines.

Hierarchical separation of machine and tape has since played an influential role in shaping the design of artificial construction and self-replication models (Mange and Sipper, 1998; McMullin, 2000), yet notable alternatives have been proposed. Hofstadter (Hofstadter, 1979), drawing inspiration from "the molecular logic of the living state" (Lehninger, 1976), blurred this separation in a typographical system he called "Typogenetics". The players in this system are "strands": strings of characters acting both as data to be manipulated and as an active "typographical enzyme" to be applied to other strands. Hofstadter called this mixing of levels a "Tangled Hierarchy" (Fig. 1), contrasting it with the case, typified by formal systems, in which there is a clear distinction between rules and the strings they apply to. Along similar lines, Laing (Laing, 1976) devised a system in which a pair of tapes undergo local sliding and state changes leading to self-reproduction via self-inspection. By means of transfer primitives, active and passive roles are arbitrarily exchangeable in this process, exemplifying a uniquely mixed-level style of execution.

All of these models assign roles in a construction process employing the intuitive concept of *levels*. Describing his system, Hofstadter asserts that: "The two way street which links 'upper' and 'lower' levels of Typogenetics shows that, in fact, neither strand nor enzyme can be thought of as being on a higher level than the other." (Hofstadter, 1979, p.513) The word "level" is used here in the sense of *containing the same information*, interpreted as passive *data* in one case (strand), and as active *process* in the other (enzyme). von Neumann's machine contains a related type of "levels". As do strand and enzyme, the tape and machine it codes for, interpreted according to a set of transition rules[1], contain the same information: one in a passive form, the other in an active one. By analogy, they are thus — according to Hofstadter's use of the word — on the same "level".

---

[1] Strictly speaking an embedded universal construction machine is also required in the initial configuration to carry out translation.

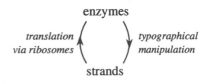

Fig. 1: The "Central Dogma of Typogenetics", an example of a Tangled Hierarchy (Hofstadter, 1979, p.513).

This concept of levels is, however, elusive: it defines two things as being on the same level if they differ only by interpretation, in one case by the human interpreter, in the other by a complicated set of CA rules. Yet this interpretation implicitly *assigns* levels: enzyme acts on strand, machine acts on tape, and never the other way around. Even in Laing's model, in which these roles are exchangeable, at any one time we interpret one tape as *active* (machine) and the other as *passive* (tape). If ultimately we aim to understand complex level-crossings and mixings of the living state, then such hierarchical interpretation is potentially misleading. Thus comes the question, the underlying theme of this paper: can we separate the functionally active *state* — that which depends on interpretation — from the structurally embedding *medium*? This requires that we replace specialized rules of *interpretation* by symmetric rules of *engagement*, and that we operate on a space that is sufficiently flexible to simultaneously accommodate both active (machine/enzyme) and passive (tape/strand) topologies.

In the sections that follow we demonstrate that these constraints are jointly satisfiable in a space of locally-interacting, collaboratively constructing graph structures in which the status of machine or tape, or the non-applicability of such labels, is an emergent and dynamic property of an *engagement* rather than an appointed *role*. By translating the output of such "level-less" engagement into the creation of new graph structures, we close the loop of the Tangled Hierarchy (Fig. 1), yet in a strictly *non-hierarchical* way. Exploiting its flexibility, we show a simple representation of a Turing machine, the quintessential top-down system, in the proposed framework. This example is used to illustrate that it is the *structure* of the graph, and not the *rules* of our formulation, that now defines machine and tape, and hence that we are working at a deeper level than any "level"-based system.

Formulation

The formulation we present in this section is conceptually divided into three parts:

(i) A *medium* of directed graph structures with input/output labels from a finite alphabet.

(ii) *Rules of engagement* according to which graph structures can read and be read by one another, producing as output a string of symbols from the alphabet of (i).

(iii) *Rules of translation* transforming the output of (ii) into an active process of growth and folding on the same graph structure (i) from which it was produced.

We introduce the medium (i) and rules (ii) by reformulating top-down input-reading of a finite-state machine as a non-hierarchical, symmetrical engagement between pointers to nodes of a directed, labeled graph. This one-on-one engagement is subsequently generalized to an arbitrary number of pointers. Finally, the translation process (iii), constituting the upward arrow of Fig. 1, is introduced, completing our formulation which we call a "Graph-Constructing Graph".

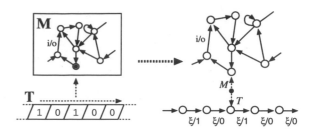

Fig. 2: Machine (**M**) and input tape (**T**) before and after symmetry is imposed.

Machines and tapes

We begin with the standard conceptual abstraction of a physical machine, that of a finite-state machine (FSM). In this paper we focus on a particular deterministic FSM called a *Mealy machine*, defined formally as a 5-tuple $\mathbf{M} = (S, I, O, \delta, \varphi)$, where S is a set of states, I an input alphabet, O an output alphabet, $\delta : I \times S \rightarrow S$ a transition function and $\varphi : I \times S \rightarrow O$ an output function. We simplify this machine by assuming a general input/output alphabet $A = I = O$, allowing us to construct composite functions from δ and φ; this will be essential for the steps that follow.

The machine **M** is a finite automaton, and in automata theory there is a clear distinction between the input/output (strings of characters from A) and the machine (a labeled directed graph). Converting input-reading, the active process in this system, from interpretation to engagement requires firstly that we reformulate the input string as a special case of a graph. For this purpose we introduce a special symbol 'ξ', inspired by the ε-transition of the nondeterministic FSM that requires no input for transition, as an element of $\Sigma = A \cup \{\xi\}$. ξ serves as a "default path" symbol, allowing us to map a linear tape to a chain of ξ/A-transitions, shown in Fig. 2 for $A = \{0, 1\}$. The set S now contains nodes (states) of both machine *and* tape, their hierarchy replaced by a *relation* between two graph structures, henceforth referred to via pointers M and T indexing states $s_M \in S$ and $s_T \in S$ respectively. Using the notation $\delta_X(\sigma) = \delta(\sigma, s_X)$ and $\varphi_X(\sigma) = \varphi(\sigma, s_X)$ for $X \in \{M, T\}$ and $\sigma \in \Sigma$, the tape-read/state-transition cycle can thus be described as:

1. T sends $\varphi_T(\xi)$ ("data") to M and advances along $\delta_T(\xi)$ (moves tape head forward).

2. M sends $\varphi_M(\varphi_T(\xi))$ to the output string and advances along $\delta_M(\varphi_T(\xi))$ (state-transition).

The above steps can be represented in terms of the function composition sequence $\xi \rightarrow T \rightarrow M$, understood as the operation: advance T on $\delta_T(\xi)$ and M on $\delta_M(\varphi_T(\xi))$ and output $\varphi_M(\varphi_T(\xi))$. This process is easily generalized to arbitrary orderings of elements from the set $\{M, T\}$, which we call *flows*. There are four of these: $(M \rightarrow T)$, (M), $(T \rightarrow M)$ and (T), drawn in Fig. 3 as directed paths on an "M/T loop". The second flow, for instance, reads as: advance M on $\delta_M(\xi)$ and output $\varphi_M(\xi)$. In general, there is a uniform procedure for each participant in these flows:

a. **Receive** output $\sigma \in \Sigma$ from the last element.

b. **Send** $\varphi_X(\sigma)$ to the next element.

c. **Advance** along the transition $\delta_X(\sigma)$.

where $X \in \{M, T\}$, $\sigma = \xi$ for the first element, and the last element's "next element" is the output string. While encompassing top-down machine/tape hierarchies in the flows $(T \rightarrow M)$ and (M) (flows that terminate with M), this procedure generalizes to more unconventional structures such as, for example, branching (graph-based) tapes. All that remains is to define, at any time, *which* flow to execute, and to do so in a way that preserves the top-down hierarchy as a special case. We do this using the following symmetric tree:

$$\xi \xrightarrow{\delta_X} \varphi_X(\xi) \xrightarrow{\delta_Y} \varphi_Y(\varphi_X(\xi))$$
$$\searrow^{\delta_Y} \varphi_Y(\xi) \xrightarrow{\delta_X} \varphi_X(\varphi_Y(\xi))$$

where $\{X, Y\}$ is a permutation of the set $\{M, T\}$. Each node in this tree corresponds to the output of a possible flow; links are prioritized from top to bottom and followed if the corresponding transition function, applied to the previous output, is defined. For example, if $\delta_Y(\xi)$ and $\delta_X(\varphi_Y(\xi))$ are defined but not $\delta_X(\xi)$, then we choose the lower branch and return the flow $(Y \rightarrow X)$ with output $\varphi_X(\varphi_Y(\xi))$. If no ξ-transition exists for any $X \in \{M, T\}$ then we say that the system has reached a *stopping configuration*. In any other case the flow returned is called the ξ-*flow*.

The steps described so far maintain symmetry in engagement: both machine and tape structures are treated in the same way. To complete the formulation and maintain this symmetry, the permutation $\{X, Y\}$ in the preceding tree must be fixed in an accordingly non-hierarchical way. We could, for instance, stipulate that $X \equiv M$, yet this would explicitly distinguish M and T by prioritizing flows $(M \rightarrow T)$ and (M) over $(T \rightarrow M)$ and (T). Instead, we opt to treat X as a dynamic *lead element*, defined as the *last* element in the *previous* ξ-flow. The ordering $\{X, Y\}$ of branches in the tree is thus also dynamic: following ξ-flows (M) and $(T \rightarrow M)$ the order is $\{M, T\}$, whereas following (T) and $(M \rightarrow T)$ the order is $\{T, M\}$. Applied to an engagement between two structures (call them "machine" and "tape") with transitions exclusively of the form σ/a and ξ/a, respectively, for $a \in A$ and $\sigma \in \Sigma$, and initiated with M pointing to the "machine" and T to the "tape", this scheme preserves top-down hierarchy since only the first two flows ever occur (Fig. 4). Yet rather than being a result of *rules*, this hierarchy results from the *structure* we have imposed. We could, alternatively, have allowed branching on the tape; output would then result from collaborative, role-changing interaction, a kind of *co-reading* process. In the latter context, "machine" and "tape", and the tape-reading process that distinguishes them, are hard to find: their structures are reading *each other*.

Loops and flows

The idea of co-reading can quite readily be generalized to an arbitrary number of elements. Consider the 2-element loop of Fig. 3 on the set $\{M, T\}$ a special case of an N-element loop (or N-loop) $L(t)$ on the sequence $\{s_i(t)\}_{i=1}^{N} \in S$, with t representing a discrete time parameter. As for the M/T system, define a lead element $l(t) \in \{1, \ldots, N\}$ and denote as L_0 the cyclic permutation of L with s_l as first element. A flow F on an N-loop L is a sequence of indices $\{f_i\}_{i=1}^{J} \in \{1, \ldots, N\}$ with output given by the composite function:

$$\Phi_F = \varphi_{f_J} \cdot \varphi_{f_{J-1}} \cdot \ldots \cdot \varphi_{f_2} \cdot \varphi_{f_1} \quad (1)$$

As an example, the set of 15 possible flows for a 3-loop is shown in Fig. 5. Execution of these flows proceeds as defined earlier for the M/T system. Beginning with $\sigma = \xi$ and ending with the output string, each participant $X \in L$

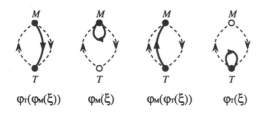

$\varphi_T(\varphi_M(\xi))$ $\varphi_M(\xi)$ $\varphi_M(\varphi_T(\xi))$ $\varphi_T(\xi)$

Fig. 3: Possible ξ-flows (solid arrows) and their output functions on the 2-element M/T loop (dashed arrows).

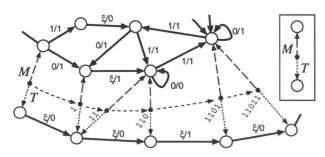

Fig. 4: Tape-reading process for machine/tape structures. Motion of M/T pointers drawn as dashed horizontal arrows. Trailing binary sequence indicates current output string at each step.

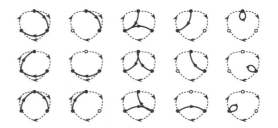

Fig. 5: Possible flows (solid arrows) on a 3-loop (dashed arrows).

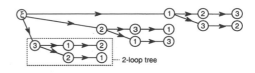

Fig. 6: Possible flows for a 3-loop $\{s_1, s_2, s_3\}$ with lead $l = 1$ as nodes of a decision tree.

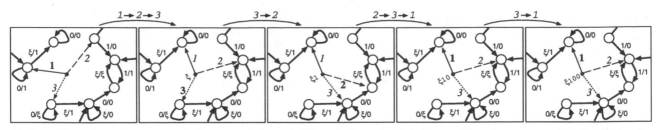

Fig. 7: Co-reading of graph structures via flows on a 3-loop. Boldface indicates lead element, trailing character sequence is output string at each step. ξ-flows drawn above frames as arrows and index sequences.

receives $\sigma \in \Sigma$ from the previous element, *sends* $\varphi_X(\sigma)$ to the next element, and *advances* along $\delta_X(\sigma)$. The decision tree of the M/T system also generalizes, in a recursive way: an N-loop tree is composed of N permutations of $(N\text{-}1)$-loop trees, is composed of $(N\text{-}1)$ permutations of $(N\text{-}2)$-loop trees, etc. To give s_l priority in this tree, branches from the root are ordered according to L_0. As an example, the tree for the 3-loop of Fig. 5 is shown in Fig. 6. Each node in this tree represents a possible flow, the ξ-node being the stopping configuration. A recursive flow-finding procedure searches depth-first, prioritizing higher branches, until it reaches a terminal branch or has no further transitions available. The branch it returns is the ξ-flow, referred to as $F_\xi(L)$.

With dynamics described by F_ξ we can express the time-dependence of $L(t) = \{s_i(t)\}_{i=1}^N$. For each $f_j \in F_\xi(L(t))$, the state $s(f_j)(t) = s_{f_j}(t)$ indexed by f_j is updated as:

$$
\begin{aligned}
s(f_1)(t+1) &= \delta_{s(f_1)(t)}(\xi) \\
s(f_j)(t+1) &= \delta_{s(f_j)(t)}(\Phi_{\{f_k\}_{k=1}^{j-1}}(\xi))
\end{aligned}
\tag{2}
$$

All elements not encountered in the ξ-flow are stationary, and the lead element is updated as $l(t+1) = f_J$ (the last element of the flow). An example of an execution process occurring via a 3-loop is shown in Fig. 7. The output string generated by this process is the non-terminating sequence '$\xi100000\ldots$', an example of *non-stopping* co-reading.

Translation

In leveling the rules of engagement between machine and tape, we have twisted the downward arrow of Hofstadter's "Central Dogma" (Fig. 1), the manipulation of strand (tape) by enzyme (machine), onto itself. The status of either *reading* (machine) or of *being read* (tape) is now relative, not to the rules of the system, but to the participants in an engagement. Translation, formerly the basis of equivalence between these two forms, is now something subtly different. It is a tranformation of the output of an engagement, as we have phrased it a string of symbols from Σ, *back* into the medium itself, in this case graph structure, as a growth and modification process. This can be abstractly represented as:

$$\mathbf{T}(\text{engagement}) = \text{process on structure}$$

This collaborative cycle of structure creation and modification, based on Hofstadter's Tangled Hierarchy, we call a "Tangled Construction Hierarchy". Though conceptually inspired by complex manipulation and translation processes

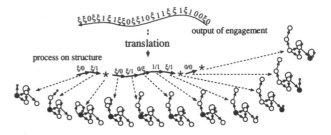

Fig. 8: Translation mapping output from engagement into process of graph growth and folding.

of the living cell, unlike machine-tape models it is not *intrinsically* hierarchical: engagement is not a structure, but a process *between* structures. In the same way, the output of $\mathbf{T}$ is not exclusively structure but a process *on* structure (Fig. 8). We draw on the analogy of complex folding from amino acid sequences into proteins, represented rather simplistically in Typogenetics as 90° "kinks" in an otherwise directionless strand. In contrast, and also differing from formal graph grammars (Rozenberg, 1997) and the recently-developed graph automata (Tomita et al., 2002), our goal is that growth and folding of the graph, as of the amino acid sequence, be locally-propagating and incremental. The following properties are thus demanded of $\mathbf{T}$: it should (1) act locally, (2) generate a minimal set of "graph-manipulation" instructions, (3) function on a minimal set of rules and (4) be construction universal in the sense that, given an appropriate output string, it should generate any arbitrary graph structure. The framework and set of rules we propose in this section achieves a desired balance between these constraints. Other possibilities, including non-Turing-computable transformations, are also being considered.

We assume to start only singly-labeled links; incorporation of input/output labels follows. Such that our graph-modification process is local (constraint (1)), we employ a dynamic *construction head*, loosely related to the write-head of machine-tape formalisms, acting in conjunction with a *root* pointer. The head and root can thought of as the "hands" of construction, with the root referencing a re-assignable position on the graph to which the head, moving about, carrying and affixing chains of nodes and links, can return during construction. Neither head nor root makes use of static global referencing; to modify a portion of the graph, the head, starting from either its current position or the root node, must travel along intervening links to get to it.

Only two types of graph-manipulation instructions (constraint (2)) are understood by the head: "insert/traverse link" (denoted by any link label) and "return" (denoted by ' * '). A minimal set of nine rules (constraint (3)) are required for universality (constraint (4)), pictured in Fig. 9 in terms of the current instruction (link or return) and neighbourhood of the head node (node pointed to by the construction head). Depending on local topology of the graph, an instruction is interpreted by the head in one of three ways: * takes it back to the root (column 1 rules), whereas a link '*a*' inserts/sprouts a new link (column 2) if the *a*-transition does not yet exist, or traverses the link (column 3) if it does. Note that the head can directly "hold" at most one link at a time, but can append complete *chains* of links by inserting and dragging them. Specifically, there is always a succession of *move, insert, return, drag* and *affix* rules that suffices to create a path connecting any accessible potions of the graph. Hence any graph having at least one node with paths to all other nodes is constructible; this is the particular form of construction universality that is achieved. Examples of strings translated and applied to a single node are given in Fig. 10.

We extend the above framework to input/output pairs by treating the pair (σ_i, σ_o) in the rules of Fig. 9 as the link instruction σ_i, with the exception that any existing output σ_o is overwritten each time a link is traversed. In addition to augmenting structure, the head can thus also modify existing links, with the exception of removing them.

Lastly, we require a mapping from the space of symbol strings (strings from the alphabet Σ) to the space of instruc-

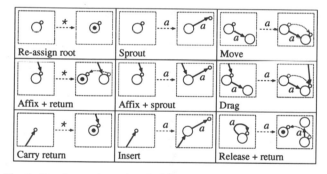

Fig. 9: Graph-growing rules. Solid arrows are links, dashed arrows are rules, '*a*' is any link label, * is the return symbol, ◯ is a node, ∘ is the head position, and ⊙ is the root node. Dotted arrows indicate change in position of construction head. A link pointing to directly to the head (→∘) is currently being held by the head. In the "re-assign root" rule, the root pointer is re-assigned to the current position of the head (the head itself does not move).

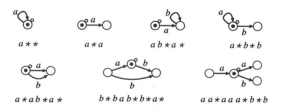

Fig. 10: Graphs produced by translating output strings and applying the resulting construction process to a single starting node.

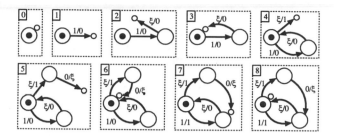

Fig. 11: Transformation of the output string 10_1 $\xi\xi0_2$ $\xi0_3$ $\xi\xi1_4$ $0\xi_5$ $\xi1_6$ 11_7 $\xi0_8$ via codons of eq. (3) into graph growth and folding. Subscript references frame index of sequence.

tions (links (σ_i, σ_o) plus *), analogous to Hofstadter's "Typogenetic Code" (Hofstadter, 1979) inspired by the Genetic Code contained in tRNA molecules of a biological cell. This mapping, shown as the downward arrow in Fig. 8, can also be thought of as the translation between "machine code" (symbol string) and "assembly code" (growth and folding instructions). Its specification determines the type of topologies that are likely to grow. To be *universally expressible*, all possible link pairs plus * must be represented; this amounts to requiring $|\Sigma|^2 + 1$ different "codons" (substrings) in our coding scheme. One possibility is the following mapping:

$$a\sigma \rightarrow a/\sigma \qquad \xi a \rightarrow * \qquad \xi\xi\sigma \rightarrow \xi/\sigma \qquad (3)$$

where $a \in A$ and $\sigma \in \Sigma$. A transformation process from output string to construction process via this mapping is shown in Fig. 11 for the alphabet $A = \{0, 1\}$. Note that ambiguity arises in the translation process if the construction head modifies structure indexed by an element of a loop (e.g. *M* or *T* in the 2-loop), or, in the case of concurrent construction, if two heads work on the same structure. Such issues are resolvable by stipulating rules for conflict resolution, for instance by disallowing construction on indexed nodes or traversal of links on a free construction thread. Neither issue is essential to the basic formalism we have described, hence we leave open the choice of how to resolve them.

The formulation we have finished describing we call a "Graph-Constructing Graph" (GCG), pictured in its entirety in Fig. 12. As does Hofstadter's "Central Dogma" (Fig. 1) and the molecular biology from which derives its inspiration, the GCG has at its core an information creation/translation cycle: interaction of structures (graphs) produces an output (strings) that is translated to the creation (growth) and modification (folding) of these same structures. Yet whereas in Typogenetics the interaction is between passive "data" and active "process", here such dis-

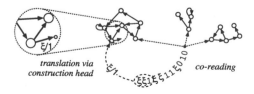

Fig. 12: The Graph-Constructing Graph, an example of a "Tangled Construction Hierarchy".

tinctions, if they exist, are entirely context-dependent. In that level-crossing and mixings are central to dynamical processes in a cell, we are thus conceptually closer to the constructing lifeforms we ultimately aim to understand.

Turing machine as GCG

The architecture of a Turng Machine as a GCG, using the codons of eq. (3) for translation, is shown in Fig. 13. Since we are only concerned with computation (reading and overwriting) and not construction (modification of graph connectivity), the return symbol '*' is not required; the root pointer thus plays no role, and is not included. We consider TM transitions of the form (a, b, c, d), where $\{a, b\} \in \{0, 1\}$ are the old and new values on the tape, respectively, $c \in \{L, R\}$ is the direction to advance the write head, and $d \in \{L, R\}$ is the direction to advance the read head.[2] We use the alphabet $A = \{0, 1\}$, with left and right symbols of the TM mapping to 0 and 1, respectively. Initially M points to a branch node on the machine, and T (read head) and H (write head) point to nodes on the tape. T begins as the lead element.

The key feature of this representation, when compared with Fig. 4, is that we have enabled 2-way motion on the tape: from a given node on the tape structure, the read head (T) may either read the data value (ξ/a), follow the left branch ($0/\xi$), or follow the right branch ($1/\xi$). ξ-transitions initiate the flow, so T (the lead element) begins by following the ξ/a transition, which moves M on the a/ξ branch. Since M has no ξ-transition, T leads again and the $\xi/a \to a/\xi$ flow is repeated. T follows ξ/a a third time, this time advancing M on the a/b branch. At this point the construction head (denoted as H) has received the codon $\xi\xi b$, instructing it to overwrite the ξ/a link with the new data value ξ/b.

T and M have not yet advanced, however. M still has no ξ-transition, so T follows ξ/b and M follows b/c, outputting c to H. Now M has a ξ-transition, ξ/d, which it follows. The value of d determines in which direction T will advance (read head move), but in both cases ξ is sent to H. The codon $d\xi$ translates to d/ξ, which advances H along d (write head move). This completes the basic read/write cycle of the Turing Machine. Assuming an infinite tape, we have achieved TM-equivalence and hence computation universality.

---

[2]The option of moving the read and write head in different directions is included as it emerges naturally from the reformulation.

Note that we have significantly handicapped the GCG to show TM-equivalence. It is quite straightforward, for instance, to build a TM with alternate structures for data storage (e.g. 2D/3D tape) or one with finite but extendible tape (as done in (Tomita et al., 2002)). Such topics will be discussed with more detail in a follow-up publication.

Conclusion

In this paper we introduced a formulation motivated by von Neumann's original search for the logic of construction. Using as medium a space of directed graph structures, a system of Graph-Constructing Graphs was described. This system, an example of what we call a "Tangled Construction Hierarchy", is ultimately aimed at abstracting tangled level-crossings of hierarchy in biological systems to a finite space of logical structures (i.e. graphs) with discrete dynamics. The underlying theme we are tackling is that of the essential or non-essential nature of machine-tape dualities in construction, and of the possibility for potentially revealing alternatives. In so doing, we are striving for a new perspective on information creation and manipulation in constructing systems. That hierarchical segregation is so patently violated by biology's most central constructing system, the cell, makes such a perspective relevant and necessary.

An immediate future goal of this work is to explore the space of graph structures and their interactions to search for fixed points in cycles of construction. Evolutionary dynamics in a "soup" of GCGs is one possible environment for this exploration; others are being considered.

Acknowledgments

C.S. acknowledges financial support by grants from the Netherlands Organization for International Cooperation in Higher Education (Nuffic) and the VSB Funds.

References

Hofstadter, D. R. (1979). *Gödel, Escher, Bach: an Eternal Golden Braid*, chapter XVI, pages 504–513. Basic Books, Inc.

Laing, R. (1976). Automaton introspection. *Journal of Computer and System Sciences*, 13(2):172–183.

Lehninger, A. (1976). *Biochemistry*. Worth Publishers, New York.

Mange, D. and Sipper, M. (1998). Von Neumann's quintessential message: genotype + ribotype = phenotype. *Artificial Life*, 4:225–227.

McMullin, B. (2000). John von Neumann and the evolutionary growth of complexity: Looking backward, looking forward... *Artificial Life*, 6:347–361.

Rozenberg, G., editor (1997). *Handbook of Graph Grammars and Computing by Graph Transformation*, volume 1 of *Foundations*. World Scientific, Singapore.

Sipper, M. (1998). Fifty years of research on self-replication: An overview. *Artificial Life*, 4:237–257.

Tomita, K., Kurokawa, H., and Murata, S. (2002). Graph automata: natural expression of self-reproduction. *Physica D*, 171(4):197–210.

Turing, A. M. (1936). On computable numbers with an application to the entscheidungsproblem. *Proc. London Math. Soc. Ser.*, 42:230–265.

von Neumann, J. (1966). *Theory of Self-Reproducing Automata*. University of Illinois Press, Urbana, Illinois. Edited and completed by A. W. Burks.

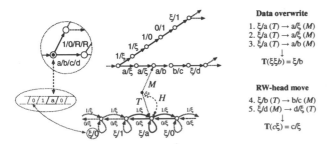

Fig. 13: Turing machine as GCG (left) and flows comprising a single read/write operation (right). a, b, c, d represent symbols from $\{0, 1\} \equiv \{L, R\}$. In this picture the TM has just completed flow 2.

Graded Artificial Chemistry in Restricted Boundaries

Barak Shenhav, Ran Kafri and Doron Lancet

Dept.of Molecular Genetics and the Crown Human Genome Center,
The Weizmann Institute of Science, Rehovot 76100, Israel
{barak.shenhav, ron.kafri, doron.lancet}@weizmann.ac.il

Abstract

The question of the origin of life is addressed by artificial life research, particularly in the realm of artificial chemistry. Such artificial chemistry is described by our Graded Autocatalysis Replication Domain (GARD) model. GARD depicts an unorthodox scenario suggested for emergence of life – the 'lipid world'. The model concerns molecular assemblies with mutual catalysis in an environment containing a plethora of molecular species. Many aspects of GARD were amply discussed. Here we concentrate on the importance of size constraints as depicted by the basic model and several of its variants. Occasional fission of a GARD assembly, which restricts the assembly size, is crucial for generating compositional quasi-stationary states ('composomes'). In a spatial version of GARD, bounded environments yield spontaneous emergence of different ecologies. Limiting the size of a population of GARD assemblies gives rise to a complex population dynamics. The last example, with possible wider impact to chemistry and nano-technology, suggests that size limit can give rise to spontaneous symmetry breaking. This latter result is compared to the classic Frank's model for homo-chirality, which requires explicit inhibition. We conclude that size restrictions are fundamental in the field of origin of life and artificial life, not only in order to facilitate evolutionary processes, as previously suggested, but also, for augmenting the dynamics portrayed by different scenarios and models.

Introduction

The question "How does life arise from the nonliving?" is first on the list of open problems in artificial life [1]. The study of the origin of life is traditionally considered to be related to classic chemistry. Starting from the seminal experiments of Oparin and Miller, a large body of knowledge was gathered on the synthesis of organic matter under possible prebiotic conditions. Combined with later insight regarding the likely supply of compounds from extraterrestrial sources, it seems that the question of availability of basic building blocks for early life is largely resolved, though many of the details are still missing, such as the exact nature of the molecules, their quantities or the environmental parameters (e.g. temperature, pressure, pH).

The assembly or emergence of life, given that the right building blocks do exist, is still out of the scope of experimental chemistry. The question of organization in a seemingly random chemical scenario was tackled by theoretical models, in the realm of artificial chemistry [2],

such as the Quasi-species and the hyper-cycle models [3, 4]. Strongly related to the work presented hereafter are the models depicting the *metabolism first* [5] approach concerning the emergence of autocatalytic sets [6, 7] and their capacity to evolve [8, 9]. These models use very simplified artificial chemical rules, usually with binary (all or nothing) parameters.

The Graded Autocatalysis Replication Domain (GARD) model takes an intermediate approach. While due to the unavailability of details, GARD is comprised of abstract molecules, it employs rigorous chemical kinetics equations and employs parameters from a realistic chemical probability distribution.

The Model

GARD entails non-covalent assemblies of mutually catalytic molecules. The molecules are usually thought to be amphiphiles (lipids), forming spontaneous molecular assemblies, e.g. micelles or liposomes, as suggested by the 'lipid world' scenario for origin of life [10]. Under the choppy conditions on early earth, such assemblies are expected to undergo occasional fission, which can serve as a primordial progeny-generation mechanism. This process presents the transfer of *compositional information* from a parent assembly to daughter assemblies as discussed in details in previous work [11, 12].

Model Description

The basic GARD model describes a single assembly in an environment containing a finite molecular repertoire of size N_G. The external concentration of the molecules, ρ, is generally taken to be equal for all species. The internal molecular counts are n_1, n_2 …, (or in vector form $\mathbf{n} = [n_1 \ \ldots \ n_{N_G}]$). Thus, the assembly size is $N = \sum n_i$. Molecules join and leave the assembly with spontaneous rates k_f and k_b, typically taken to be equal for all molecular species. The crux of the model lies in the introduction of catalytic rate enhancements, exerted by molecules within the assembly on the join/leave reactions. Using a statistical chemistry approach, the values of the rate enhancement exerted by molecules of type j on molecules of type i (β_{ij}) are drawn from a lognormal distribution. This distribution is based of the Receptor Affinity Distribution (RAD) [13] modified for rate enhancement [12, 14].

The characteristic equations that describe an assembly's growth, in terms resembling continuous concentrations, are [11]:

$$\frac{dn_i}{dt} = \left(k_f \rho N - k_b n_i\right)\left(1 + \frac{1}{N}\sum_{j=1}^{N_G}\beta_{ij}n_j\right) \qquad (1)$$

For most computer simulations of GARD, discrete stochastic chemistry algorithms were used. Stochastic chemistry is suitable for small molecular systems because it reflects stochastic dynamic noise and granularity as opposed to the common differential equation approach. Early works used the 'τ-leap' algorithm [15]. Later papers [16, 17] employed the older, yet (for the special case of GARD) much faster, 'first reaction method' [18, 19]. These calculated the reaction rates by:

$$J_i = k_f e_i\left(1 + \sum_{j=1}^{N_G}\beta_{ij}n_j\right) \qquad (2a)$$

$$L_i = k_b n_i\left(1 + \sum_{j=1}^{N_G}\beta_{ij}n_j - \beta_{ii}\right) \qquad (2b)$$

where J_i and L_i are, respectively, the join and leave rates of molecular species i, and e_i is its external count near the assembly. The subtraction of β_{ii} in eq. 2b depicts that a molecule cannot exert catalysis on itself, an idea generally ignored when concentrations are considered.

Under conditions of unlimited external resources, the molar fractions n_i/N reach a single stationary state $\mathbf{n}^*$, which corresponds to the eigen-vector with the highest eigen-value of a linear form of eq. 1 [11, 20]. Moreover, if the external supply of molecules is limited, the system will reach equilibrium for which n_i corresponds to the ratio of k_f and k_b. However, when the growth is interrupted by occasional fission, (or split), the dynamics observed may be altered dramatically. A split is modeled by random removal of molecules from the assembly, once it reaches a critical size N_{MAX}.

The Major Result - Composomes

In a system consisting of a single assembly, when a split is introduced, the effect of limited external resources is generally not observed, as the environment is generally not depleted. Thus, a GARD system is expected to approach the canonic stationary state $\mathbf{n}^*$. Indeed, for the case where the size of the assembly at the split is large ($N_{MAX} \gg N_G$) the assembly assumes a composition near the canonic one (figure 1a).

When $N_{MAX} \approx N_G$, often several quasi-stationary states – *composomes* are observed (figure 1b). Typically, one of the composomes corresponds to $\mathbf{n}^*$, though this does not have to be the most frequent one. Each composome, in this case, arises from a different underlying catalytic network [11]. These networks show mutual catalytic closure [6] to some extent, which gives them profound stability against destructive splits.

For the extreme case where $N_{MAX} \ll N_G$ often the assembly drifts most of the time in a non-stationary state. Yet, numerous short lasting quasi-stationary states are perceived, corresponding to the temporary existence of a relative strong catalyst in the assembly.

To summarize, GARD assemblies do not converge to a single steady state when a size limit is imposed on them by random fission.

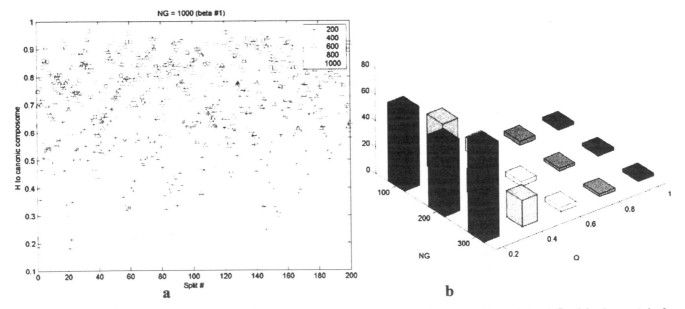

Figure 1 – Behavior of basic GARD assemblies. (a) Similarity to the canonic composition $\mathbf{n}^*$ (as defined in the text) in 3 different runs of GARD for 200 splits each. All runs use the same values for k_f, k_b, β and ρ. N_G=1000 and N_{MAX} differ (200, 600, 1000). (b) The average number of composomes observed in runs of GARD for 2000 splits each with different N_G and different Q ($= N_G/N_{MAX}$) values. Each bin was computed by averaging 30 runs. Composomes were found using the 'on-the-fly' algorithm described in [17].

GARD Extensions

Recently, we have begun to explore more realistic, and generally more computationally demanding, extensions of the GARD models.

Polymer GARD (P-GARD) introduces additional reactions, which allow the formation of more complex molecules. These reactions mimic the formation of chemical covalent bonds between two molecules within the assembly. In a first study of the model [17], with solely monomers (basic molecular species) and dimers (a concatenation of two monomers), we have observed composomes that have appreciable dimer content, and the appearance of novel metabolism-like networks for internal dimer synthesis.

The Computational Origin of Life Endeavor (CORE) [21] aims, by means of extensive computational power, to turn GARD simulations as realistic as feasible. This includes, as a first step, large molecular repertoires and P-GARD simulations. A further step would be to use exact molecular dynamics simulations, rather than the statistical chemistry approach, for obtaining parameters values. This will allow GARD simulations with concrete molecules in place of the abstract ones used contemporary.

Chiral GARD (C-GARD) (Ron Kafri *et al.* – in preparation) imposes restrictions on the values of β_{ij} to reflect that most molecular species have two chiral enantiomers (D, L). According to 'Wigner's rule', D-enatiomers of a molecular species i interact with D-enatiomers of molecular species j in the same manner that L-enatiomers of molecular species i interact with L-enatiomers of molecular species j, that is:

$$\beta_{i_D j_D} = \beta_{i_L j_L} \qquad (3a)$$

The rule also postulates that:

$$\beta_{i_L j_D} = \beta_{i_D j_L} \qquad (3b)$$

With such β values, when composomes are observed, they tend to be either symmetric (racemic), containing equal quantity of both ennatiomers for each molecular species, that is:

$$n_{i_L} \approx n_{i_D} \quad i = 1, ..., N_G \qquad (4a)$$

or asymmetric, e.g.

$$n_{i_L} \approx 0 , n_{i_D} > 0 \qquad (4b)$$

GARD Populations explores the spatio-temporal development of populations of assemblies. This is investigated in a direct manner with a computer simulation consisting of many assemblies scattered in a spatial grid [16] and with more abstract formalism [17] (further detailed in Arren Bar-Even *et al.* – in preparation). This

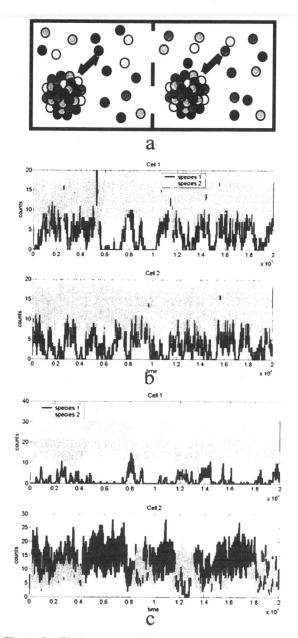

Figure 2 – The two assemblies two environment model. (a) A schematic description of the model consisting of N_G different molecular species (different shaded circle), two assemblies each dwelling in its own environment (different squares). Molecules in each environment can join/leave the assembly (depicted by the arrows). (b) The dynamics of the two assemblies with no diffusion between the environments. Light gray indicates the number of molecules supporting composome 1 (the dominant one) and Black the number of molecules supporting composome 2 (the second best composome). At the beginning, there are 20 copies of each molecular species at each environment . (c) same as case b with diffusion.

quest sheds light on emergence of rudimentary ecologies of mesobiotic entities, intermediates on the path between inanimate prebiotic chemical entities to full-fledged (biotic) living cells [22].

Results

While the basic GARD model governs the framework of our efforts, we have considered several simple variants of it, which provide insights on different aspects of the behaviors observed.

Two Environments – Two Assemblies

An example for such GARD variant is a simple population model consisting of only two assemblies, each dwelling in its own external environment (figure 2a). The rules governing the dynamics of the assemblies are similar to the rules for a single assembly described above (eqs. 2a, 2b). If the environments are independent, then the dynamics of each assembly resembles that of a single assembly.

A more interesting case is when the two environments do interact, for example, by allowing diffusion between them. In many cases, such as when the environments consist of large amounts of each molecular species or when the diffusion is fast, both environments tend to portray similar dynamics (figure 2b). Yet, if the environment size is small compared to the size of the assemblies, an initial symmetric distribution of molecules between the environments may be spontaneously broken. To be exact, one of the assemblies assumes a certain composome and molecules constituting this composome concentrate in one compartment, while the other assembly assumes another composome, leading to the concentration of other molecular species in the other compartment (figure 2c).

This result are akin, to some extent, to the results reported for many assemblies in a single environment (Lipidia) [16] showing that limited resources lead to deviation from typical dynamics. Yet, there are differences in mechanism and in outcome: spatial organization rather than speciation in Lipidia. The phenomena observed amount to induced environmental inhomogeneity, rather than merely a stochastic one.

Population Catastrophes

The formalism developed to investigate populations of GARD assemblies treats composomes as abstract entities with emergent properties such as the time elapsing between assembly divisions, the probability of an assembly to preserve its composomal state following a split or the emergence likelihood of the composome [17, 23]. The use of such abstract entities, rather than detailed molecular assemblies, has enabled easy computer simulations of populations of assemblies. In the case where the population size was not bounded, a single stationary population distribution was observed, in similarity to the canonic stationary state **n*** described above. In contrast, several quasi-stationary population distributions were observed in cases where the constant population size was imposed [24] or where populations went through occasional

'catastrophes' ('killing' half of the population once it reached a threshold size).

Detailed analysis of this result and other insights coming up from this model are out of the scope of the current manuscript and are described elsewhere (Arren Bar-Even *et al.* – in preparation). Yet, it is another example that restricting the dynamics to a limited size alters the system behavior.

Chiral toy model

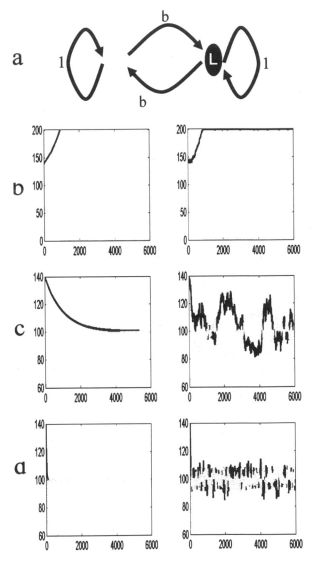

Figure 3 – The chiral toy model. (a) A schematic description of the model consisting of an L and D molecules with auto-catalysis of unit size and cross-catalysis of b. (b-d) The fraction of L molecules (Dark) and D molecules (Light) as function of time with different b values. Left shows the continuous case with no size limit (Frank's model) and right shows the stochastic dynamics with split (the chiral toy model). The values of b are: -0.05, 0.05 and 2 respectively.

We have explored the simplest possible chiral GARD, one with two molecular species, a single L molecular species and its corresponding D molecular species. For this special case, there are two independent rate enhancement values – autocatalysis (β_{LL}) and cross-catalysis (β_{LD}) since from eq. (3a) $\beta_{DD} = \beta_{LL}$ and from eq. (3b) $\beta_{DL} = \beta_{LD}$. Selecting the right set of units, we can take the autocatalysis to be 1, without losing any generality, denoting the cross-catalysis with b (figure 3a). The growth equations of this abstract GARD model are thus reduced to:

$$\frac{dn_L}{dt} = (k_f e_L - k_b n_L)(1 + n_L + bn_D) \qquad (5a)$$

$$\frac{dn_D}{dt} = (k_f e_D - k_b n_D)(1 + n_D + bn_L) \qquad (5b)$$

Eqs. (5a, 5b) resemble those described in Frank's model for the spontaneous emergence of homo-chiral systems [25], whichcontain solely L molecules or solely D molecules. In order to observe such spontaneous symmetry breaking, Frank's model requires the cross-catalysis to be negative, i.e. inhibitory. Indeed, for the cases where $b < 0$ or $b > 1$ the current model and Frank's model show the same behavior up to stochastic noise (figure 3b, 3d). However, for $0 < b < 1$ the two models display different behaviors. In Frank's model a racemic system emerges, containing similar amounts of L and D molecules. In our toy chiral GARD model, which includes a size limit (N_{MAX}), temporary biases towards L or D (enantio excesses) are observed (figure 3c).

This model may serve as a basis for studies of enantio – selection in larger molecular repertoires. For example, an asymmetric multi-component composome may be analogous to a bimolecular case with dominant autocatalysis (b < 1), while a symmetric multi-component composome could correspond to a bi-molecular system in which cross-catalysis is stronger ($b > 1$). This appears to be valid only under limited size constraints.

Whether the model relates to a single D and L molecular species or to a multi-molecular composome, it highlights the impact of imposing size boundaries on the system, leading to temporal symmetry breaking without requiring explicit inhibition.

Discussion

The basic GARD model sheds light on many aspects usually not concerned in traditional origin of life studies. We have concentrated here on the impact of imposing size restriction on the system. We have indicated that in the basic GARD model, as well as in several of its derivatives, such restrictions significantly alter the ensuing dynamics. Generally, the dynamics observed is more multifaceted, e.g. showing more quasi-stationary states or spontaneous symmetry breaking. Thus, we suggest size limiting as a general mechanism for turning seemingly lifeless systems, ones that portray a random walk or convergence to a single steady state, into more elaborate systems with several or many quasi-stationary states, capable of manifesting more lifelike faculties.

Other studies have highlighted the importance of restricted boundaries for the facilitation of evolutionary processes. For example, compartmentalization was suggested in order to maintain hyper-cycles against possible molecular parasites [26]. Another example is the requirement of encapsulating molecular replication mechanisms (replicases) by lipid vesicles. This was suggested to serve a crucial rule in allowing the evolution and takeover of better replicases [27]. We propose that augmenting the dynamics of the systems is another important aspect of closure, which requires further study.

Acknowledgments

We thank Barak Naveh for critically reading the manuscript. The research was partially supported by the European Commission through grant number HPRI-CT-1999-00026 (the TRACS Programme at EPCC) and by the Crown Human Genome Center.

References

[1] Bedau, M.A., McCaskill, J.S., Packard, N.H., Rasmussen, S., Adami, C., Green, D.G., Ikegami, T., Kaneko, K. and Ray, T.S., (2000), Open Problems in Artificial Life. Artif Life 6:363-376.

[2] Dittrich, P., Ziegler, J. and Banzhaf, W., (2001). Artificial Chemistries - a Review. Artif Life 7:225 –275.

[3] Eigen, M., (1971). Selforganization of Matter and the Evolution of Biological Macromolecules. Naturwissenchaften 58:465-523.

[4] Eigen, M. and Schuster, P., (1979). The Hypercycle. Berlin: Springer Verlag.

[5] Morowitz, H.J., (2002). The Emergence of Metabolism. In The Emergence of Everything: How the World Became Complex. Oxford Press.

[6] Farmer, J.D., Kauffman, S.A. and Packard, N.H., (1986). Autocatalytic Replication of Polymers. Physica 22D:50-67.

[7] Bagley, R.J. and Farmer, J.D., (1991). Spontaneous Emergence of a Metabolism. In Langton, C.G., Taylor, C., Farmer, J.D. and Rasmussen, S., Artificial Life II. Addison-Wesley.

[8] Jain, S. and Krishna, S., (2002). Large Extinctions in an Evolutionary Model: The Role of Innovation and Keystone Species. Proc. Natl. Acad. Sci. U. S. A. 99:2055-2060.

[9]   Bagley, R.J., Farmer, J.D. and Fontana, W., (1991). Evolution of a Metabolism. In Langton, C.G., Taylor, C., Farmer, J.D. and Rasmussen, S., Artificial Life II. Addison-Wesley.

[10]   Segre, D., Ben-Eli, D., Deamer, D.W. and Lancet, D., (2001). The Lipid World. Origins Life Evol B 31:119-145.

[11]   Segre, D., Ben-Eli, D. and Lancet, D., (2000). Compositional Genomes: Prebiotic Information Transfer in Mutually Catalytic Noncovalent Assemblies. Proc Natl Acad Sci U S A 97:4112-7.

[12]   Segre, D., Shenhav, B., Kafri, R. and Lancet, D., (2001). The Molecular Roots of Compositional Inheritance. J Theor Biol 213:481-491.

[13]   Lancet, D., Sadovsky, E. and Seidemann, E., (1993). Probability Model for Molecular Recognition in Biological Receptor Repertoires - Significance to the Olfactory System. Proc. Natl. Acad. Sci. U. S. A. 90:3715-3719.

[14]   Lancet, D., Kedem, O. and Pilpel, Y., (1994). Emergence of Order in Small Autocatalytic Sets Maintained Far from Equilibrium - Application of a Probabilistic Receptor Affinity Distribution (Rad) Model. Ber. Bunsen-Ges. Phys. Chem. Chem. Phys. 98:1166-1169.

[15]   Gillespie, D.T., (2001). Approximate Accelerated Stochastic Simulation of Chemically Reacting Systems. J. Chem. Phys. 115:1716-1733.

[16]   Naveh, B., Sipper, M., Lancet, D. and Shenhav, B., (submitted). Lipidia: An Artificial Chemistry of Self-Replicating Assemblies of Lipid-Like Molecules. In Proceedings of Artificial Life IX, Boston, MA, USA. MIT press

[17]   Shenhav, B., Bar-Even, A., Kafri, R. and Lancet, D., (in press). Polymer Gard: Computer Simulation of Covalent Bond Formation in Reproducing Molecular Assemblies. Origins Life Evol B

[18]   Gillespie, D.T., (1976). General Method for Numerically Simulating Stochastic Time Evolution of Coupled Chemical-Reactions. J. Comput. Phys. 22:403-434.

[19]   Gillespie, D.T., (1977). Master Equations for Random Walks with Arbitrary Pausing Time Distributions. Physics Letters A 64:22-24.

[20]   Jain, S. and Krishna, S., (1998). Autocatalytic Sets and the Growth of Complexity in an Evolutionary Model. Phys. Rev. Lett. 81:5684-5687.

[21]   Shenhav, B. and Lancet, D., (2004). Prospects of a Computational Origin of Life Endeavor. Origins Life Evol B 34:181-194.

[22]   Shenhav, B., Segre, D. and Lancet, D., (2003). Mesobiotic Emergence: Molecular and Ensemble Complexity in Early Evolution. Adv Complex Syst 6:15-35.

[23]   Bar-Even, A., Shenhav, B., Kafri, R. and Lancet, D., (in press). The Lipid World: From Catalytic and Informational Headgroups to Micelle Replication and Evolution without Nucleic Acids. In Seckbach, J., Cellular Origin, Life in Extreme Habitats and Astrobiology (Cole).

[24]   Kuppers, B.O., (1983). Molecular Theory of Evolution. Berlin-Heidelberg: Springer-Verlag.

[25]   Frank, F.C., (1953). On Spontaneous Assymetric Synthesis. Biochem Biophys Acta 11:459-463

[26]   Eigen, M., Gardiner, W.C. and Schuster, P., (1980). Hypercycles and Compartments. J Theor Biol 85:407-411.

[27]   Szostak, J.W., Bartel, D.P. and Luisi, P.L., (2001). Synthesizing Life. Nature 409:387-390.

Spacial Representation for Artificial Chemistry Based on Small-World Networks

Hideaki Suzuki

ATR Network Informatics Laboratories, Kyoto 619-0288 Japan

hsuzuki@atr.jp

Abstract

A method to simulate molecular reactions with a graph whose nodes represent molecules or atomic modules and whose edges represent the bond/contact relation is proposed. The graph is updated by two actions: passive rewiring of contact edges that gives the graph a small-world property and active rewiring by active nodes' programs that rewire bond and contact edges. As examples, the replication of a chain molecule and the partitioning of a network are successfully simulated.

Introduction

One of the most important properties of artificial life models is spatial representation. The information space of a modern computer – the sequence of data words addressed with integer numbers – is far from the Euclidean liquid space wherein biological molecules move and interact. To compensate for this difference and realize molecular interactions in a computational medium, several artificial information space models have been proposed: one dimensional space of Tierra (Ray 1992; 1997) or SeMar (Suzuki 1999a; 1999b; 2000), two dimensional space of Avida (Adami and Brown 1994; Adami et al. 2000) or Amoeba (Pargellis 1996a; 1996b; 2001), lattice models by Ono, Madina et al. (Ono and Ikegami 1999; 2001; Madina et al. 2003), and Speroni's planar graph (Speroni di Fenizio et al. 2001). Despite having merits, these one, two, or three dimensional artificial spaces do not satisfy all of the conditions for the emergence of complex interaction between computational objects (Suzuki et al. 2003).

In a living cell, the movement of biological molecules is based on diffusion processes. After accomplishing a reaction, a molecule moves after being influenced by the forces of other molecules, collides with molecules, and finally meets its next molecule with which it will react. Such molecular movement can be modeled with mathematical expression that specifies the interval and order of molecular collisions; hence, if we were able to devise a mathematical model that has the same properties as molecular collisions, then, we would be able to simulate molecular interaction more abstractly.

Based upon this notion, we propose a new model of artificial chemistry called "Network-based Artificial Chemistry" (NAC) that represents molecular interaction by a graph whose nodes express molecules or atomic modules in macromolecules and whose edges express contact (collisions) or bonds between molecules. The graph is modeled using 'small-world' networks which include many more clusters than random networks (Watts and Strogatz 1998). While the contact edges are passively rewired according to a local rewiring rule that emulates molecular collisions in liquid, the contact and bond edges are actively rewired by the active nodes' functional programs. To test the ability of the NAC, we conducted two experiments: the replication of molecular chains by active molecules, *polymerase* and *helicase*, and the partitioning of a hydrophilic cluster by passive rewiring of hydrophilic and hydrophobic contact edges and active molecule *splitase*. The resultant graphs are visualized in figures.

After presenting the local rewiring rule for updating contact edges, the basic model for the NAC is given. Then, experimental results, closing remarks, and some conclusions are given.

Passive Rewiring Rule

Collisions between biological molecules are strongly influenced by the dimensionality and the structure of the space in which the molecules are moving. Generally speaking, the closer molecules are to each other, the more likely they are to collide with each other; hence, we can naturally assume that after colliding with molecule B, molecule A will make contact with molecules that have been in contact with B or other molecules that have been in contact with A.

The rewiring rule used in this paper is based on this conjecture. We describe the molecules' contact relation by undirected edges. At every time step, a starting node and its adjacent node, stopping node, is randomly chosen from the network, and the edge between them is eliminated. Before eliminating it, a new stopping node is randomly chosen from nodes that have a distance of two to the starting node, creating a new undirected edge. When repeated, this rewiring process increases the cluster coefficient C of a graph while keeping the average path length L constant, creating a small-world network (Watts and Strogatz 1998). A similar rewiring procedure has already been proposed which focused on an acquaintance network in human society (Davidsen et al. 2002). A typical change of C and L by the present rule is shown in Fig. 1.

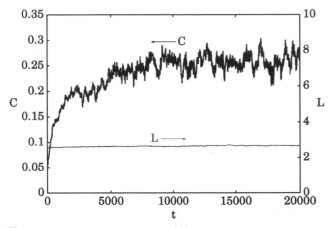

Figure 1: Cluster coefficient (C) and average path length (L) as a function of the rewiring number (t). The initial graph is random where the node number $N = 200$ and the average degree $\overline{K} = 10$. After t exceeds the edge number $N\overline{K}/2 = 1000$, C grows much larger than the random graph's theoretical value $\overline{K}/N = 0.05$, while L is kept equal to the random graph's value $\log N / \log \overline{K} = 2.3$.

Molecular Interaction in the NAC

In an actual biological cell, molecular activities are governed by four different kinds of forces: covalent, ionic, hydrogen, and van der Waals, listed in descending order of strength (Alberts et al. 1994). (Roughly speaking, a covalent bond is ten times stronger than an ionic bond, an ionic bond is ten times stronger than an hydrogen bond, and so on.) Specifically, the hydrogen bond is a key force for various interactions between molecules. The establishment of many hydrogen bonds between two macromolecules produces strong adhesion. Moreover, hydrogen bonds between water molecules and chemicals cause a hydrophilic/hydrophobic interaction and a partitioning of the solvent space.

On the analogy of these forces, we prepared four kinds of edges in the network: two kinds of directed edges for bonds, covalent (cv[ ]) and hydrogen (hy[ ]), and two kinds of undirected edges for contact, hydrophilic (il[ ])

and hydrophobic (ob[ ]). (il and ob were named after particular letters to distinguish from 'hydrophILic' and 'hydrophOBic'.) Here, cv[ ], etc. are node lists owned by each node. In the following, the rewiring of edges is always represented by the local modification of the content of the node lists. In this paper, the maximum sizes of cv[ ] and hy[ ] are always fixed at Cp = 2 and Hp = 3, respectively. (Cp is an abbreviation of 'cv pointer' whose first letter is capitalized to indicate that it stores the maximum value.) On the other hand, the maximum sizes of il[ ] and ob[ ], Ip and Op, are set to Ip = ∞ and Op = 0 for a hydrophilic node and Ip = 0 and Op = ∞ for a hydrophobic node, reflecting the 'hydro-property' of a node.

We prepared active and passive nodes in NAC. A passive node has the node lists plus a string representing its atomic constitution (sa) and an integer representing molecular status (ta). An active node has the node lists, sa and ta, plus several working registers for strings, integers, or nodes and its own program.

At every time step, the molecular graph is updated by the operation of active nodes and passive rewiring. In the active operation, every active node's program is decoded and executed. These operations not only modify its own working registers but also change the status (integer) or bond/contact edges of nearby nodes, which causes a molecular reaction or the transportation of nodes. In passive rewiring, we applied the local rewiring rule described in the previous section to the contact edges in il[ ] and ob[ ]. Although a deleted edge is randomly chosen regardless of the hydro-property of the node pair, if the hydro-properties of a created edge's node pair are in conflict with each other, the rewiring is canceled. (A hydrophilic node cannot have node information in ob[ ] and vice versa.) This forces the network to obey the rules for hydrophilic/hydrophobic interaction between molecules.

Through these operations, the bond relation in cv[ ] and hy[ ] is modified only by active operations, whereas the contact relation in il[ ] and ob[ ] is modified by both active and passive operations. This can be compared to biomolecular interaction in which molecular bonds cannot be rewired without enzymes, whereas molecular contact can be updated probabilistically by Brownian motions without catalysts.

Experiments

Replication of Molecular Chains

The elementary process of information replication in a living cell is done by complementary matching between nucleotides of DNA and RNA. In imitation, we prepared a network with a number of free hydrophilic nucleotide nodes, a polynucleotide (a chain of nucleotides) bonded in covalent bonds cv[ ], and two kinds of hydrophilic enzyme nodes, named *polymerase* and *helicase*. Polymerase's function is to read a single strand of polynucleotide with hy[0] and polymerize nucleotides with hy[1], making a double strand composed of identical nucleotide pairs. Helicase splits a double strand of polynucleotide into single ones. See Appendices A and B for the detailed programs for polymerase and helicase. Every nucleotide node has one out of four different 'base strings' in sa. The head and tail nucleotide nodes have special status numbers in ta by which polymerase and helicase identify the head/tail of a chain.

Starting from a random network which includes these molecules, we simulated the NAC reactions using the update rules presented in the previous section. The replication process of the chain is basically the same as that of Squirm3 in (Hutton 2002); however, NAC is unique because it propels replication not by rules prepared as physical laws outside but by the programs stored in active enzyme nodes. See Fig. 2 for a typical run of the replication; after 7360 time steps, the replication of a nucleotide chain is successfully carried out.

Partitioning of the Network

A lipid is an amphipathic molecule composed of hydrophilic and hydrophobic parts, and when it agglomerates, it constitutes a sheet so that hydrophobic parts will not be in contact with water (solvent). According to recent discoveries in molecular biology (Alberts et al. 1994), the inside and outside sheets of a lipid bilayer have different phospholipids (hydrophilic parts); hence, the biological membrane is asymmetrical.

In this experiment, we described a lipid with a pair of nodes, hydrophilic and hydrophobic, bonded by cv[ ]. According to strings stored in the sa, the hydrophilic nodes of the pairs are classified as inner or outer, by which a particular hydrophobic enzyme named *splitase* actively cuts hydrophilic contact bonds between a hydrophilic node that might belong to the inside of the membrane and a hydrophilic node that might belong to the outside of the membrane. See Appendix C for the detailed algorithm of the splitase program.

To test the possibility that lipid pairs might constitute a membrane in NAC, we conducted a number of experiments with a network including free nucleotide, lipid pairs, and splitase. Starting from a random network, we updated the network with the rules described in the previous section while visualizing the structure of the network. From these experiments, we observed that the complete partitioning of a hydrophilic cluster cannot take place unless passive rewiring works on the network, the splitase program is properly designed, and the numbers of lipid pairs and splitase are appro-

priately adjusted. Figure 3 shows a typical successful run of these experiments where the initially-connected hydrophilic subgraph is partitioned into two isolated subgraphs after about ten thousand time steps. We can conclude from these results that under appropriate conditions, NAC can represent the partitioning of a hydrophilic region by lipid molecules through the passive rewiring that facilitates the occurrence of clusters and the active rewiring by the splitase.

Discussion

A new method of spatial representation for artificial chemistry was proposed. Using a local rewiring rule that increases clusters, NAC can describe the passive transportation of molecules by diffusion as well as molecular reactions or active transportation by enzymic molecules. The ability of NAC was demonstrated using two experiments: template replication and domain partitioning.

We discuss several conclusions and possible future extensions of NAC in the following section.

Flexibility to represent molecular movement: A graph is a purely mathematical framework that can represent a variety of molecular interactions. Being obviated from the limitation of the rigidness of previous physical space models, NAC expresses both strong and weak interactions between artificial molecules or atomic modules in a unifying manner. This might enable NAC to represent the following molecular movement or activities in a cell: partitioning a reaction domain by semi-permeable membrane, freely changing the number of symbols/molecules in a cell, mingling between the partitioned compartments, transportation of signal molecules in or between compartments, random rearrangement of symbols, and selective transportation of symbols by active molecules. These activities are equivalent to the latter half of the ten necessary conditions for an artificial environment presented in (Suzuki et al. 2003), suggesting that NAC might be a promising approach for constructing an artificial environment for the evolution of complex forms of life.

High-dimensional space representation: In 2000, the author argued that "the dimension of a computational medium is of vital importance for the richness of the functions possibly emerging in the computational system. ... The dimension determines the number of operator objects that can cooperate simultaneously. ... We cannot expect the emergence of higher functions in a computational system when only a small number of operator objects can cooperate for a single task" (Suzuki 2000). Using the graph representation, NAC allows a

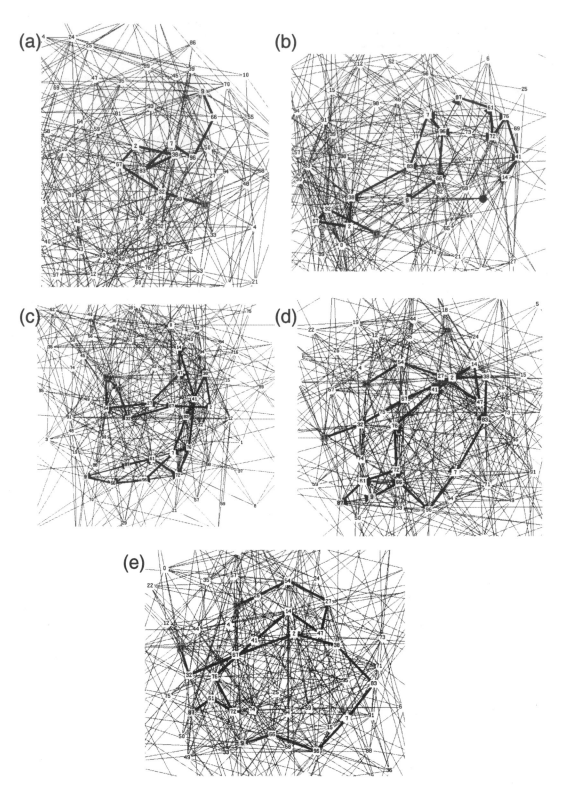

Figure 2: A typical run of the replication of a polynucleotide in a NAC cell with node number 100 and average degree 10 for hydrophilic edges. The network includes one polymerase and one helicase. (a) part of a snapshot of the initial random network at t=0, (b) during polymerization at t=3350, (c) one double strand at t=5000, (d) during the unzipping at t=7350, and (e) two unzipped strands at t=7360. Here and in the next figure, bold black edges represent covalent bond, thin black edges represent hydrogen bond, and thin grey edges represent hydrophilic contact. A white node is a nucleotide, a dark grey node in (b) is polymerase, and a node #28 in (d) is helicase. The head/tail of a polynucleotide is colored grey. The graphs are visualized with (aiSee, software).

(a)

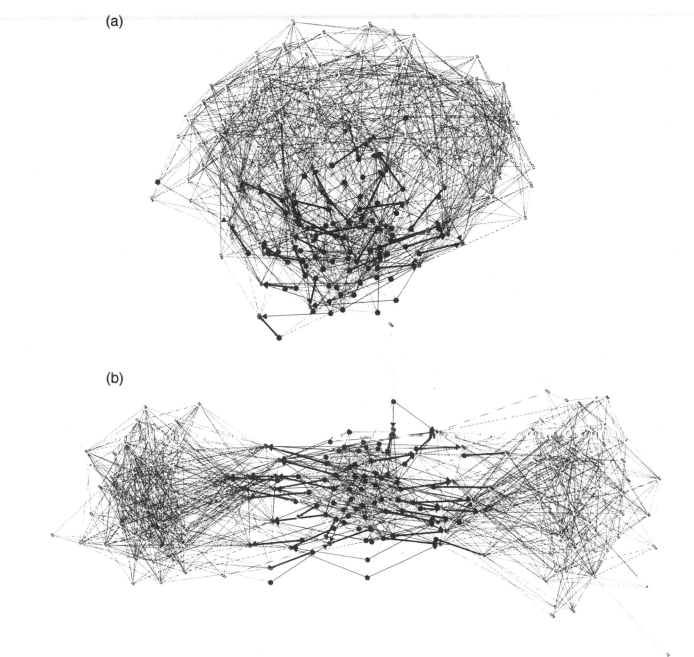

(b)

Figure 3: A typical run of the partitioning of the hydrophilic cluster in a NAC cell that has a total of 200 nodes and an average hydrophilic/hydrophobic degree of 10/20, respectively. The network includes 30 lipid pairs (15 inside and 15 outside), 40 nodes of splitase, and 100 free nucleotides. (a) snapshot at t=3000 and (b) t=10000. Thin black edges represent hydrophilic contact, and thin red edges represent hydrophobic contact. White nodes are hydrophilic nucleotides, gray nodes are hydrophilic parts of lipid, and black nodes are hydrophobic parts of lipid. Dark gray nodes which are located in the inner cluster in (b) are splitase.

node to be adjacent to any number of nodes in the network. Though particular size limitations are present for such adjacent node lists as cv[ ] or hy[ ], if several active nodes are properly connected, they can simultaneously work on a single task at the same time, suggesting that NAC can emulate such complex molecular cooperation as protein-protein interaction. The author believes that the replication in Fig. 2 demonstrates this ability of NAC, even as simple as it is. Combined with the flexibility mentioned above, this striking property of NAC might provide us with a method to simulate the organization and movement of a 'protein complex,' a highly-functional cluster of active nodes in a cell.

Future extensions: The following research topics are left for future studies: implementing a 'folding' mechanism which changes a polynucleotide into a single active molecule and makes an active node program from the string sequence of a polynucleotide; enabling a NAC cell to self-reproduce by putting the replicating and partitioning molecules into a single cellular network and synchronizing their activities; mathematically analyzing molecular diffusion in a liquid space, and examining the plausibility of the passive rewiring rule of NAC.

Dr. Ono of ATR labs provided useful comments for this work. Dr. K. Shimohara of ATR labs actively encouraged the study. This study was supported by the National Institute of Information and Communications Technology (NICT) and Doshisha University's Research Promotion Funds, Japan.

Appendix A: Pseudocode of Polymerase

The Network Cell is implemented in Java (Cadenhead and Lemay 2004). A node is prepared as an *object* in Java, and cv[ ], hy[ ], il[ ], ob[ ], sa, ta, and other node variables (na, nb, and nc) are prepared as *instance variables*. The program is implemented as a *method* of the object for an active node. Here and in the subsequent appendices, '.' is a specific Java operator representing a class member: for example, na.ta means ta of na.

```
if(hy[0] is null) // Template head is not yet found
{
  na <- (random adjacent node);
  if(na is a nucleotide AND na.ta=3)
  {
    hy[0] <- na; // hy[0] = template head
    na.ta <- 5;
  }
}
if(hy[0] is not null AND hy[1] is null)
// The first nucleotide is not yet found.
{
  na <- (random adjacent node);
```

```
  if(na is a nucleotide AND na.ta=-1 AND
     hy[0].sa=na.sa) // checking nucleo. identity
  {
    hy[1] <- na; // hy[1] = first nucleotide
    na.ta <- 5;
  }
}
if(hy[0] is not null AND hy[1] is not null AND
   hy[0].ta is not 4) // Under polymerization
{
  na <- (random adjacent node);
  if(na is a nucleotide AND na.ta=-1 AND
     hy[0].cv[0].sa=na.sa)
                      // checking nucleo. identity
  {
    hy[0].hy[0] <- hy[1]; // creates a new bond
    hy[1].hy[0] <- hy[0]; // creates a new bond
    hy[1].cv[0] <- na;    // creates a new bond
    na.cv[1] <- hy[1];    // creates a new bond
    hy[0] <- hy[0].cv[0]; // walks along chain
    hy[1] <- na;          // walks along chain
    hy[1].ta <- 2;
  }
}
if(hy[0] is not null AND hy[1] is not null AND
   hy[0].ta=4) // Polymerization is finished.
{
  hy[0].hy[0] <- hy[1]; // creates the final bond
  hy[1].hy[0] <- hy[0]; // creates the final bond
  hy[0].ta <- 6;  hy[1].ta <- 6;
  hy[0] <- null;        // releases the chain
  hy[1] <- null;        // releases the chain
}
```

Appendix B: Pseudocode of Helicase

```
if(hy[0] is null)
               // Double strand is not yet found
{
  na <- (random adjacent node);
  if(na is a nucleotide AND na.ta=6)
  {
    hy[0] <- na;        // hy[0] = strand tail
    hy[1] <- na.hy[0];  // hy[1] = strand tail
    hy[0].ta <- 4;  hy[1].ta <- 4;
  }
}
if(hy[0] is not null AND hy[0].ta is not 5)
                            // Under unzipping
{
  hy[0].hy[0] <- null;  // cuts the bond
  hy[1].hy[0] <- null;  // cuts the bond
  hy[0] <- hy[0].cv[1]; // walks along chain
  hy[1] <- hy[1].cv[1]; // walks along chain
}
if(hy[0] is not null AND hy[0].ta=5)
                            // Unzipping is finished.
{
  hy[0].hy[0] <- null;  // cuts final bond
  hy[1].hy[0] <- null;  // cuts final bond
  hy[0].ta <- 3;  hy[1].ta <- 3;
  hy[0] <- null;        // releases the chain
```

```
hy[1] <- null;          // releases the chain
}
```

Appendix C: Pseudocode of Splitase

```
na <- ob[random];
na <- na.cv[1];
if(na is a hydrophilic part of inner lipid)
{
  hy[0] <- na; // hy[0] = inner lipid surface
}else{
  hy[1] <- na; // hy[1] = outer lipid surface
}
if(hy[0] is not null AND hy[1] is not null)
                        // lipid is found
{
  nb <- hy[0].il[random];
  nc <- hy[1].il[random];
  cuts il[]-edge between hy[0] and hy[1];
  cuts il[]-edge between hy[1] and nb;
  cuts il[]-edge between hy[0] and nc;
  cuts il[]-edge between nb and nc;
}
```

References

Adami, C., Brown, C.T.: Evolutionary learning in the 2D artificial life system "Avida." In: Brooks, R., Maes, P. (eds.): Artificial Life IV: Proceedings of an Interdisciplinary Workshop on the Synthesis and Simulation of Living Systems, MIT Press, Cambridge (1994) 377–381

Adami, C., Ofria, C., Collier, T.C.: Evolution of biological complexity. Proc. Natl. Acad. Sci. USA **97** (2000) 4463-4468

aiSee is a commercial software for visualizing graphs with various algorithms such as the spring model. http://www.aisee.com/

Alberts, B., Bray, D., Lewis, J., Raff, M., Roberts, K., Watson, J.D.: Molecular Biology of the Cell, The Third Edition. Garland Publishing, New York (1994)

Cadenhead, R., Lemay, L.: Sams teach yourself Java 2 in 21 days. Macmillan Computer Pub. (2004)

Davidsen, J., Ebel, H., Bornholdt, S.: Emergence of a small world from local interactions - modeling acquaintance networks. Physical Review Letters **88**(12) (2002) 128701

Hutton, T.J.: Evolvable self-replicating molecules in an artificial chemistry. Artificial Life **8** (2002) 341-356

Madina, D., Ono, N., Ikegami, T.: Cellular evolution in a 3D lattice artificial chemistry. In: Banzhaf, W., Christaller, T., Dittrich, P., Kim, J.T., Ziegler, J. (eds.): Advances in Artificial Life (7th European Conference on Artificial Life Proceedings), Springer-Verlag, Berlin (2003) 59–68

Ono, N., Ikegami, T.: Model of Self-Replicating Cell Capable of Self-Maintenance. In: Floreano, D. et al. (eds.): Advances in Artificial Life (5th European Conference on Artificial Life Proceedings), Springer-Verlag, Berlin (1999) 399-406

Ono, N., Ikegami, T.: Artificial chemistry: computational studies on the emergence of self-reproducing units. In: Kelemen, J., Sosik, P. (eds.): Advances in Artificial Life (6th European Conference on Artificial Life Proceedings), Springer-Verlag, Berlin (2001) 186-195

Pargellis, A.N.: The spontaneous generation of digital "Life". Physica D **91** (1996) 86-96

Pargellis, A.N.: The evolution of self-replicating computer organisms. Physica D **98** (1996) 111-127

Pargellis, A.N.: Digital life behavior in the Amoeba world. Artificial Life **7** (2001) 63-75

Ray, T.S.: An approach to the synthesis of life. In: Langton, C.G. et al. (eds.): Artificial Life II: Proceedings of an Interdisciplinary Workshop on the Synthesis and Simulation of Living Systems (Santa Fe Institute Studies in the Sciences of Complexity, Vol. 10). Addison-Wesley (1992) 371–408

Ray, T.S.: Selecting Naturally for Differentiation. In: Koza, J.R. et al. (eds.): Genetic Programming 1997: Proceedings of the Second Annual Conference. Morgan Kaufmann, San Francisco (1997) 414–419

Speroni di Fenizio, P., Dittrich, P., Banzhaf, W.: Spontaneous formation of proto-cells in an universal artificial chemistry on a planar graph. In: Kelemen, J., Sosik, P. (eds.): Advances in Artificial Life (6th European Conference on Artificial Life Proceedings), Springer-Verlag, Berlin (2001) 206-215

Suzuki, H.: An approach to biological computation: unicellular core-memory creatures evolved using genetic algorithms. Artificial Life **5** N.4 (2000) 367–386

Suzuki, H.: Evolution of self-reproducing programs in a core propelled by parallel protein execution. Artificial Life **6** N.2 (2000) 103–108

Suzuki, H.: Core memory objects with address registers representing high-dimensional interaction. In: Calude, C.S. et al. (eds.): Pre-Proceedings of the Workshop on Multiset Processing (WMP-CdeA 2000), Centre for Discrete Mathematics and Theoretical Computer Science Research Report 140 (2000) 249-264

Suzuki, H., Ono, N., Yuta, K.: Several necessary conditions for the evolution of complex forms of life in an artificial environment. Artificial Life **9**(2) (2003) 537-558

Watts, D.J., Strogatz, S.H.: Collective dynamics of 'small-world' networks. Nature **393** (1998) 440-442

Behavioral Adaptive Autonomy. A milestone on the Alife route to AI?

Xabier Barandiaran[1]

[1]Department of Logic and Philosophy of Science, FICE,
UPV-EHU (University of the Basque Country)
PO BOX 1249 / 20080 Donostia - San Sebastian / Spain
barandi@sf.ehu.es
http://www.ehu.es/ias-research/barandiaran

Abstract

While central to robotics, biology and cognitive science, the concept of *autonomy* remains still difficult to make operative in the realm of Alife simulation models of cognitive agents. Its deep significance as a transition concept between life and cognition (a milestone on the Alife route to AI) remains obscured in the intricate relation between metabolic/constructive processes and behavioral adaptive processes in living systems. Within a naturalized and biologically inspired dynamical approach to cognition a definition of *behavioral adaptive autonomy* is provided: homeostatic maintenance of essential variables under viability constraints through self-modulating behavioral coupling with the environment, hierarchically decoupled from metabolic (constructive) processes. This definition allows for a naturalized notion of behavioral adaptive functionality (that defines a proper level of modelling within Alife), structurally and interactively emergent: the mapping of the agent-environment system's state space trajectories into the viability subspace of the essential variables of the organism.

Introduction

While central to Alife (Varela and Bourgine, 1992; Ruiz-Mirazo and Moreno, 2004), robotics (Maes, 1991), biology (Varela, 1979) and cognitive science (Christensen and Hooker, 2000), the concept of *autonomy* remains still difficult to make operative in the realm of Alife approaches to cognition[1]. In particular it is not yet completely clear whether an artificial cognitive autonomous agent could be built without the underlaying autopoietic autonomy being implemented; and it is not clear how could the cognitive autonomy of an artificial or simulated agent be measured or implemented. The notion of autonomy and its deep significance remains obscured in the intricate relation between metabolic/constructive processes and behavioral/cognitive processes in living beings and in the highly abstract conceptual framework in which it has been developed by Francisco Varela.

---

[1] We won't discuss here the notion of autonomy in relation to the origin and synthesis of life, this issue has long being discussed elsewhere (Ruiz-Mirazo and Moreno, 2004). Our main concern is the notion of autonomy in the cognitive domain and its interaction with basic (autopoietic) autonomy.

But far from being a neglectable term, the notion of autonomy has inspired a whole range of research projects within the Alife community and it does in fact capture the core of the conceptual shift behind most Alife research (biological grounding, self-organization, emergence, embodiment and situatedness). In fact, the subtitle of the First European Conference on Artificial Life ("towards a practice of autonomous systems") reflects the significance of the concept.

The main goal of this paper is to provide an specific definition of *behavioral adaptive autonomy* that can be implemented in Alife scientific practices and used to model adaptive behavior. The main thesis is that the hierarchical decoupling of the nervous system from metabolic constraints specifies the domain of behavioral adaptive dynamics. In this domain autonomy is defined as the capacity of the system to interactively maintain its essential variables under viability constraints.

We start (section 2) by briefly analyzing the variety of uses that the term *autonomy* has had in the literature and compiling (section 3) a set of key notions around autonomous approaches. Section 4 reconstructs the concept of basic (autopoietic) autonomy and the way it relates to functionality. We then move to specify the organization of the nervous system in the context of the whole organism (section 5) to end up defining autonomy and functionality in the dynamical framework of the behavioral adaptive domain (section 6). Finally section 7 discusses some implications of the present approach for Alife simulation models.

A quick overview of the literature

The term *autonomous robotics* has been used since the 90s (Maes, 1991) to refer to a set of engineering constraints on the construction and testing of robots, thus labeling a style of robotic research in cognitive science and engineering. Such constraints include conditions like no remote control of the agent, no external energy supply, mobility in the robot, no human intervention in robot task solving or real-time response in real-world environments. Close to situated robotics (Brooks, 1991), autonomous robotics highlights physicality, embodiment, situatedness and dynamicism versus abstract, virtual and formal approaches to artificial intelligence (in which agents operate in controlled formal or virtual environments or in toy like worlds without dynamical constraints). As a consequence of the real-world interaction of the robot the emphasis is often put on the *viability* of the robot as a task achieving agent: a self-generated and robust capacity to respond to environmental changes.

The practice of *autonomous robotics* has forced some engineers to go beyond the specification of a list of engineering constraints and to develop a more elaborated notion of autonomy that specifies the kind of interaction process that is established between the robot and its physical environment, the dynamic structure and properties of the control mechanisms and the underlaying consequences for cognitive science and epistemology. That is the case of engineers like Tim Smithers (1997) or Randall Beer (1995; 1997) who have strongly criticized computational information processing approaches to cognition highlighting dynamicism, embodiment and situatedness or Eric Prem (1997) who has put the emphasis on *epistemic autonomy* "the system's own ability to decide upon the validity of measurements" (Prem, 1997) a process that cannot be reduced to formal aspects, given the physicality of the measuring process, its preformal nature.

These authors have been greatly influenced by the biologist Francisco Varela whose definition of autonomy is much more abstract and encompassing than its robotic application[2]. Autonomy is defined by Varela as an abstract systemic kind of organization; a kind of self-maintained, self-reinforced and self-regulated system dynamics resulting from a highly recursive network of processes that generates and maintains internal invariants in the face of internal and external perturbations. A process that defines its own identity; i.e. its unity as a system distinguishable from the surrounding processes. This abstract notion of autonomy is realized at different biological scales and domains. It is precisely the autonomy of each domain what defines its specificity. As a paradigmatic example "life" is defined as a special kind of autonomy: *autopoiesis* or autonomy in the physical space. In turn "adaptive and cognitive behavior" is the result of a higher level of autonomy: that of the nervous system, producing invariant patterns of sensorimotor correlations and defining the behaving organism as a mobile unit in space (Varela, 1979; Varela and Bourgine, 1992; Varela, 1992).

---

[2] "Autonomous systems are mechanistic (dynamic) systems defined as a unity by their organization. *We shall say that autonomous systems are organizationally closed. That is, their organization is characterized by processes such that (1) the processes are related as a network, so that they recursively depend on each other in the generation and realization of the processes themselves, and (2) they constitute the system as a unity recognizable in the space (domain) in which the processes exist.*" (Varela, 1979, p.55, italics in the original).

But Varela's perspective on autonomy (although highly influential) has been recently criticized by its emphasis on closure[3] and the secondary role that system-environment interactions play in the definition and constitution of autonomous systems. Introducing ideas from complexity theory and thermodynamics authors like Bickhard (2000), Christensen and Hooker (2000), Collier (2002), and Ruiz-Mirazo and Moreno (2004), have defended a more specific notion of autonomy as a recursively self-maintaining far-from-equilibrium and thermodynamically open system. The interactive side of autonomy is essential in the definition: autonomous systems must interact continuously to assure the necessary flow of matter and energy for their self-maintenance. The philosophical consequences derived from the nature of autonomous systems are highlighted by these authors and summarized by Collier in the slogan: "No meaning without intention; no intention without function; no function without autonomy." (Collier, 2002). Autonomy is made the naturalized basis for functionality, intentionality, meaning and normativity. But it is not always clear what the relation is between this basic thermodynamic or constructive autonomy and neurally guided adaptive behavior. It is even argued that dynamical system theory (and thus computational simulation models in Alife) cannot capture the kind of organization that autonomy is (Christensen and Hooker, 2000) or that robots should be self-constructive in order to be "truly" autonomous (Ruiz Mirazo and Moreno, 2000).

Key notions covered by autonomous approaches to cognition

In general and across the differences between the uses of the term autonomy and the consequences that (more or less explicitly) are derived from it, the notion of autonomy subsumes a set of key notions in cognitive science that have been pushing towards a paradigmatic shift. Among this key notions we find:

- **Biological grounding**: the idea that the understanding of cognition and adaptive behavior must be approached bottom-up at two levels: in terms of the evolution of cognitive capacities in natural history and in terms of the biological mechanisms (neural networks, bodies, neuroendocrine systems, etc.) that produce cognitive behavior.

---

[3] It is precisely this emphasis on operational closure and its algebraic definition in Varela (1979) what makes controversial to apply Varela's notion of autonomy to a dynamical modelling of biological systems. Nonetheless Varela addresses several times the issue of a dynamical modelling of autonomy (pages 56, 86, 201 and 264) and concludes: "(...) I see these tools [dynamical system theory and computer simulations] as one way in which properties of systems, autonomous or allonomous, can be expressed. Differentiable dynamics represent, in practice, the most workable framework in which these two points of view can actually coexist and be seen as complementary in an effective way." (Varela, 1979, p.164)

- **Self-organization, complexity, emergence**: the idea that there is no central processor or homunculi that controls behavior but a distributed and functionally integrated network of recursive processes from which a coherent behavior emerges as a global product of the system. The notion of autonomy assumes a high degree of complexity in the system introducing constraints on the possible analysis and functional localization and decomposition of structures.

- **Interactivism, embodiment, situatedness, dynamicism**: Cognition is a process whose development and realization cannot be decoupled from the embodied interaction processes in which it is situated. An autonomous approach assumes a dialectics between independence and structural coupling: an interactive construction of meaning and behavior in which embodiment and situatedness are taken to be essential features of cognition that are best captured by dynamical (rather than traditional computational) notions, thus introducing time and space dependant constraints as essential features on the generation of behavior.

- **Critics to GOFAI**: The use of the notion of autonomy is often associated with a profound critique to what has been the mainstrain paradigm in cognitive science: the view that cognitive processes are logical transformations of computational states bearing a representational relation with observer independent "states of affairs" in the world. A view where the representational relation is taken to be the mark of the mental and the program-like transformation rules between representational states the causally effective mechanisms in the production of behavior. From autonomous robotics to the philosophy of biology and cognition, the approaches focused on autonomy have taken a different starting point, different theoretical primitives from which theories of cognition and adaptation have been built (complex dynamic networks, physically and thermodynamically embodied interactions, decentralized control systems, biologically grounded subsymbolic processes) to specifically address some of the problems that GOFAI approaches suffered at both practical-engineering and theoretical-philosophical levels[4].

So far so good, the concept of autonomy subsumes a set of new approaches to cognitive science... But what else? Is *autonomy* just an umbrella label to cover an undetermined set of general constraints in robotic and cognitive science? Is it just a heavy-weighted metaphysical concept that only

makes sense under the conceptual framework developed by Maturana and Varela? Or can it be conceptually and methodologically tunned in order to be introduced as a scientifically productive concept in empirical and synthetic research? The remaining of the paper will try provide an explicit and positive answer to this question.

Basic autonomy: the root for normative functionality

The origin of the word autonomy comes from the Greek *auto-nomos* (self-law). We can thus provide an intuitive first notion of autonomous systems as those producing their own laws[5]. But this notion requires a previous notion of self: autonomous systems must first produce their own identity; i.e. autonomous systems are primarily those whose basic organization is that of a self-sustaining, self-constructing entity over time and space. Their being is a process of recursive production of their constituting structure: a recursivity that generates a self. It is on top of this sense of *basic autonomy* that other levels of autonomy will appear in natural systems.

Basic Autonomy

Basic autonomy (Ruiz-Mirazo and Moreno, 2000) is the organization by which far from equilibrium and thermodynamically open systems adaptively generate internal and interactive constraints to modulate the flow of matter and energy required for their self-maintenance. Two equally fundamental but distinct aspects of basic autonomy can be distinguished:

a) *constructive*: generation of *internal* constraints to control the internal flow of matter and energy for self-maintenance. In this sense the autonomous (autopoietic) system can be understood as a highly recursive network of processes that produces the components that constitute the network itself (Maturana and Varela, 1980). Metabolism is the expression of this constructive aspect.

b) *interactive*: the generation of *interactive* constraints modulating the boundary conditions of the system to assure the necessary flow of energy and matter between the system and its environment[6]. Active transport through the membrane of a cell, control of behavior or breathing are characteristic examples of this interactive constraint generation.

On this basis we can define *constructive closure* as the satisfaction of constructive constraint generation and *interactive closure* as the satisfaction of interactive constraint generation for self maintenance.

---

[4] In this sense autonomy refers to explanations and design principles grounded on the internally driven interactive organization of the system; and not on representational or causally correlated relations between agent and environment (and often heteronomously interpreted or designed by and external observer-engineer).

[5] Strictly speaking new physical *laws* will never be created by an organism (or any other macroscopic system) but constraints can be generated that specify and govern its behavior.

[6] Unlike dissipative structures which hold their organization only under a restricted set of external conditions that the system cannot modify.

In general autonomy, at any level, will always present a twofold dialectics between internal recursive process and the necessary interactions to maintain them. In autonomous systems internal dynamics are more cohesive and integrated (more complex) than the interactive dynamics it sustains, thus producing a dynamic control asymmetry laden to the side of the autonomous agent.

The origin of functionality and normativity

What defines functionality in autonomous systems is the satisfaction of closure conditions (Collier, 2002) of internal and interactive processes. A process (internal or interactive) is functional if it contributes to the global self-maintenance of the system.

In turn functions become *normative*[7] by means of the *dynamic presupposition* of that process in the overall organization of the system (Christensen and Bickhard, 2002) since constructive and interactive functional processes are *the condition of possibility* of autonomous systems (as far from equilibrium and recursively self-maintained systems). The strength of an autonomous perspective resides in the fact that it is the very system who determines and specifies it. It is not an external observer who attributes functions to structures imposing a normative criteria according to its correspondence with states of affairs in the world. Nor is it on the basis of the agents evolutionary history Millikan (1989) or its structural matching with the environment that processes or structures acquire a function.

The organization of the nervous system

Following Moreno and Lasa (2003) if an autonomous system needs to recruit the same infrastructure to achieve both constructive and interactive closure then the space of possible biological organization becomes highly constrained. This happens because metabolic reactions (constructive processes) are slower than the reaction times required for available interactive closure opportunities, specially those available for fast body movements (motility) in big organisms (where the relative difference in velocity between metabolic reactions and body movement increases). Thus if a subset

---

[7] Normativity refers to the value attribution that is given to a process or object; e.g. adaptive or maladaptive to an interaction or structure in an organism, true or false to a cognitive state or believe, beautiful or ugly to a work of art, etc. Normativity challenges physicalist scientific approaches to the understanding of our world because it introduces a value asymmetry (good/bad, true/false, adapted/maladapted) in the description of nature, an asymmetry that is not present in any of the fundamental laws of physics. But, although alien to fundamental physics, *normativity* is an essential component of biology and cognitive science (and consequently for Alife and AI): whether an structure or interaction is adaptive or maladaptive for an organism is a value judgment that a scientist engaged in the study of living and intelligent systems must do. A judgment that must be justified in naturalistic terms; i.e. in the very organization of the system under study and not from a set of value preferences in the observer scientist.

of the interactive closure is achieved and controlled by a structure that instantiates processes which are dynamically decoupled from the constructive ones, the space of viable system organization is expanded. That's precisely the origin of the nervous system: the new opportunities for survival offered by the hierarchical decoupling of the nervous system, i.e. behavioral control decoupled from metabolic (constructive) constraints. The relation between metabolic constructive processes (M) and the nervous system (NS) is characterized by four properties that specify the organization[8] of the nervous system in the context of the whole organism:

1. **Hierarchical decoupling of the NS from M:** The NS is hierarchically decoupled from M by the:

   (a) **Bottom-up, local, constructive causation of the NS by M:** constructive/metabolic processes produce and maintain the architecture of the nervous system (neural cells, synapses, myelin, etc.) thus sustaining a new dynamical domain, new variables and relations between variables: the NS. The constructive nature of this causation establishes the *hierarchical* aspect of the decoupling.

   (b) **Dynamic underdetermination of NS by M:** the dynamic state of the NS is underdetermined by metabolic dynamics, i.e. neural dynamics are enabled but not determined by the metabolic production of the neural architecture. This underdetermination specifies the *decoupling* side of the relation.

2. **Downward causal dependency of M on NS:** Because the NS performs interactive functionality for the self-maintenance of the system, M depends on the proper functioning of NS; i.e. the organism's survival depends on neurally controlled behavior.

3. **Global and dynamic meta-regulation of NS by M:** Although the NS is dynamically underdetermined by M, M establishes the metaestability condition for the NS because the NS's functionality is defined by its interactive contribution to self-maintenance (and this must ultimately be evaluated by M). M does not directly evaluate the NS's dynamics but the interactive closure: i.e. the input of matter and energy it gets from the environment. But this meta-regulation, again, underdetermines the dynamics of the NS. Metabolism only indicates if a particular coupling is successful or not in the satisfaction of interactive closure conditions, but does not determine which one of all the possible viable/adaptive couplings should the NS undergo.

4. **Internal cohesive dynamics of the NS:** The other side of the metabolic constructive and meta-regulatory underdetermination of the NS's dynamic state is the recursive

---

[8] The identity characterizing properties of a system, i.e. the set of properties that identify a system as being a member of class.

capacity of the NS to maintain invariant patterns under internal and external perturbations; i.e. its capacity for self-generated cohesion, the degree in which the system's internal dynamics are more complex than the interactive flow so that the former can control the later to compensate for internal and external perturbations[9].

We can now abstract a second domain in biological systems (hierarchically decoupled from basic autonomy): *the domain of the organism's behavioral adaptive dynamics*, specified by the organization of the nervous system. This new dynamic domain, decoupled from local metabolic processes, provides a qualitative lower level (epistemological) boundary for the characterization of the specificity of cognition and allows for specific dynamical modelling of adaptive behavior. It is in this modelling domain that we will be able to define behavioral adaptive autonomy and thus a new level of functionality (properly cognitive but still biologically grounded).

Dynamical modelling of autonomy and functionality in the behavioral domain

Dynamically considered metabolism only acts as a set of control parameters for the nervous system; the behavioral domain is dynamically blind to metabolism's constructive functioning (although it has to be sensible to global metabolic conditions). Thus the constructive processes of basic autonomy can be modelled as a set of essential variables which tend to stay away from equilibrium; representing the cohesive limits of constructive processes and their interactive closure conditions. A similar approach was already taken by Ashby (1952) half a century ago (from whom we have taken the term essential variables) and recently recovered by Beer (1997) and Di Paolo (2003) in (evolutionary) simulation modelling of adaptive behavior. The dynamical autonomy of the behavioral domain allows for a naturalistically justified assumption of dynamical system theory (DST) as the proper conceptual framework to think about autonomy and cognition in this domain. If we model: a) the agent's NS and the environment as coupled dynamical systems (situatedness), b) coupled through sensory and motor transfer functions (embodiment), and c) the metabolic processes as essential (far from equilibrium) variables only controllable from the environment and signalling the NS; we get that functionality and autonomy can be redefined in the behavioral domain.

Behavioral adaptive autonomy

In the behavioral domain thus considered, a new level of autonomy can be described, hierarchically decoupled but interlocked with basic (metabolic) autonomy: *behavioral adaptive autonomy*.

We can now, in dynamical terms, explicitly define *behavioral adaptive autonomy* as:

homeostatic maintenance of essential variables under viability constraints [**adaptivity**] through a self-modulating behavioral coupling with the environment [**agency**], hierarchically decoupled from metabolic (constructive) processes [**domain specificity**].

This definition highlights three main aspects of behavioral adaptive autonomy:

Adaptivity: The "homeostatic maintenance of essential variables under viability constraints" condition assures a naturalized and autonomous criteria for (adaptive) functionality. Adaptivity is thus defined from the perspective of the maintenance of the organism, not from the perspective of structural adequation between the organism and the environment. It is not the organism that matches the environment in a given prespecified way. On the contrary it is through the particular way in which the agent satisfies the homeostatic maintenance of essential variables that an adaptive environment (a world) is specified cut out from a background of unspecific physical surroundings. Next section will further analyze the the nature of behavioral adaptive functionality thus considered.

Agency: The "self-modulating behavioral coupling" condition for behavioral adaptive autonomy specifies the *agency* of the organism in the adaptive process. "Self-modulation" is the consequence of the cohesive dynamics of the nervous system by which its dynamics are more complex than the interactive ones in the generation of the internal invariants (the homeostatic maintenance of essential variables under viability constraints). The notion of self-modulation refers to this control asymmetry in the production of behavior and that's precisely what we call agency. It can't be otherwise, if the state of essential variables is only accessible for the agent (through internal sensors: level of glucose, feeling of hot, pain, etc.) the homeostatic regulation must be guided by the agent's nervous system and not by the environment. Thus the NS needs to evaluate it's structural coupling through value signals from the essential variables. This way a *value system* guided by the state of essential variables and acting as metaestability condition for structural plasticity of sensorimotor transformations becomes a fundamental component of behavioral autonomy, and a defining component of agency. The higher the agent's capacity for adaptively guided self re-structuring (plasticity) the higher it's behavioral adaptive autonomy and hence its agency[10].

---

[9] This is close to what Varela refers to as "operational closure" although we believe that internal cohesion is achieved through interaction processes rather than through internal recursivity alone: i.e. closed sensorimotor loops are integrated in the recursive functioning of neural dynamics.

---

[10] By this condition external contributions to adaptation (such as parents care or artificially induced constraints in toy-like worlds), although functional for the agent, would be excluded from the domain of autonomous adaptation.

Domain specificity: The hierarchical decoupling of the nervous system from metabolic processes provides a naturalized criteria for the domain specificity of behavioral autonomy, distinct from other adaptive domains in nature (bacterian networks, plants, etc.). This domain specificity should not be considered as independency but as hierarchical decoupling (explained above), which allows for a justified specific modelling of behavioral autonomy separated from local constructive aspects. Nonetheless it should be noted that two kinds of autonomy are interlocked here: basic autonomy and behavioral autonomy. Both domains are mutually required, the behavioral domain satisfies interactive closure of basic autonomy and basic autonomy constructs the bodily and neural variables defining the NS's architecture. At the same time basic autonomy acts as a meta-regulator of the NS's dynamics.

Behavioral adaptive functionality

Functionality, in the behavioral domain thus considered, can be defined as the homeostatic effect of an interaction process on the maintenance of essential variables under viability constraints and, more specifically, as the mapping of the interactive trajectories (in the agent-environment coupled dynamic space) into the state space of the essential variables. Normativity is transitive from basic autonomy to the behavioral domain through the maintenance of essential variables under viability constraints. Thus normative functionality (adaptivity) is *the mapping of the agent-environment coupled system's state space trajectories into the viability subspace of the essential variables.*

Because this definition of function does not compromise any structural decomposition in functional primitives (unlike traditional functionalism), a dynamical approach to behavioral functionality can hold two kinds of emergence[11]:

a) *Internal emergence*: It appears when the agent's internal structure is causally integrated (and the NS often is), i.e. interactions between components are non-linear and components are highly inter-connected. Functional decomposition of components (localization) is not possible. The functionality of the system *emerges* from local non-linear and recursive interactions between components.

b) *Interactive emergence*: Because essential variables are non-controlled variables for the agent, functionality is interactively emergent (Steels, 1991; Hendriks-Jansen, 1996), not in the trivial sense that essential variables need external input, but in the sense that achieving this often requires closed sensorimotor loops for the agent to enact

the necessary sensorimotor invariants to control essential variables.[12]

What this double emergent condition shows is that the way the specific adaptive function is achieved involves a dynamic coupling between agent and environment where no particular decomposable structure of the agent can be mapped into functional components: functionality is the outcome of an interaction *process* (that can be modified by the cognitive agent according to its perceived satisfaction of closure conditions).

Discussion

Now, the problem with behavioral adaptive autonomy is the problem of a higher characterization and development of its understanding, specially in relation to its self-regulating, emergent and complex nature which does not allow for a localizationist program to succeed: i.e. functional and structural decomposition of components and aggregative causal abstraction of mutual relations (Bechtel and Richardson, 1993). When localizationist strategies are thrown away the locus of the research enquiry regarding the nature and origin of cognition and adaptation is displaced towards: a) the specification of the dynamic structure of lower level mechanisms capable of implementing behavioral adaptive autonomy (i.e. capable of self-restructuring cohesive and recursive dynamics); and b) the search for the nature of intermediate explanatory patterns between the agent-environment structural coupling and the maintenance of essential variables under viability constraints: traditional explanatory concepts (such as information, representation, memory, processing, etc.) should be dynamically grounded if introduced at all in the proposed framework. In this sense the view on *behavioral adaptive autonomy* presented here is closer to highly integrated and functionally unspecific models (such as those of evolutionary robotics) than action selection modelling techniques (Humphrys, 1996), where possible actions are pre-specified and the agents internal structure is unable to reconceptualize an interactive domain to achieve novel functionality. Behavioral adaptive autonomy is neither something to be achieved just by introducing energetic constraints on robot task solving (Kelly et al., 1999).

A-life and, more specifically, evolutionary simulation modelling[13] becomes a mayor research tool here through the synthesis and experimental manipulation and analysis of the behavior generated by embodied and situated DRNNs (dynamic recurrent neural networks). The simulation model

---

[11] We are here talking of weak emergence in the sense of an holistic, recursive and distributed causal structure that produces a global ordered/invariant pattern. We are not arguing for a strong or ontological emergence that defends the appearance of a new property or object non reducible to the underlying processes.

[12] Very often interactive emergence reinforces internal emergence because "interactions between separate sub-systems are not limited to directly visible connecting links between them, but also include interactions mediated via the environment" (Harvey et al., 1997, p.205)

[13] Evolutionary robotics (Harvey et al., 1997; Nolfi and Floreano, 2000) and Randall Beer's minimally cognitive behavior program (Beer, 2004) being the major exponents here.

acts as an artifactual blending between lower level neural mechanistic concepts and the global functional conceptualization of behavior (Barandiaran and Feltrero, 2003).

An interesting line of research has recently been proposed by Di Paolo (2003) in this direction. Di Paolo argues that behavior itself is underdetermined by survival conditions and proposes *habit formation* as the origin of intentionality. Habits are self sustaining dynamic structures of behavioral patterns, sensorimotor invariants homeostatically maintained by neural organization. Homeostatically controlled synaptic plasticity (Turrigliano, 1999) could be a relevant neural organization leading to such autonomy of behavioral patterns; as demonstrated by Di Paolo (2000). Although habit formation does not necesarily address the issue of the relation between metabolic and nervous autonomy Di Paolo points towards a fundamental step forward in current research trends: that structural coupling (and closed sensorimotor loops) is not all there is in a dynamical and situated approach and that a robust capacity of the agent to evaluate and restructure its coupling is the way to follow to achieve progresively higher levels of cognitive autonomy and intentionality.

If Alife is to throw some light on the origin of cognition and adaptive behavior, far from equilibrium essential variables and value systems capable of specifying stability conditions for a given dynamical coupling with the environment should be introduced in the simulation models. In particular essential variable based fitness functions in evolutionary simulation modelling are a particular instantiation of behavioral, internal and implicit fitness functions which (according to Floreano and Urlezai —2000) shall produce highly self-organized control systems. This principles for evolutionary simulation modelling of autonomous agents were succesfully implemented in a foraging task with alternate profitability sources (Barandiaran, 2002).

In addition to this synthetic bottom-up methodology other analytic tools should be theoretically tunned. Complexity measures to understand functional integration in neural processes (Tononi et al., 1998) are producing interesting results, an could be used to better characterize the cohesive nature of the NS. An early exploratory example of such methodology is provided by Seth (2002), fusioning both evolutionary simulation modelling and complexity measures of neural network dynamics (using dynamical graph theory) to analyze the relation between behavioral (interactive) and neural (internal) complexity.

Conclusion

A wide use of the term autonomy is found in the Alife literature: from a set of engineering constraints in robotics to a fundamental organizational principle in biology. In relation to behavior and cognition it is not clear how to operationalize the term and what the relation is between behavioral autonomy and basic (autopoietic) autonomy. We have seen that the particular organization of the nervous system allows for a specific modelling domain of cognition and adaptive behavior: the domain of behavioral adaptive dynamics. Interlocked with basic/metabolic autonomy (through the requirement to actively maintain essential variables under viability constraints) *behavioral adaptive autonomy* becomes a process of cohesive maintenance of internal invariants through continuous interaction loops, which requires, in turn, a functionally integrated and plastic neural organization with a higher internal dynamic complexity than that established between the organism and its environment.

By providing an explicit definition of behavioral autonomy and functionality in dynamical terms we hope to have contributed something to the simulation modelling approach that traces the Alife route to AI; to the understanding of the transition that goes from basic forms of life to adaptive behavior and cognition.

References

Ashby, W. (1952). *Design for a Brain. The origin of adaptive behaviour*. Chapman and Hall, 1978 edition.

Barandiaran, X. (2002). Adaptive Behaviour, Autonomy and Value Systems. Master's thesis, COGS, University of Sussex, Brighton, UK. URL: http://www.ehu.es/ias-research/doc/$2002_{bam}sth_fin.pdf$.

Barandiaran, X. and Feltrero, R. (2003). Conceptual and methodological blending in cognitive science. The role of simulated and robotic models in scientific explanation. In *Volume of abstracts of the 12th International Congress of Logic, Methodology and Philosophy of Science, Oviedo (Spain).*, page 171.

Bechtel, W. and Richardson, R. (1993). *Discovering Complexity. Decomposition and Localization as strategies in scientific research*. Princeton University Press.

Beer, R. (1995). A dynamical systems perspective on autonomous systems. *Artificial Intelligence*, (72):173–215.

Beer, R. D. (1997). The Dynamics of Adaptive Behavior: A research program. *Robotics and Autonomous Systems*, 20:257–289.

Beer, R. D. (2004). The dynamics of active categorical perception in an evolved model agent. *Adaptive Behaviour*. in press.

Bickhard, M. H. (2000). Autonomy, Function, and Representation. In Etxeberria et al. (2000), pages 111–131.

Brooks, R. A. (1991). Intelligence without representation. *Artificial Intelligence Journal*, 47:139–160.

Christensen, W. and Bickhard, M. (2002). The process dynamics of normative function. *Monist*, 85 (1):3–28.

Christensen, W. and Hooker, C. (2000). Autonomy and the emergence of intelligence: Organised interactive construction. In Etxeberria et al. (2000), pages 133–157.

Collier, J. (2002). What is autonomy? In Dubois, D., editor, *International Journal of Computing Anticipatory Systems. Partial proceedings of the Fifth International Conference CASYS'01 on Computing Anticipatory Systems, Lige, Belgium, August 13–18, 2001.*

Di Paolo, E. (2000). Homeostatic adaptation to inversion of the visual field and other sensorimotor disruptions. In Meyer, J.-A., Berthoz, A., Floreano, D., Roitblat, H., and Wilson, S., editors, *From Animals to Animats 6: Proceedings of the Sixth International Conference on Simulation of Adaptive Behavior*, pages 440–449. Harvard, MA: MIT Press.

Di Paolo, E. (2003). Organismically inspired robotics. In Murase, K. and Asakura, T., editors, *Dynamical Systems Approach to Embodiment and Sociality*, pages 19–42. Advanced Knowledge International, Adelaide, Australia.

Etxeberria, A., Umerez, J., and Moreno, A., editors (2000). *Communication and Cognition – Artificial Intelligence. Special issue on "The contribution of artificial life and the sciences of complexity to the understanding of autonomous systems"*, volume 17 (3–4).

Floreano, D. and Urzelai, J. (2000). Evolutionary robots with online self-organization and behavioural fitness. *Robotics and Autonomous Systems*, 13:431–443.

Harvey, I., Husbands, P., Cliff, D., Thompson, A., and Jakobi, N. (1997). Evolutionary Robotics: the Sussex Approach. *Robotics and Autonomous Systems*, 20:205–224.

Hendriks-Jansen, H. (1996). In praise of interactive emergence, or why explanations don't have to wait for implementations. In Boden, M., editor, *The Philosophy of Artificial Life*, pages 282–299. Oxford University Press, Oxford.

Humphrys, M. (1996). Action Selection methods using Reinforcement Learning. In Pattie Maes, Maja J. Mataric, J.-A. M. J. P. and Wilson, S. W., editors, *From Animals to Animats 4: Proceedings of the Fourth International Conference on Simulation of Adaptive Behavior (SAB-96)*, pages 135–144. MIT Press.

Kelly, I., Holland, O., Scull, M., and McFarland, D. (1999). Artificial Autonomy in the Natural World: Building a Robot Predator. In Floreano, D., Nicoud, J.-D., and Mondada, F., editors, *Advances in Artificial Life. Proc. of ECAL'99*, pages 289–293. Springer-Verlag.

Maes, P., editor (1991). *Designing Autonomous Agents*. MIT Press.

Maturana, H. and Varela, F. (1980). Autopoiesis. The realization of the living. In Maturana, H. and Varela, F., editors, *Autopoiesis and Cognition. The realization of the living*, pages 73–138. D. Reidel Publishing Company, Dordrecht, Holland.

Millikan, R. G. (1989). In defense of proper functions. *Philosophy of Science*, 56:288–302.

Moreno, A. and Lasa, A. (2003). From Basic Adaptivity to Early Mind. *Evolution and Cognition*, 9(1):12–24.

Nolfi, S. and Floreano, D. (2000). *Evolutionary Robotics: The Biology, Intelligence and Technology of Self-Organizing Machines*. MIT Press.

Prem, E. (1997). Epistemic autonomy in models of living systems. In *Proceedings of the Fourth European Conference on Artificial Life*. MIT Press, Bradford Books.

Ruiz-Mirazo, K. and Moreno, A. (2000). Searching for the Roots of Autonomy: the natural and artificial paradigms revisited. In Etxeberria et al. (2000), pages 209–228.

Ruiz-Mirazo, K. and Moreno, A. (2004). Basic Autonomy as a Fundamental Step in the Synthesis of Life. *Artificial Life*, 10:235–259.

Seth, A. K. (2002). Using dynamical graph theory to relate behavioral and mechanistic complexity in evolved neural networks. Unpublished. Url: http://www.nsi.edu/users/seth/Papers/nips2002.pdf.

Smithers, T. (1997). Autonomy in Robots and Other Agents. *Brain and Cognition*, (34):88–106.

Steels, L. (1991). Towards a Theory of Emergent Functionality. In Meyer, J. and Wilson, R., editors, *Simulation of Adaptive Behaviour*, pages 451–461. MIT Press.

Tononi, G., Edelman, G., and Sporns, O. (1998). Complexity and coherency: integrating information in the brain. *Behavioural and Brain Sciences*, 2(12):474–484.

Turrigliano, G. (1999). Homeostatic plasticity in neuronal networks: The more things change, the more they stay the same. *Trends in Neuroscience*, 22:221–227.

Varela, F. (1979). *Principles of Biologicall Autonomy*. North-Holland, New York.

Varela, F. (1992). Autopoiesis and a biology of intentionality. In McMullin, B., editor, *Proceedings of a workshop on Autopoiesis and Perception*, pages 4–14.

Varela, F. and Bourgine, P. (1992). Towards a Practice of Autonomous Systems. In Varela, F., editor, *Towards a Practice of Autonomous Systems. Proceedings of the First European Conference on Artificial Life*, pages xi–xvi.

Acknowledgments

I want to acknowledge the Basque Government for financial support through the doctoral fellowship BFI03371-AE and the University of the Basque Country for the research project grant 9/UPV 00003.230-13707/2001. I want to thank Alvaro Moreno and Ezequiel Di Paolo for usefull comments and discussion on the topics presented here. Thanks also to Jesus Siqueiros and Tomas Garcia for corrections on the final manuscript.

A Physiological Approach to the Generation of Artificial Life Forms

Marc Cavazza[1], Simon Hartley[1], Louis Bec [2], François Mourre [2], Gonzague Defos du Rau [2], Remy Lalanne [2], Mikael Le Bras[1] and Jean-Luc Lugrin[1].

[1]School of Computing, University of Teesside, Middlesbrough, TS1 3BA United Kingdom
[2]CYPRES, Friche de la Belle de Mai, 41 rue Jobin F-13003, Marseille
m.o.cavazza@tees.ac.uk

Abstract

Artificial Life has been an inspiration for Virtual Reality Art, which has proposed the design of imaginary life forms and has also used Artificial Life techniques in various installations. The creation of imaginary life forms has been so far limited to the external appearance of these creatures. We introduce a novel approach to the design of artificial life forms, which enables the description of their internal physiology from first principles, by using a simulation method known as qualitative physiology, derived from a well described Artificial Intelligence technique. We illustrate this framework by revisiting early work in Artificial Life and providing these virtual life forms with a corresponding physiology, so as to obtain complete living organism in virtual worlds.

Introduction

In this paper, we present the creation of artificial life forms described by their appearance as well as their internal physiology. These can be simulated in real-time in a virtual environment in which elementary physical phenomena are also simulated, so as to recreate part of an artificial ecosystem.

We have revisited early work in Artificial Life, originally presented at the second "ALife" conference [Bec, 1991; 1998], which proposed imaginary creatures, by extending the description of these life forms to incorporate a description of their internal anatomy as well as the associated physiology. Artistic work, with a strong inspiration from Biology [Risan, 1997], is one important aspect of investigating "life as it could be" [Langton, 1989]. In the next sections, after an introduction to qualitative simulation and a presentation of the system's architecture, we describe the modelling of the artificial creature *Diaphaplanomena* that belongs to the Upokrinomena, as well as early results obtained from the simulation of the organism behaviour in its environment.

System Overview and Architecture

The system comprises a visualisation environment, which displays the real-time motion of the Diaphaplanomena, and a qualitative simulation engine controlling the simulation of both the internal physiological processes of the creature, and of physical processes in its liquid environment (such as currents, heat flows, etc.).

The visualisation component supports the real-time display of the creature in its environment and the interaction between the Diaphaplanomena and the physical world. This includes, the creature's locomotion in the 3D environment, the visualisation of its internal or external motion as real-time animations and dynamic changes in its external appearance (colour and textures).

The visualisation engine is based on a game engine, Unreal® Tournament 2003 (UT 2003). The rationale for using a game engine in research is that it supports advanced graphic rendering and animation control as well as the integration of external software modules, which makes it an ideal development environment [Lewis and Jacobson, 2002].

In addition, UT 2003 provides a sophisticated event system, which supports the real-time interaction between the Diaphaplanomena and its environment. Physical processes simulated in the environment, such as currents, turbulences and heat flows create objects (vortices, heat sources, etc.) whose direct interaction with the creature can be processed through the event system and be interpreted by the qualitative simulation module.

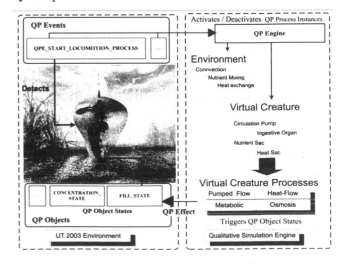

Figure 1: System Architecture

The software architecture is based on UDP communication between the qualitative simulation engine and the visualisation engine. The simulation engine passes landmark values reached by qualitative variables to the visualisation engine. These are interpreted by various scripts that control the creature's behaviour (physiological processes). In a similar fashion, physical processes in the environment such as currents, turbulences and heat flows are controlled by the simulation engine. On the other hand, dynamic interactions between the Diaphaplanomena in its environment (such as the creature entering a cold current or being hit by turbulence) generate events that are passed to the simulation engine and alter the current simulation. For instance, if the creature enters a cold current, this will eventually trigger a heat-transfer process between the current and the Diaphaplanomena, which will change its internal state (and trigger further thermal regulation mechanisms).

Artificial Physiology at the System Level: Qualitative Physiology

Modelling physiological processes has always been a major endeavour of biological modelling as it would support detailed simulations of the organism behaviour. Most models have concentrated on the numerical simulation of physiological phenomena expressed through a set of differential equations or through control theory, because of the importance of regulatory and homeostatic processes in physiology.

However, symbolic modelling is another approach which has developed more recently, mainly in medical knowledge-based systems. One advantage of symbolic reasoning is that it enables the aggregation of multiple physiological systems (even across different levels of description) and has better explanatory capabilities.

Most importantly, in the case of artificial life systems, it brings the prospect of defining artificial physiology from first principles, which opens new ways for the experimentation of artificial, alternative life forms. Previous work on Artificial Life has mostly considered molecular physiology, rather than higher level systems, with a few exceptions [Grand et al., 1997].

One of the authors has developed a technique for qualitative simulation of physiological systems [Cavazza and Simo, 2003], derived from Qualitative Process Theory (QPT), an Artificial Intelligence formalism used in qualitative physics [Forbus, 1984]. Qualitative Physiology uses the process-based formalism of QPT to represent physiological processes governed by physiological laws, and supports the real-time simulation of physiological sub-systems.

The QPT formalism is based on the concept of a qualitative variable, a discrete variable whose values correspond to the orders of magnitude of a given physical parameter. For instance, the qualitative values can be used to form a discrete set for the concentration of solute within a fluid, or for the pressure within a container. The formalism supports two basic mechanisms for updating the variables' values in the course of a process execution. Influence equations are the central mechanisms for updating variables: they correspond to abstract expressions of physical (in some cases physiological) laws. For example, in osmotic systems the transfer of solute through the semi-permeable membrane can be represented by the following influence equations.

> I+ (Amount-of-Solute(? Substance dst)(A Osmotic-rate))
> I- (Amount-of- Solute(?Substance src)(A Osmotic-rate))

Where I+ represent a positive influence and I- a negative influence upon the first quantity by the second. The Influence equation specifies what can cause a quantity to change so we have specified that the quantity amount of solute in substance src will be decreasing with the osmotic rate and the amount of solute in substance dst will be increasing with the osmotic rate.

On the other hand, qualitative proportionalities relate variables values outside causal processes, for example stating that the mass of a liquid is proportional to its volume (hence updating the volume variable whenever the mass variable is modified, to enforce that constraint).

A qualitative process is triggered whenever its pre-conditions are satisfied; for instance, when the creature is entering a cold area (such as a cold current), the conditions for triggering a heat-exchange process between itself and the current are satisfied.

Naturally, Qualitative Simulation can also be used to simulate physical phenomena in the creature's environment, which in our case provides a unified principle for simulation.

An example process that relates the behaviour of the Diaphaplanomena to its environment is the *Propulsion* process:

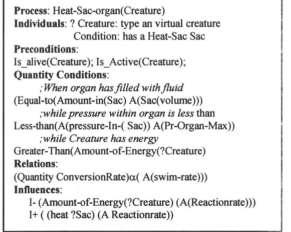

Process: Heat-Sac-organ(Creature)
Individuals: ? Creature: type an virtual creature
                 Condition: has a Heat-Sac Sac
Preconditions:
Is_alive(Creature); Is_Active(Creature);
Quantity Conditions:
    ; *When organ has filled with fluid*
(Equal-to(Amount-in(Sac) A(Sac(volume)))
    ; *while pressure within organ is less* than
Less-than(A(pressure-In-(Sac)) A(Pr-Organ-Max))
    ; *while Creature has energy*
Greater-Than(Amount-of-Energy(?Creature)
Relations:
(Quantity ConversionRate)α(A(swim-rate)))
Influences:
    I- (Amount-of-Energy(?Creature) (A(Reactionrate)))
    I+ ((heat ?Sac) (A Reactionrate))

In this process, the conversion rate is dependent upon the swim rate information received from the UT 2003 environment about the detection of danger within the environment. Since the "heat sac" is used in the

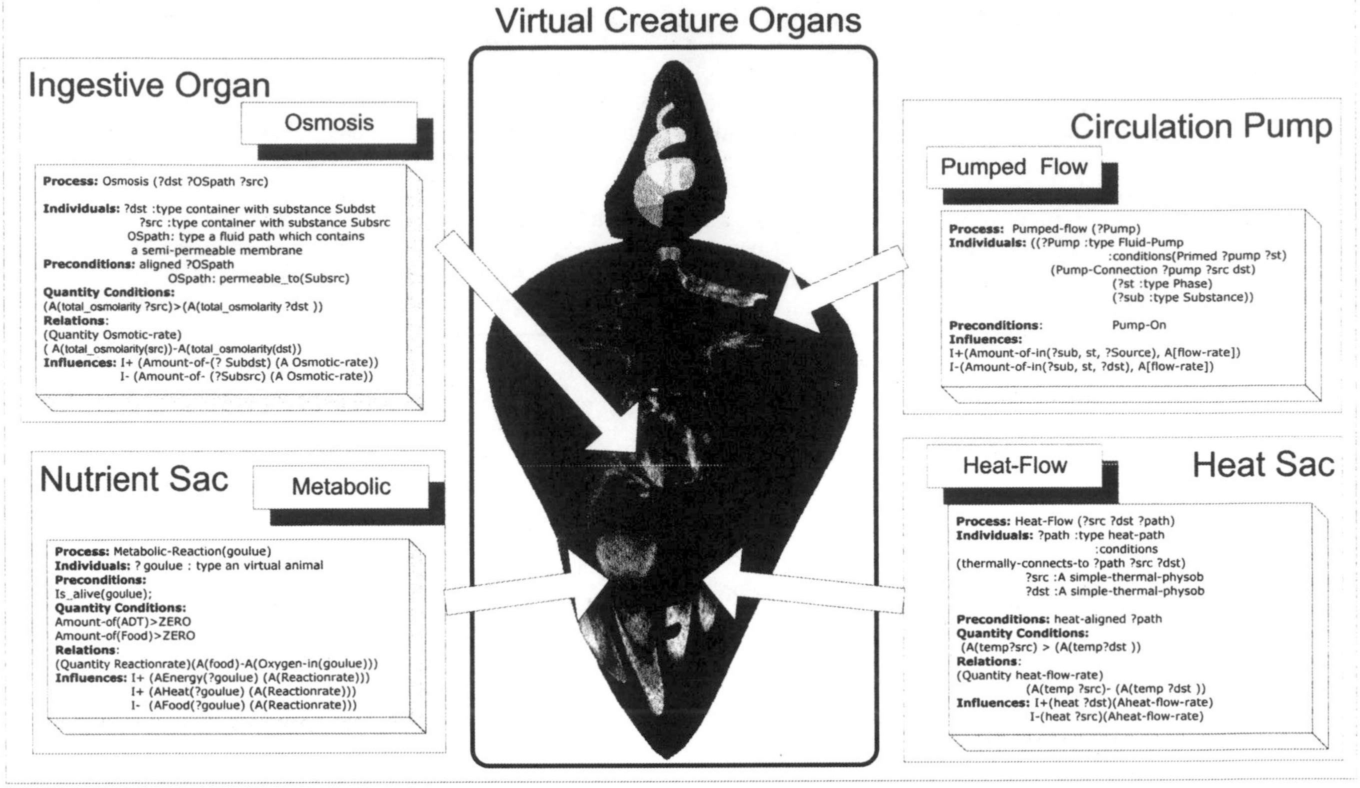

Virtual Creature Organs

Ingestive Organ

Osmosis

Process: Osmosis (?dst ?OSpath ?src)

Individuals: ?dst :type container with substance Subdst
　　　　　　?src :type container with substance Subsrc
　　　　　　OSpath: type a fluid path which contains
　　　　　　　　a semi-permeable membrane
Preconditions: aligned ?OSpath
　　　　　　　　OSpath: permeable_to(Subsrc)
Quantity Conditions:
(A(total_osmolarity ?src)>(A(total_osmolarity ?dst))
Relations:
(Quantity Osmotic-rate)
(A(total_osmolarity(src))-A(total_osmolarity(dst))
Influences: I+ (Amount-of-(? Subdst) (A Osmotic-rate))
　　　　　　I- (Amount-of- (?Subsrc) (A Osmotic-rate))

Circulation Pump

Pumped Flow

Process: Pumped-flow (?Pump)
Individuals: ((?Pump :type Fluid-Pump
　　　　　　　　:conditions(Primed ?pump ?st)
　　　　　　　　(Pump-Connection ?pump ?src dst)
　　　　　　　　(?st :type Phase)
　　　　　　　　(?sub :type Substance))

Preconditions:　　　　Pump-On
Influences:
I+(Amount-of-in(?sub, st, ?Source), A[flow-rate])
I-(Amount-of-in(?sub, st, ?dst), A[flow-rate])

Nutrient Sac

Metabolic

Process: Metabolic-Reaction(goulue)
Individuals: ? goulue : type an virtual animal
Preconditions:
Is_alive(goulue);
Quantity Conditions:
Amount-of(ADT)>ZERO
Amount-of(Food)>ZERO
Relations:
(Quantity Reactionrate)(A(food)-A(Oxygen-in(goulue)))
Influences: I+ (AEnergy(?goulue) (A(Reactionrate)))
　　　　　　I+ (AHeat(?goulue) (A(Reactionrate)))
　　　　　　I- (AFood(?goulue) (A(Reactionrate)))

Heat-Flow　　Heat Sac

Process: Heat-Flow (?src ?dst ?path)
Individuals: ?path :type heat-path
　　　　　　　　:conditions
(thermally-connects-to ?path ?src ?dst)
　　　　　　?src :A simple-thermal-physob
　　　　　　?dst :A simple-thermal-physob

Preconditions: heat-aligned ?path
Quantity Conditions:
(A(temp?src) > (A(temp?dst))
Relations:
(Quantity heat-flow-rate)
　　　　　　(A(temp ?src)- (A(temp ?dst))
Influences: I+(heat ?dst)(Aheat-flow-rate)
　　　　　　I-(heat ?src)(Aheat-flow-rate)

Figure 2: Diaphaplanomena Anatomy and Physiology

locomotion, the increased conversion rate allows the Diaphaplanomena to swim faster away from closer dangers.

Creating the Diaphaplanomena

The creation of the Diaphaplanomena includes the description of its anatomy and its physiology. The anatomical structure of the Diaphaplanomena is briefly outlined in Figure 2.

The visual contents have been produced using 3D modelling and animation packages such as 3D Studio Max™ and XSI™. The native 3D models as well as animations have been imported into the UT 2003 engine. This engine supports the use of key-framed animation to describe certain actions (in our case movements of internal organs, changes in shape of the creature, as well as locomotion), while retaining the interactive nature of the simulation. The actions are required to simulate interaction with the environment and/or other creatures forming a potential ecosystem.

We have modelled several physiological processes for the Diaphaplanomena, dealing with elementary physiological functions such as nutrition, locomotion, and certain homeostatic processes such as thermal regulation. An essential aspect is that these are defined altogether as an integrated system, which can further be refined through experimentation, as the process description is highly

modular and processes only connect through characteristic physiological variables (e.g., concentration in nutrients, temperature in certain organs, etc.). Defining the creature using this level of description is largely a functionalist approach [Bedau, 1992], although a top-down, non-emergent one.

As far as locomotion is concerned, the Diaphaplanomena is a wandering organism, which inhabits a viscous but heterogeneous space traversed by turbulences and currents. Its locomotion is mostly dedicated to maintain stability against external perturbations as well as supporting basic chemotactic mechanisms. Its locomotion is closely related to thermal generation processes and is based on the generation of turbulences due to internal temperature gradients between "hot" and "cool" organs.

Results

This approach enables various forms of experimentation in virtual worlds. In addition, because physical processes and physiological processes are described and simulated independently of one another, it is actually possible to experiment upon the adequacy of these mechanisms for the long-term survival of the Diaphaplanomena in its environment.

This process of top-down creation of a life form has also implications for the relations between Artificial Life and real living systems [Keeley, 1997]. We have implemented a first prototype of the system in which we have a basic set of physiological and environmental processes. The processes allow our experimentation with, and simulation of, the Diaphaplanomena and its environment.

In this section, we illustrate the system behaviour by describing some specific results from the simulation.

The Diaphaplanomena's Digestive System

A number of processes are active between the creature and the environment, when the creature is free floating. Active instances of the fluid flow and heat flow processes exist between the internal media and the environment. These two processes continuously refresh the internal media of nutrients and deprive the creature of heat respectively. The organs interact with this internal media.

The behaviour of an organ is determined by its representation. In our sample case, the organs are represented by containers and pumps which operate by using the bodily-fluids contained within them. To create the special behaviours desired to represent a specific organ, the container is given special properties. For instance, the absorption of nutrients through the nutrient sac by osmosis is enabled due to the nutrient sac's semi-permeable membrane property. In this way each organ has a direct correlation to a process or sequence of processes.

The primary behaviour processes that are active within each of the main organs are detailed in Figure 2. However, this is not a complete description of all of the processes within an organ. For instance, the osmosis process is the primary behaviour of the ingestive organ. The osmosis process transfers nutrients between the internal media and the bodily fluid in the organ. This transfer is balanced by the pumped flow from the circulation pump which is also transferring these required nutrients in bodily fluid to the attached receptive organs.

In our example creature, the nutrient rich bodily fluid in the ingestive organ is transferred to the nutrient sac by the fluid flow process created by the circulation pump. The primary behaviour of the nutrient sac is the metabolic process that converts the nutrients in the bodily fluid to heat and stored energy. During this conversion the metabolic process decreases the concentration of nutrients in the bodily fluid, this nutrient deficient fluid in the nutrient sac is then pumped back into the ingestive organ by the circulation pump completing the cycle. In this way the processes are combined to produce the physiology for the creature.

In our example creature, the nutrient sac is given an additional behaviour of being nutrient sensitive. Changes in the nutrient concentration correspond directly to the physical colour for the organ. This mapping is caused by triggering object states in the UT 2003 engine by communication with the qualitative simulation engine. Figure 3 illustrates the representation of nutrient concentration in an organ by using colour. The triggering of the colour object states of the "nutrient sac" organ by the qualitative simulation engine uses three qualitative process effects; these are QP_Concentration_Min, QP_Concentration_Max and QP_Concentration_Update.

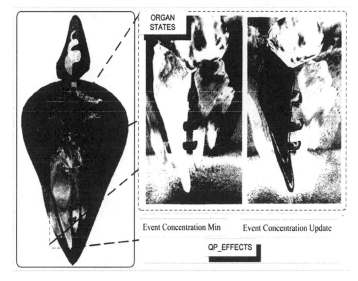

Figure 3: Changes in Internal Organs during Simulation

In the qualitative simulation engine the discretised qualitative variable that represents the concentration of nutrients in the bodily fluid that are within the "nutrient sac" is given a set of landmark values. The minimum and maximum of the landmark values are used for the respective QP_Effects, while the rest communicate using the effect QP_Concentration_Update. The distribution of the landmark values determines the responses of the organ. However, since the value of the nutrient concentration depends on the physiological processes and their associated influence equations, it is these which ultimately control the response of the organ.

The motion of the virtual creature is governed by the convection current in the environment when the creature is not under its own direct motion. As the creature is not using locomotion, the creatures Heat-sac has no active processes. The heat sac is used only to power the locomotion for the creature. In this free floating case, the creature is conserving energy and building its reserves in the nutrient sac.

The Diaphaplanomena Locomotion System

During the simulation the virtual creature may encounter the active vortices that are generated by the environment's convection process. In this case, the Touch-Vortex event will support the detection of that environmental phenomenon by the Diaphaplanomena. This will be interpreted as a danger signal and trigger its locomotion behaviour.

Figure 4: Interaction with the Creature's Liquid Medium

During locomotion, the internal processes affecting the virtual creature are mostly unchanged. An exception is that during locomotion, the rate at which the bodily fluids are carried around the organs by the circulatory pump, is raised due to an increase in the pump-rate. The main change is that the heat sac becomes active during the locomotion cycle.

The activation of the heat sac causes the release of the intake valve allowing it to draw fluid, via a fluid flow process. The fluid flow from the internal media into the heat sac generates the fill heat sac event, QP_Effect_Fill_Heat_Sac, which is sent to the UT 2003 engine. The fill heat sac effect specifies the rate at which the heat sac is expanded depending upon the quantity fluid flow rate of the fluid flow process into the heat sac. The primary behaviour of the heat sac is to convert stored energy from metabolic processes into heat; this conversion process creates a new heat source within the heat sac and consequently activates a heat flow from the new heat source. As temperature within the heat sac increases, the internal fluid contained also begins to heat. Qualitative proportionalities give the relationship that "heat sac" pressure increases with the heat. So, as within the heat sac increases the pressure increases. This increase in the qualitative pressure quantity triggers the landmark value within the expansion valve on the heat sac when the pressure reaches a maximum value, causing the contained media to escape. The activation of the expansion valve between the heat sac and the environment provides a fluid path between the heat sac and the environment. This fluid flow sends the event "empty heat sac", QP_Effect_Empty_Heat_Sac, to the UT 2003 engine, which changes the state for the heat sac to empting and triggers a particle effect whose rate depends upon the rate of the fluid flow (see figure 5. Propulsion effects).

The start of the event "empty heat sac" for the heat sac organ state activates the propulsion process for the creature which calculates the thrust and drag for the creature from the rate of the fluid flow, the viscosity of the fluid and the convection in the environment. The propulsion process sends the calculated qualitative thrust data to the UT 2003 engine via event "start process propulsion". This changes the velocity of the creature within the UT 2003 system.

Conclusion and Perspectives

We have presented a new method for the principled creation of artificial life forms and their real-time simulation inside their virtual environment. Its central feature is to provide a high-level representation for the creatures' physiology, which supports a more integrated description of the virtual life forms.

The use of qualitative simulation also provides a unified framework in which to simulate both the creature's physiology and the physical phenomena that affect it in its environment. This is of particular interest when exploring life forms populating alternative worlds in which laws of physics themselves could be different (a concept we have introduced as Alternative Reality [Cavazza et al., 2003]).

This "creationist" approach, which is in-line with the artistic perspective from which this work was originally developed, can however be reconciled with evolutionary approaches. One potential research direction is to link the symbolic description of physiological processes to some kind of evolutionary computation principles that would transform the creature's physiological processes. The important aspect would be to retain a link with Theoretical Biology [Noble, 1997] in the Physiological description rather than delegating it entirely to emergent processes.

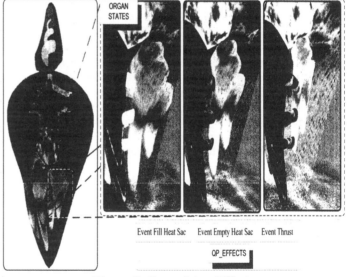

Figure 5: Propulsion Effects

Acknowledgements

This work has been funded in part by the European Commission through the ALTERNE project (IST-38575).

References

Bec, L., 1991. Elements d'Epistemologie Fabulatoire, in: C. Langton, C. Taylor, J.D. Farmer, & S. Rasmussen (Eds.), *Artificial Life II*, SFI Studies in the Sciences of Complexity, Proc. Vol. X. Redwood City, CA: Addison-Wesley (in French).

Bec, L., 1998. Artificial Life Under Tension: A Lesson in Epistemological Fabulation, In: C. Sommerer and L. Mignonneau (Eds.), Art @ Science, New York: Springer Verlag.

Bedau, M.A., 1992. Philosophical Aspects of Artificial Life, in: F.Varela and P. Bourgine (Eds.), Towards a Practice of Autonomous Systems, Cambridge: MIT Press, pp. 494-503.

Cavazza, M. and Simo, A., 2003. A Virtual Patient Based on Qualitative Simulation. In ACM Intelligent User Interfaces 2003, Miami, FL, USA, pp. 19-25

Cavazza, M., Hartley, S., Lugrin, J.-L. and Le Bras, M., 2002. Alternative Reality: Qualitative Physics for Digital Arts, Proceedings of the 17th international workshop on Qualitative Reasoning 2003, Brasilia, Brasil.

Cavazza, M., Hartley, S., Lugrin, J.-L. and Le Bras, M., 2002. Alternative Reality: A New Platform for Digital Arts, ACM Symposium on Virtual Reality Software and Technology (VRST2003), pp. 100-108, Osaka, Japan, October 2003.

Dorin, A., 2003. Artefact and Artifice: Views on Life, Artificial Life, vol. 9, issue 1, MIT Press.

Forbus, K.D., 1984 Qualitative Process Theory, Artificial Intelligence, 24, 1-3, pp. 85-168.

Keeley, B.L., 1997. Evaluating artificial life and artificial organisms, In *Artificial Life V: Proceedings of the 5th International Workshop on the Synthesis and Simulation of Living Systems*, C. G. Langton & K. Shimohara (eds.). Cambridge, MA: MIT Press, 264-271.

Grand, S., Cliff, D. and Malhotra, A., 1997. Creatures: artificial life autonomous software agents for home entertainment, In: *Proceedings of the first international conference on Autonomous agents*, pp. 22-29. ACM Press.

Langton, C. G., 1989. Artificial life. In Langton, C. G., editor, *Artificial Life* (Santa Fe Institute Studies in the Sciences of Complexity, Vol VI), pages 1-47, Reading, MA. Addison-Wesley.Lewis, M and Jacobson, Games Engines in Scientific Research. In Communications of ACM, Vol. 45, No. I, January 2002. pp 27-31.

Lewis, M and Jacobson, Games Engines in Scientific Research. Communications of ACM, Vol. 45, No. I, January 2002. pp27-31.

Noble, J. (1997). The scientific status of artificial life, *Fourth European Conference on Artificial Life (ECAL97)*, Brighton, UK, 28--31 July

Risan, L. (1997). Why are there so few biologists here? Artificial life as a theoretical biology of artistry. In Husbands, P., & Harvey, I. (Eds.), Proceedings of the Fourth European Conference on Artificial Life (ECAL'97). MIT Press / Bradford Books, Cambridge, MA.

Mechanistic and ecological explanations in agent-based models of cognition

Jason Noble[1] and Manuel de Pinedo[2]
[1]School of Computing, University of Leeds, UK
[2]Department of Philosophy, University of Granada, Spain
jasonn@comp.leeds.ac.uk

Abstract

We argue that two styles of explanation—mechanistic and ecological—are needed in accounting for the behaviour of synthetic agents. An emphasis on mechanistic explanation in some current ALife models is identified, and parallels are drawn with issues in the philosophy of mind literature. We conclude that ecological or agent-level explanation does not come with representational baggage, and that mechanistic explanations cannot stand alone.

Explaining the behaviour of artificial agents

This paper is concerned with the ways in which artificial life researchers explain the behaviour of their evolved agents. We see a role for both mechanistic explanations, and ecological or agent-level explanations. In a mechanistic explanation, the agent is accounted for by showing how its component parts interact with each other and the environment to result in the production of the behaviour in question. In an ecological or agent-level explanation, the mechanical details are abstracted away and a behaviour is explained by saying what its purpose or function is *for the agent*—there may be explicit or implicit reference to a history of learning or selection.

We are interested in exploring the use of these two types of explanation for two reasons. First, we see fruitful parallels between artificial life and philosophy of mind in this respect. Behaviour and its explanation are central to both disciplines, and we hope to show that each area has something to learn from the other. Second, we are worried about a tendency for mechanistic explanations in artificial life to be regarded as somehow more basic or more primary. This tendency mirrors the rise in philosophy of mind of an idea known as *eliminativism*, which seeks to replace agent-level talk about behaviour with a mechanistic vocabulary. The eliminativist holds that explaining a person's behaviour at the agent level (e.g., explaining John's trip to the supermarket by saying that he wanted to buy milk) is pre-scientific, and that such accounts will one day be replaced by complex but precise neurological explanations. We believe that eliminativism is a deeply flawed view. We will argue that

artificial life research should not only avoid the temptations of eliminativism, but can actually help in showing what is wrong with the idea.

Artificial life research covers a broad territory, but it can reasonably be summarized as seeking to understand life and mind through synthetic means. We want to focus on a particular strand of artificial life work in which mechanistic explanations are sought for the behaviour of evolved agents. In a typical paper of the type we have in mind, an evolutionary algorithm is applied to simulated or robotic agents with a flexible control architecture (often a recurrent neural network). The fitness function measures performance on some task, and eventually the agents achieve high levels of performance, perhaps using an unanticipated strategy. An elite or typical agent is then analyzed in detail and a mechanistic explanation of the behaviour is presented (e.g., an elaborate description of just how the weights and connections in the neural network facilitate the successful behaviour). The authors of the paper typically conclude that the behaviour is now fully explained: the artificial agent's capacity to perform behaviour X in environment Y is accounted for by having a control architecture tuned just so. The inference is often drawn that naturally occurring agents (animals) performing a similar behaviour might use a similar mechanism.

This tradition in artificial life research can be traced back to the seminal work of Walter (1950), and was revived by Braitenberg (1984), who demonstrated that behaviour seemingly requiring complex agent-level explanations could be produced by some very simple mechanisms. (Braitenberg was playfully agnostic, however, on the question of whether his models meant that agent-level explanation could be dispensed with entirely.) Braitenberg's work has been extremely influential in artificial life. Research that has followed on in a similar vein includes that of Beer (1990, 1996) who is involved in a long-term program to produce models of "minimally cognitive behavior" and to analyze the evolved architectures using a dynamical systems approach. Other work that fits our template includes Cliff, Harvey, and Husbands (1993), Floreano and Mondada (1994), Husbands, Harvey, and Cliff (1995), Quinn, Smith, Mayley, and Husbands (2002), and others too numerous to mention. Webb's (1994) paper is an interesting example in that the evolved agents to be analyzed are not synthetic but real insects; the work uses a robotic analogue of the cricket to demonstrate the plausibility of a simple mechanical explanation for sonotactic behaviour.

The work of Randall Beer epitomizes the approach we are discussing, and so from here on we will focus on his most recent paper (Beer, 2003) which presents a highly detailed dynamical systems analysis of the workings of an evolved agent. The agent is controlled by a continuous-time recurrent neural network, it is equipped with a fan-like array of simple range sensors, and it can move along a one-dimensional track while two kinds of objects fall from the sky above. The agent is the elite member of a population that has been selected for the ability to catch falling circles and to avoid falling diamonds. Beer's analysis demonstrates that the catching and avoiding behaviour can be understood as the result of a complex dynamical system that includes the agent's neural control architecture, its simulated body, and its environment. Beer presents this type of analysis as a potential new paradigm for explanation in cognitive science.

We should reassure the reader at this point that we are admirers of Beer's work and of that of the other authors cited above. Mechanistic explanations of artificially evolved agents (or of naturally occurring agents) are valuable. Our problem is with the implication that mechanistic explanations are independent and exhaustive explanations, containing all the conceptual resources necessary for understanding behaviour. We will expand on this point below, but it seems to us that such explanations are only interesting inasmuch as they shed light on broader questions about life, cognition, and the way in which an agent is situated in its environment.

One of the reasons for artificial life research to have focussed on this sort of mechanistic explanation is because of the way in which the field has defined itself in opposition to the classical artificial intelligence tradition ("GOFAI"). In classical AI, an agent's behaviour is explained as the result of it planning a course of action based on its internal model of the world. The agent is supposed to be using sensory input to update its representation of what is going on in the world, and then to be manipulating these internal representations in order to plan and re-plan the optimal way to achieve its explicit goals. ALife evangelists such as Brooks (1991) and Harvey (1996) were quick to point out problems with this picture, and to supply counter-examples in the form of simple reactive agents that were capable of highly competent behaviour without possessing anything resembling an internal representation. If you could completely describe the workings of an agent's neural network in simple mechanistic terms, as did Cliff et al. (1993) for example, you could then ask, "Where are the internal representations?" Nowhere, of course.

Beer's (2003) paper is also driven by this ongoing rejec-

tion of representationalism: Beer argues that dynamical systems explanations (i.e., mechanistic explanations) raise "important questions about the very necessity of notions of representation and computation in cognitive theorizing" (p. 3). But the rejection of representationalism leads Beer to take for granted a false dilemma. He associates explanations cast at the whole-agent level—such as an explanation of the agent's movement in terms of a belief that the falling object was a diamond—with a representational perspective. In other words, if the agent can be sensibly described as believing that a diamond is falling, then there had better be a diamond-detecting routine and an internal switch or memory register to represent the detected diamond. Alternatively, it might be possible to explain the agent's behaviour in terms of mechanistic properties of the coupled brain-body-environment system. This is the false dilemma. Obviously Beer makes a strong case for the second option: after close examination of the circle-catching, diamond-avoiding agent, we find no circle or diamond detectors and nothing that resembles a representation of a circle or a diamond. Beer concludes that internal representations are not needed to explain the agent's behaviour, whereas the right set of dynamical equations allows us to understand what is going on.

Representationalism takes something of a battering in Beer's paper. We end up with a bold new vision for the explanation of behaviour in terms of dynamical systems theory, with none of the familiar agent-level explanatory concepts—beliefs, desires, intentions—in sight. This is clearly at odds with our own view that both kinds of explanation are desirable and necessary. We will therefore question Beer's assumption that agent-level explanations and internal representations go hand in hand. In order to do so, we must first look at why representationalist views held any appeal in the first place.

Representationalism is just one aspect of a bigger and older idea, namely *internalism*: roughly speaking, this is the belief that cognition is something that goes on inside the head. And as so often happens in issues to do with the mind, the problem is all down to Descartes. Internalism is attractive because it matches the Cartesian intuition that thoughts, cognition, and knowledge are processes or things located in a special, non-physical place called the mind. Putting it another way, representations are theoretical devices to give flesh to the intuitive appeal of the idea that thought happens inside the agent—in its head, soul, brain, or wherever. The initial mistake is to suppose that a lonely homunculus inhabits the mind, and the mistake is compounded by bringing in representations for the homunculus to calculate over. Thus intuitive internalism gives rise to representationalism and not the other way around. It follows that representationalism is best dealt with at its source: that is, by challenging the notion that thought is internal.

There is some irony here. In an important sense, artificial life is the last place to expect internalism. Unlike artificial intelligence, which smacks of internalism in its resolve to keep mind and world separate, artificial life researchers have been quick to appreciate that cognition can be distributed across the agent-environment divide. Witness the popularity of the key phrases "situatedness" and "embeddedness" in the ALife literature. So we need to be clear about what we are claiming: that the diagnosis of representationalism's flaws has not been radical enough. A last vestige of internalist thinking has resulted in an incorrect association between representationalism and agent-level explanations, and this in turn has produced an unwarranted emphasis on low-level mechanistic explanations of artificial agents.

Parallels in philosophy of mind

We want to pause momentarily in our analysis of Beer (2003) and move to the philosophy of mind literature. Philosophers of mind are much concerned with the explanation of behaviour, and their nearest equivalent to the agent-level / mechanistic distinction is that between personal and sub-personal explanations. This was informally introduced by Ryle (1949) and later popularized by his student Dennett (1969). In *The Concept of Mind* (1949) Ryle argued that there are two very different classes of human behaviour requiring explanation. When we get something right, our behaviour is best explained at the personal level, in the ordinary language of beliefs and desires: John drove to the supermarket because he desired milk and believed that it was available there. On the other hand, when we get something wrong, a sub-personal explanation, phrased in terms of physical interactions between our component parts, may be called for. Suppose John crashes his car on the way to the supermarket, because he suffers a mild stroke. The relevant explanation is obviously sub-personal.

An important point to appreciate about the two kinds of explanation is that personal-level stories stop quite early. Once we have granted that John is a rational agent, there is not much more to say about his trip to the supermarket. Ryle was content with this short chain of reasons for rational acts, and was extremely sceptical about the idea that psychology or any other science could somehow supplant personal-level explanations and supply the 'real' causal story behind a person's actions. Sub-personal explanation, on the other hand, can go very deep: accounting for John's stroke might involve considerations of diet, physiology, genetics, biochemistry, and ultimately physics. Sub-personal explanation is also part of the story in explaining John's competencies: how is it that he has the sensorimotor coordination needed to drive a car, or to remember the way to the supermarket? It was in these sorts of questions that Ryle saw a role for cognitive science.

Furthermore, Ryle held that the category errors generated by confusion between the two different levels of explanation were responsible for most of the apparent mysteries about cognition. Ryle's most famous example of a category error involves a visitor being shown all of the buildings in Ox-

ford and then insisting "Yes, but where is the university?" A similar error is easily committed when thinking about agents and their component parts. For example, John is like the university: he is in a sense constituted by his component parts, but he is not the same type of thing as one of his components. Forgetting this, and imagining that the components of John's brain and John the person are on a par with each other, leads to conceptual disasters such as the 'mystery' of how John could possibly be conscious, for example. (If John is equated with the mere matter of his brain then certainly there is a mystery about how the-matter-that-is-John achieves conscious awareness, but if we recognize that John and his brain matter are concepts at different levels, then wondering about how John could possibly be conscious is properly recognized as being a bit like asking how it is that a university can make decisions despite being constructed from stone.)

Regarding the issue of internalism, it is worth noting that Ryle also embraced a radical *externalism*. Ryle argued that cognition is spread out across the agent and the world, and wanted to banish the Cartesian image of mind as a private place (whether non-physical or physical) where thoughts happen. A similar externalist view has been adopted by many more recent authors in the philosophy of mind (see, e.g., Morris, 1992; McDowell, 1998) and seems in obvious harmony with the concern for situatedness and embodiment in artificial life.

Perhaps because of his accessible style and engagement with the sciences, Dennett has been an influential philosopher in the artificial life community and so a further word on his position is in order. Dennett (1969) first espouses a view very similar to Ryle's, but in later work (Dennett, 1987) he retreated somewhat and claimed that personal-level or agent-level explanation is just a stance that can be taken when it is expedient to do so. However, Morris (1992) points out that this is plainly self-defeating, as the idea that there is anyone around to take stances toward anything presupposes the existence of persons or agents. Our own view is that Ryle's original distinction is worth defending.

We do not wish to give the impression that the personal / sub-personal distinction is universally accepted in the literature. Many authors have opposed it; the argument usually goes as follows. Even if personal-level explanations are taken to be somehow autonomous, as in Ryle's account, then more or less often we will still need sub-personal explanations to fill in the gaps, i.e., to explain what went wrong when rational action fails. If we can be successful in explaining *some* behaviour at the sub-personal level, then surely there is a prospect of explaining all behaviour this way. It would therefore be parsimonious to eliminate reference to persons (as well as to beliefs, desires, reasons, etc.) altogether. Behaviour is then to be explained by something like neuroscience. This is of course the eliminativist view: see Churchland (1981) for the classic account. In the next section we will explain why we believe eliminativism to be untenable.

Philosophers of mind have tended to focus on the minds and behaviour of normal adult humans. As a result, the status of actors occupying the broad space between thermostats and rational adults (e.g., animals, infants, people with neurological disorders, etc.) can be problematic for otherwise excellent accounts of agency. This seems to be an area where philosophy of mind could look to artificial life for richer ways of categorizing the various possible classes of agent (see Dennett, 1996, for an example of this). And although the personal / sub-personal distinction is a useful one, McDowell (1998) points out that its use in conjunction with a focus on adult humans has led to a confusion between what should be two separate distinctions. Personal-level explanation seems to belong to a realm of rational, normative agents, whereas sub-personal explanation concerns the disenchanted physical world. McDowell draws on the landmark neuroscience paper "What the frog's eye tells the frog's brain" (Lettvin, Maturana, McCulloch, & Pitts, 1959) and suggests that to fully understand the frog, we would also need a "froggy / sub-froggy" distinction. On the one hand we need to consider the frog as a whole agent in its environment or *Umwelt*, and look for example at the significance of different environmental features *for the frog*. This corresponds to what we have called agent-level or ecological explanation, and McDowell aptly cites the ecological psychology of Gibson (1979) to illustrate the approach. On the other hand we will also want to employ familiar styles of mechanistic explanation in looking at how various components of the frog (the visual system and the motor system for example) interact with each other. McDowell references Marr's (1982) work on computational models of vision as an example of the latter approach.

In none of this does McDowell (1998) argue that frogs are rational, normative agents. McDowell's conclusion is that there are really two distinctions at work: one between agents and their component parts, and one between persons and non-rational agents. We endorse McDowell's argument, and stress that we are interested here in the first of the two distinctions. We see the split between personal and sub-personal explanation in philosophy of mind as a special case of the more general distinction between agent-level or ecological explanation, and mechanistic explanation. What of the second distinction? It certainly points to a profound philosophical question about what it might be that distinguishes a person from a preying mantis. Philosophers such as Morris (1992) and many others have attempted to answer that question in terms of qualities such as moral responsibility, but we plan to remain silent on the issue. It seems to be a problem about which artificial life does not yet have anything sensible to say.

Explanatory pluralism needed in artificial life

We believe that artificial life requires both mechanistic and ecological explanations in order to make sense of the behaviour of synthetic agents. Our argument, in brief, is that agent-level explanations are necessary because without them mechanistic explanations are incomprehensible, that any suspicion that agent-level explanations might be reducible to mechanistic explanations is founded on a metaphysical error, and that an externalist perspective means there is no danger of ecological or agent-level talk being linked with representationalism.

We start by returning to our critique of Beer (2003). Beer seems to believe that because he has rejected representationalism, he is left with an eliminativist position with respect to agent-level concepts such as knowledge, meaning, belief, and desire. In other words, there is no place for such concepts in cognitive science, and the only proper explanations will be mechanistic, dynamical systems accounts of agents coupled with their environments. However, this would only follow if the representationalist perspective was the only way to make sense of knowledge and meaning. A sufficiently radical externalist perspective on these concepts does away with representationalism without disposing of all agent-level talk. Consider John and his trip to the supermarket for milk. To say that John intends to drive to the supermarket simply does not mean being committed to the notion that John has an explicit mental representation of a planned route mapped out in his head. Presumably John will get there through some distributed combination of driving habits, consulting a street directory, reactive strategies at particular intersections, environmental features such as street signs prompting the right behaviour, etc. In terms of the explanatory utility of the agent-level description of his action, the one phrased in terms of his desire for milk and his belief that the supermarket sells milk, it does not really matter. It would be interesting to study John's sub-personal capacities and find out how they constrained the sorts of milk-buying journeys he might be capable of, but such mechanistic explanations will not lead to a revision of the agent-level story: on this occasion, he went to the supermarket to buy milk.

The dilemma identified by Beer (2003) is thus a false one, and externalism means that it is possible to save the baby of agent-level explanation while throwing out the representational bathwater. Artificial life and cognitive science generally do not need to give up on talking about agent-level concepts such as knowledge or meaning: practitioners in both fields just need to recognize that knowledge and meaning, as much as perception and action, are features of the coupled agent-environment system and not something internal.

Why do we feel that agent-level description is of value? Why are we convinced that the impressive analytical vocabulary of dynamical systems theory, for example, is not the only vocabulary needed by the artificial life researcher? We refer the reader to the deceptively obvious fact that Beer (2003) needs to describe his agent as a circle catcher and a diamond avoider. Indeed, these are the propensities that his agent was selected for over many generations of evolution. This description is admittedly simple, but it is agent-level talk, and clearly of a different explanatory level than a description of the agent / environment system in terms of differential equations. As a quick thought experiment of our own, we ask whether anyone could possibly make sense of the behaviour of the agent given only the dynamical systems description so carefully developed in Beer's paper, and not the brief but enormously helpful agent-level description. Looking only at the mechanistic level, it would be extremely difficult and perhaps impossible to see that all of this complexity was in the service of circle catching and diamond avoidance.

McDowell (1998) makes the same point in slightly different language. At the agent or ecological level we can pose and answer "why?" questions. Why did the frog stick out its tongue? In order to catch what it believed was a fly. These questions and answers can in turn inspire "how?" questions at the mechanistic ("sub-froggy") level. How did the visual input lead to the appropriate motor output? When we have answered the mechanistic how-question in terms of some sort of neural circuitry diagram, we have described what McDowell calls an enabling condition for the agent-level behaviour. If we were to then insist that this mechanistic explanation could stand alone, we are mistaking an enabling condition for a constitutive one. As Davidson (1980, p. 247) puts it "... it is one thing for developments in one field to affect changes in a related field, and another thing for knowledge gained in one area to constitute knowledge of another." Even the best mechanistic explanation will be incomprehensible without an agent-level framework. If systems as simple as the one analyzed by Beer (2003) require on the one hand agent-level explanations and on the other hand a mechanistic description in terms of dynamical systems, then clearly more ambitious targets such as advanced ALife agents, frogs, and human beings will also require both levels of description.

We have said that we would show why eliminativism is wrong. Eliminativism argues that agent-level descriptions are no better than myths, and stand in need of elimination in favour of mechanistic explanations. It is in the kind of practical interplay between agent-level and mechanistic explanations described above that we think artificial life demonstrates why the former cannot be eliminated in favour of the latter. But where did anyone ever get the idea that an ecological or agent-level explanation would be eliminable by a mechanistic one? This is an instance of a widely held and often unquestioned belief that all forms of explanation will eventually be replaced by or reduced to one privileged explanatory basis, usually assumed to be the language of physics. Morris (1992) identifies this belief as *scientism*, and demolishes it in short order. Morris is in no way anti-

scientific, but he disagrees with a movement in philosophy known as *naturalism*, which seeks to use the methods of natural science as a basis for metaphysics, i.e., as a basis for thinking about what exists and about how we could know about it. The first move of the naturalist is to propose the sciences (and ultimately physics) as the only basis for knowledge, and to declare "There are only scientific facts." Morris points out that this move is immediately fatal: what kind of fact is the declaration itself? It is clearly not a testable proposition from the natural sciences, and the declaration thus perversely renders itself false.

We see eliminativism as one of the faces of scientism. The onus of proof is not on the user of agent-level explanations to say why they are autonomous with respect to mechanistic explanations. Rather the burden runs the other way: given the failure of scientism, the eliminativist must show why and how an agent-level explanation could be dispensed with. It is not valid to simply assume that mechanistic explanations are primary.

If our argument holds, then artificial life should be content to deal in multiple levels of explanation for the behaviour of synthetic agents. Undoubtedly some researchers will react to this assertion with horror, whereas others will shrug as they are already committed to such pluralism. By way of reassurance for the first group, we want to point out that there is a strong precedent for the peaceful coexistence of multiple levels of explanation in a scientific discipline: we refer to Tinbergen's (1963) seminal paper on the aims and methods of ethology. Tinbergen introduced four types of explanation for ethology, and arguably for biology in general: two of them were explanation in terms of function and in terms of mechanism, which obviously correspond to the two types of explanation we have been discussing. Tinbergen's two additional types of explanation were both historical: ontogenetic and phylogenetic explanation. The good news is therefore that one of artificial life's parent disciplines appears to be able to cope with a plurality of explanatory projects.

In conclusion, we feel that Beer (2003) should not be concerned that endorsing an agent-level description will commit him to the follies of old-fashioned representational AI. One can say that Beer's agent catches circles and avoids diamonds without conceiving of cognition as a series of rule-governed operations over internal symbols. We encourage Beer (and others in artificial life) to go all the way with the externalism exhibited in his analyses of perception and decision-making. On the view we are urging, mind is not internal, it is all over the place; indeed, "mind" is just a very abstract way of describing the agent / environment interaction. The debate over representationalism loses all urgency once the Cartesian image of the mind as a place of internal knowings has been properly dispelled.

Anticipated objections

We have anticipated two of the more likely objections to our argument and have attempted to answer them in advance.

Some might see our argument about the need for two kinds of explanation as a pragmatic move related to the difficulty of understanding the messy architectures of evolved agents. In contrast, when dealing with a traditionally engineered system (a hand-coded AI robot for example) perhaps only the mechanistic level of explanation will be necessary, as a complete and accurate blueprint of the agent's architecture is available. We disagree: such an agent will still need to be understood in terms of the mechanical interactions between its components, and at the agent-level in terms of the designer's intentions. Random wandering in a vacuuming robot, for instance, might be intended to clean the carpet, independently of the way in which that movement is implemented at a lower level of description.

Accusing your opponents of the hangovers of Cartesian thinking is a popular sport in cognitive science and philosophy, and we too might be faced with the accusation that in proposing two levels of explanation for agents we are resurrecting some sort of dualism between a physical, mechanistic, sub-agent domain and a mysterious, cognitive agent-level domain. The important thing to emphasize here is that we are not positing two kinds of thing, physical-stuff and mind-stuff. The insistence that to each variety of explanation corresponds a variety of stuff is itself a Cartesian idea. Our externalist, Rylean, perspective precludes seeing the mind as a place or as a special sort of thing. For us, mind is shorthand for a set of complex interactions in a unified world that happen to demand their own sort of explanation.

References

Beer, R. D. (1996). Toward the evolution of dynamical neural networks for minimally cognitive behavior. In Maes, P., Matarić, M., Meyer, J.-A., Pollack, J., & Wilson, S. W. (Eds.), *From Animals to Animats 4: Proceedings of the Fourth International Conference on Simulation of Adaptive Behavior*, pp. 421–429. MIT Press / Bradford Books, Cambridge, MA.

Beer, R. D. (2003). The dynamics of active categorical perception in an evolved model agent. *Adaptive Behavior*, *11*(4), 209–244.

Beer, R. D. (1990). *Intelligence as Adaptive Behavior: An Experiment in Computational Neuroethology*. Academic Press, Boston.

Braitenberg, V. (1984). *Vehicles: Experiments in Synthetic Psychology*. MIT Press, Cambridge, MA.

Brooks, R. A. (1991). Intelligence without representation. *Artificial Intelligence*, *47*, 139–159.

Churchland, P. M. (1981). Eliminative materialism and the propositional attitudes. *Journal of Philosophy*, *78*(2), 67–90.

Cliff, D., Harvey, I., & Husbands, P. (1993). Explorations in evolutionary robotics. *Adaptive Behavior*, *2*(1), 73–110.

Cliff, D., Husbands, P., Meyer, J.-A., & Wilson, S. W. (Eds.). (1994). *From Animals to Animats 3: Proceedings of the Third International Conference on Simulation of Adaptive Behavior*. MIT Press / Bradford Books, Cambridge, MA.

Davidson, D. (1980). The material mind. In *Essays on Actions and Events*, pp. 245–259. Clarendon Press, Oxford.

Dennett, D. C. (1969). *Content and Consciousness*. Routledge & Kegan Paul, London.

Dennett, D. C. (1987). *The Intentional Stance*. MIT Press / Bradford Books, Cambridge, MA.

Dennett, D. C. (1996). *Kinds of Minds: Towards an Understanding of Consciousness*. Weidenfeld & Nicolson, London.

Floreano, D., & Mondada, F. (1994). Automatic creation of an autonomous agent: Genetic evolution of a neural-network driven robot. In Cliff et al. (1994), pp. 421–430.

Gibson, J. J. (1979). *The Ecological Approach to Visual Perception*. Houghton Mifflin, Boston.

Harvey, I. (1996). Untimed and misrepresented: Connectionism and the computer metaphor. *AISB Quarterly*, *96*, 20–27.

Husbands, P., Harvey, I., & Cliff, D. (1995). Circle in the round: State space attractors for evolved sighted robots. *Robotics and Autonomous Systems*, *15*, 83–106.

Lettvin, J. Y., Maturana, H. R., McCulloch, W. S., & Pitts, W. H. (1959). What the frog's eye tells the frog's brain. *Proceedings of the Institute of Radio Engineers*, *47*, 1940–1955.

Marr, D. (1982). *Vision*. W. H. Freeman & Co., San Francisco.

McDowell, J. (1998). The content of perceptual experience. In *Mind, Value and Reality*, pp. 341–358. Harvard University Press, Cambridge, MA.

Morris, M. (1992). *The Good and the True*. Clarendon Press, Oxford.

Quinn, M., Smith, L., Mayley, G., & Husbands, P. (2002). Evolving teamwork and role allocation with real robots. In Standish, R. K., Bedau, M. A., & Abbass, H. A. (Eds.), *Artificial Life VIII: Proceedings of the Eighth International Conference on Artificial Life*, pp. 302–311. MIT Press, Cambridge, MA.

Ryle, G. (1949). *The Concept of Mind*. Hutchinson, London.

Tinbergen, N. (1963). On aims and methods of ethology. *Zeitschrift für Tierpsychologie*, *20*, 410–433.

Walter, W. G. (1950). An imitation of life. *Scientific American*, May, 42–45.

Webb, B. (1994). Robotic experiments in cricket phonotaxis. In Cliff et al. (1994), pp. 45–54.

Empiricism in Artificial Life

Eric Silverman and Seth Bullock

School of Computer Studies, University of Leeds, UK
[eric|seth]@comp.leeds.ac.uk

Abstract

Strong artificial life research is often thought to rely on Alife systems as sources of novel empirical data. It is hoped that by augmenting our observations of natural life, this novel data can help settle empirical questions, and thereby separate fundamental properties of living systems from those aspects that are merely contingent on the idiosyncrasies of terrestrial evolution. Some authors have questioned whether this approach can be pursued soundly in the absence of a prior, agreed-upon definition of life. Here we compare Alife's position to that of more orthodox empirical tools that nevertheless suffer from strong theory-dependence. Drawing on these examples, we consider what kind of justification might be needed to underwrite artificial life as empirical enquiry.

In the title of the first international artificial life conference, held over a decade and a half ago, two streams of Alife research were identified—the *synthesis* of life-like or living systems versus their *simulation*. This distinction was perhaps intended to echo that made between strong and weak artificial intelligence, a division that has been readily adopted by the Alife community. While strong AI or Alife is concerned with building *bona fide* examples of real intelligence or real life, the weak strand of research is concerned with improving our understanding of intelligence and life via the construction of models or replicas of natural systems. While the latter branch of research is understood as an orthodox type of scientific or engineering methodology, the former has often been regarded as more problematic. How can we create genuine instances of intelligence and life without a prior understanding of what constitutes valid membership of either category?

In this paper we will not consider what constitutes a definition of life, or what it is to be alive. We will assume that such questions are being pursued by others. Rather, we will be interested in a related question: what would have to be true of the relationship between artificial and natural life for it to be the case that observations of (computational) Alife systems could be legitimately employed to settle empirical questions regarding the nature of life?

The root of the problem is hinted at by the suggestion of deception, falsity or unreality that can sometimes be detected in the meanings of both *synthetic* and *simulated*. We can clarify this hint by considering two distinct meanings of the term "artificial".

First, the word artificial can be used to describe *a man-made example of something natural* (hereafter denoted $Artificial_1$). Artificial light, for example, is typically real, actual light that is *manufactured* rather than generated naturally by the sun, or forest fires, or bioluminescence of some kind—but these are all instances of the same physical phenomenon. By contrast, the word artificial can also be used to describe *something that has been designed, perhaps through artifice of some kind, to closely resemble something else* (hereafter denoted $Artificial_2$). For example, an artificial lime flavouring, E555, might have been designed to taste like real lime, but is in fact not an instance of the real taste of lime even if it tastes indistinguishable from the real thing. Note that whether or not artificial light or lime can be used to settle empirical questions regarding real light or lime hinges critically upon which definition of artificial is understood to apply.

Which meaning of the word artificial is justified in a particular case depends upon what criteria we feel must be met in order for something to count as a true instance of a particular category. In the case of light, we have a physical theory that allows us to lump together sunlight, firelight, torchlight, electric light-bulb light, etc., into one category: real light. Given this, we are in a position to pursue "strong artificial light" research, discovering new ways of generating new kinds of light. Weak artificial light research could investigate ways of producing the *appearance* of light via constructing models of light (which involve no light themselves) or perhaps fake lighting (e.g., ways of painting a room or a picture to suggest the presence of light). In doing so, we might learn about how light works and why it appears to us in the ways that it does.

In the case of lime flavours, we have a less well-specified but widely agreed upon notion that the flavour of real lime fruit cannot be lumped together with some flavouring E555 in a category called "real lime flavour", no matter how convincingly limey E555 happens to taste (unless perhaps E555 is derived directly and straightforwardly from real limes, or can be proven to be chemically identical to real lime). In this situation, "weak artificial lime" researchers could legitimately concern themselves with producing flavourings that resemble real lime in some way. Through this type of research, we might discover a lot of the chemistry, biology and psychology of what makes things taste like lime. Strong artificial lime research is a little more problematic, however, and we will return to why this is a little later.

What is important to notice about this example is not simply that some categories, such as tastes, are subjective, but rather that some are as yet unsupported by some scientific account. The appearance of light is extremely subjective, yet we are able to objectively account for both the perceived variety of different forms of light and their unity through recourse to a physical theory. By contrast, the absence of such a theoretical account of lime flavour prevents us from achieving a similar understanding.

In this paper we consider what kind of framework could underwrite strong artificial life as an empirical pursuit, since it is through the promise of generating useful empirical data that strong artificial life typically gains its strongest support.

Empirical Alife

In their seminal paper, Alan Newell and Herb Simon (1976) provide an account intended to underwrite artificial intelligence as a kind of empirical enquiry into the nature of intelligence. This account reduces to a pair of working hypotheses:

1. A physical symbol system has the necessary and sufficient means for intelligent action.
2. A suitably programmed computer is an example of a physical symbol system.

The extent to which Newell and Simon were successful is debatable, but it remains the case that no comparable, widely adopted framework has been forthcoming within artificial life. However, the absence of such an account has not prevented Alife practitioners from arguing that their work is a form of empirical enquiry. In particular, it has been suggested that computer simulations of living systems could be used to settle empirical questions concerning life itself.

For example, Bedau (1998) considers a thought experiment proposed by Gould (1989). What would happen, Gould asks, if we could rewind the universe, to a point in time before the advent of life on earth. If we gave the primordial soup a quick stir and then let time run forward again to the year 2004, what would we see on the face of the planet? Would we see creatures much the same as ourselves, and dogs and cats and lice and lichen? Or, as he suggests, is there no reason to suppose that we would not see utterly different life-forms? For Gould, this is a thought experiment about contingency. For Bedau it is an opportunity for Alife to set-

tle an empirical question. By formulating a suitable simulation, one that "manifests" important, fundamental properties of evolution and life, and repeatedly running this simulation, we would be able to observe and record the variety of artificial life-forms that evolve, their regularities and diversity. We would then be able to answer Gould without having to rely on our, rather puny, imaginations.

This type of use for Alife systems has been argued for by a number of authors (Bonabeau & Theraulaz, 1994; Ray, 1994; Taylor & Jefferson, 1994; Miller, 1995) who claim that evolutionary biologists face a problem in that they possess scant evidence with which to reconstruct the evolution of life on earth. Such evidence includes the fossil record, and our limited observations of the species that currently surround us. Compounding this claimed paucity of raw data, the authors point out that any evidence that biologists do possess can only be the result of one evolutionary sequence, since life has (presumably) only evolved once on earth. With only terrestrial life to draw upon in constructing theories to explain phenomena associated with life, biologists are unable to distinguish between aspects of life which are contingent upon the particular historical development of life on earth, and those aspects which are fundamental to life in general.

For these authors, then, one promise of artificial life is to offer whole new datasets which can be examined alongside the one provided by natural evolution. But this approach to empirical Alife is not without its problems. Consider an anecdote intended to reveal the empirical power of computer simulation, as presented by Casti (1997) in his book *Would-be Worlds: How Simulation is Changing the Frontiers of Science.*

Casti describes a particular real sporting event (a Super Bowl game) in which the unfancied American football team happened to beat the favourite. Was the underdog simply lucky, Casti wonders? If the same teams had played again, say 100 times, would the same team have won? Pundits and fans will of course never know for sure since there is only a single data point upon which to base any argument (the actual game as it was played). The data massively underdetermines the hypotheses. But wait! Casti introduces the reader to the existence of an American football computer game in which one can force the computer to control both teams (the user merely watches the game unfold). By rerunning the Super Bowl many times, with the same teams and same weather conditions, etc., Casti generates a series of new data points. On the basis of this data, he confirms that, indeed, the underdog had not merely been lucky, since under the highly controlled conditions created by the simulation, it won statistically more often than it lost. This story encapsulates the proposed role for simulation as an empirical tool, filling in missing or incomplete data sets and thereby helping to settle empirical questions. But how valid are these new data points? Can they even be classed as data at all?

Imagine that Casti's computer game had been released after the Super Bowl that he is re-creating. Could it not be possible, perhaps even highly probable, that the game programmers had noted the outcome of the Super Bowl, and tinkered with the computer program such that it tended to replicate this important result? If that were the case, through his controlled experiment, Casti would have discovered something about the ideas of the computer game programmers, not about the teams "simulated" by the computer game. Even if the computer game had been released *before* the Super Bowl in question had been played, ensuring that the programmers could not have been influenced by its result, the programmers could (must?) have been influenced by other results. How could Casti control for this influence? The programmers themselves may not have understood the idiosyncratic biases that they had perhaps inadvertently included in the game. Surely these biases and suchlike render the simulation moribund, at least as a source of empirical data?

Some authors (e.g., Di Paolo, Noble, & Bullock, 2000) have argued that Alife simulations cannot be considered to be sources of empirical data precisely because they are so loaded with the tacit pre-theoretical biases of their creators, coloured by their ideas, and polluted by their opinions. But does philosophy of science not tell us that all observations are theory-loaded in this way? When considering this problem with empirical science, Chalmers (1999) reaches the following conclusion:

> "...however informed by theory an experiment is, there is a strong sense in which the results of an experiment are determined by the world and not by the theories", and "we cannot make [the] outcomes conform to our theories." (Chalmers, 1999, pp.39–40).

Can these claims really be made of computer programs? In this paper we will explore what would have to be true for these constraints to hold in the case of strong artificial life. But first, we will consider two examples of apparently orthodox empirical science in order to discover the ways in which "artificially generated" data take part in regular science.

Trans-Cranial Magnetic Stimulation

Research into brain function often employs patients who have suffered brain damage through strokes or head injuries. Brains are examined to determine which areas are damaged, and associations between these damaged areas and the functional deficits exhibited by the patients are postulated. The technique of trans-cranial magnetic stimulation, known as TCMS or TMS, has allowed psychology researchers to extend the scope of this approach through generating temporary "artificial strokes" in normal, healthy patients.

TMS machinery consists of a set of electrodes that are placed on the outside of the skull. The researcher begins by mapping the major areas of a subject's brain using an MRI (magnetic resonance imaging) scan, and then proceeds to selectively shut off very small areas of the subject's brain using magnetic pulses. For example, TMS studies have replicated the effects of certain types of seizure (Fujiki & Steward, 1997), and have examined the effects of stimulation of the occipital cortex in patients with early-onset blindness, finding that sensory areas deprived of input begin to function in other sensory modalities (Kujala, Alho, & Naatanen, 2000). TMS can even trigger anomalous emotional responses; after inhibition of the prefrontal cortex via TMS, visual stimuli that might normally trigger a sad response were much more likely to cause a happy reaction, even laughing (Padberg, 2001).

Such methods provide a way for neuroscientists and psychologists to circumvent the lack of sufficient data from lesion studies. Much of cognitive neuropsychology is restricted in this way, forcing researchers to search through lengthy hospital records and medical journals for patients suffering from appropriate injuries. Even worse, appropriate patients may continue to go unnoticed, as some studies may require finely differentiated neurological deficits that would not normally be tested for by hospital staff. By using TMS in an attempt to mimic the effects of brain damage, researchers gain the ability to manufacture new case studies, and use this new data to establish or undermine theories regarding the functional architecture of the human brain.

TMS success depends on the precise functioning of the machinery; if subjects undergoing the procedure do not actually experience appropriate inhibition of the brain area under consideration, then the results of a study may be useless. Unfortunately there are few guarantees in TMS research. The machinery inhibits neural activity through bursts of electromagnetism which send the neurons into such a frenzy of activity that normal firing patterns are impossible. This effectively disrupts activity in the area under the pulse, but may not actually mimic the effect of a conventional lesion. Similarly, varying the frequency of the electromagnetic pulse may change the resultant effect on the subject. Finally, the pulse is only intended to affect brain areas which are near to the skull surface; however, the pulse penetrates beyond these areas, and the effect of this excess inhibition is not fully understood. Despite these shortcomings, however, use of TMS to simulate brain dysfunction has become a rapidly growing area of research within psychology and neuroscience.

The data derived from TMS is clearly strongly theory-dependent. Using this data as a way of settling empirical questions regarding the brain function of normal people requires researchers to sign up to a "backstory". This backstory is an account that lumps TMS data and regular lesion data together as examples of real brain-damage data. Despite the fact that TMS brain damage data is artificial, it will remain admissible if neuroscientists read this artificiality in the sense of Artificial$_1$: "a man-made example of something natural". Neuroscientists would reject TMS if they only regarded TMS brain damage as Artificial$_2$: "something that has been designed to closely resemble something else".

Neuroscience Studies of Rats

Studies of rats are very common within the field of neuroscience, given that rat brains are much less problematic to examine and analyse than human brains. Ideally, researchers would be able to make non-invasive, *in situ* recordings of the neural activity of free-living rats as they go about their normal everyday behaviour. Unfortunately, such recordings are currently beyond the state of the art. In their place, neuroscientists must often rely upon studies of artificially prepared rat brains or portions of rat brain, e.g., to determine the neural pathways that are used during various cognitive functions.

A study of neurons in the medial geniculate body of the rat provides a useful example (Peruzzi, Bartlett, Smith, & Oliver, 1997). Rats in this study were anaesthetised and dissected, then slices of the cortex were stimulated directly after preparation in this way. This particular study aimed to determine the possible varieties of connections within the main auditory pathway of the rat, and by extension the possible structure of the human auditory pathway.

Many neuroscientists would not see any problem with the empirical validity of this type of procedure, and it is certainly the case that one cannot currently identify specific connections between neurons without some sort of similar intervention. However, the behaviour of cortical cells in a preserved culture is certainly not the same as the behaviour that the same cells would exhibit during their normal functioning. The whole brain provides an array of stimulation to the area of cortex in question, and this stimulation is modified by a structured external environment which is itself influenced by rat behaviour. Adaptive behaviour research places a great deal of importance on this notion of embodiment and situatedness, arguing that an organism's coupling to its environment is vital to that organism's cognition and behaviour (Brooks, 1991). If this argument holds, then in some sense neuroscience studies of the kind described above could be accused of generating and recording artificial neurological data. In what way do these data apply to real rat behaviour?

In fact, research in this vein proceeds on the assumption that removal of the rat from the environment may actually increase the experimental validity of the study, since in this way environmental or observer interference can be minimised. The absorption or expulsion of chemicals by individual neurons can be monitored and catalogued with high precision, very accurate readings of neuronal activity can be taken very easily, and experimenter mistakes will likely be more apparent. Additionally, the neuroscience community must hold, perhaps tacitly, that any *artificiality* introduced by their experimental procedures (the extent to which the

behaviour of rat neurons changes when the rat is removed from its normal environment, or, in this extreme case, the rat's brain is removed from the rat) is only Artificiality$_1$, and hence acceptable.

Discussion

The two examples above demonstrate that relatively orthodox empirical tools are never-the-less theory dependent, and that this dependence need be neither straightforward nor explicitly understood. The grounds upon which the TMS and rat studies described above are considered valid methods of collecting data on real brain function are neither formal nor conclusive, but rather constitute a kind of working hypothesis supported by a tacit framework of assumptions. The fact that empirical tools can be employed in the absence of a strict account of their validity is welcome news for strong artificial life, since a workable definition of what is to count as living and what is not appears to be some way off. But the fact that empirical tools do appear to rely on some agreed-upon "backstory" begs the question: what sort of backstory might similarly support strong artificial life research?

First and foremost, this backstory must offer a convincing (to the Alife community, at least) account of why artificial life deserves to be understood as a source of Artificial$_1$ data, rather than merely Artificial$_2$ data. That is, artificial life must count as man-made life, rather than merely resembling life to some extent.

How might this be achieved? Both TMS and rat neuroscience offer the same suggestion. Propose an empirical procedure that starts with a non-controversial example of the class of systems one wishes to explore (e.g., brains) and prepare it in a manner that makes it amenable to empirical investigation. Then argue that any effects of the preparation procedure (e.g., zapping with electromagnetism; anaesthetising, slicing, and shocking; etc.) are neutral, or benign or can be controlled for. Hence, despite the fact that preparation introduces a gap between the object of enquiry and the subject of experimentation, researchers can still claim that the latter offers a window onto the former, although one that is somewhat *indirect*.

In fact, this is an approach to preparing artificial life already taken within some fields, e.g., AI (Artificial Insemination): start by getting hold of some real male and female gametes from the species you wish to artificially synthesise (uncontroversially, we typically take these gametes to be capable of becoming alive) and artificially bring them together in a way that encourages a new living creature to develop. A similar story underpins other forms of manufactured life, e.g., clones, or mutant lifeforms such as various experimental strains of drosophila.

So far, this approach appears analogous to the manufacture of lime flavour through processing real limes—if this processing is not regarded as somehow debasing or polluting, one might feel justified in using the term "real lime flavour" to describe such a product. Unfortunately, this approach will not satisfy a central goal of strong artificial life: the generation of entirely new datasets. Any artificial life produced in this way will be inherently related to existing life. We will not be able to significantly augment the dataset upon which theoretical biology can draw through this approach.[1] Indeed regular biology regards synthesised life of this type (cloned sheep, mutant flies) as straightforwardly falling within its standard remit.

However, the "preparation" that is typically involved in Alife computer simulation appears to be of an entirely different character. Alife researchers typically begin with a system that appears to be entirely removed from the living systems that they seek to investigate—an unprogrammed computer. The "preparation" that this machine undergoes is some kind of programming, which takes place in a largely idiosyncratic and informal fashion, by contrast with standardised practices in experimental neuroscience. Rather than merely being intended to make the system amenable to controlled observation and recording, this kind of preparation appears to bear the entire burden of "animating" the computer—bringing it to life. As such it would be hard to argue that this type of preparation is neutral or benign. Rather, it is substantive, and hence suspicious.

Thus, the gap between the class of living things and the Alife systems built to generate empirical data on them looks pretty wide. Rather than being a window onto life-as-it-could-be, from this perspective artificial life might appear to be quintessentially artificial$_2$—an attempt to make one thing look like something else. Can this be remedied?

One example of an alternative route is offered by Newell & Simon (1976). Rather than deriving man-made intelligence from uncontroversial examples of natural intelligence, they commence from a more radical position, arguing that computers meet the necessary and sufficient conditions for intelligence by virtue of being, along with brains, physical symbol systems. Once this is established, any computer can be considered to be only a suitable programming away from being a *bona fide* example of real intelligence. Notice that Newell and Simon do not rely on particular similarities between the behaviour or structure of AI systems and natural intelligences. Rather, they attempt to establish first principles from which to demonstrate a fundamental equivalence between computation and intelligence. On this view, programming a computer is just the same kind of preparation employed in orthodox empirical science—it is a way of subjecting computation/intelligence to controlled, critical examination. How might artificial life follow this lead?

Langton (1992) posits that life is some kind of informa-

---

[1] An alternative tack also fails for a similar reason. Real lime flavour might be successfully manufactured through building an exact copy of real lime molecules and mixing them together in the right way. But this type of synthesis will not produce new data, it simply reproduces existing data.

tion ecology—"in living systems, a dynamics of information has gained control over the dynamics of energy, which determines the behavior of most non-living systems" (Langton, 1992, p.41). Perhaps this notion (or a previous version of it, Dupuy, 2000) could be elaborated upon to form an appropriate framework for strong artificial life?

1. An information ecology provides the necessary and sufficient conditions for life.
2. A suitably programmed computer is an example of an information ecology.

Under such an account, an Alife simulation would be regarded as an instance of an information ecology, and thus a manifestation of life, rather than merely a facsimile of it. The unprogrammed computer becomes a system that is potentially alive, given the appropriate conditions. Just as Newell and Simon view the computer as simply needing the right type of program in order to realise its potential for intelligent action, strong artificial life proponents might view the computer as similar to an unfertilised egg, needing only the correct stimulation (i.e., appropriate information) to realise its potential for life. From this perspective, the gap between the object of enquiry and the subject of experimentation not only narrows, but entirely disappears; Alife becomes a means for gathering data on real digital life, rather than a means for producing behaviours similar to biological life.

This paper should not be read as necessarily endorsing this particular account or indeed the wider project of empirical artificial life. Rather, we hope to have demonstrated that if empirical artificial life is to proceed, it will require some set of agreed-upon assumptions that perform the same role as Langton's. They must motivate the notion that a computer is more than merely *capable* of being a life-supporting medium (like organic molecules), but rather the far more problematic assertion that computation is *intrinsic* to life.

Summary

Here, we have considered the potential for strong artificial life to settle empirical questions regarding the nature of life despite the absence of a unifying theory of life itself. Unfortunately, any "data" drawn from an artificial life study would appear to be strongly theory dependent, i.e., whether or not one considers the behaviour of an artificial life system to constitute Artificial$_1$ life or Artificial$_2$ life hinges critically on whether one signs up to an appropriate "backstory" or rationale.

We have seen that less controversial empirical tools also suffer strong theory-dependence, yet are in common use despite the absence of an explicit, formal, conclusive account of their validity. What underwrites the use of these tools is an informal set of tacit assumptions convincing experimentalists that, while the tool may not be a direct window onto the phenomena of interest, any attendant indirectness is not problematic. However, it appears that the kind of frameworks that support the scientific use of these tools will not help support strong artificial life. The validity of TMS and rat neuroscience studies stems from the fact that the starting point for these empirical procedures are uncontroversial examples of the systems they are intended to generate data on. (Although the procedures that are applied could be accused of introducing artificiality, the community that employ these types of technique regard this kind of "preparation" as non-problematic.) As a result, any data generated from these procedures is intimately linked to these uncontroversial examples—it is this link from which the validity of the procedure emanates. If artificial life is intended to generate new datasets that are independent of, or distinct from, life-as-we-know-it, this will not do. Indeed, when "artificial life" is derived from living things, as in cloning, the results are regarded as part of terrestrial biology, not as some distinct kind of life.

In light of this, it appears that strong artificial life will require some theory of life to be in place before it can commence. This theory might follow Newell & Simon (1976) in taking the form of an argument that computation is a kind of life. From this position, it would be possible to generate artificial life that was independent from terrestrial life, yet real in the sense of Artificial$_1$. The equation of life and computation is a formidable undertaking, yet artificial life is in possession of a few candidate theories that could act as the seeds of such an account. For instance, Langton's notion of living systems as those in which "a dynamics of information has gained control over the dynamics of energy" (Langton, 1992, p.41) may be one such seed. However, such Alife ideas must develop and mature considerably before we are in the enviable position of, say, artificial light researchers, who rely upon a well-founded theory that allows them to generate light in entirely novel ways, yet feel secure in claiming this to be real light, not Artificial$_2$ light. There is of course no stipulation that such a theory of computation/life need be correct. Ultimately, it will be the artificial life community who will decide whether or not any candidate theory is sufficiently compelling.

Acknowledgements

Thanks to Ezequiel Di Paolo, Jason Noble, an editor of a British neuroscience journal, and members of the Biosystems group for useful conversation, and to three anonymous reviewers for their comments.

References

Bedau, M. A. (1998). Philosophical content and method of artificial life. In Bynum, T. W., & Moor, J. H. (Eds.), *The Digital Phoenix: How Computers are Changing Philosophy*, pp. 135–152. Blackwell, Oxford.

Bonabeau, E. W., & Theraulaz, G. (1994). Why do we need artificial life? *Artificial Life, 1*, 303–325.

Brooks, R. A. (1991). Intelligence without reason. In Myopoulos, J., & Reiter, R. (Eds.), *Proceedings of the Twelfth International Joint Conference on Artificial Intelligence*, pp. 569–595. Morgan Kaufmann, San Mateo, CA.

Casti, J. L. (1997). *Would-be Worlds: How Simulation is Changing the Frontiers of Science.* Wiley, New York.

Chalmers, A. F. (1999). *What Is This Thing Called Science?* (3rd edition). Open University Press, Buckingham.

Di Paolo, E. A., Noble, J., & Bullock, S. (2000). Simulation models as opaque thought experiments. In Bedau, M. A., McCaskill, J. S., Packard, N. H., & Rasmussen, S. (Eds.), *Artificial Life VII*, pp. 497–506. MIT Press, Cambridge, MA.

Dupuy, J.-P. (2000). *The Mechanization of the Mind: On the Origins of Cognitive Science.* Princeton University Press, Princeton.

Fujiki, M., & Steward, O. (1997). High frequency transcranial magnetic stimulation mimics the effects of ECS in upregulating astroglial gene expression in the murine CNS. *Molecular Brain Research, 44*, 301–308.

Gould, S. J. (1989). *Wonderful Life: The Burgess Shale and the Nature of History.* W. W. Norton, New York.

Kujala, T., Alho, K., & Naatanen, R. (2000). Cross-modal reorganization of human cortical functions. *Trends in Neuroscience, 23*, 115–120.

Langton, C. G. (1992). Life at the edge of chaos. In Langton, C. G., Taylor, C., Farmer, J. D., & Rasmussen, S. (Eds.), *Artificial Life II*, pp. 41–91. Addison-Wesley, Redwood City, CA.

Miller, G. F. (1995). Artificial life as theoretical biology: How to do real science with computer simulation. Cognitive Science Research Paper 378, School of Cognitive and Computing Sciences, University of Sussex, Brighton, UK.

Newell, A., & Simon, H. A. (1976). Computer science as empirical enquiry: Symbols and search. *Communications of the Association for Computing Machinery, 19*, 113–126.

Padberg, F. (2001). Prefrontal cortex modulation of mood and emotionally induced facial expressions: A transcranial magnetic stimulation study. *Journal of Neuropsychiatry and Clinical Neurosciences, 13*, 206–212.

Peruzzi, D., Bartlett, E. L., Smith, P. H., & Oliver, D. L. (1997). A monosynaptic GABAergic input from the inferior colliculus to the medial geniculate body in rat. *Journal of Neuroscience, 17*, 3766–3777.

Ray, T. S. (1994). An evolutionary approach to synthetic biology: Zen and the art of creating life. *Artificial Life, 1*, 179–209.

Taylor, C. E., & Jefferson, D. R. (1994). Artificial life as a tool for biological inquiry. *Artificial Life, 1*, 1–13.

Using the Universal Similarity Metric to Model Artificial Creativity and Predict Human Listeners Response to Evolutionary Music

Nils Svangård[1], Jon Klein[1,2] and Peter Nordin[1]

[1]Physical Resource Theory,
Chalmers University of Technology and University of Gothenburg
SE-412 96 Gothenburg, Sweden
[2]Hampshire College
Amherst, MA 01002
nils@svangard.org

Abstract

In this paper we present a new technique for modeling Artificial Creativity in Evolutionary Music (EM) systems and predicting how appealing musical pieces are to human listeners. We use a k-Nearest Neighbor classifier where we approximate the Information Distance between the new, unclassified, musical piece and a corpus of observed musical pieces rated by the user with the Universal Similarity Metric. We approximate the Information Distance with two different methods, using standard binary compression of MIDI files, and using MP3 encoding of raw audio streams.

Our experiments indicate that the universal similarity metric can be used to discriminate between music that do and do not appeal to human listeners. Even though classification results is not perfect, it performs significantly better than the random baseline and when we combine the predictions made independently by the MIDI and MP3 classifiers, we obtain an even higher classification accuracy, ranging up to 77% on the test set. These results is in the same range as our results in predicting the aesthetic value of visual art, which indicates that the Universal Similarity Metric is a very general and versatile approach to modeling Artificial Creativity.

Introduction

Creativity is usually associated with the ability to create art, and the study of Artificial Creativity can provide significant insights into both Artificial Intelligence and Life. However, the vast majority of computational systems developed in the past few years for generating artwork require human interaction to be creative. They have neither the ability to perceive the artworks they, or other artists produce, nor are they able to perform aesthetic judgments during their creative process. In other words, they are completely blind and deaf to the outside world.

This is a well-known limitation and several attempts to solve it have been made, in both visual arts and music. In particular Rob Saunders and John Gero (2001) gives a good introduction and motivation to the study of Artificial Creativity, and they use Artificial Neural Networks (ANN) to measure the aesthetic value of computer-generated images. Other similar attempts had been made earlier by

Shumeet Baluja et al. (1994), who trained a set of artificial neural networks with low resolution images as input in order to predict whether an image looks good or not. However, both had limited success with their predictions and their systems did not generalize well. Lately, Penousal Machado et al. (1998) have made other attempts using "expert knowledge" about the artworks to make aesthetic judgments with better results, even though they where unable to guide the process of an evolutionary artist using their measures.

There have also been numerous attempts in the domain of evolutionary music. For example Spector and Alpern (1994, 1995) used a combination of Genetic Programming to evolve Jazz tunes, and an Artificial Neural Network trained on combinations of Charlie Parker's as a critic, with promising, but somewhat "unsatisfactory" results. Recently Machado et al. (2003) has developed a framework for Artificial Art Critics using feature extraction, evaluators and validation processes as a foundation for further research in this area. In their paper they raise an interesting issue, stating that:

> The amount of information contained in some artworks is huge. In visual arts, for instance, even a relatively small picture can fill a lot of memory. Taking into account the current state-of-the-art in adaptive systems (e.g., neural networks, genetic algorithms) it is clear that these techniques cannot currently handle such vast amounts of information.

They also empathize the point that an Artificial Art Critics should be general enough to easily adapt to new domains and should base it's assumptions on only the artwork itself, no other information about the artwork should be required. However, the need for generality clashes with the need to handle the particularities of each domain, which requires specialized techniques for each domain. Moreover, to the best of our knowledge there has not been any single method presented that theoretically can understand and model the aesthetic values of all possible forms of art.

However, there are new general techniques that require no expert knowledge that seem promising; recently there have been great advances in the mathematical theory of similarity where a new universal metric for calculating the similarity of any two objects has emerged. We have successfully applied this technique to the visual arts domain (Svangård et al. 2004), where we successfully used it to predict what images produced by an Evolutionary Art system would be best appreciated by the observer (having the highest "aesthetical fitness"). The Universal Similarity Metric has previously been used to categorize MIDI music (Cilibrasi et al. 2003), where it was successfully used to distinguish between music genres and could even cluster pieces by composer.

Instead of requiring detailed knowledge about the objects, and problem domain, to compute the similarity, this metric can detect all similarities between the objects that any other effective metrics could detect. This metric was developed by Li et al. in (Li et al. 2001, 2002 & 2003) and is based on the "information distance" described in (Li and Vitányi 1997) and (Bennet et al. 1998). Roughly speaking, two objects are said to be similar if we can significantly compress one given the information in the other, the idea being that if two objects are more similar then we can more easily describe one given the other. In this context, compression is based on the ideal mathematical notion of Kolmogorov complexity, which unfortunately is not effectively computable, but can be approximated with good results using existing compression algorithms.

The idea we present in this paper is to use the universal similarity metric to compute how similar pieces produced by an Evolutionary Music (EM) system is to a corpus of musical pieces previously rated by how good it sound to the listener. Then we use this information to predict how the listeners will respond to the new piece, i.e. whether they will like it or not.

We will approximate the Information Distance using both compression of the MIDI files produced by the EM system, similar to what Cilibrasi et al. did, but also using MP3 compression of the raw audio streams produced when the MIDI files is played. Using MP3 encoding to compress the songs is particularly interesting, compared to using a standard binary compressor, since it is designed to exploit known limitations of the human ear, notably the fact that there are certain harmonics in audio signals that humans cannot perceive, as well as harmonics that obscure the presence of other frequencies. Thus, MP3 is intended for compressing music that will be listened to by humans, which is exactly what we try to model.

This paper is organized as follows; the first section will give a brief introduction to our Evolutionary Music system, followed by a section that give a detailed introduction to Kolmogorov complexity and the Universal Similarity Metric. Finally, we discuss our implementation and experiments as well as an analysis of our results.

Evolutionary Music

A major obstacle to constructing interactive evolutionary music is user interaction. Especially as pieces become longer and longer, and as it takes more and more generations to produce appealing results, it becomes unreasonable to expect that users will want to patiently listen and rate music through an entire evolutionary process. This is another area where the study of Artificial Creativity could be used to speed up the interactive process and only present the user with new, interesting pieces removing those who will sound bad based on the users previous actions.

Another way that our system, "iEvolve", tries to address this problem is by allowing the user to rate music at their own pace, using software and hardware they may already have, exactly as the software and hardware are intended to be used. We thus take advantage of an existing, user-friendly process of rating music, and use it to drive the creation of evolutionary music. We do so by tapping into Apple's popular iPod music player and the freely available iTunes jukebox software. The iPod and its software companion iTunes allow users to rate songs on a scale from one to five stars. If the ratings are made using an iPod, the rating information is automatically synchronized when the iPod is plugged in. This rating information is stored in the iTunes database and can be accessed by the user to, among other things, sort the music and create playlists.

iEvolve works by automatically retrieving rating information from the iTunes database, using this information to breed and mutate music and then to upload the new music into the user's iTunes library and iPod. From the user's perspective, the entire process is completely transparent. They simply use their iPod or iTunes software to listen to and rank the music, just as they might do with any other non-evolving music.

The software relies on a hodgepodge of different technologies provided by the Mac OS X operating system in order to make the evolutionary process completely transparent to the listener. Because there are several steps required, each requiring access to different technologies, iEvolve is implemented as a series of simple programs and scripts that are connected together by a Perl script. These individual programs are described below, in the order in which they are called from the Perl script:

> GetRatings.scpt: an AppleScript that retrieves ratings information from the iTunes library.
> Trainer: a script that stores the observations about the musical pieces and their rating in a database for later use by our Artificial Art Critic.
> Breeder: a C program which loads in a set of raw-data genomes and the current song ratings. The song ratings are then used to drive the genetic operators and to generate the next generation of genomes.
> Translate: a C program which translates the raw-data genomes into their corresponding phenotypes, in the form of MIDI files. We have

experimented with several different genome-to-MIDI translation schemes.

Miditoaiff: a C program which uses QuickTime to convert MIDI files into the AIFF files that the iPod can play.

Classifier: a script that takes a set of genomes, MIDI and AIFF files and makes a prediction how well the user will like them based on the previous observed rating of musical pieces.

UpdateMusic.scpt: an AppleScript that removes the previous generation of songs from the iTunes library and replaces them with a new generation. If an iPod is connected to the machine, it also synchronizes the new music onto the iPod.

We use a simple Genetic Algorithm (GA) that converts the genomes into MIDI files. It is a simple byte-code language interpreted by a register based virtual machine that supports the basic algebraic operations, as well as some mathematical functions (sin, cos, exp, log), conditional statements, and loops, but the two most important features is that it can create tracks in the MIDI file, and play notes. The purpose of this paper is not to go into detail how this system works, but a good introduction to evolving music using GA is (Papadopoulos 1998).

Aesthetic Selection

The question about how aesthetic selection work is in hot debate, and has been for hundreds of years, but recently there have appeared a number of interesting theories, and computer models using machine learning, explaining how it might work. One of the most popular way to approach this problem seem to be using prior knowledge of images and genomes that produce visually interesting images. Since most genetic artists usually work with a large quantity of genomes that looks good to seed new runs, there exist plenty of such information. Unfortunately there's no given method to assign aesthetic fitness values to genomes based on this information, and most techniques for classification, such as artificial neural networks or statistical analysis, usually requires expert knowledge about the features of the problem and does not provide any universal method for solving this problem.

However, since the introduction of the universal similarity metric there is a very powerful tool that can be applied to this tasks without any prior knowledge about aesthetics and which conforms to the principle of Occam's Razor; "Less is more."

Kolmogorov Complexity

Given an object, x, that can be encoded as a string over a finite alphabet, e.g. the binary alphabet, the Kolmogorov complexity, $K(x)$, is the length of the shortest compressed binary version from which x can be fully reproduced. Here

"shortest" means the minimum taken over every possible decompression program, both real and imaginary. In fact, there does not even have to be a program that can compress the original object to the compressed form, but if there is one so much the better.

Technically the definition of Kolmogorov complexity is as follows: First, we fix a syntax for expressing all computable functions. The usual form is as an enumeration of all Turing machines, but an enumeration of all syntactically correct programs in some universal programming language like Java, Lisp or C is also possible. Then we define the Kolmogorov complexity of a finite binary string as the length of the shortest Turing machine that can generate it in our chosen syntax. Which syntax we use is not important, but we have to use the same syntax in all calculations. With this definition, the Kolmogorov complexity of any finite string will be a definite positive integer.

Even though Kolmogorov complexity is defined in terms of a particular machine model, it is actually machine-independent up to an additive constant. This means it is asymptotically universal and absolute through Church's thesis, and from the ability that universal machines can simulate one another and execute any effective process. The Kolmogorov complexity of an object can then be seen as an absolute and objective quantification of the amount of information it contains.

So $K(x)$ gives the length of the ultimately compressed version x^* of x which can be thought of as the amount of information, in number of bits, contained in the string. Similarly, $K(x|y)$ is the minimal number of bits required to reconstruct x given y. In a way $K(x)$ expressed the minimum amount of information required to communicate x when the sender and receiver has no knowledge where x comes from. For more information on individual information content, see (Li and Vitányi 1997).

The Similarity Metric

As mentioned our approach is based on a new and very general similarity distance that can categorize objects depending on how much information they share. In mathematics, there are many different types of distances, but they are usually required to be 'metric' in order to avoid undesired effects. A metric is a distance function, $D(\cdot,\cdot)$, that assigns a non-negative distance, $D(a,b)$, to any two objects a and b under the following conditions:

1. $D(a,b) = 0$ only when $a = b$
2. $D(a,b) = D(b,a)$ (symmetry)
3. $D(a,b) \leq D(a,c)+D(c,b)$ (triangle inequality)

In (Li et al. 2003) a new theoretical approach to a wide class of similarity metrics was proposed: the "normalized information distance." This distance is a metric and universal in the sense that this single metric uncovers all similarities simultaneously that the any metric in the class uncovers separately. This can be understood in the sense that if two objects are similar (that is, close) according to a particular feature described by a particular metric, they are also similar in the sense of the normalized information distance metric, which justifies calling the latter *the* similarity metric. Oblivious to the problem area concerned, simply using the distance according to the similarity metric can be used to automatically classify objects of any kind. Mathematically the similarity metric is defined as the distance of any pair of objects, x and y, where:

$$d(x,y) = \frac{\max\{K(x|y), K(y|x)\}}{\max\{K(x), K(y)\}}$$

It is clear that $d(x,y)$ is symmetric and in (Li et al. 2003) it is shown to be metric. It is also universal in the sense that every metric expressing some similarity that can be computed from the objects concerned is comprised in $d(x,y)$.

To compute the conditional measure, $K(x|y)$, we use a sophisticated theorem from (Li and Vitányi 1997) known as the "symmetry of algorithmic information":

$$K(x|y) \approx K(xy) - K(x)$$

So in order to compute the conditional complexity, $K(x|y)$, we can just take the difference of the unconditional complexities, $K(xy)$ and $K(x)$, which allows us to easily approximate $K(x|y)$ for any pair of objects.

Distance Based Classification

Given a set of objects with unknown aesthetic fitness, and a corpus of objects that we known are attractive to the user, we have to figure out which of the new genomes the user will find most interesting. In order to do this we have to figure out which objects shares the most information with the library of prior knowledge using the similarity metric. One method that is suited for this kind of classification tasks, which also happens to be one of the simplest ones, is k-Nearest Neighbors (kNN) (Mitchell 1997).

The kNN algorithm works by first calculating the distance, $D(a,b)$, between the object we want to classify, x, and all observations in the corpus. It then takes the 'k' observations nearest to x and sums the weight for each class and observation, $W(a,b)$. The class with the highest weight is thus the class most likely to be the correct guess for x.

Method

Since aesthetic selection is probably as far from an objective system you can get, there is no given way to measure how well our method really work. However, we wanted to get some statistical indication on the predictive power of our system so we decided to test the classification

accuracy of our system by letting the user label a new set of musical pieces and then having our system trying to predict which were the most interesting ones.

First, we let our test subject use our application to build a database of pieces (s)he find interesting. Then we test the system by repeatedly presenting the user with two new pieces and the user has to pick the one he thinks sounds best. Afterwards we let our system process all such 2-tuples of tunes and predict which sounded the best, and with the users choice as a label we can evaluate the classification accuracy of our system as a binary classification task.

Since the tunes are categorized in six numeric classes, corresponding to the rating of how good they sound, we use a slightly modified version of the kNN classifier to predict which sounds best. Instead of picking the class having the most weight, as you usually do, we calculate the average class by multiplying the weight of the class with the numeric value and dividing by the total weight for all observations. This gives us a less discreet prediction that lets us consider the relation to all classes.

To calculate the information distance between the MIDI files we use LZ77 (a.k.a. Lempel-Ziv, as used in gzip), and to compute the conditional complexities we simply concatenate the two MIDI files and compress it. For the AIFF files we use a combination of MP3 encoding, with compression set to max, using Variable Bit Rate (VBR) and psycho-acoustic models, followed by LZ77 to get the maximum compression of the files. To compute the conditional complexities of the audio files we create a new AIFF file with correct headers and concatenate all channels and frames from the two files, and then use the MP3 encoding scheme as before.

One drawback with MP3 compression is that it is lossy, compared to LZ77, which is not, and the approximation of Kolmogorov complexity requires the compression algorithm to be able to completely reproduce the object. However, this only manifests itself as a linear error term in the model, which we think is a reasonable cost for the drastically improved compression ratios. After all, the MP3 algorithm is designed to remove only the frequencies the human ear cannot perceive, so in a way it is lossless to the listener even though the binary representation is not identical.

Results

In order to test our system we carried out two independent experiments, one where we use genomes with 256 genes, and one with 1024 genes. In both experiments we use 'zlib' for the LZ77 compression, and 'LAME' for the MP3 encoding[1]. Table 1 shows the number of observations in

---

[1] LAME is an LGPL MP3 encoder: http://lame.sourceforge.net/

each class for the two experiments, and the number of genomes in the test set.

Rating (Class)	256 genes	1024 genes
0%	49	33
20% ★	8	18
40% ★★	8	15
60% ★★★	10	8
80% ★★★★	4	11
100% ★★★★★	1	3
Total:	80	88
Test set:	22	11

Table 1 The number of observed musical pieces for the different ratings in the training corpus for both experiments and the total count of observations in the training and test sets

In both experiments we use our improved kNN classifier (with k set to 21) to guess which piece sounded the best to the user with our two distance measures, d_{MIDI} for compression of the MIDI phenomes, and d_{MP3} for compression of the audio genomes. The weight function was simply one minus the distance between the two objects, since the distance is in the range [0,1] and with our limited training set we wanted the distance between objects to count more than the number of observations. Table 2 shows the classification accuracy we obtained using these parameters.

Classifier	256 genes	1024 genes
d_{MIDI}	59%	64%
d_{MP3}	68%	64%

Table 2 The accuracy of the kNN-classifier using different distance measure when predicting which musical piece sounded best to the user

We noticed that the MIDI and MP3 classifiers did not give the same answer for all the examples in the test set, so we believed we could improve the classification performance by using the information from both of them simultaneously. In Table 3 you can see the *overlap* of the two classifiers, which simply is the number of times both classifiers gave the same answer divided by the number of all answers. This led us to create a new distance measure, d_{both}, which is simply the length of the vector (d_{MIDI}, d_{MP3}):

$$d_{both}(x,y) = \sqrt{d_{MIDI}(x,y)^2 + d_{MP3}(x,y)^2}$$

The classification results using this measure can also be found in Table 3, and it is apparent that this measure significantly improved our classification performance.

The performance of the musical classification is in the same range as when we applied it to estimating image aesthetics, where we had 75% accuracy when using JPEG compression of the images and 65% when we compressed genomes using LZ77 (Svangård et al. 2004). That system

was designed a little differently though, since we computed the conditional probability between the test object and a concatenation of all the training objects, but in one experiment similar to this kNN approach we calculated the mean distance to all training genomes we got an accuracy of 59%, so it seems like our methods perform equally well on both visual art and music.

Classifier	256 genes	1024 genes
Overlap:	55%	45%
d_{both}	77%	73%

Table 3 The classification results obtained when combining the MIDI and the MP3 distance into a single distance measure. Overlap represents the percentage of independent answers that where identical between the MIDI and AIFF classifiers, while d_{both} represents the accuracy our classifier had when we considered both the MIDI and AIFF classification simultaneously

Future Work

The classifier based on the MIDI compression is very fast. Every distance measure usually takes less than three milliseconds on a modern computer and with our training corpus it takes less than a second to classify an unknown musical piece. The MP3 compression however, is as expected much slower; a single distance measure can take anything from 1 second up to several minutes depending on how long the piece lasts, and classifying an unknown genome usually takes over an hour. Therefore, the MP3 distance measure have to be optimized to be feasible to use in this context, in particular when we are going to work with larger example sets. Either by finding a better compression algorithm, smarter kNN algorithms that prune some of the observations, or in worst case parallelizing the algorithm and using clusters for the computations.

One of the first extensions to this system that springs to mind is completely automatic evolution without any user interaction (short of listening to the final piece). Therefore, instead of letting the user choose which piece to pick as parents as the next generation, we would take those that our aesthetic system believes the user would have picked. However, previous attempts using Artificial Art Critics that performed well in theory did seem to have some trouble when used to control the evolution autonomously, so this probably requires some further improvements to the model to be made.

Our experiments had a quite small training and test set, so in order to get better statistical significance how well this method works we need to make larger experiments, using many test subjects. In addition to this, one interesting extension would be to use a large library of previously rated real music instead of our evolutionary produced tunes. Since our method is general, there is no real limitation to what kind of music you could use.

Conclusion

Our experiments indicate that the universal similarity metric can be used to successfully discriminate between music that do and do not appeal to human listeners. Even though classification results is not perfect, it performs significantly better than the random baseline and in the same range as our results in predicting the aesthetic value of visual art.

The best distance measure we used seems to be the MP3 information distance, having an average accuracy of over 64% on both our test sets, but when combined with the MIDI distance the accuracy is improved by almost 10%.

We have shown that the Universal Similarity Metric can be used for Artificial Art Critics in both visual arts and for music. Even though it can be somewhat computationally expensive, the general nature of this method, and that it doesn't require any particular expert knowledge about the target domain, makes it a very promising technique to model Artificial Creativity and will be investigated further.

References

Baluja, S., Pomerlau, D. and Todd, J., 1994: Towards Automated Artificial Evolution for Computer-Generated Images. Connection Science, 6 (1994), 325-354

Bennet, C., Gács, P., Li, M., Vitányi, P., Zurek, W., 1998: Information Distance, IEEE Transactions on Information Theory, 44:4(1998), 1407-1423

Cilibrasi, R., de Wolf, R., Vitanyi, P., 2003: Algorithmic clustering of music. http://arxiv.org/archive/cs/0303025

Li, M., Vitányi, P., 1997: An Introduction To Kolmogorov Complexity and its Applications, Springer-Verlag, New York, 2nd Edition (1997)

Li, M., Badger, J. H., Chen, X., Kwong, S., Kearney, P., Zhang, H., 2001: An information-based sequence distance and its application to whole mitochondrial genome phylogeny, Bioinformatics, 17:2(2001), 149-154

Li, M., Vitányi, P., 2002: Algorithmic Complexity, International Encyclopedia of the Social & Behavioral Sciences, Smelser, N, Baltes, P., Eds., Pergamon, Oxford (2001/2002), 376-382

Li., M., Chen, X., Li, X., Ma, B., Vitányi, P., 2003: The Similarity Metric, in proceedings at the 14th ACM-SIAM Symposium on Discrete Algorithms (2003)

Machado, P., Cardoso, A., 1998: Computing Aesthetics. In: Oliveira, F. (Ed.), Procs. XIVth Brazilian Symposium on Artificial Intelligence SBIA'98, Porto Alegre, Brazil, Springer-Verlag, LNAI Series (1998), 219-229, ISBN 3-540-65190-X

Machado, P.,. Romero, J,. Manaris, B,. Santos, A., and Cardoso, A., 2003: "Power to the Critics – A Framework for the Development of Artificial Critics," in Proceedings of 3rd Workshop on Creative Systems, 18th International Joint Conference on Artificial Intelligence (IJCAI 2003), Acapulco, Mexico (2003), 55-64

Mitchell, T.M., 1997: Machine Learning, Boston MA: McGraw-Hill, 1997, 230-247

Papadopoulos, G., Phon-Amnuaisuk, S., Wiggins, G., and Tuson, A.L., 1998: Evolutionary Methods for Music Composition. Submitted to the International Conference on Computer Music (ICMC 98)

Saunders, R. and Gero J. S., 2001: Artificial Creativity: A Synthetic Approach to the Study of Creative Behaviour, in J. S. Gero (ed.), Proceedings of the Fifth Conference on Computational and Cognitive Models of Creative Design, Key Centre of Design Com-putting and Cognition. (2001)

Spector, L., and A. Alpern, 1994: Criticism, Culture, and the Automatic Generation of Artworks. In Proceedings of the Twelfth National Conference on Artificial Intelligence, AAAI-94, pp. 3-8. Menlo Park, CA and Cambridge, MA: AAAI Press/The MIT Press

Spector, L., and A. Alpern, 1995: Induction and Recapitulation of Deep Musical Structure. In Working Notes of the IJCAI-95 Workshop on Artificial Intelligence and Music . pp. 41-48

Svangård, N. and Nordin, P., 2004: Automated Aesthetic Selection of Evolutionary Art by Distance Based Classification of Genomes and Phenomes using the Universal Similarity Metric, To Appear in Proceedings of the 2nd European Workshop on Evolutionary Music and Art (EvoMUSART 2004)

Playing Music by Conducting BOID Agents
– a Style of Interaction in the Life with A-Life

Tatsuo Unemi[1] and Daniel Bisig[2]
[1]Soka University, 1-236 Tangi-machi, Hachiōji, Tokyo, 192-8577 Japan
[2]University of Zurich, Andreasstrasse 15, CH-8050 Zürich, Switzerland
unemi@iss.soka.ac.jp

Abstract

This paper presents an interactive installation that employs flocking algorithms to produce music and visuals. The user's motions are captured by a video camera and influence the flocks behaviour. Each agent moving in a virtual 3D space controls a MIDI instrument whose playing style depends on the agent's state. In this system, the user acts as a conductor influencing the flock's musical activity. In addition to gestural interaction, the acoustic properties of the system can be modified on the fly by using an intuitive GUI. The acoustical and visual output of the system results from the combination of the flock's and user's behaviour. It therefore creates on the behavioural level a mixing of natural and artificial reality. The system has been designed to run an a variety of different computational configuration ranging from small laptops to exhibition scale installations.

Introduction

Life is beautiful, stimulating, and dramatic.

It is very natural that ALife research keeps inspiring many artists. ALife reveals both on a conceptual and practical level how complex life-like forms can be produced artificially.

In the closing speech in A-Life VI in the year 1998, Chris Langton mentioned that mixing real and artificial life might become an important trend in the future. Of course, such a development raises both technical and ethical issues. At the same time the existence of various robotic and virtual reality projects reveal, that this trend has already started. From an artist's point of view the interaction between artificial agents and humans holds tremendous potential for interactive art. Depending on the complexity and adaptivity of the agent's behavioural response to interaction, interesting relations and impressive experiences might be produced for the viewer. A-Volve (Sommerer and Mignonneau, 1996) utilising evolutionary computation and MIC & MUSE (Tosa and Nakatsu, 1996) tuned by artificial neural networks are typical art works toward this direction.

Boids (Reynolds, 1987) impressively demonstrated how a particular computer simulation can produce complex phenomena from simple mechanisms. This and similar software allows the simulation of various forms of collective behaviour such as a flock of birds, a herd of herbivores, a school of fish etc. Such software has become an important and well established technique in computer animation.

Apart from using flocking algorithms to produce visual effects, some researchers and artists have applied these algorithms to create music. For example the Breve environment (Klein, 2002) served as basis to produce a musical flocking system of evolving agents (Spector and Klein, 2002). In another project, a flock of agents moves through a three dimensional space which is segmented into different acoustical regions (Tang and Shiffman, 2003). Whenever an agent enters a particular region, a predefined musical pattern is rendered audible. In both examples the flock behaviour and the resulting acoustical output is completely autonomous. The system presented in this paper differs fundamentally from this approach in that allows the user's behaviour to influences the flock. A project by Rowe and Singer (Rowe and Singer, 1997) employs the Boids concept for an interactive musical performance. In this system, the acoustical output of a several musicians modifies the behaviour of a flock controlling the visual appearance of words on a projected screen. Contrary to our system, the flock in this project serves as a visualisation of a purely human acoustic performance but doesn't produce any sound itself. In another interesting project, a group of natural and artificial musicians collaborate in a music performance. The artificial musicians are implemented as multi-swarms, moving towards musical targets produced by the participating musicians (Blackwell and Bentley, 2002). This project differs from the one presented in this paper in that the flock acts as a musician rather than a musical instrument. The flock in our system can be regarded as a set of virtual instruments which are conducted by one or several users. It therefore possesses certain similarities to a variety of projects in which performers control the behaviour of a virtual instrument by means of their gestures (Roads, 1996). Our system differs from these approaches in that the flock as a musical instrument possesses a certain amount of autonomy and can only be influenced indirectly by the user's actions. The system therefore mixes the artificial behaviour of a simulated flock with the behaviour of

Figure 1: An example of screen-shot.

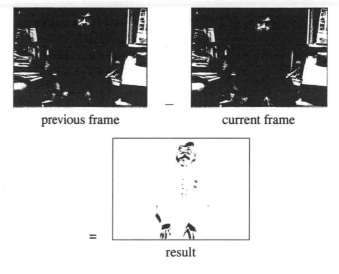

previous frame   current frame

result

Figure 2: Example of a difference image obtained from two consecutive images.

users. The relationship between the flock and a user mimics the relationship between an orchestra and a conductor. On the other hand, the flock acts completely autonomously when left on its own and the resulting musing serves as a means of catching attention and motivating users to engage into an interaction with the system.

Our system has been designed in a way that it scales very well with available computational power. Users can interact with the system running either their personal laptop or desktop computer or as an exhibition scale installation that employs high performance computer equipment and video beamers. The only particular requirement consists of a video capture device such as a web cam that captures the users gestures.

Throughout the remainder of this paper we describe the implementation of the interactive flocking system, its means of producing music, and different system set-ups for presentation. We end the paper with a short section of concluding remarks.

Interaction with Flocking Agents

Presented in Figure 1 is a screen-shot of the simulation in which depicts 64 agents flocking within a rectangular bounding box. Each agent is controlled by a set of forces. Some of these forces implement standard boids type of flocking rules whereas other forces result from the user's interaction. The flocking forces cause the following behaviour:

1. collision avoidance between agents,

2. velocity alignment with neighbouring agents,

3. flock cohesion by attraction to the neighbouring agents centre, and

4. collision avoidance with the faces of the bounding box.

The repelling forces for collision avoidance are proportional to the inverse of the square of the distance between the agent and the obstacle. The sum of all the forces affecting an agent results in its goal angle and goal speed. The agent tries to meet these goal values by modifying its current orientation and velocity within the allowed limitations of steering angle and acceleration.

Interaction forces realise the flock's behavioural respond to user input. They cause the following to behaviours;

1. movement towards a particular target position on the front plane when user motion is detected, and

2. movement away from the front plane in absence of user motion.

The target position is calculated in the following way (see equation 1 and figure 3 and 4):

1. a difference image is calculated by summing over the absolute differences of all pixel RGB values between the current and previous captured images, and

2. for each agent an individual attractor position is calculated.

This position is derived by multiplying RGB difference values of all pixels that lie within the neighbourhood circle of an agent with their corresponding x and y position in the camera image.

$$u_t = \frac{1}{|C|} \sum_{p \in C} (|r_p^t - r_p^{t-1}| + |g_p^t - g_p^{t-1}| + |b_p^t - b_p^{t-1}|) \, x_p, \quad (1)$$

The position of attraction for a particular agent at time t is denoted by $(u_t, v_t, 0)$. The pixel co-ordinates in the two consecutive frames are denoted by x and y. C is the set of all

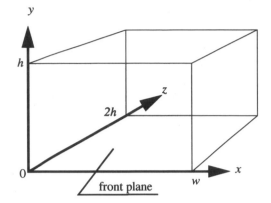

Figure 3: Co-ordinates system in flocking space.

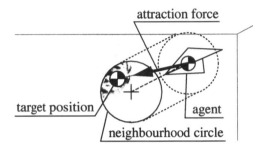

Figure 4: Illustration of the attraction force.

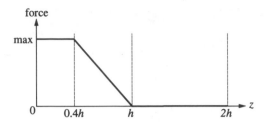

Figure 5: Distribution of the strength of the agent repulsion force. The position $z = 0$ corresponds to the front plane.

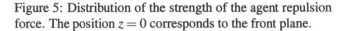

Figure 6: Graphical user interface for setting up various acoustical parameters.

pixels within a fixed distance from the x, y co-ordinates of the agent, $r_p^t, r_p^{t-1}, g_p^t, g_p^{t-1}, b_p^t,$ and b_p^{t-1} are the RGB colour values of pixel p of frame t and $t - 1$, respectively. x_p is the image co-ordinate of pixel p. v_t is defined in the similar manner.

The second interaction force causes the agents to flock freely in a certain distance from the front plane if no user motion is detected. The strength of this force is at its maximum right at the front plane up to a distance of one fifth of the bounding box depth. At larger distances the force linearly decreases until it reaches a value of zero at half the bounding box depth. In the remainder region within the bounding this force stays zero (see figure 5).

This force causes the agents to flock in the back of the bounding box in the absence of any interaction and to quickly move towards the front when user motion is detected.

To give users an idea of what the agents are seeing and how they respond to his interactive cues the captured image is blended onto the visualisation of the flocking agents.

How Agents Play Instruments

Agents play instruments by sending MIDI (MIDI, 1995) messages representing note pitch and velocity. In order to increase the amount of feedback, each agents responds to user motion not only by moving towards the front but also

by changing its MIDI instrument. For this reason, each agent controls a primary and secondary MIDI instrument. It plays its primary instrument whenever it is attracted to the user's interaction. Otherwise it plays its secondary instrument. The secondary instrument is played very quietly whereas the primary instrument produces much louder acoustic output.

Each agent possesses a certain probability of playing a note. The pitch of this note is derived from the agent's y co-ordinate with larger y values corresponding to higher pitches. The balance between left and right speaker output corresponds to the x co-ordinate of the agent. By this way, the stereo position of its acoustic output matches its visual position in flocking space. The loudness of the sound is controlled by the agent's z co-ordinate. The closer the agent is to the front plane of the bounding box the louder the sound it produces.

In a situation as depicted in figure 1 the number of agents

exceeds the number of available MIDI channels. For this reason, it is usually impossible for all agents to play their sounds simultaneously. The lowest common denominator are twelve MIDI channels which allows to use this system in conjunction with very simple general MIDI devices. In this system. the primary and secondary instruments occupy six MIDI channels each. In order to handle the MIDI channel issue efficiently, each agent possesses virtual MIDI channels. Each of these virtual MIDI channels corresponds to two real MIDI channels, ones for the left and right stereo position. One pair of real MIDI channels is shared by a maximum of four virtual MIDI channels in case these channels play notes at different pitches. Therefore, a maximum of twelve virtual MIDI channels is available to each primary and secondary instrument. In case an agent would like to play a note but all virtual channels are occupied by other agents the longest playing note is shut off and freed again. An agent gains access to a virtual MIDI channel according to one of several selection mechanisms:

1. each agent is selected in turn according to a fixed order,

2. each agent possesses the same probability of being selected, and

3. the selection probability of an agent is proportional to its movement speed.

The first selection mechanism produces repetitive phrases which are easy to recognise if relatively few agents are producing music. The second mechanisms produces similar results but both the melody and rhythm are arranged more randomly. The third mechanism is particularly good in produced feedback to the user's motion. Slow and quiet sounds are interrupted by fast and loud acoustic output as soon as user motion is detected.

The system can easily be configured to play only a restricted range of pitches and scales. These settings can be accessed via an intuitive GUI (see figure 6). Another GUI allows the user to specify particular sequence of chord progressions (see figure 7). Employing this GUI, the user arranged chord segments represented as labelled rectangles. The duration of each chord segment can be modified by dragging the right edge of its corresponding rectangle. These arrangements can be modified by a variety of standard editing commands such as copy, paste, and cut. New chord segments can be added either by importing an external chord file or by dragging chords from another table containing a number of pre-defined scales. These scales don't represent chords themselves but make it easy to create chords as long as one is aware of the fact that octave intervals produce very harmonic chords whereas other intervals such as the 9th, 11th or 13th produce highly dissonant chords. Clicking the play button automatically adapts the scale and set of pitches required to play the chord sequence. These new val-

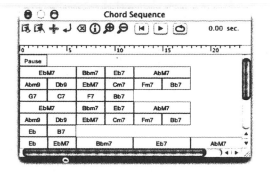

Figure 7: Graphical user interface for setting up the sequence of chord progression.

ues serve as bounding conditions for the system to improvise its flocking music.

System Configuration

The system is designed for Macintosh computers (MacOS X 10.2 or higher) and equipped with a digital movie camera. The software was written in Objective-C. In order to take advantage of the AltiVec parallel processing capabilities the computer should possess either a G4 or G5 processor. Within these technical limitations the program works well on systems ranging from laptops to very fast desktop computers. For an exhibition context, the system can be expanded with a projector producing a large scale visualisation of the program. In this context two types of users would interact with the system. Visitors can influence the flocking and music production via their movements. On the other hand, operators can control the global acoustic parameters of the system through the graphical user interfaces. These changes can be done while the simulation is running as long as the GUI windows are displayed on a secondary screen. The authors are convinced that this system would be particularly interesting to explore in a dance-performance set-up.

The response time of the system depends on the number of agents and the resolution of the camera and screen image. At VGA resolution, the system should run on a computer system possessing either dual G4 or single G5 processor in order the get an acceptable frame rate (above 10 frames per second). On slower computers, the resolution of the camera image should be reduced to 320×240 pixels whereas the visualisation can still be run at VGA resolution. Such a configuration works well on a single G4 CPU system with a clock rate of 667 MHz. Since our system scales well over a relatively wide range of computing power, it is both suitable for display on exhibitions as well as in personal environments such as at home or at the office. This flexibility should help to bring the experience of an interactive ALife system to a broad audience.

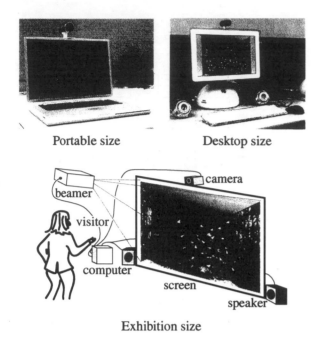

Portable size   Desktop size

Exhibition size

Figure 8: Typical system configurations in various sizes.

Concluding Remarks

In this paper, a musical system for real-time interaction with flocking agents was presented. The user can conduct the musical activity of the flock by means of his/her motions. The interaction takes the form of a collaboration between flock and user rather than an unidirectional exertion of complete control.

This system has been exposed only experimentally to some persons, several students and laboratory staffs, so far. Almost all of them were interested in it, and we observed that some members of a student jazz band were more strongly enjoying their own play using predefined chord progression of standard tunes, relatively than members of chorus group and symphony orchestra. It suggests that the interaction has a type of similar characteristics with improvisational play of jazz music.

In participating in the trend towards hybrid ALife systems, it is important to consider various aspects of interaction between real and artificial life. In our system, the interaction takes place on the behavioural level. This level is highly intuitive for most users. This approach will become even more interesting if the behavioural repertoire of the agents is made much more complex that in this simulation. Other forms of interaction which could be explored in combination with our approach could range from a genetic or chemical level up to the evolutionary and ecological level. It is interesting to think about how the fitness of humans and artificial system mutually influence each other in an interactive simulation. In hybrid systems of collaborate activity issues of competition and symbiosis between natural and artificial system become relevant as well.

According to the principle of ecological balance, it is important to match the behavioural complexity of agents with a complex artificial environment and a complex technical set-up for an interactive installation. Computer Grids and multi-camera systems will become very useful when trying to develop more complex behavioural simulations.

The authors hope that the system described in this paper forms part in creating an ever increasing diversity of interactive experiences with ALife systems.

Acknowledgement

The authors would like to thank Prof. Rolf Pfeifer at University of Zurich and Mr. Akio Yasumoto at Mama Top Co. for their continuing support during this collaboration.

References

Blackwell, T. M. and Bentley, P. J. (2002). Improvised music with swarms. In *Proceedings of the Congress on Evolutionary Computation*, pages 1462–1467.

Klein, J. (2002). Breve: a 3D simulation environment for the simulation of decentralized systems and artificial life. In *Proceedings of A-Life VIII*, Sydney, NSW, Australia.

MIDI (1995). *The Complete MIDI 1.0 Detailed Specification*. Midi Manufactures Association, La Habra, CA.

Reynolds, C. W. (1987). Flocks, herds, and schools: A distributed behavioral model. *Computer Graphics*, 21(4):25–34. (SIGGRAPH '87 Conference Proceedings).

Roads, C. (1996). *The Computer Music Tutorial*. MIT Press.

Rowe, R. and Singer, E. L. (1997). Two highly integrated real-time music and graphics performance systems. In *Proceedings of the International Computer Music Conference*, pages 133–140.

Sommerer, C. and Mignonneau, L. (1996). "A-Volve" an evolutionary artificial life environment. In *Proceedings of A-Life V*, pages 167–175, Nara, Japan.

Spector, L. and Klein, J. (2002). Complex adaptive music systems in the breve simulation environment. In Bilotta, E. e. a., editor, *Workshop Proceedings of A-Life VIII*, pages 17–23, Sydney, NSW, Australia.

Tang, E. and Shiffman, D. (2003). Musical flocking box. http://www.antiexperience.com/edtang/works/flockingbox.html.

Tosa, N. and Nakatsu, R. (1996). The esthetics of artificial life – human-like communication character, "MIC" & feeling improvisation character, "MUSE". In *Proceedings of A-Life V*, pages 143–151, Nara, Japan.

An Evolutionary Approach to Complex System Regulation Using Grammatical Evolution

Saoirse Amarteifio and Michael O'Neill
Biocomputing and Developmental Systems
University Limerick
Ireland
amartey@eircom.net, michael.oneill@ul.ie

Abstract

Motivated by difficulties in engineering adaptive distributed systems, we consider a method to evolve cooperation in swarms to model dynamical systems. We present an information processing swarm model that we find to be useful in studying control methods for adaptive distributed systems. We attempt to evolve systems that form consistent patterns through the interaction of constituent agents or particles. This model considers artificial ants as walking sensors in an information-rich environment. Grammatical Evolution is combined with this swarming model as we evolve an ant's response to information. The fitness of the swarm depends on information processing by individual ants, which should lead to appropriate macroscopic spatial and/or temporal patterns. We discuss three primary issues, which are tractability, representation and fitness evaluation of dynamical systems and show how Grammatical Evolution supports a promising approach to addressing these concerns.

Introduction

Nature clearly sets the standard on complex system regulation. However her principles can be difficult to apply forcing us to ask difficult questions about how to allow open-ended evolvability, how to reach high levels of adaptability or how to regulate developmental processes. We take an *evolutionary automatic programming* approach in addressing some early issues in using natural principles to program artificial complex systems. Specifically we begin by considering how synergy in swarms can be evolved to produce consistent patterns from the interaction of simple agents. Using Grammatical Evolution (O'Neill and Ryan, 2003) we evolve templates for simple transducers that describe how environmental information should be modelled by ants to produce responses that favour a swarm's fitness.

Background to evolving adaptive behaviour is outlined in section 2 and we discuss issues in engineering distributed adaptive systems. We provide background on grammatical evolution in section 3. In Section 4 we describe information processing swarms as an interesting model for regulating complex dynamical systems and describe a new model for evolving swarms. We outline three issues which we consider crucial; tractability, representation, and fitness evalu-

ation. Section 5 describes the swarm model. Results comparing two experiments can be found in section 6. We close with a discussion of results.

Background

Swarm Intelligence, which is comprised of Particle Swarm (Kennedy and Eberhart, 1995; Kennedy and Eberhart, 2001) and Ant Algorithms (Colorini et al., 1991; Dorigo et al., 1996; Bonabeau et al., 1999), has the potential to be used as a model for regulating distributed adaptive systems. It has been said in the context of dynamical systems and morphogenesis that finding a solution to a particular [distributed] problem is equivalent to finding a specific pattern in space or time (Bonabeau, 1997). It is of interest to understand how patterns unfold through random processes and interactions among elementary constituents or agents (Bonabeau, 1997). Swarms of interacting, information processing agents can yield the emergence of a computational power not present at the level of individual organisms. Examples of artificial systems that possess this computational power are dynamical systems such as Swarms and Cellular Automata (CA) (Wolfram, 1983; Crutchfield et al., 2003). Dynamical systems are often considered cognitive or purposeful and considered in light of achievements such as crop harvesting in ant colonies or pathogen detection in the immune system. The current study is interested in the system regulation *per se*, considering these achievements or patterns simply as by-products of system dynamics.

Dynamical systems are characterised by a continual coupling between systems and their environment. They are spatially extended, consisting of a large number of simple components each with limited communication to other components and each following simple transition rules that depend on their inputs. Properties of such systems although attractive are difficult to formalize for engineering purposes (Rosen, 1985; Kubrik, 2003). These principles are likely to be of great importance to adaptive distributed systems design over the coming years in fields such as amorphous computing (Abelson et al., 2000; Nagpal, 2001). Researchers in this field prefer to understand the construction of these systems so that they function as intended not as they evolve. In support of the direct engineering approach, (Nagpal, 2002; Nagpal, 2001) observes that in CA research, local rules are constructed empirically without providing a framework for construction of local rules to obtain any desired goal and that these approaches are difficult to generalize. On the other hand Nagpal observes that evolutionary computing, while generalising well, uses local rules that are evolved without any understanding of how they work. This problem is compounded when evolving dynamical systems as the emergent computation performed is determined by space-time behaviour (Crutchfield et al., 2003). Designing an appropriate fitness function can be as difficult as designing a control algorithm from scratch (Nagpal, 2001).

Yet we ask how evolution can be used to produce a better learning system, which could be called adaptive, dynamical or emergent. Similar work has been done in the fields of Artificial Neural Networks, which has been termed Evolutionary Artificial Neural Networks (EANN) (e.g. see (Yao, 1999)), evolving Cellular Automata (e.g. see (Crutchfield et al., 2003)). Very recently similar work has been carried out in the field of Swarm Intelligence (Williams, 2002) and the evolution of mulitcelluar programs using genetic programming (Schmutter, 2002).

Our work bears only principle similarities of varying degrees with past work in EANNs and evolving CA. CA are more akin to swarms as both are dynamical systems unlike ANNs, which are computational and have different properties. We are interested in evolving dynamical systems which are characterized by continual change, have less imposition of structure giving rise to an enabling substrate whereby higher level functionality can emerge. See (Mitchell, 1998) for discussion on dynamical versus computational systems. We deliberately avoid discussion of cooperation and the evolution of cooperation stemming from Robert Axelrod's work (Axelrod, 1987) as this work is largely concerned with cooperation among self-interested agents.

Grammatical Evolution

Grammatical Evolution (GE) is an evolutionary algorithm that can evolve computer programs in any language (O'Neill and Ryan, 2003; O'Neill, 2001; O'Neill and Ryan, 2001; Ryan et al., 1998), and can be considered a form of grammar-based genetic programming. Rather than representing the programs as parse trees, as in GP (Koza, 1992; Koza, 1994; Banzhaf et al., 1998; Koza et al., 1999; Koza et al., 2003), a linear genome representation is used. A genotype-phenotype mapping is employed such that each individual's variable length binary string, contains in its codons (groups of 8 bits) the information to select production rules from a Backus Naur Form (BNF) grammar. The grammar allows the generation of programs in an arbitrary language that are guaranteed to be syntactically correct, and as such it is used as a generative grammar, as opposed to the

classical use of grammars in compilers to check syntactic correctness of sentences. The user can tailor the grammar to produce solutions that are purely syntactically constrained, or they may incorporate domain knowledge by biasing the grammar to produce very specific forms of sentences.

BNF is a notation that represents a language in the form of production rules. It is comprised of a set of non-terminals that can be mapped to elements of the set of terminals (the primitive symbols that can be used to construct the output program or sentence(s)), according to the production rules. A simple example BNF grammar is given below, where <expr> is the start symbol from which all programs are generated. These productions state that <expr> can be replaced with either one of <expr><op><expr> or <var>. An <op> can become either +, -, or *, and a <var> can become either x, or y.

```
<expr> ::= <expr><op><expr>   (0)
         | <var>              (1)
<op>   ::= +                  (0)
         | -                  (1)
         | *                  (2)
<var>  ::= x                  (0)
         | y                  (1)
```

The grammar is used in a developmental process to construct a program by applying production rules, selected by the genome, beginning from the start symbol of the grammar. In order to select a production rule in GE, the next codon value on the genome is read, interpreted, and placed in the following formula:

$$Rule = Codon\ Value\ \%\ Num.\ Rules$$

where % represents the modulus operator.

Beginning from the the left hand side of the genome, codon integer values are generated and used to select appropriate rules for the left-most non-terminal in the developing program from the BNF grammar, until one of the following situations arise: (a) A complete program is generated. This occurs when all the non-terminals in the expression being mapped are transformed into elements from the terminal set of the BNF grammar. (b) The end of the genome is reached, in which case the *wrapping* operator is invoked. This results in the return of the genome reading frame to the left hand side of the genome once again. The reading of codons will then continue unless an upper threshold representing the maximum number of wrapping events has occurred during this individuals mapping process. (c) In the event that a threshold on the number of wrapping events has occurred and the individual is still incompletely mapped, the mapping process is halted, and the individual assigned the lowest possible fitness value. A full description of GE can be found in (O'Neill and Ryan, 2003).

Swarm Evolution

We consider three primary concerns in evolving dynamical systems.

Tractability The problem tends to be intractable in terms of evolutionary time and search. Emergent computation performed by CA is determined by its overall space-time behavior (Crutchfield et al., 2003). This is also true of swarms. Individual experiments evaluated by the evolutionary system may run for several minutes. The problem contains non-linearities and therefore mirrors a rugged search space. On a desktop PC it could take weeks to evolve a solution, if a solution is to be found at all.

Representation As an automatic programming problem we must consider how to encode the problem, what terminals to use and how to map the genetic material onto the swarm program. This representation should contribute to a Language that allows us to relate microscopic and macroscopic phenomena (Kubrik, 2003).

Fitness Evaluation A Method to determine the fitness of the swarm can be difficult to produce for dynamical systems problems (Williams, 2002). Methods to identify patterns in space or time are required. For example Spatial Entropy values (Bonabeau et al., 1999), Hough Transforms (Williams, 2002) and techniques from computational mechanics (Crutchfield et al., 2003) each describe patterns and hence fitness of dynamical pattern-forming systems. This is not to say there is a one-size-fits-all fitness evaluation method for complex systems but that dynamical systems might be evaluated based on the recognition of patterns.

In previous work we ran a number of experiments that considered the information processing capability of swarms on clustering tasks and considered the impact of different information usage. We concluded that different functions of information did effect the swarm's ability to model its environment although it was difficult to determine appropriate information and functions. Each dynamical environment has its own properties and information - where information is the meaning of events. In complex environments there are consistent patterns in space and time. If an entity can recognise relevant patterns it has the potential to be adaptive and anticipatory (Rosen, 1985). The information-theoretic notion of relevant information and the potential of taking an agent-centered theoretic approach to the design of distributed adaptive systems is discussed in (Nehaniv et al., 2002).

Every creature in nature survives by modelling its environment as it senses it and through a transducer system makes a response that is fit for that environment. Using GE, we evolve templates for simple transducers that describe how environmental information should be modelled by an ant to produce a response that favours the swarm's fitness.

Many of the distributed systems envisioned are data and event-centric and closely tied to their environments. Agents will make appropriate responses in the presence of certain environmental information. Grammar-based Genetic Programming approaches such as GE are a powerful means to describe legal interpretations of such terminal information yet still allow the open-ended evolution of novel solutions. GE's distinction between genotype and phenotype aids in representation of the problem and provides a substrate for processes that will regulate swarm construction.

Experimental Setup

We describe a multi-agent or swarm simulation model where information processing ants cooperate to solve an abstract clustering problem. Ants cluster identical objects which is a distinct problem from sorting (where a similarity measure is used to sort like objects). The ant's world is a square toroidal grid of 177*177 pixels (similar area to the circular world used in (Bonabeau et al., 1999))and uses a Moore neighbourhood. We use the Repast simulation tool and the Colt math library (Repast, 2004). Ants will move one pixel per time step. Ants can move in random directions based on a Gaussian probability distribution centered around the forward direction. A 'left antenna' at the north-west Moore pixel surrounding the ant and a 'right antenna' at the north-east are used to sense the concentration gradient. An ant will move deterministically in the direction of highest concentration or will continue moving straight if there is no difference in concentrations. These design choices are based on natural phenomena (Wilson, 1971; Hölldobler and Wilson, 1990). An ant will deposit a single pheromone signal of a certain concentration as it moves. This pheromone signal diffuses and evaporates at a constant rate. Diffusion and evaporation is implemented by the repast simulation tool (Repast, 2004). An ant may pickup or deposit objects in it's environment. Ants have a 'browsing' function whereby objects and other ants are observed and 'remembered'. The ant has a limited memory map where each object type encounter over a certain period is stored.

Ants use non-deterministic threshold functions (Bonabeau et al., 1999) to determine action. Equations 1 and 2 give the probability to pick up and drop an object respectively. $k1$ is the threshold value for picking up and object and $k2$ is the threshold value for dropping an object. f is the stimulus that the ant perceives - here it is the fraction of objects perceived over some period i.e. the ant's memory length.

$$\rho_p = \left(\frac{k1}{k1+f}\right)^2 \qquad (1)$$

$$\rho_d = \left(\frac{k2}{k2+f}\right)^2 \qquad (2)$$

The parameters used in the simulation model are shown in Table 1. We experimentally decided the simulation time.

Parameter	Value
World Dimensions	177 X 177
Ant Count	400
Object Count	1000
Memory Length	50
Decay factor	80 timesteps
Evaporation Rate	0.87
Diffusion Rate	0.42
Simulation Duration	5000 timesteps

Table 1: Swarm simulation parameters.

This time is a short duration that can be used to effectively evaluate the clustering performance. Evaporation and diffusion rates were chosen to represent 'recruitment' signals with medium spatial effect and short temporal effect. Other parameters were chosen over a number of trials. In the clustering task, ants have a number of behaviours. There is a behaviour for picking up objects, dropping objects, sensing stimulus, dropping pheromone and depositing 'pheromone traces' on objects. Each of these has a corresponding *gene*, which is the evolved template mentioned above. These templates are equivalent to GP S-Expressions, arranged in sequences that make up a complex of S-Expressions or genes. These genes regulate ant responses using environmental information as inputs. Each ant individual is encoded by a complex of genes mapped to behaviours. These genes regulate the values used for k1, k2 and f in equations 1 and 2. When objects are deposited, they emit pheromone traces of concentration specified by the appropriate gene. The concentration of pheromone used in ant trails is also regulated by a gene. Pheromone is emitted for a period of time specified by the decay factor in Table 1. GE uses the following grammar to map a genome to an S-Expression, or a complex of S-Expressions called GeneComplex. Start Symbol, S, Non-terminals, N, Production rules, P, and Terminals, T are shown below.

```
S = GeneComplex
P = GeneComplex> ::=     <expr> <expr>
                      <expr> <expr>
                      <expr>    (0)

<expr> ::= <expr> <op> <expr> (0)
         | <var>              (1)

<op> ::=  + (0)
        | - (1)
        | / (2)
        | * (3)

<var> ::= 10                  (0)
        | 100                 (1)
        | ants                (2)
        | working_ants        (3)
        | current_pheromone   (4)
        | pheromone           (5)
        | objects             (6)
```

Notice the terminal symbols of the grammar correspond to environmental information. Production rule 1 describes a complex of 5 S-Expressions. These expressions are mapped onto ant behaviours. A complex of S-Expressions are mapped in an arbitrary but consistent manner from a single genome using the grammar above. This complex is passed to a swarm simulation and used to construct genes in ants using a sequential mapping of S-Expressions onto the behaviour to be encoded. During the simulation at each time step, the ant senses its environment and passes in a table of all sensed variables corresponding to terminal symbols. Terminal symbols represent the number of ants, working ants, objects etc. (see grammar) encountered at each time step. Each behaviour class then computes a response value based on the environmental information supplied. These computed values are used for example in equations 1 and 2 to decide when to pick up and drop objects. In this way, each S-expression determines the value for a gene or parameter that determines the expression of a behaviour based on information inputs.

After the specified experiment length, the fitness of the swarm is computed and returned to the evolutionary engine. Spatial Entropy (Gutowiz, 1993; Bonabeau et al., 1999) was used as a measure for clustering performance in a lattice divided into a number of grids. The equation below gives the spatial entropy E, at a certain grid scale, s. P is the fraction of objects in one grid of total objects. In our experiments s = 6, there are 36 grids.

$$E_s = - \sum_{I \in s-patches} P_I log P_I \qquad (3)$$

Spatial Entropy (Gutowiz, 1993) is a macroscopic measure that corresponds to the individual ant's microscopic goals. As the ants 'work', spatial entropy values tend to decrease as the world becomes more ordered. In time-dependant problems it is advantageous to have fitness evaluation measures where the fitness value tends towards the optimum. In previous work we ran experiments for 50,000 time steps. This was a reasonable 'convergence' time over different clustering models. As mentioned, we run simulations for only 5,000. However, this time reasonably approximates clustering behaviour over 50,000 time steps.

It is the nature of the swarm clustering task that it should be easy to find a good solution (approximate ratio between stimulus and threshold values) while it can be difficult to identify the features of the multi-parameter problem. Consequently we focus on showing evidence of progressive search rather than on finding an optimal solution. GE genome individuals are evaluated only once (assuming they are not mutated). We use a variable-length generational GA with tournament selection, one-point crossover and integer mutation (as opposed to bit mutation). See evolutionary parameters table 2.

Parameter	Value
Mutation rate	.1
Crossover Rate	.7
Population size	200
Generations	5
Fitness Measure	Spatial Entropy

Table 2: GE parameters.

Generations	1	2	3	4	5
GE(mean)	1.79	1.79	1.70	1.43	1.70
GE(sd)	0.47	0.11	0.11	0.35	0.33
Rand(mean)	2.02	1.99	2.42	1.65	1.98
Rand(sd)	0.60	0.45	0.49	0.47	0.15

Table 3: Comparison between random run (Rand) and Grammatical Evolution (GE) showing mean Spatial Entropy results and standard deviation. (Initial entropy values approx. 2.7 on average)

Results

Results showing best solutions found in each generation for both GE search and random search are shown in Table 3. Random search generates and evaluates a random population on each generation. Due to computational time requirements, we have only taken 5 samples of each. We hoped to observe evidence of evolvability (Altenberg, 1994) rather than find optimal solutions. In evolving complex systems it would seem more important to enable *scaffolding*, where we maintain good sub-structures in a population and build on them. Although not discernable from the tables above, GE did find the most favourable spatial entropy value of 1.059 although only marginally better than the best value in the random search which was 1.127. However GE made improvements over successive generations in most samples. Over many clustering experiments, spatial entropy values ranged from average 2.7 (worst) to 0.9 (best) using given parameters.

We observed a number of clustering patterns. All experiments used the same agent models and differed only in the information-processing templates used. Clustering models in the literature show consistent formation of several small clusters, gradually becoming three or four large clusters. We observed these similar patterns but also observed patterns where objects seemed to be 'swept' into regions of the ant's world. The regions first contained sparse clusters that were gradually swept into dense compact clusters. We observed clusters that formed stripe-like patterns in addition to 'spots' although this may have been as a result of agent trajectories.

Conclusions

The purpose of this article was to demonstrate evolutionary pattern-forming swarms using Grammatical Evolution, with the result that the ant colony successfully evolved templates that exhibited clustering behaviour based on a spatial entropy measure. We have focused on the representation of evolving dynamical systems. We feel our information-theoretic representation could be easily generalized. GE provides independence between the evolutionary aspects and the program representation. Many complex systems can be considered in terms of swarms of information processing particles. The use of grammars provides a means to describe transducers for information processing in complex system nodes. Grammars provide a powerful means to describe legal interpretations of information yet still allow the open-ended evolution of novel solutions.

Fitness evaluation methods that evaluate patterns are an interesting way to evolve dynamical systems. The choice of fitness function is important as depending on how well it approximates or anticipates performance of the dynamical system in the early stages of a simulation, one can use shorter simulation times and less runs in the evolutionary stages.

The use of templates in a homogenous colony leads to behaviorally heterogenous ants based on their environmental information context. In a sense it also realises a type of ontogeny in that over time, the ant has features that may have variable fitness. This is an important aspect to exploit given the temporal development of simulations and makes the colony more adaptive. We have observed this through comparisons between models using static parameters and those using template-based parameters. In all areas of complex system research, a bridge of understanding between microscopic and macroscopic phenomena is required. Some research perspectives focus more on one or the other of these suffering the critique of others. Templates represent for us loci at which to study this connection. The evolutionary search implicitly identifies these templates as features of complex systems where for some global task, we can see how individuals process information. For example, even on the simple clustering task we observed several patterns of clustering. Given that we can observe these macroscopic patterns, can we analyze the templates and find out what the ants were thinking?

References

Abelson, H., Allen, D., Coore, D., Hanson, C., Homsy, G., Knight, T. F., Nagpal, R., Rauch, E., Sussman, G. J., and Weiss, R. (2000). Amorphous computing. *Communications of the ACM*, 43(5):74–82.

Altenberg, L. (1994). The evolution of evolvability in genetic programming. In Kinnear, K. E., editor, *Advances in Genetic Programming*, pages 47–74. MIT Press, Cambridge, MA.

Axelrod, R. (1987). The evolution of strategies in the iterated prisoner's dilemma. In Davis, L., editor, *Ge-*

netic Algorithms and Simulated Annealing, pages 32–41. Princeton University Press.

Banzhaf, W., Nordin, P., Keller, R., and Francone, F. (1998). *Genetic Programming – An Introduction; On the Automatic Evolution of Computer Programs and its Applications*. Morgan Kaufmann.

Bonabeau, E. (1997). From classical models of morphogenesis to agent-based models of pattern formation. *Artificial Life 3*, pages 191–209.

Bonabeau, E., Theraulaz, G., and Dorigo, M. (1999). *Swarm Intelligence*. Oxford Press.

Colorini, A., Dorigo, M., and Maniezzo, V. (1991). Distributed optimsation by ant colonies. In Varela, F. and Bourgine, P., editors, *European Conference on Artifical Life*, pages 134–142. MIT-Press.

Crutchfield, J., Mitchell, M., and Das, R. (2003). Evolutionary design of collective computation in cellular automata. In Crutchfield, J. and Schuster, P., editors, *Evolutionary Dynamics*. Oxford University Press.

Dorigo, M., Maniezzo, V., and Colorini, A. (1996). Ant system: Optimization by a colony of cooperating agents. *IEEE Transactions*, pages 29–41.

Gutowiz, H. (1993). Complexity seeking ants: (Unpublished).

Hölldobler, B. and Wilson, E. O. (1990). *The Ants*. Sprinter-Verlag.

Kennedy, J. and Eberhart, R. (1995). Particle swarm optimzation. In *IEEE International Conference On Neural Networks*, pages 1942–1948.

Kennedy, J. and Eberhart, R. (2001). *Swarm Intelligence*. Morgan Kaufmann.

Koza, J. (1992). *Genetic Programming: On the Programming of Computers by Means of Natural Selection*. MIT Press.

Koza, J. (1994). *Genetic Programming II: Automatic Discovery of Reusable Programs*. MIT Press.

Koza, J., Andre, D., Bennett III, F., and Keane, M. (1999). *Genetic Programming 3: Darwinian Invention and Problem Solving*. Morgan Kaufmann.

Koza, J., Keane, M., Streeter, M., Mydlowec, W., Yu, J., and Lanza, G. (2003). *Genetic Programming IV: Routine Human-Competitive Machine Intelligence*. Kluwer Academic Publishers.

Kubrik, A. (2003). Towards a formalisation of emergence. *Artifical Life*, 9:41–65.

Mitchell, M. (1998). A complex-systems perspective on the 'computation vs. dynamics' debate in cognitive science. In *Twentieth Annual Conference of the Cognitive Science Society*.

Nagpal, R. (2001). *Progammable Self-Assembly: Constructing Global Shape using Biologically-Inspired Local Interactions and Oragami Mathematics*. PhD thesis, Massachusetts Institute of Technology.

Nagpal, R. (2002). Programmable self-assembly using biologically-inspired multiagent control. *First International Joint Conference on Autonomous Agents and Multi-Agent Systems*, pages 418–425.

Nehaniv, C. L., Polani, D., and Dautenhahn, K. (2002). Meaningful information, sensor evolution, and the temporal horizon of embodied organisms. pages 345–349. MIT Press.

O'Neill, M. (2001). *Automatic Programming in an Arbitrary Language: Evolving Programs in Grammatical Evolution*. PhD thesis, University of Limerick.

O'Neill, M. and Ryan, C. (2001). Grammatical Evolution. *IEEE Transactions on Evolutionary Computation*.

O'Neill, M. and Ryan, C. (2003). *Grammatical Evolution: Evolutionary Automatic Programming in an Arbitrary Language*. KluwerAcademic Publishers.

Repast (2004). http://repast.sourceforge.net, Social Science Research Computing University of Chicago.

Rosen, R. (1985). *Anticipatory Systems: Philosophical, Mathematical and Methodological Foundations*. Pergamon Press.

Ryan, C., Collins, J., and O'Neill, M. (1998). Grammatical evolution: Evolving programs for an arbitrary language. In *Proceedings of the First European Workshop on Genetic Programming*, pages 83–95.

Schmutter, P. (2002). Object-oriented ontogenetic programming: Breeding computer programs that work like multicellual creatures. Master's thesis, University Dortmund.

Williams, H. (2002). Spatial organisation of a homogenous agent population using diffusive signalling and role diffrenciation. Master's thesis, University Sussex.

Wilson, E. O. (1971). *The Social Insects*. Harvard University Press.

Wolfram, S. (1983). Statistcal mechanics of cellular automata. *Review of Modern Physics*, pages 601–644.

Yao, X. (1999). Evolving artificial neural networks. In *Proceedings of the IEEE Vol 87 No 9*, pages 1423–1447.

Analyzing Evolved Fault-Tolerant Neurocontrollers

Alon Keinan

School of Computer Science, Tel-Aviv University, Tel-Aviv, Israel
keinanak@post.tau.ac.il

Abstract

Evolutionary autonomous agents whose behavior is determined by a neurocontroller "brain" are a promising model for studying neural processing. Nevertheless, they are missing an important quality prevalently found in all levels of natural systems, *fault-tolerance*, the lack of which results in overly simplistic neurocontrollers. We present a way of modifying a given evolutionary process for encouraging the creation of neurocontrollers that manifest high levels of fault-tolerance, using both direct and incremental evolutions. The evolved neurocontrollers are more robust not only against the faults introduced during the evolutionary process, but also against much more extreme ones. This robustness poses a great challenge for an analysis of the workings of the neurocontrollers, the latter being the focus of this paper: We utilize the Multi perturbation Shapley value Analysis (MSA) to uncover the important neurons, as well as the interactions between them, revealing the mechanisms underlying the evolved fault-tolerance.

Introduction

Neurally-driven Evolved Autonomous Agents (EAAs) are software programs embedded in a simulated virtual environment, performing tasks such as navigating and gathering food. An agent's behavior is determined by a neurocontroller which receives and processes sensory inputs from the surrounding environment and governs the activation of its motors. In recent years, much progress has been made in finding ways to evolve agents that successfully cope with diverse behavioral tasks (see Kodjabachian and Meyer (1998); Yao (1999); Guillot and Meyer (2001) for reviews). Furthermore, numerous EAA studies have yielded networks which manifest interesting biological-like characteristics (e.g. Cangelosi and Parisi (1997); Ijspeert et al. (1999); Aharonov-Barki et al. (2001)). Hence, EAAs are a very promising model for studying neural processing and developing methods for its analysis (Ruppin, 2002). Being abstractions, EAAs are missing many qualities of natural systems. This paper focuses on one such important quality, *fault-tolerance*, which is prevalently found in all levels of living organisms. Fault-tolerance emerges as a consequence of evolutionary pressure which favors organisms which are more resilient to harm. In this paper we emulate such a pressure by modifying a given evolutionary fitness function, encouraging the creation of neurocontrollers that manifest high levels of fault-tolerance. Those neurocontrollers are more plausible, both biologically and for hardware implementation.

A large amount of work has been invested in enhancing the fault-tolerance of feedforward neural networks, either by explicitly adding redundant hidden nodes or by modifying the back-propagation training algorithm. Evolutionary techniques for producing fault-tolerant systems have been mainly developed in the context of evolvable hardware, starting from Thompson (1996) who suggested to deliberately subject an evolving system to faults during its fitness evaluations. Canham and Tyrrell (2002) have used Thompson's method, while investigating what causes an evolved circuit to be tolerant to faults. In a recent paper (Zhou and Chen, 2003), a genetic algorithm based method is applied for improving the fault-tolerance of feedforward, back-propagation trained, neural networks. All these studies first produce a processing system that perform the required function and only then aim at producing fault-tolerance. In this paper, we evolve fault-tolerant embedded neurocontrollers both using a direct evolution and an incremental one. Further, we deal with fully-recurrent networks, without an explicit error function, rather than restricting the scope to feedforward networks trained via back-propagation.

One of the major challenges in the field of EAAs is understanding the way neurocontrollers operate. Lesion studies, where functional performance is measured after lesioning different elements of the system, have been employed in neuroscience for localizing function in a causal manner. However, most of the lesion studies employ single-lesions, in which only one element is lesioned at a time. Such approaches are limited in their ability to reveal the significance of interacting elements. One obvious example is provided by two elements that exhibit a high degree of overlap in their function, as is likely to be the case in fault-tolerant systems: Lesioning either element alone will not reveal its significance. Acknowledging that single lesions are insufficient for localizing functions in neural systems, we have previously presented the Multi-perturbation Shapley value Analysis (MSA) (Keinan et al., 2004b). The MSA processes a data set composed of numerous multiple lesions that are afflicted upon a neural system, together with the corresponding system performance level in each. It quantifies the contribution of the different system elements to the successful performance of the system's functions, as well as the functional interactions between groups of elements. The MSA was first developed for deciphering the mechanisms underlying EAAs' behavior. In comparison to previous analyzed neurocontrollers (Keinan et al., 2004a; Saggie et al., 2003; Ganon et al., 2003), the fault-tolerant agents evolved in this study pose a much greater analysis challenge, serving as a more biologically plausible testbed for such analysis methods. In this paper, we utilize the MSA for the analysis of the evolved neurocontrollers, examining the fault-tolerance mechanisms. We also present a new variant, *MSA K-limited contributions*, introduced in this paper in order to examine, in the presence of high levels of fault-tolerance, the gap between the single-lesion approach and the full multi-lesion analysis.

The Evolutionary Environment

The EAA environment is described in detail in Aharonov-Barki et al. (2001). The agents live in a discrete 2D grid "world" surrounded by walls. Poison items are scattered all around the world, while food items are scattered only in a "food zone" in one corner. The agent's goal is to find and eat as many food items as possible during its life, while avoiding the poison items. The agent is equipped with a set of sensors, motors, and a fully recurrent neurocontroller containing n McCulloch-Pitts neurons ($n = 10$ in all simulations). The four sensors encode the presence of a wall, a resource (food or poison, without distinction between the two), or a vacancy in the cell the agent occupies and in the three cells directly in front of it. A fifth sensor is a "smell" sensor which can differentiate between food and poison underneath the agent and gives a random reading if the agent is in an empty cell. The four motor neurons dictate movement forward (neuron 1), a turn left (neuron 2) or right (neuron 3), and control the state of the mouth (open or closed, neuron 4).

As in previous studies (Aharonov-Barki et al., 2001), a *genetic algorithm* is used to evolve the synaptic weights by directly encoding them in the genome as real valued numbers. In this study, in order to encourage the creation of fault-tolerant neurocontrollers, phenotypic faults are introduced while an agent's fitness is being evaluated throughout the evolutionary process, thus making the fault-tolerance an integral part of the task specification. Particularly, the fitness of an agent is determined by evaluating it n times, in addition to the regular evaluation, with a different neuron being lesioned (removed from the neurocontroller) in each

of these evaluations.[1] The *mean lesion fitness* function is then defined by

$$\frac{1}{n+1}\left(f(N) + \sum_{i=1}^{n} f(N \setminus \{i\})\right), \qquad (1)$$

where $N = \{1, \ldots, n\}$ is the set of neurons and f is the performance. $f(N)$ denotes the performance level of the intact neurocontroller, which is the *standard fitness* used to evolve the agents in Aharonov-Barki et al. (2001), while $f(N \setminus \{i\})$ denotes the performance level of the neurocontroller when neuron i is lesioned. The mean lesion fitness function is utilized to create fault-tolerant neurocontrollers in two ways: One is incremental evolution, in which an evolutionary run is first conducted using the standard fitness function, starting from random neurocontrollers. Then, in the incremental stage, starting from the most successful agent in this evolutionary run, agents are evolved using the mean lesion fitness function. The second one consists of a direct evolutionary run using the mean lesion fitness function, starting with random neurocontrollers. Ten evolutionary runs of each of the two types were performed, all with the same genetic algorithm parameters as in Aharonov-Barki et al. (2001).

Analysis of Evolutionary Results

We begin by testing whether the introduced change to the fitness function results in more fault-tolerant agents. Figure 1 presents both the standard and mean lesion fitness of the successful agents. The difference between the two fitness functions serves to quantify the level of fault-tolerance. As evident, the agents evolved using the mean lesion fitness are much more fault-tolerant compared with the agents evolved using the standard fitness (the ones serving as the basis for the incremental stage), while reaching almost the same level of intact performance. This testifies that **the modified evolutionary pressure indeed encourages the creation of fault-tolerance, when using either direct or incremental evolutions.**

In order to further quantify the level of fault-tolerance of a neurocontroller, we measure the degradation in the agent's performance level as it is subjected to more and more concurrent lesions. Figure 2 shows the average performance level of agent F, an incrementally evolved fault-tolerant agent, as a function of the lesioning depth (the number of concurrently lesioned neurons). Obviously, the agent after the incremental evolution maintains, for all lesion depths, a much higher level of performance than its predecessor, before applying the incremental stage of the evolution. Interestingly, **though during the evolution only**

---

[1] When lesioning motor neurons the activity transmitted to the motors themselves in not altered, but only the activity transmitted to other neurons, thus hindering only the role they play in the recurrent neurocontroller's computations.

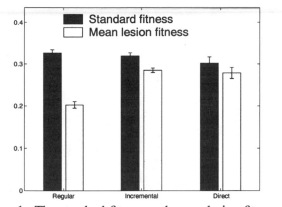

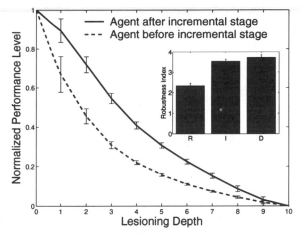

Figure 1: The standard fitness and mean lesion fitness for the agents evolved with the standard fitness (*Regular*), for the agents evolved using incremental evolution (*Incremental*) and for the agents evolved using direct evolution (*Direct*). All are mean (and standard deviation of the mean) of the most successful agent from each of the 10 evolutionary runs of that type.

Figure 2: The mean normalized performance level as a function of the lesioning depth, for agent F (solid line) and for its predecessor, serving as the basis for the incremental stage in which F was evolved (dashed line). The robustness index equals 3.87 for the former and 2.56 for the latter. The inset depicts the mean robustness index of the agents of each type, in the same order of bars as in Figure 1.

single-lesions are afflicted, the evolutionary pressure encourages a higher level of fault-tolerance, e.g., the agent's performance level when four of its neurons are lesioned is about the same as the performance level of the agent evolved without the mean lesion fitness when only two of its neurons are lesioned. We define the *robustness index* to be the area below the curve of the type introduced in Figure 2. Evidently, based on this measure, **the evolution with the mean lesion fitness yields agents which are much more robust** (Figure 2).

Lesioning a neuron that plays no part in carrying out the neurocontroller's function might have no effect on the performance level. Hence, as suggested by Canham and Tyrrell (2002) in the context of evolvable hardware, systems whose function is carried out by a small fraction of its elements would be regarded as more robust. We wish to test whether the above results are indeed a manifestation of the evolution of backup mechanisms or whether they are merely an outcome of smaller effective sizes of the neurocontrollers evolved with the mean lesion fitness. An intuitive way to attack this question, analogous to the one suggested by Canham and Tyrrell (2002), is to test, for a neuron that has no effect on the performance level when single-lesioned, whether is has an effect when lesioned together with any other neuron. Indeed, there are incidents where this suffices to expose the effect of neurons. For instance, in agent F, lesioning either neuron number 5 or neuron number 10 by themselves does not change the performance level considerably, while lesioning them together does cause a large decrease of 62%. Nevertheless, since the evolved neurocontrollers exhibit an improved fault-tolerance to deep levels of lesioning, as shown above, double-lesions might not be enough to reveal all the neurons playing a true part in carrying out the

function. Indeed, Canham and Tyrrell (2002) report that using this kind of approach still gives only an impression of what is happening, while missing elements which must be of importance. To overcome the drawbacks of finding the important neurons based on single-lesions and/or double-lesions, we turn to utilize the MSA.

The MSA framework (Keinan et al., 2004b) addresses the challenge of defining and calculating the contributions of system elements from a data set of multiple lesions that are afflicted upon the system. In this framework, we view a set of multiple lesion experiments as a *coalitional game*, borrowing concepts and analytical approaches from the field of Game Theory. Specifically, we define the set of contributions to be the *Shapley value* (Shapley, 1953), which stands for the unique fair division of the game's worth (the system's performance level when all elements are intact) among the different players (the system elements). Applying the MSA for the analysis of the neurocontrollers presented in this paper, the *marginal importance* of neuron i to a set of neurons S, with $i \notin S$, is defined by

$$\Delta_i(S) = f(S \cup \{i\}) - f(S), \qquad (2)$$

where $f(S)$ is the performance level when only the neurons in the set S are intact, while the rest are lesioned. Then, the Shapley value defines the contribution

$$\gamma_i = \frac{1}{n!} \sum_{R \in \mathcal{R}} \Delta_i(S_i(R)) \qquad (3)$$

of each neuron $i \in N$, where $\mathcal{R}$ is the set of all $n!$ orderings of the set of neurons N and $S_i(R)$ is the set of neurons preceding i in the ordering R. It can be interpreted as follows:

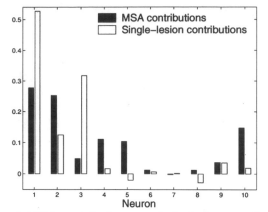

Figure 3: MSA contributions vs. single-lesion contributions of agent *F*. The single-lesion contribution of a neuron is the decrease in the performance level when the neuron is lesioned. In order to be comparable, both are normalized such that the sum across all neurons equals 1.

Suppose that all the neurons are arranged in some order, all orders being equally likely. Then γ_i is the expected marginal importance of neuron i to the set of neurons who precede him. *This unique and fair contribution measures the part the neuron plays in successfully performing the neurocontroller's function.*

Based on all the possible multi-lesion experiments, the MSA reveals the true contribution of neurons whose importance has been missed by the single-lesion approach (Figure 3). For instance, arbitrarily defining an important neuron as one with a normalized contribution greater then 0.03, the MSA reveals 7 important neurons while the single-lesion approach reveals only 4 of them. The performance level even slightly increases when lesioning neuron number 5, resulting in a negative single-lesion contribution, while actually this neuron is an important one. Generally, **the disagreement between the single-lesion approach and the MSA is greater for the agents evolved using the mean lesion fitness, with single-lesion contributions being as far from the MSA contributions as random normalized vectors are** (Figure 4). Armed with the MSA contributions, we can finally return to test whether the neurocontrollers evolved with the mean lesion fitness are of smaller effective sizes. The results testify that this in not the case as agents of all three evolutionary types have, on the average, the same number of important neurons (mean of 7.1 neurons), with no significant difference.[2] This new finding leads to the conclusion that the agents evolved with the mean lesion fitness are **more robust due to the evolution of some backup mechanisms, rather than the mere evolution of smaller solutions**. We return to further investigate these mechanisms in the next section.

------

[2]This conclusion still holds for other importance thresholds, though the number of important neurons varies (data not shown).

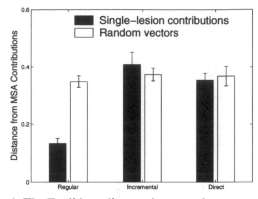

Figure 4: The Euclidean distance between the vector of normalized single-lesion contributions and the vector of normalized MSA contributions (dark bars). For comparison, the distance between random normalized vectors and the MSA contributions is also plotted (light bars). Both present the mean of all agents of each type.

To examine the wide gap between the contributions yielded by single-lesions and the contributions yielded by the MSA, the approach should be generalized. We seek a variant of the MSA which, for a given lesioning depth K, yields the neuronal contributions based on all multi-lesions up to that depth. Hence, we define $\Delta_i(S)$ (eq. (2)) to be zero for all S such that $|S| \geq K$. Based on this definition, the contribution of neuron i may be defined according to eq. (3), properly normalized by changing the denominator to be $K \cdot (n-1)!$ instead of $n!$. These *MSA K-limited contributions* coincide with the actual MSA contributions for $K = N$ and with the single-lesion contributions for $K = 1$. Figure 5 plots the distance between the MSA K-limited contributions and the actual ones as a function of K. Obviously, **the K-limited contributions approach the actual contributions much slower for the fault-tolerant agent, testifying that one must use a deep level of multi-lesions in order to gain true insights into the working of such systems**. For instance, an analysis of the regular neurocontroller based on single-lesions solely is as accurate as an analysis of the fault-tolerant neurocontroller based on all multi-lesions of up to depth 4.

Underlying Fault-Tolerance Mechanisms

The MSA contributions introduced in the previous section serve as a summary of the system's functionality, indicating the average marginal importance of each element over all possible orderings of the elements. For complex systems, where the importance of an element strongly depends on the state (lesioned or intact) of other elements, the MSA further suggests a higher order description (Keinan et al., 2004b). Focusing here on a two-dimensional description, the MSA defines the functional interaction between each pair of elements. This functional interaction quantifies how much the

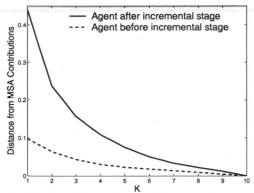

Figure 5: The Euclidean distance between the normalized MSA contributions and the normalized MSA K-limited contributions as a function of the lesioning depth K, for agent F (solid line) and for its predecessor (dashed line).

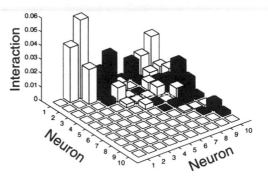

Figure 6: The symmetric MSA interactions between each pair of neurons of agent F. The figure presents the absolute values, where dark bars denote negative interactions and light bars denote positive ones.

average marginal importance of the two elements together is larger (or smaller) than the sum of the average marginal importance of each of them when the other one is lesioned. Intuitively, this interaction measures how much "the whole is greater than the sum of its parts" (*synergism*), where the whole is the pair of elements. In cases where the whole is smaller than the sum of its parts, that is, when the two elements exhibit functional overlap (*antagonism*), the interaction is negative. Clearly, single-lesion approaches cannot uncover such functional interactions.

Observing the MSA interactions between all pairs of neurons of agent F reveals many negative ones, pointing to pairs of neurons which backup each other's function (Figure 6). Obviously, the backup scenario is not a clear-cut case as one might have expected, according to which each redundant neuron has another redundant one completely backing it up. Rather, several neurons backup several others to some extent, e.g., each of neurons 4, 9 and 10 backup each of the two others. These results exemplify the multiplicity of negative interactions in the agents evolved with the mean lesion fitness. While those agents have, on the average, 15.1 negative interactions of meaningful magnitude, the agents evolved with the standard fitness have only 7.8, with this difference being significant (p-value < 0.05). Comparing the number of positive interactions, the former have much less than the latter (mean of 17.8, compared with mean of 27.1; p-value < 0.05). These results testify to the fact that **the evolutionary pressure introduced by the mean lesion fitness function encourages the formation of functional overlap between the neurons, at the expense of the formation of cooperation**, which results in the evolution of neurocontrollers that are much more fault-tolerant.

Since, as shown above, the backup scenario is not a clear-cut case, we turn to more closely analyze the fault-tolerance mechanisms by focusing on the level of the individual synaptic connections. The Evolutionary Network Minimization (ENM) algorithm (Ganon et al., 2003) is a

genetic algorithm with an additional step in which synaptic connections are irreversibly eliminated. Ganon et al. (2003) have shown that, given an agent, the ENM tends to largely minimize its neurocontroller, while keeping the performance level high and the principal functional characteristics intact. Here, we utilize the ENM to minimize the successful neurocontrollers, while examining the number of remaining synaptic connections in the resulted backbones. The agents evolved with the mean lesion fitness have more synaptic connections in their backbones (mean of 20.4, compared with mean of 17.1; p-value < 0.05), implying that **those fault-tolerant agents perform a more complicated function**. Bearing in mind that, as shown in the previous section, both the fault-tolerant and the regular neurocontrollers consist of the same number of important neurons, we conjecture that the additional 3.3 synapses found, on the average, are due to synaptic connections playing a role in the backup mechanisms between those neurons. A further direct inspection of the small backbones yielded by the ENM helps in understanding the way the neurocontrollers operate and the redundancies between synapses (omitted due to space considerations).

Discussion

We have introduced a modification to evolutionary processes for evolving neurocontrollers that, while reaching a good performance level, exhibit high fault-tolerance to neuronal lesioning. The modification was shown to be successful both when starting from a successful, but not fault-tolerant, neurocontroller (incremental evolution) and when starting from random ones (direct evolution). The evolved neurocontrollers exhibit a high level of fault-tolerance, not only to the faults introduced during the evolutionary process, but also to much more extreme ones. Aiming to understand the workings of those evolved neurocontrollers, their robustness poses a great challenge. We have utilized the MSA to uncover the important neurons, as well as the interactions between them, while overcoming the inherent disadvantages of single-lesion approaches, showing that these are amplified in the face of fault-tolerance. Based on the MSA, the depth of lesioning required to get a good insight into the workings of a neurocontroller has been quantified. Furthermore, the analysis reveals that the robustness of the neurocontrollers is due to the actual evolution of backup mechanisms, captured by the functional interactions, rather than the evolution of smaller solutions. Lastly, while the fault-tolerant neurocontrollers utilize the same number of important neurons to perform their function, the underlying synaptic networks, captured by the ENM, are more complicated than the ones of regular neurocontrollers.

The development and study of the MSA has been first done within the framework of EAAs. Nevertheless, the MSA is geared toward general experimental biological applications. In Keinan et al. (2004b), the applicability of the MSA to the analysis of neurophysiological models was demonstrated, as well as to the analysis of behavioral data from experimental deactivation studies of the cat brain. The MSA is further applicable to the analysis of Transcranial Magnetic Stimulation (TMS) experiments. In biology in general, the recent development of RNA interference (RNAi) has made the possibility of multiple concomitant gene knockouts a reality, allowing the utilization of the MSA to the analysis of genetic and metabolic networks. Surely, in such biological applications only a limited number of multi-perturbation experiments can be performed. Hence, the MSA encompasses prediction and estimation variants which approximates the Shapley value in an accurate, scalable and efficient manner (Keinan et al., 2004b). Robustness, being a common quality of biological systems, must be correctly handled by any analysis method applied to biological experiments. This paper has demonstrated the applicability of the MSA to the analysis of robust systems, overcoming the disadvantages of the single-lesion approach, and has established that a deep level of lesioning should be used in order to correctly identify the important elements in such systems.

References

Aharonov-Barki, R., Beker, T., and Ruppin, E. (2001). Emergence of memory-driven command neurons in evolved artificial agents. *Neural Computation*, 13(3):691–716.

Cangelosi, A. and Parisi, D. (1997). A neural network model of Caenorhabditis Elegans: The circuit of touch sensitivity. *Neural Processing Letters*, 6:91–98.

Canham, R. O. and Tyrrell, A. M. (2002). Evolved fault tolerance in evolvable hardware. In *proceedings of the Congress on Evolutionary Computation 2002 (CEC2002)*, pages 1267–1272.

Ganon, Z., Keinan, A., and Ruppin, E. (2003). Evolution-

ary network minimization: Adaptive implicit pruning of successful agents. In Banzhaf, W., Christaller, T., Dittrich, P., Kim, J. T., and Ziegler, J., editors, *Advances in Artificial Life - Proceedings of the 7th European Conference on Artificial Life (ECAL)*, volume 2801 of *Lecture Notes in Artificial Intelligence*, pages 319–327. Springer Verlag Berlin, Heidelberg.

Guillot, A. and Meyer, J. A. (2001). The animat contribution to cognitive systems research. *Journal of Cognitive Systems Research*, 2(2):157–165.

Ijspeert, A. J., Hallam, J., and Willshaw, D. (1999). Evolving swimming controllers for a simulated lamprey with inspiration from neurobiology. *Adaptive Behavior*, 7(2):151–172.

Keinan, A., Hilgetag, C. C., Meilijson, I., and Ruppin, E. (2004a). Causal localization of neural function: The Shapley value method. *Neurocomputing, to appear*.

Keinan, A., Sandbank, B., Hilgetag, C. C., Meilijson, I., and Ruppin, E. (2004b). Fair attribution of functional contribution in artificial and biological networks. *Neural Computation, to appear*.

Kodjabachian, J. and Meyer, J. A. (1998). Evolution and development of neural controllers for locomotion, gradient-following and obstacle-avoidance in artificial insects. *IEEE Transactions on Neural Networks*, 9(5):796–812.

Ruppin, E. (2002). Evolutionary autonomous agents: A neuroscience perspective. *Nature Reviews Neuroscience*, 3:132–141.

Saggie, K., Keinan, A., and Ruppin, E. (2003). Solving a delayed response task with spiking and Mcculloch-Pitts agents. In Banzhaf, W., Christaller, T., Dittrich, P., Kim, J. T., and Ziegler, J., editors, *Advances in Artificial Life - Proceedings of the 7th European Conference on Artificial Life (ECAL)*, volume 2801 of *Lecture Notes in Artificial Intelligence*, pages 199–208. Springer Verlag Berlin, Heidelberg.

Shapley, L. S. (1953). A value for n-person games. In Kuhn, H. W. and Tucker, A. W., editors, *Contributions to the Theory of Games*, volume II of *Annals of Mathematics Studies 28*, pages 307–317. Princeton University Press, Princeton.

Thompson, A. (1996). Evolutionary techniques for fault tolerance. In *Proceedings UKACC Int. Conf. on Control (CONTROL 96)*, pages 693–698. IEE Conference Publication No. 427.

Yao, X. (1999). Evolving artificial neural networks. *Proceedings of the IEEE*, 87(9):1423–1447.

Zhou, Z. H. and Chen, S. F. (2003). Evolving fault-tolerant neural networks. *Neural Computing and Applications*, 11(3-4): 156-160.

Tracking Information Flow through the Environment: Simple Cases of Stigmergy

Alexander S. Klyubin, Daniel Polani, and Chrystopher L. Nehaniv

Adaptive Systems Research Group
School of Computer Science, Faculty of Engineering and Information Sciences
University of Hertfordshire, College Lane, Hatfield Herts AL10 9AB, UK
{A.Kljubin,D.Polani,C.L.Nehaniv}@herts.ac.uk

Abstract

Recent work in sensor evolution aims at studying the perception-action loop in a formalized information-theoretic manner. By treating sensors as extracting information and actuators as having the capability to "imprint" information on the environment we can view agents as creating, maintaining and making use of various information flows. In our paper we study the perception-action loop of agents using Shannon information flows. We use information theory to track and reveal the important relationships between agents and their environment. For example, we provide an information-theoretic characterization of stigmergy and evolve finite-state automata as agent controllers to engage in stigmergic communication. Our analysis of the evolved automata and the information flow provides insight into how evolution organizes sensoric information acquisition, implicit internal and external memory, processing and action selection.

1 Introduction

We approach the study of information flows through the perception-action loop and the environment using classical information theory. We believe that a formalized approach to the perception-action loop may bring us closer to finding principles underlying adaptive behavior. Such principles could be used both for guiding the construction of systems with desired information flows and for studying their behavior. The use of information theory provides us with a universal framework which minimizes the influences of a particular implementation.

Consider an agent with sensors and actuators. Sensors capture some information, the information gets processed, and based on the results the actuators act upon the environment. If sensors are seen as taking information in, it seems also reasonable to see actuators as modifying the environment informationally. Surprisingly, it seems that little research has been done to quantitatively treat perception-action in terms of information – an observation also made by Touchette and Lloyd in the context of control [15].

In [14, 15] the problem of control is quantitatively treated in terms of Shannon information which is seen as flowing from a controlled system into a controller and then back. An important information-theoretic bound is obtained for the usefulness of any sensor for control. [9] introduces an information-theoretic view of perception and actuation and discusses usefulness as a means to attribute agent-specific meaning to information. This is further formalized in the context of relevant information [12] measured in bits. Relevant information "flows" from the environment via sensors to actuators, thus connecting them.

1.1 Shannon Information Flow & Environment

One of the motivations for our study is sensory evolution [5]. Originally the idea was focused on sensors evolving to capture more useful or relevant information. However, using sensors is often inseparable from what an agent needs to *do* in its environment. It makes more sense to consider perception and action together as a single entity, a loop. Actions in the environment can influence the sensors, creating a loop. This is an important link which the quantitative approaches above have not addressed directly and which this paper does address in a quantitative manner (Fig. 1).

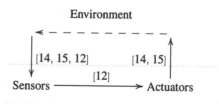

Figure 1: Information flow within an agent (box) and between the agent and its environment addressed by quantitative information-theoretic approaches. Solid line – prior work. Dashed and solid – this paper.

The perception-action *loop* is important for understanding the behavior of adapted agents. If we view sensors as capturing Shannon information and actuators as being capable of "imprinting" information on the environment, we can then treat agents as creating, maintaining and using various information flows, both internal and external. The view may be quite useful since there are strong indications that biological agents are partly driven by the necessity of acquiring, exchanging and also concealing information.

Treating the perception-action loop in terms of Shannon information flows enables us to quantitatively capture certain phenomena, like "imprinting" information onto the environment, offloading and later reacquisition of information, and active perception. An agent imprints information onto its environment by changing the environment in different ways depending on its internal state.

Uncovering hidden information using the perception-action loop is demonstrated by Kirsh and Maglio in [8]. There advanced Tetris game players are shown to quickly rotate a falling block while it still is not visible completely. This active modification of the environment allows the players to discover the shape of the block before it actually becomes completely visible on the screen. *Active perception* in general is seen as a powerful technique [10].

Offloading of information into the environment can be illustrated by the fact that we often write notes or reminders, which we then later look at to reacquire some information. A good account of such information flow (not in the Shannon sense) between several people is given in the analysis of how members of an airliner crew indirectly communicate using cockpit controls as a medium [7]. For example, long before landing one of the pilots calculates proper flap settings for various speeds and then based on the results sets special markers on the airspeed indicator. Later, during the landing phase, the markers allow the crew to quickly and reliably find out what flap settings to use for the momentary speed. In a wider context, indirect communication via the environment is of high importance in distributed systems [6, 2].

Stigmergy is usually considered as the indirect communication between agents via the environment without targeting a specific recipient. Stigmergy has been used to explain nest building, sorting, and foraging in social insects [13, 3, 2].

To our knowledge this work is the first information-theoretic characterization of stigmergy. It allows us to refine the understanding of stigmergic communication by including the *relationship* between agents and the environment (cf. [1]). For example, as a reminder to do something one could go to sleep in one of several rooms. Later, using the room one wakes up in as a cue, one could engage in different corresponding activities. Information may thus be communicated via one's relation with the environment.

This paper is structured as follows. Sec. 2 introduces the relevant concepts of information theory. Sec. 3 presents the model of the perception-action loop we are using. Sec. 4 describes our experimental approach. In Sec. 5 we evolve an agent to stigmergically communicate with itself through time. In Sec. 6 we evolve pairs of agents to engage in classical stigmergic communication. The work is summarized and discussed in Sec. 7.

2   Information Theory

We denote random variables with uppercase letters, e.g., X, their sets of values with calligraphic letters, e.g., $\mathcal{X}$, and the values with lowercase letters, e.g., x. In this paper we deal exclusively with discrete variables. By abuse of notation we denote the probability that $X = x$ as $p(x)$ and the conditional probability of $X = x$ given that $Y = y$ with $p(x|y)$.

The *entropy* of X, denoted by $H(X)$, is defined as a measure of uncertainty of the probability distribution of X:

$$H(X) := -\sum_{x \in \mathcal{X}} p(x) \log_2 p(x)$$

The *conditional entropy* of X given Y, denoted $H(X|Y)$, is defined as uncertainty of X knowing Y weighted by the probability of a particular realization of Y occurring:

$$H(X|Y) := \sum_{y \in \mathcal{Y}} p(y) H(X|Y = y)$$
$$= -\sum_{y \in \mathcal{Y}} p(y) \sum_{x \in \mathcal{X}} p(x|y) \log_2 p(x|y)$$

The *mutual information* between X and Y, denoted $I(X;Y)$, is defined as reduction in the uncertainty of X given Y:

$$I(X;Y) := H(X) - H(X|Y)$$

All of the above quantities are *nonnegative*. We always calculate them using the binary logarithm, hence they are measured in *bits*. An important property of information-theoretic measures is that they do not depend on the particular values of the variables – *the measures only depend on the probability distributions* of the values. For a detailed introduction to information theory consult, for example, [4].

3   A Model of the Perception-Action Loop

In this section we present a model of the perception-action loop of an agent. The constituents of the loop, which are modeled as random variables, are sensors S, actuators A, the memory M of the controller, and R – the rest of the world. We need R to formally account for the effects of actuation and the environment on the sensors.

As in [15] we interpret the perception-action loop in terms of a communication channel-like model. In order to model the temporal aspects we *unroll* the loop in time by introducing a time variable t. To account for the complete loop we model the dynamics of arbitrary number of time steps, as opposed to [15] where only one time step is modeled.

We model the relations between the variables as a causal Bayesian network [11] which is a directed acyclic graph where any node, given its parents, is conditionally independent from any other node which is not its parent or successor (any node directly or indirectly reachable from the node). In our model this property results in conditional independence from the past.

We show the pattern of relations between variables at two consecutive time steps on Fig. 2. We assume that the pattern

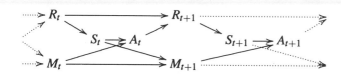

Figure 2: A temporal section of the unrolled perception-action loop modeled a Bayesian network. The dotted lines show how the section connects to adjacent sections. S – sensoric input, A – action performed, M – memory of the controller, R – the rest of the world.

of relations is time-invariant and thus holds for any t. Thus, the graph on Fig. 2 is just a section of the network.

The diagram can be read as follows: A_t is picked given S_t and M_t. Furthermore, S_t is obtained from R_t.

The perception-action loop is created by information flowing from sensors to actuators, and from actuators via the rest of the world to future sensoric input. If these flows extend over more than one time step, they may be mediated by a combination of internal and external variables potentially enabling complex behavior, such as filtering based on the state of the environment or the memory.

4 Experimental Approach

We study simple stigmergic scenarios and track information flows through the environment and the perception-action loop. We measure the flows using information-theoretic tools. We evolve agent controllers to maximize various information flows passing via the environment.

4.1 The Testbed

We base our experiments on a model which despite its simplicity captures important features of systems employing gradient sensors.

The environment consists of a two-dimensional grid of infinite size. A source is located at the center of the grid. The source emits a signal, the strength P of which in any cell of the grid is $P(d) = d^{-2}$ ($P(0) = 2$), where d is the distance to the source. The exact relation is not important for our experiments – it is only important that the decrease is strictly monotonic with distance.

An agent is situated in a single cell at a time. The agent has a gradient *sensor*. The gradient points to the cell with highest signal strength among the four adjacent cells (north, east, south, west). If there are several cells with highest signal strength, the gradient randomly points to one of these with equal probability (see Fig. 3). The agent also has an *actuator* – at each time step the agent performs one of the four available actions: move north, east, south or west.

Following the model of perception-action loop presented in Sec. 3 and its notation, we denote the sensoric input (the gradient) with the random variable S with values in set $S = \{s_N, s_E, s_S, s_W\}$; the action with the random variable A

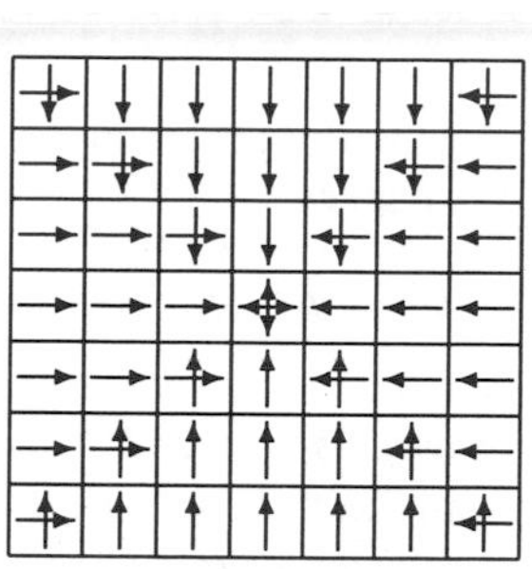

Figure 3: Sensoric input S vs. the position R. In cells with only one arrow the input is constant. In cells with multiple arrows it is randomly chosen from the listed options. The source is located at the center of the grid.

with values in set $A = \{a_N, a_E, a_S, a_W\}$; the internal state or the memory of the agent's controller with the random variable M with values in $M = \{1, 2, \ldots, N\}$, where N is the number of states; the two-dimensional position of the agent with the random variable R with values in $\mathbb{Z}^2$. S, A, M, and R completely describe the state of the world at any instant.

According to the model in Sec. 3 A_t and M_{t+1} depend directly only on (S_t, M_t). The agent's controller can thus be seen as implementing the mapping: $(S_t, M_t) \mapsto (A_t, M_{t+1})$. We assume that the mapping is time-invariant.

The mapping can be implemented by a finite state automaton operating with input set S, output set A, and state set M. Here we use deterministic automata hence allowing only for deterministic mappings. However, without any loss of generality our approach can be used with stochastic mappings implemented by nondeterministic finite-state automata. Determinism of the controller is an experimental choice, not a limitation of the model.

To summarize, the world is an infinite two-dimensional grid. A symmetric signal field is created around a source located at the center of the grid. The strength of the field decreases strictly monotonically with distance. The environment is static. An agent moves on the grid, one cell at a time, in one of the four adjacent cells (north, east, south, west). The agent has a nondeterministic sensor which can distinguish between four directions of the local signal gradient. The agent is controlled by a controller, which is modeled as a deterministic finite-state automaton taking current sensoric input and producing an action. All the controller has access to is its own memory and the momentary sensoric input.

4.2 Information Flow

In the experiments presented in this paper we use a special case of information flow. We "inject" information independent of the past and present state of the system into a variable X (e.g., M_0) by making its distribution independent and entropic. The *information flow* from X to any variable Y in the system is then $I(X; Y)$. This gives us a characterization of stigmergy if X is from one agent and Y is from another.

In general, it is possible to measure information flow without injecting independent information. However, this is out of scope of this paper.

4.3 Measuring Information Flows

We use Monte Carlo simulations for estimating the information flows. Accordingly, to spot and avoid undersampling, for each of the possible initial states of the system we produce 32, 256, or more samples depending on the particular quantities measured. We then increase the number of samples by at least a factor of 16 to check whether the quantities of interest remain stable.

4.4 Evolution

We use evolution as a search in the space of controllers. A minimal setup is used. This is to emphasize that nothing in our general approach is specific to the particular model or the search tool employed. To evaluate the fitness of a controller, it is allowed to control the agent in the particular setup of the experiment. The fitness is expressed as the information flow specific to the experiment.

We initialize the population with five randomly generated controllers. In every generation five best controllers are selected into the next generation and also produce five offspring each. Thus the size of the population is between 5 and 30.

An offspring is produced from its parent by mutation. To speed up the search and make it more efficient we have incorporated ideas from simulated annealing and tabu search: (1) the number of mutations performed is uniformly distributed between 1 and $1 + (G \bmod 20)$, where G is the generation; and (2) we do not add offspring controllers which have been evaluated before or are already present in the population. On our problems these adjustments do improve the efficiency of the search. Additionally, we perform at least five separate evolutionary runs to sample different solutions.

The transition matrix of a controller is represented by a mapping $S \times \mathcal{M} \rightarrow \mathcal{A} \times \mathcal{M}$. This mapping can represent any finite-state automaton. We limit our search to deterministic automata only, hence the mapping can be represented as an array of length $|S| \cdot |\mathcal{M}|$ with each element containing a value between 1 and $|\mathcal{A}| \cdot |\mathcal{M}|$ corresponding to the action to perform and the next state to go into. A mutation is performed by setting a randomly chosen element of the array to a randomly chosen value in the range.

5   Stigmergy for One

Here we study a special case of stigmergy where an agent has to pass information to itself through the environment over a fixed number of time steps. We initialize the agent with information, let the agent run for a while, then erase its memory and let the agent reacquire the lost information.[1]

---

[1]The source code of experiments is available on request.

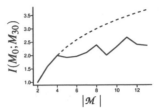

Figure 4: Fitness of best evolved controllers. Solid line – fitness measured as the information flow $I(M_0; M_{30})$ for controllers with $|\mathcal{M}|$ states. Dashed line – theoretical upper bound $\log_2 |\mathcal{M}|$.

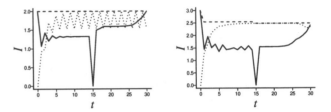

Figure 5: Information flows constructed by the best evolved 4-state (left) and 8-state (right) controller. Solid line – $I(M_0; M_t)$, dotted – $I(M_0; R_t)$, dashed – $I(M_0; M_t, R_t)$.

The agent starts at the center of the grid. We inject $\log_2 |\mathcal{M}|$ bits of information into its controller's internal state M_0 by making its probability distribution uniform. At time step 15 we erase the controller's memory by setting it into state 1. We then measure how much of the information "injected" into M_0 is contained in M_{30}. In terms of information flow we want to find a controller which maximizes the information flow from M_0 to M_{30}. As we erase M_{15}, at time step 15 the flow can pass only via the environment. The amount of flow is measured as $I(M_0; M_{30})$.

5.1 Results and Discussion

We have evolved separate populations of controllers with 2 to 13 states for 1000 generations. Evolution does indeed find controllers capable of *self-stigmergy* (Fig. 4).

To analyze the behavior of evolved controllers we track how the injected information "diffuses" through the memory and the position of the agent by measuring $I(M_0; M_t)$, $I(M_0; R_t)$, and $I(M_0; M_t, R_t)$ (Fig. 5).

The environment used in this experiment is static. The agent offloads information into its own position R, that is into own *relation* with the environment. This phenomenon can be explained by the fact that the environment, as perceived by the controller, is dynamic. The evolved controllers employ this to offload and reacquire information.

6   Stigmergy for Two

In this section we study a classical case of stigmergy where one agent (the *sender*) indirectly communicates with another agent (the *receiver*) by changing the environment.

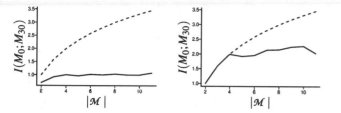

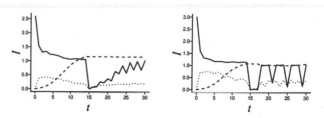

Figure 6: Fitness of best evolved controllers. Left: initial position of the sender is distributed over 11×11 cells, Right: the sender always starts at the source. Dashed line – theoretical upper bound $\log_2 |\mathcal{M}|$.

Figure 7: Information flows constructed by the best evolved 6-state (left) and 8-state (right) controller pairs. Solid line – $I(M_0; M_t)$, dashed – $I(M_0; B_t)$, dotted – $I(R_0; M_t)$.

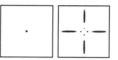

Figure 8: Implicit representations of states in M_0 by box position B_{15} as used by most controllers. Each frame shows a mapping from an initial state onto B_{15}. Intensity of cells reflects the probability of finding the box there: white – zero, black – highest.

To enable the environment to be modified we introduce a pushable box. We denote the position of the box with a random variable B with values in $\mathbb{Z}^2$. When an agent hits the box it is moved one cell in the direction the agent last moved. For example, if the agent hits the box having moved north, the box is moved north one cell. There is *no special sensor for the box*. To facilitate the perception of the box with minimal changes to the model, we make the box emit a signal same way as the source does. Signals from the source and from the box are summed. Therefore, the gradient field depends on the position of the box and in principle enables an agent to capture some information about the position of the box using the gradient sensor.

The sender starts with its initial position R_0 distributed uniformly over a square of 11×11 cells centered at the source. We inject $\log_2 |\mathcal{M}|$ bits of information into its controller's internal state M_0. At time step 15 we remove the sender and introduce the receiver by placing it at the source. The receiver's controller is set to state 1. At time step 30 we measure how much of the information about the sender's memory M_0 is contained in the receiver's memory M_{30}. As the receiver's initial state and position are independent of those of the sender, the information flow from the sender's memory into the recipient's memory at time step 15 can pass only via the position of the box B_{15}.

We use evolution to find *pairs* of controllers maximizing the flow $I(M_0; M_{30})$. Instead of individual controllers the evolutionary algorithm operates on sender-receiver pairs.

6.1 Results and Discussion

We have evolved pairs of controllers with 2 to 11 states for 1000 generations. At most 1 bit of information is stigmergically communicated (Fig. 6, left). More information gets communicated if we perform the same experiment but with the sender always starting at the source (Fig. 6, right).

The information flow from the sender's memory M_0 to the receiver's memory M_{30} at time step 15 goes exclusively via the position of the box B_{15}. Thus the amount of information about M_0 in B_{15} is the maximum the receiver could in principle recover. The flow through the memory ($I(M_0; M_t)$) and the position of the box ($I(M_0, B_t)$) is shown on Fig. 7.

Occasionally the receiver reacquires some information about the sender's initial position R_0. The corresponding information flow from R_0 into memory M_t is visualized on Fig. 7. This finding parallels the fact that one's activity usually leaves traces which might be perceived and exploited by others, potentially resulting in a race between strategies for leaving fewer traces and for discovering them.

To understand how the injected information is represented in the position of the box B_{15} we visualize the mapping $M_0 \mapsto B_{15}$ (Fig. 8). It turns out that virtually all evolved controllers use the same representation with minor differences: depending on M_0 the box either remains in the center of the grid, or is pushed away in one of the four straight lines heading north, east, west or south.

7 Conclusions

We have presented an information-theoretic approach to quantifying information flows in agent-environment interactions. In addition to quantification, the approach allows us to "inject" a piece of information into the system and then track how the information diffuses. An advantage of the information-theoretic quantities is that they ignore representation and deal only with the underlying information. This creates a versatile, powerful and flexible view where information is treated as a measurable commodity.

To show the approach in action we have evolved agent controllers for stigmergic behavior. We have characterized stigmergy as the offloading and reacquisition of Shannon information in the relationship between agents and the environment. Stigmergy thus naturally lends itself to being treated in terms of Shannon information flow between agents. As a testbed we have used a two-dimensional grid world where agents have access to a gradient sensor only. The agents are controlled by finite-state automata controllers with limited amount of memory. Although the model is very simple we believe it pertains to a range of Artificial Life models, especially those concerned with stigmergy.

In the first experiment an agent was provided with information which it had to offload from and later reacquire back into its memory, resulting in a kind of self-stigmergy. At first sight a surprising finding is that the agent was able to perform this task without modifying the environment in the common sense. The environment was static and the agent offloaded information *into its own position* in the environment. In other words, the information was offloaded into the *relation* between the agent and the environment.

In the second experiment one agent (the sender) was provided with information to stigmergically communicate to another agent (the receiver) *by modifying the environment*. Interestingly, in addition to the required information the sender offloaded some extra information about itself, some of which was later occasionally acquired by the recipient. This parallels the fact that one's activity usually leaves traces which might be perceived and exploited by others, potentially resulting in a race between strategies for leaving fewer traces and for discovering them.

The sender could only communicate indirectly by pushing a box. The communicated information was thus stored in the position of the box. Neither the sender nor the receiver had explicit sensors for the position or the proximity of the box. The agents did not evolve any "concept" of the box to stigmergically communicate either, rather they managed to create a suitable information flow by interacting with the environment.[2] This emphasizes the fact that the information-theoretic approach enables us to avoid imposing our own biases or assumptions on the agents and thus potentially allows more efficient solutions to be found.

We use the information-theoretic approach based on tracking information flows in order to understand stigmergy and ultimately more general phenomena in a *quantitative* manner. There is strong evidence that interactions of biological agents are partly due to the need for acquiring, exchanging and concealing information. Therefore it is important that the approach enables us to measure and also construct various information flows without much bias, without imposing our own models on the interactions. We believe this may provide novel insights into adaptive systems.

Acknowledgments

We would like to thank the Condor Team from the University of Wisconsin, whose High Throughput Computing system Condor (http://www.cs.wisc.edu/condor) enabled us to conveniently run large numbers of simulations on ordinary workstations.

---

[2]This approach is consistent with the *enactive* view [16] that considers agents as creating the conditions for their interactions.

References

[1] Mark A. Bedau. Dynamics of the environment for adaptation in static resource models. In J. Kelemen and P. Sośik, editors, *Advances in Artificial Life*, pages 76–85. Springer-Verlag, 2001.

[2] Eric Bonabeau, Marco Dorigo, and Guy Theraulaz. *Swarm Intelligence: From Natural to Artificial Systems*. Oxford University Press Inc., 1999.

[3] Eric Bonabeau, Guy Theraulaz, Jean-Louis Deneubourg, Serge Aron, and Scott Camazine. Self-organization in social insects. *Trends in Ecology & Evolution*, 12(5):188–193, May 1997.

[4] Thomas M. Cover and Joy A. Thomas. *Elements of Information Theory*. John Wiley & Sons, Inc., 1991.

[5] Kerstin Dautenhahn, Daniel Polani, and Thomas Uthmann. Guest editors' introduction: Special issue on sensor evolution. *Artificial Life*, 7(2):95–98, 2001.

[6] James Hollan, Edwin Hutchins, and David Kirsh. Distributed cognition: Toward a new foundation for human-computer interaction research. *ACM Transactions on Computer-Human Interaction*, 7(2):174–196, Jun 2000.

[7] Edwin Hutchins. How a cockpit remembers its speeds. *Cognitive Science*, 19(3):265–288, 1995.

[8] David Kirsh and Paul Maglio. On distinguishing epistemic from pragmatic action. *Cognitive Science*, 18(4):513–549, 1994.

[9] Chrystopher L. Nehaniv. Meaning for observers and agents. In *IEEE International Symposium on Intelligent Control / Intelligent Systems and Semiotics, ISIC/ISAS'99*, pages 435–440. IEEE Press, 1999.

[10] Stefano Nolfi and Davide Marocco. Active perception: A sensorimotor account of object categorization. In B. Hallam, D. Floreano, J. Hallam, G. Hayes, and J-A. Meyer, editors, *From Animals to Animats 7*, pages 266–271. MIT Press, 2002.

[11] Judea Pearl. *Causality: Models, Reasoning, and Inference*. Cambridge University Press, 2001.

[12] Daniel Polani, J. T. Kim, and T. Martinetz. An information-theoretic approach for the quantification of relevance. In J. Kelemen and P. Sośik, editors, *Advances in Artificial Life*, LNCS, pages 704–713. Springer, 2001.

[13] Guy Theraulaz and Eric Bonabeau. Modelling the collective building of complex architectures in social insects with lattice swarms. *Journal of Theoretical Biology*, 177(4):381–400, Dec 1995.

[14] Hugo Touchette and Seth Lloyd. Information-theoretic limits of control. *Phys. Rev. Lett.*, 84:1156, 2000.

[15] Hugo Touchette and Seth Lloyd. Information-theoretic approach to the study of control systems. *Physica A*, 331(1–2):140–172, January 2004.

[16] Francisco J. Varela, Evan Thompson, and Eleanor Rosch. *The Embodied Mind: Cognitive Science and Human Experience*. The MIT Press, 1993.

Ant Foraging Revisited

Liviu A. Panait and Sean Luke
George Mason University, Fairfax, VA 22030
lpanait@cs.gmu.edu, sean@cs.gmu.edu

Abstract

Most previous artificial ant foraging algorithms have to date relied to some degree on *a priori* knowledge of the environment, in the form of explicit gradients generated by the nest, by hard-coding the nest location in an easily-discoverable place, or by imbuing the artificial ants with the knowledge of the nest direction. In contrast, the work presented solves ant foraging problems using two pheromones, one applied when searching for food and the other when returning food items to the nest. This replaces the need to use complicated nest-discovery devices with simpler mechanisms based on pheromone information, which in turn reduces the ant system complexity. The resulting algorithm is orthogonal and simple, yet ants are able to establish increasingly efficient trails from the nest to the food in the presence of obstacles. The algorithm replaces the blind addition of new amounts of pheromones with an adjustment mechanism that resembles dynamic programming.

Introduction

Swarm-behavior algorithms are increasingly popular approaches to clustering and foraging tasks multi-agent and robotics research (Beckers et al., 1994; Deneubourg et al., 1991). We are interested in communication among swarm agents, and the focus of this paper is the application of pheromones to swarm foraging. Ants and other social insects use pheromones very successfully to mark trails connecting the nest to food sources (Holldobler and Wilson, 1990) and to recruit other ants to the foraging task. For example, leafcutter ants are able to organize trails connecting their nest with food sources located as far as hundreds of meters away (Holldobler and Wilson, 1990).

We are interested in the capabilities of pheromones as a guide for artificial agents and robots. In some sense pheromones may be viewed as a mechanism for inter-agent communication that can help reduce the complexity of individual agents. But pheromones cannot be viewed as essentially a blackboard architecture. While pheromones are *global* in that they may be deposited and read by any ant, and generally last long periods of time, they are *local* in that ants may only read or change the pheromone concentrations local to themselves in the environment.

However, nearly all existing pheromone models for artificial agents have eschewed a formalism such as this, concentrating instead on approaches inspired by biology if not biologically plausible. These models assume a single pheromone in order to mark locations to food, and use an ad-hoc mechanism for getting back home (for example, compass information). However, this problem is very easily represented in a symmetric fashion under the assumption of *two pheromones*, one for finding the food and one leading back to the nest. The use of multiple pheromones not only eliminates the ad-hoc nature of previous work, but also permits obstacles and other environment features.

Previous Work

The problem discussed in this paper is known as "central place food foraging", and consists of two phases: leaving the nest to search for food, and returning to the nest laden with food (Sudd and Franks, 1987). There are several ant foraging applications already suggested in the literature. One of the earliest ones (Deneubourg et al., 1990) concerns modeling the path selection decision making process of the Argentine ant *Linepithema humile*. Given two paths of identical lengths connecting the nest and the food source, Deneubourg *et al* show that the ants collectively choose one of the two paths by depositing pheromones over time. When paths of different lengths connect the two sites, the ant system gradually learns to forage along the shortest one. Bonabeau *et al* suggest that this happens because larger amounts of pheromones accumulate on the shorter paths more quickly (Bonabeau, 1996; Bonabeau and Cogne, 1996). Deneubourg *et al* also present a model of the foraging patterns of army ants (Deneubourg et al., 1989). Monte Carlo simulations of their model produces patterns strikingly similar to the ones observed in nature.

All models presented thus far have used a single type of pheromone. A foraging model that uses two kinds of pheromones is presented in (Resnick, 1994): ants deposit one pheromone to mark trails to the food sources, but a second type of pheromone is released by the nest itself. This second pheromone diffuses in the environment and creates a gradient that the ants can follow to locate the nest. Resnick's ants learn to forage from the closest food source. Moreover, when food sources deplete, the ants learn to search for other more distant food sources and establish paths connecting them to the nest, "forgetting" about the previously established trails to depleted sites. Further investigation of this model is presented in (Nakamura and Kurumatani, 1997).

The models discussed so far all make the assumption that the ants have an ad-hoc oracle — essentially a compass — which leads them back to the nest. This assumption is either stated explicitly, or it is adopted implicitly by using specific environment models and transition functions. The main justifications are not computational, but biological: ants are believed to employ sophisticated navigational techniques (including orientation based on landmark memorization and the position of the sun) to return to the nest (Holldobler and Wilson, 1990).

We know of two papers which do not rely on ad-hoc methods to return to the nest. The first such work (Wodrich and Bilchev, 1997) is similar to our own, using two pheromones to establish gradients to the nest and to a food source in the presence of an obstacle. The ants use simple pheromone addition to create trails and so the authors rely on high rates of pheromone evaporation to maintain the gradients. Another work (Vaughan et al., 2000) proposes agent behaviors using *directed pheromones* (pheromones that indicate a direction), and show successful foraging in simulation and on real robots. Aside from using directed pheromones, their work also assumes that the agents can deposit pheromones at any location in the environment, even if they are located far away from it (similar to a random-access memory).

Method

Our goal is to adapt the *notion* of pheromones (if not biologically plausible pheromones per se) as a communication mode for artificial foraging agents. Hence the method described in this paper applies two different pheromones, one used to locate the food source, and the other used to locate the nest. These pheromones are deposited by the ants and may evaporate and diffuse. The algorithm assumes that both types of pheromones can co-exist at the same location.

Upon leaving the nest, ants follow food-pheromone gradients to the food source, but also leave "home" (toward-nest) pheromones behind to indicate the path back home. When the ants are at the food source, they pick up a food item and start heading back to the nest. To do so, they now follow the trail of home pheromones back to the nest, while leaving a trail of food pheromones behind to mark the trail toward the food source.

Ants have one of eight possible orientations: N, NE, E, SE, S, SW, W, NW. When determining where to move, the ants first look at three nearby locations: directly in front, to the front-left, and to the front-right. We will refer to these three locations as the *forward locations*. For example, an

Parameter	Value
Duration of simulation	5000 time steps
Environment	100x100, Non-Toroidal
Nest location and size	(70,70), single location
Food source location and size	(20,20), single location
Maximum ants in simulation	1000
Maximum ants per location	10
Life time for each ant	500
Number of initial ants at nest	2
Ants borne per time step	2
Min amount of pheromone	0.0
Max amount of pheromone	1000.0
Evaporation ratio	0.1%
Diffusion ratio	0.1%
Obstacles	none
K	0.001
N	10.0

Table 1: Parameters for the experiments

ant with orientation N considers its NE, N and NW neighbors. Each location receives a weight based on the amounts of pheromones it contains. Ants searching for food are more likely to transition to locations with more food pheromones, while ants returning to the nest are attracted by more home pheromones. Transition decisions are stochastic to add exploration. An ant does not consider moving to a neighboring location that is too crowded (more than ten ants) or that is covered by an obstacle. If none of the three forward locations are valid, the ant then considers moving to one of the five remaining (non-forward) locations. If none of these locations are valid either, then the ant stays in its current location until the next time step.

At each location visited, ants "top off" the current pheromone level to a desired value. If there is already more pheromone than the desired value, the ant deposits nothing. The desired value is defined as the maximum amount of pheromones in the surrounding eight neighbors, minus a constant. Thus as the ant wanders away from the nest (or food), its desired level of deposited nest (or food) pheromone drops, establishing a gradient. When reaching a goal (nest or food) the desired concentration is set to a maximum value.

The Algorithm

Foraging ants execute the `Ant-Forage` procedure at each time step. The pseudocode is shown next:

```
Ant-Forage
    If HasFoodItem
        Ant-Return-To-Nest
    Else
        Ant-Find-Food-Source
```

```
Ant-Return-To-Nest
    If ant located at food source
        Orientation ← neighbor location with max home
                      pheromones
    X ← forward location with max home pheromones
    If X = NULL
        X ← neighbor location with max home
              pheromones
    If X ≠ NULL
        Drop-Food-Pheromones
        Orientation ← heading to X from current location
        Move to X
        If ant located at nest
            Drop food item
            HasFoodItem ← FALSE

Ant-Find-Food-Source
    If ant located at nest
        Orientation ← neighbor location with max food
                      pheromones
    X ← Select-Location(forward locations)
    If X = NULL
        X ← Select-Location(neighbor locations)
    If X ≠ NULL
        Drop-Home-Pheromones
        Orientation ← heading to X from current location
        Move to X
        If ant located at food source
            Pick up food item
            HasFoodItem ← TRUE

Select-Location(LocSet)
    LocSet ← LocSet - Obstacles
    LocSet ← LocSet - LocationsWithTooManyAnts
    If LocSet = NULL
        Return NULL
    Else
        Select a location from LocSet, where each
            location is chosen with probability
            (K + FoodPheromonesAtLocation)^N

Drop-Home-Pheromones
    If ant located at nest
        Top off home pheromones to maximum level
    Else
        MAX ← max home pheromones of neighbor
               locations
        DES ← MAX - 2
        D ← DES - home pheromones at current location
        If D > 0
            Deposit D home pheromones at current location

Drop-Food-Pheromones
```
Same as Drop-Home-Pheromones, *but modify*
"home"→"food" and "nest"→"food source".

	Ph. Adjust	Ph. Increment
Mean	**10450.70**	4396.04
Std. Dev.	1850.69	2835.90

Table 2: Mean and standard deviation of the food items foraged when adjusting or incrementing the amount of pheromone at the current location.

Evaporation Rate	Mean	Std. Dev.
0.0	9783.38	2477.07
0.001	**10450.70**	1850.69
0.01	1203.98	409.62

Table 3: Mean and standard deviation of the food items foraged when different evaporation rates are used

Experiments

The experiments were performed using the MASON multi-agent system simulator (Luke et al., 2003). Unless stated otherwise, the experiments use the settings described in Table 1. Claims of better or worse performance are verified with a Welch's two-sample statistical test at 95% confidence over samples of 50 runs each.

The first experiment compares the algorithm using the "top off" mechanism with one that simply deposits a fixed amount of pheromones, incrementing the amount existing at the current location. Table 2 shows the means and standard deviations of the amounts of food items collected with the two algorithms. The pheromone adjustment algorithm is significantly better.

Next, we analyzed the sensitivity of the "top off" algorithm to evaporation and diffusion rates (Tables 3 and 4). The results show that the amount of food foraged is relatively sensitive to both parameters. Lower evaporation rates do not significantly affect the results, while higher rates deplete information too rapidly for successful foraging. Diffusion is different: a zero diffusion rate significantly decreases performance. This suggests that without diffusion, pheromone information can spread to neighboring locations only through ant propagation. As expected, too much diffusion again depletes information.

We then investigated the effectiveness of the foraging behaviors at different exploration-exploitation settings. In the Selection-Location procedure, N is an exploitation factor, where larger values influence greedier decisions. The results shown in Table 5 compare the impact of the N parameter on the performance. We found that increasing the greediness led to a direct improvement in performance. We then replaced Select-Location with Greedy-Select-Location to examine extreme greediness ($N \rightarrow \infty$). The revised algorithm is shown here:

Timestep: 0   1100   1600   2600   6600

Figure 1: Foraging sequence with predefined suboptimal path, showing local path optimization.

Diffusion Rate	Mean	Std. Dev.
0.0	6133.2	1159.17
0.001	**10450.70**	1850.69
0.01	6162.52	1145.94

Table 4: Mean and standard deviation of the food items foraged when different diffusion rates are used

N	Mean	Std. Dev.
2	4111.3	648.02
5	8435.36	1514.27
10	**10450.70**	1850.69

Table 5: Mean and standard deviation of the food items foraged when different values for the N parameter are used in the Select-Location procedure. Results are significantly improving with increasing N from 2 to 5 and then 10.

```
Greedy-Select-Location(LocSet)
  LocSet ← LocSet - Obstacles
  LocSet ← LocSet - LocationsWithTooManyAnts
  If LocSet = NULL
    Return NULL
  Select from LocSet the location with max food
      pheromones
```

This increased the total number of food items returned to the nest by about three times (Table 6). However, if the algorithm uses the fixed-incrementing method, greedy selection method shows no statistically significant improvement (a mean of 4926.88, standard deviation of 2425.68 — compare to the fixed-incrementing results in Table 2). Not only does a "topping-off" method do better than more traditional approaches, but it improves with more greedy behavior.

We also observed that the foraging trails for the greedy foraging are usually optimally short and straight. Our final experiment marked a clearly suboptimal path before the simulation, and adjusted the neighborhood transition probabilities to reflect the fact that diagonal transitions are longer than horizontal and vertical ones (this adjustment was used only for this experiment). The progress of the foraging process is presented in Figure 1. The ants collectively change the path

	$(N = 10)$ Select-Neighbor	Greedy-Select-Neighbor
Mean	10450.70	**28369.98**
Std. Dev.	1850.69	849.16

Table 6: Increasing the greediness of the algorithm. The Greedy-Select-Neighbor method shows an increased performance by 171%

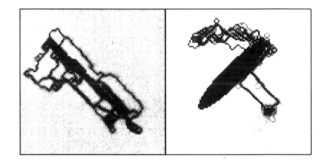
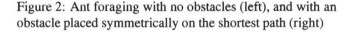

Figure 2: Ant foraging with no obstacles (left), and with an obstacle placed symmetrically on the shortest path (right)

until it becomes optimal, which suggests that our foraging technique also exhibits emergent path optimization.

We have also tested the behavior of the pheromone adjustment algorithm in more difficult domains in our simulator. Two examples are shown in Figure 2. Ants are represented by black dots. Shades of gray represent pheromones: the lighter the shade, the smaller the amount of pheromones. Nest is located in the lower right region, while the food source is in the upper left area.

The left snapshot in Figure 2 shows the foraging process after some 1250 time steps. The nest and food source locations are located close to the two ends of the dense center trail. There are many ants on the shortest path connecting the nest and the food source. Additionally, most space has a light gray color: ants have previously performed an extensive exploration for the food source.

The right snapshot in Figure 2 shows the foraging task in the presence of an obstacle centered on the path. The ants discover a path on one side of the obstacle and concentrate the foraging in that direction. Taken later in the run, the image shows most pheromones have already evap-

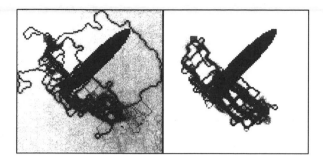

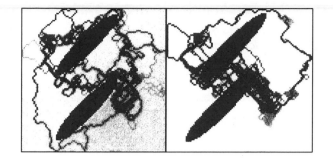

Figure 3: Ant foraging with an obstacle placed non-symmetrically: early (left) and late (right) snapshots of the simulation

Figure 4: Ant foraging with two obstacles: early (left) and late (right) snapshots of the simulation

orated throughout the environment, suggesting that the ants are mainly exploiting the existing food source and they are not exploring for alternatives.

Figure 3 shows the same algorithm applied in an environment where the only obstacle is slightly modified such that one of its sides offers a shorter path than the other does. In an early (left) snapshot, the ants discover several paths on both sides of the obstacle. Later, the ants concentrate on the shorter path and the pheromones disappear in most of the space (right snapshot). It is important to note that, because in this environment, diagonal transitions cost the same as others, most of the paths shown are in fact optimal.

Figure 4 shows the exploration with two obstacles placed such that the shortest path goes between them. Several paths are initially found, but their number decreases with time. If more time were allowed, the longer trail close to the top-right corner of the image would probably disappear.

Conclusions and Future Work

We proposed a new algorithm for foraging inspired from ant colonies which uses multiple pheromones. This model does not rely on ad-hoc methods to return to the nest, and is essentially symmetric. We showed its efficacy at solving basic foraging tasks and its ability to perform path optimization. We then experimented with the effects of different parameter settings (degrees of evaporation and diffusion; greediness) on the rate of food transfer in the model.

The pheromone-adjustment procedure detailed earlier bears some resemblance to Dijkstra's single source shortest path graph algorithm in that both perform essentially the same relaxation procedure. However in the pheromone model, the *selection* of the state transition to relax is not greedily chosen but is determined by the gradient of the *other* pheromone. We also note a relationship between this approach and reinforcement learning methods. In some sense pheromone concentrations may be viewed as state utility values. However we note that reinforcement learning performs backups, iteratively painting the "pheromone" back one step each time the ant traverses the entire length of the

path. In contrast our method paints the pheromone forward every step the ant takes, resulting in a significant speedup. We further explore the similarities between reinforcement learning and pheromone-based foraging in (Panait and Luke, 2004).

For future work we plan to extend the method and apply it to other problem domains involving defense of foraging trails and nest, competition for resources with other ant colonies, and depleting and moving food sources. We are also interested in formal investigations of the properties of the model. In particular, we wish to cast the model in a form based on dynamic programming.

Acknowledgements

We would like to thank Dr. Larry Rockwood, Elena Popovici, Gabriel Balan, Zbigniew Skolicki, Jeff Bassett, Paul Wiegand and Marcel Barbulescu for comments and suggestions.

References

Beckers, R., Holland, O. E., and Deneubourg, J.-L. (1994). From local actions to global tasks: Stigmergy and collective robotics. In *Artificial Life IV: Proceedings of the International Workshop on the Synthesis and Simulation of Living Systems , third edition*. MIT Press.

Bonabeau, E. (1996). Marginally stable swarms are flexible and efficient. *Phys. I France*, pages 309–320.

Bonabeau, E. and Cogne, F. (1996). Oscillation-enhanced adaptability in the vicinity of a bifurcation: The example of foraging in ants. In Maes, P., Mataric, M., Meyer, J., Pollack, J., and Wilson, S., editors, *Proceedings of the Fourth International Conference on Simulation of Adaptive Behavior: From Animals to Animats 4*, pages 537–544. MIT Press.

Deneubourg, J. L., Aron, S., Goss, S., and Pasteels, J. M. (1990). The self-organizing exploratory pattern of the argentine ant. *Insect Behavior*, 3:159–168.

Deneubourg, J. L., Goss, S., Franks, N., Sendova-Franks, A., Detrain, C., and Chretien, L. (1991). The dynamics of collective sorting: Robot-like ants and ant-like robots. In *From Animals to Animats: Proceedings of the First International Conference on Simulation of Adaptive Behavior*, pages 356–363. MIT Press.

Deneubourg, J. L., Goss, S., Franks, N. R., and Pasteels, J. (1989). The blind leading the blind: Modeling chemically mediated army ant raid patterns. *Insect Behavior*, 2:719–725.

Holldobler, B. and Wilson, E. O. (1990). *The Ants*. Harvard University Press.

Luke, S., Balan, G. C., Panait, L. A., Cioffi-Revilla, C., and Paus, S. (2003). MASON: A Java multi-agent simulation library. In *Proceedings of Agent 2003 Conference on Challenges in Social Simulation*.

Nakamura, M. and Kurumatani, K. (1997). Formation mechanism of pheromone pattern and control of foraging behavior in an ant colony model. In Langton, C. G. and Shimohara, K., editors, *Artificial Life V: Proceedings of the Fifth International Workshop on the Synthesis and Simulation of Living Systems*, pages 67–76. MIT Press.

Panait, L. and Luke, S. (2004). A pheromone-based utility model for collaborative foraging. In *Proceedings of the Third International Joint Conference on Autonomous Agents and Multi Agent Systems (AAMAS-2004)*.

Resnick, M. (1994). *Turtles, Termites and Traffic Jams*. MIT Press.

Sudd, J. H. and Franks, N. R. (1987). *The Behavioral Ecology of Ants*. Chapman & Hall, New York.

Vaughan, R. T., Støy, K., Sukhatme, G. S., and Mataric, M. J. (2000). Whistling in the dark: Cooperative trail following in uncertain localization space. In Sierra, C., Gini, M., and Rosenschein, J. S., editors, *Proceedings of the Fourth International Conference on Autonomous Agents*, pages 187–194. ACM Press.

Wodrich, M. and Bilchev, G. (1997). Cooperative distributed search: The ants' way. *Control and Cybernetics*, 26.

Learning Ant Foraging Behaviors

Liviu A. Panait and Sean Luke
George Mason University, Fairfax, VA 22030
lpanait@cs.gmu.edu, sean@cs.gmu.edu

Abstract

Insects are good at cooperatively solving many complex tasks. For example, foraging for food far away from a nest can be solved through relatively simple behaviors in combination with pheromones. As task complexity increases, however, it may become difficult to find individual agent rules which yield a desired emergent cooperative behavior, or to know if any such rules exist at all. For such tasks, machine learning techniques like evolutionary computation (EC) may prove a valuable approach to searching the space of possible rule combinations. This paper presents an application of genetic programming to search for foraging behaviors. The learned foraging behaviors use only pheromone information to find the path to the nest and to the food source.

Introduction

Artificial Intelligence has drawn many ideas from biology: evolutionary computation, neural networks, robotics, vision, and cooperative problem solving all steal liberally from Mother Nature. One such area of particular recent interest in AI has been in algorithms inspired from social insects such as ants, termites and bees. The interest stems from the capacity of such simple organisms to collaboratively work together to solve problems no one individual could. Some social-insect-inspired AI literature has focused on foraging and related tasks through the use of pheromones (Bonabeau et al., 1999). The social memory mechanism of pheromones is an inviting paradigm for designing multiagent systems with blackboards, joint utility tables, and other global memory mechanisms. However, hand-coding of agent behaviors using this paradigm can prove problematic given the unexpected emergent group behaviors that arise.

While previous work has applied machine learning methods to the *use* of pheromone information, they have still tended to hard-code the pheromone depositing procedure. In contrast, this paper shows that it is possible to have the entire foraging behavior discovered by the learning system.

The paper proceeds with a description of previous work in learning foraging behaviors, and a description of an evolutionary computation approach to the learning task. A set of three experiments in increasingly difficult environments shows that good foraging behaviors can be discovered. A later experiment shows that behaviors learned for complex domains are robust to simpler environments.

Previous Work

The specific problem at hand is called *central place food foraging*, and it consists of two main phases: an initial exploration for food, followed by carrying it back to the nest (Sudd and Franks, 1987). When an ant reaches a food source, the ant automatically becomes laden with food; and when the ant reaches the nest, it automatically drops the food off at the nest.

Various learning algorithms have been used to attack this problem. Some algorithms related to reinforcement learning adopt a fixed pheromone laying procedure, then use the sensed pheromone information to explore the space or to update state-action utility estimates (Leerink et al., 1995; Monekosso et al., 2002). Evolutionary computation techniques have also been applied to learn exploration/exploitation strategies using pheromones deposited by hardcoded mechanisms. For example, Sauter et al show how EC can be used to tune the action-selection behavior in an application involving multiple "digital" pheromones (Sauter et al., 2002). A similar idea applied to network routing is presented in (White et al., 1998).

AntFarm (Collins and Jefferson, 1992) is another system that combines communication via pheromones and evolutionary computation, and it is the closest work to the algorithm presented in this paper. AntFarm uses multiple colonies of homogeneous ants, with each colony in a separate 16x16 grid environment. The ants use a single pheromone to mark trails to food sources, but use a compass to point themselves along the shortest path back to the nest. The system uses evolutionary computation to search for foraging behaviors represented as neural networks.

A trend common to all previously described algorithms is that the ants *know* how to return to the nest. This assumption is mainly based on observations that ants use sophisticated navigational techniques for this task, including orientation based on landmark memorization or using the position of

Pheromone Depositing Tree Function	Description
scalar ←CurFoodPhLevel()	Food pheromone at my location
scalar ←CurHomePhLevel()	Home pheromone at my location
scalar ←LastDeposited()	How much pheromone I deposited last time
scalar ←DistanceFromSite()	Number of time steps elapsed since I last visited the nest (or food, depending on state)
scalar ←MaxDistanceFromSite()	Max possible distance from site (depends on the maximum lifetime of ants)
scalar ←MaxLocalFoodPheromone()	Max food pheromone at my eight neighboring locations
scalar ←MinLocalFoodPheromone()	Min food pheromone at my eight neighboring locations
scalar ←MaxLocalHomePheromone()	Max home pheromone at my eight neighboring locations
scalar ←MinLocalHomePheromone()	Min home pheromone at my eight neighboring locations
scalar ←MaxPheromone()	Max amount of pheromone possible
scalar ←MaxPhomoneDividedByMaxDistanceFromSite()	MaxPheromone() / DistanceFromSite()
scalar ←MaxPheromoneDividedByMaxDistanceFromSite()	MaxPheromone() / MaxDistanceFromSite()
scalar ←Add(*scalar, scalar*)	Add two scalars
scalar ←Sub(*scalar, scalar*)	Subtract two scalars
scalar ←Max(*scalar, scalar*)	Maximum of two scalars
scalar ←Min(*scalar, scalar*)	Minimum of two scalars

Behavior Selection Tree Function	Description
vector ←FoodPheromones()	Amounts of food pheromones at the eight neighboring locations
vector ←HomePheromones()	Amounts of home pheromones at the eight neighboring locations
vector ←AddV(*vector, vector*)	Add two vectors
vector ←SubV(*vector, vector*)	Subtract two vectors
vector ←Mul2V(*vector*)	Multiply each component of a vector by 2
vector ←Div2V(*vector*)	Divide each component of a vector by 2
vector ←SqrV(*vector*)	Square each component of a vector
vector ←Sqrt(*vector*)	Take the square root of each component of a vector
direction ←MinO(*vector*)	Return the index of the smallest component of a vector
direction ←MaxO(*vector*)	Return the index of the largest component of a vector
direction ←ProbO(*vector*)	Return a random index, chosen using the normalized component sizes as probabilities (+ .001)

Table 1: Function set for an ant's GP pheromone-depositing and behavior-selection trees. Functions depicted take the form of *returnType*←functionName(*argumentTypes*). Leaf nodes have no arguments.

the sun (Hölldobler and Wilson, 1990). However, we argue thzat most current robotics applications are still far from that level of sophistication. Moreover, the discovery of pure-pheromone behaviors is appealing in that its analysis seems more likely to lead to useful applications of pheromone-like global memories to problems for which such "hand-coded" hacks are of less utility. Last, by using *only* pheromone functions, we hope to move towards a formal description of the system as a variation of dynamic programming.

The work in this paper is concerned with learning foraging behaviors that can find food and nest locations in relatively simple, obstacle-free environments. In an accompanying paper (Panait and Luke, 2004a), we present a hard-coded ant foraging algorithm for environments with obstacles.

Evolving Foraging Strategies

To evolve ant behaviors, we used a form of EC known as "strongly-typed" genetic programming (GP) (Koza, 1992; Montana, 1995). In the common form of genetic programming, which we adopted, evolutionary individuals (candidate solutions) use a parse tree structure representation. Leaf nodes in the parse tree return various external state values

for the ant. Internal nodes in the tree are passed by their children and return the result of some function applied to those values. Crossover swaps subtrees among individuals. In strongly-typed GP, type constraints specify which nodes may be children to various other nodes: we used strong typing to enable a large set of available functions operating on vector, scalar, and directional information. Even so, the representational complexity available to the GP learner was significantly less than that afforded in the hand-coded design presented in our accompanying poster paper.

The GP system uses three data types. The first data type is *scalar*, representing any real valued information (for example, the level of food pheromones at the current location). A second data type is *vector*, representing a collection of related scalar values (such as the food pheromone levels in neighboring locations). The third data type is *orientation*, used by ants to decide to which neighboring location (of the possible eight) to move next.

In our approach, a GP individual consists of two trees: the pheromone-depositing tree and the behavior-selection tree. An ant is in one of two states: either he is laden with food, or he is not. The pheromone-depositing tree tells the ant

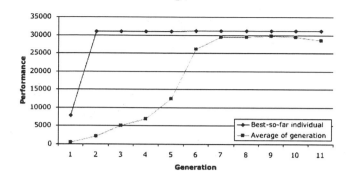

Figure 1: Evolution of performance in the 10x10 grid world.

Figure 2: The emergent foraging pattern for `LearnedBehavior10x10`. Circles represent ants, and squares represent the locations of the nest and the food source.

`LearnedBehavior10x10`
    If carrying a food item
        Set the amount of food pheromones to *MaxDistanceFromSite − DistanceFromSite*
        Move to the neighboring location with most home pheromones
    Else
        Set the amount of home pheromones to *MaxDistanceFromSite − DistanceFromSite*
        Move to the neighboring location with most food pheromones

Figure 3: The learned behavior in the 10x10 environment.

how much pheromone to deposit; but the ant's state tells it *which* pheromone to deposit. Additionally, the trees consist of nodes labeled by a given pheromone name (for example, *MaxLocalHomePheromone* for the pheromone to the "home", or nest). These labels are correct when the ant is *not* laden with food, but when the ant has food, the pheromones actually dealt with by these nodes are swapped[1]. Thus for example, when the ant is laden with food *MaxLocalHomePheromone* actually returns the max value of the local pheromone to the *food*, and not the nest.

The root node of the pheromone-depositing tree returns a scalar value (the pheromone to deposit); the absolute value is always used. The root node of the behavior-selection tree returns a direction. Accordingly, these two trees are constructed out of two different sets of nodes. The sets are shown in Table 1. The functions shown are admittedly simple; but for a first attempt we felt this was reasonable.

The algorithm the learning ants followed is:

---

[1]We used this symmetry in the foraging task to reduce the size of the search space.

`Foraging-Behavior`
    Call the first tree to select the desired level of pheromones
    Call the second tree to select where to move next
    Deposit pheromones and move to desired location

The experiments were implemented using the MASON (Luke et al., 2003) multi-agent simulator and the ECJ (Luke, 2002) evolutionary computation framework. The parameters for the EC system were: elitism of size 2, 100 individuals per population, minimum/maximum depth for Ramped Half-and-Half tree generation of 2/4, minimum/maximum depth for Grow tree generation of 3/3, and re-attempting unsuccessful crossover operations 100 times before giving up and returning the original parents. All other parameters have default values as specified in (Koza, 1992). The fitness of an individual is computed as the average performance of three trials. The performance in each trial is calculated as $FoodPickedUp + 10 * FoodCarriedToNest$. The parameters for the multiagent foraging simulation are: minimum/maximum amount of a given type of pheromone per location of 0/100, evaporation rate of 0.1%, and diffusion rate of 0.1%. Demanding simulations resulted in an extremely slow evolutionary process, which limited the cur-

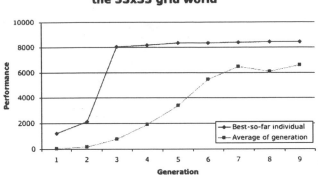

Figure 4: Evolution of performance in the 33x33 grid world.

Figure 5: The emergent foraging pattern for `LearnedBehavior33x33`. Dots represent ants, and shades of grey indicate concentration of pheromones.

`LearnedBehavior33x33`

   If carrying a food item

      Set the amount of food pheromones to *MaxPheromoneDividedByDistanceFromSite*

      Move to the neighboring location with minimum value for $FoodPheromones - 3*HomePheromones$

   Else

      Set the amount of home pheromones to *MaxPheromoneDividedByDistanceFromSite*

      Move to the neighboring location with minimum value for $HomePheromones - 3*FoodPheromones$

Figure 6: The learned behavior in the 33x33 environment.

rent experiment to only a single run consisting of few generations. Additional runs are required to make statistically significant conclusions; so this work should be considered proof-of-concept only.

Likewise, the proof-of-concept experimentation presented in this paper relies on three assumptions primarily as a simplification of the search space. Other work of ours (Panait and Luke, 2004a,b) has eliminated these assumptions for hand-coded ant algorithms, and we have no reason to doubt that such elimination would be problematic for an evolutionary method in future work. The first assumption is that the agents can move to any of the eight neighboring locations (this eliminates obstacles and requires the world to be toroidal). Second, ants die and new ants are created at the nest. Third, ants cannot only add, but can also *remove* pheromones from the environment (the concept of anti-pheromones was previously used in (Montgomery and Randall, 2002) to improve exploration and help the system escape from local optima).

Experiments

The first experiment concerned learning foraging behaviors in a small 10x10 toroidal grid world. Other parameters for

the ant foraging simulation were as follows: 501 simulation steps, food source located at (5,3), nest located at (7,7), ant lifespan of 50 simulation steps, one initial ant, one new ant per time step, and maximum 50 ants in simulation at each time step.

The performance of the best-so-far individuals and the average performance per generation are plotted in Figure 1. The graph shows that a good solution is discovered relatively easily, within two generations. The behavior of a well performing forager, as well as an emergent foraging trail it creates, are shown in Figures 2 and 3.

The second experiment concerned learning foraging behaviors in a larger 33x33 grid world. Other parameters for the ant foraging simulation were as follows: 1001 simulation steps, food source located at (17,10), nest located at (23,23), ant lifespan of 50 simulation steps, one initial ant, one new ant per time step, and maximum 50 ants in simulation at each time step.

The performance of the best-so-far individuals and the average performance per generation are plotted in Figure 4. The graph shows that a good solution is still discovered relatively easily, within three generations. The new individual contains a simple, but useful formula for exploring: when

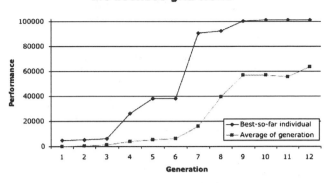

Figure 7: Evolution of performance in the 100x100 grid world.

Figure 8: The emergent foraging pattern for `LearnedBehavior100x100`. Dots represent ants, and shades of grey indicate concentration of pheromones.

`LearnedBehavior100x100`

    If carrying a food item

        Set the amount of food pheromones to *Max(MinLocalFoodPheromone, MaxPheromoneDividedByDistanceFromSite)*

        Move to the neighboring location with minimum value for *FoodPheromones−2∗HomePheromones*

    Else

        Set the amount of home pheromones to *Min(MinLocalHomePheromone, MaxPheromoneDividedByDistanceFromSite)*

        Move to the neighboring location with minimum value for *HomePheromones−2∗FoodPheromones*

Figure 9: The learned behavior in the 100x100 environment.

searching for the food source, advance towards more food pheromones and also less nest pheromones. This improves the initial search process by guiding the ants away from the nest, and it represents an interesting alternative to the exploration strategy in our accompanying paper (Panait and Luke, 2004a). A behavior of a well performing ant forager is shown in Figure 6, and an emergent foraging trail in one application of the specific learned foraging behavior is presented in Figure 5.

The third experiment concerned learning foraging behaviors in a 100x100 grid world. Other parameters for the ant foraging simulation were as follows: 2501 simulation steps, food source located at (50,30), nest located at (70,70), ant lifespan of 500 simulation steps, one initial ant, one new ant per time step, and maximum 500 ants in simulation at each time step.

The performance of the best-so-far individuals and the average performance per generation are plotted in Figure 7. The graph shows continual improvement over the first nine generations, suggesting incrementally more complex foraging strategies are discovered. A relatively similar, but somewhat more complex individual was discovered; its behavioral algorithm is presented in Figure 9. Additionally, Figure

8 presents an emergent foraging trail obtained when using this behavior for the ants, and it shows how most of the ants have converged on a straight trail connecting the nest to the food source.

In the fourth experiment, we took the best evolved individuals from each of the three previous experiments and tested them in all three environments. For each individual and each grid size, we performed 10 runs. The results are shown in Table 2.

As can be seen, the more difficult the training problem domain, the more general the solutions (they perform equally well to other solutions specifically evolved for simpler domains). This suggests that for simpler problems, there is no learning gradient toward more sophisticated foraging behaviors. Rather, simple enough such strategies perform as well as more advanced strategies, and the learning system is not capable to distinguish among them. However, as the problem domain becomes more and more challenging, increasingly general foraging strategies are discovered. Additional experiments are required to support this hypothesis.

	10x10 environment	33x33 environment	100x100 environment
LearnedBehavior10x10	**2801.00** (27.91)	113.80 (281.78)	2.20 (3.52)
LearnedBehavior33x33	**2800.50** (38.65)	**929.80** (17.67)	3958.50 (2808.00)
LearnedBehavior100x100	**2802.90** (26.77)	**931.90** (18.10)	**7098.90** (1636.22)

Table 2: The performance of evolved foraging behaviors across the three grid worlds. Numbers (mean performance, standard deviation in parentheses) summarize food items returned to the nest in 10 runs. Bold numbers represent statistically significantly better performance (95% confidence) for a given grid size (down a column).

Conclusions and Future Work

This paper presented a successful application of evolutionary learning to search the space of foraging behaviors. The results of the learning approach are agent behaviors capable of learning by themselves: they use pheromones to mark trails connecting the nest to the food source such that they can navigate faster between the two sites.

Additionally, the results suggest that behaviors learned in more difficult domains also have good performance in easier ones. However, the opposite does not hold: behaviors learned with for easier settings have significantly worse performance when tested on more difficult problems.

Our future work will analyze representational bias, eliminate some of the simplifying assumptions in our model, and apply the learning approach to more complex domains possibly involving obstacles, multiple possibly-decaying food sources, and predator agents, all of which may require the system to develop specialized behaviors for different ants.

Acknowledgments

The authors would like to thank Elena Popovici, Gabriel Balan, Zbigniew Skolicki, Jeff Bassett, Paul Wiegand and Marcel Barbulescu for discussions and suggestions related to the ant algorithms.

References

Bonabeau, E., Dorigo, M., and Theraulaz, G. (1999). *Swarm Intelligence: From Natural to Artificial Systems*. Santa Fe Institute Studies in the Sciences of Complexity. Oxford University Press.

Collins, R. J. and Jefferson, D. R. (1992). Antfarm : Towards simulated evolution. In Langton, C. G., Taylor, C., Farmer, J. D., and Rasmussen, S., editors, *Artificial Life II*, pages 579–601. Addison-Wesley, Redwood City, CA.

Hölldobler, B. and Wilson, E. O. (1990). *The Ants*. Harvard University Press.

Koza, J. (1992). *Genetic Programming: on the Programming of Computers by Means of Natural Selection*. MIT Press.

Leerink, L. R., Schultz, S. R., and Jabri, M. A. (1995). A reinforcement learning exploration strategy based on ant foraging mechanisms. In *Proceedings of the Sixth Australian Conference on Neural Networks*, Sydney, Australia.

Luke, S. (2002). ECJ 9: A Java EC research system. http://www.cs.umd.edu/projects/plus/ec/ecj/.

Luke, S., Balan, G. C., Panait, L. A., Cioffi-Revilla, C., and Paus, S. (2003). MASON: A Java multi-agent simulation library. In *Proceedings of Agent 2003 Conference on Challenges in Social Simulation*.

Monekosso, N., Remagnino, P., and Szarowicz, A. (2002). An improved q-learning algorithm using synthetic pheromones. In B. Dunin-Keplicz, E. N., editor, *From Theory to Practice in Multi-Agent Systems*, Lecture Notes in Artificial Intelligence LNAI-2296. Springer-Verlag.

Montana, D. J. (1995). Strongly typed genetic programming. *Evolutionary Computation*, 3:199–230.

Montgomery, J. and Randall, M. (2002). Anti-Pheromone as a Tool for Better Exploration of Search Spaces. In et al, M. D., editor, *Ant Algorithms: Third International Workshop (ANTS 2002)*, Lecture Notes in Computer Science LNCS 2463. Springer-Verlag.

Panait, L. and Luke, S. (2004a). Ant foraging revisited. In *Proceedings of the Ninth International Conference on the Simulation and Synthesis of Living Systems (ALIFE9)*.

Panait, L. and Luke, S. (2004b). A pheromone-based utility model for collaborative foraging. In *Proceedings of the Third International Joint Conference on Autonomous Agents and Multi Agent Systems (AAMAS-2004)*.

Sauter, J., Matthews, R. S., Parunak, H. V. D., and Brueckner, S. (2002). Evolving adaptive pheromone path planning mechanisms. In *Proceedings of First International Joint Conference on Autonomous Agents and Multi-Agent Systems (AAMAS-02)*, pages 434–440.

Sudd, J. H. and Franks, N. R. (1987). *The Behavioral Ecology of Ants*. Chapman & Hall, New York.

White, T., Pagurek, B., and Oppacher, F. (1998). ASGA : Improving the ant system by integration with genetic algorithms. In et al, J. R. K., editor, *Genetic Programming 1998: Proceedings of the Third Annual Conference*, pages 610–617. Morgan Kaufmann.

Systems Biology Thought Experiments in Human Genetics Using Artificial Life and Grammatical Evolution

Bill C. White and Jason H. Moore

Center for Human Genetics Research, Vanderbilt University, Nashville, TN 37323-0700
{bwhite, moore}@chgr.mc.vanderbilt.edu

Abstract

A goal of systems biology and human genetics is to understand how DNA sequence variations impact human health through a hierarchy of biochemical, metabolic, and physiological systems. We present here a proof-of-principle study that demonstrates how artificial life in the form of agent-based simulation can be used to generate hypothetical systems biology models that are consistent with pre-defined genetic models of disease susceptibility. Here, an evolutionary computing strategy called grammatical evolution is utilized to discover artificial life models. The goal of these studies is to perform thought experiments about the nature of complex biological systems that are consistent with genetic models of disease susceptibility. It is anticipated that the utility of this approach will be the generation of biological hypotheses that can then be tested using experimental systems.

Introduction

Human genetics is largely concerned with the relationship between DNA sequence variations and measures of human health. Genotypes influence phenotypes through a hierarchical network of biochemical, metabolic, and physiological systems in the context of environmental exposure. Systems biology is an emerging discipline focused on developing comprehensive laboratory and analytical strategies for understanding the complex biological systems that complete the genotype to phenotype mapping relationship (Ideker et al. 2001). The promise of a joint human genetics and systems biology approach is improved human health through advances in disease diagnosis, prevention, and treatment.

Understanding the genetic architecture of common human diseases such as essential hypertension requires a research strategy that embraces, rather than ignores, complexity due to nonlinear gene-gene interactions or epistasis (Moore and Williams 2002). The biological definition of epistasis is one gene standing upon or masking the effects of another gene (Bateson 1909) while the statistical definition is a deviation from additivity in a linear model (Fisher 1918). We have presented the working hypothesis that epistasis is a ubiquitous component of the genetic architecture of common human diseases (Moore 2003).

There are several general systems biology strategies that can be utilized to study the genetics of susceptibility to common human diseases. One strategy relies on the collection and integrated analysis of genetic, genomic, and proteomic data from complex biological systems (e.g. Reif et al. 2004). A second strategy relies on perturbation of complex biological systems in model organisms (Jansen 2003). A third strategy relies on computer simulations for carrying out thought experiments that can be used to generate testable hypotheses (e.g. Di Paolo et al. 2000). In the present study, we explore the utility of using artificial life in the form of agent-based modeling for carrying out systems biology thought experiments for generating hypotheses about the mapping relationship between genotype and susceptibility to common disease.

Our previous work in this area has focused on the use of a discrete dynamical systems modeling tool called Petri nets. Petri nets are a type of directed graph (Desel and Juhas 2001) that have been used to model biochemical systems (Goss and Pecoud 1998). We utilized an evolutionary computing strategy called grammatical evolution to discover Petri net models of hypothetical biochemical systems that are consistent with a fixed genetic model of disease susceptibility (Moore and Hahn 2003a, 2003b, 2004a, 2004b). These proof-of-principle studies provided evidence that it is possible to routinely generate discrete dynamic systems models that are consistent with genetic models in which disease susceptibility is dependent on nonlinear interactions among genotypes from two or three DNA sequence variations.

While the Petri net approach has been very successful, it is our conjecture that agent-based modeling will provide more flexibility for carrying out thought experiments. That is, we anticipate a system comprised of agents and rules for their physical interaction will provide a wider range of possible system behaviors than that afforded by the Petri nets. To this end, we provide here an initial proof-of-principle study that demonstrates an artificial life modeling strategy is capable of generating systems biology models that are consistent with a defined genetic model. This interdisciplinary study brings together concepts from several disparate fields that are summarized in the next several sections. We first review the nonlinear gene-gene interaction models from which the systems biology models are derived. Next, we review agent-based simulation and then our agent-based modeling strategy. Finally, we review our evolutionary computing strategy for model discovery that is based on grammatical evolution. The final sections of the paper present the research results and a discussion with future directions.

The Nonlinear Gene-Gene Interaction Models

Our two high-order, nonlinear, gene-gene interaction (i.e. epistasis) models are based on penetrance functions. Penetrance functions represent one approach to modeling the relationship between genetic variations and risk of disease. Penetrance is simply the probability (P) of disease (D) given a particular combination of genotypes (G) that was inherited (i.e. P[D|G]). Figure 1 illustrates the penetrance functions used for Models 1 and 2, respectively. Each model has been described previously (Frankel and Schork 1996; Li and Reich 2000; Moore et al. 2002). What makes these models interesting is that disease risk is dependent on each particular combination of all three genotypes inherited. Each single genotype has effectively no main effect on disease risk.

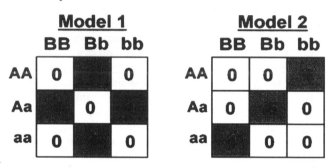

Figure 1: Gene-gene interaction Models 1 and 2.

Agent Based Simulation

Agent based simulation (ABS) refers to a branch of artificial intelligence in computer science that is concerned with multiple, autonomous, interacting computing elements and their emergent behavior (d'Inverno 2001, Woolridge 2002). While the idea of what an agent is and does is not well defined, agents used in this study follow the notion of "weak agency" as put forth by Wooldridge and Jennings (1995) in which agents are defined as being autonomous, interacting with other agents, reacting to their environment and pursuing their own goals in self interest.

Agents in our simulations move and collide with each other on a grid of fixed size. Boundary conditions are handled by allowing wraparound at the grid edges, that is, the world is a toroidial grid. Agents begin in a random spatial configuration with predefined move and collision behaviors and end in a final state after a specified number of time steps. Some agents move in fixed ways, while others are dependent on global environmental conditions specified at the beginning of the simulation, specifically whether agent behavior is dependent on characteristics of the system being modeled. Agent interaction (therefore communication) happens via collisions. Agents can sense whether the location they move to has an agent already

occupying that location. If so, the agent is said to have collided with the existing agent, and it reacts using its collision rule. Agents are allowed to occupy the same space; therefore, infinite move-collision recursions are avoided. In cases where no collision is detected, agents continue to move according to their move rules, which define their goals. The expected emergent behavior relates agent interaction dynamics to the system being modeled.

Our Agent Based Simulation Modeling Strategy

Moore and Hahn (2003a, 2003b, 2004a, 2004b) developed a strategy for identifying discrete, dynamic models of biochemical systems that are consistent with observed gene-gene interactions that define disease susceptibility. The specific systems used to model the biochemical pathways were Petri nets with time (Merlin 1974, Ramchandani 1974).

The goal of identifying discrete models of biochemical systems that are consistent with observed population-level gene-gene interactions is accomplished here by developing agent based simulations that are dependent on specific genotypes from two DNA sequence variations. We allow movement and collision behavior to be genotype-dependent; therefore, simulations can yield different ending configurations. Each agent based simulation model is related to the genetic model using a discrete version of the threshold model from population genetics (Falconer and Mackay, 1996). With a classic threshold or liability model, it is the concentration of a biochemical or environmental substance that is related to the risk of disease, under the hypothesis that risk of disease is greatly increased once a particular substance exceeds some threshold concentration. Conversely, the risk of disease may increase in the absence of a particular factor or with any significant deviation from a reference level. In such cases, high or low levels are associated with high risk while an intermediate level is associated with low risk. Here, we use a discrete version of this model for our deterministic ABS. For each model, the number of agents at a particular space of the simulation world is recorded, and if they exceed a certain threshold, the appropriate risk assignment is made. If the number of agents does not exceed the threshold, the alternative risk assignment is made. The high-risk and low-risk assignments made by the discrete threshold from the output of the ABS can then be compared to the high-risk and low-risk genotypes from the genetic model. A perfect match indicates the ABS model is consistent with the gene-gene interactions observed in the genetic model. The ABS then becomes a model that relates the DNA sequence variations to risk of disease through an intermediate biochemical network.

Identifying ABS models that are consistent with the genotype-dependent distribution of risk is challenging by trial and error. Therefore, we developed an evolutionary computing approach to the discovery of ABS models. This approach is described in the next section.

A Grammatical Evolution Approach to Discovering Agent Based Simulation Models

Overview of Grammatical Evolution

Evolutionary computation arose from early work on evolutionary programming (Fogel 1962, Fogel et al. 1966) and evolution strategies (Rechenberg, 1964, Schwefel 1965) that used simulated evolution for artificial intelligence. The focus on representations at the genotypic level lead to the development of genetic algorithms by Holland (1962, 1975) and others. Genetic algorithms have become a popular machine intelligence strategy because they can be effective for implementing parallel searches of rugged fitness landscapes (Goldberg 1989). Briefly, this is accomplished by generating a random population of models or solutions, evaluating their ability to solve the problem at hand, selecting the best models or solutions, and generating variability in these models by exchanging model components among different models. The process of selecting models and introducing variability is iterated until an optimal model is identified or some termination criteria are satisfied. Koza (1992) developed a variation on genetic algorithms called genetic programming where the models or solutions are represented by tree structures that are in turn executed as computer programs. Koza (2001) and others (Kitagawa 2003) have applied genetic programming to modeling metabolic networks.

Grammatical evolution has been described by O'Neill and Ryan (2001, 2003) as a variation on genetic programming. Here, a grammar is specified that allows a computer program or model to be constructed by a simple genetic algorithm operating on an array of numbers called "codons." The evolved codons select grammar elements in the derivation of a valid sentence in the language specified by the grammar. For our purposes the language L is all valid ABS configurations. This approach is appealing because only a text file specifying the grammar needs to be altered for different simulations, that is, as long as the grammar specifies valid sentences of the language L(G). There is no need to modify and recompile source code during development once the fitness function is specified. The end result is a decrease in development time and an increase in computational flexibility.

A Grammar for Agent Based Simulation Models in Backus-Naur Form

Backus-Naur Form (BNF) is a formal notation for describing the syntax of a context-free grammar as a set of production rules that consist of terminals and nonterminals (Hopcroft and Ulman 1979). Nonterminals form the left-hand side of production rules while both terminals and nonterminals form the right-hand side. A terminal is a simulation/model element or parameter, and a nonterminal is the name of a production rule. Use of nonterminals in the right-hand side of production rules allows for recursion, deriving more complex sentences, thus

simulations, by expanding these nonterminals recursively. For the ABS models, the terminal set includes, for example, the basic building blocks of an ABS: agents and their movement and collision behavior rules. The nonterminal set includes the names of production rules that construct the ABS. For example, a nonterminal might name a production rule for determining whether an agent has movement and collision behavior that is fixed or genotype-dependent. We show in (1) below the production rule that is the start symbol to begin the derivation and thus the ABS configuration. In the grammar rules, variables shown in all capital letters represent terminals and variables contained within angle brackets represent nonterminals.

<simulation>	::=	NUM_TIMESTEPS = <constant > GRID_SIZE = <constant> <constant> <statement_list>	(1)

The <simulation> production rule specifies that a valid simulation configuration must have a number of timesteps, a grid/world size (fixed here to 12 by 12) and a statement list. The nonterminal <statement_list> is a production rule that allows the ABS to grow. The production rule for <statement_list> is shown below in (2).

<statement_list>	::=	<statement>	<statement> <statement_list>	(2)

Here, the BNF symbol "|" separates choices in the substitution for <statement_list> in the derivation of valid ABS configurations. The specific choice is determined by applying the modulus operator to the current genetic algorithm codon. In the case of <statement_list>, the genetic algorithm codon would be applied modulo two, making the choice equally probable. If the second of the alternatives is chosen, the derived configuration is extended by substituting an instance of the <statement_list> nonterminal itself. This process can repeat recursively until the first alternative is chosen and a <statement> nonterminal ends the recursion.

A <statement>, shown in (3) below, defines an agent in the ABS; therefore, rule (2) above allows the simulation to have one or more agents.

<statement>	::=	AGENT MOVE <move_rule> COLLIDE <collision_rule>	(3)

Agent move and collision behavior is then defined by the grammar to allow for many types of movement and interaction, including becoming stuck, that is, an agent can become an obstacle. The grammar for all behaviors is too large to reproduce here. A full grammar can be obtained from the authors upon request.

The Fitness Function

Once an ABS model is constructed using the BNF grammar, as dictated by the genetic algorithm chromosome (vector of codons), the model fitness is determined. Similar to the Petri net approach described by Moore and Hahn (2003a, 2003b, 2004a, 2004b), this is carried out by executing the ABS model for each combination of genotypes in the genetic model and comparing the final agent counts at a defined quadrant of the grid world to a threshold constant to determine the risk assignment. Let G be the set of $i = 1$ to n possible genotype combinations where n = 9 when there are two DNA sequence variations, each with three genotypes. Let Z_i be the final number of agents from the designated ABS quadrant for the ith genotype combination and let c be the threshold constant. Let $d(G_i)$ be the risk assignment for the ith genotype combination in the genetic model and let $f(G_i)$ be the risk assignment made by the ABS. If $Z_i \geq c$ then $f(G_i)$ = "high risk" else if $Z_i < c$ then $f(G_i)$ = "low risk". The dichotomous risk assignment is consistent with epidemiological study designs in which subjects with the disease (cases) and subjects without the disease (controls) are used to identify genetic risk factors. Genotypes more common in cases than controls can be thought of as high risk. Fitness (E) of the ABS model is determined by comparing the high risk and low risk assignments made by the ABS to those from the given nonlinear gene-gene interaction model. Calculation of the fitness value, E, is given by the classification error function E shown in (4) below. In the present study, max(E) = 9 and min(E) = 0. The goal is to minimize E.

$$E = \sum_{i=1}^{|G|} e_i \quad e_i = 0 \ \text{if} \ f(G_i) = d(G_i)$$

$$e_i = 1 \ \text{if} \ f(G_i) \neq d(G_i) \quad (4)$$

Genetic Algorithm Parameters

Grammatical evolution works by decoding genetic algorithm chromosomes. Our GA chromosome consisted of 250 integer codons. In implementing the grammar, it is possible to reach the end of a chromosome with an incomplete instantiation of the grammar. To complete the instance, chromosome wraparound was used (O'Neill and Ryan 2001, 2003). In other words, the instance of the grammar was completed be reusing the chromosome. Wraparound imperfectly solves this problem by simply returning to the beginning of the chromosome to acquire more codons to expand the production rules. The solution is imperfect because there is no guarantee that the grammar is finite; therefore, it is possible for the derivation to recurse endlessly, so there must be a stopping point. In the present study, we used a stopping point of 10 wraparounds.

Genetic algorithms require the setting of many parameters. Table 1 below summarizes the genetic algorithm parameter settings used in this study. These initial settings were selected based on Goldberg's simple GA (Goldberg 1989) and our previous experience in this domain (Moore and Hahn 2003a, 2003b, 2004a, 2004b). We ran the genetic algorithm a total of 100 times with different random seeds for each gene-gene interaction model. The genetic algorithm was stopped when a model with a classification error of zero was discovered (i.e. $E = 0$) or when 90% of the population converged, where convergence was measured by the genetic algorithm library used, GAlib (Wall 2003). We used a parallel search strategy (Cantu-Paz 2000) of ten demes, each with a population size of 500, for a total population size of 5000. A best chromosome migrated from each deme to all other demes every 25 generations.

Number of runs	100
Stopping criteria	Classification error = 0 or population convergence at 90% as defined by GAlib
Population size	5000
Number of demes	10
Generations (max.)	n/a
Selection	Roulette wheel
Crossover	Single point
Crossover probability	0.90
Mutation probability	0.01

Table 1: Summary of the genetic algorithm parameters used.

Results

The grammatical evolution algorithm was run a total of 100 times for each of the two nonlinear gene-gene interaction models. For both Model 1 and Model 2 (see Figure 1), the grammatical evolution strategy yielded an ABS model that was perfectly consistent with the high-risk and low-risk assignments for each combination of genotypes. Thus, the ABS model discovery method routinely found perfect models.

Figure 2 below illustrates the starting and ending configurations for an ABS that is consistent with genetic Model 2. Note that the ABS for genotypes *AAbb*, *AaBb*, and *aaBB* all have at least three agents in the upper left quadrant while there are less than three agents in the upperleft quadrant for all the other genotype combinations. This pattern of high-risk and low-risk agent counts is perfectly consistent with the high-risk and low-risk genotype combinations in genetic Model 2 and thus has a maximum fitness. This ABS was executed for 91 timesteps and had nine total agents. Six agents had movements that were dependent on genotype while five had collisions that were genotype-dependent. The agents that finished in the upper left quadrant were all dependent on genotype.

Tables 2 and 3 below summarize the mode (i.e. most common) and range of the number of agents, collisions and types of moves and collision behaviors that define the genotype-dependencies of the elements in the best ABS models found across the 100 runs for each model. Agents are broken into categories according to their movement and collision behaviors. Locus independent (LI) and locus dependent (LD) move (M) and collision (C) behaviors are shown. As expected, every ABS had at least two agents that had moves or collisions that were locus dependent. For Model 1 and Model 2, the most frequent numbers of agents were six and five, respectively.

	Agents	LIM	LDM	LIC	LDC	C
Avg	4.5	1.3	3.2	2.3	2.2	49.3
Min	2	0	1	0	0	2
Max	12	5	9	9	6	533
Mode	6	0	2	2	2	2

Table 2: Summary of results from 100 runs for Model 1.

	Agents	LIM	LDM	LIC	LDC	C
Avg	5.1	1.3	3.7	2.5	2.5	67.3
Min	2	0	2	0	0	4
Max	13	5	8	8	7	521
Mode	5	0	2	3	3	6

Table 3: Summary of results from 100 runs for Model 2.

Discussion

The primary conclusion of this study is that it is possible to use artificial life in the form of agent-based simulation (ABS) to generate hypothetical systems biology models that are consistent with genetic models of disease susceptibility. This study represent the first step towards the use of artificial life to carry out thought experiments about the nature of the genotype to phenotype mapping relationship in the context of human health and disease. We anticipate a combined computational approach to both systems biology and human genetics will lead to a better understanding of human health which in turn will lead to better disease diagnosis, prevention, and treatment strategies.

We acknowledge that the approach presented here is only the first step and that additional changes to the overall modeling strategy will be needed before this prototype will be useful for routine thought experiments. For example, it will be important to incorporate measures of agent dynamics into the system. The current approach measures a static agent endpoint. It will also be important to incorporate measures of systems complexity such as

entropy (reviewed by Adami 1998). A wider range of agent behaviors such as birth and death will need to be explored in addition to optimization of the grammatical evolution strategy for higher-order genetic models as has been done for the Petri net approach (Moore and Hahn 2004b). Further, higher-level measures of the system will need to be implemented to capture more interesting patterns of behavior such as dynamical hierarchies (e.g. Dorin and McCormack 2003). Finally, interpretation of ABS models will be necessary if useful biological hypotheses are to be generated from these thought experiments. We anticipate that this study will provide a useful starting point for those hoping to use artificial life models as hypothesis-generating thought experiments.

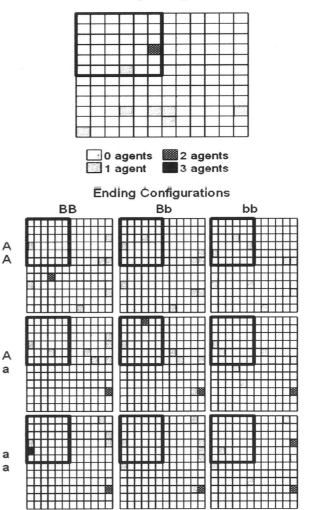

Figure 2: Starting (top) and ending (bottom) configurations for an ABS that is consistent with Model 2. In this ABS, each square in the 12x12 grid is occupied by either zero (white), one (grey), two (checkerboard), or three (black) agents. For simplicity, the types of agents are not illustrated. The upper left quadrant of the grid outlined in black is used to determine risk.

Acknowledgements

This work was supported by National Institutes of Health grants HL65234, HL65962, GM31304, AG19085, and AG20135.

References

Adami, C. 1998. *Introduction to Artificial Life*. TELOS Springer-Verlag, Santa Clara.

Bateson, W. 1909. *Mendel's Principles of Heredity*. Cambridge University Press, Cambridge.

Cantu-Paz, E., 2000. *Efficient and Accurate Parallel Genetic Algorithms*. Kluwer Academic Publishers.

Desel, J., Juhas, G. 2001. What is a Petri net? Informal answers for the informed reader. In H. Ehrig, G. Juhas, J. Padberg, and G. Rozenberg (Eds.), *Unifying Petri Nets, Lecture Notes in Computer Science* 2128, pp 1-27. Springer.

D'Inverno, M. and Luck, M. 2001. *Understanding Agent Systems*. Berlin, Springer.

Di Paolo, E.A., Noble, J., Bullock, S. 2000. Simulation models as opaque thought experiments. In: Dedau, M.A. et al. (Eds.), *Artificial Life VI*. The MIT Press, Cambridge.

Dorin, A., McCormack, J. Self-assembling dynamical hierarchies. In: Standish, R.K., Bedau, M.A., Abbass, H.A. (Eds.), *Artificial Life VIII*. The MIT Press, Cambridge.

Falconer, D.S., Mackay, T.F.C. 1996. *Introduction to Quantitative Genetics*. Longman, Essex.

Fisher, R.A. 1918. The correlation between relatives on the supposition of Mendelian inheritance. *Transactions of the Royal Society of Edinburgh* 52, 399-433.

Fogel, L.J. 1962. Autonomous automata. *Industrial Research* 4, 14-19.

Fogel, L.J., Owens, A.J., Walsh, M.J. 1966. *Artificial Intelligence through Simulated Evolution*. John Wiley, New York.

Frankel, W.N., Schork, N.J. 1996. Who's afraid of epistasis. *Nature Genetics* 14, 371-3.

Goldberg, D.E. 1989. *Genetic Algorithms in Search, Optimization, and Machine Learning*. Addison-Wesley.

Goss, P.J., Peccoud, J. 1998. Quantitative modeling of stochastic systems in molecular biology by using stochastic Petri nets. *Proceedings of the National Academy of Sciences USA* 95, 6750-5.

Holland, J.H. 1969. Adaptive plans optimal for payoff-only environments. In: *Proceedings of the 2nd Hawaii International Conference on Systems Sciences*, pp 917-920.

Holland, J.H. 1975. *Adaptation in Natural and Artificial Systems*. University of Michigan Press, Ann Arbor.

Hopcroft, J. and Ullman, J. 1979. *Introduction to Automata Theory, Languages and Computation*. Reading, MA: Addison-Wesley.

Ideker, T., Galitski, T., Hood, L. 2001. A new approach to decoding life: systems biology. *Annual Review of Genomics and Human Genetics* 2, 343-72.

Jansen, R.C. 2003. Studying complex biological systems using multifactorial perturbation. *Nature Reviews Genetics* 4, 145-51.

Kitagawa, J., Iba, H. 2003. Identifying metabolic pathways and gene regulation networks with evolutionary algorithms. In:

Fogel, G.B., Corne, D.W. (Eds.), *Evolutionary Computation and Bioinformatics*. Morgan Kaufmann Publishers.

Koza, J.R. 1992. *Genetic Programming: On the Programming of Computers by Means of Natural Selection*. The MIT Press.

Koza, J.R., Mydlowec, W., Lanza, G., Yu, J., Keane, M.A. 2001. Reverse engineering of metabolic pathways from observed data using genetic programming. *Pacific Symposium on Biocomputing* 6, 434-45.

Li, W., Reich, J. 2000. A complete enumeration and classification of two-locus disease models. Human Heredity 50, 334-49.

Merlin, P. 1974. *A Study of the Recoverability of Computer Systems*. Ph.D. Thesis, University of California, Irvine (1974).

Moore, J.H. 2003. The ubiquitous nature of epistasis in determining susceptibility to common human diseases. *Human Heredity* 56, 73-82.

Moore, J.H., Hahn, L.W. 2003a. Grammatical evolution for the discovery of Petri net models of complex genetic systems. In: Cantu-Paz et al. (Eds.), *Genetic and Evolutionary Computation – GECCO 2003. Lecture Notes in Computer Science* 2724, 2412-13.

Moore, J.H., Hahn, L.W. 2003b. Petri net modeling of high-order genetic systems using grammatical evolution. *BioSystems* 72, 177-86.

Moore, J.H., Hahn, L.W. 2004a. Evaluation of a discrete dynamic systems approach for modeling the hierarchical relationship between genes, biochemistry, and disease susceptibility. *Discrete and Continuous Dynamical Systems: Series B* 4, 275-87.

Moore, J.H., Hahn, L.W. 2004b. An improved grammatical evolution strategy for Petri net modeling of complex genetic systems. In: *Applications of Evolutionary Computing, Lecture Notes in Computer Science*, in press.

Moore, J.H., Hahn, L.W., Ritchie, M.D., Thornton, T.A., White, B.C. 2002. Application of genetic algorithms to the discovery of complex genetic models for simulation studies in human genetics. In: Langdon, W.B., et al. (Eds.), *Proceedings of the Genetic and Evolutionary Computation Conference*. Morgan Kaufmann Publishers, San Francisco.

Moore, J.H., Williams, S.M. 2002. New strategies for identifying gene-gene interactions in hypertension. *Annals of Medicine* 34, 88-95.

O'Neill, M., Ryan, C. 2001. Grammatical evolution. *IEEE Transactions on Evolutionary Computation* 5, 349-358.

O'Neill, M., Ryan, C. 2003. *Grammatical Evolution: Evolutionary Automatic Programming in an Arbitrary Language*. Kluwer Academic Publishers, Boston.

Rechenberg, I. 1965. *Cybernetic solution path of an experimental problem*. Royal Aircraft Establishment, Farnborough, U.K., Library Translation No. 1122, August.

Reif, D.M., White, B.C., Moore, J.H. 2004. Integrated analysis of genetic, genomic, and proteomic data. *Expert Review of Proteomics*, in press.

Schwefel, H.-P. 1965. *Kybernetische Evolution als Strategie der experimentellen Forschung in der Stromungstechnik*. Diploma Thesis, Technical University of Berlin.

Wall, M. 2003. GAlib. http://lancet.mit.edu/ga/

Woolridge, M. 2002. *An Introduction to MultiAgent Systems*. Sussex, England: Wiley.

Wooldridge, M., Jennings, N.R. 1995. Intelligent agents: theory and practice. *The Knowledge Engineering Review* 10, 115-52.

Author Index

Ackley, David: 274
Adami, Christoph: 438
Amarteifio, Saoirse: 551
Antony, Antony: 262
Arita, Takaya: 369
Au, Randy: 244

Bankhead, Armand III: 220
Banzhaf, Wolfgang: 404
Barandiaran, Xabier: 514
Bardeen, Matthew: 292
Bec, Louis: 522
Bedau, Mark: 297
Beer, Randall: 381, 415
Bersini, Hugues: 226, 472, 478
Bisig, Daniel: 546
Bonabeau, Eric: 232
Bongard, Josh: 57
Borenstein, Elhanan: 146
Bosque Moran, Martin: 75
Braynen, William: 244
Buchanan, Andrew: 297
Buchrmann, Thomas: 63
Bullock, Seth: 534

Cañamero, Lola: 176
Capcarrere, Mathieu: 21
Capi, Genci: 69
Cavazza, Marc: 522
Chu, Dominique: 433
Clune, Jeff: 303
Commeyras, Auguste: 478
Connolly, John: 244
Cornish-Bowden, Athel: 450
Craft, Michael: 88
Curran, Dara: 152
Cárdenas, María-Luz: 450

Defos du Rau, Gonzague: 522
de Pinedo, Manuel: 528
Desnoyer, Mark: 15
Di Paolo, Ezequiel: 1, 63, 94, 139
Doya, Kenji: 69, 170

Drennan, Barry: 381

Eggenberger Hotz, Peter: 387
Eriksson, Anders: 456
Eriksson, Karl-Erik: 456

Fernando, Chrisantha: 1
Floreano, Dario: 9
Fontana, Alessandro: 393
Fraccaro, Walter Steven: 393
Funes, Pablo: 232

Garcia Carbajal, Santi: 75
Gershenson, Carlos: 238
Goings, Sherri: 303
Gong, Tao: 158
Gonzalez Martinez, Fermin: 75
Grim, Patrick: 164, 244
Guíñez-Abarzúa, Flavio: 450
Gulyas, Laszlo: 328

Hallinan, Jennifer: 399
Hampton, Alan: 438
Hartley, Simon: 522
Harvey, Inman: 139, 309, 352
Heckendorn, Robert B.: 220
Hénaff, Patrick: 81
Hoinville, Thierry: 81
Howsman, Thomas: 88
Hsiao, Kaijen: 113
Huang, Chien-Feng: 489
Huang, Wei: 315
Husbands, Phil: 106
Hutton, Tim: 444

Ikegami, Takashi: 188, 194, 421, 461, 495
Izquierdo-Torres, Eduardo: 322

James, Minett: 158

Kafri, Ran: 501
Kamp, Hans: 170

Kampis, George: 328
Ke, Jinyun: 158
Keinan, Alon: 557
Kennedy, Paul: 427
Kim, Chang Hun: 125
Kim, Sun Jeong: 125
Klein, Jon: 540
Klein, Michael: 170
Klyubin, Alexander: 563
Kokalis, Trina: 164
Kuo, P. Dwight: 404

Lalanne, Remy: 522
Lancet, Doron: 466, 501
Le Bras, Mikael: 522
Lenaerts, Tom: 472
Lenski, Richard: 340
Letelier, Juan-Carlos: 450
Lindgren, Kristian: 456
Lipson, Hod: 15, 57, 100
Louie, Nancy: 244
Lowe, Robert: 176
Lugrin, Jean-Luc: 522
Luke, Sean: 569, 575

Macinnes, Ian: 94
Madina, Duraid: 461
Magnuson, Nancy: 220
Malone, Evan: 100
Marcus, David: 15
Marshall, James: 334
Mattiussi, Claudio: 9
McHale, Gary: 106
Minett, James: 158
Mirolli, Marco: 182
Misevic, Dusan: 340
Mitteldorf, Joshua: 346
Moore, Jason: 581
Morimoto, Gentaro: 188
Mourre, François: 522
Mueller, Philip: 250
Mytilinaios, Efstathios: 15

Naveh, Barak: 466
Nehaniv, Chrystopher: 119, 176, 563
Nishimura, Shin: 410

Noble, Jason: 528
Nordin, Peter: 540

O'Kelly, Michael: 113
O'Neil, Daniel: 88
O'Neill, Michael: 551
O'Riordan, Colm: 152
Ofria, Charles: 303, 315, 340
Olsson, Lars: 119
Orme, Belinda: 232
Özturkeri, Can: 21

Palm, Guenther: 170
Panait, Liviu: 569, 575
Parisi, Domenico: 182
Penn, Alexandra: 352
Pennock, Robert T.: 303
Philippe, Pierre: 472
Plasson, Raphaël: 478
Polani, Daniel: 119, 176, 563
Pollack, Jordan: 45
Portegys, Thomas: 484
Prokopenko, Mikhail: 27
Psujek, Sean: 415

Rocha, Luis Mateus: 256, 489
Roggen, Daniel: 33
Rosenberger, Robert: 244
Ruppin, Eytan: 146

Salzberg, Chris: 262, 495
Sanchez, Eduardo: 33
Sasahara, Kazutoshi: 194
Sasai, Maaski: 410
Sayama, Hiroki: 262, 495
Selinger, Evan: 244
Shenhav, Barak: 466, 501
Shim, Yoon Sik: 125
Silverman, Eric: 534
Sipper, Moshe: 466
Skusa, Andre: 427
Soto-Andrade, Jorge: 450
Spier, Emmet: 133
Standish, Russell: 358, 364
Steels, Luc: 200
Stevens, William: 39

Suzuki, Hideaki: 507
Suzuki, Keisuke: 421
Suzuki, Reiji: 369
Svangård, Nils: 540

Taylor, Charles E.: 208
Taylor, Tim: 268, 375
Thoma, Yann: 33
Tokumine, Simon: 334
Torng, Eric: 315
Triant, Mark: 297

Unemi, Tatsuo: 546

Vabø, Rune: 433
Vallejo, Edgar E.: 208
Van Belle, Terry: 274
Vaughan, Eric: 139
Viswanathan, Shivakumar: 45
Vogt, Paul: 214

Wait, Alexander: 280
Wang, Peter: 27
Wang, William S-Y: 158
Weiss Solís, David: 472
White, Bill: 581
Whitehead, Dion: 427
Wiles, Janet: 399
Wills, Peter: 51
Wuensche, Andrew: 286